CREATING A VIEWING RECTANGLE

Each of the viewing rectangles shows portions of the graph of $y = 0.1x^4 - x^3 + 2x^2$. The first viewing rectangle shows the most complete graph of the equation.

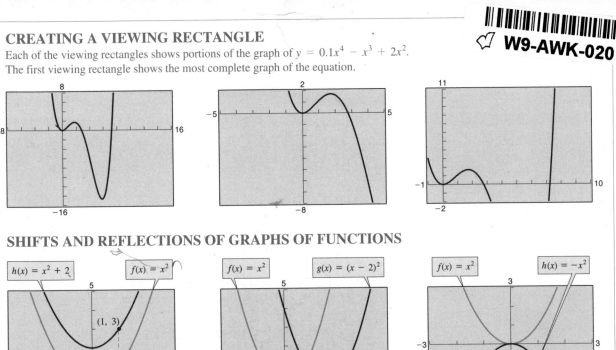

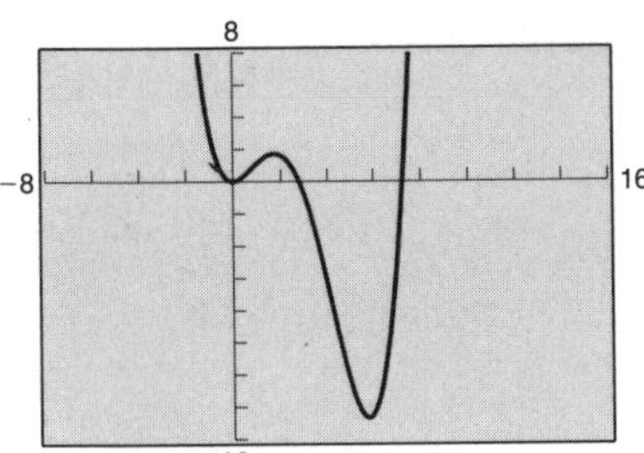

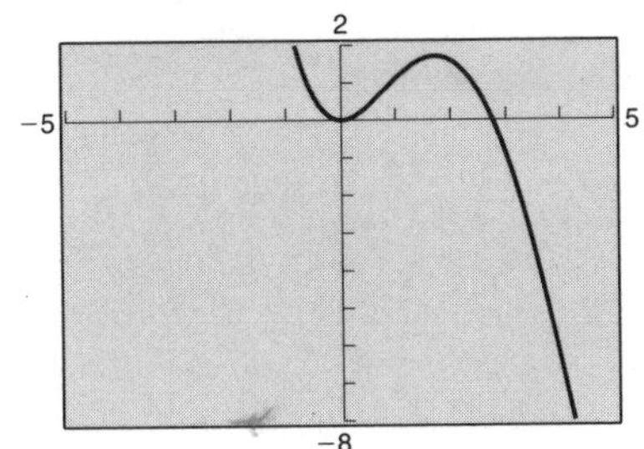

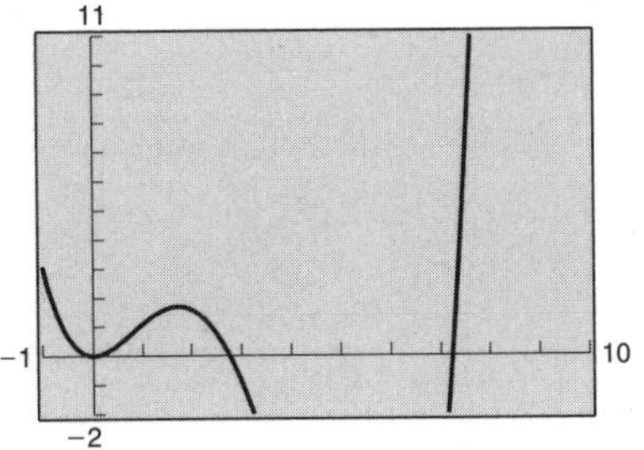

SHIFTS AND REFLECTIONS OF GRAPHS OF FUNCTIONS

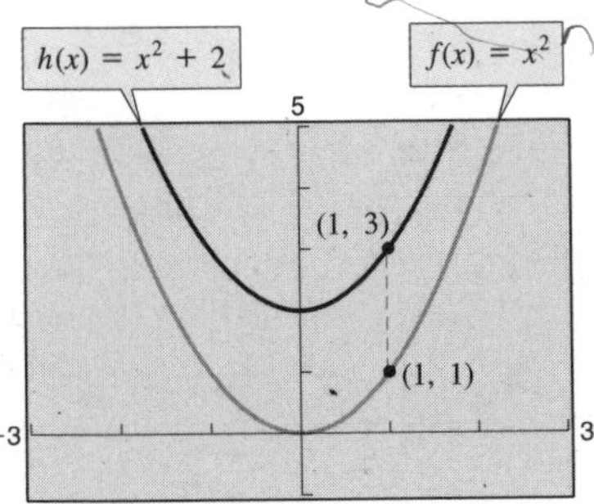

Vertical shift upward: 2 units

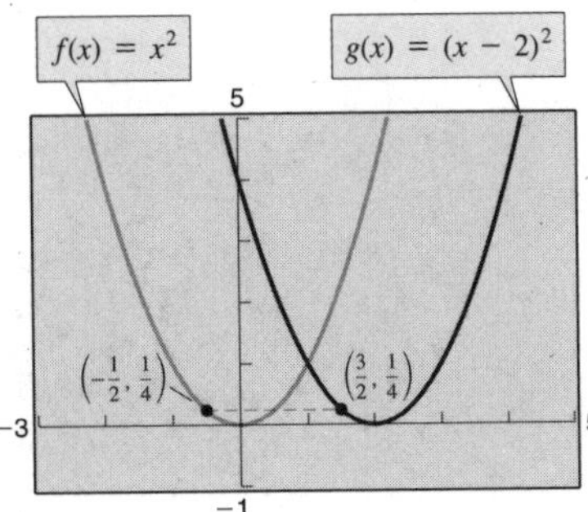

Horizontal shift to the right: 2 units

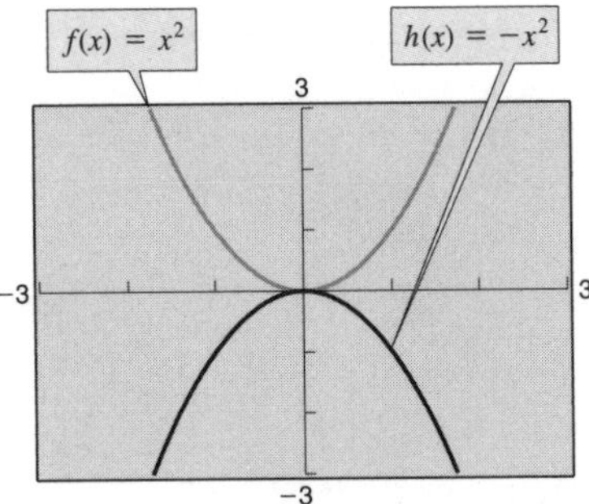

Reflection

MAINTAINING GEOMETRIC PERSPECTIVE WITH A SQUARE VIEWING RECTANGLE

To maintain a proper geometric perspective for graphs such as perpendicular lines and circles, use a viewing rectangle with a "square setting" in which the ticks on the horizontal and vertical axes are equally spaced.

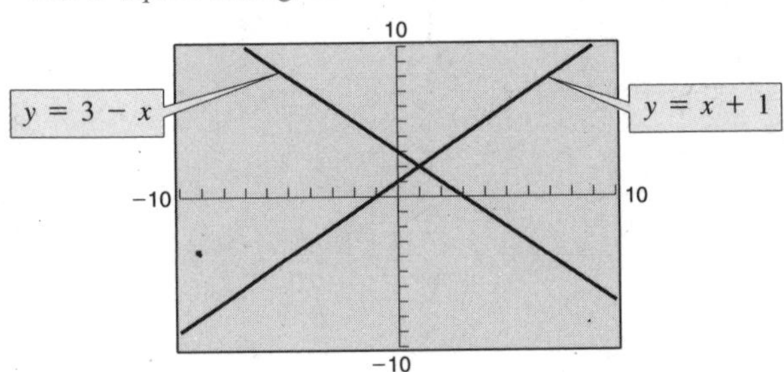

Nonsquare setting: Lines do not appear to be perpendicular

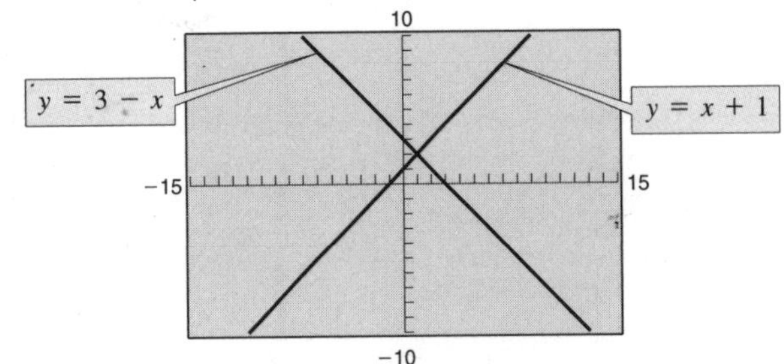

Square setting: Lines appear perpendicular

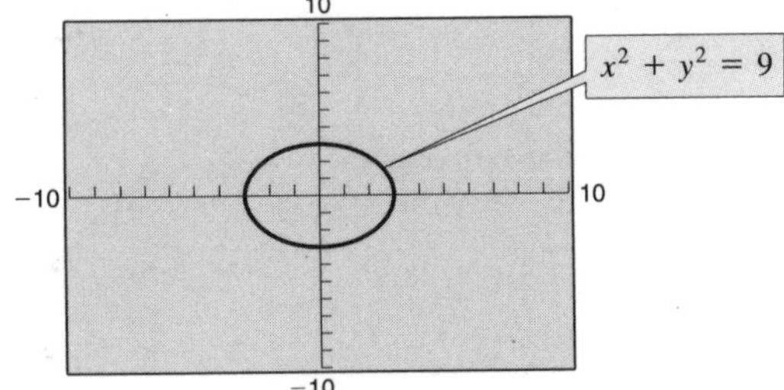

Nonsquare setting: Circle does not appear to be circular

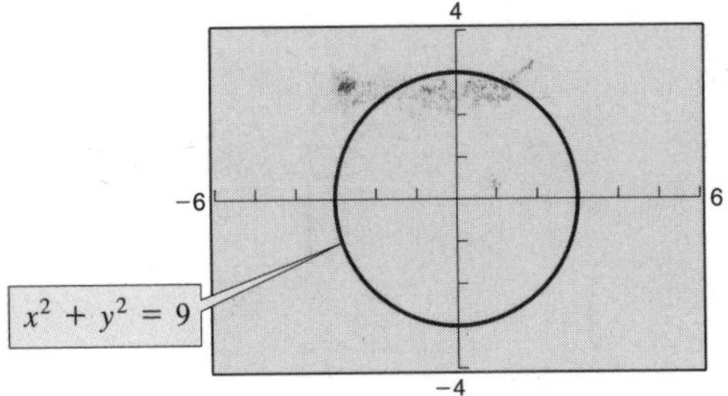

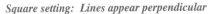

Square setting: Circle appears circular

Kevin Chilcoat P.C. 2007-3

College Algebra

College Algebra

A Graphing Approach

SECOND EDITION

Roland E. Larson
The Pennsylvania State University
The Behrend College

Robert P. Hostetler
The Pennsylvania State University
The Behrend College

Bruce H. Edwards
University of Florida

WITH THE ASSISTANCE OF

David E. Heyd
The Pennsylvania State University
The Behrend College

HOUGHTON MIFFLIN COMPANY Boston New York

Sponsoring Editor: Christine B. Hoag
Senior Associate Editor: Maureen Brooks
Managing Editor: Catherine B. Cantin
Senior Project Editor: Karen Carter
Associate Project Editor: Rachel D'Angelo Wimberly
Production Supervisor: Lisa Merrill
Art Supervisor: Gary Crespo
Marketing Manager: Charles Cavaliere

Cover design by Harold Burch Design, NYC

Composition: Meridian Creative Group

Printed in the U.S.A.

Library of Congress Catalog Card Number: 96-076660

ISBN: 0-669-41732-7

23456789–DC–00-99-98-97-96

Preface

College Algebra: A Graphing Approach, Second Edition, is the premier text for a reform-oriented course. Designed to build a strong foundation in algebra, the text encourages students to develop a firm grasp of the underlying mathematical concepts while using algebra as a tool for solving real-life problems. The comprehensive text presentation invites discovery and exploration, while the integrated technology and consistent problem-solving strategies help the student develop strong college algebra skills.

College Algebra Reform

The college algebra course has changed over the past few years in response to the growing discussion of reform in mathematics education. Generally speaking, these changes have focused on the following areas: technology, real-life applications, problem-solving, and communicating about mathematics. The Second Edition embodies the spirit of these reform ideals without compromising the mathematical integrity of the course presentation. All text elements from the previous edition were considered for revision and many new examples, exercises, and applications were added.

Technology Graphing technology is consistently incorporated throughout the Second Edition. The visualization and exploration capabilities of technology encourage the student to participate actively in the learning process, to develop their intuitive understanding of mathematical concepts, and to solve problems using actual data. Thus, students learn how algebra functions as a modeling language for real-life problems. Technology is used as a tool, drawn into the discussion whenever it offers a useful perspective on the topic at hand. For example, the power of graphing technology may be used to guide the students through thought-provoking explorations or to show alternative problem-solving techniques. Where appropriate, situations in which the results obtained through the use of technology may be misleading are also noted.

The Second Edition assumes that the student will use a graphing calculator on a daily basis in the course. Integrated throughout the text at point of use are many opportunities for investigation using technology (e.g., see page 318) and exercises that require the use of a graphing utility (e.g., see page 188). The text also carefully shows how to use graphing technology to best advantage (e.g., see page 185).

Whenever possible, references to graphing technology are generic. In a few cases, however, the text includes programs that will enable the student to investigate particular mathematical concepts (e.g., see page 134). Comparable programs for a wide variety of Texas Instruments, Casio, Sharp, and Hewlett-

Packard graphing calculators—including the most current models—are given in the appendix.

To accommodate a variety of teaching and learning styles, *College Algebra: A Graphing Approach,* Second Edition, is also available in a multimedia, CD-ROM format. *Interactive College Algebra: A Graphing Approach* offers students a variety of additional tutorial assistance, including examples and exercises with detailed solutions; pre-, post-, and self-tests with answers; and *TI-82* and *TI-83* graphing calculator emulators. (See pages xviii–xx for more detailed information.)

Real-Life Applications To emphasize for students the connection between mathematical concepts and real-world situations, up-to-date, real-life applications are integrated throughout the text. These applications appear as chapter introductions with related exercises (e.g., see pages 75 and 115), examples (e.g., see page 360), exercises (e.g., see page 87), Group Activities (e.g., see page 97), and Chapter Projects (e.g., see page 455).

Students have many opportunities to collect and interpret data, to make conjectures, and to construct mathematical models in the examples, exercises, Group Activities, and Chapter Projects. Students work on modeling problems with experimental and theoretical probabilities (e.g., see page 591), use mathematical models to make predictions or draw conclusions from real data (e.g., see page 354), compare models (e.g., see page 307), and apply curve-fitting techniques to create their own models from data (e.g., see page 230). In the process, the Second Edition gives students many more opportunities to use charts, tables, scatter plots, and graphs to summarize, analyze, and interpret data.

Problem Solving The primary goal of any mathematics textbook is to encourage students to become competent and confident problem solvers. Many aspects of this revision focused on this goal—including the addition of new features such as Chapter Projects, Explorations, and Group Activities, as well as extensive and careful revision of the examples and exercise sets. Students are asked to use numerical, graphical, and algebraic techniques, and the use of graphing technology as a problem-solving tool is encouraged as appropriate (e.g., see page 215). Throughout, students are encouraged to follow a consistent approach to solving applied problems: Construct a verbal model, label terms, construct an algebraic model, solve the problem using the model, and check the answer in the original statement of the problem.

Like the previous edition, the Second Edition has an abundance of exercises that are designed to develop skills. The text also includes many other types of exercises that offer students the opportunity to refine their problem-solving skills, such as exercises that require interpretations (e.g., see page 329), those having many correct answers (e.g., see page 288), and multipart exercises designed to lead the student through problem-solving strategies (e.g., see page 226).

Communicating about Mathematics Each section in the Second Edition ends with a Group Activity. Designed to be completed in class or as homework assignments, the Group Activities give students the opportunity to work cooperatively as they think, talk, and write about mathematics. Students' understanding is reinforced through interpretation of mathematical concepts and results (e.g., see page 250), problem posing and error analysis (e.g., see page 428), and constructing mathematical models, tables and graphs (e.g., see page 418).

Making connections between algebra and real-world situations also helps students understand the underlying theory. Other connections are emphasized in this text as well, including those to probability (e.g., see Chapter 7), geometry (e.g., see page 397), and statistics (see the Sections P.6 and 2.6).

Improved Coverage

Chapter P, Prerequisites, is streamlined in the Second Edition. All or part of this review material may be covered or it may be omitted, offering greater flexibility in designing the course syllabus.

Several topics are now covered earlier in the Second Edition. Linear modeling and scatter plots are covered much earlier, in Chapter 2. Complex numbers also are now in Chapter 2. Rational functions are covered with polynomials in Chapter 3. The discussion of exponential and logarithmic functions appears a chapter earlier, in Chapter 4. Coverage of systems of equations and inequalities was moved to Chapter 5 and now includes partial fractions and linear programming.

New sections on exploring data have been added to Chapters P, 2, and 4. The coverage of matrices in Chapter 6 was expanded to include determinants of matrices and applications. Conics and translations of conics have been moved to a new chapter, Chapter 8, along with new coverage of parametric equations.

Features of the Second Edition

Chapter Opener Each chapter opens with a look at a real-life application. Real data is presented using graphical, numerical, and algebraic techniques.

Notes Notes anticipate students' needs by offering additional insights, pointing out common errors, and describing generalizations.

Theorems, Definitions, and Guidelines All of the important rules, formulas, theorems, guidelines, properties, definitions, and summaries are highlighted for emphasis. Each is also titled for easy reference.

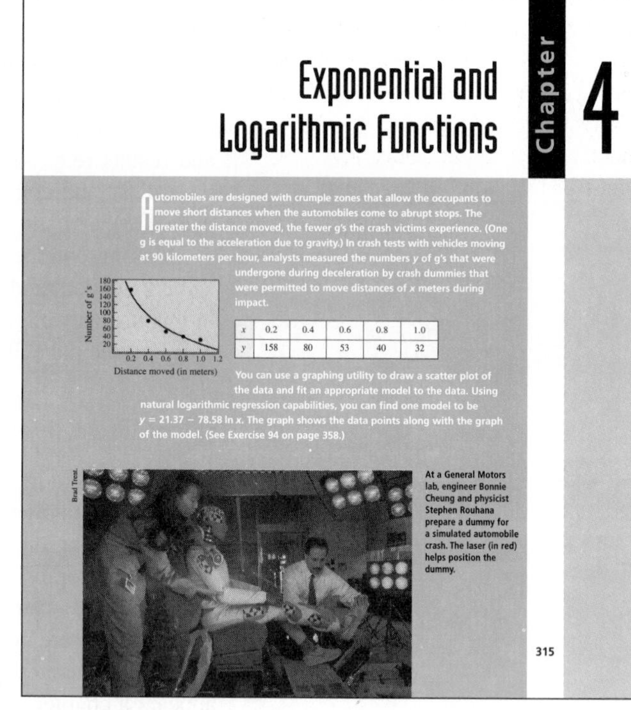

Chapter 4

Exponential and Logarithmic Functions

Automobiles are designed with crumple zones that allow the occupants to move short distances when the automobiles come to abrupt stops. The greater the distance moved, the fewer g's the crash victims experience. (One g is equal to the acceleration due to gravity.) In crash tests with vehicles moving at 90 kilometers per hour, analysts measured the numbers y of g's that were undergone during deceleration by crash dummies that were permitted to move distances of x meters during impact.

x	0.2	0.4	0.6	0.8	1.0
y	158	80	53	40	32

You can use a graphing utility to draw a scatter plot of the data and fit an appropriate model to the data. Using natural logarithmic regression capabilities, you can find one model to be $y = 21.37 - 78.58 \ln x$. The graph shows the data points along with the graph of the model. (See Exercise 94 on page 358.)

Brad Trent.

At a General Motors lab, engineer Bonnie Cheung and physicist Stephen Rouhana prepare a dummy for a simulated automobile crash. The laser (in red) helps position the dummy.

315

Section Outline Each section begins with a list of the major topics covered in the section. These topics are also the subsection titles and can be used for easy reference and review by students. In addition, an exercise application that uses a skill or illustrates a concept covered in the section is highlighted to emphasize the connection between mathematical concepts and real-life situations.

Library of Functions The concept of the function is introduced in Chapter 1. In the material that follows, the icon appears each time a new type of function is described in detail.

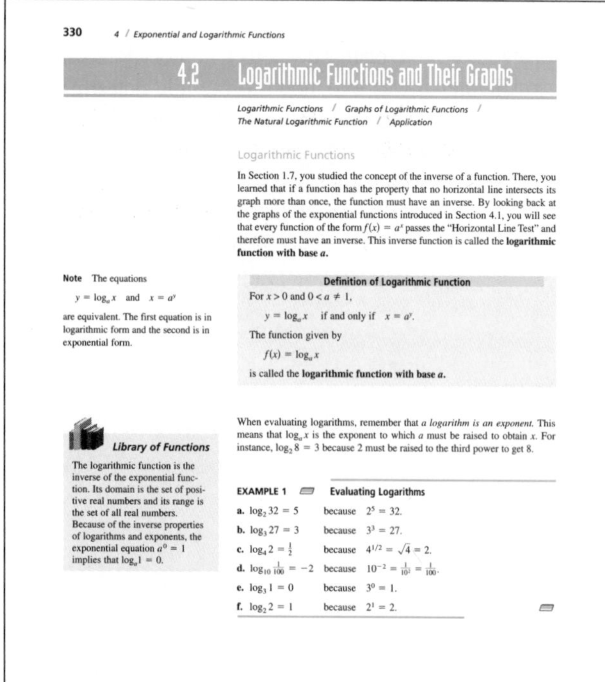

330 4 / Exponential and Logarithmic Functions

4.2 Logarithmic Functions and Their Graphs

Logarithmic Functions / Graphs of Logarithmic Functions / The Natural Logarithmic Function / Application

Logarithmic Functions

In Section 1.7, you studied the concept of the inverse of a function. There, you learned that if a function has the property that no horizontal line intersects its graph more than once, the function must have an inverse. By looking back at the graphs of the exponential functions introduced in Section 4.1, you will see that every function of the form $f(x) = a^x$ passes the "Horizontal Line Test" and therefore must have an inverse. This inverse function is called the **logarithmic function with base *a*.**

Note The equations

$$y = \log_a x \quad \text{and} \quad x = a^y$$

are equivalent. The first equation is in logarithmic form and the second is in exponential form.

Definition of Logarithmic Function

For $x > 0$ and $0 < a \neq 1$,

$$y = \log_a x \quad \text{if and only if} \quad x = a^y.$$

The function given by

$$f(x) = \log_a x$$

is called the **logarithmic function with base *a*.**

Library of Functions

The logarithmic function is the inverse of the exponential function. Its domain is the set of positive real numbers and its range is the set of all real numbers. Because of the inverse properties of logarithms and exponents, the exponential equation $a^0 = 1$ implies that $\log_a 1 = 0$.

When evaluating logarithms, remember that *a logarithm is an exponent.* This means that $\log_a x$ is the exponent to which a must be raised to obtain x. For instance, $\log_2 8 = 3$ because 2 must be raised to the third power to get 8.

EXAMPLE 1 Evaluating Logarithms

a. $\log_2 32 = 5$ because $2^5 = 32$.

b. $\log_3 27 = 3$ because $3^3 = 27$.

c. $\log_4 2 = \frac{1}{2}$ because $4^{1/2} = \sqrt{4} = 2$.

d. $\log_{10} \frac{1}{100} = -2$ because $10^{-2} = \frac{1}{10^2} = \frac{1}{100}$.

e. $\log_3 1 = 0$ because $3^0 = 1$.

f. $\log_2 2 = 1$ because $2^1 = 2$.

Intuitive Foundation for Calculus Special emphasis is given to the algebraic skills that are needed in calculus. Many examples in the Second Edition discuss algebraic techniques or graphically show concepts that are used in calculus, providing an intuitive foundation for future work.

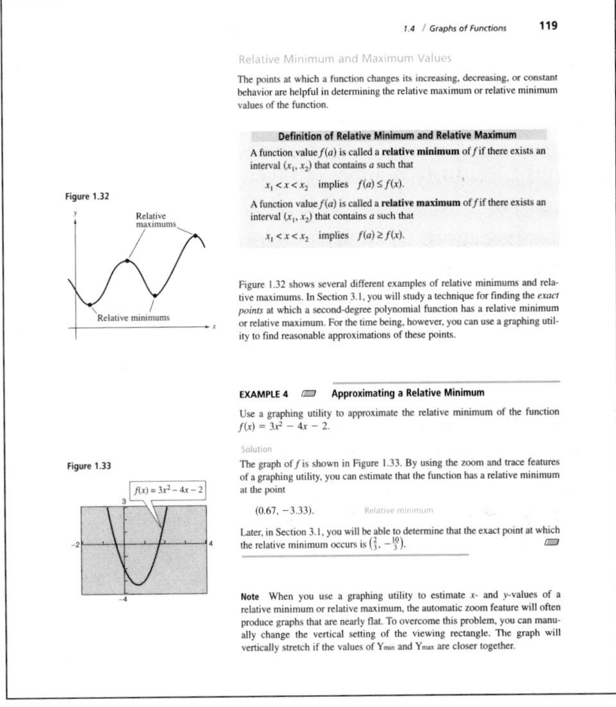

1.4 / Graphs of Functions 119

Relative Minimum and Maximum Values

The points at which a function changes its increasing, decreasing, or constant behavior are helpful in determining the relative maximum or relative minimum values of the function.

Definition of Relative Minimum and Relative Maximum

A function value $f(a)$ is called a **relative minimum** of f if there exists an interval (x_1, x_2) that contains a such that

$x_1 < x < x_2$ implies $f(a) \le f(x)$.

A function value $f(a)$ is called a **relative maximum** of f if there exists an interval (x_1, x_2) that contains a such that

$x_1 < x < x_2$ implies $f(a) \ge f(x)$.

Figure 1.32

Figure 1.32 shows several different examples of relative minimums and relative maximums. In Section 3.1, you will study a technique for finding the *exact points* at which a second-degree polynomial function has a relative minimum or relative maximum. For the time being, however, you can use a graphing utility to find reasonable approximations of these points.

EXAMPLE 4 **Approximating a Relative Minimum**

Use a graphing utility to approximate the relative minimum of the function $f(x) = 3x^2 - 4x - 2$.

Solution

The graph of f is shown in Figure 1.33. By using the zoom and trace features of a graphing utility, you can estimate that the function has a relative minimum at the point

$(0.67, -3.33)$. Relative minimum

Later, in Section 3.1, you will be able to determine that the exact point at which the relative minimum occurs is $\left(\frac{2}{3}, -\frac{10}{3}\right)$.

Figure 1.33

Note When you use a graphing utility to estimate x- and y-values of a relative minimum or relative maximum, the automatic zoom feature will often produce graphs that are nearly flat. To overcome this problem, you can manually change the vertical setting of the viewing rectangle. The graph will vertically stretch if the values of Ymin and Ymax are closer together.

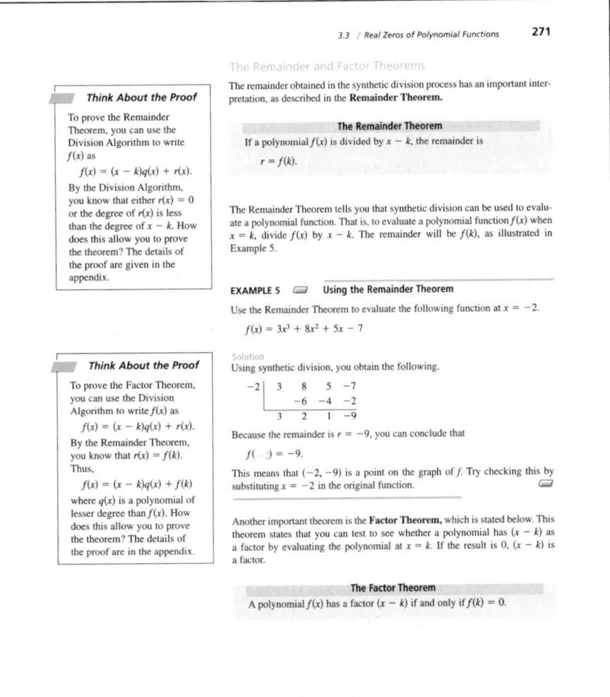

3.3 / Real Zeros of Polynomial Functions 271

The Remainder and Factor Theorems

The remainder obtained in the synthetic division process has an important interpretation, as described in the **Remainder Theorem.**

Think About the Proof

To prove the Remainder Theorem, you can use the Division Algorithm to write $f(x)$ as

$f(x) = (x - k)q(x) + r(x)$.

By the Division Algorithm, you know that either $r(x) = 0$ or the degree of $r(x)$ is less than the degree of $x - k$. How does this allow you to prove the theorem? The details of the proof are given in the appendix.

The Remainder Theorem

If a polynomial $f(x)$ is divided by $x - k$, the remainder is

$r = f(k)$.

The Remainder Theorem tells you that synthetic division can be used to evaluate a polynomial function. That is, to evaluate a polynomial function $f(x)$ when $x = k$, divide $f(x)$ by $x - k$. The remainder will be $f(k)$, as illustrated in Example 5.

EXAMPLE 5 **Using the Remainder Theorem**

Use the Remainder Theorem to evaluate the following function at $x = -2$.

$f(x) = 3x^3 + 8x^2 + 5x - 7$

Solution

Using synthetic division, you obtain the following.

-2	3	8	5	-7
		-6	-4	-2
	3	2	1	-9

Because the remainder is $r = -9$, you can conclude that

$f(-2) = -9$.

This means that $(-2, -9)$ is a point on the graph of f. Try checking this by substituting $x = -2$ in the original function.

Think About the Proof

To prove the Factor Theorem, you can use the Division Algorithm to write $f(x)$ as

$f(x) = (x - k)q(x) + r(x)$.

By the Remainder Theorem, you know that $r(x) = f(k)$. Thus,

$f(x) = (x - k)q(x) + f(k)$

where $q(x)$ is a polynomial of lesser degree than $f(x)$. How does this allow you to prove the theorem? The details of the proof are in the appendix.

Another important theorem is the **Factor Theorem,** which is stated below. This theorem states that you can test to see whether a polynomial has $(x - k)$ as a factor by evaluating the polynomial at $x = k$. If the result is 0, $(x - k)$ is a factor.

The Factor Theorem

A polynomial $f(x)$ has a factor $(x - k)$ if and only if $f(k) = 0$.

Think About the Proof Located in the margin adjacent to the corresponding theorem, each Think About the Proof feature offers strategies for proving the theorem. Detailed proofs for all theorems are given in Appendix B.

Technology Technology is integrated throughout the text at point of use as a tool for visualization, investigation, and verification. Instructions for using graphing utilities are given as necessary.

Study Tips Study Tips appear in the margin at point of use and offer students specific suggestions for studying algebra.

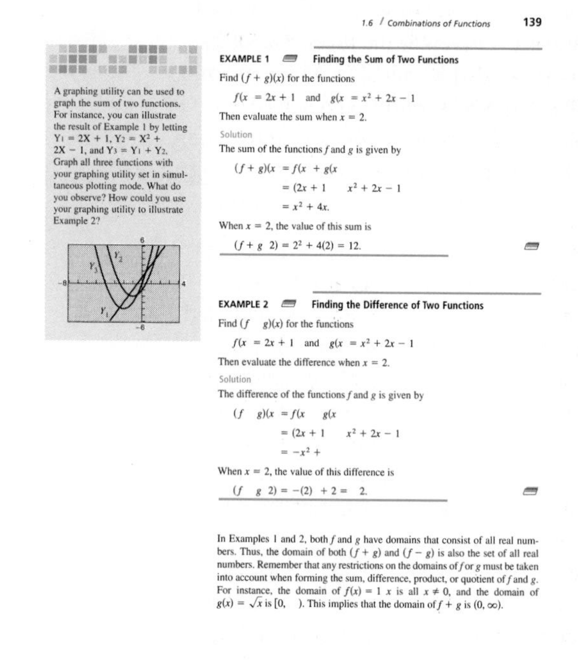

1.6 / Combinations of Functions 139

A graphing utility can be used to graph the sum of two functions. For instance, you can illustrate the result of Example 1 by letting $Y_1 = 2X + 1$, $Y_2 = X^2 + 2X - 1$, and $Y_3 = Y_1 + Y_2$. Graph all three functions with your graphing utility set in simultaneous plotting mode. What do you observe? How could you use your graphing utility to illustrate Example 2?

EXAMPLE 1 Finding the Sum of Two Functions

Find $(f + g)(x)$ for the functions

$f(x = 2x + 1$ and $g(x = x^2 + 2x - 1$

Then evaluate the sum when $x = 2$.

Solution

The sum of the functions f and g is given by

$(f + g)(x = f(x + g(x$

$= (2x + 1 \quad x^2 + 2x - 1$

$= x^2 + 4x.$

When $x = 2$, the value of this sum is

$(f + g \; 2) = 2^2 + 4(2) = 12.$

EXAMPLE 2 Finding the Difference of Two Functions

Find $(f \quad g)(x)$ for the functions

$f(x = 2x + 1$ and $g(x = x^2 + 2x - 1$

Then evaluate the difference when $x = 2$.

Solution

The difference of the functions f and g is given by

$(f \quad g)(x = f(x \quad g(x$

$= (2x + 1 \quad x^2 + 2x - 1$

$= -x^2 +$

When $x = 2$, the value of this difference is

$(f \quad g \; 2) = -(2) + 2 = \; 2.$

In Examples 1 and 2, both f and g have domains that consist of all real numbers. Thus, the domain of both $(f + g)$ and $(f - g)$ is also the set of all real numbers. Remember that any restrictions on the domains of f or g must be taken into account when forming the sum, difference, product, or quotient of f and g. For instance, the domain of $f(x) = 1 \; x$ is all $x \neq 0$, and the domain of $g(x) = \sqrt{x}$ is $[0, \quad)$. This implies that the domain of $f + g$ is $(0, \infty)$.

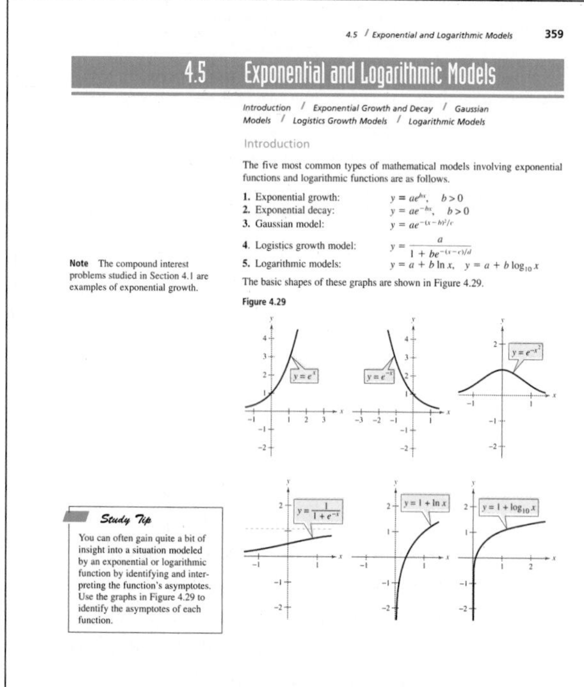

4.5 / Exponential and Logarithmic Models 359

4.5 Exponential and Logarithmic Models

Introduction / Exponential Growth and Decay / Gaussian Models / Logistics Growth Models / Logarithmic Models

Introduction

The five most common types of mathematical models involving exponential functions and logarithmic functions are as follows.

1. Exponential growth: $y = ae^{bx}, \quad b > 0$
2. Exponential decay: $y = ae^{-bx}, \quad b > 0$
3. Gaussian model: $y = ae^{-(x-b)^2/c}$
4. Logistics growth model: $y = \dfrac{a}{1 + be^{-(x-c)/d}}$
5. Logarithmic models: $y = a + b \ln x, \quad y = a + b \log_{10} x$

Note The compound interest problems studied in Section 4.1 are examples of exponential growth.

The basic shapes of these graphs are shown in Figure 4.29.

Figure 4.29

Study Tip

You can often gain quite a bit of insight into a situation modeled by an exponential or logarithmic function by identifying and interpreting the function's asymptotes. Use the graphs in Figure 4.29 to identify the asymptotes of each function.

2.3 / Complex Numbers 193

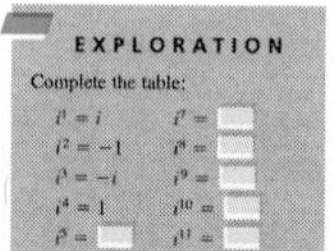

EXPLORATION

Complete the table:

$i^1 = i$	$i^7 =$
$i^2 = -1$	$i^8 =$
$i^3 = -i$	$i^9 =$
$i^4 = 1$	$i^{10} =$
$i^5 =$	$i^{11} =$
$i^6 =$	$i^{12} =$

What pattern do you see? Write a brief description of how you would find i raised to any positive integer power.

Many of the properties of real numbers are valid for complex numbers as well. Here are some examples.

Associative Property of Addition and Multiplication
Commutative Property of Addition and Multiplication
Distributive Property of Multiplication Over Addition

Notice how these properties are used when two complex numbers are multiplied.

$(a + bi)(c + di) = a(c + di) + bi(c + di)$ Distributive
$= ac + (ad)i + (bc)i + (bd)i^2$ Distributive
$= ac + (ad)i + (bc)i + (bd)(-1)$ Definition of i
$= ac - bd + (ad)i + (bc)i$ Commutative
$= (ac - bd) + (ad + bc)i$ Associative

Rather than trying to memorize this multiplication rule, we suggest that you simply remember how the distributive property is used to multiply two complex numbers. The procedure is similar to multiplying two polynomials and combining like terms (as in the FOIL Method).

EXAMPLE 2 Multiplying Complex Numbers

a. $(i)(-3i) = -3i^2$ Multiply.
$= -3(-1)$ Definition of i
$= 3$ Simplify.

b. $(2 - i)(4 + 3i) = 8 + 6i - 4i - 3i^2$ Product of binomials
$= 8 + 6i - 4i - 3(-1)$ Definition of i
$= 8 + 3 + 6i - 4i$ Group like terms.
$= 11 + 2i$ Standard form

c. $(3 + 2i)(3 - 2i) = 9 - 6i + 6i - 4i^2$ Product of binomials
$= 9 - 4(-1)$ Definition of i
$= 9 + 4$ Simplify.
$= 13$ Standard form

d. $(3 + 2i)^2 = 9 + 6i + 6i + 4i^2$ Product of binomials
$= 9 + 4(-1) + 12i$ Definition of i
$= 9 - 4 + 12i$ Simplify.
$= 5 + 12i$ Standard form

Exploration Throughout the text, the Exploration features encourage active participation by students, strengthening their intuition and critical thinking skills by exploring mathematical concepts and discovering mathematical relationships. Using a variety of approaches—including visualization, verification, use of graphing utilities, pattern recognition, and modeling—students are encouraged to develop a conceptual understanding of theoretical topics.

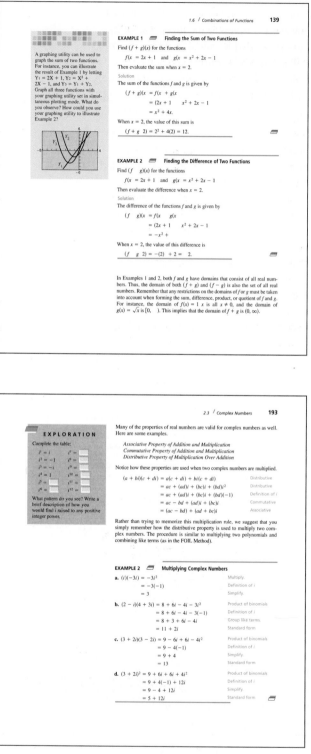

Historical Notes To help students understand that algebra has a past, historical notes featuring mathematicians and their work and mathematical artifacts are included in each chapter.

Graphics Visualization is a critical problem-solving skill. To encourage the development of this ability, the text has nearly 1500 figures in examples, exercises, and answers to exercises. Included are graphs of equations and functions, geometric figures, displays of statistical information, scatter plots, and numerous screen outputs from graphing technology. All graphs of equations and functions are computer- or calculator-generated for accuracy, and they are designed to resemble students' actual screen outputs as closely as possible. Graphics are also used to emphasize graphical interpretation, comparison, and estimation.

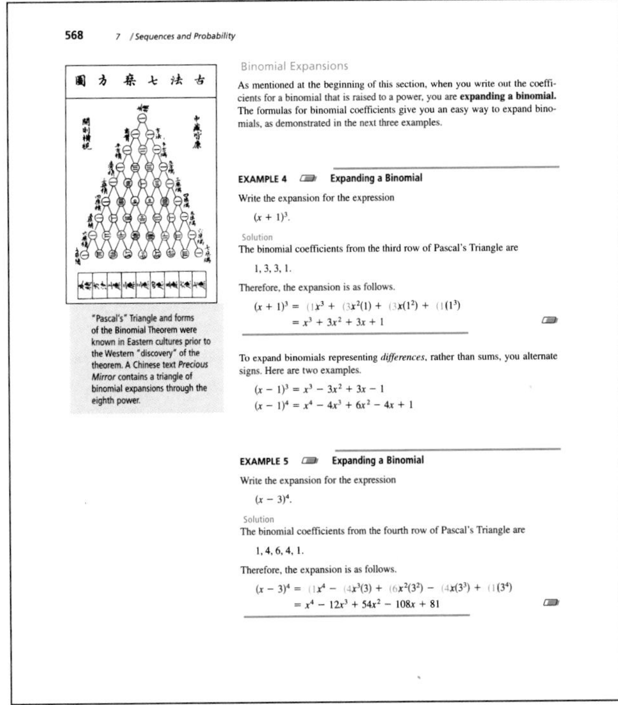

568 7 / Sequences and Probability

Binomial Expansions

As mentioned at the beginning of this section, when you write out the coefficients for a binomial that is raised to a power, you are **expanding a binomial.** The formulas for binomial coefficients give you an easy way to expand binomials, as demonstrated in the next three examples.

EXAMPLE 4 Expanding a Binomial

Write the expansion for the expression

$(x + 1)^3$.

Solution

The binomial coefficients from the third row of Pascal's Triangle are

1, 3, 3, 1.

Therefore, the expansion is as follows.

$$(x + 1)^3 = (1)x^3 + (3)x^2(1) + (3)x(1^2) + (1)(1^3) = x^3 + 3x^2 + 3x + 1$$

"Pascal's" Triangle and forms of the Binomial Theorem were known in Eastern cultures prior to the Western "discovery" of the theorem. A Chinese text *Precious Mirror* contains a triangle of binomial expansions through the eighth power.

To expand binomials representing *differences*, rather than sums, you alternate signs. Here are two examples.

$$(x - 1)^3 = x^3 - 3x^2 + 3x - 1$$
$$(x - 1)^4 = x^4 - 4x^3 + 6x^2 - 4x + 1$$

EXAMPLE 5 Expanding a Binomial

Write the expansion for the expression

$(x - 3)^4$.

Solution

The binomial coefficients from the fourth row of Pascal's Triangle are

1, 4, 6, 4, 1.

Therefore, the expansion is as follows.

$$(x - 3)^4 = (1)x^4 - (4)x^3(3) + (6)x^2(3^2) - (4)x(3^3) + (1)(3^4) = x^4 - 12x^3 + 54x^2 - 108x + 81$$

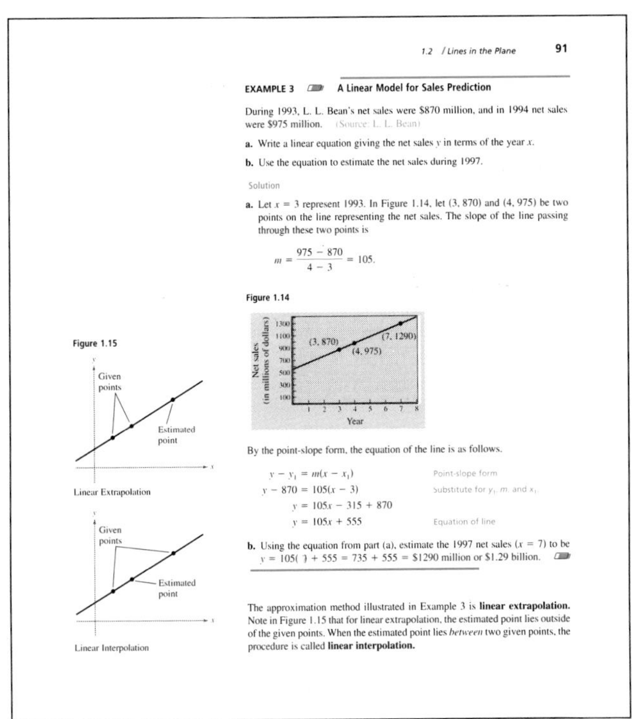

1.2 / Lines in the Plane 91

EXAMPLE 3 A Linear Model for Sales Prediction

During 1993, L. L. Bean's net sales were \$870 million, and in 1994 net sales were \$975 million. (Source: L. L. Bean)

a. Write a linear equation giving the net sales y in terms of the year x.

b. Use the equation to estimate the net sales during 1997.

Solution

a. Let $x = 3$ represent 1993. In Figure 1.14, let (3, 870) and (4, 975) be two points on the line representing the net sales. The slope of the line passing through these two points is

$$m = \frac{975 - 870}{4 - 3} = 105.$$

Figure 1.14

By the point-slope form, the equation of the line is as follows.

$$y - y_1 = m(x - x_1) \quad \text{Point-slope form}$$
$$y - 870 = 105(x - 3) \quad \text{Substitute for } y_1, m, \text{ and } x_1.$$
$$y = 105x - 315 + 870$$
$$y = 105x + 555 \quad \text{Equation of line}$$

b. Using the equation from part (a), estimate the 1997 net sales ($x = 7$) to be $y = 105(7) + 555 = 735 + 555 = \$1290$ million or \$1.29 billion.

The approximation method illustrated in Example 3 is **linear extrapolation.** Note in Figure 1.15 that for linear extrapolation, the estimated point lies outside of the given points. When the estimated point lies *between* two given points, the procedure is called **linear interpolation.**

Figure 1.15

Applications Real-life applications are integrated throughout the text in examples and exercises. These applications offer students constant review of problem-solving skills, and they emphasize the relevance of the mathematics. Many of the applications use recent, real data, and all are titled for easy reference. Photographs with captions in the introduction to the chapter also encourage students to see the link between mathematics and real life.

Examples Each of the more than 400 text examples was carefully chosen to illustrate a particular mathematical concept, problem-solving approach, or computational technique, and to enhance students' understanding. The examples in the text cover a wide variety of problem types, including theoretical problems, real-life applications (many with real data), and problems requiring the use of graphing technology. Each example is titled for easy reference, and real-life applications are labeled. Many examples include side comments in color that clarify the steps of the solution.

Problem Solving The text provides ample opportunity for students to hone their problem-solving skills. In both the exercises and the examples in the Second Edition, students are asked to apply verbal, analytical, graphical, and numerical approaches to problem solving. Students are also encouraged to use a graphing utility as a tool for solving problems. Students are taught the following approach to solving applied problems: (1) construct a verbal model; (2) label variable and constant terms; (3) construct an algebraic model; (4) using the model, solve the problem; and (5) check the answer in the original statement of the problem.

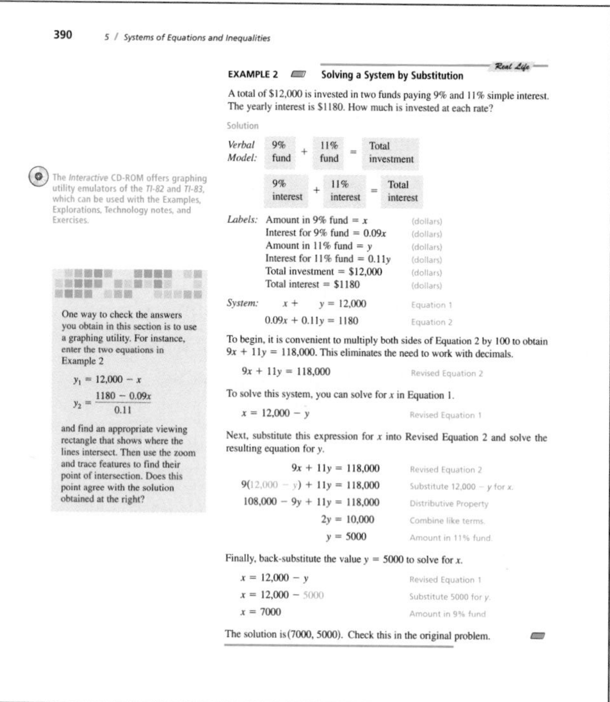

390 5 / Systems of Equations and Inequalities

The *Interactive* CD-ROM offers graphing utility emulators of the *TI-82* and *TI-83*, which can be used with the Examples, Explorations, Technology notes, and Exercises.

One way to check the answers you obtain in this section is to use a graphing utility. For instance, enter the two equations in Example 2

$y_1 = 12{,}000 - x$

$y_2 = \dfrac{1180 - 0.09x}{0.11}$

and find an appropriate viewing rectangle that shows where the lines intersect. Then use the zoom and trace features to find their point of intersection. Does this point agree with the solution obtained at the right?

Real Life

EXAMPLE 2 Solving a System by Substitution

A total of \$12,000 is invested in two funds paying 9% and 11% simple interest. The yearly interest is \$1180. How much is invested at each rate?

Solution

Verbal Model: 9% fund + 11% fund = Total investment

9% interest + 11% interest = Total interest

Labels:
Amount in 9% fund $= x$ (dollars)
Interest for 9% fund $= 0.09x$ (dollars)
Amount in 11% fund $= y$ (dollars)
Interest for 11% fund $= 0.11y$ (dollars)
Total investment $= \$12{,}000$ (dollars)
Total interest $= \$1180$ (dollars)

System:
$x + y = 12{,}000$ Equation 1
$0.09x + 0.11y = 1180$ Equation 2

To begin, it is convenient to multiply both sides of Equation 2 by 100 to obtain $9x + 11y = 118{,}000$. This eliminates the need to work with decimals.

$9x + 11y = 118{,}000$ Revised Equation 2

To solve this system, you can solve for x in Equation 1.

$x = 12{,}000 - y$ Revised Equation 1

Next, substitute this expression for x into Revised Equation 2 and solve the resulting equation for y.

$9x + 11y = 118{,}000$ Revised Equation 2
$9(12{,}000 - y) + 11y = 118{,}000$ Substitute 12,000 − y for x.
$108{,}000 - 9y + 11y = 118{,}000$ Distributive Property
$2y = 10{,}000$ Combine like terms.
$y = 5000$ Amount in 11% fund

Finally, back-substitute the value $y = 5000$ to solve for x.

$x = 12{,}000 - y$ Revised Equation 1
$x = 12{,}000 - 5000$ Substitute 5000 for y.
$x = 7000$ Amount in 9% fund

The solution is (7000, 5000). Check this in the original problem.

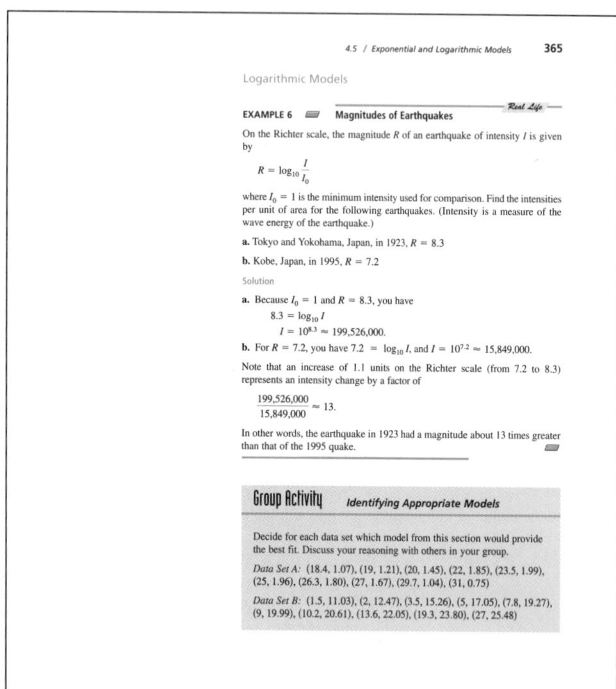

4.5 / Exponential and Logarithmic Models 365

Logarithmic Models

Real Life

EXAMPLE 6 Magnitudes of Earthquakes

On the Richter scale, the magnitude R of an earthquake of intensity I is given by

$$R = \log_{10} \frac{I}{I_0}$$

where $I_0 = 1$ is the minimum intensity used for comparison. Find the intensities per unit of area for the following earthquakes. (Intensity is a measure of the wave energy of the earthquake.)

a. Tokyo and Yokohama, Japan, in 1923, $R = 8.3$

b. Kobe, Japan, in 1995, $R = 7.2$

Solution

a. Because $I_0 = 1$ and $R = 8.3$, you have

$8.3 = \log_{10} I$

$I = 10^{8.3} \approx 199{,}526{,}000.$

b. For $R = 7.2$, you have $7.2 = \log_{10} I$, and $I = 10^{7.2} \approx 15{,}849{,}000.$

Note that an increase of 1.1 units on the Richter scale (from 7.2 to 8.3) represents an intensity change by a factor of

$$\frac{199{,}526{,}000}{15{,}849{,}000} \approx 13.$$

In other words, the earthquake in 1923 had a magnitude about 13 times greater than that of the 1995 quake.

Group Activity *Identifying Appropriate Models*

Decide for each data set which model from this section would provide the best fit. Discuss your reasoning with others in your group.

Data Set A: (18.4, 1.07), (19, 1.21), (20, 1.45), (22, 1.85), (23.5, 1.99), (25, 1.96), (26.3, 1.80), (27, 1.67), (29.7, 1.04), (31, 0.75)

Data Set B: (1.5, 11.03), (2, 12.47), (3.5, 15.26), (5, 17.05), (7.8, 19.27), (9, 19.99), (10.2, 20.61), (13.6, 22.05), (19.3, 23.80), (27, 25.48)

CD-ROM The icon refers to additional features of *Interactive College Algebra: A Graphing Approach* that enhance the text presentation, such as exercises, computer animations, examples, tests, and graphing calculator emulators.

Group Activities The Group Activities that appear at the ends of sections reinforce students' understanding by studying mathematical concepts in a variety of ways, including talking and writing about mathematics, creating and solving problems, analyzing errors, and developing and using mathematical models. Designed to be completed as group projects in class or as homework assignments, the Group Activities give students opportunities to do interactive learning and to think, talk, and write about mathematics.

Exercises The exercise sets were completely revised for the Second Edition. More than 5000 exercises with a broad range of conceptual, computational, and applied problems accommodate a variety of teaching and learning styles. Included in the section and review exercise sets are multipart, writing, and more challenging problems with extensive graphics that encourage exploration and discovery, enhance students' skills in mathematical modeling, estimation, and data interpretation and analysis, and encourage the use of graphing technology for conceptual understanding. Applications are labeled for easy reference. The exercise sets are designed to build competence, skill, and understanding; each exercise set is graded in difficulty to allow students to gain confidence as they progress. Detailed solutions to all odd-numbered exercises are given in the *Study and Solutions Guide;* answers to all odd-numbered exercises appear in the back of the text.

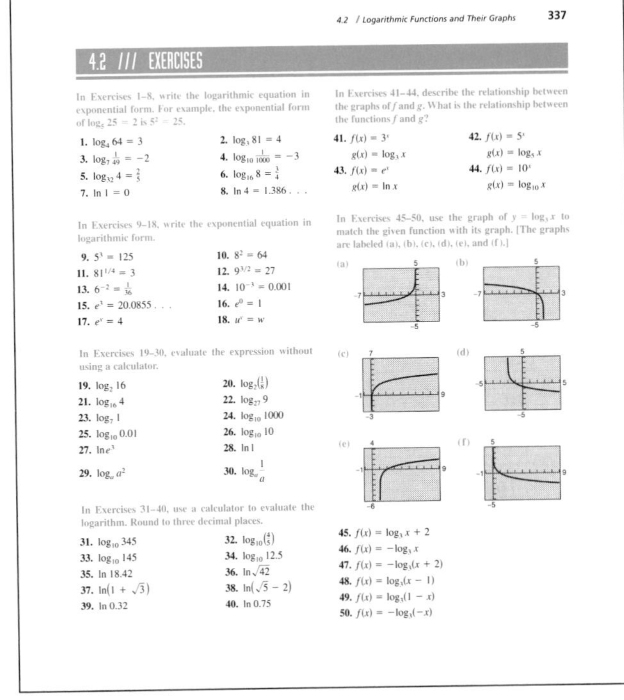

4.2 / *Logarithmic Functions and Their Graphs* 337

4.2 /// EXERCISES

In Exercises 1–8, write the logarithmic equation in exponential form. For example, the exponential form of $\log_5 25 = 2$ is $5^2 = 25$.

1. $\log_4 64 = 3$
2. $\log_3 81 = 4$
3. $\log_7 \frac{1}{49} = -2$
4. $\log_{10} \frac{1}{1000} = -3$
5. $\log_{32} 4 = \frac{2}{5}$
6. $\log_{16} 8 = \frac{3}{4}$
7. $\ln 1 = 0$
8. $\ln 4 = 1.386\ldots$

In Exercises 9–18, write the exponential equation in logarithmic form.

9. $5^3 = 125$
10. $8^2 = 64$
11. $81^{1/4} = 3$
12. $9^{3/2} = 27$
13. $6^{-2} = \frac{1}{36}$
14. $10^{-3} = 0.001$
15. $e^3 = 20.0855\ldots$
16. $e^0 = 1$
17. $e^x = 4$
18. $u^v = w$

In Exercises 19–30, evaluate the expression without using a calculator.

19. $\log_2 16$
20. $\log_2(\frac{1}{8})$
21. $\log_{16} 4$
22. $\log_{27} 9$
23. $\log_7 1$
24. $\log_{10} 1000$
25. $\log_{10} 0.01$
26. $\log_{10} 10$
27. $\ln e^3$
28. $\ln 1$
29. $\log_a a^2$
30. $\log_a \frac{1}{a}$

In Exercises 31–40, use a calculator to evaluate the logarithm. Round to three decimal places.

31. $\log_{10} 345$
32. $\log_{10}(\frac{4}{5})$
33. $\log_{10} 145$
34. $\log_{10} 12.5$
35. $\ln 18.42$
36. $\ln\sqrt{42}$
37. $\ln(1 + \sqrt{3})$
38. $\ln(\sqrt{5} - 2)$
39. $\ln 0.32$
40. $\ln 0.75$

In Exercises 41–44, describe the relationship between the graphs of f and g. What is the relationship between the functions f and g?

41. $f(x) = 3^x$, $g(x) = \log_3 x$
42. $f(x) = 5^x$, $g(x) = \log_5 x$
43. $f(x) = e^x$, $g(x) = \ln x$
44. $f(x) = 10^x$, $g(x) = \log_{10} x$

In Exercises 45–50, use the graph of $y = \log_3 x$ to match the given function with its graph. [The graphs are labeled (a), (b), (c), (d), (e), and (f).]

45. $f(x) = \log_3 x + 2$
46. $f(x) = -\log_3 x$
47. $f(x) = -\log_3(x + 2)$
48. $f(x) = \log_3(x - 1)$
49. $f(x) = \log_3(1 - x)$
50. $f(x) = -\log_3(-x)$

2.4 / *Solving Equations Algebraically* 215

Exploration In Exercises 111 and 112, find x such that the distance between the points is 13.

111. $(1, 2), (x, -10)$
112. $(-8, 0), (x, 5)$

113. *Oxygen Consumption* The metabolic rate of ectothermic organisms increases with increasing temperature within a certain range. Experimental data for oxygen consumption (microliters per gram per hour) of a beetle for certain temperatures yielded the model

$$C = 0.45x^2 - 1.65x + 50.75, \qquad 10 \le x \le 25$$

where x is the air temperature in degrees Celsius.

(a) Use a graphing utility to graph the consumption function over the specified domain.

(b) Use the graph to approximate the air temperature resulting in oxygen consumption of 150 microliters per gram per hour.

(c) If the temperature is increased from 10 to 20 degrees, the oxygen consumption is increased by approximately what factor?

114. *Saturated Steam* The temperature T (in degrees Fahrenheit) of saturated steam increases as pressure increases. This relationship is approximated by the model

$$T = 75.82 - 2.11x + 43.51\sqrt{x}, \qquad 5 \le x \le 40$$

where x is the absolute pressure in pounds per square inch.

(a) Use a graphing utility to graph the temperature function over the specified domain.

(b) The temperature of steam at sea level ($x = 14.696$) is 212°F. Evaluate the model at this pressure, and verify the result graphically.

(c) Use the model to approximate the pressure for a steam temperature of 240°F.

115. *Fuel Efficiency* The distance d a car can travel on one tank of fuel is approximated by the model

$$d = -0.024s^2 + 1.455s + 431.5, \quad 0 < s \le 75$$

where s is the average speed of the car.

(a) Use a graphing utility to graph the distance function over the specified domain.

(b) Use the graph to determine the greatest distance that can be traveled on a tank of fuel. How long will the trip take?

(c) Determine the greatest distance that can be traveled in this car in 8 hours with no refueling. How fast should the car be driven? [*Hint:* The distance traveled in 8 hours is $8s$. Graph this expression in the same viewing rectangle as the graph in part (a) and approximate the point of intersection.]

116. *Solving Graphically, Numerically, and Algebraically* A meteorologist is positioned 100 feet from the point where a weather balloon is launched. When the balloon is at height h, the distance d between the meteorologist and the balloon is given by $d = \sqrt{100^2 + h^2}$.

(a) Use a graphing utility to graph the equation. Use the trace feature to approximate the value of h when $d = 200$.

(b) Complete the table. Use the table to approximate the value of h when $d = 200$.

h	160	165	170	175	180	185
d						

(c) Find h algebraically when $d = 200$.

(d) Compare the results of each method. In each case, what information did you gain that wasn't revealed by another solution method?

In Exercises 117 and 118, solve for the variable.

117. *Surface Area of a Cone*
Solve for h: $S = \pi r\sqrt{r^2 + h^2}$

118. *Inductance*
Solve for Q: $i = \pm\sqrt{\frac{1}{LC}}\sqrt{Q^2 - q}$

In Exercises 119 and 120, consider an equation of the form $x + \sqrt{x - a} = b$, where a and b are constants.

119. *Exploration* Find a and b if the solution to the equation is $x = 20$. (There are many correct answers.)

120. *Essay* Write a short paragraph listing the steps required for solving an equation involving radicals.

Geometry Geometric formulas and concepts are reviewed throughout the text in examples, Group Activities, and exercises. For reference, common formulas are listed inside the back cover of this text.

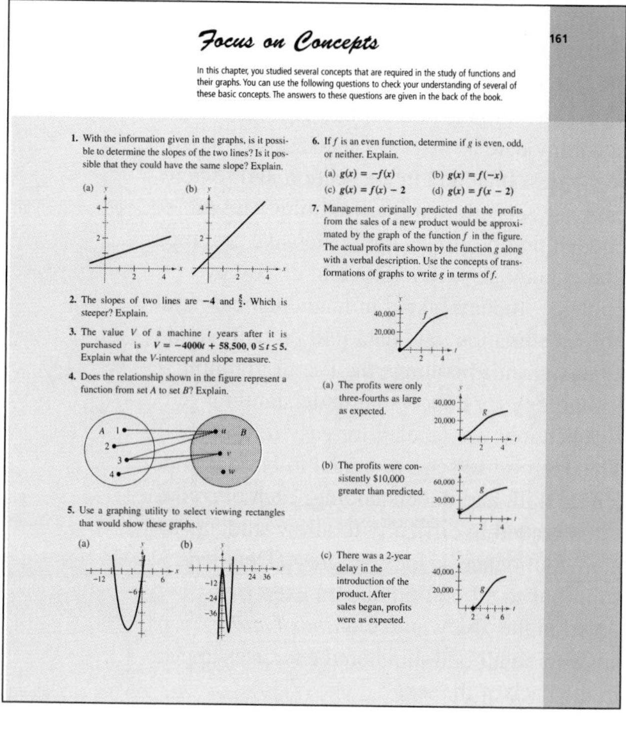

Focus on Concepts 161

In this chapter, you studied several concepts that are required in the study of functions and their graphs. You can use the following questions to check your understanding of several of these basic concepts. The answers to these questions are given in the back of the book.

1. With the information given in the graphs, is it possible to determine the slopes of the two lines? Is it possible that they could have the same slope? Explain.

2. The slopes of two lines are -4 and $\frac{5}{2}$. Which is steeper? Explain.

3. The value V of a machine t years after it is purchased is $V = -4000t + 58{,}500,\ 0 \le t \le 5$. Explain what the V-intercept and slope measure.

4. Does the relationship shown in the figure represent a function from set A to set B? Explain.

5. Use a graphing utility to select viewing rectangles that would show these graphs.

6. If f is an even function, determine if g is even, odd, or neither. Explain.
   (a) $g(x) = -f(x)$ (b) $g(x) = f(-x)$
   (c) $g(x) = f(x) - 2$ (d) $g(x) = f(x - 2)$

7. Management originally predicted that the profits from the sales of a new product would be approximated by the graph of the function f in the figure. The actual profits are shown by the function g along with a verbal description. Use the concepts of transformations of graphs to write g in terms of f.
   (a) The profits were only three-fourths as large as expected.
   (b) The profits were consistently $10,000 greater than predicted.
   (c) There was a 2-year delay in the introduction of the product. After sales began, profits were as expected.

Focus on Concepts Each Focus on Concepts feature is a set of exercises that test students' understanding of the basic concepts covered in the chapter. Answers to all questions are given in the back of the text.

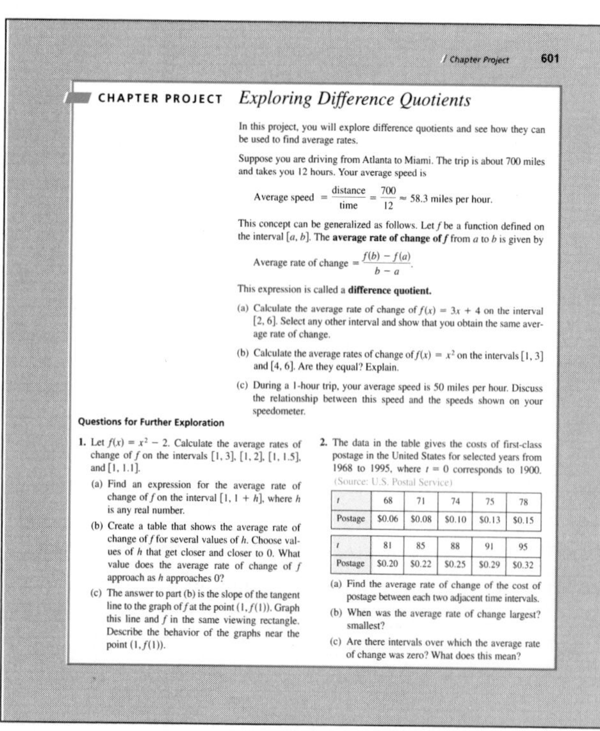

/ Chapter Project 601

CHAPTER PROJECT *Exploring Difference Quotients*

In this project, you will explore difference quotients and see how they can be used to find average rates.

Suppose you are driving from Atlanta to Miami. The trip is about 700 miles and takes you 12 hours. Your average speed is

$$\text{Average speed} = \frac{\text{distance}}{\text{time}} = \frac{700}{12} \approx 58.3 \text{ miles per hour.}$$

This concept can be generalized as follows. Let f be a function defined on the interval $[a, b]$. The **average rate of change of f** from a to b is given by

$$\text{Average rate of change} = \frac{f(b) - f(a)}{b - a}.$$

This expression is called a **difference quotient.**

(a) Calculate the average rate of change of $f(x) = 3x + 4$ on the interval $[2, 6]$. Select any other interval and show that you obtain the same average rate of change.

(b) Calculate the average rates of change of $f(x) = x^2$ on the intervals $[1, 3]$ and $[4, 6]$. Are they equal? Explain.

(c) During a 1-hour trip, your average speed is 50 miles per hour. Discuss the relationship between this speed and the speeds shown on your speedometer.

Questions for Further Exploration

1. Let $f(x) = x^2 - 2$. Calculate the average rates of change of f on the intervals $[1, 3]$, $[1, 2]$, $[1, 1.5]$, and $[1, 1.1]$.
   (a) Find an expression for the average rate of change of f on the interval $[1, 1 + h]$, where h is any real number.
   (b) Create a table that shows the average rate of change of f for several values of h. Choose values of h that get closer and closer to 0. What value does the average rate of change of f approach as h approaches 0?
   (c) The answer to part (b) is the slope of the tangent line to the graph of f at the point $(1, f(1))$. Graph this line and f in the same viewing rectangle. Describe the behavior of the graphs near the point $(1, f(1))$.

2. The data in the table gives the costs of first-class postage in the United States for selected years from 1968 to 1995, where $t = 0$ corresponds to 1900. (Source: U.S. Postal Service)

t	68	71	74	75	78
Postage	$0.06	$0.08	$0.10	$0.13	$0.15

t	81	85	88	91	95
Postage	$0.20	$0.22	$0.25	$0.29	$0.32

   (a) Find the average rate of change of the cost of postage between each two adjacent time intervals.
   (b) When was the average rate of change largest? smallest?
   (c) Are there intervals over which the average rate of change was zero? What does this mean?

Chapter Projects Chapter Projects are extended applications that use real data, graphs, and modeling to enhance students' understanding of mathematical concepts. Designed as individual or group projects, they offer additional opportunities to think, discuss, and write about mathematics. Many projects give students the opportunity to collect, analyze, and interpret data.

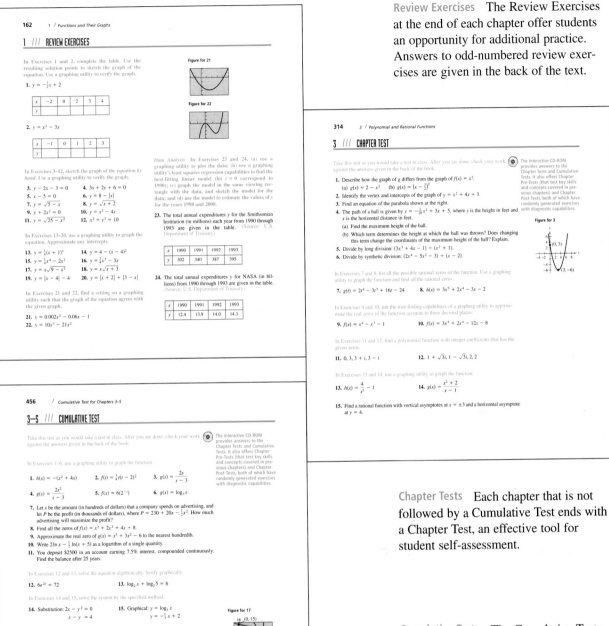

Review Exercises The Review Exercises at the end of each chapter offer students an opportunity for additional practice. Answers to odd-numbered review exercises are given in the back of the text.

162 1 / Functions and Their Graphs

1 /// REVIEW EXERCISES

In Exercises 1 and 2, complete the table. Use the resulting solution points to sketch the graph of the equation. Use a graphing utility to verify the graph.

1. $y = -\frac{1}{2}x + 2$

x	-2	0	2	3	4
y					

2. $y = x^2 - 3x$

x	-1	0	1	2	3
y					

In Exercises 3–12, sketch the graph of the equation *by hand*. Use a graphing utility to verify the graph.

3. $y - 2x - 3 = 0$
4. $3x + 2y + 6 = 0$
5. $x - 5 = 0$
6. $y = 8 - |x|$
7. $y = \sqrt{5 - x}$
8. $y = \sqrt{x + 2}$
9. $y + 2x^2 = 0$
10. $y = x^2 - 4x$
11. $y = \sqrt{25 - x^2}$
12. $x^2 + y^2 = 10$

In Exercises 13–20, use a graphing utility to graph the equation. Approximate any intercepts.

13. $y = \frac{1}{4}(x + 1)^3$
14. $y = 4 - (x - 4)^2$
15. $y = \frac{1}{4}x^4 - 2x^2$
16. $y = \frac{1}{4}x^3 - 3x$
17. $y = x\sqrt{9 - x^2}$
18. $y = x\sqrt{x + 3}$
19. $y = |x - 4| - 4$
20. $y = |x + 2| + |3 - x|$

In Exercises 21 and 22, find a setting on a graphing utility such that the graph of the equation agrees with the given graph.

21. $y = 0.002x^2 - 0.06x - 1$
22. $y = 10x^3 - 21x^2$

Figure for 21

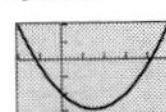

Figure for 22

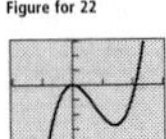

Data Analysis In Exercises 23 and 24, (a) use a graphing utility to plot the data; (b) use a graphing utility's least squares regression capabilities to find the best-fitting linear model (let $t = 0$ correspond to 1990); (c) graph the model in the same viewing rectangle with the data, and sketch the model for the data; and (d) use the model to estimate the values of y for the years 1998 and 2000.

23. The total annual expenditures y for the Smithsonian Institution (in millions) each year from 1990 through 1993 are given in the table. (Source: U.S. Department of Treasury)

x	1990	1991	1992	1993
y	302	340	387	395

24. The total annual expenditures y for NASA (in billions) from 1990 through 1993 are given in the table. (Source: U.S. Department of Treasury)

x	1990	1991	1992	1993
y	12.4	13.9	14.0	14.3

Chapter Tests Each chapter that is not followed by a Cumulative Test ends with a Chapter Test, an effective tool for student self-assessment.

314 3 / Polynomial and Rational Functions

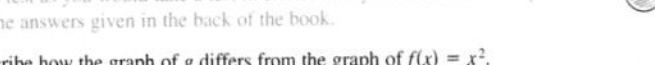

3 /// CHAPTER TEST

Take this test as you would take a test in class. After you are done, check your work against the answers given in the back of the book.

The *Interactive* CD-ROM provides answers to the Chapter Tests and Cumulative Tests. It also offers Chapter Pre-Tests (that test key skills and concepts covered in previous chapters) and Chapter Post-Tests, both of which have randomly generated exercises with diagnostic capabilities.

1. Describe how the graph of g differs from the graph of $f(x) = x^2$.
   (a) $g(x) = 2 - x^2$ (b) $g(x) = \left(x - \frac{3}{2}\right)^2$
2. Identify the vertex and intercepts of the graph of $y = x^2 + 4x + 3$.
3. Find an equation of the parabola shown at the right.
4. The path of a ball is given by $y = -\frac{1}{20}x^2 + 3x + 5$, where y is the height in feet and x is the horizontal distance in feet.
   (a) Find the maximum height of the ball.
   (b) Which term determines the height at which the ball was thrown? Does changing this term change the coordinates of the maximum height of the ball? Explain.
5. Divide by long division: $(3x^3 + 4x - 1) \div (x^2 + 1)$.
6. Divide by synthetic division: $(2x^4 - 5x^2 - 3) \div (x - 2)$.

Figure for 3

In Exercises 7 and 8, list all the possible rational zeros of the function. Use a graphing utility to graph the function and find all the rational zeros.

7. $g(t) = 2t^4 - 3t^3 + 16t - 24$
8. $h(x) = 3x^5 + 2x^4 - 3x - 2$

In Exercises 9 and 10, use the root-finding capabilities of a graphing utility to approximate the real zeros of the function accurate to three decimal places.

9. $f(x) = x^4 - x^3 - 1$
10. $f(x) = 3x^5 + 2x^4 - 12x - 8$

In Exercises 11 and 12, find a polynomial function with integer coefficients that has the given zeros.

11. $0, 3, 3 + i, 3 - i$
12. $1 + \sqrt{3}i, 1 - \sqrt{3}i, 2, 2$

In Exercises 13 and 14, use a graphing utility to graph the function.

13. $h(x) = \frac{4}{x^2} - 1$
14. $g(x) = \frac{x^2 + 2}{x - 1}$

15. Find a rational function with vertical asymptotes at $x = \pm 3$ and a horizontal asymptote at $y = 4$.

Cumulative Tests The Cumulative Tests that follow Chapters 2, 5, and 8 help students judge their mastery of previously covered material as well as reinforce the knowledge they have been accumulating throughout the text—preparing them for other exams and for future courses.

456 / Cumulative Test for Chapters 3–5

3–5 /// CUMULATIVE TEST

Take this test as you would take a test in class. After you are done, check your work against the answers given in the back of the book.

The *Interactive* CD-ROM provides answers to the Chapter Tests and Cumulative Tests. It also offers Chapter Pre-Tests (that test key skills and concepts covered in previous chapters) and Chapter Post-Tests, both of which have randomly generated exercises with diagnostic capabilities.

In Exercises 1–6, use a graphing utility to graph the function.

1. $h(x) = -(x^2 + 4x)$
2. $f(t) = \frac{1}{4}t(t - 2)^2$
3. $g(s) = \frac{2s}{s - 3}$
4. $g(s) = \frac{2s^2}{s - 3}$
5. $f(x) = 6(2^{-x})$
6. $g(x) = \log_3 x$
7. Let x be the amount (in hundreds of dollars) that a company spends on advertising, and let P be the profit (in thousands of dollars), where $P = 230 + 20x - \frac{1}{2}x^2$. How much advertising will maximize the profit?
8. Find all the zeros of $f(x) = x^3 + 2x^2 + 4x + 8$.
9. Approximate the real zero of $g(x) = x^3 + 3x^2 - 6$ to the nearest hundredth.
10. Write $2 \ln x - \frac{1}{2} \ln(x + 5)$ as a logarithm of a single quantity.
11. You deposit \$2500 in an account earning 7.5% interest, compounded continuously. Find the balance after 25 years.

In Exercises 12 and 13, solve the equation algebraically. Verify graphically.

12. $6e^{2x} = 72$
13. $\log_2 x + \log_2 5 = 6$

In Exercises 14 and 15, solve the system by the specified method.

14. Substitution: $2x - y^2 = 0$, $x - y = 4$
15. Graphical: $y = \log_3 x$, $y = -\frac{1}{3}x + 2$
16. Find the equation of the parabola $y = ax^2 + bx + c$ passing through the points $(0, 6)$, $(-2, 2)$, and $\left(3, \frac{9}{2}\right)$.
17. Derive a set of inequalities to describe the region shown in the figure.
18. A merchant plans to sell two models of compact disc players. One model sells for \$275 and yields a profit of \$55, and the other model sells for \$400 and yields a profit of \$75. The merchant estimates that the total monthly demand will not exceed 300 units. The merchant does not want to invest more than \$100,000 in inventory for these products. Find the number of units of each model that should be stocked in order to maximize profit.

Figure for 17

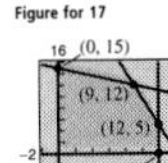

Supplements

College Algebra: A Graphing Approach, Second Edition, by Larson, Hostetler, and Edwards is accompanied by a comprehensive supplements package. Most items are keyed to the text.

Printed Resources

For the student

Study and Solutions Guide by Bruce Edwards, University of Florida, and Dianna L. Zook, Indiana University—Purdue University at Fort Wayne

- Section summaries of key concepts
- Detailed, step-by-step solutions to all odd-numbered exercises
- Key solution steps for Chapter Tests and Cumulative Tests
- Practice tests with solutions
- Study strategies

Graphing Technology Keystroke Guide: Precalculus

- Keystroke instructions for a wide variety of Texas Instruments, Casio, Sharp, and Hewlett-Packard graphing calculators—including the most current models.
- Examples with step-by-step solutions
- Extensive graphics screen output
- Technology tips

For the instructor

Instructor's Annotated Edition

- Includes the entire student edition of the text, with the student answers section
- Instructor's Answers section: Answers to all even-numbered exercises, and answers to all Explorations, Technology exercises, Group Activities, and Chapter Project exercises
- Annotations at point of use offer specific teaching strategies and suggestions for implementing Group Activities, point out common student errors, and give additional examples, exercises, class activities, and group activities.

Solutions to Even-Numbered Exercises

- Detailed, step-by-step solutions to even-numbered exercises

Test Item File and Instructor's Resource Guide

- Printed test bank with approximately 2000 test items (multiple-choice, open-ended, and writing) coded by level of difficulty
- Technology-required test items coded for easy reference
- Bank of chapter test forms with answer keys

- Two final exam test forms
- Notes to the instructor, including materials for alternative assessment and managing the multicultural and cooperative-learning classrooms

Problem Solving, Modeling, and Data Analysis Labs by Wendy Metzger, Palomar College

- Multipart, guided discovery activities and applications
- Keystroke instructions for Derive and *TI-82*
- Keyed to the text by topic
- Funded in part by NSF (National Science Foundation, Instrumentation and Laboratory Improvement) and California Community College Fund for Instructional Improvement

Media Resources

For the student

Interactive College Algebra: A Graphing Approach (See pages xviii–xx for a description, or visit the Houghton Mifflin home page at http://www.hmco.com for a preview.)

- Interactive, multimedia CD-ROM format
- IBM-PC for Windows

Tutor software

- Interactive tutorial software keyed to the text by section
- Diagnostic feedback
- Chapter self-tests
- Guided exercises with step-by-step solutions
- Glossary

Videotapes by Dana Mosely

- Comprehensive, text-specific coverage keyed to the text by section
- Real-life application vignettes introduced where appropriate
- Computer-generated animation
- For media/resource centers
- Additional explanation of concepts, sample problems, and applications
- Instructional graphing calculator videotape also available

For the instructor

Computerized Testing (IBM, Macintosh, Windows)

- New on-line testing
- New grade-management capabilities
- Algorithmic test-generating software provides an unlimited number of tests
- Approximately 2000 test items
- Also available as a printed test bank

Transparency Package

- 50 color transparencies color-coded by topic

Interactive College Algebra: A Graphing Approach

To accommodate a variety of teaching and learning styles, *College Algebra: A Graphing Approach* is also available in a multimedia, CD-ROM format. In this interactive format, the text offers the student additional tutorial assistance with

- Complete solutions to all odd-numbered text exercises.
- Chapter pre-tests, self-tests, and post-tests.
- *TI-82* and *TI-83* emulators.
- Guided examples with step-by-step solutions.
- Editable graphs.
- Animations of mathematical concepts.
- Section and tutorial exercises.
- Glossary of key terms.

These and other pedagogical features of the CD-ROM are illustrated by the screen dumps shown below.

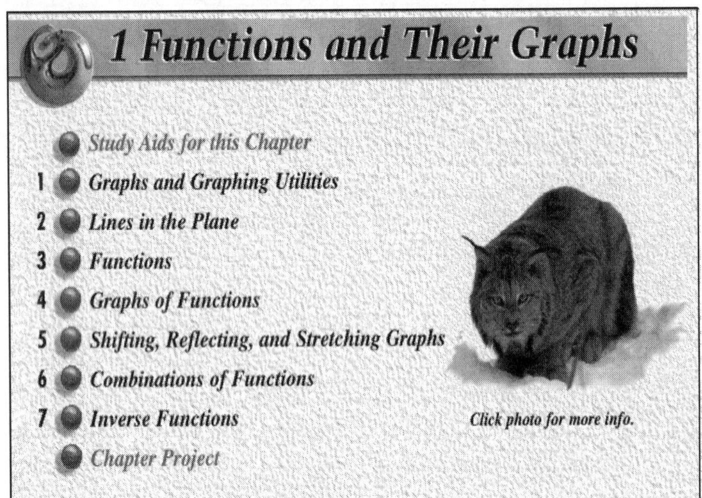

Chapter Topics Each chapter begins with an outline of the topics to be covered. Using the buttons at the bottom of the screen, the student can quickly move to the appropriate section.

Introductory Chapter Application Each chapter opens with a real-data application that illustrates the key concepts and techniques to be covered. Clicking on the photo, the student can access additional data and background information that frames the real-world context for a mathematical concept.

Chapter Project Each chapter is accompanied by a Chapter Project. This offers the student the opportunity to synthesize the algebraic techniques and concepts studied in the chapter. Many projects use real data and emphasize data analysis and mathematical modeling.

Study Aids Each section offers the student an array of additional study aids, including Chapter Pre-, Post-, and Self-Tests, Review Exercises, and Focus on Concepts. With diagnostics, complete solutions, or answers, these helpful features promote the focused practice needed to master mathematical concepts. Short, informative video segments are also included.

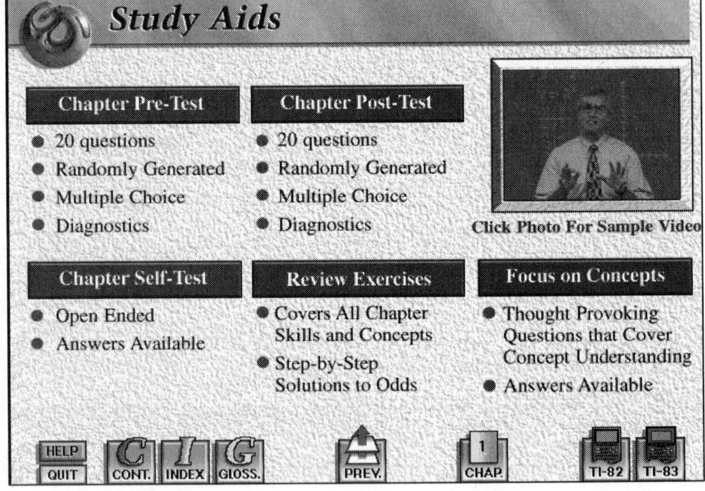

Examples *Interactive College Algebra: A Graphing Approach* illustrates mathematical concepts by featuring all of the Examples found in the text. Guided Examples with step-by-step solutions that appear one line at a time offer additional opportunities for practice and skill development. Group Activities, Exercises, and Integrated Examples help synthesize the concepts in the section.

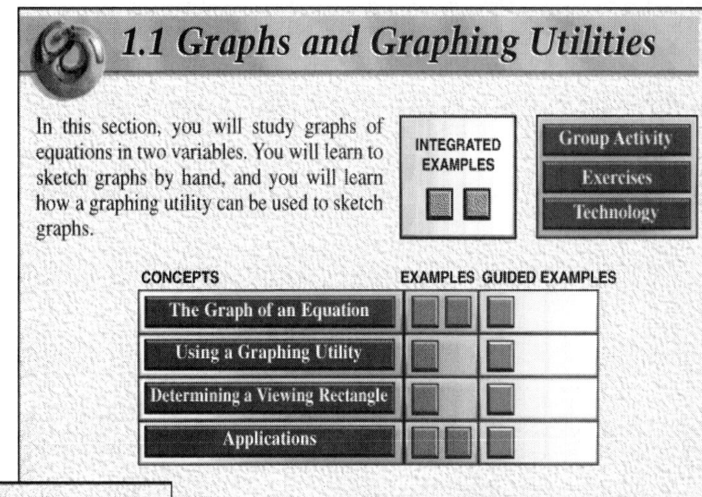

Example ***Sketching the Graph of an Equation***

Sketch the graph of $y = 7 - 3x$.

Solution

The simplest way to sketch the graph of an equation is the *point-plotting method.* With this method, you construct a **table of values** that consists of several solution points of the equation. For instance, when $x = 0$,

$$y = 7 - 3(0) = 7,$$

which implies that $(0, 7)$ is a solution point of the graph.

After plotting the points given in the table, you can see that they appear to lie on a line, as shown in the figure. The graph of the equation is the line that passes through the five plotted points.

(0, 7) (1, 4) (2, 1) (3, −2) (4, −5)

EDIT WINDOW ZOOM TRACE

Try It!

HELP QUIT CONT. INDEX GLOSS. PREV. CHAP. SECT. TI-82 TI-83

Graphs Some examples are accompanied by editable graphs for exploration and discovery. Using the keys below the graph, the student can change the function, the graphing window, and the x- and y-scales, and can trace and zoom.

Try It! After studying the worked-out example, the student can use the Try It! button to access similar examples—with solutions following on separate screens—to test his or her mastery of mathematical concepts and techniques.

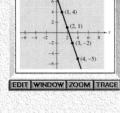

TI-82 and *TI-83* Emulators Accessible on every screen, the *TI-82* and *TI-83* emulators give instant access to graphing utilities as tools for computation and exploration. They are also available for working exercises in the text that require the use of a graphing utility. Instruction on using the emulators is also included at the click of a button.

These emulators were developed and copyrighted by Meridian Creative Group with the prior written permission of Texas Instruments.

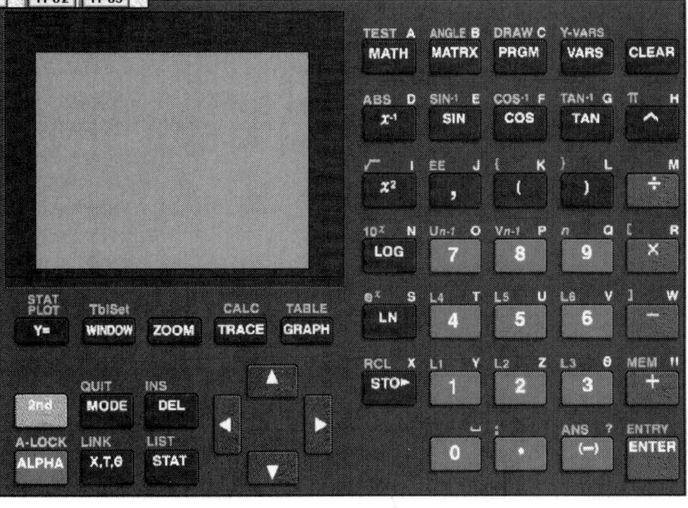

In Exercises 1 and 2, complete a table of values. Use the solution points to sketch the graph of the equation.

1. $y = -\frac{1}{2}x + 2$
2. $y = x^2 - 3x$

In Exercises 3–12, sketch the graph *by hand.*

3. $y - 2x - 3 = 0$
4. $3x + 2y + 6 = 0$
5. $x - 5 = 0$
6. $y = 8 - |x|$
7. $y = \sqrt{5 - x}$
8. $y = \sqrt{x + 2}$
9. $y + 2x^2 = 0$
10. $y = x^2 - 4x$
11. $y = \sqrt{25 - x^2}$
12. $x^2 + y^2 = 10$

HELP QUIT CONT. INDEX GLOSS.

PREV.

Section Exercises Each section is accompanied by a comprehensive set of exercises promoting skills mastery and conceptual understanding. Solutions to all odd-numbered exercises are available for instant feedback.

Tutorial Exercises Every section has a set of exercises in a multiple-choice format that offer students additional practice. Examples and diagnostics enhance this guided practice.

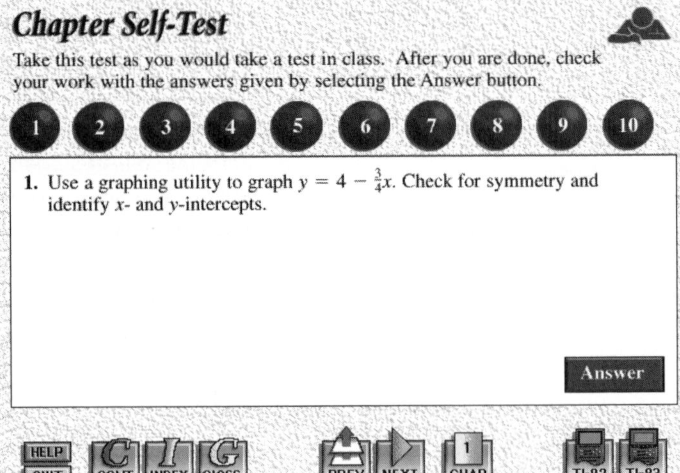
Chapter Self-Test
Take this test as you would take a test in class. After you are done, check your work with the answers given by selecting the Answer button.
1 2 3 4 5 6 7 8 9 10
1. Use a graphing utility to graph $y = 4 - \frac{3}{4}x$. Check for symmetry and identify x- and y-intercepts.
Answer
HELP QUIT CONT. INDEX GLOSS. PREV. NEXT CHAP. TI-82 TI-83

Tests Every chapter of the interactive text includes tests that are different from those in the textbook: Chapter Pre-Tests (testing key skills and concepts covered in previous chapters) and Chapter Post-Tests test mastery of the material covered in the textbook. The Chapter Self-Tests from the text are also included. Answers to all tests are included.

The *Interactive* CD-ROM shows every example with its solution; clicking on the *Try It!* button brings up similar problems. Guided Examples and Integrated Examples show step-by-step solutions to additional examples. Integrated Examples are related to several concepts in the section.

Length of
Height of
Length of

Equation: $\frac{x}{170.25} =$

$x =$

Thus, the World Trade C

Interactive College Algebra: A Graphing Approach supports the mathematical presentation in the text *College Algebra: A Graphing Approach,* Second Edition, with a variety of tutorial, diagnostic, and demonstration features. Throughout both the student text and the Instructor's Annotated Edition, CD-ROM icons identify these additional functions of the interactive text, as illustrated by the sample text page (at left).

Acknowledgments

We would like to thank the many people who have helped us at various stages of this project to prepare the text and supplements package. Their encouragement, criticisms, and suggestions have been invaluable to us.

Second Edition Reviewers: Daniel D. Anderson, University of Iowa; Anne E. Brown, Indiana University–South Bend; Eunice Everett, Seminole Community College; Jeff Frost, Johnson County Community College; Khadiga H. Gamgoum, Northern Virginia Community College; Michele Greenfield, Middlesex County College; Zenas Hartvigson, University of Colorado at Denver; Allen Hesse, Rochester Community College; Jean M. Horn, Northern Virginia Community College; Bill Huston, Missouri Western State College; Francine Winston Johnson, Howard Community College; Peter A. Lappan, Michigan State University; Judy McInerney, Sandhills Community College; Roger B. Nelsen, Lewis and Clark College; Jon Odell, Richland Community College; Wing M. Park, College of Lake County; Robert Pearce, South Plains College; George W. Shultz, St. Petersburg Junior College; Cathryn U. Stark, Collin County Community College; G. Bryan Stewart, Tarrant County Junior College; Mahbobeh Vezvaei, Kent State University; Joel E. Wilson, Eastern Kentucky University; and Karl M. Zilm, Lewis and Clark Community College.

Second Edition Survey Respondents: Marwan A. Abu-Sawwa, Florida Community College at Jacksonville; Barbara C. Armenta, Pima Community College—East; Gladwin E. Bartel, Otero Junior College; Carole A. Bauer, Triton College; Joyce M. Becker, Luther College; Marybeth Beno, South Suburban College; Charles M. Biles, Humboldt State University; Ruthane Bopp, Lake Forest College; Tim Chappell, North Central Missouri College; Michael Davidson, Cabrillo College; Diane L. Doyle, Adirondack Community College; Donna S. Fatheree, University of Louisiana at Lafayette; John R. Formsma, Los Angeles City College; John S. Frohliger, Saint Norbert College; Gary Glaze, Eastern Washington University; Irwin S. Goldfine, Truman College; Elise M. Grabner, Slippery Rock University; Donnie Hallstone, Green River Community College; Lois E. Higbie, Brookdale Community College; Susan S. Hollar, Kalamazoo Valley Community College; Fran Hopf, Hillsborough Community College; Margaret D. Hovde, Grossmont College; John F. Keating, Massasoit Community College; John G. LaMaster, Indiana University–Purdue University at Fort Wayne; Giles Wilson Maloof, Boise State University, Kenneth Mangels, Concordia State University; Peggy I. Miller, University of Nebraska at Kearney; Gilbert F. Orr, University of Southern Colorado; Jim Paige, Wayne State College; Elise Price, Tarrant County Junior College; Doris Schraeder, McLennan Community College; Linda Schultz, McHenry County College; Fay Sewell, Montgomery County Community

College; Patricia G. Shelton, North Carolina A & T State University; Hazel Shows, Hinds Community College; Joseph F. Stokes, Western Kentucky University; Diane Van Nostrand, University of Tulsa; and Raymond D. Wuco, San Joaquin Delta College.

Thanks to all of the people at Houghton Mifflin Company who worked with us in the development and production of the text, especially Chris Hoag, Sponsoring Editor; Cathy Cantin, Managing Editor; Maureen Brooks, Senior Associate Editor; Carolyn Johnson, Assistant Editor; Karen Carter, Senior Project Editor; Rachel Wimberly, Associate Project Editor; Gary Crespo, Art Supervisor; Lisa Merrill, Production Supervisor; Ros Kane, Marketing Associate; and Carrie Lipscomb, Editorial Assistant.

We would also like to thank the staff at Larson Texts, Inc. who assisted with proofreading the manuscript, preparing and proofreading the art package, and checking and typesetting the supplements.

On a personal level, we are grateful to our wives, Deanna Gilbert Larson, Eloise Hostetler, and Consuelo Edwards, for their love, patience, and support. Also, special thanks go to R. Scott O'Neil.

If you have suggestions for improving the text, please feel free to write to us. Over the past two decades, we have received many useful comments from both instructors and students, and we value these very much.

Roland E. Larson
Robert P. Hostetler
Bruce H. Edwards

Contents

P Prerequisites 1

1 Functions and Their Graphs 75

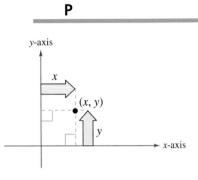

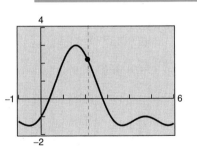

2 Intercepts, Zeros, and Solutions 167

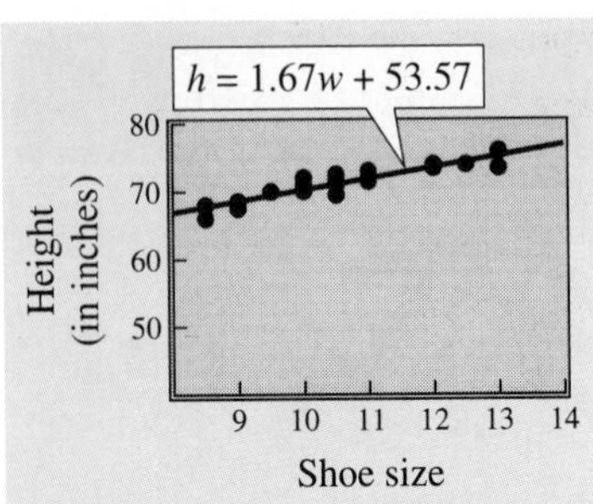

3 Polynomial and Rational Functions 243

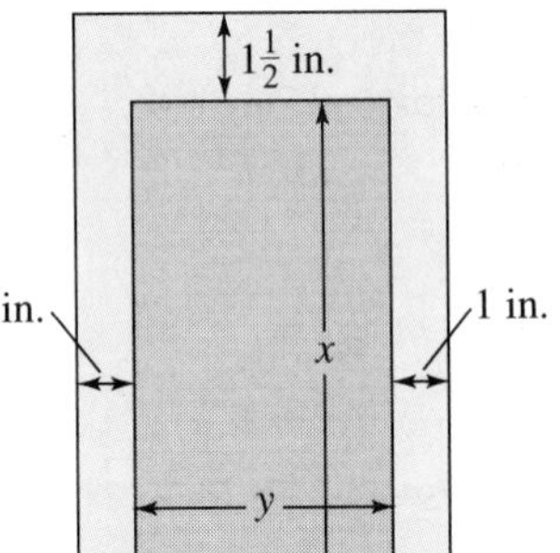

4 Exponential and Logarithmic Functions 315

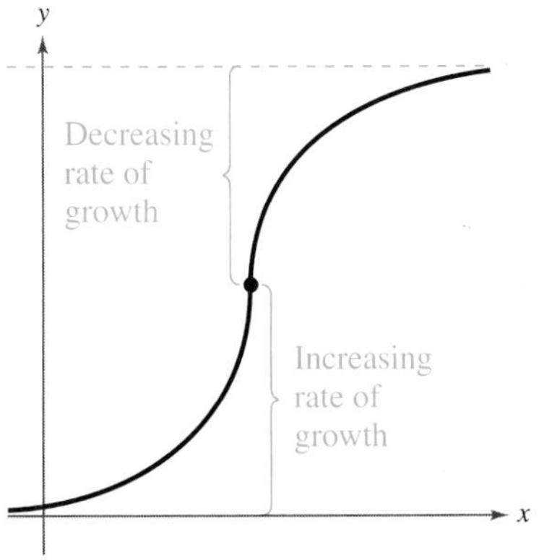

5 Systems of Equations and Inequalities 387

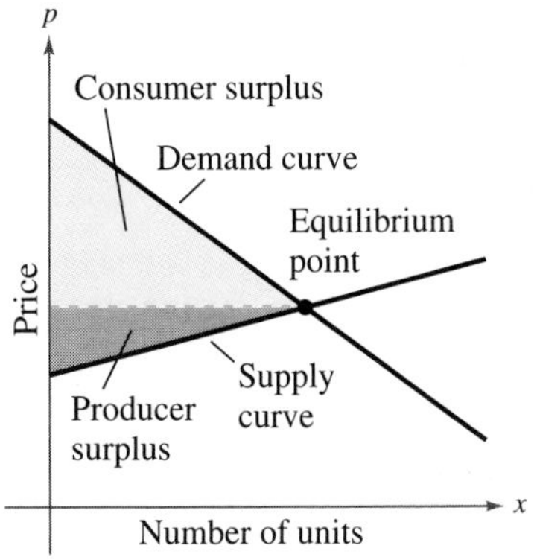

6 Matrices and Determinants 457

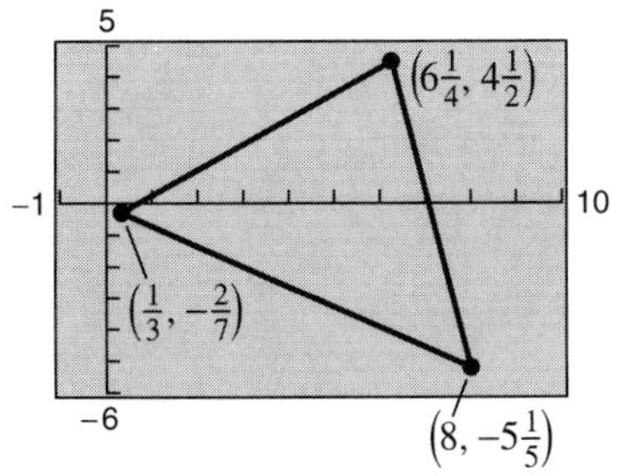

7 Sequences and Probability 523

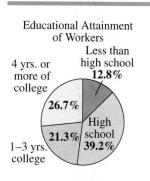

8 Conics and Parametric Equations 603

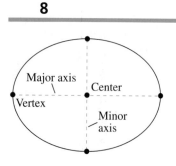

Appendices

Answers

Indices

College Algebra

Chapter P

Prerequisites

The concept for a museum dedicated to the heritage of rock and roll began in 1983 with a group of people from the music industry. This group created the Rock and Roll Hall of Fame Foundation.

The following data shows the numbers of recording artists who were elected to the Rock and Roll Hall of Fame from 1986 through 1995.

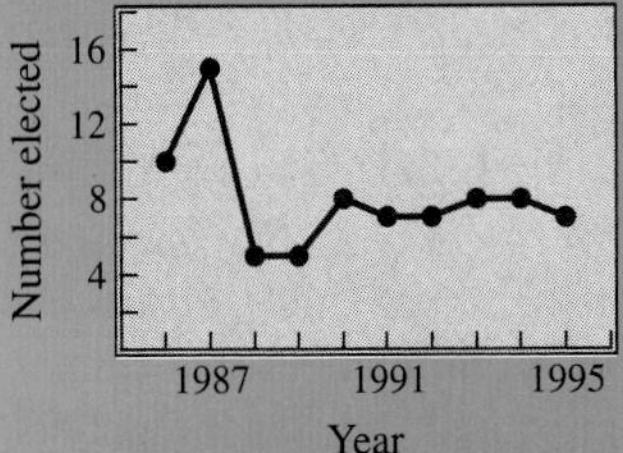

1986	(10)	1987	(15)	1988	(5)
1989	(5)	1990	(8)	1991	(7)
1992	(7)	1993	(8)	1994	(8)
1995	(7)				

You can use a graphing utility to represent this data visually as a *line graph*, as shown at the left, which is useful for showing trends in data over periods of time. (See Exercise 89 on page 58.)

The Plain Dealer, Cleveland, Ohio; (inset) Ingbet Grittner.

The Rock and Roll Hall of Fame and Museum is located in Cleveland, Ohio. The 150,000 square foot building was designed by I. M. Pei. The museum opened in September 1995.

P.1 Real Numbers

Real Numbers / Ordering Real Numbers / Absolute Value and Distance / Algebraic Expressions / Basic Rules of Algebra

Real Numbers

Real numbers are used in everyday life to describe quantities such as age, miles per gallon, container size, and population. To represent real numbers, you can use symbols such as 9, 0, $\frac{4}{3}$, 0.666 . . . , 28.21, $\sqrt{2}$, π, and $\sqrt[3]{-32}$.

Here are some important subsets of the real numbers.

$\{1, 2, 3, 4, \ldots\}$ Set of natural numbers

$\{0, 1, 2, 3, 4, \ldots\}$ Set of whole numbers

$\{\ldots -3, -2, -1, 0, 1, 2, 3, \ldots\}$ Set of integers

A real number is **rational** if it can be written as the ratio p/q of two integers, where $q \neq 0$. For instance, the numbers

$$\frac{1}{3} = 0.3333 \ldots, \quad \frac{1}{8} = 0.125, \quad \text{and} \quad \frac{125}{111} = 1.126126 \ldots$$

are rational. The decimal representation of a rational number either *repeats* (as in 3.1454545 . . .) or *terminates* (as in $\frac{1}{2} = 0.5$). A real number that cannot be written as the ratio of two integers is called **irrational.** Irrational numbers have infinite *nonrepeating* decimal representations. For instance, the numbers

$$\sqrt{2} \approx 1.4142136 \quad \text{and} \quad \pi \approx 3.1415927$$

are irrational. (The symbol $\approx$ means "is approximately equal to.")

Real numbers are represented graphically by a **real number line.** The point 0 on the real number line is the **origin.** Numbers to the right of 0 are positive, and numbers to the left of 0 are negative, as shown in Figure P.1. The term **nonnegative** describes a number that is either positive or zero.

Figure P.1 The Real Number Line

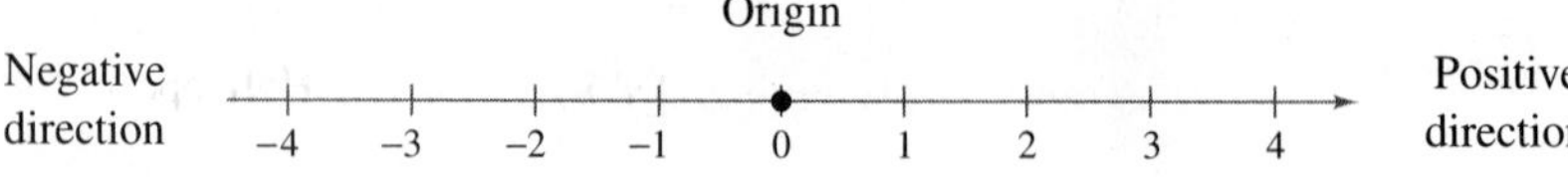

As illustrated in Figure P.2, there is a *one-to-one correspondence* between real numbers and points on the real number line.

Figure P.2 One-to-One Correspondence

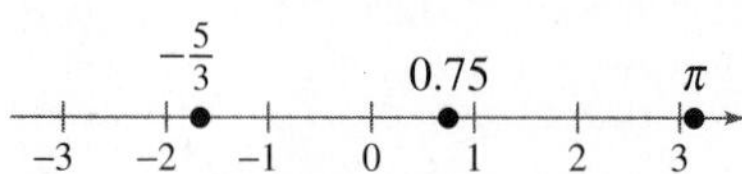

Every real number corresponds to exactly one point on the real number line.

Every point on the real number line corresponds to exactly one real number.

Ordering Real Numbers

One important property of real numbers is that they are **ordered.**

> **Definition of Order on the Real Number Line**
>
> If a and b are real numbers, a is **less than** b if $b - a$ is positive. This order is denoted by the **inequality**
>
> $$a < b.$$
>
> This can also be described by saying that b is **greater than** a and writing $b > a$. The inequality $a \leq b$ means that a is **less than or equal to** b, and the inequality $b \geq a$ means that b is **greater than or equal to** a. The symbols $<$, $>$, $\leq$, and $\geq$ are **inequality symbols.**

Figure P.3 $a < b$ if and only if a lies to the left of b.

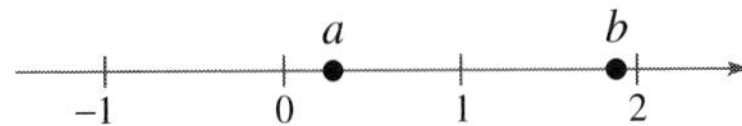

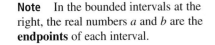

Geometrically, this definition implies that $a < b$ if and only if a lies to the *left* of b on the real number line, as shown in Figure P.3.

Figure P.4

(a) $x \leq 2$

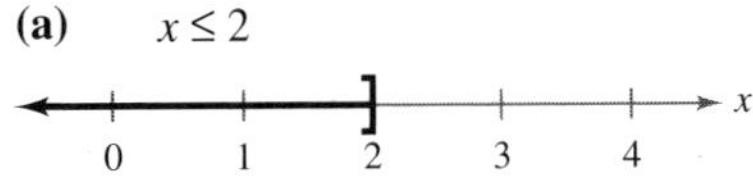

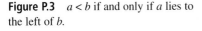

(b) $-2 \leq x < 3$

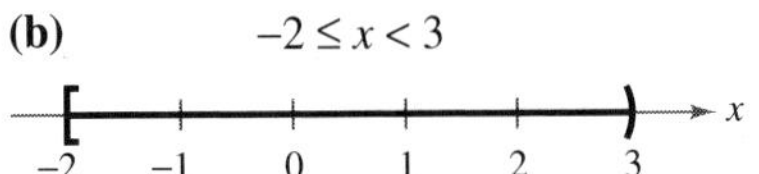

EXAMPLE 1 **Interpreting Inequalities**

a. The inequality $x \leq 2$ denotes all real numbers less than or equal to 2, as shown in Figure P.4(a).

b. The inequality $-2 \leq x < 3$ means that $x \geq -2$ *and* $x < 3$. The "double inequality" denotes all real numbers between -2 and 3, including -2 but *not* including 3, as shown in Figure P.4(b).

Inequalities can be used to describe subsets of real numbers called **intervals.**

Bounded Intervals on the Real Number Line

Notation	*Interval Type*	*Inequality*	*Graph*
$[a, b]$	Closed	$a \leq x \leq b$	a b
(a, b)	Open	$a < x < b$	a b
$[a, b)$	Half-open	$a \leq x < b$	a b
$(a, b]$	Half-open	$a < x \leq b$	a b

Note In the bounded intervals at the right, the real numbers a and b are the **endpoints** of each interval.

Unbounded Intervals on the Real Number Line

Notation	*Interval Type*	*Inequality*	*Graph*
$[a, \infty)$	Half-open	$x \geq a$	
(a, ∞)	Open	$x > a$	
$(-\infty, b]$	Half-open	$x \leq b$	
$(-\infty, b)$	Open	$x < b$	
$(-\infty, \infty)$	Entire real line		

Note The symbols ∞, **positive infinity,** and $-\infty$, **negative infinity,** do not represent real numbers. They are simply convenient symbols used to describe the unboundedness of an interval such as $(1, \infty)$ or $(-\infty, 3]$.

The *Interactive* CD-ROM shows every example with its solution; clicking on the *Try It!* button brings up similar problems. Guided Examples and Integrated Examples show step-by-step solutions to additional examples. Integrated Examples are related to several concepts in the section.

EXAMPLE 2 Using Inequalities to Represent Intervals

Use inequality notation to describe each of the following.

a. c is at most 2.

b. All x in the interval $(-3, 5]$

Solution

a. The statement "c is at most 2" can be represented by $c \leq 2$.

b. "All x in the interval $(-3, 5]$" can be represented by $-3 < x \leq 5$.

EXAMPLE 3 Interpreting Intervals

Give a verbal description of each interval.

a. $(-1, 0)$ **b.** $[2, \infty)$ **c.** $(-\infty, 0)$

Solution

a. This interval consists of all real numbers that are greater than -1 and less than 0.

b. This interval consists of all real numbers that are greater than or equal to 2.

c. This interval consists of all negative real numbers.

The **Law of Trichotomy** states that for any two real numbers a and b, *precisely* one of three relationships is possible:

$$a = b, \quad a < b, \quad \text{or} \quad a > b. \qquad \text{Law of Trichotomy}$$

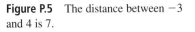

Absolute Value and Distance

The **absolute value** of a real number is its *magnitude*.

Definition of Absolute Value

If a is a real number, the **absolute value** of a is

$$|a| = \begin{cases} a, & \text{if } a \geq 0 \\ -a, & \text{if } a < 0. \end{cases}$$

Notice from this definition that the absolute value of a real number is never negative. For instance, if $a = -5$, then $|-5| = -(-5) = 5$.

EXPLORATION

Absolute value expressions can be evaluated on a graphing utility. To evaluate $|-4|$ with a *TI-82*, use the keystrokes below:

[ABS] [(−)] 4 [ENTER]

To evaluate $|-4|$ with a *TI-83*, use these keystrokes:

[MATH] (NUM) (1:abs( ) [(−)] 4 [ENTER]

When evaluating an expression such as $|3 - 8|$, parentheses should surround the entire expression. Evaluate each expression below. What can you conclude?

a. $|6|$ b. $|-1|$

c. $|5 - 2|$ d. $|2 - 5|$

Note The absolute value of a real number is either positive or zero. Moreover, 0 is the only real number whose absolute value is 0. Thus, $|0| = 0$.

EXAMPLE 4 Evaluating the Absolute Value of a Number

Evaluate $\dfrac{|x|}{x}$ for (a) $x > 0$ and (b) $x < 0$.

Solution

a. If $x > 0$, then $|x| = x$ and $\dfrac{|x|}{x} = \dfrac{x}{x} = 1$.

b. If $x < 0$, then $|x| = -x$ and $\dfrac{|x|}{x} = \dfrac{-x}{x} = -1$.

Properties of Absolute Value

1. $|a| \geq 0$ 2. $|-a| = |a|$

3. $|ab| = |a||b|$ 4. $\left|\dfrac{a}{b}\right| = \dfrac{|a|}{|b|}, \quad b \neq 0$

Absolute value can be used to define the distance between two numbers on the real number line. For instance, the distance between -3 and 4 is $|-3 - 4| = |-7| = 7$, as shown in Figure P.5.

Figure P.5 The distance between -3 and 4 is 7.

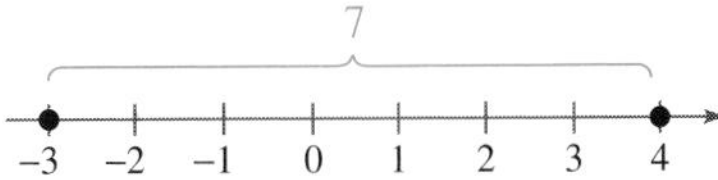

Distance Between Two Points on the Real Line

Let a and b be real numbers. The **distance between a and b** is

$$d(a, b) = |b - a| = |a - b|.$$

Algebraic Expressions

Study Tip

When an algebraic expression is evaluated, the Substitution Principle is used. It states, "If $a = b$, then a can be replaced by b in any expression involving a." In the first evaluation shown at the right, for instance, 3 is *substituted* for x in the expression $-3x + 5$.

One characteristic of algebra is the use of letters to represent numbers. The letters are **variables,** and combinations of letters and numbers are **algebraic expressions.** Here are a few examples of algebraic expressions.

$$5x, \quad 2x - 3, \quad \frac{4}{x^2 + 2}, \quad 7x + y$$

Definition of an Algebraic Expression

A collection of letters (**variables**) and real numbers (**constants**) combined using the operations of addition, subtraction, multiplication, division, and exponentiation is an **algebraic expression.**

The French mathematician Nicolas Chuquet (ca. 1500) wrote *Triparty en la science des nombres,* in which a form of exponent notation was used. Our expressions $6x^3$ and $10x^2$ were written as $.6.^3$ and $.10.^2$. Zero and negative exponents were also represented, so x^0 would be written as $.1.^0$ and $3x^{-2}$ as $.3.^{2.m}$. Chuquet wrote that $.72.^1$ divided by $.8.^3$ is $.9.^{2.m}$. That is, $72x \div 8x^3 = 9x^{-2}$.

The **terms** of an algebraic expression are those parts that are separated by *addition.* For example,

$$x^2 - 5x + 8 = x^2 + (-5x) + 8$$

has three terms: x^2 and $-5x$ are the **variable terms** and 8 is the **constant term.** The numerical factor of a variable term is the **coefficient** of the variable term. For instance, the coefficient of $-5x$ is -5, and the coefficient of x^2 is 1.

To **evaluate** an algebraic expression, substitute numerical values for each of the variables in the expression. Here are two examples.

Expression	*Value of Variable*	*Substitute*	*Value of Expression*
$-3x + 5$	$x = 3$	$-3(3) + 5$	$-9 + 5 = -4$
$3x^2 + 2x - 1$	$x = -1$	$3(-1)^2 + 2(-1) - 1$	$3 - 2 - 1 = 0$

Basic Rules of Algebra

There are four arithmetic operations with real numbers: **addition, multiplication, subtraction,** and **division,** denoted by the symbols $+$, $\times$ or $\cdot$, $-$, and $\div$. Of these, addition and multiplication are the two primary operations. Subtraction and division are the inverse operations of addition and multiplication, respectively.

Subtraction

$$a - b = a + (-b)$$

Division

$$\text{If } b \neq 0, \text{ then } a \div b = a\left(\frac{1}{b}\right) = \frac{a}{b}.$$

In these definitions, $-b$ is the **additive inverse** (or opposite) of b, and $1/b$ is the **multiplicative inverse** (or reciprocal) of b. In the fractional form a/b, a is the **numerator** of the fraction and b is the **denominator.**

Be sure you see that the following **basic rules of algebra** are true for variables and algebraic expressions as well as for real numbers. Try to formulate a verbal description of each property. For instance, the first property states that *the order in which two real numbers are added does not affect their sum.*

Basic Rules of Algebra

Let a, b, and c be real numbers, variables, or algebraic expressions.

Property		*Example*
Commutative Property of Addition:	$a + b = b + a$	$4x + x^2 = x^2 + 4x$
Commutative Property of Multiplication:	$ab = ba$	$(4 - x)x^2 = x^2(4 - x)$
Associative Property of Addition:	$(a + b) + c = a + (b + c)$	$(x + 5) + x^2 = x + (5 + x^2)$
Associative Property of Multiplication:	$(ab)c = a(bc)$	$(2x \cdot 3y)(8) = (2x)(3y \cdot 8)$
Distributive Properties:	$a(b + c) = ab + ac$	$3x(5 + 2x) = 3x \cdot 5 + 3x \cdot 2x$
	$(a + b)c = ac + bc$	$(y + 8)y = y \cdot y + 8 \cdot y$
Additive Identity Property:	$a + 0 = a$	$5y^2 + 0 = 5y^2$
Multiplicative Identity Property:	$a \cdot 1 = a$	$(4x^2)(1) = 4x^2$
Additive Inverse Property:	$a + (-a) = 0$	$5x^3 + (-5x^3) = 0$
Multiplicative Inverse Property:	$a \cdot \frac{1}{a} = 1, \quad a \neq 0$	$(x^2 + 4)\left(\frac{1}{x^2 + 4}\right) = 1$

Note Because subtraction is defined as "adding the opposite," the Distributive Properties are also true for subtraction. For instance, the "subtraction form" of $a(b + c) = ab + ac$ is

$$a(b - c) = ab - ac.$$

As well as formulating a verbal description for each of the following basic properties of negation, zero, and fractions, try to gain an *intuitive sense* for the validity of each.

Properties of Negation

Let a and b be real numbers, variables, or algebraic expressions.

	Property	*Example*
1.	$(-1)a = -a$	$(-1)7 = -7$
2.	$-(-a) = a$	$-(-6) = 6$
3.	$(-a)b = -(ab) = a(-b)$	$(-5)3 = -(5 \cdot 3) = 5(-3)$
4.	$(-a)(-b) = ab$	$(-2)(-x) = 2x$
5.	$-(a + b) = (-a) + (-b)$	$-(x + 8) = (-x) + (-8)$
		$= -x - 8$

Note Be sure you see the difference between the *opposite of a number* and a *negative number.* If a is already negative, then its opposite, $-a$, is positive. For instance, if $a = -5$, then $-a = -(-5) = 5$.

Note The "or" in the Zero-Factor Property includes the possibility that either or both factors may be zero. This is an **inclusive or,** and it is the way the word "or" is generally used in mathematics.

Properties of Zero

Let a and b be real numbers, variables, or algebraic expressions.

1. $a + 0 = a$ and $a - 0 = a$
2. $a \cdot 0 = 0$
3. $\dfrac{0}{a} = 0, \quad a \neq 0$
4. $\dfrac{a}{0}$ is undefined.
5. **Zero-Factor Property:** If $ab = 0$, then $a = 0$ or $b = 0$.

Note In Property 1, the phrase "if and only if" implies two statements. One statement is: If $a/b = c/d$, then $ad = bc$. The other statement is: If $ad = bc$, where $b \neq 0$ and $d \neq 0$, then $a/b = c/d$.

Properties of Fractions

Let a, b, c, and d be real numbers, variables, or algebraic expressions such that $b \neq 0$ and $d \neq 0$.

1. **Equivalent Fractions:** $\dfrac{a}{b} = \dfrac{c}{d}$ if and only if $ad = bc$.
2. **Rules of Signs:** $-\dfrac{a}{b} = \dfrac{-a}{b} = \dfrac{a}{-b}$ and $\dfrac{-a}{-b} = \dfrac{a}{b}$
3. **Generate Equivalent Fractions:** $\dfrac{a}{b} = \dfrac{ac}{bc}, \quad c \neq 0$
4. **Add or Subtract with Like Denominators:** $\dfrac{a}{b} \pm \dfrac{c}{b} = \dfrac{a \pm c}{b}$
5. **Add or Subtract with Unlike Denominators:** $\dfrac{a}{b} \pm \dfrac{c}{d} = \dfrac{ad \pm bc}{bd}$
6. **Multiply Fractions:** $\dfrac{a}{b} \cdot \dfrac{c}{d} = \dfrac{ac}{bd}$
7. **Divide Fractions:** $\dfrac{a}{b} \div \dfrac{c}{d} = \dfrac{a}{b} \cdot \dfrac{d}{c} = \dfrac{ad}{bc}, \quad c \neq 0$

EXAMPLE 5 Properties of Fractions

a. $\dfrac{x}{5} = \dfrac{3 \cdot x}{3 \cdot 5} = \dfrac{3x}{15}$ Generate equivalent fractions.

b. $\dfrac{x}{3} + \dfrac{2x}{5} = \dfrac{5 \cdot x + 3 \cdot 2x}{15}$ Add fractions with unlike denominators.

c. $\dfrac{7}{x} \div \dfrac{3}{2} = \dfrac{7}{x} \cdot \dfrac{2}{3} = \dfrac{14}{3x}$ Divide fractions.

Properties of Equality

Let a, b, and c be real numbers, variables, or algebraic expressions.

1. If $a = b$, then $a + c = b + c$. — Add c to both sides.
2. If $a = b$, then $ac = bc$. — Multiply both sides by c.
3. If $a + c = b + c$, then $a = b$. — Subtract c from both sides.
4. If $ac = bc$ and $c \neq 0$, then $a = b$. — Divide both sides by c.

If a, b, and c are integers such that $ab = c$, then a and b are **factors** or **divisors** of c. A **prime number** is a positive integer that has exactly two positive factors: itself and 1. For example, 2, 3, 5, 7, and 11 are prime numbers. The numbers 4, 6, 8, 9, and 10 are **composite** because they can be written as the product of two or more prime numbers. The number 1 is neither prime nor composite. The **Fundamental Theorem of Arithmetic** states that every positive integer greater than 1 can be written as the product of prime numbers in precisely one way (disregarding order). For instance, the *prime factorization* of 24 is $24 = 2 \cdot 2 \cdot 2 \cdot 3$.

When adding or subtracting fractions with unlike denominators, you have two options. You can use Property 5 of fractions as in Example 5(b), or you can rewrite the fractions with like denominators. Here is an example.

$$\frac{2}{15} - \frac{5}{9} + \frac{4}{5} = \frac{2(3)}{15(3)} - \frac{5(5)}{9(5)} + \frac{4(9)}{5(9)} \qquad \text{The LCD is 45.}$$
$$= \frac{6 - 25 + 36}{45}$$
$$= \frac{17}{45}$$

Study Tip

An important use of technology is in verifying your hand calculations. For instance, try evaluating $\frac{2}{15} - \frac{5}{9} + \frac{4}{5}$ with your calculator. You should obtain 0.3777777778, which is a decimal approximation of the answer obtained at the right, $\frac{17}{45}$.

Group Activity

Finding Prime Numbers

Discuss how you could decide if a given integer is a prime number. Are any of the following numbers prime? Explain how you were able to decide.

381 773 1741 2043

Work together to make a list of all the prime numbers less than 300. Describe the search method you used. You might want to consider using technology to facilitate your search. (Check with your instructor for hints on how to use technology in this way.)

The *Interactive* CD-ROM contains step-by-step solutions to all odd-numbered Section and Review Exercises. It also provides Tutorial Exercises, which link to Guided Examples for additional help.

P.1 /// EXERCISES

In Exercises 1–6, determine which numbers are (a) natural numbers, (b) integers, (c) rational numbers, and (d) irrational numbers.

1. $-9, -\frac{7}{2}, 5, \frac{2}{3}, \sqrt{2}, 0, 1$
2. $\sqrt{5}, -7, -\frac{7}{3}, 0, 3.12, \frac{5}{4}$
3. $2.01, 0.666 \ldots, -13, 0.010110111 \ldots$
4. $2.30300030003 \ldots, 0.7575, -4.63, \sqrt{10}$
5. $-\pi, -\frac{1}{3}, \frac{6}{3}, \frac{1}{2}\sqrt{2}, -7.5$
6. $25, -17, -\frac{12}{5}, \sqrt{9}, 3.12, \frac{1}{2}\pi$

In Exercises 7–10, use a calculator to find the decimal form of the rational number. If it is a nonterminating decimal, write the repeating pattern.

7. $\frac{5}{8}$
8. $\frac{1}{3}$
9. $\frac{41}{333}$
10. $\frac{6}{11}$

In Exercises 11 and 12, approximate the numbers and place the correct symbol (< or >) between them.

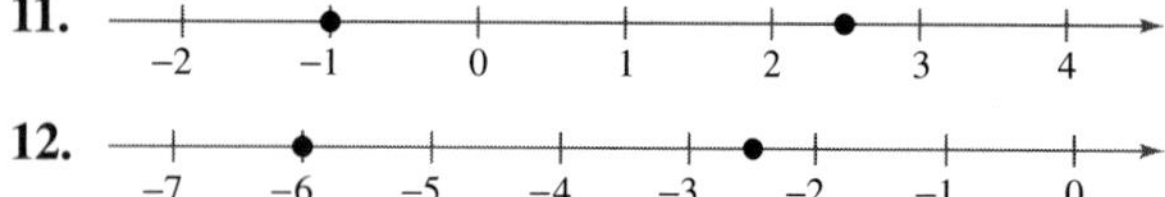

In Exercises 13–18, plot the two real numbers on the real number line. Then place the appropriate inequality symbol (< or >) between them.

13. $\frac{3}{2}, 7$
14. $-3.5, 1$
15. $-4, -8$
16. $1, \frac{16}{3}$
17. $\frac{5}{6}, \frac{2}{3}$
18. $-\frac{8}{7}, -\frac{3}{7}$

In Exercises 19–28, verbally describe the subset of real numbers represented by the inequality. Then sketch the subset on the real number line. State whether the interval is bounded or unbounded.

19. $x \leq 5$
20. $x \geq -2$
21. $x < 0$
22. $x > 3$
23. $x \geq 4$
24. $x < 2$
25. $-2 < x < 2$
26. $0 \leq x \leq 5$
27. $-1 \leq x < 0$
28. $0 < x \leq 6$

In Exercises 29 and 30, use a calculator to order the numbers from smallest to largest.

29. $\frac{7071}{5000}, \frac{584}{413}, \sqrt{2}, \frac{47}{33}, \frac{127}{90}$
30. $\frac{26}{15}, \sqrt{3}, 1.7320, \frac{381}{220}, \sqrt{10} - \sqrt{2}$

In Exercises 31–36, use inequality and interval notation to describe the set.

31. x is negative.
32. z is at least 10.
33. y is nonnegative.
34. y is no more than 25.
35. The person's age A is at least 30.
36. The annual rate of inflation r is expected to be at least 2.5%, but no more than 5%.

In Exercises 37–46, evaluate the expression.

37. $|-10|$
38. $|0|$
39. $|3 - \pi|$
40. $|4 - \pi|$
41. $\dfrac{-5}{|-5|}$
42. $-3 - |-3|$
43. $-3|-3|$
44. $|-1| - |-2|$
45. $-|16.25| + 20$
46. $2|33|$

In Exercises 47–52, place the correct symbol ($<$, $>$, or $=$) between the pair of real numbers.

47. $|-3|$ ▇ $-|-3|$ **48.** $|-4|$ ▇ $|4|$

49. -5 ▇ $-|5|$ **50.** $-|-6|$ ▇ $|-6|$

51. $-|-2|$ ▇ $-|2|$ **52.** $-(-2)$ ▇ -2

In Exercises 53–60, find the distance between a and b.

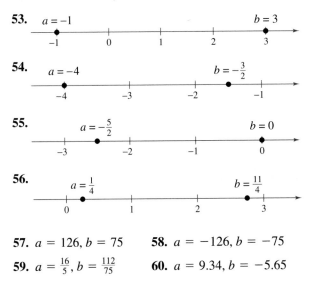

53. $a = -1$, $b = 3$

(number line: −1, 0, 1, 2, 3)

54.

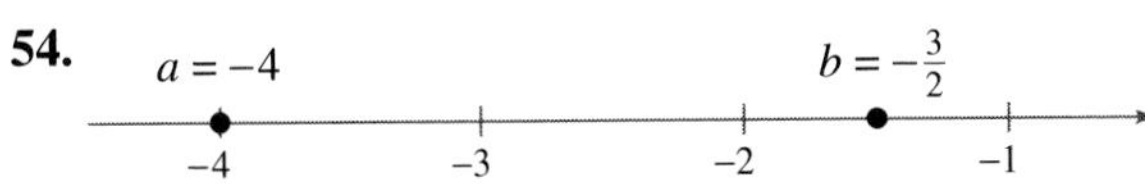

55. $a = -\frac{5}{2}$, $b = 0$

(number line: −3, −2, −1, 0)

56. $a = \frac{1}{4}$, $b = \frac{11}{4}$

(number line: 0, 1, 2, 3)

57. $a = 126, b = 75$ **58.** $a = -126, b = -75$

59. $a = \frac{16}{5}, b = \frac{112}{75}$ **60.** $a = 9.34, b = -5.65$

In Exercises 61–66, use absolute value notation to describe the situation.

61. The distance between x and 5 is no more than 3.

62. The distance between x and -10 is at least 6.

63. While traveling, you pass milepost 7, then milepost 18. How far do you travel during that time period?

64. While traveling, you pass milepost 103, then milepost 86. How far do you travel during that time period?

65. y is at least six units from 0.

66. y is at most two units from a.

Budget Variance **In Exercises 67–70, the accounting department of a company is checking to see whether the actual expenses of a department differ from the budgeted expenses by more than \$500 or by more than 5%. Fill in the missing parts of the table, and determine whether the actual expense passes the "budget variance test."**

	Budgeted Expense, b	*Actual Expense, a*	$\lvert a - b\rvert$	$0.05b$
67. Wages	\$112,700	\$113,356	▇	▇
68. Utilities	\$9400	\$9772	▇	▇
69. Taxes	\$37,640	\$37,335	▇	▇
70. Insurance	\$2575	\$2613	▇	▇

Federal Deficit **In Exercises 71–74, use the bar graph, which shows the receipts of the federal government (in billions of dollars) for selected years from 1960 through 1993. In each exercise you are given the outlay of the federal government. Find the magnitude of the surplus or deficit for the year.** (Source: U.S. Treasury Department)

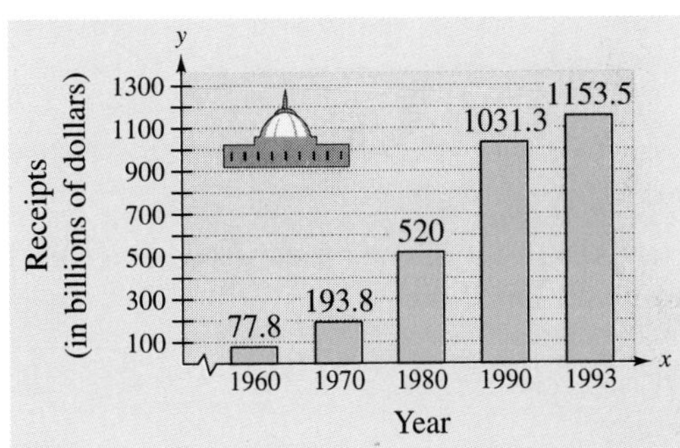

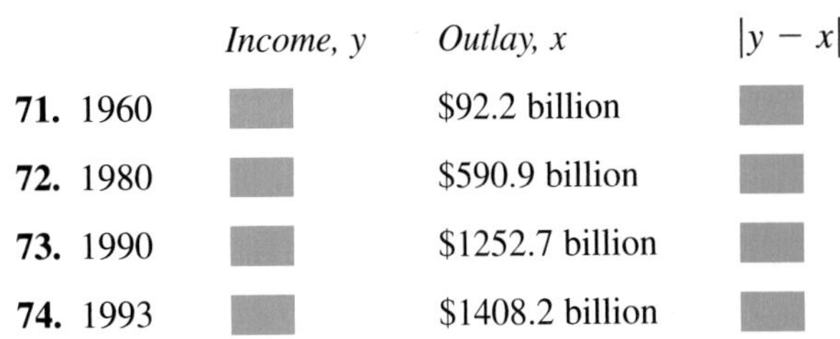

	Income, y	*Outlay, x*	$\lvert y - x\rvert$
71. 1960	▇	\$92.2 billion	▇
72. 1980	▇	\$590.9 billion	▇
73. 1990	▇	\$1252.7 billion	▇
74. 1993	▇	\$1408.2 billion	▇

75. *Exploration* Consider $|u + v|$ and $|u| + |v|$.

(a) Are the values of the expressions always equal? If not, under what conditions are they unequal?

(b) If the two expressions are not equal for certain values of u and v, is one of the expressions always greater than the other? Explain.

76. *Think About It* Is there a difference between saying that a real number is positive and saying that a real number is nonnegative? Explain.

In Exercises 77–80, identify the terms of the expression.

77. $7x + 4$ **78.** $3x^2 - 8x - 11$

79. $4x^3 + x - 5$ **80.** $3x^4 + 3x^3$

In Exercises 81–86, evaluate the expression for the values of x. (If not possible, state the reason.)

	Expression	*Values*	
81.	$4x - 6$	(a) $x = -1$	(b) $x = 0$
82.	$9 - 7x$	(a) $x = -3$	(b) $x = 3$
83.	$x^2 - 3x + 4$	(a) $x = -2$	(b) $x = 2$
84.	$-x^2 + 5x - 4$	(a) $x = -1$	(b) $x = 1$
85.	$\dfrac{x+1}{x-1}$	(a) $x = 1$	(b) $x = -1$
86.	$\dfrac{x}{x+2}$	(a) $x = 2$	(b) $x = -2$

In Exercises 87–96, identify the rule(s) of algebra illustrated by the equation.

87. $x + 9 = 9 + x$ **88.** $2\left(\frac{1}{2}\right) = 1$

89. $\dfrac{1}{h+6}(h + 6) = 1, \quad h \neq -6$

90. $(x + 3) - (x + 3) = 0$

91. $2(x + 3) = 2x + 6$

92. $(z - 2) + 0 = z - 2$

93. $1 \cdot (1 + x) = 1 + x$

94. $x + (y + 10) = (x + y) + 10$

95. $x(3y) = (x \cdot 3)y = (3x)y$

96. $\frac{1}{7}(7 \cdot 12) = \left(\frac{1}{7} \cdot 7\right)12 = 1 \cdot 12 = 12$

In Exercises 97–100, evaluate the expression. (If not possible, state the reason.)

97. $\dfrac{81 - (90 - 9)}{5}$ **98.** $10(23 - 30 + 7)$

99. $\dfrac{8 - 8}{-9 + (6 + 3)}$ **100.** $15 - \dfrac{3 - 3}{5}$

In Exercises 101–110, perform the operations. (Write fractional answers in reduced form.)

101. $(4 - 7)(-2)$ **102.** $\dfrac{27 - 35}{4}$

103. $\frac{3}{16} + \frac{5}{16}$ **104.** $\frac{6}{7} - \frac{4}{7}$

105. $\frac{5}{8} - \frac{5}{12} + \frac{1}{6}$ **106.** $\frac{10}{11} + \frac{6}{33} - \frac{13}{66}$

107. $\frac{4}{5} \cdot \frac{1}{2} \cdot \frac{3}{4}$ **108.** $\frac{11}{16} \div \frac{3}{4}$

109. $12 \div \frac{1}{4}$ **110.** $\left(\frac{3}{5} \div 3\right) - \left(6 \cdot \frac{4}{8}\right)$

In Exercises 111–114, use a calculator to evaluate the expression. (Round your answer to two decimal places.)

111. $-3 + \frac{3}{7}$ **112.** $3\left(-\frac{5}{12} + \frac{3}{8}\right)$

113. $\dfrac{11.46 - 5.37}{3.91}$ **114.** $\dfrac{\frac{1}{5}(-8 - 9)}{-\frac{1}{3}}$

115. Use a calculator to complete the table.

n	1	0.5	0.01	0.0001	0.000001
$5/n$					

116. *Think About It* Use the result of Exercise 115 to make a conjecture about the value of $5/n$ as n approaches 0.

117. Use a calculator to complete the table.

n	1	10	100	10,000	100,000
$5/n$					

118. *Think About It* Use the result of Exercise 117 to make a conjecture about the value of $5/n$ as n increases without bound.

P.2 Exponents and Radicals

Exponents / Scientific Notation / Radicals and Their Properties / Simplifying Radicals / Rationalizing Denominators and Numerators / Rational Exponents / Radicals and Calculators

EXPLORATION

Use the definition of exponential form to write each of the expressions as a single power of 2. From the results, can you find a general rule for simplifying expressions of the forms

$$a^m a^n \text{ and } \frac{a^m}{a^n}?$$

a. $2^3 \cdot 2^2$ **b.** $2^5 \cdot 2^1$

c. $\frac{2^3}{2^2}$ **d.** $\frac{2^5}{2^1}$

Exponents

Repeated *multiplications* can be written in **exponential form.**

Repeated Multiplication	*Exponential Form*
$a \cdot a \cdot a \cdot a \cdot a$	a^5
$(-4)(-4)(-4)$	$(-4)^3$
$(2x)(2x)(2x)(2x)$	$(2x)^4$

In general, if a is a real number, variable, or algebraic expression and n is a positive integer, then

$$a^n = \underbrace{a \cdot a \cdot a \cdots a}_{n \text{ factors}}$$

where n is the **exponent** and a is the **base.** The expression a^n is read "a to the nth **power.**"

Properties of Exponents

Let a and b be real numbers, variables, or algebraic expressions, and let m and n be integers. (All denominators and bases are nonzero.)

Property	*Example*
1. $a^m a^n = a^{m+n}$	$3^2 \cdot 3^4 = 3^{2+4} = 3^6 = 729$
2. $\frac{a^m}{a^n} = a^{m-n}$	$\frac{x^7}{x^4} = x^{7-4} = x^3$
3. $a^{-n} = \frac{1}{a^n} = \left(\frac{1}{a}\right)^n$	$y^{-4} = \frac{1}{y^4} = \left(\frac{1}{y}\right)^4$
4. $a^0 = 1, \quad a \neq 0$	$(x^2 + 1)^0 = 1$
5. $(ab)^m = a^m b^m$	$(5x)^3 = 5^3x^3 = 125x^3$
6. $(a^m)^n = a^{mn}$	$(y^3)^{-4} = y^{3(-4)} = y^{-12} = \frac{1}{y^{12}}$
7. $\left(\frac{a}{b}\right)^m = \frac{a^m}{b^m}$	$\left(\frac{2}{x}\right)^3 = \frac{2^3}{x^3} = \frac{8}{x^3}$
8. $\lvert a^2 \rvert = \lvert a \rvert^2 = a^2$	$\lvert(-2)^2\rvert = \lvert -2 \rvert^2 = (-2)^2 = 4$

Note It is important to recognize the difference between expressions such as $(-2)^4$ and -2^4. In $(-2)^4$, the parentheses indicate that the exponent applies to the negative sign as well as to the 2, but in $-2^4 = -(2^4)$, the exponent applies only to the 2. Hence, $(-2)^4 = 16$, whereas $-2^4 = -16$.

> **Study Tip**
>
> Rarely in algebra is there only one way to solve a problem. Don't be concerned if the steps you use to solve a problem are not exactly the same as the steps presented in this text. The important thing is to use steps that you understand *and,* of course, are justified by the rules of algebra. For instance, you might prefer the following steps for Example 2(d).
>
> $$\left(\frac{3x^2}{y}\right)^{-2} = \left(\frac{y}{3x^2}\right)^2 = \frac{y^2}{9x^4}$$

The properties of exponents listed on the previous page apply to *all* integers m and n, not just positive integers. For instance, by Property 2, you can write

$$\frac{3^4}{3^{-5}} = 3^{4-(-5)} = 3^{4+5} = 3^9.$$

EXAMPLE 1 **Using Properties of Exponents**

a. $(-3ab^4)(4ab^{-3}) = -12(a)(a)(b^4)(b^{-3}) = -12a^2b$

b. $(2xy^2)^3 = 2^3(x)^3(y^2)^3 = 8x^3y^6$

c. $3a(-4a^2)^0 = 3a(1) = 3a, \qquad a \neq 0$

d. $\left(\frac{5x^3}{y}\right)^2 = \frac{5^2(x^3)^2}{y^2} = \frac{25x^6}{y^2}$

EXAMPLE 2 **Rewriting with Positive Exponents**

a. $x^{-1} = \frac{1}{x}$ Property 3: $a^{-n} = \frac{1}{a^n}$

b. $\frac{1}{3x^{-2}} = \frac{1(x^2)}{3} = \frac{x^2}{3}$ -2 exponent does not apply to 3.

c. $\frac{12a^3b^{-4}}{4a^{-2}b} = \frac{12a^3 \cdot a^2}{4b \cdot b^4} = \frac{3a^5}{b^5}$

d. $\left(\frac{3x^2}{y}\right)^{-2} = \frac{3^{-2}(x^2)^{-2}}{y^{-2}} = \frac{3^{-2}x^{-4}}{y^{-2}} = \frac{y^2}{3^2x^4} = \frac{y^2}{9x^4}$

e. $\frac{x^{-2}}{y^{-2}} = \frac{y^2}{x^2}$

Note The graphing calculator keystrokes given in this text correspond to the *TI-82* and *TI-83* graphing calculators from Texas Instruments. For other graphing calculators, the keystrokes may differ. Be sure you are familiar with the use of the keys on your own calculator.

EXAMPLE 3 **Calculators and Exponents**

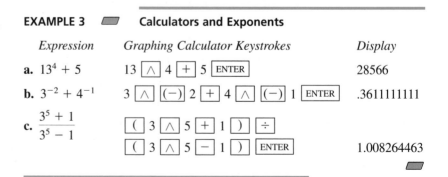

Expression	*Graphing Calculator Keystrokes*	*Display*
a. $13^4 + 5$	13 [∧] 4 [+] 5 [ENTER]	28566
b. $3^{-2} + 4^{-1}$	3 [∧] [(−)] 2 [+] 4 [∧] [(−)] 1 [ENTER] 	.3611111111
c. $\frac{3^5 + 1}{3^5 - 1}$	[(] 3 [∧] 5 [+] 1 [)] [÷] [(] 3 [∧] 5 [−] 1 [)] [ENTER]	1.008264463

Scientific Notation

Exponents provide an efficient way of writing and computing with very large (or very small) numbers. For instance, a drop of water contains more than 33 billion billion molecules—that is, 33 followed by 18 zeros.

$$33{,}000{,}000{,}000{,}000{,}000{,}000$$

It is convenient to write such numbers in **scientific notation.** This notation has the form $\pm c \times 10^n$, where $1 \le c < 10$ and n is an integer. Thus, the number of molecules in a drop of water can be written in scientific notation as

$$3.3 \times 10{,}000{,}000{,}000{,}000{,}000{,}000 = 3.3 \times 10^{19}.$$

The *positive* exponent 19 indicates that the number is *large* (10 or more) and that the decimal point has been moved 19 places. A *negative* exponent indicates that the number is *small* (less than 1). For instance, the mass (in grams) of one electron is approximately

$$9.0 \times 10^{-28} = 0.0000000000000000000000000009.$$

28 decimal places

EXAMPLE 4 Scientific Notation

a. $1.345 \times 10^2 = 134.5$

b. $0.0000782 = 7.82 \times 10^{-5}$

c. $-9.36 \times 10^{-6} = -0.00000936$

d. $836{,}100{,}000 = 8.361 \times 10^8$

EXAMPLE 5 Using Scientific Notation with a Calculator

Use a calculator to evaluate $65{,}000 \times 3{,}400{,}000{,}000$.

Solution

Because $65{,}000 = 6.5 \times 10^4$ and $3{,}400{,}000{,}000 = 3.4 \times 10^9$, you can multiply the two numbers using the following graphing calculator steps.

6.5 [EE] 4 [×] 3.4 [EE] 9 [ENTER]

After entering these keystrokes, the calculator display should read [2.21 E 14]. Therefore, the product of the two numbers is

$$(6.5 \times 10^4)(3.4 \times 10^9) = 2.21 \times 10^{14}$$
$$= 221{,}000{,}000{,}000{,}000.$$

Note Most calculators switch to scientific notation when they are showing large (or small) numbers that exceed the display range. Try evaluating $86{,}500{,}000 \times 6000$. If your calculator follows standard conventions, its display should be

[5.19 11] or [5.19 E 11]

which is 5.19×10^{11}.

Radicals and Their Properties

A **square root** of a number is one of its two equal factors. For example, 5 is a square root of 25 because 5 is one of the two equal factors of $25 = 5 \cdot 5$. In a similar way, a **cube root** of a number is one of its three equal factors.

Definition of the *n*th Root of a Number

Let a and b be real numbers and let $n \geq 2$ be a positive integer. If

$$a = b^n$$

then b is an ***n*th root of *a*.** If $n = 2$, the root is a **square root.** If $n = 3$, the root is a **cube root.**

Some numbers have more than one *n*th root. For example, both 5 and -5 are square roots of 25. The **principal *n*th root** of a number is defined as follows.

Principal *n*th Root of a Number

Let a be a real number that has at least one *n*th root. The **principal *n*th root of *a*** is the *n*th root that has the same sign as a. It is denoted by a **radical symbol**

$$\sqrt[n]{a}. \qquad \text{Principal } n\text{th root}$$

The positive integer n is the **index** of the radical, and the number a is the **radicand.** If $n = 2$, we omit the index and write $\sqrt{a}$ rather than $\sqrt[2]{a}$. (The plural of index is *indices.*)

Study Tip

Try using your calculator to verify the calculations shown in Example 6.

EXAMPLE 6 **Evaluating Expressions Involving Radicals**

a. $\sqrt{36} = 6$ because $6^2 = 36$.

b. $-\sqrt{36} = -6$ because $-(\sqrt{36}) = -(6) = -6$.

c. $\sqrt[3]{\dfrac{125}{64}} = \dfrac{5}{4}$ because $\left(\dfrac{5}{4}\right)^3 = \dfrac{5^3}{4^3} = \dfrac{125}{64}$.

d. $\sqrt[5]{-32} = -2$ because $(-2)^5 = -32$.

e. $\sqrt[4]{-81}$ is not a real number because there is no real number that can be raised to the fourth power to produce -81.

Here are some generalizations about the nth roots of a real number.

1. If a is a positive real number and n is a positive *even* integer, then a has exactly two real nth roots denoted by $\sqrt[n]{a}$ and $-\sqrt[n]{a}$. See Examples 6(a) and 6(b).
2. If a is any real number and n is an *odd* integer, then a has only one real nth root denoted by $\sqrt[n]{a}$. See Examples 6(c) and 6(d).
3. If a is a negative real number and n is an *even* integer, then a has no real nth root. See Example 6(e).
4. $\sqrt[n]{0} = 0$.

Integers such as 1, 4, 9, 16, 25, and 36 are called **perfect squares** because they have integer square roots. Similarly, integers such as 1, 8, 27, 64, and 125 are called **perfect cubes** because they have integer cube roots.

Properties of Radicals

Let a and b be real numbers, variables, or algebraic expressions such that the indicated roots are real numbers, and let m and n be positive integers.

Property	*Example*
1. $\sqrt[n]{a^m} = \left(\sqrt[n]{a}\right)^m$	$\sqrt[3]{8^2} = \left(\sqrt[3]{8}\right)^2 = (2)^2 = 4$
2. $\sqrt[n]{a} \cdot \sqrt[n]{b} = \sqrt[n]{ab}$	$\sqrt{5} \cdot \sqrt{7} = \sqrt{5 \cdot 7} = \sqrt{35}$
3. $\dfrac{\sqrt[n]{a}}{\sqrt[n]{b}} = \sqrt[n]{\dfrac{a}{b}}, \quad b \neq 0$	$\dfrac{\sqrt[4]{27}}{\sqrt[4]{9}} = \sqrt[4]{\dfrac{27}{9}} = \sqrt[4]{3}$
4. $\sqrt[m]{\sqrt[n]{a}} = \sqrt[mn]{a}$	$\sqrt[3]{\sqrt{10}} = \sqrt[6]{10}$
5. $\left(\sqrt[n]{a}\right)^n = a$	$\left(\sqrt{3}\right)^2 = 3$
6. For n even, $\sqrt[n]{a^n} = \lvert a \rvert$.	$\sqrt{(-12)^2} = \lvert -12 \rvert = 12$
For n odd, $\sqrt[n]{a^n} = a$.	$\sqrt[3]{(-12)^3} = -12$

Note A common special case of Property 6 is $\sqrt{a^2} = \lvert a \rvert$.

EXAMPLE 7 **Using Properties of Radicals**

a. $\sqrt{8} \cdot \sqrt{2} = \sqrt{8 \cdot 2} = \sqrt{16} = 4$

b. $\left(\sqrt[3]{5}\right)^3 = 5$

c. $\sqrt[3]{x^3} = x$

Simplifying Radicals

An expression involving radicals is in **simplest form** when the following conditions are satisfied.

1. All possible factors have been removed from the radical.
2. All fractions have radical-free denominators (accomplished by a process called *rationalizing the denominator*).
3. The index of the radical is reduced.

To simplify a radical, factor the radicand into factors whose exponents are multiples of the index. The roots of these factors are written outside the radical, and the "leftover" factors make up the new radicand.

EXAMPLE 8 Simplifying Even Roots

Perfect 4th power — Leftover factor

a. $\sqrt[4]{48} = \sqrt[4]{16 \cdot 3} = \sqrt[4]{2^4 \cdot 3} = 2\sqrt[4]{3}$

Perfect square — Leftover factor

b. $\sqrt{75x^3} = \sqrt{25x^2 \cdot 3x}$ Find largest square factor.

$= \sqrt{(5x)^2 \cdot 3x}$

$= 5x\sqrt{3x}$ Find root of perfect square.

c. $\sqrt[4]{(5x)^4} = |5x| = 5|x|$

Note In Example 8(b), the expression $\sqrt{75x^3}$ makes sense only for nonnegative values of x.

EXAMPLE 9 Simplifying Odd Roots

Perfect cube — Leftover factor

a. $\sqrt[3]{24} = \sqrt[3]{8 \cdot 3} = \sqrt[3]{2^3 \cdot 3} = 2\sqrt[3]{3}$

Perfect cube — Leftover factor

b. $\sqrt[3]{24a^4} = \sqrt[3]{8a^3 \cdot 3a}$ Find largest cube factor.

$= \sqrt[3]{(2a)^3 \cdot 3a}$

$= 2a\sqrt[3]{3a}$ Find root of perfect cube.

c. $\sqrt[3]{-40x^6} = \sqrt[3]{(-8x^6) \cdot 5} = \sqrt[3]{(-2x^2)^3 \cdot 5} = -2x^2\sqrt[3]{5}$

Rationalizing Denominators and Numerators

To rationalize a denominator or numerator of the form $a - b\sqrt{m}$ or $a + b\sqrt{m}$, multiply both numerator and denominator by a **conjugate:** $a + b\sqrt{m}$ and $a - b\sqrt{m}$ are conjugates of each other. If $a = 0$, the rationalizing factor for $\sqrt{m}$ is itself, $\sqrt{m}$.

EXAMPLE 10 **Rationalizing Single-Term Denominators**

a. $\dfrac{5}{2\sqrt{3}} = \dfrac{5}{2\sqrt{3}} \cdot \dfrac{\sqrt{3}}{\sqrt{3}} = \dfrac{5\sqrt{3}}{2(3)} = \dfrac{5\sqrt{3}}{6}$

b. $\dfrac{2}{\sqrt[3]{5}} = \dfrac{2}{\sqrt[3]{5}} \cdot \dfrac{\sqrt[3]{5^2}}{\sqrt[3]{5^2}} = \dfrac{2\sqrt[3]{5^2}}{\sqrt[3]{5^3}} = \dfrac{2\sqrt[3]{25}}{5}$

Note In Example 10(b) the numerator and denominator are multiplied by $\sqrt[3]{5^2}$ to produce a perfect cube radicand.

EXAMPLE 11 **Rationalizing a Denominator with Two Terms**

$$\frac{2}{3+\sqrt{7}} = \frac{2}{3+\sqrt{7}} \cdot \frac{3-\sqrt{7}}{3-\sqrt{7}}$$ Multiply numerator and denominator by conjugate.

$$= \frac{2(3-\sqrt{7})}{(3)^2 - (\sqrt{7})^2}$$

$$= \frac{2(3-\sqrt{7})}{9-7}$$

$$= \frac{2(3-\sqrt{7})}{2}$$

$$= 3 - \sqrt{7}$$ Cancel like factors.

Note Do not confuse the expression $\sqrt{5} + \sqrt{7}$ with the expression $\sqrt{5+7}$. In general, $\sqrt{x+y}$ does not equal $\sqrt{x} + \sqrt{y}$. Similarly, $\sqrt{x^2 + y^2}$ does not equal $x + y$.

EXAMPLE 12 **Rationalizing the Numerator**

$$\frac{\sqrt{5}-\sqrt{7}}{2} = \frac{\sqrt{5}-\sqrt{7}}{2} \cdot \frac{\sqrt{5}+\sqrt{7}}{\sqrt{5}+\sqrt{7}}$$ Multiply numerator and denominator by conjugate.

$$= \frac{5-7}{2(\sqrt{5}+\sqrt{7})}$$

$$= \frac{-2}{2(\sqrt{5}+\sqrt{7})}$$

$$= \frac{-1}{\sqrt{5}+\sqrt{7}}$$ Simplify.

Rational Exponents

Definition of Rational Exponents

If a is a real number and n is a positive integer such that the principal nth root of a exists, we define $a^{1/n}$ to be

$$a^{1/n} = \sqrt[n]{a}.$$

Moreover, if m is a positive integer that has no common factor with n, then

$$a^{m/n} = (a^{1/n})^m = \left(\sqrt[n]{a}\right)^m \quad \text{and} \quad a^{m/n} = (a^m)^{1/n} = \sqrt[n]{a^m}.$$

The numerator of a rational exponent denotes the *power* to which the base is raised, and the denominator denotes the *index* or the *root* to be taken, as shown below.

Power

Index

$$b^{m/n} = \left(\sqrt[n]{b}\right)^m = \sqrt[n]{b^m}$$

When you are working with rational exponents, the properties of integer exponents still apply. For instance,

$$2^{1/2}2^{1/3} = 2^{(1/2)+(1/3)} = 2^{5/6}.$$

EXAMPLE 13 **Changing from Radical to Exponential Form**

a. $\sqrt{3} = 3^{1/2}$

b. $\sqrt{(3xy)^5} = \sqrt[2]{(3xy)^5} = (3xy)^{(5/2)}$

c. $2x\sqrt[4]{x^3} = (2x)(x^{3/4}) = 2x^{1+(3/4)} = 2x^{7/4}$

Note Rational exponents can be tricky, and you must remember that the expression $b^{m/n}$ is not defined unless $\sqrt[n]{b}$ is a real number. This restriction produces some unusual-looking results. For instance, the number $(-8)^{1/3}$ is defined because $\sqrt[3]{-8} = -2$, but the number $(-8)^{2/6}$ is undefined because $\sqrt[6]{-8}$ is not a real number.

EXAMPLE 14 **Changing from Exponential to Radical Form**

a. $(x^2 + y^2)^{3/2} = \left(\sqrt{x^2 + y^2}\right)^3 = \sqrt{(x^2 + y^2)^3}$

b. $2y^{3/4}z^{1/4} = 2(y^3z)^{1/4} = 2\sqrt[4]{y^3z}$

c. $a^{-3/2} = \dfrac{1}{a^{3/2}} = \dfrac{1}{\sqrt{a^3}}$

d. $x^{0.2} = x^{1/5} = \sqrt[5]{x}$

Rational exponents are particularly useful for evaluating roots of numbers on a calculator, for reducing the index of a radical, and for simplifying expressions encountered in calculus.

EXAMPLE 15 **Simplifying with Rational Exponents**

a. $(27)^{2/6} = (27)^{1/3} = \sqrt[3]{27} = 3$

b. $(-32)^{-4/5} = \left(\sqrt[5]{-32}\right)^{-4} = (-2)^{-4} = \dfrac{1}{(-2)^4} = \dfrac{1}{16}$

c. $(-5x^{5/3})(3x^{-3/4}) = -15x^{(5/3)-(3/4)} = -15x^{11/12}, \qquad x \neq 0$

d. $\sqrt[9]{a^3} = a^{3/9} = a^{1/3} = \sqrt[3]{a}$

e. $\sqrt[3]{\sqrt{125}} = \sqrt[6]{125} = \sqrt[6]{(5)^3} = 5^{3/6} = 5^{1/2} = \sqrt{5}$

f. $(2x-1)^{4/3}(2x-1)^{-1/3} = (2x-1)^{(4/3)-(1/3)}$

$$= 2x - 1, \qquad x \neq \frac{1}{2}$$

g. $\dfrac{x-1}{(x-1)^{-1/2}} = (x-1)^{1-(-1/2)}$

$$= (x-1)^{3/2}, \qquad x \neq 1$$

Radical expressions can be combined (added or subtracted) if they are **like radicals**—that is, if they have the same index and radicand. For instance, $\sqrt{2}$, $3\sqrt{2}$, and $\frac{1}{2}\sqrt{2}$ are like radicals, but $\sqrt{3}$ and $\sqrt{2}$ are unlike radicals. To determine whether two radicals are like radicals, you should first simplify each radical.

Note Try using your calculator to check the result of Example 16(a). You should obtain -1.732050808, which is the same as the calculator's approximation for $-\sqrt{3}$.

EXAMPLE 16 **Combining Radicals**

a.
$$\begin{aligned} 2\sqrt{48} - 3\sqrt{27} &= 2\sqrt{16 \cdot 3} - 3\sqrt{9 \cdot 3} && \text{Find square factors.} \\ &= 8\sqrt{3} - 9\sqrt{3} && \text{Find square roots.} \\ &= (8-9)\sqrt{3} && \text{Combine like terms.} \\ &= -\sqrt{3} \end{aligned}$$

b.
$$\begin{aligned} \sqrt[3]{16x} - \sqrt[3]{54x^4} &= \sqrt[3]{8 \cdot 2x} - \sqrt[3]{27 \cdot x^3 \cdot 2x} \\ &= 2\sqrt[3]{2x} - 3x\sqrt[3]{2x} \\ &= (2 - 3x)\sqrt[3]{2x} \end{aligned}$$

Radicals and Calculators

There are four methods of evaluating radicals on most graphing calculators. For square roots, you can use the *square root key* $\boxed{\sqrt{\ }}$. For cube roots, you can use the *cube root key* $\boxed{\sqrt[3]{\ }}$ (or menu choice). For other roots, you can first convert the radical to exponential form and then use the *exponential key* $\boxed{\wedge}$, or you can use the *nth root key* $\boxed{\sqrt[x]{\ }}$.

EXAMPLE 17 **Evaluating Radicals with a Calculator**

Use a calculator to evaluate $\sqrt[4]{56} = 56^{1/4}$.

Graphing Calculator Keystrokes

a. 56 $\boxed{\wedge}$ $\boxed{(}$ 1 $\boxed{\div}$ 4 $\boxed{)}$ $\boxed{\text{ENTER}}$

b. 4 $\boxed{\text{MATH}}$ (MATH) $(5: \sqrt[x]{\ })$ $\boxed{\text{ENTER}}$ 56 $\boxed{\text{ENTER}}$

For each of these two keystroke sequences, the display is $\sqrt[4]{56} \approx 2.7355648$.

Group Activity

A Famous Mathematical Discovery

Johannes Kepler (1571–1630), a well-known German astronomer, discovered a relationship between the average distance of a planet from the sun and the time (or period) it takes the planet to orbit the sun. People then knew that planets that are closer to the sun take less time to complete an orbit than planets that are farther from the sun. Kepler discovered that the distance and period are related by an exact mathematical formula. The table shows the average distance x (in astronomical units) and period y (in years) for the six planets that are closest to the sun. By completing the table, can you rediscover Kepler's relationship? Discuss your conclusions.

Planet	Mercury	Venus	Earth	Mars	Jupiter	Saturn
x	0.387	0.723	1.0	1.523	5.203	9.541
$\sqrt{x}$						
y	0.241	0.615	1.0	1.881	11.861	29.457
$\sqrt[3]{y}$						

P.2 /// EXERCISES

In Exercises 1–6, evaluate the expression.

1. (a) $4^2 \cdot 3$ (b) $3 \cdot 3^3$

2. (a) $\dfrac{5^5}{5^2}$ (b) $\dfrac{3^2}{3^4}$

3. (a) $(3^3)^2$ (b) -3^2

4. (a) $(2^3 \cdot 3^2)^2$ (b) $\left(-\frac{3}{5}\right)^3\left(\frac{5}{3}\right)^2$

5. (a) $\dfrac{3}{3^{-4}}$ (b) $24(-2)^{-5}$

6. (a) $\dfrac{4 \cdot 3^{-2}}{2^{-2} \cdot 3^{-1}}$ (b) $(-2)^0$

In Exercises 7–10, use a calculator to evaluate the expression. (Round to three decimal places.)

7. $(-4)^3(5^2)$ 8. $(8^{-4})(10^3)$

9. $\dfrac{3^6}{7^3}$ 10. $\dfrac{4^3}{3^{-4}}$

In Exercises 11–14, evaluate the expression for the value of x.

	Expression	*Value*
11.	$-3x^3$	2
12.	$7x^{-2}$	4
13.	$6x^0 - (6x)^0$	10
14.	$5(-x)^3$	3

In Exercises 15–24, simplify the expression.

15. (a) $(-5z)^3$ (b) $5x^4(x^2)$

16. (a) $(3x)^2$ (b) $(4x^3)^2$

17. (a) $\dfrac{7x^2}{x^3}$ (b) $\dfrac{12(x + y)^3}{9(x + y)}$

18. (a) $\dfrac{r^4}{r^6}$ (b) $\left(\dfrac{4}{y}\right)^3\left(\dfrac{3}{y}\right)^4$

19. (a) $(x + 5)^0, \quad x \neq -5$ (b) $(2x^2)^{-2}$

20. (a) $(2x^5)^0, \quad x \neq 0$ (b) $(z + 2)^{-3}(z + 2)^{-1}$

21. (a) $\left(\dfrac{x}{10}\right)^{-1}$ (b) $\left(\dfrac{x^{-3}y^4}{5}\right)^{-3}$

22. (a) $[(x^2y^{-2})^{-1}]^{-1}$ (b) $(5x^2z^6)^3(5x^2z^6)^{-3}$

23. (a) $3^n \cdot 3^{2n}$ (b) $\left(\dfrac{a^{-2}}{b^{-2}}\right)\left(\dfrac{b}{a}\right)^3$

24. (a) $\dfrac{x^2 \cdot x^n}{x^3 \cdot x^n}$ (b) $\left(\dfrac{a^{-3}}{b^{-3}}\right)\left(\dfrac{a}{b}\right)^3$

In Exercises 25–34, fill in the missing description.

	Radical Form	*Rational Exponent Form*
25.	$\sqrt{9} = 3$	
26.	$\sqrt[3]{64} = 4$	
27.		$32^{1/5} = 2$
28.		$-(144^{1/2}) = -12$
29.		$196^{1/2} = 14$
30.	$\sqrt[3]{614.125} = 8.5$	
31.	$\sqrt[3]{-216} = -6$	
32.		$(-243)^{1/5} = -3$
33.	$\sqrt[4]{81^3} = 27$	
34.		$16^{5/4} = 32$

In Exercises 35–44, evaluate each expression. (Do not use a calculator.)

35. $\sqrt{9}$ 36. $\sqrt{49}$

37. $-\sqrt[3]{-27}$ 38. $\dfrac{\sqrt[4]{81}}{3}$

39. $\left(\sqrt[3]{-125}\right)^3$ 40. $\sqrt[4]{562^4}$

41. $32^{-3/5}$ 42. $\left(\frac{9}{4}\right)^{-1/2}$

43. $\left(-\dfrac{1}{64}\right)^{-1/3}$ 44. $-\left(\dfrac{1}{125}\right)^{-4/3}$

In Exercises 45–48, use a calculator to approximate the number. (Round to three decimal places.)

45. $\sqrt[5]{-27^3}$

46. $\sqrt[3]{45^2}$

47. $(1.2^{-2})\sqrt{75} + 3\sqrt{8}$

48. $(3.4)^{2.5}$

In Exercises 49–54, simplify by removing all possible factors from the radical.

49. (a) $\sqrt{8}$ (b) $\sqrt[3]{24}$

50. (a) $\sqrt[3]{\frac{16}{27}}$ (b) $\sqrt{\frac{75}{4}}$

51. (a) $\sqrt{72x^3}$ (b) $\sqrt{\dfrac{18^2}{z^3}}$

52. (a) $\sqrt{54xy^4}$ (b) $\sqrt{\dfrac{32a^4}{b^2}}$

53. (a) $\sqrt[3]{16x^5}$ (b) $\sqrt{75x^2y^{-4}}$

54. (a) $\sqrt[4]{(3x^2)^4}$ (b) $\sqrt[5]{96x^5}$

In Exercises 55–60, perform the operations and simplify.

55. $5^{4/3} \cdot 5^{8/3}$

56. $\dfrac{8^{12/5}}{8^{2/5}}$

57. $\dfrac{(2x^2)^{3/2}}{2^{1/2}x^4}$

58. $\dfrac{x^{4/3}y^{2/3}}{(xy)^{1/3}}$

59. $\dfrac{x^{-3} \cdot x^{1/2}}{x^{3/2} \cdot x^{-1}}$

60. $\dfrac{5^{-1/2} \cdot 5x^{5/2}}{(5x)^{3/2}}$

In Exercises 61–64, rationalize the denominator. Then simplify your answer.

61. (a) $\dfrac{1}{\sqrt{3}}$ (b) $\dfrac{8}{\sqrt[3]{2}}$

62. (a) $\dfrac{5}{\sqrt{10}}$ (b) $\dfrac{5}{\sqrt[3]{(5x)^2}}$

63. (a) $\dfrac{2x}{5 - \sqrt{3}}$ (b) $\dfrac{3}{\sqrt{5} + \sqrt{6}}$

64. (a) $\dfrac{5}{\sqrt{14} - 2}$ (b) $\dfrac{5}{2\sqrt{10} - 5}$

In Exercises 65–68, rationalize the numerator. Then simplify your answer.

65. $\dfrac{\sqrt{8}}{2}$

66. $\dfrac{\sqrt{2}}{3}$

67. $\dfrac{\sqrt{5} + \sqrt{3}}{3}$

68. $\dfrac{\sqrt{3} - \sqrt{2}}{2}$

In Exercises 69 and 70, reduce the index of the radical.

69. (a) $\sqrt[4]{3^2}$ (b) $\sqrt[6]{(x + 1)^4}$

70. (a) $\sqrt[6]{x^3}$ (b) $\sqrt[4]{(3x^2)^4}$

In Exercises 71 and 72, write as a single radical. Then simplify your answer.

71. (a) $\sqrt{\sqrt{32}}$ (b) $\sqrt{\sqrt[4]{2x}}$

72. (a) $\sqrt{\sqrt{243(x + 1)}}$ (b) $\sqrt{\sqrt[3]{10a^7b}}$

In Exercises 73–76, simplify the expression.

73. (a) $2\sqrt{50} + 12\sqrt{8}$ (b) $10\sqrt{32} - 6\sqrt{18}$

74. (a) $4\sqrt{27} - \sqrt{75}$ (b) $\sqrt[3]{16} + 3\sqrt[3]{54}$

75. (a) $5\sqrt{x} - 3\sqrt{x}$ (b) $-2\sqrt{9y} + 10\sqrt{y}$

76. (a) $3\sqrt{x + 1} + 10\sqrt{x + 1}$
(b) $7\sqrt{80x} - 2\sqrt{125x}$

In Exercises 77–80, fill in the blank with $<$, $=$, or $>$.

77. $\sqrt{5} + \sqrt{3}$ ▭ $\sqrt{5 + 3}$

78. $\sqrt{\dfrac{3}{11}}$ ▭ $\dfrac{\sqrt{3}}{\sqrt{11}}$

79. 5 ▭ $\sqrt{3^2 + 2^2}$

80. 5 ▭ $\sqrt{3^2 + 4^2}$

In Exercises 81–84, write the number in scientific notation.

81. Land Area of Earth: 57,500,000 square miles

82. Light Year: 9,461,000,000,000,000 kilometers

83. Relative Density of Hydrogen: 0.0000899 gram per cm^3

84. One Micron (Millionth of a Meter): 0.00003937 inch

In Exercises 85–88, write the number in decimal form.

85. U.S. Daily Coca-Cola Consumption: 5.24×10^8 servings

86. Interior Temperature of Sun: 1.3×10^7 degrees Celsius

87. Charge of Electron: 4.8×10^{-10} electrostatic unit

88. Width of Human Hair: 9.0×10^{-4} meter

In Exercises 89–92, use a calculator to evaluate the expression. (Round to three decimal places.)

89. (a) $750\left(1 + \frac{0.11}{365}\right)^{800}$

(b) $\frac{67{,}000{,}000 + 93{,}000{,}000}{0.0052}$

90. (a) $(9.3 \times 10^6)^3(6.1 \times 10^{-4})$

(b) $\frac{(2.414 \times 10^4)^6}{(1.68 \times 10^5)^5}$

91. (a) $\sqrt{4.5 \times 10^9}$ (b) $\sqrt[3]{6.3 \times 10^4}$

92. (a) $(2.65 \times 10^{-4})^{1/3}$ (b) $\sqrt{9 \times 10^{-4}}$

93. *Exploration* List all possible unit digits of the square of a positive integer. Use that list to determine whether $\sqrt{5233}$ is an integer.

94. *Think About It* Square the real number $2/\sqrt{5}$ and note that the radical is eliminated from the denominator. Is this equivalent to rationalizing the denominator? Why or why not?

95. *Period of a Pendulum* The period T in seconds of a pendulum is

$$T = 2\pi\sqrt{\frac{L}{32}}$$

where L is the length of the pendulum in feet. Find the period of a pendulum whose length is 2 feet.

96. *Mathematical Modeling* A funnel is filled with water to a height of h centimeters. The time t (in seconds) for the funnel to empty is

$$t = 0.03[12^{5/2} - (12 - h)^{5/2}], \quad 0 \le h \le 12.$$

Find t for $h = 7$ centimeters.

97. *Declining Balances Depreciation* For the given data, find the annual depreciation rate r by the **declining balances method,** which uses the formula

$$r = 1 - \left(\frac{S}{C}\right)^{1/n}$$

where n is the useful life of the item (in years), S is the salvage value (in dollars), and C is the original cost (in dollars).

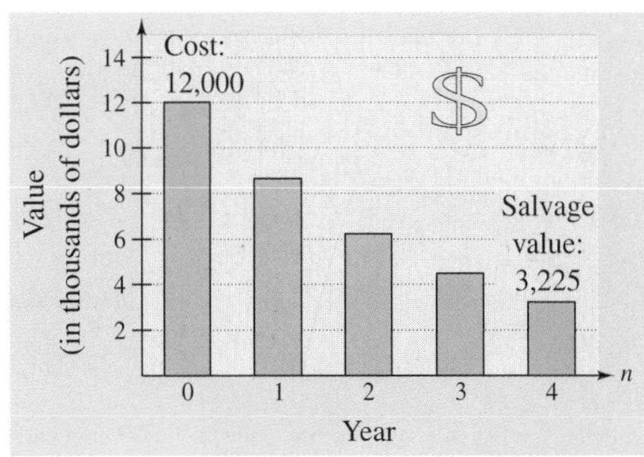

98. *Erosion* A stream of water moving at the rate of v feet per second can carry particles of size $0.03\sqrt{v}$ inches. Find the size of the particle that can be carried by a stream flowing at the rate of $\frac{3}{4}$ foot per second.

99. *Speed of Light* The speed of light is 11,160,000 miles per minute. The distance from the sun to the earth is 93,000,000 miles. Find the time for light to travel from the sun to the earth.

100. *Organizing Data* There were 1.957×10^8 million tons of municipal waste generated in 1990. Find the number of tons for each of the categories in the figure. (Source: U.S. Environmental Protection Agency)

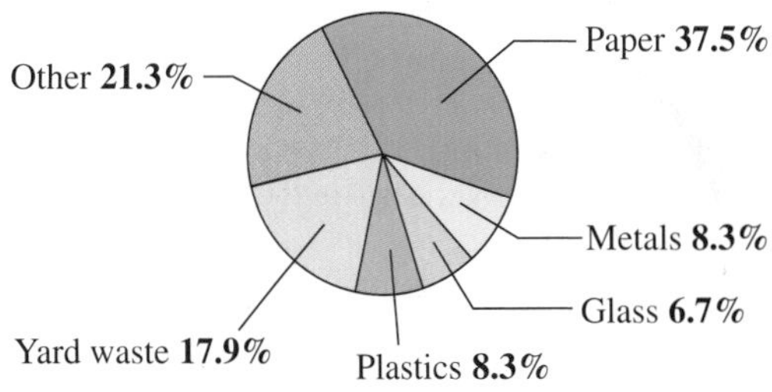

P.3 Polynomials and Factoring

Polynomials / Operations with Polynomials / Special Products / Factoring / Factoring Special Polynomial Forms / Trinomials with Binomial Factors / Factoring by Grouping

Polynomials

An algebraic expression is a collection of variables and real numbers. The most common type of algebraic expression is the **polynomial.** Some examples are

$$2x + 5, \quad 3x^4 - 7x^2 + 2x + 4, \quad \text{and} \quad 5x^2y^2 - xy + 3.$$

The first two are *polynomials in x* and the third is a *polynomial in x and y.* The terms of a polynomial in x have the form ax^k, where a is the **coefficient** and k is the **degree** of the term. For instance, the third-degree polynomial

$$2x^3 - 5x^2 + 1 = 2x^3 + (-5)x^2 + (0)x + 1$$

has coefficients 2, -5, 0, and 1.

Note Polynomials with one, two, and three terms are called **monomials, binomials,** and **trinomials,** respectively.

Definition of a Polynomial in x

Let $a_0, a_1, a_2, \ldots, a_n$ be *real numbers* and let n be a *nonnegative integer.* A **polynomial in x** is an expression of the form

$$a_nx^n + a_{n-1}x^{n-1} + \cdots + a_1x + a_0$$

where $a_n \neq 0$. The polynomial is of **degree** n, a_n is the **leading coefficient,** and a_0 is the **constant term.**

In **standard form,** a polynomial is written with descending powers of x.

Note A polynomial that has all zero coefficients is called the **zero polynomial,** denoted by 0. No degree is assigned to this particular polynomial. For polynomials in more than one variable, the degree of a *term* is the sum of the exponents of the variables in the term. The degree of the *polynomial* is the highest degree of its terms.

EXAMPLE 1 **Writing Polynomials in Standard Form**

	Polynomial	*Standard Form*	*Degree*
a.	$4x^2 - 5x^7 - 2 + 3x$	$-5x^7 + 4x^2 + 3x - 2$	7
b.	$4 - 9x^2$	$-9x^2 + 4$	2
c.	8	$8 \; (8 = 8x^0)$	0

Operations with Polynomials

You can **add** and **subtract** polynomials in much the same way you add and subtract real numbers. Simply add or subtract the *like terms* (terms having the same variables to the same powers) by adding their coefficients. For instance, $-3xy^2$ and $5xy^2$ are like terms and their sum is

$$-3xy^2 + 5xy^2 = (-3 + 5)xy^2 = 2xy^2.$$

Note A common mistake is to fail to change the sign of *each* term inside parentheses preceded by a negative sign. For instance, note that

$$-(x^2 - x + 3) = -x^2 + x - 3$$

and

$$-(x^2 - x + 3) \neq -x^2 - x + 3.$$

EXAMPLE 2 **Sums and Differences of Polynomials**

a. $(5x^3 - 7x^2 - 3) + (x^3 + 2x^2 - x + 8)$

$= (5x^3 + x^3) + (2x^2 - 7x^2) - x + (8 - 3)$ Group like terms.

$= 6x^3 - 5x^2 - x + 5$ Combine like terms.

b. $(7x^4 - x^2 - 4x + 2) - (3x^4 - 4x^2 + 3x)$

$= 7x^4 - x^2 - 4x + 2 - 3x^4 + 4x^2 - 3x$

$= (7x^4 - 3x^4) + (4x^2 - x^2) + (-3x - 4x) + 2$ Group like terms.

$= 4x^4 + 3x^2 - 7x + 2$ Combine like terms.

To find the **product** of two polynomials, use the left and right Distributive Properties.

EXAMPLE 3 **Multiplying Polynomials: The FOIL Method**

Multiply $(3x - 2)$ by $(5x + 7)$.

Solution

$$(3x - 2)(5x + 7) = 3x(5x + 7) - 2(5x + 7)$$

$$= (3x)(5x) + (3x)(7) - (2)(5x) - (2)(7)$$

$$= 15x^2 + 21x - 10x - 14$$

Product of First terms | Product of Outer terms | Product of Inner terms | Product of Last terms

$$= 15x^2 + 11x - 14$$

Note in this **FOIL Method** that for binomials the outer (O) and inner (I) terms are alike and can be combined into one term.

Special Products

Special Products

Let u and v be real numbers, variables, or algebraic expressions.

Special Product	*Example*
Sum and Difference of Same Terms	
$(u + v)(u - v) = u^2 - v^2$	$(x + 4)(x - 4) = x^2 - 4^2 = x^2 - 16$
Square of a Binomial	
$(u + v)^2 = u^2 + 2uv + v^2$	$(x + 3)^2 = x^2 + 2(x)(3) + 3^2 = x^2 + 6x + 9$
$(u - v)^2 = u^2 - 2uv + v^2$	$(3x - 2)^2 = (3x)^2 - 2(3x)(2) + 2^2 = 9x^2 - 12x + 4$
Cube of a Binomial	
$(u + v)^3 = u^3 + 3u^2v + 3uv^2 + v^3$	$(x + 2)^3 = x^3 + 3x^2(2) + 3x(2^2) + 2^3 = x^3 + 6x^2 + 12x + 8$
$(u - v)^3 = u^3 - 3u^2v + 3uv^2 - v^3$	$(x - 1)^3 = x^3 - 3x^2(1) + 3x(1^2) - 1^3 = x^3 - 3x^2 + 3x - 1$

EXAMPLE 4 **The Product of Two Trinomials**

Find the product of $(x + y - 2)$ and $(x + y + 2)$.

Solution

By grouping $x + y$ in parentheses, you can write

$$\begin{aligned}(x + y - 2)(x + y + 2) &= [(x + y) - 2][(x + y) + 2]\\ &= (x + y)^2 - 2^2\\ &= x^2 + 2xy + y^2 - 4.\end{aligned}$$

The *Interactive* CD-ROM offers graphing utility emulators of the *TI-82* and *TI-83*, which can be used with the Examples, Explorations, Technology notes, and Exercises.

```
PROGRAM:EVALUATE
:Lbl 1
:Prompt X
:Disp Y1
:Goto 1
```

There are several ways to use a graphing utility to evaluate a function. Here is one way that we like to use for evaluating functions on a *TI-82* or a *TI-83*. Begin by entering the program EVALUATE shown at the left. Now enter the expression $x^2 - 3x + 2$ into Y1. Then run the program for several values of x. Organize your results in a table. Programs for other graphing calculator models can be found in the Appendix.

Factoring

The process of writing a polynomial as a product is called **factoring.** It is an important tool for solving equations and for reducing fractional expressions.

Unless noted otherwise, when you are asked to factor a polynomial, you can assume that you are hunting for factors with integer coefficients. If a polynomial cannot be factored using integer coefficients, it is **prime** or **irreducible over the integers.** For instance, the polynomial $x^2 - 3$ is irreducible over the integers. Over the real numbers, this polynomial can be factored as

$$x^2 - 3 = \left(x + \sqrt{3}\right)\left(x - \sqrt{3}\right).$$

A polynomial is **completely factored** when each of its factors is prime. For instance,

$$x^3 - x^2 + 4x - 4 = (x - 1)(x^2 + 4)$$

is completely factored, but

$$x^3 - x^2 - 4x + 4 = (x - 1)(x^2 - 4)$$

is not completely factored. Its complete factorization would be

$$x^3 - x^2 - 4x + 4 = (x - 1)(x + 2)(x - 2).$$

The simplest type of factoring involves a polynomial that can be written as the product of a monomial and another polynomial. The technique used here is the Distributive Property, $a(b + c) = ab + ac$, in the *reverse* direction.

$$ab + ac = a(b + c)$$ a is a common factor.

Removing (factoring out) a common factor is the first step in completely factoring a polynomial.

EXAMPLE 5 **Removing Common Factors**

Factor each polynomial.

a. $3x^3 + 9x^2$ **b.** $6x^3 - 4x$ **c.** $(x - 2)(2x) + (x - 2)(3)$

Solution

a. $3x^3 + 9x^2 = 3x^2(x) + 3x^2(3)$ $3x^2$ is a common factor.

$= 3x^2(x + 3)$

b. $6x^3 - 4x = 2x(3x^2) - 2x(2)$ $2x$ is a common factor.

$= 2x(3x^2 - 2)$

c. $(x - 2)(2x) + (x - 2)(3) = (x - 2)(2x + 3)$ $x - 2$ is a common factor.

Factoring Special Polynomial Forms

Factored Form	Example
Difference of Two Squares	
$u^2 - v^2 = (u + v)(u - v)$	$9x^2 - 4 = (3x)^2 - 2^2 = (3x + 2)(3x - 2)$
Perfect Square Trinomial	
$u^2 + 2uv + v^2 = (u + v)^2$	$x^2 + 6x + 9 = x^2 + 2(x)(3) + 3^2 = (x + 3)^2$
$u^2 - 2uv + v^2 = (u - v)^2$	$x^2 - 6x + 9 = x^2 - 2(x)(3) + 3^2 = (x - 3)^2$
Sum or Difference of Two Cubes	
$u^3 + v^3 = (u + v)(u^2 - uv + v^2)$	$x^3 + 8 = x^3 + 2^3 = (x + 2)(x^2 - 2x + 4)$
$u^3 - v^3 = (u - v)(u^2 + uv + v^2)$	$27x^3 - 1 = (3x)^3 - 1^3 = (3x - 1)(9x^2 + 3x + 1)$

One of the easiest special polynomial forms to factor is the difference of two squares. Think of the form as follows.

$$u^2 - v^2 = (u + v)(u - v)$$

Difference (pointing to the minus in $u^2 - v^2$)   Opposite signs (pointing to the signs in $(u + v)(u - v)$)

To recognize perfect square terms, look for coefficients that are squares of integers and variables raised to *even powers.*

Note In Example 6, note that the first step in factoring a polynomial is to check for common factors. Once the common factor is removed, it is often possible to recognize patterns that were not immediately obvious.

EXAMPLE 6 **Removing a Common Factor First**

$$3 - 12x^2 = 3(1 - 4x^2) = 3[1^2 - (2x)^2] = 3(1 + 2x)(1 - 2x)$$

EXAMPLE 7 **Factoring the Difference of Two Squares**

a. $(x + 2)^2 - y^2 = [(x + 2) + y][(x + 2) - y]$

$= (x + 2 + y)(x + 2 - y)$

b. $16x^4 - 81 = (4x^2)^2 - 9^2$

$= (4x^2 + 9)(4x^2 - 9)$ Difference of two squares

$= (4x^2 + 9)[(2x)^2 - 3^2]$

$= (4x^2 + 9)(2x + 3)(2x - 3)$ Difference of two squares

A perfect square trinomial is the square of a binomial, and it has the following form.

$$u^2 + 2uv + v^2 = (u + v)^2 \qquad \text{or} \qquad u^2 - 2uv + v^2 = (u - v)^2$$

Like signs   Like signs

Note that the first and last terms are squares and the middle term is twice the product of u and v.

EXAMPLE 8 **Factoring Perfect Square Trinomials**

a. $16x^2 + 8x + 1 = (4x)^2 + 2(4x)(1) + 1^2$
$= (4x + 1)^2$

b. $x^2 - 10x + 25 = x^2 - 2(x)(5) + 5^2$
$= (x - 5)^2$

EXPLORATION

Find a formula for completely factoring $u^6 - v^6$ using the formulas from this section. Use your formula to completely factor $x^6 - 1$ and $x^6 - 64$.

The next two formulas show the sums and differences of cubes. Pay special attention to the signs of the terms.

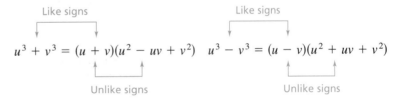

Like signs   Like signs

$$u^3 + v^3 = (u + v)(u^2 - uv + v^2) \qquad u^3 - v^3 = (u - v)(u^2 + uv + v^2)$$

Unlike signs   Unlike signs

EXAMPLE 9 **Factoring the Difference of Cubes**

$x^3 - 27 = x^3 - 3^3$ — Rewrite 27 as 3^3.
$= (x - 3)(x^2 + 3x + 9)$ — Factor.

EXAMPLE 10 **Factoring the Sum of Cubes**

a. $y^3 + 8 = y^3 + 2^3$ — Rewrite 8 as 2^3.
$= (y + 2)(y^2 - 2y + 4)$ — Factor.

b. $3(x^3 + 64) = 3(x^3 + 4^3)$ — Rewrite 64 as 4^3.
$= 3(x + 4)(x^2 - 4x + 16)$ — Factor.

Trinomials with Binomial Factors

To factor a trinomial of the form $ax^2 + bx + c$, use the following pattern.

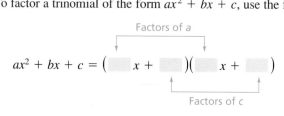

Factors of a

$$ax^2 + bx + c = (\square x + \square)(\square x + \square)$$

Factors of c

The goal is to find a combination of factors of a and c so that the outer and inner products add up to the middle term bx. For instance, in the trinomial $6x^2 + 17x + 5$, you can write

F O I L

$$(2x + 5)(3x + 1) = 6x^2 + 2x + 15x + 5 = 6x^2 + 17x + 5.$$

Note that the outer (O) and inner (I) products add up to $17x$.

> **Study Tip**
>
> Factoring a trinomial can involve trial and error. However, once you have produced the factored form, it is an easy matter to check your answer. For instance, you can verify the factorization in Example 11 by multiplying out the expression $(x - 3)(x - 4)$ to see that you obtain the original trinomial, $x^2 - 7x + 12$.

EXAMPLE 11 Factoring a Trinomial: Leading Coefficient Is 1

Factor $x^2 - 7x + 12$.

Solution

The possible factorizations are

$$(x - 2)(x - 6), \quad (x - 1)(x - 12), \quad \text{and} \quad (x - 3)(x - 4).$$

Testing the middle term, you will find the correct factorization to be

$$x^2 - 7x + 12 = (x - 3)(x - 4).$$

EXAMPLE 12 Factoring a Trinomial: Leading Coefficient Is Not 1

Factor $2x^2 + x - 15$.

Solution

The eight possible factorizations are as follows.

$(2x - 1)(x + 15)$ $\quad$ $(2x + 1)(x - 15)$

$(2x - 3)(x + 5)$ $\quad$ $(2x + 3)(x - 5)$

$(2x - 5)(x + 3)$ $\quad$ $(2x + 5)(x - 3)$

$(2x - 15)(x + 1)$ $\quad$ $(2x + 15)(x - 1)$

Testing the middle term, you will find the correct factorization to be

$$2x^2 + x - 15 = (2x - 5)(x + 3).$$

Factoring by Grouping

Sometimes polynomials with more than three terms can be factored by a method called **factoring by grouping.** It is not always obvious which terms to group, and sometimes several different groupings will work.

EXAMPLE 13 **Factoring by Grouping**

$$\begin{aligned} x^3 - 2x^2 - 3x + 6 &= (x^3 - 2x^2) - (3x - 6) && \text{Group terms.} \\ &= x^2(x - 2) - 3(x - 2) && \text{Factor groups.} \\ &= (x - 2)(x^2 - 3) && \text{Distributive Property} \end{aligned}$$

Factoring a trinomial can involve quite a bit of trial and error. Some of this trial and error can be lessened by using factoring by grouping.

EXAMPLE 14 **Factoring a Trinomial by Grouping**

Use factoring by grouping to factor $2x^2 + 5x - 3$.

Solution

In the trinomial $2x^2 + 5x - 3$, we have $a = 2$ and $c = -3$, which implies that the product ac is -6. Now, because -6 factors as $(6)(-1)$ and $6 - 1 = 5 = b$, we rewrite the middle term as $5x = 6x - x$. This produces the following.

$$\begin{aligned} 2x^2 + 5x - 3 &= 2x^2 + 6x - x - 3 && \text{Rewrite middle term.} \\ &= (2x^2 + 6x) - (x + 3) && \text{Group terms.} \\ &= 2x(x + 3) - (x + 3) && \text{Factor groups.} \\ &= (x + 3)(2x - 1) && \text{Distributive Property} \end{aligned}$$

Therefore, the trinomial factors as

$$2x^2 + 5x - 3 = (x + 3)(2x - 1).$$

Guidelines for Factoring Polynomials

1. Factor out any common factors by the Distributive Property.
2. Factor according to one of the special polynomial forms.
3. Factor as $ax^2 + bx + c = (mx + r)(nx + s)$.
4. Factor by grouping.

Group Activity

A Three-Dimensional View of a Special Product

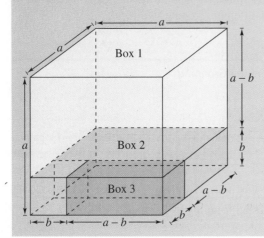

The figure at the left shows two cubes.

a. The large cube has a volume of a^3.

b. The small cube has a volume of b^3.

If the smaller cube is removed from the larger, the remaining solid has a volume of $a^3 - b^3$ and is composed of three rectangular boxes labeled Box 1, Box 2, and Box 3. Find the volume of each box and describe how these results are related to the special product formula.

$$a^3 - b^3 = (a - b)(a^2 + ab + b^2)$$
$$= (a - b)a^2 + (a - b)ab + (a - b)b^2$$

P.3 /// EXERCISES

In Exercises 1–12, perform the operations and write the result in standard form.

1. $(6x + 5) - (8x + 15)$
2. $(2x^2 + 1) - (x^2 - 2x + 1)$
3. $-(x^3 - 2) + (4x^3 - 2x)$
4. $-(5x^2 - 1) - (-3x^2 + 5)$
5. $(15x^2 - 6) - (-8x^3 - 14x^2 - 17)$
6. $(15x^4 - 18x - 19) - (13x^4 - 5x + 15)$
7. $3x(x^2 - 2x + 1)$
8. $y^2(4y^2 + 2y - 3)$
9. $-5z(3z - 1)$
10. $-4x(3 - x^3)$
11. $(1 - x^3)(4x)$
12. $(-2x)(-3x)(5x + 2)$

In Exercises 13–36, find the product.

13. $(x + 3)(x + 4)$
14. $(x - 5)(x + 10)$
15. $(3x - 5)(2x + 1)$
16. $(7x - 2)(4x - 3)$
17. $(2x - 5y)^2$
18. $(5 - 8x)^2$
19. $[(x - 3) + y]^2$
20. $[(x + 1) - y]^2$
21. $(x + 10)(x - 10)$
22. $(2x + 3)(2x - 3)$
23. $(x + 2y)(x - 2y)$
24. $(2x + 3y)(2x - 3y)$
25. $[(m - 3) + n][(m - 3) - n]$
26. $[(x + y) + 1][(x + y) - 1]$
27. $(2r^2 - 5)(2r^2 + 5)$
28. $(3a^3 - 4b^2)(3a^3 + 4b^2)$
29. $(x + 1)^3$
30. $(x - 2)^3$
31. $(2x - y)^3$
32. $(3x + 2y)^3$
33. $5x(x + 1) - 3x(x + 1)$
34. $(2x - 1)(x + 3) + 3(x + 3)$
35. $(u + 2)(u - 2)(u^2 + 4)$
36. $(x + y)(x - y)(x^2 + y^2)$
37. *Think About It* Must the sum of two second-degree polynomials be a second-degree polynomial? If not, give an example.
38. *Think About It* Is the product of two binomials always a binomial? Explain.

39. *Compound Interest* After 2 years, an investment of \$500 compounded annually at an interest rate r will yield an amount of

$500(1 + r)^2.$

(a) Write this polynomial in standard form.

(b) Use a calculator to evaluate the expression for the values of r in the table.

r	$5\frac{1}{2}\%$	7%	8%	$8\frac{1}{2}\%$	9%
$500(1 + r)^2$					

(c) What conclusion can you make from the table?

40. *Compound Interest* After 3 years, an investment of \$1200 compounded annually at an interest rate r will yield an amount of

$1200(1 + r)^3.$

(a) Write this polynomial in standard form.

(b) Use a calculator to evaluate the expression for the values of r in the table.

r	6%	7%	$7\frac{1}{2}\%$	8%	$8\frac{1}{2}\%$
$1200(1 + r)^3$					

(c) What conclusion can you make from the table?

41. *Volume of a Box* A closed box is constructed by cutting along the solid lines and folding along the broken lines on the rectangular piece of metal shown in the figure. The length and width of the rectangle are 45 centimeters and 15 centimeters, respectively. Find the volume of the box in terms of x. Find the volume when $x = 3$, $x = 5$, and $x = 7$.

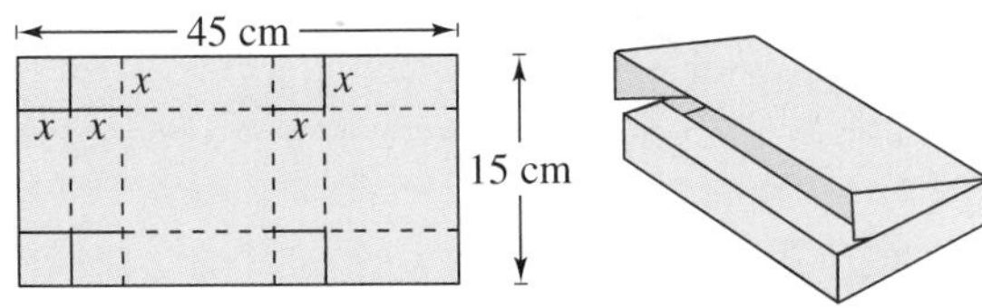

42. *Volume of a Box* An open box is made by cutting squares out of the corners of a piece of metal that is 18 centimeters by 26 centimeters (see figure). If the edge of each cut-out square is x inches, find the volume when $x = 1$, $x = 2$, and $x = 3$.

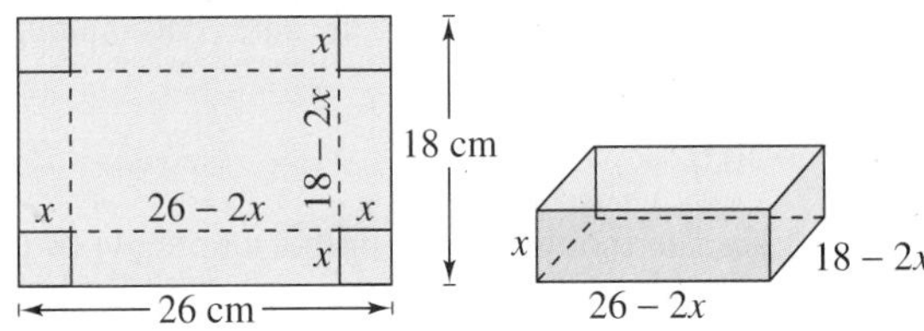

43. *Stopping Distance* The stopping distance of an automobile is the distance traveled during the driver's reaction time plus the distance traveled after the brakes are applied. In an experiment, these distances were measured (in feet) when the automobile was traveling at a speed of x miles per hour (see figure). The distance traveled during the reaction time was $R = 1.1x$, and the braking distance was $B = 0.14x^2 - 4.43x + 58.40$.

(a) Determine the polynomial that represents the total stopping distance.

(b) Use the result of part (a) to estimate the total stopping distance when $x = 30$, $x = 40$, and $x = 55$.

(c) Use the bar graph to make a statement about the total stopping distance required for increasing speeds.

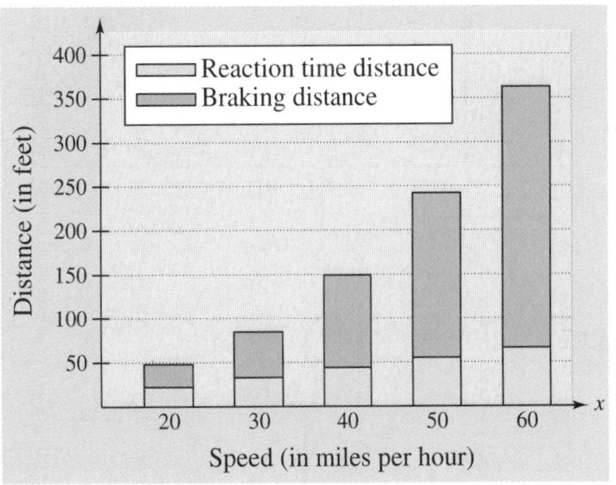

44. *Safe Beam Load* A uniformly distributed load is placed on a 1-inch-wide steel beam. When the span of the beam is x feet and its depth is 6 inches, the safe load is approximated by

$$S_6 = (0.06x^2 - 2.42x + 38.71)^2.$$

When the depth is 8 inches, the safe load is approximated by

$$S_8 = (0.08x^2 - 3.30x + 51.93)^2.$$

(a) Estimate the difference in the safe loads of these two beams when the span is 12 feet (see figure).

(b) How does the difference in safe load change as the span increases?

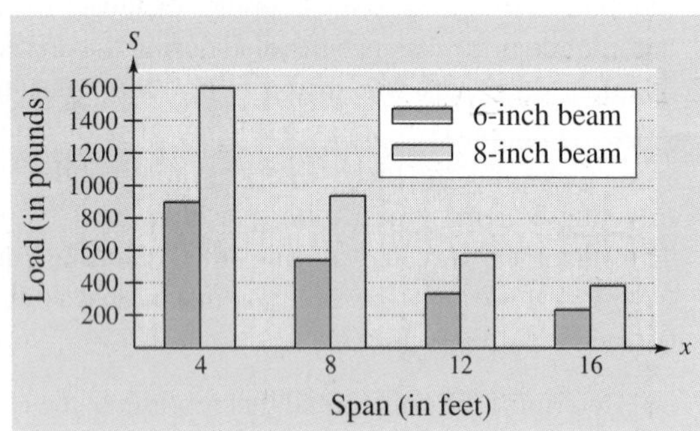

Geometrical Modeling **In Exercises 45 and 46, use the area model to write two different expressions for the area. Then equate the two expressions and name the algebraic property that is illustrated.**

45.

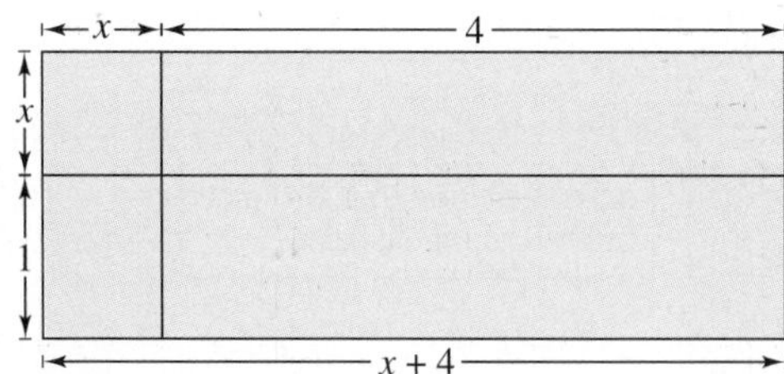

46.

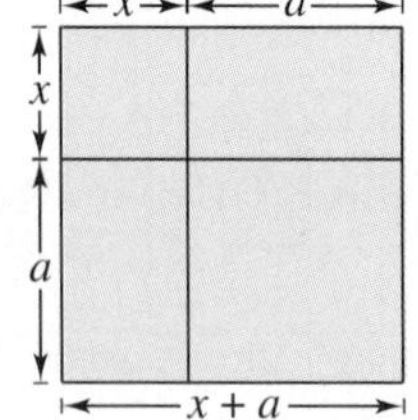

In Exercises 47–50, factor out the common factor.

47. $3x + 6$ **48.** $5y - 30$

49. $2x^3 - 6x$ **50.** $4x^3 - 6x^2 + 12x$

In Exercises 51–56, factor the difference of two squares.

51. $x^2 - 36$ **52.** $x^2 - \frac{1}{4}$

53. $16y^2 - 9$ **54.** $49 - 9y^2$

55. $(x - 1)^2 - 4$ **56.** $25 - (z + 5)^2$

In Exercises 57–60, factor the perfect square trinomial.

57. $x^2 - 4x + 4$

58. $x^2 + 10x + 25$

59. $4t^2 + 4t + 1$

60. $9x^2 - 12x + 4$

In Exercises 61–70, factor the trinomial.

61. $x^2 + x - 2$ **62.** $x^2 + 5x + 6$

63. $s^2 - 5s + 6$ **64.** $t^2 - t - 6$

65. $20 - y - y^2$ **66.** $24 + 5z - z^2$

67. $3x^2 - 5x + 2$

68. $2x^2 - x - 1$

69. $5x^2 + 26x + 5$

70. $-5u^2 - 13u + 6$

In Exercises 71–74, factor the sum or difference of cubes.

71. $x^3 - 8$ **72.** $x^3 - 27$

73. $y^3 + 64$ **74.** $z^3 + 125$

In Exercises 75–78, factor by grouping.

75. $x^3 - x^2 + 2x - 2$

76. $x^3 + 5x^2 - 5x - 25$

77. $2x^3 - x^2 - 6x + 3$

78. $5x^3 - 10x^2 + 3x - 6$

In Exercises 79–110, completely factor the expression.

79. $x^3 - 9x$

80. $12x^2 - 48$

81. $x^3 - 4x^2$

82. $6x^2 - 54$

83. $x^2 - 2x + 1$

84. $16 + 6x - x^2$

85. $1 - 4x + 4x^2$

86. $-9x^2 + 6x - 1$

87. $2x^2 + 4x - 2x^3$

88. $2y^3 - 7y^2 - 15y$

89. $9x^2 + 10x + 1$

90. $13x + 6 + 5x^2$

91. $3x^3 + x^2 + 15x + 5$

92. $5 - x + 5x^2 - x^3$

93. $x^4 - 4x^3 + x^2 - 4x$

94. $3u - 2u^2 + 6 - u^3$

95. $25 - (z + 5)^2$

96. $(t - 1)^2 - 49$

97. $(x^2 + 1)^2 - 4x^2$

98. $(x^2 + 8)^2 - 36x^2$

99. $2t^3 - 16$

100. $5x^3 + 40$

101. $4x(2x - 1) + (2x - 1)^2$

102. $5(3 - 4x)^2 - 8(3 - 4x)(5x - 1)$

103. $2(x + 1)(x - 3)^2 - 3(x + 1)^2(x - 3)$

104. $7(3x + 2)^2(1 - x)^2 + (3x + 2)(1 - x)^3$

105. $7x(2)(x^2 + 1)(2x) - (x^2 + 1)^2(7)$

106. $3(x - 2)^2(x + 1)^4 + (x - 2)^3(4)(x + 1)^3$

107. $2x(x - 5)^4 - x^2(4)(x - 5)^3$

108. $5(x^6 + 1)^4(6x^5)(3x + 2)^3 + 3(3x + 2)^2(3)(x^6 + 1)^5$

109. $\frac{x^2}{2}(x^2 + 1)^4 - (x^2 + 1)^5$

110. $5w^3(9w + 1)^4(9) + (2w + 1)^5(3w^2)$

Geometric Modeling **In Exercises 111–114, match the factoring formula with the correct "geometric factoring model." [The models are labeled (a), (b), (c), and (d).]**

111. $a^2 - b^2 = (a + b)(a - b)$

112. $a^2 + 2ab + b^2 = (a + b)^2$

113. $a^2 + 2a + 1 = (a + 1)^2$

114. $ab + a + b + 1 = (a + 1)(b + 1)$

(a)

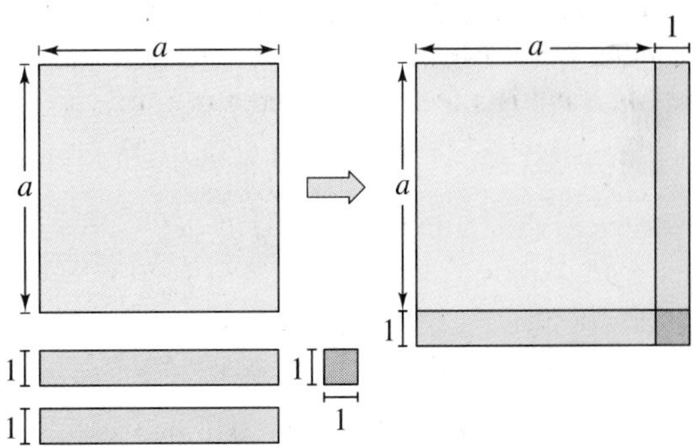

(b)

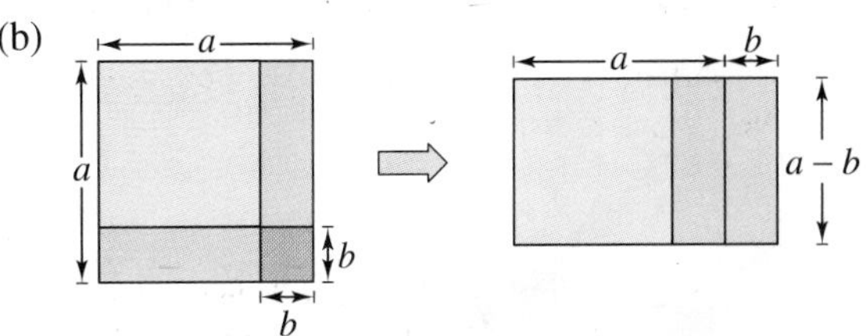

(c)

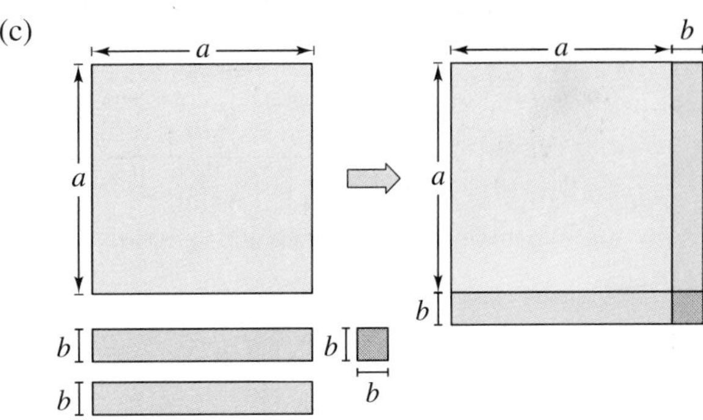

(d)

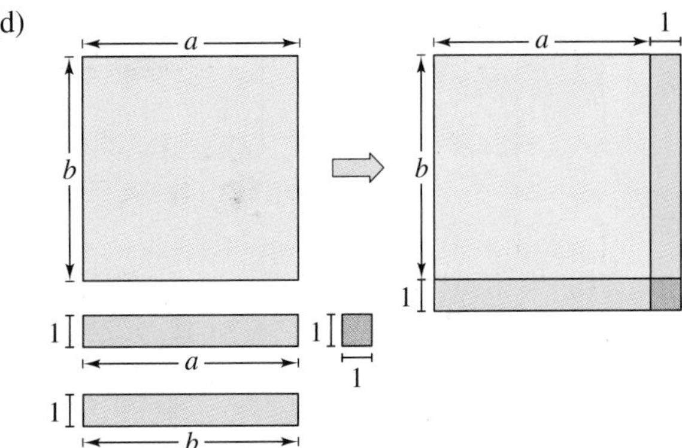

Geometric Modeling **In Exercises 115–118, draw a "geometric factoring model" to represent the factorization. For instance, a factoring model for**

$$2x^2 + 3x + 1 = (2x + 1)(x + 1)$$

is shown in the figure.

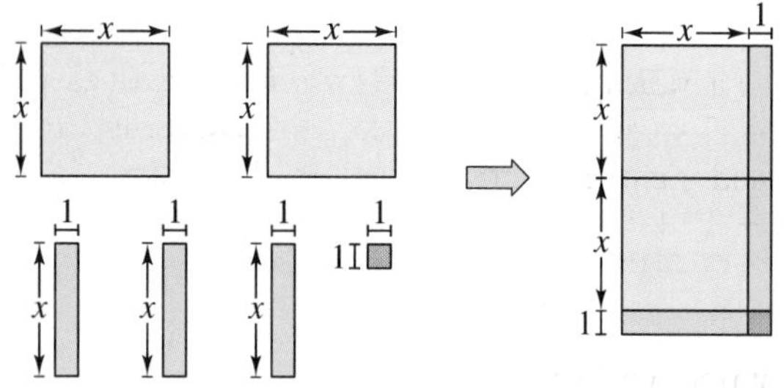

115. $3x^2 + 7x + 2 = (3x + 1)(x + 2)$

116. $x^2 + 4x + 3 = (x + 3)(x + 1)$

117. $2x^2 + 7x + 3 = (2x + 1)(x + 3)$

118. $x^2 + 3x + 2 = (x + 2)(x + 1)$

Geometry **In Exercises 119–122, write, in factored form, an expression for the shaded portion of the figure.**

119.

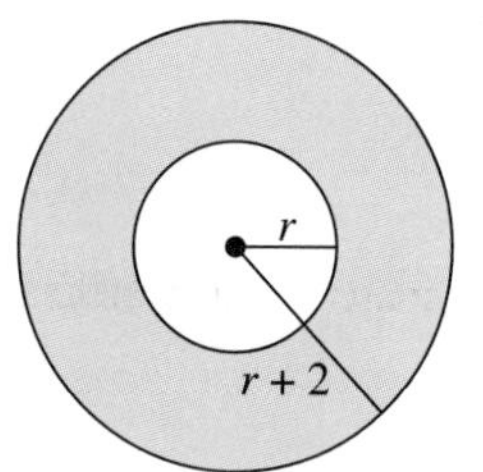

120.

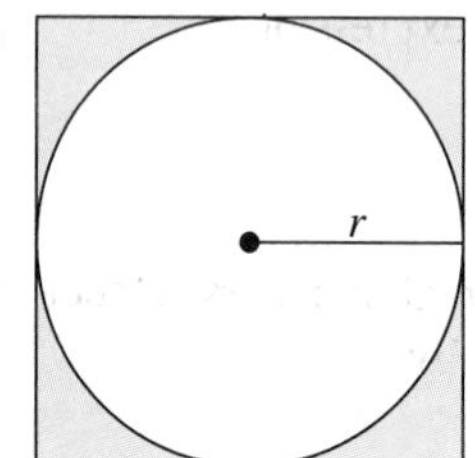

121.

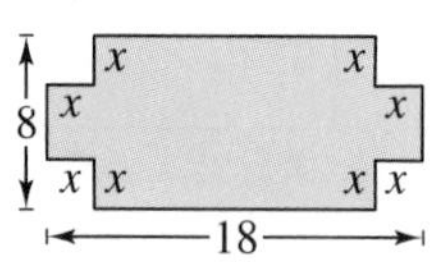

122.

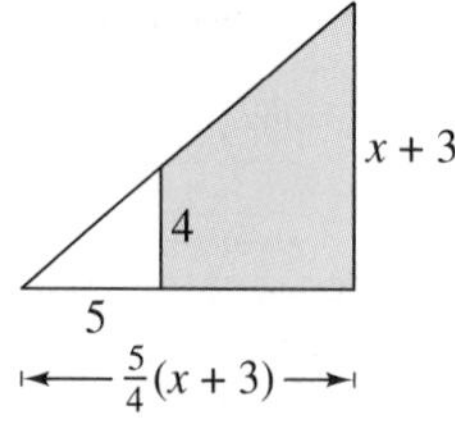

In Exercises 123 and 124, find all values of b for which the trinomial can be factored.

123. $x^2 + bx - 15$

124. $x^2 + bx + 50$

In Exercises 125 and 126, find two integers c such that the trinomial can be factored. (There are many correct answers.)

125. $2x^2 + 5x + c$

126. $3x^2 - 10x + c$

127. *Error Analysis* Describe the error.

$$9x^2 - 9x - 54 = (3x + 6)(3x - 9)$$
$$= 3(x + 2)(x - 3)$$

128. *Think About It* Is $(3x - 6)(x + 1)$ completely factored? Explain.

129. *Geometry* The cylindrical shell shown in the figure has a volume of

$$V = \pi R^2 h - \pi r^2 h.$$

(a) Factor the expression for the volume.

(b) From the result of part (a), show that the volume is 2π (average radius)(thickness of the shell)h.

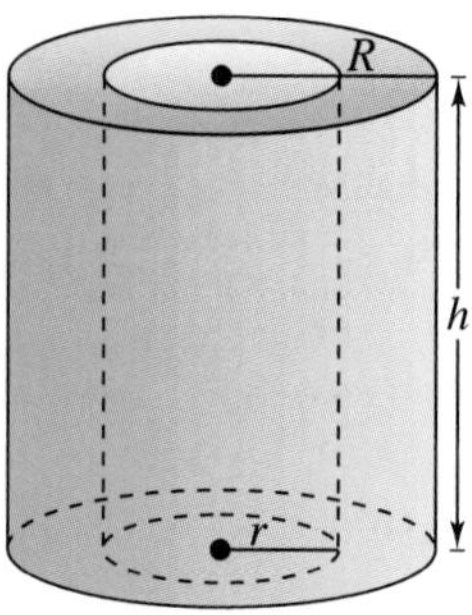

130. *Chemistry* The rate of change of an autocatalytic chemical reaction is $kQx - kx^2$, where Q is the amount of the original substance, x is the amount of substance formed, and k is a constant of proportionality. Factor the expression.

P.4 Fractional Expressions

Domain of an Algebraic Expression / Simplifying Rational Expressions / Operations with Rational Expressions / Compound Fractions

Domain of an Algebraic Expression

The set of real numbers for which an algebraic expression is defined is the **domain** of the expression. Two algebraic expressions are **equivalent** if they have the same domain and yield the same values for all numbers in their domain. For instance, $(x + 1) + (x + 2)$ and $2x + 3$ are equivalent.

EXAMPLE 1 **Finding the Domain of an Algebraic Expression**

a. The domain of the polynomial

$$2x^3 + 3x + 4$$

is the set of all real numbers. In fact, the domain of any polynomial is the set of all real numbers, *unless* the domain is specifically restricted.

b. The domain of the radical expression

$$\sqrt{x - 2}$$

is the set of real numbers greater than or equal to 2, because the square root of a negative number is not a real number.

c. The domain of the expression

$$\frac{x + 2}{x - 3}$$

is the set of all real numbers except $x = 3$, which would produce an undefined division by zero.

The quotient of two algebraic expressions is a **fractional expression.** Moreover, the quotient of two *polynomials* such as

$$\frac{1}{x}, \quad \frac{2x - 1}{x + 1}, \quad \text{or} \quad \frac{x^2 - 1}{x^2 + 1}$$

is a **rational expression.** Recall that a fraction is in reduced form if its numerator and denominator have no factors in common aside from ± 1. To write a fraction in reduced form, apply the following rule.

$$\frac{a \cdot \cancel{c}}{b \cdot \cancel{c}} = \frac{a}{b}, \qquad b \neq 0 \quad \text{and} \quad c \neq 0.$$

The key to success in simplifying rational expressions lies in your ability to *factor* polynomials.

EXAMPLE 2 **Reducing a Rational Expression**

Write $\dfrac{x^2 + 4x - 12}{3x - 6}$ in reduced form.

Solution

$$\frac{x^2 + 4x - 12}{3x - 6} = \frac{(x + 6)(x - 2)}{3(x - 2)} \qquad \text{Factor completely.}$$

$$= \frac{x + 6}{3}, \quad x \neq 2 \qquad \text{Cancel common factors.}$$

Note that the original expression is undefined when $x = 2$ (because division by zero is undefined). To make sure that the reduced expression is *equivalent* to the original expression, you must restrict the domain of the reduced expression by excluding the value $x = 2$.

Study Tip

In Example 2, do not make the mistake of trying to reduce further by dividing *terms.*

$$\frac{x + 6}{3} \neq \frac{x + \overset{2}{\cancel{6}}}{\cancel{3}} = x + 2$$

Remember that to reduce fractions, divide common *factors*, not terms.

Simplifying Rational Expressions

When simplifying rational expressions, be sure to factor each polynomial completely before concluding that the numerator and denominator have no factors in common. Moreover, changing the sign of a factor may allow further reduction, as shown in part (b) of the next example.

EXAMPLE 3 **Reducing Rational Expressions**

a. $$\frac{x^3 - 4x}{x^2 + x - 2} = \frac{x(x^2 - 4)}{(x + 2)(x - 1)}$$

$$= \frac{x(x + 2)(x - 2)}{(x + 2)(x - 1)} \qquad \text{Factor completely.}$$

$$= \frac{x(x - 2)}{x - 1}, \quad x \neq -2 \qquad \text{Cancel common factors.}$$

b. $$\frac{12 + x - x^2}{2x^2 - 9x + 4} = \frac{(4 - x)(3 + x)}{(2x - 1)(x - 4)} \qquad \text{Factor completely.}$$

$$= \frac{-(x - 4)(3 + x)}{(2x - 1)(x - 4)} \qquad (4 - x) = -(x - 4)$$

$$= -\frac{3 + x}{2x - 1}, \quad x \neq 4 \qquad \text{Cancel common factors.}$$

Operations with Rational Expressions

To multiply or divide rational expressions, we use the properties of fractions discussed in Section P.1. Recall that to divide fractions we invert the divisor and multiply.

EXAMPLE 4 **Multiplying Rational Expressions**

$$\frac{2x^2 + x - 6}{x^2 + 4x - 5} \cdot \frac{x^3 - 3x^2 + 2x}{4x^2 - 6x} = \frac{(2x - 3)(x + 2)}{(x + 5)(x - 1)} \cdot \frac{x(x - 2)(x - 1)}{2x(2x - 3)}$$

$$= \frac{(x + 2)(x - 2)}{2(x + 5)}, \qquad x \neq 0, x \neq 1, x \neq \tfrac{3}{2}$$

EXAMPLE 5 **Dividing Rational Expressions**

$$\frac{x^3 - 8}{x^2 - 4} \div \frac{x^2 + 2x + 4}{x^3 + 8} = \frac{x^3 - 8}{x^2 - 4} \cdot \frac{x^3 + 8}{x^2 + 2x + 4} \qquad \text{Invert and multiply.}$$

$$= \frac{(x - 2)(x^2 + 2x + 4)}{(x + 2)(x - 2)} \cdot \frac{(x + 2)(x^2 - 2x + 4)}{x^2 + 2x + 4}$$

$$= x^2 - 2x + 4, \qquad x \neq \pm 2$$

To add or subtract rational expressions, you can use the LCD (least common denominator) method or the basic definition

$$\frac{a}{b} \pm \frac{c}{d} = \frac{ad \pm bc}{bd}, \qquad b \neq 0 \text{ and } d \neq 0. \qquad \text{Basic definition}$$

This definition provides an efficient way of adding or subtracting *two* fractions that have no common factors in their denominators.

EXAMPLE 6 **Subtracting Rational Expressions**

$$\frac{x}{x - 3} - \frac{2}{3x + 4} = \frac{x(3x + 4) - 2(x - 3)}{(x - 3)(3x + 4)} \qquad \text{Basic definition}$$

$$= \frac{3x^2 + 4x - 2x + 6}{(x - 3)(3x + 4)} \qquad \text{Remove parentheses.}$$

$$= \frac{3x^2 + 2x + 6}{(x - 3)(3x + 4)} \qquad \text{Combine like terms.}$$

For three or more fractions, or for fractions with a repeated factor in the denominators, the LCD method works well. Recall that the least common denominator of several fractions consists of the product of all prime factors in the denominators, with each factor given the highest power of its occurrence in any denominator. Here is a numerical example.

$$\begin{aligned}\frac{1}{6}+\frac{3}{4}-\frac{2}{3} &= \frac{1 \cdot 2}{6 \cdot 2}+\frac{3 \cdot 3}{4 \cdot 3}-\frac{2 \cdot 4}{3 \cdot 4} \\ &= \frac{2}{12}+\frac{9}{12}-\frac{8}{12} \\ &= \frac{3}{12} \\ &= \frac{1}{4}\end{aligned}$$

The LCD is 12.

Note Sometimes the numerator of the answer has a factor in common with the denominator. In such cases the answer should be reduced. For instance, in the example above, $\frac{3}{12}$ was reduced to $\frac{1}{4}$.

EXAMPLE 7 **Combining Rational Expressions: The LCD Method**

Perform the operations and simplify.

$$\frac{3}{x-1}-\frac{2}{x}+\frac{x+3}{x^2-1}$$

Solution

Using the factored denominators $(x-1)$, x, and $(x+1)(x-1)$, you can see that the LCD is $x(x+1)(x-1)$.

$$\begin{aligned}\frac{3}{x-1}-\frac{2}{x}+\frac{x+3}{(x+1)(x-1)} &= \frac{3(x)(x+1)}{x(x+1)(x-1)}-\frac{2(x+1)(x-1)}{x(x+1)(x-1)}+\frac{(x+3)(x)}{x(x+1)(x-1)} \\ &= \frac{3(x)(x+1)-2(x+1)(x-1)+(x+3)(x)}{x(x+1)(x-1)} \\ &= \frac{3x^2+3x-2x^2+2+x^2+3x}{x(x+1)(x-1)} \\ &= \frac{2x^2+6x+2}{x(x+1)(x-1)} \\ &= \frac{2(x^2+3x+1)}{x(x+1)(x-1)}\end{aligned}$$

Compound Fractions

Fractional expressions with separate fractions in the numerator, denominator, or both, are called **compound** or **complex fractions.** Here are two examples.

$$\frac{\left(\frac{1}{x}\right)}{x^2+1} \quad \text{and} \quad \frac{\left(\frac{1}{x}\right)}{\left(\frac{1}{x^2+1}\right)}$$

A compound fraction can be simplified by first combining both its numerator and its denominator into single fractions, then inverting the denominator and multiplying.

Note In Example 8, the disclaimer $x \neq 1$ is added to the final expression to make its domain agree with the domain of the original expression.

EXAMPLE 8 **Simplifying a Compound Fraction**

$$\frac{\left(\frac{2}{x}-3\right)}{\left(1-\frac{1}{x-1}\right)} = \frac{\left[\frac{2-3(x)}{x}\right]}{\left[\frac{1(x-1)-1}{x-1}\right]} \qquad \text{Combine fractions.}$$

$$= \frac{\left(\frac{2-3x}{x}\right)}{\left(\frac{x-2}{x-1}\right)} \qquad \text{Simplify.}$$

$$= \frac{2-3x}{x} \cdot \frac{x-1}{x-2} \qquad \text{Invert and multiply.}$$

$$= \frac{(2-3x)(x-1)}{x(x-2)}, \quad x \neq 1$$

Another way to simplify a compound fraction is to multiply each term in its numerator and denominator by the LCD of all fractions in its numerator and denominator. This method is applied to the fraction in Example 8 as follows.

$$\frac{\left(\frac{2}{x}-3\right)}{\left(1-\frac{1}{x-1}\right)} = \frac{\left(\frac{2}{x}-3\right)}{\left(1-\frac{1}{x-1}\right)} \cdot \frac{x(x-1)}{x(x-1)}$$

$$= \frac{2(x-1)-3x(x-1)}{x(x-1)-x}$$

$$= \frac{-3x^2+5x-2}{x^2-2x}$$

$$= \frac{(2-3x)(x-1)}{x(x-2)}, \quad x \neq 1$$

The next three examples illustrate some methods for simplifying fractional expressions involving radicals and negative exponents. These types of expressions occur frequently in calculus.

EXAMPLE 9 **Simplifying an Expression with Negative Exponents**

Simplify

$$x(1 - 2x)^{-3/2} + (1 - 2x)^{-1/2}.$$

Solution

By rewriting the expression with positive exponents, you obtain

$$\frac{x}{(1 - 2x)^{3/2}} + \frac{1}{(1 - 2x)^{1/2}}$$

which can then be combined by the LCD method. However, the process can be simplified by first removing the common factor with the *smaller exponent.*

$$\begin{aligned} x(1 - 2x)^{-3/2} + (1 - 2x)^{-1/2} &= (1 - 2x)^{-3/2}[x + (1 - 2x)^{(-1/2)-(-3/2)}] \\ &= (1 - 2x)^{-3/2}[x + (1 - 2x)^1] \\ &= \frac{1 - x}{(1 - 2x)^{3/2}} \end{aligned}$$

Note In Example 9, note that when factoring, you subtract exponents.

EXAMPLE 10 **Simplifying a Compound Fraction**

Simplify

$$\frac{(4 - x^2)^{1/2} + x^2(4 - x^2)^{-1/2}}{4 - x^2}.$$

Solution

$$\begin{aligned} \frac{(4 - x^2)^{1/2} + x^2(4 - x^2)^{-1/2}}{4 - x^2} &= \frac{(4 - x^2)^{1/2} + x^2(4 - x^2)^{-1/2}}{4 - x^2} \cdot \frac{(4 - x^2)^{1/2}}{(4 - x^2)^{1/2}} \\ &= \frac{(4 - x^2)^1 + x^2(4 - x^2)^0}{(4 - x^2)^{3/2}} \\ &= \frac{4 - x^2 + x^2}{(4 - x^2)^{3/2}} \\ &= \frac{4}{(4 - x^2)^{3/2}} \end{aligned}$$

Some graphing utilities have a table feature that can be used to create tables of values. For instance, to evaluate the expression $x^2 - 4$ for $x = 1, 2, 3, 4, 5, 6$, and 7 on a *TI-83,* you can use the following keystrokes.

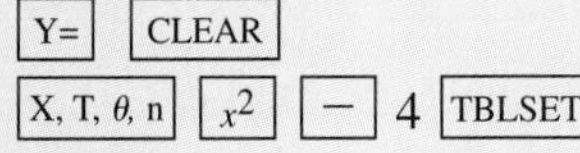

Y= CLEAR

X, T, θ, n x^2 − 4 TBLSET

TblStart=1
ΔTbl=1
Indpnt: Auto
Depend: Auto

TABLE

The table produced by these keystrokes is shown below.

X	Y1	
1	-3	
2	0	
3	5	
4	12	
5	21	
6	32	
7	45	
X = 1		

EXAMPLE 11 **Simplifying a Compound Fraction**

The expression from calculus

$$\frac{\sqrt{x+h} - \sqrt{x}}{h}$$

is an example of a *difference quotient.* Rewrite this expression by rationalizing its numerator.

Solution

$$\begin{aligned}
\frac{\sqrt{x+h} - \sqrt{x}}{h} &= \frac{\sqrt{x+h} - \sqrt{x}}{h} \cdot \frac{\sqrt{x+h} + \sqrt{x}}{\sqrt{x+h} + \sqrt{x}} \\
&= \frac{\left(\sqrt{x+h}\right)^2 - \left(\sqrt{x}\right)^2}{h\left(\sqrt{x+h} + \sqrt{x}\right)} \\
&= \frac{h}{h\left(\sqrt{x+h} + \sqrt{x}\right)} \\
&= \frac{1}{\sqrt{x+h} + \sqrt{x}}, \qquad h \neq 0
\end{aligned}$$

Notice that the original expression is meaningless when $h = 0$, but the final expression *could* be evaluated when $h = 0$.

Group Activity — *Comparing Domains of Two Expressions*

Complete the following table by evaluating the expressions

$$\frac{x^2 - 3x + 2}{x - 2} \quad \text{and} \quad x - 1$$

for the values of x. If your graphing utility has a *table feature,* use it to help create the table. Write a short paragraph describing the equivalence or nonequivalence of the two expressions.

x	−3	−2	−1	0	1	2	3
$\frac{x^2 - 3x + 2}{x - 2}$							
$x - 1$							

P.4 /// EXERCISES

In Exercises 1–10, find the domain of the expression.

1. $3x^2 - 4x + 7$
2. $2x^2 + 5x - 2$
3. $4x^3 + 3, \quad x \geq 0$
4. $6x^2 - 9, \quad x > 0$
5. $\dfrac{1}{x - 2}$
6. $\dfrac{x + 1}{2x + 1}$
7. $\dfrac{x - 1}{x^2 - 4x}$
8. $\dfrac{2x + 1}{x^2 - 9}$
9. $\sqrt{x + 1}$
10. $\dfrac{1}{\sqrt{x + 1}}$

In Exercises 11 and 12, find the missing factor in the numerator so that the two fractions will be equivalent.

11. $\dfrac{5}{2x} = \dfrac{5(\quad)}{6x^2}$
12. $\dfrac{3}{4} = \dfrac{3(\quad)}{4(x + 1)}$

In Exercises 13–26, write the rational expression in reduced form.

13. $\dfrac{15x^2}{10x}$
14. $\dfrac{18y^2}{60y^5}$
15. $\dfrac{3xy}{xy + x}$
16. $\dfrac{9x^2 + 9x}{2x + 2}$
17. $\dfrac{x - 5}{10 - 2x}$
18. $\dfrac{x^2 - 25}{5 - x}$
19. $\dfrac{x^3 + 5x^2 + 6x}{x^2 - 4}$
20. $\dfrac{x^2 + 8x - 20}{x^2 + 11x + 10}$
21. $\dfrac{y^2 - 7y + 12}{y^2 + 3y - 18}$
22. $\dfrac{3 - x}{x^2 + 11x + 10}$
23. $\dfrac{2 - x + 2x^2 - x^3}{x - 2}$
24. $\dfrac{x^2 - 9}{x^3 + x^2 - 9x - 9}$
25. $\dfrac{z^3 - 8}{z^2 + 2z + 4}$
26. $\dfrac{y^3 - 2y^2 - 3y}{y^3 + 1}$

In Exercises 27 and 28, complete the table. What can you conclude?

27.

x	0	1	2	3	4	5	6
$\dfrac{x^2 - 2x - 3}{x - 3}$							
$x + 1$							

28.

x	0	1	2	3	4	5	6
$\dfrac{x - 3}{x^2 - x - 6}$							
$\dfrac{1}{x + 2}$							

29. *Error Analysis* Describe the error.

$$\frac{5x^3}{2x^3 + 4} = \frac{5x^3}{2x^3 + 4} = \frac{5}{2 + 4} = \frac{5}{6}$$

30. *Think About It* Is the statement $(ax - b)/(b - ax) = -1$ true for all nonzero real numbers a and b? Explain.

In Exercises 31 and 32, find the ratio of the area of the shaded portion of the figure to the total area of the figure.

31.

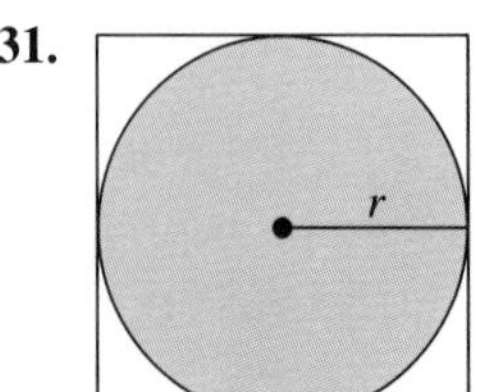

32.

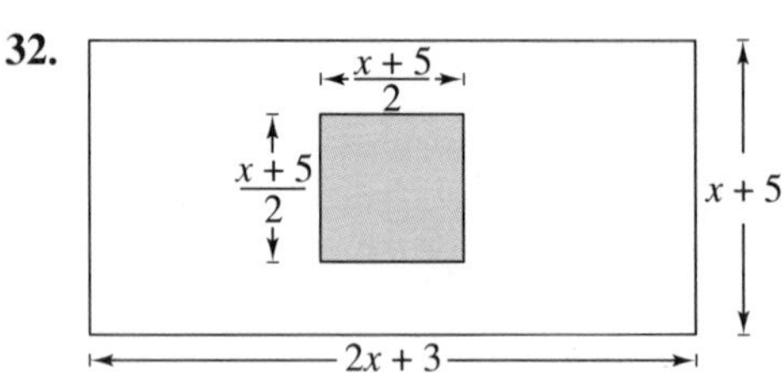

In Exercises 33–42, perform the multiplication or division and simplify.

33. $\dfrac{5}{x-1} \cdot \dfrac{x-1}{25(x-2)}$

34. $\dfrac{x+13}{x^3(3-x)} \cdot \dfrac{x(x-3)}{5}$

35. $\dfrac{r}{r-1} \cdot \dfrac{r^2-1}{r^2}$

36. $\dfrac{4y-16}{5y+15} \cdot \dfrac{2y+6}{4-y}$

37. $\dfrac{t^2-t-6}{t^2+6t+9} \cdot \dfrac{t+3}{t^2-4}$

38. $\dfrac{y^3-8}{2y^3} \cdot \dfrac{4y}{y^2-5y+6}$

39. $\dfrac{3(x+y)}{4} \div \dfrac{x+y}{2}$

40. $\dfrac{x+2}{5(x-3)} \div \dfrac{x-2}{5(x-3)}$

41. $\dfrac{\left[\dfrac{x^2}{(x+1)^2}\right]}{\left[\dfrac{x}{(x+1)^3}\right]}$

42. $\dfrac{\left(\dfrac{x^2-1}{x}\right)}{\left[\dfrac{(x-1)^2}{x}\right]}$

In Exercises 43–54, perform the addition or subtraction and simplify.

43. $\dfrac{5}{x-1} + \dfrac{x}{x-1}$

44. $\dfrac{2x-1}{x+3} + \dfrac{1-x}{x+3}$

45. $6 - \dfrac{5}{x+3}$

46. $\dfrac{3}{x-1} - 5$

47. $\dfrac{3}{x-2} + \dfrac{5}{2-x}$

48. $\dfrac{2x}{x-5} - \dfrac{5}{5-x}$

49. $\dfrac{1}{x^2-x-2} - \dfrac{x}{x^2-5x+6}$

50. $\dfrac{2}{x^2-x-2} + \dfrac{10}{x^2+2x-8}$

51. $-\dfrac{1}{x} + \dfrac{2}{x^2+1} + \dfrac{1}{x^3+x}$

52. $\dfrac{2}{x+1} + \dfrac{2}{x-1} + \dfrac{1}{x^2-1}$

53. $x^2(x^2+1)^{-5} - (x^2+1)^{-4}$

54. $2x(x-5)^{-3} - 4x^2(x-5)^{-4}$

In Exercises 55–62, simplify the compound fraction.

55. $\dfrac{\left(\dfrac{x}{2}-1\right)}{(x-2)}$

56. $\dfrac{(x-4)}{\left(\dfrac{x}{4}-\dfrac{4}{x}\right)}$

57. $\dfrac{\left[\dfrac{1}{(x+h)^2}-\dfrac{1}{x^2}\right]}{h}$

58. $\dfrac{\left(\dfrac{x+h}{x+h+1}-\dfrac{x}{x+1}\right)}{h}$

59. $\dfrac{\left(\sqrt{x}-\dfrac{1}{2\sqrt{x}}\right)}{\sqrt{x}}$

60. $\dfrac{3x^{1/3}-x^{-2/3}}{3x^{-2/3}}$

61. $\dfrac{\left(\dfrac{t^2}{\sqrt{t^2+1}}-\sqrt{t^2+1}\right)}{t^2}$

62. $\dfrac{-x^3(1-x^2)^{-1/2}-2x(1-x^2)^{1/2}}{x^4}$

In Exercises 63 and 64, rationalize the numerator of the expression.

63. $\dfrac{\sqrt{x+2}-\sqrt{x}}{2}$

64. $\dfrac{\sqrt{z-3}-\sqrt{z}}{3}$

65. *Rate* A photocopier copies at a rate of 16 pages per minute.

(a) Find the time required to copy one page.

(b) Find the time required to copy x pages.

(c) Find the time required to copy 60 pages.

66. *Rate* After working together for t hours on a common task, two workers have done fractional parts of the job equal to $t/3$ and $t/5$, respectively. What fractional part of the task has been completed?

67. *Average* Determine the average of the two real numbers $x/3$ and $2x/5$.

68. *Partition into Equal Parts* Find three real numbers that divide the real number line between $x/3$ and $3x/4$ into four equal parts.

Monthly Payment **In Exercises 69 and 70, use the formula that gives the approximate annual interest rate r of a monthly installment loan:**

$$r = \frac{\left[\frac{24(NM - P)}{N}\right]}{\left(P + \frac{NM}{12}\right)}$$

where N is the total number of payments, M is the monthly payment, and P is the amount financed.

69. (a) Approximate the annual interest rate for a 4-year car loan of $15,000 that has monthly payments of $400.

(b) Simplify the expression for the annual interest rate r, and then rework part (a).

70. (a) Approximate the annual interest rate for a 5-year car loan of $18,000 that has monthly payments of $400.

(b) Simplify the expression for the annual interest rate r, and then rework part (a).

71. *Refrigeration* When food (at room temperature) is placed in the refrigerator, the time required for the food to cool depends on the amount of food, the air circulation in the refrigerator, the original temperature of the food, and the temperature of the refrigerator. Consider the model that gives the temperature of the food that is at 75°F and is placed in a 40°F refrigerator

$$T = 10\left(\frac{4t^2 + 16t + 75}{t^2 + 4t + 10}\right)$$

where T is the temperature in degrees Fahrenheit and t is the time in hours.

(a) Complete the table.

t	0	1	2	3	4	5
T						

(b) Create a bar graph showing the temperatures at the times given in the table in part (a).

72. *Precious Metals* The costs per fine ounce of gold and silver for the years 1988 through 1992 are given in the table. (Source: U.S. Bureau of Mines)

Year	1988	1989	1990	1991	1992
Gold	$438	$383	$385	$363	$345
Silver	$6.54	$5.50	$4.82	$4.04	$3.94

Mathematical models for this data are

$$\text{Cost of gold} = \frac{5301t + 37{,}498}{19t + 100}$$

and

$$\text{Cost of silver} = \frac{237t + 4734}{176t + 1000}$$

where $t = 0$ corresponds to the year 1990.

(a) Create a table using the models to estimate the prices of each metal for the given years. Compare the estimates given by the models with the actual prices.

(b) Determine a model for the ratio of the price of gold to the price of silver. Use the model to find this ratio over the given years. Over this period of time, did the price of gold increase or decrease relative to the price of silver?

Probability **In Exercises 73 and 74, consider an experiment in which a marble is tossed into a box whose base is shown in the figure. The probability that the marble will come to rest in the shaded portion of the box is equal to the ratio of the shaded area to the total area of the figure. Find the probability.**

73.

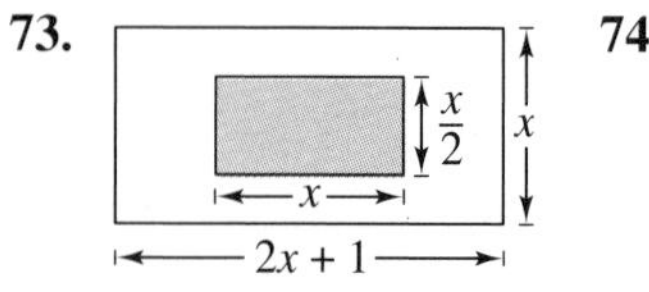

74.

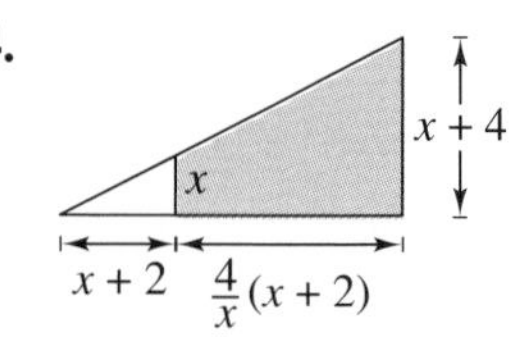

P.5 The Cartesian Plane

The Cartesian Plane / The Distance Formula / The Midpoint Formula / The Equation of a Circle / Application

Figure P.6

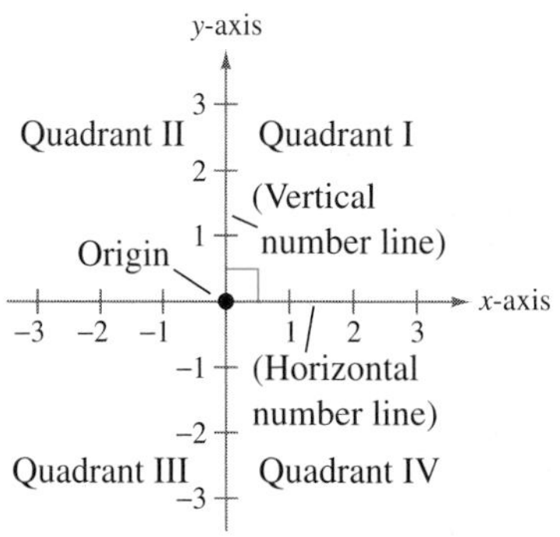

The Cartesian Plane

Just as you can represent real numbers by points on a real number line, you can represent ordered pairs of real numbers by points in a plane called the **rectangular coordinate system,** or the **Cartesian plane,** after the French mathematician René Descartes (1596–1650).

The Cartesian plane is formed by using two real lines intersecting at right angles, as shown in Figure P.6. The horizontal real line is usually called the ***x*-axis,** and the vertical real line is usually called the ***y*-axis.** The point of intersection of these two axes is the **origin,** and the two axes divide the plane into four parts called **quadrants.**

Figure P.7

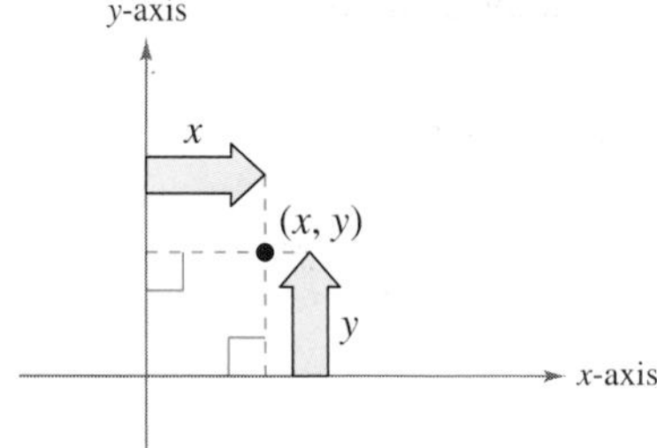

Each point in the plane corresponds to an **ordered pair** (x, y) of real numbers x and y, called **coordinates** of the point. The ***x*-coordinate** represents the directed distance from the y-axis to the point, and the ***y*-coordinate** represents the directed distance from the x-axis to the point, as shown in Figure P.7.

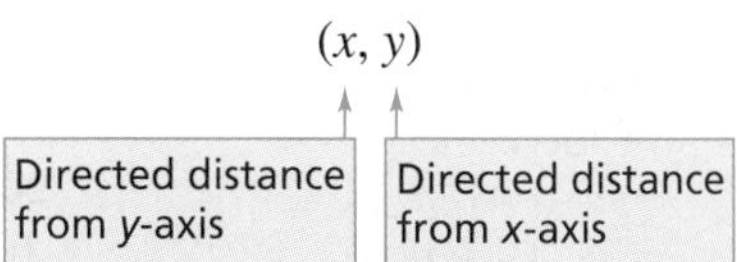

Note The notation (x, y) denotes both a point in the plane and an open interval on the real line. The context will tell you which meaning is intended.

Figure P.8

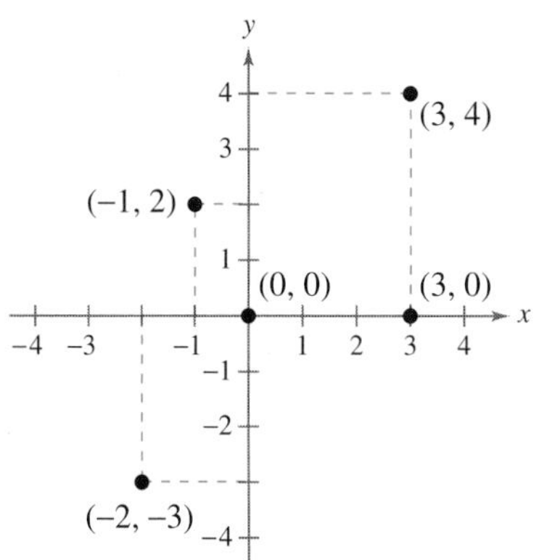

EXAMPLE 1 Plotting Points in the Cartesian Plane

Plot the points $(-1, 2)$, $(3, 4)$, $(0, 0)$, $(3, 0)$, and $(-2, -3)$.

Solution

To plot the point

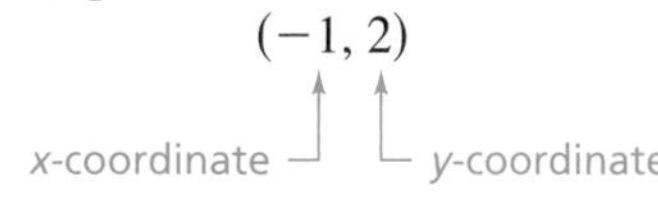

imagine a vertical line through -1 on the x-axis and a horizontal line through 2 on the y-axis. The intersection of these two lines is the point $(-1, 2)$. The other four points can be plotted in a similar way, and are shown in Figure P.8.

Figure P.9

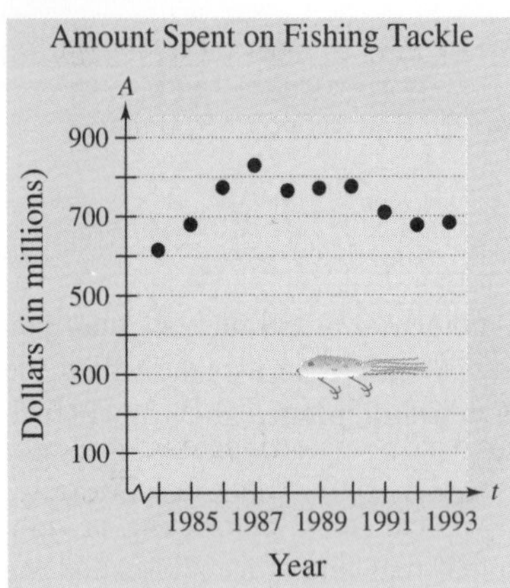

The beauty of a rectangular coordinate system is that it allows you to see relationships between two variables. It would be difficult to overestimate the importance of Descartes's introduction of coordinates to the plane. Today, his ideas are in common use in virtually every scientific and business-related field.

EXAMPLE 2 Sketching a Scatter Plot

The amounts A (in millions of dollars) spent on fishing tackle in the United States for the years 1984 to 1993 are given in the table, where t represents the year. Sketch a scatter plot of the data. (Source: National Sporting Goods Association)

t	1984	1985	1986	1987	1988	1989	1990	1991	1992	1993
A	616	681	773	830	766	769	776	711	678	685

Solution

To sketch a *scatter plot* of the data given in the table, you simply represent each pair of values by an ordered pair (t, A) and plot the resulting points, as shown in Figure P.9. For instance, the first pair of values is represented by the ordered pair (1984, 616). Note that the break in the t-axis indicates that the numbers between 0 and 1984 have been omitted.

Note In Example 2, you could have let $t = 1$ represent the year 1984. In that case, the horizontal axis would not have been broken, and the tick marks would have been labeled 1 through 10 (instead of 1984 through 1993).

The scatter plot in Example 2 is only one way to represent the data graphically. Two other techniques are shown at the right. The first is a *bar graph* and the second is a *line graph*. All three graphical representations were created with a computer. Try using a graphing utility to graphically represent the data given in Example 2.

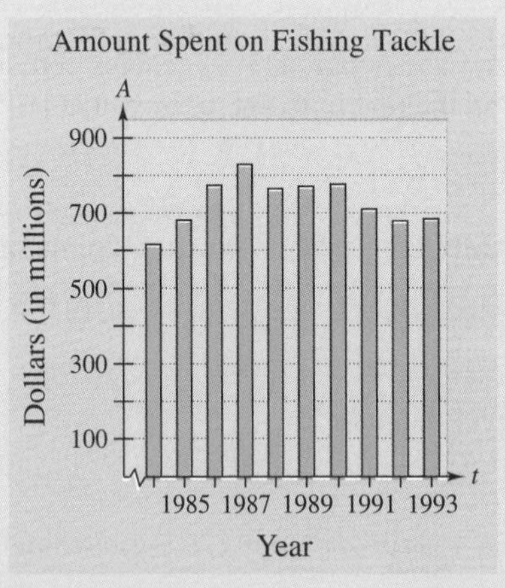

Bar graph

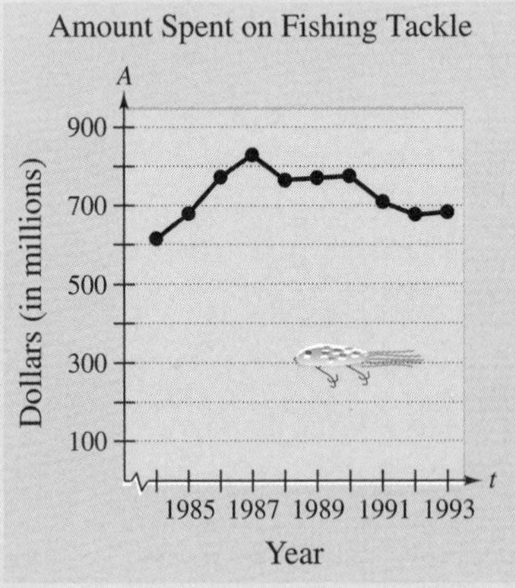

Line graph

The Distance Formula

Figure P.10

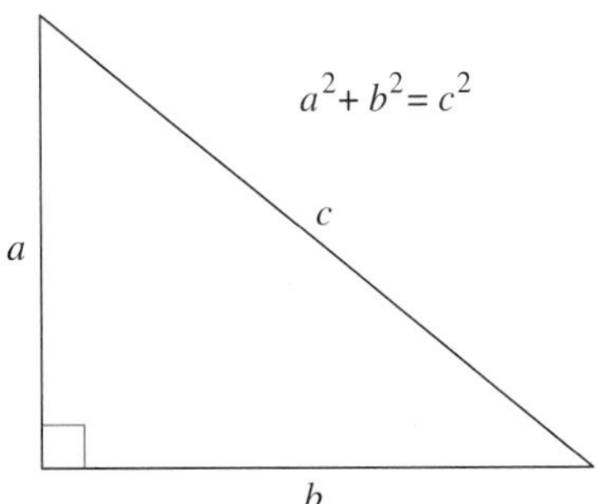

Recall from the Pythagorean Theorem that, for a right triangle with hypotenuse of length c and sides of lengths a and b, you have

$$a^2 + b^2 = c^2 \qquad \text{Pythagorean Theorem}$$

as shown in Figure P.10. (The converse is also true. That is, if $a^2 + b^2 = c^2$, the triangle is a right triangle.)

Suppose you want to determine the distance d between two points (x_1, y_1) and (x_2, y_2) in the plane. With these two points, a right triangle can be formed, as shown in Figure P.11. The length of the vertical side of the triangle is $|y_2 - y_1|$, and the length of the horizontal side is $|x_2 - x_1|$. By the Pythagorean Theorem, you can write

Figure P.11

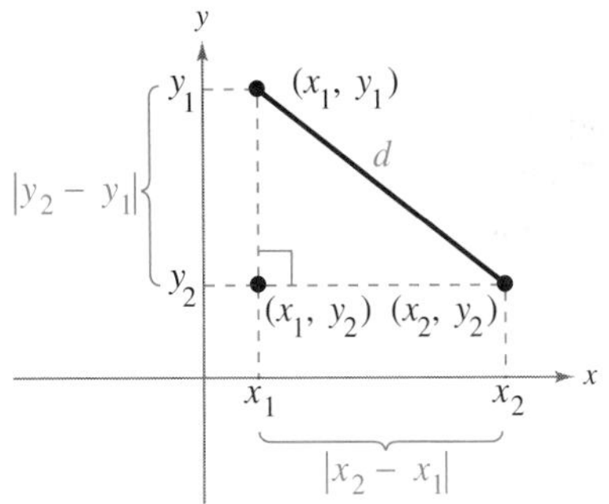

$$d^2 = |x_2 - x_1|^2 + |y_2 - y_1|^2$$
$$d = \sqrt{|x_2 - x_1|^2 + |y_2 - y_1|^2}$$
$$d = \sqrt{(x_2 - x_1)^2 + (y_2 - y_1)^2}.$$

This result is the **Distance Formula.**

The Distance Formula

The distance d between the points (x_1, y_1) and (x_2, y_2) in the plane is

$$d = \sqrt{(x_2 - x_1)^2 + (y_2 - y_1)^2}.$$

EXAMPLE 3 Finding a Distance

Find the distance between the points $(-2, 1)$ and $(3, 4)$.

Figure P.12

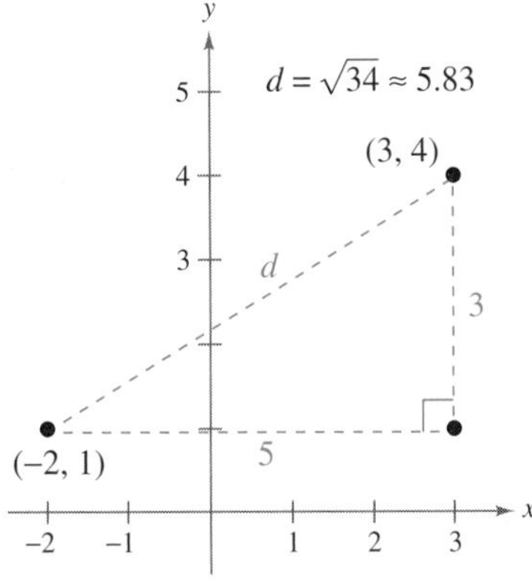

Solution

Let $(x_1, y_1) = (-2, 1)$ and $(x_2, y_2) = (3, 4)$. Then apply the Distance Formula as follows.

$$\begin{aligned} d &= \sqrt{(x_2 - x_1)^2 + (y_2 - y_1)^2} && \text{Distance Formula} \\ &= \sqrt{[3 - (-2)]^2 + (4 - 1)^2} && \text{Substitute for } x_1, y_1, x_2, \text{ and } y_2. \\ &= \sqrt{(5)^2 + (3)^2} && \text{Simplify.} \\ &= \sqrt{34} \\ &\approx 5.83 && \text{Use a calculator.} \end{aligned}$$

Note in Figure P.12 that a distance of 5.83 looks about right.

The Midpoint Formula

To find the **midpoint** of the line segment that joins two points in a coordinate plane, you can simply find the average values of the respective coordinates of the two endpoints.

The Midpoint Formula

The midpoint of the segment joining the points (x_1, y_1) and (x_2, y_2) is

$$\text{Midpoint} = \left(\frac{x_1 + x_2}{2}, \frac{y_1 + y_2}{2}\right).$$

EXAMPLE 4 Finding a Segment's Midpoint

Find the midpoint of the line segment joining the points $(-5, -3)$ and $(9, 3)$, as shown in Figure P.13.

Figure P.13

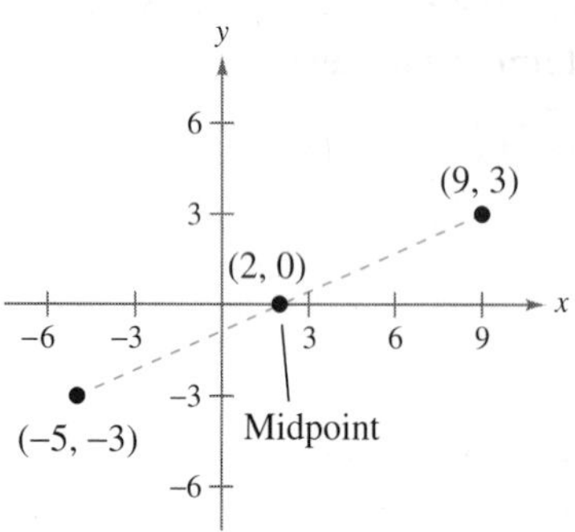

Solution

Let $(x_1, y_1) = (-5, -3)$ and $(x_2, y_2) = (9, 3)$.

$$\text{Midpoint} = \left(\frac{x_1 + x_2}{2}, \frac{y_1 + y_2}{2}\right) \qquad \text{Midpoint Formula}$$

$$= \left(\frac{-5 + 9}{2}, \frac{-3 + 3}{2}\right) \qquad \text{Substitute for } x_1, y_1, x_2, \text{ and } y_2.$$

$$= (2, 0) \qquad \text{Simplify.}$$

Real Life

EXAMPLE 5 Estimating Annual Sales

Ben and Jerry's had annual sales of $132.0 million in 1992 and $148.8 million in 1994. Without knowing any additional information, what would you estimate the 1993 sales to have been? (Source: Ben and Jerry's, Inc.)

Figure P.14

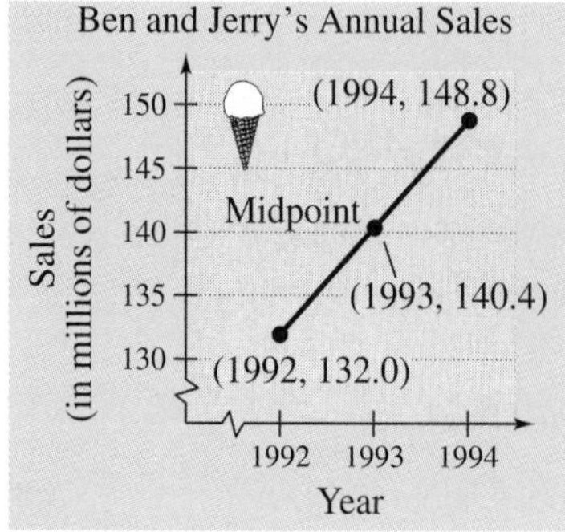

Solution

One solution to the problem is to assume that sales followed a linear pattern. With this assumption, you can estimate the 1993 sales by finding the midpoint of the segment connecting the points (1992, 132.0) and (1994, 148.8).

$$\text{Midpoint} = \left(\frac{1992 + 1994}{2}, \frac{132.0 + 148.8}{2}\right) = (1993, 140.4)$$

Hence, you would estimate the 1993 sales to have been about $140.4 million, as shown in Figure P.14. (The actual 1993 sales were $140.3 million.)

Figure P.15

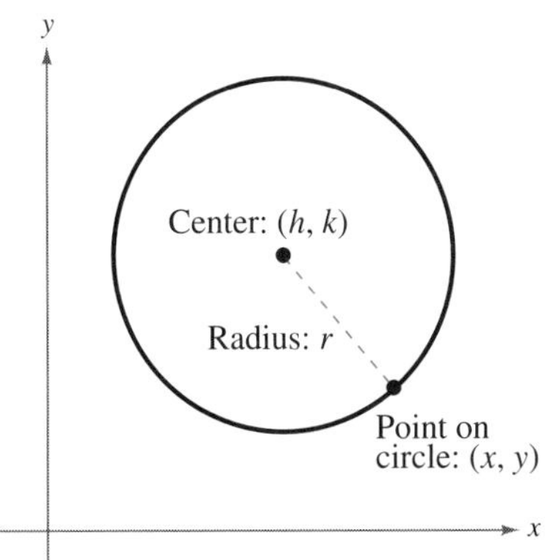

The Equation of a Circle

The Distance Formula provides a convenient way to define circles. A **circle of radius r** with **center** at the point (h, k) is shown in Figure P.15. The point (x, y) is on this circle if and only if its distance from the center (h, k) is r. This means that a **circle** in the plane consists of all points (x, y) that are a given positive distance r from a fixed point (h, k). Using the Distance Formula, you can express this relationship by saying that the point (x, y) lies on the circle if and only if

$$\sqrt{(x-h)^2 + (y-k)^2} = r.$$

By squaring both sides of this equation, you can obtain the **standard form of the equation of a circle.**

Standard Form of the Equation of a Circle

The **standard form of the equation of a circle** is

$$(x-h)^2 + (y-k)^2 = r^2.$$

The point (h, k) is the **center** of the circle, and the positive number r is the **radius** of the circle. The standard form of the equation of a circle whose center is the *origin* is $x^2 + y^2 = r^2$.

EXAMPLE 6 Finding an Equation of a Circle

The point $(3, 4)$ lies on a circle whose center is at $(-1, 2)$, as shown in Figure P.16. Find an equation for the circle.

Figure P.16

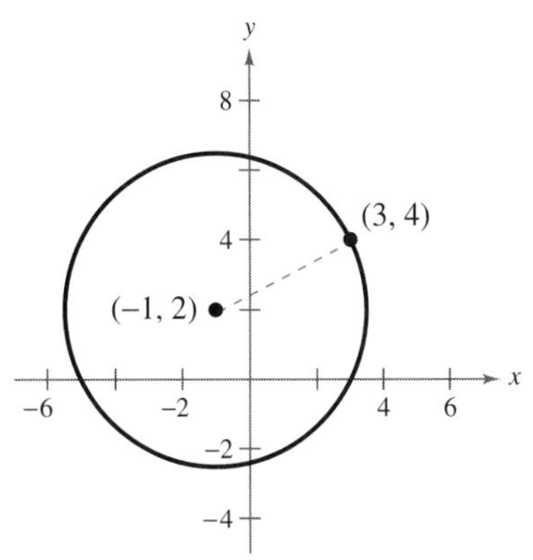

Solution

The radius r of the circle is the distance between $(-1, 2)$ and $(3, 4)$.

$$\begin{aligned} r &= \sqrt{[3-(-1)]^2 + (4-2)^2} \\ &= \sqrt{16+4} \\ &= \sqrt{20} \end{aligned}$$

Thus, the center of the circle is $(h, k) = (-1, 2)$ and the radius is $r = \sqrt{20}$, and you can write the standard form of the equation of the circle as follows.

$$(x-h)^2 + (y-k)^2 = r^2 \qquad \text{Standard form}$$

$$[x-(-1)]^2 + (y-2)^2 = \left(\sqrt{20}\right)^2 \qquad \text{Substitute for } h, k, \text{ and } r.$$

$$(x+1)^2 + (y-2)^2 = 20 \qquad \text{Equation of circle}$$

Application

EXAMPLE 7 **Translating Points in the Plane**

The triangle in Figure P.17(a) has vertices at the points $(-1, 2)$, $(1, -4)$, and $(2, 3)$. Shift the triangle three units to the right and two units up and find the vertices of the shifted triangle, as shown in Figure P.17(b).

Figure P.17

(a)

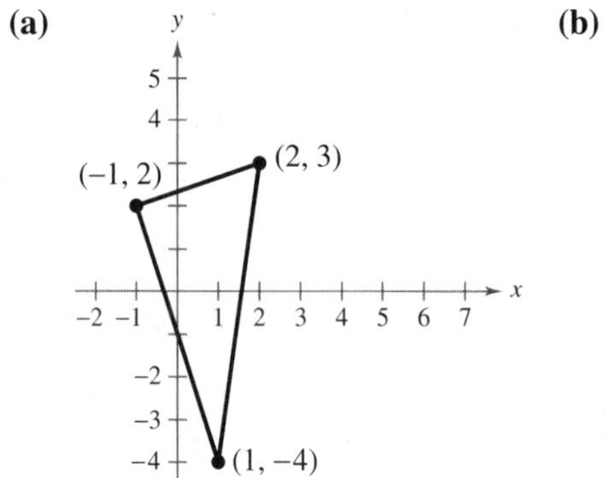

(b)

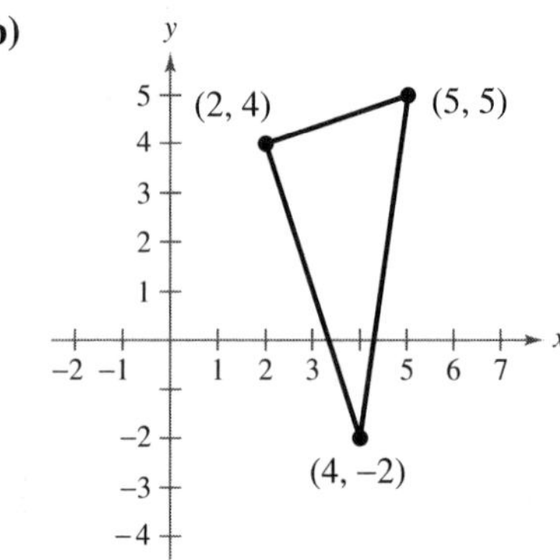

Solution

To shift the vertices three units to the right, add 3 to each x-coordinate. To shift the vertices two units up, add 2 to each y-coordinate.

Original Point	*Translated Point*
$(-1, 2)$	$(-1 + 3, 2 + 2) = (2, 4)$
$(1, -4)$	$(1 + 3, -4 + 2) = (4, -2)$
$(2, 3)$	$(2 + 3, 3 + 2) = (5, 5)$

Group Activity — *Extending the Example*

Example 7 shows how to translate points in a coordinate plane. How are the following transformed points related to the original points?

Original Point	*Transformed Point*
(x, y)	$(-x, y)$
(x, y)	$(x, -y)$
(x, y)	$(-x, -y)$

P.5 /// EXERCISES

In Exercises 1–4, sketch the polygon with the indicated vertices.

1. Triangle: $(-1, 1), (2, -1), (3, 4)$
2. Triangle: $(0, 3), (-1, -2), (4, 8)$
3. Square: $(2, 4), (5, 1), (2, -2), (-1, 1)$
4. Parallelogram: $(5, 2), (7, 0), (1, -2), (-1, 0)$

In Exercises 5–8, approximate the coordinates of the points.

5.

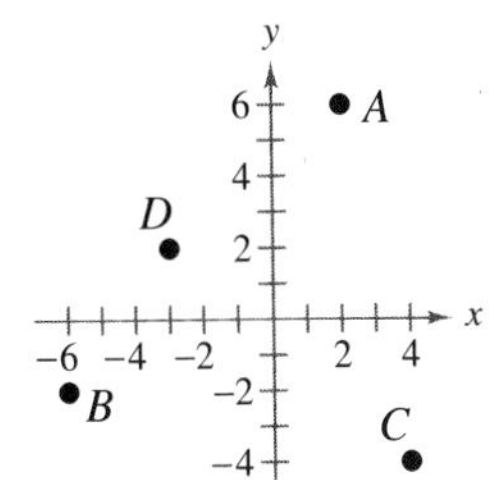

6.

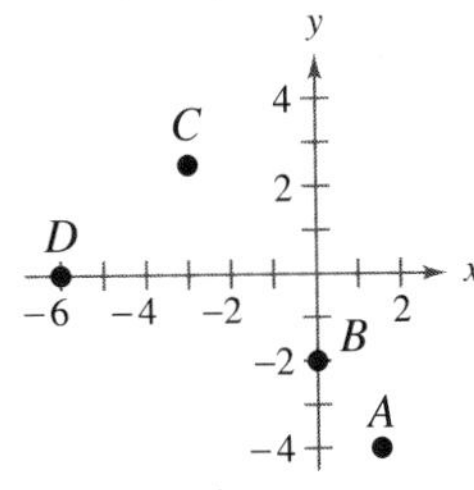

7.

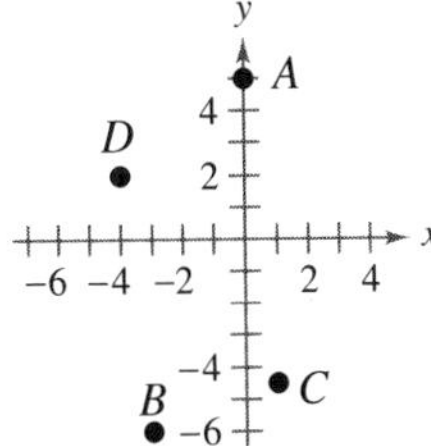

8.

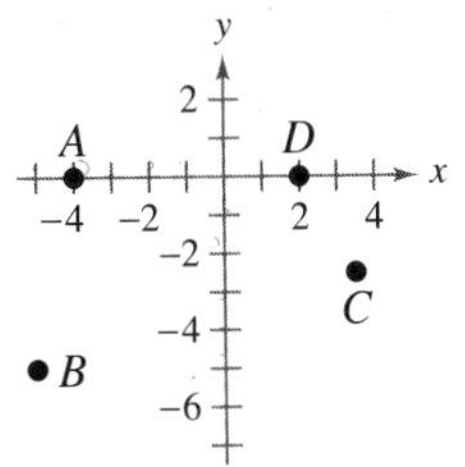

In Exercises 9–12, find the coordinates of the point.

9. The point is located three units to the left of the y-axis and four units above the x-axis.
10. The point is located eight units below the x-axis and four units to the right of the y-axis.
11. The point is located five units below the x-axis and the coordinates of the point are equal.
12. The point is on the x-axis and twelve units to the left of the y-axis.
13. *Think About It* What is the y-coordinate of any point on the x-axis? What is the x-coordinate of any point on the y-axis?
14. *Think About It* When plotting points on the rectangular coordinate system, is it true that the scales on the x- and y-axes must be the same? Explain.

In Exercises 15–24, determine the quadrant(s) in which (x, y) is located so that the condition(s) is (are) satisfied.

15. $x > 0$ and $y < 0$
16. $x < 0$ and $y < 0$
17. $x = -4$ and $y > 0$
18. $x > 2$ and $y = 3$
19. $y < -5$
20. $x > 4$
21. $(x, -y)$ is in the second quadrant.
22. $(-x, y)$ is in the fourth quadrant.
23. $xy > 0$
24. $xy < 0$

In Exercises 25 and 26, sketch a scatter plot of the data given in the table.

25. *Normal Temperatures* The normal temperature y (in degrees Fahrenheit) in Duluth, Minnesota, for each month x, where $x = 1$ represents January, is given in the table. (Source: NOAA)

x	1	2	3	4	5	6
y	6	12	23	38	50	59

x	7	8	9	10	11	12
y	65	63	54	44	28	14

26. *Wal-Mart* The number y of Wal-Mart stores for each year x from 1985 through 1994 is given in the table. (Source: Wal-Mart Annual Report for 1994)

x	1985	1986	1987	1988	1989
y	745	859	980	1114	1259

x	1990	1991	1992	1993	1994
y	1399	1568	1714	1850	1953

In Exercises 27 and 28, the polygon is shifted to a new position in the plane. Find the coordinates of the vertices of the polygon in its *new* position.

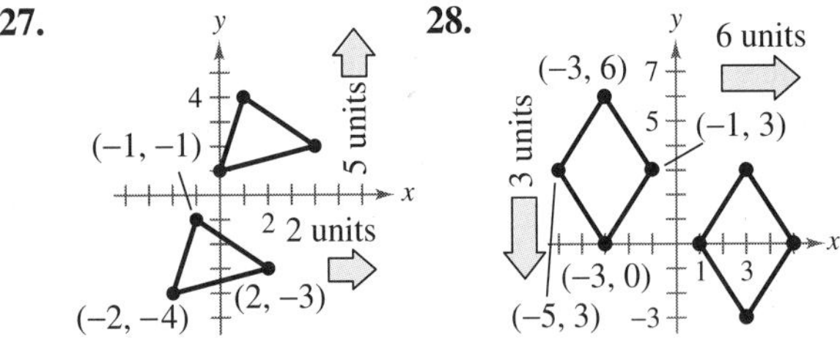

In Exercises 29 and 30, make a table of values for $x = -2, -1, -\frac{1}{2}, 0, \frac{1}{2}, 1$, and 2. Then plot the points on a rectangular coordinate system.

29. $y = 2 - \frac{1}{2}x$ **30.** $y = 2 - \frac{1}{2}x^2$

TV Advertising **In Exercises 31 and 32, refer to the figure.** (Source: Nielson Media Research)

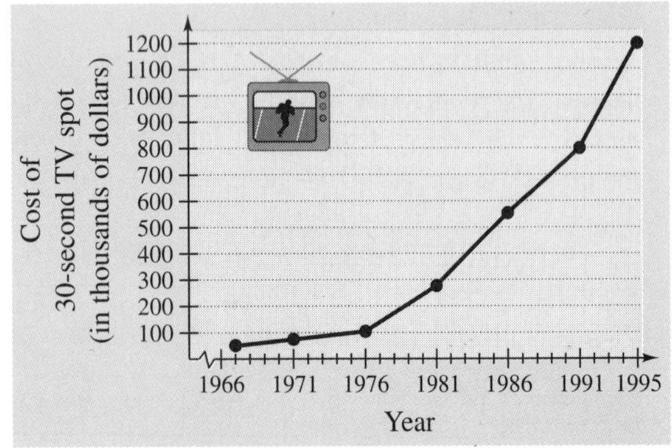

31. Approximate the percent increase in cost of a TV spot from Super Bowl I in 1967 to Super Bowl XXIX in 1995.

32. Estimate the increase in cost of a TV spot (a) from Super Bowl V to Super Bowl XV, and (b) from Super Bowl XV to Super Bowl XXV.

Milk Prices **In Exercises 33 and 34, refer to the figure, which shows the average price paid to farmers for milk.** (Source: U.S. Department of Agriculture and the National Milk Producers Federation)

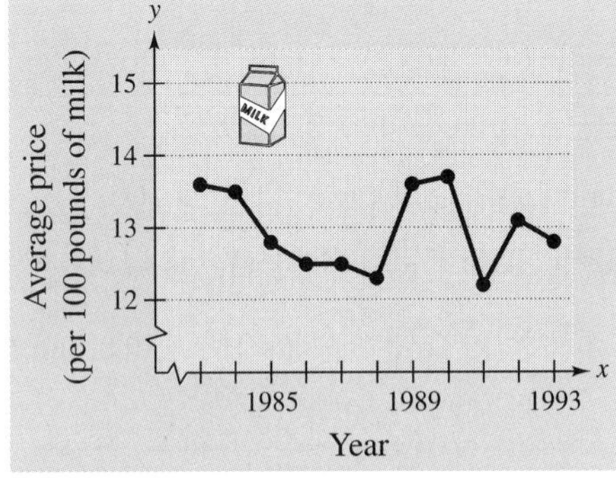

33. Approximate the highest price of milk shown in the graph. When did this occur?

34. Approximate the percent drop in the price of milk from the highest price shown in the graph to the price paid to farmers in January 1993.

Minimum Wage **In Exercises 35 and 36, refer to the figure.** (Source: U.S. Department of Labor)

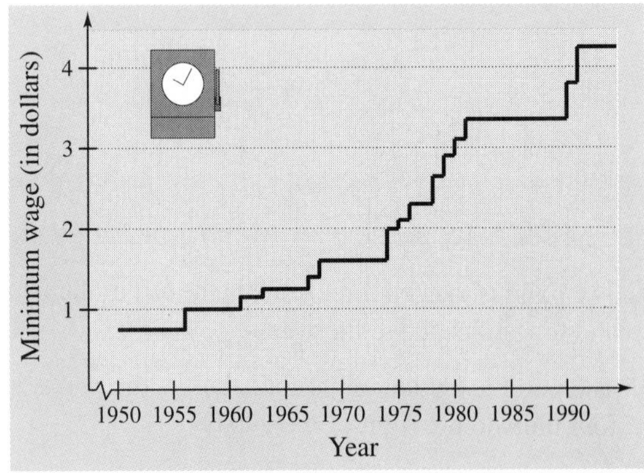

35. During which decade did the minimum wage increase most rapidly?

36. Approximate the percent increase in the minimum wage from 1990 to 1994.

Analyzing Data In Exercises 37 and 38, refer to the figure, which shows the mathematics entrance test scores x, and the final examination scores y, in an algebra course for a sample of 10 students.

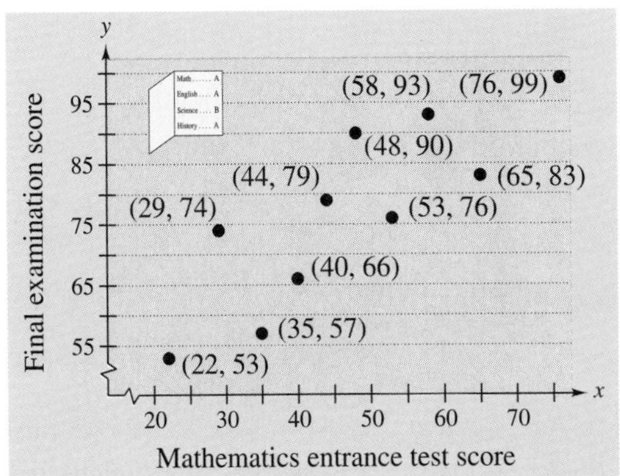

37. Find the entrance exam score of any student with a final exam score in the 80's.

38. Does a higher entrance exam score necessarily imply a higher final exam score? Explain.

In Exercises 39–42, (a) find the length of each side of a right triangle, and (b) show that these lengths satisfy the Pythagorean Theorem.

39.

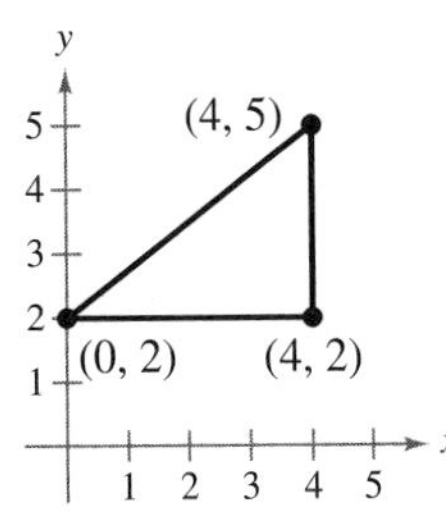

40.

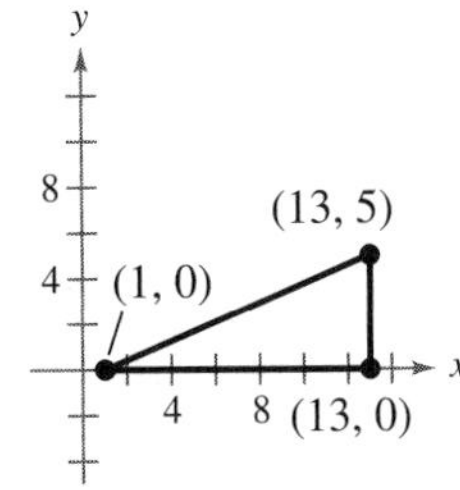

41.

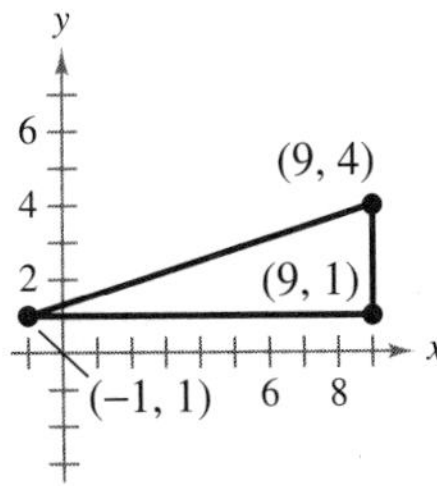

42.

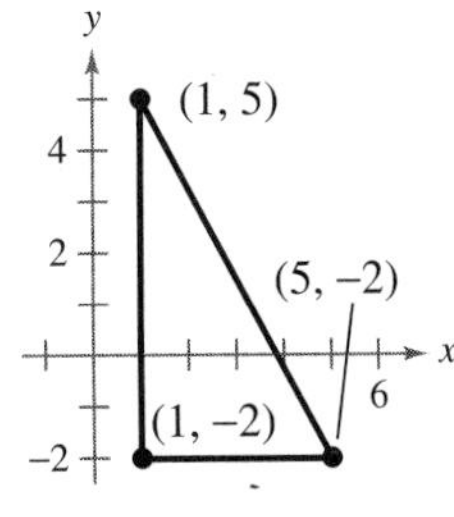

In Exercises 43–46, find the distance between the points. (*Note*: In each case the two points lie on the same horizontal or vertical line.)

43. $(6, -3), (6, 5)$ **44.** $(1, 4), (8, 4)$

45. $(-3, -1), (2, -1)$ **46.** $(-3, -4), (-3, 6)$

In Exercises 47–58, (a) plot the points, (b) find the distance between the points, and (c) find the midpoint of the line segment joining the points.

47. $(1, 1), (9, 7)$ **48.** $(1, 12), (6, 0)$

49. $(-4, 10), (4, -5)$ **50.** $(-7, -4), (2, 8)$

51. $(-1, 2), (5, 4)$ **52.** $(2, 10), (10, 2)$

53. $\left(\frac{1}{2}, 1\right), \left(-\frac{5}{2}, \frac{4}{3}\right)$

54. $\left(-\frac{1}{3}, -\frac{1}{3}\right), \left(-\frac{1}{6}, -\frac{1}{2}\right)$

55. $(6.2, 5.4), (-3.7, 1.8)$

56. $(-16.8, 12.3), (5.6, 4.9)$

57. $(-36, -18), (48, -72)$

58. $(1.451, 3.051), (5.906, 11.360)$

In Exercises 59 and 60, use the Midpoint Formula to estimate the sales of a company in 1993, given the sales in 1991 and 1995. Assume the sales followed a linear pattern.

59.

Year	1991	1995
Sales	\$520,000	\$740,000

60.

Year	1991	1995
Sales	\$4,200,000	\$5,650,000

In Exercises 61–66, show that the points form the vertices of the polygon.

61. Right triangle: $(4, 0), (2, 1), (-1, -5)$

62. Isosceles triangle: $(1, -3), (3, 2), (-2, 4)$

63. Rhombus: $(0, 0), (1, 2), (2, 1), (3, 3)$

(A rhombus is a parallelogram whose sides are all the same length.)

64. Rhombus: $(4, 0)$, $(0, 6)$, $(-4, 0)$, $(0, -6)$

65. Parallelogram: $(2, 5)$, $(0, 9)$, $(-2, 0)$, $(0, -4)$

66. Parallelogram: $(0, 1)$, $(3, 7)$, $(4, 4)$, $(1, -2)$

In Exercises 67–74, find the standard form of the equation of the specified circle.

67. Center: $(0, 0)$; radius: 3 **68.** Center: $(0, 0)$; radius: 5

69. Center: $(2, -1)$; radius: 4

70. Center: $\left(0, \frac{1}{3}\right)$; radius: $\frac{1}{3}$

71. Center: $(-1, 2)$; solution point: $(0, 0)$

72. Center: $(3, -2)$; solution point: $(-1, 1)$

73. Endpoints of a diameter: $(0, 0)$, $(6, 8)$

74. Endpoints of a diameter: $(-4, -1)$, $(4, 1)$

In Exercises 75–80, find the center and radius, and sketch the circle.

75. $x^2 + y^2 = 4$ **76.** $x^2 + y^2 = 16$

77. $(x - 1)^2 + (y + 3)^2 = 4$

78. $x^2 + (y - 1)^2 = 4$ **79.** $\left(x - \frac{1}{2}\right)^2 + \left(y - \frac{1}{2}\right)^2 = \frac{9}{4}$

80. $(x - 2)^2 + (y + 1)^2 = 2$

81. A line segment has (x_1, y_1) as one endpoint and (x_m, y_m) as its midpoint. Find the other endpoint (x_2, y_2) of the line segment in terms of x_1, y_1, x_m, and y_m.

82. Use the result of Exercise 81 to find the coordinates of the endpoint of a line segment if the coordinates of the other endpoint and midpoint are, respectively,

(a) $(1, -2)$, $(4, -1)$ (b) $(-5, 11)$, $(2, 4)$

83. Use the Midpoint Formula three times to find the three points that divide the line segment joining (x_1, y_1) and (x_2, y_2) into four parts.

84. Use the result in Exercise 83 to find the points that divide the line segment joining the given points into four equal parts.

(a) $(1, -2)$, $(4, -1)$ (b) $(-2, -3)$, $(0, 0)$

85. *Football Pass* In a football game, a quarterback throws a pass from the 15-yard line, 10 yards from the sideline (see figure). The pass is caught on the 40-yard line, 45 yards from the same sideline. How long is the pass?

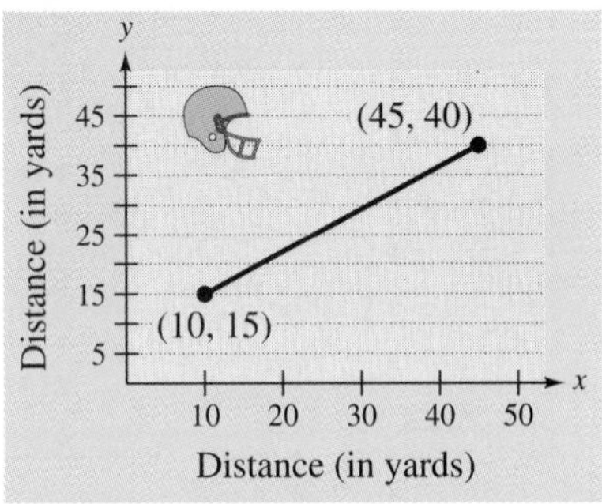

86. *Flying Distance* A plane flies in a straight line to a city that is 100 kilometers east and 150 kilometers north of the point of departure. How far does it fly?

87. *Make a Conjecture* Plot the points $(2, 1)$, $(-3, 5)$, and $(7, -3)$ on a rectangular coordinate system. Then change the sign of the x-coordinate of each point and plot the three new points on the same rectangular coordinate system. What conjecture can you make about the location of a point when the sign of the x-coordinate is changed?

88. Prove that the diagonals of the parallelogram in the figure bisect each other.

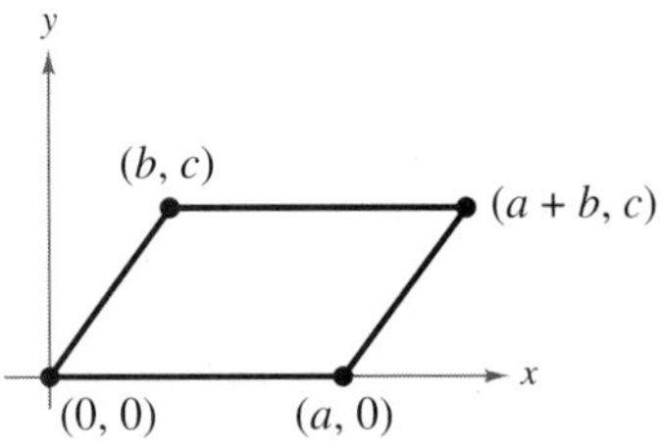

89. *Chapter Opener* Use the graph on page 1.

(a) Describe any trends in the data. From these trends, predict the number of artists elected in 1996.

(b) Why do you think the numbers elected in 1986 and 1987 were greater than in other years?

P.6 Exploring Data: Representing Data Graphically

Line Plots / Stem-and-Leaf Plots / Histograms and Frequency Distributions / Line Graphs

Line Plots

Statistics is the branch of mathematics that studies techniques for collecting, organizing, and interpreting data. In this section, you will study several ways to organize data. The first is a **line plot,** which uses a portion of a real number line to order numbers. Line plots are especially useful for ordering small sets of numbers (about 50 or less) by hand.

EXAMPLE 1 Constructing a Line Plot

Use a line plot to organize the following test scores. Which number occurs with the greatest frequency?

93, 70, 76, 67, 86, 93, 82, 78, 83, 86, 64, 78, 76, 66, 83
83, 96, 74, 69, 76, 64, 74, 79, 76, 88, 76, 81, 82, 74, 70

Solution

Begin by scanning the data to find the smallest and largest numbers. For this data, the smallest number is 64 and the largest is 96. Next, draw a portion of a real number line that includes the interval [64, 96]. To create the line plot, start with the first number, 93, and enter an $\times$ above 93 on the number line. Continue recording $\times$'s for each number in the list until you obtain the line plot shown in Figure P.18. From the line plot, you can see that 76 had the greatest frequency.

Figure P.18

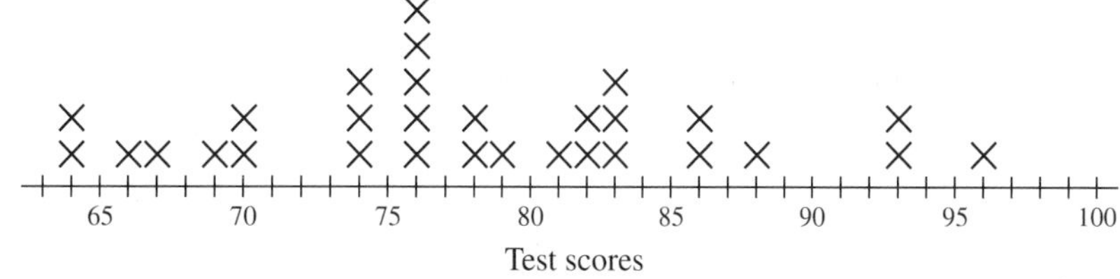

Many computer programs and calculators will sort data. Try using a computer or calculator to sort the data in Example 1.

Stem-and-Leaf Plots

Another type of plot that is used to organize sets of numbers is a **stem-and-leaf plot.** A stem-and-leaf plot for the test scores in Example 1 is shown below.

Stems	*Leaves*
6	4 4 6 7 9
7	0 0 4 4 4 6 6 6 6 6 8 8 9
8	1 2 2 3 3 3 6 6 8
9	3 3 6

Note that the *leaves* represent the units digits of the numbers and the *stems* represent the tens digits. Stem-and-leaf plots can also be used to compare two sets of data, as shown in the next example.

EXAMPLE 2 Comparing Two Sets of Data

Use a stem-and-leaf plot to compare the test scores in Example 1 with the following test scores. Which set of test scores is better?

90, 81, 70, 62, 64, 73, 81, 92, 73, 81, 92, 93, 83, 75, 76
83, 94, 96, 86, 77, 77, 86, 96, 86, 77, 86, 87, 87, 79, 88

Solution

Begin by ordering the second set of scores.

62, 64, 70, 73, 73, 75, 76, 77, 77, 77, 79, 81, 81, 81, 83
83, 86, 86, 86, 86, 87, 87, 88, 90, 92, 92, 93, 94, 96, 96

Now that the data is ordered, you can construct a *double* stem-and-leaf plot by letting the leaves to the right of the stem represent the units digits for the first group of test scores and the leaves to the left of the stem represent the units digits for the second group of test scores.

Leaves (Second Group)	*Stems*	*Leaves (First Group)*
4 2	6	4 4 6 7 9
9 7 7 7 6 5 3 3 0	7	0 0 4 4 4 6 6 6 6 6 8 8 9
8 7 7 6 6 6 6 3 3 1 1 1	8	1 2 2 3 3 3 6 6 8
6 6 4 3 2 2 0	9	3 3 6

From the two sets of leaves, you can see that the second group of test scores is better than the first group.

EXAMPLE 3 Using a Stem-and-Leaf Plot

The table shows the percent of the population of each state (and the District of Columbia) that was 65 or older in 1993. Use a stem-and-leaf plot to organize the data. (Source: U.S. Bureau of Census)

AK 4.4	**AL** 13.0	**AR** 15.0	**AZ** 13.4	**CA** 10.6	**CO** 10.0
CT 14.1	**D.C.** 13.3	**DE** 12.4	**FL** 18.6	**GA** 10.1	**HI** 11.7
IA 15.5	**ID** 11.8	**IL** 12.6	**IN** 12.7	**KS** 13.9	**KY** 12.7
LA 11.3	**MA** 14.0	**MD** 11.1	**ME** 13.7	**MI** 12.4	**MN** 12.6
MO 14.2	**MS** 12.5	**MT** 13.4	**NC** 12.5	**ND** 14.8	**NE** 14.2
NH 11.9	**NJ** 13.6	**NM** 11.0	**NV** 11.1	**NY** 13.1	**OH** 13.3
OK 13.6	**OR** 13.8	**PA** 15.8	**RI** 15.5	**SC** 11.7	**SD** 14.7
TN 12.8	**TX** 10.2	**UT** 8.9	**VA** 11.0	**VT** 12.0	**WA** 11.6
WI 13.4	**WV** 15.3	**WY** 10.9			

Solution

Begin by ordering the numbers, as shown on the left. Next construct the stem-and-leaf plot using the leaves to represent the digits to the right of the decimal points. The stem-and-leaf plot is shown on the right.

4.4, 8.9, 10.0, 10.1, 10.2,
10.6, 10.9, 11.0, 11.0, 11.1,
11.1, 11.3, 11.6, 11.7, 11.7,
11.8, 11.9, 12.0, 12.4, 12.4,
12.5, 12.5, 12.6, 12.6, 12.7,
12.7, 12.8, 13.0, 13.1, 13.3,
13.3, 13.4, 13.4, 13.4, 13.6,
13.6, 13.7, 13.8, 13.9, 14.0,
14.1, 14.2, 14.2, 14.7, 14.8,
15.0, 15.3, 15.5, 15.5, 15.8,
18.6

Stems	Leaves	
4.	4	Alaska has the lowest percent.
5.		
6.		
7.		
8.	9	
9.		
10.	0 1 2 6 9	
11.	0 0 1 1 3 6 7 7 8 9	
12.	0 4 4 5 5 6 6 7 7 8	
13.	0 1 3 3 4 4 4 6 6 7 8 9	
14.	0 1 2 2 7 8	
15.	0 3 5 5 8	
16.		
17.		
18.	6	Florida has the highest percent.

Histograms and Frequency Distributions

With data such as that given in Example 3, it is useful to group the numbers into intervals and plot the frequency of the data in each interval. For instance, the **frequency distribution** and **histogram** shown in Figure P.19 represent the data given in Example 3.

Try using a computer or graphing calculator to create a bar histogram for the data at the right. How does the histogram change when the intervals change?

Figure P.19

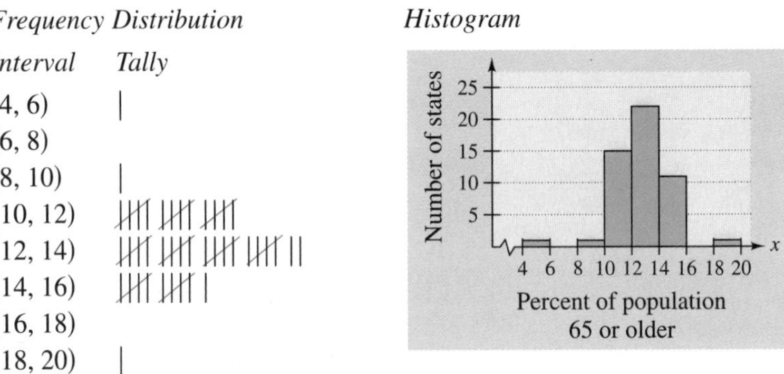

A histogram has a portion of the real number line as its horizontal axis. A **bar graph** is similar to a histogram, except that the rectangles (bars) can be either horizontal or vertical and the labels of the bars are not necessarily numbers. Another difference between a bar graph and a histogram is that the bars in a bar graph are usually separated by spaces, whereas the bars in a histogram are not separated by spaces.

EXAMPLE 4 Constructing a Bar Graph

The data below shows the average monthly precipitation (in inches) in Houston, Texas. Construct a bar graph for this data. What can you conclude? (Source: PC USA)

January	3.2	*February*	3.3	*March*	2.7
April	4.2	*May*	4.7	*June*	4.1
July	3.3	*August*	3.7	*September*	4.9
October	3.7	*November*	3.4	*December*	3.7

Solution

To create a bar graph, begin by drawing a vertical axis to represent the precipitation and a horizontal axis to represent the month. The bar graph is shown in Figure P.20. From the graph, you can see that Houston receives a fairly consistent amount of rain throughout the year—the driest month tends to be March and the wettest month tends to be September.

Figure P.20

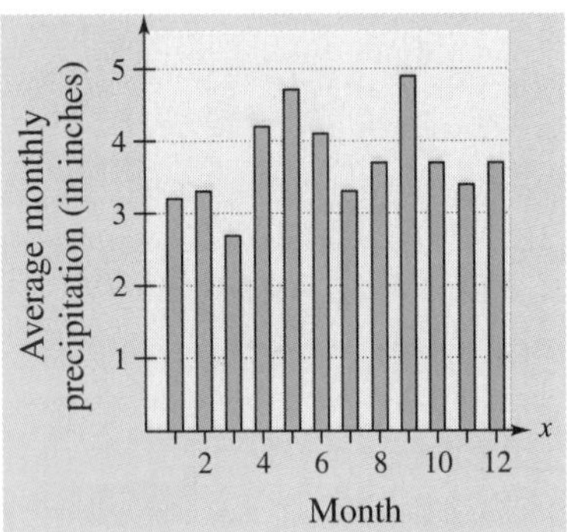

EXAMPLE 5 **Constructing a Double Bar Graph**

The table shows the percents of bachelor's degrees awarded to males and females for selected majors in the United States in 1991. Construct a double bar graph for this data. (Source: U.S. National Center for Education Statistics)

Field of Study	% Female	% Male
Agriculture and Natural Resources	31.5	68.5
Area and Ethnic Studies	80.4	19.6
Education	73.1	26.9
Engineering	8.7	91.3
Home Economics	73.7	26.3
Liberal/General Studies	59.4	40.6
Mathematics	39.0	61.0
Military Sciences	9.4	90.6
Physical Sciences	38.1	61.9
Social Sciences	59.6	40.4

Solution

For this data, a horizontal bar graph seems to be appropriate. Such a graph is shown in Figure P.21.

Figure P.21

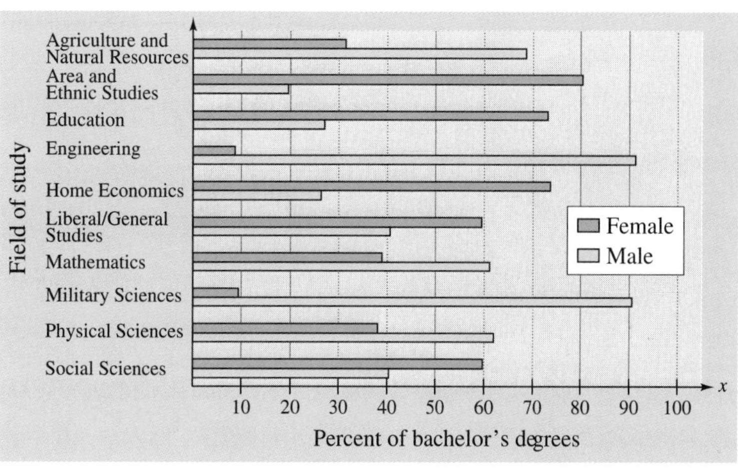

Line Graphs

EXAMPLE 6 Constructing a Line Graph

Decade	*Number*
1851–1860	2598
1861–1870	2315
1871–1880	2812
1881–1890	5247
1891–1900	3688
1901–1910	8795
1911–1920	5736
1921–1930	4107
1931–1940	528
1941–1950	1035
1951–1960	2515
1961–1970	3322
1971–1980	4493
1981–1990	6447

The table at the left shows the number of immigrants (in thousands) entering the United States in each decade from 1851 to 1990. Construct a line graph of this data. What can you conclude? (Source: U.S. Bureau of Census)

Solution

Begin by drawing a vertical axis to represent the number of immigrants in thousands. Then label the horizontal axis with decades and plot the points given in the table. Finally, connect the points with line segments, as shown in Figure P.22. From the line graph, you can see that the number of immigrants hit a low point during the depression of the 1930s. Since then the number has steadily increased.

Figure P.22

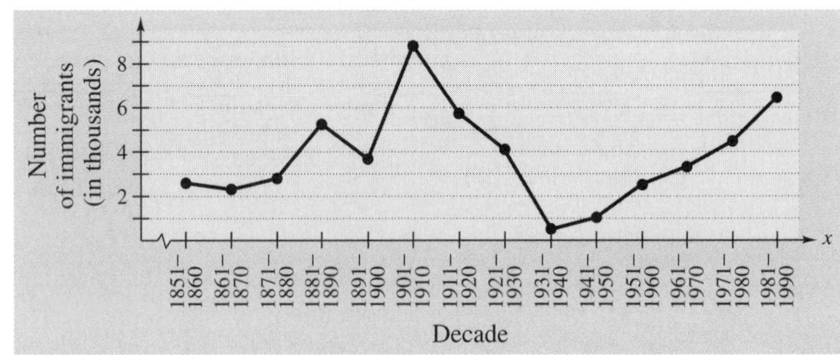

Note A **line graph** is similar to a standard coordinate graph. Line graphs are usually used to show trends over periods of time.

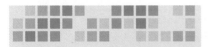

To sketch a line graph using a *TI-82* or *TI-83,* use the following steps.

1. Use the STAT key to enter the data into L_1 and L_2.
2. Use the STAT PLOT key to select Plot 1 and the line graph icon.
3. Use the ZOOM key to select ZoomStat, which will set an appropriate viewing rectangle.

Group Activity *Organizing Data*

Listed below are the winning times (in seconds) for the 110-meter hurdles in the summer Olympics from 1896 through 1992. With others in your group, decide which type of graph can best represent this data. Sketch the graph and discuss any patterns or trends that you see.

1896 (17.60)	1900 (15.40)	1904 (16.00)	1908 (15.00)	1912 (15.10)
1920 (14.80)	1924 (15.00)	1928 (14.80)	1932 (14.60)	1936 (14.20)
1948 (13.90)	1952 (13.70)	1956 (13.50)	1960 (13.80)	1964 (13.60)
1968 (13.30)	1972 (13.24)	1976 (13.30)	1980 (13.39)	1984 (13.20)
1988 (12.98)	1992 (13.12)			

P.6 /// EXERCISES

1. *Gasoline Prices* The line plot shows a sample of prices of unleaded regular gasoline from 25 different cities.

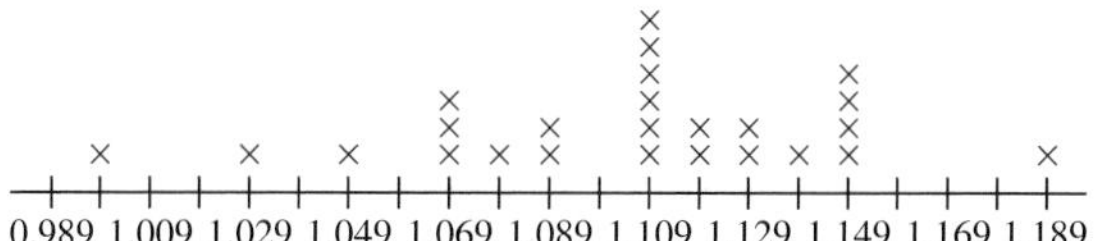

(a) What price occurred with the greatest frequency?

(b) What is the range of prices?

2. *Livestock Weights* The line plot shows the weights (to the nearest hundred pounds) of 30 head of cattle sold by a rancher.

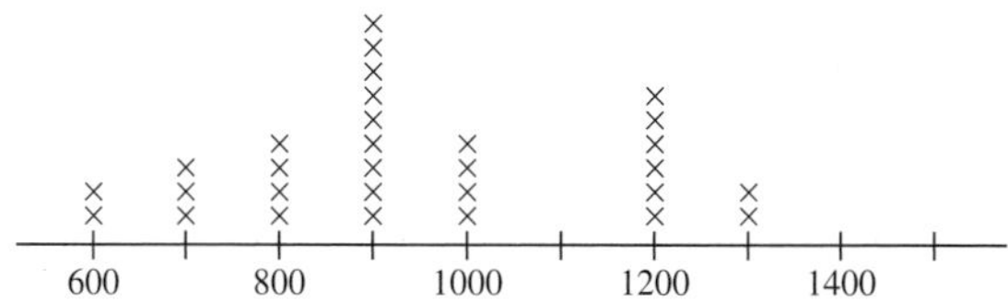

(a) What weight occurred with the greatest frequency?

(b) What is the range of weights?

Quiz and Exam Scores In Exercises 3–8, use the following scores from a math class of 30 students. The scores are for two 25-point quizzes and two 100-point exams.

Quiz #1 **20, 15, 14, 20, 16, 19, 10, 21, 24, 15, 15, 14, 15, 21, 19, 15, 20, 18, 18, 22, 18, 16, 18, 19, 21, 19, 16, 20, 14, 12**

Quiz #2 **22, 22, 23, 22, 21, 24, 22, 19, 21, 23, 23, 25, 24, 22, 22, 23, 23, 23, 23, 22, 24, 23, 22, 24, 21, 24, 16, 21, 16, 14**

Exam #1 **77, 100, 77, 70, 83, 89, 87, 85, 81, 84, 81, 78, 89, 78, 88, 85, 90, 92, 75, 81, 85, 100, 98, 81, 78, 75, 85, 89, 82, 75**

Exam #2 **76, 78, 73, 59, 70, 81, 71, 66, 66, 73, 68, 67, 63, 67, 77, 84, 87, 71, 78, 78, 90, 80, 77, 70, 80, 64, 74, 68, 68, 68**

3. Construct a line plot for Quiz #1. Which score occurred with the greatest frequency?

4. Construct a line plot for Quiz #2. Which score occurred with the greatest frequency?

5. Construct a line plot for Exam #1. Which score occurred with the greatest frequency?

6. Construct a line plot for Exam #2. Which score occurred with the greatest frequency?

7. Construct a stem-and-leaf plot for Exam #1.

8. Construct a double stem-and-leaf plot to compare the scores for Exam #1 and Exam #2. Which set of scores is higher?

9. *Educational Expenses* The list gives the per capita expenditures for public elementary and secondary education in the 50 states and the District of Columbia in 1993. Use a stem-and-leaf plot to organize the data. (Source: National Education Association)

AK 1800	AL 692	AR 762	AZ 915
CA 919	CO 974	CT 1252	D.C. 1061
DE 974	FL 864	GA 834	HI 932
IA 943	ID 898	IL 865	IN 1029
KS 1024	KY 841	LA 804	MA 913
MD 1003	ME 1098	MI 1155	MN 1124
MO 764	MS 676	MT 1056	NC 818
ND 863	NE 964	NH 916	NJ 1351
NM 919	NV 964	NY 1265	OH 1024
OK 850	OR 1138	PA 1096	RI 903
SC 848	SD 885	TN 663	TX 1049
UT 834	VA 940	VT 1197	WA 1131
WI 1123	WV 1045	WY 1336	

10. *Land Value* The list gives the average value of land and buildings per acre in the 48 contiguous states in 1993. Use a stem-and-leaf plot to organize the data. (Source: U.S. Department of Agriculture)

AL 863	AR 759	AZ 305	CA 1722
CO 383	CT 4299	DE 2362	FL 2074
GA 964	IA 1245	ID 691	IL 1503
IN 1366	KS 494	KY 1084	LA 945
MA 3662	MD 2521	ME 992	MI 1130
MN 896	MO 715	MS 757	MT 270
NC 1319	ND 388	NE 580	NH 2178
NJ 4536	NM 225	NV 215	NY 1119
OH 1267	OK 512	OR 657	PA 1747
RI 4894	SC 871	SD 370	TN 1049
TX 471	UT 464	VA 1295	VT 1158
WA 782	WI 932	WV 696	WY 149

11. *Cellular-Phone Fraud* The bar graph gives the industry losses (in millions of dollars) from cellular-phone fraud for the years 1991 through 1994 and the projected loss for 1995. Determine the percent increase in losses from 1991 to 1995. (Source: Cellular Telecommunications Industry Association)

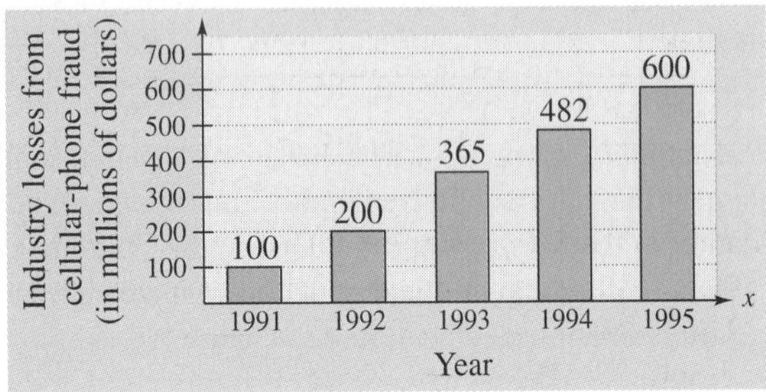

12. *Food Production* The double bar graph gives the production and exports (in millions of metric tons) of corn, soybeans, and wheat for the year 1992. Determine the percent of each product that is exported. (Source: U.S. Department of Agriculture)

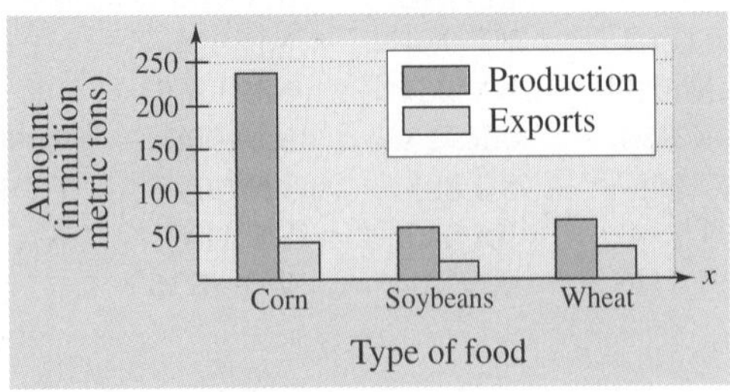

13. *Snowfall* The list gives the seasonal snowfall (in inches) at Erie, Pennsylvania, starting with the 1960/61 winter and ending with the 1994/95 winter. The amounts are listed in order by year. Organize the data graphically. (Source: National Oceanic and Atmospheric Association)

69.6, 42.5, 75.9, 115.9, 92.9, 84.8, 68.6, 107.9, 79.7, 85.6, 120.0, 92.3, 53.7, 68.6, 66.7, 66.0, 111.5, 142.8, 76.5, 55.2, 89.4, 71.3, 41.2, 110.0, 106.3, 124.9, 68.2, 103.5, 76.5, 114.9, 59.6, 104.8, 108.5, 131.3, 53.7

14. *Fruit Crops* The list gives farmers' cash receipts (in millions of dollars) from fruit crops in 1992. Construct a bar graph for the data. (Source: U.S. Department of Agriculture)

Apples	1159	Oranges	1616
Cherries	235	Peaches	373
Grapefruit	403	Pears	276
Grapes	1713	Plums and Prunes	245
Lemons	235	Strawberries	685

15. *Travel to the United States* The places of origin and numbers of travelers (in millions) to the United States in 1992 are as follows: Canada, 18.6; Mexico, 8.3; Europe, 8.3; Latin America, 3.3; Other, 6.3. Construct a horizontal bar graph for this data. (Source: U.S. Travel and Tourism Administration)

16. *Sports Participants* The list gives the numbers of males and females (in millions) over the age of seven that participated in popular sports activities in the year 1992 in the United States. Construct a double bar graph for the data. (Source: National Sporting Goods Association)

Activity	*Male*	*Female*
Exercise walking	23.2	44.6
Swimming	29.8	33.3
Bicycling	28.0	26.6
Camping	25.7	21.7
Bowling	21.8	20.7
Basketball	20.6	7.6
Running	12.7	9.2
Aerobic exercising	5.1	22.8

17. *Personal Savings* The line graph shows the percent of disposable personal income saved in the United States in selected years from 1970 to 1993. (Source: U.S. Bureau of Economic Analysis)

(a) Determine the percent decrease in the rate of saving from 1975 to 1993.

(b) Is the trend shown in the line graph good for the country? Explain your reasoning.

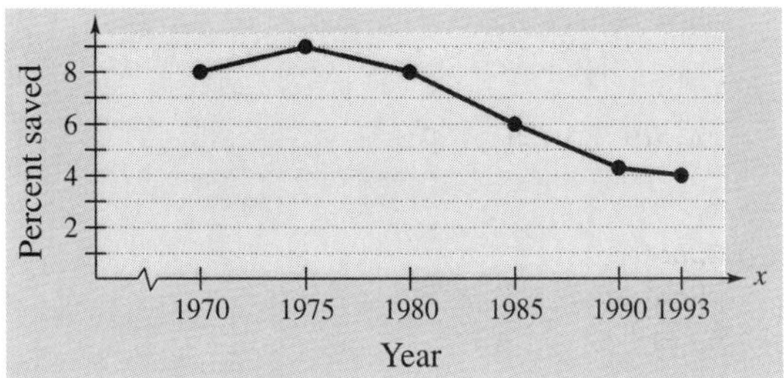

18. *Unemployment Rate* The double line graph shows the percent of the labor force in mining and manufacturing that were unemployed in the years 1987 to 1993. (Source: U.S. Bureau of Labor Statistics)

(a) Which industry showed the greater variability in unemployment rate?

(b) During which year was there the greatest difference between the two unemployment rates? During which year was the difference least?

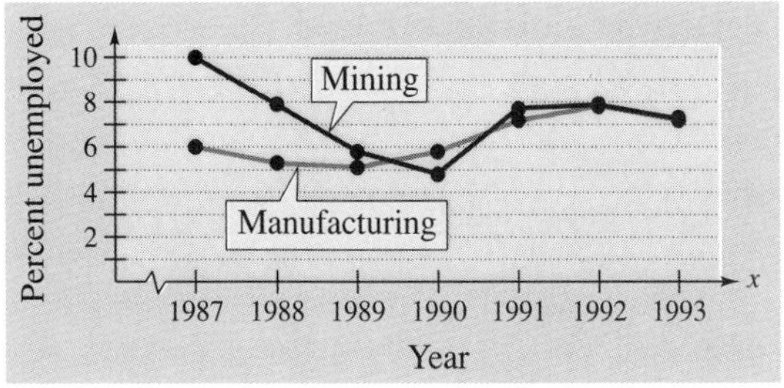

19. *College Attendance* The table shows the enrollment in a liberal arts college for the years 1988 to 1995. Construct a line graph for the data.

Year	1988	1989	1990	1991
Enrollment	1675	1704	1710	1768

Year	1992	1993	1994	1995
Enrollment	1833	1918	1967	1972

20. *Oil Imports* The table shows the amount of crude oil imported into the United States (in millions of barrels) for the years 1984 through 1993. Construct a line graph for the data and state what information the graph reveals. (Source: Energy Information Administration)

Year	1984	1985	1986	1987	1988
Imports	1254	1168	1525	1706	1869

Year	1989	1990	1991	1992	1993
Imports	2133	2145	2110	2226	2457

21. *Federal Income* The list gives the receipts (in billions of dollars) for the federal government of the United States from 1972 to 1993. Construct a line graph for the data. What can you conclude? (Source: U.S. Office of Management and Budget)

1972	207.3	1983	600.6
1973	230.8	1984	666.5
1974	263.2	1985	734.1
1975	279.1	1986	769.1
1976	298.1	1987	854.1
1977	355.6	1988	909.0
1978	399.6	1989	990.7
1979	463.3	1990	1031.3
1980	517.1	1991	1054.3
1981	599.3	1992	1090.5
1982	617.8	1993	1153.5

Focus on Concepts

In this chapter, you studied several concepts that are required in the study of algebra. You can use the following questions to check your understanding of several of these basic concepts. The answers to these questions are given in the back of the book.

1. Describe the differences among the sets of natural numbers, integers, rational numbers, and irrational numbers.

2. Three real numbers are shown on the real number line. Determine the sign of each expression.

(a) $-A$ (b) $-C$
(c) $B - A$ (d) $A - C$

3. You may hear it said that to take the absolute value of a real number you simply remove any negative sign and make the number positive. Can it ever be true that $|a| = -a$ for a real number a? Explain.

4. Explain why each of the following is *not* an equality.

(a) $(3x)^{-1} \neq \dfrac{3}{x}$ (b) $y^3 \cdot y^2 \neq y^6$
(c) $(a^2b^3)^4 \neq a^6b^7$ (d) $(a + b)^2 \neq a^2 + b^2$
(e) $\sqrt{4x^2} \neq 2x$ (f) $\sqrt{2} + \sqrt{3} \neq \sqrt{5}$

5. Is the real number 52.7×10^5 written in scientific notation? Explain.

6. A third-degree polynomial and a fourth-degree polynomial are added.

(a) Can the sum be a fourth-degree polynomial? Explain or give an example.

(b) Can the sum be a second-degree polynomial? Explain or give an example.

(c) Can the sum be a seventh-degree polynomial? Explain or give an example.

7. Explain what is meant when it is said that a polynomial is in factored form.

8. How do you determine whether a rational expression is in reduced form?

In Exercises 9–12, use the plot of the point (x_0, y_0) in the figure. Match the transformation of the point with the correct plot. [The plots are labeled (a), (b), (c), and (d).]

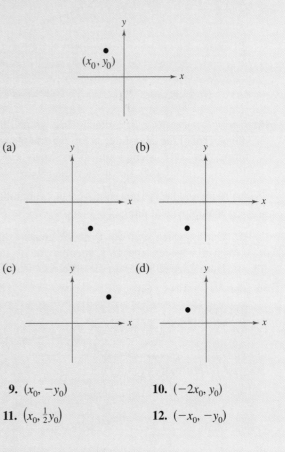

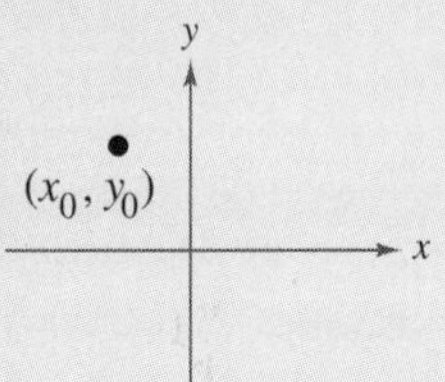

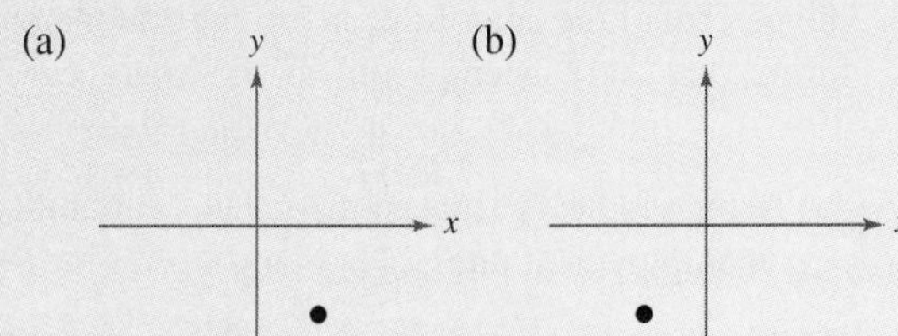

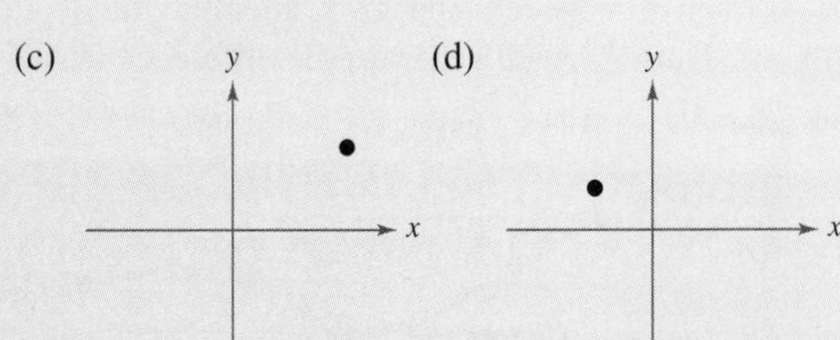

9. $(x_0, -y_0)$ **10.** $(-2x_0, y_0)$

11. $\left(x_0, \frac{1}{2}y_0\right)$ **12.** $(-x_0, -y_0)$

P /// REVIEW EXERCISES

In Exercises 1 and 2, determine which numbers in the set are (a) natural numbers, (b) integers, (c) rational numbers, and (d) irrational numbers.

1. $\left\{11, -14, -\frac{8}{9}, \frac{5}{2}, \sqrt{6}, 0.4\right\}$

2. $\left\{\sqrt{15}, -22, -\frac{10}{3}, 0, 5.2, \frac{3}{7}\right\}$

In Exercises 3 and 4, use a calculator to find the decimal form of the rational number. If it is a nonterminating decimal, write the repeating pattern.

3. (a) $\frac{5}{6}$ (b) $\frac{7}{8}$ **4.** (a) $\frac{9}{25}$ (b) $\frac{5}{7}$

In Exercises 5 and 6, give a verbal description of the real numbers that are represented by the inequality. Then sketch the inequality on the real number line.

5. $x \leq 7$ **6.** $x > 1$

In Exercises 7–10, use absolute value notation to describe the expression.

7. The distance between x and 7 is at least 4.

8. The distance between x and 25 is no more than 10.

9. The distance between y and -30 is less than 5.

10. The distance between y and $\frac{1}{2}$ is more than 2.

In Exercises 11 and 12, perform the operations.

11. $|-3| + 4(-2) - 6$ **12.** $\dfrac{|-10|}{-10}$

In Exercises 13–16, identify the rule of algebra illustrated by the equation.

13. $2x + (3x - 10) = (2x + 3x) - 10$

14. $\dfrac{2}{y+4} \cdot \dfrac{y+4}{2} = 1, \quad y \neq -4$

15. $(t + 4)(2t) = (2t)(t + 4)$

16. $0 + (a - 5) = a - 5$

In Exercises 17–20, simplify the expression.

17. (a) $(-2z)^3$ (b) $(a^2b^4)(3ab^{-2})$

18. (a) $\dfrac{(8y)^0}{y^2}$ (b) $\dfrac{40(b-3)^5}{75(b-3)^2}$

19. (a) $\dfrac{6^2u^3v^{-3}}{12u^{-2}v}$ (b) $\dfrac{3^{-4}m^{-1}n^{-3}}{9^{-2}mn^{-3}}$

20. (a) $(x + y^{-1})^{-1}$ (b) $\left(\dfrac{x^{-3}}{y}\right)\left(\dfrac{x}{y}\right)^{-1}$

In Exercises 21 and 22, write the number in scientific notation.

21. *1994 Net Sales of Procter and Gamble Company:* \$30,296,000,000 (Source: 1994 Annual Report)

22. *Number of Meters in One Foot:* 0.3048

In Exercises 23 and 24, write the number in decimal form.

23. *Miles from Sun to Jupiter:* 4.833×10^8

24. *Ratio of Day to Year:* 2.74×10^{-3}

In Exercises 25 and 26, use a calculator to evaluate the expression. (Round your answer to three decimal places.)

25. (a) $1800(1 + 0.08)^{24}$

(b) $0.0024(7{,}658{,}400)$

26. (a) $50{,}000\left(1 + \dfrac{0.075}{12}\right)^{48}$

(b) $\dfrac{28{,}000{,}000 + 34{,}000{,}000}{87{,}000{,}000}$

In Exercises 27 and 28, fill in the missing description.

	Radical Form	*Rational Exponent Form*
27.	$\sqrt{16} = 4$	$\square = 4$
28.	$\square = 2$	$16^{1/4} = 2$

In Exercises 29 and 30, simplify by removing all possible factors from the radical.

29. $\sqrt{4x^4}$ **30.** $\sqrt[5]{64x^6}$

In Exercises 31 and 32, rewrite the expression by rationalizing the denominator. Simplify your answer.

31. $\dfrac{1}{2 - \sqrt{3}}$ **32.** $\dfrac{1}{\sqrt{x} - 1}$

In Exercises 33 and 34, simplify the expression.

33. $\sqrt{50} - \sqrt{18}$ **34.** $\sqrt{8x^3} + \sqrt{2x}$

35. *Strength of a Wooden Beam* The rectangular cross section of a wooden beam cut from a log of diameter 24 inches (see figure) will have a maximum strength if its width w and height h are given by

$w = 8\sqrt{3}$ and $h = \sqrt{24^2 - (8\sqrt{3})^2}$.

Find the area of the rectangular cross section and express the answer in simplest form.

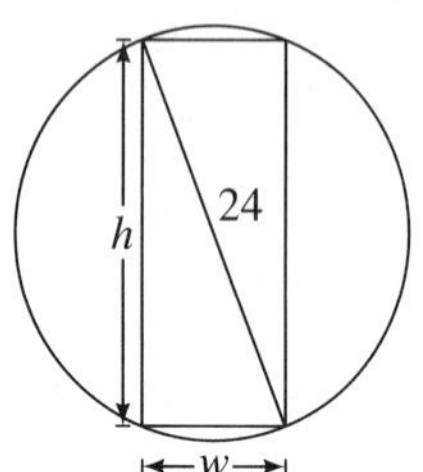

36. *Essay* Explain why $\sqrt{5u} + \sqrt{3u} \neq 2\sqrt{2u}$.

In Exercises 37–42, describe the error and then make the necessary correction.

37. ~~$\dfrac{x - 1}{1 - x} = 1$~~

38. ~~$-x^2(-x^2 + 3) = x^4 + 3x^2$~~

39. ~~$(2x)^4 = 2x^4$~~ **40.** ~~$(-x)^6 = -x^6$~~

41. ~~$\sqrt{3^2 + 4^2} = 3 + 4$~~ **42.** ~~$\sqrt{10x} = 10\sqrt{x}$~~

In Exercises 43–50, perform the operations and write the result in standard form.

43. $-(3x^2 + 2x) + (1 - 5x)$

44. $8y - [2y^2 - (3y - 8)]$

45. $(2x - 3)^2$ **46.** $(3\sqrt{5} + 2)(3\sqrt{5} - 2)$

47. $(x^2 - 2x + 1)(x^3 - 1)$

48. $(x^3 - 3x)(2x^2 + 3x + 5)$

49. $(y^2 - y)(y^2 + 1)(y^2 + y + 1)$

50. $\left(x - \dfrac{1}{x}\right)(x + 2)$

51. *Geometric Modeling* Use the area model to write two different expressions for the total area. Then equate the two expressions and name the algebraic property illustrated.

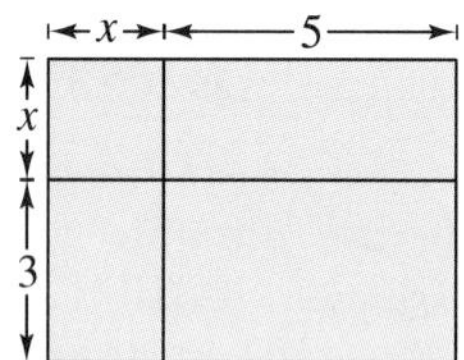

52. *Geometry* Write an expression for the area of the region and simplify the result.

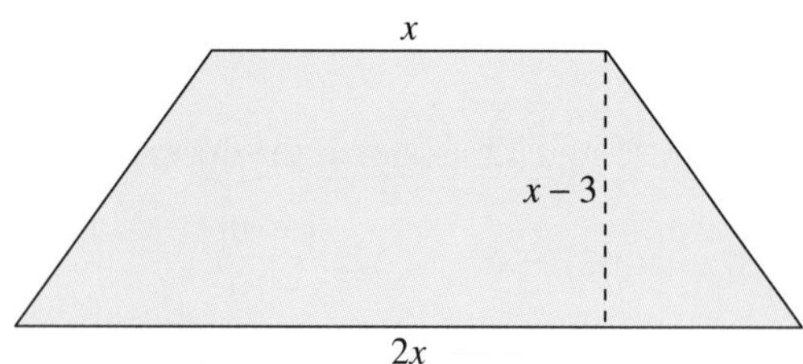

In Exercises 53–58, factor completely.

53. $x^3 - x$ **54.** $x(x - 3) + 4(x - 3)$

55. $2x^2 + 21x + 10$ **56.** $3x^2 + 14x + 8$

57. $x^3 - x^2 + 2x - 2$ **58.** $x^3 - 1$

59. *Exploration* The surface area of a right circular cylinder is $S = 2\pi r^2 + 2\pi rh$.

(a) Draw a right circular cylinder of radius r and height h. Use the figure to explain how the surface area formula was obtained.

(b) Factor the expression for surface area.

60. *Revenue* The revenue for selling x units of a product at a price of p dollars per unit is $R = xp$. For a particular product the revenue is

$$R = 1600x - 0.50x^2.$$

Factor the expression and determine an expression that gives the price in terms of x.

In Exercises 61–68, perform the operation and simplify the answer.

61. $\dfrac{x^2 - 4}{x^4 - 2x^2 - 8} \cdot \dfrac{x^2 + 2}{x^2}$

62. $\dfrac{2x - 1}{x + 1} \cdot \dfrac{x^2 - 1}{2x^2 - 7x + 3}$

63. $\dfrac{x^2(5x - 6)}{2x + 3} \div \dfrac{5x}{2x + 3}$

64. $\dfrac{4x - 6}{(x - 1)^2} \div \dfrac{2x^2 - 3x}{x^2 + 2x - 3}$

65. $x - 1 + \dfrac{1}{x + 2} + \dfrac{1}{x - 1}$

66. $2x + \dfrac{3}{2(x - 4)} - \dfrac{1}{2(x + 2)}$

67. $\dfrac{1}{x} - \dfrac{x - 1}{x^2 + 1}$

68. $\dfrac{1}{x - 1} + \dfrac{1 - x}{x^2 + x + 1}$

In Exercises 69 and 70, simplify the compound fraction.

69. $\dfrac{\left(\dfrac{1}{x} - \dfrac{1}{y}\right)}{(x^2 - y^2)}$

70. $\dfrac{\left(\dfrac{1}{x} - \dfrac{1}{y}\right)}{\left(\dfrac{1}{x} + \dfrac{1}{y}\right)}$

Geometry **In Exercises 71 and 72, plot the points and verify that the points form the polygon.**

71. *Right Triangle:* (2, 3), (13, 11), (5, 22)

72. *Parallelogram:* (1, 2), (8, 3), (9, 6), (2, 5)

In Exercises 73 and 74, determine the quadrant(s) in which (x, y) is located so that the conditions are satisfied.

73. $x > 0$ and $y = -2$

74. (x, y), $xy = 4$

In Exercises 75 and 76, (a) plot the points, (b) find the distance between the points, and (c) find the midpoint of the line segment joining the points.

75. $(-3, 8)$, $(1, 5)$

76. $(5.6, 0)$, $(0, 8.2)$

77. *Weather* The normal daily maximum and minimum temperatures for each month for the city of Chicago are given in the table. Make a double line graph for the data. (Source: NOAA)

Month	Jan	Feb	Mar	Apr	May	Jun
Max	29.0	33.5	45.8	58.6	70.1	79.6
Min	12.9	17.2	28.5	38.6	47.7	57.5

Month	Jul	Aug	Sep	Oct	Nov	Dec
Max	83.7	81.8	74.8	63.3	48.4	34.0
Min	62.6	61.6	53.9	42.2	31.6	19.1

78. *CompuServe Revenues* The revenues (in millions of dollars) for CompuServe, owned by H & R Block, Inc., for the years 1990 through 1994 are given in the table. Create a bar graph for the data. (Source: H & R Block 1994 Annual Report)

Year	1990	1991	1992	1993	1994
Revenue	206.7	251.6	280.9	315.4	429.9

CHAPTER PROJECT *Modeling the Volume of a Box*

Many mathematical results are discovered experimentally by calculating examples and looking for patterns. Prior to the 1950s, this mode of discovery was very time-consuming because the calculations had to be done by hand. The introduction of computer and calculator technology has removed much of this drudgery. In the following project, you are asked to model a real-life situation and solve a problem by looking for patterns in the corresponding data.

Consider a rectangular box with a square base and a surface area of 216 square inches. Let x represent the length (in inches) of each side of the base and let h represent the height (in inches) of the box, as shown at the left. Your goal is to answer the question "Of all rectangular boxes with square bases and surface area of 216 square inches, which has the greatest volume?"

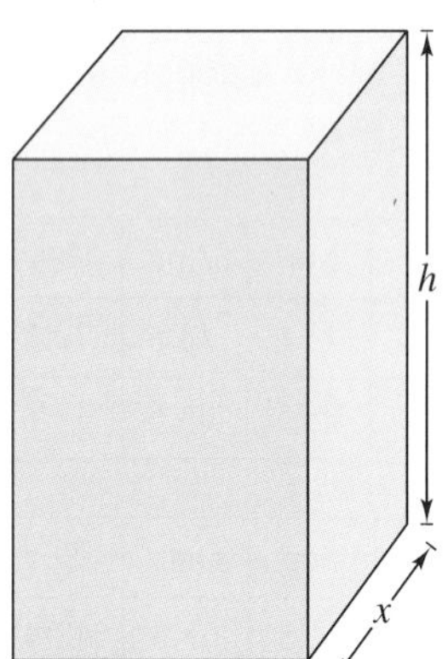

(a) Express the areas of the base, top, and sides in terms of x and h.

(b) Find an expression in terms of x and h for the surface area of the box.

(c) Use the fact that the surface area is 216 square inches to express the variable h in terms of x.

(d) Find an expression for the volume of the box in terms of x alone.

(e) Use the expression in part (d) and a graphing utility to complete the table. Then use the results to decide which box has the greatest volume.

Base, x	*Height*	*Surface Area*	*Volume*
1.0	53.5	216.0	53.5
1.5	35.3	216.0	79.3
2.0	26.0	216.0	104.0
⋮	⋮	⋮	⋮
10.0	0.4	216.0	40.0

Questions for Further Exploration

1. What happens to the height of the box as x gets closer and closer to 0? Of all boxes with square base and a surface area of 216 square inches, is there a tallest? Explain your reasoning.

2. What is the maximum value of x? What happens to the height of the box as x gets closer and closer to this maximum value? Is there a shortest box that has a square base and a surface area of 216 square inches? Explain your reasoning.

3. Complete the table. Does it lend further support to your answer to part (e)? Explain.

x	5.9	5.99	5.999	6.001	6.01	6.1
V						

4. Of all rectangular boxes with surface area of 216 square inches and a base that is x inches by $2x$ inches, which has the maximum volume? Explain your reasoning.

P /// CHAPTER TEST

Take this test as you would take a test in class. After you are done, check your work against the answers given in the back of the book.

The *Interactive* CD-ROM provides answers to the Chapter Tests and Cumulative Tests. It also offers Chapter Pre-Tests (that test key skills and concepts covered in previous chapters) and Chapter Post-Tests, both of which have randomly generated exercises with diagnostic capabilities.

1. Place the correct symbol (< or >) between $-\frac{10}{3}$ and $-|-4|$.
2. Find the distance between the real numbers -5.4 and $3\frac{3}{4}$.

In Exercises 3–6, evaluate the quantity without the aid of a calculator.

3. (a) $27\left(-\dfrac{2}{3}\right)$ (b) $\dfrac{5}{18} \div \dfrac{15}{8}$
4. (a) $\left(-\dfrac{3}{5}\right)^3$ (b) $\left(\dfrac{3^2}{2}\right)^{-3}$
5. (a) $\sqrt{5} \cdot \sqrt{125}$ (b) $\dfrac{\sqrt{72}}{\sqrt{2}}$
6. (a) $\dfrac{5.4 \times 10^8}{3 \times 10^3}$ (b) $(3 \times 10^4)^3$

In Exercises 7–9, simplify the expression.

7. (a) $3z^2(2z^3)^2$ (b) $(u-2)^{-4}(u-2)^{-3}$
8. (a) $\left(\dfrac{x^{-2}y^2}{3}\right)^{-1}$ (b) $\sqrt[3]{\dfrac{16}{v^5}}$
9. (a) $9z\sqrt{8z} - 3\sqrt{2z^3}$ (b) $-5\sqrt{16y} + 10\sqrt{y}$

In Exercises 10–13, perform the operations and simplify.

10. $(x^2 + 3) - [3x + (8 - x^2)]$
11. $\left(x + \sqrt{5}\right)\left(x - \sqrt{5}\right)$
12. $\dfrac{8x}{x-3} + \dfrac{24}{3-x}$
13. $\left(\dfrac{2}{x} - \dfrac{2}{x+1}\right) \div \left(\dfrac{4}{x^2-1}\right)$
14. Factor (a) $2x^4 - 3x^3 - 2x^2$ and (b) $x^3 + 2x^2 - 4x - 8$ completely.
15. Rationalize the denominators of (a) $16/\sqrt[3]{16}$ and (b) $6/\left(1 - \sqrt{3}\right)$.
16. Plot the points $(-2, 5)$ and $(6, 0)$. Find the coordinates of the midpoint of the line segment joining the points and the distance between the points.
17. The numbers (in millions) of votes cast for the Democratic candidates for president in 1980, 1984, 1988, and 1992 were 35.5, 37.6, 41.8, and 44.9, respectively. Create a bar graph for this data.

Library of Functions

In Chapter 1, you will be introduced to the concept of a *function.* As you proceed through the text, you will see that functions play a primary role in modeling real-life situations.

Over the past few hundred years, many different types of functions have been introduced and studied. Those that have proven to be most important in modeling real life have come to be known as *elementary functions.* There are three basic types of elementary functions: algebraic functions, exponential and logarithmic functions, and trigonometric and inverse trigonometric functions.

You will also encounter other types of functions in this text, such as functions defined by real-life data and piecewise-defined functions.

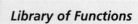

Library of Functions

This "library icon" will appear in the text each time a new type of function is studied in detail.

For instance, the icon appears in Section 1.2 because that section is about lines, and the graph of a linear function is a line.

Algebraic Functions		
Polynomial Functions	Linear or First-Degree Polynomial Functions	Section 1.2
	Quadratic or Second-Degree Polynomial Functions	Section 3.1
	Cubic or Third-Degree Polynomial Functions	Section 3.2
	Fourth- and Higher-Degree Polynomial Functions	Section 3.2
Rational Functions		Section 3.5
Radical Functions		Section 1.3
Exponential and Logarithmic Functions		
Exponential Functions		Section 4.1
Logarithmic Functions		Section 4.2
Trigonometric and Inverse Trigonometric Functions		
Trigonometric Functions		Not covered in this text.
Inverse Trigonometric Functions		Not covered in this text.
Nonelementary Functions		
Piecewise-Defined Functions		Section 1.3
Greatest Integer Functions		Section 1.4
Absolute Value Functions		Section 1.4
Functions Defined by Real-Life Data		Section 1.3

Functions and Their Graphs

Chapter 1

Many wildlife populations follow a cyclical "predator-prey" pattern. One example is the populations of snowshoe hare and lynx in the Yukon Territory. The researchers shown in the photo kept track of the lynx and hare populations from 1988 through 1995. Lynx numbers in a 350-square-kilometer region of the Yukon are shown below.

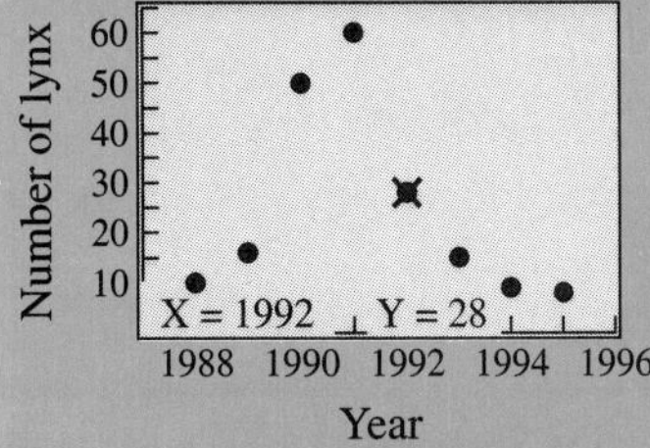

1988	(10)	1991	(60)	1994	(9)
1989	(16)	1992	(28)	1995	(8)
1990	(50)	1993	(15)		

The hare population was low in 1986, increased to a high in 1990, and then decreased to a low again in 1992.

The number of lynx is a function of the year. You can use a graphing utility to create a scatter plot that depicts the lynx population as a function of the year, as shown above. Use the *trace feature* to identify the coordinates of each point. The data was supplied by Mark O'Donoghue, as part of the Kluane Boreal Forest Ecosystem Project. (See Exercise 87 on page 115.)

Alejandro Frid/Biological Photo Service.

Husband and wife researchers, Elizabeth Hofer and Peter Upton, are measuring a lynx that has been trapped and sedated. The researchers work in the Yukon Territory, Canada.

1.1 Graphs and Graphing Utilities

The Graph of an Equation / Using a Graphing Utility / Determining a Viewing Rectangle / Applications

The Graph of an Equation

News magazines often show graphs comparing the rate of inflation, the federal deficit, wholesale prices, or the unemployment rate to the time of year. Industrial firms and businesses use graphs to report their monthly production and sales statistics. Such graphs provide geometric pictures of the way one quantity changes with respect to another. Frequently, the relationship between two quantities is expressed as an equation. This section introduces the basic procedure for determining the geometric picture associated with an equation.

For an equation in variables x and y, a point (a, b) is a **solution point** if the substitution of $x = a$ and $y = b$ satisfies the equation. Most equations have *infinitely* many solution points. For example, the equation

$$3x + y = 5$$

has solution points $(0, 5)$, $(1, 2)$, $(2, -1)$, $(3, -4)$, and so on. The set of all solution points of an equation is the **graph** of the equation.

Study Tip

To sketch the graph of an equation by point plotting, use the following procedure.

1. If possible, rewrite the equation so that one of the variables is isolated on one side of the equation.
2. Make a table of several solution points.
3. Plot these points in the coordinate plane.
4. Connect the points with a smooth curve.

EXAMPLE 1 Sketching a Graph by Point Plotting

Use point plotting and graph paper to sketch the graph of

$$3x + y = 6.$$

Solution

In this case you can isolate the variable y to obtain

$$y = 6 - 3x. \qquad \text{Solve equation for } y.$$

Using negative, zero, and positive values for x, you can obtain the following table of values (solution points).

x	-1	0	1	2	3
$y = 6 - 3x$	9	6	3	0	-3

Next, plot these points and connect them, as shown in Figure 1.1. It appears that the graph is a straight line. You will study lines extensively in Section 1.2.

Figure 1.1

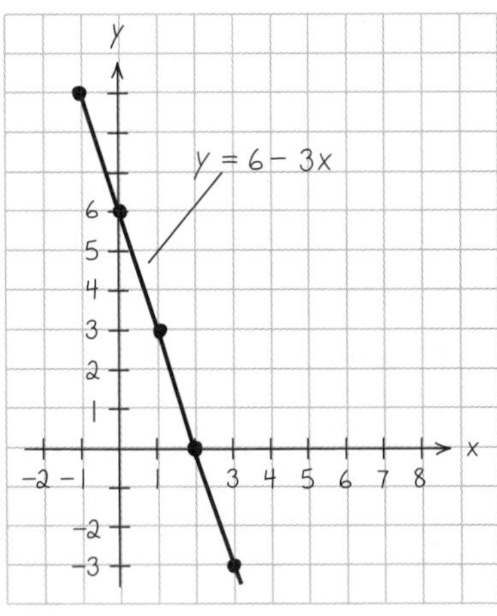

The points at which a graph touches or crosses an axis are the **intercepts** of the graph. For instance, in Example 1 the point (0, 6) is the y-intercept of the graph because the graph crosses the y-axis at that point. The point (2, 0) is the x-intercept of the graph because the graph crosses the x-axis at that point.

EXAMPLE 2 Sketching a Graph by Point Plotting

The *Interactive* CD-ROM shows every example with its solution; clicking on the *Try It!* button brings up similar problems. Guided Examples and Integrated Examples show step-by-step solutions to additional examples. Integrated Examples are related to several concepts in the section.

Use point plotting and graph paper to sketch the graph of $y = x^2 - 2$.

Solution

First, make a table of values by choosing several convenient values of x and calculating the corresponding values of y.

x	-2	-1	0	1	2	3
$y = x^2 - 2$	2	-1	-2	-1	2	7

Next, plot the corresponding solution points, as shown in Figure 1.2(a). Finally, connect the points with a smooth curve, as shown in Figure 1.2(b).

A computer animation of this concept appears in the *Interactive* CD-ROM.

Figure 1.2

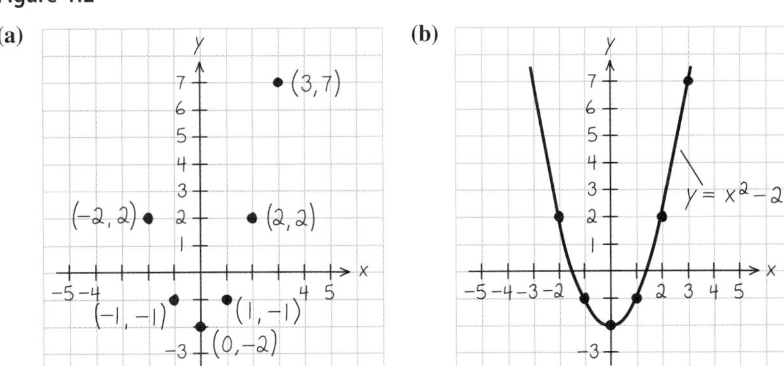

This graph is called a **parabola.** You will study parabolas in Section 3.1.

Note In this text, you will study two basic ways to create graphs: *by hand* and *using a graphing utility.* For instance, the graphs in Figures 1.1 and 1.2 were sketched by hand and the graph in Figure 1.4 was sketched using a graphing utility.

Using a Graphing Utility

One of the disadvantages of the point-plotting method is that to get a good idea about the shape of a graph you need to plot *many* points. With only a few points, you could badly misrepresent the graph. For instance, consider the equation

$$y = \frac{1}{30}x(x^4 - 10x^2 + 39).$$

Suppose you plotted only five points: $(-3, -3)$, $(-1, -1)$, $(0, 0)$, $(1, 1)$, and $(3, 3)$, as shown in Figure 1.3(a). From these five points, you might assume that the graph of the equation is a straight line. That, however, is not correct. By plotting several more points, you can see that the actual graph is not straight at all. See Figure 1.3(b).

Figure 1.3

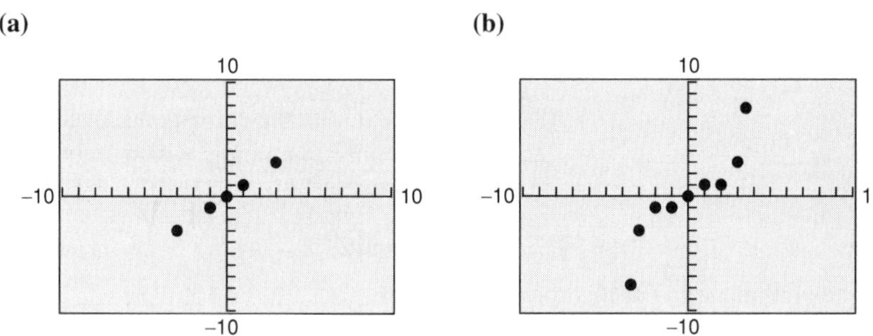

From this, you can see that the point-plotting method leaves you with a dilemma. On the one hand, the method can be very inaccurate if only a few points are plotted. But on the other hand, it is very time-consuming to plot a dozen (or more) points. Technology can help solve this dilemma. Plotting several (even several hundred) points on a rectangular coordinate system is something that a computer or calculator can do easily.

Note The point-plotting method is the method used by *all* graphing utilities. Each computer or calculator screen is made up of a grid of hundreds or thousands of small areas called **pixels.** Screens that have many pixels per square inch are said to have a higher **resolution** than screens with fewer pixels.

Using a Graphing Utility to Graph an Equation

To graph an equation involving x and y on a graphing utility, use the following procedure.

1. Rewrite the equation so that y is isolated on the left side.
2. Enter the equation into a graphing utility.
3. Determine a **viewing rectangle** that shows all important features of the graph. For some graphing utilities, the standard viewing rectangle ranges between -10 and 10 for both x- and y-values.
4. Activate the graphing utility.

The following table, which can be generated on a *TI-82* or a *TI-83*, shows some points of the graph in Figure 1.4.

X	Y1	
−3	7.5	
−2	0	
−1	−1.5	
0	0	
1	1.5	
2	0	
3	−7.5	
X = −3		

EXAMPLE 3 **Using a Graphing Utility**

Use a graphing utility to graph $2y + x^3 = 4x$.

Solution

To begin, solve the equation for y in terms of x.

$$2y + x^3 = 4x \qquad \text{Original equation}$$

$$2y = -x^3 + 4x \qquad \text{Subtract } x^3 \text{ from both sides.}$$

$$y = -\frac{1}{2}x^3 + 2x \qquad \text{Divide both sides by 2.}$$

Now, by entering this equation into a graphing utility (using a standard viewing rectangle), you can obtain the graph shown in Figure 1.4.

Figure 1.4

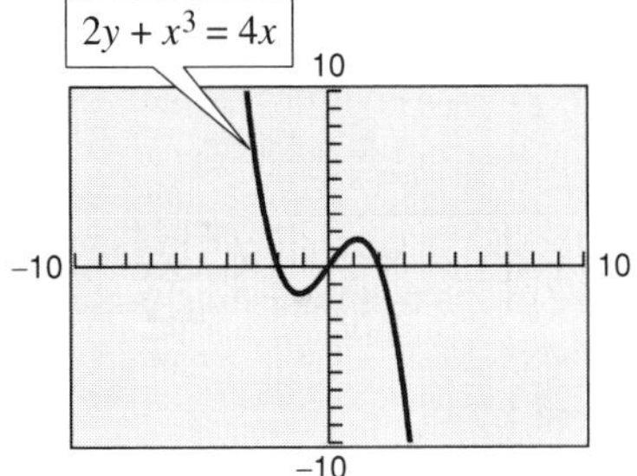

EXPLORATION

Use your graphing utility to duplicate the graphs shown below and state the viewing rectangle. Work with a partner. If possible, choose a partner who has the same model of graphing utility as you do.

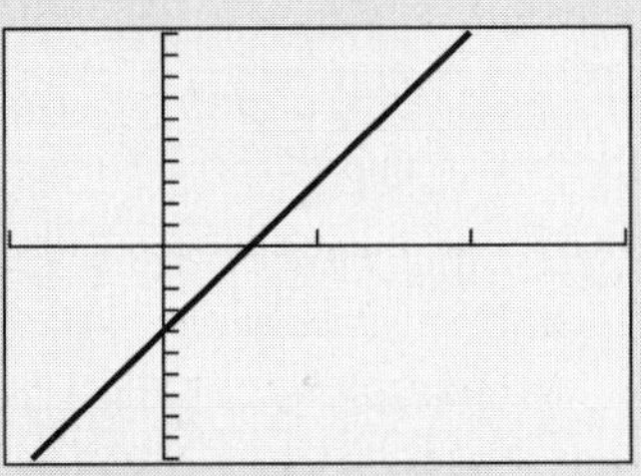

Graph of $y = 7x - 4$

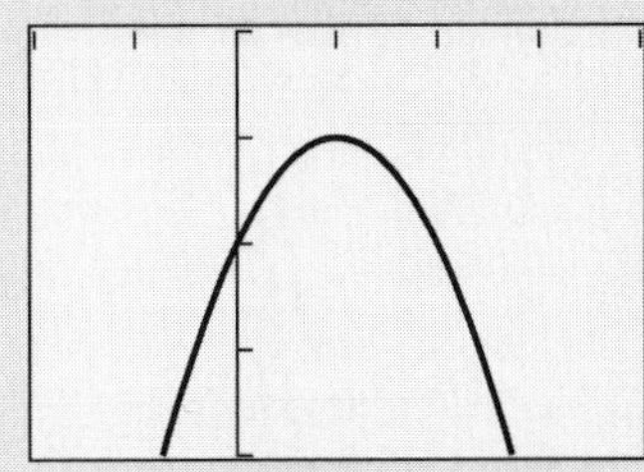

Graph of $y = -x^2 + 2x - 8$

Determining a Viewing Rectangle

A **viewing rectangle** for a graph is a rectangular portion of the Cartesian plane. A viewing rectangle is determined by six values: the minimum x-value, the maximum x-value, the x-scale, the minimum y-value, the maximum y-value, and the y-scale. The **standard** viewing rectangle for some graphing utilities uses the following values.

EXPLORATION

It is important to choose a viewing rectangle that shows all the important features of a graph. For example, consider the problem of graphing the equation

$$y = \frac{x^2 + 12x}{x - 21}.$$

Does the standard viewing rectangle show all the important features of the graph? Can you find a better viewing rectangle? Try using the zoom feature of your graphing utility several times.

```
Xmin = -10
Xmax = 10
Xscl = 1
Ymin = -10
Ymax = 10
Yscl = 1
```

By choosing different viewing rectangles for a graph, it is possible to obtain very different impressions of the graph's shape. For instance, Figure 1.5 shows four different viewing rectangles for the graph of

$$y = 0.1x^4 - x^3 + 2x^2.$$

Of these, the view shown in Figure 1.5(a) is the most complete because it shows more of the distinguishing portions of the graph.

Figure 1.5

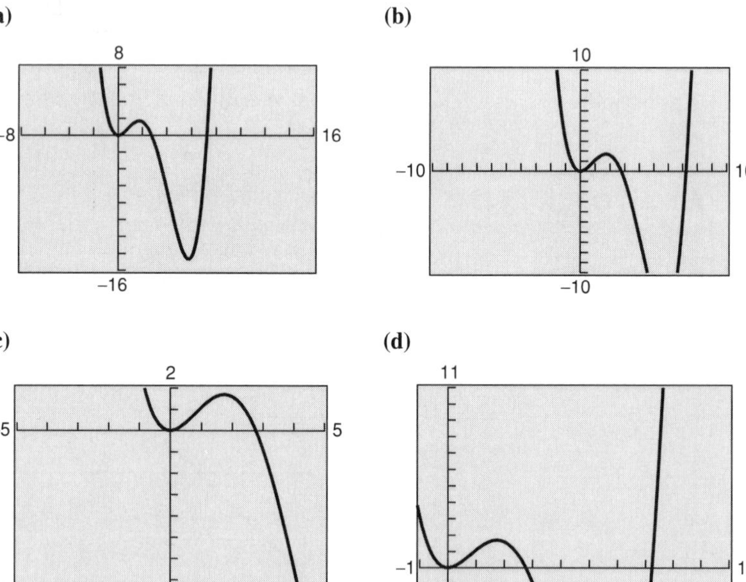

EXAMPLE 4 Sketching a Circle with a Graphing Utility

Use a graphing utility to graph

$$x^2 + y^2 = 9.$$

Solution

The graph of $x^2 + y^2 = 9$ is a circle whose center is the origin and whose radius is 3. (See Section P.5 in preceding chapter.) To graph the equation, begin by solving the equation for y.

$$x^2 + y^2 = 9 \qquad \text{Original equation}$$
$$y^2 = 9 - x^2 \qquad \text{Solve for } y^2.$$
$$y = \pm\sqrt{9 - x^2} \qquad \text{Solve for } y.$$

Note The standard viewing rectangle on many graphing utilities does not give a true geometric perspective. That is, perpendicular lines will not appear to be perpendicular and circles will not appear to be circular. To overcome this, you can use a square setting, as demonstrated in Example 4.

The graph of

$$y = \sqrt{9 - x^2} \qquad \text{Upper semicircle}$$

is the upper semicircle. The graph of

$$y = -\sqrt{9 - x^2} \qquad \text{Lower semicircle}$$

is the lower semicircle. Enter *both* equations in your graphing utility and generate the resulting graphs. In Figure 1.6(a), note that if you use a standard viewing rectangle, the two graphs do not appear to form a circle. You can overcome this problem by using a **square setting,** in which the horizontal and vertical tick marks have equal spacing, as shown in Figure 1.6(b). On many graphing utilities, a square setting can be obtained by using the ratio

$$\frac{\text{Ymax} - \text{Ymin}}{\text{Xmax} - \text{Xmin}} = \frac{2}{3}.$$

Figure 1.6

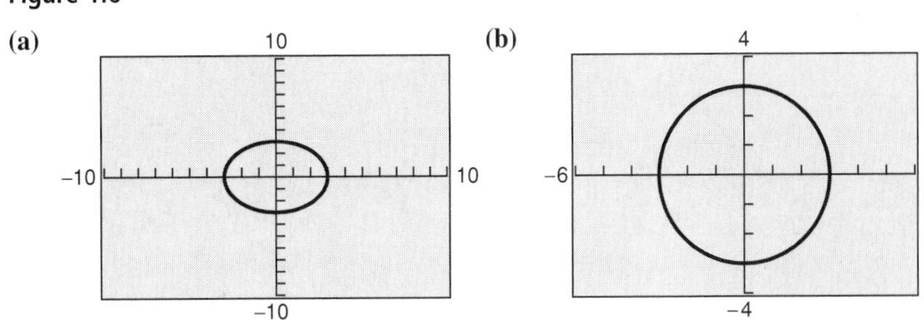

In applications, it is convenient to use variable names that suggest real-life quantities: d for distance, t for time, and so on. Most graphing utilities, however, require the variable names to be x and y.

Applications

The following two applications show how to develop mathematical models to represent real-world situations. You will see that both a graphing utility and algebra can be used to understand and resolve the problems posed.

Once an appropriate viewing rectangle is chosen for a particular graph, the *zoom* and *trace* features of a graphing utility are useful for approximating specific values from the graph.

EXAMPLE 5 Using the Zoom and Trace Features

A runner runs at a constant rate of 4.8 miles per hour. The verbal model and algebraic equation relating distance run and elapsed time are as follows.

Verbal Model: Distance = Rate · Time

Equation: $d = 4.8t$

a. Use a graphing utility and an appropriate viewing rectangle to graph the equation $d = 4.8t$. (Represent d by y and t by x.)

b. Estimate how far the runner can run in 3.2 hours.

c. Estimate how long it will take to run a 26-mile marathon.

Solution

Note The viewing rectangle on your graphing utility may differ from those shown in parts (b) and (c) of Figure 1.7.

a. An appropriate viewing rectangle and graph are shown in Figure 1.7(a).

b. Figure 1.7(b) shows the viewing rectangle after zooming in (near $x = 3.2$) once by a factor of 4. Using the trace feature, you can determine that for $x = 3.2$, the distance is $y \approx 15.36$ miles.

c. Figure 1.7(c) shows the viewing rectangle after zooming in (near $y = 26$) twice by a factor of 4. Using the trace feature, you can determine that for $y = 26$, the time is $x \approx 5.42$ hours.

Figure 1.7

(a)

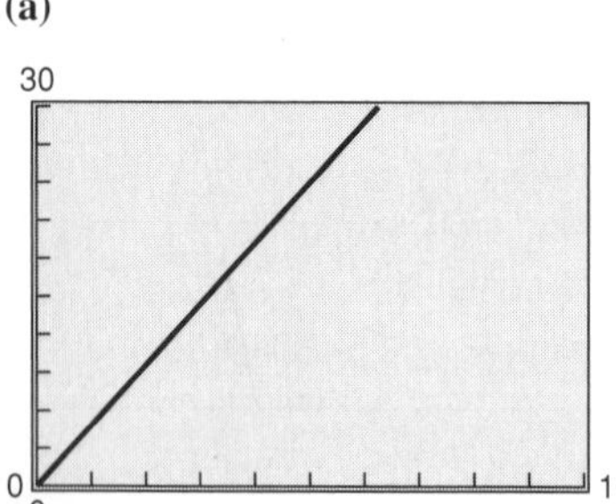

(b)

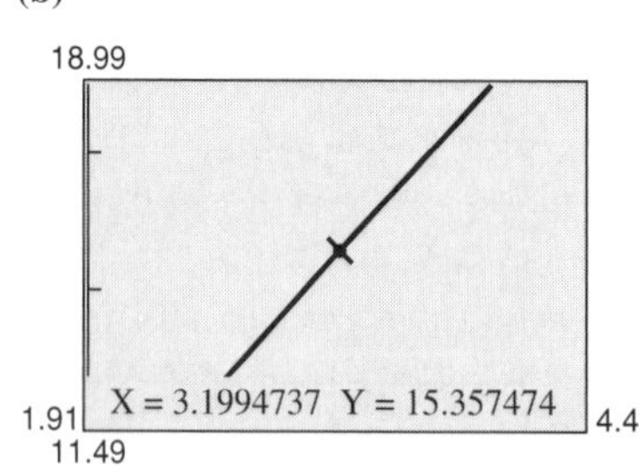

(c)

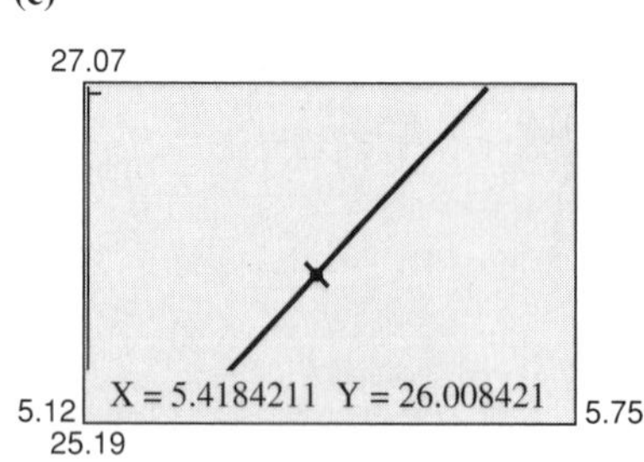

EXAMPLE 6 **An Application: Monthly Wages**

You receive a monthly salary of \$2000 plus a commission of 10% of sales.

a. Find an equation expressing the monthly wages y in terms of the sales x.

b. If sales are $x = 1480$ in August, what are your wages for that month?

c. If you receive \$2225 in wages for September, what were your sales for that month?

Figure 1.8

(a)

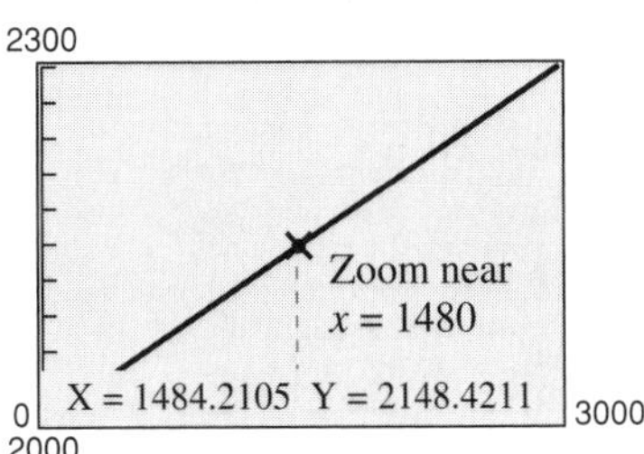

(b)

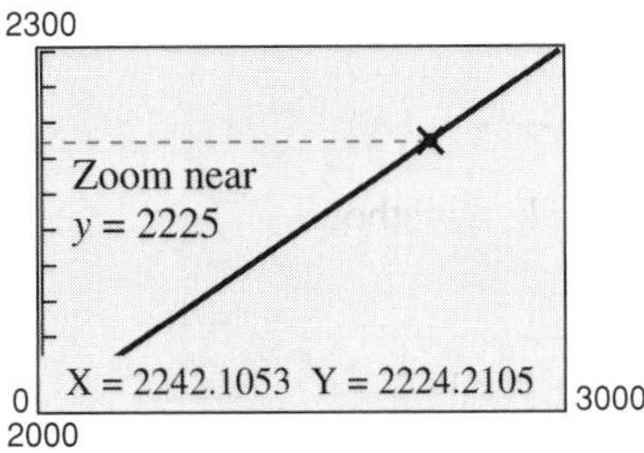

Solution

a. The monthly wages are the sum of the fixed \$2000 salary and the 10% commission on sales x.

Verbal Model: Wages = Salary + Commission on Sales

Equation: $y = 2000 + 0.1x$

b. If $x = 1480$, the corresponding wages are

$$y = 2000 + 0.1(1480) = 2000 + 148 = \$2148.$$

You can confirm this result using a graphing utility. Because $x \geq 0$ and the monthly wages are at least \$2000, a reasonable viewing rectangle is the one shown in Figure 1.8(a). Using the zoom and trace features near $x = 1480$ shows that the wages are about \$2148.

c. To answer the third question, you can use the graphing utility to find the value along the x-axis (sales) that corresponds to a y-value of 2225 (wages). Beginning with Figure 1.8(b) and using the zoom and trace features, you can estimate x to be $x \approx 2250$. You can verify this answer by observing that

$$\text{Wages} = 2000 + 0.1(2250) = 2000 + 225 = \$2225.$$

Group Activity

Comparison of Wages

Your employer offers you a choice of wage scales: a monthly salary of \$3000 plus commission of 7% of sales or a salary of \$3400 plus a 5% commission. Discuss how you would choose your option. At what sales level would the options yield the same salary? Be sure to take advantage of your graphing utility.

1.1 /// EXERCISES

In Exercises 1–8, determine whether the points lie on the graph of the equation.

Equation	*Points*	
1. $y = \sqrt{x+4}$	(a) $(0, 2)$	(b) $(5, 3)$
2. $y = x^2 - 3x + 2$	(a) $(2, 0)$	(b) $(-2, 8)$
3. $y = 4 - \lvert x - 2 \rvert$	(a) $(1, 5)$	(b) $(6, 0)$
4. $y = \frac{1}{3}x^3 - 2x^2$	(a) $\left(2, -\frac{16}{3}\right)$	(b) $(-3, 9)$
5. $2x - y - 3 = 0$	(a) $(1, 2)$	(b) $(1, -1)$
6. $x^2 + y^2 = 20$	(a) $(3, -2)$	(b) $(-4, 2)$
7. $x^2y - x^2 + 4y = 0$	(a) $\left(1, \frac{1}{5}\right)$	(b) $\left(2, \frac{1}{2}\right)$
8. $y = \dfrac{1}{x^2 + 1}$	(a) $(0, 0)$	(b) $(3, 0.1)$

In Exercises 9–16, complete the table. Use the resulting solution points to sketch the graph of the equation. Use a graphing utility to verify the graph.

9. $y = -2x + 3$

x	-1	0	1	$\frac{3}{2}$	2
y					

10. $y = \frac{3}{2}x - 1$

x	-2	0	$\frac{2}{3}$	1	2
y					

11. $y = x^2 - 2x$

x	-1	0	1	2	3
y					

12. $y = 4 - x^2$

x	-2	-1	0	1	2
y					

13. $y = \frac{1}{4}x - 3$

x	-2	-1	0	1	2
y					

14. $y = 3 - \lvert x - 2 \rvert$

x	0	1	2	3	4
y					

15. $y = \sqrt{x - 1}$

x	1	2	5	10	17
y					

16. $y = \dfrac{6x}{x^{-2} + 1}$

x	-2	-1	0	1	2
y					

17. *Think About It* Repeat Exercise 13 for the equation $y = -\frac{1}{4}x - 3$. Use the result to describe any differences in the graphs.

18. *Think About It* Continue the table in Exercise 16 for x-values of 5, 10, 20, and 40. What is the value of y approaching? Can y be negative for positive values of x? Explain.

In Exercises 19–26, use a graphing utility to graph the equation. Use a standard setting. Approximate any x- or y-intercepts of the graph.

19. $y = x - 5$

20. $y = (x + 1)(x - 3)$

21. $y = x^2 + x - 2$

22. $y = 9 - x^2$

23. $y = x\sqrt{x + 6}$

24. $y = (6 - x)\sqrt{x}$

25. $y = \dfrac{2x}{x - 1}$

26. $y = \dfrac{4}{x}$

In Exercises 27–32, match the equation with its graph. [The graphs are labeled (a), (b), (c), (d), (e), and (f).]

(a)
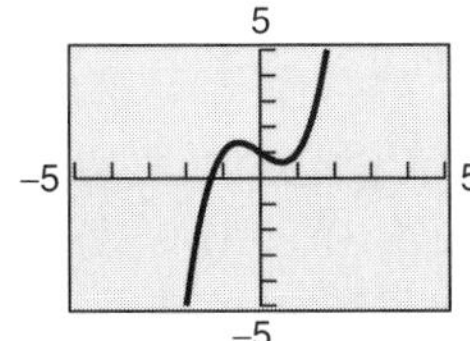

(b)
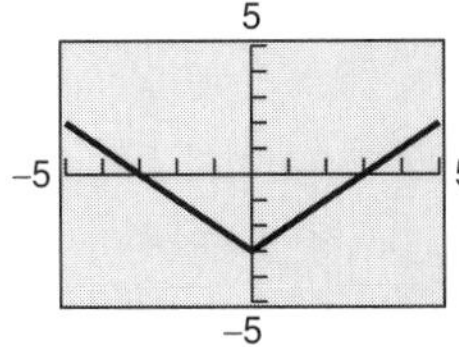

(c)
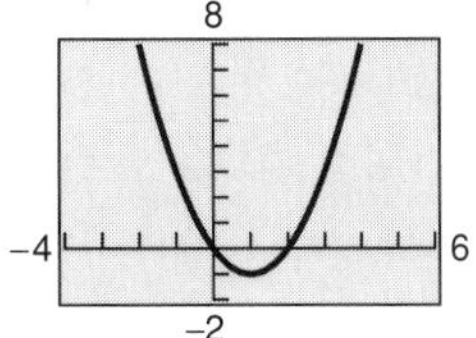

(d)
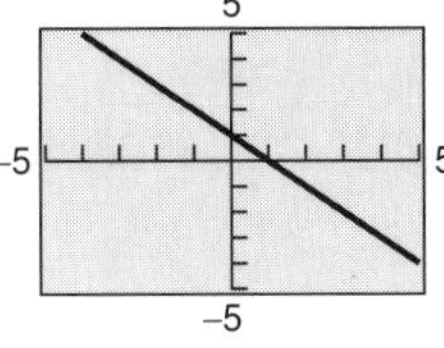

(e)
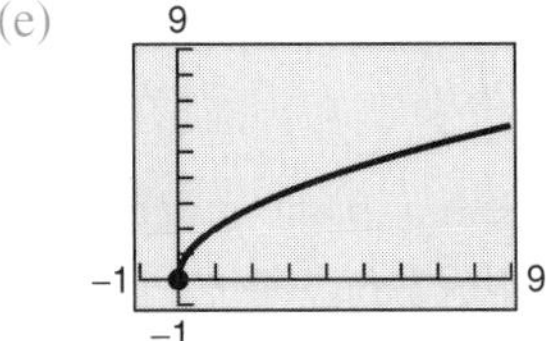

(f)
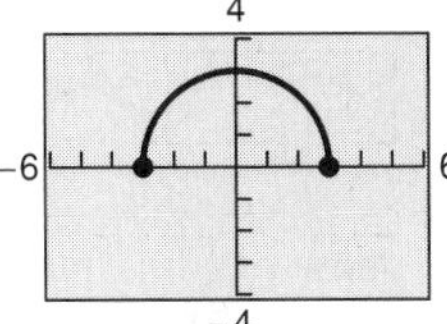

27. $y = 1 - x$

28. $y = x^2 - 2x$

29. $y = \sqrt{9 - x^2}$

30. $y = 2\sqrt{x}$

31. $y = x^3 - x + 1$

32. $y = |x| - 3$

In Exercises 33–46, sketch the graph of the equation.

33. $y = -3x + 2$

34. $y = 2x - 3$

35. $y = 1 - x^2$

36. $y = x^2 - 1$

37. $y = x^2 - 3x$

38. $y = -x^2 - 4x$

39. $y = x^3 + 2$

40. $y = x^3 - 1$

41. $y = \sqrt{x - 3}$

42. $y = \sqrt{1 - x}$

43. $y = |x - 2|$

44. $y = 4 - |x|$

45. $x = y^2 - 1$

46. $x = y^2 - 4$

The *Interactive* CD-ROM contains step-by-step solutions to all odd-numbered Section and Review Exercises. It also provides Tutorial Exercises, which link to Guided Examples for additional help.

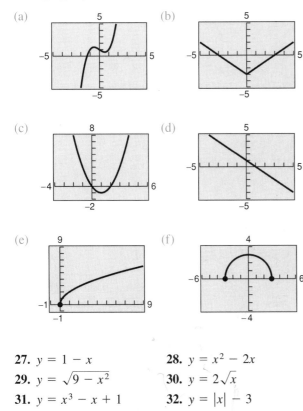

In Exercises 47–54, use a graphing utility to graph the equation. Use a standard setting. Approximate any intercepts.

47. $y = 3 - \frac{1}{2}x$

48. $y = \frac{2}{3}x - 1$

49. $y = x^2 - 4x + 3$

50. $y = \frac{1}{2}(x + 4)(x - 2)$

51. $y = x(x - 2)^2$

52. $y = \dfrac{4}{x^2 + 1}$

53. $y = \sqrt[3]{x}$

54. $y = \sqrt[3]{x + 1}$

In Exercises 55–58, use a graphing utility to sketch the graph of the equation. Begin by using a standard setting. Then graph the equation a second time using the specified setting. Which setting is better? Explain.

55. $y = \frac{5}{2}x + 5$

Xmin = 0
Xmax = 6
Xscl = 1
Ymin = 0
Ymax = 10
Yscl = 1

56. $y = -3x + 50$

Xmin = -1
Xmax = 4
Xscl = 1
Ymin = -5
Ymax = 60
Yscl = 5

57. $y = -x^2 + 10x - 5$

Xmin = -1
Xmax = 11
Xscl = 1
Ymin = -5
Ymax = 25
Yscl = 2

58. $y = 4(x + 5)\sqrt{4 - x}$

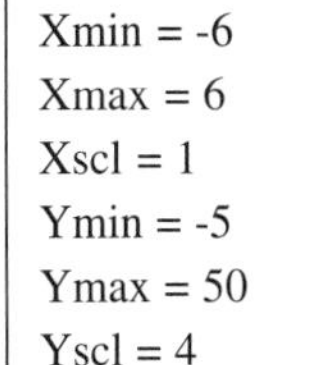
Xmin = -6
Xmax = 6
Xscl = 1
Ymin = -5
Ymax = 50
Yscl = 4

In Exercises 59–62, describe the viewing rectangle.

59. $y = 4x^2 - 25$

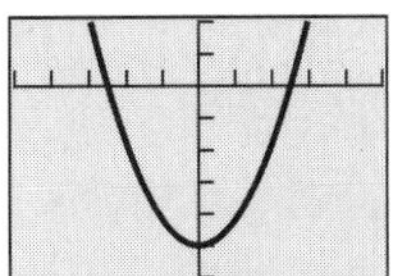

60. $y = x^3 - 3x^2 + 4$

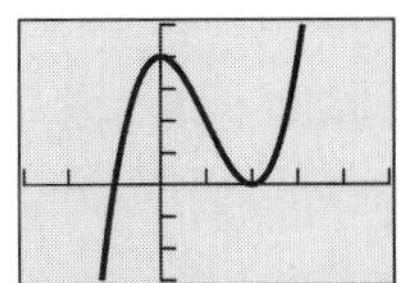

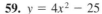

61. $y = |x| + |x - 10|$

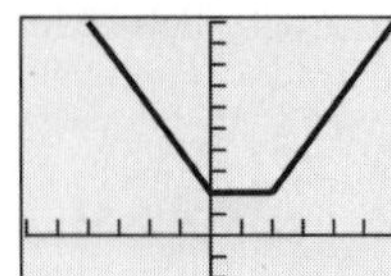

62. $y = 8\sqrt[3]{x - 6}$

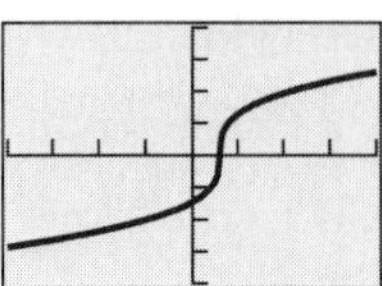

In Exercises 63 and 64, use a graphing utility to graph the equation using each of the suggested viewing rectangles. Assume that the equation gives the profit y when x units of a product are sold. Note that a graph can distort the information presented simply by changing the viewing rectangle. Which viewing rectangle would be selected by a person who wishes to argue that profits will increase dramatically with increased sales?

63. $y = 0.25x - 50$

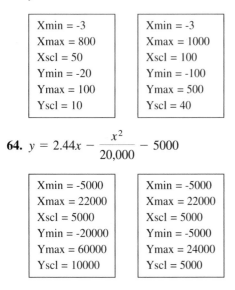

Xmin = -3	Xmin = -3
Xmax = 800	Xmax = 1000
Xscl = 50	Xscl = 100
Ymin = -20	Ymin = -100
Ymax = 100	Ymax = 500
Yscl = 10	Yscl = 40

64. $y = 2.44x - \dfrac{x^2}{20{,}000} - 5000$

Xmin = -5000	Xmin = -5000
Xmax = 22000	Xmax = 22000
Xscl = 5000	Xscl = 5000
Ymin = -20000	Ymin = -5000
Ymax = 60000	Ymax = 24000
Yscl = 10000	Yscl = 5000

In Exercises 65–68, solve for y and use a graphing utility to graph each of the resulting equations in the same viewing rectangle. Adjust the viewing rectangle so that a circle really does appear circular.

65. $x^2 + y^2 = 64$

66. $(x - 1)^2 + (y - 2)^2 = 16$

67. $x^2 + y^2 = 49$

68. $(x - 3)^2 + (y - 1)^2 = 25$

In Exercises 69–72, explain how to use a graphing utility to verify that $y_1 = y_2$. Identify the rule of algebra that is illustrated.

69. $y_1 = \frac{1}{4}(x^2 - 8)$
$y_2 = \frac{1}{4}x^2 - 2$

70. $y_1 = \frac{1}{2}x + (x + 1)$
$y_2 = \frac{3}{2}x + 1$

71. $y_1 = \frac{1}{5}[10(x^2 - 1)]$
$y_2 = 2(x^2 - 1)$

72. $y_1 = (x - 3) \cdot \dfrac{1}{x - 3}$
$y_2 = 1$

73. *Depreciation* A manufacturing plant purchases a new molding machine for $225,000. The depreciated value y after t years is given by

$$y = 225{,}000 - 20{,}000t, \qquad 0 \le t \le 8.$$

(a) Use the constraints of the model to determine an appropriate viewing rectangle.

(b) Use a graphing utility to graph the equation.

74. *Dimensions of a Rectangle* A rectangle of length x and width w has a perimeter of 12 meters.

(a) Draw a rectangle that gives a visual representation of the problem. Use the specified variables to label the sides of the rectangle.

(b) Show that $w = 6 - x$ is the width of the rectangle and that $A = x(6 - x)$ is its area.

(c) Use a graphing utility to graph the area equation.

(d) From the graph in part (c), estimate the dimensions of the rectangle that yield a maximum area.

75. *Think About It* Suppose you correctly enter an expression for the variable y on a graphing utility. However, no graph appears on the display when you graph the equation. Give a possible explanation and the steps you could take to remedy the problem. Illustrate your explanation with an example.

Data Analysis In Exercises 76 and 77, (a) sketch a graph comparing the data and the model for the data, (b) use the model to estimate the value of y for the year 1998, and (c) repeat part (b) for the year 2000.

76. *Federal Debt* The table gives the per capita federal debt for the United States for several years. (Source: U.S. Treasury Department)

Year	1950	1960	1970
Per Capita Debt	\$1688	\$1572	\$1807

Year	1980	1990	1994
Per Capita Debt	\$3981	\$12,848	\$15,750

A mathematical model for the per capita debt during this period is

$$y = 0.255t^3 - 4.096t^2 + 1570.417$$

where y represents the per capita debt and t is the time in years, with $t = 0$ corresponding to 1950.

77. *Life Expectancy* The table gives the life expectancy of a child (at birth) in the United States for selected years from 1920 to 1990. (Source: Department of Health and Human Services)

Year	1920	1930	1940	1950
Life Expectancy	54.1	59.7	62.9	68.2

Year	1960	1970	1980	1990
Life Expectancy	69.7	70.8	73.7	75.4

A mathematical model for the life expectancy during this period is

$$y = \frac{t + 66.93}{0.01t + 1}$$

where y represents the life expectancy and t is the time in years, with $t = 0$ corresponding to 1950.

78. *Think About It* Find a and b if the x-intercepts of the graph of $y = (x - a)(x - b)$ are $(-2, 0)$ and $(5, 0)$.

79. *Dividends Per Share* The dividends per common share of Procter and Gamble Company from 1990 through 1994 can be approximated by the mathematical model

$$y = 0.086t + 0.872, \qquad 0 \le t \le 4$$

where y is the dividend and t is the calendar year, with $t = 0$ corresponding to 1990. Sketch a graph of this equation. (Source: Procter and Gamble Company 1994 Annual Report)

80. *Copper Wire* The resistance y in ohms of 1000 feet of solid copper wire at 77°F can be approximated by the mathematical model

$$y = \frac{10{,}770}{x^2} - 0.37, \qquad 5 \le x \le 100$$

where x is the diameter of the wire in mils (0.001 in.). Use the model to estimate the resistance when $x = 50$. (Source: American Wire Gage)

In Exercises 81–84, use a graphing utility to graph the equation. Move the cursor along the curve to approximate the unknown coordinate of each given solution point accurate to two decimal places. (_Hint:_ You may need to use the zoom feature of the graphing utility to obtain the required accuracy.)

81. $y = \sqrt{5 - x}$
(a) $(2, y)$
(b) $(x, 3)$

82. $y = x^3(x - 3)$
(a) $(2.25, y)$
(b) $(x, 20)$

83. $y = x^5 - 5x$
(a) $(-0.5, y)$
(b) $(x, -4)$

84. $y = |x^2 - 6x + 5|$
(a) $(2, y)$
(b) $(x, 1.5)$

Review Solve Exercises 85–88 as a review of the skills and problem-solving techniques you learned in previous sections.

85. Identify the terms: $9x^5 + 4x^3 - 7$.

86. Write the expression using exponential notation.

$$-(7 \times 7 \times 7 \times 7)$$

87. True or False? $(3 + 4)^2 \stackrel{?}{=} 3^2 + 4^2$

88. Simplify: $\sqrt{18x} - \sqrt{2x}$.

1.2 Lines in the Plane

The Slope of a Line / The Point-Slope Form of the Equation of a Line / Sketching Graphs of Lines / Changing the Viewing Rectangle / Parallel and Perpendicular Lines

Figure 1.9

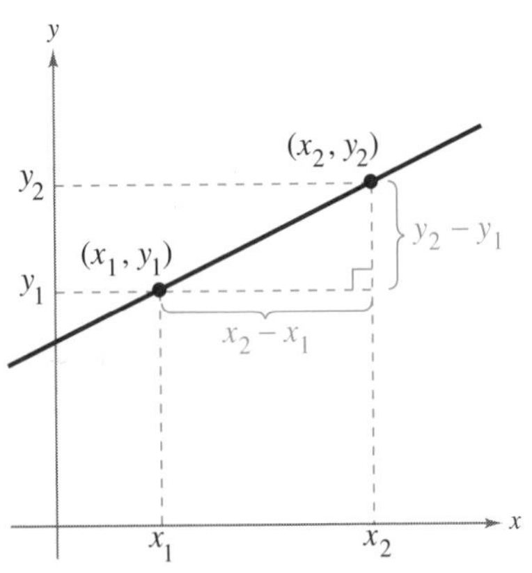

The Slope of a Line

In this section, you will study lines and their equations. The **slope** of a nonvertical line represents the number of units a line rises or falls vertically for each unit of horizontal change from left to right. For instance, consider the two points (x_1, y_1) and (x_2, y_2) on the line shown in Figure 1.9. As you move from left to right along this line, a change of $(y_2 - y_1)$ units in the vertical direction corresponds to a change of $(x_2 - x_1)$ units in the horizontal direction. That is,

$$y_2 - y_1 = \text{the change in } y$$

and

$$x_2 - x_1 = \text{the change in } x.$$

The slope of the line is given by the ratio of these two changes.

Definition of the Slope of a Line

The **slope** m of the nonvertical line through (x_1, y_1) and (x_2, y_2) is

$$m = \frac{y_2 - y_1}{x_2 - x_1} = \frac{\text{change in } y}{\text{change in } x}$$

where $x_1 \neq x_2$.

EXPLORATION

Use a graphing utility to compare the slopes of the lines $y = 0.5x$, $y = x$, $y = 2x$, and $y = 4x$. What do you observe about the slopes of the lines? Compare the slopes of the lines $y = -0.5x$, $y = -x$, $y = -2x$, and $y = -4x$. What do you observe about the slopes of these lines? (*Hint:* Use a square setting to guarantee a true geometric perspective.)

When this formula is used, the *order of subtraction* is important. Given two points on a line, you are free to label either one of them as (x_1, y_1) and the other as (x_2, y_2). However, once this has been done, you must form the numerator and denominator using the same order of subtraction.

$$m = \frac{y_2 - y_1}{x_2 - x_1} \qquad m = \frac{y_1 - y_2}{x_1 - x_2} \qquad m = \frac{y_2 - y_1}{x_1 - x_2}$$

Correct — Correct — Incorrect

Note Throughout this text, the term **line** always means a *straight* line.

EXAMPLE 1 Finding the Slope of a Line

Find the slope of the line passing through each pair of points.

a. $(-2, 0)$ and $(3, 1)$ **b.** $(-1, 2)$ and $(2, 2)$ **c.** $(0, 4)$ and $(1, -1)$

Solution

$$\textbf{a. } m = \overbrace{\frac{y_2 - y_1}{x_2 - x_1}}^{\text{Difference in } y\text{-values}}_{\text{Difference in } x\text{-values}} = \frac{1 - 0}{3 - (-2)} = \frac{1}{3 + 2} = \frac{1}{5}$$

$$\textbf{b. } m = \frac{2 - 2}{2 - (-1)} = \frac{0}{3} = 0$$

$$\textbf{c. } m = \frac{-1 - 4}{1 - 0} = \frac{-5}{1} = -5$$

The graphs of the three lines are shown in Figure 1.10.

Note In Figure 1.10, note that the square setting gives the correct "steepness" of the lines.

Figure 1.10

(a)

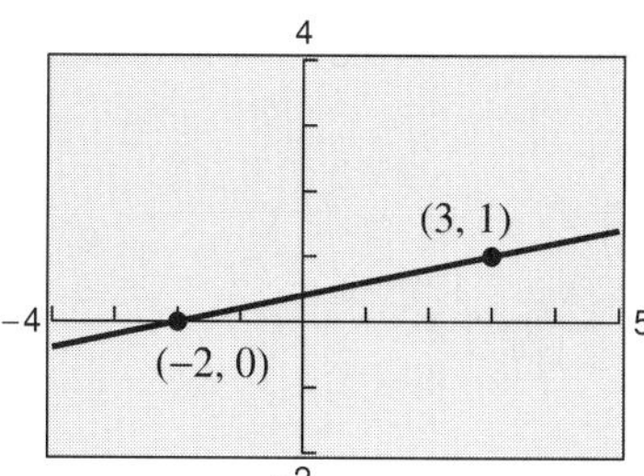

(b)

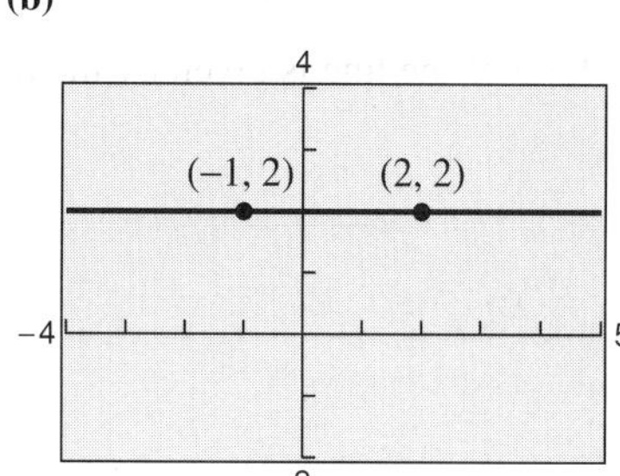

(c)

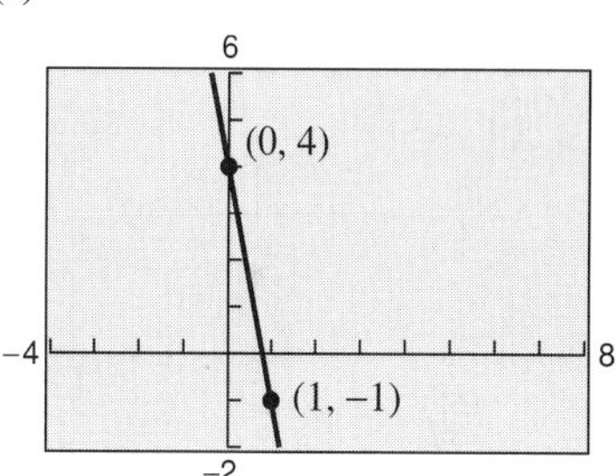

Figure 1.11

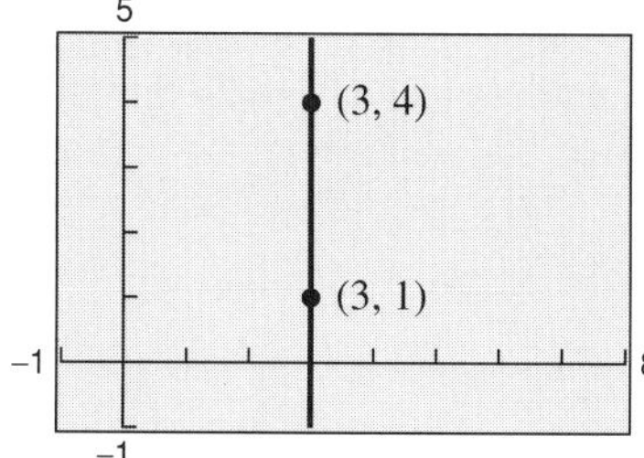

The definition of slope does not apply to vertical lines. For instance, consider the points $(3, 4)$ and $(3, 1)$ on the vertical line shown in Figure 1.11. Applying the formula for slope, you obtain $m = (4 - 1)/(3 - 3)$. Because division by zero is undefined, the slope of a vertical line is undefined.

From the slopes of the lines shown in Figures 1.10 and 1.11, you can make the following generalizations about the slope of a line.

1. A line with positive slope $(m > 0)$ *rises* from left to right.
2. A line with negative slope $(m < 0)$ *falls* from left to right.
3. A line with zero slope $(m = 0)$ is *horizontal.*
4. A line with undefined slope is *vertical.*

The Point-Slope Form of the Equation of a Line

Figure 1.12

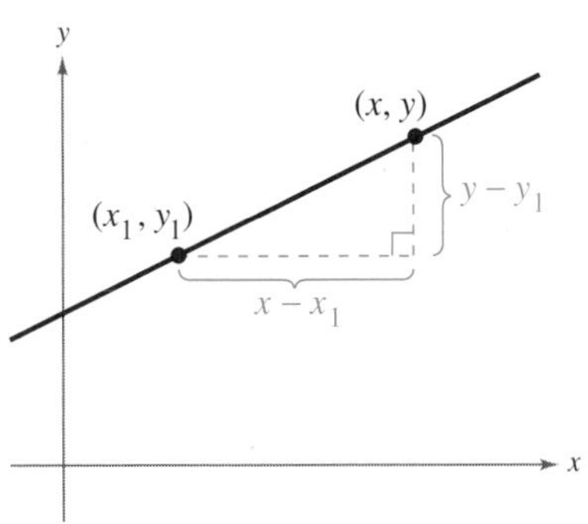

If you know the slope of a line *and* you also know the coordinates of one point on the line, you can find an equation for the line. For instance, in Figure 1.12, let (x_1, y_1) be a given point on the line whose slope is m. If (x, y) is any *other* point on a line, it follows that

$$\frac{y - y_1}{x - x_1} = m.$$

This equation in the variables x and y can be rewritten in the **point-slope form** of the equation of a line.

Point-Slope Form of the Equation of a Line

The **point-slope form** of the equation of the line that passes through the point (x_1, y_1) and has a slope of m is

$$y - y_1 = m(x - x_1).$$

EXAMPLE 2 The Point-Slope Form of the Equation of a Line

Find an equation of the line that passes through the point $(1, -2)$ and has a slope of 3.

Figure 1.13

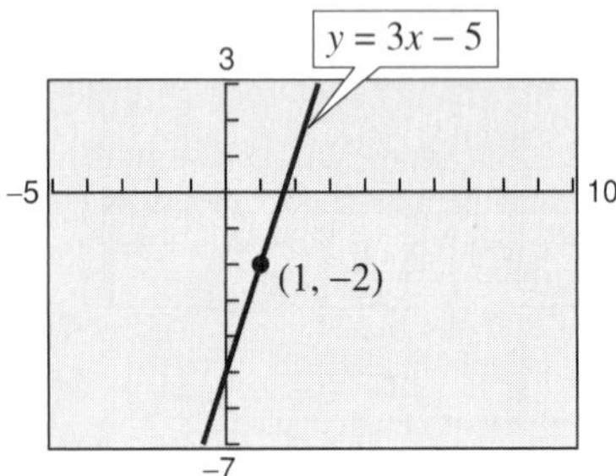

Solution

$$y - y_1 = m(x - x_1) \quad \text{Point-slope form}$$
$$y - (-2) = 3(x - 1) \quad \text{Substitute for } y_1, m, \text{ and } x_1.$$
$$y + 2 = 3x - 3$$
$$y = 3x - 5 \quad \text{Equation of line}$$

This line is shown in Figure 1.13.

The point-slope form can be used to find an equation of a nonvertical line passing through two points (x_1, y_1) and (x_2, y_2). First, use the formula for the slope of the line passing through two points.

$$m = \frac{y_2 - y_1}{x_2 - x_1}$$

Then, once you know the slope, use the point-slope form to obtain the equation

$$y - y_1 = \frac{y_2 - y_1}{x_2 - x_1}(x - x_1) = m(x - x_1).$$

This is sometimes called the **two-point form** of the equation of a line.

EXAMPLE 3 A Linear Model for Sales Prediction

During 1993, L. L. Bean's net sales were \$870 million, and in 1994 net sales were \$975 million. (Source: L. L. Bean)

a. Write a linear equation giving the net sales y in terms of the year x.

b. Use the equation to estimate the net sales during 1997.

Solution

a. Let $x = 3$ represent 1993. In Figure 1.14, let $(3, 870)$ and $(4, 975)$ be two points on the line representing the net sales. The slope of the line passing through these two points is

$$m = \frac{975 - 870}{4 - 3} = 105.$$

Figure 1.14

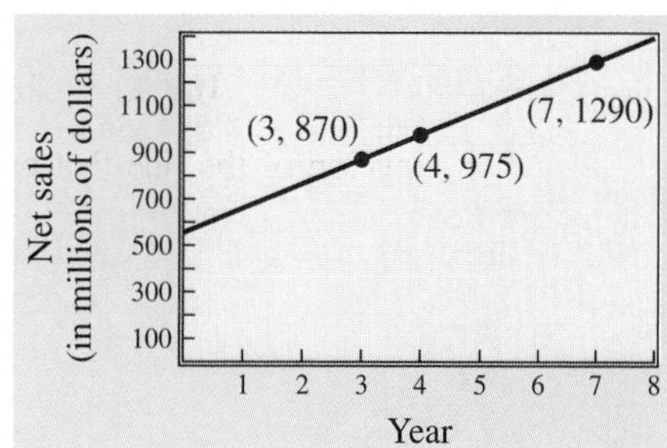

By the point-slope form, the equation of the line is as follows.

$$y - y_1 = m(x - x_1) \quad \text{Point-slope form}$$
$$y - 870 = 105(x - 3) \quad \text{Substitute for } y_1, m, \text{ and } x_1.$$
$$y = 105x - 315 + 870$$
$$y = 105x + 555 \quad \text{Equation of line}$$

b. Using the equation from part (a), estimate the 1997 net sales ($x = 7$) to be $y = 105(7) + 555 = 735 + 555 = \$1290$ million or \$1.29 billion.

Figure 1.15

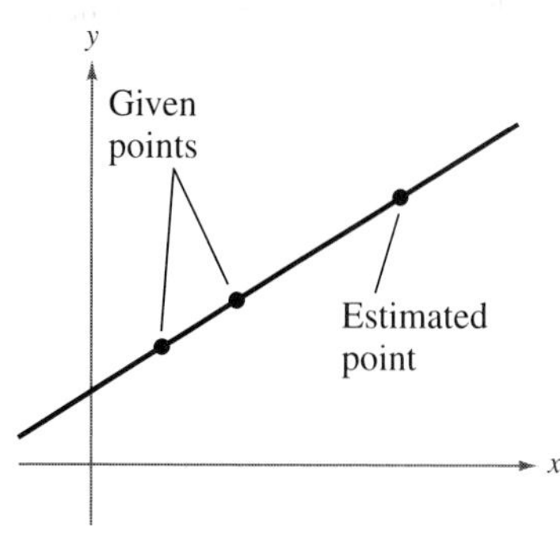

Linear Extrapolation

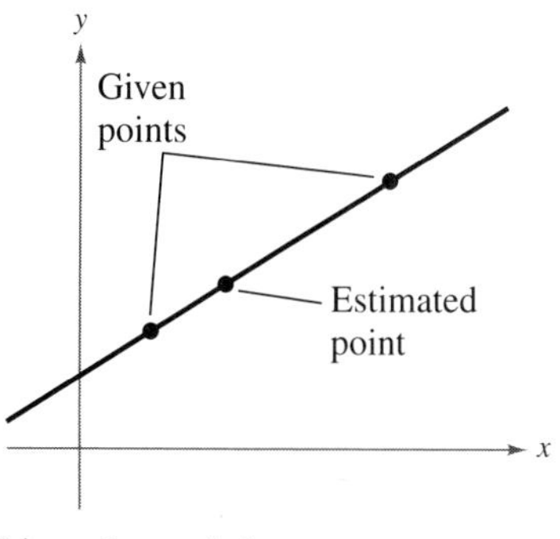

Linear Interpolation

The approximation method illustrated in Example 3 is **linear extrapolation.** Note in Figure 1.15 that for linear extrapolation, the estimated point lies outside of the given points. When the estimated point lies *between* two given points, the procedure is called **linear interpolation.**

Sketching Graphs of Lines

Many problems in coordinate geometry can be classified as follows.

1. Given a graph (or parts of it), find its equation.
2. Given an example, find its graph.

For lines, the first problem is solved easily by using the point-slope form. This formula, however, is not particularly useful for solving the second type of problem. The form that is better suited to graphing linear equations is the **slope-intercept form** $y = mx + b$ of the equation of a line.

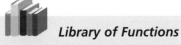

Library of Functions

In the next section, you will be introduced to the precise meaning of the term *function*. The simplest type of function is a linear function and has the form

$$y = ax + b.$$

As its name implies, the graph of a linear function is a line that has a slope of a and a y-intercept at $(0, b)$.

EXAMPLE 4 Determining the Slope and y-Intercepts

a. Graph the lines $y = 2x + 1$, $y = \frac{1}{2}x + 1$, and $y = -2x + 1$ in the same viewing rectangle. What can you observe?

b. Graph the lines $y = 2x + 1$, $y = 2x$, and $y = 2x - 1$ in the same viewing rectangle. What can you observe?

Solution

a. In Figure 1.16(a), you can see that all three lines have the same y-intercept $(0, 1)$, but slopes of 2, $\frac{1}{2}$, and -2, respectively.

b. In Figure 1.16(b), you can see that all three lines have a slope of 2, but y-intercepts of $(0, 1)$, $(0, 0)$, and $(0, -1)$, respectively.

Figure 1.16

(a)

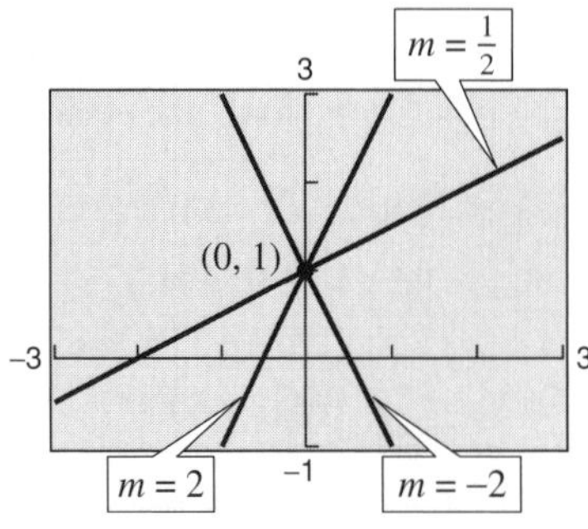

(b)

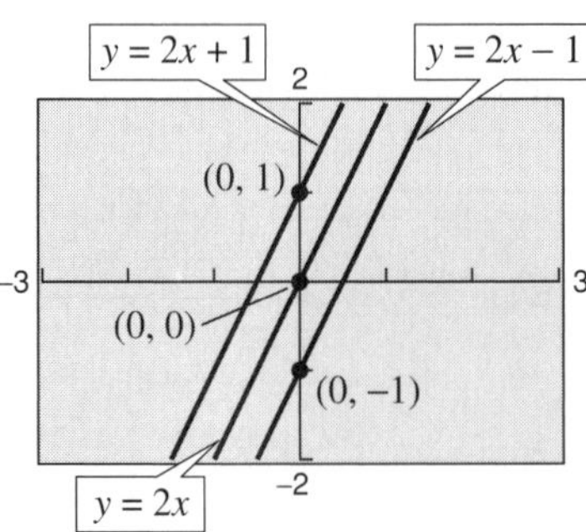

Slope-Intercept Form of the Equation of a Line

The graph of the equation

$$y = mx + b$$

is a line whose slope is m and whose y-intercept is $(0, b)$.

Figure 1.17

(a)

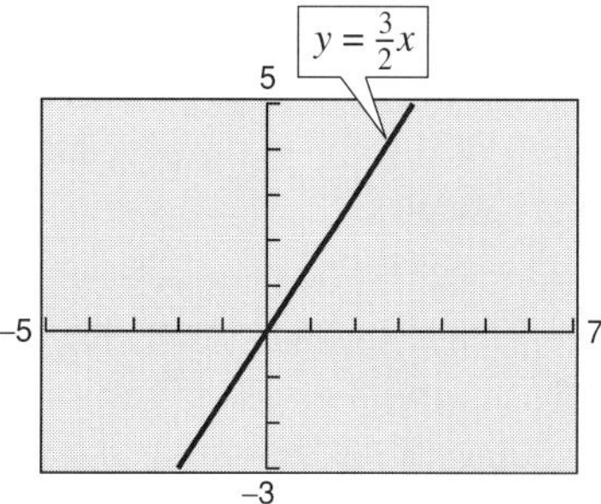

(b)

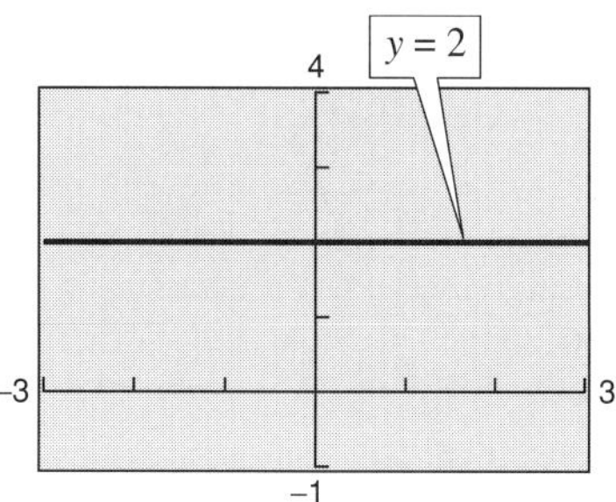

(c)

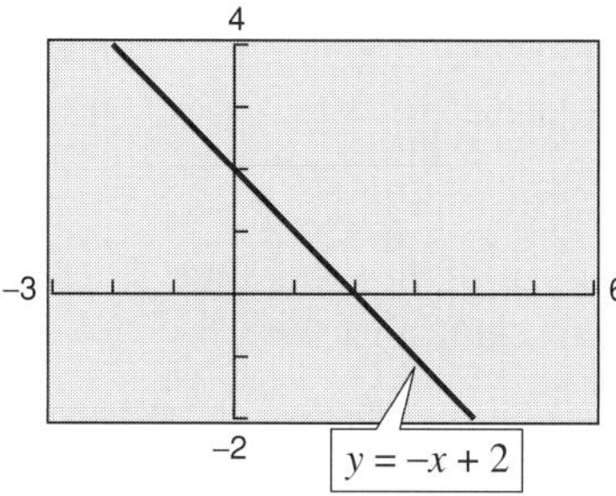

EXAMPLE 5 **Using the Slope-Intercept Form**

Describe the graph of each linear equation.

a. $y = \dfrac{3}{2}x$ **b.** $y = 2$ **c.** $x + y = 2$

Solution

a. Because $b = 0$, the y-intercept is $(0, 0)$. Moreover, because the slope is $m = \frac{3}{2}$, this line *rises* three units for every two units it moves to the right, as shown in Figure 1.17(a).

b. By writing the equation $y = 2$ in the form $y = (0)x + 2$, you can see that the y-intercept is $(0, 2)$ and the slope is zero. A zero slope implies that the line is horizontal, as shown in Figure 1.17(b).

c. By writing the equation $x + y = 2$ in slope-intercept form, $y = -x + 2$, you can see that the y-intercept is $(0, 2)$. Moreover, because the slope is $m = -1$, this line *falls* one unit for every unit it moves to the right, as shown in Figure 1.17(c).

From the slope-intercept form of the equation of a line, you can see that a horizontal line ($m = 0$) has an equation of the form $y = b$. This is consistent with the fact that each point on a horizontal line through $(0, b)$ has a y-coordinate of b.

Similarly, each point on a vertical line through $(a, 0)$ has an x-coordinate of a. Hence, a vertical line has an equation of the form $x = a$. This equation cannot be written in the slope-intercept form, because the slope of a vertical line is undefined. However, *every* line has an equation that can be written in the **general form**

$$Ax + By + C = 0$$ General form of the equation of a line

where A and B are not *both* zero.

Summary of Equations of Lines

1. General form: $Ax + By + C = 0$
2. Vertical line: $x = a$
3. Horizontal line: $y = b$
4. Slope-intercept form: $y = mx + b$
5. Point-slope form: $y - y_1 = m(x - x_1)$

Changing the Viewing Rectangle

When a graphing utility is used to sketch a straight line, it is important to realize that the graph of the line may not visually appear to have the slope indicated by its equation. This occurs because of the viewing rectangle used for the graph. For instance, Figure 1.18 shows graphs of $y = 2x + 1$ produced on a graphing utility using three different viewing rectangles.

Notice that the slopes in Figure 1.18(a) and (b) do not visually appear to be equal to 2. However, if you use the *square* viewing rectangle, as in Figure 1.18(c), the slope visually appears to be 2. In general, two graphs of the same equation can appear to be quite different depending on the viewing rectangle selected.

Figure 1.18

(a)

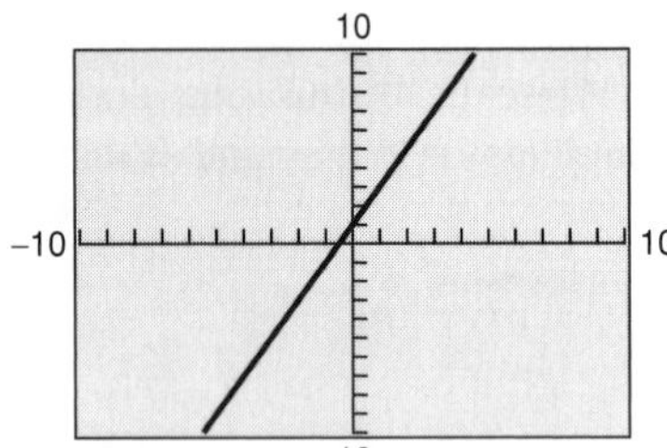

(b)

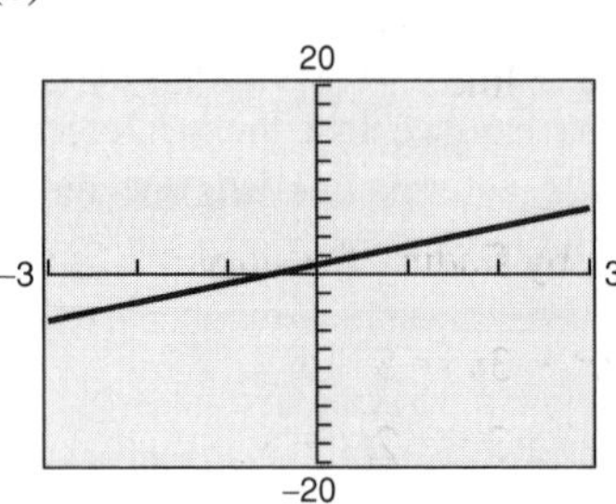

(c)

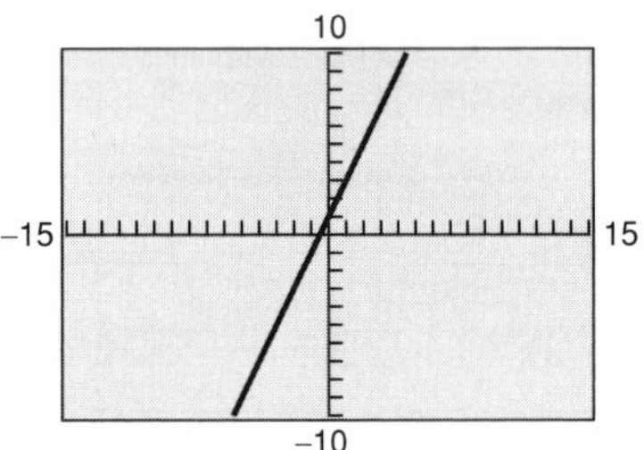

EXAMPLE 6  **Different Viewing Rectangles**

The graphs of the two lines

$$y = -x - 1 \qquad \text{and} \qquad y = -10x - 1$$

are shown in Figure 1.19. Even though the slopes of these lines are different (-1 and -10, respectively), the graphs seem similar because the viewing rectangles are different.

Figure 1.19

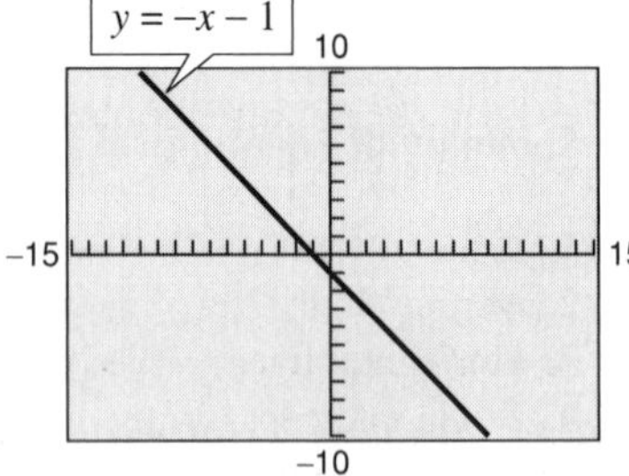

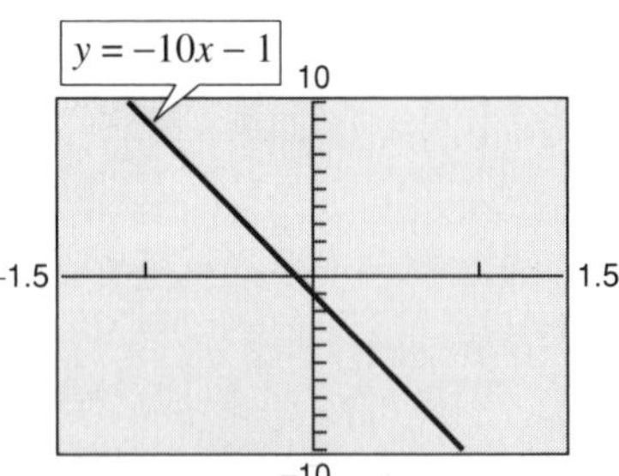

> **Study Tip**
>
> Be careful when you graph equations such as $y = \frac{2}{3}x - \frac{7}{3}$ on your graphing utility. A common mistake is to type it in as
>
> Y1 = 2/3X − 7/3,
>
> which is not the original equation. You should use one of the following formulas.
>
> Y1 = 2X/3 − 7/3 or
> Y1 = (2/3)X − 7/3.
>
> Do you see why?

Parallel and Perpendicular Lines

The slope of a line is a convenient tool for determining whether two lines are parallel or perpendicular. Example 4(b) suggests the following property of parallel lines.

> **Parallel Lines**
>
> Two distinct nonvertical lines are **parallel** if and only if their slopes are equal.

EXAMPLE 7 Equations of Parallel Lines

Find an equation of the line that passes through the point $(2, -1)$ and is parallel to the line $2x - 3y = 5$, as shown in Figure 1.20.

Figure 1.20

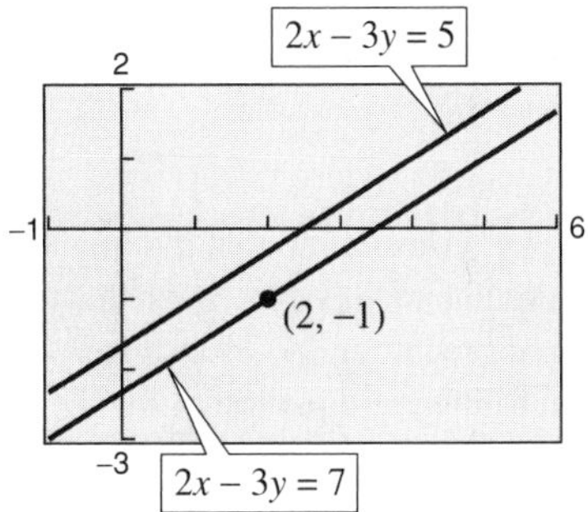

Solution

Begin by finding the slope of the given line.

$$2x - 3y = 5 \qquad \text{Original equation}$$

$$3y = 2x - 5$$

$$y = \frac{2}{3}x - \frac{5}{3} \qquad \text{Slope-intercept form}$$

Therefore, the given line has a slope of $m = \frac{2}{3}$. Because any line parallel to the given line must also have a slope of $\frac{2}{3}$, the required line through $(2, -1)$ has the following equation.

$$y - (-1) = \frac{2}{3}(x - 2) \qquad \text{Point-slope form}$$

$$y = \frac{2}{3}x - \frac{4}{3} - 1$$

$$y = \frac{2}{3}x - \frac{7}{3} \qquad \text{Slope-intercept form}$$

Notice the similarity between the slope-intercept form of the original equation and the slope-intercept form of the parallel equation.

Two nonvertical lines are *perpendicular* if and only if their slopes are negative reciprocals of each other. For instance, the lines $y = 2x$ and $y = -\frac{1}{2}x$ are perpendicular because one has a slope of 2 and the other has a slope of $-\frac{1}{2}$.

Perpendicular Lines

Two nonvertical lines are **perpendicular** if and only if their slopes are negative reciprocals of each other. That is,

$$m_1 = -\frac{1}{m_2}.$$

Figure 1.21

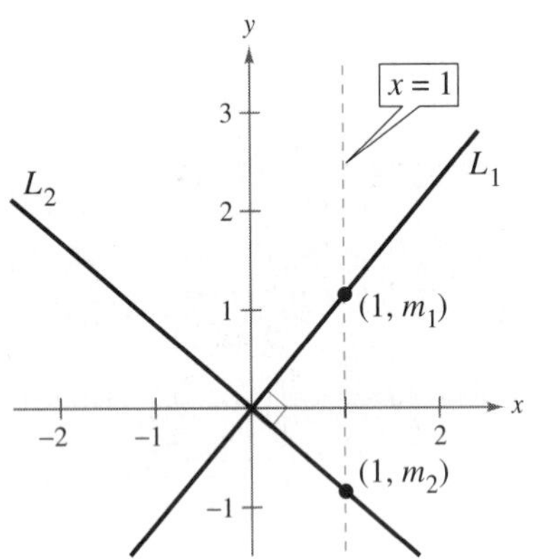

Proof /// The phrase "if and only if" is a way of stating two rules in one. One rule says "If two nonvertical lines are perpendicular, their slopes are negative reciprocals." The other rule, which is the converse, says "If two lines have slopes that are negative reciprocals, the lines must be perpendicular." The proof of the first rule is outlined as follows.

Assume you are given two nonvertical perpendicular lines L_1 and L_2, with slopes of m_1 and m_2. For simplicity's sake, let these two lines intersect at the origin, as shown in Figure 1.21. The vertical line $x = 1$ will intersect L_1 and L_2 at the points $(1, m_1)$ and $(1, m_2)$. Because L_1 and L_2 are perpendicular, the triangle formed by these two points and the origin is a right triangle. Using the Pythagorean Theorem, it follows that

$$\left(\sqrt{1 + m_1^2}\right)^2 + \left(\sqrt{1 + m_2^2}\right)^2 = \left(\sqrt{0^2 + (m_1 - m_2)^2}\right)^2.$$

By simplifying this equation, you can conclude that $m_1 = -1/m_2$. ///

EXAMPLE 8 Equations of Perpendicular Lines

Find an equation of the line that passes through the point $(2, -1)$ and is perpendicular to the line $2x - 3y = 5$.

Figure 1.22

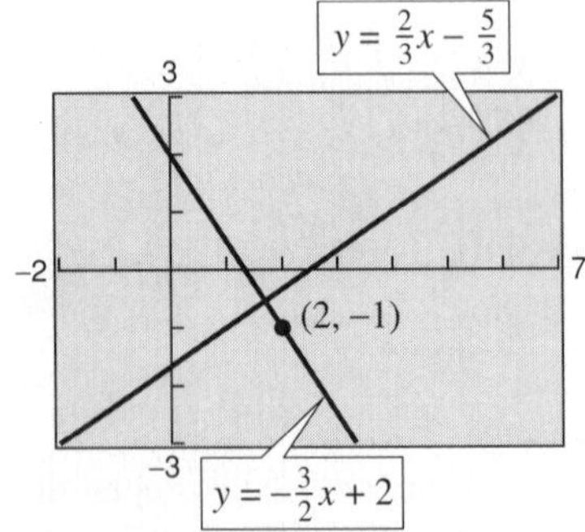

Solution

By writing the given line in the form $y = \frac{2}{3}x - \frac{5}{3}$, you can see that the line has a slope of $\frac{2}{3}$. Hence, any line that is perpendicular to this line must have a slope of $-\frac{3}{2}$ (because $-\frac{3}{2}$ is the negative reciprocal of $\frac{2}{3}$). Therefore, the required line through the point $(2, -1)$ has the following equation.

$$y - (-1) = -\frac{3}{2}(x - 2) \qquad \text{Point-slope form}$$
$$y = -\frac{3}{2}x + 3 - 1$$
$$y = -\frac{3}{2}x + 2 \qquad \text{Slope-intercept form}$$

The graphs of both equations are shown in Figure 1.22.

EXAMPLE 9 Graphs of Perpendicular Lines

Use a graphing utility to graph the lines given by $y = x + 1$ and $y = -x + 3$. Display *both* graphs in the same viewing rectangle. The lines are supposed to be perpendicular (they have slopes of $m_1 = 1$ and $m_2 = -1$). Do they appear to be perpendicular on the display?

Solution

If the viewing rectangle is nonsquare, as in Figure 1.23(a), the two lines will not appear perpendicular. If, however, the viewing rectangle is square, as in Figure 1.23(b), the lines will appear perpendicular.

Figure 1.23

(a)

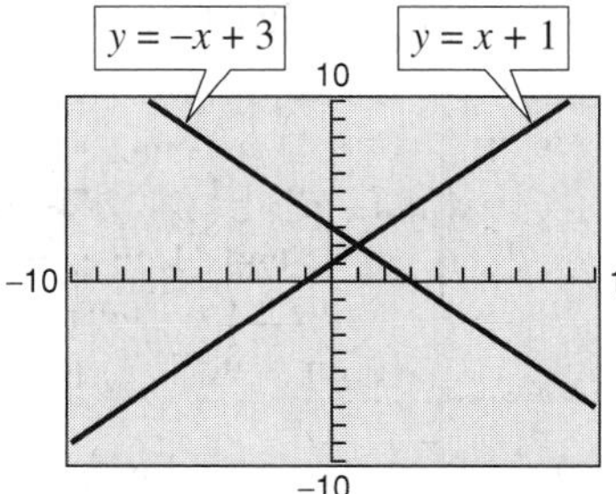

(b)

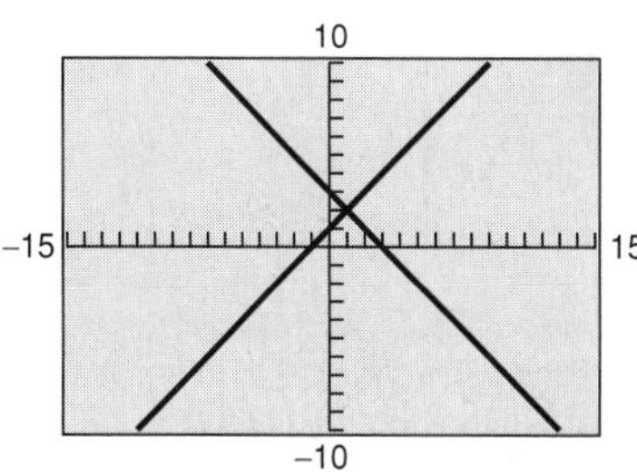

Group Activity

An Application of Slope

In 1985, a college had an enrollment of 5500 students. By 1995, the enrollment had increased to 7000 students.

a. What was the average annual change in enrollment from 1985 to 1995?

b. Use the average annual change in enrollment to estimate the enrollments in 1989, 1993, and 1997.

c. Write the equation of the line that represents the data given in part (b). What is the slope of this line? Interpret the slope in the context of the problem.

d. Discuss the concepts of *slope* and *average rate of change* in a group or in a short paragraph.

1.2 /// EXERCISES

In Exercises 1 and 2, identify the line that has the specified slope.

1. (a) $m = \frac{2}{3}$ (b) m is undefined. (c) $m = -2$

2. (a) $m = 0$ (b) $m = -\frac{3}{4}$ (c) $m = 1$

Figure for 1

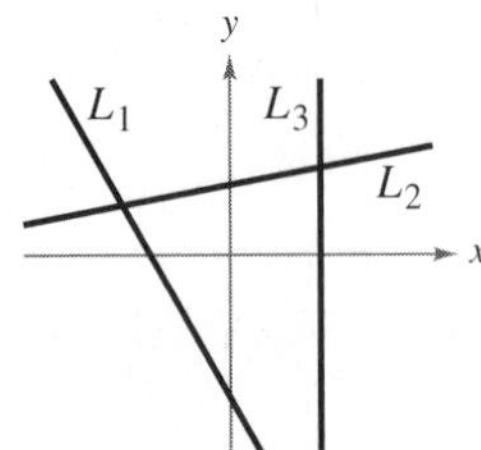

Figure for 2

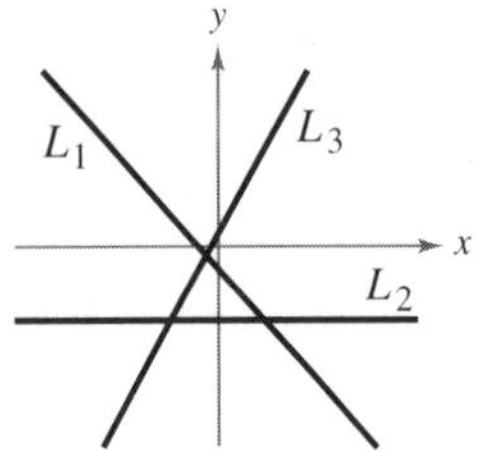

In Exercises 3–8, estimate the slope of the line.

3.

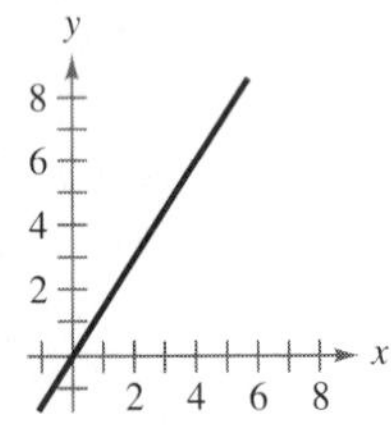

4.

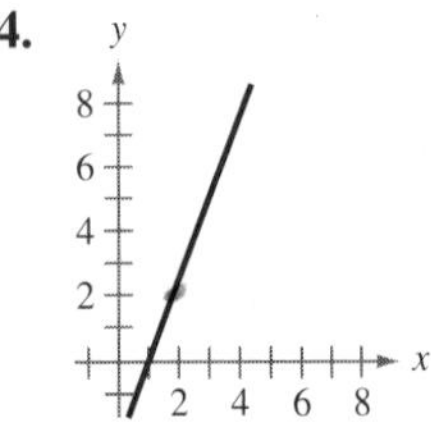

5.

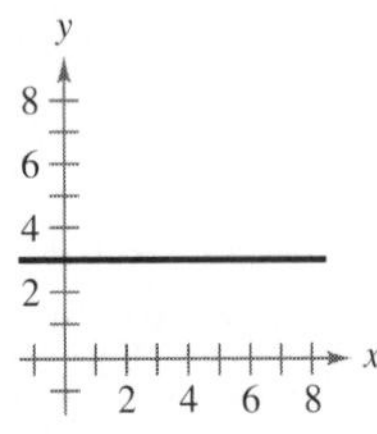

6.

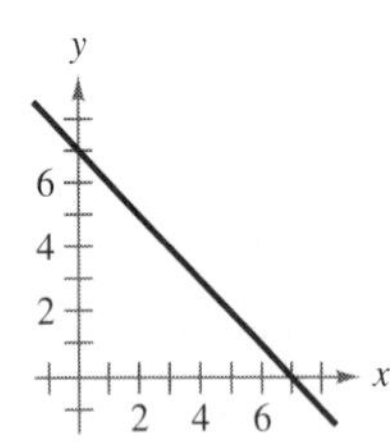

7.

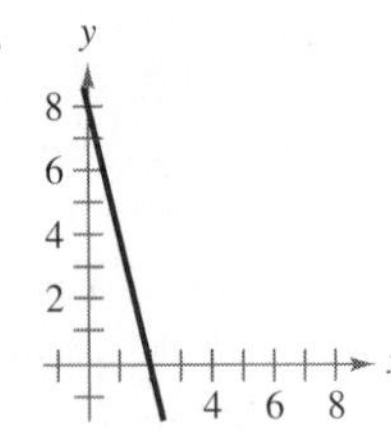

8.

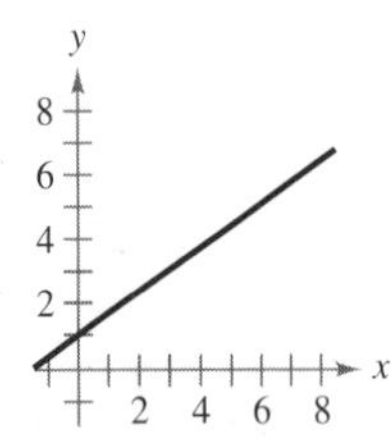

In Exercises 9 and 10, sketch the lines with the given slopes through the given point.

	Point	*Slopes*			
9.	(2, 3)	(a) 0	(b) 1	(c) 2	(d) −3
10.	(−4, 1)	(a) 3	(b) −3	(c) $\frac{1}{2}$	(d) undefined

In Exercises 11–16, plot the points and find the slope of the line passing through the points. Use a graphing utility to graph the line segment connecting the two points. (Use the square setting.)

11. $(-3, -2), (1, 6)$ **12.** $(2, 4), (4, -4)$

13. $(-6, -1), (-6, 4)$ **14.** $(0, -10), (-4, 0)$

15. $(1, 2), (-2, -2)$ **16.** $\left(\frac{7}{8}, \frac{3}{4}\right), \left(\frac{5}{4}, -\frac{1}{4}\right)$

In Exercises 17–22, you are given the slope of the line and a point on the line. Find three additional points through which the line passes. (The solution is not unique.)

	Point	*Slope*
17.	(2, 1)	$m = 0$
18.	(−4, 1)	m is undefined.
19.	(5, −6)	$m = 1$
20.	(10, −6)	$m = -1$
21.	(−8, 1)	m is undefined.
22.	(−3, −1)	$m = 0$

In Exercises 23–26, determine if the lines L_1 and L_2 passing through the pairs of points are parallel, perpendicular, or neither. Use a graphing utility to graph the line segments connecting the pairs of points on the respective lines. (Use a square setting.)

23. L_1: $(0, -1), (5, 9)$
L_2: $(0, 3), (4, 1)$

24. L_1: $(-2, -1), (1, 5)$
L_2: $(1, 3), (5, -5)$

25. L_1: $(3, 6), (-6, 0)$
L_2: $(0, -1), \left(5, \frac{7}{3}\right)$

26. L_1: $(4, 8), (-4, 2)$
L_2: $(3, -5), \left(-1, \frac{1}{3}\right)$

27. *Essay* Write a brief paragraph explaining whether or not any pair of points on a line can be used to calculate the slope of the line.

28. *Think About It* Is it possible for two lines with positive slopes to be perpendicular? Explain.

29. *Rate of Change* The following are the slopes of lines representing annual sales y in terms of time x in years. Use the slopes to interpret any change in annual sales for a 1-year increase in time.

(a) $m = 135$ (b) $m = 0$ (c) $m = -40$

30. *Rate of Change* The following are the slopes of lines representing daily revenues y in terms of time x in days. Use the slopes to interpret any change in daily revenues for a 1-day increase in time.

(a) $m = 400$ (b) $m = 100$ (c) $m = 0$

31. *Data Analysis* The data gives the earnings per share of common stock for General Mills for the years 1987 through 1994. Time in years is represented by t, with $t = 0$ corresponding to 1990, and the earnings per share are represented by y. (Source: General Mills)

$(-3, 1.25), (-2, 1.63), (-1, 2.53), (0, 2.32)$
$(1, 2.87), (2, 2.99), (3, 3.10), (4, 2.95)$

(a) Use a graphing utility to create a line graph of the data.

(b) Use the slope to determine the years when earnings decreased most rapidly and increased most rapidly.

32. *Data Analysis* The data gives the declared dividend per share of common stock for the Procter and Gamble Company for the years 1987 through 1994. Time in years is represented by t, with $t = 0$ corresponding to 1990, and the dividends per share are represented by y. (Source: Procter and Gamble)

$(-3, 0.675), (-2, 0.688), (-1, 0.750), (0, 0.875)$
$(1, 0.975), (2, 1.025), (3, 1.100), (4, 1.240)$

(a) Use a graphing utility to create a line graph representing the data.

(b) Use the slope to determine the year in which earnings increased most rapidly.

In Exercises 33 and 34, use a graphing utility to graph the equation using each of the suggested viewing rectangles. Describe the difference between the two views.

33. $y = 0.5x - 3$

Viewing rectangle 1	Viewing rectangle 2
Xmin = -5	Xmin = -2
Xmax = 10	Xmax = 10
Xscl = 1	Xscl = 1
Ymin = -1	Ymin = -4
Ymax = 10	Ymax = 1
Yscl = 1	Yscl = 1

34. $y = -8x + 5$

Viewing rectangle 1	Viewing rectangle 2
Xmin = -5	Xmin = -5
Xmax = 5	Xmax = 10
Xscl = 1	Xscl = 1
Ymin = -10	Ymin = -80
Ymax = 10	Ymax = 80
Yscl = 1	Yscl = 20

In Exercises 35–40, find the slope and y-intercept (if possible) of the equation of the line. Sketch a graph of the line by hand. Use a graphing utility to verify your sketch.

35. $5x - y + 3 = 0$ **36.** $2x + 3y - 9 = 0$

37. $5x - 2 = 0$ **38.** $3y + 5 = 0$

39. $7x + 6y - 30 = 0$ **40.** $x - y - 10 = 0$

In Exercises 41–48, find an equation for the line passing through the points. Use a graphing utility to sketch a graph of the line.

41. $(5, -1), (-5, 5)$ **42.** $(4, 3), (-4, -4)$

43. $\left(2, \frac{1}{2}\right), \left(\frac{1}{2}, \frac{5}{4}\right)$ **44.** $(-1, 4), (6, 4)$

45. $(-8, 1), (-8, 7)$ **46.** $(1, 1), \left(6, -\frac{2}{3}\right)$

47. $(1, 0.6), (-2, -0.6)$ **48.** $(-8, 0.6), (2, -2.4)$

49. *Mountain Driving* When driving down a mountain road, you notice warning signs indicating that it is a "12% grade." This means that the slope of the road is $-\frac{12}{100}$. Approximate the amount of horizontal change in your position if you note from elevation markers that you have descended 2000 feet vertically.

50. *Attic Height* The "rise to run" ratio that determines the steepness of the roof on a house is 3 to 4. Determine the maximum height in the attic if the house is 32 feet wide (see figure).

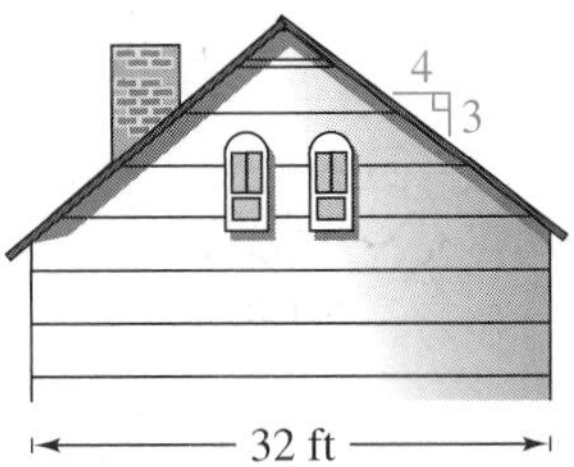

In Exercises 51–60, find an equation of the line that passes through the given point and has the indicated slope. Sketch a graph of the line by hand. Use a graphing utility to verify your sketch.

	Point	*Slope*
51.	$(0, -2)$	$m = 3$
52.	$(0, 10)$	$m = -1$
53.	$(-3, 6)$	$m = -2$
54.	$(0, 0)$	$m = 4$
55.	$(4, 0)$	$m = -\frac{1}{3}$
56.	$(-2, -5)$	$m = \frac{3}{4}$
57.	$(6, -1)$	m is undefined.
58.	$(-10, 4)$	$m = 0$
59.	$\left(4, \frac{5}{2}\right)$	$m = \frac{4}{3}$
60.	$\left(-\frac{1}{2}, \frac{3}{2}\right)$	$m = -3$

Conjecture **In Exercises 61 and 62, use the values of a and b and a graphing utility to graph the equation of the line given by**

$$\frac{x}{a} + \frac{y}{b} = 1, \qquad a \neq 0, b \neq 0.$$

Use the graphs to make a conjecture about what a and b represent. Verify your conjecture.

61. $a = 5, \quad b = -3$

62. $a = -6, \quad b = 2$

In Exercises 63 and 64, use the results of Exercises 61 and 62 to write an equation of the line that passes through the points.

63. x-intercept: $(2, 0)$
y-intercept: $(0, 3)$

64. x-intercept: $\left(-\frac{1}{6}, 0\right)$
y-intercept: $\left(0, -\frac{2}{3}\right)$

In Exercises 65–68, write equations of the lines through the given point (a) parallel to the given line and (b) perpendicular to the given line.

	Point	*Line*
65.	$(2, 1)$	$4x - 2y = 3$
66.	$\left(\frac{7}{8}, \frac{3}{4}\right)$	$5x + 3y = 0$
67.	$(-1, 0)$	$y = -3$
68.	$(2, 5)$	$x = 4$

Graphical Analysis **In Exercises 69–72, use a graphing utility to graph the three equations in the same viewing rectangle. Adjust the viewing rectangle so that the slope appears visually correct. Identify any lines that are parallel or perpendicular.**

69. L_1: $y = 2x$
L_2: $y = -2x$
L_3: $y = \frac{1}{2}x$

70. L_1: $y = \frac{2}{3}x$
L_2: $y = -\frac{3}{2}x$
L_3: $y = \frac{2}{3}x + 2$

71. L_1: $y = -\frac{1}{2}x$
L_2: $y = -\frac{1}{2}x + 3$
L_3: $y = 2x - 4$

72. L_1: $y = x - 8$
L_2: $y = x + 1$
L_3: $y = -x + 3$

Rate of Change **In Exercises 73–76, you are given the dollar value of an item in 1996 *and* the rate at which the value of the item is expected to change during the next 5 years. Use this information to write a linear equation that gives the dollar value V of the item in terms of the year t. (Let $t = 6$ represent 1996.)**

	1996 Value	*Rate*
73.	\$2540	\$125 increase per year
74.	\$156	\$4.50 increase per year
75.	\$20,400	\$2000 decrease per year
76.	\$245,000	\$5600 decrease per year

Graphical Interpretation In Exercises 77–80, match the description with its graph. Also determine the slope and how it is interpreted in the situation. [The graphs are labeled (a), (b), (c), and (d).]

(a)
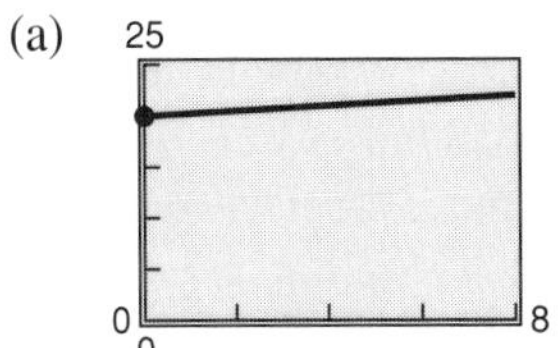

(b)
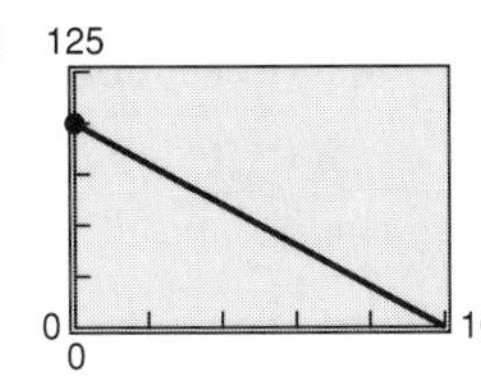

(c)
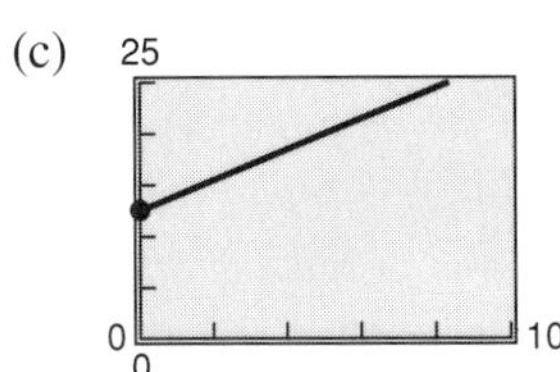

(d)
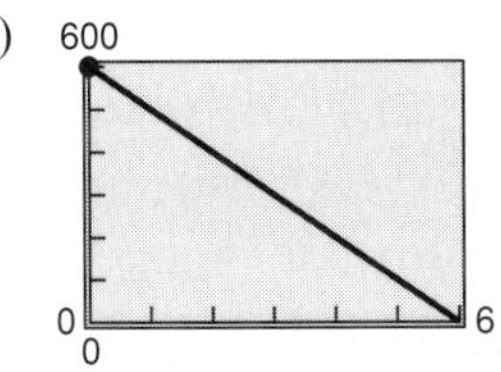

77. A person is paying \$10 per week to a friend to repay a \$100 loan.

78. An employee is paid \$12.50 per hour plus \$1.50 for each unit produced per hour.

79. A sales representative receives \$20 per day for food plus \$0.25 for each mile traveled.

80. A word processor that was purchased for \$600 depreciates \$100 per year.

In Exercises 81 and 82, find a relationship between x and y such that (x, y) is equidistant from the two points.

81. $(4, -1), (-2, 3)$ **82.** $\left(3, \frac{5}{2}\right), (-7, 1)$

83. *Temperature* Find the equation of the line giving the relationship between the temperature in degrees Celsius C and degrees Fahrenheit F. Remember that water freezes at 0°C (32°F) and boils at 100°C (212°F).

84. *Temperature* Use the result of Exercise 83 to complete the table.

C		$-10°$	$10°$			$177°$
F	$0°$			$68°$	$90°$	

85. *Annual Salary* Suppose that your salary was \$28,500 in 1994 and \$32,900 in 1996. If your salary follows a linear growth pattern, what will your salary be in 2000?

86. *College Enrollment* A small college had 2546 students in 1994 and 2702 students in 1996. If the enrollment follows a linear growth pattern, how many students will the college have in 2000?

87. *Straight-Line Depreciation* A small business purchases a piece of equipment for \$875. After 5 years, the equipment will be outdated and have no value.

(a) Write a linear equation giving the value V of the equipment during the 5 years it will be used.

(b) Use a graphing utility to graph the linear equation representing the depreciation of the equipment, and use the trace key to complete the table.

t	0	1	2	3	4	5
V						

88. *Straight-Line Depreciation* A small business purchases a piece of equipment for \$25,000. After 10 years, the equipment will have to be replaced. Its value at that time is expected to be \$2000.

(a) Write a linear equation giving the value V of the equipment during the 10 years it will be used.

(b) Use a graphing utility to graph the linear equation representing the depreciation of the equipment and find when $V = \$13,000$.

89. *Sales Price and List Price* A store is offering a 15% discount on all items. Write a linear equation giving the sale price S for an item with a list price L.

90. *Hourly Wages* A manufacturer pays its assembly line workers \$11.50 per hour. In addition, workers receive a piecework rate of \$0.75 per unit produced. Write a linear equation for the hourly wages W in terms of the number of units x produced per hour.

91. *Contracting Purchase* A contractor purchases a piece of equipment for \$36,500. The equipment requires an average expenditure of \$5.25 per hour for fuel and maintenance, and the operator is paid \$11.50 per hour.

(a) Write a linear equation giving the total cost C of operating this equipment for t hours. (Include the purchase cost of the equipment.)

(b) Assuming that customers are charged \$27 per hour of machine use, write an equation for the revenue R derived from t hours of use.

(c) Use the formula ($P = R - C$) to write an equation for the profit derived from t hours of use.

(d) *Break-Even Point* Use the result of part (c) to find the number of hours this equipment must be used to yield a profit of 0 dollars.

92. *Perimeter* The length and width of a rectangular garden are 15 meters and 10 meters, respectively. A walkway of width x surrounds the garden.

(a) Write the outside perimeter y of the walkway in terms of x.

(b) Use a graphing utility to graph the equation for the perimeter.

(c) Determine the slope of the graph in part (b). For each additional 1-meter increase in the width of the walkway, determine the increase in its outside perimeter.

93. *Data Analysis* The average annual salaries y of major league baseball players (in thousands of dollars) from 1983 to 1992 are given as ordered pairs (t, y), where t is the time in years, with $t = 0$ corresponding to 1980. The ordered pairs are (3, 289), (4, 329), (5, 371), (6, 413), (7, 412), (8, 439), (9, 497), (10, 598), (11, 851), and (12, 1029). (Source: Major League Baseball Players Association)

(a) Use the regression capabilities of a graphing utility to find the least squares regression line.

(b) Use a graphing utility to plot the points and graph the regression line in the same viewing rectangle.

(c) Use the regression line to predict the average salary in the year 2000.

(d) Interpret the meaning of the slope of the regression line.

94. *Data Analysis* An instructor gives regular 20-point quizzes and 100-point exams in a mathematics course. Average scores for six students, given as ordered pairs (x, y) where x is the average quiz score and y is the average exam score, are (18, 87), (10, 55), (19, 96), (16, 79), (13, 76), and (15, 82).

(a) Use the regression capabilities of a graphing utility to find the least squares regression line.

(b) Use a graphing utility to plot the points and graph the regression line in the same viewing rectangle.

(c) Use the regression line to predict the average test score for a student whose average quiz score is 17.

(d) Interpret the meaning of the slope of the regression line.

(e) If the instructor added 4 points to the average test score of everyone in the class, describe the changes in the positions of the plotted points and the change in the equation of the line.

95. *Real Estate Purchase* A real estate office handles an apartment complex with 50 units. When the rent per unit is \$580 per month, all 50 units are occupied. However, when the rent is \$625 per month, the average number of occupied units drops to 47. Assume that the relationship between the monthly rent p and the demand x is linear.

(a) Write the equation of the line giving the demand x in terms of the rent p.

(b) Use a graphing utility to graph the demand equation and use the trace feature to predict the number of units occupied if the rent is raised to \$655.

(c) Use the demand equation to predict the number of units occupied if the rent is lowered to \$595. Verify graphically.

96. *Simple Interest* An inheritance of \$12,000 is invested in two different mutual funds. One fund pays $5\frac{1}{2}\%$ simple interest and the other pays 8% simple interest.

(a) If x dollars is invested in the fund paying $5\frac{1}{2}\%$, how much is invested in the fund paying 8%?

(b) Write the annual interest y in terms of x.

(c) Use a graphing utility to graph the function in part (b) over the interval $0 \le x \le 12{,}000$.

(d) Explain why the slope of the line in part (c) is negative.

1.3 Functions

Introduction to Functions / Function Notation / The Domain of a Function / Applications

Library of Functions

Many functions do not have simple mathematical formulas but are defined by real-life data. Such functions arise when you are using collections of data to model real-life applications. You will see that it is often convenient to *approximate* the data using a mathematical model or formula.

Introduction to Functions

Many everyday phenomena involve pairs of quantities that are related to each other by some rule of correspondence. Here are some examples.

1. The simple interest I earned on \$1000 for 1 year is related to the annual interest rate r by the formula $I = 1000r$.
2. The distance d traveled on a bicycle in 2 hours is related to the speed s of the bicycle by the formula $d = 2s$.
3. The area A of a circle is related to its radius r by the formula $A = \pi r^2$.

Not all correspondences between two quantities have simple mathematical formulas. For instance, people commonly match up NFL starting quarterbacks with touchdown passes, and hours of the day with temperature. In each of these cases, however, there is some rule of correspondence that matches each item from one set with exactly one item from a different set. Such a rule of correspondence is called a **function.**

Definition of a Function

A **function** f from a set A to a set B is a rule of correspondence that assigns to each element x in the set A exactly one element y in the set B. The set A is the **domain** (or set of inputs) of the function f, and the set B contains the **range** (or set of outputs).

Figure 1.24

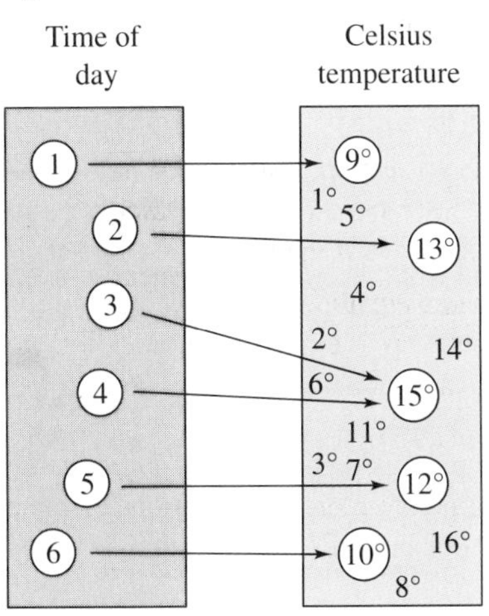

To help understand this definition, look at the function illustrated in Figure 1.24. This function can be represented by the following ordered pairs.

$$\{(1, 9°), (2, 13°), (3, 15°), (4, 15°), (5, 12°), (6, 10°)\}$$

In each ordered pair, the first coordinate is the input and the second coordinate is the output. In this example, note the following characteristics of a function.

1. Each element in A must be matched with an element of B.
2. Some elements in B may not be matched with any element in A.
3. Two or more elements of A may be matched with the same element of B.

The converse of the third statement is not true. That is, an element of A (the domain) cannot be matched with two different elements of B.

Note Be sure you see that the *range* of a function is not the same as the use of *range* relating to the viewing rectangle.

In the following example, you are asked to decide whether different correspondences are functions. To do this, you must decide whether each element in the domain A is matched with exactly one element in the range B. If any element in A is matched with two or more elements in B, the correspondence is not a function.

EXAMPLE 1 Testing for Functions

Let $A = \{a, b, c\}$ and $B = \{1, 2, 3, 4, 5\}$. Which of the following sets of ordered pairs or figures represent functions from set A to set B?

a. $\{(a, 2), (b, 3), (c, 4)\}$ **b.** $\{(a, 4), (b, 5)\}$

c.

a b c → 1 2 3 4 5

d.

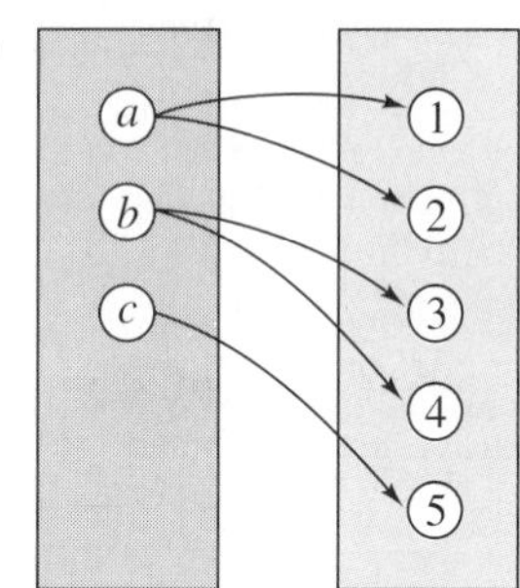

Solution

a. This collection of ordered pairs *does* represent a function from A to B. Each element of A is matched with exactly one element of B.

b. This collection of ordered pairs *does not* represent a function from A to B. Not every element of A is matched with an element of B.

c. This figure *does* represent a function from A to B. It does not matter that each element of A is matched with the same element of B.

d. This figure *does not* represent a function from A to B. The element a in A is matched with *two* elements, 1 and 2, of B. This is also true of the element b.

Leonhard Euler (1707–1783), a Swiss mathematician, is considered to have been the most prolific and productive mathematician in history. One of his greatest influences on mathematics was his use of symbols, or notation. The function notation $y = f(x)$ was introduced by Euler.

Representing functions by sets of ordered pairs is common in *discrete mathematics.* In algebra, however, it is more common to represent functions by equations or formulas involving two variables. For instance, the equation

$$y = x^2 \qquad y \text{ is a function of } x.$$

represents the variable y as a function of the variable x. In this equation, x is the **independent variable** and y is the **dependent variable.** The domain of the function is the set of all values taken on by the independent variable x, and the range of the function is the set of all values taken on by the dependent variable y.

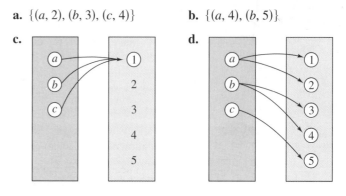

The *Interactive* CD-ROM offers graphing utility emulators of the *TI-82* and *TI-83*, which can be used with the Examples, Explorations, Technology notes, and Exercises.

EXPLORATION

Use a graphing utility to graph $x^2 + y = 1$. Then use the graph to write a convincing argument that each x-value has at most one y-value.

Use a graphing utility to graph $-x + y^2 = 1$. (*Hint:* You will need to use two equations.) Then use the graph to find an x-value that corresponds to two y-values. Why does the graph not represent y as a function of x?

EXAMPLE 2 **Testing for Functions Represented by Equations**

Which of the equations represents y as a function of x?

a. $x^2 + y = 1$ **b.** $-x + y^2 = 1$

Solution

To determine whether y is a function of x, try to solve for y in terms of x.

a. Solving for y yields the following.

$$x^2 + y = 1 \qquad \text{Original equation}$$
$$y = 1 - x^2 \qquad \text{Solve for } y.$$

To each value of x there corresponds exactly one value of y. Thus, y *is* a function of x.

b. Solving for y yields the following.

$$-x + y^2 = 1 \qquad \text{Original equation}$$
$$y^2 = 1 + x \qquad \text{Add } x \text{ to both sides.}$$
$$y = \pm\sqrt{1 + x} \qquad \text{Solve for } y.$$

The $\pm$ indicates that to a given value of x there correspond two values of y. Thus, y *is not* a function of x.

Function Notation

When an equation is used to represent a function, it is convenient to name the function so that it can be referenced easily. For example, you know that the equation $y = 1 - x^2$ describes y as a function of x. Suppose you give this function the name "f." Then you can use the following **function notation.**

Input	*Output*	*Equation*
x	$f(x)$	$f(x) = 1 - x^2$

The symbol $f(x)$ is read as the **value of f at x** or simply **f of x.** The symbol $f(x)$ corresponds to the y-value for a given x. Thus, you can write $y = f(x)$. Keep in mind that f is the *name* of the function, whereas $f(x)$ is the *value* of the function at x. For instance, the function given by

$$f(x) = 3 - 2x$$

has *function values* denoted by $f(-1), f(0), f(2)$, and so on. To find these values, substitute the specified input values into the given equation.

For $x = -1$, $\quad f(-1) = 3 - 2(-1) = 3 + 2 = 5.$

For $x = 0$, $\quad f(0) = 3 - 2(0) = 3 - 0 = 3.$

For $x = 2$, $\quad f(2) = 3 - 2(2) = 3 - 4 = -1.$

Library of Functions

The function in Example 4 is a *piecewise-defined* function. This means that the function is defined by two or more equations over a specified domain. In Example 4, you use the top equation for all x-values less than 0, and the bottom equation for all x-values greater than or equal to 0.

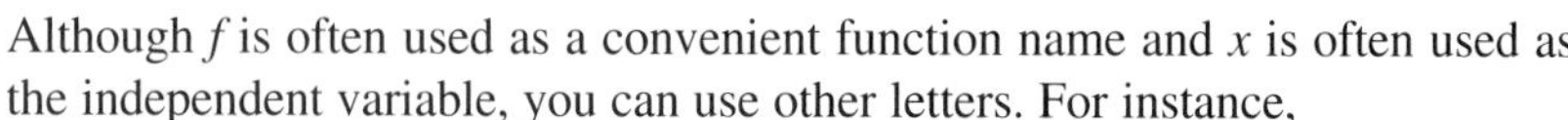

Although f is often used as a convenient function name and x is often used as the independent variable, you can use other letters. For instance,

$$f(x) = x^2 - 4x + 7, \quad f(t) = t^2 - 4t + 7, \quad \text{and} \quad g(s) = s^2 - 4s + 7$$

all define the same function. In fact, the role of the independent variable is that of a "placeholder." Consequently, the function could be described by

$$f(\quad) = (\quad)^2 - 4(\quad) + 7.$$

EXAMPLE 3 **Evaluating a Function**

Let $g(x) = -x^2 + 4x + 1$ and find the following.

a. $g(2)$ **b.** $g(t)$ **c.** $g(x + 2)$

Solution

a. Replacing x with 2 in $g(x) = -x^2 + 4x + 1$ yields the following.

$$g(2) = -(2)^2 + 4(2) + 1 = -4 + 8 + 1 = 5$$

b. Replacing x with t yields the following.

$$g(t) = -(t)^2 + 4(t) + 1 = -t^2 + 4t + 1$$

c. Replacing x with $x + 2$ yields the following.

$$\begin{aligned} g(x + 2) &= -(x + 2)^2 + 4(x + 2) + 1 \\ &= -(x^2 + 4x + 4) + 4x + 8 + 1 \\ &= -x^2 - 4x - 4 + 4x + 8 + 1 \\ &= -x^2 + 5 \end{aligned}$$

Note In Example 3, note that $g(x + 2)$ is not equal to $g(x) + g(2)$. In general, $g(u + v) \neq g(u) + g(v)$.

EXAMPLE 4 **A Piecewise-Defined Function**

Evaluate the function when $x = -1, 0$, and 1.

$$f(x) = \begin{cases} x^2 + 1, & x < 0 \\ x - 1, & x \geq 0 \end{cases}$$

Solution

Because $x = -1$ is less than 0, use $f(x) = x^2 + 1$ to obtain

$$f(-1) = (-1)^2 + 1 = 2.$$

For $x = 0$, use $f(x) = x - 1$ to obtain

$$f(0) = (0) - 1 = -1.$$

For $x = 1$, use $f(x) = x - 1$ to obtain $f(1) = (1) - 1 = 0$.

Most graphing utilities can graph functions that are defined piecewise. For example, on the *TI-82* or *TI-83*, you can obtain the graph of the function in Example 4 as follows.

Y1 = (X² + 1)(X < 0) + (X − 1)(X ≥ 0).

EXPLORATION

Use a graphing utility to graph $y = \sqrt{4 - x^2}$. What is the domain of this function? Then graph $y = \sqrt{x^2 - 4}$. What is the domain of this function? Do the domains of these two functions overlap? If so, for what values?

The Domain of a Function

The domain of a function can be described explicitly or it can be *implied* by the expression used to define the function. The **implied domain** is the set of all real numbers for which the expression is defined. For instance, the function given by

$$f(x) = \frac{1}{x^2 - 4}$$

has an implied domain that consists of all real x other than $x = \pm 2$. These two values are excluded from the domain because division by zero is undefined. Another common type of implied domain is that used to avoid even roots of negative numbers. For example, the function given by

$$f(x) = \sqrt{x}$$

is defined only for $x \geq 0$. Hence, its implied domain is the interval $[0, \infty)$. In general, the domain of a function *excludes* values that would cause division by zero *or* result in the even root of a negative number.

EXAMPLE 5 **Finding the Domain of a Function**

Find the domain of each function.

a. f: $\{(-3, 0), (-1, 4), (0, 2), (2, 2), (4, -1)\}$

b. $g(x) = \dfrac{1}{x + 5}$

c. Volume of a sphere: $V = \frac{4}{3}\pi r^3$

d. $h(x) = \sqrt{4 - x}$

Solution

a. The domain of f consists of all first coordinates in the set of ordered pairs.

$$\text{Domain} = \{-3, -1, 0, 2, 4\}.$$

b. Excluding x-values that yield zero in the denominator, the domain of g is the set of all real numbers $x \neq -5$.

c. Because this function represents the volume of a sphere, the values of the radius r must be positive. Thus, the domain is the set of all real numbers r such that $r > 0$.

d. This function is defined only for x-values for which $4 - x \geq 0$. The domain is all real numbers that are less than or equal to 4.

Library of Functions

The *square root* function $f(x) = \sqrt{x}$ is not defined for $x < 0$. This means that you must be careful when analyzing the domain of complicated functions involving the square root symbol.

Note In Example 5(c), note that the domain of a function may be implied by the physical context. For instance, from the equation $V = \frac{4}{3}\pi r^3$, you would have no reason to restrict r to positive values, but the physical context implies that a sphere cannot have a negative radius.

Applications

EXAMPLE 6 The Dimensions of a Container

Real Life

Figure 1.25

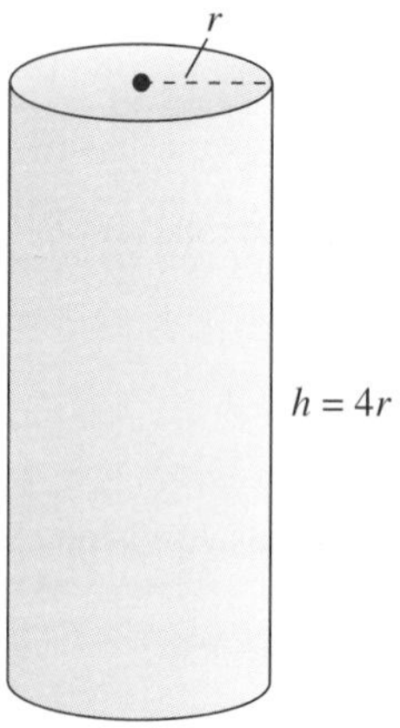

r

$h = 4r$

You work in the marketing department of a soft-drink company and are experimenting with a new soft-drink can that is slightly narrower and taller than a standard can. For your experimental can, the ratio of the height to the radius is 4, as shown in Figure 1.25.

a. Express the volume of the can as a function of the radius r.

b. Express the volume of the can as a function of the height h.

Solution

The volume of a right circular cylinder is given by the formula

$$V = \pi(\text{radius})^2(\text{height}) = \pi r^2 h.$$

Because the ratio of the height to the radius is 4, you can write $h = 4r$.

a. To write the volume as a function of the radius, use the fact that $h = 4r$.

$$V = \pi r^2 h = \pi r^2(4r) = 4\pi r^3$$

b. To write the volume as a function of the height, use the fact that $r = h/4$.

$$V = \pi\left(\frac{h}{4}\right)^2 h = \frac{\pi h^3}{16}$$

EXAMPLE 7 The Path of a Baseball

Real Life

Figure 1.26

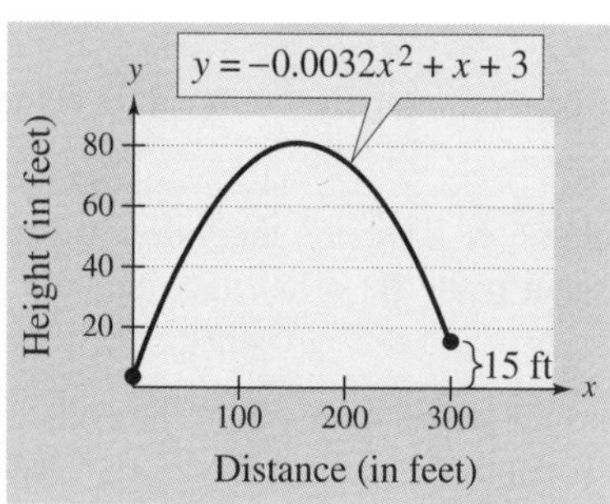

A baseball is hit at a point 3 feet above ground at a velocity of 100 feet per second and an angle of 45°. The path of the baseball is given by the function

$$y = -0.0032x^2 + x + 3$$

where y and x are measured in feet, as shown in Figure 1.26. Will the baseball clear a 10-foot fence located 300 feet from home plate?

Solution

When $x = 300$, the height of the baseball is given by

$$y = -0.0032(300)^2 + 300 + 3 = 15 \text{ feet}.$$

Thus, the ball will clear the fence.

Note In the equation in Example 7, the height of the baseball is a function of the horizontal distance from home plate.

Real Life

EXAMPLE 8 **U.S. Travelers Abroad**

Figure 1.27

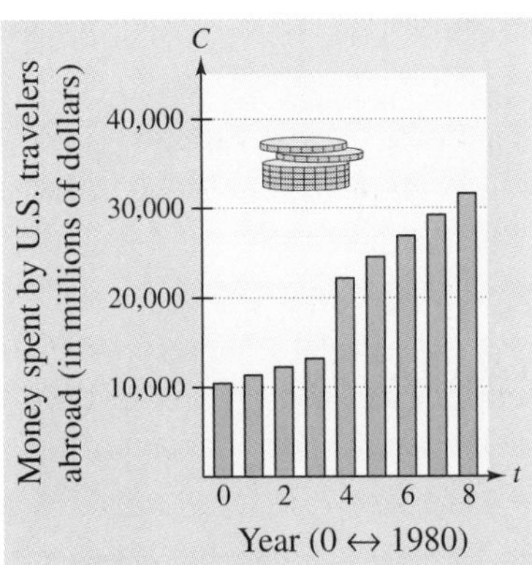

The money C (in millions of dollars) spent by U.S. travelers in other countries increased in a linear pattern from 1980 to 1983, as shown in Figure 1.27. Then, in 1984, the money spent took a sharp jump and until 1988 increased in a *different* linear pattern. These two patterns can be approximated by the piecewise-defined function

$$C = \begin{cases} 10{,}479 + 917.1t, & 0 \le t \le 3 \\ 12{,}808 + 2350.4t, & 4 \le t \le 8 \end{cases}$$

where $t = 0$ represents 1980. Use this function to approximate the total amount spent by U.S. travelers abroad between 1980 and 1988. (Source: U.S. Bureau of Economic Analysis)

Solution

From 1980 to 1983, use the formula $C = 10{,}479 + 917.1t$.

$$\underbrace{\$10{,}479,}_{1980} \quad \underbrace{\$11{,}396,}_{1981} \quad \underbrace{\$12{,}313,}_{1982} \quad \underbrace{\$13{,}230}_{1983}$$

From 1984 to 1988, use the formula $C = 12{,}808 + 2350.4t$.

$$\underbrace{\$22{,}210,}_{1984} \quad \underbrace{\$24{,}560,}_{1985} \quad \underbrace{\$26{,}910,}_{1986} \quad \underbrace{\$29{,}261,}_{1987} \quad \underbrace{\$31{,}611}_{1988}$$

The total of these nine amounts is \$181,970, which implies that the total amount spent was approximately \$181,970,000,000.

EXAMPLE 9 **From Calculus: Evaluating a Difference Quotient**

For $f(x) = x^2 - 4x + 7$, find $\dfrac{f(x+h) - f(x)}{h}$.

Solution

$$\begin{aligned} \frac{f(x+h) - f(x)}{h} &= \frac{[(x+h)^2 - 4(x+h) + 7] - (x^2 - 4x + 7)}{h} \\ &= \frac{x^2 + 2xh + h^2 - 4x - 4h + 7 - x^2 + 4x - 7}{h} \\ &= \frac{2xh + h^2 - 4h}{h} \\ &= \frac{h(2x + h - 4)}{h} \\ &= 2x + h - 4, \quad h \neq 0 \end{aligned}$$

Note One of the basic definitions in calculus employs the ratio

$$\frac{f(x+h) - f(x)}{h}, \quad h \neq 0$$

which is called a **difference quotient,** as illustrated in Example 9.

Summary of Function Terminology

Function: A **function** is a relationship between two variables such that to each value of the independent variable there corresponds exactly one value of the dependent variable.

Function Notation: $y = f(x)$

f is the **name** of the function.
y is the **dependent variable.**
x is the **independent variable.**
$f(x)$ is the **value of the function at x.**

Domain: The **domain** of a function is the set of all values (inputs) of the independent variable for which the function is defined. If x is in the domain of f, we say that f is **defined** at x. If x is not in the domain of f, we say that f is **undefined** at x.

Range: The **range** of a function is the set of all values (outputs) assumed by the dependent variable (that is, the set of all function values).

Implied Domain: If f is defined by an algebraic expression and the domain is not specified, the **implied domain** consists of all real numbers for which the expression is defined.

Group Activity

Modeling with Piecewise-Defined Functions

x	y
1	5.2
2	5.6
3	6.6
4	8.3
5	11.5
6	15.8
7	12.8
8	10.1
9	8.6
10	6.9
11	4.5
12	2.7

The table at the left shows the monthly revenue y (in thousands of dollars) for one year of a landscaping business, with $x = 1$ representing January.

A mathematical model that represents this data is

$$f(x) = \begin{cases} -1.97x + 26.33 \\ 0.51x^2 - 1.47x + 6.31. \end{cases}$$

For what values of x is each part of the piecewise-defined function defined? How can you tell? Explain your reasoning.

Find $f(5)$ and $f(11)$, and interpret your results in the context of the problem. How do these model values compare with the actual data values?

1.3 /// EXERCISES

In Exercises 1–4, is the relationship a function?

1. *Domain* *Range*

−2, −1, 0, 1, 2 → 5, 6, 7, 8

2. *Domain* *Range*

−2, −1, 0, 1, 2 → 3, 4, 5

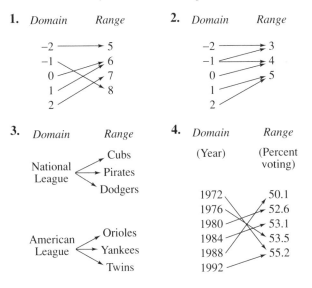

3. *Domain* *Range*

National League → Cubs, Pirates, Dodgers

American League → Orioles, Yankees, Twins

4. *Domain* (Year) *Range* (Percent voting)

1972, 1976, 1980, 1984, 1988, 1992 → 50.1, 52.6, 53.1, 53.5, 55.2

In Exercises 5–8, does the table describe a function? Explain your reasoning.

5.

Input Value	−2	−1	0	1	2
Output Value	−8	−1	0	1	8

6.

Input Value	0	1	2	1	0
Output Value	−4	−2	0	2	4

7.

Input Value	10	7	4	7	10
Output Value	3	6	9	12	15

8.

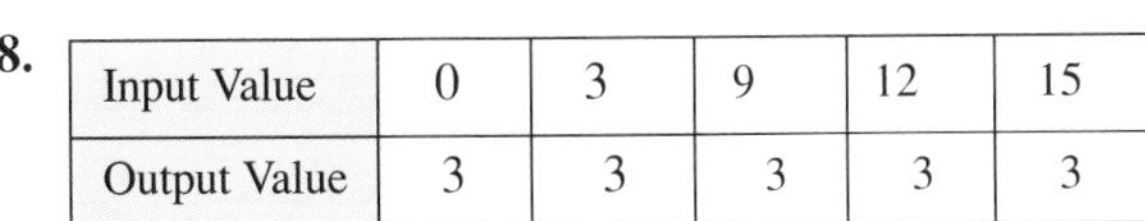

Input Value	0	3	9	12	15
Output Value	3	3	3	3	3

In Exercises 9 and 10, which sets of ordered pairs represent a function from A to B? Explain.

9. $A = \{0, 1, 2, 3\}$ and $B = \{-2, -1, 0, 1, 2\}$

(a) $\{(0, 1), (1, -2), (2, 0), (3, 2)\}$

(b) $\{(0, -1), (2, 2), (1, -2), (3, 0), (1, 1)\}$

(c) $\{(0, 0), (1, 0), (2, 0), (3, 0)\}$

10. $A = \{a, b, c\}$ and $B = \{0, 1, 2, 3\}$

(a) $\{(a, 1), (c, 2), (c, 3), (b, 3)\}$

(b) $\{(a, 1), (b, 2), (c, 3)\}$

(c) $\{(1, a), (0, a), (2, c), (3, b)\}$

Circulation of Newspapers **In Exercises 11 and 12, use the graph, which shows the circulation (in millions) of daily newspapers in the United States.** (Source: Editor & Publisher Company)

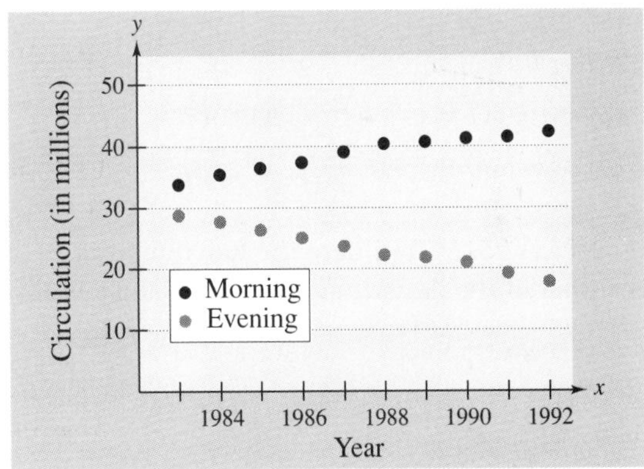

11. Is the circulation of morning newspapers a function of the year? Is the circulation of evening newspapers a function of the year? Explain.

12. Let $f(x)$ represent the circulation of evening newspapers in year x. Find $f(1988)$.

In Exercises 13–22, determine if the equation represents y as a function of x.

13. $x^2 + y^2 = 4$

14. $x = y^2$

15. $x^2 + y = 4$

16. $x + y^2 = 4$

17. $2x + 3y = 4$

18. $(x - 2)^2 + y^2 = 4$

19. $y^2 = x^2 - 1$

20. $y = \sqrt{x + 5}$

21. $y = |4 - x|$

22. $|y| = 4 - x$

In Exercises 23 and 24, fill in the blanks using the specified function and the given values of the independent variable. Simplify the result.

23. $f(s) = \dfrac{1}{s + 1}$

(a) $f(4) = \dfrac{1}{(\quad) + 1}$

(b) $f(0) = \dfrac{1}{(\quad) + 1}$

(c) $f(4x) = \dfrac{1}{(\quad) + 1}$

(d) $f(x + c) = \dfrac{1}{(\quad) + 1}$

24. $g(x) = x^2 - 2x$

(a) $g(2) = (\quad)^2 - 2(\quad)$

(b) $g(-3) = (\quad)^2 - 2(\quad)$

(c) $g(t + 1) = (\quad)^2 - 2(\quad)$

(d) $g(x + c) = (\quad)^2 - 2(\quad)$

In Exercises 25–36, evaluate the function at the specified values of the independent variable and simplify.

25. $f(x) = 2x - 3$

(a) $f(1)$ (b) $f(-3)$ (c) $f(x - 1)$

26. $g(y) = 7 - 3y$

(a) $g(0)$ (b) $g(\frac{7}{3})$ (c) $g(s + 2)$

27. $h(t) = t^2 - 2t$

(a) $h(2)$ (b) $h(1.5)$ (c) $h(x + 2)$

28. $V(r) = \frac{4}{3}\pi r^3$

(a) $V(3)$ (b) $V(\frac{3}{2})$ (c) $V(2r)$

29. $f(y) = 3 - \sqrt{y}$

(a) $f(4)$ (b) $f(0.25)$ (c) $f(4x^2)$

30. $f(x) = \sqrt{x + 8} + 2$

(a) $f(-8)$ (b) $f(1)$ (c) $f(x - 8)$

31. $q(x) = \dfrac{1}{x^2 - 9}$

(a) $q(0)$ (b) $q(3)$ (c) $q(y + 3)$

32. $q(t) = \dfrac{2t^2 + 3}{t^2}$

(a) $q(2)$ (b) $q(0)$ (c) $q(-x)$

33. $f(x) = \dfrac{|x|}{x}$

(a) $f(2)$ (b) $f(-2)$ (c) $f(x - 1)$

34. $f(x) = |x| + 4$

(a) $f(2)$ (b) $f(-2)$ (c) $f(x^2)$

35. $f(x) = \begin{cases} 2x + 1, & x < 0 \\ 2x + 2, & x \geq 0 \end{cases}$

(a) $f(-1)$ (b) $f(0)$ (c) $f(2)$

36. $f(x) = \begin{cases} x^2 + 2, & x \leq 1 \\ 2x^2 + 2, & x > 1 \end{cases}$

(a) $f(-2)$ (b) $f(1)$ (c) $f(2)$

In Exercises 37–42, complete the table.

37. $f(x) = x^2 - 3$

x	-2	-1	0	1	2
$f(x)$					

38. $g(x) = \sqrt{x - 3}$

x	3	4	5	6	7
$g(x)$					

39. $h(t) = \frac{1}{2}|t + 3|$

t	-5	-4	-3	-2	-1
$h(t)$					

40. $f(s) = \dfrac{|s - 2|}{s - 2}$

s	0	1	$\frac{3}{2}$	$\frac{5}{2}$	4
$f(s)$					

41. $f(x) = \begin{cases} -\frac{1}{2}x + 4, & x \le 0 \\ (x-2)^2, & x > 0 \end{cases}$

x	-2	-1	0	1	2
$f(x)$					

42. $h(x) = \begin{cases} 9 - x^2, & x < 3 \\ x - 3, & x \ge 3 \end{cases}$

x	1	2	3	4	5
$h(x)$					

In Exercises 43–46, find all real values of x such that $f(x) = 0$.

43. $f(x) = 15 - 3x$

44. $f(x) = \dfrac{3x - 4}{5}$

45. $f(x) = x^2 - 9$

46. $f(x) = x^3 - x$

In Exercises 47–50, find the value(s) of x for which $f(x) = g(x)$.

47. $f(x) = x^2, \quad g(x) = x + 2$

48. $f(x) = x^2 + 2x + 1, \quad g(x) = 3x + 3$

49. $f(x) = \sqrt{3x} + 1, \quad g(x) = x + 1$

50. $f(x) = x^4 - 2x^2, \quad g(x) = 2x^2$

In Exercises 51–60, find the domain of the function.

51. $f(x) = 5x^2 + 2x - 1$

52. $g(x) = 1 - 2x^2$

53. $h(t) = \dfrac{4}{t}$

54. $s(y) = \dfrac{3y}{y + 5}$

55. $g(y) = \sqrt{y - 10}$

56. $f(t) = \sqrt[3]{t + 4}$

57. $f(x) = \sqrt[4]{1 - x^2}$

58. $h(x) = \dfrac{10}{x^2 - 2x}$

59. $g(x) = \dfrac{1}{x} - \dfrac{3}{x + 2}$

60. $f(s) = \dfrac{\sqrt{s - 1}}{s - 4}$

In Exercises 61–64, assume that the domain of f is the set $A = \{-2, -1, 0, 1, 2\}$. Determine the set of ordered pairs representing the function f.

61. $f(x) = x^2$

62. $f(x) = \dfrac{2x}{x^2 + 1}$

63. $f(x) = \sqrt{x + 2}$

64. $f(x) = |x + 1|$

65. *Essay* In your own words, explain the meaning of *domain* and *range*.

66. *Think About It* Describe an advantage of function notation.

Exploration **In Exercises 67–70, select a function from $f(x) = cx$, $g(x) = cx^2$, $h(x) = c\sqrt{|x|}$, or $r(x) = c/x$ and determine the value of the constant c such that the function fits the data given in the table.**

67.

x	-4	-1	0	1	4
y	-32	-2	0	-2	-32

68.

x	-4	-1	0	1	4
y	-1	$-\frac{1}{4}$	0	$\frac{1}{4}$	1

69.

x	-4	-1	0	1	4
y	-8	-32	Undef.	32	8

70.

x	-4	-1	0	1	4
y	6	3	0	3	6

In Exercises 71–76, find the difference quotient and simplify your answer.

71. $f(x) = x^2 - x + 1, \quad \dfrac{f(2 + h) - f(2)}{h}, h \neq 0$

72. $f(x) = 5x - x^2, \quad \dfrac{f(5 + h) - f(5)}{h}, h \neq 0$

73. $f(x) = x^3, \quad \dfrac{f(x + c) - f(x)}{c}, c \neq 0$

74. $f(x) = 2x, \quad \dfrac{f(x+c) - f(x)}{c}, c \neq 0$

75. $g(x) = 3x - 1, \quad \dfrac{g(x) - g(3)}{x - 3}, x \neq 3$

76. $f(t) = \dfrac{1}{t}, \quad \dfrac{f(t) - f(1)}{t - 1}, t \neq 0$

77. *Area of a Circle* Express the area A of a circle as a function of its circumference C.

78. *Area of a Triangle* Express the area A of an equilateral triangle as a function of the length s of its sides.

79. *Exploration* An open box of maximum volume is to be made from a square piece of material, 24 centimeters on a side, by cutting equal squares from the corners and turning up the sides (see figure).

(a) Use the table feature of a graphing utility to complete six rows of the table. Use the result to guess the maximum volume.

Height, x	Width	Volume, V
1	$24 - 2(1)$	$1[24 - 2(1)]^2 = 484$
2	$24 - 2(2)$	$2[24 - 2(2)]^2 = 800$

(b) Use a graphing utility to plot the points (x, V). Is V a function of x? If yes, write the volume V as a function of x, and determine its domain.

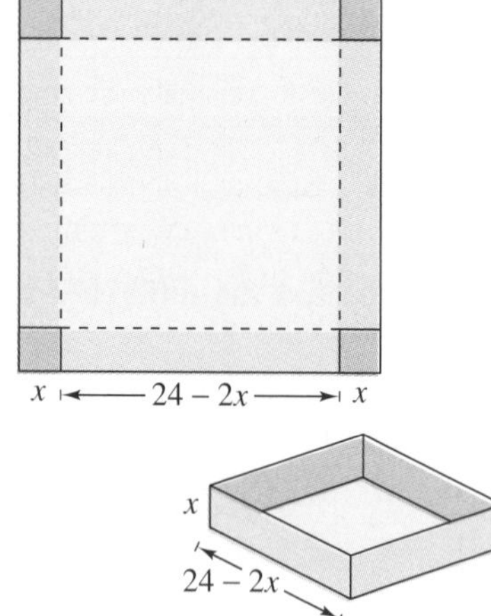

80. *Exploration* The cost per unit in the production of a certain radio model is \$60. The manufacturer charges \$90 per unit for orders of 100 or less. To encourage large orders, the manufacturer reduces the charge by \$0.15 per radio for each unit ordered in excess of 100.

(a) Use the table feature of a graphing utility to complete six rows of the table. Use the result to estimate the maximum profit.

Units, x	Price, p	Profit, P
102	$90 - 2(0.15)$	$xp - 102(60)$
104	$90 - 4(0.15)$	$xp - 104(60)$

(b) Use a graphing utility to plot the points (x, P). Is P a function of x? If yes, write the profit P as a function of x, and determine its domain.

81. *Area of a Triangle* A right triangle is formed in the first quadrant by the x and y axes and a line through the point $(2, 1)$ (see figure). Write the area of the triangle as a function of x, and determine the domain of the function.

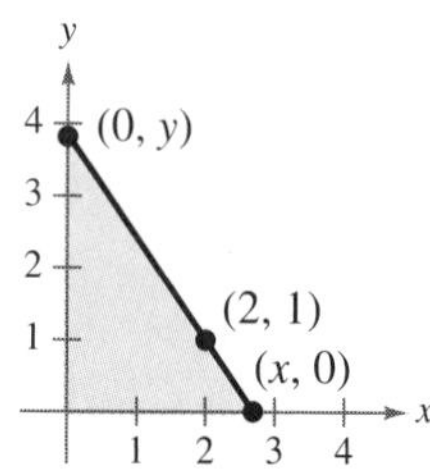

82. *Area of a Rectangle* A rectangle is bounded by the x-axis and the semicircle $y = \sqrt{36 - x^2}$ (see figure). Write the area of the rectangle as a function of x, and determine the domain of the function.

Figure for 82

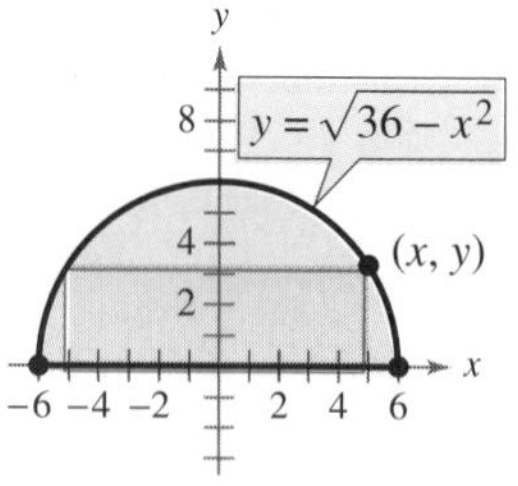

Figure for 83

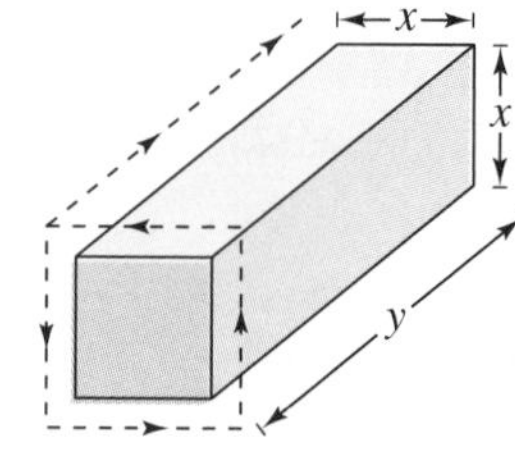

83. *Volume of a Package* A rectangular package to be sent by a postal service can have a maximum combined length and girth (perimeter of a cross section) of 108 inches (see figure). Write the volume of the package as a function of x. What is the domain of the function?

84. *Price of Mobile Homes* The average price p (in thousands of dollars) of a new mobile home in the United States from 1974 to 1993 can be approximated by the piecewise-defined model

$$p(t) = \begin{cases} 19.247 + 1.694t, & -6 \le t \le -1 \\ 19.305 + 0.427t + 0.033t^2, & 0 \le t \le 13 \end{cases}$$

where $t = 0$ represents 1980 (see figure). Use a graphing utility to graph the model and find the average price of a mobile home in 1978, 1988, and 1993. (Source: U.S. Bureau of Census, Construction Reports)

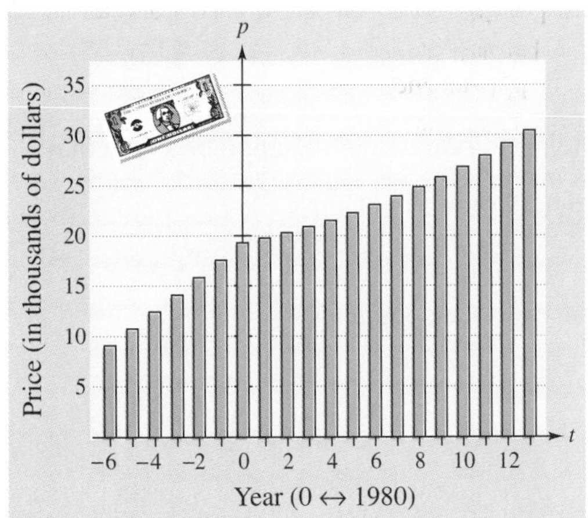

85. *Cost, Revenue, and Profit* A company produces a product for which the variable cost is \$12.30 per unit and the fixed costs are \$98,000. The product sells for \$17.98. Let x be the number of units produced and sold.

(a) Write the total cost C as a function of the number of units produced.

(b) Write the revenue R as a function of the number of units sold.

(c) Write the profit P as a function of the number of units sold. (*Note*: $P = R - C$.)

86. *Charter Bus Fares* For groups of 80 or more people, a charter bus company determines the rate per person according to the formula

$$\text{Rate} = 8 - 0.05(n - 80), \qquad n \ge 80$$

where the rate is given in dollars and n is the number of people.

(a) Express the revenue R for the bus company as a function of n.

(b) Use the function from part (a) to complete the table. What can you conclude?

n	90	100	110	120	130	140	150
$R(n)$							

(c) Use a graphing utility to graph R and determine the number of people that will produce a maximum revenue. Compare the result with your conclusion from part (b).

87. *Chapter Opener* Use the data on page 75. Let $f(t)$ represent the number of lynx in year t.

(a) Find $f(1992)$.

(b) Find $\dfrac{f(1994) - f(1991)}{1994 - 1991}$

and interpret the result in the context of the problem.

(c) An approximate model for the function is

$$N(t) = \frac{434t + 4387}{45t^2 - 55t + 100}$$

where N is the number of lynx and t is the time in years, with $t = 0$ corresponding to 1990. Complete the table and compare the result with the data. Use a graphing utility to graph the model and data in the same viewing rectangle. Comment on the validity of the model.

t	1988	1989	1990	1991
N				

t	1992	1993	1994	1995
N				

1.4 Graphs of Functions

The Graph of a Function / Increasing and Decreasing Functions / Relative Minimum and Maximum Values / Step Functions / Even and Odd Functions

The Graph of a Function

In Section 1.3 you studied functions from an algebraic point of view. In this section, you will study functions from a geometric perspective. The **graph of a function** f is the collection of ordered pairs $(x, f(x))$ such that x is in the domain of f. As you study this section, remember the following geometrical interpretation of x and $f(x)$.

$$x = \text{the directed distance from the } y\text{-axis}$$
$$f(x) = \text{the directed distance from the } x\text{-axis}$$

Example 1 shows how to use the graph of a function to find the domain and range of the function.

Figure 1.28

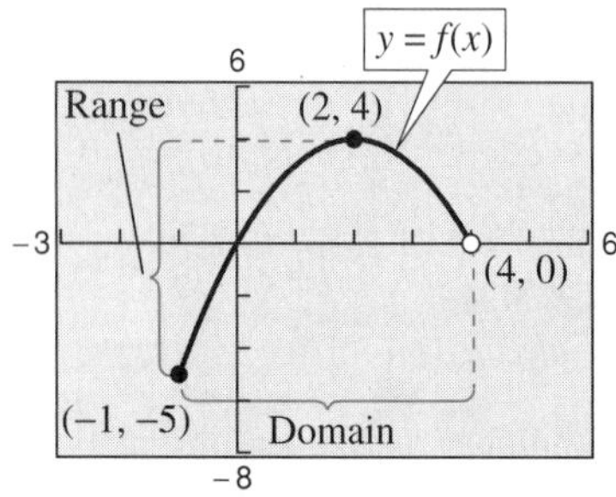

EXAMPLE 1 **Finding the Domain and Range of a Function**

Use the graph of the function f, shown in Figure 1.28.

a. Find the domain of f.

b. Find the function values $f(-1)$ and $f(2)$.

c. Find the range of f.

Solution

a. The closed dot (on the left) indicates that $x = -1$ is in the domain of f, whereas the open dot (on the right) indicates that $x = 4$ is not in the domain. Thus, the domain of f is all x in the interval $[-1, 4)$.

b. Because $(-1, -5)$ is a point on the graph of f, it follows that

$$f(-1) = -5.$$

Similarly, because $(2, 4)$ is a point on the graph of f, it follows that

$$f(2) = 4.$$

c. Because the graph does not extend below $f(-1) = -5$ or above $f(2) = 4$, the range of f is the interval $[-5, 4]$.

By the definition of a function, at most one y-value corresponds to a given x-value. It follows, then, that a vertical line can intersect the graph of a function at most once. This observation provides a convenient visual test for functions.

Vertical Line Test for Functions

A set of points in a coordinate plane is the graph of y as a function of x if and only if no vertical line intersects the graph at more than one point.

EXAMPLE 2 Vertical Line Test for Functions

Which of the graphs in Figure 1.29 represent y as a function of x?

Solution

a. This *is not* a graph of y as a function of x because you can find a vertical line that intersects the graph twice.

b. This *is* a graph of y as a function of x because every vertical line intersects the graph at most once.

c. This *is* a graph of y as a function of x. (Note that if a vertical line does not intersect the graph, it simply means that the function is undefined for that particular value of x.)

Figure 1.29

(a)

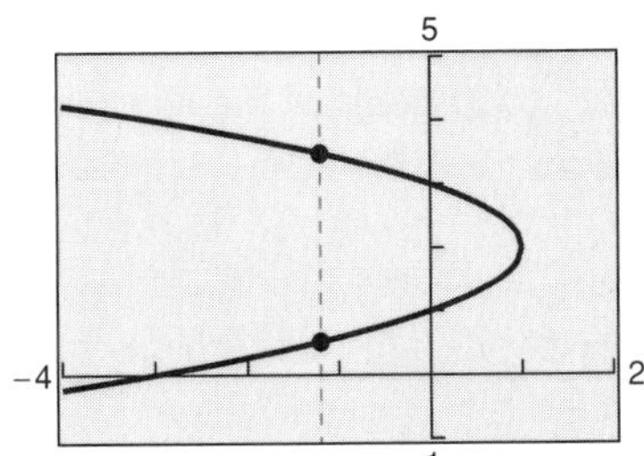

(b)

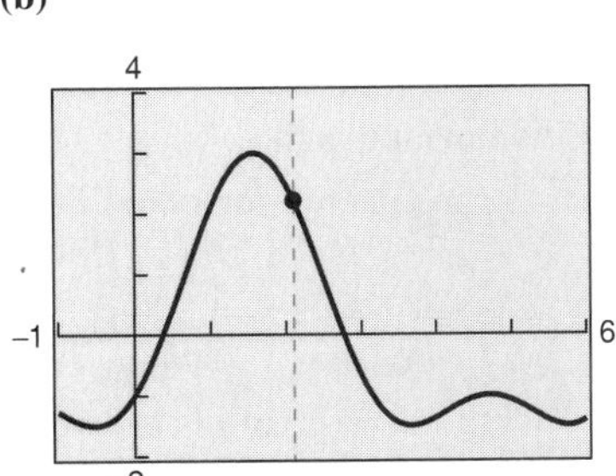

(c)

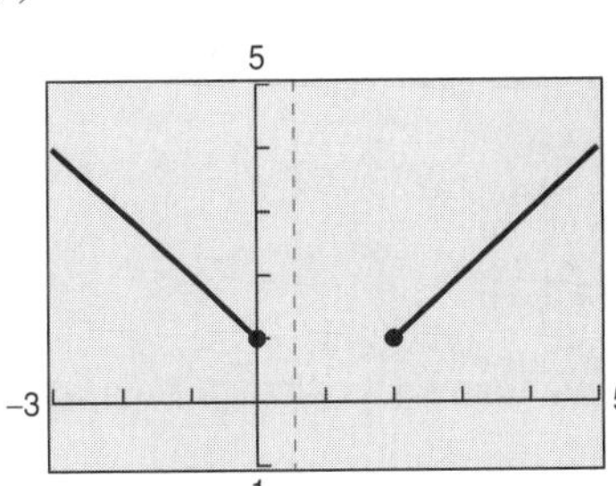

Note Most graphing utilities are designed to graph functions of x more easily than other types of equations. For instance, the graph shown in Figure 1.29(a) represents the equation $x + y^2 - 4y + 3 = 0$. Try using a graphing utility to duplicate this graph. To sketch the graph, did you need to use *two* equations?

Increasing and Decreasing Functions

Figure 1.30

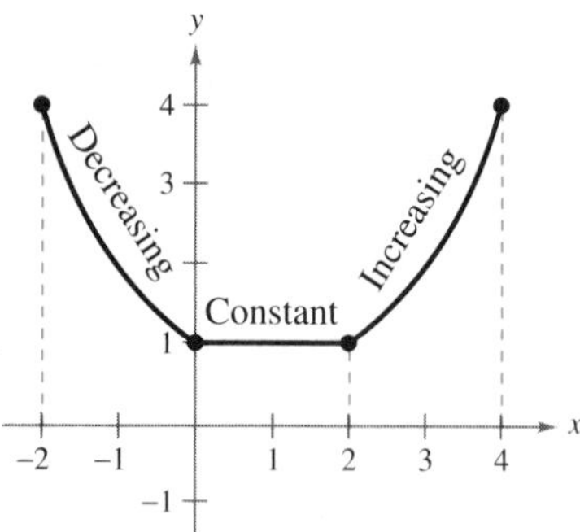

The more you know about the graph of a function, the more you know about the function itself. Consider the graph shown in Figure 1.30. Moving from *left to right,* this graph falls from $x = -2$ to $x = 0$, is constant from $x = 0$ to $x = 2$, and rises from $x = 2$ to $x = 4$.

Increasing, Decreasing, and Constant Functions

A function f is **increasing** on an interval if, for any x_1 and x_2 in the interval, $x_1 < x_2$ implies $f(x_1) < f(x_2)$.

A function f is **decreasing** on an interval if, for any x_1 and x_2 in the interval, $x_1 < x_2$ implies $f(x_1) > f(x_2)$.

A function f is **constant** on an interval if, for any x_1 and x_2 in the interval, $f(x_1) = f(x_2)$.

EXAMPLE 3 **Increasing and Decreasing Functions**

In Figure 1.31, determine the open intervals on which each function is increasing, decreasing, or constant.

Solution

a. Although it might appear that there is an interval in which this function is constant, you can see that if $x_1 < x_2$, then $f(x_1) = x_1^3 < x_2^3 = f(x_2)$. Thus, the function is increasing over the entire real line.

b. This function is increasing on the interval $(-\infty, -1)$, decreasing on the interval $(-1, 1)$, and increasing on the interval $(1, \infty)$.

c. This function is increasing on the interval $(-\infty, 0)$, constant on the interval $(0, 2)$, and decreasing on the interval $(2, \infty)$.

Figure 1.31

(a)

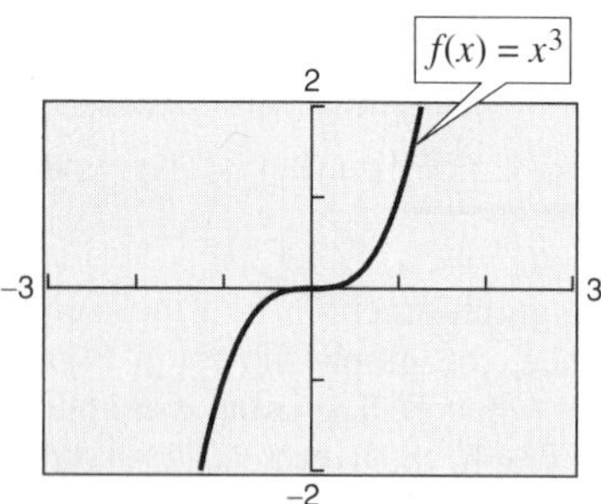

(b)

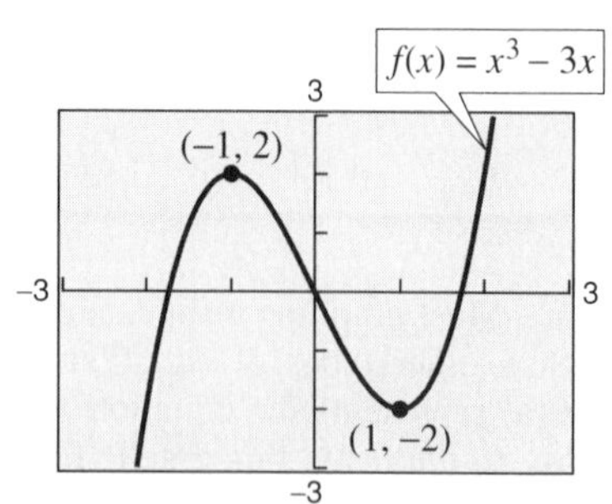

(c)

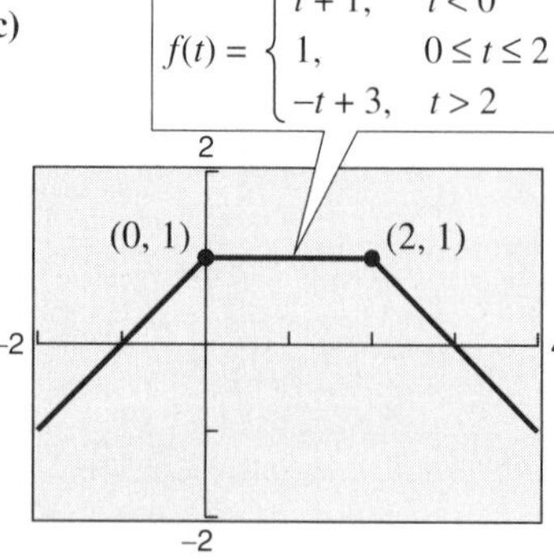

Relative Minimum and Maximum Values

The points at which a function changes its increasing, decreasing, or constant behavior are helpful in determining the relative maximum or relative minimum values of the function.

Definition of Relative Minimum and Relative Maximum

A function value $f(a)$ is called a **relative minimum** of f if there exists an interval (x_1, x_2) that contains a such that

$$x_1 < x < x_2 \quad \text{implies} \quad f(a) \le f(x).$$

A function value $f(a)$ is called a **relative maximum** of f if there exists an interval (x_1, x_2) that contains a such that

$$x_1 < x < x_2 \quad \text{implies} \quad f(a) \ge f(x).$$

Figure 1.32

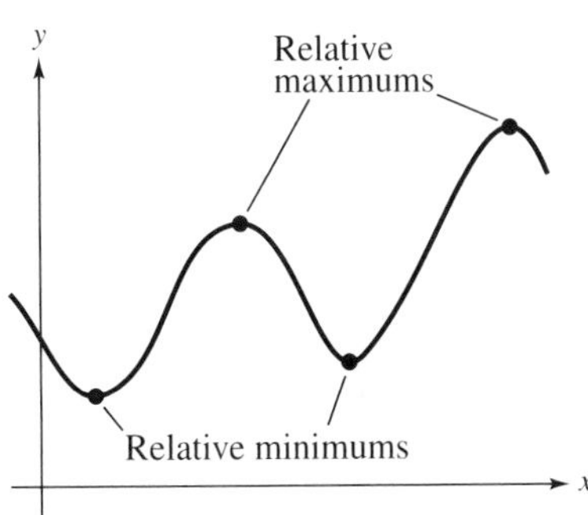

Figure 1.32 shows several different examples of relative minimums and relative maximums. In Section 3.1, you will study a technique for finding the *exact points* at which a second-degree polynomial function has a relative minimum or relative maximum. For the time being, however, you can use a graphing utility to find reasonable approximations of these points.

EXAMPLE 4 **Approximating a Relative Minimum**

Use a graphing utility to approximate the relative minimum of the function $f(x) = 3x^2 - 4x - 2$.

Solution

The graph of f is shown in Figure 1.33. By using the zoom and trace features of a graphing utility, you can estimate that the function has a relative minimum at the point

$(0.67, -3.33)$. Relative minimum

Later, in Section 3.1, you will be able to determine that the exact point at which the relative minimum occurs is $\left(\frac{2}{3}, -\frac{10}{3}\right)$.

Figure 1.33

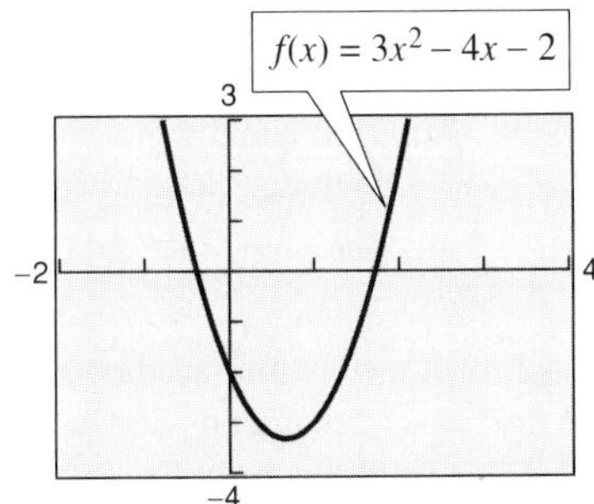

Note When you use a graphing utility to estimate x- and y-values of a relative minimum or relative maximum, the automatic zoom feature will often produce graphs that are nearly flat. To overcome this problem, you can manually change the vertical setting of the viewing rectangle. The graph will vertically stretch if the values of Ymin and Ymax are closer together.

EXAMPLE 5 Approximating Relative Minimums and Maximums

Use a graphing utility to approximate the relative minimum and relative maximum of the function $f(x) = -x^3 + x$.

Solution

A sketch of the graph of f is shown in Figure 1.34. By using the zoom and trace features of the graphing utility, you can estimate that the function has a relative minimum at the point

$(-0.58, -0.38)$ Relative minimum

and a relative maximum at the point

$(0.58, 0.38).$ Relative maximum

If you go on to take a course in calculus, you will learn a technique for finding the exact points at which this function has a relative minimum and a relative maximum.

Figure 1.34

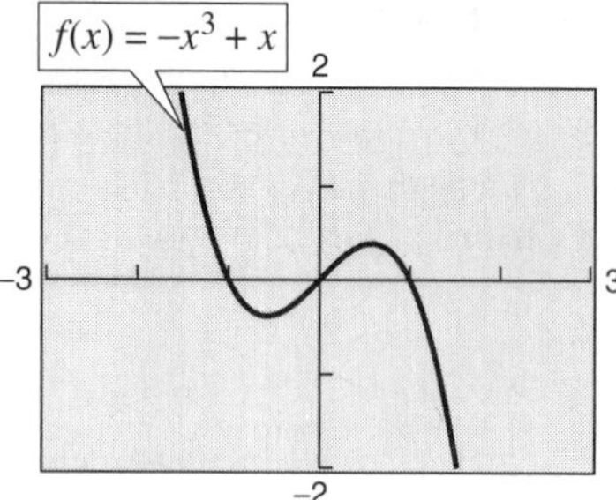

Note Some graphing utilities have built-in programs that will find minimum or maximum values. For instance, on a *TI-82* or *TI-83*, the minimum program can be accessed by CALC (3:minimum). If your graphing utility has such features, try using them to rework Example 5.

EXAMPLE 6 The Price of Diamonds

During the 1980s, the average price of a 1-carat polished diamond decreased and then increased according to the model

$$C = -0.7t^3 + 16.25t^2 - 106t + 388, \qquad 2 \le t \le 10,$$

where C is the average price in dollars (on the Antwerp Index) and t represents the calendar year, with $t = 2$ corresponding to January 1, 1982. According to this model, during which years was the price of diamonds decreasing? During which years was the price of diamonds increasing? Approximate the minimum price of a 1-carat diamond between 1982 and 1990. (Source: Diamond High Council)

Solution

To solve this problem, sketch an accurate graph of the function, as shown in Figure 1.35. From the graph, you can see that the price of diamonds decreased from 1982 until late 1984. Then, from late 1984 to 1990, the price increased. The minimum price during the 8-year period was approximately \$175.

Figure 1.35

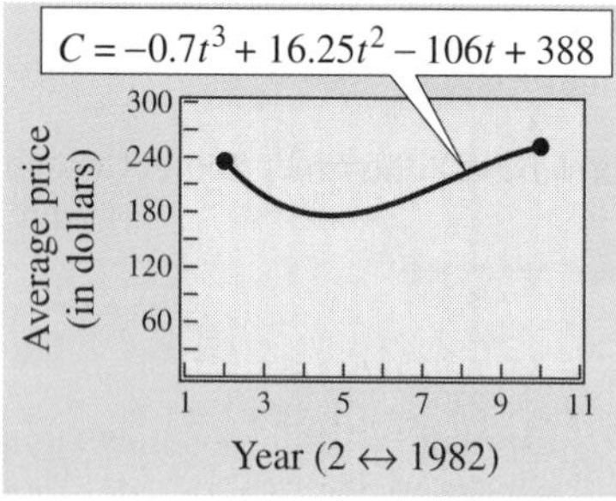

Step Functions

Figure 1.36

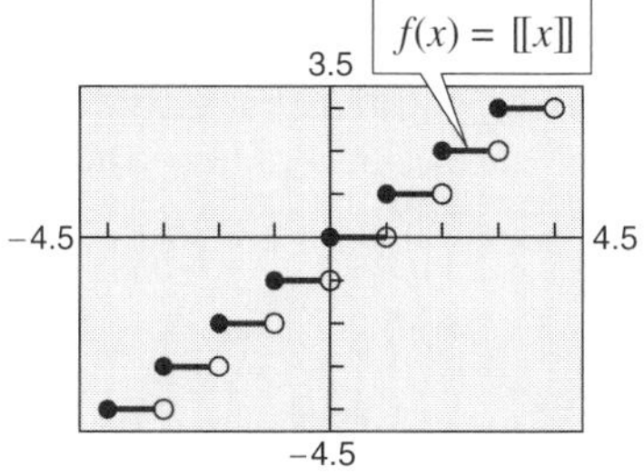

EXAMPLE 7 The Greatest Integer Function

The **greatest integer function** is denoted by $[\![x]\!]$ and is defined by

$$f(x) = [\![x]\!] = \text{the greatest integer less than or equal to } x.$$

The graph of this function is shown in Figure 1.36. Note that the graph of the greatest integer function jumps vertically one unit at each integer and is constant (a horizontal line segment) between each pair of consecutive integers. Because of the jumps in its graph, the greatest integer function is an example of a **step function.** Some values of the greatest integer function are as follows.

$$[\![-1]\!] = -1 \qquad [\![-0.5]\!] = -1$$
$$[\![0]\!] = 0 \qquad [\![0.5]\!] = 0$$
$$[\![1]\!] = 1 \qquad [\![1.5]\!] = 1$$

The range of the greatest integer function is the set of all integers.

Library of Functions

The *greatest integer* function has an infinite number of breaks or steps—one at each integer value in its domain. Could you describe the greatest integer function using a piecewise-defined function? How does the graph of the greatest integer function differ from the graph of a line with zero slope?

EXAMPLE 8 The Cost of a Telephone Call

Suppose the cost of a telephone call between Los Angeles and San Francisco is \$0.50 for the first minute and \$0.36 for each additional minute. The greatest integer function can be used to create a model for the cost of this call.

$$C = 0.50 + 0.36[\![t]\!], \qquad t > 0,$$

where C is the total cost of the call in dollars and t is the length of the call in minutes. Sketch the graph of this function.

Solution

For calls up to 1 minute, the cost is \$0.50. For calls between 1 and 2 minutes, the cost is \$0.86, and so on.

Length of Call:	$0 < t < 1$	$1 \le t < 2$	$2 \le t < 3$	$3 \le t < 4$	$4 \le t < 5$
Cost of Call:	\$0.50	\$0.86	\$1.22	\$1.58	\$1.94

Using these values, you can sketch the graph shown in Figure 1.37.

Figure 1.37

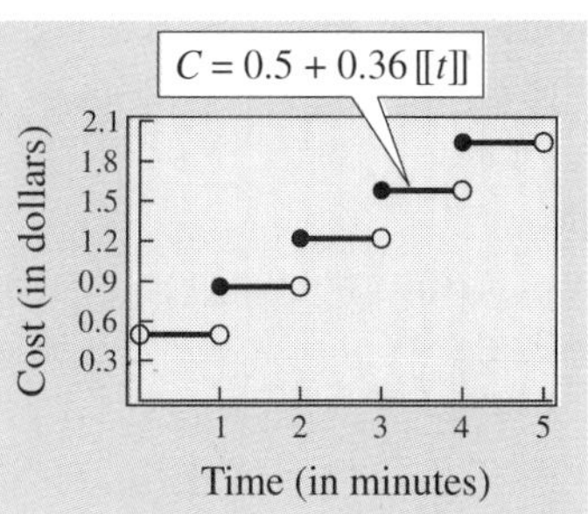

Note Most graphing utilities display graphs in *connected mode*, which means that the graph has no breaks. When you are sketching graphs that do have breaks, it is better to use *dot mode*. Graph the greatest integer function [often called Int (x)] in connected and dot modes, and compare the two results.

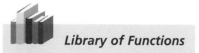

Library of Functions

The *absolute value* function can be expressed as a piecewise-defined function.

$$|x| = \begin{cases} -x, & x < 0 \\ x, & x \geq 0 \end{cases}$$

Use this definition to show that $|-5| = 5$ and $|0| = 0$.

EXPLORATION

Use a graphing utility to graph the functions $y_1 = x^2$, $y_2 = x^4$, and $y_3 = 1/x^2$ in the standard viewing rectangle. What kind of symmetry do you observe? Then graph and discuss the symmetry of $y_1 = x^3$, $y_2 = x^5$, and $y_3 = 1/x$. Can you predict the symmetry of $y = x^8$ and $y = 1/x^5$ without graphing them?

Even and Odd Functions

A computer animation of this concept appears in the *Interactive* CD-ROM.

A graph has **symmetry with respect to the y-axis** if, whenever (x, y) is on the graph, so is the point $(-x, y)$. A graph has **symmetry with respect to the origin** if, whenever (x, y) is on the graph, so is the point $(-x, -y)$. A graph has **symmetry with respect to the x-axis** if, whenever (x, y) is on the graph, so is the point $(x, -y)$. These three types of symmetry are illustrated in Figure 1.38.

Figure 1.38

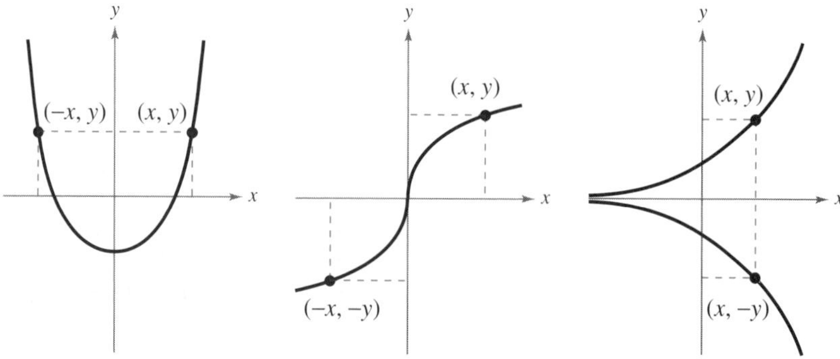

A function whose graph is symmetric with respect to the y-axis is an **even** function. A function whose graph is symmetric with respect to the origin is an **odd** function. The graph of a (nonzero) function cannot be symmetric with respect to the x-axis.

Test for Even and Odd Functions

A function f is **even** if, for each x in the domain of f, $f(-x) = f(x)$.

A function f is **odd** if, for each x in the domain of f, $f(-x) = -f(x)$.

EXAMPLE 9 Testing for Evenness and Oddness

Figure 1.39 shows the $\bigvee$-shape of the **absolute value function,** $f(x) = |x|$. This was produced using the "abs" key and the standard viewing rectangle. Is the function even, odd, or neither?

Figure 1.39

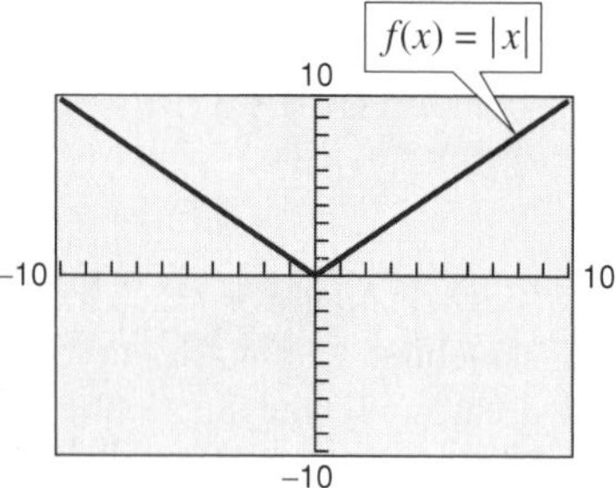

Solution

Because the graph is symmetric about the y-axis, it is an even function. You can verify this algebraically by observing that

$$\begin{aligned} f(-x) &= |-x| \\ &= |x| \\ &= f(x). \end{aligned}$$

EXAMPLE 10 Even and Odd Functions

Determine whether each function is even, odd, or neither.

a. $g(x) = x^3 - x$ **b.** $h(x) = x^2 + 1$ **c.** $f(x) = x^3 - 1$

The graphs of the three functions are shown in Figure 1.40.

Solution

a. This function is odd because

$$g(-x) = (-x)^3 - (-x) = -x^3 + x = -(x^3 - x) = -g(x).$$

b. This function is even because

$$h(-x) = (-x)^2 + 1 = x^2 + 1 = h(x).$$

c. Substituting $-x$ for x produces $f(-x) = (-x)^3 - 1 = -x^3 - 1$. Because $f(x) = x^3 - 1$ and $-f(x) = -x^3 + 1$, you can conclude that $f(-x) \neq f(x)$ and $f(-x) \neq -f(x)$. Hence, the function is neither even nor odd.

Figure 1.40

(a)

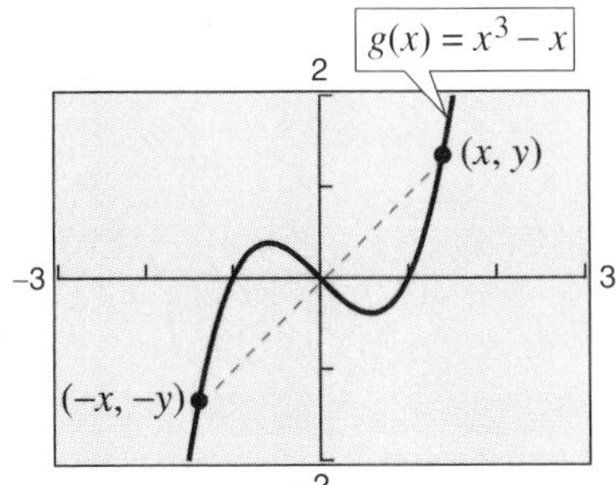

(b)

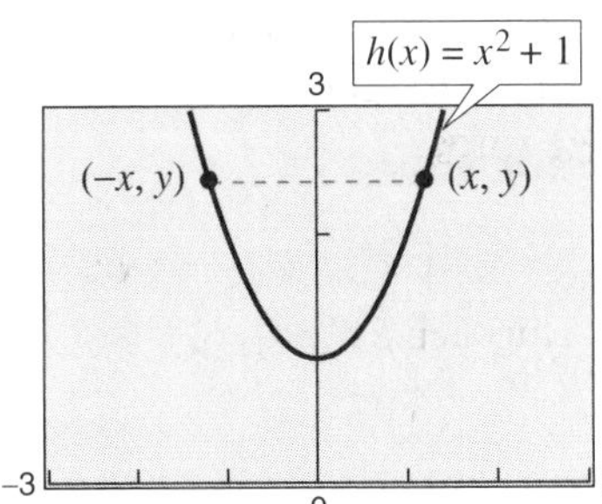

(c)

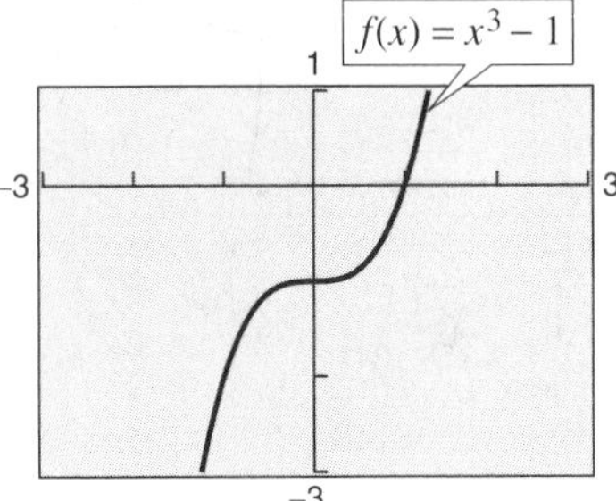

Group Activity

Increasing and Decreasing Functions

Describe three different functions that represent quantities between 1980 and 1995. Describe one that decreased during this time, one that increased, and one that was constant. For instance, the value of the dollar decreased, the cost of first-class postage increased, and the land size of the United States remained constant. Present your results graphically.

1.4 /// EXERCISES

In Exercises 1–4, find the domain and range of the function.

1. $f(x) = \sqrt{x^2 - 1}$

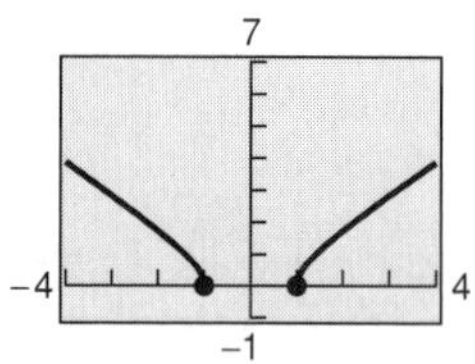

2. $f(x) = \frac{1}{2}|x - 2|$

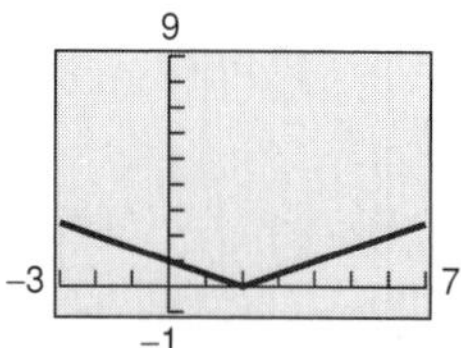

3. $h(x) = \sqrt{16 - x^2}$

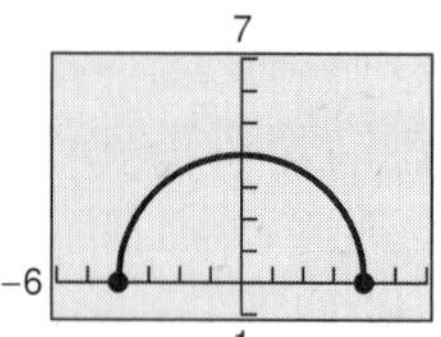

4. $g(x) = \dfrac{|x - 1|}{x - 1}$

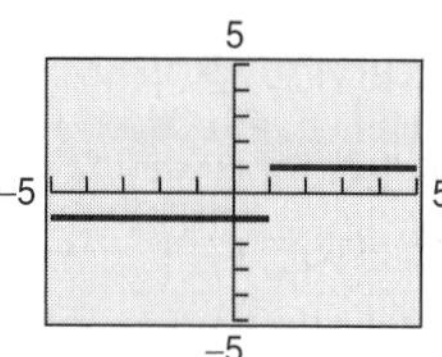

In Exercises 5–8, use a graphing utility to graph the function and find its domain and range.

5. $g(x) = 1 - x^2$

6. $f(x) = \sqrt{x - 1}$

7. $f(x) = |x + 3|$

8. $h(t) = \sqrt{4 - t^2}$

In Exercises 9–14, use the Vertical Line Test to determine if *y* is a function of *x*. Describe how you can use a graphing utility to produce the given graph.

9. $y = \frac{1}{2}x^2$

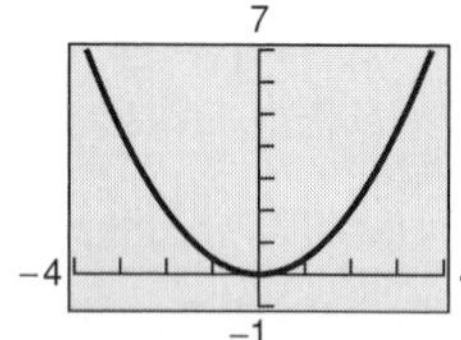

10. $y = \frac{1}{4}x^3$

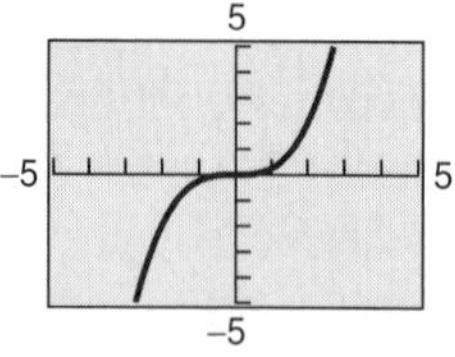

11. $x - y^2 = 1$

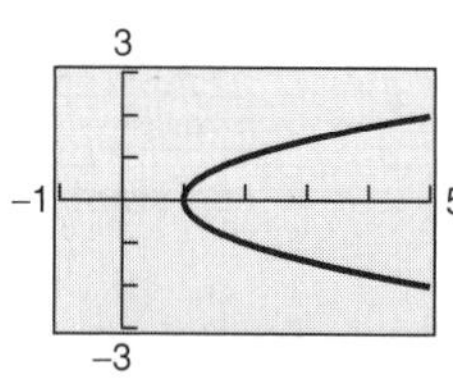

12. $x^2 + y^2 = 25$

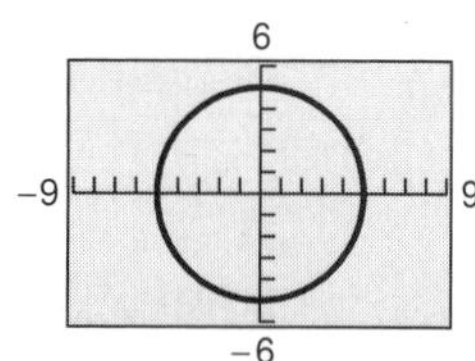

13. $x^2 = 2xy - 1$

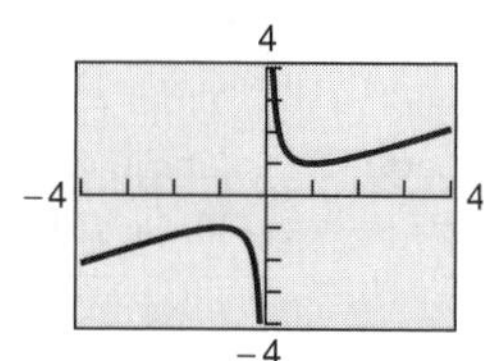

14. $x = |y + 2|$

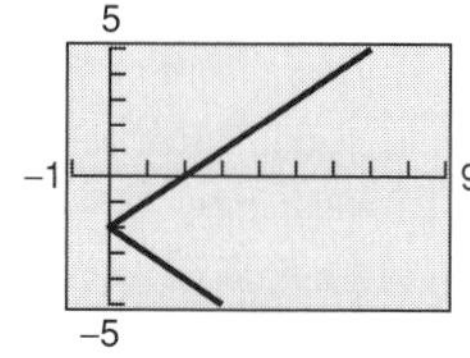

In Exercises 15–18, select the viewing rectangle on a graphing utility that shows the most complete graph of the function.

15. $f(x) = -0.2x^2 + 3x + 32$

Xmin = -2	Xmin = -10	Xmin = 0
Xmax = 20	Xmax = 30	Xmax = 10
Xscl = 1	Xscl = 5	Xscl = 0.5
Ymin = -10	Ymin = -5	Ymin = 0
Ymax = 30	Ymax = 50	Ymax = 200
Yscl = 4	Yscl = 5	Yscl = 25

16. $f(x) = 6[x - (0.1x)^5]$

Xmin = -500	Xmin = -25	Xmin = -20
Xmax = 5000	Xmax = 25	Xmax = 20
Xscl = 50	Xscl = 5	Xscl = 5
Ymin = -500	Ymin = -25	Ymin = -100
Ymax = 500	Ymax = 25	Ymax = 100
Yscl = 50	Yscl = 5	Yscl = 20

17. $f(x) = 4x^3 - x^4$

Xmin = -2	Xmin = -50	Xmin = 0
Xmax = 6	Xmax = 50	Xmax = 2
Xscl = 1	Xscl = 5	Xscl = 0.2
Ymin = -10	Ymin = -50	Ymin = -2
Ymax = 30	Ymax = 50	Ymax = 2
Yscl = 4	Yscl = 5	Yscl = 0.5

18. $f(x) = 10x\sqrt{400 - x^2}$

Xmin = -5	Xmin = -20	Xmin = -25
Xmax = 50	Xmax = 20	Xmax = 25
Xscl = 5	Xscl = 2	Xscl = 5
Ymin = -5000	Ymin = -500	Ymin = -2000
Ymax = 5000	Ymax = 500	Ymax = 2000
Yscl = 500	Yscl = 50	Yscl = 200

In Exercises 19–22, (a) determine the intervals over which the function is increasing, decreasing, or constant, and (b) determine if the function is even, odd, or neither.

19. $f(x) = \frac{3}{2}x$

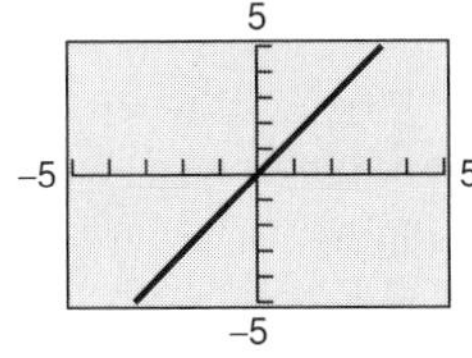

20. $f(x) = x^2 - 4x$

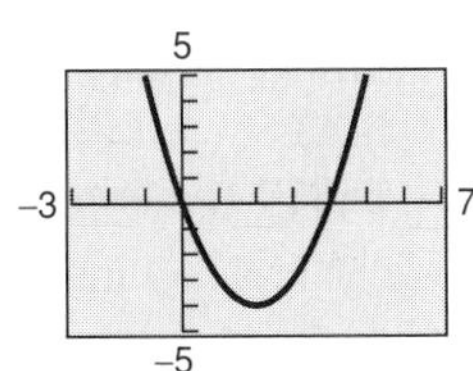

21. $f(x) = x^3 - 3x^2 + 2$

22. $f(x) = \sqrt{x^2 - 1}$

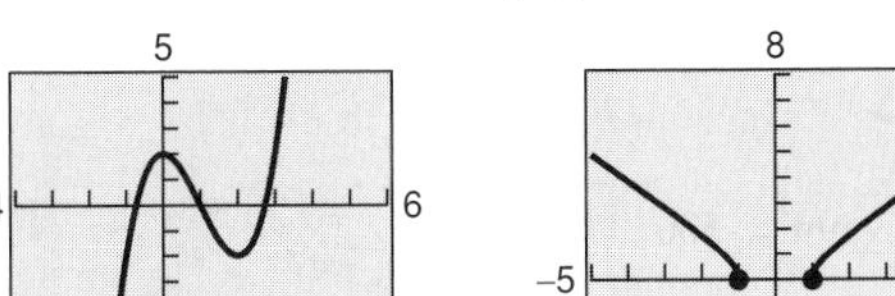

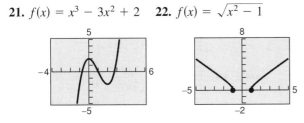

23. *Think About It* Does the graph in Exercise 11 represent x as a function of y? Explain.

24. *Think About It* Does the graph in Exercise 12 represent x as a function of y? Explain.

In Exercises 25–28, (a) use a graphing utility to graph the function, (b) determine the intervals over which the function is increasing, decreasing, or constant, and (c) determine if the function is even, odd, or neither.

25. $f(x) = 3x^4 - 6x^2$

26. $f(x) = x^{2/3}$

27. $f(x) = x\sqrt{x + 3}$

28. $f(x) = |x + 1| + |x - 1|$

In Exercises 29–34, determine if the function is even, odd, or neither.

29. $f(x) = x^6 - 2x^2 + 3$

30. $h(x) = x^3 - 5$

31. $g(x) = x^3 - 5x$

32. $f(x) = x\sqrt{1 - x^2}$

33. $f(t) = t^2 + 2t - 3$

34. $g(s) = 4s^{2/3}$

Think About It **In Exercises 35 and 36, find the coordinates of a second point on the graph of a function f if the given point is on the graph and the function is (a) even and (b) odd.**

35. $\left(-\frac{3}{2}, 4\right)$

36. $(4, 9)$

In Exercises 37–46, sketch the graph of the function and determine if the function is even, odd, or neither. Use a graphing utility to verify your sketch.

37. $f(x) = 3$

38. $g(x) = x$

39. $f(x) = 5 - 3x$

40. $h(x) = x^2 - 4$

41. $f(x) = \sqrt{1 - x}$

42. $f(x) = x^{3/2}$

43. $g(t) = \sqrt[3]{t - 1}$

44. $f(x) = |x + 2|$

45. $f(x) = \begin{cases} x + 3, & x \le 0 \\ 3, & 0 < x \le 2 \\ 2x - 1, & x > 2 \end{cases}$

46. $f(x) = \begin{cases} 2x + 1, & x \le -1 \\ x^2 - 2, & x > -1 \end{cases}$

In Exercises 47–50, use a graphing utility to graph the piecewise-defined function.

47. $f(x) = \begin{cases} 2x + 3, & x < 0 \\ 3 - x, & x \ge 0 \end{cases}$

48. $f(x) = \begin{cases} \sqrt{4 + x}, & x < 0 \\ \sqrt{4 - x}, & x \ge 0 \end{cases}$

49. $f(x) = \begin{cases} x^2 + 5, & x \le 1 \\ -x^2 + 4x + 3, & x > 1 \end{cases}$

50. $f(x) = \begin{cases} 1 - (x - 1)^2, & x \le 2 \\ \sqrt{x - 2}, & x > 2 \end{cases}$

In Exercises 51–56, use a graphing utility to approximate (to two-decimal-place accuracy) any relative minimum or maximum values of the function.

51. $f(x) = x^2 - 6x$

52. $f(x) = (x - 1)^2(x + 2)$

53. $y = 2x^3 + 3x^2 - 12x$

54. $y = x^3 - 6x^2 + 15$

55. $h(x) = (x - 1)\sqrt{x}$

56. $g(x) = x\sqrt{4 - x}$

57. *Maximum Area* The perimeter of a rectangle (see figure) is 100 meters.

(a) Show that the area of the rectangle is given by $A = x(50 - x)$, where x is its length.

(b) Use a graphing utility to graph the area function.

(c) Use a graphing utility to approximate the maximum area of the rectangle and the dimensions that yield the maximum area.

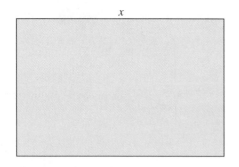

58. *Maximum Profit* The marketing department of a company estimates that the demand for a product is given by

$$p = 100 - 0.0001x$$

where p is the price per unit and x is the number of units. The cost of producing x units is given by

$$C = 350{,}000 + 30x$$

and the profit for producing and selling x units is given by

$$P = R - C = xp - C.$$

Use a graphing utility to graph the profit function and estimate the number of units that would produce a maximum profit.

In Exercises 59 and 60, use a graphing utility to graph the function. State the domain and range of the function. Describe the pattern of the graph.

59. $s(x) = 2\left(\frac{1}{4}x - [\![\frac{1}{4}x]\!]\right)$

60. $g(x) = 2\left(\frac{1}{4}x - [\![\frac{1}{4}x]\!]\right)^2$

61. *Cost of a Telephone Call* The cost of a telephone call between two cities is \$0.65 for the first minute and \$0.40 for each additional minute.

(a) It is required that a model be created for the cost C of a telephone call between the two cities lasting t minutes. Which of the following is the appropriate model? Explain.

$$C_1(t) = 0.65 + 0.4[\![t - 1]\!]$$
$$C_2(t) = 0.65 - 0.4[\![-(t - 1)]\!]$$

(b) Use a graphing utility to graph the appropriate model and use the zoom and trace features to determine the cost of an 18-minute, 45-second call.

62. *Cost of Overnight Delivery* Suppose that the cost of sending an overnight package from New York to Atlanta is \$9.80 for under one pound and \$2.50 for each additional pound. Use the greatest integer function to create a model for the cost C of overnight delivery of a package weighing x pounds where $x > 0$. Sketch the graph of the function.

In Exercises 63–70, graph the function and determine the interval(s) (if any) on the real axis for which $f(x) \geq 0$. Use a graphing utility to verify your results.

63. $f(x) = 4 - x$

64. $f(x) = 4x + 2$

65. $f(x) = x^2 - 9$

66. $f(x) = x^2 - 4x$

67. $f(x) = 1 - x^4$

68. $f(x) = \sqrt{x + 2}$

69. $f(x) = x^2 + 1$

70. $f(x) = -(1 + |x|)$

In Exercises 71–74, write the height h of the rectangle as a function of x.

71.

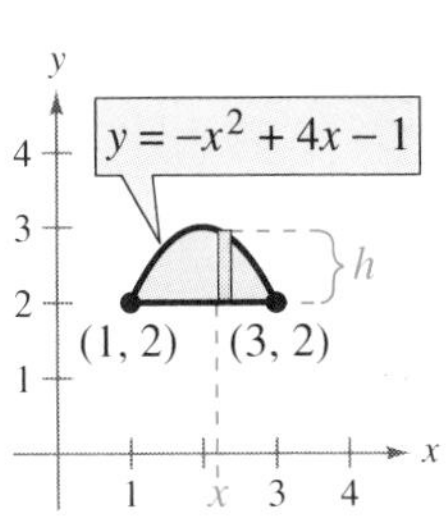

72.

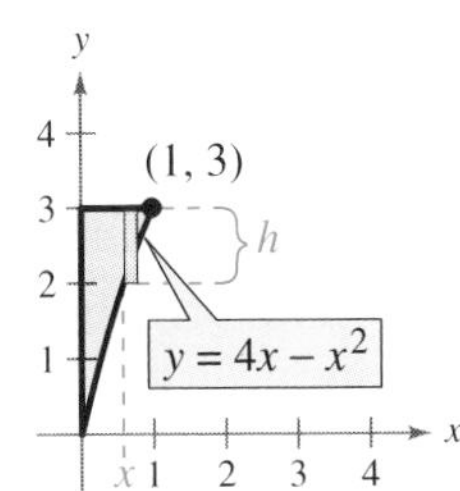

73.

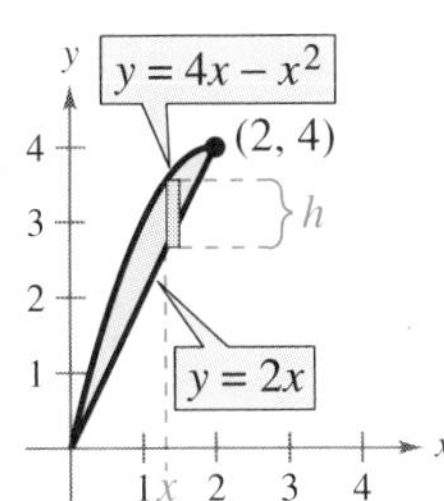

74.

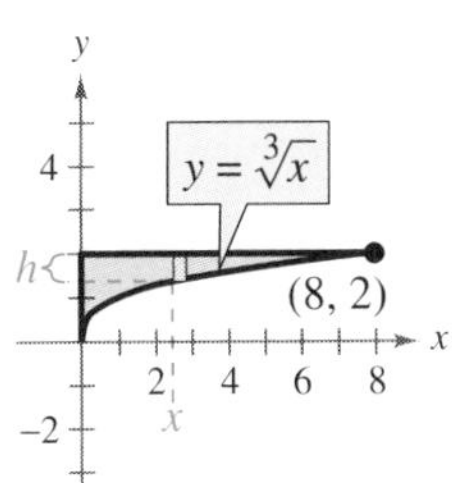

In Exercises 75 and 76, write the length L of the rectangle as a function of y.

75.

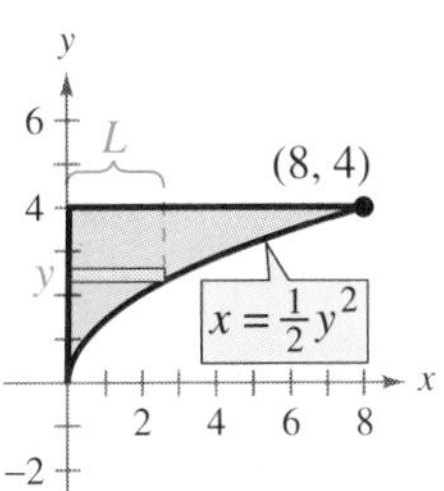

76.

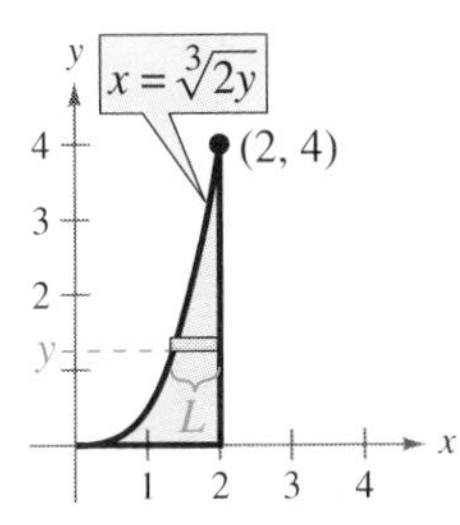

77. *Data Analysis* The amounts y of the merchandise trade balance of the United States in billions of dollars for the years 1986 through 1993 were as follows: 1986(−152.7), 1987(−152.1), 1988(−118.6), 1989 (−109.6), 1990(−101.7), 1991(−65.4), 1992 (−85.4), 1993(−115.8). (Source: U.S. Bureau of the Census)

(a) Let t be the time in years, with $t = 0$ corresponding to 1990. Use the regression capabilities of a graphing utility to find a cubic model for the data. What is the domain of the model?

(b) Use a graphing utility to graph the data and model.

(c) For which year does the model most accurately estimate the actual data? During which year is it least accurate?

(d) Why would economists be concerned if this model remained valid in the future?

78. *Fluid Flow* The intake pipe of a 100-gallon tank has a flow rate of 10 gallons per minute, and two drain pipes have a flow rate of 5 gallons per minute each. The figure shows the volume V of fluid in the tank as a function of time t. Determine the pipes in which the fluid is flowing in specific subintervals of the 1 hour of time shown on the graph. (There is more than one correct answer.)

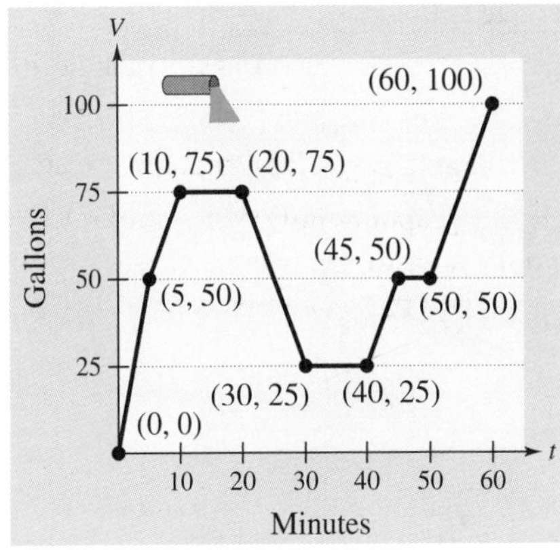

79. Prove that a function of the following form is odd.

$$y = a_{2n+1}x^{2n+1} + a_{2n-1}x^{2n-1} + \cdots + a_3x^3 + a_1x$$

80. Prove that a function of the following form is even.

$$y = a_{2n}x^{2n} + a_{2n-2}x^{2n-2} + \cdots + a_2x^2 + a_0$$

1.5 Shifting, Reflecting, and Stretching Graphs

Summary of Graphs of Common Functions / Vertical and Horizontal Shifts / Reflections / Nonrigid Transformations

Summary of Graphs of Common Functions

One of the goals of this text is to enable you to build your intuition for the basic shapes of the graphs of different types of functions. For instance, from your study of lines in Section 1.2, you can determine the basic shape of the graph of the linear function

$$f(x) = ax + b.$$

Specifically, you know that the graph of this function is a line whose slope is a and whose y-intercept is b.

The six graphs shown in Figure 1.41 represent the most commonly used functions in algebra. Familiarity with the basic characteristics of these simple graphs will help you analyze the shapes of more complicated graphs. Try using a graphing utility to verify these graphs.

Figure 1.41

(a) Constant Function

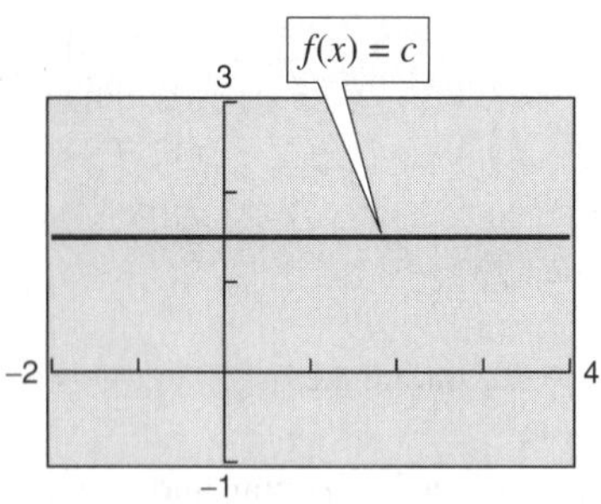

(b) Identity Function

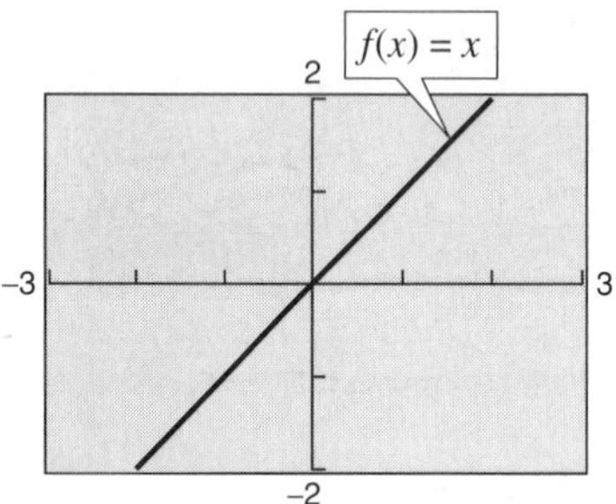

(c) Absolute Value Function

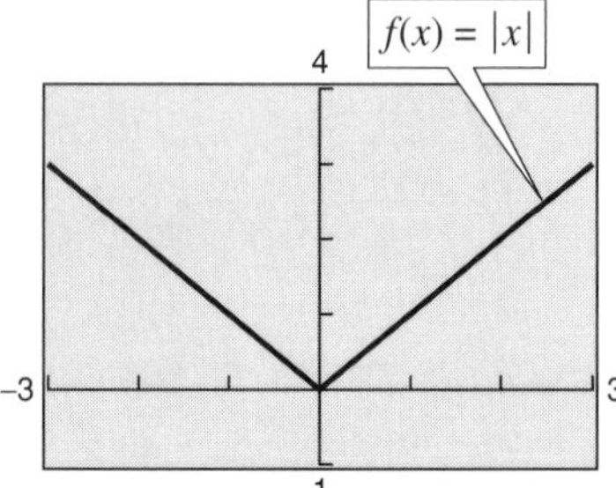

(d) Square Root Function

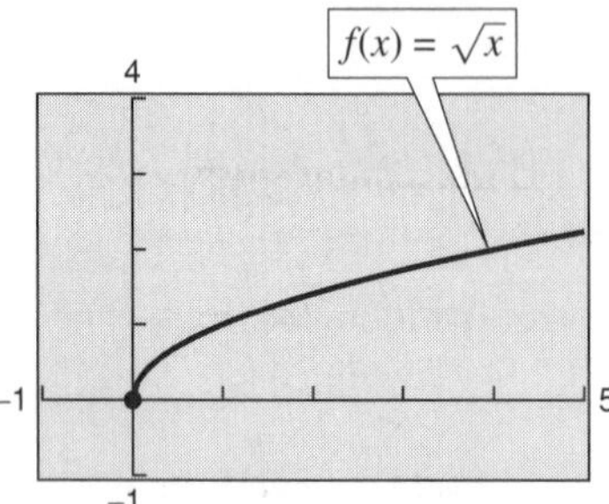

(e) Squaring Function

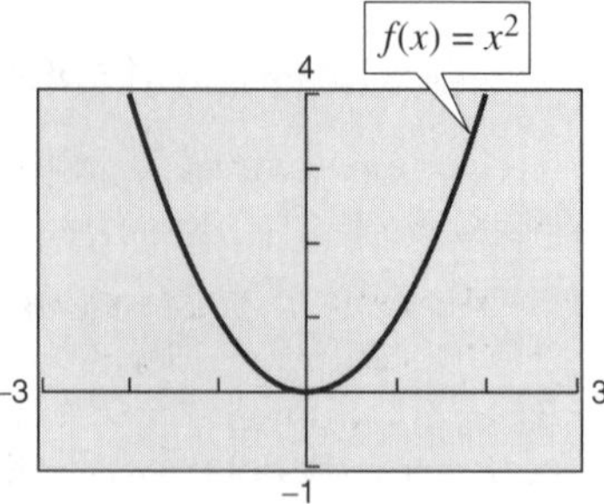

(f) Cubing Function

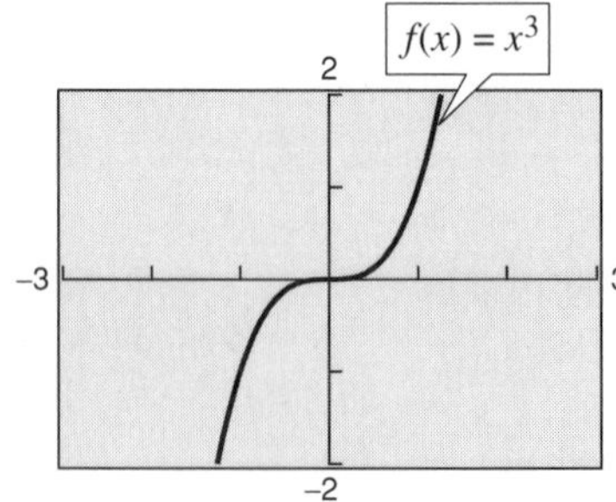

Vertical and Horizontal Shifts

 A computer animation of this concept appears in the *Interactive* CD-ROM.

Many functions have graphs that are simple transformations of the common graphs summarized in Figure 1.41. For example, you can obtain the graph of

$$h(x) = x^2 + 2$$

by shifting the graph of $f(x) = x^2$ *up* two units, as shown in Figure 1.42. In function notation, h and f are related as follows.

$$\begin{aligned} h(x) &= x^2 + 2 \\ &= f(x) + 2 \end{aligned} \qquad \text{Upward shift of 2}$$

Similarly, you can obtain the graph of

$$g(x) = (x - 2)^2$$

by shifting the graph of $f(x) = x^2$ to the *right* two units, as shown in Figure 1.43. In this case, the functions g and f have the following relationship.

$$\begin{aligned} g(x) &= (x - 2)^2 \\ &= f(x - 2) \end{aligned} \qquad \text{Right shift of 2}$$

> **EXPLORATION**
>
> Use a graphing utility to display the graphs of $y = x^2 + c$ where $c = -2, 0, 2,$ and 4. Use the result to describe the effect that c has on the graph.
>
> Use a graphing utility to display the graphs of $y = (x + c)^2$ where $c = -2, 0, 2,$ and 4. Use the result to describe the effect that c has on the graph.

Figure 1.42

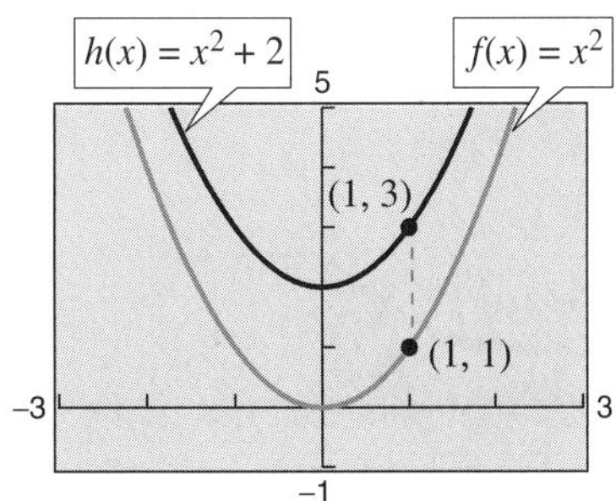

Vertical shift upward: two units

Figure 1.43

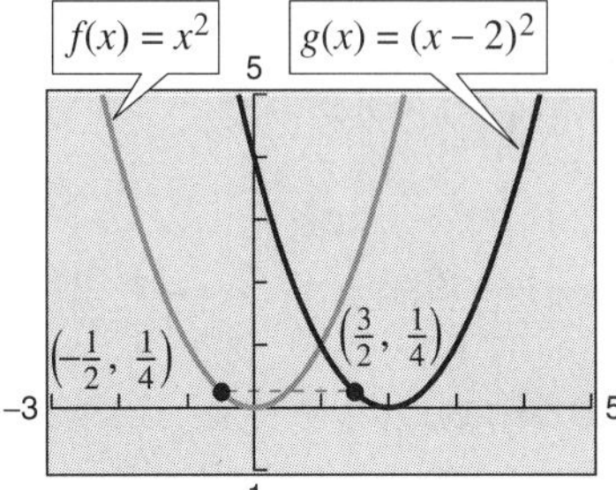

Horizontal shift to the right: two units

The following list summarizes this discussion about horizontal and vertical shifts.

> **Vertical and Horizontal Shifts**
>
> Let c be a positive real number. **Vertical and horizontal shifts** in the graph of $y = f(x)$ are represented as follows.
>
> 1. Vertical shift c units **upward:** $h(x) = f(x) + c$
> 2. Vertical shift c units **downward:** $h(x) = f(x) - c$
> 3. Horizontal shift c units to the **right:** $h(x) = f(x - c)$
> 4. Horizontal shift c units to the **left:** $h(x) = f(x + c)$

EXAMPLE 1 Shifts in the Graph of a Function

Compare the graph of each function with the graph of $f(x) = x^3$.

a. $g(x) = x^3 + 1$ **b.** $h(x) = (x - 1)^3$ **c.** $k(x) = (x + 2)^3 + 1$

Solution

a. Begin by graphing $f(x) = x^3$ and $g(x) = x^3 + 1$ in the same viewing rectangle, as shown in Figure 1.44(a). From the graph, you can see that you can obtain the graph of g by shifting the graph of f one unit up.

b. Begin by graphing $f(x) = x^3$ and $h(x) = (x - 1)^3$ in the same viewing rectangle, as shown in Figure 1.44(b). From the graph, you can see that you can obtain the graph of h by shifting the graph of f one unit to the right.

c. Begin by graphing $f(x) = x^3$ and $k(x) = (x + 2)^3 + 1$ in the same viewing rectangle, as shown in Figure 1.44(c). From the graph, you can see that you can obtain the graph of k by shifting the graph of f two units to the left and then one unit up.

Figure 1.44

(a) Vertical shift: one unit up

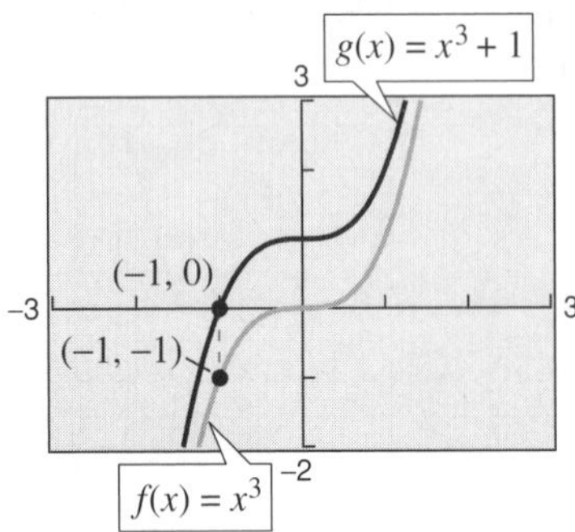

(b) Horizontal shift: one unit right

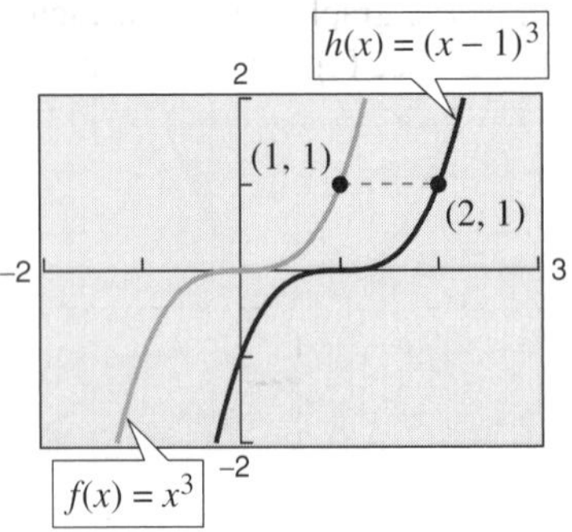

(c) Two units left and one unit up

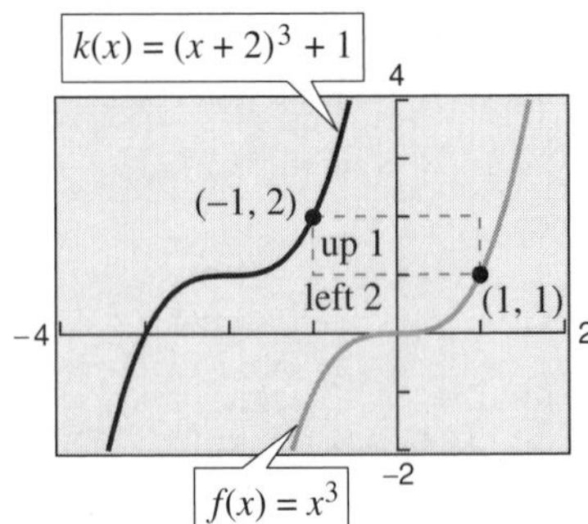

Figure 1.45

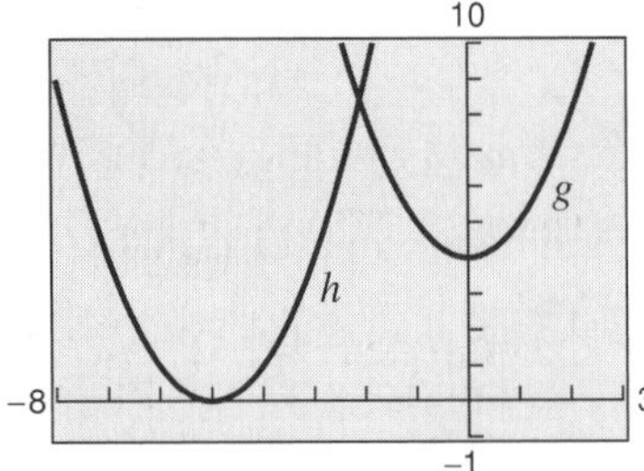

EXAMPLE 2 Finding Equations from Graphs

Each of the graphs shown in Figure 1.45 is a vertical or horizontal shift of the graph of $f(x) = x^2$. Find an equation for each function.

Solution

a. The graph of g is a vertical shift of four units upward of the graph of $f(x) = x^2$. Thus, the equation for g is $g(x) = x^2 + 4$.

b. The graph of h is a horizontal shift of five units to the left of the graph of $f(x) = x^2$. Thus, the equation for h is $h(x) = (x + 5)^2$.

Figure 1.46

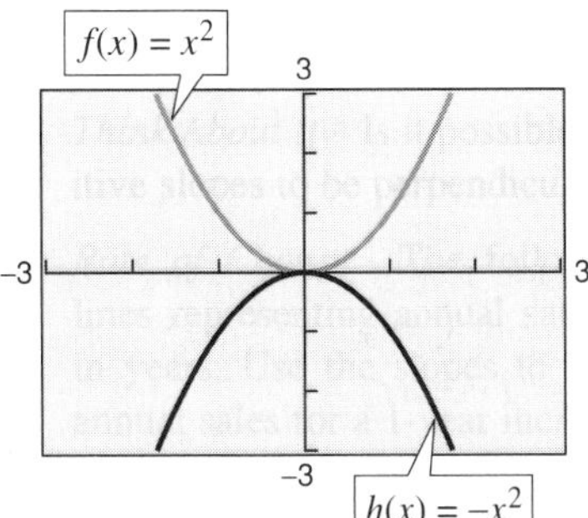

A computer animation of this concept appears in the *Interactive* CD-ROM.

Reflections

The second common type of transformation is called a **reflection.** For instance, if you consider the x-axis to be a mirror, the graph of

$$h(x) = -x^2$$

is the mirror image (or reflection) of the graph of $f(x) = x^2$, as shown in Figure 1.46.

Reflections in the Coordinate Axes

Reflections in the coordinate axes of the graph of $y = f(x)$ are represented as follows.

1. Reflection in the x-axis: $h(x) = -f(x)$
2. Reflection in the y-axis: $h(x) = f(-x)$

EXAMPLE 3 Finding Equations from Graphs

Each of the graphs shown in Figure 1.48 is a transformation of the graph of $f(x) = x^4$ (see Figure 1.47). Find an equation for each function.

Figure 1.47

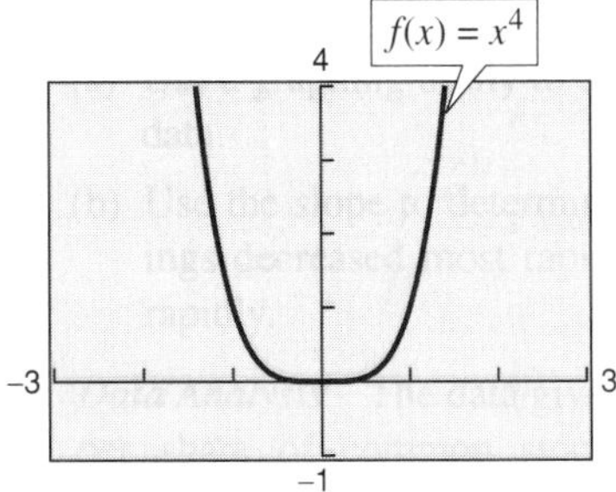

Figure 1.48

(a)

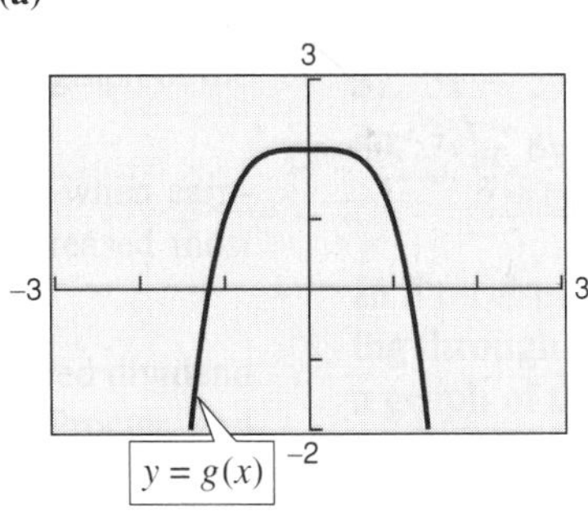

(b)

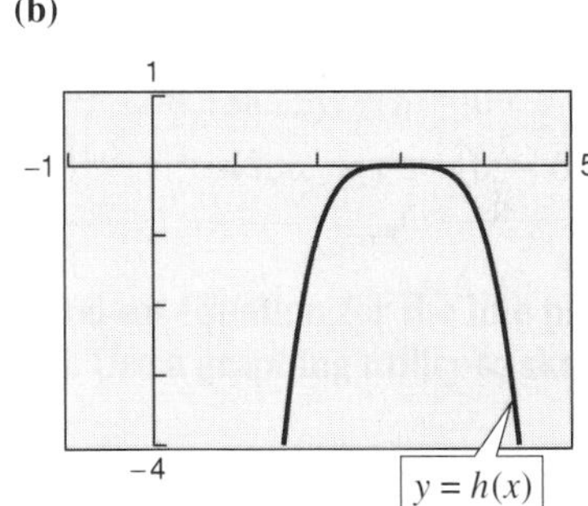

Solution

a. The graph of g is a reflection in the x-axis *followed* by an upward shift of two units of the graph of $f(x) = x^4$. Thus, the equation for g is $g(x) = -x^4 + 2$.

b. The graph of h is a horizontal shift of three units to the right *followed* by a reflection in the x-axis of the graph of $f(x) = x^4$. Thus, the equation for h is $h(x) = -(x - 3)^4$.

EXAMPLE 4 Reflections and Shifts

Compare the graph of each function with the graph of $f(x) = \sqrt{x}$.

a. $g(x) = -\sqrt{x}$ **b.** $h(x) = \sqrt{-x}$ **c.** $k(x) = -\sqrt{x+2}$

Solution

a. Relative to the graph of $f(x) = \sqrt{x}$, the graph of g is a reflection in the x-axis because

$$g(x) = -\sqrt{x} = -f(x).$$

b. The graph of h is a reflection of the graph of $f(x) = \sqrt{x}$ in the y-axis because

$$h(x) = \sqrt{-x} = f(-x).$$

c. From the equation

$$k(x) = -\sqrt{x+2} = -f(x+2)$$

you can conclude that the graph of k is a left shift of two units, followed by a reflection in the x-axis.

The graphs of all three functions are shown in Figure 1.49.

Figure 1.49

(a)

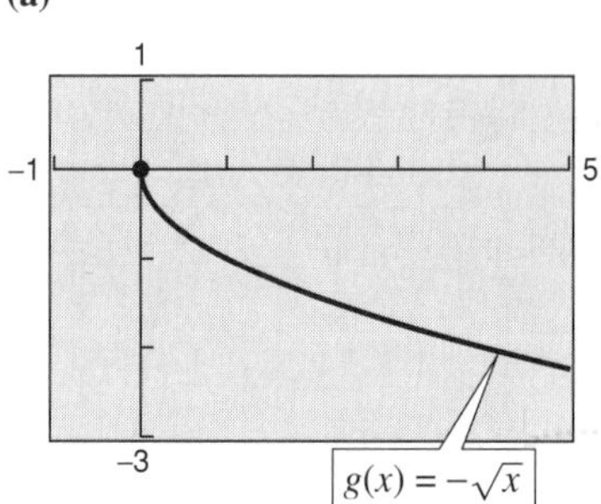

(b)

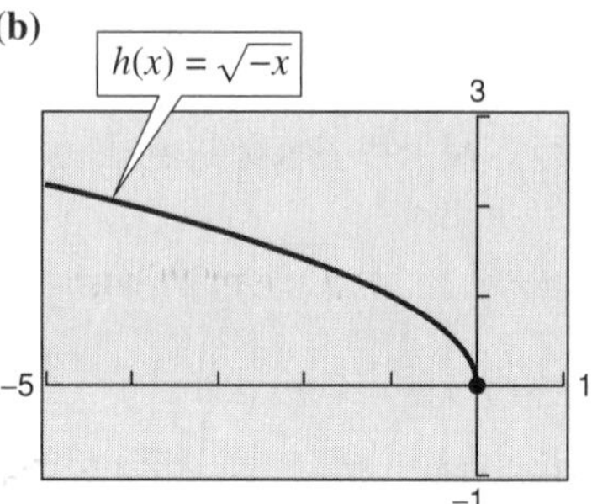

(c)

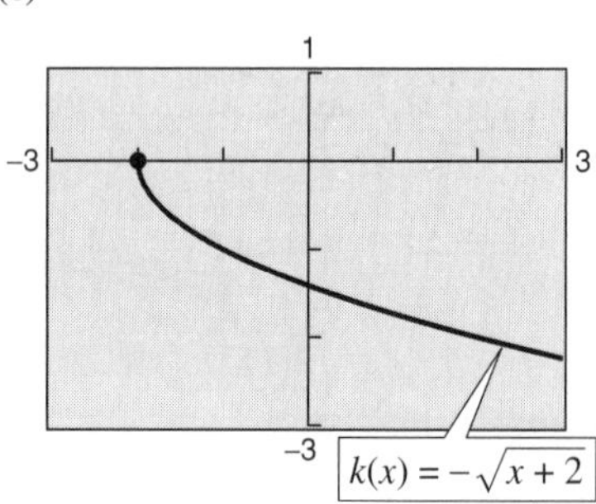

Note When sketching the graph of a function involving square roots, remember that the domain must be restricted to exclude negative numbers inside the radical. For instance, here are the domains of the functions in Example 4.

Domain of $g(x) = -\sqrt{x}$: $0 \le x$
Domain of $h(x) = \sqrt{-x}$: $0 \ge x$
Domain of $k(x) = -\sqrt{x+2}$: $-2 \le x$

Nonrigid Transformations

Horizontal shifts, vertical shifts, and reflections are called **rigid** transformations because the basic shape of the graph is unchanged. These transformations change only the *position* of the graph in the xy-plane. **Nonrigid** transformations are those that cause a *distortion*—a change in the shape of the original graph. For instance, a nonrigid transformation of the graph of $y = f(x)$ is represented by $y = cf(x)$, where the transformation is a **vertical stretch** if $c > 1$ and a **vertical shrink** if $0 < c < 1$.

Figure 1.50

(a)

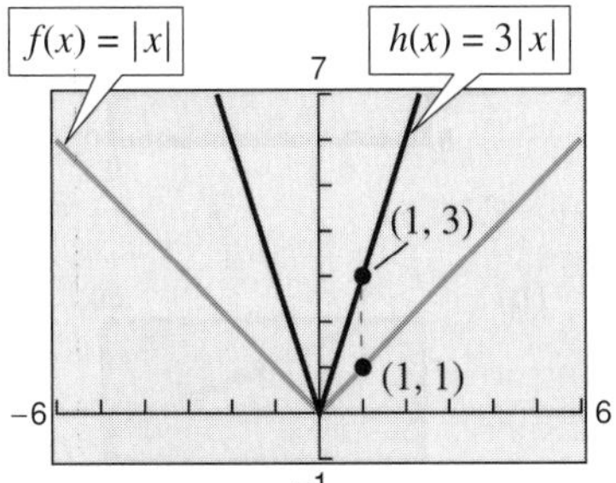

(b)

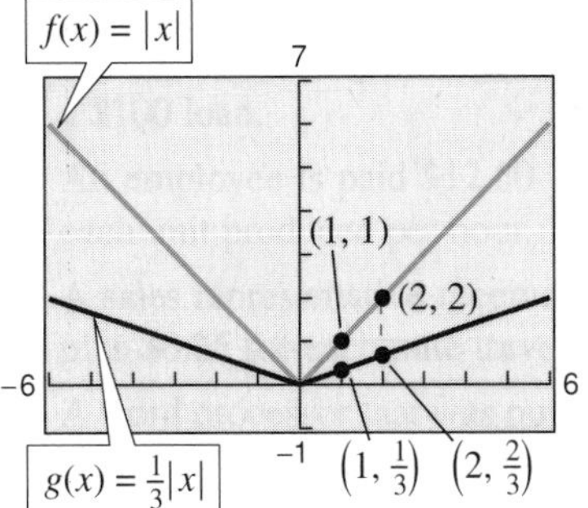

EXAMPLE 5 **Nonrigid Transformations**

Compare the graph of each function with the graph of $f(x) = |x|$.

a. $h(x) = 3|x|$ **b.** $g(x) = \dfrac{1}{3}|x|$

Solution

a. Relative to the graph of $f(x) = |x|$, the graph of

$$h(x) = 3|x| = 3f(x)$$

is a vertical stretch (multiply each y-value by 3) of the graph of f.

b. Similarly, the equation

$$g(x) = \frac{1}{3}|x| = \frac{1}{3}f(x)$$

indicates that the graph of g is a vertical shrink (multiply each y-value by $\frac{1}{3}$) of the graph of f.

The graphs of all three functions are shown in Figure 1.50.

Figure 1.51

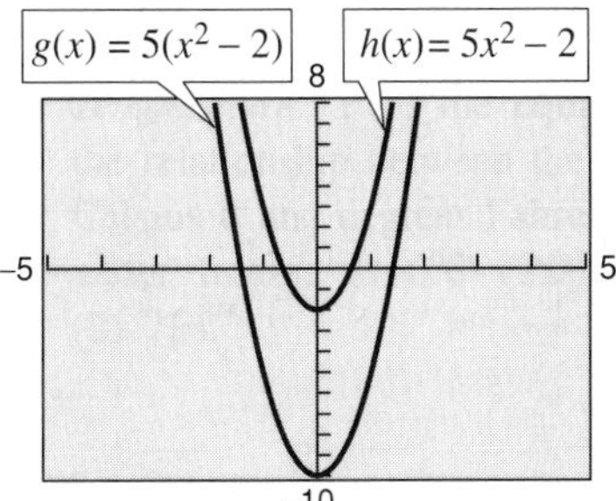

EXAMPLE 6 **Sequences of Transformations**

Use a graphing utility to graph the two functions

$$g(x) = 5(x^2 - 2) \qquad \text{and} \qquad h(x) = 5x^2 - 2$$

in the same viewing rectangle. Describe how each function was obtained from $f(x) = x^2$ as a sequence of shifts and stretches.

Solution

Notice that the two graphs in Figure 1.51 are different. The graph of g is a downward shift of two units followed by a vertical stretch, whereas h is a vertical stretch followed by a downward shift of two units. Hence, the order of applying the transformations is important.

Group Activity — *A Program for Practice*

If you have a programmable calculator, try entering the following program. (These program steps are for a Texas Instruments *TI-82* or *TI-83*. Programs for other graphing calculator models may be found in the appendix.)

```
PROGRAM:PARABOLA
:-6+int (12rand)→H
:-3+int (6rand)→V
:rand→R
:If R<.5
:Then
:-1→R
:Else
:1→R
:End
:"R(X+H)²+V"→Y1
:-9→Xmin
:9→Xmax
:1→Xscl
:-6→Ymin
:6→Ymax
:1→Yscl
:DispGraph
:Pause
:Disp "Y = R(X+H)²+V"
:Disp "R = ",R
:Disp "H = ",H
:Disp "V = ",V
:Pause
```

This program will sketch a graph of the function

$$y = R(x + H)^2 + V$$

where $R = \pm 1$, H is an integer between -6 and 6, and V is an integer between -3 and 3. (Each time you run the program, different values of R, H, and V are possible.) From the graph, you should be able to determine the values of R, H, and V. After you have determined the values, press ENTER to see the answer. (To look at the graph again, press GRAPH.)

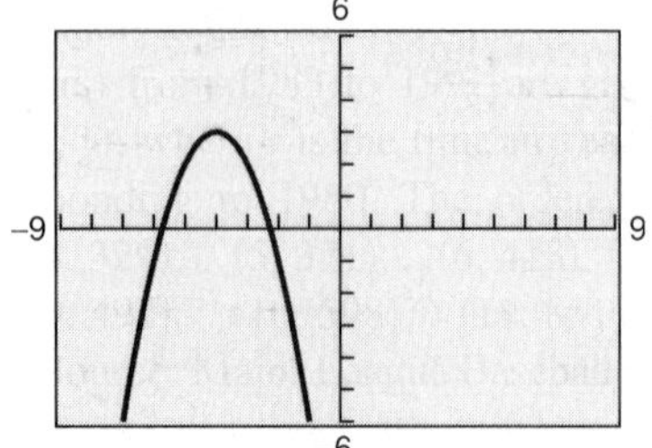

For example, in the graph shown at the left, you can make the following conclusions. Because the graph of $f(x) = x^2$ has been reflected about the x-axis, you know that $R = -1$. Because the graph of $f(x) = x^2$ has been shifted four units to the left, you know that $H = 4$. Because the graph of $f(x) = x^2$ has been shifted three units up, you know that $V = 3$. Thus, the equation of the graph shown at the left must be

$$y = -(x + 4)^2 + 3.$$

Try running this program several times. It will give you practice in working with reflections, horizontal shifts, and vertical shifts.

1.5 /// EXERCISES

In Exercises 1–10, sketch the graphs of the three functions *by hand* on the same rectangular coordinate system. Verify your result with a graphing utility.

1. $f(x) = x$
$g(x) = x - 4$
$h(x) = 3x$

2. $f(x) = \frac{1}{2}x$
$g(x) = \frac{1}{2}x + 2$
$h(x) = \frac{1}{2}(x - 2)$

3. $f(x) = x^2$
$g(x) = x^2 + 2$
$h(x) = (x - 2)^2$

4. $f(x) = x^2$
$g(x) = x^2 - 4$
$h(x) = (x + 2)^2 + 1$

5. $f(x) = -x^2$
$g(x) = -x^2 + 1$
$h(x) = -(x - 2)^2$

6. $f(x) = (x - 2)^2$
$g(x) = (x - 2)^2 + 2$
$h(x) = -(x - 2)^2 + 4$

7. $f(x) = x^2$
$g(x) = \left(\frac{1}{2}x\right)^2$
$h(x) = (2x)^2$

8. $f(x) = x^2$
$g(x) = \left(\frac{1}{4}x\right)^2 + 2$
$h(x) = -\left(\frac{1}{4}x\right)^2$

9. $f(x) = |x|$
$g(x) = |x| - 1$
$h(x) = |x - 3|$

10. $f(x) = \sqrt{x}$
$g(x) = \sqrt{x + 1}$
$h(x) = \sqrt{x - 2} + 1$

11. Use the graph of f to sketch the graphs.
(a) $y = f(x) + 2$ (b) $y = -f(x)$
(c) $y = f(x - 2)$ (d) $y = f(x + 3)$
(e) $y = f(2x)$ (f) $y = f(-x)$

Figure for 11

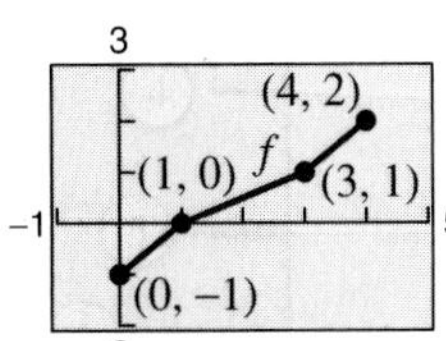

Figure for 12

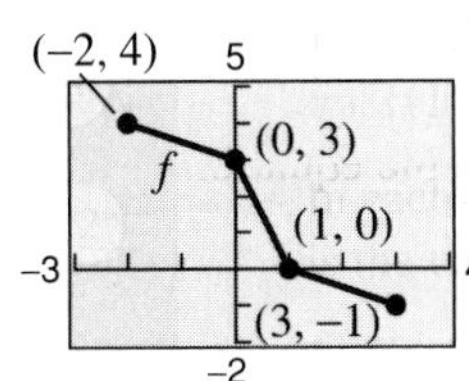

12. Use the graph of f to sketch the graphs.
(a) $y = f(x) - 1$ (b) $y = f(x + 1)$
(c) $y = f(x - 1)$ (d) $y = -f(x - 2)$
(e) $y = f(-x)$ (f) $y = \frac{1}{2}f(x)$

In Exercises 13–24, identify the common function and the transformation shown in the graph. Write the formula for the graphed function.

13.

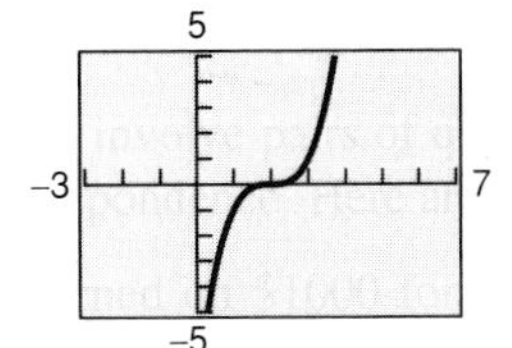

14.

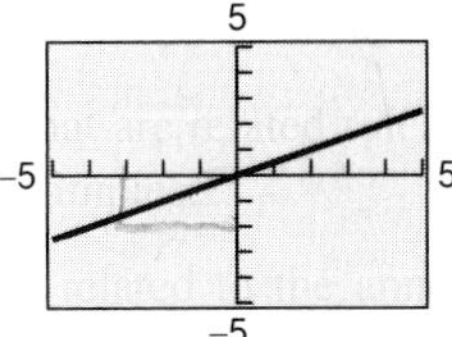

15.

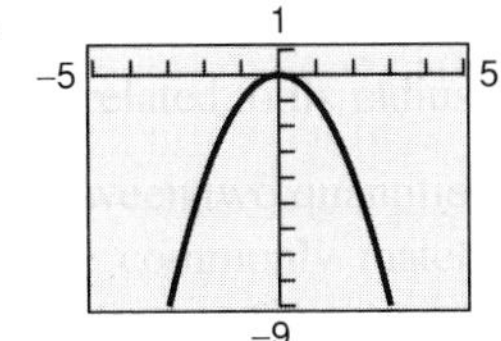

16.

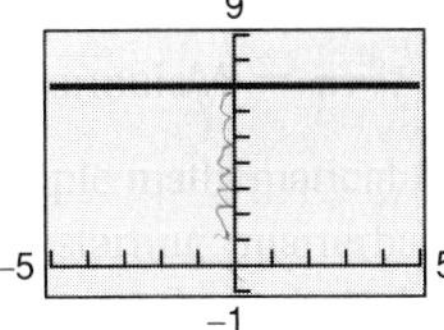

17.

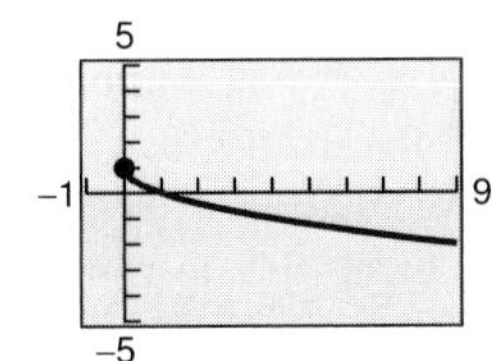

18.

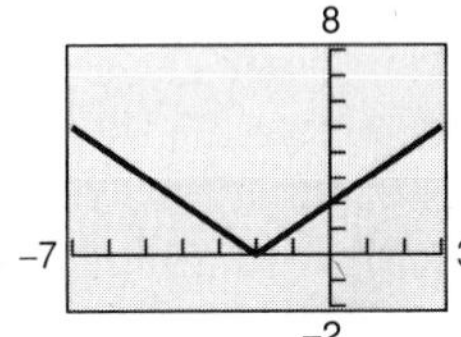

19.

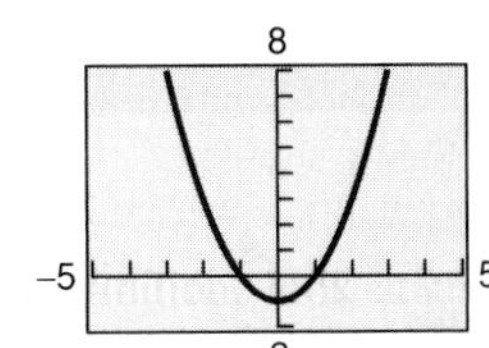

20.

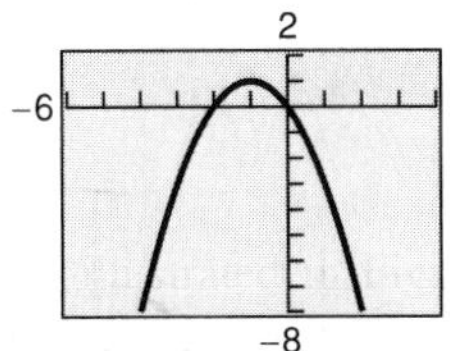

21.

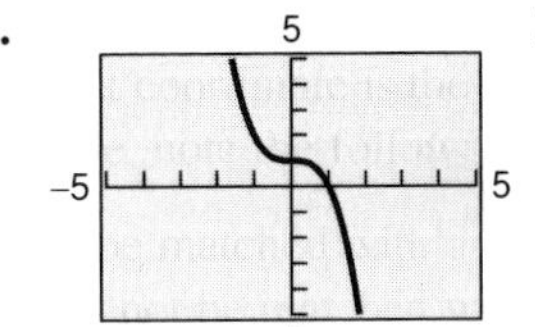

22.

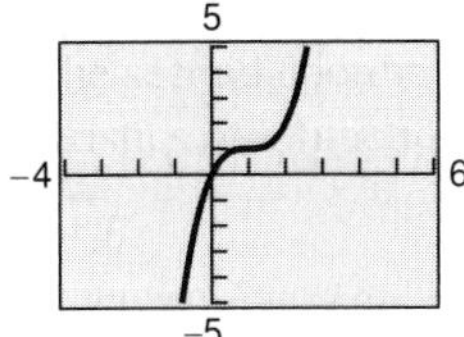

23.

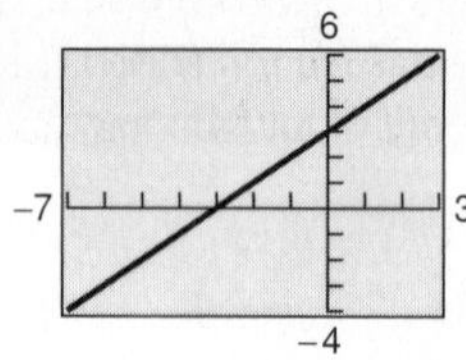

24.

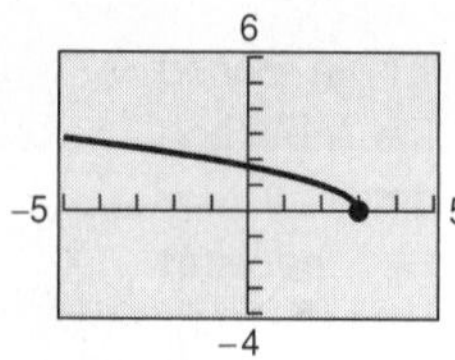

In Exercises 25–30, compare the graph of the function with the graph of $f(x) = \sqrt{x}$.

25. $y = \sqrt{x} + 2$

26. $y = -\sqrt{x}$

27. $y = \sqrt{x - 2}$

28. $y = \sqrt{x + 3}$

29. $y = \sqrt{2x}$

30. $y = \sqrt{-x}$

In Exercises 31–36, compare the graph of the function with the graph of $f(x) = |x|$.

31. $y = |x + 2|$

32. $y = |x| - 3$

33. $y = -|x|$

34. $y = |-x|$

35. $y = \frac{1}{3}|x|$

36. $y = \left|\frac{1}{2}x\right|$

In Exercises 37–40, use a graphing utility to graph the three functions in the same viewing rectangle. Describe the graphs of g and h relative to the graph of f.

37. $f(x) = x^3 - 3x^2$
$g(x) = f(x + 2)$
$h(x) = f\left(\frac{1}{2}x\right)$

38. $f(x) = x^3 - 3x^2 + 2$
$g(x) = f(x - 1)$
$h(x) = f(2x)$

39. $f(x) = x^3 - 3x^2$
$g(x) = -\frac{1}{3}f(x)$
$h(x) = f(-x)$

40. $f(x) = x^3 - 3x^2 + 2$
$g(x) = -f(x)$
$h(x) = f(-x)$

In Exercises 41–46, specify the sequence of transformations that will yield the graph of the given function from the graph of the function $f(x) = x^3$.

41. $g(x) = 4 - x^3$

42. $g(x) = -(x - 4)^3$

43. $h(x) = \frac{1}{4}(x + 2)^3$

44. $h(x) = -2(x - 1)^3 + 3$

45. $p(x) = \left(\frac{1}{3}x\right)^3 + 2$

46. $p(x) = [3(x - 2)]^3$

In Exercises 47 and 48, use the graph of $f(x) = x^3 - 3x^2$ (see Exercise 37) to write a formula for the function g shown in the graph.

47.

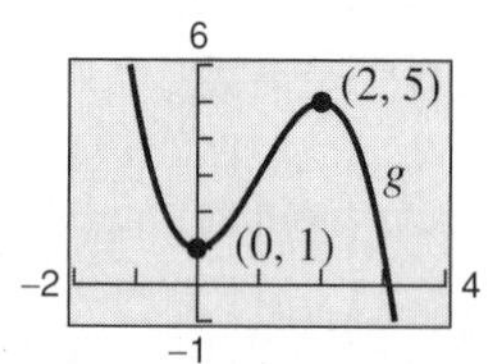

48.

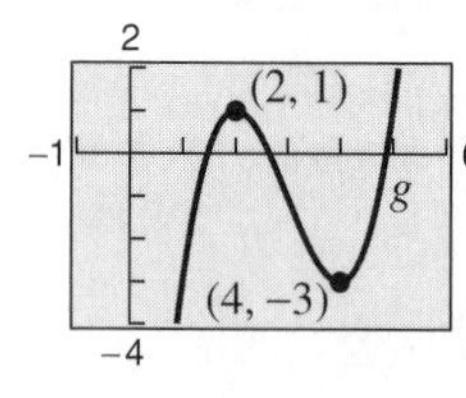

49. *Profit* The profit P per week on a certain product is given by the model

$$P(x) = 80 + 20x - 0.5x^2, \qquad 0 \le x \le 20$$

where x is the amount spent on advertising. In this model, x and P are both measured in hundreds of dollars.

(a) Use a graphing utility to graph the profit function.

(b) The business estimates that taxes and operating costs will increase by an average of $2500 per week during the next year. Rewrite the profit equation to reflect this expected decrease in profits. Identify the type of transformation applied to the graph of the equation.

(c) Rewrite the profit equation so that x measures advertising expenditures in dollars. [Find $P\left(\frac{x}{100}\right)$.] Identify the type of transformation applied to the graph of the profit function.

50. *Automobile Aerodynamics* The number of horsepower H required to overcome wind drag on a certain automobile is approximated by

$$H(x) = 0.002x^2 + 0.005x - 0.029, \qquad 10 \le x \le 100$$

where x is the speed of the car in miles per hour.

(a) Use a graphing utility to graph the power function.

(b) Rewrite the power function so that x represents the speed in kilometers per hour. [Find $H(1.6x)$.] Identify the type of transformation applied to the graph of the power function.

51. *Exploration* Use a graphing utility to graph each function. Describe any similarities and differences you observe among the graphs.

(a) $y = x$ (b) $y = x^2$

(c) $y = x^3$ (d) $y = x^4$

(e) $y = x^5$ (f) $y = x^6$

52. *Conjecture* Use the results of Exercise 51 to make a conjecture about the shapes of the graphs of $y = x^7$ and $y = x^8$. Use a graphing utility to verify your conjecture.

53. Use the results of Exercise 51 to sketch the graph of $y = (x - 3)^3$ by hand. Use a graphing utility to verify your graph.

54. Use the results of Exercise 51 to sketch the graph of $y = (x + 1)^2$ by hand. Use a graphing utility to verify your graph.

Conjecture **In Exercises 55–58, use the results of Exercise 51 to make a conjecture about the shape of the graph of the function. Use a graphing utility to verify your conjecture.**

55. $f(x) = x^2(x - 6)^2$

56. $f(x) = x^3(x - 6)^2$

57. $f(x) = x^2(x - 6)^3$

58. $f(x) = x^3(x - 6)^3$

59. *Graphical Reasoning* An electronically controlled thermostat in a home is programmed to automatically lower the temperature during the night (see figure). The temperature in the house T, in degrees Fahrenheit, is given in terms of t, the time in hours on a 24-hour clock.

(a) Explain why T is a function of t.

(b) Approximate $T(4)$ and $T(15)$.

(c) Suppose the thermostat were reprogrammed to produce a temperature H where $H(t) = T(t - 1)$. How would this change the temperature in the house? Explain.

(d) Suppose the thermostat were reprogrammed to produce a temperature H where $H(t) = T(t) - 1$. How would this change the temperature in the house? Explain.

Figure for 59

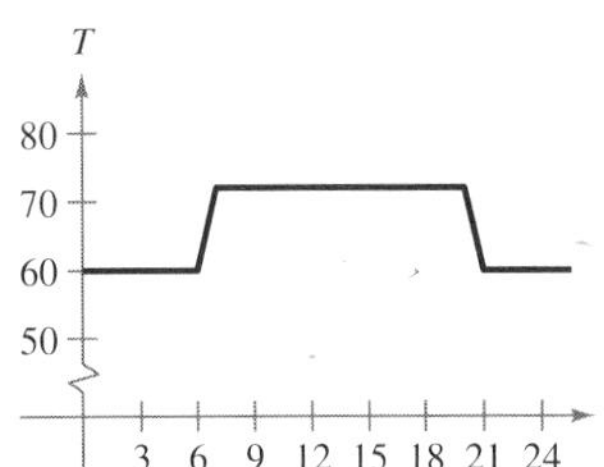

60. *Coordinate Axis Scale* It is necessary to graph the function $f(t)$, which models the specified data for the years 1980 through 1996, with $t = 0$ corresponding to the year 1980. State a possible scale for the vertical axis (e.g., hundreds, thousands, millions, etc.) of the graph and give a reason for your answer.

(a) $f(t)$ represents the average salary of college professors.

(b) $f(t)$ represents the population of the United States.

(c) $f(t)$ represents the percent of the civilian work force that is unemployed.

In Exercises 61–66, use the graph of $f(x)$ (see figure) to sketch the graph of the specified function.

61. $f(x - 4)$

62. $f(x + 2)$

63. $f(x) + 4$

64. $f(x) - 1$

65. $2f(x)$

66. $\frac{1}{2}f(x)$

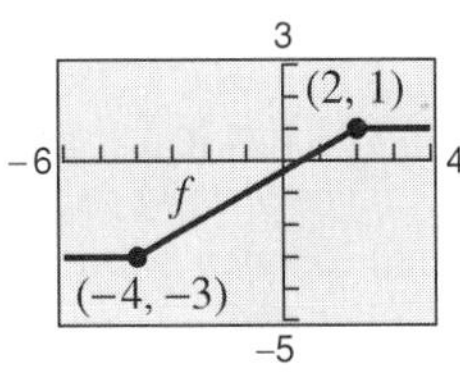

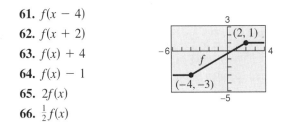

1.6 Combinations of Functions

Arithmetic Combinations of Functions / Compositions of Functions / Application

Arithmetic Combinations of Functions

Just as two real numbers can be combined by the operations of addition, subtraction, multiplication, and division to form other real numbers, two *functions* can be combined to create new functions. For example, if

$$f(x) = 2x - 3 \quad \text{and} \quad g(x) = x^2 - 1$$

you can form the sum, difference, product, and quotient of f and g as follows.

$$\begin{aligned} f(x) + g(x) &= (2x - 3) + (x^2 - 1) \\ &= x^2 + 2x - 4 \end{aligned} \qquad \text{Sum}$$

$$\begin{aligned} f(x) - g(x) &= (2x - 3) - (x^2 - 1) \\ &= -x^2 + 2x - 2 \end{aligned} \qquad \text{Difference}$$

$$\begin{aligned} f(x) \cdot g(x) &= (2x - 3)(x^2 - 1) \\ &= 2x^3 - 3x^2 - 2x + 3 \end{aligned} \qquad \text{Product}$$

$$\frac{f(x)}{g(x)} = \frac{2x - 3}{x^2 - 1}, \quad x \neq \pm 1 \qquad \text{Quotient}$$

The domain of an arithmetic combination of functions f and g consists of all real numbers that are common to the domains of f and g. In the case of the quotient $f(x)/g(x)$, there is the further restriction that $g(x) \neq 0$.

Sum, Difference, Product, and Quotient of Functions

Let f and g be two functions with overlapping domains. Then, for all x common to both domains, the **sum, difference, product,** and **quotient** of f and g are defined as follows.

1. Sum: $(f + g)(x) = f(x) + g(x)$
2. Difference: $(f - g)(x) = f(x) - g(x)$
3. Product: $(fg)(x) = f(x) \cdot g(x)$
4. Quotient: $\left(\frac{f}{g}\right)(x) = \frac{f(x)}{g(x)}, \quad g(x) \neq 0$

A graphing utility can be used to graph the sum of two functions. For instance, you can illustrate the result of Example 1 by letting $Y_1 = 2X + 1$, $Y_2 = X^2 + 2X - 1$, and $Y_3 = Y_1 + Y_2$. Graph all three functions with your graphing utility set in simultaneous plotting mode. What do you observe? How could you use your graphing utility to illustrate Example 2?

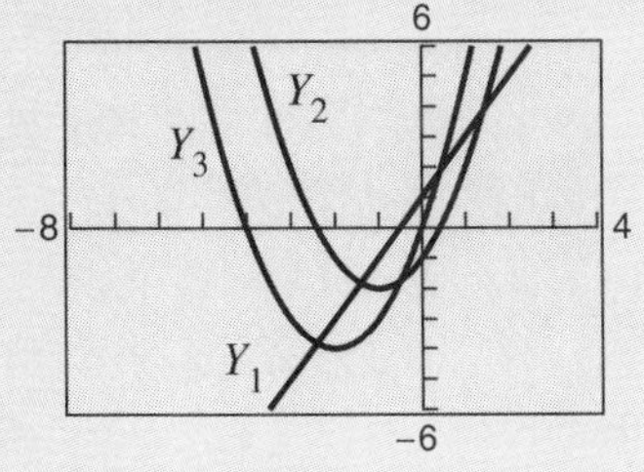

EXAMPLE 1 **Finding the Sum of Two Functions**

Find $(f + g)(x)$ for the functions

$$f(x) = 2x + 1 \quad \text{and} \quad g(x) = x^2 + 2x - 1.$$

Then evaluate the sum when $x = 2$.

Solution

The sum of the functions f and g is given by

$$\begin{aligned}(f + g)(x) &= f(x) + g(x)\\ &= (2x + 1) + (x^2 + 2x - 1)\\ &= x^2 + 4x.\end{aligned}$$

When $x = 2$, the value of this sum is

$$(f + g)(2) = 2^2 + 4(2) = 12.$$

EXAMPLE 2 **Finding the Difference of Two Functions**

Find $(f - g)(x)$ for the functions

$$f(x) = 2x + 1 \quad \text{and} \quad g(x) = x^2 + 2x - 1.$$

Then evaluate the difference when $x = 2$.

Solution

The difference of the functions f and g is given by

$$\begin{aligned}(f - g)(x) &= f(x) - g(x)\\ &= (2x + 1) - (x^2 + 2x - 1)\\ &= -x^2 + 2.\end{aligned}$$

When $x = 2$, the value of this difference is

$$(f - g)(2) = -(2)^2 + 2 = -2.$$

In Examples 1 and 2, both f and g have domains that consist of all real numbers. Thus, the domain of both $(f + g)$ and $(f - g)$ is also the set of all real numbers. Remember that any restrictions on the domains of f or g must be taken into account when forming the sum, difference, product, or quotient of f and g. For instance, the domain of $f(x) = 1/x$ is all $x \neq 0$, and the domain of $g(x) = \sqrt{x}$ is $[0, \infty)$. This implies that the domain of $f + g$ is $(0, \infty)$.

You can confirm the results of Example 3 with your graphing utility. Enter the three functions

Y1 = $\sqrt{X}$

Y2 = $\sqrt{(4 - X^2)}$

Y3 = Y1/Y2

and graph them using a zoom decimal screen (ZDecimal on the *TI-82* or *TI-83*). The zoom decimal screen displays the graph in a new viewing rectangle that sets the change in the x-values and the change in the y-values to 0.1. Trace along the third curve between X = 0 and X = 2. What happens when you arrive at X = 2? Repeat this experiment for Y3 = Y2/Y1 and observe any changes in the domain.

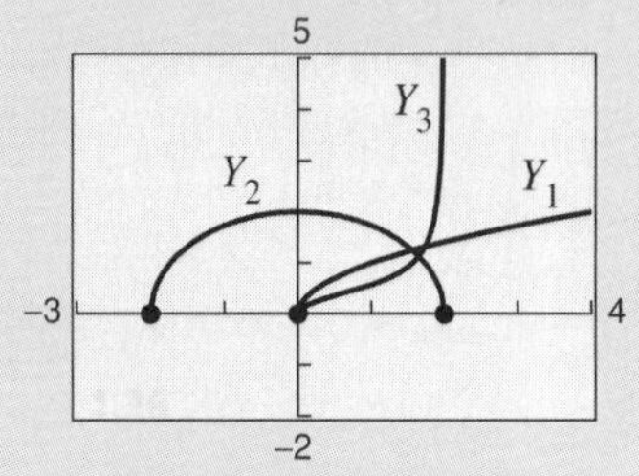

EXAMPLE 3 Finding the Quotient of Two Functions

Find $(f/g)(x)$ and $(g/f)(x)$ for the functions

$$f(x) = \sqrt{x} \quad \text{and} \quad g(x) = \sqrt{4 - x^2}.$$

Then find the domains of f/g and g/f.

Solution

The quotient of f and g is given by

$$\left(\frac{f}{g}\right)(x) = \frac{f(x)}{g(x)} = \frac{\sqrt{x}}{\sqrt{4 - x^2}},$$

and the quotient of g and f is given by

$$\left(\frac{g}{f}\right)(x) = \frac{g(x)}{f(x)} = \frac{\sqrt{4 - x^2}}{\sqrt{x}}.$$

The domain of f is $[0, \infty)$ and the domain of g is $[-2, 2]$. The intersection of these domains is $[0, 2]$. Thus, the domains for f/g and g/f are as follows.

$$\text{Domain of } \frac{f}{g}: [0, 2) \qquad \text{Domain of } \frac{g}{f}: (0, 2]$$

Can you see why these two domains differ slightly?

Compositions of Functions

Another way of combining two functions is to form the **composition** of one with the other. For instance, if

$$f(x) = x^2 \qquad \text{and} \qquad g(x) = x + 1,$$

the composition of f with g is

$$f(g(x)) = f(x + 1) = (x + 1)^2.$$

This composition is denoted as $f \circ g$.

Figure 1.52

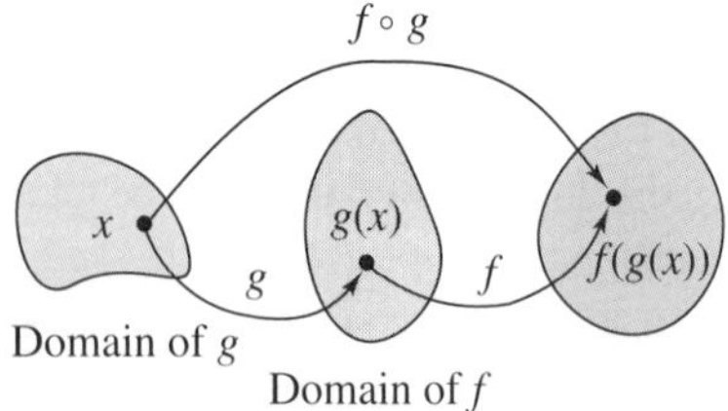

Definition of Composition of Two Functions

The **composition** of the function f with g is

$$(f \circ g)(x) = f(g(x)).$$

The domain of $f \circ g$ is the set of all x in the domain of g such that $g(x)$ is in the domain of f. (See Figure 1.52.)

EXPLORATION

Let $f(x) = x + 2$ and $g(x) = 4 - x^2$. Are the compositions $f \circ g$ and $g \circ f$ equal? You can use your graphing utility to answer this question by entering and graphing the following functions.

$$Y_1 = (4 - x^2) + 2$$

$$Y_2 = 4 - (x + 2)^2$$

What do you observe? Which function represents $f \circ g$ and which represents $g \circ f$?

EXAMPLE 4 **Forming the Composition of *f* with *g***

Find $(f \circ g)(x)$ for $f(x) = \sqrt{x}$, $x \geq 0$, and $g(x) = x - 1$, $x \geq 1$. If possible, find $(f \circ g)(2)$ and $(f \circ g)(0)$.

Solution

$$(f \circ g)(x) = f(g(x)) \qquad \text{Definition of } f \circ g$$

$$= f(x - 1) \qquad \text{Definition of } g(x)$$

$$= \sqrt{x - 1}, \quad x \geq 1 \qquad \text{Definition of } f(x)$$

The domain of $f \circ g$ is $[1, \infty)$. Thus,

$$(f \circ g)(2) = \sqrt{2 - 1} = 1$$

is defined, but $(f \circ g)(0)$ is not defined because 0 is not in the domain of $f \circ g$.

The composition of f with g is generally not the same as the composition of g with f. This is illustrated in Example 5.

EXAMPLE 5 **Compositions of Functions**

Given $f(x) = x + 2$ and $g(x) = 4 - x^2$, find the following.

a. $(f \circ g)(x)$

b. $(g \circ f)(x)$

Solution

a. $(f \circ g)(x) = f(g(x))$ Definition of $f \circ g$

$$= f(4 - x^2) \qquad \text{Definition of } g(x)$$

$$= (4 - x^2) + 2 \qquad \text{Definition of } f(x)$$

$$= -x^2 + 6$$

b. $(g \circ f)(x) = g(f(x))$ Definition of $g \circ f$

$$= g(x + 2) \qquad \text{Definition of } f(x)$$

$$= 4 - (x + 2)^2 \qquad \text{Definition of } g(x)$$

$$= 4 - (x^2 + 4x + 4)$$

$$= -x^2 - 4x$$

Note in this case that $(f \circ g)(x) \neq (g \circ f)(x)$.

Note The following tables help illustrate numerically the composition $f \circ g$ of the functions f and g given in Example 5.

x	0	1	2	3
$g(x)$	4	3	0	−5

$g(x)$	4	3	0	−5
$f(g(x))$	6	5	2	−3

x	0	1	2	3
$f(g(x))$	6	5	2	−3

Note that the first two tables can be combined (or "composed") to produce the values given in the third table.

EXAMPLE 6 **Finding the Domain of a Composite Function**

Find the composition $(f \circ g)(x)$ for the functions

$$f(x) = x^2 - 9 \quad \text{and} \quad g(x) = \sqrt{9 - x^2}.$$

Then find the domain of $(f \circ g)$.

Solution

The composition of the functions is as follows.

$$\begin{aligned}(f \circ g)(x) &= f(g(x)) \\ &= f\left(\sqrt{9 - x^2}\right) \\ &= \left(\sqrt{9 - x^2}\right)^2 - 9 \\ &= 9 - x^2 - 9 \\ &= -x^2\end{aligned}$$

Figure 1.53

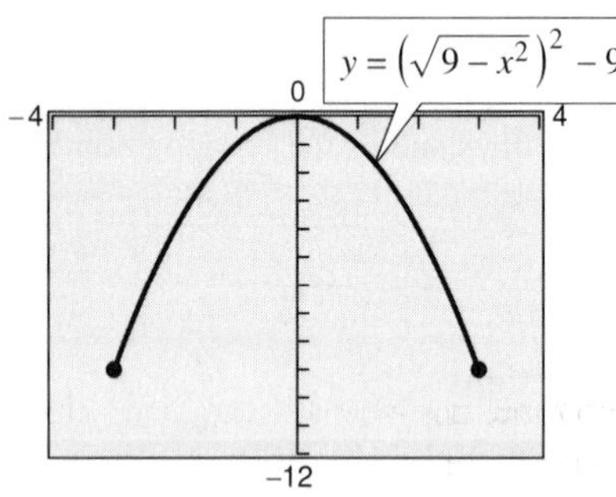

From this, it might appear that the domain of the composition is the set of all real numbers. This, however, is not true because the domain of g is $-3 \le x \le 3$. To convince yourself of this, use a graphing utility to graph

$$y = \left(\sqrt{9 - x^2}\right)^2 - 9$$

as shown in Figure 1.53. Notice that the graphing utility does not extend the graph to the left of $x = -3$ or to the right of $x = 3$. Thus, the domain of $f \circ g$ is $-3 \le x \le 3$.

EXAMPLE 7 **A Case in Which $f \circ g = g \circ f$**

Given $f(x) = 2x + 3$ and $g(x) = \frac{1}{2}(x - 3)$, find the following.

a. $(f \circ g)(x)$ **b.** $(g \circ f)(x)$

Solution

a.
$$\begin{aligned}(f \circ g)(x) &= f(g(x)) \\ &= f\left(\frac{1}{2}(x - 3)\right) \\ &= 2\left[\frac{1}{2}(x - 3)\right] + 3 \\ &= x - 3 + 3 \\ &= x\end{aligned}$$

b.
$$\begin{aligned}(g \circ f)(x) &= g(f(x)) \\ &= g(2x + 3) \\ &= \frac{1}{2}[(2x + 3) - 3] \\ &= \frac{1}{2}(2x) \\ &= x\end{aligned}$$

Note In Example 7, note that the two composite functions $f \circ g$ and $g \circ f$ are equal, and both represent the identity function. That is,

$$(f \circ g)(x) = x$$
$$(g \circ f)(x) = x.$$

You will study this special case in the next section.

In Examples 4, 5, 6, and 7 you formed the composition of two given functions. In calculus, it is also important to be able to identify two functions that make up a given composite function. For instance, the function h given by

$$h(x) = (3x - 5)^3$$

is the composition of f with g, where $f(x) = x^3$ and $g(x) = 3x - 5$. That is,

$$h(x) = (3x - 5)^3 = [g(x)]^3 = f(g(x)).$$

Basically, to "decompose" a composite function, look for an "inner" and an "outer" function. In the function h above, $g(x) = 3x - 5$ is the inner function and $f(x) = x^3$ is the outer function.

EXAMPLE 8 Identifying a Composite Function

Express the function

$$h(x) = \frac{1}{(x - 2)^2}$$

as a composition of two functions.

Solution

One way to write h as a composition of two functions is to take the inner function to be $g(x) = x - 2$ and the outer function to be

$$f(x) = \frac{1}{x^2} = x^{-2}.$$

Then you can write

$$h(x) = \frac{1}{(x - 2)^2} = (x - 2)^{-2} = f(x - 2) = f(g(x)).$$

EXPLORATION

The function in Example 8 can be decomposed in other ways. For which of the following pairs of functions is $h(x)$ equal to $f(g(x))$?

a. $g(x) = \dfrac{1}{x - 2}$ and $f(x) = x^2$

b. $g(x) = x^2$ and $f(x) = \dfrac{1}{x - 2}$

c. $g(x) = \dfrac{1}{x}$ and $f(x) = (x - 2)^2$

EXPLORATION

Use a graphing utility to graph $y = 320t^2 + 420$ and $y = 2000$ in the same viewing rectangle. (Use a viewing rectangle in which $0 \le x \le 3$ and $420 \le y \le 4000$.) Explain how the graphs can be used to answer the question asked in Example 9(c). Compare your answer with that given in part (c). When will the bacteria count reach 3200?

Notice that the model for this bacteria count situation is valid only for a span of 3 hours. Now suppose the minimum number of bacteria in the food is reduced from 420 to 100. Will the number of bacteria still reach a level of 2000 within the 3-hour time span? Will the number of bacteria reach a level of 3200 within 3 hours?

Application

EXAMPLE 9 Bacteria Count

The number of bacteria in a refrigerated food is given by

$$N(T) = 20T^2 - 80T + 500, \qquad 2 \le T \le 14$$

where T is the Celsius temperature of the food. When the food is removed from refrigeration, the temperature is given by

$$T(t) = 4t + 2, \qquad 0 \le t \le 3$$

where t is the time in hours. Find the following.

a. The composite $N(T(t))$. What does this function represent?

b. The number of bacteria in the food when $t = 2$ hours

c. The time when the bacteria count reaches 2000

Solution

a.
$$\begin{aligned} N(T(t)) &= 20(4t + 2)^2 - 80(4t + 2) + 500 \\ &= 20(16t^2 + 16t + 4) - 320t - 160 + 500 \\ &= 320t^2 + 320t + 80 - 320t - 160 + 500 \\ &= 320t^2 + 420 \end{aligned}$$

This composite function represents the number of bacteria as a function of the amount of time the food has been out of refrigeration.

b. When $t = 2$, the number of bacteria is

$$N = 320(2)^2 + 420 = 1280 + 420 = 1700.$$

c. The bacteria count will reach $N = 2000$ when $320t^2 + 420 = 2000$. You can solve this equation for t algebraically as follows.

$$\begin{aligned} 320t^2 + 420 &= 2000 \\ 320t^2 &= 1580 \\ t^2 &= \frac{1580}{320} = \frac{79}{16} \\ t &= \frac{\sqrt{79}}{4} \approx 2.2 \text{ hours} \end{aligned}$$

Or you can use a graphing utility to approximate the solution, as shown in Figure 1.54.

Figure 1.54

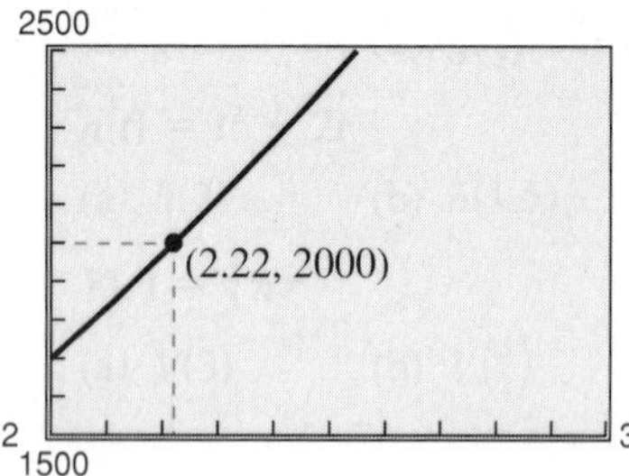

Group Activity *The Composition of Two Functions*

You are considering buying a new sport-utility vehicle for which the regular price is \$20,800. The dealership has advertised a factory rebate of \$2000 *and* a 12% discount. Using function notation, the rebate and the discount can be represented as

$$f(x) = x - 2000 \qquad \text{Rebate of \$2000}$$

and

$$g(x) = 0.88x \qquad \text{Discount of 12\%}$$

where x is the price of the vehicle, $f(x)$ is the price after subtracting the rebate, and $g(x)$ is the price after taking the 12% discount. Compare the sale price obtained by subtracting the rebate first and then taking the discount with the sale price obtained by taking the discount first and then subtracting the rebate.

1.6 /// EXERCISES

In Exercises 1–4, use the graphs to graph $h(x) = (f + g)(x)$.

1.

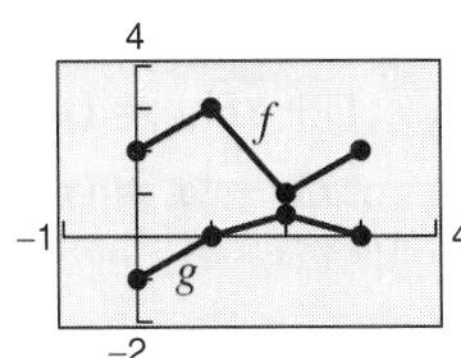

2.

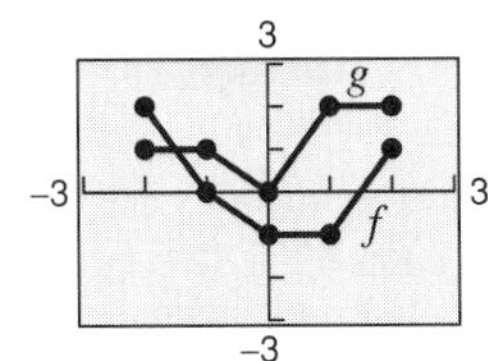

3.

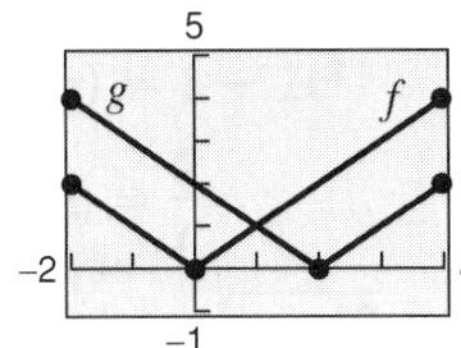

4.

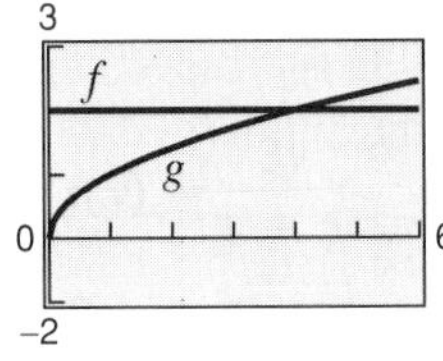

In Exercises 5–12, find the following.

(a) $(f + g)(x)$ (b) $(f - g)(x)$

(c) $(fg)(x)$ (d) $(f/g)(x)$

What is the domain of f/g?

5. $f(x) = x + 1, \quad g(x) = x - 1$

6. $f(x) = 2x - 5, \quad g(x) = 1 - x$

7. $f(x) = x^2, \quad g(x) = 1 - x$

8. $f(x) = 2x - 5, \quad g(x) = 5$

9. $f(x) = x^2 + 5, \quad g(x) = \sqrt{1 - x}$

10. $f(x) = \sqrt{x^2 - 4}, \quad g(x) = \dfrac{x^2}{x^2 + 1}$

11. $f(x) = \dfrac{1}{x}, \quad g(x) = \dfrac{1}{x^2}$

12. $f(x) = \dfrac{x}{x + 1}, \quad g(x) = x^3$

In Exercises 13–24, evaluate the indicated function for $f(x) = x^2 + 1$ and $g(x) = x - 4$.

13. $(f + g)(3)$
14. $(f - g)(-2)$
15. $(f - g)(0)$
16. $(f + g)(1)$
17. $(f - g)(2t)$
18. $(f + g)(t - 1)$
19. $(fg)(4)$
20. $(fg)(-6)$
21. $\left(\frac{f}{g}\right)(5)$
22. $\left(\frac{f}{g}\right)(0)$
23. $\left(\frac{f}{g}\right)(-1) - g(3)$
24. $(2f)(5)$

In Exercises 25–28, graph the functions f, g, and $f + g$ in the same viewing rectangle.

25. $f(x) = \frac{1}{2}x, \quad g(x) = x - 1$
26. $f(x) = \frac{1}{3}x, \quad g(x) = -x + 4$
27. $f(x) = x^2, \quad g(x) = -2x$
28. $f(x) = 4 - x^2, \quad g(x) = x$

In Exercises 29 and 30, use a graphing utility to sketch the graphs of f, g, and $f + g$ in the same viewing rectangle. Which function contributes most to the magnitude of the sum when $0 \le x \le 2$? Which function contributes most to the magnitude of the sum when $x > 6$?

29. $f(x) = 3x, \quad g(x) = -\dfrac{x^3}{10}$
30. $f(x) = \dfrac{x}{2}, \quad g(x) = \sqrt{x}$

31. *Stopping Distance* A car traveling x miles per hour stops quickly. The distance a car travels during the driver's reaction time is given by $R(x) = \frac{3}{4}x$. The distance traveled while braking is given by $B(x) = \frac{1}{15}x^2$.
    (a) Find the stopping-distance function T.
    (b) Use a graphing utility to graph the functions R, B, and T in the interval $0 \le x \le 60$.
    (c) Which function contributes most to the magnitude of the sum at higher speeds? Explain.

32. *Comparing Sales* You own two restaurants. From 1990 to 1995, the sales R_1 (in thousands of dollars) for one restaurant can be modeled by

$$R_1 = 480 - 8t - 0.8t^2, \qquad t = 0, 1, 2, 3, 4, 5$$

where $t = 0$ represents 1990. During the same 6-year period, the sales R_2 (in thousands of dollars) for the other restaurant can be modeled by

$$R_2 = 254 + 0.78t, \qquad t = 0, 1, 2, 3, 4, 5.$$

(a) Write a function that represents the total sales for the two restaurants. Use a graphing utility to graph the total sales function.
(b) Use the *stacked bar graph* in the figure, which represents the total sales during the 6-year period, to determine whether the total sales have been increasing or decreasing. Explain.

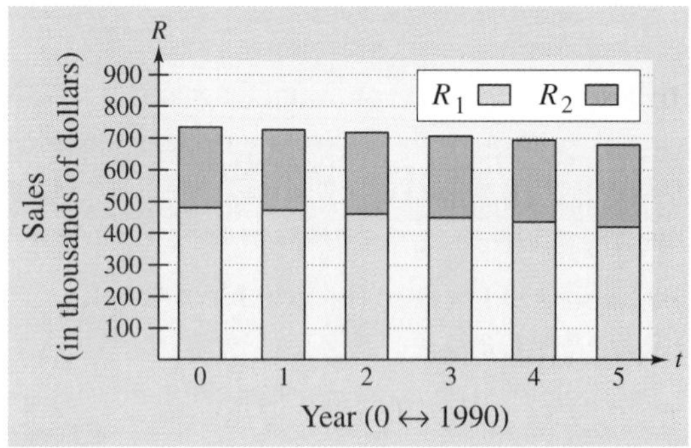

Data Analysis **In Exercises 33 and 34, use the data in the table, which gives the variable costs for operating an automobile in the United States for the years 1985 through 1991. The functions y_1, y_2, and y_3 represent the costs in cents per mile for gas and oil, maintenance, and tires, respectively.** (Source: American Automobile Manufacturers Association)

Year	1985	1986	1987	1988	1989	1990	1991
y_1	6.16	4.48	4.80	5.20	5.20	5.40	6.70
y_2	1.23	1.37	1.60	1.60	1.90	2.10	2.20
y_3	0.65	0.67	0.80	0.80	0.80	0.90	0.90

33. Create a stacked bar graph for the data.

34. Mathematical models for the data are given by

$$y_1 = 0.16t^2 - 2.43t + 13.96$$
$$y_2 = 0.17t + 0.38$$
$$y_3 = 0.04t + 0.44$$

where $t = 5$ represents 1985. Use a graphing utility to graph y_1, y_2, y_3, and $y_1 + y_2 + y_3$ in the same viewing rectangle. Use the model to estimate the total variable cost per mile in 1995.

In Exercises 35–38, find (a) $f \circ g$, (b) $g \circ f$, and (c) $f \circ f$.

35. $f(x) = x^2, \quad g(x) = x - 1$

36. $f(x) = \sqrt[3]{x - 1}, \quad g(x) = x^3 + 1$

37. $f(x) = 3x + 5, \quad g(x) = 5 - x$

38. $f(x) = x^3, \quad g(x) = \dfrac{1}{x}$

In Exercises 39–44, (a) find $f \circ g$ and $g \circ f$. (b) Use a graphing utility to graph $f \circ g$ and $g \circ f$. Determine if $f \circ g = g \circ f$.

39. $f(x) = \sqrt{x + 4}, \quad g(x) = x^2$

40. $f(x) = \sqrt[3]{x + 1}, \quad g(x) = x^3 - 1$

41. $f(x) = \frac{1}{3}x - 3, \quad g(x) = 3x + 1$

42. $f(x) = \sqrt{x}, \quad g(x) = \sqrt{x}$

43. $f(x) = x^{2/3}, \quad g(x) = x^6$

44. $f(x) = |x|, \quad g(x) = x + 6$

In Exercises 45–48, use the graphs of f and g (see figures) to evaluate the functions.

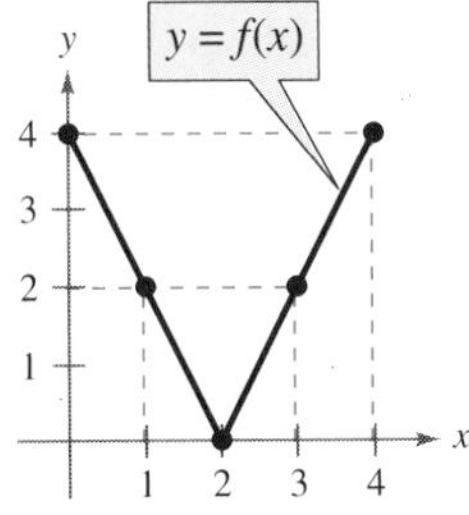

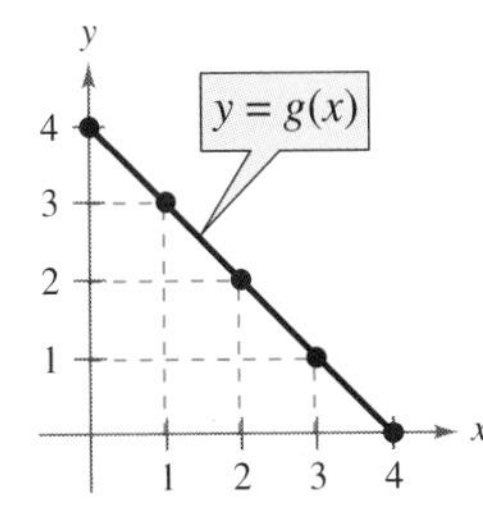

45. (a) $(f + g)(3)$ (b) $(f/g)(2)$

46. (a) $(f - g)(1)$ (b) $(fg)(4)$

47. (a) $(f \circ g)(2)$ (b) $(g \circ f)(2)$

48. (a) $(f \circ g)(1)$ (b) $(g \circ f)(3)$

In Exercises 49–56, find two functions f and g such that $(f \circ g)(x) = h(x)$. (There are many correct answers.)

49. $h(x) = (2x + 1)^2$

50. $h(x) = (1 - x)^3$

51. $h(x) = \sqrt[3]{x^2 - 4}$

52. $h(x) = \sqrt{9 - x}$

53. $h(x) = \dfrac{1}{x + 2}$

54. $h(x) = \dfrac{4}{(5x + 2)^2}$

55. $h(x) = (x + 4)^2 + 2(x + 4)$

56. $h(x) = (x + 3)^{3/2}$

In Exercises 57–60, determine the domains of (a) f, (b) g, and (c) $f \circ g$.

57. $f(x) = \sqrt{x}, \quad g(x) = x^2 + 1$

58. $f(x) = \dfrac{1}{x}, \quad g(x) = x + 3$

59. $f(x) = \dfrac{3}{x^2 - 1}, \quad g(x) = x + 1$

60. $f(x) = 2x + 3, \quad g(x) = \dfrac{x}{2}$

Average Rate of Change **In Exercises 61–64, find the difference quotient**

$$\frac{f(x + h) - f(x)}{h}$$

and simplify your answer.

61. $f(x) = 3x - 4$

62. $f(x) = 1 - x^2$

63. $f(x) = \dfrac{4}{x}$

64. $f(x) = \sqrt{2x + 1}$

65. *Ripples* A pebble is dropped into a calm pond, causing ripples in the form of concentric circles (see figure). The radius (in feet) of the outer ripple is given by $r(t) = 0.6t$, where t is the time in seconds after the pebble strikes the water. The area of the circle is given by the function $A(r) = \pi r^2$. Find and interpret $(A \circ r)(t)$.

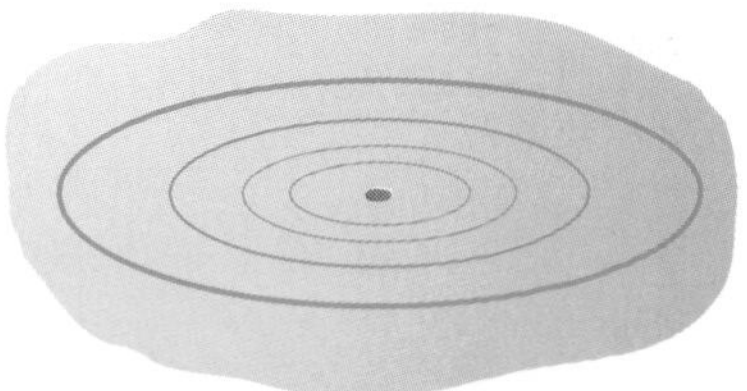

66. *Area* A square concrete foundation was prepared as a base for a large cylindrical gasoline tank (see figure).

(a) Express the radius r of the tank as a function of the length x of the sides of the square.

(b) Express the area A of the circular base of the tank as a function of the radius r.

(c) Find and interpret $(A \circ r)(x)$.

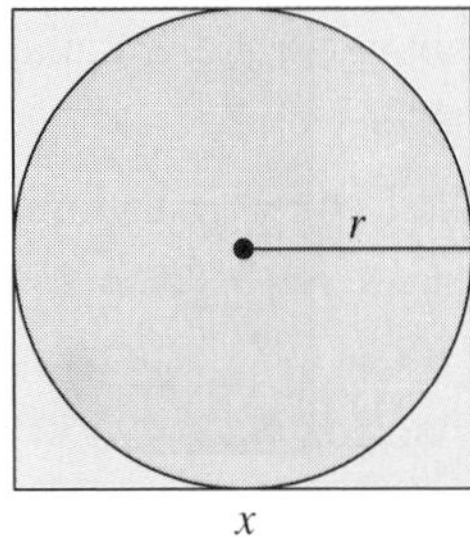

67. *Cost* The weekly cost of producing x units in a manufacturing process is given by the function $C(x) = 60x + 750$. The number of units produced in t hours is given by $x(t) = 50t$.

(a) Find and interpret $(C \circ x)(t)$.

(b) Use a graphing utility to graph the cost as a function of time. Move the cursor along the curve to estimate (to two-decimal-place accuracy) the time that must elapse until the cost increases to \$15,000.

68. *Air Traffic Control* An air traffic controller spots two planes at the same altitude flying toward each other (see figure). Their flight paths form a right angle at point P. One plane is 150 miles from point P and is moving at 450 miles per hour. The other plane is 200 miles from point P and is moving at 450 miles per hour. Write the distance s between the planes as a function of time t.

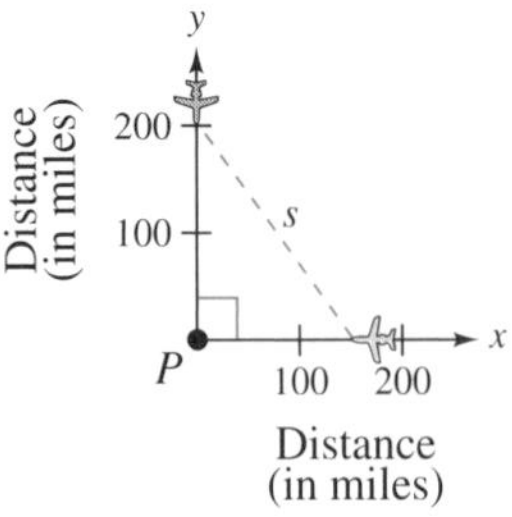

69. *Think About It* You are a sales representative for an automobile manufacturer. You are paid an annual salary plus a bonus of 3% of your sales over \$500,000. Consider the two functions $f(x) = x - 500{,}000$ and $g(x) = 0.03x$. If x is greater than \$500,000, which of the following represents your bonus? Explain.

(a) $f(g(x))$ (b) $g(f(x))$

70. *Exploration* The suggested retail price of a new car is p dollars. The dealership advertised a factory rebate of \$1200 and an 8% discount.

(a) Write a function R in terms of p, giving the cost of the car after receiving the rebate from the factory.

(b) Write a function S in terms of p, giving the cost of the car after receiving the dealership discount.

(c) Form the composite functions $(R \circ S)(p)$ and $(S \circ R)(p)$, and interpret each.

(d) Find $(R \circ S)(18{,}400)$ and $(S \circ R)(18{,}400)$. Which yields the lower cost for the car? Explain.

71. Prove that the product of two odd functions is an even function and the product of two even functions is an even function.

72. *Conjecture* Use examples to hypothesize whether the product of an odd function and an even function is even or odd. Then prove your hypothesis.

73. Given a function f, prove that $g(x)$ is even and $h(x)$ is odd where

$$g(x) = \frac{1}{2}[f(x) + f(-x)] \quad \text{and}$$

$$h(x) = \frac{1}{2}[f(x) - f(-x)].$$

74. Use the result of Exercise 73 to prove that any function can be written as a sum of even and odd functions. (*Hint:* Add the two equations in Exercise 73.)

75. Use the result of Exercise 74 to write each function as a sum of even and odd functions.

(a) $f(x) = x^2 - 2x + 1$ (b) $f(x) = \dfrac{1}{x + 1}$

76. *Exploration* Consider the repeated composition of $f(x) = \sqrt{x + k}$ with itself defined by

$$R_n(x) = \underbrace{f(f(f(\cdot \cdot \cdot f(x) \cdot \cdot \cdot)))}_{n}.$$

(*Hint:* If you use a *TI-82* or *TI-83*, use the following procedure. Let Y1 = $\sqrt{x + k}$ and store 0 in x. Using the Y-VARS menu, enter Y1 → x on the home screen. Pressing the ENTER key repeatedly will yield the repeated composition of f.)

(a) Use a graphing utility to complete the table for $k = \frac{1}{9}$.

n	1	2	3	4	5
$R_n(0)$					

n	6	7	8	9	10
$R_n(0)$					

(b) Use a graphing utility to complete the table for $k = 9$.

n	1	2	3	4	5
$R_n(0)$					

n	6	7	8	9	10
$R_n(0)$					

(c) Describe what appears to be happening for the repeated composition of the function f.

Data Analysis **In Exercises 77 and 78, use the data in the table, which gives the circulations of morning and evening newspapers in the United States for the years 1983 through 1992. The variables y_1 and y_2 represent the circulations in millions of the morning and evening papers, respectively.** (Source: Editor & Publisher Co.)

Year	1983	1984	1985	1986	1987
y_1	33.8	35.4	36.4	37.4	39.1
y_2	28.8	27.7	26.4	25.1	23.7

Year	1988	1989	1990	1991	1992
y_1	40.4	40.7	41.3	41.5	42.4
y_2	22.2	21.8	21.0	19.2	17.8

77. Create a stacked bar graph for the data.

78. Use a graphing utility to create a scatter plot of each data set. What type of model would best fit the data? Use the least squares regression capabilities of a graphing utility to find models for y_1 and y_2 (use $t = 3$ to represent 1983). Graph the models for y_1, y_2, and $y_1 - y_2$ in the same viewing rectangle. What does the graph of the difference of the functions indicate about newspaper circulation in general?

Review **Solve Exercises 79–82 as a review of the skills and problem-solving techniques you learned in previous sections.**

79. Sketch the polygon with vertices (1, 0), (6, 3), (8, 5), and (3, 2).

80. Find the coordinates of a point that is eight units to the left of the y-axis and three units above the x-axis.

81. Determine the quadrant(s) of a point (x, y) where $xy > 0$.

82. Find the distance between the points (0, 8) and (5, 4).

1.7 Inverse Functions

The Inverse of a Function / The Graph of an Inverse Function / The Existence of an Inverse Function / Finding Inverse Functions

The Inverse of a Function

Recall from Section 1.3 that a function can be represented by a set of ordered pairs. For instance, the function $f(x) = x + 4$ from the set $A = \{1, 2, 3, 4\}$ to the set $B = \{5, 6, 7, 8\}$ can be written as follows.

$$f(x) = x + 4: \ \{(1, 5), (2, 6), (3, 7), (4, 8)\}$$

By interchanging the first and second coordinates of each of these ordered pairs, you can form the **inverse function** of f, which is denoted by f^{-1}. It is a function from the set B to the set A, and can be written as follows.

$$f^{-1}(x) = x - 4: \ \{(5, 1), (6, 2), (7, 3), (8, 4)\}$$

Note that the domain of f is equal to the range of f^{-1}, and vice versa, as shown in Figure 1.55. Also note that the functions f and f^{-1} have the effect of "undoing" each other. In other words, when you form the composition of f with f^{-1} or the composition of f^{-1} with f, you obtain the identity function.

$$f(f^{-1}(x)) = f(x - 4) = (x - 4) + 4 = x$$
$$f^{-1}(f(x)) = f^{-1}(x + 4) = (x + 4) - 4 = x$$

Figure 1.55

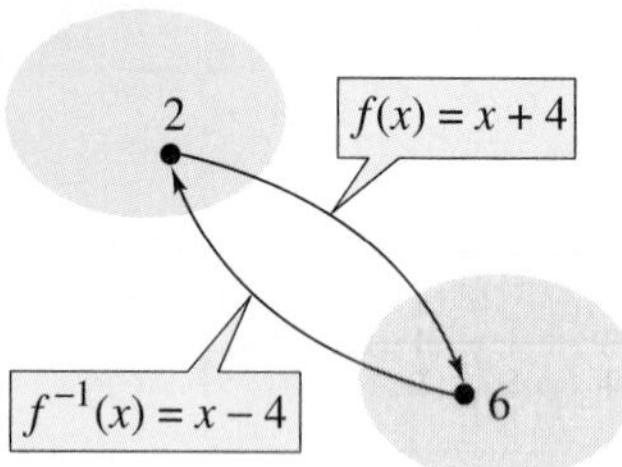

Note A table of values can help you understand inverse functions. For instance, the following table shows several values of the function f in Example 1.

x	-2	-1	0	1	2
$f(x)$	-8	-4	0	4	8

By interchanging the rows of the table, you obtain values of the inverse function f^{-1}.

x	-8	-4	0	4	8
$f^{-1}(x)$	-2	-1	0	1	2

In the first table, each output is 4 times the input, and in the second table each output is $\frac{1}{4}$ the input.

EXAMPLE 1 **Finding Inverse Functions Informally**

Find the inverse of $f(x) = 4x$. Then verify that both $f(f^{-1}(x))$ and $f^{-1}(f(x))$ are equal to the identity function.

Solution

The given function *multiplies* each input by 4. To "undo" this function, you need to *divide* each input by 4. Thus, the inverse function of $f(x) = 4x$ is

$$f^{-1}(x) = \frac{x}{4}.$$

You can verify that both $f(f^{-1}(x))$ and $f^{-1}(f(x))$ are equal to the identity function as follows.

$$f(f^{-1}(x)) = f\left(\frac{x}{4}\right) = 4\left(\frac{x}{4}\right) = x$$

$$f^{-1}(f(x)) = f^{-1}(4x) = \frac{4x}{4} = x$$

EXAMPLE 2 **Finding Inverse Functions Informally**

Find the inverse of $f(x) = x - 6$. Then verify that both $f(f^{-1}(x))$ and $f^{-1}(f(x))$ are equal to the identity function.

Solution

The given function *subtracts* 6 from each input. To "undo" this function, you need to *add* 6 to each input. Thus, the inverse function of $f(x) = x - 6$ is

$$f^{-1}(x) = x + 6.$$

You can verify that both $f(f^{-1}(x))$ and $f^{-1}(f(x))$ are equal to the identity function as follows.

$$f(f^{-1}(x)) = f(x + 6) = (x + 6) - 6 = x$$
$$f^{-1}(f(x)) = f^{-1}(x - 6) = (x - 6) + 6 = x$$

The formal definition of the inverse of a function is as follows.

Definition of the Inverse of a Function

Let f and g be two functions such that

$f(g(x)) = x$ for every x in the domain of g

and

$g(f(x)) = x$ for every x in the domain of f.

Under these conditions, the function g is the **inverse** of the function f. The function g is denoted by f^{-1} (read "f-inverse"). Thus,

$$f(f^{-1}(x)) = x$$

and

$$f^{-1}(f(x)) = x.$$

The domain of f must be equal to the range of f^{-1}, and the range of f must be equal to the domain of f^{-1}.

Note Don't be confused by the use of -1 to denote the inverse function f^{-1}. In this text, whenever we write f^{-1}, we will *always* be referring to the inverse of the function f and *not* to the reciprocal of $f(x)$.

If the function g is the inverse of the function f, it must also be true that the function f is the inverse of the function g. For this reason, you can say that the functions f and g are *inverses of each other.*

Most graphing utilities can graph $y = x^{1/3}$ in two ways:

Y1 = X ∧(1/3) or

Y1 = $\sqrt[3]{X}$.

You may not be able to obtain the complete graph of $y = x^{2/3}$ by entering Y1 = X ∧ (2/3). If not, you should use

Y1 = (X ∧(1/3))^2 or

Y1 = $\sqrt[3]{X^2}$.

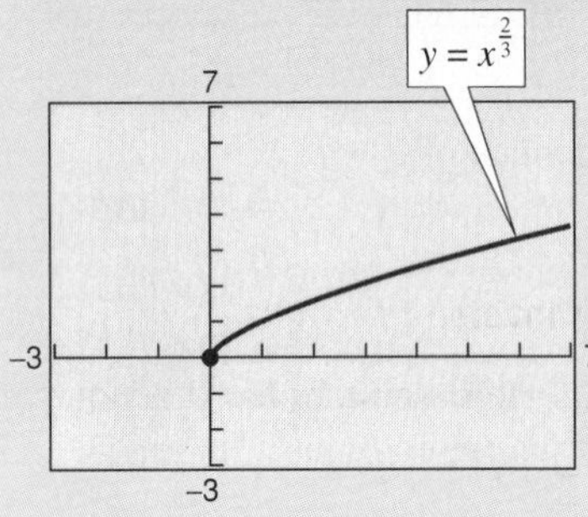

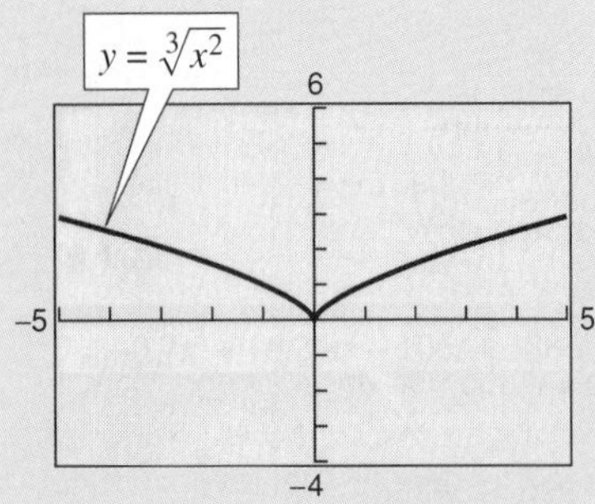

EXAMPLE 3 **Verifying Inverse Functions**

Show that the functions are inverses of each other.

$$f(x) = 2x^3 - 1 \qquad \text{and} \qquad g(x) = \sqrt[3]{\frac{x+1}{2}}$$

Solution

$$\begin{aligned} f(g(x)) = f\left(\sqrt[3]{\frac{x+1}{2}}\right) &= 2\left(\sqrt[3]{\frac{x+1}{2}}\right)^3 - 1 \\ &= 2\left(\frac{x+1}{2}\right) - 1 \\ &= x + 1 - 1 \\ &= x \end{aligned}$$

$$\begin{aligned} g(f(x)) = g(2x^3 - 1) &= \sqrt[3]{\frac{(2x^3 - 1) + 1}{2}} \\ &= \sqrt[3]{\frac{2x^3}{2}} \\ &= \sqrt[3]{x^3} \\ &= x \end{aligned}$$

EXAMPLE 4 **Verifying Inverse Functions**

Which of the functions is the inverse of $f(x) = \dfrac{5}{x-2}$?

$$g(x) = \frac{x-2}{5} \qquad \text{and} \qquad h(x) = \frac{5}{x} + 2$$

Solution

By forming the composition of f with g, you have

$$f(g(x)) = f\left(\frac{x-2}{5}\right) = \frac{5}{[(x-2)/5] - 2} = \frac{25}{x - 12} \neq x.$$

Because this composition is not equal to the identity function x, it follows that g *is not* the inverse of f. By forming the composition of f with h, you have

$$f(h(x)) = f\left(\frac{5}{x} + 2\right) = \frac{5}{[(5/x) + 2] - 2} = \frac{5}{5/x} = x.$$

Thus, it appears that h *is* the inverse of f. You can confirm this by showing that the composition of h with f is also equal to the identity function.

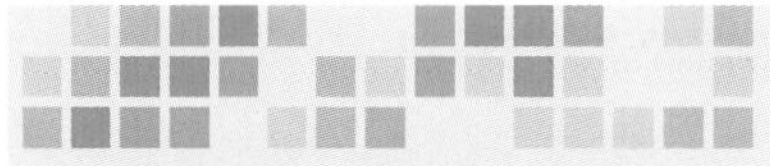

In Examples 3 and 4, inverse functions were verified algebraically. A graphing utility can also be helpful in checking to see whether one function is the inverse of another function. Use the Graph Reflection Program found in the appendix to verify graphically Example 4.

The Graph of an Inverse Function

The graphs of f and f^{-1} are related to each other in the following way. If the point (a, b) lies on the graph of f, then the point (b, a) lies on the graph of f^{-1} and vice versa. This means that the graph of f^{-1} is a reflection of the graph of f in the line $y = x$, as shown in Figure 1.56.

Figure 1.56

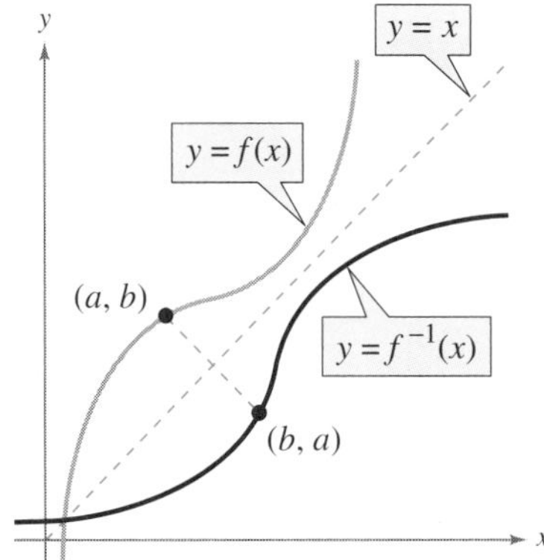

Figure 1.57

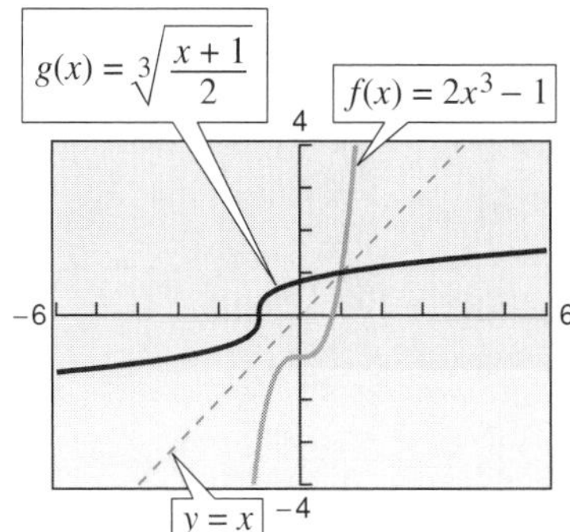

x	-2	-1	0	1	2
$f(x)$	-17	-3	-1	1	15

x	-17	-3	-1	1	15
$g(x)$	-2	-1	0	1	2

EXAMPLE 5 **Graphical Check for Inverse Functions**

Example 3 shows how to verify *algebraically* that the functions

$$f(x) = 2x^3 - 1 \quad \text{and} \quad g(x) = \sqrt[3]{\frac{x+1}{2}}$$

are inverses of each other. Verify that f and g are inverses of each other *graphically* by graphing them in the same viewing rectangle and observing that the graph of g is the reflection of the graph of f in the line $y = x$. (Use a square setting.)

Solution

Using a graphing utility, you can obtain the graphs shown in Figure 1.57. From this figure you can see that the graph of g is the reflection of the graph of f in the line $y = x$, which confirms graphically that g is the inverse of f.

Note You can verify *numerically* that the functions f and g are inverses by creating two tables of values, as shown at the left. Note that the entries in the two tables are the same except that their rows are interchanged.

The **parametric mode** of a graphing utility allows you to graph the inverse of a function without having to determine the actual equation of the inverse. The key idea is to express both x and y as functions of a third variable t. For instance, you could graph the function $y = 2x^3 - 1$ by setting your graphing utility to parametric mode and graphing the following equations.

X1T = T

Y1T = 2T ∧ 3 − 1.

Use the viewing window

Tmin = −4

Tmax = 4

Tstep = .05

with $-6 \le x \le 6$ and $-4 \le y \le 4$.

To graph the inverse of $y = 2x^3 - 1$, all you need to do is reverse the roles of x and y:

X2T = 2T ∧ 3 − 1

Y2T = T.

The Existence of an Inverse Function

A function need not have an inverse function. For instance, the function

$$f(x) = x^2$$

has no inverse [assuming a domain of $(-\infty, \infty)$]. To have an inverse, a function must be **one-to-one,** which means that no two elements in the domain of f correspond to the same element in the range of f.

Definition of a One-to-One Function

A function f is **one-to-one** if, for a and b in its domain,

$$f(a) = f(b) \quad \text{implies that} \quad a = b.$$

Existence of an Inverse Function

A function f has an inverse function f^{-1} if and only if f is one-to-one.

EXAMPLE 6 Testing for One-to-One Functions

Which functions are one-to-one and have inverse functions?

a. $f(x) = x^3 + 1$ **b.** $g(x) = x^2 - x$ **c.** $h(x) = \sqrt{x}$

Solution

a. Let a and b be real numbers with $f(a) = f(b)$.

$$a^3 + 1 = b^3 + 1 \qquad \text{Set } f(a) = f(b).$$
$$a^3 = b^3$$
$$a = b$$

Therefore, $f(a) = f(b)$ implies that $a = b$. From this, it follows that f is one-to-one and has an inverse function.

b. Because $g(-1) = (-1)^2 - (-1) = 2$ and $g(2) = 2^2 - 2 = 2$, you have two distinct inputs matched with the same output. Thus, g *is not* a one-to-one function and has no inverse function.

c. Let a and b be nonnegative real numbers with $h(a) = h(b)$.

$$\sqrt{a} = \sqrt{b} \qquad \text{Set } h(a) = h(b).$$
$$a = b$$

Therefore, $h(a) = h(b)$ implies that $a = b$. Thus, h *is* one-to-one and has an inverse function.

From its graph, it is easy to tell whether a function of x is one-to-one. Simply check to see that every *horizontal* line intersects the graph of the function at most once. For instance, Figure 1.58 shows the graphs of the three functions given in Example 6. On the graph of $g(x) = x^2 - x$, you can find a horizontal line that intersects the graph twice.

Figure 1.58

(a) f is one-to-one.

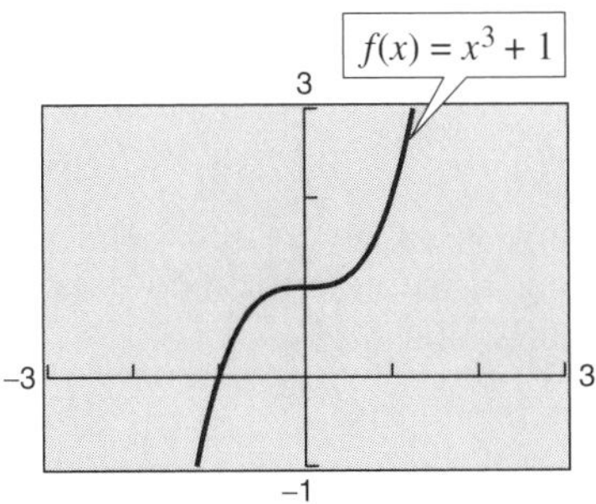

(b) g is not one-to-one.

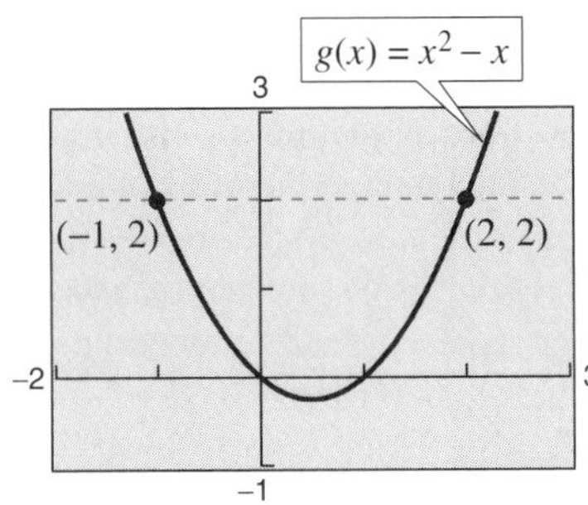

(c) h is one-to-one.

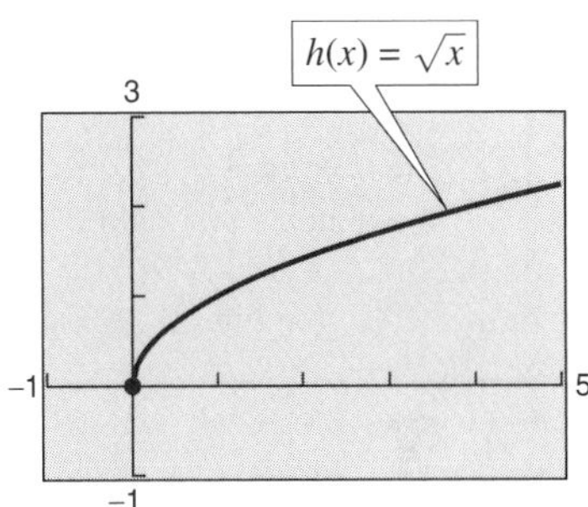

Two special types of functions that pass the **horizontal line test** are those that are increasing or decreasing on their entire domains.

1. If f is *increasing* on its entire domain, f is one-to-one.
2. If f is *decreasing* on its entire domain, f is one-to-one.

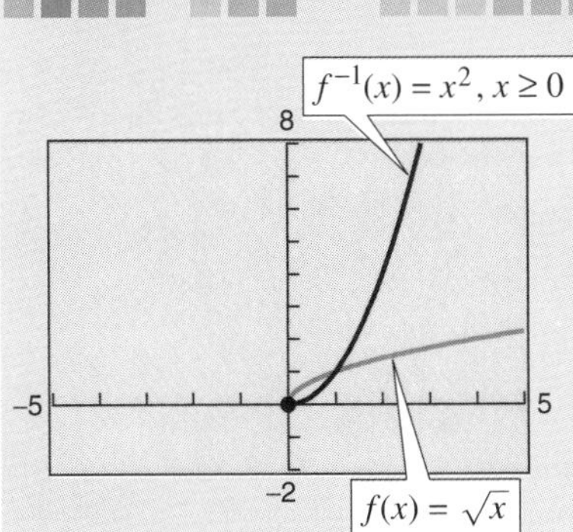

Many graphing utilities, such as the *TI-82* or *TI-83*, have a built-in feature to draw the inverse of a function. To see how this works, consider the function $f(x) = \sqrt{x}$. The inverse of f is $f^{-1}(x) = x^2, x \geq 0$. Enter the function

$$Y_1 = \sqrt{X}$$

and graph it in the standard viewing rectangle. Now go to the DRAW menu and select item 8 (DrawInv), and then select Y_1 from the Y-VARS menu.

DrawInv Y_1

You should obtain the figure on the left, which shows both f and its inverse f^{-1}.

Finding Inverse Functions

For simple functions (such as the ones in Examples 1 and 2) you can find inverse functions by inspection. For instance, the inverse of $f(x) = 8x$ is $f^{-1}(x) = x/8$. For more complicated functions, however, it is best to use the following procedure for finding the inverse of a function.

Note The problem of finding the inverse of a function can be difficult (or even impossible) for two reasons. First, given $y = f(x)$, it may be algebraically difficult to solve for x in terms of y. Second, if f is not one-to-one, then f^{-1} does not exist.

Finding the Inverse of a Function

To find the inverse of f, use the following steps.

1. In the equation for $f(x)$, replace $f(x)$ by y.
2. Interchange the roles of x and y.
3. If the new equation does not represent y as a function of x, the function f does not have an inverse function. If the new equation does represent y as a function of x, solve the new equation for y.
4. Replace y by $f^{-1}(x)$.
5. Verify that f and f^{-1} are inverses of each other by showing that $f(f^{-1}(x)) = x$ and $f^{-1}(f(x)) = x$.

EXAMPLE 7 **Finding the Inverse of a Function**

Find the inverse (if it exists) of

$$f(x) = \frac{5 - 3x}{2}.$$

Solution

From Figure 1.59, you can see that f is one-to-one, and therefore has an inverse.

Figure 1.59

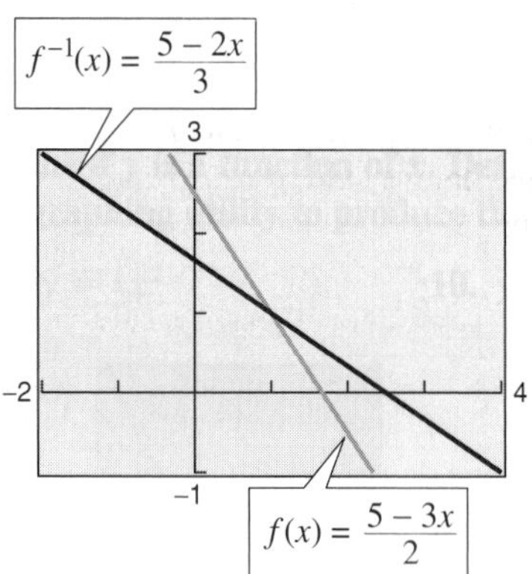

$$y = \frac{5 - 3x}{2} \quad \text{Replace } f(x) \text{ by } y.$$

$$x = \frac{5 - 3y}{2} \quad \text{Interchange } x \text{ and } y.$$

$$2x = 5 - 3y$$

$$3y = 5 - 2x$$

$$y = \frac{5 - 2x}{3} \quad \text{Solve for } y.$$

$$f^{-1}(x) = \frac{5 - 2x}{3} \quad \text{Replace } y \text{ by } f^{-1}(x).$$

The domain and range of both f and f^{-1} consist of all real numbers.

Figure 1.60

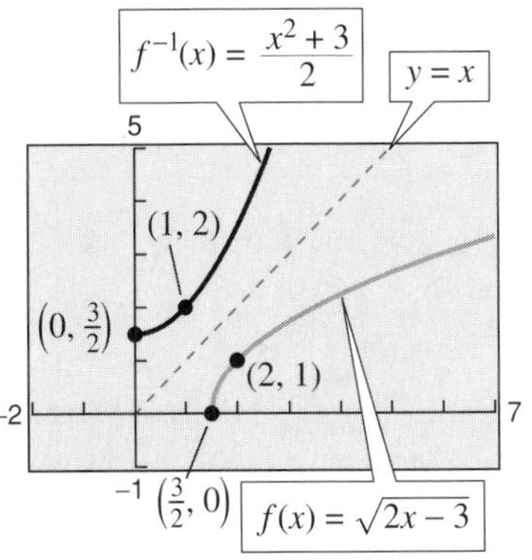

EXAMPLE 8 Finding the Inverse of a Function

Find the inverse of $f(x) = \sqrt{2x - 3}$ and sketch the graphs of f and f^{-1}.

Solution

$$y = \sqrt{2x - 3} \qquad \text{Replace } f(x) \text{ by } y.$$

$$x = \sqrt{2y - 3} \qquad \text{Interchange } x \text{ and } y.$$

$$x^2 = 2y - 3$$

$$2y = x^2 + 3$$

$$y = \frac{x^2 + 3}{2} \qquad \text{Solve for } y.$$

$$f^{-1}(x) = \frac{x^2 + 3}{2}, \quad x \geq 0 \qquad \text{Replace } y \text{ by } f^{-1}(x).$$

The graph of f^{-1} is the reflection of the graph of f in the line $y = x$, as shown in Figure 1.60. Note that the domain of f is the interval $\left[\frac{3}{2}, \infty\right)$ and the range of f is the interval $[0, \infty)$. Moreover, the domain of f^{-1} is the interval $[0, \infty)$ and the range of f^{-1} is the interval $\left[\frac{3}{2}, \infty\right)$.

Group Activity — *The Existence of an Inverse Function*

Write a short paragraph describing why the following functions do or do not have inverse functions. Give a numerical example.

a. Your hourly wage is \$7.50 plus \$0.90 for each unit x produced per hour. Let $f(x)$ represent your weekly wage for 40 hours of work. Does this function have an inverse?

b. Let x represent the retail price of an item (in dollars), and let $f(x)$ represent the sales tax on the item. Assume that the sales tax is 7% of the retail price *and* that the sales tax is rounded to the nearest cent. Does this function have an inverse? (*Hint:* Can you undo this function? For instance, if you know that the sales tax is \$0.14, can you determine *exactly* what the retail price is?)

1.7 /// EXERCISES

In Exercises 1–4, match the graph of the function with the graph of its inverse. [The graphs of the inverse functions are labeled (a), (b), (c), and (d).]

(a)

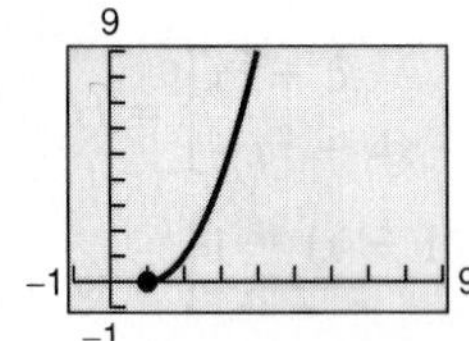

(b)

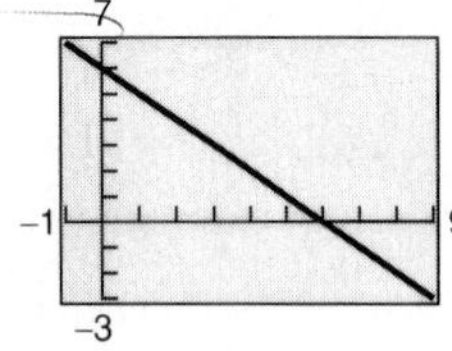

(c)

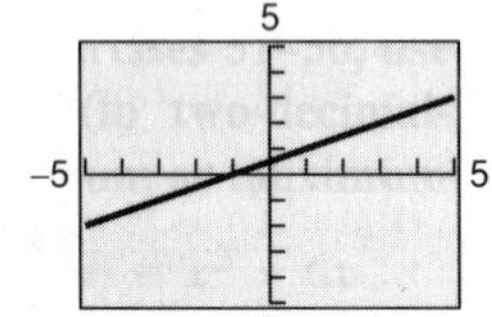

(d)

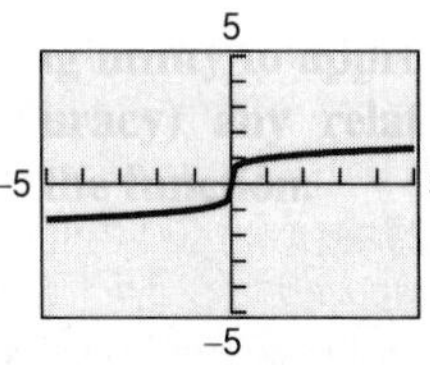

1.

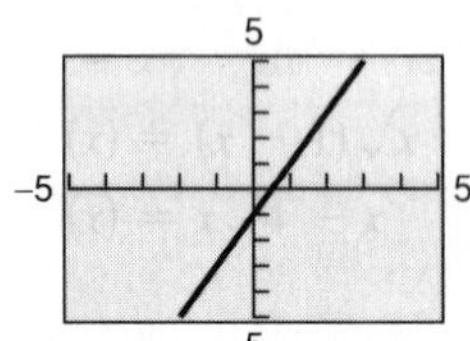

2.

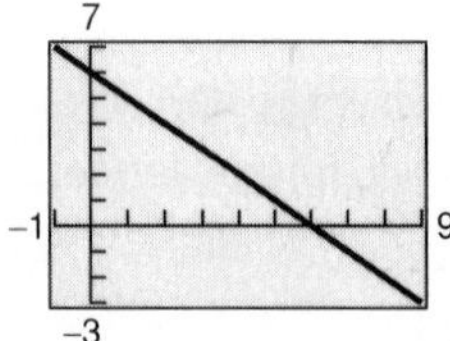

3.

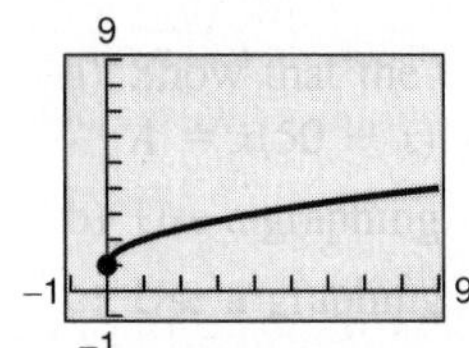

4.

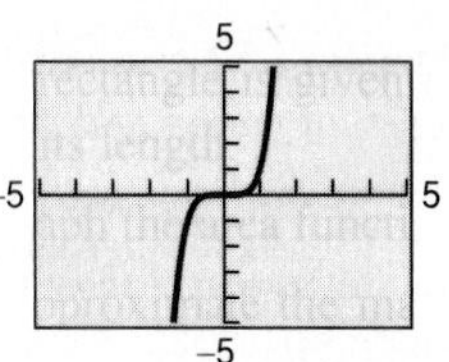

In Exercises 5–10, find the inverse of f informally. Verify that $f(f^{-1}(x)) = x$ and $f^{-1}(f(x)) = x$.

5. $f(x) = 8x$

6. $f(x) = \frac{1}{5}x$

7. $f(x) = x + 10$

8. $f(x) = x - 5$

9. $f(x) = \sqrt[3]{x}$

10. $f(x) = x^5$

In Exercises 11–14, show that f and g are inverse functions (a) algebraically and (b) graphically.

11. $f(x) = 2x, \quad g(x) = \dfrac{x}{2}$

12. $f(x) = x - 5, \quad g(x) = x + 5$

13. $f(x) = 5x + 1, \quad g(x) = \dfrac{x - 1}{5}$

14. $f(x) = 3 - 4x, \quad g(x) = \dfrac{3 - x}{4}$

In Exercises 15–20, show that f and g are inverse functions algebraically. Use a graphing utility to graph f and g in the same viewing rectangle. Describe the relationship between the graphs.

15. $f(x) = x^3, \quad g(x) = \sqrt[3]{x}$

16. $f(x) = \dfrac{1}{x}, \quad g(x) = \dfrac{1}{x}$

17. $f(x) = \sqrt{x - 4}, \quad g(x) = x^2 + 4, \ x \geq 0$

18. $f(x) = 9 - x^2, \quad x \geq 0$

$g(x) = \sqrt{9 - x}, \quad x \leq 9$

19. $f(x) = 1 - x^3, \quad g(x) = \sqrt[3]{1 - x}$

20. $f(x) = \dfrac{1}{1 + x}, \quad x \geq 0$

$g(x) = \dfrac{1 - x}{x}, \quad 0 < x \leq 1$

In Exercises 21–30, use a graphing utility to graph the function and use the Horizontal Line Test to determine whether the function is one-to-one.

21. $f(x) = 3 - \frac{1}{2}x$

22. $h(x) = \sqrt{16 - x^2}$

23. $h(x) = \dfrac{x^2}{x^2 + 1}$

24. $f(x) = \sqrt{x - 2}$

25. $g(x) = \dfrac{4 - x}{6}$

26. $f(x) = 10$

27. $h(x) = |x + 4| - |x - 4|$

28. $g(x) = (x + 5)^3$

29. $f(x) = -2x\sqrt{16 - x^2}$

30. $f(x) = \frac{1}{8}(x + 2)^2 - 1$

In Exercises 31–40, find the inverse of the function f. Use a graphing utility to graph both f and f^{-1} in the same viewing rectangle. Describe the relationship between the graphs.

31. $f(x) = 2x - 3$

32. $f(x) = 3x$

33. $f(x) = x^5$

34. $f(x) = x^3 + 1$

35. $f(x) = \sqrt{x}$

36. $f(x) = x^2, \quad x \geq 0$

37. $f(x) = \sqrt{4 - x^2}, \quad 0 \leq x \leq 2$

38. $f(x) = \dfrac{4}{x}$

39. $f(x) = \sqrt[3]{x - 1}$

40. $f(x) = x^{3/5}$

In Exercises 41–56, determine whether the function is one-to-one. If it is, find its inverse.

41. $f(x) = x^4$

42. $f(x) = \dfrac{1}{x^2}$

43. $g(x) = \dfrac{x}{8}$

44. $f(x) = 3x + 5$

45. $p(x) = -4$

46. $f(x) = \dfrac{3x + 4}{5}$

47. $f(x) = (x + 3)^2, \quad x \geq -3$

48. $q(x) = (x - 5)^2$

49. $h(x) = \dfrac{4}{x^2}$

50. $f(x) = |x - 2|, \quad x \leq 2$

51. $f(x) = \sqrt{2x + 3}$

52. $f(x) = \sqrt{x - 2}$

53. $g(x) = x^2 - x^4$

54. $f(x) = \dfrac{x^2}{x^2 + 1}$

55. $f(x) = 25 - x^2, \quad x \leq 0$

56. $f(x) = ax + b, \quad a \neq 0$

In Exercises 57–60, delete part of the graph of the function so that the part that remains is one-to-one. Find the inverse of the remaining part and give the domain of the inverse. (There is more than one correct answer.)

57. $f(x) = (x - 2)^2$

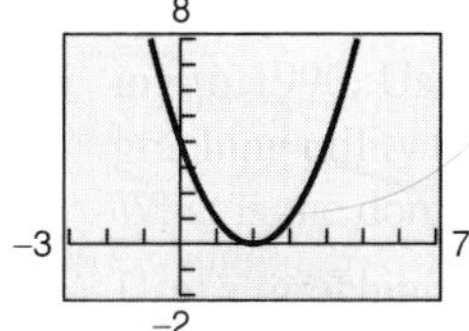

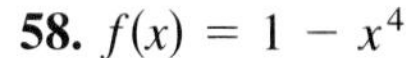

58. $f(x) = 1 - x^4$

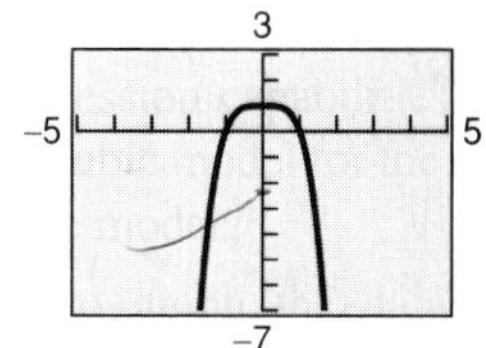

59. $f(x) = |x + 2|$

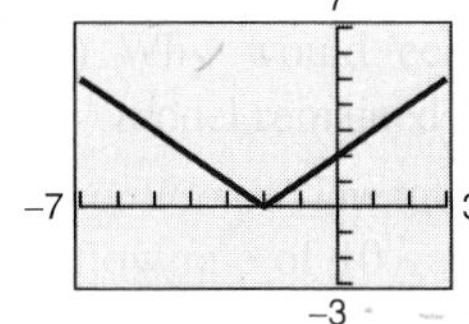

60. $f(x) = |x - 2|$

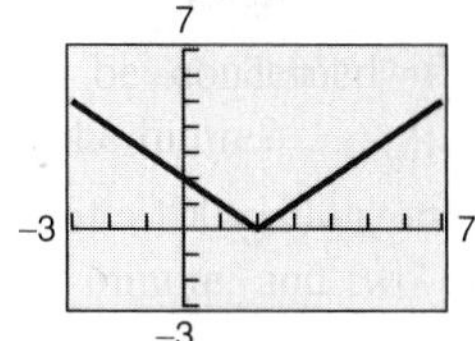

In Exercises 61 and 62, use the graph of the function f to complete the table and sketch the graph of f^{-1}.

61.

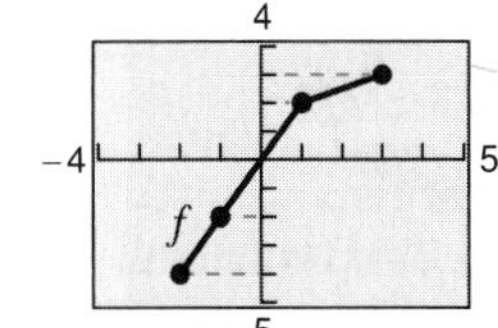

x	$f^{-1}(x)$
-4	
-2	
2	
3	

62.

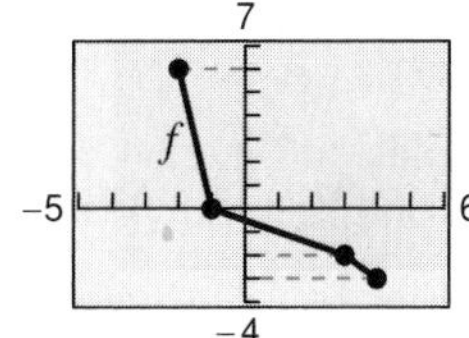

x	$f^{-1}(x)$
-3	
-2	
0	
6	

Graphical Reasoning In Exercises 63–66, (a) use a graphing utility to graph the function, (b) use the DrawInv feature of the graphing utility to draw the inverse of the function, and (c) determine if the graph of the inverse relation is an inverse function (explain).

63. $f(x) = x^3 + x + 1$

64. $h(x) = x\sqrt{4 - x^2}$

65. $g(x) = \dfrac{3x^2}{x^2 + 1}$

66. $f(x) = \dfrac{4x}{\sqrt{x^2 + 15}}$

True or False? In Exercises 67–70, determine if the statement is true or false. If it is false, give an example to show why.

67. If f is an even function, f^{-1} exists.

68. If the inverse of f exists, the y-intercept of f is an x-intercept of f^{-1}.

69. If $f(x) = x^n$ where n is odd, f^{-1} exists.

70. There exists no function f such that $f = f^{-1}$.

In Exercises 71–76, use the functions $f(x) = \frac{1}{8}x - 3$ and $g(x) = x^3$ to find the indicated value or function.

71. $(f^{-1} \circ g^{-1})(1)$

72. $(g^{-1} \circ f^{-1})(-3)$

73. $(f^{-1} \circ f^{-1})(6)$

74. $(g^{-1} \circ g^{-1})(-4)$

75. $(f \circ g)^{-1}$

76. $g^{-1} \circ f^{-1}$

In Exercises 77–80, use the functions $f(x) = x + 4$ and $g(x) = 2x - 5$ to find the specified functions.

77. $g^{-1} \circ f^{-1}$

78. $f^{-1} \circ g^{-1}$

79. $(f \circ g)^{-1}$

80. $(g \circ f)^{-1}$

81. Prove that if f and g are one-to-one functions, $(f \circ g)^{-1}(x) = (g^{-1} \circ f^{-1})(x)$.

82. Prove that if f is a one-to-one odd function, f^{-1} is an odd function.

83. *Hourly Wage* Your wage is \$8.00 per hour plus \$0.75 for each unit produced per hour. Thus, your hourly wage y in terms of the number of units produced is given by $y = 8 + 0.75x$.

(a) Determine the inverse of the function. What does each variable in the inverse function represent?

(b) Use a graphing utility to graph the function and its inverse.

(c) Use the trace key to find the hourly wage if 10 units are produced per hour.

(d) Use the trace key to find the number of units produced when your hourly wage is \$22.25.

84. *Cost* Suppose you need 50 pounds of two commodities costing \$1.25 and \$1.60 per pound, respectively.

(a) Verify that the total cost is $y = 1.25x + 1.60(50 - x)$, where x is the number of pounds of the less expensive commodity.

(b) Find the inverse of the cost function. What does each variable in the inverse function represent?

(c) Use the context of the problem to determine the domain of the inverse function.

(d) Determine the number of pounds of the less expensive commodity purchased if the total cost is \$73.

85. *Diesel Engine* The function

$$y = 0.03x^2 + 254.50, \quad 0 < x < 100$$

approximates the exhaust temperature y of a diesel engine in degrees Fahrenheit, where x is the percent load on the engine.

(a) Determine the inverse of the function. What does each variable in the inverse function represent?

(b) Use a graphing utility to graph the inverse function.

(c) Determine the percent load interval if the exhaust temperature of the engine must not exceed 500° F.

86. *Think About It* The function

$$f(x) = k(2 - x - x^3)$$

is one-to-one and $f^{-1}(3) = -2$. Find k.

Focus on Concepts

In this chapter, you studied several concepts that are required in the study of functions and their graphs. You can use the following questions to check your understanding of several of these basic concepts. The answers to these questions are given in the back of the book.

1. With the information given in the graphs, is it possible to determine the slopes of the two lines? Is it possible that they could have the same slope? Explain.

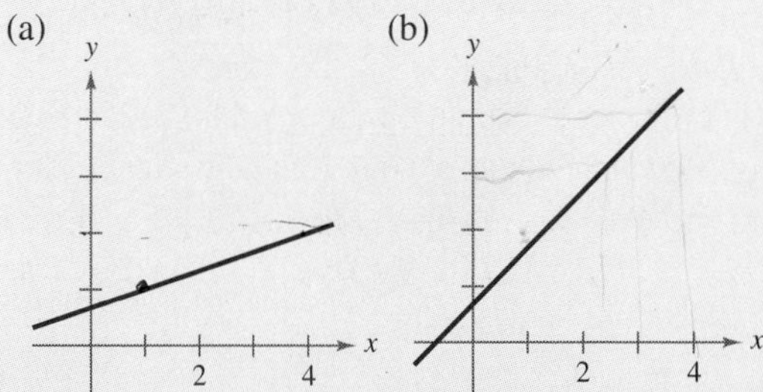

2. The slopes of two lines are -4 and $\frac{5}{2}$. Which is steeper? Explain.

3. The value V of a machine t years after it is purchased is $V = -4000t + 58{,}500, 0 \le t \le 5$. Explain what the V-intercept and slope measure.

4. Does the relationship shown in the figure represent a function from set A to set B? Explain.

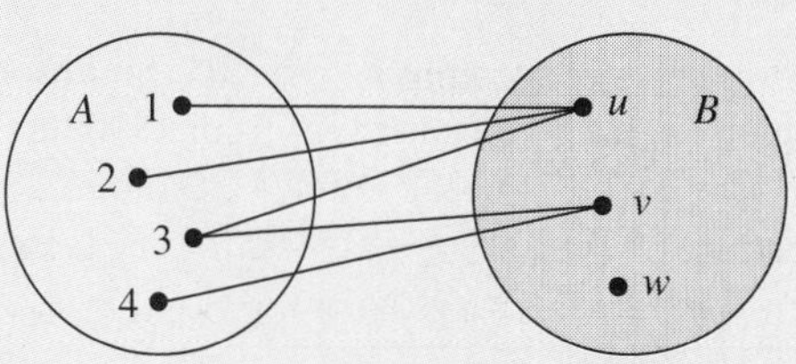

5. Use a graphing utility to select viewing rectangles that would show these graphs.

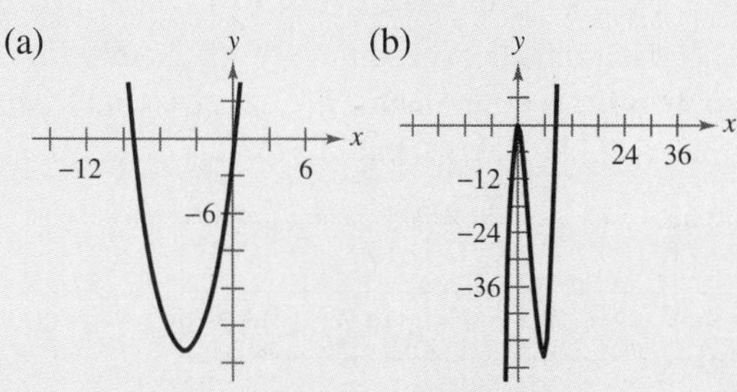

6. If f is an even function, determine if g is even, odd, or neither. Explain.

(a) $g(x) = -f(x)$ (b) $g(x) = f(-x)$

(c) $g(x) = f(x) - 2$ (d) $g(x) = f(x - 2)$

7. Management originally predicted that the profits from the sales of a new product would be approximated by the graph of the function f in the figure. The actual profits are shown by the function g along with a verbal description. Use the concepts of transformations of graphs to write g in terms of f.

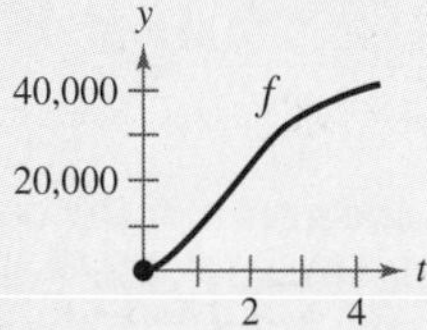

(a) The profits were only three-fourths as large as expected.

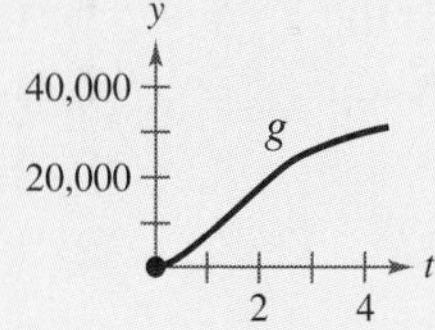

(b) The profits were consistently \$10,000 greater than predicted.

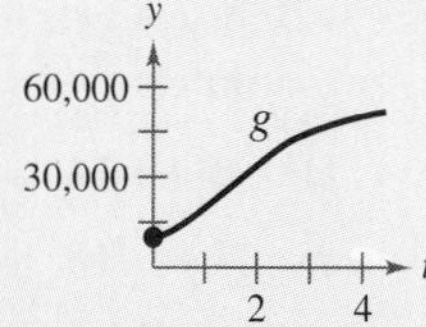

(c) There was a 2-year delay in the introduction of the product. After sales began, profits were as expected.

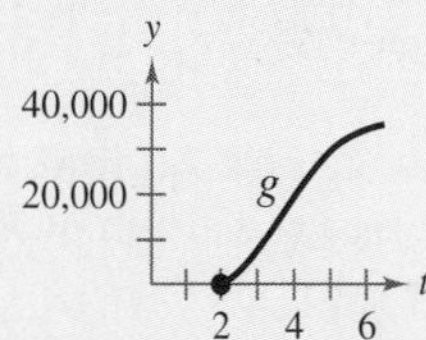

1 /// REVIEW EXERCISES

In Exercises 1 and 2, complete the table. Use the resulting solution points to sketch the graph of the equation. Use a graphing utility to verify the graph.

1. $y = -\frac{1}{2}x + 2$

x	-2	0	2	3	4
y					

2. $y = x^2 - 3x$

x	-1	0	1	2	3
y					

In Exercises 3–12, sketch the graph of the equation *by hand*. Use a graphing utility to verify the graph.

3. $y - 2x - 3 = 0$ **4.** $3x + 2y + 6 = 0$

5. $x - 5 = 0$ **6.** $y = 8 - |x|$

7. $y = \sqrt{5 - x}$ **8.** $y = \sqrt{x + 2}$

9. $y + 2x^2 = 0$ **10.** $y = x^2 - 4x$

11. $y = \sqrt{25 - x^2}$ **12.** $x^2 + y^2 = 10$

In Exercises 13–20, use a graphing utility to graph the equation. Approximate any intercepts.

13. $y = \frac{1}{4}(x + 1)^3$ **14.** $y = 4 - (x - 4)^2$

15. $y = \frac{1}{4}x^4 - 2x^2$ **16.** $y = \frac{1}{4}x^3 - 3x$

17. $y = x\sqrt{9 - x^2}$ **18.** $y = x\sqrt{x + 3}$

19. $y = |x - 4| - 4$ **20.** $y = |x + 2| + |3 - x|$

In Exercises 21 and 22, find a setting on a graphing utility such that the graph of the equation agrees with the given graph.

21. $y = 0.002x^2 - 0.06x - 1$

22. $y = 10x^3 - 21x^2$

Figure for 21

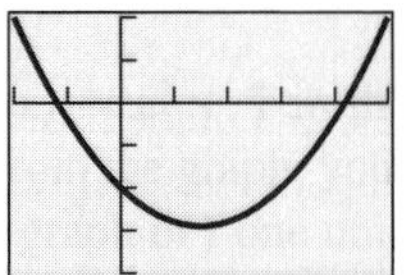

Figure for 22

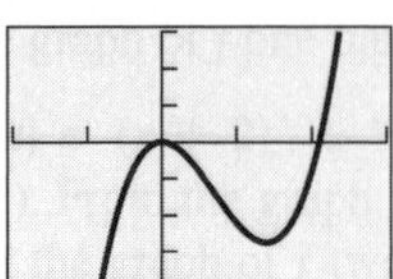

Data Analysis **In Exercises 23 and 24, (a) use a graphing utility to plot the data; (b) use a graphing utility's least squares regression capabilities to find the best-fitting linear model (let $t = 0$ correspond to 1990); (c) graph the model in the same viewing rectangle with the data, and sketch the model for the data; and (d) use the model to estimate the values of y for the years 1998 and 2000.**

23. The total annual expenditures y for the Smithsonian Institution (in millions) each year from 1990 through 1993 are given in the table. (Source: U.S. Department of Treasury)

x	1990	1991	1992	1993
y	302	340	387	395

24. The total annual expenditures y for NASA (in billions) from 1990 through 1993 are given in the table. (Source: U.S. Department of Treasury)

x	1990	1991	1992	1993
y	12.4	13.9	14.0	14.3

In Exercises 25–28, plot the two points and find the slope of the line that passes through the points.

25. $(-4.5, 6), (2.1, 3)$ **26.** $(-3, 2), (8, 2)$

27. $\left(\frac{3}{2}, 1\right), \left(5, \frac{5}{2}\right)$ **28.** $(7, -1), (7, 12)$

In Exercises 29–32, use the concept of slope to find t such that the three points are collinear.

29. $(-2, 5), (0, t), (1, 1)$

30. $(-6, 1), (1, t), (10, 5)$

31. $(1, -4), (t, 3), (5, 10)$

32. $(-3, 3), (t, -1), (8, 6)$

In Exercises 33–36, use the point on the line and the slope of the line to find three additional points through which the line passes. (The solution is not unique.)

	Point	*Slope*
33.	$(2, -1)$	$m = \frac{1}{4}$
34.	$(-3, 5)$	$m = -\frac{3}{2}$
35.	$(-6, -5)$	$m = -2$
36.	$(10, -6)$	m is undefined.

In Exercises 37–42, find an equation of the line that passes through the points.

37. $(0, 0), (0, 10)$ **38.** $(-1, 4), (2, 0)$

39. $(2, 1), (14, 6)$ **40.** $(-2, 2), (3, -10)$

41. $(-1, 0), (6, 2)$ **42.** $(1, 6), (4, 2)$

In Exercises 43–46, find an equation of the line that passes through the given point and has the specified slope. Sketch the graph of the line.

43. $(0, -5), \quad m = \frac{3}{2}$ **44.** $(-2, 6), \quad m = 0$

45. $(3, 0), \quad m = -\frac{2}{3}$ **46.** $(5, 4), \quad m$ is undefined.

In Exercises 47 and 48, write equations of the lines through the point (a) parallel to the given line and (b) perpendicular to the given line. Verify your result with a graphing utility (use a square setting).

	Point	*Line*
47.	$(3, -2)$	$5x - 4y = 8$
48.	$(-8, 3)$	$2x + 3y = 5$

Rate of Change In Exercises 49 and 50, you are given the dollar value of a product in 1996 *and* the rate at which the value of the item is expected to change during the next 5 years. Use this information to write a linear equation that gives the dollar value V of the product in terms of the year t. (Let $t = 6$ represent 1996.)

	1996 Value	*Rate*
49.	\$12,500	\$850 increase per year
50.	\$72.95	\$5.15 increase per year

Exploration In Exercises 51 and 52, find a relationship between x and y such that (x, y) is equidistant from the two points.

51. $(-2, -5), (6, 3)$ **52.** $\left(1, \frac{7}{2}\right), (5, 0)$

53. *Fourth-Quarter Sales* During the second and third quarters of the year, a business had sales of \$160,000 and \$185,000, respectively. If the growth of sales follows a linear pattern, estimate sales during the fourth quarter.

54. *Dollar Value* The dollar value of a product in 1995 is \$85 and the product will increase in value at an expected rate of \$3.75 per year.

(a) Write a linear equation that gives the dollar value V of the product in terms of the year t. (Let $t = 5$ represent 1995.)

(b) Use a graphing utility to graph the sales equation.

(c) Move the cursor along the graph of the sales model to estimate the dollar value of the product in 2000.

In Exercises 55–58, identify the equations that determine y as a function of x.

55. $16x - y^4 = 0$

56. $2x - y - 3 = 0$

57. $y = \sqrt{1 - x}$

58. $|y| = x + 2$

In Exercises 59 and 60, evaluate the function at the specified values of the independent variable. Simplify your answers.

59. $f(x) = x^2 + 1$

(a) $f(2)$ (b) $f(t^2)$ (c) $-f(x)$

60. $g(x) = x^{4/3}$

(a) $g(8)$ (b) $g(t + 1)$ (c) $\dfrac{g(8) - g(1)}{8 - 1}$

In Exercises 61–64, determine the domain of the function. Verify your result with a graphing utility.

61. $f(x) = \sqrt{25 - x^2}$

62. $f(x) = 3x + 4$

63. $g(s) = \dfrac{5}{3s - 9}$

64. $f(x) = \sqrt{x^2 + 8x}$

In Exercises 65 and 66, select the viewing rectangle on a graphing utility that shows the most complete graph of the function.

65. $f(x) = \dfrac{3x}{2(3 - x)}$

Xmin = -4	Xmin = -5	Xmin = 0
Xmax = 4	Xmax = 10	Xmax = 20
Xscl = 1	Xscl = 1	Xscl = 2
Ymin = -3	Ymin = -8	Ymin = 0
Ymax = 3	Ymax = 6	Ymax = 10
Yscl = 1	Yscl = 1	Yscl = 2

66. $f(x) = 4[(0.3x)^3 - 5x]$

Xmin = -200	Xmin = -10	Xmin = -15
Xmax = 200	Xmax = 10	Xmax = 15
Xscl = 50	Xscl = 2	Xscl = 5
Ymin = -500	Ymin = -20	Ymin = -150
Ymax = 500	Ymax = 20	Ymax = 150
Yscl = 50	Yscl = 4	Yscl = 50

Graphical Analysis **In Exercises 67 and 68, use a graphing utility to (a) approximate the intervals in which the function is increasing, decreasing, or constant; (b) approximate (to two-decimal-place accuracy) any relative maximum or minimum values of the function; and (c) determine if the function is even, odd, or neither.**

67. $f(x) = (x^2 - 4)^2$

68. $h(x) = 4x^3 - x^4$

In Exercises 69–72, (a) find f^{-1}, (b) graph f and f^{-1} in the same viewing rectangle, and (c) verify that $f^{-1}(f(x)) = x = f(f^{-1}(x))$.

69. $f(x) = \frac{1}{2}x - 3$

70. $f(x) = 5x - 7$

71. $f(x) = \sqrt{x + 1}$

72. $f(x) = x^3 + 2$

In Exercises 73 and 74, restrict the domain of the function f to an interval over which the function is increasing, and determine f^{-1} over that interval. Use a graphing utility to graph f and f^{-1} in the same viewing rectangle.

73. $f(x) = 2(x - 4)^2$

74. $f(x) = |x - 2|$

In Exercises 75–82, let $f(x) = 3 - 2x$, $g(x) = \sqrt{x}$, and $h(x) = 3x^2 + 2$, and find the indicated value.

75. $(f - g)(4)$

76. $(f + h)(5)$

77. $(fh)(1)$

78. $\left(\dfrac{g}{h}\right)(1)$

79. $(h \circ g)(7)$

80. $(g \circ f)(-2)$

81. $g^{-1}(3)$

82. $(h \circ f^{-1})(1)$

83. *Exploration* A wire 24 inches long is to be cut into four pieces to form a rectangle whose shortest side has a length of x.

(a) Express the area A of the rectangle as a function of x.

(b) Determine the domain of the function and use a graphing utility to graph the function over that domain.

(c) Use the graph of the function to approximate the maximum area of the rectangle. Make a conjecture about the dimensions that yield the maximum area of the rectangle.

CHAPTER PROJECT *Modeling the Area of a Plot*

Many real-life problems can be analyzed from a *graphical,* a *numerical,* and an *analytical* perspective. In this project, you will use all three strategies to determine the maximum size of a rectangular plot that can be enclosed by a fixed amount of fencing.

You have 100 meters of fencing material to enclose a rectangular plot. Your goal is to determine the dimensions of the plot such that you enclose the maximum area possible.

(a) Express the area $A(x)$ of the rectangular plot as a function of the length x of one side, as shown in the figure at the left.

(b) Analyze the problem *numerically* by completing the table.

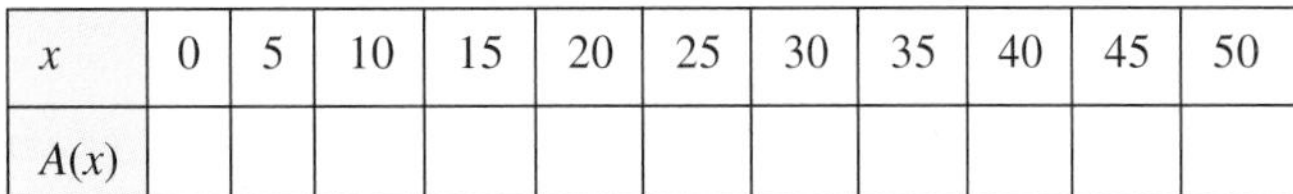

x	0	5	10	15	20	25	30	35	40	45	50
$A(x)$											

According to this table, what do you think the dimensions of the plot should be to enclose the maximum area? Explain.

(c) Use a graphing utility to graph the area function. What is the domain of the function? Solve the problem *graphically* by using the trace feature of your graphing utility to find the value of x that yields the maximum area.

(d) Solve the problem *analytically* by showing that the area function can be written as $A(x) = 625 - (x - 25)^2$. How does this form of the function allow you to find the dimensions that produce a maximum area?

(e) Discuss the strengths and weaknesses of the three strategies used in parts (b), (c), and (d).

Questions for Further Exploration

1. Suppose you were not restricted to a rectangular plot. Would you be able to use 100 meters of fencing to enclose a greater area? Explain.

2. In the project above, you found the maximum area that can be enclosed in a rectangular plot using 100 meters of fencing. If you doubled the amount of fencing, could you enclose twice as much area? Use numerical, graphical, and analytical approaches and explain your reasoning.

3. Suppose the rectangular plot runs along a building, so that you need to fence only three sides. What dimensions will now yield a maximum area?

1 /// CHAPTER TEST

Take this test as you would take a test in class. After you are done, check your work against the answers given in the back of the book.

The *Interactive* CD-ROM provides answers to the Chapter Tests and Cumulative Tests. It also offers Chapter Pre-Tests (that test key skills and concepts covered in previous chapters) and Chapter Post-Tests, both of which have randomly generated exercises with diagnostic capabilities.

In Exercises 1–4, graph the equation and identify any intercepts.

1. $y = 4 - \frac{3}{4}|x|$ **2.** $y = 4 - (x - 2)^2$

3. $y = x - x^3$ **4.** $y = \sqrt{3 - x}$

5. A line passes through the point $(3, -1)$ with slope $m = \frac{3}{2}$. List three additional points on the line.

6. Find the x- and y-intercepts of the graph of $3x - 2y - 9 = 0$.

7. Find an equation of the line that passes through the point $(0, 4)$ and is perpendicular to the line $5x + 2y = 3$.

8. The graph of $y^2(4 - x) = x^3$ is shown at the right. Does the graph represent y as a function of x? Explain.

Figure for 8

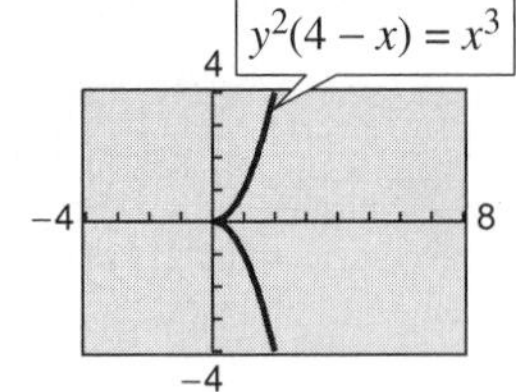

In Exercises 9–12, use the function $f(x) = 10 - \sqrt{3 - x}$.

9. Evaluate: $f(-6)$ **10.** Simplify: $f(t - 3)$

11. Simplify: $\dfrac{f(x) - f(2)}{x - 2}$ **12.** Determine the domain of f.

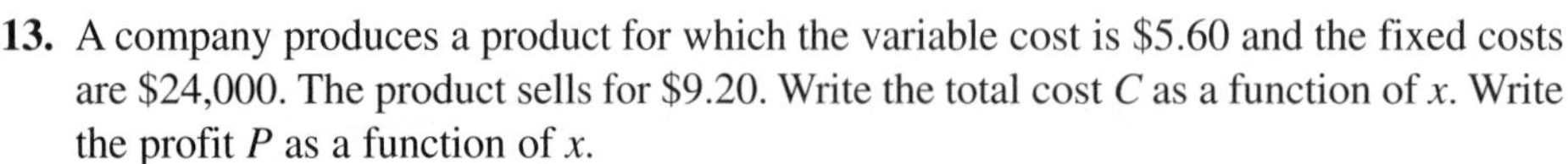

13. A company produces a product for which the variable cost is \$5.60 and the fixed costs are \$24,000. The product sells for \$9.20. Write the total cost C as a function of x. Write the profit P as a function of x.

14. The graph of a function g is shown at the right. Sketch graphs of (a) $\frac{1}{2}g(x - 2)$ and (b) $g\left(\frac{1}{2}x\right) - 1$.

Figure for 14

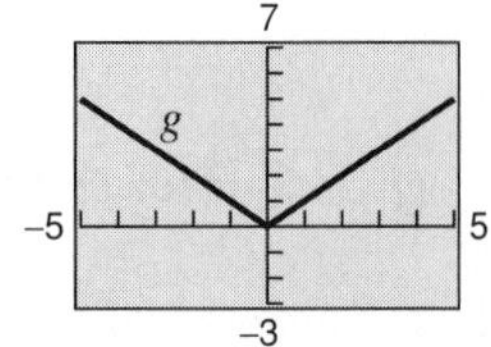

In Exercises 15 and 16, find the intervals for which the function is increasing, decreasing, and/or constant.

15. $h(x) = \frac{1}{4}x^4 - 2x^2$ **16.** $g(t) = |t + 2| - |t - 2|$

17. Use the functions $f(x) = x^2$ and $g(x) = \sqrt{2 - x}$ to find the specified function and its domain.

(a) $(f - g)(x)$ (b) $\left(\dfrac{f}{g}\right)(x)$ (c) $(f \circ g)(x)$ (d) $g^{-1}(x)$

18. Explain how to determine whether a function has an inverse. What is the relationship between the graph of a function and its inverse?

Chapter 2

Intercepts, Zeros, and Solutions

Many classical guitarists complain about the sound made by the guitar—the volume is too low, the notes are difficult to sustain, the treble is feeble, and the sound is uneven. Yet, in spite of advances in physics and acoustics, the design of classical guitars has remained unchanged for over 150 years.

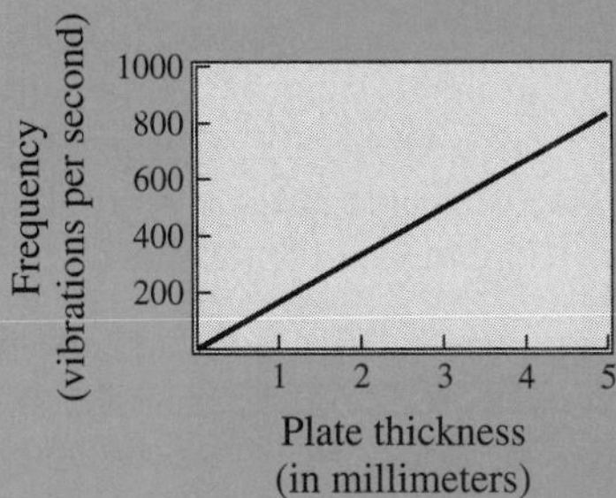

To correct these complaints, Michael Kasha, of Florida State University, used physics and mathematics to design a new classical guitar. As part of his work, he used the equation for the frequency of the vibrations in a circular plate

$$v = \frac{2.6t}{d^2}\sqrt{\frac{E}{\rho}}.$$

In this equation, v is the frequency, t is the plate thickness, d is the diameter of the plate, E is the elasticity of the plate material, and ρ is the density of the plate material. For fixed values of d, E, and ρ, the graph of the line is $v = \frac{500}{3}t$. (See Exercises 74–77 on page 227.)

This classical guitar was designed by Michael Kasha of Florida State University. The hole was moved to increase the surface area for bass response, and interior bracing is used for greater resonant range.

Richard Schneider; (inset) Ruth Goodman (Dr. Kasha/B&W).

2.1 Linear Equations and Modeling

Equations and Solutions of Equations / Linear Equations / Using Mathematical Models to Solve Problems / Common Formulas

Equations and Solutions of Equations

An **equation** is a statement that two algebraic expressions are equal. For example, $3x - 5 = 7$, $x^2 - x - 6 = 0$, and $\sqrt{2x} = 4$ are equations. To **solve** an equation in x means to find all values of x for which the equation is true. Such values are **solutions.** For instance, $x = 4$ is a solution of the equation $3x - 5 = 7$, because $3(4) - 5 = 7$ is a true statement.

The solutions of an equation depend on the kinds of numbers being considered. For instance, in the set of rational numbers $x^2 = 10$ has no solution because there is no rational number whose square is 10. However, in the set of real numbers the equation has the two solutions $\sqrt{10}$ and $-\sqrt{10}$.

An equation that is true for *every* real number in the domain of the variable is called an **identity.** For example, $x^2 - 9 = (x + 3)(x - 3)$ is an identity because it is a true statement for any real value of x, and $x/(3x^2) = 1/(3x)$, where $x \neq 0$, is an identity because it is true for any nonzero real value of x.

An equation that is true for just *some* (or even none) of the real numbers in the domain of the variable is called a **conditional equation.** For example, the equation $x^2 - 9 = 0$ is conditional because $x = 3$ and $x = -3$ are the only values in the domain that satisfy the equation. Learning to solve conditional equations is the primary focus of this chapter.

Linear Equations

An ancient Egyptian papyrus, discovered in 1858, contains one of the earliest examples of mathematical writing in existence. The papyrus itself dates back to around 1650 B.C., but it is actually a copy of writings from two centuries earlier. The algebraic equations on the papyrus were written in words. Diophantus, a Greek who lived around A.D. 250, is often called the Father of Algebra. He was the first to use abbreviated word forms in equations.

Definition of Linear Equation

A **linear equation** in one variable x is an equation that can be written in the standard form

$$ax + b = 0$$

where a and b are real numbers with $a \neq 0$.

A linear equation has exactly one solution. To see this, consider the following steps. (Remember that $a \neq 0$.)

$$ax + b = 0 \qquad \text{Original equation}$$

$$ax = -b \qquad \text{Subtract } b \text{ from both sides.}$$

$$x = -\frac{b}{a} \qquad \text{Divide both sides by } a.$$

To solve a conditional equation in x, isolate x on one side of the equation by a sequence of **equivalent** (and usually simpler) equations, each having the same solution(s) as the original equation. The operations that yield equivalent equations come from the Substitution Principle and the simplification techniques studied in Chapter P.

Generating Equivalent Equations

An equation can be transformed into an *equivalent equation* by one or more of the following steps.

	Original Equation	*Equivalent Equation*
1. Remove symbols of grouping, combine like terms, or reduce fractions on one or both sides of the equation.	$2x - x = 4$	$x = 4$
2. Add (or subtract) the same quantity to (from) *both* sides of the equation.	$x + 1 = 6$	$x = 5$
3. Multiply (or divide) *both* sides of the equation by the same *nonzero* quantity.	$2x = 6$	$x = 3$
4. Interchange the two sides of the equation.	$2 = x$	$x = 2$

EXAMPLE 1 **Solving a Linear Equation**

Solve $3x - 6 = 0$.

Solution

$3x - 6 = 0$ Original equation

$3x = 6$ Add 6 to both sides.

$x = 2$ Divide both sides by 3.

Check: After solving an equation, you should **check each solution** in the *original* equation.

$3x - 6 = 0$ Original equation

$3(2) - 6 \stackrel{?}{=} 0$ Substitute 2 for x.

$0 = 0$ Solution checks. ✓

EXPLORATION

Use a graphing utility to graph the equation $y = 3x - 6$. Use the result to estimate the x-intercept of the graph. Explain how the x-intercept is related to the solution of the equation $3x - 6 = 0$, as shown in Example 1.

Study Tip

Students sometime tell us that a solution looks easy when we work it out in class, but that they don't see where to begin when trying it alone. Keep in mind that no one—not even great mathematicians—can expect to look at every mathematical problem and know immediately where to begin. Many problems involve some trial and error before a solution is found. To make algebra work for you, put in a lot of time, expect to try solution methods that end up not working, and learn from both your successes and your failures.

To solve an equation involving fractional expressions, find the least common denominator of all terms in the equation and multiply every term by this LCD. This procedure clears the equation of fractions.

EXAMPLE 2 Solving an Equation Involving Fractions

$$\frac{x}{3} + \frac{3x}{4} = 2 \qquad \text{Original equation}$$

$$(12)\frac{x}{3} + (12)\frac{3x}{4} = (12)2 \qquad \text{Multiply by the LCD.}$$

$$4x + 9x = 24 \qquad \text{Reduce and multiply.}$$

$$13x = 24 \qquad \text{Combine like terms.}$$

$$x = \frac{24}{13} \qquad \text{Divide both sides by 13.}$$

The solution is $\frac{24}{13}$. Check this in the original equation.

When multiplying or dividing an equation by a *variable* expression, it is possible to introduce an **extraneous** solution—one that does not satisfy the original equation. The next example demonstrates the importance of checking your solution when you have multiplied or divided by a variable expression.

EXAMPLE 3 An Equation with an Extraneous Solution

Solve the equation for x.

$$\frac{1}{x-2} = \frac{3}{x+2} - \frac{6x}{x^2-4}$$

Solution

In this case, the LCD is $x^2 - 4 = (x + 2)(x - 2)$. Multiplying every term by the LCD and reducing produces the following.

$$\frac{1}{x-2}(x+2)(x-2) = \frac{3}{x+2}(x+2)(x-2) - \frac{6x}{x^2-4}(x+2)(x-2)$$

$$x + 2 = 3(x - 2) - 6x, \quad x \neq \pm 2$$

$$x + 2 = 3x - 6 - 6x$$

$$4x = -8$$

$$x = -2$$

A check of $x = -2$ in the original equation shows that it yields a denominator of zero. Thus, $x = -2$ is extraneous, and the equation has *no solution.*

EXPLORATION

Use a graphing utility to graph the two functions

$y_1 = 1/(x - 2)$

$y_2 = 3/(x + 2) - 6x/(x^2 - 4)$

in the same viewing rectangle. What are the domains of these functions? Do the graphs of the functions intersect each other? What does this tell you about the solution to Example 3?

Using Mathematical Models to Solve Problems

One of the primary goals of this text is to learn how algebra can be used to solve problems that occur in real-life situations. This procedure is called **mathematical modeling.**

A good approach to mathematical modeling is to use two stages. Begin by using the verbal description of the problem to form a *verbal model.* Then, after assigning labels to the unknown quantities in the verbal model, form a *mathematical model* or *algebraic equation.*

Verbal description → Verbal model → Algebraic equation

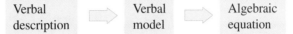

When you are trying to construct a verbal model, it is helpful to look for a *hidden equality*—a statement that two algebraic expressions are equal. These two expressions might be explicitly stated as being equal, or they might be known to be equal (based on prior knowledge or experience).

Real Life

EXAMPLE 4 Finding the Dimensions of a Room

A rectangular family room is twice as long as it is wide, and its perimeter is 84 feet. Find the dimensions of the family room.

Solution

For this problem, it helps to sketch a picture, as shown in Figure 2.1.

Figure 2.1

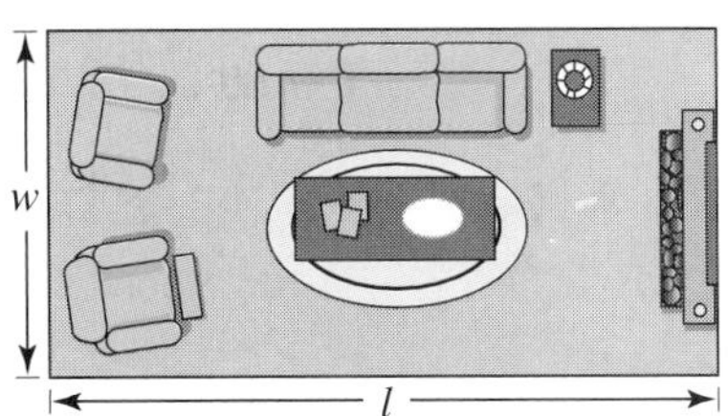

Verbal Model: $2 \cdot \text{Length} + 2 \cdot \text{Width} = \text{Perimeter}$

Labels:
Perimeter $= 84$ (feet)
Width $= w$ (feet)
Length $= l = 2w$ (feet)

Equation:

$$2(2w) + 2w = 84$$
$$6w = 84$$
$$w = 14$$
$$l = 2w = 28$$

The dimensions of the room are 14 feet by 28 feet.

EXAMPLE 5 **A Distance Problem** *Real Life*

Figure 2.2

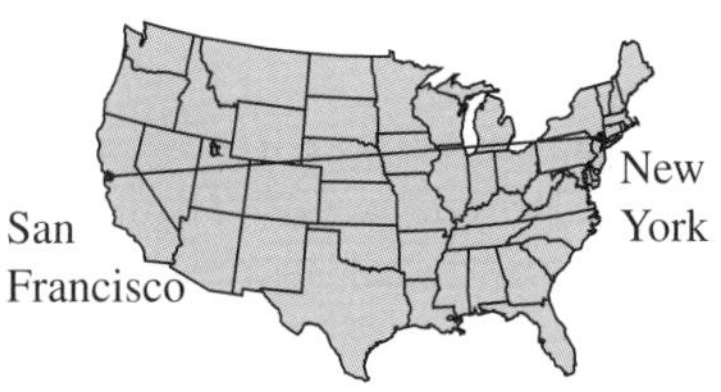

A plane is flying nonstop from New York to San Francisco, a distance of about 2700 miles, as shown in Figure 2.2. After $1\frac{1}{2}$ hours in the air, the plane flies over Chicago (a distance of 800 miles from New York). Estimate the time it will take the plane to fly from New York to San Francisco.

Solution

Verbal Model: Distance = Rate · Time

Labels:
Distance = 2700 (miles)
Time = t (hours)
$\text{Rate} = \frac{\text{Distance to Chicago}}{\text{Time to Chicago}} = \frac{800}{1.5}$ (miles per hour)

Equation:
$$2700 = \frac{800}{1.5}t$$
$$5.06 \approx t$$

The trip will take about 5.06 hours or 5 hours and 4 minutes.

EXAMPLE 6 **An Application Involving Similar Triangles** *Real Life*

Figure 2.3

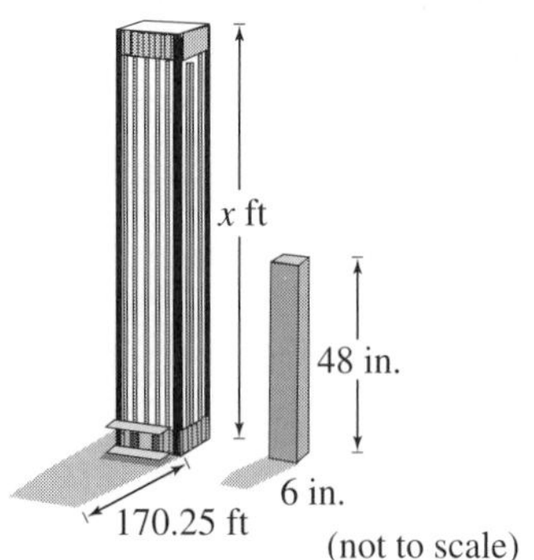

To measure the height of the twin towers of the World Trade Center, you measure the shadow cast by one of the buildings and find it to be 170.25 feet long, as shown in Figure 2.3. Then you measure the shadow cast by a 4-foot post and find it to be 6 inches long. Estimate the building's height.

Solution

To solve this problem, you use a result from geometry that states that the ratios of corresponding sides of similar triangles are equal.

Verbal Model:
$$\frac{\text{Height of building}}{\text{Length of building's shadow}} = \frac{\text{Height of post}}{\text{Length of post's shadow}}$$

Labels:
Height of building = x (feet)
Length of building's shadow = 170.25 (feet)
Height of post = 4 feet = 48 inches (inches)
Length of post's shadow = 6 (inches)

Equation:
$$\frac{x}{170.25} = \frac{48}{6}$$
$$x = 1362$$

Thus, the World Trade Center is about 1362 feet high.

The *Interactive* CD-ROM shows every example with its solution; clicking on the *Try It!* button brings up similar problems. Guided Examples and Integrated Examples show step-by-step solutions to additional examples. Integrated Examples are related to several concepts in the section.

Real Life

EXAMPLE 7 **A Simple Interest Problem**

You invested \$10,000 at $9\frac{1}{2}\%$ and 11% simple interest. During one year, the two accounts earned \$1038.50. How much did you invest at each rate?

Note Example 7 uses the simple interest formula $I = Prt$, where I is the interest, P is the principal, r is the annual interest rate (in decimal form), and t is the time in years.

Solution

Verbal Model: [Interest from $9\frac{1}{2}\%$] + [Interest from 11%] = [Total interest]

Labels:

Amount invested at $9\frac{1}{2}\% = x$ (dollars)
Amount invested at $11\% = 10{,}000 - x$ (dollars)
Interest from $9\frac{1}{2}\% = Prt = (x)(0.095)(1)$ (dollars)
Invested from $11\% = Prt = (10{,}000 - x)(0.11)(1)$ (dollars)
Total interest $= 1038.50$ (dollars)

Equation:

$$0.095x + 0.11(10{,}000 - x) = 1038.50$$
$$-0.015x = -61.5$$
$$x = \$4100 \text{ at } 9\tfrac{1}{2}\%$$
$$10{,}000 - x = \$5900 \text{ at } 11\%$$

Study Tip

Notice in Examples 7 and 8 that percents are expressed as decimals. For instance, $r = 11\%$ is written as $r = 0.11$ in Example 7, and 22% is written as 0.22 in Example 8.

Real Life

EXAMPLE 8 **An Inventory Problem**

A store has \$30,000 of inventory in 12-inch and 19-inch color televisions. The profit on a 12-inch set is 22% and the profit on a 19-inch set is 40%. The profit for the entire stock is 35%. How much was invested in each type of television?

Solution

Verbal Model: [Profit from 12-inch sets] + [Profit from 19-inch sets] = [Total profit]

Labels:

Inventory of 12-inch sets $= x$ (dollars)
Inventory of 19-inch sets $= 30{,}000 - x$ (dollars)
Profit from 12-inch sets $= 0.22x$ (dollars)
Profit from 19-inch sets $= 0.40(30{,}000 - x)$ (dollars)
Total profit $= 0.35(30{,}000) = 10{,}500$ (dollars)

Equation:

$$0.22x + 0.40(30{,}000 - x) = 10{,}500$$
$$-0.18x = -1500$$
$$x \approx \$8333.33 \text{ in 12-inch sets}$$
$$30{,}000 - x \approx \$21{,}666.67 \text{ in 19-inch sets}$$

Common Formulas

Many common types of geometric, scientific, and investment problems use ready-made equations, called **formulas.** Knowing these formulas will help you translate and solve a wide variety of real-life applications.

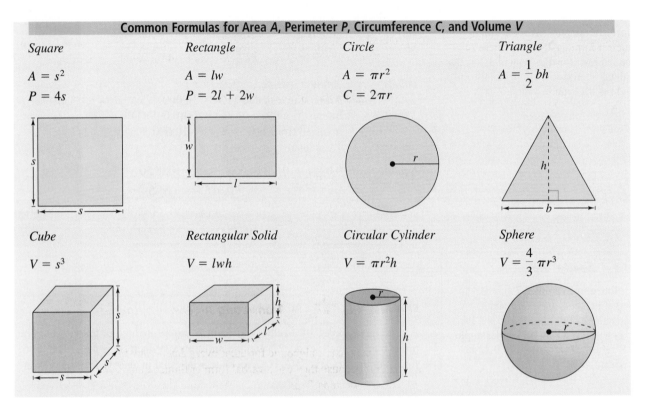

Common Formulas for Area A, Perimeter P, Circumference C, and Volume V

Square

$A = s^2$

$P = 4s$

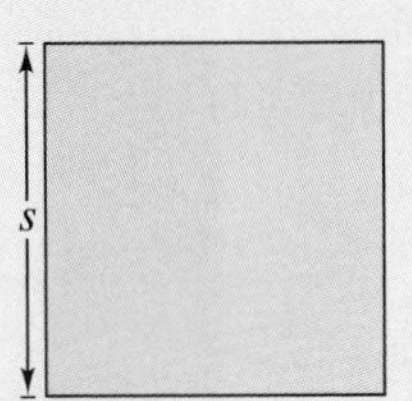

Rectangle

$A = lw$

$P = 2l + 2w$

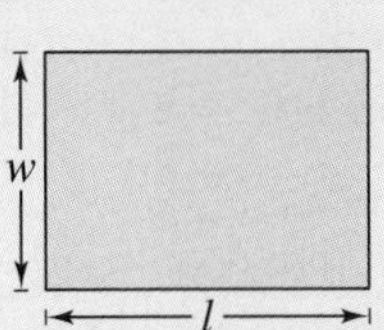

Circle

$A = \pi r^2$

$C = 2\pi r$

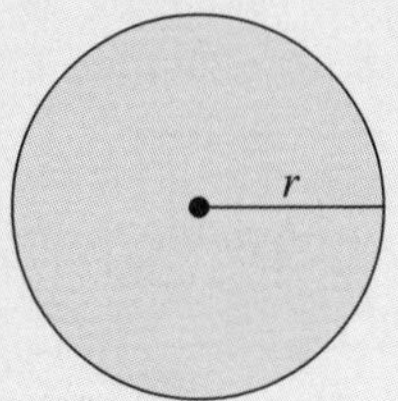

Triangle

$A = \frac{1}{2}bh$

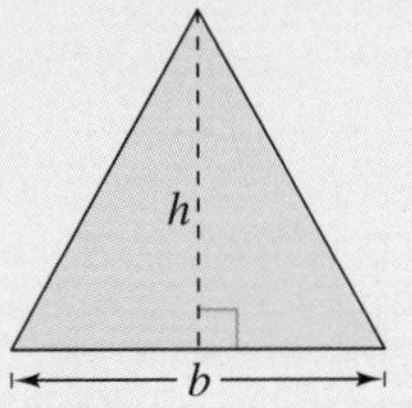

Cube

$V = s^3$

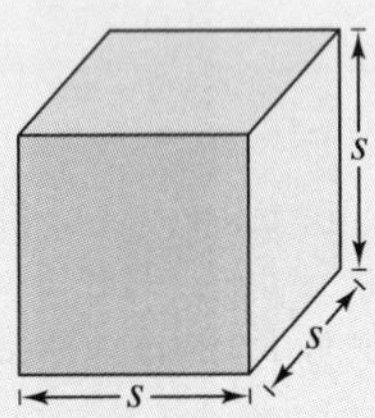

Rectangular Solid

$V = lwh$

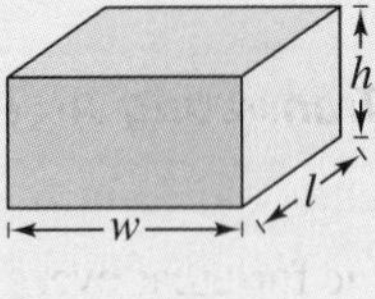

Circular Cylinder

$V = \pi r^2 h$

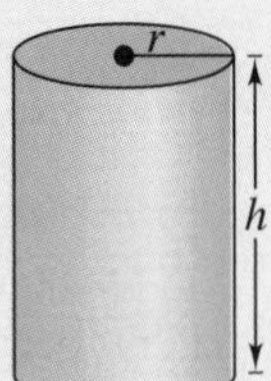

Sphere

$V = \frac{4}{3}\pi r^3$

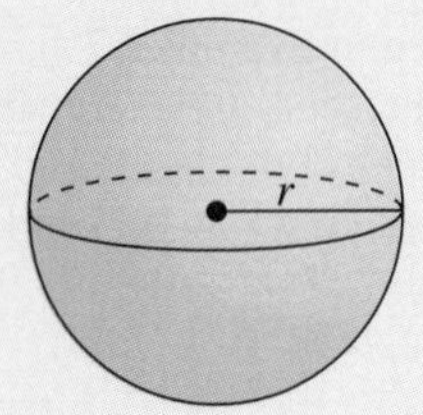

Miscellaneous Common Formulas

Temperature:	$F = \frac{9}{5}C + 32$	F = degrees Fahrenheit, C = degrees Celsius
Simple Interest:	$I = Prt$	I = interest, P = principal, r = annual interest rate, t = time in years
Compound Interest:	$A = P\left(1 + \frac{r}{n}\right)^{nt}$	A = balance, P = principal, r = annual interest rate, n = compoundings per year, t = time in years
Distance:	$d = rt$	d = distance traveled, r = rate, t = time

When working with applied problems you often need to rewrite one of the common formulas. For instance, the formula $P = 2l + 2w$, for the perimeter of a rectangle, can be rewritten or solved for w as $w = \frac{1}{2}(P - 2l)$.

Figure 2.4

Real Life

EXAMPLE 9 **Using a Formula**

A cylindrical can has a volume of 300 cubic centimeters (cm^3) and a radius of 3 centimeters (cm), as shown in Figure 2.4. Find the height of the can.

Solution

The formula for the *volume of a cylinder* is $V = \pi r^2 h$. To find the height of the can, solve for h.

$$h = \frac{V}{\pi r^2}$$

Then, using $V = 300 \text{ cm}^3$ and $r = 3$ cm, find the height.

$$h = \frac{300 \text{ cm}^3}{\pi(3 \text{ cm})^2} = \frac{300 \text{ cm}^3}{9\pi \text{ cm}^2} \approx 10.61 \text{ cm}$$

Group Activity *Translating Algebraic Formulas*

Most people use algebraic formulas every day—sometimes without realizing it because they use a verbal form or think of an often-repeated calculation in steps. Translate each of the following verbal descriptions into an algebraic formula, and demonstrate the use of each formula.

a. *The Christmas Tree Rule* "To find out how many lights your Christmas tree needs, multiply the tree height times the tree width times three."—Michael Spence, lawyer (Source: *Rules of Thumb* by Tom Parker)

b. *Percent of Calories from Fat* "To calculate percent of calories from fat, multiply grams of total fat per serving by 9, then divide by the number of calories per serving." (Source: *Good Housekeeping*)

c. *Building Stairs* "A set of steps will be comfortable to use if two times the height of one riser plus the width of one tread is equal to 26 inches."—Alice Lukens Bachelder, gardener (Source: *Rules of Thumb* by Tom Parker)

The *Interactive* CD-ROM contains step-by-step solutions to all odd-numbered Section and Review Exercises. It also provides Tutorial Exercises, which link to Guided Examples for additional help.

2.1 /// EXERCISES

In Exercises 1–6, determine whether the given values of x are solutions of the equation.

Equation	*Values*	
1. $5x - 3 = 3x + 5$	(a) $x = 0$	(b) $x = -5$
	(c) $x = 4$	(d) $x = 10$
2. $7 - 3x = 5x - 17$	(a) $x = -3$	(b) $x = 0$
	(c) $x = 8$	(d) $x = 3$
3. $\frac{5}{2x} - \frac{4}{x} = 3$	(a) $x = -\frac{1}{2}$	(b) $x = 4$
	(c) $x = 0$	(d) $x = \frac{1}{4}$
4. $3 + \frac{1}{x+2} = 4$	(a) $x = -1$	(b) $x = -2$
	(c) $x = 0$	(d) $x = 5$
5. $(x + 5)(x - 3) = 20$	(a) $x = 3$	(b) $x = -2$
	(c) $x = 0$	(d) $x = -7$
6. $\sqrt[3]{x - 8} = 3$	(a) $x = 2$	(b) $x = -5$
	(c) $x = 35$	(d) $x = 8$

In Exercises 7–14, determine whether the equation is an identity or a conditional equation.

7. $2(x - 1) = 2x - 2$

8. $3(x + 2) = 5x + 4$

9. $-6(x - 3) + 5 = -2x + 10$

10. $-7(x - 3) + 4x = 3(7 - x)$

11. $x^2 - 8x + 5 = (x - 4)^2 - 11$

12. $x^2 + 2(3x - 2) = x^2 + 6x - 4$

13. $3 + \frac{1}{x+1} = \frac{4x}{x+1}$

14. $\frac{5}{x} + \frac{3}{x} = 24$

15. *Think About It* What is meant by equivalent equations? Give an example of two equivalent equations.

16. *Essay* In your own words, describe the steps used to transform an equation into an equivalent equation.

In Exercises 17 and 18, justify each step of the solution.

17.
$$4x + 32 = 83$$
$$4x + 32 - 32 = 83 - 32$$
$$4x = 51$$
$$\frac{4x}{4} = \frac{51}{4}$$
$$x = \frac{51}{4}$$

18.
$$3(x - 4) + 10 = 7$$
$$3x - 12 + 10 = 7$$
$$3x - 2 = 7$$
$$3x - 2 + 2 = 7 + 2$$
$$3x = 9$$
$$\frac{3x}{3} = \frac{9}{3}$$
$$x = 3$$

In Exercises 19–22, solve the equation mentally.

19. $3x = 15$

20. $\frac{1}{2}t = 7$

21. $s + 12 = 18$

22. $2u - 3 = 25$

In Exercises 23 and 24, solve the equation in two ways. Then explain which way was easier for you.

23. $3(x - 1) = 4$

24. $\frac{3}{4}(z - 4) = 6$

In Exercises 25–36, solve the equation and use a graphing utility to verify your solution.

25. $8x - 5 = 3x + 10$

26. $7x + 3 = 3x - 13$

27. $2(x + 5) - 7 = 3(x - 2)$

28. $2(13t - 15) + 3(t - 19) = 0$

29. $6[x - (2x + 3)] = 8 - 5x$

30. $3(x + 3) = 5(1 - x) - 1$

31. $\frac{5x}{4} + \frac{1}{2} = x - \frac{1}{2}$

32. $\frac{x}{5} - \frac{x}{2} = 3$

33. $\frac{3}{2}(z + 5) - \frac{1}{4}(z + 24) = 0$

34. $\frac{3x}{2} + \frac{1}{4}(x - 2) = 10$

35. $0.25x + 0.75(10 - x) = 3$

36. $0.60x + 0.40(100 - x) = 50$

In Exercises 37–54, solve the equation (if possible) and use a graphing utility to verify your solution.

37. $\frac{100 - 4u}{3} = \frac{5u + 6}{4} + 6$

38. $\frac{17 + y}{y} + \frac{32 + y}{y} = 100$

39. $\frac{5x - 4}{5x + 4} = \frac{2}{3}$

40. $\frac{15}{x} - 4 = \frac{6}{x} + 3$

41. $\frac{1}{x - 3} + \frac{1}{x + 3} = \frac{10}{x^2 - 9}$

42. $\frac{1}{x - 2} + \frac{3}{x + 3} = \frac{4}{x^2 + x - 6}$

43. $\frac{x}{x + 4} + \frac{4}{x + 4} + 2 = 0$

44. $\frac{2}{(x - 4)(x - 2)} = \frac{1}{x - 4} + \frac{2}{x - 2}$

45. $\frac{7}{2x + 1} - \frac{8x}{2x - 1} = -4$

46. $\frac{4}{u - 1} + \frac{6}{3u + 1} = \frac{15}{3u + 1}$

47. $\frac{1}{x} + \frac{2}{x - 5} = 0$

48. $\frac{6}{x} - \frac{2}{x + 3} = \frac{3(x + 5)}{x(x + 3)}$

49. $\frac{3}{x(x - 3)} + \frac{4}{x} = \frac{1}{x - 3}$

50. $3 = 2 + \frac{2}{z + 2}$

51. $(x + 2)^2 + 5 = (x + 3)^2$

52. $(x + 1)^2 + 2(x - 2) = (x + 1)(x - 2)$

53. $(x + 2)^2 - x^2 = 4(x + 1)$

54. $(2x + 1)^2 = 4(x^2 + x + 1)$

In Exercises 55 and 56, find an equation of the form $ax + b = cx$ that has the given solution. (There are many correct answers.)

55. $x = -3$

56. $x = \frac{1}{4}$

Human Height In Exercises 57 and 58, use the following information. The relationship between the length of an adult's thigh bone and the height of the adult can be approximated by the linear equations

$y = 0.432x - 10.44$ Female

$y = 0.449x - 12.15$ Male

where y is the length of the thigh bone in inches and x is the height in inches (see figure).

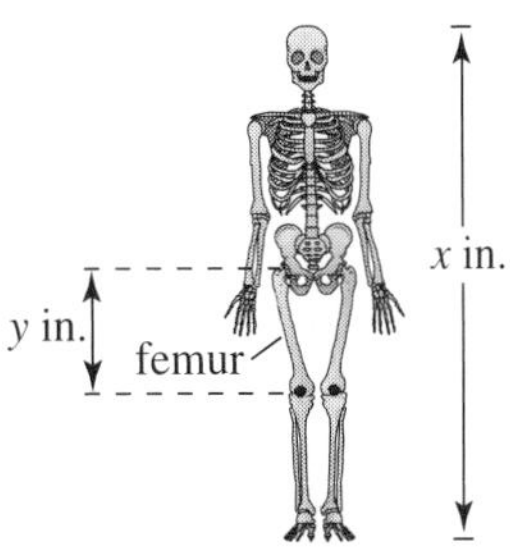

57. An anthropologist discovers a thigh bone belonging to an adult human female. The bone is 16 inches long. Estimate the height of the female.

58. From the foot bones of an adult human male, an anthropologist estimates that the height of the male was 69 inches. A few feet away from the site where the foot bone was discovered, the anthropologist discovered a male adult thigh bone that was 19 inches long. Is it possible that both the foot bone and the thigh bone came from the same person?

59. *Geometry* The surface area S of the rectangular solid in the figure is given by

$$S = 2(24) + 2(4x) + 2(6x).$$

Find the length of the box x if the surface area is 248 square centimeters.

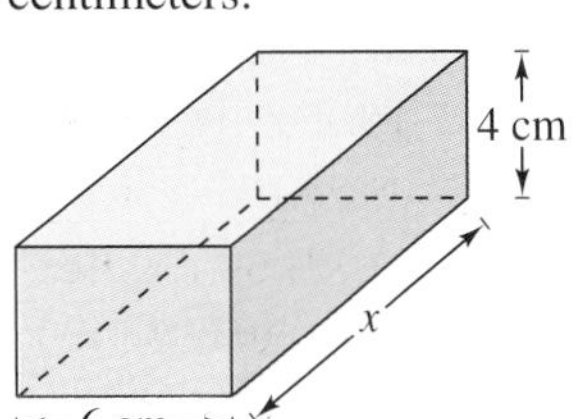

60. *Operating Cost* A delivery company has a fleet of vans. The annual operating cost per van is

$C = 0.32m + 2500$

where m is the number of miles traveled by a van in a year. What number of miles will yield an annual operating cost of $10,000?

Then and Now **In Exercises 61–64, the values or prices of different items are given for 1980 and 1992. Find the percent change for each item.** (Source: Statistical Abstract of the U.S.)

	Item	*1980*	*1992*
61.	Median weekly earnings/family	$400.00	$688.00
62.	Cable TV monthly basic rate	$7.85	$19.08
63.	An ounce of gold	$613.00	$350.00
64.	One acre of farmland	$737.00	$685.00

65. *Travel Time* Suppose you are driving on a Canadian freeway to a town that is 300 kilometers from your home. After 30 minutes you pass a freeway exit that you know is 50 kilometers from your home. Assuming that you continue at the same constant speed, how long will it take for the entire trip?

66. *Travel Time* On the first part of a 317-mile trip, a salesman averaged 58 miles per hour. He averaged only 52 miles per hour on the last part of the trip because of an increased volume of traffic. Find the amount of time at each of the speeds if the total time was 5 hours and 45 minutes.

67. *Travel Time* Two families meet at a park for a picnic. At the end of the day one family travels east at an average speed of 42 miles per hour and the other travels west at an average speed of 50 miles per hour. Both families have approximately 160 miles to travel.

(a) Find the time it takes each family to get home.

(b) Find the time that will have elapsed when they are 100 miles apart.

(c) Find the distance the eastbound family has to travel after the westbound family has arrived home.

68. *Average Speed* A truck driver traveled at an average speed of 55 miles per hour on a 200-mile trip to pick up a load of freight. On the return trip (with the truck fully loaded), the average speed was 40 miles per hour. Find the average speed for the round trip.

69. *Wind Speed* An executive flew in the corporate jet to a meeting in a city 1500 kilometers away. After traveling the same amount of time on the return flight, the pilot mentioned that they still had 300 kilometers to go. If the air speed of the plane was 600 kilometers per hour, how fast was the wind blowing? (Assume that the wind direction was parallel to the flight path and constant all day.)

70. *Speed of Light* Light travels at the speed of 3.0×10^8 meters per second. Find the time in minutes required for light to travel from the sun to the earth (a distance of 1.5×10^{11} meters).

71. *Radio Waves* Radio waves travel at the same speed as light, 3.0×10^8 meters per second. Find the time required for a radio wave to travel from mission control in Houston to NASA astronauts on the surface of the moon 3.86×10^8 meters away.

72. *Height of a Tree* To obtain the height of a tree, you measure the tree's shadow and find that it is 8 meters long. You also measure the shadow of a 2-meter lamppost and find that it is 75 centimeters long (see figure). How tall is the tree?

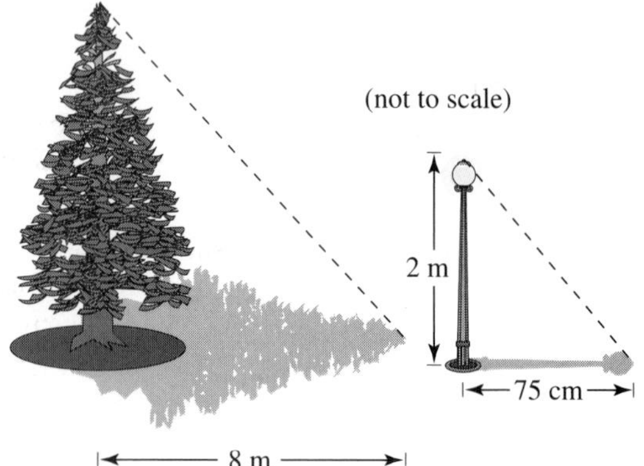

73. *Height of a Building* To obtain the height of a building, you measure the building's shadow and find that it is 80 feet long. You also measure the shadow of a 4-foot stake and find that it is $3\frac{1}{2}$ feet long.

(a) Create a figure that gives a visual representation of the problem. Let h represent the height of the building.

(b) Find the height of the building.

74. *Height of a Flagpole* A person who is 6 feet tall walks away from a flagpole toward the tip of the shadow of the pole. When the person is 30 feet from the pole, the tips of the person's shadow and the shadow cast by the pole coincide at a point 5 feet in front of the person.

(a) Create a figure that gives a visual representation of the problem. Let h represent the height of the pole.

(b) Find the height of the pole.

75. *Investment Mix* You plan to invest \$12,000 in two funds paying $7\frac{1}{2}\%$ and 10% simple interest. (There is more risk in the 10% fund.) Your goal is to obtain a total annual interest income of \$1000 from the investments. What is the smallest amount you can invest in the 10% fund in order to meet your objective?

76. *Investment Mix* You plan to invest \$25,000 in two funds paying 11% and $12\frac{1}{2}\%$ simple interest. (There is more risk in the $12\frac{1}{2}\%$ fund.) Your goal is to obtain a total annual interest income of \$3000 from the investments. What is the smallest amount you can invest in the $12\frac{1}{2}\%$ fund in order to meet your objective?

77. *Comparing Investment Returns* Suppose you invested \$12,000 in a fund paying $9\frac{1}{2}\%$ simple interest and \$8000 in a fund with a variable interest rate. At the end of the year you were notified that the total interest for both funds was \$2054.40. Find the equivalent simple interest rate on the variable-rate fund.

78. *Comparing Investment Returns* Suppose you have \$10,000 on deposit earning simple interest with the interest rate linked to the *prime rate*. Because of a drop in the prime rate, the rate on your investment dropped by $1\frac{1}{2}\%$ for the last quarter of the year. Your annual earnings on the fund were \$1112.50. Find the interest rate for the first three quarters of the year and the interest rate for the last quarter.

79. *Mixture Problem* A grocer mixes two kinds of nuts that cost \$2.49 per pound and \$3.89 per pound, respectively, to make 100 pounds of a mixture that costs \$3.19 per pound. How much of each kind of nut is put into the mixture?

80. *Production Limit* A company has fixed costs of \$10,000 per month and variable costs of \$8.50 per unit manufactured. The company has \$85,000 available to cover the monthly costs. How many units can the company manufacture? (*Fixed costs* are those that occur regardless of the level of production. *Variable costs* depend on the level of production.)

Statics Problems In Exercises 81 and 82, suppose you have a uniform beam of length L with a fulcrum x feet from one end (see figure). If objects with weights W_1 and W_2 are placed at opposite ends of the beam, the beam will balance if

$$W_1x = W_2(L - x).$$

Find x such that the beam will balance.

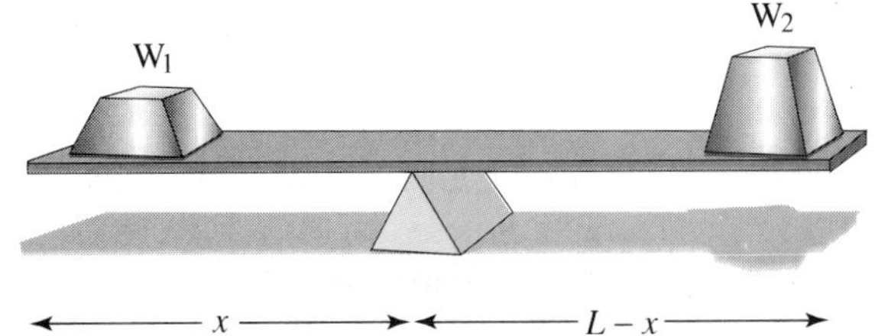

81. Two children weighing 50 pounds and 75 pounds are going to play on a seesaw that is 10 feet long.

82. A person weighing 200 pounds is attempting to move a 550-pound rock with a bar that is 5 feet long.

In Exercises 83–86, solve for the indicated variable.

83. *Area of a Triangle*

Solve for h: $A = \frac{1}{2}bh$

84. *Investment at Compound Interest*

Solve for P: $A = P\left(1 + \frac{r}{n}\right)^{nt}$

85. *Area of a Trapezoid*

Solve for b: $A = \frac{1}{2}(a + b)h$

86. *Geometric Progression*

Solve for r: $S = \dfrac{rL - a}{r - 1}$

2.2 Solving Equations Graphically

Intercepts, Zeros, and Solutions / Finding Solutions Graphically / Viewing Rectangles, Scale, and Accuracy / Points of Intersection of Two Graphs

Intercepts, Zeros, and Solutions

In Section 1.1, you learned that the **intercepts** of a graph are the points at which the graph intersects the x- or y-axis.

Definition of Intercepts

1. The point $(a, 0)$ is called an ***x*-intercept** of the graph of an equation if it is a solution point of the equation. To find the x-intercept(s), let $y = 0$ and solve the equation for x.
2. The point $(0, b)$ is called a ***y*-intercept** of the graph of an equation if it is a solution point of the equation. To find the y-intercept(s), let $x = 0$ and solve the equation for y.

EXPLORATION

Use a graphing utility to graph each function. How many times does each of the graphs intersect the x-axis and how many times does each intersect the y-axis?

$y_1 = x^4 - 5x^2 + 4$

$y_2 = 3 + x^2$

$y_3 = x^2 - 3$

$y_4 = 2 + \sqrt{4x - x^2 - 3}$

$y_5 = 2 - \sqrt{4x - x^2 - 3}$

In general, how many times can a function's graph intersect the x-axis? How many times can it intersect the y-axis?

Note Sometimes it is convenient to denote the x-intercept as simply the x-coordinate of the point $(a, 0)$ rather than the point itself. Unless it is necessary to make a distinction, we will use "intercept" to mean either the point or the coordinate.

The *Interactive* CD-ROM offers graphing utility emulators of the *TI-82* and *TI-83*, which can be used with the Examples, Explorations, Technology notes, and Exercises.

It is possible that a particular graph will have no intercepts or several intercepts. For instance, consider the three graphs in Figure 2.5.

Figure 2.5

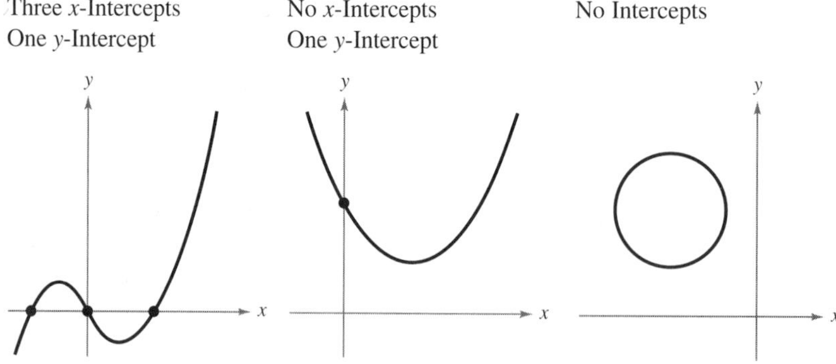

Figure 2.6

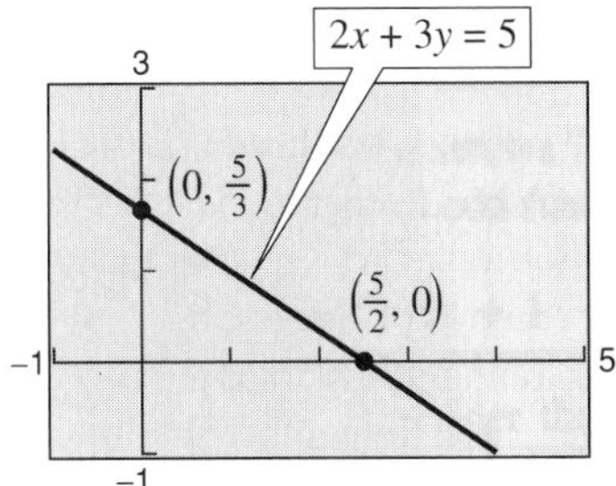

EXAMPLE 1 Finding x- and y-Intercepts

Find the x- and y-intercepts of the graph of $2x + 3y = 5$.

Solution

To find the x-intercept, let $y = 0$. This produces

$$2x = 5 \quad \Rightarrow \quad x = \frac{5}{2}$$

which implies that the graph has one x-intercept: $\left(\frac{5}{2}, 0\right)$. To find the y-intercept, let $x = 0$. This produces

$$3y = 5 \quad \Rightarrow \quad y = \frac{5}{3}$$

which implies that the graph has one y-intercept: $\left(0, \frac{5}{3}\right)$. See Figure 2.6.

A **zero** of a function $y = f(x)$ is a number a such that $f(a) = 0$. Thus, to find the zeros of a function, you must solve the equation $f(x) = 0$.

EXAMPLE 2 Verifying Zeros of Functions

a. The real number 3 is a zero of the function $f(x) = 4x - 12$. Check that $f(3) = 0$.

b. The real numbers -2 and 1 are zeros of the function $f(x) = x^2 + x - 2$. Check that $f(-2) = 0$ and $f(1) = 0$.

The concepts of x-intercepts, zeros of functions, and solutions of equations are closely related. In fact, the following statements are equivalent.

1. The point $(a, 0)$ is an *x-intercept* of the graph of $y = f(x)$.
2. The number a is a *zero* of the function f.
3. The number a is a *solution* of the equation $f(x) = 0$.

Figure 2.7

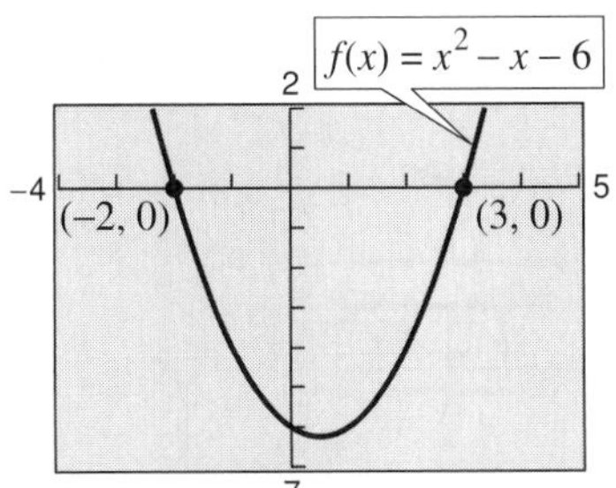

Figure 2.7 shows the graph of $f(x) = x^2 - x - 6$. Note that the graph has *x-intercepts* of $(3, 0)$ and $(-2, 0)$ because the equation $0 = x^2 - x - 6$ has the *solutions* $x = 3$ and $x = -2$.

This close connection among x-intercepts, zeros, and solutions is crucial to our study of algebra, and you can take advantage of this connection in two basic ways. You can use your algebraic "equation-solving skills" to find the x-intercepts of a graph, and you can use your "graphing skills" to approximate the solutions of an equation.

Finding Solutions Graphically

Polynomial equations of degree 1 or 2 can be solved in relatively straightforward ways. Polynomial equations of higher degrees can, however, be quite difficult, especially if you rely only on algebraic techniques. For such equations, a graphing utility can be very helpful.

Note In Chapter 3 you will learn techniques for determining the number of solutions of a polynomial equation. For now, you should know that a polynomial equation of degree n cannot have more than n different solutions.

Graphical Approximations of Solutions of an Equation

1. Write the equation in *standard form,* $f(x) = 0$, with the nonzero terms on one side of the equation and zero on the other side.
2. Use a graphing utility to graph the function $y = f(x)$. Be sure the viewing rectangle shows all the relevant features of the graph.
3. Use the zoom and trace features of the graphing utility to approximate each of the x-intercepts of the graph of f. Remember that a graph can have more than one x-intercept, so you may need to change the viewing rectangle a few times.

EXPLORATION

Use a graphing utility to graph

$$y = 24x^3 - 36x + 17.$$

Describe a viewing rectangle that allows you to determine the number of real solutions of the equation

$$24x^3 - 36x + 17 = 0.$$

Use the same technique to determine the number of real solutions of

$$97x^3 - 102x^2 - 200x - 63 = 0.$$

EXAMPLE 3 Finding Solutions of an Equation Graphically

Use a graphing utility to approximate the solutions of $2x^3 - 3x + 2 = 0$.

Solution

Begin by graphing the function $y = 2x^3 - 3x + 2$, as shown in Figure 2.8(a). You can see from the graph that there is only one x-intercept. It lies between -1 and -2 and is approximately -1.5. By using the zoom feature of a graphing utility you can improve the approximation, as shown in the graph in Figure 2.8(b). To three-decimal-place accuracy, the solution is $x \approx -1.476$. Check this approximation on your calculator. You will find that the value of y is $y = 2(-1.476)^3 - 3(-1.476) + 2 \approx -0.003$.

Figure 2.8

(a)

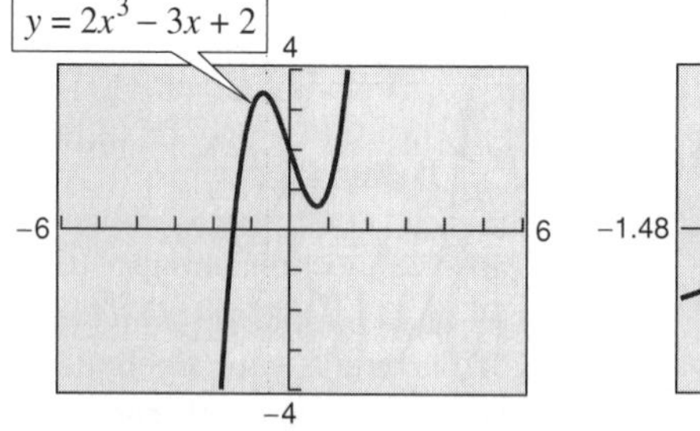

(b)

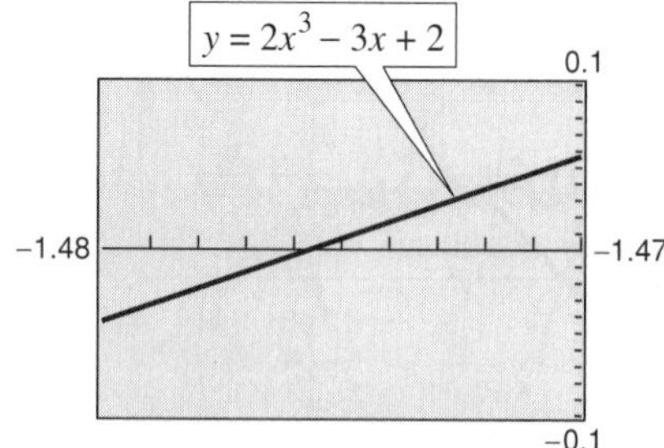

Viewing Rectangles, Scale, and Accuracy

Here are some suggestions for using the *zoom-in* feature of a graphing utility.

1. With each successive zoom-in, adjust the x-scale (if necessary) so that the resulting viewing rectangle shows at least the two scale marks between which the solution lies.
2. The accuracy of the approximation will always be such that the error is less than the distance between two scale marks.
3. If you have a *trace* feature on your graphing utility, you can generally add one more decimal place of accuracy without changing the viewing rectangle.

Unless stated otherwise, this book will approximate all real solutions with an error of *at most* 0.01.

Some graphing utilities have built-in programs that will approximate solutions of equations or approximate x-intercepts of graphs. If your graphing utility has such features, try using them to approximate the solutions in Example 4.

EXAMPLE 4 **Rewriting in Standard Form First**

Use a graphing utility to approximate the solutions of $x^2 + 3 = 5x$.

Solution

In standard form, this equation is

$$x^2 - 5x + 3 = 0. \quad \text{Equation in standard form}$$

Thus, you can begin by graphing

$$y = x^2 - 5x + 3 \quad \text{Function to be graphed}$$

as shown in Figure 2.9(a). This graph has two x-intercepts, and by using the zoom and trace features you can approximate the corresponding solutions to be $x \approx 0.70$ and $x \approx 4.30$, as shown in Figures 2.9(b) and 2.9(c).

Figure 2.9

(a)

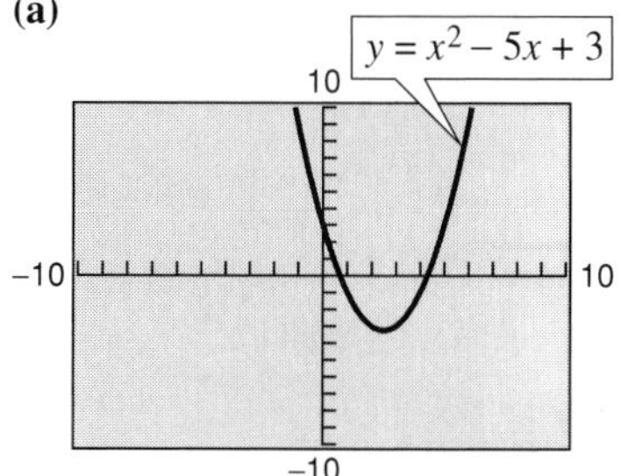

(b)

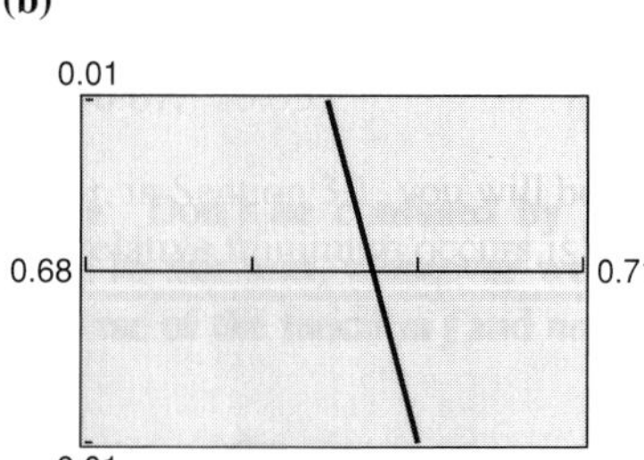

(c)

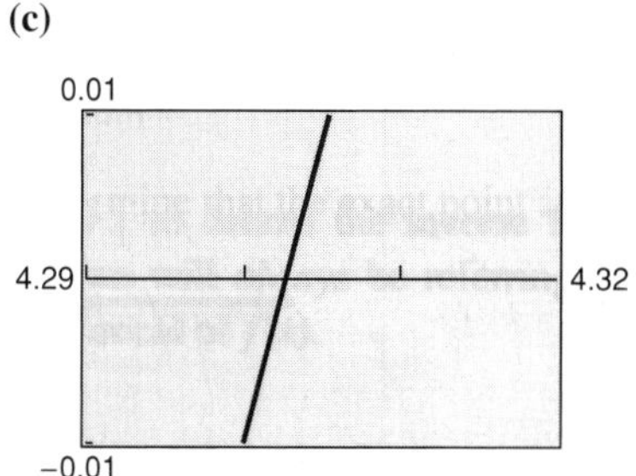

Points of Intersection of Two Graphs

An ordered pair that is a solution of two different equations is called a **point of intersection** of the graphs of the two equations. For instance, in Figure 2.10 you can see that the graphs of the following equations have two points of intersection.

Figure 2.10

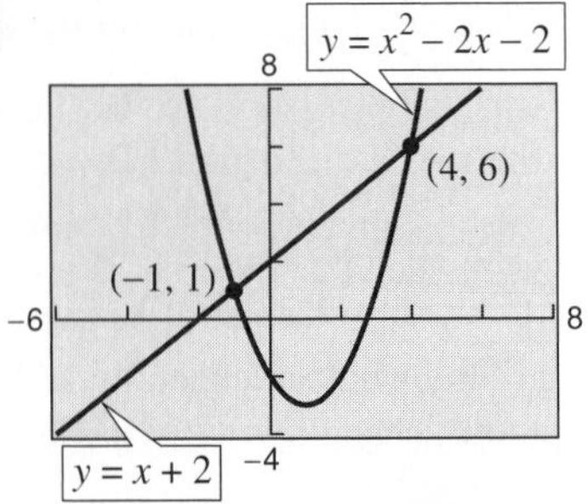

$y = x + 2$ Equation 1

$y = x^2 - 2x - 2$ Equation 2

The point $(-1, 1)$ is a solution of both equations, and the point $(4, 6)$ is a solution of both equations. To check this algebraically, substitute -1 and 4 into each equation.

Check that $(-1, 1)$ *is a solution.*

Equation 1: $y = -1 + 2 = 1$

Equation 2: $y = (-1)^2 - 2(-1) - 2 = 1$ ✓

Check that $(4, 6)$ *is a solution.*

Equation 1: $y = 4 + 2 = 6$

Equation 2: $y = (4)^2 - 2(4) - 2 = 6$ ✓

The following table, which can be generated on a *TI-82* or a *TI-83*, shows some points of the graphs of the equations at the right. Explain how the table can be used to find the points of intersection of the graphs.

X	Y1	Y2
−2	0	6
−1	1	1
0	2	−2
1	3	−3
2	4	−2
3	5	1
4	6	6
X = −2		

To find the points of intersection of the graphs of two equations, solve each equation for y (or x) and set the two results equal to each other. The resulting equation will be an equation in one variable, which can be solved using standard procedures, as shown in Example 5.

EXAMPLE 5 Finding Points of Intersection

Find the points of intersection of the graphs of $2x - 3y = -2$ and $4x - y = 6$.

Solution

To begin, solve each equation for y to obtain $y = \frac{2}{3}x + \frac{2}{3}$ and $y = 4x - 6$. Next, set the two expressions for y equal to each other and solve the resulting equation for x, as follows.

$$\frac{2}{3}x + \frac{2}{3} = 4x - 6$$

$$2x + 2 = 12x - 18$$

$$-10x = -20$$

$$x = 2$$

When $x = 2$, the y-value of each of the given equations is 2. Thus, the graphs have one point of intersection, $(2, 2)$, as shown in Figure 2.11.

Figure 2.11

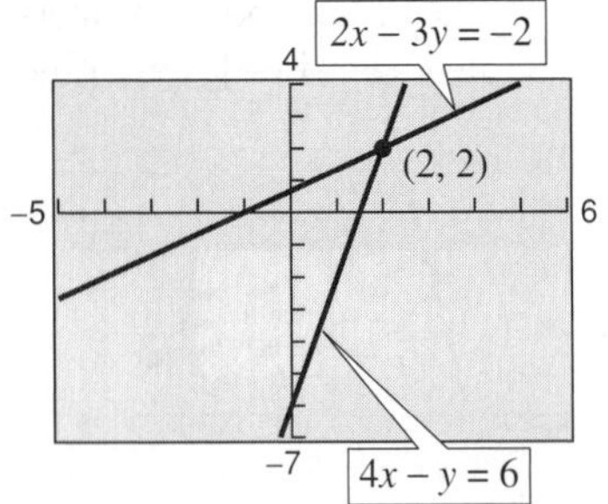

Another way to approximate points of intersection of two graphs is to graph both equations and use the zoom and trace features (or the intersect feature) to find the point or points at which the two graphs intersect.

EXAMPLE 6 **Approximating Points of Intersection**

Approximate the point(s) of intersection of the graphs of the following equations.

$y = x^2 - 3x - 4$ Equation 1 (quadratic function)

$y = x^3 + 3x^2 - 2x - 1$ Equation 2 (cubic function)

Solution

Figure 2.12

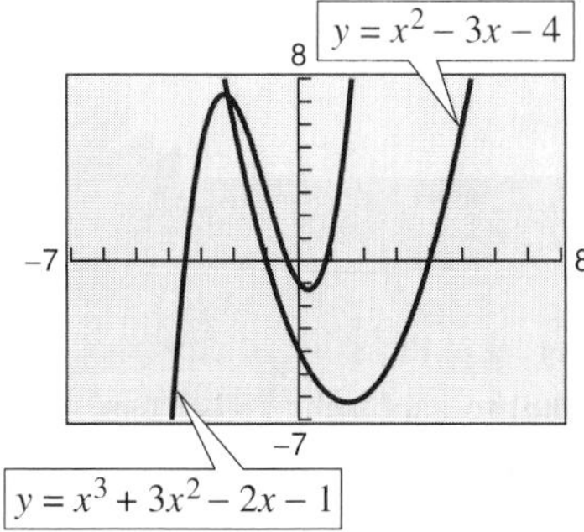

Begin by using a graphing utility to graph both functions, as shown in Figure 2.12. From this display, you can see that the two graphs have only one point of intersection. Then, using the zoom and trace features, approximate the point of intersection to be $(-2.17, 7.25)$. To test the reasonableness of this approximation, you can evaluate both functions when $x = -2.17$.

Quadratic Function:

$$y = (-2.17)^2 - 3(-2.17) - 4$$
$$\approx 7.22$$

Cubic Function:

$$y = (-2.17)^3 + 3(-2.17)^2 - 2(-2.17) - 1$$
$$\approx 7.25$$

Because both functions yield approximately the same y-value, you can conclude that the approximate coordinates of the point of intersection are $x \approx -2.17$ and $y \approx 7.25$.

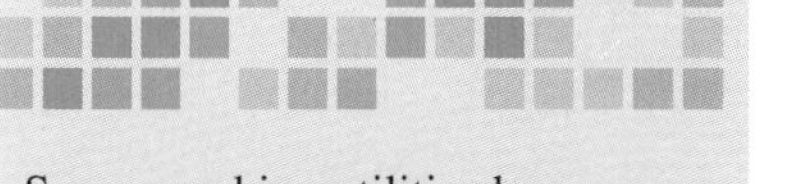

Some graphing utilities have built-in programs that approximate the point of intersection of two graphs. If your graphing utility has such a feature, try using it to approximate the result in Example 6.

The method shown in Example 6 gives a nice graphical picture of points of intersection of two graphs. However, for actual approximation purposes, it is better to use the procedure described in Example 5. That is, the point of intersection of $y = x^2 - 3x - 4$ and $y = x^3 + 3x^2 - 2x - 1$ coincides with the solution of the equation

$x^3 + 3x^2 - 2x - 1 = x^2 - 3x - 4$ Equate y-values.

$x^3 + 2x^2 + x + 3 = 0.$ Write in standard form.

By graphing $y = x^3 + 2x^2 + x + 3$ on a graphing utility and using the zoom and trace features (or the root feature), you can approximate the solution of this equation to be $x \approx -2.17$. The corresponding y-value for *both* of the functions given in Example 6 is $y \approx 7.25$.

EXAMPLE 7 **A Historical Look at Stereo Equipment**

Between 1983 and 1987, the number of compact disc players sold each year in the United States was *increasing* and the number of turntables was *decreasing*. Two models that approximate the sales S are

$$S = -1700 + 496t \qquad \text{Compact disc players}$$

$$S = 1972 - 82t \qquad \text{Turntables}$$

where $t = 3$ represents 1983. According to these two models, when would you expect the sales of compact disc players to have exceeded the sales of turntables? (Source: Dealerscope Merchandising)

Solution

Because the first equation has already been solved for S in terms of t, substitute this value into the second equation and solve for t, as follows.

$$-1700 + 496t = 1972 - 82t$$

$$496t + 82t = 1972 + 1700$$

$$578t = 3672$$

$$t \approx 6.35$$

Thus, from the given models, you would expect that the sales of compact disc players exceeded the sales of turntables sometime during 1986. The graphs of $S = -1700 + 496t$ and $S = 1972 - 82t$ are shown in Figure 2.13, confirming that they intersect at approximately $t = 6.35$.

Figure 2.13

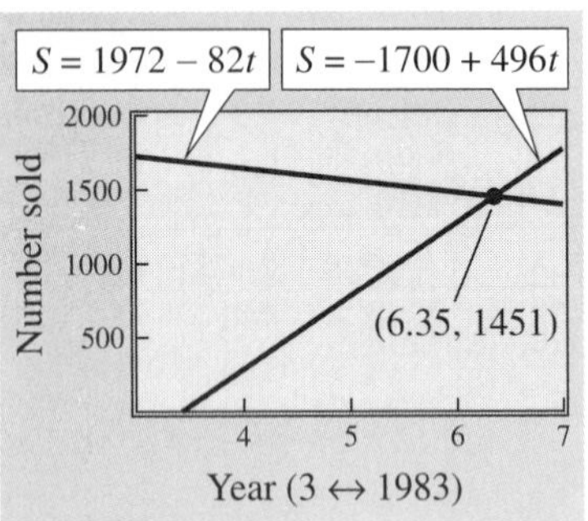

Group Activity

Judging the Accuracy of an Approximate Solution

Suppose you are solving the equation

$$\frac{x}{x-1} - \frac{99}{100} = 0$$

for x, and you obtain $x = -99.1$ as your solution. Substituting this value back into the equation produces

$$\frac{-99.1}{-99.1-1} - \frac{99}{100} = 0.00000999 = 9.99 \times 10^{-6} \approx 0.$$

Does this mean that -99.1 is a good approximation to the solution? Explain your reasoning.

2.2 /// EXERCISES

In Exercises 1–10, find the x- and y-intercepts of the graph of the equation.

1. $y = x - 5$

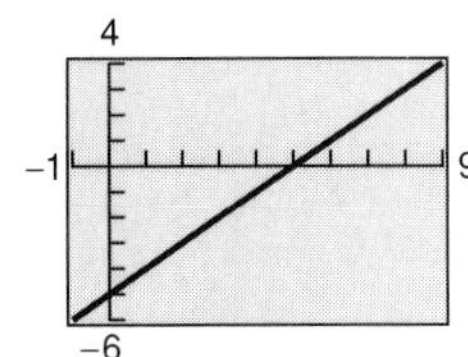

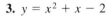

2. $y = (x - 1)(x - 3)$

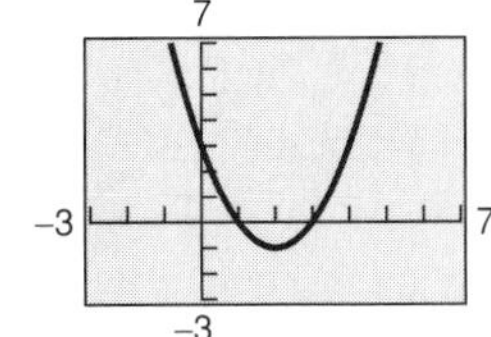

3. $y = x^2 + x - 2$

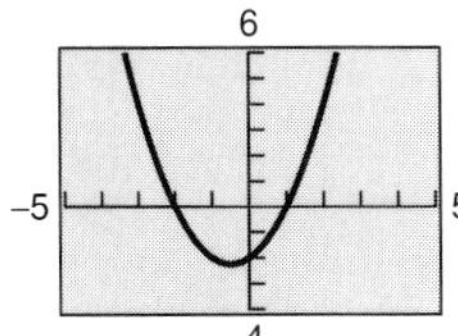

4. $y = 4 - x^2$

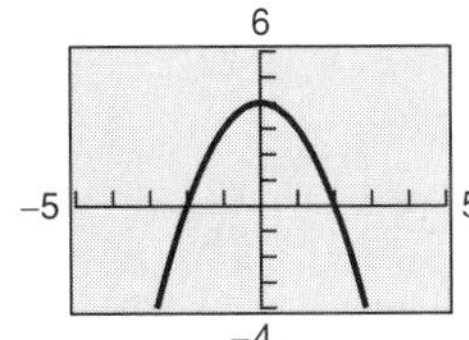

5. $y = x\sqrt{x + 2}$

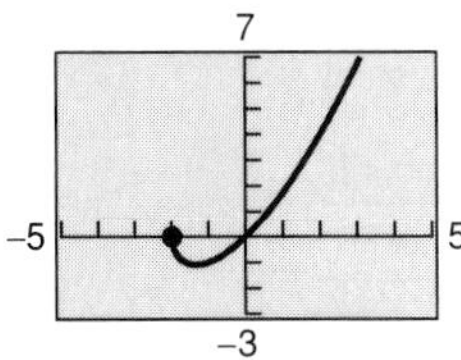

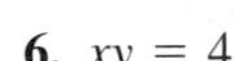

6. $xy = 4$

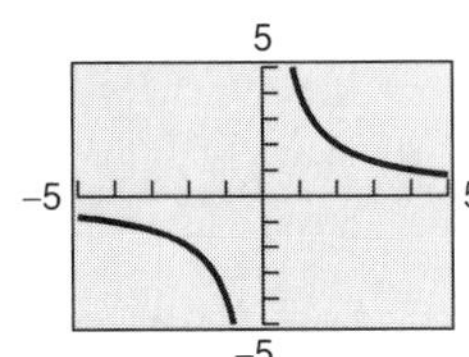

7. $y = |x - 2| - 3$

8. $y = 4 - \frac{1}{2}|x + 1|$

9. $xy - 2y - x + 1 = 0$

10. $x^2y - x^2 + 4y = 0$

In Exercises 11–16, use a graphing utility to graph the function and verify its zero(s).

Function	*Zero(s)*
11. $f(x) = 12 - 4x$	$x = 3$
12. $f(x) = 3(x - 5) + 9$	$x = 2$
13. $f(x) = x^2 - 2.5x - 6$	$x = -1.5, 4$
14. $f(x) = x^3 - 9x^2 + 18x$	$x = 0, 3, 6$
15. $f(x) = \dfrac{x + 2}{3} - \dfrac{x - 1}{5} - 1$	$x = 1$
16. $f(x) = x - 3 - \dfrac{10}{x}$	$x = -2, 5$

Graphical Analysis **In Exercises 17–20, use a graphing utility to graph the equation and approximate any x-intercepts. Set $y = 0$ and solve the resulting equation. Compare the results with the x-intercepts of the graph.**

17. $y = 2(x - 1) - 4$

18. $y = \frac{4}{3}x + 2$

19. $y = 20 - (3x - 10)$

20. $y = 10 + 2(x - 2)$

In Exercises 21–26, write the equation in the form $f(x) = 0$. (There are many correct answers.)

21. $25(x - 3) = 12(x + 2) - 10$

22. $1200 = 300 + 2(x - 500)$

23. $\dfrac{2x}{3} = 10 - \dfrac{1}{x}$

24. $\dfrac{x - 3}{25} = \dfrac{x - 5}{12}$

25. $\dfrac{3}{x + 2} - \dfrac{4}{x - 2} = 5$

26. $\dfrac{6}{x} + \dfrac{8}{x + 5} = 10$

In Exercises 27–32, solve the equation algebraically. Then write the equation in the form $f(x) = 0$ and use a graphing utility to verify the algebraic solution.

27. $27 - 4x = 12$

28. $3.5x - 8 = 0.5x$

29. $\dfrac{3x}{2} + \dfrac{1}{4}(x - 2) = 10$

30. $0.60x + 0.40(100 - x) = 50$

31. $3(x + 3) = 5(1 - x) - 1$

32. $(x + 1)^2 + 2(x - 2) = (x + 1)(x - 2)$

In Exercises 33–44, use a graphing utility to approximate any solutions (accurate to three decimal places) of the equation. [Remember to write the equation in the form $f(x) = 0$.]

33. $\frac{1}{4}(x^2 - 10x + 17) = 0$

34. $-2(x^2 - 6x + 6) = 0$

35. $x^3 + x + 4 = 0$

36. $\frac{1}{9}x^3 + x + 4 = 0$

37. $2x^3 - x^2 - 18x + 9 = 0$

38. $4x^3 + 12x^2 - 26x - 24 = 0$

39. $x^4 = 2x^3 + 1$

40. $x^5 = 3 + 2x^3$

41. $\dfrac{2}{x+2} = 3$

42. $\dfrac{5}{x} = 1 + \dfrac{3}{x+2}$

43. $|x - 3| = 4$

44. $\sqrt{x-2} = 3$

45. *Exploration*

(a) Use a graphing utility to complete the table.

x	-1	0	1	2	3	4
$3.2x - 5.8$						

(b) Use the table to determine the interval in which the solution to the equation $3.2x - 5.8 = 0$ is located. Explain your reasoning.

(c) Use a graphing utility to complete the table.

x	1.5	1.6	1.7	1.8	1.9	2
$3.2x - 5.8$						

(d) Use the table to determine the interval in which the solution to the equation $3.2x - 5.8 = 0$ is located. Explain how this process can be used to approximate the solution to any desired degree of accuracy.

(e) Use a graphing utility to verify graphically the solution to $3.2x - 5.8 = 0$ found in part (d).

46. *Exploration* Use the procedure of Exercise 45 to approximate the solution of the equation

$0.3(x - 1.5) - 2 = 0$

accurate to two decimal places.

In Exercises 47–60, use a graphing utility to approximate any points of intersection (accurate to three decimal places) of the graphs of the equations.

47. $y = 2 - x$
$y = 2x - 1$

48. $y = 7 - x$
$y = \frac{3}{2} - \frac{11}{2}x$

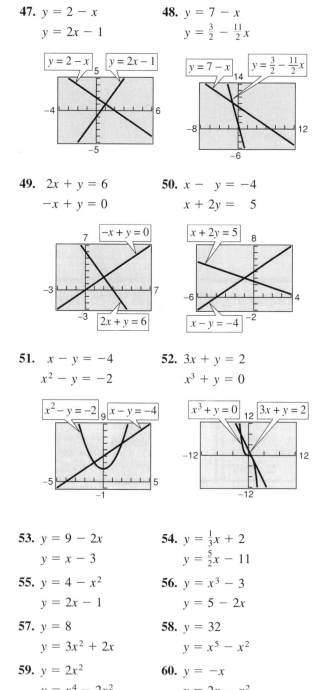

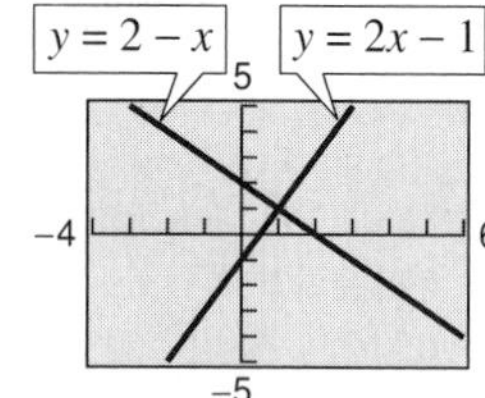

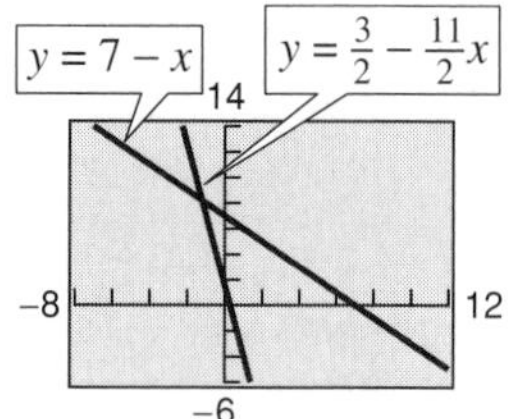

49. $2x + y = 6$
$-x + y = 0$

50. $x - y = -4$
$x + 2y = 5$

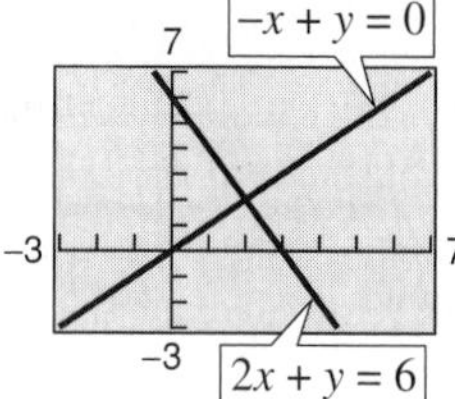

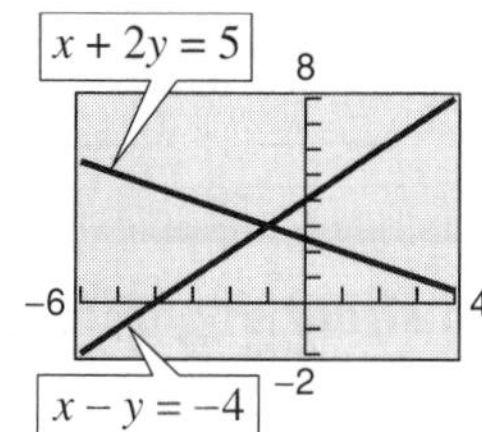

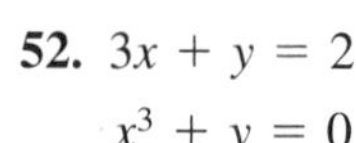

51. $x - y = -4$
$x^2 - y = -2$

52. $3x + y = 2$
$x^3 + y = 0$

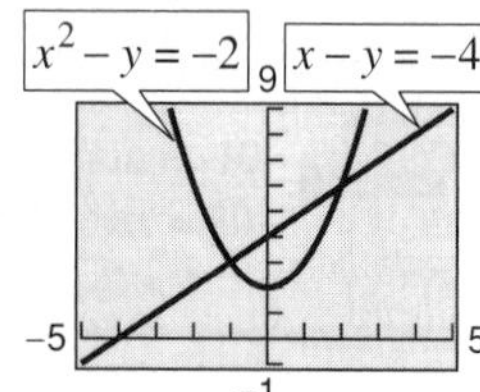

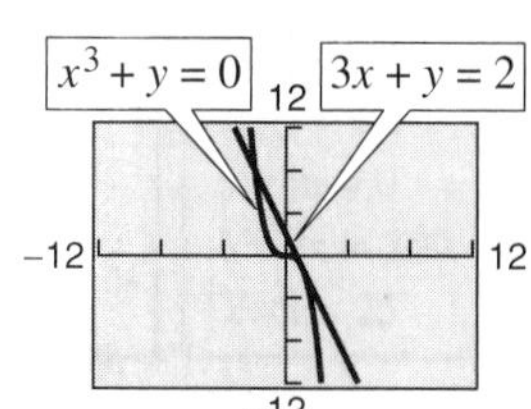

53. $y = 9 - 2x$
$y = x - 3$

54. $y = \frac{1}{3}x + 2$
$y = \frac{5}{2}x - 11$

55. $y = 4 - x^2$
$y = 2x - 1$

56. $y = x^3 - 3$
$y = 5 - 2x$

57. $y = 8$
$y = 3x^2 + 2x$

58. $y = 32$
$y = x^5 - x^2$

59. $y = 2x^2$
$y = x^4 - 2x^2$

60. $y = -x$
$y = 2x - x^2$

In Exercises 61 and 62, evaluate the expression in two ways. (a) Calculate entirely on your calculator by storing intermediate results and then rounding the final answer to two decimal places. (b) Round both the numerator and denominator to two decimal places before dividing, and then round the final answer to two decimal places. Does the method in part (b) decrease the accuracy? Explain.

61. $\dfrac{1 + 0.73205}{1 - 0.73205}$ **62.** $\dfrac{1 + 0.86603}{1 - 0.86603}$

63. *Travel Time* On the first part of a 280-mile trip a salesman averaged 63 miles per hour. He averaged only 54 miles per hour on the last part of the trip because of an increased volume of traffic.

(a) Express the total time for the trip as a function of the distance x traveled at an average speed of 63 miles per hour.

(b) Use a graphing utility to graph the time function. What is the domain of the function?

(c) Approximate the number of miles traveled at 63 miles per hour if the total time was 4 hours and 45 minutes.

64. *Production Limit* A company has fixed costs of $25,000 per month and a variable cost of $18.65 per unit manufactured. (*Fixed costs* are those that occur regardless of the level of production.)

(a) Write the total monthly costs C as a function of the number of units x produced.

(b) Use a graphing utility to graph the cost function. Approximate the number of units that can be produced per month if total costs cannot exceed $200,000. Verify algebraically. Is this better solved algebraically or graphically? Explain.

65. *Mixture Problem* A 55-gallon barrel contains a mixture with a concentration of 33%. You remove x gallons of this mixture and replace it with 100% concentrate.

(a) Write the amount of concentrate in the final mixture as a function of x.

(b) Use a graphing utility to graph the concentration function. What is the domain of the function?

(c) Approximate (accurate to one decimal place) the value of x if the final mixture is 60% concentrate.

66. *Dimensions of a Rectangle* A rectangular region with a perimeter of 230 meters has a length of x.

(a) Create a figure that gives a visual representation of the problem.

(b) Express the rectangle's area as a function of x.

(c) Use a graphing utility to graph the area function. Because area is nonnegative, what is the domain of the function?

(d) Approximate (accurate to one decimal place) the dimensions of the region if its area is 2000 square feet.

Geometry **In Exercises 67 and 68, (a) write a function for the area of the region, (b) use a graphing utility to graph the function, and (c) approximate the value of x if the area of the region is 200 square units.**

67.

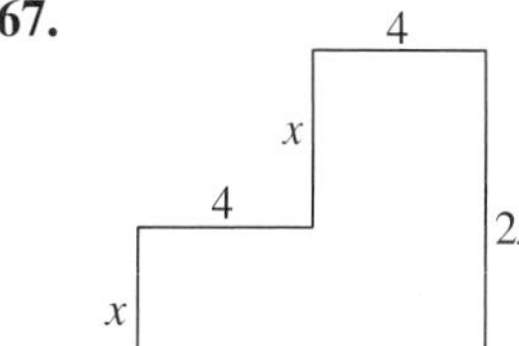

68.

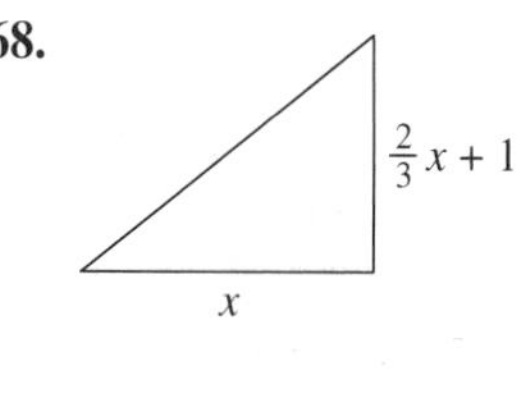

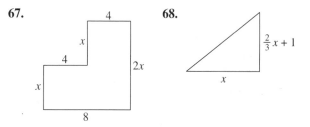

69. *Exploration* The graph of the function

$f(x) = \frac{1}{5}(x^5 - 20x^3 + 64x) + k$

for $k = 0$ is given in the figure.

(a) Determine the number of zeros of the function when $k = 0$.

(b) Use a graphing utility to graph the function for different values of k. Find a value of k such that the function has one zero. Find k so there are three distinct zeros.

(c) Is there a value of k such that the function has no zero? Explain.

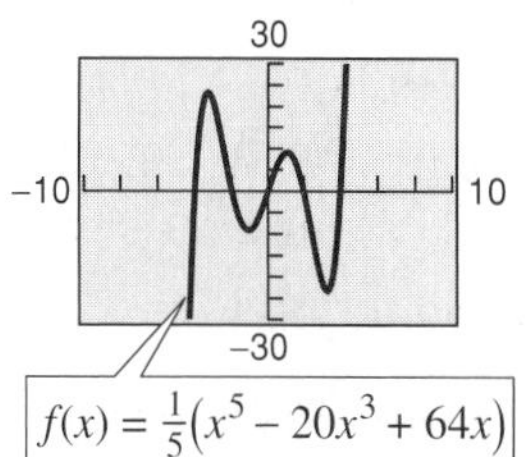

70. *Volume* Consider the swimming pool in the figure. (When finding its volume, use the fact that the volume is the area of the region on the vertical sidewall times the width of the pool.)

(a) Find the volume of the pool.

(b) Find an equation of the line representing the base of the pool.

(c) If the depth of the water at the deep end of the pool is d, show that the volume of water is given by

$$V(d) = \begin{cases} 80d^2, & 0 \le d \le 5 \\ 800d - 2000, & 5 < d \le 9 \end{cases}.$$

(d) Graph the volume function.

(e) Use a graphing utility to complete the table.

d	3	5	7	9
V				

(f) Approximate the depth of the water if the volume is 4800 cubic feet.

(g) How many gallons of water are in the pool? (There are 7.48 gallons of water in 1 cubic foot.)

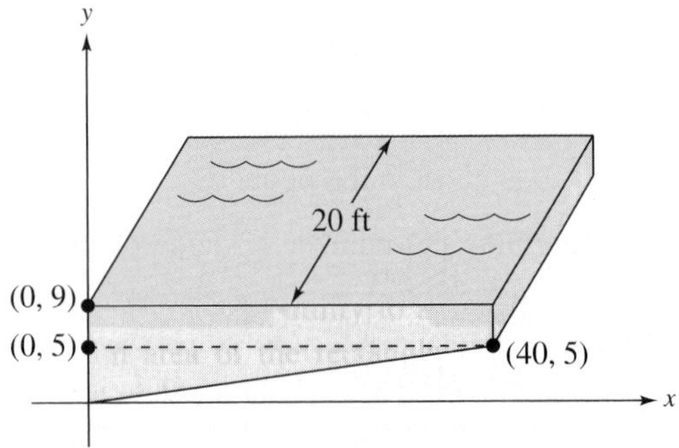

Negative Income Tax In Exercises 71–74, use the following information about a possible negative income tax for a family consisting of two adults and two children. The plan would guarantee the poor a minimum income while encouraging a family to increase its private income ($0 \le x \le 20{,}000$).

Family's earned income: $I = x$

Subsidy: $S = 10{,}000 - \frac{1}{2}x$

Total income: $T = I + S$

71. Express the total income T in terms of x.

72. Use a graphing utility to find the earned income x if the subsidy is \$6600. Verify your answer algebraically.

73. Use a graphing utility to find the earned income x if the total income is \$13,800. Verify your answer algebraically.

74. Find the subsidy S graphically if the total income is \$12,500.

75. *Using a Model* The number of married women y in the civilian work force (in millions) in the United States from 1988 to 1992 can be approximated by the model

$$y = 0.43t + 30.86$$

where $t = 0$ represents 1990 (see figure). According to this model, during which year did this number reach 31 million? Explain how to answer the question graphically and algebraically. (Source: U.S. Bureau of Labor Statistics)

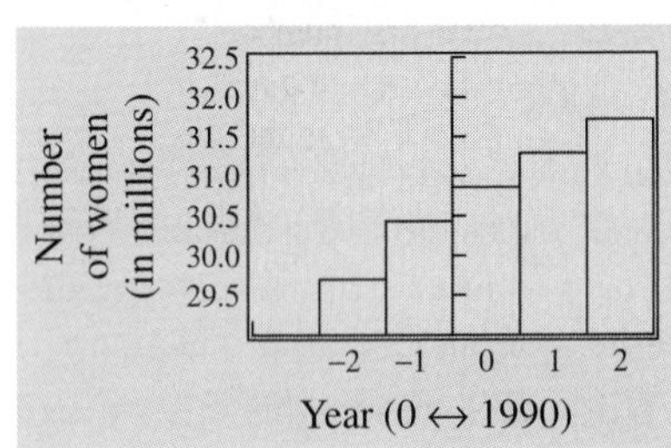

76. *Recommended Weight* The median recommended weight of women of small frame who are 25 to 59 years old can be approximated by the mathematical model

$$y = 0.041x^2 - 2.525x + 113.639, \qquad 58 \le x \le 72$$

where y is the median recommended weight in pounds and x is the height in inches. Suppose you are an insurance agent. After taking information from a female client, you compute her recommended weight to be 123 pounds. Later, when you are filling out her paperwork, you are unable to remember her height. Use the model and a graphing utility to estimate your client's height based on her recommended weight. (Source: Metropolitan Life Insurance Company)

2.3 Complex Numbers

The Imaginary Unit *i* / Operations with Complex Numbers / Complex Conjugates and Division / Applications

Carl Friedrich Gauss (1777–1855) proved that all the roots of any algebraic equation are "numbers" of the form $a + bi$, where a and b are real numbers and i is the square root of -1. These "numbers" were called complex.

The Imaginary Unit *i*

Some quadratic equations have no real solutions. For instance, the quadratic equation

$$x^2 + 1 = 0 \qquad \text{Equation with no real solution}$$

has no real solution because there is no real number x that can be squared to produce -1. To overcome this deficiency, mathematicians created an expanded system of numbers using the **imaginary unit *i*,** defined as

$$i = \sqrt{-1} \qquad \text{Imaginary unit}$$

where $i^2 = -1$. By adding real numbers to real multiples of this imaginary unit, you obtain the set of **complex numbers.** Each complex number can be written in the **standard form, $a + bi$.**

Definition of a Complex Number

For real numbers a and b, the number

$$a + bi$$

is a **complex number.** If $a = 0$ and $b \neq 0$, the complex number bi is an **imaginary number.**

The set of real numbers is a subset of the set of complex numbers, as shown in Figure 2.14. This is true because every real number a can be written as a complex number using $b = 0$. That is, for every real number a, we can write $a = a + 0i$.

Figure 2.14

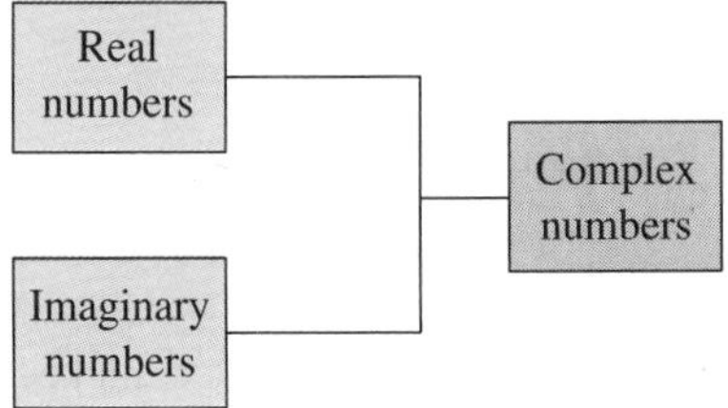

Equality of Complex Numbers

Two complex numbers $a + bi$ and $c + di$, written in standard form, are **equal** to each other

$$a + bi = c + di \qquad \text{Equality of two complex numbers}$$

if and only if $a = c$ and $b = d$.

Operations with Complex Numbers

To add (or subtract) two complex numbers, you add (or subtract) the real and imaginary parts of the numbers separately.

Addition and Subtraction of Complex Numbers

If $a + bi$ and $c + di$ are two complex numbers written in standard form, their sum and difference are defined as follows.

Sum: $(a + bi) + (c + di) = (a + c) + (b + d)i$

Difference: $(a + bi) - (c + di) = (a - c) + (b - d)i$

The **additive identity** in the complex number system is zero (the same as in the real number system). Furthermore, the **additive inverse** of the complex number $a + bi$ is

$$-(a + bi) = -a - bi. \quad \text{Additive inverse}$$

Thus, you have

$$(a + bi) + (-a - bi) = 0 + 0i = 0.$$

EXAMPLE 1 **Adding and Subtracting Complex Numbers**

a. $(3 - i) + (2 + 3i) = 3 - i + 2 + 3i$ Remove parentheses.

$= 3 + 2 - i + 3i$ Group like terms.

$= (3 + 2) + (-1 + 3)i$

$= 5 + 2i$ Standard form

b. $2i + (-4 - 2i) = 2i - 4 - 2i$ Remove parentheses.

$= -4 + 2i - 2i$ Group like terms.

$= -4$ Standard form

c. $3 - (-2 + 3i) + (-5 + i) = 3 + 2 - 3i - 5 + i$

$= 3 + 2 - 5 - 3i + i$

$= 0 - 2i$

$= -2i$

Note In Example 1(b) the sum of two complex numbers can be a real number.

EXPLORATION

Complete the table:

$i^1 = i$	$i^7 =$
$i^2 = -1$	$i^8 =$
$i^3 = -i$	$i^9 =$
$i^4 = 1$	$i^{10} =$
$i^5 =$	$i^{11} =$
$i^6 =$	$i^{12} =$

What pattern do you see? Write a brief description of how you would find i raised to any positive integer power.

Many of the properties of real numbers are valid for complex numbers as well. Here are some examples.

Associative Property of Addition and Multiplication
Commutative Property of Addition and Multiplication
Distributive Property of Multiplication Over Addition

Notice how these properties are used when two complex numbers are multiplied.

$$\begin{aligned}(a + bi)(c + di) &= a(c + di) + bi(c + di) && \text{Distributive}\\ &= ac + (ad)i + (bc)i + (bd)i^2 && \text{Distributive}\\ &= ac + (ad)i + (bc)i + (bd)(-1) && \text{Definition of } i\\ &= ac - bd + (ad)i + (bc)i && \text{Commutative}\\ &= (ac - bd) + (ad + bc)i && \text{Associative}\end{aligned}$$

Rather than trying to memorize this multiplication rule, we suggest that you simply remember how the distributive property is used to multiply two complex numbers. The procedure is similar to multiplying two polynomials and combining like terms (as in the FOIL Method).

EXAMPLE 2 **Multiplying Complex Numbers**

a.
$$\begin{aligned}(i)(-3i) &= -3i^2 && \text{Multiply.}\\ &= -3(-1) && \text{Definition of } i\\ &= 3 && \text{Simplify.}\end{aligned}$$

b.
$$\begin{aligned}(2 - i)(4 + 3i) &= 8 + 6i - 4i - 3i^2 && \text{Product of binomials}\\ &= 8 + 6i - 4i - 3(-1) && \text{Definition of } i\\ &= 8 + 3 + 6i - 4i && \text{Group like terms.}\\ &= 11 + 2i && \text{Standard form}\end{aligned}$$

c.
$$\begin{aligned}(3 + 2i)(3 - 2i) &= 9 - 6i + 6i - 4i^2 && \text{Product of binomials}\\ &= 9 - 4(-1) && \text{Definition of } i\\ &= 9 + 4 && \text{Simplify.}\\ &= 13 && \text{Standard form}\end{aligned}$$

d.
$$\begin{aligned}(3 + 2i)^2 &= 9 + 6i + 6i + 4i^2 && \text{Product of binomials}\\ &= 9 + 4(-1) + 12i && \text{Definition of } i\\ &= 9 - 4 + 12i && \text{Simplify.}\\ &= 5 + 12i && \text{Standard form}\end{aligned}$$

Complex Conjugates and Division

Notice in Example 2(c) that the product of two complex numbers can be a real number. This occurs with pairs of complex numbers of the form $a + bi$ and $a - bi$, called **complex conjugates.**

$$\begin{aligned}(a + bi)(a - bi) &= a^2 - abi + abi - b^2i^2\\ &= a^2 - b^2(-1)\\ &= a^2 + b^2\end{aligned}$$

To find the quotient of $a + bi$ and $c + di$ where c and d are not both zero, multiply the numerator and denominator by the conjugate of the denominator to obtain

$$\frac{a + bi}{c + di} = \frac{a + bi}{c + di}\left(\frac{c - di}{c - di}\right) = \frac{(ac + bd) + (bc - ad)i}{c^2 + d^2}.$$

EXAMPLE 3 **Dividing Complex Numbers**

$$\begin{aligned}\frac{1}{1 + i} &= \frac{1}{1 + i}\left(\frac{1 - i}{1 - i}\right) && \text{Multiply by conjugate.}\\ &= \frac{1 - i}{1^2 - i^2} && \text{Expand.}\\ &= \frac{1 - i}{1 - (-1)} && \text{Definition of } i\\ &= \frac{1 - i}{2} && \text{Simplify.}\\ &= \frac{1}{2} - \frac{1}{2}i && \text{Standard form}\end{aligned}$$

Some graphing utilities, such as the *TI-92* or *TI-83* from Texas Instruments, can perform operations with complex numbers. For instance, to divide $2 + 3i$ by $4 - 2i$, enter

[(] 2 [+] 3 [i] [)] [÷]
[(] 4 [−] 2 [i] [)] [ENTER].

The display is

1/10 + 4/5*i*.

EXAMPLE 4 **Dividing Complex Numbers**

$$\begin{aligned}\frac{2 + 3i}{4 - 2i} &= \frac{2 + 3i}{4 - 2i}\left(\frac{4 + 2i}{4 + 2i}\right) && \text{Multiply by conjugate.}\\ &= \frac{8 + 4i + 12i + 6i^2}{16 - 4i^2} && \text{Expand.}\\ &= \frac{8 - 6 + 16i}{16 + 4} && \text{Definition of } i\\ &= \frac{1}{20}(2 + 16i) && \text{Simplify.}\\ &= \frac{1}{10} + \frac{4}{5}i && \text{Standard form}\end{aligned}$$

Figure 2.15

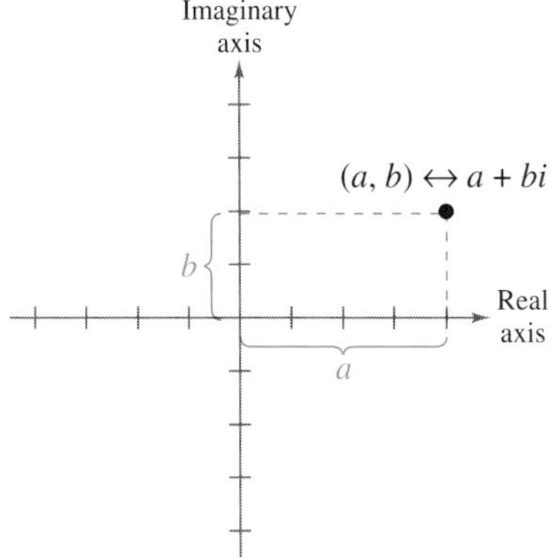

Applications

Most applications involving complex numbers are either theoretical or very technical, and are therefore not appropriate for inclusion in this text. However, to give you some idea of how complex numbers can be used in applications, we give a general description of their use in **fractal geometry.**

To begin, consider a coordinate system called the **complex plane.** Just as every real number corresponds to a point on the real line, every complex number corresponds to a point in the complex plane, as shown in Figure 2.15. In this figure, note that the vertical axis is the **imaginary axis** and the horizontal axis is the **real axis.** The point that corresponds to the complex number $a + bi$ is (a, b).

Figure 2.16

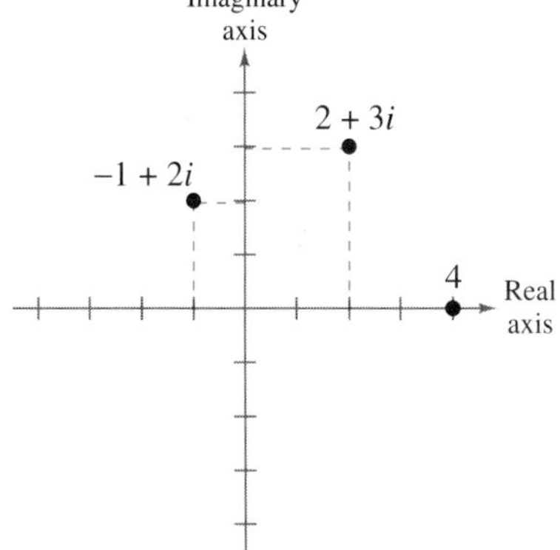

EXAMPLE 5 Plotting Complex Numbers

Plot each complex number in the complex plane.

a. $2 + 3i$ **b.** $-1 + 2i$ **c.** 4

Solution

a. To plot the complex number $2 + 3i$, move (from the origin) two units to the right on the real axis and then three units up, as shown in Figure 2.16. In other words, plotting the complex number $2 + 3i$ in the complex plane is comparable to plotting the point $(2, 3)$ in the Cartesian plane.

b. The complex number $-1 + 2i$ corresponds to the point $(-1, 2)$, as shown in Figure 2.16.

c. The complex number 4 corresponds to the point $(4, 0)$, as shown in Figure 2.16.

In the hands of a person who understands "fractal geometry," the complex plane can become an easel on which stunning pictures, called **fractals,** can be drawn. The most famous such picture is called the **Mandelbrot Set,** named after the Polish-born mathematician Benoit Mandelbrot. To draw the Mandelbrot Set, consider the following sequence of numbers.

$$c, \quad c^2 + c, \quad (c^2 + c)^2 + c, \quad [(c^2 + c)^2 + c]^2 + c, \quad \ldots$$

The behavior of this sequence depends on the value of the complex number c. For some values of c this sequence is **bounded,** and for other values it is **unbounded.** If the sequence is bounded, the complex number c is in the Mandelbrot Set, and if the sequence is unbounded, the complex number c is not in the Mandelbrot Set.

EXAMPLE 6 **Members of the Mandelbrot Set**

a. The complex number -2 is in the Mandelbrot Set because for $c = -2$, the corresponding Mandelbrot sequence is $-2, 2, 2, 2, 2, 2, \ldots$, which is bounded.

b. The complex number i is also in the Mandelbrot Set because for $c = i$, the corresponding Mandelbrot sequence is

$$i, \quad -1 + i, \quad -i, \quad -1 + i, \quad -i, \quad -1 + i, \quad \ldots$$

which is bounded.

c. The complex number $1 + i$ is *not* in the Mandelbrot Set because for $c = 1 + i$, the corresponding Mandelbrot sequence is

$$1 + i, \quad 1 + 3i, \quad -7 + 7i, \quad 1 - 97i, \quad -9407 - 193i,$$
$$88454401 + 3631103i, \quad \ldots$$

which is unbounded.

With this definition, a picture of the Mandelbrot Set would have only two colors. One color for points that are in the set (the sequence is bounded) and one color for points that are outside the set (the sequence is unbounded). Figure 2.17 shows a black and blue picture of the Mandelbrot Set. The points that are black are in the Mandelbrot Set and the points that are blue are not.

Figure 2.18

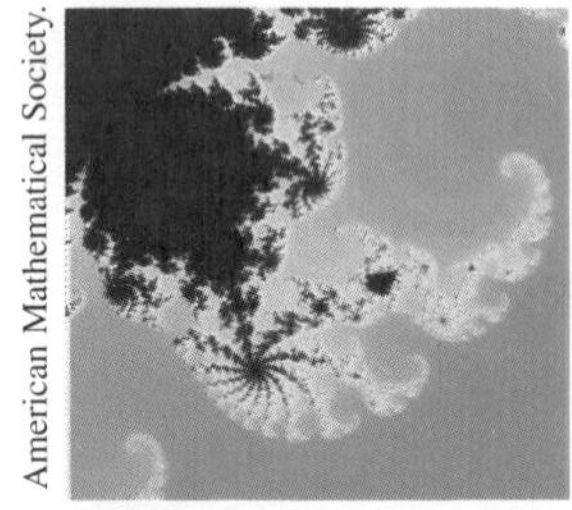

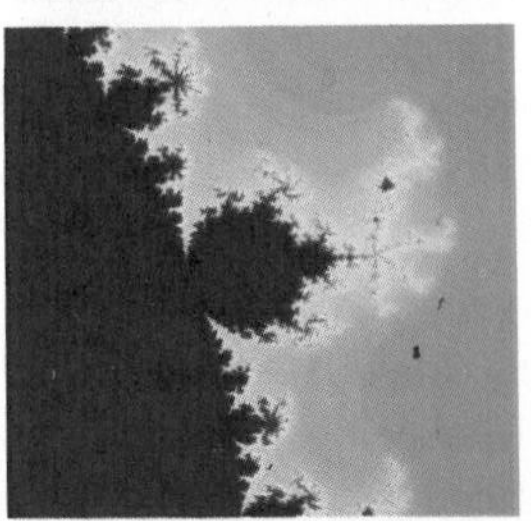

American Mathematical Society.

Figure 2.17

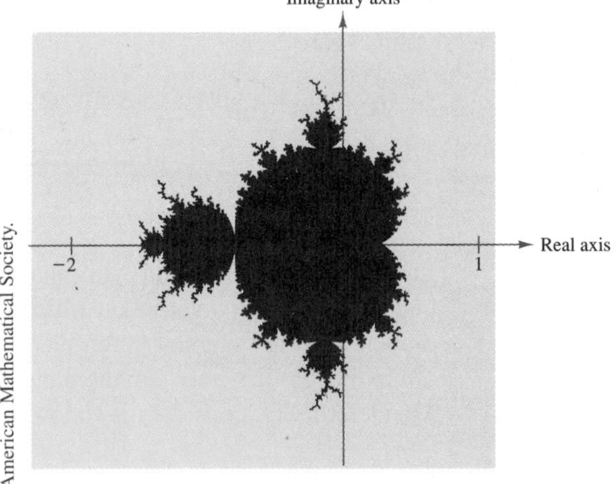

American Mathematical Society.

To add more interest to the picture, computer scientists discovered that the points that are not in the Mandelbrot Set can be assigned a variety of colors, depending on "how quickly" their sequences diverge. Figure 2.18 shows three different appendages of the Mandelbrot Set. (The black portions of the picture represent points that are in the Mandelbrot Set.)

Figures 2.19, 2.20, and 2.21 show other types of fractals. From these pictures, you can see why fractals have fascinated people since their discovery (around 1980). The fractals shown were produced on a graphing calculator.

Figure 2.19 The Sierpinski Gasket

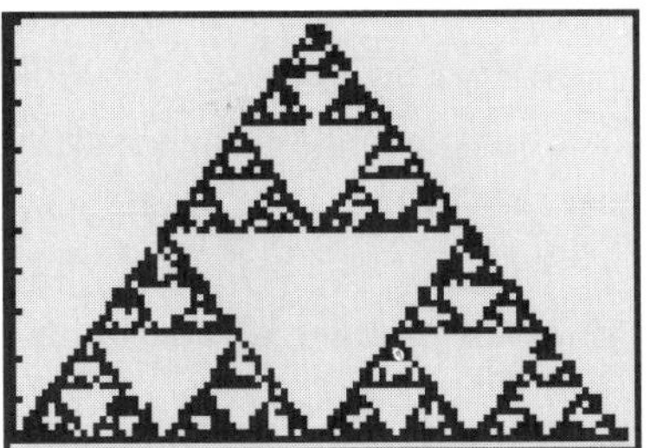

Figure 2.20 The Fractal "Dragon"

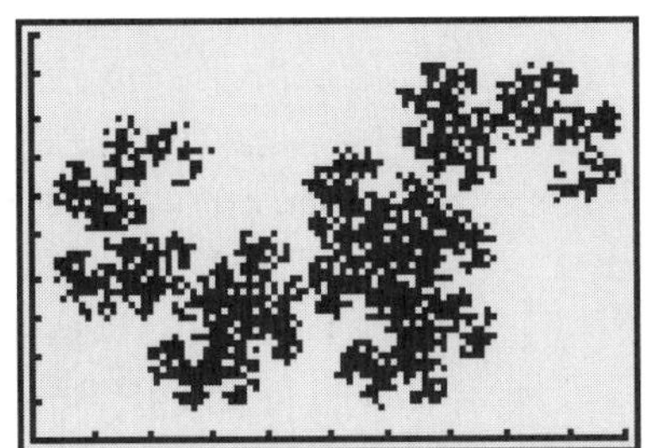

Figure 2.21 A Fractal Fern

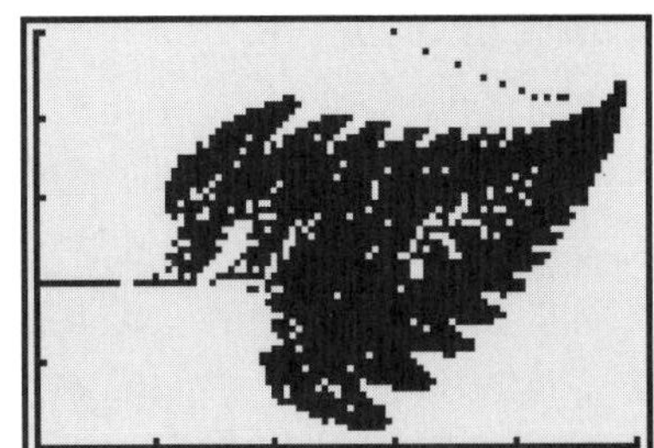

Group Activity

Error Analysis

Suppose you are a math instructor, and one of your students has handed in the following quiz. Find the error(s) in each solution and discuss how to explain each error to your student.

Question	*Answer*
1. Write $\dfrac{5}{3-2i}$ in standard form.	$\dfrac{5}{3-2i} \cdot \dfrac{3+2i}{3+2i} = \dfrac{15+10i}{9-4} = 3+2i$
2. Multiply: $\left(\sqrt{-4}+3\right)\left(i-\sqrt{-3}\right)$.	$\left(\sqrt{-4}+3\right)\left(i-\sqrt{-3}\right) = i\sqrt{-4} - \sqrt{-4}\sqrt{-3} + 3i - 3\sqrt{-3}$ $= -2i - \sqrt{12} + 3i - 3i\sqrt{3}$ $= \left(1-3\sqrt{3}\right)i - 2\sqrt{3}$
3. Sketch the graph of $y = -x^2 + 2$.	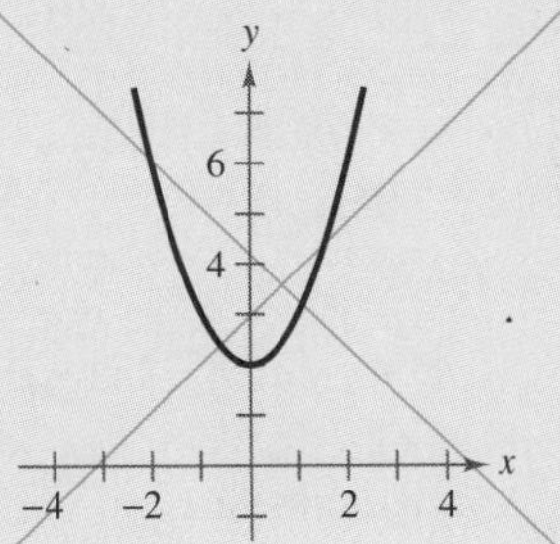

2.3 /// EXERCISES

In Exercises 1–4, solve for a and b.

1. $a + bi = -10 + 6i$
2. $a + bi = 13 + 4i$
3. $(a - 1) + (b + 3)i = 5 + 8i$
4. $(a + 6) + 2bi = 6 - 5i$

In Exercises 5–16, write in standard form.

5. $4 + \sqrt{-9}$
6. $3 + \sqrt{-16}$
7. $2 - \sqrt{-27}$
8. $1 + \sqrt{-8}$
9. $\sqrt{-75}$
10. 45
11. $-6i + i^2$
12. $-4i^2 + 2i$
13. 8
14. $(\sqrt{-4})^2 - 5$
15. $\sqrt{-0.09}$
16. $\sqrt{-0.0004}$

In Exercises 17–26, perform the addition or subtraction and write the result in standard form.

17. $(5 + i) + (6 - 2i)$
18. $(13 - 2i) + (-5 + 6i)$
19. $(8 - i) - (4 - i)$
20. $(3 + 2i) - (6 + 13i)$
21. $(-2 + \sqrt{-8}) + (5 - \sqrt{-50})$
22. $(8 + \sqrt{-18}) - (4 + 3\sqrt{2}i)$
23. $13i - (14 - 7i)$
24. $22 + (-5 + 8i) + 10i$
25. $-(\frac{3}{2} + \frac{5}{2}i) + (\frac{5}{3} + \frac{11}{3}i)$
26. $(1.6 + 3.2i) + (-5.8 + 4.3i)$

In Exercises 27–40, perform the operation and write the result in standard form.

27. $\sqrt{-6} \cdot \sqrt{-2}$
28. $\sqrt{-5} \cdot \sqrt{-10}$
29. $(\sqrt{-10})^2$
30. $(\sqrt{-75})^2$
31. $(1 + i)(3 - 2i)$
32. $(6 - 2i)(2 - 3i)$
33. $6i(5 - 2i)$
34. $-8i(9 + 4i)$
35. $(\sqrt{14} + \sqrt{10}i)(\sqrt{14} - \sqrt{10}i)$
36. $(3 + \sqrt{-5})(7 - \sqrt{-10})$
37. $(4 + 5i)^2$
38. $(2 - 3i)^3$
39. $(2 + 3i)^2 + (2 - 3i)^2$
40. $(1 - 2i)^2 - (1 + 2i)^2$

41. *Error Analysis* Describe the error.

    $\sqrt{-6}\sqrt{-6} = \sqrt{(-6)(-6)} = \sqrt{36} = 6$

42. *True or False?* There is no complex number that is equal to its conjugate. Explain.

In Exercises 43–50, find the product of the number and its conjugate.

43. $5 + 3i$
44. $9 - 12i$
45. $-2 - \sqrt{5}i$
46. $-4 + \sqrt{2}i$
47. $20i$
48. $\sqrt{-15}$
49. $\sqrt{8}$
50. $1 + \sqrt{8}$

In Exercises 51–64, perform the operation and write the result in standard form.

51. $\dfrac{6}{i}$
52. $-\dfrac{10}{2i}$
53. $\dfrac{4}{4 - 5i}$
54. $\dfrac{3}{1 - i}$
55. $\dfrac{2 + i}{2 - i}$
56. $\dfrac{8 - 7i}{1 - 2i}$
57. $\dfrac{6 - 7i}{i}$
58. $\dfrac{8 + 20i}{2i}$
59. $\dfrac{1}{(4 - 5i)^2}$
60. $\dfrac{(2 - 3i)(5i)}{2 + 3i}$
61. $\dfrac{2}{1 + i} - \dfrac{3}{1 - i}$
62. $\dfrac{2i}{2 + i} + \dfrac{5}{2 - i}$
63. $\dfrac{i}{3 - 2i} + \dfrac{2i}{3 + 8i}$
64. $\dfrac{1 + i}{i} - \dfrac{3}{4 - i}$

65. Express the first 16 positive integer powers of i as i, $-i$, 1, or -1. Describe any pattern you see.
66. Use your result from Exercise 65 to express each of the following powers of i as i, $-i$, 1, or -1: (a) i^{40} (b) i^{25} (c) i^{50} (d) i^{67}

In Exercises 67–74, simplify the complex number and write it in standard form.

67. $-6i^3 + i^2$

68. $4i^2 - 2i^3$

69. $-5i^5$

70. $(-i)^3$

71. $(\sqrt{-75})^3$

72. $(\sqrt{-2})^6$

73. $\dfrac{1}{i^3}$

74. $\dfrac{1}{(2i)^3}$

In Exercises 75–78, determine the complex number shown in the complex plane.

75.

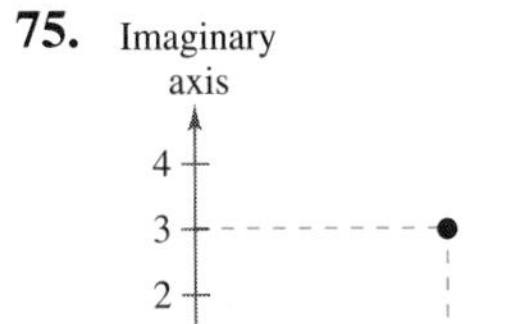

76.

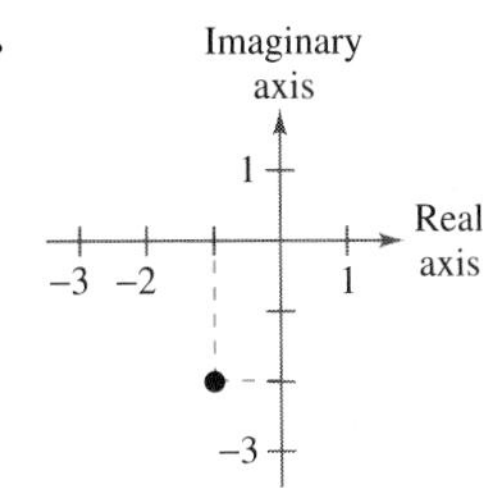

77.

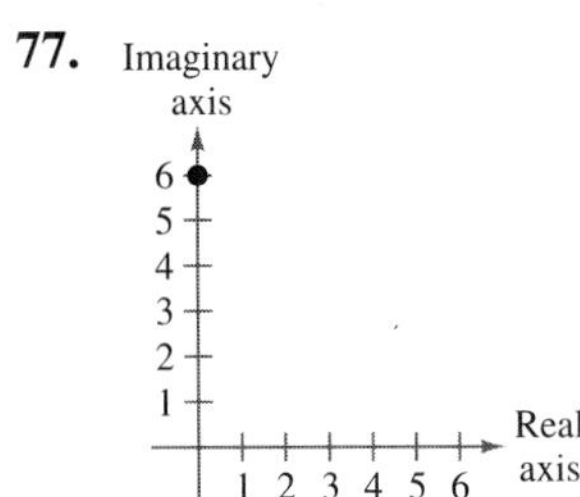

78.

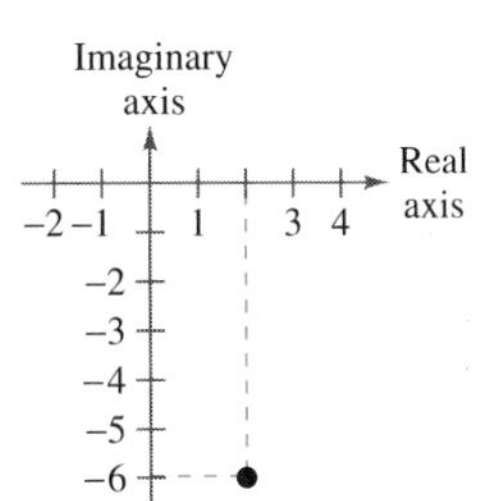

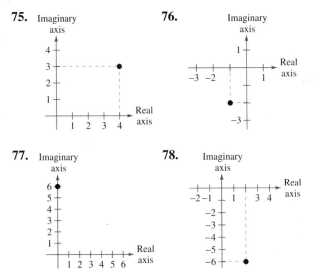

In Exercises 79–82, plot the complex number in the complex plane.

79. $3 - 4i$

80. $2i$

81. -5

82. $-5 + 3i$

Fractals **In Exercises 83–88, find the first six terms in the following sequence.**

$$c,\ c^2 + c,\ (c^2 + c)^2 + c,\ [(c^2 + c)^2 + c]^2 + c,\ \ldots$$

From these terms, do you think the given complex number is in the Mandelbrot Set?

83. $c = 0$

84. $c = 2$

85. $c = \frac{1}{2}i$

86. $c = -i$

87. $c = 1$

88. $c = -1$

89. Cube the complex numbers. What do you notice?

$2, \quad -1 + \sqrt{3}i, \quad -1 - \sqrt{3}i$

90. Raise the numbers to the fourth power.

$2, \quad -2, \quad 2i, \quad -2i$

91. Prove that the sum of a complex number $a + bi$ and its conjugate is a real number, and that the difference of a complex number $a + bi$ and its conjugate is an imaginary number.

92. Prove that the product of a complex number $a + bi$ and its conjugate is a real number.

93. Prove that the conjugate of the product of two complex numbers $a_1 + b_1i$ and $a_2 + b_2i$ is the product of their conjugates.

94. Prove that the conjugate of the sum of two complex numbers $a_1 + b_1i$ and $a_2 + b_2i$ is the sum of their conjugates.

Review **Solve Exercises 95–103 as a review of the skills and problem-solving techniques you learned in previous sections.**

95. Subtract: $(x^3 - 3x^2) - (6 - 2x - 4x^2)$

96. Add: $(4 + 3x) + (8 - 6x - x^2)$

97. Multiply: $(3x - \frac{1}{2})(x + 4)$

98. Expand: $(2x - 5)^2$

99. Expand: $[(x + y) + 3]^2$

100. *Volume of an Oblate Spheroid*

Solve for a: $V = \frac{4}{3}\pi a^2 b$

101. *Newton's Law of Universal Gravitation*

Solve for r: $F = \alpha \dfrac{m_1 m_2}{r^2}$

102. *Mixture Problem* A 5-liter container contains a mixture with a concentration of 50%. How much of this mixture must be withdrawn and replaced by 100% concentrate to bring the mixture up to 60% concentration?

103. *Average Speed* A business executive traveled at an average speed of 100 kilometers per hour on a 200-kilometer trip. Because of heavy traffic, the average speed on the return trip was only 80 kilometers per hour. Find the average speed for the round trip.

2.4 Solving Equations Algebraically

Quadratic Equations / Polynomial Equations of Higher Degree / Equations Involving Radicals / Equations Involving Fractions or Absolute Values / Applications

Quadratic Equations

A **quadratic equation** in x is an equation that can be written in the standard form

$$ax^2 + bx + c = 0$$

where a, b, and c are real numbers with $a \neq 0$. Another name for a quadratic equation in x is a **second-degree polynomial equation in x**. You should be familiar with the following four methods for solving quadratic equations.

Solving a Quadratic Equation

Method	*Example*
Factoring: If $ab = 0$, then $a = 0$ or $b = 0$.	$x^2 - x - 6 = 0$ $(x - 3)(x + 2) = 0$ $x - 3 = 0 \Rightarrow x = 3$ $x + 2 = 0 \Rightarrow x = -2$
Square Root Principle: If $u^2 = c$, where $c > 0$, then $u = \pm\sqrt{c}$.	$(x + 3)^2 = 16$ $x + 3 = \pm 4$ $x = -3 \pm 4$ $x = 1$ or $x = -7$
Completing the Square: If $x^2 + bx = c$, then $x^2 + bx + \left(\frac{b}{2}\right)^2 = c + \left(\frac{b}{2}\right)^2$ $\left(x + \frac{b}{2}\right)^2 = c + \frac{b^2}{4}$.	$x^2 + 6x = 5$ $x^2 + 6x + 3^2 = 5 + 3^2$ $(x + 3)^2 = 14$ $x + 3 = \pm\sqrt{14}$ $x = -3 \pm \sqrt{14}$
Quadratic Formula: If $ax^2 + bx + c = 0$, then $x = \frac{-b \pm \sqrt{b^2 - 4ac}}{2a}$.	$2x^2 + 3x - 1 = 0$ $x = \frac{-3 \pm \sqrt{3^2 - 4(2)(-1)}}{2(2)}$ $= \frac{-3 \pm \sqrt{17}}{4}$

EXAMPLE 1 Solving a Quadratic Equation by Factoring

Solve each quadratic equation.

a. $6x^2 = 3x$ **b.** $9x^2 - 6x + 1 = 0$

Solution

a.

$6x^2 = 3x$		Original equation
$6x^2 - 3x = 0$		Standard form
$3x(2x - 1) = 0$		Factored form
$3x = 0$	$\Rightarrow x = 0$	Set 1st factor equal to 0.
$2x - 1 = 0$	$\Rightarrow x = \frac{1}{2}$	Set 2nd factor equal to 0.

b.

$9x^2 - 6x + 1 = 0$		Original equation
$(3x - 1)^2 = 0$		Factored form
$3x - 1 = 0$	$\Rightarrow x = \frac{1}{3}$	Set repeated factor equal to 0.

Try programming the Quadratic Formula into a computer or graphing calculator. For instance, the following program is for the *TI-82* or *TI-83*. Programs for other graphing calculator models may be found in the Appendix.

```
PROGRAM:QUADRAT
:Disp "AX²+BX+C = 0"
:Prompt A
:Prompt B
:Prompt C
:B²-4AC→D
:If D≥0
:Then
:(-B+√(D))/(2A)→M
:Disp M
:(-B-√(D))/(2A)→N
:Disp N
:Else
:Disp "NO REAL SOLUTION"
:End
```

To use this program, you must first write the equation in standard form. Then enter the values of *a, b,* and *c*. After the final value has been entered, the program will display either two real solutions *or* the words "NO REAL SOLUTION."

Throughout the text, when solving equations, be sure to check your solutions *algebraically* by substituting in the original equation or *graphically.* For instance, you can check the solution in Example 1(b) algebraically as follows.

Check

$9x^2 - 6x + 1 = 0$	Original equation
$9\left(\frac{1}{3}\right)^2 - 6\left(\frac{1}{3}\right) + 1 \stackrel{?}{=} 0$	Substitute $\frac{1}{3}$ for x.
$1 - 2 + 1 \stackrel{?}{=} 0$	Simplify.
$0 = 0$	Solution checks. ✓

Similarly, the graphs in Figure 2.22 provide a graphical check.

Figure 2.22

(a)

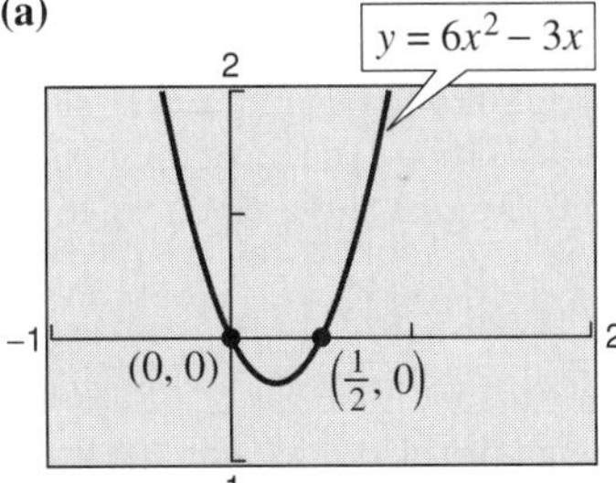

(b)

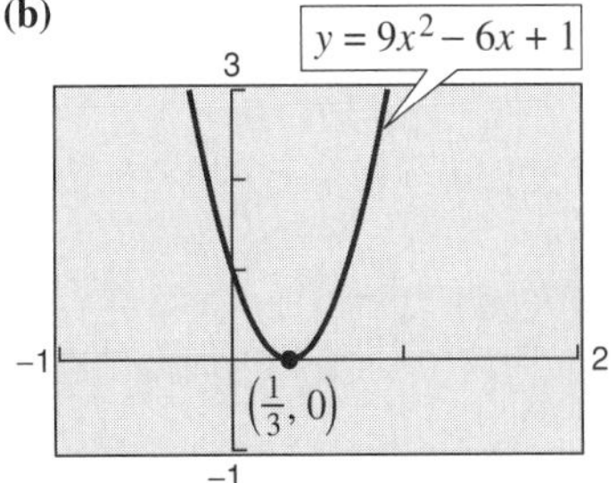

EXAMPLE 2 **Extracting Square Roots**

Solve each quadratic equation.

a. $4x^2 = 12$ **b.** $(x - 3)^2 = 7$

Solution

a. $4x^2 = 12$ Original equation

$x^2 = 3$ Divide both sides by 4.

$x = \pm\sqrt{3}$ Extract square roots.

b. $(x - 3)^2 = 7$ Original equation

$x - 3 = \pm\sqrt{7}$ Extract square roots.

$x = 3 \pm \sqrt{7}$ Add 3 to both sides.

The graphs of $y = 4x^2 - 12$ and $y = (x - 3)^2 - 7$, as shown in Figure 2.23, verify the solutions graphically.

Figure 2.23

(a)

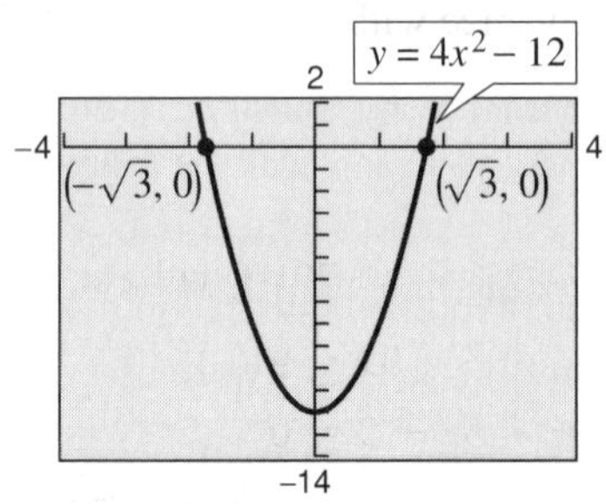

(b)

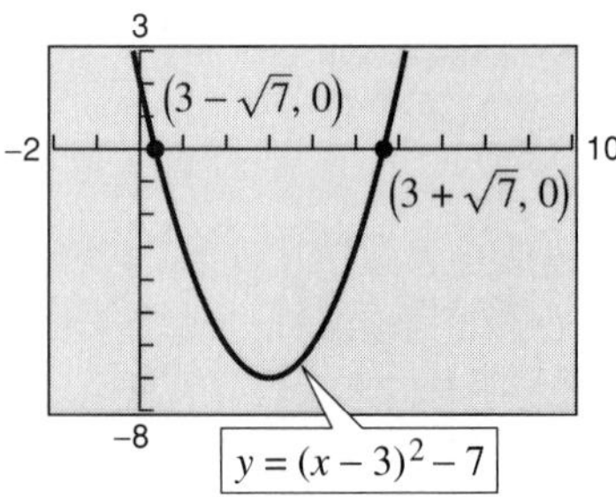

The solutions shown in Example 2 are listed in *exact* form. Most graphing utilities produce decimal approximations of solutions rather than exact forms. For instance, if you solve the equations in Example 2 with a *TI-82* or *TI-83*, you will obtain $x \approx \pm 1.732$ in part (a) and $x \approx 5.646$ and $x \approx 0.354$ in part (b).

Some graphing utilities, such as *Derive* and the *TI-92,* have symbolic algebra programs that *can* list the exact form of a solution.

EXAMPLE 3 Completing the Square: Leading Coefficient Is Not 1

$$3x^2 - 4x - 5 = 0 \quad \text{Original equation}$$

$$3x^2 - 4x = 5 \quad \text{Add 5 to both sides.}$$

$$x^2 - \frac{4}{3}x = \frac{5}{3} \quad \text{Divide both sides by 3.}$$

$$x^2 - \frac{4}{3}x + \left(\frac{2}{3}\right)^2 = \frac{5}{3} + \left(\frac{2}{3}\right)^2 \quad \text{Add } \left(\tfrac{2}{3}\right)^2 \text{ to both sides.}$$

(Half)^2

$$\left(x - \frac{2}{3}\right)^2 = \frac{19}{9}$$

$$x - \frac{2}{3} = \pm\frac{\sqrt{19}}{3} \quad \text{Extract square roots.}$$

$$x = \frac{2}{3} \pm \frac{\sqrt{19}}{3} \quad \text{Solutions}$$

Using a calculator, the two solutions are approximately 2.11963 and −0.78630, which agree with the graphical solution shown in Figure 2.24.

Figure 2.24

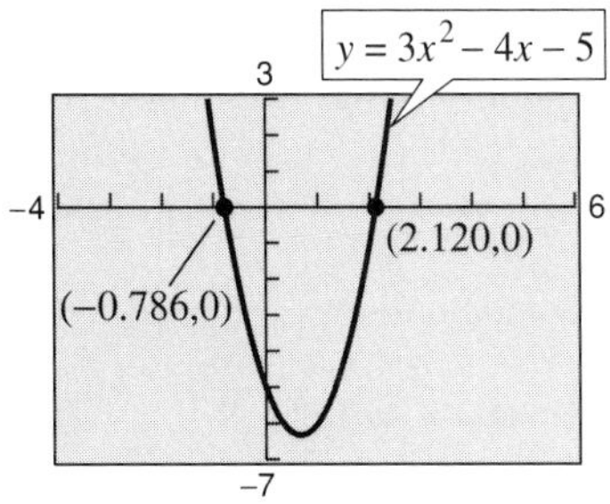

EXAMPLE 4 Quadratic Formula: Two Distinct Solutions

$$x^2 + 3x = 9 \quad \text{Original equation}$$

$$x^2 + 3x - 9 = 0 \quad \text{Standard form}$$

$$x = \frac{-b \pm \sqrt{b^2 - 4ac}}{2a} \quad \text{Quadratic Formula}$$

$$x = \frac{-3 \pm \sqrt{3^2 - 4(1)(-9)}}{2(1)} \quad \text{Substitute.}$$

$$x = \frac{-3 \pm \sqrt{45}}{2}$$

$$x = \frac{-3 \pm 3\sqrt{5}}{2}$$

$$x \approx 1.85 \text{ or } -4.85 \quad \text{Solutions}$$

The equation has two solutions: $x \approx 1.85$ and $x \approx -4.85$. Check these solutions in the original equation. The graph of $y = x^2 + 3x - 9$, as shown in Figure 2.25, verifies the solutions graphically.

Figure 2.25

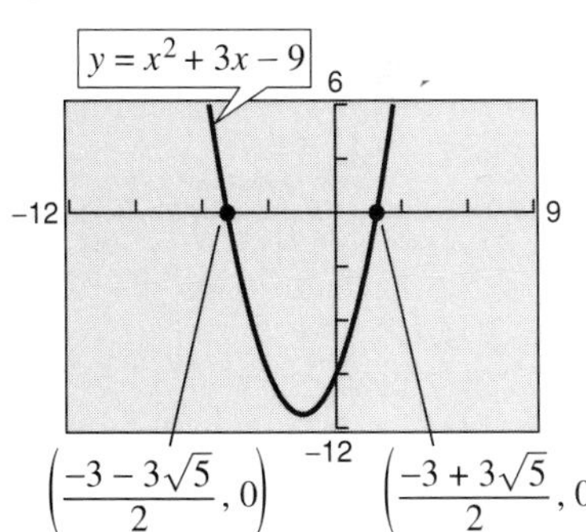

EXAMPLE 5 Quadratic Formula: One Repeated Solution

Use the Quadratic Formula to solve $8x^2 - 24x + 18 = 0$.

Solution

This equation has a common factor of 2. To simplify things, first divide both sides of the equation by 2.

$$8x^2 - 24x + 18 = 0 \qquad \text{Original equation}$$

$$4x^2 - 12x + 9 = 0 \qquad \text{Standard form}$$

$$x = \frac{-b \pm \sqrt{b^2 - 4ac}}{2a} \qquad \text{Quadratic Formula}$$

$$x = \frac{-(-12) \pm \sqrt{(-12)^2 - 4(4)(9)}}{2(4)}$$

$$x = \frac{12 \pm \sqrt{0}}{8} = \frac{3}{2} \qquad \text{Repeated solution}$$

Figure 2.26

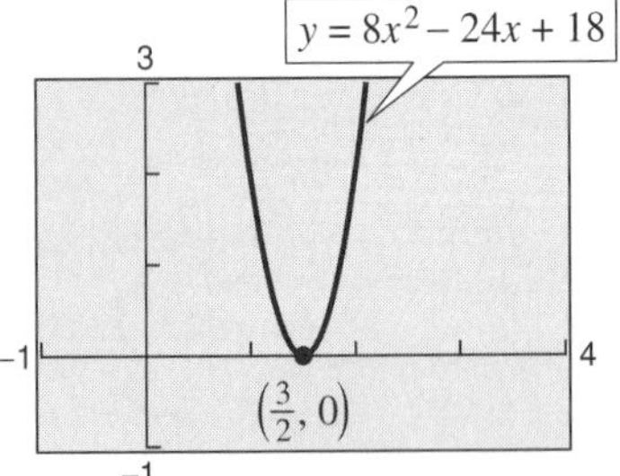

This quadratic equation has only one solution: $\frac{3}{2}$. Check this solution in the original equation. The graph of $y = 8x^2 - 24x + 18$ is shown in Figure 2.26.

EXPLORATION

Use a graphing utility to graph the three quadratic equations

$$y_1 = x^2 - 2x$$
$$y_2 = x^2 - 2x + 1$$
$$y_3 = x^2 - 2x + 2$$

in the same viewing rectangle. Compute the *discriminant* $\sqrt{b^2 - 4ac}$ for each and discuss the relationship between the discriminant and the number of zeros of the quadratic function.

EXAMPLE 6 Complex Solutions of a Quadratic Equation

Solve $3x^2 - 2x + 5 = 0$.

Solution

By the Quadratic Formula, you can write the solutions as follows.

$$x = \frac{-b \pm \sqrt{b^2 - 4ac}}{2a}$$

$$= \frac{-(-2) \pm \sqrt{(-2)^2 - 4(3)(5)}}{2(3)}$$

$$= \frac{2 \pm \sqrt{-56}}{6}$$

$$= \frac{2 \pm 2\sqrt{14}i}{6}$$

$$= \frac{1}{3} \pm \frac{\sqrt{14}}{3}i$$

Figure 2.27

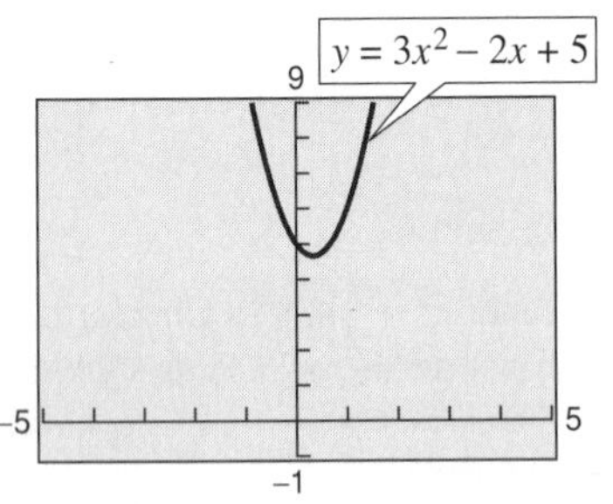

The equation has two solutions: $\frac{1}{3}(1 + \sqrt{14}i)$ and $\frac{1}{3}(1 - \sqrt{14}i)$. Because the original equation has no real solutions, the graph of $y = 3x^2 - 2x + 5$ should have no x-intercepts. This is confirmed in Figure 2.27.

Polynomial Equations of Higher Degree

The methods used to solve quadratic equations can sometimes be extended to polynomial equations of higher degree, as shown in the next two examples.

EXAMPLE 7 Solving an Equation of Quadratic Type

Solve the equation

$$x^4 - 3x^2 + 2 = 0.$$

Solution

$$x^4 - 3x^2 + 2 = 0 \quad \text{Original equation}$$
$$(x^2)^2 - 3(x^2) + 2 = 0 \quad \text{Quadratic form in } x^2$$
$$(x^2 - 1)(x^2 - 2) = 0 \quad \text{Partially factor.}$$
$$(x + 1)(x - 1)(x^2 - 2) = 0 \quad \text{Factor.}$$
$$x + 1 = 0 \Rightarrow x = -1$$
$$x - 1 = 0 \Rightarrow x = 1$$
$$x^2 - 2 = 0 \Rightarrow x = \pm\sqrt{2}$$

Figure 2.28

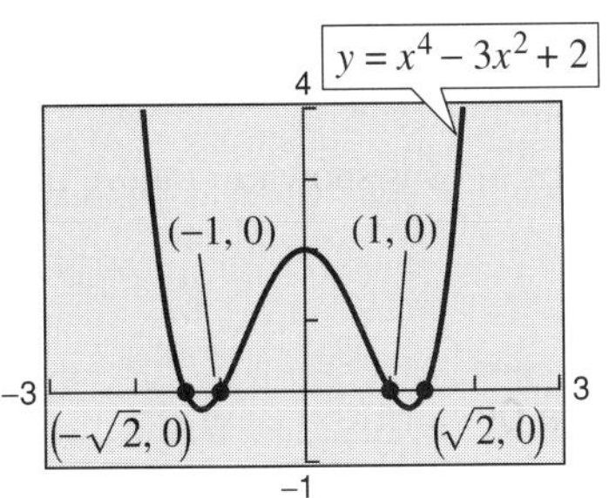

The equation has four solutions: -1, 1, $\sqrt{2}$, and $-\sqrt{2}$. Check these solutions in the original equation. The graph of $y = x^4 - 3x^2 + 2$, as shown in Figure 2.28, verifies the solutions graphically.

EXAMPLE 8 Solving a Polynomial Equation by Factoring

Solve the equation

$$x^3 - 3x^2 - 3x + 9 = 0.$$

Solution

$$x^3 - 3x^2 - 3x + 9 = 0 \quad \text{Original equation}$$
$$x^2(x - 3) - 3(x - 3) = 0 \quad \text{Group terms.}$$
$$(x - 3)(x^2 - 3) = 0 \quad \text{Factor by grouping.}$$
$$x - 3 = 0 \Rightarrow x = 3$$
$$x^2 - 3 = 0 \Rightarrow x = \pm\sqrt{3}$$

Figure 2.29

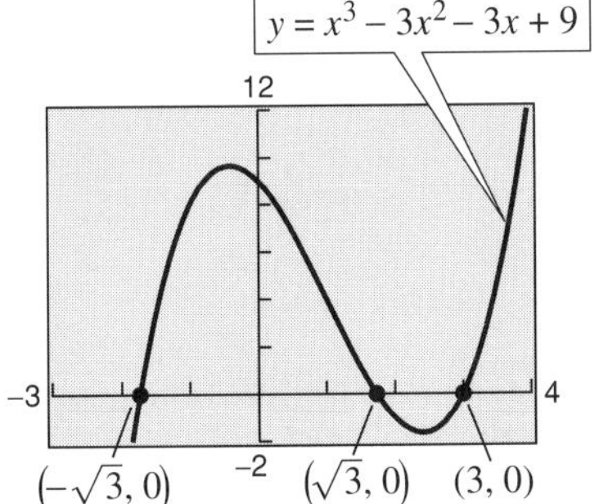

The equation has three solutions: 3, $\sqrt{3}$, and $-\sqrt{3}$. Check these solutions in the original equation. The graph of $y = x^3 - 3x^2 - 3x + 9$, as shown in Figure 2.29, verifies the solutions graphically.

Figure 2.30

(a)

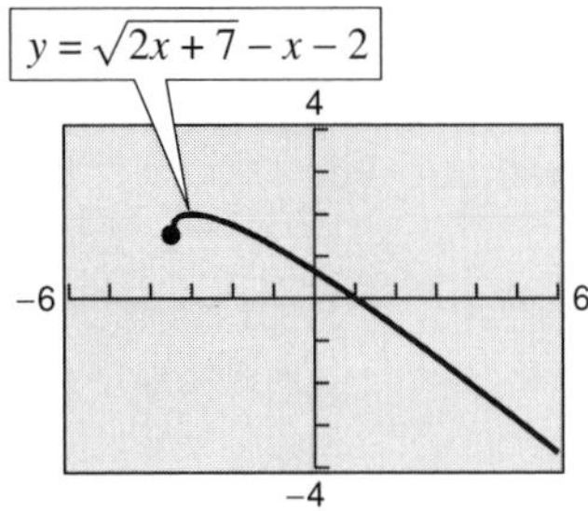

(b)

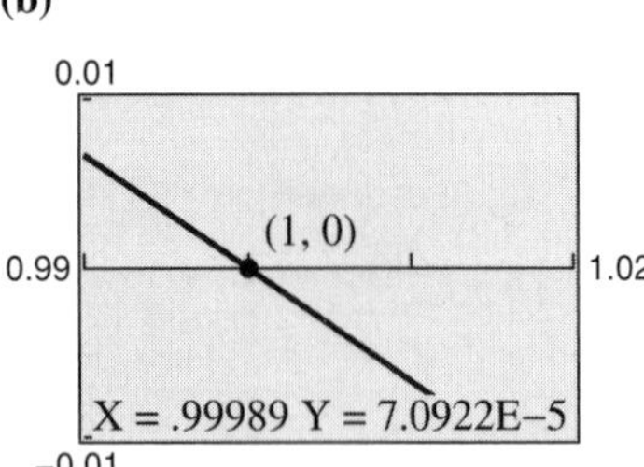

Equations Involving Radicals

An equation involving a radical expression can often be cleared of radicals by raising both sides of the equation to an appropriate power. When using this procedure, remember that it can introduce extraneous solutions—so be sure to check each solution in the original equation.

EXAMPLE 9 Solving an Equation Involving a Radical

Solve $\sqrt{2x + 7} - x = 2$.

Solution

The graph of $y = \sqrt{2x + 7} - x - 2$ is shown in Figure 2.30(a). Notice that the domain is $x \geq -\frac{7}{2}$, because the expression under the radical cannot be negative. Furthermore, there seems to be one solution near $x = 1$. Using the zoom and trace features shown in Figure 2.30(b), you can verify that $x = 1$ is the only solution. The equation can also be solved algebraically. To do this, first isolate the radical. Then eliminate the square root by squaring both sides of the equation. You will obtain two solutions: -3 and 1. By substituting into the original equation, you can determine that -3 is extraneous whereas 1 is valid. Thus, the equation has only one real solution: $x = 1$.

EXAMPLE 10 Solving an Equation Involving Two Radicals

Solve the equation

$$\sqrt{2x + 6} - \sqrt{x + 4} = 1.$$

Solution

$$\sqrt{2x + 6} - \sqrt{x + 4} = 1 \qquad \text{Original equation}$$

$$\sqrt{2x + 6} = 1 + \sqrt{x + 4} \qquad \text{Isolate radical.}$$

$$2x + 6 = 1 + 2\sqrt{x + 4} + (x + 4) \qquad \text{Square both sides.}$$

$$x + 1 = 2\sqrt{x + 4} \qquad \text{Isolate radical.}$$

$$x^2 + 2x + 1 = 4(x + 4) \qquad \text{Square both sides.}$$

$$x^2 - 2x - 15 = 0 \qquad \text{Standard form}$$

$$(x - 5)(x + 3) = 0 \qquad \text{Factor.}$$

$$x - 5 = 0 \implies x = 5 \qquad \text{Set 1st factor equal to 0.}$$

$$x + 3 = 0 \implies x = -3 \qquad \text{Set 2nd factor equal to 0.}$$

Figure 2.31

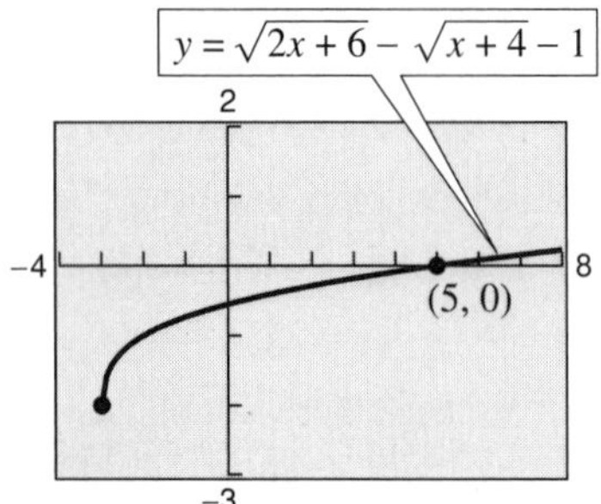

From Figure 2.31, you can conclude that $x = 5$ is the only solution.

Equations Involving Fractions or Absolute Values

As demonstrated in Section 2.1, you can algebraically solve an equation involving fractions by multiplying both sides of the equation by the least common denominator of each term in the equation. This procedure will "clear an equation of fractions." For instance, in the equation

$$\frac{2}{x^2 + 1} + \frac{1}{x} = \frac{2}{x}$$

you can multiply both sides of the equation by the LCD $x(x^2 + 1)$ to obtain

$$(2x) + (x^2 + 1) = 2(x^2 + 1).$$

Try solving this equation. You should obtain one solution: $x = 1$.

EXPLORATION

Using dot mode, graph the two functions

$$y_1 = \frac{2}{x}$$

$$y_2 = \frac{3}{x - 2} - 1$$

in the same viewing rectangle. How many times do the graphs of the functions intersect each other? What does this tell you about the solution to Example 11?

EXAMPLE 11 **Solving an Equation Involving Fractions**

Solve $\dfrac{2}{x} = \dfrac{3}{x - 2} - 1.$

Solution

For this equation, the least common denominator of the three terms is $x(x - 2)$, so you can begin by multiplying each term in the equation by this expression.

$$\begin{aligned}
\frac{2}{x} &= \frac{3}{x - 2} - 1 \\
x(x - 2)\frac{2}{x} &= x(x - 2)\frac{3}{x - 2} - x(x - 2)(1) \\
2(x - 2) &= 3x - x(x - 2), \qquad x \neq 0, 2 \\
2x - 4 &= -x^2 + 5x \\
x^2 - 3x - 4 &= 0 \\
(x - 4)(x + 1) &= 0 \\
x - 4 &= 0 \quad \Rightarrow \quad x = 4 \\
x + 1 &= 0 \quad \Rightarrow \quad x = -1
\end{aligned}$$

The equation has two solutions: 4 and -1. Check these solutions in the original equation. Use a graphing utility to verify these solutions graphically.

Note Graphs of functions involving variable denominators can be tricky because of the way graphing utilities skip over points where the denominator is zero. You will study graphs of such functions in Sections 3.5 and 3.6.

EXAMPLE 12 Solving an Equation Involving Absolute Value

Solve $|x^2 - 3x| = -4x + 6$.

Solution

Begin by writing the equation as $|x^2 - 3x| + 4x - 6 = 0$. From the graph of $y = |x^2 - 3x| + 4x - 6$ in Figure 2.32, you can estimate the solutions to be -3 and 1. These can be verified by substitution into the equation. To solve *algebraically* an equation involving an absolute value, you must consider the fact that the expression inside the absolute value symbols can be positive or negative. This consideration results in *two* separate equations, each of which must be solved.

Figure 2.32

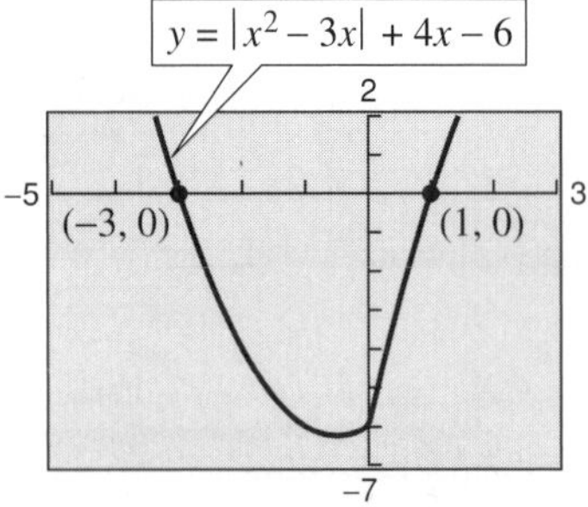

First Equation:

$x^2 - 3x = -4x + 6$		Use positive expression.
$x^2 + x - 6 = 0$		Standard form
$(x + 3)(x - 2) = 0$		Factor.
$x + 3 = 0$	$x = -3$	Set 1st factor equal to 0.
$x - 2 = 0$	$x = 2$	Set 2nd factor equal to 0.

Second Equation:

$-(x^2 - 3x) = -4x + 6$		Use negative expression.
$x^2 - 7x + 6 = 0$		Standard form
$(x - 1)(x - 6) = 0$		Factor.
$x - 1 = 0$	$x = 1$	Set 1st factor equal to 0.
$x - 6 = 0$	$x = 6$	Set 2nd factor equal to 0.

Check

$|(-3)^2 - 3(-3)| = -4(-3) + 6$	-3 checks. ✓
$|2^2 - 3(2)| \neq -4(2) + 6$	2 does not check.
$|1^2 - 3(1)| = -4(1) + 6$	1 checks. ✓
$|6^2 - 3(6)| \neq -4(6) + 6$	6 does not check.

The equation has only two solutions: -3 and 1, just as you obtained by graphing.

Note In Figure 2.32, the graph of $y = |x^2 - 3x| + 4x - 6$ appears to be a straight line to the right of the y-axis. Is it? Explain how you decided.

Applications

A common application of quadratic equations involves an object that is falling (or projected into the air). The general equation that gives the height of such an object is called a **position equation,** and on *earth's* surface it has the form

Note The position equation at the right ignores air resistance.

$$s = -16t^2 + v_0t + s_0.$$

In this equation, s represents the height of the object (in feet), v_0 represents the original velocity of the object (in feet per second), s_0 represents the original height of the object (in feet), and t represents the time (in seconds).

Real Life

EXAMPLE 13 Falling Time

Figure 2.33

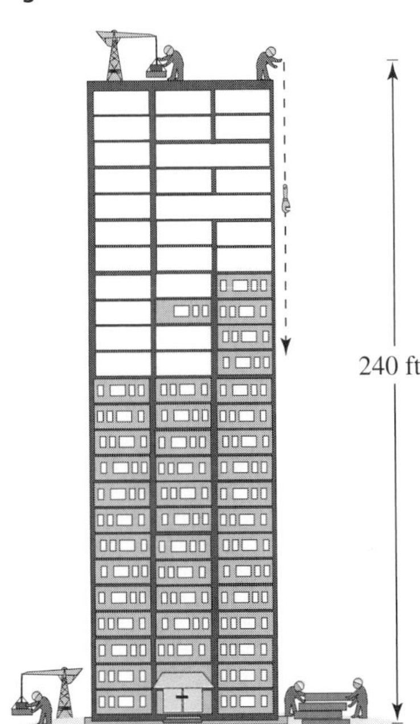

A construction worker on the 24th floor of a building project (see Figure 2.33) accidentally drops a wrench and yells "Look out below!" Could a person at ground level hear this warning in time to get out of the way?

Solution

Assume that each floor of the building is 10 feet high, so that the wrench is dropped from a height of 240 feet. Because sound travels at about 1100 feet per second, it follows that a person at ground level hears the warning within 1 second of the time the wrench is dropped. To set up a mathematical model for the height of the wrench, use the position equation

$$s = -16t^2 + v_0t + s_0.$$

Because the object is dropped rather than thrown, the initial velocity is $v_0 = 0$. Thus, with an initial height of $s_0 = 240$ feet, you have the following model.

$$s = -16t^2 + 240$$

After falling for 1 second, the height of the wrench is $-16(1)^2 + 240 = 224$. After falling for 2 seconds, the height of the wrench is $-16(2)^2 + 240 = 176$. To find the number of seconds it takes the wrench to hit the ground, let the height s be zero and solve the equation for t.

$$s = -16t^2 + 240 \quad \text{Position equation}$$

$$0 = -16t^2 + 240 \quad \text{Set height equal to 0.}$$

$$16t^2 = 240 \quad \text{Add } 16t^2 \text{ to both sides.}$$

$$t^2 = 15 \quad \text{Divide both sides by 16.}$$

$$t = \sqrt{15} \approx 3.87 \quad \text{Extract positive square root.}$$

The wrench will take about 3.87 seconds to hit the ground. If the person hears the warning 1 second after the wrench is dropped, the person still has almost 3 seconds to get out of the way.

Real Life

EXAMPLE 14 **Quadratic Modeling: Cable Television**

From 1985 to 1992, the number of hours spent annually per person watching basic cable television in the United States closely followed the quadratic model

$$\text{Hours} = 3.8t^2 - 29t + 167$$

where $t = 5$ represents 1985. The number of hours per year is shown graphically in Figure 2.34. According to this model, in which year will the number of hours spent per person reach or surpass 400? (Source: Veronis, Suhler, & Associates)

Figure 2.34

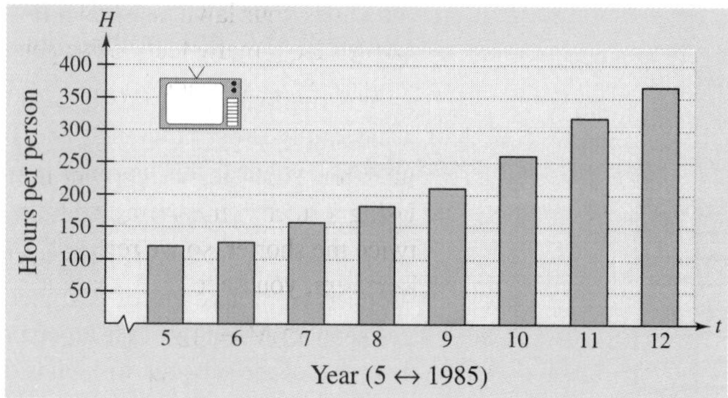

You can solve Example 14 with your graphing utility by graphing the two functions

$$y_1 = 3.8x^2 - 29x + 167$$
$$y_2 = 400$$

in the same viewing rectangle and finding their point of intersection. You should obtain $x \approx 12.53$, which verifies the answer obtained analytically.

Solution

To find when the number of hours spent per person will reach 400, you need to solve the equation

$$3.8t^2 - 29t + 167 = 400.$$

To begin, write the equation in standard form.

$$3.8t^2 - 29t - 233 = 0$$

Then apply the Quadratic Formula.

$$t = \frac{-(-29) \pm \sqrt{(-29)^2 - 4(3.8)(-233)}}{2(3.8)}$$

Choosing the positive solution, you find that

$$t = \frac{29 + \sqrt{(-29)^2 - 4(3.8)(-233)}}{2(3.8)} \approx 12.53 \approx 13.$$

Because $t = 5$ corresponds to 1985, it follows that $t = 13$ must correspond to 1993. Thus, the number of hours spent annually per person watching basic cable should have reached 400 during 1993.

Figure 2.35

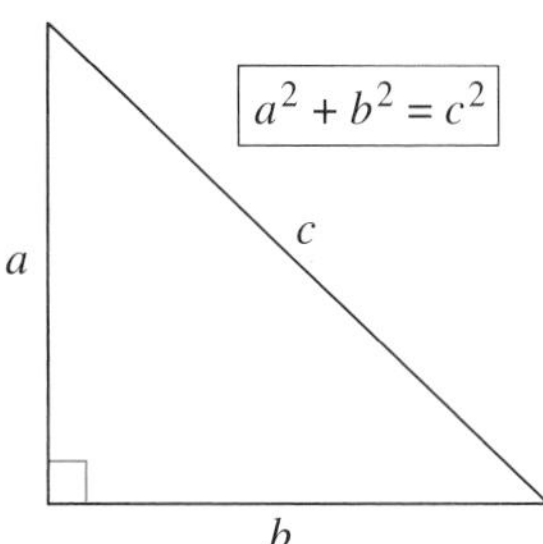

Another type of application that often involves a quadratic equation is one dealing with the hypotenuse of a right triangle. These types of applications often use the Pythagorean Theorem, which states that

$$a^2 + b^2 = c^2 \qquad \text{Pythagorean Theorem}$$

where a and b are the legs of a right triangle and c is the hypotenuse, as indicated in Figure 2.35.

Real Life

EXAMPLE 15 **Cutting Across the Lawn**

Your house is on a large corner lot. Several of the children in the neighborhood cut across your lawn, as shown in Figure 2.36. The distance across the lawn is 32 feet. How many feet does a person save by walking across the lawn instead of walking on the sidewalk?

Figure 2.36

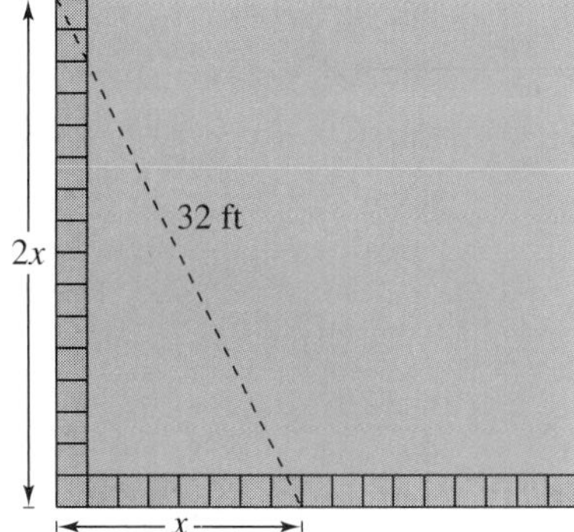

Solution

In Figure 2.36, let x represent the length of the shorter part of the sidewalk. Using a ruler you can find that the length of the longer part of the sidewalk is twice the shorter, so we represent its length by $2x$. Now, using the Pythagorean Theorem, you have

$$x^2 + (2x)^2 = 32^2 \qquad \text{Pythagorean Theorem}$$

$$5x^2 = 1024 \qquad \text{Combine like terms.}$$

$$x^2 = 204.8 \qquad \text{Divide both sides by 5.}$$

$$x = \sqrt{204.8} \qquad \text{Extract positive square root.}$$

The total distance on the sidewalk is

$$x + 2x = 3x = 3\sqrt{204.8} \approx 42.9 \text{ feet.}$$

Cutting across the lawn saves a person about $42.9 - 32$ or 10.9 feet.

Group Activity

Investigating Intercepts

On a graphing utility, store the value 5 in A, -2 in B, and 1 in C. Then use a graphing utility to sketch the graph of $y = C(x - A)(x - B)$. Explain how the values of A and B can be determined from the graph. Now store any other nonzero value in C. Does the value of C affect the x-intercepts of the graph? Find values of A, B, and C such that the graph opens downward and has x-intercepts at $(-5, 0)$ and $(0, 0)$.

2.4 /// EXERCISES

In Exercises 1–4, write the quadratic equation in standard form.

1. $2x^2 = 3 - 5x$ **2.** $x^2 = 25x$

3. $\frac{1}{5}(3x^2 - 10) = 12x$ **4.** $x(x + 2) = 3x^2 + 1$

In Exercises 5–12, solve the quadratic equation for x by factoring. Use a graphing utility to verify your solutions graphically.

5. $6x^2 + 3x = 0$ **6.** $9x^2 - 1 = 0$

7. $x^2 - 2x - 8 = 0$ **8.** $x^2 - 10x + 9 = 0$

9. $3 + 5x - 2x^2 = 0$ **10.** $2x^2 = 19x + 33$

11. $x^2 + 4x = 12$ **12.** $(x + a)^2 - b^2 = 0$

In Exercises 13–16, solve the equation by extracting square roots. List both the exact solution *and* the decimal solution rounded to two decimal places.

13. $x^2 = 7$ **14.** $9x^2 = 25$

15. $(x - 12)^2 = 18$ **16.** $(x - 5)^2 = 20$

In Exercises 17–20, solve the quadratic equation by completing the square. Verify your answer graphically.

17. $x^2 + 6x + 2 = 0$ **18.** $x^2 + 8x + 14 = 0$

19. $9x^2 - 18x + 3 = 0$ **20.** $9x^2 - 12x - 14 = 0$

Graphical Reasoning **In Exercises 21–28, use a graphing utility to graph the equation. Use the graph to approximate any x-intercepts of the graph. Set $y = 0$ and solve the resulting equation. Compare the result with the x-intercepts.**

21. $y = (x + 3)^2 - 4$ **22.** $y = 1 - (x - 2)^2$

23. $y = -4x^2 + 4x + 3$ **24.** $y = x^2 + 3x - 4$

25. $y = \frac{1}{4}(4x^2 - 20x + 25)$

26. $y = -(x^2 - 4x + 3)$

27. $y = -(x^2 - 4x + 5)$

28. $y = \frac{1}{4}(x^2 - 2x + 9)$

In Exercises 29–32, use a graphing utility to determine the number of real solutions of the quadratic equation.

29. $2x^2 - 5x + 5 = 0$ **30.** $2x^2 - x - 1 = 0$

31. $\frac{1}{5}x^2 + \frac{6}{5}x - 8 = 0$ **32.** $\frac{1}{3}x^2 - 5x + 25 = 0$

In Exercises 33–42, use the Quadratic Formula to solve the equation. Use a graphing utility to verify your solutions graphically.

33. $2 + 2x - x^2 = 0$ **34.** $x^2 - 10x + 22 = 0$

35. $x^2 + 8x - 4 = 0$ **36.** $4x^2 - 4x - 4 = 0$

37. $28x - 49x^2 = 4$ **38.** $9x^2 + 24x + 16 = 0$

39. $x^2 - 2x + 2 = 0$ **40.** $x^2 + 6x + 10 = 0$

41. $4x^2 + 16x + 15 = 0$ **42.** $9x^2 - 6x - 35 = 0$

In Exercises 43–46, solve the equation by any convenient method.

43. $x^2 - 2x - 1 = 0$ **44.** $11x^2 + 33x = 0$

45. $(x + 3)^2 = 81$ **46.** $x^2 + 3x - \frac{3}{4} = 0$

47. *True or False?* If

$$(2x - 3)(x + 5) = 8,$$

then $2x - 3 = 8$ or $x + 5 = 8$. Explain.

48. *Exploration* Solve the equation

$$3(x + 4)^2 + (x + 4) - 2 = 0$$

in two ways.

(a) Let $u = x + 4$, and solve the resulting equation for u. Then find the corresponding values of x that are solutions of the original equation.

(b) Expand and collect like terms in the original equation, and solve the resulting equation for x.

(c) Which method is easier? Explain.

49. *Think About It* Find a quadratic equation with solutions $x = -4$ and $x = 6$. (There are many correct answers.)

50. *Exploration* Given that a and b are nonzero real numbers, determine the solutions of the equations.

(a) $ax^2 + bx = 0$ (b) $ax^2 - ax = 0$

51. *Numerical, Graphical, and Analytical Analysis* A rancher has 100 meters of fencing to enclose two adjacent rectangular corrals (see figure).

(a) Write the area of the enclosed region as a function of x.

(b) Use a graphing utility to generate additional rows of the table. Use the table to estimate the dimensions that will produce a maximum area.

x	y	Area
2	$\frac{92}{3}$	$\frac{368}{3} \approx 123$
4	28	224

(c) Use a graphing utility to graph the area function, and use the graph to estimate the dimensions that will produce a maximum area.

(d) Use the graph to approximate the dimensions such that the enclosed area will be 350 square meters.

(e) Find the required dimensions of part (d) analytically.

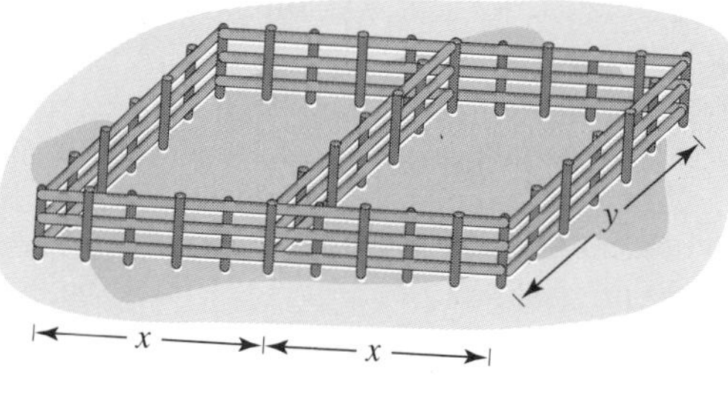

$4x + 3y = 100$

52. *Airline Traffic* The total number y of annual airline flight departures in the United States from 1984 through 1993 can be approximated by the mathematical model

$$y = -0.021t^2 + 0.524t + 3.769, \qquad 4 \le t \le 13$$

where y is in millions and t is the year, with $t = 4$ corresponding to 1984. (Source: National Transportation Safety Board)

(a) Use a graphing utility to graph the model.

(b) Use the graph to estimate the year in which the total number of departures was 7.0 million.

(c) Use the model to check analytically your estimate from part (b).

In Exercises 53–70, find all solutions of the equation. Use a graphing utility to verify the solutions graphically.

53. $4x^4 - 18x^2 = 0$

54. $20x^3 - 125x = 0$

55. $x^4 - 81 = 0$

56. $x^6 - 64 = 0$

57. $5x^3 + 30x^2 + 45x = 0$

58. $9x^4 - 24x^3 + 16x^2 = 0$

59. $x^3 - 3x^2 - x + 3 = 0$

60. $x^3 + 2x^2 + 3x + 6 = 0$

61. $x^4 - x^3 + x - 1 = 0$

62. $x^4 + 2x^3 - 8x - 16 = 0$

63. $x^4 - 4x^2 + 3 = 0$

64. $x^4 + 5x^2 - 36 = 0$

65. $4x^4 - 65x^2 + 16 = 0$

66. $36t^4 + 29t^2 - 7 = 0$

67. $\frac{1}{t^2} + \frac{8}{t} + 15 = 0$

68. $6\left(\frac{s}{s+1}\right)^2 + 5\left(\frac{s}{s+1}\right) - 6 = 0$

69. $2x + 9\sqrt{x} - 5 = 0$

70. $3x^{1/3} + 2x^{2/3} = 5$

Graphical Analysis **In Exercises 71–74, use a graphing utility to graph the equation. Use the graph to approximate any x-intercepts of the graph. Set $y = 0$ and solve the resulting equation. Compare the result with the x-intercepts of the graph.**

71. $y = x^3 - 2x^2 - 3x$

72. $y = 2x^4 - 15x^3 + 18x^2$

73. $y = x^4 - 10x^2 + 9$

74. $y = x^4 - 29x^2 + 100$

In Exercises 75–86, find all solutions of the equation. Check your solutions in the original equation.

75. $\sqrt{x - 10} - 4 = 0$

76. $\sqrt{5 - x} - 3 = 0$

77. $\sqrt[3]{2x + 5} + 3 = 0$

78. $\sqrt[3]{3x + 1} - 5 = 0$

79. $\sqrt{x + 1} - 3x = 1$

80. $\sqrt{x + 5} = \sqrt{x - 5}$

81. $\sqrt{x} - \sqrt{x - 5} = 1$

82. $\sqrt{x} + \sqrt{x - 20} = 10$

83. $(x - 5)^{2/3} = 16$

84. $(x^2 - x - 22)^{4/3} = 16$

85. $3x(x - 1)^{1/2} + 2(x - 1)^{3/2} = 0$

86. $4x^2(x - 1)^{1/3} + 6x(x - 1)^{4/3} = 0$

Graphical Analysis **In Exercises 87–90, use a graphing utility to graph the equation. Use the graph to approximate any x-intercepts of the graph. Set $y = 0$ and solve the resulting equation. Compare the result with the x-intercepts of the graph.**

87. $y = \sqrt{11x - 30} - x$

88. $y = 2x - \sqrt{15 - 4x}$

89. $y = \sqrt{7x + 36} - \sqrt{5x + 16} - 2$

90. $y = 3\sqrt{x} - \dfrac{4}{\sqrt{x}} - 4$

In Exercises 91–102, find all solutions of the equation. Use a graphing utility to verify your solutions graphically.

91. $\dfrac{20 - x}{x} = x$

92. $\dfrac{4}{x} - \dfrac{5}{3} = \dfrac{x}{6}$

93. $\dfrac{1}{x} - \dfrac{1}{x + 1} = 3$

94. $\dfrac{x}{x^2 - 4} + \dfrac{1}{x + 2} = 3$

95. $x = \dfrac{3}{x} + \dfrac{1}{2}$

96. $4x + 1 = \dfrac{3}{x}$

97. $\dfrac{4}{x + 1} - \dfrac{3}{x + 2} = 1$

98. $\dfrac{x + 1}{3} - \dfrac{x + 1}{x + 2} = 0$

99. $|2x - 1| = 5$

100. $|3x + 2| = 7$

101. $|x| = x^2 + x - 3$

102. $|x - 10| = x^2 - 10x$

Graphical Analysis **In Exercises 103–106, use a graphing utility to graph the equation. Use the graph to approximate any x-intercepts of the graph. Set $y = 0$ and solve the resulting equation. Compare the result with the x-intercepts of the graph.**

103. $y = \dfrac{1}{x} - \dfrac{4}{x - 1} - 1$

104. $y = x + \dfrac{9}{x + 1} - 5$

105. $y = |x + 1| - 2$

106. $y = |x - 2| - 3$

Think About It **In Exercises 107 and 108, find an equation having the given solutions. (There are many correct answers.)**

107. $\sqrt{2}, -\sqrt{2}, 4$

108. $-2, 2, i, -i$

109. *Sharing the Cost* A college charters a bus for \$1700 to take a group of students to a museum. When six more students join the trip, the cost per student drops by \$7.50. How many students were in the original group?

110. *Average Speed* A family drove 1080 miles to their vacation lodge. Because of increased traffic density, their average speed on the return trip was decreased by 6 miles per hour and the trip took $2\frac{1}{2}$ hours longer. Determine their average speed on the way to the lodge.

Exploration In Exercises 111 and 112, find x such that the distance between the points is 13.

111. $(1, 2), (x, -10)$ **112.** $(-8, 0), (x, 5)$

113. *Oxygen Consumption* The metabolic rate of ectothermic organisms increases with increasing temperature within a certain range. Experimental data for oxygen consumption (microliters per gram per hour) of a beetle for certain temperatures yielded the model

$$C = 0.45x^2 - 1.65x + 50.75, \qquad 10 \le x \le 25$$

where x is the air temperature in degrees Celsius.

(a) Use a graphing utility to graph the consumption function over the specified domain.

(b) Use the graph to approximate the air temperature resulting in oxygen consumption of 150 microliters per gram per hour.

(c) If the temperature is increased from 10 to 20 degrees, the oxygen consumption is increased by approximately what factor?

114. *Saturated Steam* The temperature T (in degrees Fahrenheit) of saturated steam increases as pressure increases. This relationship is approximated by the model

$$T = 75.82 - 2.11x + 43.51\sqrt{x}, \qquad 5 \le x \le 40$$

where x is the absolute pressure in pounds per square inch.

(a) Use a graphing utility to graph the temperature function over the specified domain.

(b) The temperature of steam at sea level ($x = 14.696$) is 212°F. Evaluate the model at this pressure, and verify the result graphically.

(c) Use the model to approximate the pressure for a steam temperature of 240°F.

115. *Fuel Efficiency* The distance d a car can travel on one tank of fuel is approximated by the model

$$d = -0.024s^2 + 1.455s + 431.5, \quad 0 < s \le 75$$

where s is the average speed of the car.

(a) Use a graphing utility to graph the distance function over the specified domain.

(b) Use the graph to determine the greatest distance that can be traveled on a tank of fuel. How long will the trip take?

(c) Determine the greatest distance that can be traveled in this car in 8 hours with no refueling. How fast should the car be driven? [*Hint:* The distance traveled in 8 hours is $8s$. Graph this expression in the same viewing rectangle as the graph in part (a) and approximate the point of intersection.]

116. *Solving Graphically, Numerically, and Algebraically* A meteorologist is positioned 100 feet from the point where a weather balloon is launched. When the balloon is at height h, the distance d between the meteorologist and the balloon is given by $d = \sqrt{100^2 + h^2}$.

(a) Use a graphing utility to graph the equation. Use the trace feature to approximate the value of h when $d = 200$.

(b) Complete the table. Use the table to approximate the value of h when $d = 200$.

h	160	165	170	175	180	185
d						

(c) Find h algebraically when $d = 200$.

(d) Compare the results of each method. In each case, what information did you gain that wasn't revealed by another solution method?

In Exercises 117 and 118, solve for the variable.

117. *Surface Area of a Cone*

Solve for h: $S = \pi r\sqrt{r^2 + h^2}$

118. *Inductance*

Solve for Q: $i = \pm\sqrt{\frac{1}{LC}}\sqrt{Q^2 - q}$

In Exercises 119 and 120, consider an equation of the form $x + \sqrt{x - a} = b$, where a and b are constants.

119. *Exploration* Find a and b if the solution to the equation is $x = 20$. (There are many correct answers.)

120. *Essay* Write a short paragraph listing the steps required for solving an equation involving radicals.

2.5 Solving Inequalities Algebraically and Graphically

Properties of Inequalities / Solving a Linear Inequality / Inequalities Involving Absolute Value / Polynomial Inequalities / Rational Inequalities / Applications

Properties of Inequalities

Simple inequalities were reviewed in Section P.1. There, inequality symbols $<$, $\le$, $>$, and $\ge$ were used to compare two numbers and to denote subsets of real numbers. For instance, the simple inequality $x \ge 3$ denotes all real numbers x that are greater than or equal to 3. In this section you will study inequalities that contain more involved statements such as

$$5x - 7 > 3x + 9 \quad \text{and} \quad -3 \le 6x - 1 < 3.$$

As with an equation, you **solve an inequality** in the variable x by finding all values of x for which the inequality is true. These values are **solutions** and are said to **satisfy** the inequality. For instance, the number 9 is a solution of the first inequality listed above because

$$5(9) - 7 > 3(9) + 9.$$

On the other hand, the number 7 is not a solution because

$$5(7) - 7 \not> 3(7) + 9.$$

The set of all real numbers that are solutions of an inequality is the **solution set** of the inequality.

The set of all points on the real number line that represent the solution set is the **graph** of the inequality. Graphs of many types of inequalities consist of intervals on the real number line.

The procedures for solving linear inequalities in one variable are much like those for solving linear equations. To isolate the variable you can make use of the **properties of inequalities.** These properties are similar to the properties of equality, but there are two important exceptions. When both sides of an inequality are multiplied or divided by a negative number, *the direction of the inequality symbol must be reversed.* Here is an example.

$$-2 < 5 \qquad \text{Original inequality}$$

$$(-3)(-2) > (-3)(5) \qquad \text{Multiply both sides by } -3 \text{ and reverse inequality.}$$

$$6 > -15 \qquad \text{New inequality}$$

Two inequalities that have the same solution set are **equivalent.** The properties listed at the top of the next page describe operations that can be used to create equivalent inequalities.

Properties of Inequalities

Let a, b, c, and d be real numbers.

1. *Transitive Property*

   $a < b$ and $b < c$ $\Longrightarrow$ $a < c$

2. *Addition of Inequalities*

   $a < b$ and $c < d$ $\Longrightarrow$ $a + c < b + d$

3. *Addition of a Constant*

   $a < b$ $\Longrightarrow$ $a + c < b + c$

4. *Multiplying by a Constant*

   For $c > 0$, $a < b$ $\Longrightarrow$ $ac < bc$

   For $c < 0$, $a < b$ 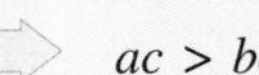 $ac > bc$

Note Each of the properties at the right is true if the symbol $<$ is replaced by $\leq$ and $>$ is replaced by $\geq$. For instance, another form of Property 3 would be as follows.

$a \leq b$ $\Longrightarrow$ $a + c \leq b + c$

EXPLORATION

Use a graphing utility to graph $f(x) = 5x - 7$ and $g(x) = 3x + 9$ in the same viewing rectangle. (Use $-1 \leq x \leq 15$ and $-5 \leq y \leq 50$.) For which values of x does the graph of f lie above the graph of g? Explain how the answer to this question can be used to solve the inequality in Example 1.

Solving a Linear Inequality

The simplest type of inequality to solve is a **linear inequality** in a single variable, such as $2x + 3 > 4$.

EXAMPLE 1 **Solving a Linear Inequality**

Solve the inequality

$$5x - 7 > 3x + 9.$$

Solution

$5x - 7 > 3x + 9$	Original inequality
$5x > 3x + 16$	Add 7 to both sides.
$5x - 3x > 16$	Subtract $3x$ from both sides.
$2x > 16$	Combine like terms.
$x > 8$	Divide both sides by 2.

Thus, the solution set consists of all real numbers that are greater than 8. The interval notation for this solution set is $(8, \infty)$. The number line graph of this solution set is shown in Figure 2.37.

Figure 2.37 Solution interval: $(8, \infty)$

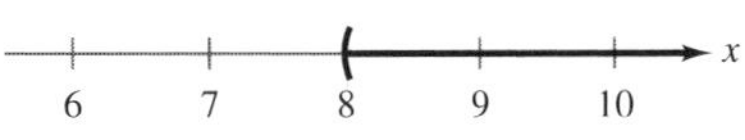

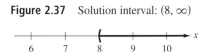

Note The five inequalities forming the solution steps of Example 1 are all **equivalent** in the sense that each has the same solution set.

EXAMPLE 2 Solving an Inequality

Solve $1 - \frac{3x}{2} \geq x - 4.$

Solution

$1 - \frac{3x}{2} \geq x - 4$	Original inequality
$2 - 3x \geq 2x - 8$	Multiply both sides by the LCD.
$-3x \geq 2x - 10$	Subtract 2 from both sides.
$-5x \geq -10$	Subtract $2x$ from both sides.
$x \leq 2$	Divide both sides by -5 and reverse inequality.

The solution set consists of all real numbers that are less than or equal to 2. The interval notation for this solution set is $(-\infty, 2]$. The number line graph of this solution set is shown in Figure 2.38. Graphically, you can solve the given inequality by sketching the graphs of the left and right sides

$$f(x) = 1 - \frac{3x}{2} \quad \text{and} \quad g(x) = x - 4.$$

In Figure 2.39, the graphs appear to intersect at the point $(2, -2)$. This is confirmed by the fact that $f(2) = -2 = g(2)$. Moreover, the graph of f lies above the graph of g to the left of their point of intersection, which implies that $f(x) \geq g(x)$ for all $x \leq 2$.

Figure 2.38 Solution interval: $(-\infty, 2]$

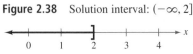

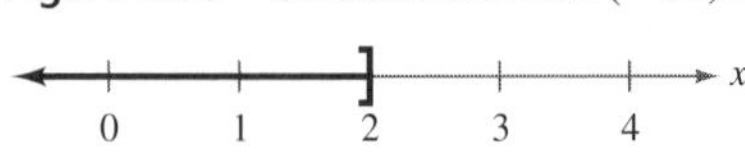

Figure 2.39

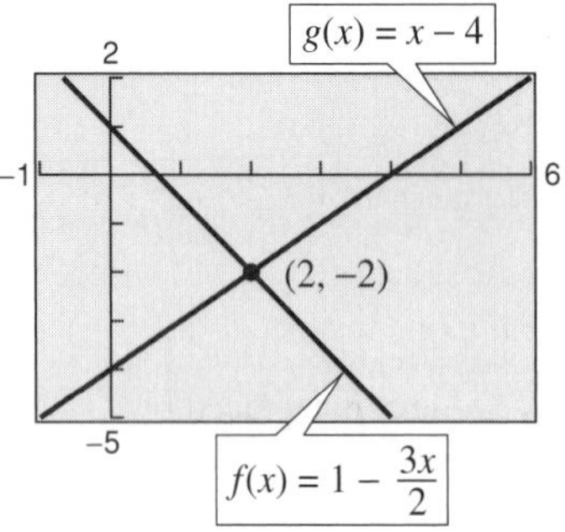

Sometimes it is possible to write two inequalities as a **double inequality.** For instance, you can write the two inequalities $-4 \leq 5x - 2$ and $5x - 2 < 7$ more simply as $-4 \leq 5x - 2 < 7$. This form allows you to solve the two inequalities together.

Study Tip

Checking the solution set of an inequality is not as simple as checking the solution of an equation because there are simply too many x-values to substitute into the original inequality. However, you can get an indication of the validity of the solution set by substituting a few convenient values of x. For instance, in Example 2 try substituting $x = 0$ and $x = 4$ into the original inequality.

EXAMPLE 3 Solving a Double Inequality

$-3 \leq 6x - 1 < 3$	Original inequality
$-2 \leq 6x < 4$	Add 1 to all three parts.
$-\frac{1}{3} \leq x < \frac{2}{3}$	Divide by 6 and reduce.

The solution set consists of all real numbers that are greater than or equal to $-\frac{1}{3}$ and less than $\frac{2}{3}$. The interval notation for this solution set is $\left[-\frac{1}{3}, \frac{2}{3}\right)$. The number line graph of this solution set is shown in Figure 2.40.

Figure 2.40 Solution interval: $\left[-\frac{1}{3}, \frac{2}{3}\right)$

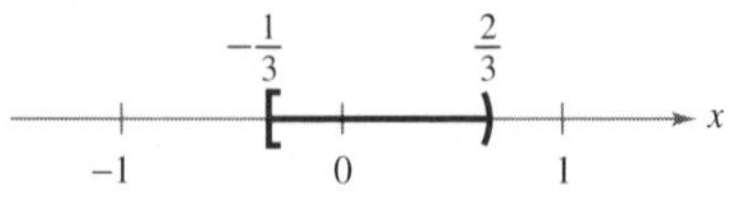

$-\frac{1}{3}$ $\frac{2}{3}$ x

−1 0 1

Inequalities Involving Absolute Value

Solving an Absolute Value Inequality

Let x be a variable of an algebraic expression and let a be a real number such that $a \geq 0$.

1. The solutions of $|x| < a$ are all values of x that lie between $-a$ and a.

$$|x| < a \quad \text{if and only if} \quad -a < x < a.$$

2. The solutions of $|x| > a$ are all values of x that are less than $-a$ or greater than a.

$$|x| > a \quad \text{if and only if} \quad x < -a \quad \text{or} \quad x > a.$$

These rules are also valid if $<$ is replaced by $\leq$ and $>$ is replaced by $\geq$.

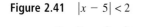

EXAMPLE 4 **Solving an Absolute Value Inequality**

$$|x - 5| < 2 \qquad \text{Original inequality}$$

$$-2 < x - 5 < 2 \qquad \text{Equivalent inequalities}$$

$$-2 + 5 < x - 5 + 5 < 2 + 5 \qquad \text{Add 5 to all three parts.}$$

$$3 < x < 7 \qquad \text{Simplify.}$$

The solution set is all real numbers that are greater than 3 and less than 7. The interval notation for this solution set is $(3, 7)$. The number line graph of this solution set is shown in Figure 2.41.

Figure 2.41 $|x - 5| < 2$

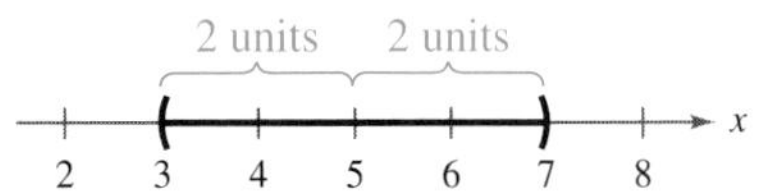

EXAMPLE 5 **Solving an Absolute Value Inequality Graphically**

Use a graphing utility to solve $\left|2 - \frac{x}{3}\right| < 0.01$.

Solution

You can solve this inequality graphically by graphing the function

$$f(x) = \left|2 - \frac{x}{3}\right| - 0.01$$

and determining the values of x for which the graph lies *below* the x-axis, as shown in Figure 2.42. Using the zoom and trace features, you can determine that the x-intercepts occur at $x \approx 5.97$ and $x \approx 6.03$. Because the graph of f lies below the x-axis between these two values, the solution interval is $(5.97, 6.03)$. Try solving this inequality algebraically to verify the solution interval.

Figure 2.42

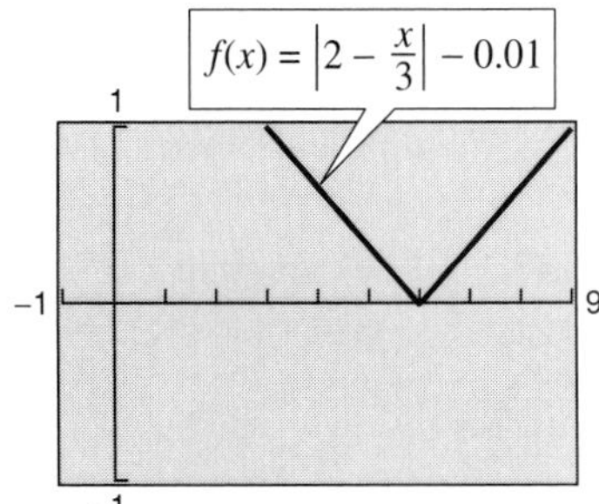

Polynomial Inequalities

To solve a polynomial inequality such as $x^2 - 2x - 3 < 0$, use the fact that a polynomial can change signs only at its zeros (the x-values that make the polynomial equal to zero). Between two consecutive zeros a polynomial must be entirely positive or entirely negative. This means that when the real zeros of a polynomial are put in order, they divide the real number line into intervals in which the polynomial has no sign changes. These zeros are the **critical numbers** of the inequality, and the resulting intervals are the **test intervals** for the inequality. For instance, the polynomial

$$x^2 - 2x - 3 = (x + 1)(x - 3)$$

has two zeros, $x = -1$ and $x = 3$, which divide the real number line into three test intervals: $(-\infty, -1)$, $(-1, 3)$, and $(3, \infty)$. To solve the inequality $x^2 - 2x - 3 < 0$, you only need to test one value from each test interval.

Finding Test Intervals for a Polynomial

To determine the intervals on which the values of a polynomial are entirely negative or entirely positive, use the following steps.

1. Find all real zeros of the polynomial, and arrange the zeros in increasing order (from smallest to largest). The zeros of a polynomial are its **critical numbers.**
2. Use the critical numbers of the polynomial to determine its **test intervals.**
3. Choose one representative x-value in each test interval and evaluate the polynomial at that value. If the value of the polynomial is negative, the polynomial will have negative values for *every* x-value in the interval. If the value of the polynomial is positive, the polynomial will have positive values for *every* x-value in the interval.

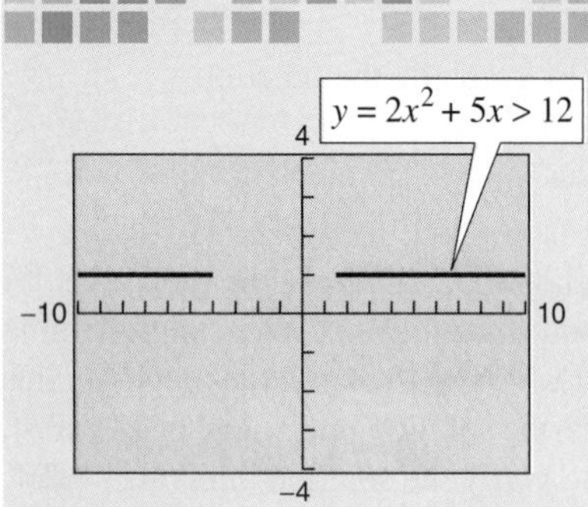

Some graphing utilities will produce graphs of inequalities. For instance, on a *TI-82* or *TI-83*, you can graph $2x^2 + 5x > 12$ (see Example 7) by entering

$$Y_1 = 2x^2 + 5x > 12$$

and pressing GRAPH. Using $-10 \le x \le 10$ and $-4 \le y \le 4$, the graph should look like that at the left. Solve the problem algebraically to verify that the solution is $(-\infty, -4) \cup \left(\frac{3}{2}, \infty\right)$.

EXAMPLE 6 Investigating Polynomial Behavior

Determine the intervals on which $x^2 - x - 6$ is entirely negative and those on which it is entirely positive.

Solution

By factoring the quadratic as $x^2 - x - 6 = (x + 2)(x - 3)$, you can see that the critical numbers occur at $x = -2$ and $x = 3$. Therefore, the test intervals for the quadratic are $(-\infty, -2)$, $(-2, 3)$, and $(3, \infty)$. In each test interval, choose a representative x-value and evaluate the polynomial, as shown in the table.

Figure 2.43

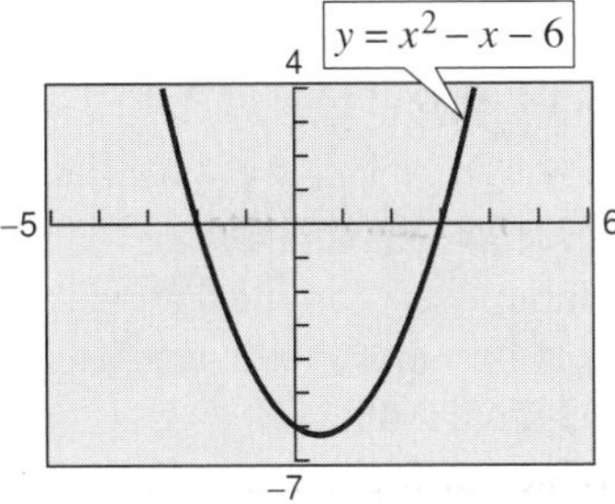

Interval	x-Value	Value of Polynomial	Sign of Polynomial
$(-\infty, -2)$	$x = -3$	$(-3)^2 - (-3) - 6 = 6$	Positive
$(-2, 3)$	$x = 0$	$(0)^2 - (0) - 6 = -6$	Negative
$(3, \infty)$	$x = 4$	$(4)^2 - (4) - 6 = 6$	Positive

The polynomial has positive values for every x in the intervals $(-\infty, -2)$ and $(3, \infty)$ and negative values for every x in the interval $(-2, 3)$. This result is shown graphically in Figure 2.43.

Note To determine the test intervals for a polynomial inequality, the inequality must first be written in standard form with the polynomial on one side and zero on the other side.

EXAMPLE 7 Solving a Polynomial Inequality

Solve $2x^2 + 5x > 12$.

Solution

$$2x^2 + 5x > 12 \quad \text{Original inequality}$$
$$2x^2 + 5x - 12 > 0 \quad \text{Standard form}$$
$$(x + 4)(2x - 3) > 0 \quad \text{Factor.}$$

Critical Numbers: $x = -4, x = \frac{3}{2}$

Test Intervals: $(-\infty, -4), \left(-4, \frac{3}{2}\right), \left(\frac{3}{2}, \infty\right)$

Test: Is $(x + 4)(2x - 3) > 0$?

Figure 2.44

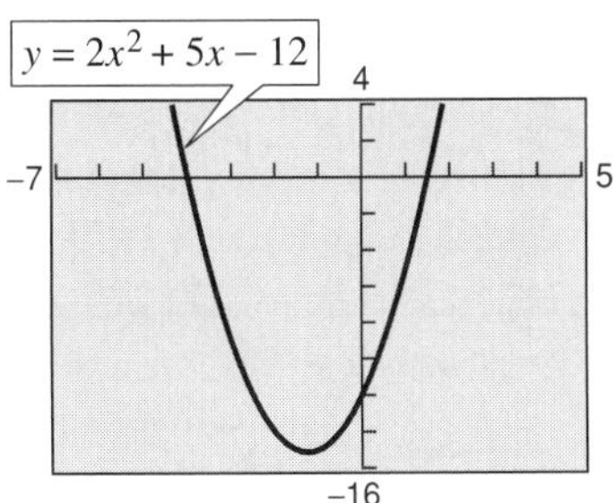

After testing these intervals, you can see that the polynomial $2x^2 + 5x - 12$ is positive in the open intervals $(-\infty, -4)$ and $\left(\frac{3}{2}, \infty\right)$. Therefore, the solution set of the inequality is $(-\infty, -4) \cup \left(\frac{3}{2}, \infty\right)$. You could have solved this example with a graphing utility by graphing the polynomial function $y = 2x^2 + 5x - 12$, and noting where the graph is *above* the x-axis. You can see in Figure 2.44 that the polynomial lies above the x-axis for $x < -4$ or $x > \frac{3}{2}$.

EXAMPLE 8 Unusual Solution Sets

a. The solution set of $x^2 + 2x + 4 > 0$ consists of the entire set of real numbers, $(-\infty, \infty)$. In other words, the quadratic $x^2 + 2x + 4$ is positive for every real value of x, as indicated in Figure 2.45(a). (Note that this quadratic inequality has *no* critical numbers. In such a case, there is only one test interval—the entire real number line.)

b. The solution set of $x^2 + 2x + 1 \le 0$ consists of the single real number -1, because the graph touches the x-axis just at -1, as shown in Figure 2.45(b).

c. The solution set of $x^2 + 3x + 5 < 0$ is empty. In other words, the quadratic $x^2 + 3x + 5$ is not less than zero for any value of x, as indicated in Figure 2.45(c).

d. The solution set of $x^2 - 4x + 4 > 0$ consists of all real numbers *except* the number 2. In interval notation, this solution set can be written as $(-\infty, 2) \cup (2, \infty)$. The graph of $x^2 - 4x + 4$ lies above the x-axis except at $x = 2$, where it touches it, as indicated in Figure 2.45(d).

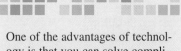

> **Study Tip**
>
> When solving a polynomial inequality, be sure you have accounted for the particular type of inequality symbol in the inequality. For instance, an inequality containing the < or > symbols should have solution sets consisting of open intervals such as (x_1, x_2), etc. Inequalities containing ≤ or ≥ should have solution sets consisting of closed intervals such as $[x_1, x_2]$, etc.

One of the advantages of technology is that you can solve complicated polynomial inequalities that might be difficult, or even impossible, to factor. For instance, explain how you could use a graphing utility to approximate the solution to the inequality

$$x^3 - 0.26x^2 - 3.1416x + 1.414 < 0.$$

Figure 2.45

(a)

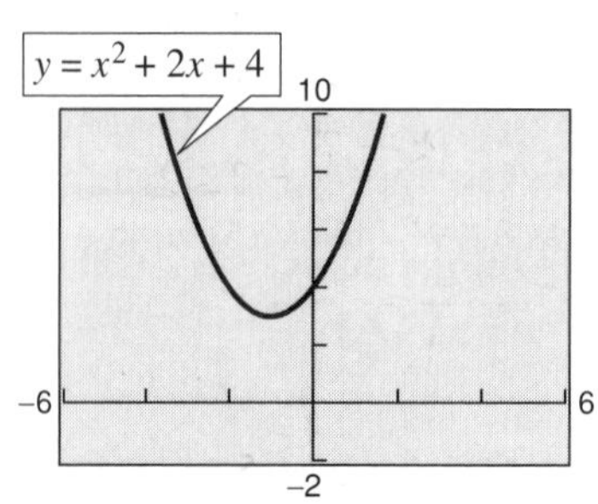

(b)

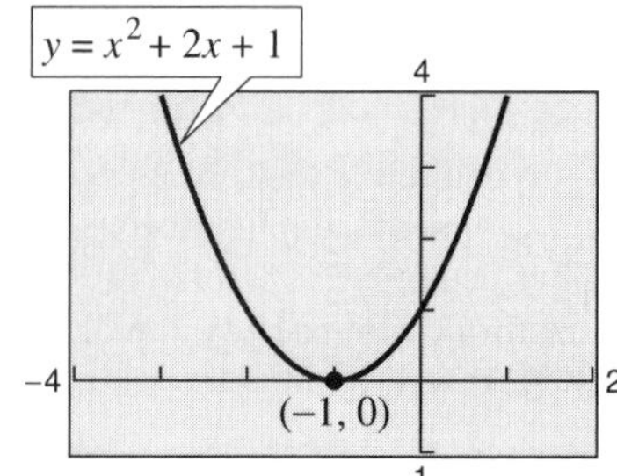

(c)

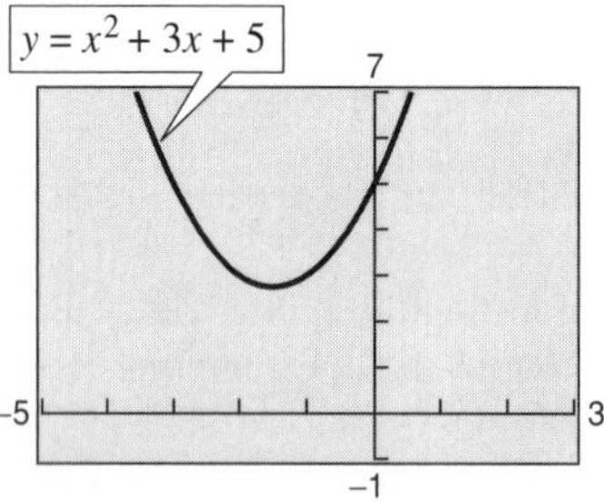

(d)

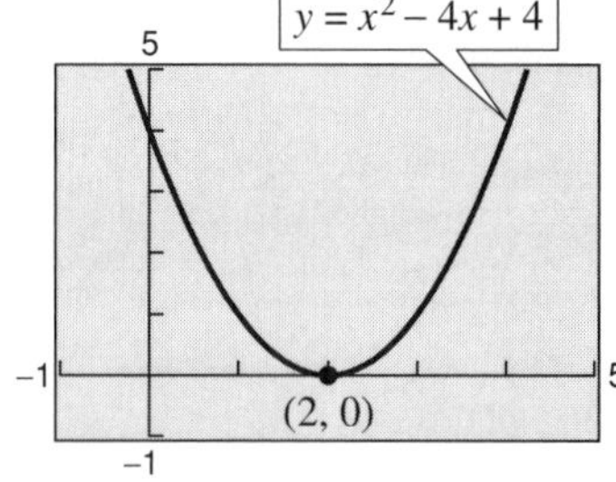

Rational Inequalities

The concepts of critical numbers and test intervals can be extended to inequalities involving rational expressions. To do this, use the fact that the value of a rational expression can change sign only at its *zeros* (the x-values for which its numerator is zero) and its *undefined values* (the x-values for which its denominator is zero). These two types of numbers make up the **critical numbers** of a rational inequality.

EXAMPLE 9 **Solving a Rational Inequality**

Solve $\dfrac{2x-7}{x-5} \le 3$.

Solution

In Figure 2.46, you can see that the graph of

$$y = \frac{2x-7}{x-5} - 3$$

is less than or equal to zero on the intervals $(-\infty, 5)$ and $[8, \infty)$. To solve the inequality algebraically, use the following steps.

Figure 2.46

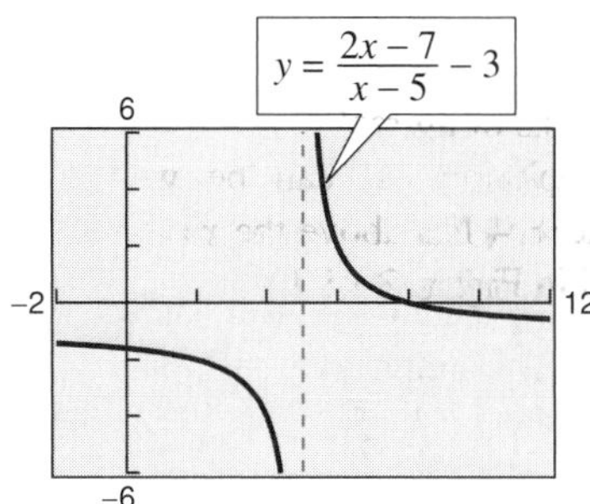

$$\frac{2x-7}{x-5} \le 3 \qquad \text{Original inequality}$$

$$\frac{2x-7}{x-5} - 3 \le 0 \qquad \text{Standard form}$$

$$\frac{2x-7-3x+15}{x-5} \le 0 \qquad \text{Add fractions.}$$

$$\frac{-x+8}{x-5} \le 0 \qquad \text{Simplify.}$$

Now, in standard form you can see that the critical numbers are 5 and 8, and you can proceed as follows.

Critical Numbers: $x = 5, x = 8$

Test Intervals: $(-\infty, 5), (5, 8), (8, \infty)$

Test: Is $\dfrac{-x+8}{x-5} \le 0$?

By testing these intervals, you can determine that the rational expression $(-x+8)/(x-5)$ is negative in the open intervals $(-\infty, 5)$ and $(8, \infty)$. Moreover, because $(-x+8)/(x-5) = 0$ when $x = 8$, you can conclude that the solution set of the inequality is $(-\infty, 5) \cup [8, \infty)$.

Study Tip

In Example 9, it is incorrect to write the solution set $(-\infty, 5) \cup [8, \infty)$ as a double inequality $8 \le x < 5$. Do you see why?

Applications

EXAMPLE 10 Finding the Domain of an Expression

Find the domain of

$$\sqrt{64 - 4x^2}.$$

Solution

Remember that the domain of an expression is the set of all x-values for which the expression is defined (has real values). Because $\sqrt{64 - 4x^2}$ is defined only if $64 - 4x^2$ is nonnegative, the domain is given by $64 - 4x^2 \geq 0$.

$$64 - 4x^2 \geq 0 \quad \text{Standard form}$$

$$16 - x^2 \geq 0 \quad \text{Divide both sides by 4.}$$

$$(4 - x)(4 + x) \geq 0 \quad \text{Factor.}$$

Figure 2.47

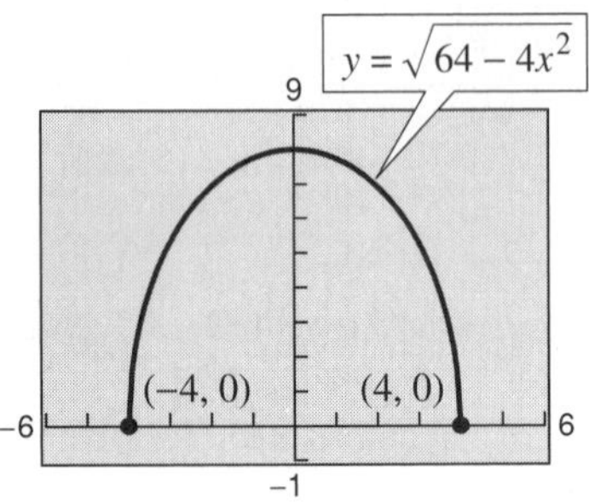

The inequality has two critical numbers: -4 and 4. A test shows that $64 - 4x^2 \geq 0$ in the *closed interval* $[-4, 4]$. The graph of $y = \sqrt{64 - 4x^2}$, shown in Figure 2.47, confirms that the domain is $[-4, 4]$.

EXAMPLE 11 The Height of a Projectile

Real Life

A projectile is fired straight upward from ground level with an initial velocity of 384 feet per second. During what time period will its height exceed 2000 feet?

Solution

The position of an object moving vertically is given by $s = -16t^2 + v_0 t + s_0$, where s is the height in feet and t is the time in seconds. In this case, $s_0 = 0$ and $v_0 = 384$. Thus, you need to solve the inequality $-16t^2 + 384t > 2000$. Using a graphing utility, graph $s = -16t^2 + 384t$ and $s = 2000$, as shown in Figure 2.48. From the graph, you can determine that $-16t^2 + 384t > 2000$ for t between 7.64 and 16.36. You can verify this result algebraically as follows.

Figure 2.48

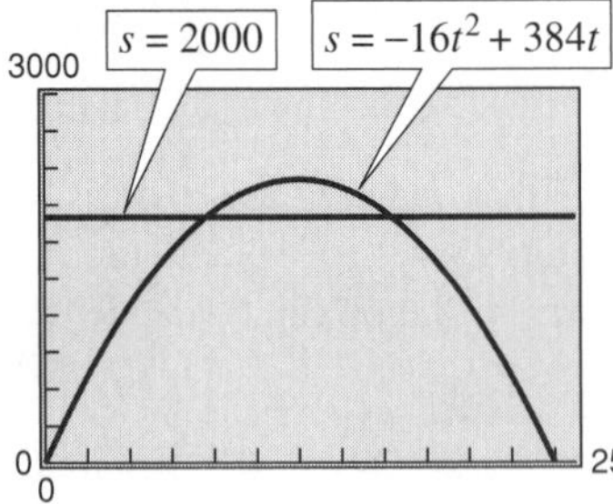

$$-16t^2 + 384t > 2000 \quad \text{Original inequality}$$

$$t^2 - 24t < -125 \quad \text{Divide by −16 and reverse inequality.}$$

$$t^2 - 24t + 125 < 0 \quad \text{Standard form}$$

By the Quadratic Formula the critical numbers are 7.64 and 16.36. A test will verify that the height of the projectile will exceed 2000 feet during the time interval 7.64 seconds $< t <$ 16.36 seconds.

Group Activity

Communicating Mathematically

Some people find that it is easier to remember a verbal statement, a numerical example, or a picture than it is to remember a mathematical formula. For instance, you can remember the factoring formula $(u + v)(u - v) = u^2 - v^2$ as "the product of the sum and difference of two terms is the difference of each term squared."

Four different properties of inequalities are listed on page 217. For each property, (a) translate the mathematical statement into a *verbal statement,* (b) compile a list of several numerical examples that demonstrate the property, and (c) construct a number line or series of number lines that graphically illustrates the property.

2.5 /// EXERCISES

In Exercises 1–4, match the inequality with its graph. [The graphs are labeled (a), (b), (c), and (d).]

1. $x < 3$

2. $x \geq 5$

3. $-3 < x \leq 4$

4. $0 \leq x \leq \frac{9}{2}$

(a) 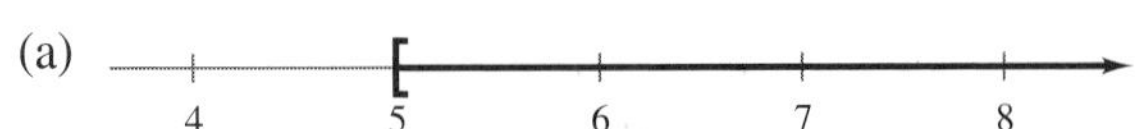

(b) 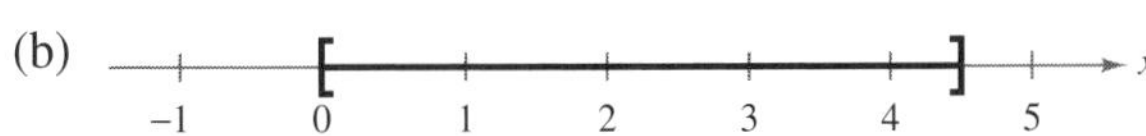

(c) 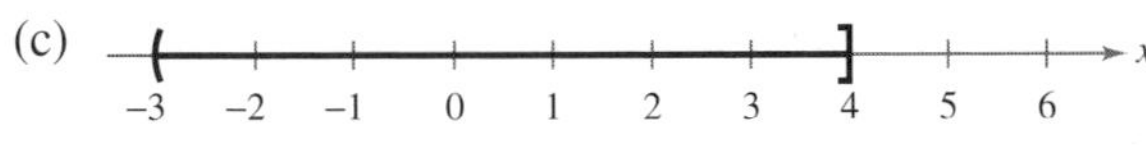

(d) 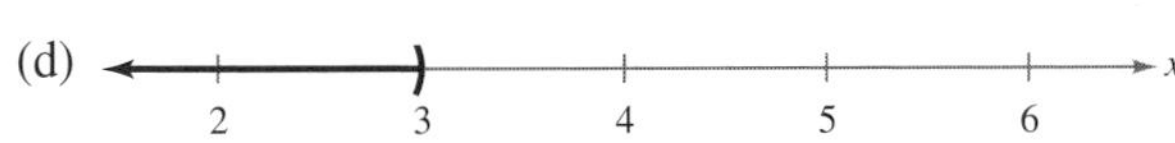

In Exercises 5–8, determine whether the given values of x are solutions of the inequality.

Inequality	*Values*	
5. $5x - 12 > 0$	(a) $x = 3$	(b) $x = -3$
	(c) $x = \frac{5}{2}$	(d) $x = \frac{3}{2}$
6. $-1 < \dfrac{3 - x}{2} \leq 1$	(a) $x = 0$	(b) $x = \sqrt{5}$
	(c) $x = 1$	(d) $x = 5$
7. $\lvert x - 10 \rvert \geq 3$	(a) $x = 13$	(b) $x = -1$
	(c) $x = 14$	(d) $x = 9$
8. $\lvert 2x - 3 \rvert < 15$	(a) $x = -6$	(b) $x = 0$
	(c) $x = 12$	(d) $x = 7$

In Exercises 9–18, solve the inequality and sketch the solution on the real number line. Use a graphing utility to verify your solution graphically.

9. $-10x < 40$

10. $2x > 3$

11. $4(x + 1) < 2x + 3$

12. $2x + 7 < 3$

13. $1 < 2x + 3 < 9$

14. $-8 \leq 1 - 3(x - 2) < 13$

15. $-4 < \dfrac{2x - 3}{3} < 4$

16. $0 \leq \dfrac{x + 3}{2} < 5$

17. $-1 < -\dfrac{x}{3} < 1$

18. $\dfrac{3}{4} > x + 1 > \dfrac{1}{4}$

Graphical Analysis **In Exercises 19–24, use a graphing utility to graph the inequality.**

19. $6x > 12$

20. $3x - 1 \leq 5$

21. $5 - 2x \geq 1$

22. $3(x + 1) < x + 7$

23. $0 \leq 2(x + 4) < 20$

24. $-2 < 3x + 1 < 10$

Graphical Analysis **In Exercises 25–28, use a graphing utility to graph the equation. Use the graph to approximate the values of x that satisfy the specified inequalities.**

Equation	*Inequalities*	
25. $y = 2x - 3$	(a) $y \geq 1$	(b) $y \leq 0$
26. $y = \frac{2}{3}x + 1$	(a) $y \leq 5$	(b) $y \geq 0$
27. $y = -\frac{1}{2}x + 2$	(a) $0 \leq y \leq 3$	(b) $y \geq 0$
28. $y = -3x + 8$	(a) $-1 \leq y \leq 3$	(b) $y \leq 0$

In Exercises 29 and 30, find the interval(s) on the real number line for which the radicand is nonnegative (greater than or equal to zero).

29. $\sqrt{x - 5}$

30. $\sqrt[4]{6x + 15}$

In Exercises 31–40, solve the inequality and sketch the solution on the real number line.

31. $\left|\frac{x}{2}\right| > 3$

32. $|5x| > 10$

33. $|x - 20| \leq 4$

34. $|x - 7| < 6$

35. $|x - 20| \geq 4$

36. $|x + 14| + 3 > 17$

37. $\left|\frac{x - 3}{2}\right| \geq 5$

38. $|1 - 2x| < 5$

39. $|x - 5| < 0$

40. $3|4 - 5x| \leq 9$

Graphical Analysis **In Exercises 41 and 42, use a graphing utility to graph the equation. Use the graph to approximate the values of x that satisfy the specified inequalities.**

Equation	*Inequalities*	
41. $y = |x - 3|$	(a) $y \leq 2$	(b) $y \geq 4$
42. $y = \left|\frac{1}{2}x + 1\right|$	(a) $y \leq 4$	(b) $y \geq 1$

In Exercises 43–48, use absolute value notation to define each interval (or pair of intervals) on the real number line.

43.
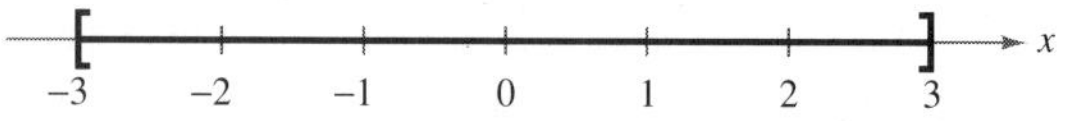

44.
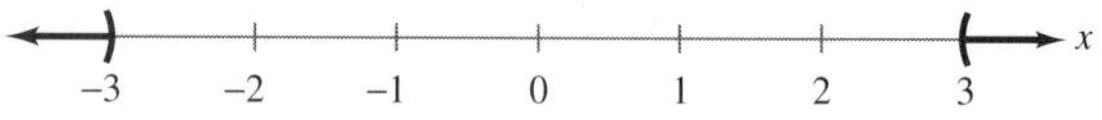

45.
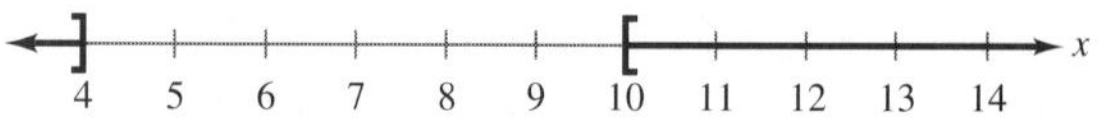

46.
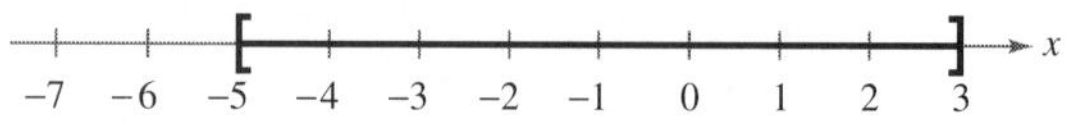

47. All real numbers within 10 units of 12

48. All real numbers whose distances from -3 are more than 5

49. *Data Analysis* The college admissions office wants to decide if there is a relationship between IQ scores x and grade-point averages y after the first year. A sample of 12 students yielded the following data.

x	118	131	125	123	133	136
y	2.2	2.4	3.2	2.4	3.5	3.0

x	128	124	116	120	134	131
y	3.0	2.8	2.2	1.8	3.4	3.6

(a) Use a graphing utility to plot the points. What shape does the plot have?

(b) Use the regression capabilities of a graphing utility to find a model appropriate to the shape of the plotted data. Graph the model in the same viewing rectangle as part (a).

(c) Which values of x predict a grade-point average of at least 3.0?

(d) Use the graph to write a statement about the accuracy of the model. If you think the graph indicates that IQ scores are not particularly good predictors of grade-point average, list other factors that may influence college performance.

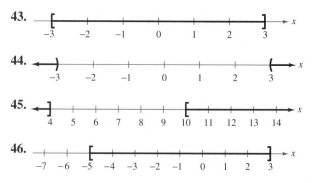

50. *Teachers' Salaries* The average salary for elementary and secondary teachers in the United States from 1984 to 1993 is approximated by the model

$$\text{Salary} = 15.812 + 1.472t$$

where the salary is given in thousands of dollars and the time t represents the calendar year, with $t = 4$ corresponding to 1984. (Source: National Education Association)

(a) Use a graphing utility to graph the model.

(b) Assuming the model is correct, when will the average salary *exceed* \$40,000?

In Exercises 51–58, solve the inequality and graph the solution on the real number line. Use a graphing utility to verify your solution graphically.

51. $(x + 2)^2 < 25$

52. $(x + 6)^2 \le 8$

53. $x^2 + 4x + 4 \ge 9$

54. $x^2 - 6x + 9 < 16$

55. $x^2 + x < 6$

56. $4x^3 - 12x^2 > 0$

57. $x^3 - 4x \ge 0$

58. $x^4(x - 3) \le 0$

Graphical Analysis **In Exercises 59–62, use a graphing utility to graph the equation. Use the graph to approximate the values of x that satisfy the specified inequalities.**

	Equation	*Inequalities*	
59.	$y = -x^2 + 2x + 3$	(a) $y \le 0$	(b) $y \ge 3$
60.	$y = \frac{1}{2}x^2 - 2x + 1$	(a) $y \le 1$	(b) $y \ge 7$
61.	$y = \frac{1}{8}x^3 - \frac{1}{2}x$	(a) $y \ge 0$	(b) $y \le 6$
62.	$y = x^3 - x^2 - 16x + 16$	(a) $y \le 0$	(b) $y \ge 36$

In Exercises 63–66, solve the inequality and graph the solution on the real number line. Use a graphing utility to verify your solution graphically.

63. $\dfrac{1}{x} - x > 0$

64. $\dfrac{1}{x} - 4 < 0$

65. $\dfrac{x + 6}{x + 1} - 2 < 0$

66. $\dfrac{x + 12}{x + 2} - 3 \ge 0$

Graphical Analysis **In Exercises 67–70, use a graphing utility to graph the equation. Use the graph to approximate the values of x that satisfy the specified inequalities.**

	Equation	*Inequalities*	
67.	$y = \dfrac{3x}{x - 2}$	(a) $y \le 0$	(b) $y \ge 6$
68.	$y = \dfrac{2(x - 2)}{x + 1}$	(a) $y \le 0$	(b) $y \ge 8$
69.	$y = \dfrac{2x^2}{x^2 + 4}$	(a) $y \ge 1$	(b) $y \le 2$
70.	$y = \dfrac{5x}{x^2 + 4}$	(a) $y \ge 1$	(b) $y \le 0$

In Exercises 71 and 72, find the domain of x in the expression.

71. $\sqrt[4]{4 - x^2}$

72. $\sqrt{x^2 - 4}$

73. *Percent of College Graduates* The percent P of the American population that graduated from college between 1950 and 1990 is approximated by

$$P = 5.9556 + 1.492t + 0.0056t^2$$

where the time t represents the calendar year, with $t = 0$ corresponding to 1950. (Source: U.S. Bureau of Census)

(a) Use a graphing utility to graph the model over the indicated years.

(b) According to this model, when will the percent of college graduates exceed 25% of the population? Solve algebraically and verify graphically.

Chapter Opener **In Exercises 74–77, graph the equation on page 167 on a graphing utility and use the trace feature.**

74. Estimate the frequency if the plate thickness is 2 millimeters.

75. Estimate the plate thickness if the frequency is 600 vibrations per second.

76. Approximate the interval for the plate thickness if the frequency is between 200 and 400 vibrations per second.

77. Approximate the interval for the frequency if the plate thickness is less than 3 millimeters.

2.6 Exploring Data: Linear Models and Scatter Plots

Scatter Plots and Correlation / Fitting a Line to Data

Scatter Plots and Correlation

Many real-life situations involve finding relationships between two variables such as the year and the number of people in the labor force. In a typical situation, data is collected and written as a set of ordered pairs. We discussed the graph of such a set, a **scatter plot,** briefly in Section P.5.

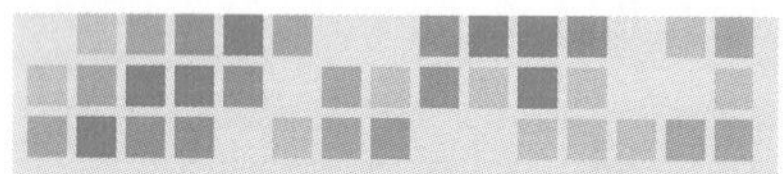

Most graphing utilities have built-in statistical programs that can create scatter plots. Use your graphing utility to plot the points given in the table at the right.

EXAMPLE 1 Constructing a Scatter Plot — *Real Life*

The data in the table shows the number of people P (in millions) in the United States who were part of the labor force from 1983 through 1993. In the table, t represents the year, with $t = 3$ corresponding to 1983. Sketch a scatter plot of the data. (Source: U.S. Bureau of Labor Statistics)

t	3	4	5	6	7	8	9	10	11	12	13
P	113	115	117	120	122	123	126	126	127	129	130

Solution

Begin by representing the data with a set of ordered pairs.

$(3, 113), (4, 115), (5, 117), (6, 120), (7, 122), (8, 123),$
$(9, 126), (10, 126), (11, 127), (12, 129), (13, 130)$

Then plot each point in a coordinate plane, as shown in Figure 2.49.

Figure 2.49

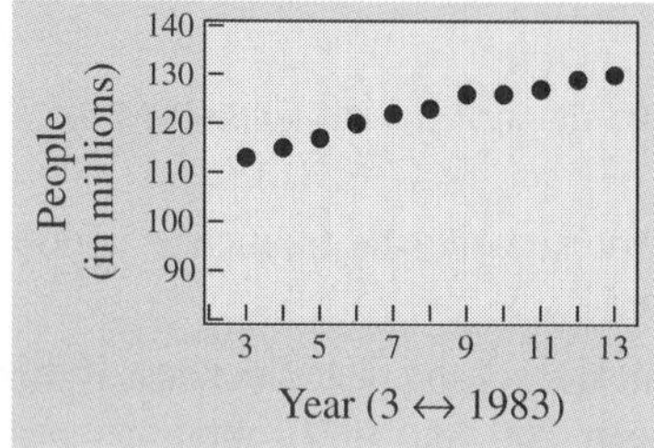

From the scatter plot in Figure 2.49, it appears that the points describe a relationship that is nearly linear. The relationship is not *exactly* linear because the labor force did not increase by precisely the same amount each year.

A mathematical equation that approximates the relationship between t and P is called a *mathematical model.* When developing a mathematical model, you strive for two (often conflicting) goals—accuracy and simplicity. For the data above, a linear model of the form $P = at + b$ appears to be best. It is simple and relatively accurate.

Consider a collection of ordered pairs of the form (x, y). If y tends to increase as x increases, the collection is said to have a **positive correlation.** If y tends to decrease as x increases, the collection is said to have a **negative correlation.** Figure 2.50 shows three examples: one with a positive correlation, one with a negative correlation, and one with no (discernible) correlation.

Figure 2.50

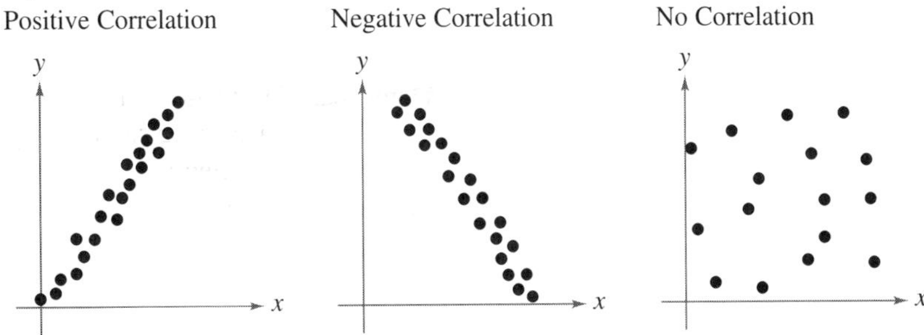

Real Life

EXAMPLE 2 Interpreting Correlation

On a Friday, 22 students in a class were asked to keep track of the number of hours they spent studying for a test on Monday and the number of hours they spent watching television. The numbers are shown below. Construct a scatter plot for each set of data. Then determine whether the points are positively correlated, are negatively correlated, or have no discernible correlation. What can you conclude? (The first coordinate is the number of hours and the second coordinate is the score obtained on Monday's test.)

Study Hours: (0, 40), (1, 41), (2, 51), (3, 58), (3, 49), (4, 48), (4, 64), (5, 55), (5, 69), (5, 58), (5, 75), (6, 68), (6, 63), (6, 93), (7, 84), (7, 67), (8, 90), (8, 76), (9, 95), (9, 72), (9, 85), (10, 98)

TV Hours: (0, 98), (1, 85), (2, 72), (2, 90), (3, 67), (3, 93), (3, 95), (4, 68), (4, 84), (5, 76), (7, 75), (7, 58), (9, 63), (9, 69), (11, 55), (12, 58), (14, 64), (16, 48), (17, 51), (18, 41), (19, 49), (20, 40)

Figure 2.51

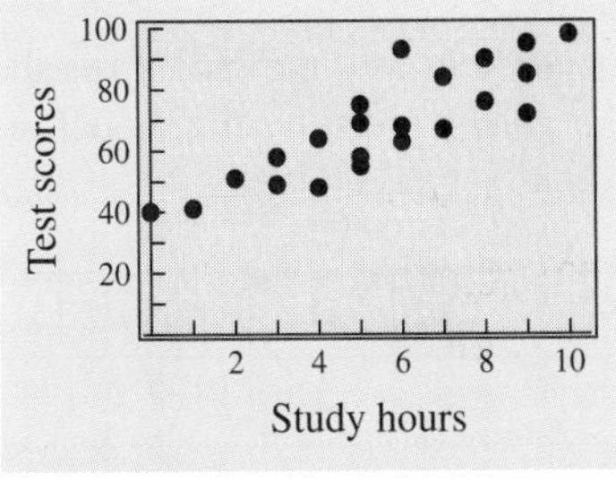

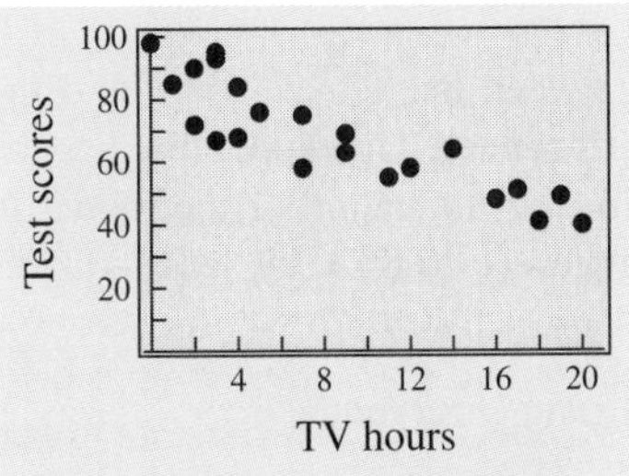

Solution

Scatter plots for the two sets of data are shown in Figure 2.51. The scatter plot relating study hours and test scores has a positive correlation. This means that the more a student studied, the higher his or her score tended to be. The scatter plot relating television hours and test scores has a negative correlation. This means that the more time a student spent watching television, the lower his or her score tended to be.

Fitting a Line to Data

Finding a linear model to represent the relationship described by a scatter plot is called **fitting a line to data.** You can do this graphically by simply sketching the line that appears to fit the points, finding two points on the line, and then finding the equation of the line that passes through the two points.

Real Life

EXAMPLE 3 Fitting a Line to Data

Find a linear model that relates the year with the number of people in the United States labor force. (See Example 1.)

t	3	4	5	6	7	8	9	10	11	12	13
P	113	115	117	120	122	123	126	126	127	129	130

Figure 2.52

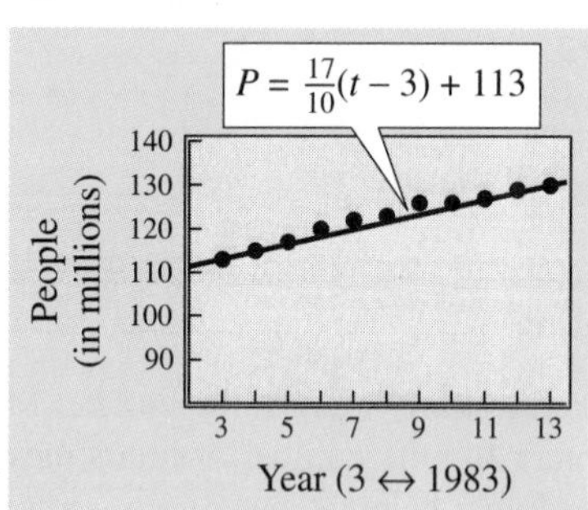

Solution

After plotting the data in the table, draw the line that you think best represents the data, as shown in Figure 2.52. Two points that lie on this line are (3, 113) and (13, 130). Using the point-slope form, you can find the equation of the line to be

$$P = \frac{17}{10}(t - 3) + 113.$$ Linear model

Once you have found a model, you can measure how well the model fits the data by comparing the actual values with the values given by the model, as shown in the following table.

	t	3	4	5	6	7	8	9	10	11	12	13
Actual	P	113	115	117	120	122	123	126	126	127	129	130
Model	P	113	114.7	116.4	118.1	119.8	121.5	123.2	124.9	126.6	128.3	130

The sum of the squares of the differences between the actual values and the model's values is the **sum of the squared differences.** The model that has the least sum is the **least squares regression line** for the data. For the model in Example 3, the sum of the squared differences is 20.85. The least squares regression line for the data is

$$P = 1.7t + 108.9.$$ Best-fitting linear model

Its sum of squared differences is 8.85.

Least Squares Regression Line

The least squares regression line, $y = ax + b$, for the points (x_1, y_1), (x_2, y_2), (x_3, y_3), . . . , (x_n, y_n) is given by

$$a = \frac{n\sum_{i=1}^{n} x_i y_i - \sum_{i=1}^{n} x_i \sum_{i=1}^{n} y_i}{n\sum_{i=1}^{n} x_i^2 - \left(\sum_{i=1}^{n} x_i\right)^2} \text{ and } b = \frac{1}{n}\left(\sum_{i=1}^{n} y_i - a\sum_{i=1}^{n} x_i\right).$$

EXAMPLE 4 **Finding a Least Squares Regression Line**

Find the least squares regression line for the points $(-3, 0)$, $(-1, 1)$, $(0, 2)$, and $(2, 3)$.

Solution

Begin by constructing a table like that shown below.

x	y	xy	x^2
-3	0	0	9
-1	1	-1	1
0	2	0	0
2	3	6	4
$\sum_{i=1}^{n} x_i = -2$	$\sum_{i=1}^{n} y_i = 6$	$\sum_{i=1}^{n} x_i y_i = 5$	$\sum_{i=1}^{n} x_i^2 = 14$

Applying the formulas for the least squares regression line with $n = 4$ produces

$$a = \frac{n\sum_{i=1}^{n} x_i y_i - \sum_{i=1}^{n} x_i \sum_{i=1}^{n} y_i}{n\sum_{i=1}^{n} x_i^2 - \left(\sum_{i=1}^{n} x_i\right)^2} = \frac{4(5) - (-2)(6)}{4(14) - (-2)^2} = \frac{32}{52} = \frac{8}{13}$$

and

$$b = \frac{1}{n}\left(\sum_{i=1}^{n} y_i - a\sum_{i=1}^{n} x_i\right) = \frac{1}{4}\left[6 - \frac{8}{13}(-2)\right] = \frac{47}{26}.$$

The least squares regression line is $y = \frac{8}{13}x + \frac{47}{26}$, as shown in Figure 2.53.

Figure 2.53

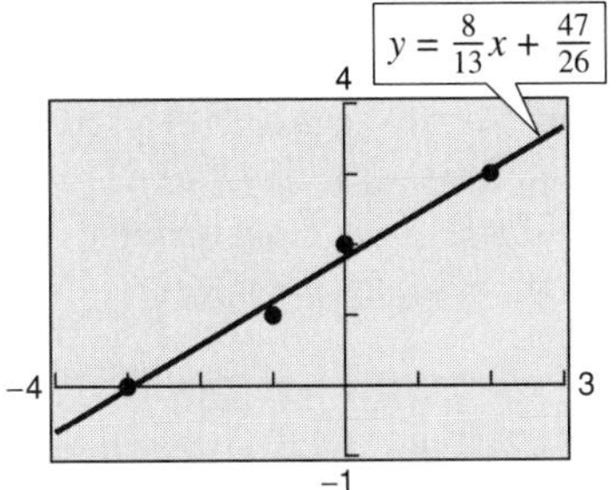

Real Life

EXAMPLE 5 A Mathematical Model

The annual amounts of advertising expenses y (in billions of dollars) in the United States from 1983 to 1992 are given in the table. (Source: McCann Erickson)

Year	1983	1984	1985	1986	1987	1988	1989	1990	1991	1992
y	75.9	88.1	94.8	102.1	109.8	118.1	125.6	128.6	126.4	131.7

A linear model that approximates this data is

$$y = 6.1703t + 63.8327, \quad 3 \leq t \leq 12$$

where $t = 3$ corresponds to 1983. Plot the actual data *and* the model on the same graph. How closely does the model represent the data?

Solution

The actual data is plotted in Figure 2.54, along with the graph of the linear model. From the figure, it appears that the model is a "good fit" for the actual data. You can see how well the model fits by comparing the actual values of y with the values of y given by the model (these are labeled y^* in the table below).

Figure 2.54

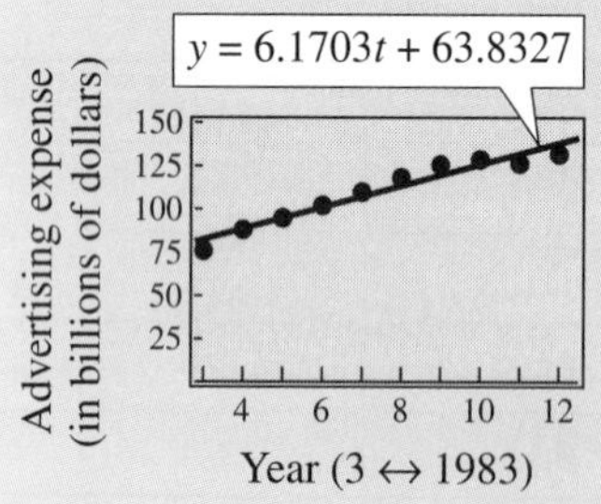

t	3	4	5	6	7	8	9	10	11	12
y	75.9	88.1	94.8	102.1	109.8	118.1	125.6	128.6	126.4	131.7
y^*	82.3	88.5	94.7	100.9	107.0	113.2	119.4	125.5	131.7	137.9

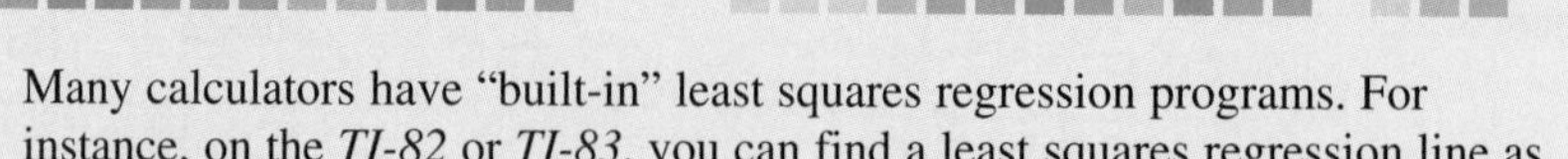

Many calculators have "built-in" least squares regression programs. For instance, on the *TI-82* or *TI-83*, you can find a least squares regression line as shown below.

1. Use the stat and edit menus to enter the data in lists L_1 and L_2.
2. For the *TI-82,* use the stat and calc menus to select SetUp. Make sure the XList is set to L_1 and the YList is set to L_2. (These are the default settings on the *TI-83*.)
3. Use the stat and calc menus to select LinReg(ax+b). After running this program, the calculator will display the values of a and b in the model $y = ax + b$. The *TI-82* will also display the correlation coefficient r.

Real Life

EXAMPLE 6 **Finding a Least Squares Regression Line**

The following ordered pairs (w, h) represent the shoe sizes w and the heights h (in inches) of 25 men. Use a computer program or a statistical calculator to find the least squares regression line for the data.

(10.0, 70.0)	(10.5, 71.0)	(9.5, 70.0)	(11.0, 72.0)	(12.0, 74.0)
(8.5, 66.0)	(9.0, 68.5)	(13.0, 76.0)	(10.5, 71.5)	(10.5, 70.5)
(10.0, 72.0)	(9.5, 70.0)	(10.0, 71.0)	(10.5, 69.5)	(11.0, 71.5)
(12.0, 73.5)	(12.5, 74.0)	(11.0, 71.5)	(9.0, 67.5)	(10.0, 70.0)
(13.0, 73.5)	(10.5, 72.5)	(10.5, 71.0)	(11.0, 73.0)	(8.5, 68.0)

Figure 2.55

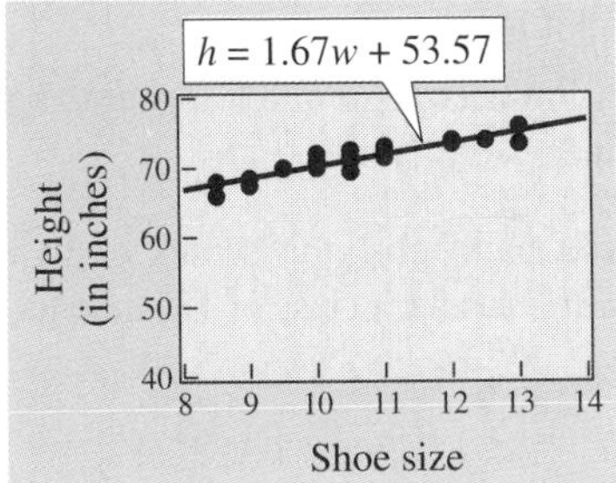

Solution

A scatter plot for the data is shown in Figure 2.55. Note that the plot does not have 25 points because some of the ordered pairs graph as the same point. After entering the data into a graphing utility, you can obtain

$$a = 1.67 \quad \text{and} \quad b = 53.57.$$

Thus, the least squares regression line for the data is

$$h = 1.67w + 53.57.$$

In Figure 2.55, this line is plotted with the data. Note that the line is a relatively good fit for the data.

If you use a graphing calculator or computer program to duplicate the results of Example 6, you will notice that the program also outputs a value of $r \approx 0.918$. This number is the **correlation coefficient** of the data. Correlation coefficients vary between -1 and 1. Basically, the closer $|r|$ is to 1, the better the points can be described by a line. Three examples are shown in Figure 2.56.

Figure 2.56

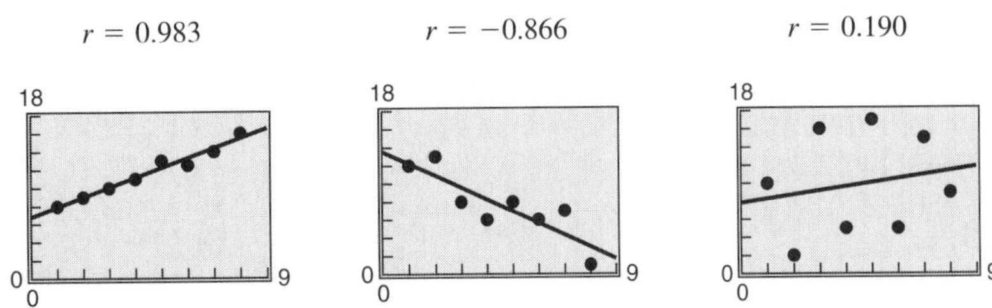

Group Activity

Research Project

Use your school's library or some other reference source to locate data that you think describes a linear relationship. Create a scatter plot of the data, and find the least squares regression line that represents the points. Interpret the slope and y-intercept in the context of the data.

2.6 /// EXERCISES

Correlation **In Exercises 1–4, the scatter plots of sets of data are given. Determine whether there is positive correlation, negative correlation, or very little correlation between the variables.**

1. 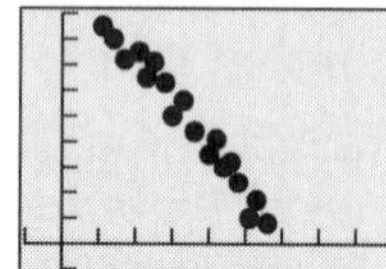

2. 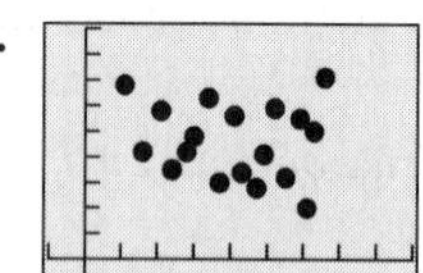

3. 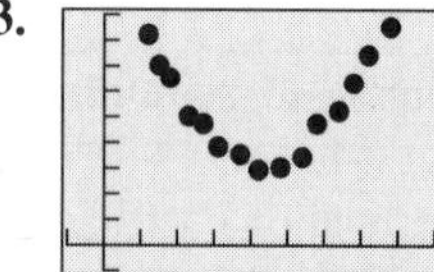

4. 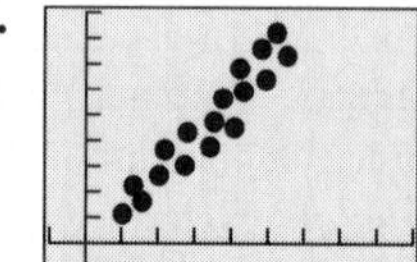

5. The following ordered pairs give the exposure index x of a carcinogenic substance and the cancer mortality y per 100,000 population. The higher the index level, the higher the level of contamination.

(3.50, 150.1), (3.58, 133.1),
(4.42, 132.9), (2.26, 116.7),
(2.63, 140.7), (4.85, 165.5),
(12.65, 210.7), (7.42, 181.0),
(9.35, 213.4)

(a) Create a scatter plot for the data.

(b) Does the relationship between x and y appear to be approximately linear? Explain.

6. The following ordered pairs give the scores of two consecutive 15-point quizzes for a class of 18 students.

(7, 13), (9, 7), (14, 14),
(15, 15), (10, 15), (9, 7),
(14, 11), (14, 15), (8, 10),
(9, 10), (15, 9), (10, 11),
(11, 14), (7, 14), (11, 10),
(14, 11), (10, 15), (9, 6)

(a) Create a scatter plot for the data.

(b) Does the relationship between consecutive quiz scores appear to be approximately linear? If not, give some possible explanations.

In Exercises 7–10, (a) find the least squares regression line by hand (see Example 4) and use a graphing utility to verify your results, (b) graph the data points and the regression line, and (c) comment on the validity of the model.

7. 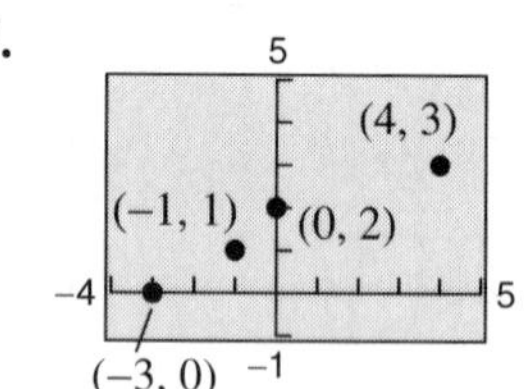

8. 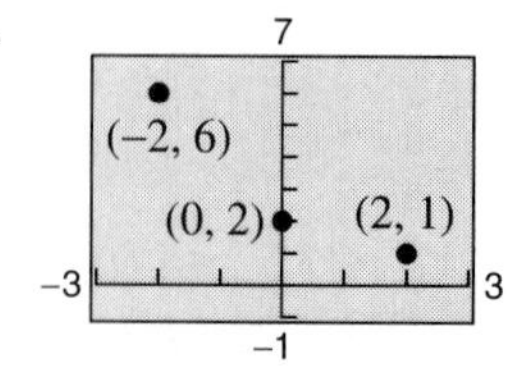

9. 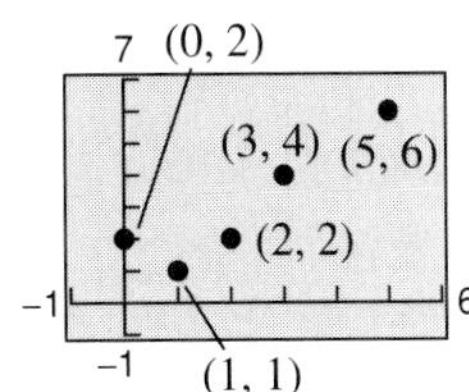

10. 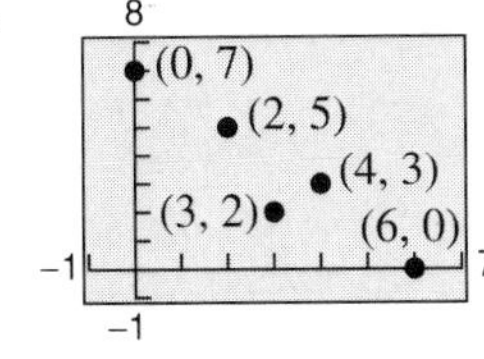

11. *Hooke's Law* Hooke's Law states that the force F required to compress or stretch a spring (within its elastic limits) is proportional to the distance d that the spring is compressed or stretched from its original length. That is, $F = kd$, where k is the measure of the stiffness of the spring and is called the *spring constant.* The table gives the elongation d in centimeters of a spring when a force of F kilograms is applied.

F	20	40	60	80	100
d	1.4	2.5	4.0	5.3	6.6

(a) Use the regression capabilities of a graphing utility to find a linear model for the data.

(b) Use a graphing utility to plot the data and graph the model. How well does the model fit the data? Explain your reasoning.

(c) Use the model to estimate the elongation of the spring when a force of 55 kilograms is applied.

12. *Falling Object* In an experiment students measured the speed s (in meters per second) of a falling object t seconds after it was released. The results are given in the table.

t	0	1	2	3	4
s	0	11.0	19.4	29.2	39.4

(a) Use the regression capabilities of a graphing utility to find a linear model for the data.

(b) Use a graphing utility to plot the data and graph the model. How well does the model fit the data? Explain your reasoning.

(c) Use the model to estimate the speed of the object after 2.5 seconds.

13. *Cable TV* The average monthly basic rate R (in dollars) for cable TV for the years 1986 through 1993 in the United States is given in the table. (Source: Paul Kagan Associates, Inc.)

Year	1986	1987	1988	1989
R	11.09	13.27	14.45	15.97

Year	1990	1991	1992	1993
R	17.58	18.61	19.08	19.39

Let t represent the year, with $t = 6$ corresponding to 1986.

(a) Use the regression capabilities of a graphing utility to find a linear model for the data.

(b) Use a graphing utility to plot the data and graph the model.

(c) Interpret the slope of the model in the context of the problem.

(d) Use the model to predict the average monthly basic rate for cable TV for the year 2000.

14. *Price of Homes* The median sales price P (in thousands of dollars) of existing one-family homes for 1986 through 1993 in the United States is given in the table. (Source: National Association of Realtors)

Year	1986	1987	1988	1989
P	80.3	85.6	89.3	93.1

Year	1990	1991	1992	1993
P	95.5	100.3	103.7	106.8

Let t represent the year, with $t = 6$ corresponding to 1986.

(a) Use the regression capabilities of a graphing utility to find a linear model for the data.

(b) Use a graphing utility to plot the data and graph the model.

(c) Interpret the slope of the model in the context of the problem.

(d) Use the model to predict the median price of existing one-family homes for the year 2000.

15. *Advertising and Sales* The table gives the advertising expenditures x and sales volume y for a company for seven randomly selected months. Both are measured in thousands of dollars.

Month	1	2	3	4	5	6	7
x	2.4	1.6	2.0	2.6	1.4	1.6	2.0
y	202	184	220	240	180	164	186

(a) Use the regression capabilities of a graphing utility to find a linear model for the data.

(b) Use a graphing utility to plot the data and graph the model.

(c) Interpret the slope of the model in the context of the problem.

(d) Use the model to predict sales for advertising expenditures of $1500.

16. *Energy Consumption* The data gives the per capita energy usage (in thousands of kilograms of coal equivalent) and the per capita gross national product (in thousands of U.S. dollars) for a sample of countries in 1990. (Source: Statistical Office of the United Nations)

Argentina (1.83, 3.7); Bangladesh (0.01, 0.2); Brazil (0.77, 2.6); Canada (10.46, 21.5); Denmark (4.70, 24.0); Finland (5.93, 26.0); France (3.87, 20.8); Greece (3.05, 6.8); India (0.31, 0.3); Italy (3.86, 19.4); Japan (4.21, 26.2); Mexico (1.75, 3.0); Pakistan (0.28, 0.4); South Korea (2.47, 6.0); Tanzania (0.04, 0.1); United States (10.32, 23.0)

(a) Use the regression capabilities of a graphing utility to find a linear model for the data.

(b) Use a graphing utility to plot the data and graph the model.

(c) Interpret the graph in part (b). Use the graph to identify any countries that appear to differ substantially from most of the others.

17. *Holders of Mortgage Debts* The table gives the amount of mortgage debt (in billions of dollars) held by savings institutions x and commercial banks y for the years 1988 through 1992 in the United States. (Source: The Federal Reserve Bulletin)

Year	1988	1989	1990	1991	1992
x	925	910	802	705	628
y	674	767	845	876	895

(a) Use the regression capabilities of a graphing utility to find a linear model for the data.

(b) Use a graphing utility to plot the data and graph the model.

(c) Interpret the slope of the model in the context of the problem. What information is given by the sign of the slope?

18. *Voter Registration and Turnout* The data shows the percent x of the voting-age population that was registered and the percent y that actually voted by state in 1992. (Source: U.S. Bureau of the Census)

AK (75.1, 68.6)	AL (77.1, 63.6)	AR (66.5, 58.0)
AZ (70.5, 64.7)	CA (57.6, 52.8)	CO (74.7, 68.8)
CT (76.8, 71.7)	D.C. (73.9, 66.1)	DE (70.7, 66.1)
FL (62.7, 55.8)	GA (62.0, 54.1)	HI (60.6, 55.2)
IA (77.9, 70.8)	ID (71.8, 66.9)	IL (72.1, 65.1)
IN (68.0, 63.0)	KS (77.1, 71.9)	KY (64.9, 57.6)
LA (77.0, 68.6)	MA (72.5, 65.8)	MD (71.7, 66.4)
ME (85.4, 74.1)	MI (74.6, 65.9)	MN (86.1, 74.2)
MO (74.2, 66.2)	MS (79.3, 66.7)	MT (77.3, 71.4)
NC (68.7, 60.0)	ND (90.8, 71.1)	NE (73.1, 66.1)
NH (69.9, 64.6)	NJ (68.0, 61.2)	NM (67.1, 62.6)
NV (63.4, 58.1)	NY (62.0, 56.8)	OH (69.6, 64.3)
OK (74.3, 67.5)	OR (74.7, 69.1)	PA (65.1, 59.9)
RI (74.0, 69.0)	SC (67.0, 58.0)	SD (80.1, 70.2)
TN (65.0, 55.6)	TX (64.9, 55.6)	UT (78.8, 72.3)
VA (65.4, 61.2)	VT (76.1, 68.6)	WA (71.5, 66.3)
WI (84.4, 75.3)	WV (64.9, 57.3)	WY (68.1, 64.2)

(a) Use the regression capabilities of a graphing utility to find a linear model for the data.

(b) Use a graphing utility to plot the data and graph the model.

(c) Interpret the graph in part (b). Use the graph to identify any states that appear to differ substantially from most of the others.

(d) Interpret the slope of the model in the context of the problem.

Focus on Concepts

In this chapter, you studied algebraic and graphical methods related to solving equations and inequalities. You can use the following questions to check your understanding of several of these basic methods. The answers to these questions are given in the back of the book.

1. In your own words, explain the difference between an identity and a conditional equation.

2. In your own words, explain what is meant by equivalent equations. Describe the steps used to transform an equation into an equivalent equation.

3. Consider the linear equation $ax + b = 0$.

(a) What is the sign of the solution if $ab > 0$?

(b) What is the sign of the solution if $ab < 0$?

4. The graphs show the solution(s) of equations plotted on the real number line. In each case, determine if the solution(s) is (are) for a linear equation, a quadratic equation, both, or neither. Explain.

(a) 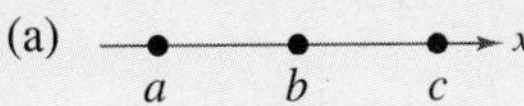

(b) x
a

(c) x
a b

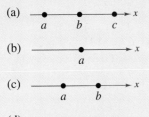

(d) x
a b c d

5. To solve the equation $2x^2 + 3x = 15x$, a student divides both sides by x and solves the equation $2x + 3 = 15$. The resulting solution $(x = 6)$ satisfies the given equation. Is there an error? Explain.

6. *True or False?* The graph of a function may have two distinct y-intercepts.

7. Describe the relationship among the x-intercepts of a graph, the zeros of a function, and the solutions of an equation.

8. *True or False?* The sum of two complex numbers cannot be a real number. Explain.

9. In your own words, explain why the equation

$$x^2 + 5 = 0$$

has no real solutions.

10. Which graph shows the solution of the inequality $|x - a| \geq 2$? Explain.

(a) 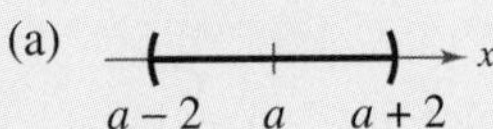

(b) 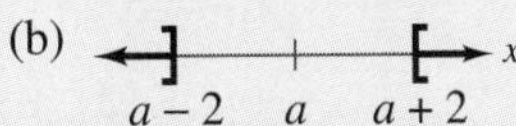

(c) x
$2 - a$ 2 $2 + a$

(d) x
$2 - a$ 2 $2 + a$

11. Consider the polynomial $(x - a)(x - b)$ and the real number line (see figure).

(a) Identify the points on the line where the polynomial is zero.

(b) In each of the three subintervals of the line, write the sign of each factor and the sign of the product.

(c) For which x-values does a polynomial change signs?

12. *True or False?* If the correlation coefficient is close to -1, the regression line cannot be used to describe the data. Explain.

13. *True or False?* The signs of the slope of the regression line and the correlation coefficient are always the same.

2 /// REVIEW EXERCISES

In Exercises 1 and 2, determine whether the equation is an identity or a conditional equation.

1. $6 - (x - 2)^2 = 2 + 4x - x^2$

2. $3(x - 2) + 2x = 2(x + 3)$

In Exercises 3 and 4, determine whether the given values of x are solutions of the equation.

Equation	*Values*	
3. $2x^2 + 7x - 4 = 0$	(a) $x = 0$	(b) $x = -4$
	(c) $x = \frac{1}{2}$	(d) $x = -1$
4. $6 + \dfrac{3}{x - 4} = 5$	(a) $x = 5$	(b) $x = 0$
	(c) $x = -2$	(d) $x = 1$

In Exercises 5–34, solve the equation (if possible) and use a graphing utility to verify your solution.

5. $3x - 2(x + 5) = 10$

6. $4x + 2(7 - x) = 5$

7. $4(x + 3) - 3 = 2(4 - 3x) - 4$

8. $\frac{1}{2}(x - 3) - 2(x + 1) = 5$

9. $3\left(1 - \dfrac{1}{5t}\right) = 0$

10. $\dfrac{1}{x - 2} = 3$

11. $6x = 3x^2$

12. $15 + x - 2x^2 = 0$

13. $(x + 4)^2 = 18$

14. $16x^2 = 25$

15. $x^2 - 12x + 30 = 0$

16. $x^2 + 6x - 3 = 0$

17. $5x^4 - 12x^3 = 0$

18. $4x^3 - 6x^2 = 0$

19. $\dfrac{4}{(x - 4)^2} = 1$

20. $\dfrac{1}{(t + 1)^2} = 1$

21. $\sqrt{x + 4} = 3$

22. $\sqrt{x - 2} - 8 = 0$

23. $2\sqrt{x} - 5 = 0$

24. $\sqrt{3x - 2} = 4 - x$

25. $\sqrt{2x + 3} + \sqrt{x - 2} = 2$

26. $5\sqrt{x} - \sqrt{x - 1} = 6$

27. $(x - 1)^{2/3} - 25 = 0$

28. $(x + 2)^{3/4} = 27$

29. $(x + 4)^{1/2} + 5x(x + 4)^{3/2} = 0$

30. $8x^2(x^2 - 4)^{1/3} + (x^2 - 4)^{4/3} = 0$

31. $|x - 5| = 10$

32. $|2x + 3| = 7$

33. $|x^2 - 3| = 2x$

34. $|x^2 - 6| = x$

In Exercises 35–40, use a graphing utility to graph the equation. Use the graph to approximate any x-intercepts of the graph. Set $y = 0$ and solve the resulting equation. Compare the result with the x-intercepts of the graph.

35. $y = 4x^3 - 12x^2 + 8x$

36. $y = 12x^3 - 84x^2 + 120x$

37. $y = \dfrac{1}{x} + \dfrac{1}{x + 1} - 2$

38. $y = \dfrac{4}{x - 3} - \dfrac{4}{x} - 1$

39. $y = \sqrt{x^2 + 1} + x - 9$

40. $y = |2x - 3| - 5$

In Exercises 41–44, solve the equation for the indicated variable.

41. Solve for r: $V = \frac{1}{3}\pi r^2 h$

42. Solve for X: $Z = \sqrt{R^2 - X^2}$

43. Solve for p: $L = \dfrac{k}{3\pi r^2 p}$

44. Solve for v: $E = 2kw\left(\dfrac{v}{2}\right)^2$

In Exercises 45 and 46, find the constant C such that the ordered pair is a solution point of the equation.

45. $y = C\sqrt{x + 1}$, $(3, 8)$

46. $x + C(y + 2) = 0$, $(4, 3)$

In Exercises 47–56, perform the operations and write the result in standard form.

47. $(7 + 5i) + (-4 + 2i)$

48. $\left(\frac{\sqrt{2}}{2} - \frac{\sqrt{2}}{2}i\right) - \left(\frac{\sqrt{2}}{2} + \frac{\sqrt{2}}{2}i\right)$

49. $5i(13 - 8i)$

50. $(1 + 6i)(5 - 2i)$

51. $(10 - 8i)(2 - 3i)$

52. $i(6 + i)(3 - 2i)$

53. $\dfrac{6 + i}{i}$

54. $\dfrac{3 + 2i}{5 + i}$

55. $\dfrac{4}{-3i}$

56. $\dfrac{1}{(2 + i)^4}$

In Exercises 57–60, use a graphing utility to graph the function. Determine the number of x-intercepts of the graph and compare these intercepts with the real zeros of the function. (Find all the zeros of the function.)

57. $y = 3x^2 + 1$

58. $y = x^3 + x$

59. $y = x^3 - 4x^2 + 5x$

60. $y = 2 + 8x^{-2}$

In Exercises 61–68, solve the inequality. Use a graphing utility to verify your solution.

61. $\frac{1}{2}(3 - x) > \frac{1}{3}(2 - 3x)$

62. $x^2 - 2x \geq 3$

63. $\dfrac{x - 5}{3 - x} < 0$

64. $\dfrac{2}{x + 1} \leq \dfrac{3}{x - 1}$

65. $|x - 2| < 1$

66. $|x| \leq 4$

67. $\left|x - \frac{3}{2}\right| \geq \frac{3}{2}$

68. $|x - 3| > 4$

In Exercises 69–72, use a graphing utility to solve the inequality.

69. $\dfrac{x}{5} - 6 \leq -\dfrac{x}{2} + 6$

70. $2x^2 + x \geq 15$

71. $(x - 4)|x| > 0$

72. $|x(x - 6)| < 5$

In Exercises 73 and 74, find the domain of the expression by finding the interval(s) on the real number line for which the radicand is nonnegative.

73. $\sqrt{2x - 10}$

74. $\sqrt{x(x - 4)}$

75. *Monthly Profit* In October, a company's total profit was 12% more than it was in September. The total profit for the two months was \$689,000. Find the profit for each month.

76. *Discount Rate* The price of a television set has been discounted \$85. The sale price is \$340. What is the percent discount?

77. *Mixture Problem* A car radiator contains 10 liters of a 30% antifreeze solution. How many liters will have to be replaced with pure antifreeze if the resulting solution is to be 50% antifreeze?

78. *Starting Positions* A fitness center has two running tracks around a rectangular playing floor. The tracks are 1 meter wide and form semicircles at the narrow ends of the rectangular floor (see figure). Determine the distance between the starting positions if two runners must run the same distance to the finish line in one lap around the track.

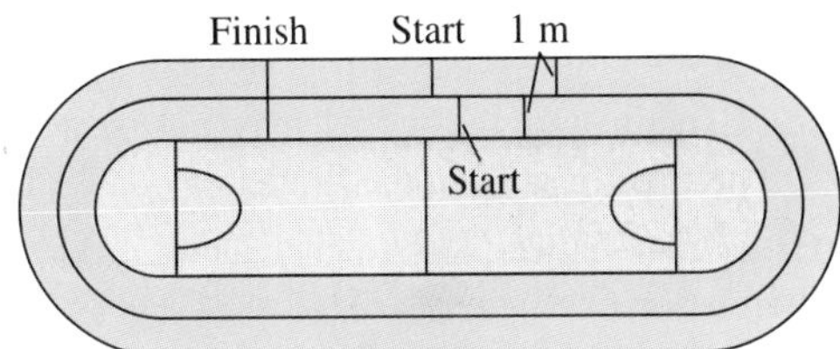

79. *Cost Sharing* A group of farmers agree to share equally in the cost of a \$48,000 piece of machinery. If they could find two more farmers to join the group, each person's share of the cost would decrease by \$4000. How many farmers are presently in the group?

80. *Venture Capital* An individual is planning on starting a small business that will require \$90,000 before any income can be generated. Because it is difficult to borrow for new ventures, the individual wants a group of friends to divide the cost equally for future shares of the profit. Some are willing, but three more are needed so that the price per person will be \$2500 less. How many investors are needed?

81. *Average Speed* You drove 56 miles one way on a service call. On the return trip, your average speed was 8 miles per hour greater and the trip took 10 fewer minutes. What was your average speed on the return trip?

82. *Data Analysis* The total sales of sporting goods in the United States from 1981 through 1992 can be approximated by the model

$$\text{Sales} = 16.8091 + 0.7151t^2 - 0.0446t^3$$

where the sales are measured in billions of dollars and the time t represents the calendar year, with $t = 1$ corresponding to 1981. The actual sales (in billions) are given in the table.

Year	1981	1982	1983	1984	1985	1986
Sales	18.7	18.7	23.1	26.4	27.4	30.6

Year	1987	1988	1989	1990	1991	1992
Sales	33.9	42.1	45.2	44.1	42.8	42.4

(a) Use a graphing utility to plot the data and graph the model.

(b) Use the model to estimate sales in 1994.

(c) Explain why the model may not be accurate in predicting sales in the future.

83. Order the scatter plots in increasing order by their correlation coefficients.

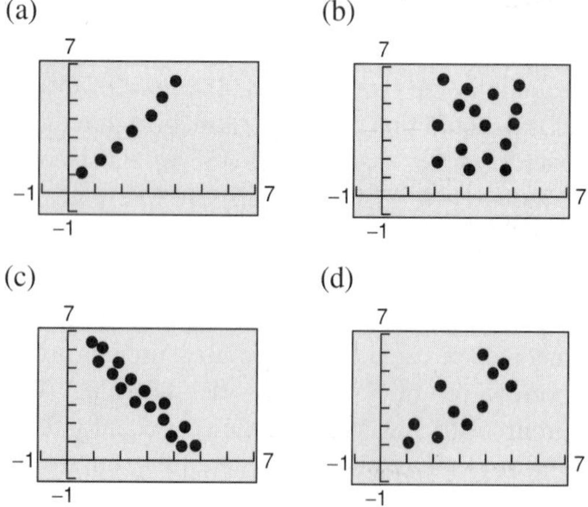

84. *Sales* The sales manager of a company wants to determine if there is a relationship between sales and years of experience for sales personnel who have been with the company 4 or fewer years. The table gives the years of experience x for eight of the firm's sales personnel and monthly sales y in thousands of dollars.

x	1.5	1.0	0.3	3.0
y	46.7	32.9	19.2	48.4

x	4.0	0.5	2.5	1.8
y	51.2	28.5	53.4	35.5

(a) Use the regression capabilities of a graphing utility to find a linear model for the data.

(b) Use a graphing utility to plot the data and graph the model.

(c) Interpret the slope of the model in the context of the problem.

(d) Use the model to predict monthly sales of a salesperson with 5 years of experience.

85. *Stress Test* A machine part was tested by bending it x centimeters 10 times per minute until it failed (y = time to failure, in hours). The results are recorded in the table.

x	3	6	9	12	15
y	61	56	53	55	48

x	18	21	24	27	30
y	35	36	33	44	23

(a) Use the regression capabilities of a graphing utility to find a linear model for the data.

(b) Use a graphing utility to plot the data and graph the model.

(c) Use the graph to determine if there may have been an error made in conducting one of the tests or in recording the results. If so, eliminate the point and find the model for the remaining data.

CHAPTER PROJECT *Mozart and the Golden Ratio*

Sonata	x	y	Sonata	x	y
279, I	62	38	310, I	84	49
279, II	46	28	311, I	73	39
279, III	102	56	330, I	92	58
280, I	88	56	330, III	103	68
280, II	36	24	332, I	136	93
280, III	113	77	332, III	155	90
281, I	69	40	333, I	102	63
281, II	60	46	333, III	50	31
282, I	18	15	457, I	93	74
282, III	63	39	533, I	137	102
283, I	67	53	533, II	76	46
283, II	23	14	545, I	45	28
283, III	171	102	547a, I	118	78
284, I	76	51	570, I	130	79
309, I	97	58			

Music and mathematics have always been closely related. In this project, you will discover a curious relationship between the structure of Mozart's (1756–1791) piano sonatas and Euclid's (c. 300 B.C.) famous golden ratio. (Source: John F. Putz, *Mathematics Magazine*, October 1995)

Sonatas can be naturally divided into two parts: the *exposition,* which introduces the musical theme, and the *development and recapitulation,* which develop and repeat the theme. The data at the left shows all of Mozart's piano sonata movements that have these two parts. The first column identifies the sonata movement. The second column, labeled x, identifies the length of the development and recapitulation in measures. The third column, labeled y, identifies the length of the exposition in measures.

(a) Enter the points (x, y) in a graphing utility and sketch a scatter plot of the points. Describe the relationship between x and y.

(b) Use the statistical capabilities of a graphing utility to find the least squares regression line for the data. How well does the line fit the data? Explain. Interpret the meaning of the slope and the y-intercept.

(c) The **golden section** was defined by Euclid as the point B on a line segment AC such that

$$\frac{AB}{BC} = \frac{BC}{AC}.$$

Assume that $AC = 1$ and $BC = r$. Then this proportion can be written as

$$\frac{1 - r}{r} = \frac{r}{1}.$$

Use a graphing utility to graph $y_1 = \dfrac{1 - x}{x}$ and $y_2 = \dfrac{x}{1}$ and approximate their point of intersection in Quadrant I. This value is called the **golden ratio.** How does it compare to the line in part (b)?

Questions for Further Exploration

1. Use the data for Mozart's sonatas to complete a table that calculates ratios of the form $\dfrac{x}{x + y}$.

How do these ratios compare with the golden ratio?

2. The Fibonacci Sequence is given by 1, 1, 2, 3, 5, 8, 13, 21, 34, ...

(a) Explain how to generate succeeding terms in the sequence.

(b) Complete and expand the table, which shows decimal approximations of ratios of adjacent terms in the sequence.

$\frac{1}{1}$	$\frac{1}{2}$	$\frac{2}{3}$	$\frac{3}{5}$	$\frac{5}{8}$	$\frac{8}{13}$	$\frac{13}{21}$	$\frac{21}{34}$
1.000	0.500	0.667					

What can you conclude about the ratios?

3. *Research Project* Obtain a copy of the article "The Golden Section and the Piano Sonatas of Mozart," John F. Putz, *Mathematics Magazine*, Vol. 68, No. 4, October 1995, pp. 275–282. Read the article and write a short paper about other aspects of Mozart's sonatas and the golden ratio.

P–2 /// CUMULATIVE TEST

Take this test as you would take a test in class. After you are done, check your work against the answers given in the back of the book.

The *Interactive* CD-ROM provides answers to the Chapter Tests and Cumulative Tests. It also offers Chapter Pre-Tests (that test key skills and concepts covered in previous chapters) and Chapter Post-Tests, both of which have randomly generated exercises with diagnostic capabilities.

In Exercises 1 and 2, simplify the expression.

1. $\dfrac{8x^2y^{-3}}{30x^{-1}y^2}$ **2.** $\sqrt{24x^4y^3}$

In Exercises 3–5, perform the operations and simplify the result.

3. $4x - [2x + 3(2 - x)]$ **4.** $(x - 2)(x^2 + x - 3)$ **5.** $\dfrac{2}{s + 3} - \dfrac{1}{s + 1}$

In Exercises 6–8, factor the expression completely.

6. $25 - (x - 2)^2$ **7.** $x - 5x^2 - 6x^3$ **8.** $54 - 16x^3$

In Exercises 9–11, graph the equation.

9. $x - 3y + 12 = 0$ **10.** $y = x^2 - 9$ **11.** $y = \sqrt{4 - x}$

12. Find an equation for the line passing through $\left(-\frac{1}{2}, 1\right)$ and $(3, 8)$.

13. Does the graph at the right represent y as a function of x? Explain.

14. Evaluate $f(x) = \dfrac{x}{x - 2}$ at $x = 6$, $x = 2$, and $x = s + 2$.

15. Compare the graphs of the functions with the graph of $y = \sqrt[3]{x}$.

(a) $r(x) = \dfrac{1}{2}\sqrt[3]{x}$ (b) $h(x) = \sqrt[3]{x} + 2$ (c) $g(x) = \sqrt[3]{x + 2}$

16. Decide whether $h(x) = 5x - 2$ is one-to-one. If it is, find its inverse.

Figure for 13

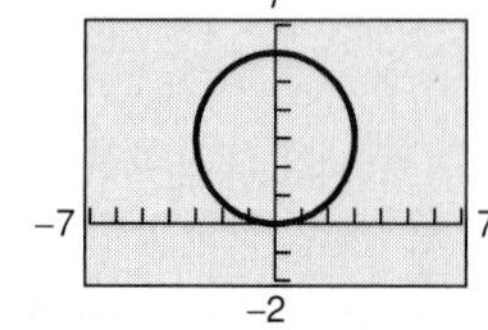

In Exercises 17–20, solve (if possible) the equation.

17. $2x - 3(x - 4) = 5$ **18.** $\dfrac{2}{t - 3} + \dfrac{2}{t - 2} = \dfrac{10}{t^2 - 5t + 6}$

19. $3y^2 + 6y + 2 = 0$ **20.** $\sqrt{x + 10} = x - 2$

21. The points (75, 2.3), (82, 3.2), (90, 3.6), and (65, 2.3) give the entrance exam scores x and the grade-point averages y after 1 year of college for four students. Find a least squares regression line for the data and use it to predict the GPA of a student with an entrance exam score of 88.

Polynomial and Rational Functions

Chapter 3

The Fundamental Theorem of Algebra implies that an nth-degree polynomial equation has precisely n solutions. This result, however, is true only if repeated and complex solutions are counted. For instance, the equation $x^4 - 2x^3 + 2x^2 - 2x + 1 = 0$ has solutions of 1, 1, i, and $-i$.

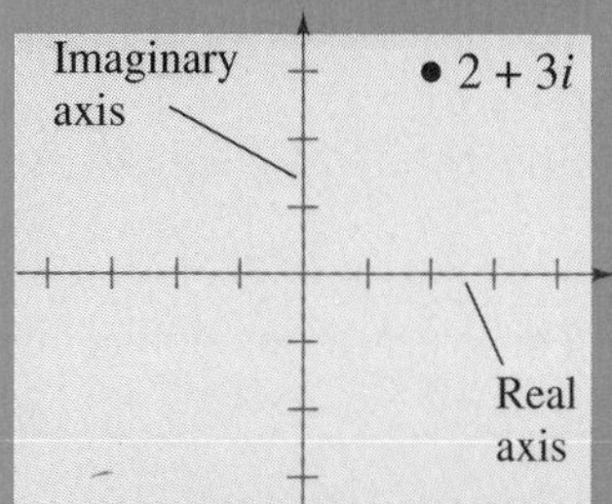

You can verify the real solutions with a graphing utility by graphing the equation $y = x^4 - 2x^3 + 2x^2 - 2x + 1$ and observing that it only touches the x-axis at $(1, 0)$.

When first developed, complex numbers were used primarily for theoretical results such as this. Today, however, they have several other uses.

One use of complex numbers is in creating fractals such as that shown in the Zuckerman photograph. To program this fractal with a computer, complex numbers are plotted in the complex plane.

In the complex plane, the point (a, b) represents the complex number $a + bi$. For instance, the number $2 + 3i$ is plotted at the left. (See Exercises 58 and 59 on page 288.)

Jim Zuckerman; (inset) Rondi Ballard.

This fractal was created on a computer by photographer Jim Zuckerman. Zuckerman owns and operates a photography and digital imaging company in Northridge, California.

3.1 Quadratic Functions

The Graph of a Quadratic Function / The Standard Form of a Quadratic Function / Applications

The Graph of a Quadratic Function

In this and the next section, you will study the graphs of polynomial functions.

Definition of Polynomial Function

Let n be a nonnegative integer and let $a_n, a_{n-1}, \ldots, a_2, a_1, a_0$ be real numbers with $a_n \neq 0$. The function given by

$$f(x) = a_n x^n + a_{n-1}x^{n-1} + \cdots + a_2x^2 + a_1x + a_0$$

is called a **polynomial function of x with degree n.**

Polynomial functions are classified by degree. For instance, the polynomial function

$$f(x) = a, \qquad a \neq 0 \qquad \text{Constant function}$$

has degree 0 and is called a **constant function.** In Chapter 1, you learned that the graph of this type of function is a horizontal line. The polynomial function

$$f(x) = ax + b, \qquad a \neq 0 \qquad \text{Linear function}$$

has degree 1 and is called a **linear function.** In Chapter 1, you learned that the graph of the linear function $f(x) = ax + b$ is a line whose slope is a and whose y-intercept is $(0, b)$. In this section you will study second-degree polynomial functions, which are called **quadratic functions.**

Library of Functions

The graph of a quadratic function is called a parabola. Graph $f(x) = x^2$ and $g(x) = |x|$ in the same viewing rectangle. Zoom in near the origin and compare the shape of the two graphs. Which graph grows faster as x gets larger and larger? Why does the definition of a quadratic function require that $a \neq 0$?

Definition of Quadratic Function

Let a, b, and c be real numbers with $a \neq 0$. The function of x given by

$$f(x) = ax^2 + bx + c \qquad \text{Quadratic function}$$

is called a **quadratic function.**

Note The graph of a quadratic function is a "∪"-shaped curve that is called a **parabola.**

All parabolas are symmetric with respect to a line called the **axis of symmetry,** or simply the **axis** of the parabola. The point where the axis intersects the parabola is the **vertex** of the parabola, as shown in Figure 3.1. If the leading coefficient is positive, the graph of $f(x) = ax^2 + bx + c$ is a parabola that opens upward, and if the leading coefficient is negative, the graph of $f(x) = ax^2 + bx + c$ is a parabola that opens downward.

Figure 3.1

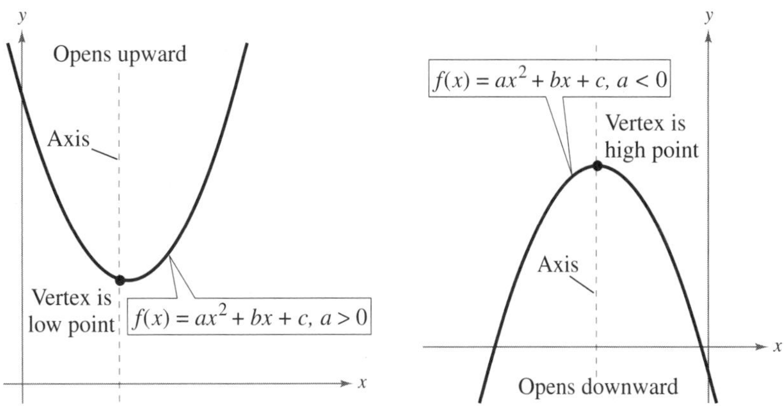

The simplest type of quadratic function is $f(x) = ax^2$. Its graph is a parabola whose vertex is (0, 0). If $a > 0$, the vertex is the *minimum* point on the graph, and if $a < 0$, the vertex is the *maximum* point on the graph, as shown in Figure 3.2.

Figure 3.2

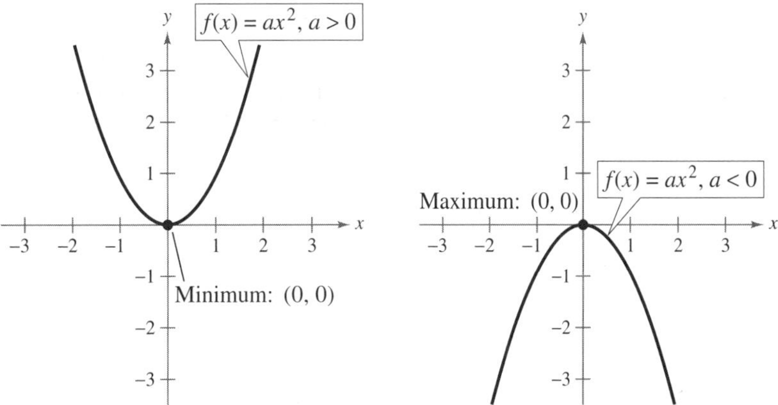

When sketching the graph of $f(x) = ax^2$, it is helpful to use the graph of $y = x^2$ as a reference, as discussed in Section 1.5.

EXAMPLE 1 Graphing Simple Quadratic Functions

Describe how the graph of each function is related to the graph of $y = x^2$.

a. $f(x) = \dfrac{1}{3}x^2$ **b.** $g(x) = 2x^2$

c. $h(x) = -x^2 + 1$ **d.** $k(x) = (x + 2)^2 - 3$

Solution

a. Compared with $y = x^2$, each output of f "shrinks" by a factor of $\frac{1}{3}$. The result is a parabola that opens upward and is broader than the parabola represented by $y = x^2$, as shown in Figure 3.3(a).

b. Compared with $y = x^2$, each output of g "stretches" by a factor of 2, creating a narrower parabola, as shown in Figure 3.3(b).

c. With respect to the graph of $y = x^2$, the negative coefficient in $h(x) = -x^2 + 1$ reflects the graph *downward* and the positive constant term shifts the vertex *up* one unit. The graph of h is shown in Figure 3.3(c).

d. With respect to the graph of $y = x^2$, the graph of $k(x) = (x + 2)^2 - 3$ is obtained by a horizontal shift two units *to the left* and a vertical shift three units *down*, as shown in Figure 3.3(d).

Note In Example 1, note that the coefficient a determines how widely the parabola given by $f(x) = ax^2$ opens. If $|a|$ is small, the parabola opens more widely than if $|a|$ is large.

The *Interactive* CD-ROM shows every example with its solution; clicking on the *Try It!* button brings up similar problems. Guided Examples and Integrated Examples show step-by-step solutions to additional examples. Integrated Examples are related to several concepts in the section.

Figure 3.3

(a)

(b)

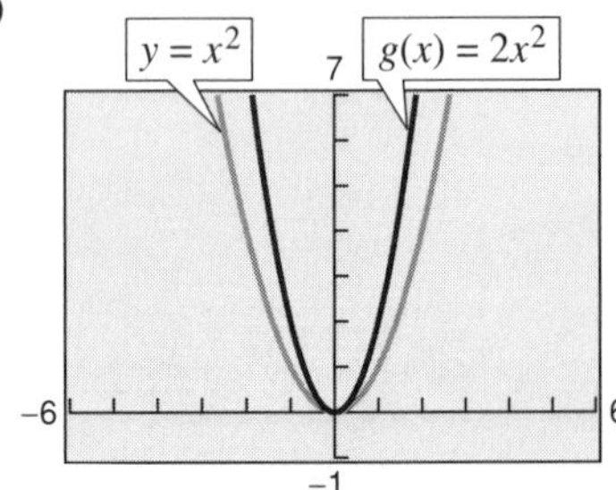

(c)

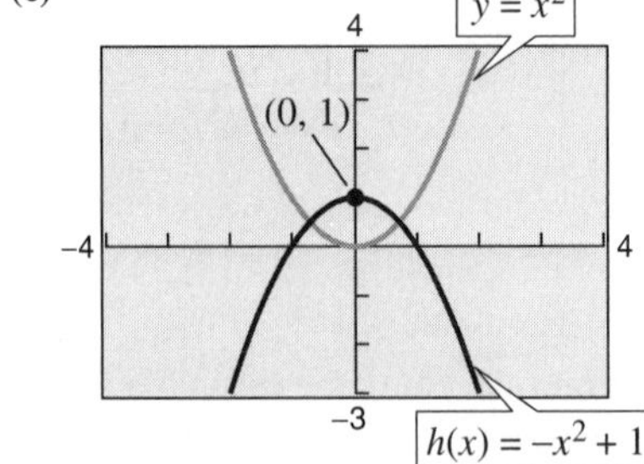

(d)

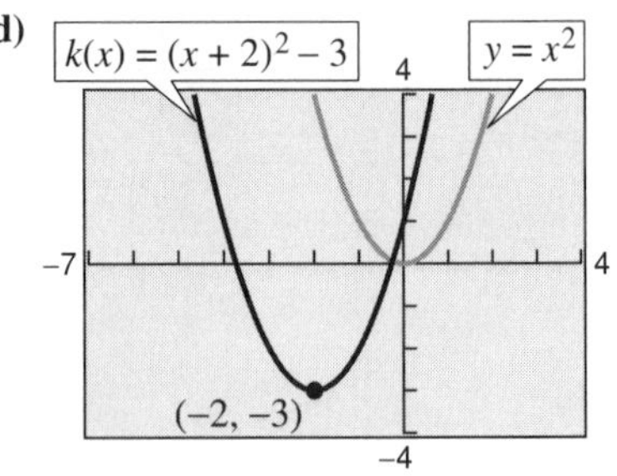

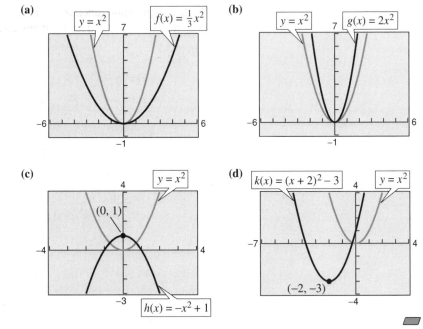

Note Recall from Section 1.5 that the graphs of $y = f(x \pm c)$, $y = f(x) \pm c$, $y = -f(x)$, and $y = f(-x)$ are rigid transformations of the graph of $y = f(x)$.

$y = f(x \pm c)$	Horizontal shift
$y = f(x) \pm c$	Vertical shift
$y = -f(x)$	Reflection in x-axis
$y = f(-x)$	Reflection in y-axis

The Standard Form of a Quadratic Function

The equation in Example 1(d) is written in the **standard form**

$$f(x) = a(x - h)^2 + k.$$

This form is especially convenient for sketching a parabola because it identifies the vertex of the parabola as (h, k).

EXPLORATION

Use a graphing utility to graph $y = ax^2$ with $a = -2, -1, -0.5, 0.5, 1$, and 2. How does the value of a affect the graph?

Use a graphing utility to graph $y = (x - h)^2$ with $h = -4, -2, 2$, and 4. How does the value of h affect the graph?

Use a graphing utility to graph $y = x^2 + k$ with $k = -4, -2, 2$, and 4. How does the value of k affect the graph?

Standard Form of a Quadratic Function

The quadratic function

$$f(x) = a(x - h)^2 + k, \qquad a \neq 0$$

is said to be in **standard form.** The graph of f is a parabola whose axis is the vertical line $x = h$ and whose vertex is the point (h, k). If $a > 0$, the parabola opens upward, and if $a < 0$, the parabola opens downward.

EXAMPLE 2 Writing a Quadratic Function in Standard Form

Describe the graph of $f(x) = 2x^2 + 8x + 7$.

Solution

Write the quadratic function in standard form by completing the square. Notice that the first step is to factor out any coefficient of x^2 that is different from 1.

$$f(x) = 2x^2 + 8x + 7 \quad \text{Original function}$$

$$= 2(x^2 + 4x) + 7 \quad \text{Factor 2 out of } x\text{-terms.}$$

$$= 2(x^2 + 4x + 4 - 4) + 7 \quad \text{Add and subtract 4 within parentheses.}$$

$(b/2)^2$

$$= 2(x^2 + 4x + 4) - 2(4) + 7 \quad \text{Regroup terms.}$$

$$= 2(x + 2)^2 - 1 \quad \text{Standard form}$$

From the standard form, you can see that the graph of f is a parabola that opens upward with vertex $(-2, -1)$, as shown in Figure 3.4.

Figure 3.4

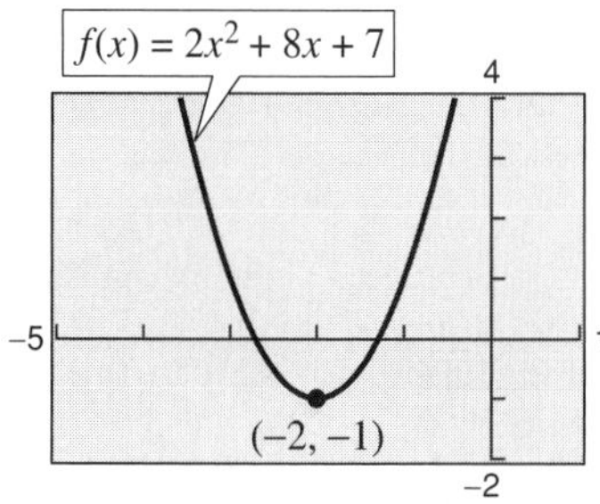

To find the x-intercepts of the graph of $f(x) = ax^2 + bx + c$, solve the equation $ax^2 + bx + c = 0$. If $ax^2 + bx + c$ does not factor, you can use the Quadratic Formula or a graphing utility to find the x-intercepts. Remember, however, that a parabola may have no x-intercept.

EXAMPLE 3 Writing a Quadratic Function in Standard Form

Describe the graph of $f(x) = -x^2 + 6x - 8$.

Solution

$$\begin{aligned} f(x) &= -x^2 + 6x - 8 && \text{Original form} \\ &= -(x^2 - 6x) - 8 && \text{Factor } -1 \text{ out of } x\text{-terms.} \\ &= -(x^2 - 6x + \underbrace{9 - 9}_{(b/2)^2}) - 8 && \text{Add and subtract 9 within parentheses.} \\ &= -(x^2 - 6x + 9) - (-9) - 8 && \text{Regroup terms.} \\ &= -(x - 3)^2 + 1 && \text{Standard form} \end{aligned}$$

The graph of f is a parabola that opens downward with vertex at (3, 1), as shown in Figure 3.5.

Figure 3.5

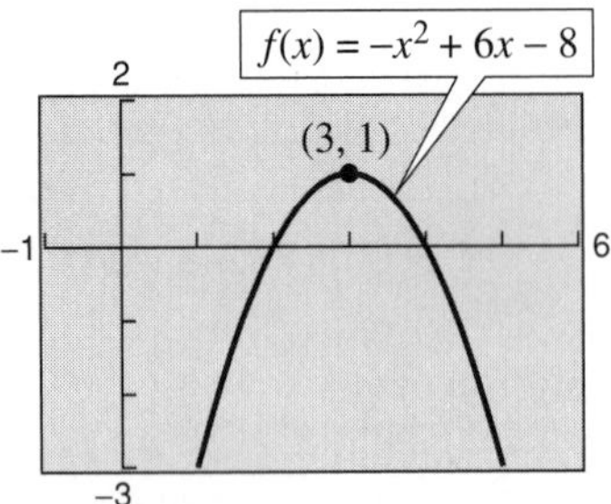

EXAMPLE 4 Finding the Equation of a Parabola

Find an equation for the parabola that has its vertex at (1, 2) and passes through the point (0, 0), as shown in Figure 3.6.

Figure 3.6

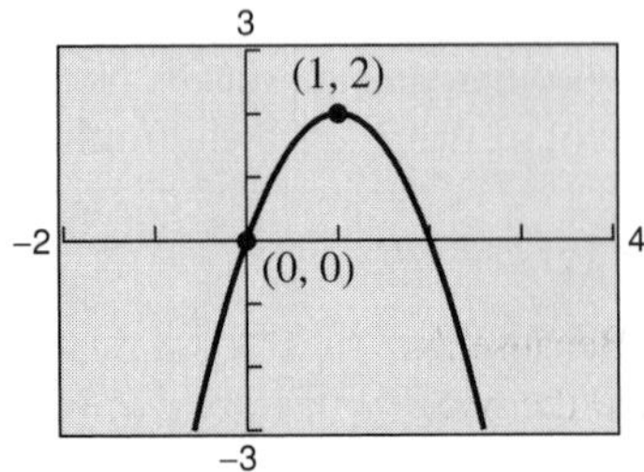

Solution

Because the parabola has a vertex at $(h, k) = (1, 2)$, the equation has the form

$$f(x) = a(x - 1)^2 + 2. \qquad \text{Standard form}$$

Because the parabola passes through the point (0, 0), it follows that $f(0) = 0$. Thus, you obtain

$$0 = a(0 - 1)^2 + 2,$$

which implies that $a = -2$. The equation is

$$\begin{aligned} f(x) &= -2(x - 1)^2 + 2 \\ &= -2x^2 + 4x. \end{aligned}$$

Try graphing $f(x) = -2x^2 + 4x$ with a graphing utility to confirm that its vertex is (1, 2) and that it passes through the point (0, 0).

Note In Example 4, there are infinitely many different parabolas that have a vertex at (1, 2). Of these, however, the only one that passes through the point (0, 0) is the one given by $f(x) = -2x^2 + 4x$.

Applications

> **Study Tip**
>
> To obtain the standard form at the right, you can complete the square of the form
>
> $$f(x) = ax^2 + bx + c.$$
>
> Try verifying this computation.

Many applications involve finding the maximum or minimum value of a quadratic function. By writing the quadratic function $f(x) = ax^2 + bx + c$ in standard form,

$$f(x) = a\left(x + \frac{b}{2a}\right)^2 + \left(c - \frac{b^2}{4a}\right)$$

you can see that the vertex occurs at $x = -b/(2a)$, which implies the following.

1. If $a > 0$, f has a *minimum* that occurs at $x = -b/(2a)$.
2. If $a < 0$, f has a *maximum* that occurs at $x = -b/(2a)$.

EXAMPLE 5 The Maximum Height of a Baseball

Real Life

A baseball is hit 3 feet above ground at a velocity of 100 feet per second and at an angle of 45 degrees with respect to level ground. The path of the baseball is given by the function

$$f(x) = -0.0032x^2 + x + 3$$

where $f(x)$ is the height of the baseball (in feet) and x is the distance from home plate (in feet). What is the maximum height reached by the baseball? (See Example 7 in Section 1.3.)

Solution

For this quadratic function, you have

$$f(x) = ax^2 + bx + c = -0.0032x^2 + x + 3$$

which implies that $a = -0.0032$ and $b = 1$. Because the function has a maximum when $x = -b/2a$, you can conclude that the baseball reaches its maximum height when

$$x = -\frac{b}{2a} = -\frac{1}{2(-0.0032)} = 156.25 \text{ feet}$$

from home plate. At this distance, the maximum height is

$$f(156.25) = -0.0032(156.25)^2 + 156.25 + 3 = 81.125 \text{ feet.}$$

Figure 3.7

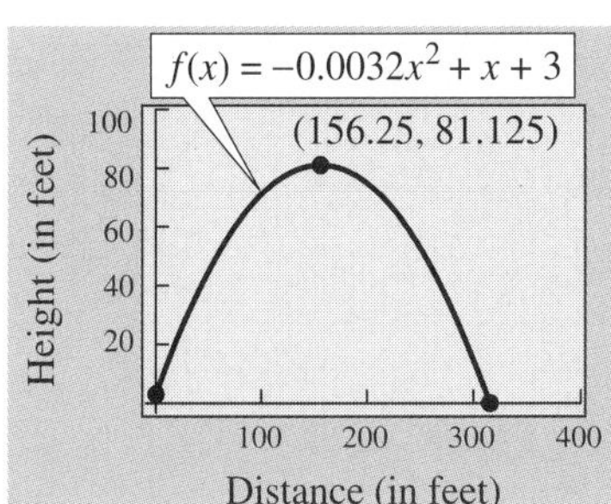

The path of the baseball is shown in Figure 3.7. You can also solve this problem by using a graphing utility. By using the zoom and trace features, you can determine that the maximum height on the graph occurs when $x \approx 156.25$. Note that you might have to change the y-scale in order to avoid a graph that is "too flat."

Real Life

EXAMPLE 6 Charitable Contributions

According to a survey conducted by *Independent Sector*, the percent of their income that Americans give to charities is related to their household income. For families with an annual income of \$100,000 or less, the percent is approximately given by

$$P = 0.0014x^2 - 0.1529x + 5.855, \qquad 5 \le x \le 100$$

where P is the percent of annual income given and x is the annual income in thousands of dollars. According to this model, what income level corresponds to the minimum percent of charitable contributions?

Figure 3.8

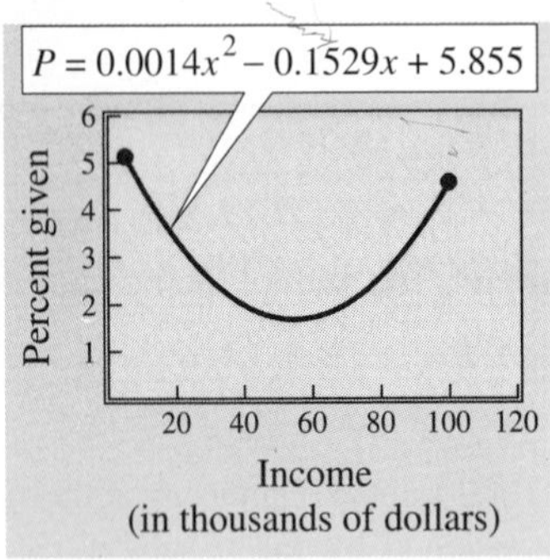

Solution

There are two ways to answer this question. One is to use a graphing utility to graph the quadratic function, as shown in Figure 3.8. From this graph, it appears that the minimum percent corresponds to an income level of about \$55,000. By using the zoom and trace features, you can improve the approximation to $x \approx 54{,}600$. The other way to answer the question is to use the fact that the minimum point of the parabola occurs when $x = -b/2a$. For this function, you have $a = 0.0014$ and $b = -0.1529$. Thus,

$$x = -\frac{b}{2a} = -\frac{-0.1529}{2(0.0014)} \approx 54.6.$$

From this x-value, you can conclude that the minimum percent corresponds to an income level of about \$54,600.

Group Activity — *Finding an Equation for a Curve*

The parabola in the figure at the right has an equation of the form

$$y = ax^2 + bx - 4.$$

Find the equation for this parabola in two different ways, by hand and with technology. Discuss the methods you used. Compare the results of the two methods and compare your results with those of your group members.

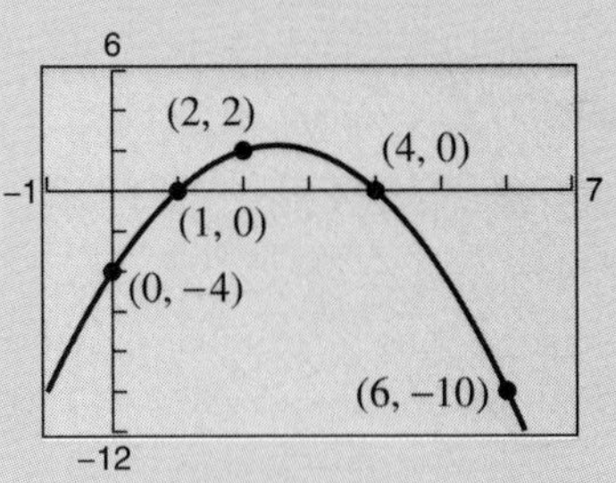

GROUP ACTIVITY SUGGESTION

Ask each group member to use a different point on the curve when finding the equation of the parabola by hand. Have them compare results and verify that their equations are equivalent no matter which point on the curve is used.

3.1 /// EXERCISES

In Exercises 1–8, match the quadratic function with the correct graph. [The graphs are labeled (a), (b), (c), (d), (e), (f), (g), and (h).]

(a)

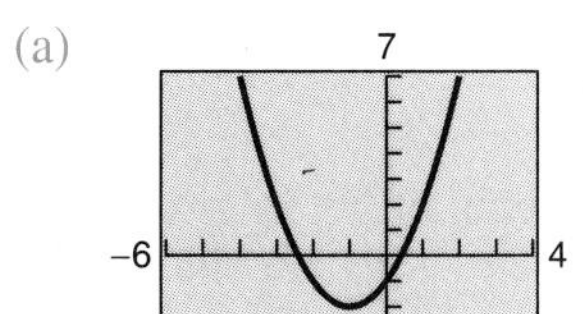

(b)

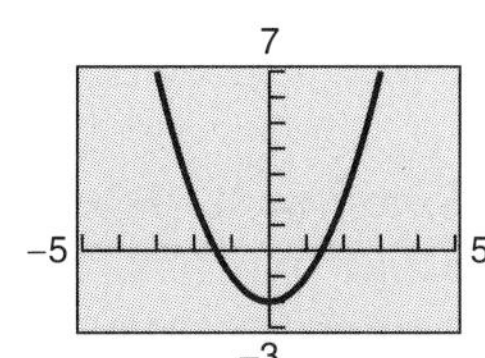

(c)

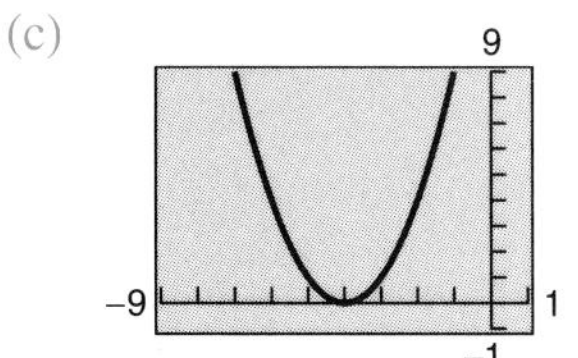

(d)

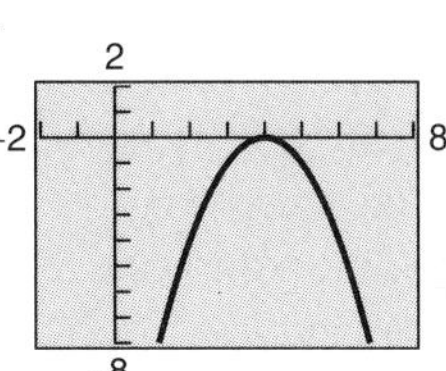

(e)

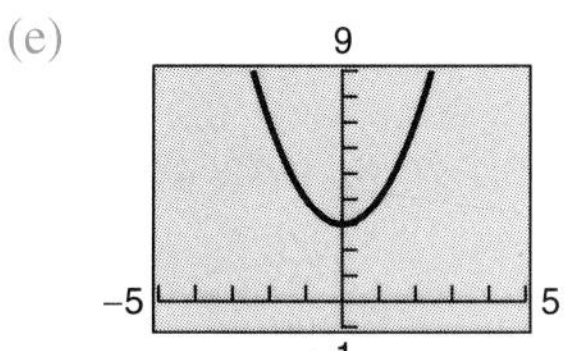

(f)

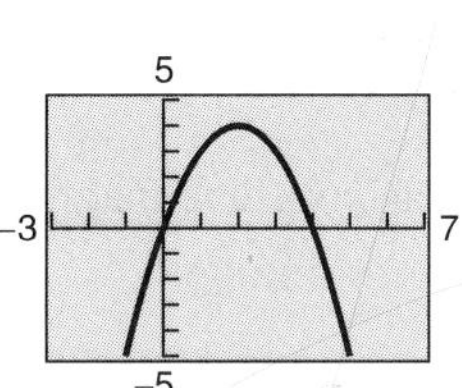

(g)

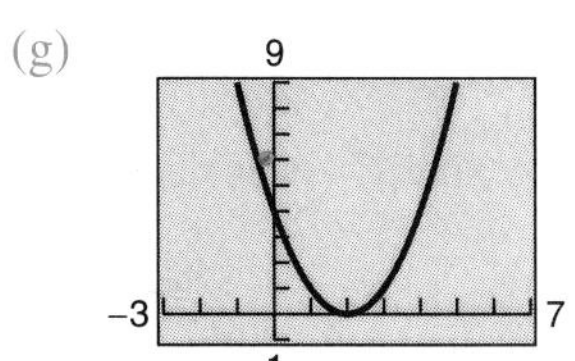

(h)

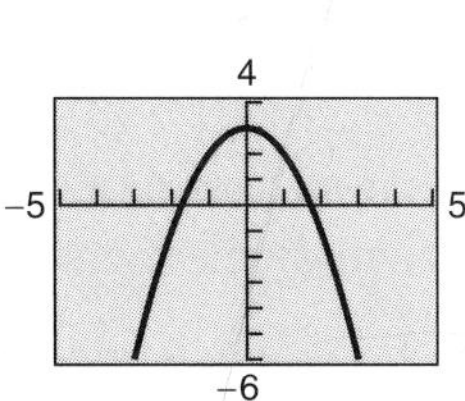

1. $f(x) = (x - 2)^2$

2. $f(x) = (x + 4)^2$

3. $f(x) = x^2 - 2$

4. $f(x) = 3 - x^2$

5. $f(x) = 4 - (x - 2)^2$

6. $f(x) = (x + 1)^2 - 2$

7. $f(x) = x^2 + 3$

8. $f(x) = -(x - 4)^2$

The *Interactive* CD-ROM contains step-by-step solutions to all odd-numbered Section and Review Exercises. It also provides Tutorial Exercises, which link to Guided Examples for additional help.

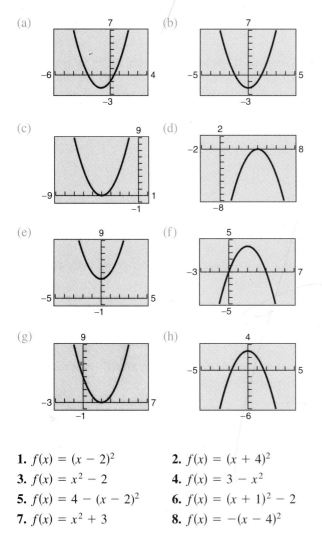

Exploration In Exercises 9–12, use a graphing utility to graph each equation. Describe how each differs from the graph of $y = x^2$.

9. (a) $y = \frac{1}{2}x^2$ (b) $y = -\frac{1}{8}x^2$
(c) $y = \frac{3}{2}x^2$ (d) $y = -3x^2$

10. (a) $y = x^2 + 1$ (b) $y = x^2 - 1$
(c) $y = x^2 + 3$ (d) $y = x^2 - 3$

11. (a) $y = (x - 1)^2$ (b) $y = (x + 1)^2$
(c) $y = (x - 3)^2$ (d) $y = (x + 3)^2$

12. (a) $y = -\frac{1}{2}(x - 2)^2 + 1$ (b) $y = \frac{1}{2}(x - 2)^2 + 1$

In Exercises 13–22, sketch the graph of the function. Identify the vertex, intercepts, and zeros of the function. Use a graphing utility to verify your results.

13. $f(x) = 16 - x^2$

14. $f(x) = \frac{1}{2}x^2 - 4$

15. $h(x) = x^2 - 8x + 16$

16. $g(x) = x^2 + 2x + 1$

17. $f(x) = x^2 - x + \frac{5}{4}$

18. $f(x) = x^2 + 3x + \frac{1}{4}$

19. $f(x) = -x^2 + 2x + 5$

20. $f(x) = -x^2 - 4x + 1$

21. $h(x) = 4x^2 - 4x + 21$

22. $f(x) = 2x^2 - x + 1$

In Exercises 23–26, use a graphing utility to graph the quadratic function. Identify the vertex, intercepts, and zeros of the function. Then check your results algebraically by completing the square.

23. $f(x) = -(x^2 + 2x - 3)$

24. $g(x) = x^2 + 8x + 11$

25. $f(x) = 2x^2 - 16x + 31$

26. $g(x) = \frac{1}{2}(x^2 + 4x - 2)$

In Exercises 27–30, find an equation for the parabola. Use a graphing utility to graph the equation and verify your result.

27.

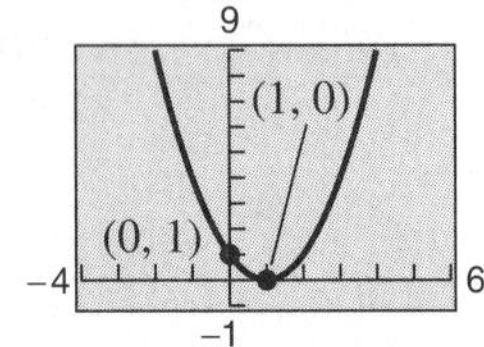

28.

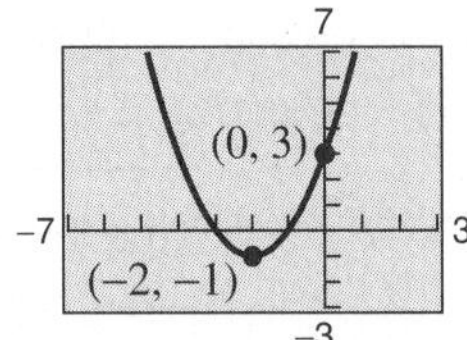

29.

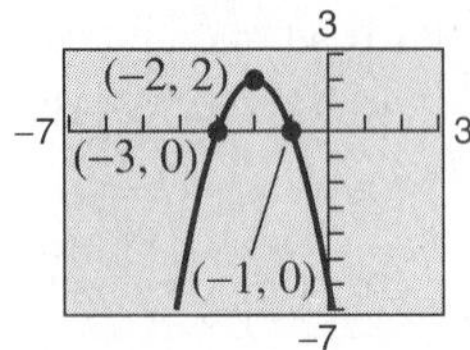

30.

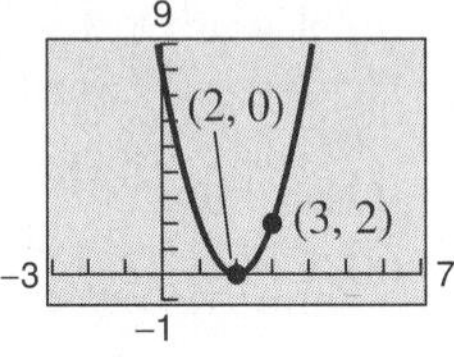

In Exercises 31–36, find the quadratic function that has the indicated vertex and whose graph passes through the given point. Confirm your result with a graphing utility.

31. Vertex: $(-2, 5)$; Point: $(0, 9)$

32. Vertex: $(4, -1)$; Point: $(2, 3)$

33. Vertex: $(3, 4)$; Point: $(1, 2)$

34. Vertex: $(2, 3)$; Point: $(0, 2)$

35. Vertex: $(5, 12)$; Point: $(7, 15)$

36. Vertex: $(-2, -2)$; Point: $(-1, 0)$

Graphical Reasoning **In Exercises 37–40, determine the x-intercepts of the graph visually. How do the x-intercepts correspond to the solutions of the quadratic equation when $y = 0$?**

37.

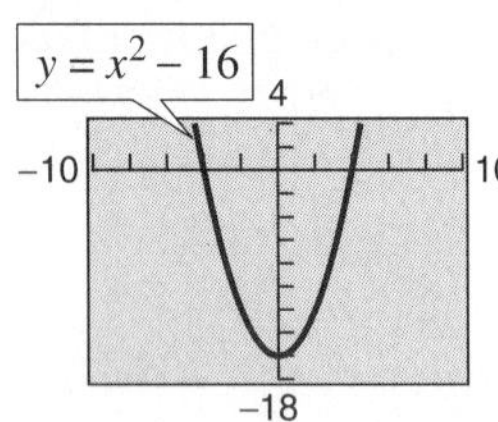

38.

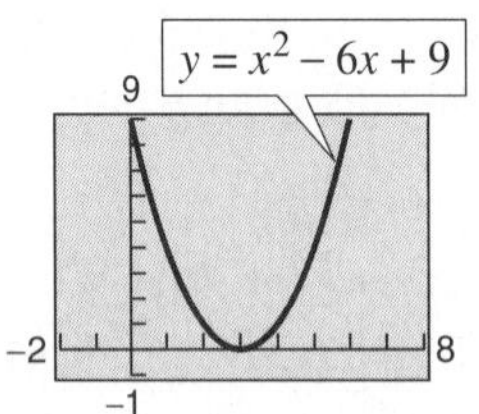

39.

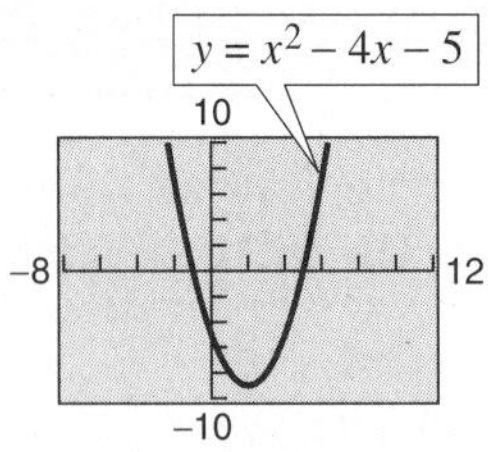

40.

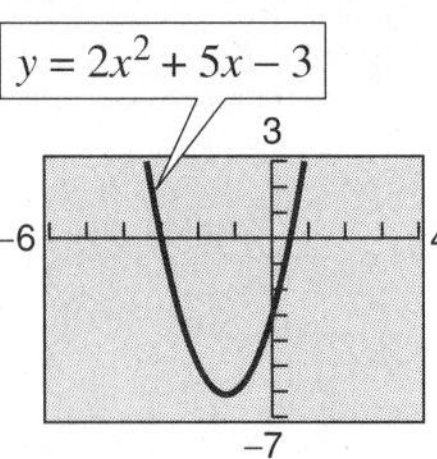

In Exercises 41–44, use a graphing utility to graph the quadratic function. Find the x-intercepts of the graph and compare them with the solutions of the corresponding quadratic equation when $y = 0$.

41. $y = x^2 - 4x$

42. $y = x^2 - 9x + 18$

43. $y = 2x^2 - 7x - 30$

44. $y = -\frac{1}{2}(x^2 - 6x - 7)$

In Exercises 45–48, find two quadratic functions whose graphs—one opening upward and the other downward—have the given x-intercepts. (The answers are not unique.)

45. $(-1, 0), (3, 0)$

46. $(4, 0), (8, 0)$

47. $(-3, 0), \left(-\frac{1}{2}, 0\right)$

48. $\left(-\frac{5}{2}, 0\right), (2, 0)$

In Exercises 49 and 50, find two positive real numbers that satisfy the requirements.

49. The sum is 110 and the product is a maximum.

50. The sum is S and the product is a maximum.

Maximum Area **In Exercises 51 and 52, consider a rectangle of length x and perimeter P.**

(a) Express the area A as a function of x and determine the domain of the function.

(b) Use a graphing utility to graph the area function.

(c) Use the graph to approximate the length and width of the rectangle of maximum area, and verify algebraically.

51. $P = 100$ feet

52. $P = 36$ meters

53. *Maximum Revenue* Find the number of units that produces a maximum revenue given by $R = 900x - 0.1x^2$, where R is the total revenue in dollars and x is the number of units sold.

54. *Maximum Profit* Let x be the amount (in hundreds of dollars) a company spends on advertising, and let P be the profit, where $P = 230 + 20x - 0.5x^2$. What expenditure for advertising results in the maximum profit?

55. *Numerical, Graphical, and Analytical Analysis* A rancher has 200 feet of fencing to enclose two adjacent rectangular corrals (see figure). Use the following methods to determine the dimensions that will produce a maximum enclosed area.

(a) Complete six rows of a table such as the one below. (The first two rows are shown.)

x	y	Area
2	$\frac{1}{3}[200 - 4(2)]$	$2xy = 256$
4	$\frac{1}{3}[200 - 4(4)]$	$2xy \approx 491$

(b) Use a graphing utility to generate additional rows of the table in part (a). Use the table to estimate the dimensions that will produce the maximum enclosed area.

(c) Write the area A as a function of x.

(d) Use a graphing utility to graph the area function. Use the graph to approximate the dimensions that will produce the maximum enclosed area.

(e) Write the area function in standard form to find analytically the dimensions that will produce the maximum area.

(f) Compare your results from parts (b), (d), and (e).

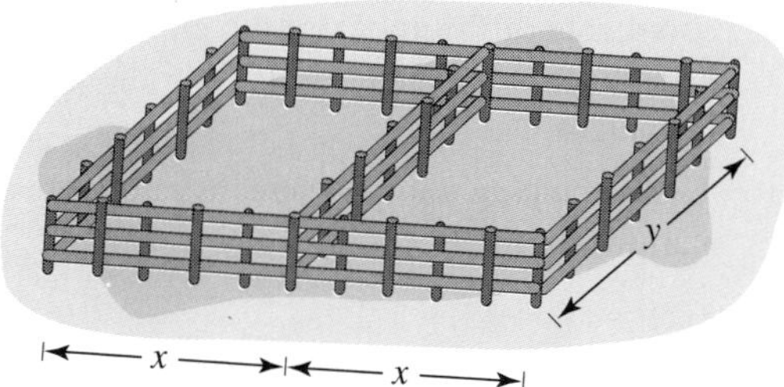

56. *Maximum Area* An indoor physical fitness room consists of a rectangular region with a semicircle on each end. The perimeter of the room is to be a 200-meter running track.

(a) Draw a figure that visually represents the problem. Let x and y represent the length and width of the rectangular region.

(b) Determine the radius of the semicircular ends of the track. Determine the distance, in terms of y, around the two semicircular parts of the track.

(c) Use the result of part (b) to write an equation, in terms of x and y, for the distance traveled in one lap around the track. Solve for y.

(d) Use the result of part (c) to write the area A of the rectangular region as a function of x.

(e) Use a graphing utility to graph the area function of part (d). Use the graph to approximate the dimensions that will produce a maximum area of the rectangle.

57. *Trajectory of a Ball* The height y (in feet) of a ball thrown by a child is given by

$$y = -\frac{1}{12}x^2 + 2x + 4$$

where x is the horizontal distance (in feet) from where the ball is thrown (see figure).

(a) Use a graphing utility to graph the path of the ball.

(b) How high is the ball when it leaves the child's hand? (*Note:* Find y when $x = 0$.)

(c) How high is the ball when it is at its maximum height?

(d) How far from the child does the ball strike the ground?

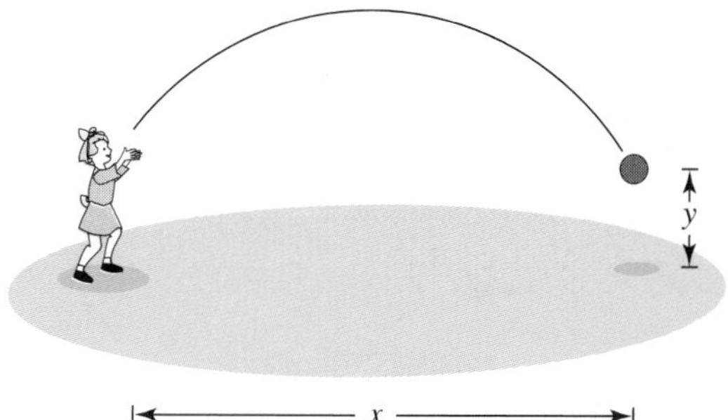

58. *Maximum Height of a Dive* The path of a diver is given by

$$y = -\frac{4}{9}x^2 + \frac{24}{9}x + 12$$

where y is the height in feet and x is the horizontal distance from the end of the diving board in feet (see figure). What is the maximum height of the dive?

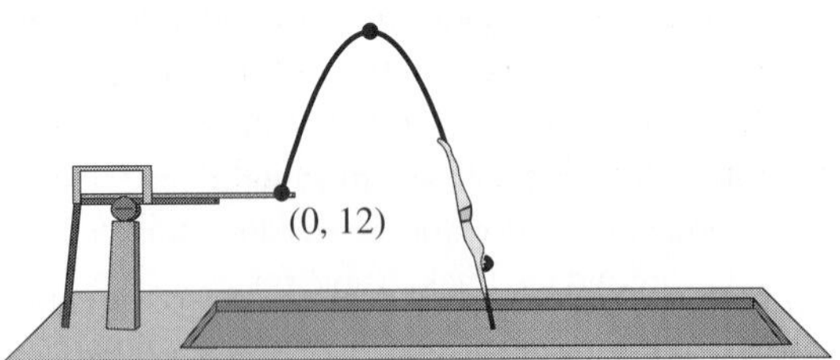

59. *Forestry* The number of board feet in a 16-foot log is approximated by the model

$$V = 0.77x^2 - 1.32x - 9.31, \qquad 5 \le x \le 40$$

where V is the number of board feet and x is the diameter (in inches) of the log at the small end. (One board foot is a measure of volume equivalent to a board that is 12 inches wide, 12 inches long, and 1 inch thick.)

(a) Use a graphing utility to graph the function.

(b) Estimate the number of board feet in a 16-foot log with a diameter of 16 inches.

(c) Estimate the diameter of a 16-foot log that scaled 500 board feet when the lumber was sold.

60. *Automobile Aerodynamics* The number of horsepower y required to overcome wind drag on a certain automobile is approximated by

$$y = 0.002s^2 + 0.005s - 0.029, \qquad 0 \le s \le 100$$

where s is the speed of the car in miles per hour.

(a) Use a graphing utility to graph the function.

(b) Estimate the maximum speed of the car if the power required to overcome wind drag is not to exceed 10 horsepower.

61. *Graphical Analysis* From 1950 to 1990, the average annual per capita consumption C of cigarettes by Americans (18 and older) can be modeled by

$$C = 4024.5 + 51.4t - 3.1t^2, \qquad -10 \le t \le 30$$

where t is the year, with $t = 0$ corresponding to 1960. (Source: U.S. Center for Disease Control)

(a) Use a graphing utility to graph the model.

(b) Use the graph of the model to approximate the maximum average annual consumption. Beginning in 1966, all cigarette packages were required by law to carry a health warning. Do you think the warning had any effect? Explain.

(c) In 1960, the U.S. population (18 and over) was 116,530,000. Of those, about 48,500,000 were smokers. What was the average annual cigarette consumption *per smoker* in 1960? What was the average daily cigarette consumption *per smoker?*

62. *Data Analysis* The number y (in millions) of VCRs in use in the United States for the years 1984 through 1993 are given below in the form (t, y). The variable t represents time in years, where $t = 4$ represents 1984. (Source: Television Bureau of Advertising, Inc.)

$(4, 9), (5, 18), (6, 31), (7, 43), (8, 51),$
$(9, 58), (10, 63), (11, 67), (12, 69), (13, 72)$

(a) Use the regression capabilities of a graphing utility to fit a quadratic model to the data.

(b) Use a graphing utility to graph the model and the data in the same viewing rectangle.

(c) Do you think the model can be used to predict VCR utilization in the year 2000? Explain.

63. (a) Assume that the function $f(x) = ax^2 + bx + c$ $(a \ne 0)$ has two real zeros. Show that the x-coordinate of the vertex of the graph is the average of the zeros of f. (*Hint:* Use the Quadratic Formula.)

(b) Use a graphing utility to demonstrate the result of part (a) for the function

$$f(x) = \frac{1}{2}(x - 3)^2 - 2.$$

3.2 Polynomial Functions of Higher Degree

Graphs of Polynomial Functions / The Leading Coefficient Test / Zeros of Polynomial Functions / The Intermediate Value Theorem

Graphs of Polynomial Functions

You should be able to sketch accurate graphs of polynomial functions of degrees 0, 1, and 2. The graphs of polynomial functions of degree greater than 2 are more difficult to sketch by hand. However, in this section you will learn how to recognize some of the basic features of the graphs of polynomial functions.

The graph of a polynomial function is **continuous.** Essentially, this means that the graph of a polynomial function has no breaks, as shown in Figure 3.9(a). Another feature of the graph of a polynomial function is that it has only smooth, rounded turns, as shown in Figure 3.10(a). It cannot have a sharp, pointed turn such as the one shown in Figure 3.10(b).

EXPLORATION

Use a graphing utility to graph $y = x^n$ with $n = 2, 4$, and 8. (Use $-1.5 \le x \le 1.5$ and $-1 \le y \le 6$.) Compare the graphs. In the interval $(-1, 1)$, which graph is on the bottom? Outside the interval $(-1, 1)$, which graph is on the bottom?

Use a graphing utility to graph $y = x^n$ with $n = 3, 5$, and 7. (Use $-1.5 \le x \le 1.5$ and $-4 \le y \le 4$.) Compare the graphs. In the interval $(-1, 1)$, which graph is on the bottom? Outside the interval $(-1, 1)$, which graph is on the bottom?

The *Interactive* CD-ROM offers graphing utility emulators of the *TI-82* and *TI-83*, which can be used with the Examples, Explorations, Technology notes, and Exercises.

Figure 3.9

(a) Continuous

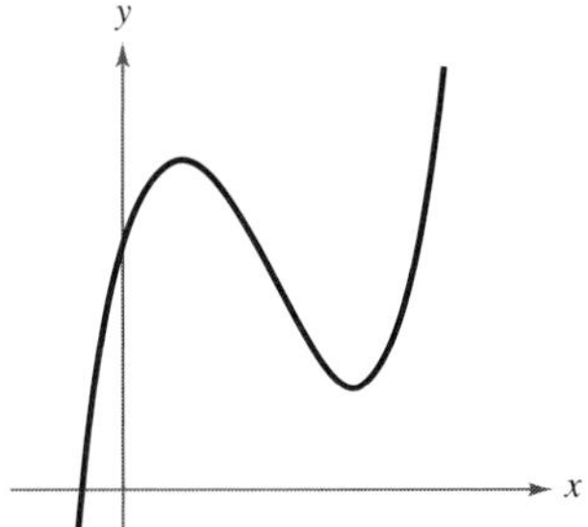

(b) Discontinuous

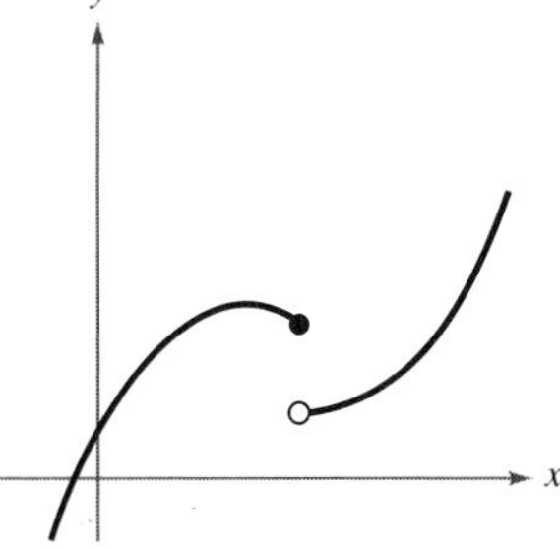

Figure 3.10

(a) Polynomial functions have smooth, rounded graphs.

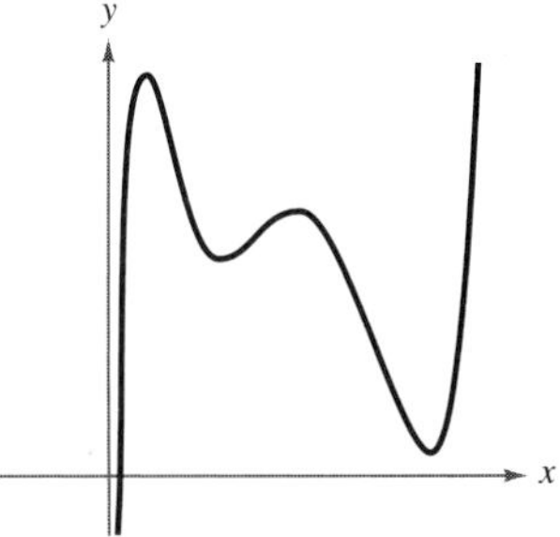

(b) Graphs of polynomial functions cannot have sharp turns.

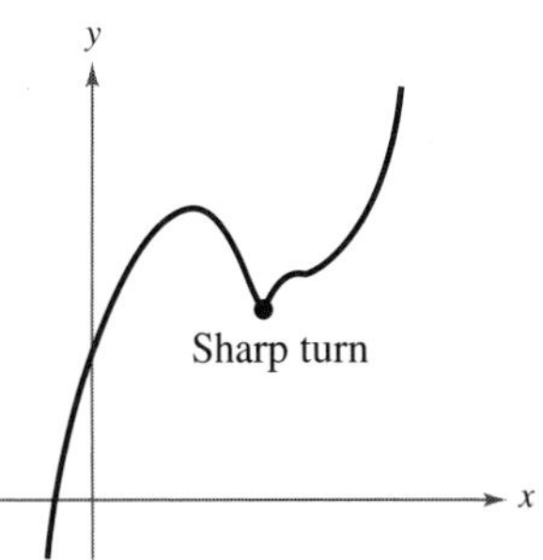

The polynomial functions that have the simplest graphs are monomials of the form $f(x) = x^n$, where n is an integer greater than zero.

Note In Figure 3.11, you can see that when n is *even* the graph is similar to the graph of $f(x) = x^2$, and when n is *odd* the graph is similar to the graph of $f(x) = x^3$. Moreover, the greater the value of n, the flatter the graph is on the interval $[-1, 1]$.

Figure 3.11

If n is even, the graph of $y = x^n$ *touches* axis at x-intercept.

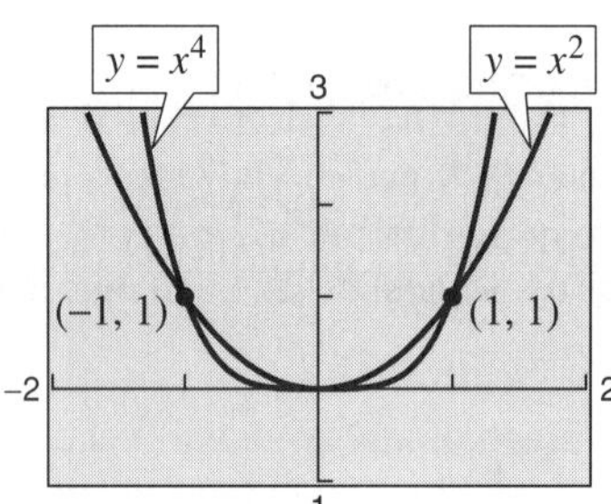

If n is odd, the graph of $y = x^n$ *crosses* axis at x-intercept.

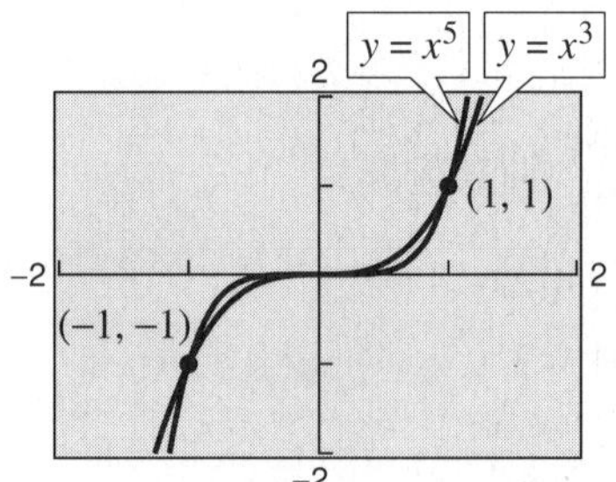

EXAMPLE 1 Sketching Transformations of Monomial Functions

Sketch the graph of each polynomial function.

a. $f(x) = -x^5$ **b.** $g(x) = x^4 + 1$ **c.** $h(x) = (x + 1)^4$

Solution

a. Because the degree of f is odd, the graph is similar to the graph of $y = x^3$. Moreover, the negative coefficient reflects the graph in the x-axis, as shown in Figure 3.12(a).

b. The graph of g is an upward shift, by one unit, of the graph of $y = x^4$, as shown in Figure 3.12(b).

c. The graph of h is a left shift, by one unit, of the graph of $y = x^4$, as shown in Figure 3.12(c).

Figure 3.12

(a)

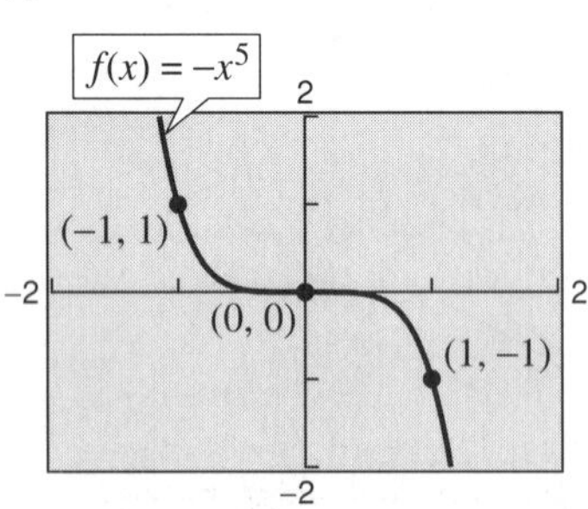

(b)

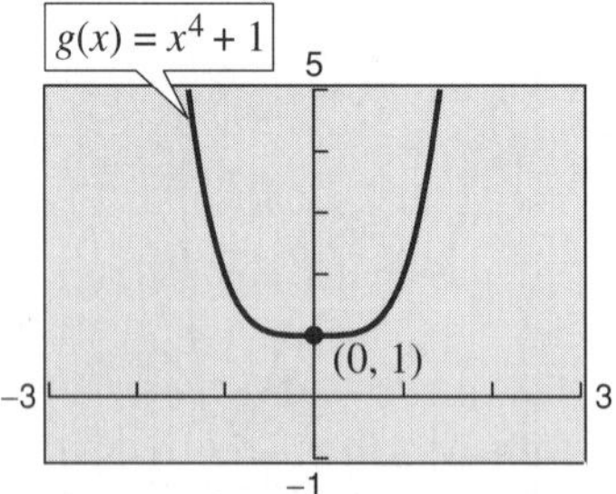

(c)

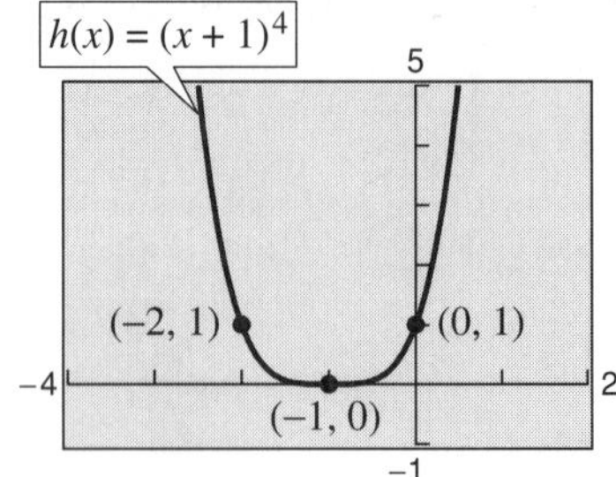

The Leading Coefficient Test

In Example 1, note that all three graphs eventually rise or fall without bound as x moves to the right. Whether the graph of a polynomial eventually rises or falls can be determined by the function's degree (even or odd) and by its leading coefficient, as indicated in the **Leading Coefficient Test.**

EXPLORATION

Use a graphing utility to investigate the behavior of the graph of

$$y = x^3 - 105x^2 + 21.$$

First use a viewing rectangle in which $-2 \le x \le 2$ and $-10 \le y \le 30$. How complete a view of the graph does this viewing rectangle show? Does the graph move down as x increases indefinitely? Find a viewing rectangle that gives a good view of the basic characteristics of the graph.

Library of Functions

The graphs of polynomials of degree 1 are lines, and those of degree 2 are parabolas. The graphs of polynomials of higher degree are smooth and continuous. The graphs eventually rise or fall without bound as x moves to the right (or left).

Note The dashed portions of the graphs indicate that the test determines *only* the right and left behavior of the graph.

Leading Coefficient Test

As x moves without bound to the left or to the right, the graph of the polynomial function $f(x) = a_nx^n + \cdot \cdot \cdot + a_1x + a_0$ eventually rises or falls in the following manner.

1. When n is *odd:*

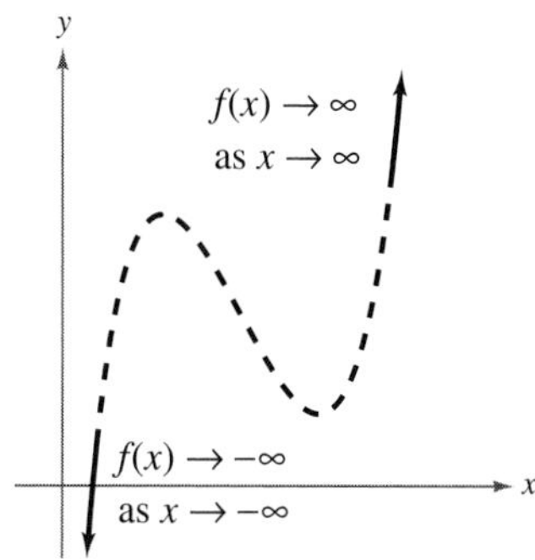

If the leading coefficient is positive ($a_n > 0$), the graph falls to the left and rises to the right.

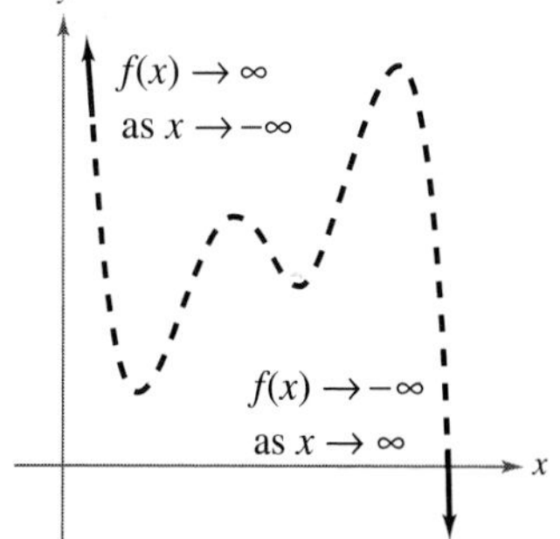

If the leading coefficient is negative ($a_n < 0$), the graph rises to the left and falls to the right.

2. When n is *even:*

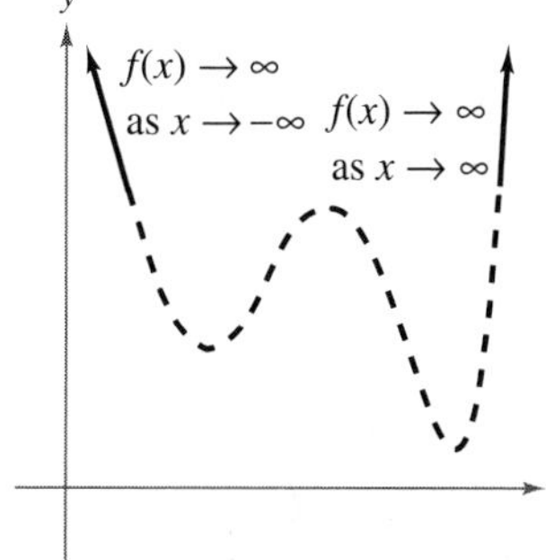

If the leading coefficient is positive ($a_n > 0$), the graph rises to the left and right.

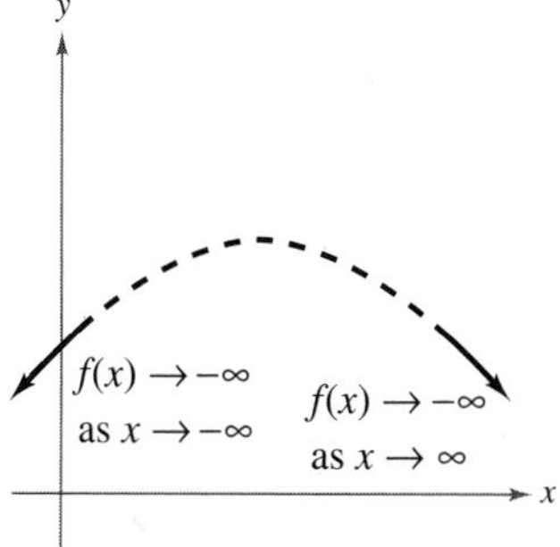

If the leading coefficient is negative ($a_n < 0$), the graph falls to the left and right.

EXAMPLE 2 Applying the Leading Coefficient Test

Use the Leading Coefficient Test to determine the right and left behavior of the graph of each polynomial function.

a. $f(x) = -x^3 + 4x$ **b.** $f(x) = x^4 - 5x^2 + 4$ **c.** $f(x) = x^5 - x$

Solution

a. Because the degree is odd and the leading coefficient is negative, the graph rises to the left and falls to the right, as shown in Figure 3.13(a).

b. Because the degree is even and the leading coefficient is positive, the graph rises to the left and right, as shown in Figure 3.13(b).

c. Because the degree is odd and the leading coefficient is positive, the graph falls to the left and rises to the right, as shown in Figure 3.13(c).

Figure 3.13

(a)

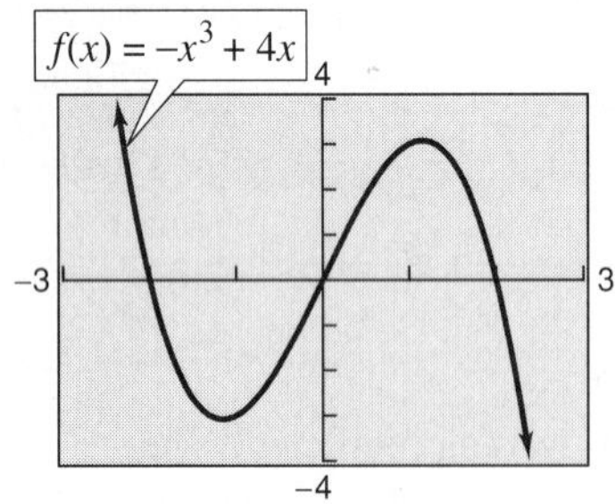

(b)

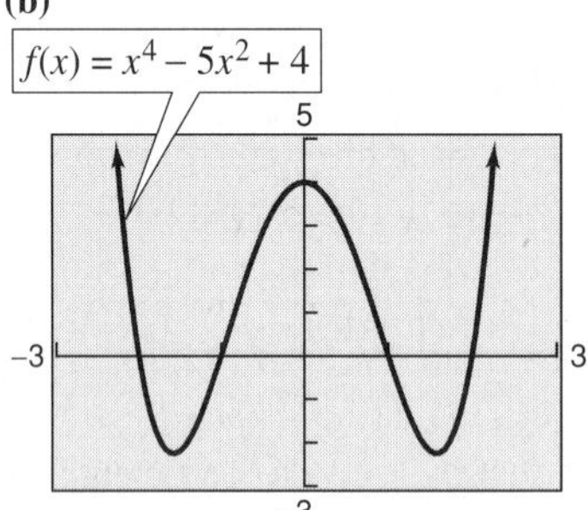

(c)

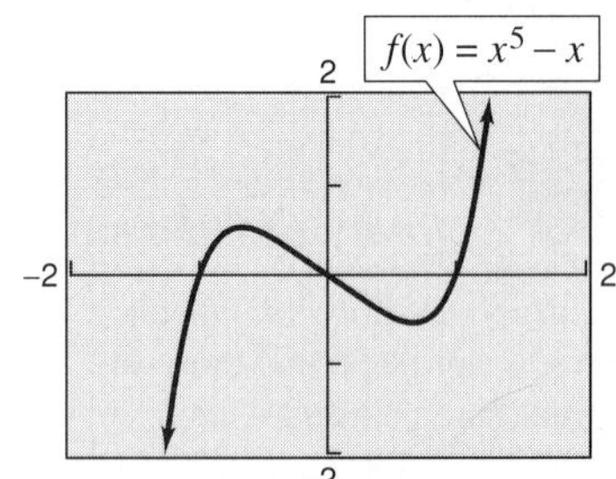

EXPLORATION

For each of the graphs in Figure 3.13, count the number of zeros of the polynomial function and the number of relative extrema, and compare these numbers with the degree of the polynomial. What do you observe?

Zeros of Polynomial Functions

It can be shown that for a polynomial function f of degree n, the following statements are true.

1. The graph of f has at most n real zeros. (This result is discussed in detail in Section 3.4.)
2. The function f has at most $n - 1$ relative **extrema** (local minimums or maximums).

Recall that a **zero** of a function f is a number x for which $f(x) = 0$. Finding the zeros of polynomial functions is one of the most important problems in algebra. You have already seen that there is a strong interplay between graphical and algebraic approaches to this problem. Sometimes you can use information about the graph of a function to help find its zeros. In other cases you can use information about the zeros of a function to find a good viewing rectangle.

Some graphing utilities, such as the *TI-82* and *TI-83*, have two features that analyze the graph of a function: (1) the *root* feature of the *TI-82* or the zero feature of the *TI-83* for finding zeros of a function, and (2) the *minimum* and *maximum* features for finding the relative extrema. If your graphing utility has these features, use them to confirm the results in Examples 3 and 4.

Real Zeros of Polynomial Functions

If f is a polynomial function and a is a real number, the following statements are equivalent.

1. $x = a$ is a *zero* of the function f.
2. $x = a$ is a *solution* of the polynomial equation $f(x) = 0$.
3. $(x - a)$ is a *factor* of the polynomial $f(x)$.
4. $(a, 0)$ is an *x-intercept* of the graph of f.

Finding zeros of polynomial functions is closely related to factoring and finding x-intercepts, as demonstrated in Examples 3, 4, and 5.

Figure 3.14

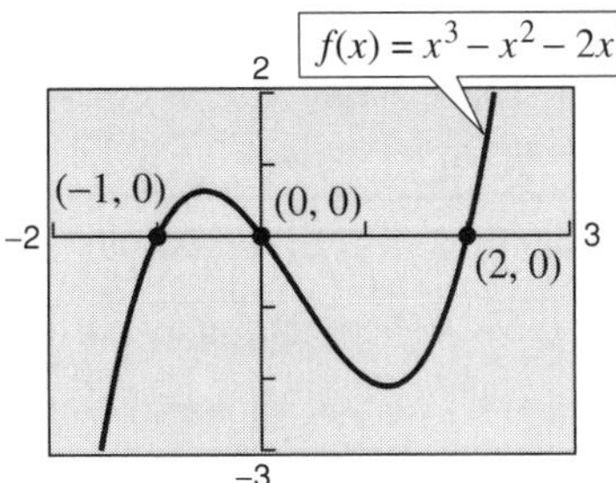

EXAMPLE 3 Finding Zeros of a Polynomial Function

Find all real zeros of $f(x) = x^3 - x^2 - 2x$.

Solution

$$\begin{aligned} f(x) &= x^3 - x^2 - 2x && \text{Original function} \\ &= x(x^2 - x - 2) && \text{Remove common monomial factor.} \\ &= x(x - 2)(x + 1) && \text{Factor completely.} \end{aligned}$$

Thus, the real zeros are $x = 0$, $x = 2$, and $x = -1$, and the corresponding x-intercepts are $(0, 0)$, $(2, 0)$, and $(-1, 0)$, as shown in Figure 3.14. Note in the figure that the graph has two relative extrema, which is consistent with the fact that a third-degree polynomial can have *at most* two relative extrema.

Figure 3.15

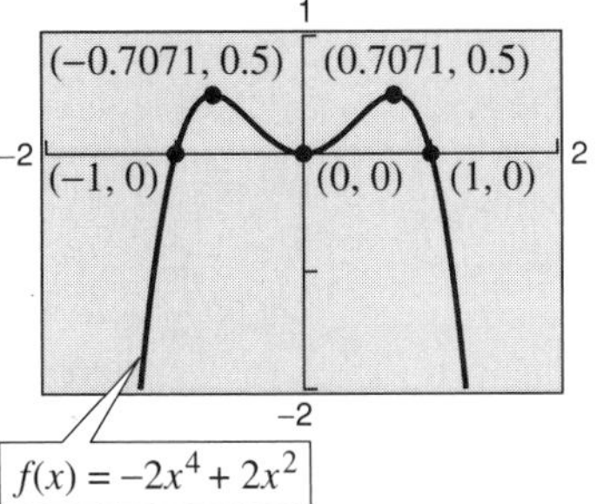

EXAMPLE 4 Analyzing a Polynomial Function

Find all real zeros and relative extrema of $f(x) = -2x^4 + 2x^2$.

Solution

$$\begin{aligned} f(x) &= -2x^4 + 2x^2 && \text{Original function} \\ &= -2x^2(x^2 - 1) && \text{Remove common monomial factor.} \\ &= -2x^2(x - 1)(x + 1) && \text{Factor completely.} \end{aligned}$$

Thus, the real zeros are $x = 0$, $x = 1$, and $x = -1$, and the corresponding x-intercepts are $(0, 0)$, $(1, 0)$, and $(-1, 0)$, as shown in Figure 3.15. Using the minimum and maximum features of a graphing utility, you can determine the three relative extrema to be $(-0.7071, 0.5)$, $(0, 0)$, and $(0.7071, 0.5)$.

In Example 4, the real zero arising from $-2x^2 = 0$ is a **repeated zero.** In general, a factor of $(x - a)^k$ yields a repeated zero $x = a$ of **multiplicity** k. If k is odd, the graph *crosses* the x-axis at $x = a$. If k is even, the graph *touches* (but does not cross) the x-axis at $x = a$, as shown in Figure 3.15.

EXAMPLE 5 **Finding Zeros of a Polynomial Function**

Find all real zeros of $f(x) = x^5 - 3x^3 - x^2 - 4x - 1$.

Solution

Use a graphing utility to obtain the graph shown in Figure 3.16. From the graph, you can see that there are three zeros. Using the root feature, you can determine that the zeros are approximately $x = -1.861$, $x = -0.254$, and $x = 2.115$. It should be noted that this fifth-degree polynomial factors as

$$\begin{aligned} f(x) &= x^5 - 3x^3 - x^2 - 4x - 1 \\ &= (x^2 + 1)(x^3 - 4x - 1). \end{aligned}$$

The three zeros obtained above are the zeros of the cubic on the right (the quadratic $x^2 + 1$ has two complex zeros and, thus, no *real* zeros).

Figure 3.16

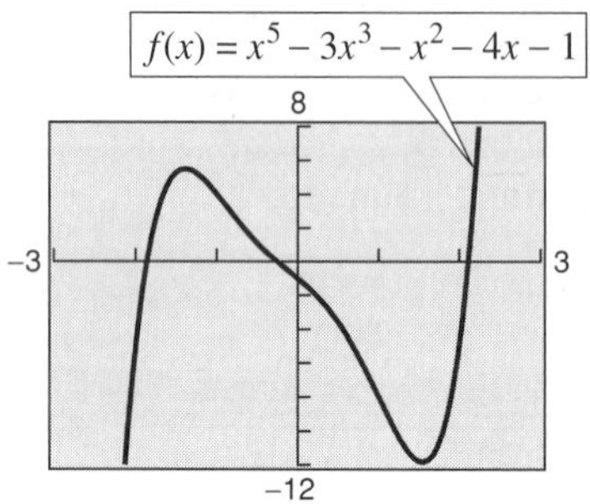

EXAMPLE 6 **Finding a Polynomial Function with Given Zeros**

Find polynomial functions with the following zeros. (There are many correct solutions.)

a. $-2, -1, 1, 2$ **b.** $-\frac{1}{2}, 3, 3$

Solution

a. For each of the given zeros, form a corresponding factor. For instance, the zero given by $x = -2$ corresponds to the factor $(x + 2)$. Thus, you can write the function as

$$\begin{aligned} f(x) &= (x + 2)(x + 1)(x - 1)(x - 2) \\ &= (x^2 - 4)(x^2 - 1) \\ &= x^4 - 5x^2 + 4. \end{aligned}$$

b. Note that the zero $x = -\frac{1}{2}$ corresponds to either $\left(x + \frac{1}{2}\right)$ or $(2x + 1)$. To avoid fractions, choose the second factor and write

$$\begin{aligned} f(x) &= (2x + 1)(x - 3)^2 \\ &= (2x + 1)(x^2 - 6x + 9) \\ &= 2x^3 - 11x^2 + 12x + 9. \end{aligned}$$

EXPLORATION

Use a graphing utility to graph

$y_1 = x + 2$

$y_2 = (x + 2)(x - 1)$.

Predict the shape of the curve $y = (x + 2)(x - 1)(x - 3)$, and verify your answer with a graphing utility.

EXAMPLE 7 **Sketching the Graph of a Polynomial Function**

Sketch the graph of $f(x) = 3x^4 - 4x^3$ by hand.

Solution

Because the leading coefficient is positive and the degree is even, you know that the graph eventually rises to the left and to the right, as shown in Figure 3.17(a). By factoring to obtain

$$f(x) = 3x^4 - 4x^3 = x^3(3x - 4),$$

you can see that the zeros of f are $x = 0$ and $x = \frac{4}{3}$ (both of odd multiplicity). Thus, the x-intercepts occur at $(0, 0)$ and $\left(\frac{4}{3}, 0\right)$. Finally, plot a few additional points, as indicated in the table, and obtain the graph shown in Figure 3.17(b).

x	-1	0.5	1	1.5
$f(x)$	7	-0.3125	-1	1.6875

Figure 3.17

(a)

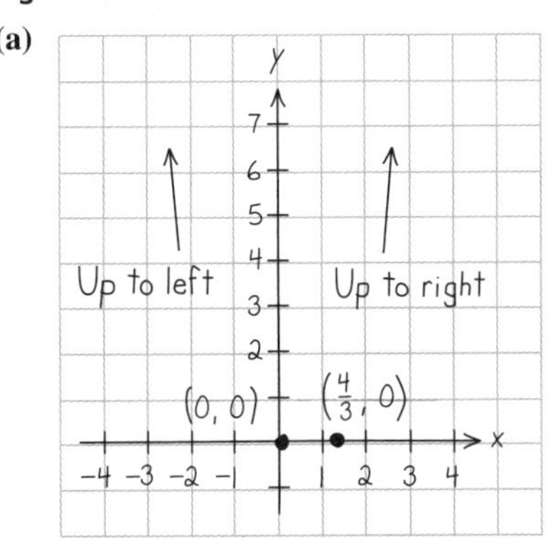

(b)

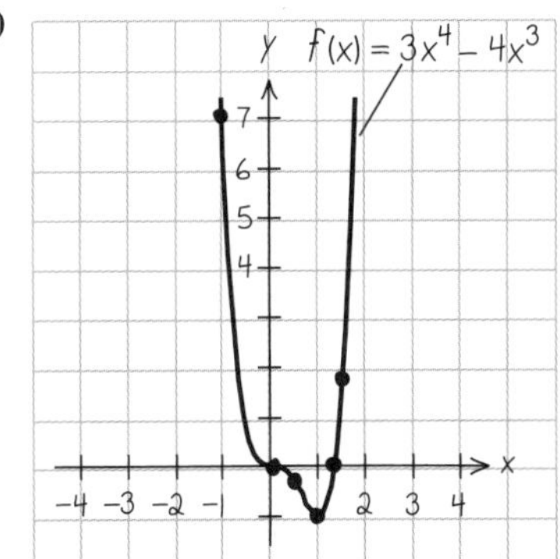

Study Tip

It is easy to make mistakes when entering functions into a graphing utility. Thus, it is important to have an understanding of the basic shapes of graphs, and to be able to graph simple polynomials *by hand*. For example, suppose you had entered the function in Example 7 as $f(x) = 3x^5 - 4x^3$. By looking at the graph, what mathematical principles would alert you to the fact that you had made a mistake?

EXPLORATION

Partner Activity Write the equation of a polynomial function that has a degree of 3, 4, or 5. Exchange equations with your partner and sketch, *by hand,* the graph of the equation that your partner wrote. When you are finished, use a graphing utility to check each other's work.

Figure 3.18

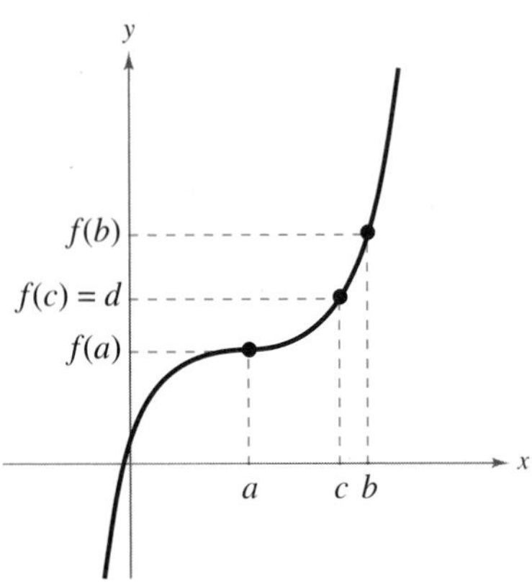

The Intermediate Value Theorem

The **Intermediate Value Theorem** concerns the existence of real zeros of polynomial functions. The theorem states that if $(a, f(a))$ and $(b, f(b))$ are two points on the graph of a polynomial function such that $f(a) \neq f(b)$, then for any number d between $f(a)$ and $f(b)$ there must be a number c between a and b such that $f(c) = d$. (See Figure 3.18.)

Intermediate Value Theorem

Let a and b be real numbers such that $a < b$. If f is a polynomial function such that $f(a) \neq f(b)$, then, in the interval $[a, b]$, f takes on every value between $f(a)$ and $f(b)$.

This theorem helps locate the real zeros of a polynomial function in the following way. If you can find a value $x = a$ where a polynomial function is positive, and another $x = b$ where it is negative, you can conclude that the function has at least one real zero between these two values. For example, the function $f(x) = x^3 + x^2 + 1$ is negative when $x = -2$ and positive when $x = -1$. Therefore, it follows from the Intermediate Value Theorem that f must have a real zero somewhere between -2 and -1, as shown in Figure 3.19.

Figure 3.19

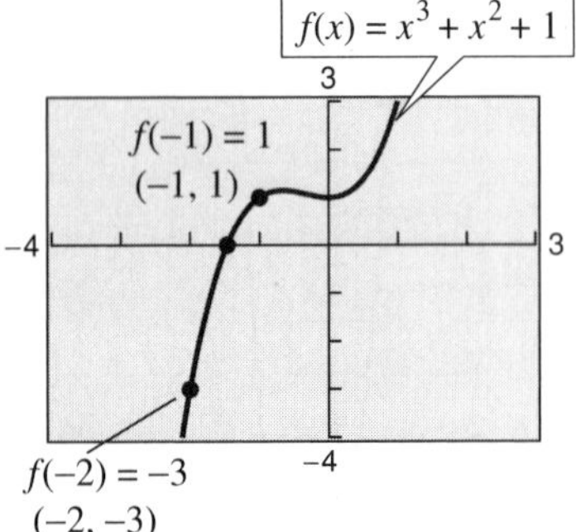

EXAMPLE 8 Approximating a Function's Zeros

Use the Intermediate Value Theorem to find three intervals of length 1 in which the polynomial $f(x) = 12x^3 - 32x^2 + 3x + 5$ is guaranteed to have a zero.

Solution

With a graphing utility, you can determine that there are three real zeros, as shown in Figure 3.20. The following table shows several function values.

Figure 3.20

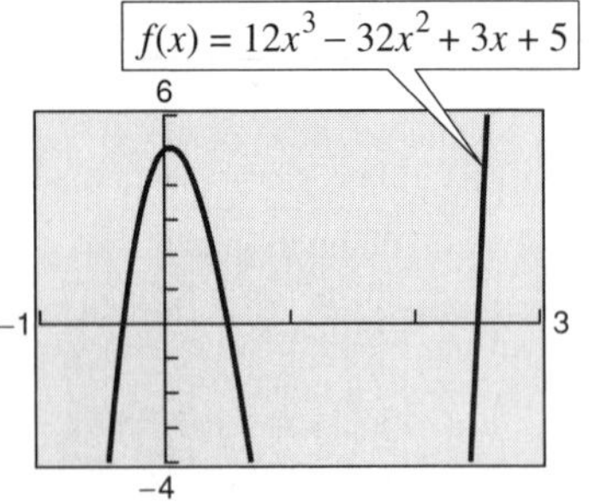

x	-1	0	1	2	3
$f(x)$	-42	5	-12	-21	50

Because $f(-1)$ is negative and $f(0)$ is positive, you can conclude from the Intermediate Value Theorem that the function has a zero between -1 and 0, as indicated in the figure. Similarly, the function has a zero between 0 and 1, and another between 2 and 3. You could obtain more accurate approximations of these zeros by using the root or zero feature of a graphing utility.

Group Activity

The Graphs of Cubic Polynomials

The graphs of cubic polynomials can be categorized according to the four basic shapes below. Match the graph of each function with one of the basic shapes and write a short paragraph describing how you reached your conclusion. Compare your conclusions with those of two classmates and resolve any differences.

1. $f(x) = -x^3$ **2.** $f(x) = -x^3 + 4x$ **3.** $f(x) = x^3$ **4.** $f(x) = x^3 - 4x$

(a)

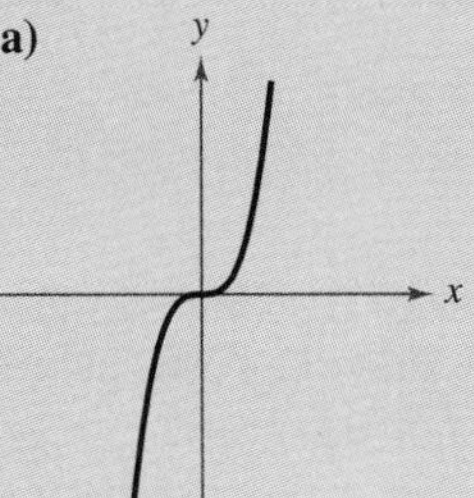

(b)

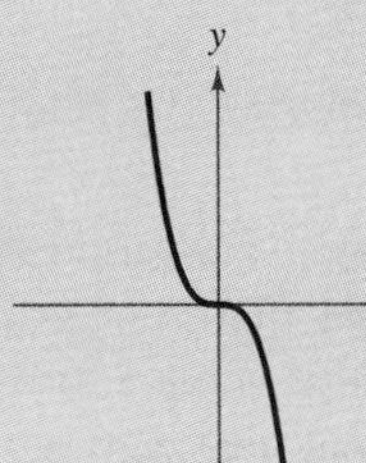

(c)

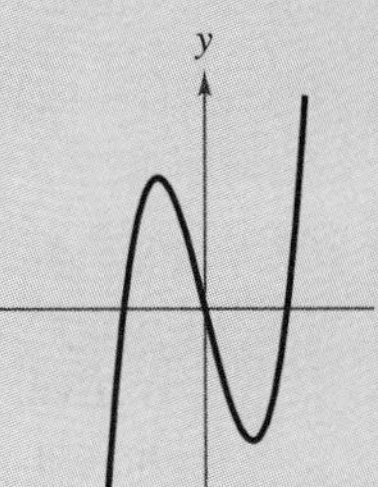

(d)

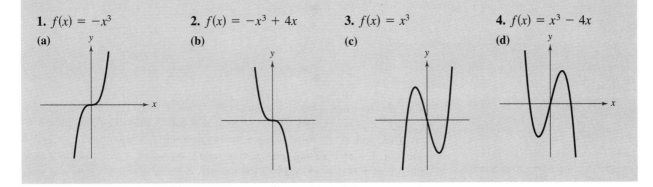

3.2 /// EXERCISES

In Exercises 1–8, match the polynomial function with its graph. [The graphs are labeled (a) through (h).]

1. $f(x) = -2x + 3$

2. $f(x) = x^2 - 4x$

3. $f(x) = -2x^2 - 5x$

4. $f(x) = 2x^3 - 3x + 1$

5. $f(x) = -\frac{1}{4}x^4 + 3x^2$

6. $f(x) = -\frac{1}{3}x^3 + x^2 - \frac{4}{3}$

7. $f(x) = x^4 + 2x^3$

8. $f(x) = \frac{1}{5}x^5 - 2x^3 + \frac{9}{5}x$

(a)

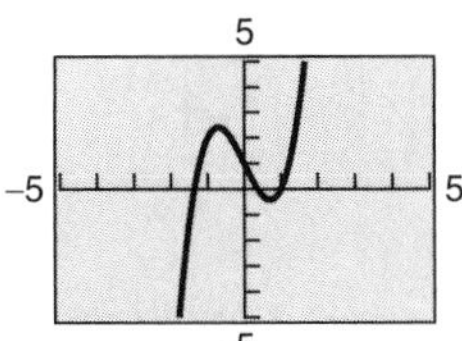

(b)

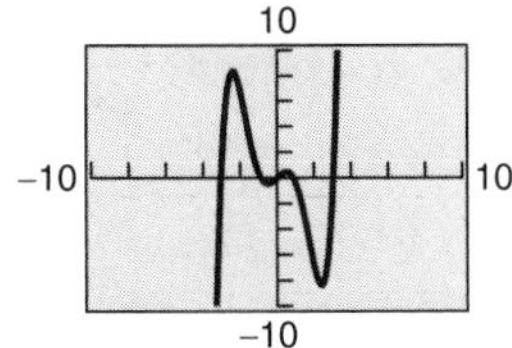

(e)

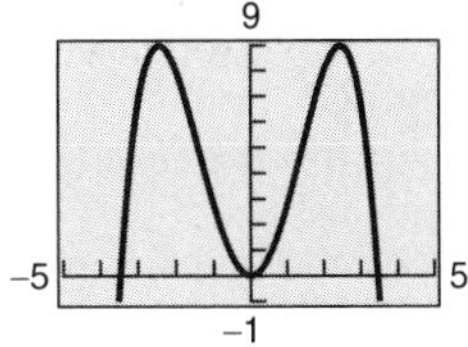

(f)

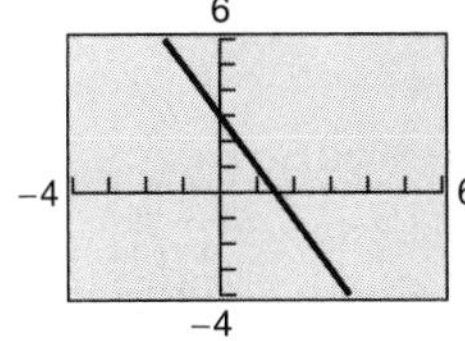

(c)

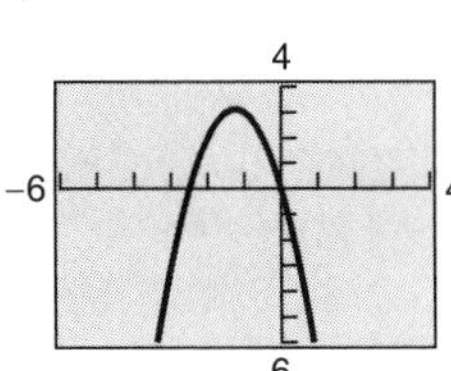

(d)

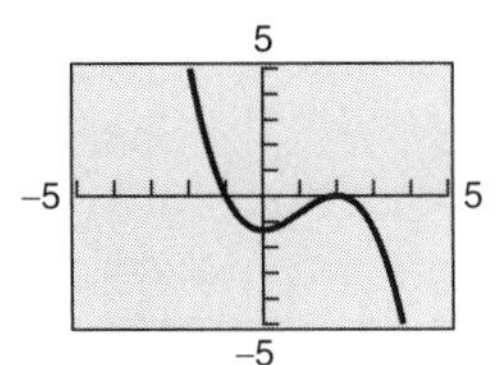

(g)

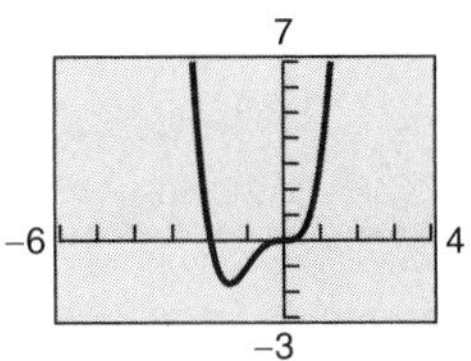

(h)

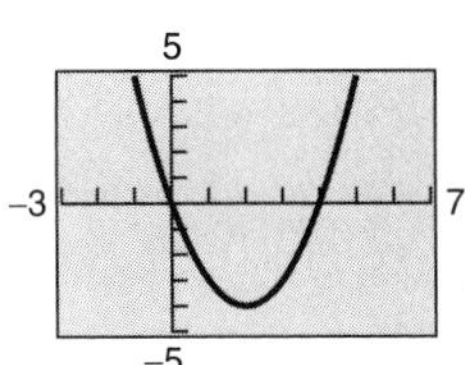

In Exercises 9–12, sketch the graphs of $y = x^n$ and the specified transformations.

9. $y = x^3$

(a) $f(x) = (x - 2)^3$ (b) $f(x) = x^3 - 2$

(c) $f(x) = -\frac{1}{2}x^3$ (d) $f(x) = (x - 2)^3 - 2$

10. $y = x^5$

(a) $f(x) = (x + 1)^5$ (b) $f(x) = x^5 + 1$

(c) $f(x) = 1 - \frac{1}{2}x^5$ (d) $f(x) = -\frac{1}{2}(x + 1)^5$

11. $y = x^4$

(a) $f(x) = (x + 3)^4$ (b) $f(x) = x^4 - 3$

(c) $f(x) = 4 - x^4$ (d) $f(x) = \frac{1}{2}(x - 1)^4$

12. $y = x^6$

(a) $f(x) = -\frac{1}{8}x^6$ (b) $f(x) = x^6 - 4$

(c) $f(x) = -\frac{1}{4}x^6 + 1$ (d) $f(x) = (x + 2)^6 - 4$

Graphical Analysis **In Exercises 13–16, use a graphing utility to graph the functions f and g in the same viewing rectangle. Zoom out sufficiently far to show that the right-hand and left-hand behavior of f and g appear identical.**

13. $f(x) = 3x^3 - 9x + 1,\quad g(x) = 3x^3$

14. $f(x) = -\frac{1}{3}(x^3 - 3x + 2),\quad g(x) = -\frac{1}{3}x^3$

15. $f(x) = -(x^4 - 4x^3 + 16x),\quad g(x) = -x^4$

16. $f(x) = 3x^4 - 6x^2,\quad g(x) = 3x^4$

In Exercises 17–26, use the Leading Coefficient Test to determine the right-hand and left-hand behavior of the graph of the polynomial function. Verify your result by using a graphing utility to graph the function.

17. $f(x) = 2x^2 - 3x + 1$

18. $f(x) = \frac{1}{3}x^3 + 5x$

19. $g(x) = 5 - \frac{7}{2}x - 3x^2$

20. $f(x) = -2.1x^5 + 4x^3 - 2$

21. $f(x) = 2x^5 - 5x + 7.5$

22. $h(x) = 1 - x^6$

23. $f(x) = 6 - 2x + 4x^2 - 5x^3$

24. $f(x) = \dfrac{3x^4 - 2x + 5}{4}$

25. $h(t) = -\frac{2}{3}(t^2 - 5t + 3)$

26. $f(s) = -\frac{7}{8}(s^3 + 5s^2 - 7s + 1)$

In Exercises 27–34, find all the real zeros of the polynomial function.

27. $f(x) = x^2 - 25$ **28.** $f(x) = 49 - x^2$

29. $h(t) = t^2 - 6t + 9$

30. $f(x) = x^2 + 10x + 25$

31. $f(x) = x^2 + x - 2$ **32.** $f(x) = \frac{1}{2}x^2 + \frac{5}{2}x - \frac{3}{2}$

33. $f(t) = t^3 - 4t^2 + 4t$

34. $f(x) = x^4 - x^3 - 20x^2$

Graphical Analysis **In Exercises 35–46, use a graphing utility to graph the function. Use the graph to approximate any x-intercepts of the graph. Set $y = 0$ and solve the resulting equation. Compare the result with any x-intercepts of the graph.**

35. $f(x) = 3x^2 - 12x + 3$

36. $g(x) = 5(x^2 - 2x - 1)$

37. $g(t) = \frac{1}{2}t^4 - \frac{1}{2}$

38. $f(x) = x^5 + x^3 - 6x$

39. $f(x) = 2x^4 - 2x^2 - 40$

40. $g(t) = t^5 - 6t^3 + 9t$

41. $f(x) = 5x^4 + 15x^2 + 10$

42. $f(x) = x^3 - 4x^2 - 25x + 100$

43. $y = 4x^3 - 20x^2 + 25x$

44. $y = 4x^3 + 4x^2 - 7x + 2$

45. $y = x^5 - 5x^3 + 4x$ **46.** $y = \frac{1}{4}x^3(x^2 - 9)$

In Exercises 47–56, find a polynomial function that has the given zeros. (There are many correct answers.)

47. 0, 10 **48.** 0, −3

49. 2, −6 **50.** −4, 5

51. 0, −2, −3 **52.** 0, 2, 5

53. 4, −3, 3, 0 **54.** −2, −1, 0, 1, 2

55. $1 + \sqrt{3}, 1 - \sqrt{3}$ **56.** $2, 4 + \sqrt{5}, 4 - \sqrt{5}$

In Exercises 57–60, (a) use the Intermediate Value Theorem and a graphing utility to find intervals of length 1 in which the polynomial function is guaranteed to have a zero. (b) Use the root feature of a graphing utility to approximate the zeros of the function.

57. $f(x) = x^3 - 3x^2 + 3$

58. $f(x) = 0.11x^3 - 2.07x^2 + 9.81x - 6.88$

59. $g(x) = 3x^4 + 4x^3 - 3$

60. $h(x) = x^4 - 10x^2 + 2$

In Exercises 61–64, use a graphing utility to graph the function. Describe a viewing rectangle that gives a good view of the basic characteristics of the graph.

61. $f(x) = -\frac{3}{2}$

62. $h(x) = \frac{1}{3}x - 3$

63. $f(t) = \frac{1}{4}(t^2 - 2t + 15)$

64. $g(x) = -x^2 + 10x - 16$

In Exercises 65–72, use a graphing utility to graph the function. Identify any symmetry with respect to the x-axis, y-axis, or origin. Determine the number of x-intercepts of the graph.

65. $f(x) = x^2(x - 4)$

66. $h(x) = \frac{1}{3}x^3(x - 4)^2$

67. $g(t) = -\frac{1}{4}(t - 2)^2(t + 2)^2$

68. $g(x) = \frac{1}{10}(x + 1)^2(x - 3)^3$

69. $f(x) = x^3 - 4x$

70. $f(x) = \frac{1}{4}x^4 - 2x^2$

71. $g(x) = \frac{1}{5}(x + 1)^2(x - 3)(2x - 9)$

72. $h(x) = \frac{1}{5}(x + 2)^2(3x - 5)^2$

73. *Graphical Reasoning* Use a graphing utility to graph the function $f(x) = x^4$. Explain how the graph of g differs (if it does) from the graph of f and confirm your result with a graphing utility. Determine whether g is odd, even, or neither.

(a) $g(x) = f(x) + 2$ (b) $g(x) = f(x + 2)$

(c) $g(x) = f(-x)$ (d) $g(x) = -f(x)$

(e) $g(x) = f\left(\frac{1}{2}x\right)$ (f) $g(x) = \frac{1}{2}f(x)$

(g) $g(x) = f\left(x^{3/4}\right)$ (h) $g(x) = (f \circ f)(x)$

74. *Exploration* Explore the transformations of the form $g(x) = a(x - h)^5 + k$.

(a) Use a graphing utility to graph the functions

$$y_1 = -\frac{1}{3}(x - 2)^5 + 1$$

and

$$y_2 = \frac{3}{5}(x + 2)^5 - 3.$$

Determine whether the graphs are increasing or decreasing. Explain.

(b) Will the graph of g always be increasing or decreasing? If so, is this behavior determined by a, h, or k? Explain.

(c) Use a graphing utility to graph the function

$$H(x) = x^5 - 3x^3 + 2x + 1.$$

Use the graph and the result of part (b) to determine whether H can be written in the form

$$H(x) = a(x - h)^5 + k.$$

Explain your reasoning.

(d) Determine the natural number exponents such that the results of part (b) are true.

75. *Volume of a Box* An open box with locking tabs is to be made from a square piece of material 24 inches on a side. This is done by cutting equal squares from the corners and folding along the dashed lines shown in the figure.

(a) Verify that the volume of the box is given by

$$V(x) = 8x(6 - x)(12 - x).$$

(b) Determine the domain of the function V.

(c) Sketch the graph of the function and estimate the value of x for which $V(x)$ is maximum.

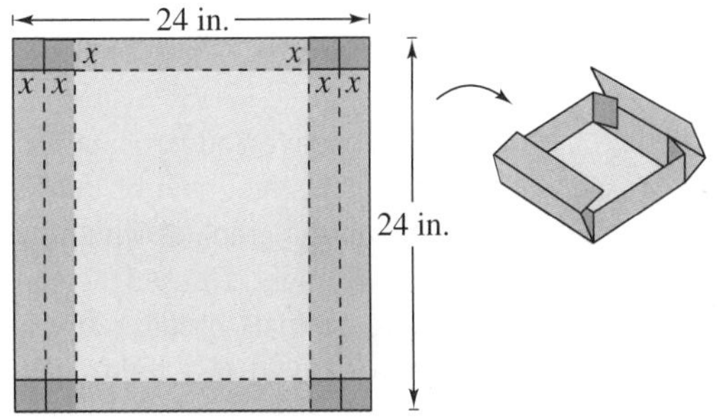

76. *Numerical and Graphical Analysis* An open box is to be made from a square piece of material 36 centimeters on a side by cutting equal squares from the corners and turning up the sides (see figure).

(a) Complete four rows of a table like the one below.

Height	Width	Volume
1	$36 - 2(1)$	$1[36 - 2(1)]^2 = 1156$
2	$36 - 2(2)$	$2[36 - 2(2)]^2 = 2048$

(b) Use a graphing utility to generate additional rows of the table. Use the table to estimate a range of dimensions within which the maximum volume is produced.

(c) Verify that the volume of the box is given by $V(x) = x(36 - 2x)^2$. Determine the domain of the function.

(d) Use a graphing utility to graph V, and use the range of dimensions from part (b) to find the x-value for which $V(x)$ is maximum.

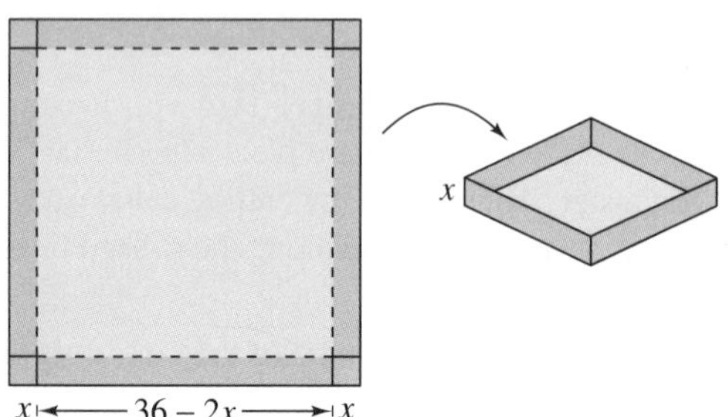

77. *Advertising Expenses* The total revenue R (in millions of dollars) for a company is related to its advertising expense by the function

$$R = \frac{1}{100{,}000}(-x^3 + 600x^2), \qquad 0 \le x \le 400$$

where x is the amount spent on advertising (in tens of thousands of dollars). Use the graph of this function to estimate the point on the graph at which the function is increasing most rapidly. This point is called the **point of diminishing returns** because any expense above this amount will yield less return per dollar invested in advertising. (*Hint:* Use a viewing rectangle in which $-200 \le x \le 600$ and $0 \le y \le 650$.)

78. *Data Analysis* The vertical deflection y of a 4-meter beam is measured in $\frac{1}{2}$-meter intervals x. The measurements are given by the following ordered pairs:

(0, 0), (0.5, 0.06), (1, 0.11), (1.5, 0.15), (2, 0.16)
(2.5, 0.15), (3, 0.11), (3.5, 0.06), (4, 0)

(a) Use the regression capabilities of a graphing utility to fit a quartic equation to the data.

(b) Use a graphing utility to graph the data and the regression equation. How do they compare?

(c) Because $y = 0$ when $x = 0$, what should the constant term in the model be? Does your answer agree with the result of part (a)? Explain.

79. *Data Analysis* The table gives the median values of privately owned U.S. homes for 1981 through 1993. In the table, t is the time in years, with $t = 1$ corresponding to 1981, and y_1 and y_2 are the median prices (in thousands of dollars) in the Northeast and the South, respectively. (Source: U.S. Department of Housing and Urban Development)

t	1	2	3	4	5
y_1	76.0	78.2	82.2	88.6	103.3
y_2	64.4	66.1	70.9	72.0	75.0

t	6	7	8	9	10
y_1	125.0	140.0	149.0	159.6	159.0
y_2	82.2	88.0	92.0	96.4	99.0

t	11	12	13
y_1	155.9	169.0	162.6
y_2	100.0	105.5	115.0

(a) Use the regression capabilities of a graphing utility to fit a cubic model to the median prices of homes in the Northeast.

(b) Use the regression capabilities of a graphing utility to fit a cubic model to the median prices of homes in the South.

(c) Use the graphs of the models to write a short paragraph about the relationship between the median prices of homes in the two regions.

3.3 Real Zeros of Polynomial Functions

Long Division of Polynomials / Synthetic Division / The Remainder and Factor Theorems / The Rational Zero Test / Bounds for Real Zeros of Polynomial Functions

Long Division of Polynomials

Figure 3.21

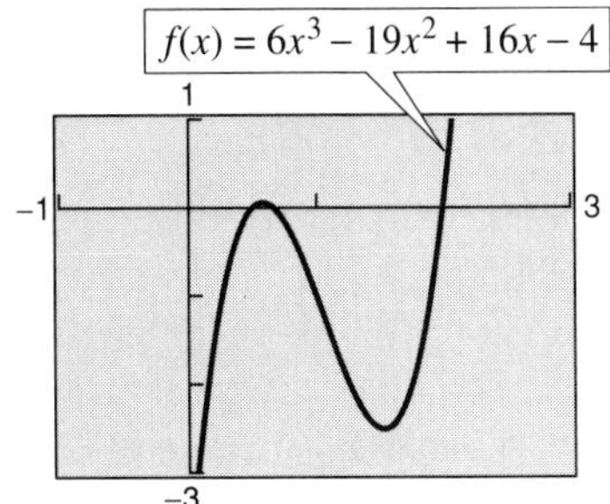

Consider the graph of

$$f(x) = 6x^3 - 19x^2 + 16x - 4.$$

Notice that a zero of f occurs at $x = 2$, as shown in Figure 3.21. Because $x = 2$ is a zero of the polynomial function f, you know that $(x - 2)$ is a factor of $f(x)$. This means that there exists a second-degree polynomial $q(x)$ such that $f(x) = (x - 2) \cdot q(x)$. To find $q(x)$, you can use **long division of polynomials.**

EXAMPLE 1 **Long Division of Polynomials**

Divide $f(x) = 6x^3 - 19x^2 + 16x - 4$ by $x - 2$, and use the result to factor the $f(x)$ completely.

Solution

Partial quotients

$$\begin{array}{r|l} x - 2 \overline{) 6x^3 - 19x^2 + 16x - 4} \quad 6x^2 - 7x + 2 & \\ \end{array}$$

Quotient	$6x^2 - 7x + 2$	
	$x - 2 \overline{)\,6x^3 - 19x^2 + 16x - 4}$	
	$6x^3 - 12x^2$	Multiply: $6x^2(x - 2)$.
	$-7x^2 + 16x$	Subtract.
	$-7x^2 + 14x$	Multiply: $-7x(x - 2)$.
	$2x - 4$	Subtract.
	$2x - 4$	Multiply: $2(x - 2)$.
	0	Subtract.

You can see that

$$\begin{aligned} 6x^3 - 19x^2 + 16x - 4 &= (x - 2)(6x^2 - 7x + 2) \\ &= (x - 2)(2x - 1)(3x - 2). \end{aligned}$$

Note that this factorization agrees with the graph of f (Figure 3.21) in that the three x-intercepts occur at $x = 2$, $x = \frac{1}{2}$, and $x = \frac{2}{3}$.

In Example 1, $x - 2$ is a factor of the polynomial $6x^3 - 19x^2 + 16x - 4$, and the long division process produces a remainder of zero. Often, long division will produce a nonzero remainder. For instance, if you divide $x^2 + 3x + 5$ by $x + 1$, you obtain the following.

$$\begin{array}{r@{}l} & x + 2 \quad \leftarrow \text{Quotient} \\ \text{Divisor} \rightarrow x + 1 \,\big) & x^2 + 3x + 5 \quad \leftarrow \text{Dividend} \\ & \underline{x^2 + x } \\ & \phantom{x^2+{}} 2x + 5 \\ & \phantom{x^2+{}} \underline{2x + 2} \\ & \phantom{x^2+2x+{}} 3 \quad \leftarrow \text{Remainder} \end{array}$$

In fractional form, you can write this result as follows.

$$\underbrace{\frac{\overbrace{x^2 + 3x + 5}^{\text{Dividend}}}{x + 1}}_{\text{Divisor}} = \overbrace{x + 2}^{\text{Quotient}} + \frac{\overset{\text{Remainder}}{3}}{\underbrace{x + 1}_{\text{Divisor}}}$$

This implies that $x^2 + 3x + 5 = (x + 1)(x + 2) + 3$, which illustrates the following well-known theorem called the **Division Algorithm.**

The Division Algorithm

If $f(x)$ and $d(x)$ are polynomials such that $d(x) \neq 0$, and the degree of $d(x)$ is less than or equal to the degree of $f(x)$, there exist unique polynomials $q(x)$ and $r(x)$ such that

$$f(x) = d(x)q(x) + r(x)$$

Dividend: $f(x)$; Divisor: $d(x)$; Quotient: $q(x)$; Remainder: $r(x)$

where $r(x) = 0$ *or* the degree of $r(x)$ is less than the degree of $d(x)$. If the remainder $r(x)$ is zero, $d(x)$ **divides evenly** into $f(x)$.

The Division Algorithm can also be written as

$$\frac{f(x)}{d(x)} = q(x) + \frac{r(x)}{d(x)}.$$

In the Division Algorithm, the rational expression $f(x)/d(x)$ is **improper** because the degree of $f(x)$ is greater than or equal to the degree of $d(x)$. On the other hand, the rational expression $r(x)/d(x)$ is **proper** because the degree of $r(x)$ is less than the degree of $d(x)$.

EXAMPLE 2 Long Division of Polynomials

Divide $x^3 - 1$ by $x - 1$.

Solution

Because there is no x^2-term or x-term in the dividend, you need to line up the subtraction by using zero coefficients (or leaving spaces) for the missing terms.

$$\begin{array}{r}
x^2 + x + 1 \\
x - 1 \overline{) x^3 + 0x^2 + 0x - 1} \\
\underline{x^3 - x^2} \qquad\qquad \\
x^2 \qquad\qquad \\
\underline{x^2 - x} \qquad \\
x - 1 \\
\underline{x - 1} \\
0
\end{array}$$

Thus, $x - 1$ divides evenly into $x^3 - 1$ and you can write

$$\frac{x^3 - 1}{x - 1} = x^2 + x + 1, \qquad x \neq 1.$$

Note You can check the result of a division problem by multiplying. For instance, in Example 2, try checking that $(x - 1)(x^2 + x + 1) = x^3 - 1$.

EXAMPLE 3 Long Division of Polynomials

Divide $2x^4 + 4x^3 - 5x^2 + 3x - 2$ by $x^2 + 2x - 3$.

Solution

$$\begin{array}{r}
2x^2 \qquad\quad + 1 \\
x^2 + 2x - 3 \overline{) 2x^4 + 4x^3 - 5x^2 + 3x - 2} \\
\underline{2x^4 + 4x^3 - 6x^2} \qquad\qquad \\
x^2 + 3x - 2 \\
\underline{x^2 + 2x - 3} \\
x + 1
\end{array}$$

Note that the first subtraction eliminated two terms from the dividend. When this happens, the quotient skips a term. Thus, you can write

$$\frac{2x^4 + 4x^3 - 5x^2 + 3x - 2}{x^2 + 2x - 3} = 2x^2 + 1 + \frac{x + 1}{x^2 + 2x - 3}.$$

Synthetic Division

There is a nice shortcut for long division of polynomials by divisors of the form $x - k$. The shortcut is called **synthetic division.** We summarize the pattern for synthetic division of a cubic polynomial as follows. (The pattern for higher-degree polynomials is similar.)

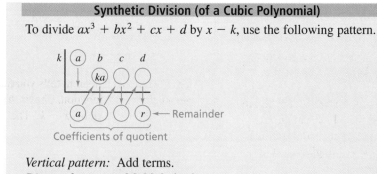

Synthetic Division (of a Cubic Polynomial)

To divide $ax^3 + bx^2 + cx + d$ by $x - k$, use the following pattern.

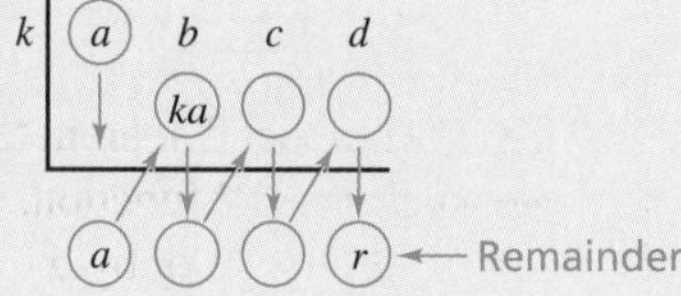

Vertical pattern: Add terms.
Diagonal pattern: Multiply by k.

Note Synthetic division works *only* for divisors of the form $x - k$. [Remember that $x + k = x - (-k)$.] You cannot use synthetic division to divide a polynomial by a quadratic such as $x^2 - 3$.

EXAMPLE 4 Using Synthetic Division

Use synthetic division to divide $x^4 - 10x^2 - 2x + 4$ by $x + 3$.

Solution

You should set up the array as follows. Note that a zero is included for each missing term in the dividend.

Divisor: $x + 3$ Dividend: $x^4 - 10x^2 - 2x + 4$

-3	1	0	-10	-2	4
		-3	9	3	-3
	1	-3	-1	1	1

Remainder: 1

Quotient: $x^3 - 3x^2 - x + 1$

Thus, you have

$$\frac{x^4 - 10x^2 - 2x + 4}{x + 3} = x^3 - 3x^2 - x + 1 + \frac{1}{x + 3}.$$

The Remainder and Factor Theorems

The remainder obtained in the synthetic division process has an important interpretation, as described in the **Remainder Theorem.**

The Remainder Theorem

If a polynomial $f(x)$ is divided by $x - k$, the remainder is

$$r = f(k).$$

> ***Think About the Proof***
>
> To prove the Remainder Theorem, you can use the Division Algorithm to write $f(x)$ as
>
> $$f(x) = (x - k)q(x) + r(x).$$
>
> By the Division Algorithm, you know that either $r(x) = 0$ or the degree of $r(x)$ is less than the degree of $x - k$. How does this allow you to prove the theorem? The details of the proof are given in the appendix.

The Remainder Theorem tells you that synthetic division can be used to evaluate a polynomial function. That is, to evaluate a polynomial function $f(x)$ when $x = k$, divide $f(x)$ by $x - k$. The remainder will be $f(k)$, as illustrated in Example 5.

EXAMPLE 5 Using the Remainder Theorem

Use the Remainder Theorem to evaluate the following function at $x = -2$.

$$f(x) = 3x^3 + 8x^2 + 5x - 7$$

Solution

Using synthetic division, you obtain the following.

$$\begin{array}{r|rrrr} -2 & 3 & 8 & 5 & -7 \\ & & -6 & -4 & -2 \\ \hline & 3 & 2 & 1 & -9 \end{array}$$

Because the remainder is $r = -9$, you can conclude that

$$f(-2) = -9.$$

This means that $(-2, -9)$ is a point on the graph of f. Try checking this by substituting $x = -2$ in the original function.

> ***Think About the Proof***
>
> To prove the Factor Theorem, you can use the Division Algorithm to write $f(x)$ as
>
> $$f(x) = (x - k)q(x) + r(x).$$
>
> By the Remainder Theorem, you know that $r(x) = f(k)$. Thus,
>
> $$f(x) = (x - k)q(x) + f(k)$$
>
> where $q(x)$ is a polynomial of lesser degree than $f(x)$. How does this allow you to prove the theorem? The details of the proof are in the appendix.

Another important theorem is the **Factor Theorem,** which is stated below. This theorem states that you can test to see whether a polynomial has $(x - k)$ as a factor by evaluating the polynomial at $x = k$. If the result is 0, $(x - k)$ is a factor.

The Factor Theorem

A polynomial $f(x)$ has a factor $(x - k)$ if and only if $f(k) = 0$.

EXAMPLE 6 **Factoring a Polynomial**

Show that $(x - 2)$ and $(x + 3)$ are factors of

$$f(x) = 2x^4 + 7x^3 - 4x^2 - 27x - 18.$$

Then find the remaining factors of $f(x)$.

Solution

Using synthetic division with 2 and -3 *successively,* you obtain the following.

$$\begin{array}{r|rrrrr} 2 & 2 & 7 & -4 & -27 & -18 \\ & & 4 & 22 & 36 & 18 \\ \hline & 2 & 11 & 18 & 9 & 0 \end{array}$$

0 remainder
$(x - 2)$ is a factor.

$$\begin{array}{r|rrrr} -3 & 2 & 11 & 18 & 9 \\ & & -6 & -15 & -9 \\ \hline & 2 & 5 & 3 & 0 \end{array}$$

0 remainder
$(x + 3)$ is a factor.

Figure 3.22

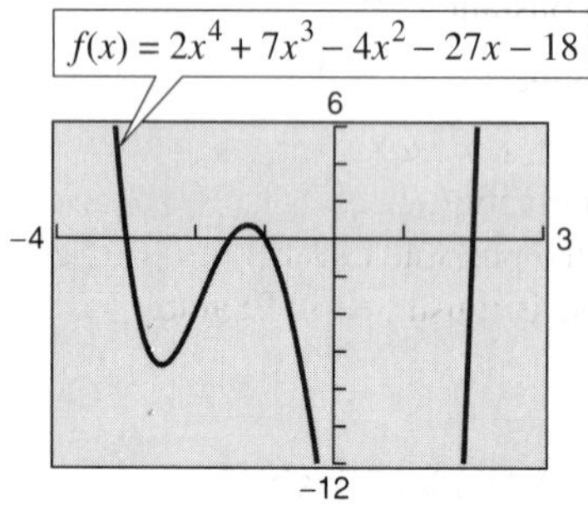

Because the resulting quadratic factors as

$$2x^2 + 5x + 3 = (2x + 3)(x + 1)$$

the complete factorization of $f(x)$ is

$$f(x) = (x - 2)(x + 3)(2x + 3)(x + 1).$$

Note that this factorization implies that f has four real zeros:

$$2, -3, -\tfrac{3}{2}, \text{ and } -1.$$

This is confirmed by the graph of f, which is shown in Figure 3.22.

In summary, the remainder r, obtained in the synthetic division of $f(x)$ by $x - k$, provides the following information.

1. The remainder r gives the value of f at $x = k$. That is, $r = f(k)$.
2. If $r = 0$, $(x - k)$ is a factor of $f(x)$.
3. If $r = 0$, $(k, 0)$ is an x-intercept of the graph of f.

Note Throughout this text, we have emphasized the importance of developing several problem-solving strategies. In the exercises for this section, try using more than one strategy to solve several of the exercises. For instance, if you find that $x - k$ divides evenly into $f(x)$, try sketching the graph of f. You should find that $(k, 0)$ is an x-intercept of the graph.

The Rational Zero Test

The **Rational Zero Test** relates the possible rational zeros of a polynomial (having integer coefficients) to the leading coefficient and to the constant term of the polynomial.

Note Graph the polynomial $f(x) = x^3 - 53x^2 + 103x - 51$ in the standard viewing rectangle. From the graph alone, you might think that there is only one zero, but the Rational Zero Test says that ± 51 might be zeros of the function. If you zoom out several times, you will see a more complete picture of the graph. What are the zeros of f?

The Rational Zero Test

If the polynomial $f(x) = a_n x^n + a_{n-1}x^{n-1} + \cdot \cdot \cdot + a_2 x^2 + a_1 x + a_0$ has *integer* coefficients, every rational zero of f has the form

$$\text{Rational zero} = \frac{p}{q}$$

where p and q have no common factors other than 1, p is a factor of the constant term a_0, and q is a factor of the leading coefficient a_n.

To use the Rational Zero Test, first list all rational numbers whose numerators are factors of the constant term and whose denominators are factors of the leading coefficient.

$$\text{Possible rational zeros} = \frac{\text{factors of constant term}}{\text{factors of leading coefficient}}$$

Now that you have formed this list of *possible rational zeros,* use a trial-and-error method to determine which, if any, are actual zeros of the polynomial. Note that when the leading coefficient is 1, the possible rational zeros are simply the factors of the constant term. This case is illustrated in Example 7.

EXAMPLE 7 Rational Zero Test with Leading Coefficient of 1

Find the rational zeros of $f(x) = x^3 + x + 1$.

Solution

Because the leading coefficient is 1, the possible rational zeros are simply the factors of the constant term.

Possible Rational Zeros: ± 1

By testing these possible zeros, you can see that neither works.

$$f(1) = (1)^3 + 1 + 1 = 3$$

$$f(-1) = (-1)^3 + (-1) + 1 = -1$$

Thus, the polynomial has *no* rational zeros. Note from the graph of f in Figure 3.23 that f does have one real zero (between -1 and 0). However, by the Rational Zero Test, you know that this real zero is *not* a rational number.

Figure 3.23

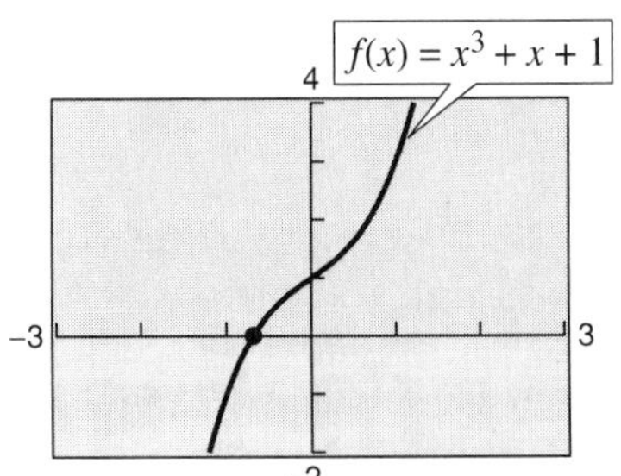

If the leading coefficient of a polynomial is not 1, the list of possible rational zeros can increase dramatically. In such cases the search can be shortened in several ways.

1. A programmable calculator can be used to speed up the calculations.
2. A graphing utility can give a good estimate of the locations of the zeros.
3. Synthetic division can be used to test the possible rational zeros.

Finding the first zero is often the most difficult part. After that, the search is simplified by working with the lower-degree polynomial obtained in synthetic division. A graphing utility can help you determine which possible rational zeros to test, as demonstrated in Example 8.

EXAMPLE 8 Using the Rational Zero Test

Find all the real zeros of $f(x) = 10x^3 - 15x^2 - 16x + 12$.

Solution

Because the leading coefficient is 10 and the constant term is 12, there is a long list of possible rational zeros.

Possible Rational Zeros:

$$\frac{\text{Factors of } 12}{\text{Factors of } 10} = \frac{\pm 1, \pm 2, \pm 3, \pm 4, \pm 6, \pm 12}{\pm 1, \pm 2, \pm 5, \pm 10}$$

Figure 3.24

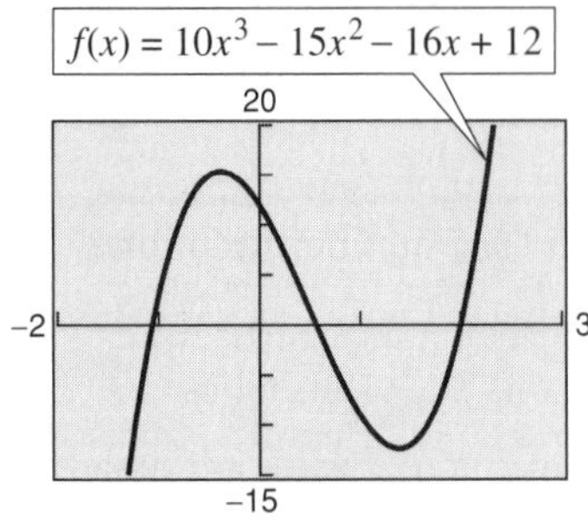

With so many possibilities (32, in fact), it is worth your time to use a graphing utility to focus on just a few. From Figure 3.24, it looks like three reasonable choices would be $x = -\frac{6}{5}$, $x = \frac{1}{2}$, and $x = 2$. Testing these by synthetic division shows that only $x = 2$ works.

$$\begin{array}{r|rrrr} 2 & 10 & -15 & -16 & 12 \\ & & 20 & 10 & -12 \\ \hline & 10 & 5 & -6 & 0 \end{array}$$

Thus, you have

$$f(x) = (x - 2)(10x^2 + 5x - 6).$$

Using the Quadratic Formula, you find that the two additional zeros are irrational numbers.

$$x = \frac{-5 + \sqrt{265}}{20} \approx 0.5639$$

and

$$x = \frac{-5 - \sqrt{265}}{20} \approx -1.0639$$

Bounds for Real Zeros of Polynomial Functions

The third test for zeros of a polynomial function is related to the sign pattern in the last row of the synthetic division tableau. This test can give you an upper or lower bound of the real zeros of f.

Note A real number b is an **upper bound** for the real zeros of f if no zeros are greater than b. Similarly, b is a **lower bound** if no real zeros of f are less than b.

Upper and Lower Bound Rule

Let $f(x)$ be a polynomial with real coefficients and a positive leading coefficient. Suppose $f(x)$ is divided by $x - c$, using synthetic division.

1. If $c > 0$ and each number in the last row is either positive or zero, c is an *upper bound* for the real zeros of f.
2. If $c < 0$ and the numbers in the last row are alternately positive and negative (zero entries count as positive or negative), c is a *lower bound* for the real zeros of f.

EXPLORATION

Use a graphing utility to graph

$$f(x) = 6x^3 - 4x^2 + 3x - 2.$$

Notice that the graph intersects the x-axis at the point $\left(\frac{2}{3}, 0\right)$. How does this relate to the real zero found in Example 9? Use a graphing utility to graph

$$g(x) = x^4 - 5x^3 + 3x^2 + x.$$

How many times does the graph intersect the x-axis? How many real zeros does g have?

EXAMPLE 9 Finding the Zeros of a Polynomial Function

Find the real zeros of $f(x) = 6x^3 - 4x^2 + 3x - 2$.

Solution

The possible real zeros are as follows.

$$\frac{\text{Factors of } 2}{\text{Factors of } 6} = \frac{\pm 1, \pm 2}{\pm 1, \pm 2, \pm 3, \pm 6}$$

$$= \pm 1, \pm\frac{1}{2}, \pm\frac{1}{3}, \pm\frac{1}{6}, \pm\frac{2}{3}, \pm 2$$

Trying $x = 1$ produces the following.

$$\begin{array}{r|rrrr} 1 & 6 & -4 & 3 & -2 \\ & & 6 & 2 & 5 \\ \hline & 6 & 2 & 5 & 3 \end{array}$$

Thus, $x = 1$ is not a zero, but because the last row has all positive entries, you know that $x = 1$ is an upper bound for the real zeros. Thus, you can restrict the search to zeros less than 1. By trial and error, you can determine that $x = \frac{2}{3}$ is a zero. Thus,

$$f(x) = \left(x - \tfrac{2}{3}\right)(6x^2 + 3).$$

Because $6x^2 + 3$ has no real zeros, it follows that $x = \frac{2}{3}$ is the only real zero.

EXPLORATION

Use a graphing utility to graph

$$f(x) = x^3 + 4.9x^2 - 126x + 382.5$$

in the standard viewing rectangle. From the graph, what do the real zeros appear to be? Discuss how the mathematical tools of this section might help you realize that the graph does not show all the important features of the polynomial function. Now use the zoom feature to find all the zeros of this function.

Before concluding this section, we list two additional hints that can help you find the real zeros of a polynomial.

1. If the terms of $f(x)$ have a common monomial factor, it should be factored out before applying the tests in this section. For instance, by writing

$$\begin{aligned} f(x) &= x^4 - 5x^3 + 3x^2 + x \\ &= x(x^3 - 5x^2 + 3x + 1) \end{aligned}$$

you can see that $x = 0$ is a zero of f and that the remaining zeros can be obtained by analyzing the cubic factor.

2. If you are able to find all but two zeros of $f(x)$, you are home free because you can always use the Quadratic Formula on the remaining quadratic factor. For instance, if you succeeded in writing

$$\begin{aligned} f(x) &= x^4 - 5x^3 + 3x^2 + x \\ &= x(x - 1)(x^2 - 4x - 1) \end{aligned}$$

you can apply the Quadratic Formula to $x^2 - 4x - 1$ to conclude that the two remaining zeros are

$$x = 2 + \sqrt{5} \quad \text{and} \quad x = 2 - \sqrt{5}.$$

Group Activity

Finding Patterns in Polynomial Division

Complete the following polynomial divisions.

a. $\dfrac{x^2 - 1}{x - 1} =$

b. $\dfrac{x^3 - 1}{x - 1} =$

c. $\dfrac{x^4 - 1}{x - 1} =$

Describe the pattern that you obtain, and use your result to find a formula for the polynomial division

$$\frac{x^n - 1}{x - 1}.$$

Create a numerical example to test your formula.

3.3 /// EXERCISES

Graphical Analysis In Exercises 1–6, use a graphing utility to graph the two equations in the same viewing rectangle. Use the graphs to verify that the expressions are equivalent. Verify the results algebraically.

1. $y_1 = \dfrac{4x}{x-1}, \quad y_2 = 4 + \dfrac{4}{x-1}$
2. $y_1 = \dfrac{3x-5}{x-3}, \quad y_2 = 3 + \dfrac{4}{x-3}$
3. $y_1 = \dfrac{x^2}{x+2}, \quad y_2 = x - 2 + \dfrac{4}{x+2}$
4. $y_1 = \dfrac{x^4 - 3x^2 - 1}{x^2+5}, \quad y_2 = x^2 - 8 + \dfrac{39}{x^2+5}$
5. $y_1 = \dfrac{x^5 - 3x^3}{x^2+1}, \quad y_2 = x^3 - 4x + \dfrac{4x}{x^2+1}$
6. $y_1 = \dfrac{x^3 - 2x^2 + 5}{x^2+x+1}, \quad y_2 = x - 3 + \dfrac{2(x+4)}{x^2+x+1}$

In Exercises 7–20, divide by long division.

7. Divide $2x^2 + 10x + 12$ by $x + 3$.
8. Divide $5x^2 - 17x - 12$ by $x - 4$.
9. Divide $4x^3 - 7x^2 - 11x + 5$ by $4x + 5$.
10. Divide $6x^3 - 16x^2 + 17x - 6$ by $3x - 2$.
11. Divide $x^4 + 5x^3 + 6x^2 - x - 2$ by $x + 2$.
12. Divide $x^3 + 4x^2 - 3x - 12$ by $x^2 - 3$.
13. Divide $7x + 3$ by $x + 2$.
14. Divide $8x - 5$ by $2x + 1$.
15. $(6x^3 + 10x^2 + x + 8) \div (2x^2 + 1)$
16. $(x^3 - 9) \div (x^2 + 1)$
17. $\dfrac{x^4 + 3x^2 + 1}{x^2 - 2x + 3}$
18. $(x^5 + 7) \div (x^3 - 1)$
19. $\dfrac{2x^3 - 4x^2 - 15x + 5}{(x-1)^2}$
20. $\dfrac{x^4}{(x-1)^3}$

Think About It In Exercises 21 and 22, perform the division by assuming that n is a positive integer.

21. $\dfrac{x^{3n} + 9x^{2n} + 27x^n + 27}{x^n + 3}$
22. $\dfrac{x^{3n} - 3x^{2n} + 5x^n - 6}{x^n - 2}$

In Exercises 23–32, divide by synthetic division.

23. $(4x^3 - 9x + 8x^2 - 18) \div (x + 2)$
24. $(9x^3 - 16x - 18x^2 + 32) \div (x - 2)$
25. $(5x^3 - 6x^2 + 8) \div (x - 4)$
26. $(5x^3 + 6x + 8) \div (x + 2)$
27. $\dfrac{x^3 + 512}{x + 8}$
28. $\dfrac{5x^3}{x + 3}$
29. $\dfrac{-3x^4}{x - 2}$
30. $\dfrac{-3x^4}{x + 2}$
31. $\dfrac{4x^3 + 16x^2 - 23x - 15}{x + \frac{1}{2}}$
32. $\dfrac{3x^3 - 4x^2 + 5}{x - \frac{3}{2}}$

33. *Think About It* What does it mean for a divisor to *divide evenly* into a dividend?
34. *Essay* Explain how you can check polynomial division. Give an example.

In Exercises 35–38, express the function in the form $f(x) = (x - k)q(x) + r(x)$ for the given value of k. Use a graphing utility to demonstrate that $f(k) = r$.

	Function	*Value of k*
35.	$f(x) = x^3 - x^2 - 14x + 11$	$k = 4$
36.	$f(x) = 15x^4 + 10x^3 - 6x^2 + 14$	$k = -\frac{2}{3}$
37.	$f(x) = x^3 + 3x^2 - 2x - 14$	$k = \sqrt{2}$
38.	$f(x) = 4x^3 - 6x^2 - 12x - 4$	$k = 1 - \sqrt{3}$

In Exercises 39 and 40, use synthetic division to find the required function values. Use a graphing utility to verify your result.

39. $f(x) = 4x^3 - 13x + 10$

(a) $f(1)$ (b) $f(-2)$

(c) $f(\frac{1}{2})$ (d) $f(8)$

40. $f(x) = 0.4x^4 - 1.6x^3 + 0.7x^2 - 2$

(a) $f(1)$ (b) $f(-2)$

(c) $f(5)$ (d) $f(-10)$

In Exercises 41–48, use synthetic division to show that x is a solution of the third-degree polynomial equation, and use the result to factor the polynomial completely. List all the real zeros of the function.

	Polynomial Equation	*Value of x*
41.	$x^3 - 7x + 6 = 0$	$x = 2$
42.	$x^3 - 28x - 48 = 0$	$x = -4$
43.	$2x^3 - 15x^2 + 27x - 10 = 0$	$x = \frac{1}{2}$
44.	$48x^3 - 80x^2 + 41x - 6 = 0$	$x = \frac{2}{3}$
45.	$x^3 + 2x^2 - 3x - 6 = 0$	$x = \sqrt{3}$
46.	$x^3 + 2x^2 - 2x - 4 = 0$	$x = \sqrt{2}$
47.	$x^3 - 3x^2 + 2 = 0$	$x = 1 + \sqrt{3}$
48.	$x^3 - x^2 - 13x - 3 = 0$	$x = 2 - \sqrt{5}$

49. *Data Analysis* The average monthly basic rates R for cable television in the United States for the years 1985 through 1993 are given in the table. The variable t represents the year, with $t = 0$ corresponding to 1990. (Source: *The Cable TV Financial Databook*)

t	-5	-4	-3	-2	-1
R	9.73	10.67	12.18	13.86	15.21

t	0	1	2	3
R	16.78	18.10	19.08	19.39

(a) Use the regression capabilities of a graphing utility to fit a cubic model to the data.

(b) Use a graphing utility to plot the data and graph the model in the same viewing rectangle.

(c) Use a graphing utility and the model to create a table of estimated values of R. Compare the model with the actual data.

(d) Use synthetic division to evaluate the model for the year 1996. Even though the model is relatively accurate for estimating the given data, do you think it is accurate for predicting future cable rates? Explain.

50. *Exploration* Use the form $f(x) = (x - k)q(x) + r$ to create a cubic function that (a) passes through the point $(2, 5)$ and rises to the right, and (b) passes through the point $(-3, 1)$ and falls to the right. (The answers are not unique.)

In Exercises 51–54, use the Rational Zero Test to list all possible rational zeros of f. Verify that the zeros of f shown on the graph are contained in the list.

51. $f(x) = x^3 + 3x^2 - x - 3$

52. $f(x) = x^3 - 4x^2 - 4x + 16$

Figure for 51

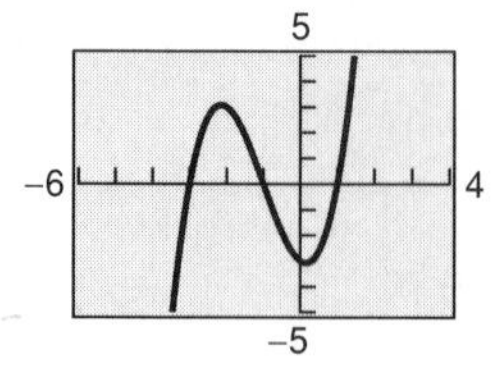

Figure for 52

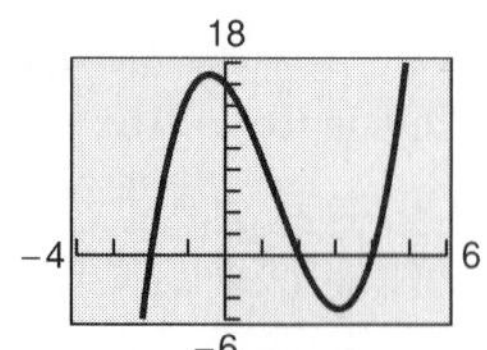

53. $f(x) = 2x^4 - 17x^3 + 35x^2 + 9x - 45$

54. $f(x) = 6x^3 - 71x^2 - 13x + 12$

Figure for 53

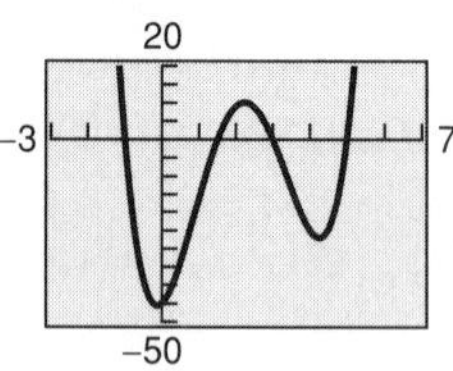

Figure for 54

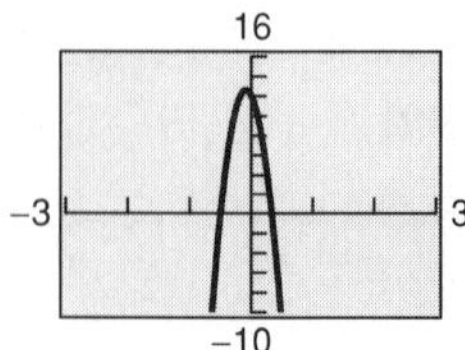

In Exercises 55–64, (a) list the possible rational zeros of f, (b) use a graphing utility to graph f so that some of the possible zeros in part (a) can be disregarded, and (c) determine all the real zeros of f.

55. $f(x) = x^3 + x^2 - 4x - 4$

56. $f(x) = -3x^3 + 20x^2 - 36x + 16$

57. $f(x) = -4x^3 + 15x^2 - 8x - 3$

58. $f(x) = 4x^3 - 12x^2 - x + 15$

59. $f(x) = -2x^4 + 13x^3 - 21x^2 + 2x + 8$

60. $f(x) = 4x^4 - 17x^2 + 4$

61. $f(x) = 32x^3 - 52x^2 + 17x + 3$

62. $f(x) = 6x^3 - x^2 - 13x + 8$

63. $f(x) = 4x^3 + 7x^2 - 11x - 18$

64. $f(x) = 2x^3 + 5x^2 - 21x - 10$

In Exercises 65–70, find all the real solutions of the polynomial equation.

65. $z^4 - z^3 - 2z - 4 = 0$

66. $x^4 - x^3 - 29x^2 - x - 30 = 0$

67. $x^4 - 13x^2 - 12x = 0$

68. $2y^4 + 7y^3 - 26y^2 + 23y - 6 = 0$

69. $2x^4 - 11x^3 - 6x^2 + 64x + 32 = 0$

70. $x^5 - x^4 - 3x^3 + 5x^2 - 2x = 0$

Graphical Analysis **In Exercises 71–74, (a) use the root-finding capabilities of a graphing utility to approximate the zeros of the function accurate to three decimal places, and (b) determine one of the exact zeros, use synthetic division to verify your result, and then factor the polynomial completely.**

71. $f(x) = x^4 - 3x^2 + 2$

72. $P(t) = t^4 - 7t^2 + 12$

73. $h(x) = x^5 - 7x^4 + 10x^3 + 14x^2 - 24x$

74. $g(x) = 6x^4 - 11x^3 - 51x^2 + 99x - 27$

In Exercises 75–78, use synthetic division to verify the upper bound and/or lower bound of the zeros of f.

75. $f(x) = x^4 - 4x^3 + 15$

Upper bound: $x = 4$; Lower bound: $x = -1$

76. $f(x) = 2x^3 - 3x^2 - 12x + 8$

Upper bound: $x = 4$; Lower bound: $x = -3$

77. $f(x) = x^4 - 4x^3 + 16x - 16$

Upper bound: $x = 5$; Lower bound: $x = -3$

78. $f(x) = 2x^4 - 8x + 3$

Upper bound: $x = 3$; Lower bound: $x = -4$

In Exercises 79–82, find the rational zeros of the polynomial function.

79. $P(x) = x^4 - \frac{25}{4}x^2 + 9 = \frac{1}{4}(4x^4 - 25x^2 + 36)$

80. $f(x) = x^3 - \frac{3}{2}x^2 - \frac{23}{2}x + 6 = \frac{1}{2}(2x^3 - 3x^2 - 23x + 12)$

81. $f(x) = x^3 - \frac{1}{4}x^2 - x + \frac{1}{4} = \frac{1}{4}(4x^3 - x^2 - 4x + 1)$

82. $f(z) = z^3 + \frac{11}{6}z^2 - \frac{1}{2}z - \frac{1}{3} = \frac{1}{6}(6z^3 + 11z^2 - 3z - 2)$

In Exercises 83–86, match the cubic function with the correct numbers of rational and irrational zeros.

(a) Rational zeros: 0; Irrational zeros: 1
(b) Rational zeros: 3; Irrational zeros: 0
(c) Rational zeros: 1; Irrational zeros: 2
(d) Rational zeros: 1; Irrational zeros: 0

83. $f(x) = x^3 - 1$

84. $f(x) = x^3 - 2$

85. $f(x) = x^3 - x$

86. $f(x) = x^3 - 2x$

Think About It **In Exercises 87–92, determine (if possible) the zeros of the function g if the function f has zeros at $x = r_1$, $x = r_2$, and $x = r_3$.**

87. $g(x) = -f(x)$

88. $g(x) = 3f(x)$

89. $g(x) = f(x - 5)$

90. $g(x) = f(2x)$

91. $g(x) = 3 + f(x)$

92. $g(x) = f(-x)$

93. *Dimensions of a Box* An open box is to be made from a rectangular piece of material 15 centimeters by 9 centimeters by cutting equal squares from the corners and turning up the sides.

(a) Let x represent the length of the sides of the squares. Draw a figure showing the squares removed from the original piece of material and the resulting dimensions of the open box.

(b) Use the figure in part (a) to write the volume V of the box as a function of x. Determine the domain of the function.

(c) Use a graphing utility to graph the function and approximate the dimensions of the box that yield maximum volume.

(d) Find values of x such that $V = 56$. Which of these values is a physical impossibility in the construction of the box? Explain.

94. *Dimensions of a Package* A rectangular package sent by a shipping service can have a maximum combined length and girth (perimeter of a cross section) of 120 inches (see figure).

(a) Show that the volume of the package is given by $V(x) = 4x^2(30 - x)$.

(b) Use a graphing utility to graph the function and approximate the dimensions of the package that yield a maximum volume.

(c) Find values of x such that $V = 13{,}500$. Which of these values is a physical impossibility in the construction of the package? Explain.

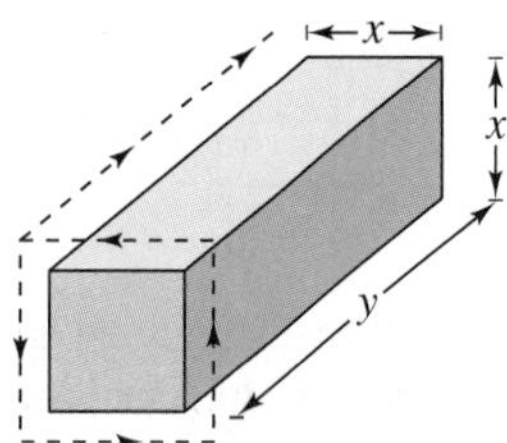

95. *Automobile Emissions* The number of parts per million of nitric oxide emissions y from a certain car engine is approximated by the model

$$y = -5.05x^3 + 3857x - 38{,}411.25, \qquad 13 \leq x \leq 18$$

where x is the air-fuel ratio.

(a) Use a graphing utility to graph the emissions function.

(b) It is observed from the graph that two air-fuel ratios produce 2400 parts per million of nitric oxide, with one being 15. Use the graph to approximate the second air-fuel ratio.

(c) Algebraically approximate the second air-fuel ratio that produces 2400 parts per million of nitric oxide. (*Hint:* Because you know that an air-fuel ratio of 15 produces the specified nitric oxide emission, you can use synthetic division.)

96. *Advertising Costs* A company that manufactures bicycles estimates that the profit for selling a particular model is given by

$$P = -45x^3 + 2500x^2 - 275{,}000, \qquad 0 \leq x \leq 50$$

where P is the profit in dollars and x is the advertising expense in tens of thousands of dollars. According to this model, find the smaller of two advertising amounts that yield a profit of \$800,000.

97. *Imports into the United States* The values of goods (in billions of dollars) imported into the United States for the years 1984 through 1993 are given in the table. (Source: U.S. General Imports and Imports for Consumption)

Year	1984	1985	1986	1987	1988
Imports	325.7	345.3	370.0	406.2	441.0

Year	1989	1990	1991	1992	1993
Imports	473.4	495.3	487.1	532.7	580.5

(a) A model for this data is given by

$$I = 0.161t^3 + 0.626t^2 + 25.646t + 484.137$$

where I is the annual value of goods imported (in billions of dollars) and t represents the calendar year, with $t = 0$ corresponding to 1990. Use a graphing utility to graph the data and the model in the same viewing rectangle. How do they compare?

(b) According to this model, when did the annual value of imports reach 525 billion dollars?

(c) According to the right-hand behavior of the model, will the value of imports continue to increase? Explain.

3.4 The Fundamental Theorem of Algebra

The Fundamental Theorem of Algebra / Conjugate Pairs / Factoring a Polynomial

The Fundamental Theorem of Algebra

You have been using the fact that an nth-degree polynomial can have at most n real zeros. In the complex number system, this statement can be improved. That is, in the complex number system, every nth-degree polynomial function has *precisely* n zeros. This important result is derived from the **Fundamental Theorem of Algebra,** first proved by the famous German mathematician Carl Friedrich Gauss (1777–1855).

The Fundamental Theorem of Algebra

If $f(x)$ is a polynomial of degree n, where $n > 0$, f has at least one zero in the complex number system.

Jean Le Rond d'Alembert (1717–1783) worked independently of Carl Gauss trying to prove the Fundamental Theorem of Algebra. His efforts were such that in France, the Fundamental Theorem of Algebra is frequently known as the Theorem of d'Alembert. (Credit: The Fogg Art Museum)

Using the Fundamental Theorem of Algebra and the equivalence of zeros and factors, you obtain the following theorem.

Linear Factorization Theorem

If $f(x)$ is a polynomial of degree n

$$f(x) = a_n x^n + a_{n-1}x^{n-1} + \cdot \cdot \cdot + a_1 x + a_0$$

where $n > 0$, f has precisely n linear factors

$$f(x) = a_n(x - c_1)(x - c_2) \cdot \cdot \cdot (x - c_n)$$

where $c_1, c_2, \ldots, c_n$ are complex numbers and a_n is the leading coefficient of $f(x)$.

Note that neither the Fundamental Theorem of Algebra nor the Linear Factorization Theorem tells you *how* to find the zeros or factors of a polynomial. Such theorems are called **existence theorems.** To find the zeros of a polynomial function, you still must rely on the techniques developed earlier in the text.

Remember that the n zeros of a polynomial function can be real or complex, and they may be repeated. Example 1 illustrates several cases.

Think About the Proof

To prove the Linear Factorization Theorem, you can use the Fundamental Theorem of Algebra to conclude that f must have at least one zero, c_1. Thus, $(x - c_1)$ is a factor of $f(x)$, and by the Factor Theorem, it follows that

$$f(x) = (x - c_1)f_1(x).$$

If the degree of $f_1(x)$ is greater than zero, you can apply the Fundamental Theorem of Algebra again to conclude that f_1 has a zero, c_2. How can you continue this reasoning to complete the proof? The details of the proof are in the appendix.

EXAMPLE 1 **Zeros of Polynomial Functions**

a. The first-degree polynomial $f(x) = x - 2$ has exactly *one* zero: $x = 2$.

b. Counting multiplicity, the second-degree polynomial function

$$f(x) = x^2 - 6x + 9 = (x - 3)(x - 3)$$

has exactly *two* zeros: $x = 3$ and $x = 3$.

c. The third-degree polynomial function

$$f(x) = x^3 + 4x = x(x^2 + 4) = x(x - 2i)(x + 2i)$$

has exactly *three* zeros: $x = 0$, $x = 2i$, and $x = -2i$.

d. The fourth-degree polynomial function

$$f(x) = x^4 - 1 = (x - 1)(x + 1)(x - i)(x + i)$$

has exactly *four* zeros: $x = 1$, $x = -1$, $x = i$, and $x = -i$.

Look at the graphs below and match them with the four polynomial functions in Example 1. Which zeros appear on the graphs?

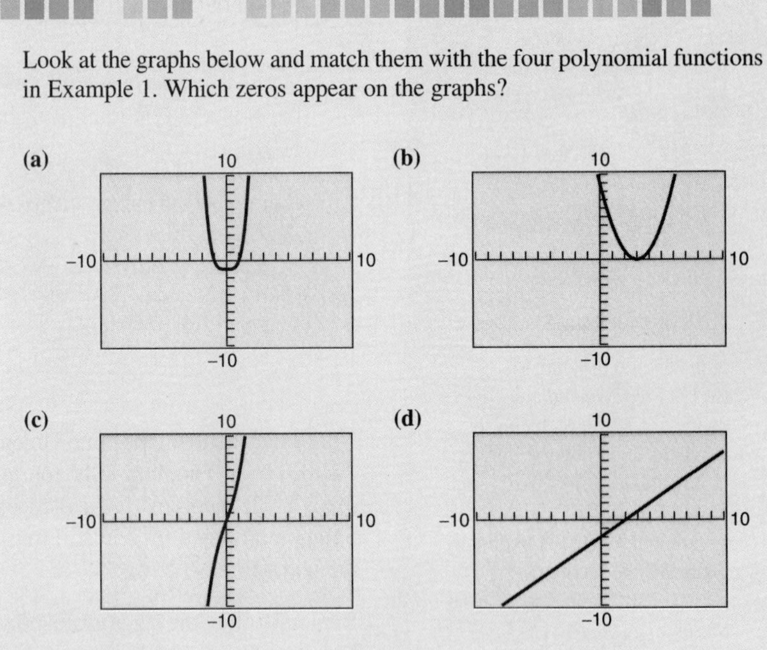

Example 2 shows how to use the methods described in Sections 3.2 and 3.3 (The Rational Zero Test, synthetic division, and factoring) to find all the zeros of a polynomial function, including complex zeros.

EXAMPLE 2 **Finding the Zeros of a Polynomial Function**

Write $f(x) = x^5 + x^3 + 2x^2 - 12x + 8$ as the product of linear factors, and list all of its zeros.

Solution

The possible rational zeros are $\pm 1, \pm 2, \pm 4$, and ± 8. The graph shown in Figure 3.25 indicates that 1 and -2 are good guesses. Synthetic division produces the following.

$$\begin{array}{r|rrrrrr} 1 & 1 & 0 & 1 & 2 & -12 & 8 \\ & & 1 & 1 & 2 & 4 & -8 \\ \hline & 1 & 1 & 2 & 4 & -8 & 0 \end{array}$$

1 is a zero.

$$\begin{array}{r|rrrrr} 1 & 1 & 1 & 2 & 4 & -8 \\ & & 1 & 2 & 4 & 8 \\ \hline & 1 & 2 & 4 & 8 & 0 \end{array}$$

1 is a repeated zero.

$$\begin{array}{r|rrrr} -2 & 1 & 2 & 4 & 8 \\ & & -2 & 0 & -8 \\ \hline & 1 & 0 & 4 & 0 \end{array}$$

-2 is a zero.

Figure 3.25

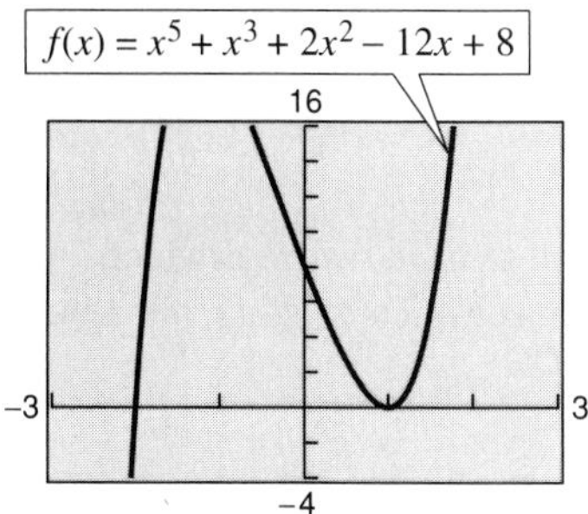

Thus, you have

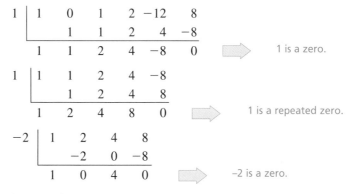

$$\begin{aligned} f(x) &= x^5 + x^3 + 2x^2 - 12x + 8 \\ &= (x-1)(x-1)(x+2)(x^2+4). \end{aligned}$$

By factoring $x^2 + 4$ as

$$x^2 - (-4) = \left(x - \sqrt{-4}\right)\left(x + \sqrt{-4}\right) = (x - 2i)(x + 2i)$$

you obtain

$$f(x) = (x-1)(x-1)(x+2)(x-2i)(x+2i)$$

which gives the following five zeros of f.

$$1, \quad 1, \quad -2, \quad 2i, \quad \text{and} \quad -2i$$

Note from the graph of f shown in Figure 3.25 that the *real* zeros are the only ones that appear as x-intercepts.

Conjugate Pairs

In Example 2, note that the two complex zeros are **conjugates.** That is, they are of the form $a + bi$ and $a - bi$.

Complex Zeros Occur in Conjugate Pairs

Let $f(x)$ be a polynomial function that has *real coefficients.* If $a + bi$, where $b \neq 0$, is a zero of the function, the conjugate $a - bi$ is also a zero of the function.

Note Be sure you see that this result is true only if the polynomial function has *real coefficients.* For instance, the result applies to the function $f(x) = x^2 + 1$, but not to the function $g(x) = x - i$.

EXAMPLE 3 **Finding a Polynomial with Given Zeros**

Find a *fourth-degree* polynomial function, with real coefficients, that has -1, -1, and $3i$ as zeros.

Solution

Because $3i$ is a zero *and* the polynomial is stated to have real coefficients, you know that the conjugate $-3i$ must also be a zero. Thus, from the Linear Factorization Theorem, $f(x)$ can be written as

$$f(x) = a(x + 1)(x + 1)(x - 3i)(x + 3i).$$

For simplicity, let $a = 1$, and obtain

$$\begin{aligned} f(x) &= (x^2 + 2x + 1)(x^2 + 9) \\ &= x^4 + 2x^3 + 10x^2 + 18x + 9. \end{aligned}$$

Factoring a Polynomial

The Linear Factorization Theorem shows that you can write any nth-degree polynomial as the product of n linear factors.

$$f(x) = a(x - c_1)(x - c_2)(x - c_3) \cdot \cdot \cdot (x - c_n)$$

However, this result includes the possibility that some of the values of c_i are complex. The boxed statement at the top of page 285 implies that even if you do not want to get involved with "complex factors," you can still write $f(x)$ as the product of linear and/or quadratic factors.

Think About the Proof

To prove the theorem at the right, begin by using the Linear Factorization Theorem to write the polynomial function as

$$f(x) = a(x - c_1) \cdots (x - c_n).$$

If each c_i is real, there is nothing to prove. If any of the c_i's are complex, how can you use the theorem on page 284 to conclude that $f(x)$ has a quadratic factor with real coefficients? The details of the proof are given in the appendix.

Factors of a Polynomial

Every polynomial of degree $n > 0$ with real coefficients can be written as the product of linear and quadratic factors with real coefficients, where the quadratic factors have no real zeros.

A quadratic factor with no real zeros is said to be **irreducible over the reals.** Be sure you see that this is not the same as being *irreducible over the rationals.* For example, the quadratic

$$x^2 + 1 = (x - i)(x + i)$$

is irreducible over the reals (and therefore over the rationals). On the other hand, the quadratic

$$x^2 - 2 = \left(x - \sqrt{2}\right)\left(x + \sqrt{2}\right)$$

is irreducible over the rationals, but *reducible* over the reals.

EXAMPLE 4 Factoring a Polynomial

Write the polynomial $f(x) = x^4 - x^2 - 20$

a. as the product of factors that are irreducible over the *rationals,*

b. as the product of linear factors and quadratic factors that are irreducible over the *reals,* and

c. in completely factored form.

Solution

a. Begin by factoring the polynomial into the product of two quadratic polynomials.

$$x^4 - x^2 - 20 = (x^2 - 5)(x^2 + 4)$$

Both of these factors are irreducible over the rationals.

b. By factoring over the reals, you have

$$x^4 - x^2 - 20 = \left(x + \sqrt{5}\right)\left(x - \sqrt{5}\right)(x^2 + 4)$$

where the quadratic factor is irreducible over the reals.

c. In completely factored form, you have

$$x^4 - x^2 - 20 = \left(x + \sqrt{5}\right)\left(x - \sqrt{5}\right)(x - 2i)(x + 2i).$$

Note In Example 4, notice from the completely factored form that the fourth-degree polynomial has four zeros.

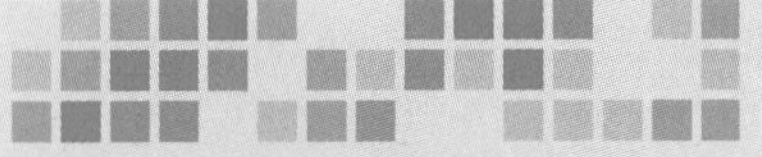

Some graphing utilities can find both real and complex zeros of a function. For instance, to find all the zeros of

$$f(x) = x^4 - 3x^3 + 6x^2 + 2x - 60$$

on a *TI–92,* you can enter the following.

F2 A 1 (cSolve

Enter $f(x)$.

= 0 , X) ENTER

The displayed solutions are

$x = 1 + 3i$ or $x = 1 - 3i$ or $x = 3$ or $x = -2$.

EXAMPLE 5 **Finding the Zeros of a Polynomial Function**

Find all the zeros of $f(x) = x^4 - 3x^3 + 6x^2 + 2x - 60$, given that $1 + 3i$ is a zero of f.

Solution

Because complex zeros occur in conjugate pairs, you know that $1 - 3i$ is also a zero of f. This means that both

$$x - (1 + 3i) \quad \text{and} \quad x - (1 - 3i)$$

are factors of $f(x)$. Multiplying these two factors produces

$$\begin{aligned}[x - (1 + 3i)][x - (1 - 3i)] &= [(x - 1) - 3i][(x - 1) + 3i] \\ &= (x - 1)^2 - 9i^2 \\ &= x^2 - 2x + 10.\end{aligned}$$

Using long division, you can divide $x^2 - 2x + 10$ into $f(x)$ to obtain the following.

$$\begin{array}{r} x^2 - x - 6 \\ x^2 - 2x + 10\overline{)\,x^4 - 3x^3 + 6x^2 + 2x - 60} \\ \underline{x^4 - 2x^3 + 10x^2} \qquad\qquad \\ -x^3 - 4x^2 + 2x \qquad \\ \underline{-x^3 + 2x^2 - 10x} \qquad \\ -6x^2 + 12x - 60 \\ \underline{-6x^2 + 12x - 60} \\ 0 \end{array}$$

Therefore, you have

$$f(x) = (x^2 - 2x + 10)(x^2 - x - 6) = (x^2 - 2x + 10)(x - 3)(x + 2)$$

and you can conclude that the zeros of f are $1 + 3i$, $1 - 3i$, 3, and -2.

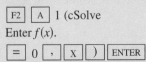

Study Tip

Throughout this chapter, we have basically stated the results and theorems in terms of zeros of polynomial functions. Be sure you see that the same results could have been stated in terms of solutions of polynomial equations. This is true because the zeros of the polynomial function

$$f(x) = a_nx^n + a_{n-1}x^{n-1} + \cdots + a_2x^2 + a_1x + a_0$$

are precisely the solutions of the polynomial equation

$$a_nx^n + a_{n-1}x^{n-1} + \cdots + a_2x^2 + a_1x + a_0 = 0.$$

Group Activity — *Factoring a Polynomial*

Compile a list of all the various techniques for factoring a polynomial that have been covered thus far in the text. Give an example illustrating each technique, and discuss when the use of each technique is appropriate.

3.4 /// EXERCISES

In Exercises 1–4, find all the zeros of the function.

1. $f(x) = x(x - 6)^2$

2. $g(x) = (x - 2)(x + 4)^3$

3. $h(t) = (t - 3)(t - 2)(t - 3i)(t + 3i)$

4. $h(m) = (m - 4)^2(m - 2 + 4i)(m - 2 - 4i)$

Graphical and Analytical Analysis **In Exercises 5–8, find all the zeros of the function. Is there a relationship between the number of real zeros and the number of x-intercepts of the graph? Explain.**

5. $f(x) = x^3 - 4x^2 + x - 4$

6. $f(x) = x^3 - 4x^2 - 4x + 16$

Figure for 5

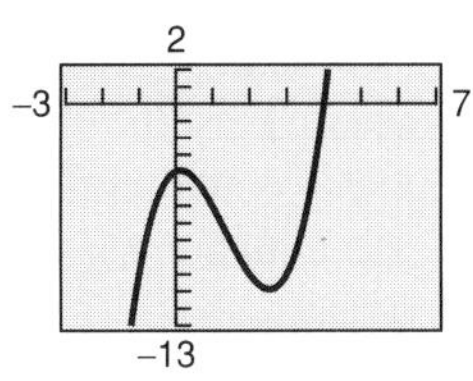

Figure for 6

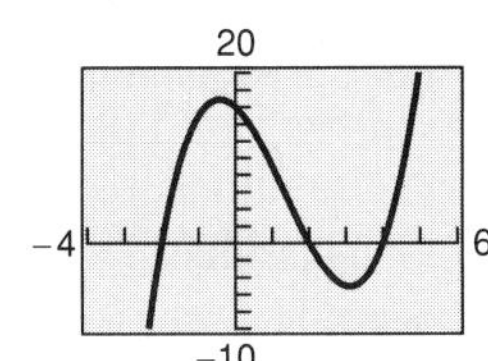

7. $f(x) = x^4 + 4x^2 + 4$

8. $f(x) = x^4 - 3x^2 - 4$

Figure for 7

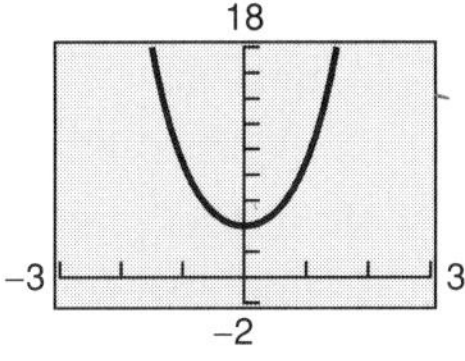

Figure for 8

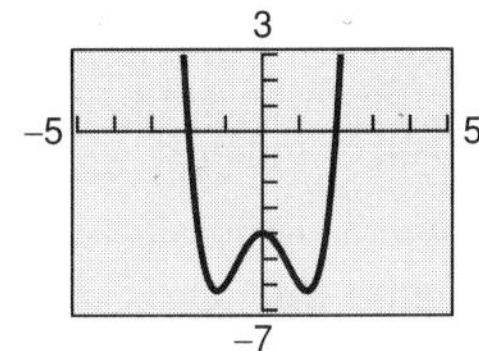

In Exercises 9–28, find all the zeros of the function and write the polynomial as a product of linear factors. Use a graphing utility to graph the function as an aid in finding the zeros and as a check of your results. (If you have a computer algebra system, use it to verify the complex zeros.)

9. $f(x) = x^2 + 25$

10. $f(x) = x^2 - x + 56$

11. $h(x) = x^2 - 4x + 1$

12. $g(x) = x^2 + 10x + 23$

13. $f(x) = x^4 - 81$

14. $f(y) = y^4 - 625$

15. $f(z) = z^2 - 2z + 2$

16. $h(x) = x^3 - 3x^2 + 4x - 2$

17. $f(t) = t^3 - 3t^2 - 15t + 125$

18. $f(x) = x^3 + 11x^2 + 39x + 29$

19. $f(x) = 16x^3 - 20x^2 - 4x + 15$

20. $f(s) = 2s^3 - 5s^2 + 12s - 5$

21. $f(x) = 5x^3 - 9x^2 + 28x + 6$

22. $g(x) = 3x^3 - 4x^2 + 8x + 8$

23. $f(x) = x^4 + 10x^2 + 9$

24. $f(x) = x^4 + 29x^2 + 100$

25. $g(x) = x^4 - 4x^3 + 8x^2 - 16x + 16$

26. $h(x) = x^4 + 6x^3 + 10x^2 + 6x + 9$

27. $f(x) = 2x^4 + 5x^3 + 4x^2 + 5x + 2$

28. $g(x) = x^5 - 8x^4 + 28x^3 - 56x^2 + 64x - 32$

In Exercises 29–36, find a polynomial function with integer coefficients that has the given zeros. (There are many correct answers.)

29. $1, 5i, -5i$

30. $4, 3i, -3i$

31. $2, 4 + i, 4 - i$

32. $6, -5 + 2i, -5 - 2i$

33. $i, -i, 6i, -6i$

34. $2, 2, 2, 4i, -4i$

35. $-5, -5, 1 + \sqrt{3}i$

36. $0, 0, 4, 1 + i$

In Exercises 37–40, write the polynomial (a) as the product of factors that are irreducible over the *rationals*, (b) as the product of linear and quadratic factors that are irreducible over the *reals*, and (c) in completely factored form.

37. $f(x) = x^4 + 6x^2 - 27$

38. $f(x) = x^4 - 2x^3 - 3x^2 + 12x - 18$
(*Hint:* One factor is $x^2 - 6$.)

39. $f(x) = x^4 - 4x^3 + 5x^2 - 2x - 6$
(*Hint:* One factor is $x^2 - 2x - 2$.)

40. $f(x) = x^4 - 3x^3 - x^2 - 12x - 20$
(*Hint:* One factor is $x^2 + 4$.)

In Exercises 41–46, use the given zero to find all the zeros of the function.

	Function	*Zero*
41.	$f(x) = 2x^3 + 3x^2 + 50x + 75$	$r = 5i$
42.	$f(x) = x^3 + x^2 + 9x + 9$	$r = 3i$
43.	$f(x) = 2x^4 - x^3 + 7x^2 - 4x - 4$	$r = 2i$
44.	$g(x) = x^3 - 7x^2 - x + 87$	$r = 5 + 2i$
45.	$g(x) = 4x^3 + 23x^2 + 34x - 10$	$r = -3 + i$
46.	$h(x) = 3x^3 - 4x^2 + 8x + 8$	$r = 1 - \sqrt{3}i$

Graphical Analysis **In Exercises 47–50, (a) use the root or zero feature of a graphing utility to approximate the zeros of the function accurate to three decimal places, and (b) determine one of the exact zeros, use synthetic division to verify your result, and find the exact values of the remaining zeros.**

47. $f(x) = x^4 + 3x^3 - 5x^2 - 21x + 22$

48. $f(x) = x^3 + 4x^2 + 14x + 20$

49. $h(x) = 8x^3 - 14x^2 + 18x - 9$

50. $f(x) = 25x^3 - 55x^2 - 54x - 18$

51. *Think About It* A zero of the function $f(x) = x^3 + ix^2 + ix - 1$ is $x = -i$.

(a) Show that the conjugate $x = i$ is not a zero of f.

(b) Does part (a) contradict the theorem that complex zeros occur in conjugate pairs? Explain.

52. *Exploration* Use a graphing utility to graph the function $f(x) = x^4 - 4x^2 + k$ for different values of k. Find values of k such that the zeros of f satisfy the specified characteristics. (Some parts do not have unique answers.)

(a) Four real zeros

(b) Two real zeros each of multiplicity 2

(c) Two real zeros and two complex zeros

(d) Four complex zeros

53. *Think About It* Will the answers to Exercise 52 change for the function g?

(a) $g(x) = f(x - 2)$ (b) $g(x) = f(2x)$

54. *Maximum Height* A baseball is thrown upward from ground level with an initial velocity of 48 feet per second, and its height h in feet is given by

$$h = -16t^2 + 48t, \quad 0 \le t \le 3$$

where t is the time in seconds. Suppose you are told that the ball reaches a height of 64 feet. Is this possible? Explain.

55. *Profit* The demand equation for a certain product is given by $p = 140 - 0.0001x$, where p is the unit price (in dollars) of the product and x is the number of units produced and sold. The cost equation for the product is $C = 80x + 150{,}000$, where C is the total cost (in dollars) and x is the number of units produced. The total profit obtained by producing and selling x units is given by $P = R - C = xp - C$. Suppose you are working in the marketing department that produces this product, and you are asked to determine a price p that would yield a profit of 9 million dollars. Is this possible? Explain.

56. Find a quadratic function f (with integer coefficients) that has $\pm\sqrt{b}i$ as zeros. Assume that b is a positive integer.

57. Find a quadratic function f (with integer coefficients) that has $a \pm bi$ as zeros. Assume that b is a positive integer.

Chapter Opener **In Exercises 58 and 59, plot the complex number.**

58. $-4 + 2i$ **59.** $\frac{3}{2} + \frac{11}{2}i$

3.5 Rational Functions and Asymptotes

Introduction to Rational Functions / Horizontal and Vertical Asymptotes / Applications

Library of Functions

A rational function $f(x)$ is the quotient of two polynomials, $f(x) = p(x)/q(x)$. A rational function is not defined at points x for which the denominator $q(x) = 0$. Near these points, the graph of the rational function may increase or decrease without bound.

Introduction to Rational Functions

A **rational function** can be written in the form

$$f(x) = \frac{p(x)}{q(x)}$$

where $p(x)$ and $q(x)$ are polynomials and $q(x)$ is not the zero polynomial. In this section we assume that $p(x)$ and $q(x)$ have no common factors.

In general, the *domain* of a rational function of x includes all real numbers except x-values that make the denominator zero. Much of our discussion of rational functions will focus on their graphical behavior near these x-values.

EXAMPLE 1 **Finding the Domain of a Rational Function**

Find the domain of

$$f(x) = \frac{1}{x}$$

and discuss the behavior of f near any excluded x-values.

Solution

Because the denominator is zero when $x = 0$, the domain of f is all real numbers except $x = 0$. To determine the behavior of f near this excluded value, evaluate $f(x)$ to the left and right of $x = 0$, as indicated in the following two tables.

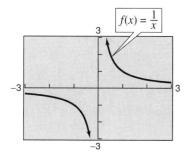

x	-1	-0.5	-0.1	-0.01	-0.001	$\to 0$
$f(x)$	-1	-2	-10	-100	-1000	$\to -\infty$

x	$0 \leftarrow$	0.001	0.01	0.1	0.5	1
$f(x)$	$\infty \leftarrow$	1000	100	10	2	1

Note that as x approaches 0 *from the left*, $f(x)$ decreases without bound, whereas as x approaches 0 *from the right*, $f(x)$ increases without bound. The graph of f is shown in Figure 3.26.

Figure 3.26

$f(x) = \frac{1}{x}$

Figure 3.27

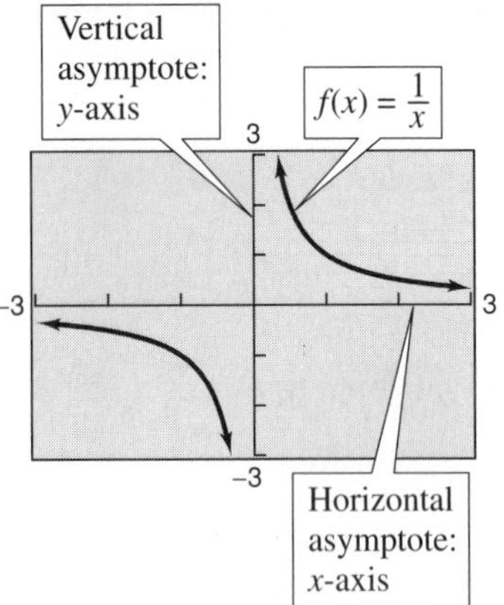

Horizontal and Vertical Asymptotes

In Example 1, the behavior of $f(x) = 1/x$ near $x = 0$ is denoted as follows.

$f(x) \to -\infty$ as $x \to 0^-$ — $f(x)$ decreases without bound as x approaches 0 from the left.

$f(x) \to \infty$ as $x \to 0^+$ — $f(x)$ increases without bound as x approaches 0 from the right.

The line $x = 0$ is a **vertical asymptote** of the graph of f, as shown in Figure 3.27. The graph of f also has a **horizontal asymptote**—the line $y = 0$. This means that the values of $f(x) = 1/x$ approach zero as x increases or decreases without bound.

$f(x) \to 0$ as $x \to -\infty$ — $f(x)$ approaches 0 as x decreases without bound.

$f(x) \to 0$ as $x \to \infty$ — $f(x)$ approaches 0 as x increases without bound.

Definition of Vertical and Horizontal Asymptotes

1. The line $x = a$ is a **vertical asymptote** of the graph of f if

$$f(x) \to \infty \qquad \text{or} \qquad f(x) \to -\infty$$

as $x \to a$, either from the right or from the left.

2. The line $y = b$ is a **horizontal asymptote** of the graph of f if

$$f(x) \to b$$

as $x \to \infty$ or $x \to -\infty$.

Eventually (as $x \to \infty$ or $x \to -\infty$), the distance between the horizontal asymptote and the points on the graph must approach zero. Figure 3.28 shows the horizontal and vertical asymptotes of the graphs of three rational functions.

Figure 3.28

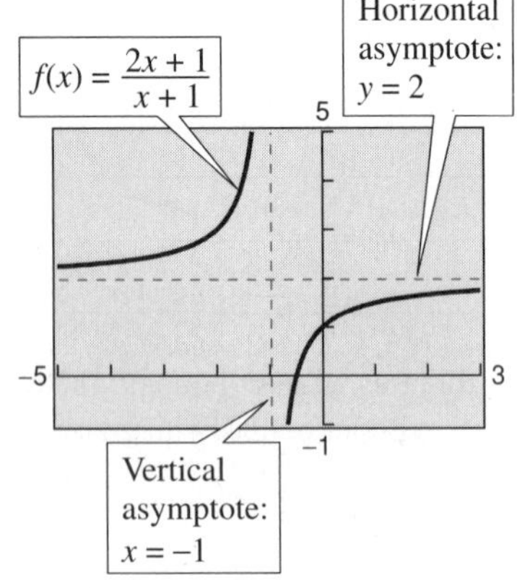

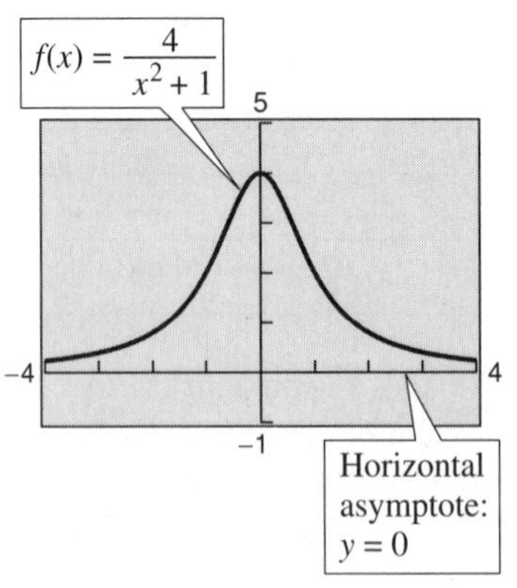

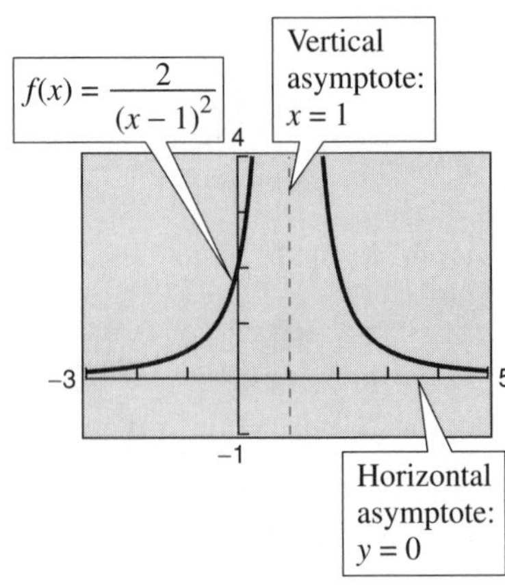

Asymptotes of a Rational Function

Let f be the rational function given by

$$f(x) = \frac{p(x)}{q(x)} = \frac{a_n x^n + a_{n-1}x^{n-1} + \cdot \cdot \cdot + a_1 x + a_0}{b_m x^m + b_{m-1}x^{m-1} + \cdot \cdot \cdot + b_1 x + b_0}$$

where $p(x)$ and $q(x)$ have no common factors.

1. The graph of f has vertical asymptotes at the zeros of $q(x)$.
2. The graph of f has at most one horizontal asymptote, as follows.

   a. If $n < m$, the x-axis $(y = 0)$ is a horizontal asymptote.

   b. If $n = m$, the line $y = a_n/b_m$ is a horizontal asymptote.

   c. If $n > m$, the graph of f has no horizontal asymptote.

EXPLORATION

Use a graphing utility to compare the graphs of

$$y = \frac{3x^3 - 5x^2 + 4x - 5}{2x^2 - 6x + 7}$$

and

$$y = \frac{3x^3}{2x^2}.$$

First, use a viewing rectangle in which $-5 \le x \le 5$ and $-10 \le y \le 10$, then zoom out. Write a convincing argument that the shape of the graph of a rational function eventually behaves like the graph of $y = a_n x^n / b_m x^m$, where $a_n x^n$ is the leading term of the numerator and $b_m x^m$ is the leading term of the denominator.

You can apply this theorem by comparing the degrees of the numerator and denominator, as illustrated in Figure 3.29.

a. The graph of

$$f(x) = \frac{2x}{3x^2 + 1}$$ See Figure 3.29(a).

has the x-axis as a horizontal asymptote.

b. The graph of

$$f(x) = \frac{2x^2}{3x^2 + 1}$$ See Figure 3.29(b).

has the line $y = \frac{2}{3}$ as a horizontal asymptote.

c. The graph of

$$f(x) = \frac{2x^3}{3x^2 + 1}$$ See Figure 3.29(c).

has no horizontal asymptote.

Figure 3.29

(a)

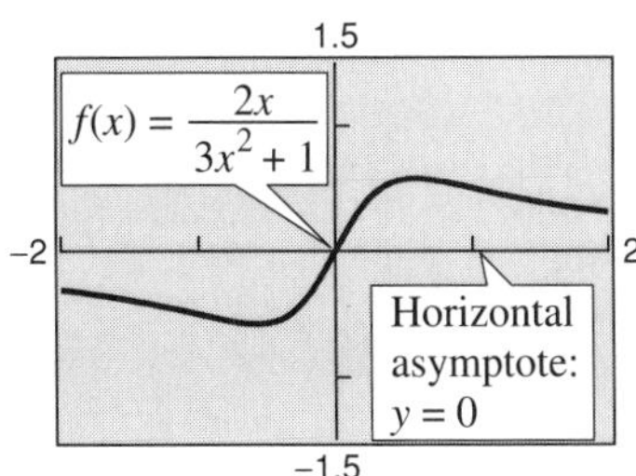

(b)

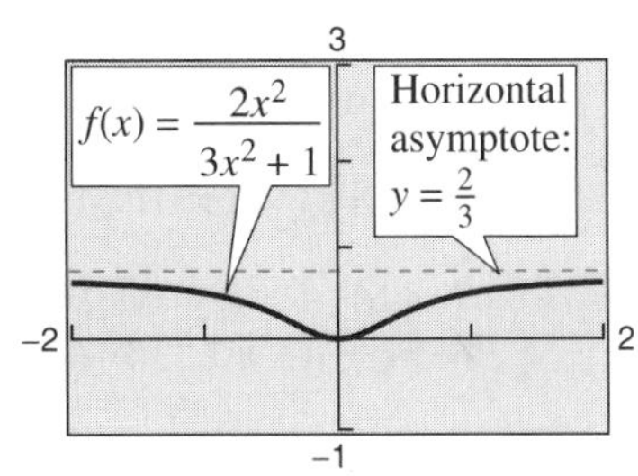

(c)

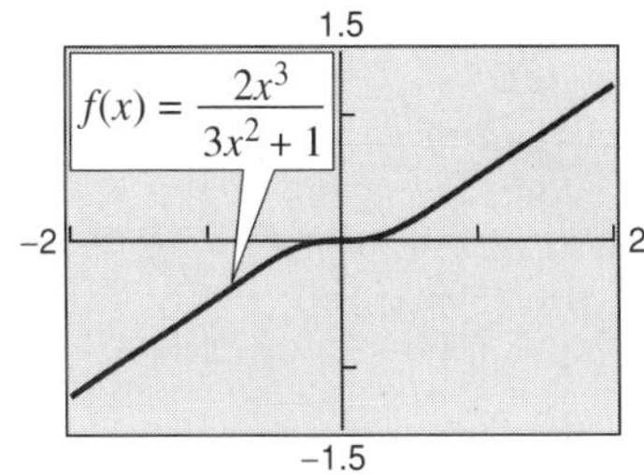

EXAMPLE 2 A Calculator Experiment

Find the horizontal asymptote of the graph of $f(x) = \dfrac{3x^3 + 7x^2 + 2}{-4x^3 + 5}$.

Solution

Because the degree of the numerator and denominator are the same, the horizontal asymptote is given by the ratio of the leading coefficients.

$$y = \frac{\text{leading coefficient of numerator}}{\text{leading coefficient of denominator}} = \frac{3}{-4} = -\frac{3}{4}$$

The tables at the right can be generated on a *TI-82* or *TI-83* using the Indpnt:Ask feature in the table setup menu.

The following tables show how the values of $f(x)$ become closer and closer to $-\frac{3}{4}$ as x becomes increasingly large or small.

x Decreases Without Bound

X	Y1	
−1	0.66667	
−10	−0.5738	
−100	−0.7325	
−1,000	−0.7482	
−10,000	−0.7498	
X =		

x Increases Without Bound

X	Y1	
1	12	
10	−0.9267	
100	−0.7675	
1,000	−0.7518	
10,000	−0.7502	
X =		

The graph is shown in Figure 3.30. Try using a graphing utility to confirm the graph.

Figure 3.30

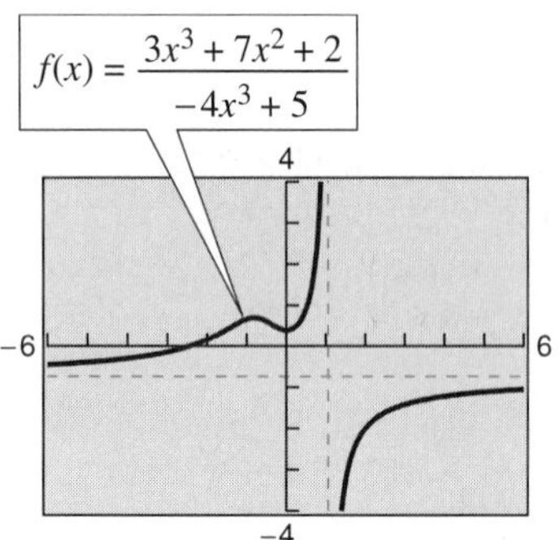

EXAMPLE 3 A Graph with Two Horizontal Asymptotes

A function that is not rational can have two horizontal asymptotes—one to the left and one to the right. For instance, the graph of

$$f(x) = \frac{x + 10}{|x| + 2}$$

is shown in Figure 3.31. It has the line $y = -1$ as a horizontal asymptote to the left and the line $y = 1$ as a horizontal asymptote to the right. You can confirm this by rewriting the function as follows.

$$f(x) = \begin{cases} \dfrac{x + 10}{-x + 2}, & x \le 0 \\ \dfrac{x + 10}{x + 2}, & x > 0 \end{cases}$$

Figure 3.31

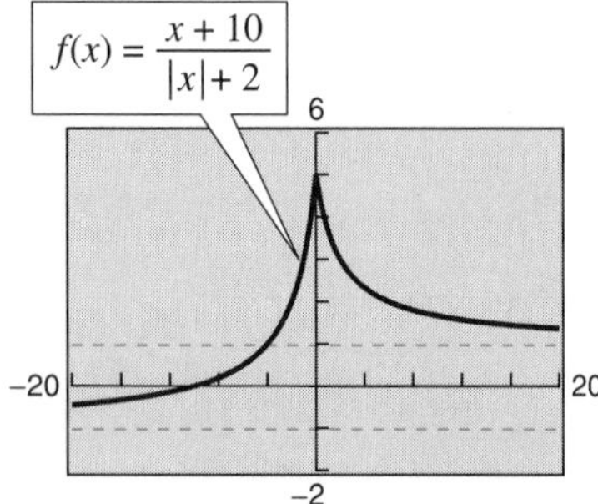

Applications

There are many examples of asymptotic behavior in real life. For instance, Example 4 shows how a vertical asymptote can be used to analyze the cost of removing pollutants from smokestack emissions.

Real Life

EXAMPLE 4 **Cost-Benefit Model**

A utility company burns coal to generate electricity. The cost of removing a certain *percent* of the pollutants from the smokestack emissions is typically not a linear function. That is, if it costs C dollars to remove 25% of the pollutants, it would cost more than $2C$ dollars to remove 50% of the pollutants. As the percent of removed pollutants approaches 100%, the cost tends to become prohibitive. Suppose that the cost C of removing $p\%$ of the smokestack pollutants is given by

$$C = \frac{80{,}000p}{100 - p}, \qquad 0 \leq p < 100.$$

Sketch the graph of this function. Suppose you are a member of a state legislature that is considering a law that would require utility companies to remove 90% of the pollutants from their smokestack emissions. If the current law requires 85% removal, how much additional expense would the new law ask the utility company to spend?

Solution

The graph of this function is shown in Figure 3.32. Note that the graph has a vertical asymptote at $p = 100$. Because the current law requires 85% removal, the current cost to the utility company is

$$C = \frac{80{,}000(85)}{100 - 85} \qquad \text{Evaluate } C \text{ at } p = 85.$$

$$\approx \$453{,}333.$$

Figure 3.32

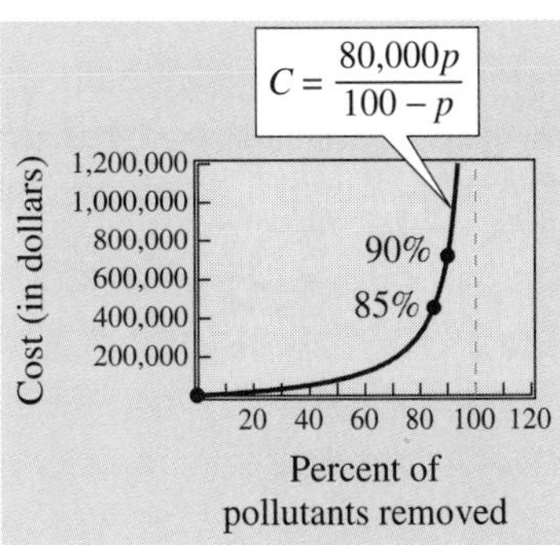

If the new law increased the percent removal to 90%, the cost to the utility company would be

$$C = \frac{80{,}000(90)}{100 - 90} \qquad \text{Evaluate } C \text{ at } p = 90.$$

$$= \$720{,}000.$$

Thus, the new law would require the utility company to spend an additional

$$720{,}000 - 453{,}333 = \$266{,}667.$$

Figure 3.33

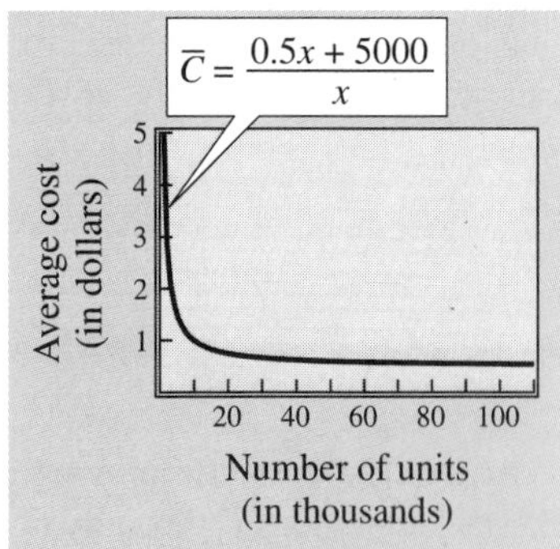

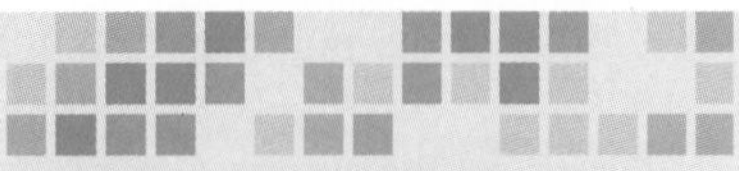

The table feature of a graphing utility can be used to find vertical and horizontal asymptotes of rational functions. For an example of how to use the table feature of a *TI-82* or *TI-83* graphing calculator, see page 45. Use the table feature to find any horizontal or vertical asymptotes of

$$f(x) = \frac{2}{x + 1}.$$

Write a statement explaining how you found the asymptote(s) using the table.

Real Life

EXAMPLE 5 Average Cost of Producing a Product

A business has a cost function of $C = 0.5x + 5000$, where C is measured in dollars and x is the number of units produced. The *average cost per unit* is

$$\overline{C} = \frac{C}{x} = \frac{0.5x + 5000}{x}.$$

Find the average cost per unit when $x = 1000$, 5000, 10,000, and 100,000. What is the horizontal asymptote for this function, and what does it represent?

Solution

When $x = 1000$, $\overline{C} = \dfrac{0.5(1000) + 5000}{1000} = \$5.50.$

When $x = 5000$, $\overline{C} = \dfrac{0.5(5000) + 5000}{5000} = \$1.50.$

When $x = 10{,}000$, $\overline{C} = \dfrac{0.5(10{,}000) + 5000}{10{,}000} = \$1.00.$

When $x = 100{,}000$, $\overline{C} = \dfrac{0.5(100{,}000) + 5000}{100{,}000} = \$0.55.$

As shown in Figure 3.33, the horizontal asymptote is given by the line $\overline{C} = 0.50$. This line represents the least possible unit cost for the product.

Note that this example points out one of the major problems of a small business. That is, it is difficult to have competitively low prices when the production level is low.

Group Activity

Common Factors in the Numerator and Denominator

When sketching the graph of a rational function, be sure that the rational function has no factor that is common to its numerator and denominator. To see why, consider the function given by

$$f(x) = \frac{x(x - 1)}{x}$$

which has a common factor of x in the numerator and denominator. Sketch the graph of this function. Does it have a vertical asymptote at $x = 0$?

3.5 /// EXERCISES

In Exercises 1–6, (a) complete each table, (b) determine the vertical and horizontal asymptotes of the function, and (c) find the domain of the function.

x	$f(x)$
0.5	
0.9	
0.99	
0.999	

x	$f(x)$
1.5	
1.1	
1.01	
1.001	

x	$f(x)$
5	
10	
100	
1000	

1. $f(x) = \dfrac{1}{x-1}$

2. $f(x) = \dfrac{5x}{x-1}$

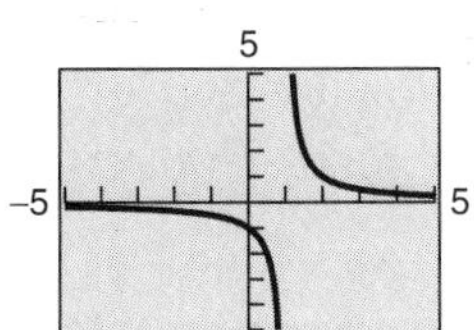

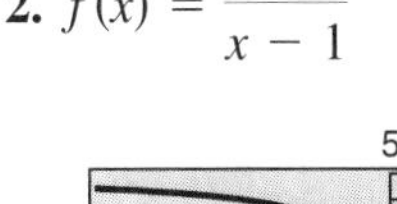

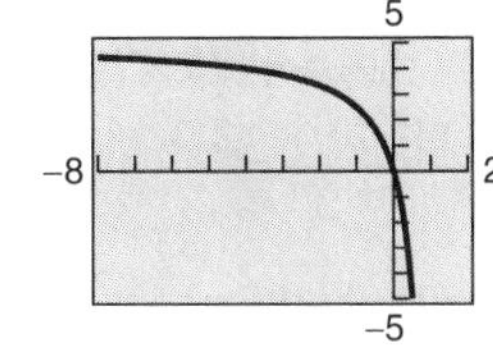

3. $f(x) = \dfrac{3x}{|x-1|}$

4. $f(x) = \dfrac{3}{|x-1|}$

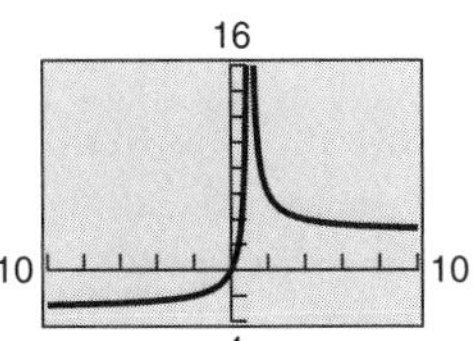

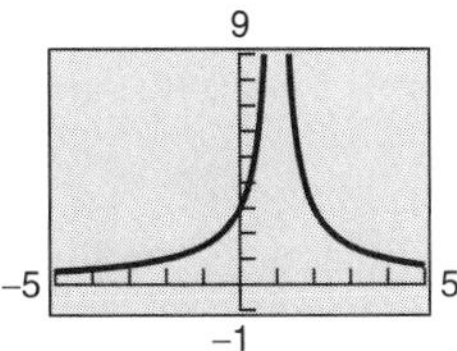

5. $f(x) = \dfrac{3x^2}{x^2-1}$

6. $f(x) = \dfrac{4x}{x^2-1}$

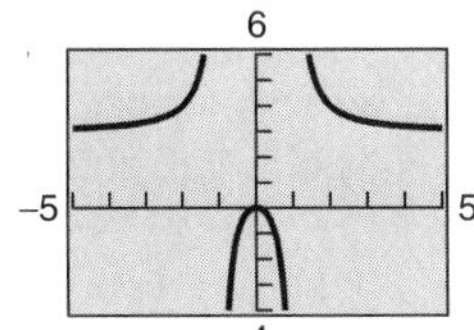

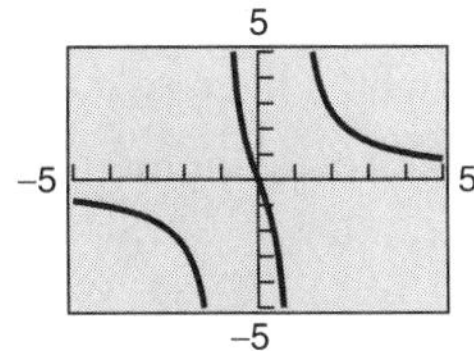

In Exercises 7–12, match the function with its graph. [The graphs are labeled (a), (b), (c), (d), (e), and (f).]

(a)

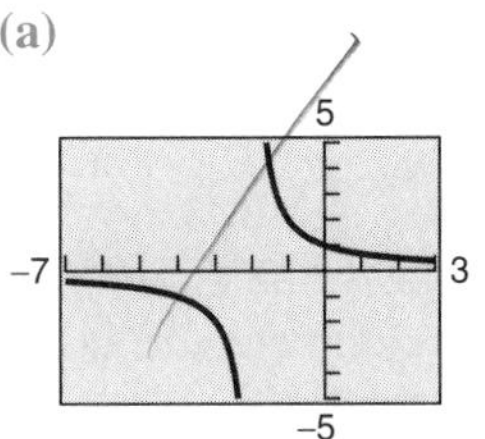

(b)

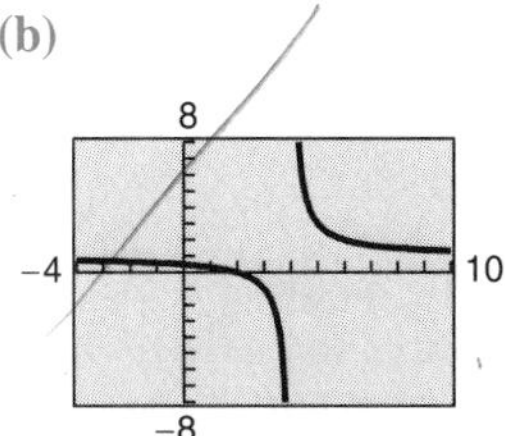

(c)

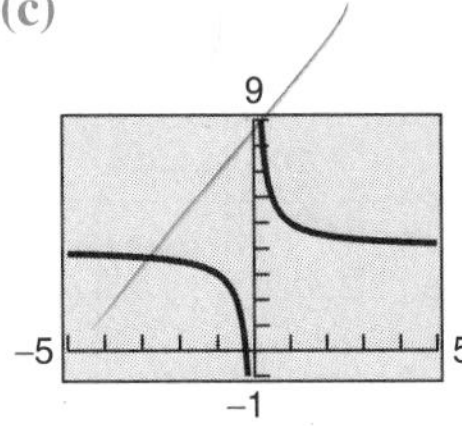

(d)

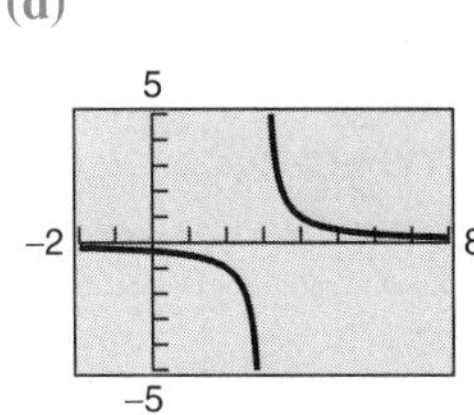

(e)

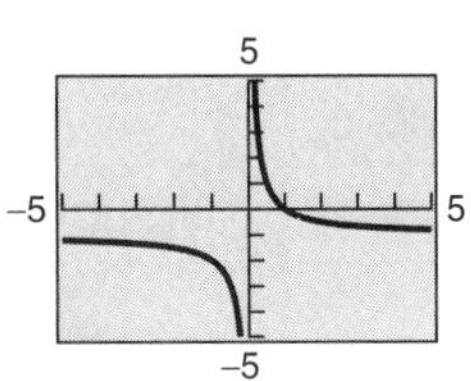

(f)

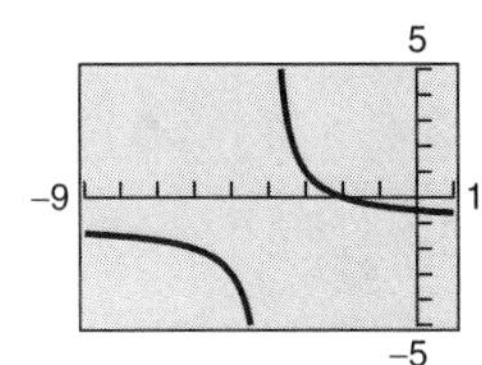

7. $f(x) = \dfrac{2}{x+2}$

8. $f(x) = \dfrac{1}{x-3}$

9. $f(x) = \dfrac{4x+1}{x}$

10. $f(x) = \dfrac{1-x}{x}$

11. $f(x) = \dfrac{x-2}{x-4}$

12. $f(x) = -\dfrac{x+2}{x+4}$

In Exercises 13–20, find the domain of the function and identify any horizontal and vertical asymptotes.

13. $f(x) = \dfrac{1}{x^2}$

14. $f(x) = \dfrac{4}{(x-2)^3}$

15. $f(x) = \dfrac{2 + x}{2 - x}$

16. $f(x) = \dfrac{1 - 5x}{1 + 2x}$

17. $f(x) = \dfrac{x^3}{x^2 - 1}$

18. $f(x) = \dfrac{2x^2}{x + 1}$

19. $f(x) = \dfrac{3x^2 + 1}{x^2 + x + 9}$

20. $f(x) = \dfrac{3x^2 + x - 5}{x^2 + 1}$

Analytical and Numerical Explanation **In Exercises 21–24, (a) determine the domain of f and g, (b) find any vertical asymptotes of f, (c) complete the table, and (d) decide what your results suggest.**

21. $f(x) = \dfrac{x^2 - 4}{x + 2}, \qquad g(x) = x - 2$

x	-4	-3	-2.5	-2	-1.5	-1	0
$f(x)$							
$g(x)$							

22. $f(x) = \dfrac{x^2(x - 3)}{x^2 - 3x}, \qquad g(x) = x$

x	-1	0	1	2	3	3.5	4
$f(x)$							
$g(x)$							

23. $f(x) = \dfrac{x - 3}{x^2 - 3x}, \qquad g(x) = \dfrac{1}{x}$

x	-1	-0.5	0	0.5	2	3	4
$f(x)$							
$g(x)$							

24. $f(x) = \dfrac{2x - 8}{x^2 - 9x + 20}, \qquad g(x) = \dfrac{2}{x - 5}$

x	0	1	2	3	4	5	6
$f(x)$							
$g(x)$							

Think About It **In Exercises 25–28, write a rational function f having the specified characteristics.**

25. Vertical asymptotes: $x = -2, x = 1$

26. Vertical asymptote: None
Horizontal asymptote: $y = 0$

27. Vertical asymptote: None
Horizontal asymptote: $y = 2$

28. Vertical asymptotes: $x = 0, x = \frac{5}{2}$
Horizontal asymptote: $y = -3$

Exploration **In Exercises 29–32, (a) determine the value the function f approaches as the magnitude of x increases. Is $f(x)$ greater than or less than this functional value when (b) x is positive and large in magnitude and (c) x is negative and large in magnitude?**

29. $f(x) = 4 - \dfrac{1}{x}$

30. $f(x) = 2 + \dfrac{1}{x - 3}$

31. $f(x) = \dfrac{2x - 1}{x - 3}$

32. $f(x) = \dfrac{2x - 1}{x^2 + 1}$

In Exercises 33–36, find the zeros (if any) of the rational function.

33. $g(x) = \dfrac{x^2 - 4}{x + 1}$

34. $h(x) = 4 + \dfrac{5}{x^2 + 2}$

35. $f(x) = 1 - \dfrac{2}{x - 3}$

36. $g(x) = \dfrac{x^3 - 8}{x^2 + 1}$

37. *Cost of Clean Water* The cost in millions of dollars for removing $p\%$ of the industrial and municipal pollutants discharged into a river is given by

$$C = \frac{255p}{100 - p}, \qquad 0 \le p < 100.$$

(a) Find the cost of removing 10% of the pollutants.
(b) Find the cost of removing 40% of the pollutants.
(c) Find the cost of removing 75% of the pollutants.
(d) According to this model, would it be possible to remove 100% of the pollutants? Explain.

38. *Recycling Costs* In a pilot project, a rural township was given recycling bins for separating and storing recyclable products. The cost in dollars for supplying bins to $p\%$ of the population is given by

$$C = \frac{25{,}000p}{100 - p}, \qquad 0 \le p < 100.$$

(a) Find the cost of giving bins to 15% of the population.

(b) Find the cost of giving bins to 50% of the population.

(c) Find the cost of giving bins to 90% of the population.

(d) According to this model, would it be possible to supply bins to 100% of the residents? Explain.

39. *Data Analysis* Consider a physics laboratory experiment designed to determine an unknown mass. A flexible metal meter stick is clamped to a table with 50 centimeters overhanging the edge (see figure). Known masses M ranging from 200 grams to 2000 grams are attached to the end of the meter stick. For each mass the meter stick is displaced vertically and then allowed to oscillate. The average time t in seconds of one oscillation for each mass is recorded in the table.

M	200	400	600	800	1000
t	0.450	0.597	0.721	0.831	0.906

M	1200	1400	1600	1800	2000
t	1.003	1.088	1.168	1.218	1.338

A model for the data is given by

$$t = \frac{38M + 16{,}965}{10(M + 5000)}.$$

(a) Use a graphing utility to create a table showing the predicted time based on the model for each of the masses shown in the table. What can you conclude?

(b) Use the model to approximate the mass of an object if the average for one oscillation is 1.056 seconds.

Figure for 39

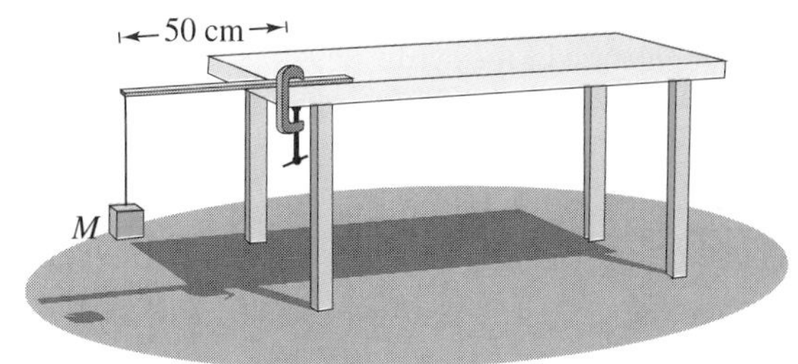

40. *Data Analysis* The endpoints of the interval over which distinct vision is possible are called the *near point* and *far point* of the eye (see figure). With increasing age these points normally change. The table gives the approximate near points y in centimeters for various ages x.

x	10	20	30	40	50
y	7	10	14	22	40

(a) Fit a rational model to the data. Take the reciprocals of the near points to generate the points $(x, 1/y)$. Use the regression capabilities of a graphing utility to fit a line to this data. The resulting line has the form

$$\frac{1}{y} = ax + b.$$

Solve for y.

(b) Use a graphing utility to create a table giving the predicted near point based on the model for each of the ages in the given table.

(c) Do you think the model can be used to predict the near point for a person who is 60 years old? Explain.

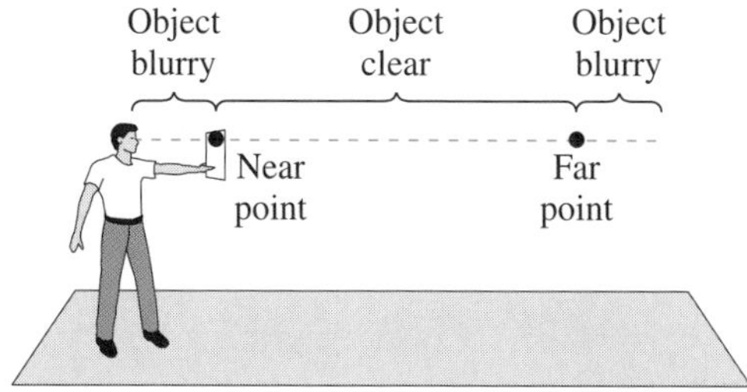

41. *Population of Deer* The game commission introduces 100 deer into newly acquired state game lands. The population of the herd is given by

$$N = \frac{20(5 + 3t)}{1 + 0.04t}, \qquad 0 \leq t$$

where t is the time in years (see figure).

(a) Find the population when t is 5, 10, and 25.

(b) What is the limiting size of the herd as time increases? Explain your reasoning.

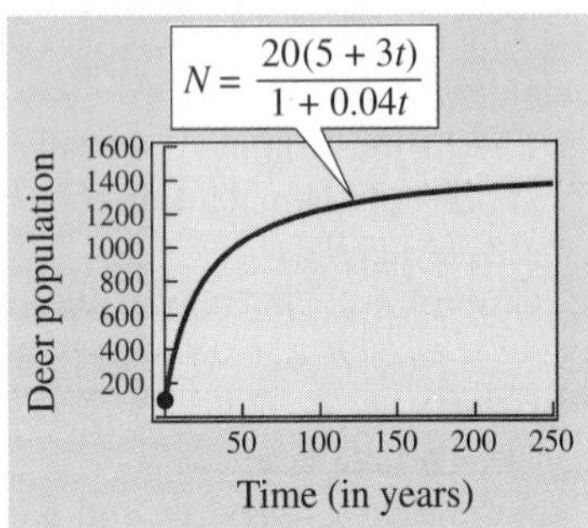

42. *Food Consumption* A biology class performs an experiment comparing the quantity of food consumed by a certain kind of moth with the quantity supplied (see figure). The model for their experimental data is given by

$$y = \frac{1.568x - 0.001}{6.360x + 1}, \qquad 0 < x$$

where x is the quantity (in milligrams) of food supplied and y is the quantity (in milligrams) eaten. At what level of consumption will the moth become satiated?

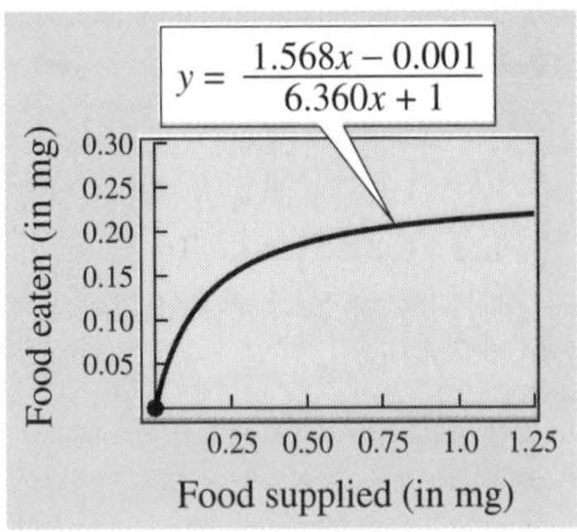

43. *Human Memory Model* Psychologists have developed mathematical models for predicting performance as a function of the number of trials n of a certain task. Consider the learning curve given by

$$P = \frac{0.5 + 0.9(n - 1)}{1 + 0.9(n - 1)}, \qquad 0 < n$$

where P is the percent of correct responses after n trials.

(a) Complete the table for this model. What does it suggest?

n	1	2	3	4	5	6	7	8	9	10
P										

(b) According to this model, what is the limiting percent of correct responses as n increases? Explain your reasoning.

44. *Human Memory Model* How would the limiting percent of correct responses change if the human memory model in Exercise 43 were changed to

$$P = \frac{0.5 + 0.6(n - 1)}{1 + 0.8(n - 1)}, \qquad 0 < n?$$

Review **Solve Exercises 45–48 as a review of the skills and problem-solving techniques you learned in previous sections.**

45. $225 - 50x = 0$

46. $t^3 - 50t = 0$

47. $2z^2 - 3z - 35 = 0$

48. $x(10 - x) = 25$

3.6 Graphs of Rational Functions

The Graph of a Rational Function / Slant Asymptotes / Application

The Graph of a Rational Function

Note Testing for symmetry can be useful, especially for simple rational functions. For example, the graph of $f(x) = 1/x$ is symmetrical with respect to the origin, and the graph of $g(x) = 1/x^2$ is symmetrical with respect to the y-axis.

Analyzing Graphs of Rational Functions

Let $f(x) = p(x)/q(x)$, where $p(x)$ and $q(x)$ are polynomials with no common factors.

1. The y-intercept (if any) is the value of $f(0)$.
2. The x-intercepts (if any) are the zeros of the numerator—that is, the solutions of the equation $p(x) = 0$.
3. The vertical asymptotes (if any) are the zeros of the denominator—that is, the solutions of the equation $q(x) = 0$.
4. The horizontal asymptote (if any) is the value that $f(x)$ approaches as x increases or decreases without bound.
5. Determining the behavior of the graph *between* and *beyond* each x-intercept and vertical asymptote is crucial for describing the complete graph of a rational function.

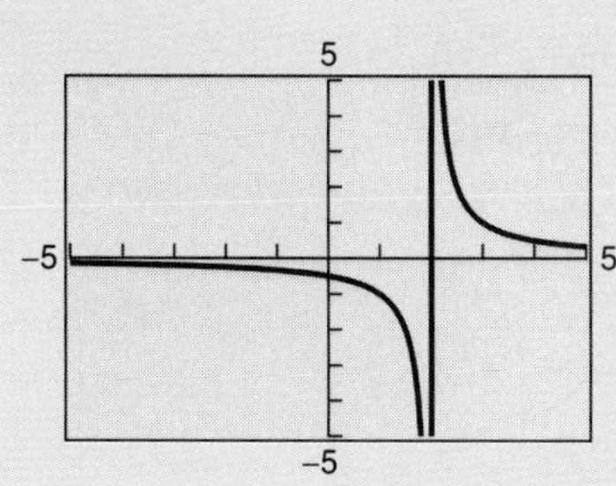

Graphing utilities have difficulty sketching graphs of rational functions that have vertical asymptotes. Often, the utility will connect parts of the graph that are not supposed to be connected. For instance, the screen at the left shows the graph of

$$f(x) = \frac{1}{x - 2}.$$

Notice that the graph should consist of two *unconnected* portions—one to the left of $x = 2$ and the other to the right of $x = 2$. To eliminate this problem, you can try changing the *mode* of the graphing utility to *dot mode.* The problem with this is that the graph is then represented as a collection of dots rather than as a smooth curve.

EXAMPLE 1 **Analyzing a Rational Function**

Analyze the function $g(x) = \dfrac{3}{x - 2}$.

Solution

y-Intercept: $\left(0, -\frac{3}{2}\right)$, because $g(0) = -\frac{3}{2}$.

x-Intercept: None, because $3 \neq 0$.

Vertical Asymptote: $x = 2$, zero of denominator

Horizontal Asymptote: $y = 0$, degree of $p(x) <$ degree of $q(x)$

Additional Points:

x	-4	1	3	5
$g(x)$	-0.5	-3	3	1

By plotting the intercepts, asymptotes, and a few additional points, you can obtain the graph shown in Figure 3.34. Confirm this with a graphing utility.

Figure 3.34

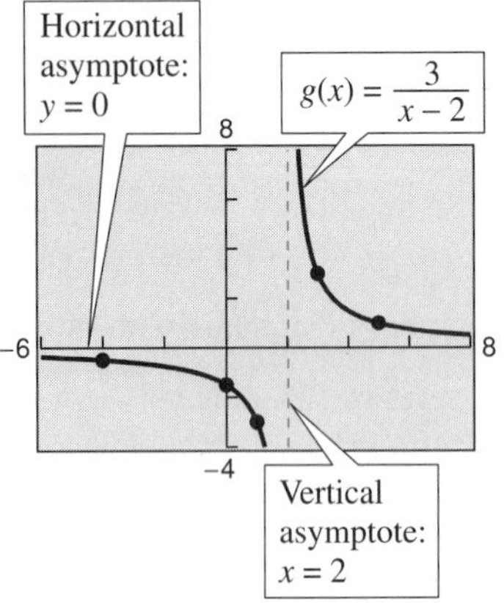

Note The graph of g in Example 1 is a vertical stretch and a right shift of the graph of $f(x) = 1/x$ because

$$g(x) = \frac{3}{x - 2} = 3\left(\frac{1}{x - 2}\right) = 3f(x - 2).$$

EXAMPLE 2 **Analyzing a Rational Function**

Analyze the function $f(x) = \dfrac{2x - 1}{x}$.

Solution

y-Intercept: None, because $x = 0$ is not in the domain.

x-Intercept: $\left(\frac{1}{2}, 0\right)$, because $2x - 1 = 0$.

Vertical Asymptote: $x = 0$, zero of denominator

Horizontal Asymptote: $y = 2$, degree of $p(x) =$ degree of $q(x)$

Additional Points:

x	-4	-1	$\frac{1}{4}$	4
$f(x)$	2.25	3	-2	1.75

By plotting the intercepts, asymptotes, and a few additional points, you can obtain the graph shown in Figure 3.35. Confirm this with a graphing utility.

Figure 3.35

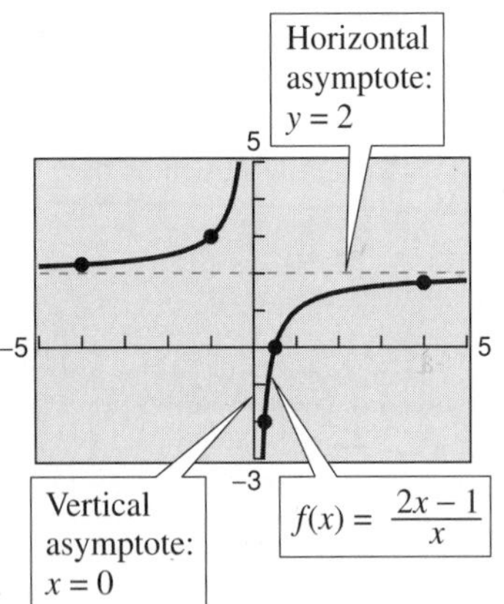

EXAMPLE 3 Analyzing a Rational Function

Analyze the function $f(x) = \dfrac{x}{x^2 - x - 2}$.

Solution

By factoring the denominator, you have

$$f(x) = \frac{x}{x^2 - x - 2} = \frac{x}{(x + 1)(x - 2)}.$$

y-Intercept: $(0, 0)$, because $f(0) = 0$.

x-Intercept: $(0, 0)$

Vertical Asymptotes: $x = -1$, $x = 2$, zeros of denominator

Horizontal Asymptote: $y = 0$, degree of $p(x) <$ degree of $q(x)$

Additional Points:

x	-3	-0.5	1	3
$f(x)$	-0.3	0.4	-0.5	0.75

The graph is shown in Figure 3.36.

Figure 3.36

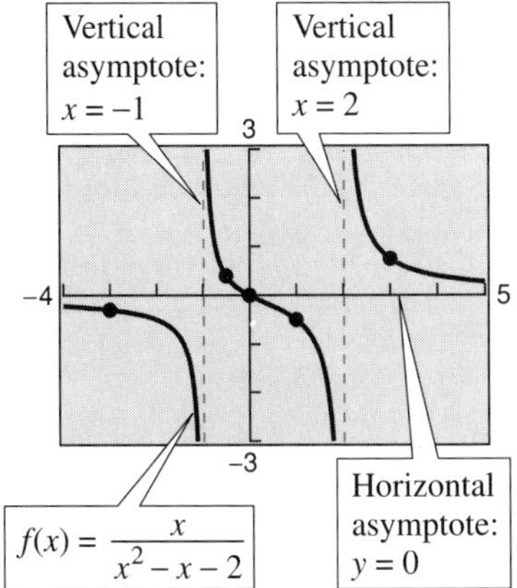

EXAMPLE 4 Analyzing a Rational Function

Analyze the function $f(x) = \dfrac{2(x^2 - 9)}{x^2 - 4}$.

Solution

By factoring the numerator and denominator, you have

$$f(x) = \frac{2(x^2 - 9)}{x^2 - 4} = \frac{2(x - 3)(x + 3)}{(x - 2)(x + 2)}.$$

y-Intercept: $\left(0, \frac{9}{2}\right)$, because $f(0) = \frac{9}{2}$.

x-Intercepts: $(-3, 0)$, $(3, 0)$

Vertical Asymptotes: $x = -2$, $x = 2$, zeros of denominator

Horizontal Asymptote: $y = 2$, degree of $p(x) =$ degree of $q(x)$

Symmetry: With respect to y-axis, because $f(-x) = f(x)$.

Additional Points:

x	0.5	2.5	6
$f(x)$	4.67	-2.44	1.69

The graph is shown in Figure 3.37.

Figure 3.37

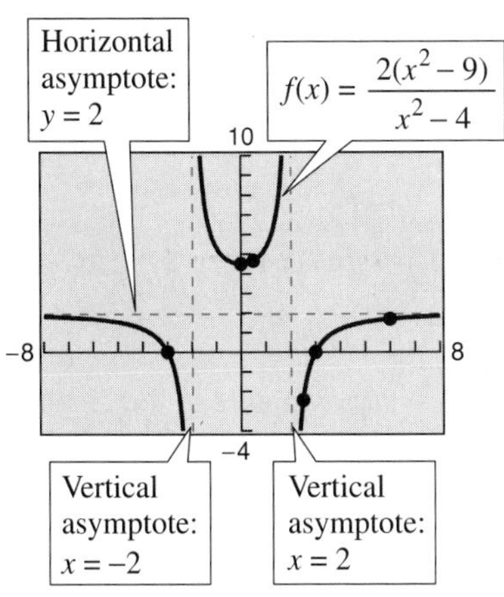

Slant Asymptotes

If the degree of the numerator of a rational function is exactly *one more* than the degree of its denominator, the graph of the function has a **slant asymptote.** For example, the graph of

$$f(x) = \frac{x^2 - x}{x + 1}$$

has a slant asymptote, as shown in Figure 3.38. To find the equation of a slant asymptote, use long division. For instance, by dividing $x + 1$ into $x^2 - x$, you have

$$f(x) = \frac{x^2 - x}{x + 1} = \underbrace{x - 2}_{\text{Slant asymptote } (y = x - 2)} + \frac{2}{x + 1}.$$

$\frac{2}{x+1} \to 0$ as $x \to \infty$

Figure 3.38

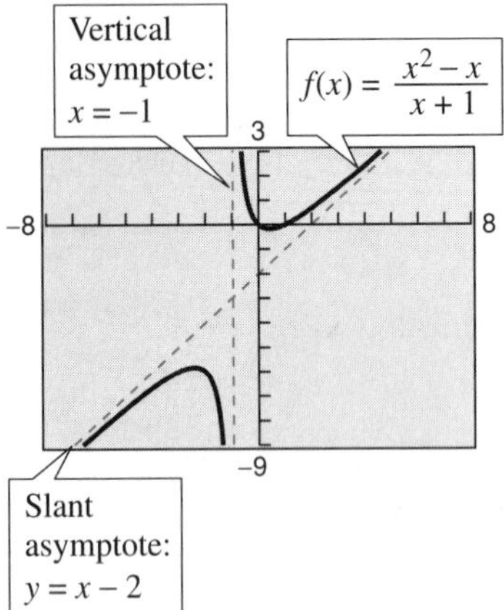

In Figure 3.38, notice that the graph of f approaches the line $y = x - 2$ as x moves to the right or left.

EXAMPLE 5 A Rational Function with a Slant Asymptote

Graph the function $f(x) = \dfrac{x^2 - x - 2}{x - 1}$.

Solution

First write $f(x)$ in two different ways. Factoring the numerator

$$f(x) = \frac{x^2 - x - 2}{x - 1} = \frac{(x - 2)(x + 1)}{x - 1}$$

allows you to recognize the x-intercepts, and long division

$$f(x) = \frac{x^2 - x - 2}{x - 1} = x - \frac{2}{x - 1}$$

$\frac{2}{x-1} \to 0$ as $x \to \infty$

allows you to recognize that the line $y = x$ is a slant asymptote of the graph.

y-Intercept: $(0, 2)$, because $f(0) = 2$.

x-Intercepts: $(-1, 0)$, $(2, 0)$

Vertical Asymptote: $x = 1$

Slant Asymptote: $y = x$

Additional Points:

x	-2	0.5	1.5	3
$f(x)$	-1.33	4.5	-2.5	2

Figure 3.39

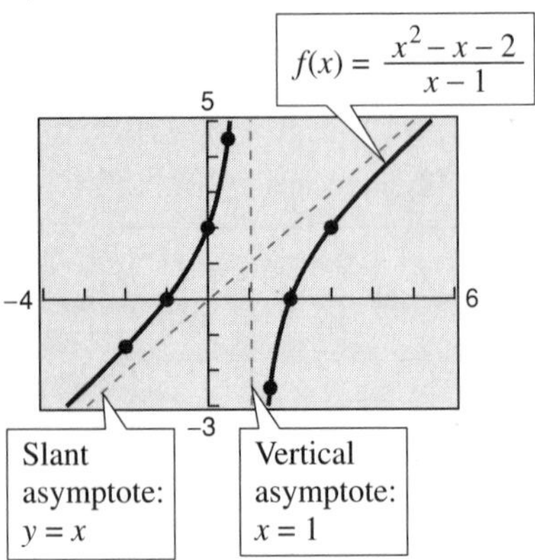

The graph is shown in Figure 3.39.

Figure 3.40

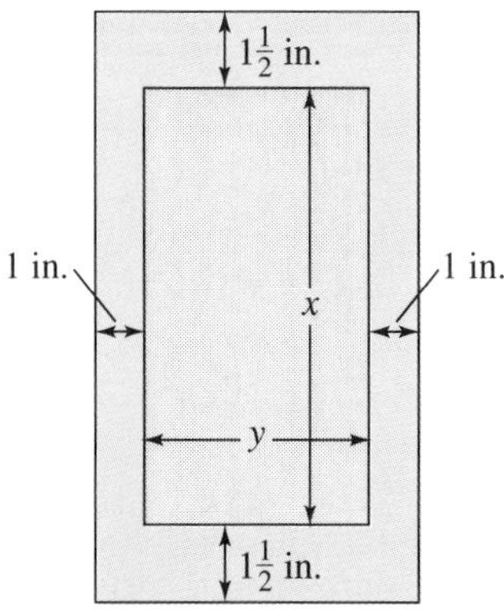

Application

EXAMPLE 6 **Finding a Minimum Area** *Real Life*

A rectangular page is to contain 48 square inches of print. The margins at the top and bottom of the page are each $1\frac{1}{2}$ inches. The margins on each side are 1 inch. What should the dimensions of the page be so that the minimum amount of paper is used?

Solution

Let A be the area to be minimized. From Figure 3.40, you can write

$$A = (x + 3)(y + 2).$$

The printed area inside the margins is given by $48 = xy$ or $y = 48/x$. To find the minimum area, rewrite the equation for A in terms of just one variable by substituting $48/x$ for y.

$$A = (x + 3)\left(\frac{48}{x} + 2\right) = \frac{(x + 3)(48 + 2x)}{x}, \qquad x > 0$$

Figure 3.41

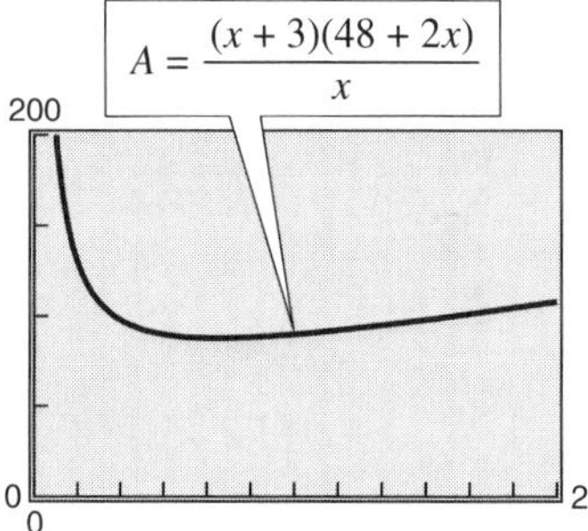

The graph of this rational function is shown in Figure 3.41. Because x represents the height of the printed area, you need consider only the portion of the graph for which x is positive. Using the zoom and trace features of a graphing utility, you can approximate the minimum value of A to occur when $x \approx 8.5$ inches. The corresponding value of y is $48/8.5 \approx 5.6$ inches. Thus, the dimensions should be

$$x + 3 \approx 11.5 \text{ inches} \qquad \text{by} \qquad y + 2 \approx 7.6 \text{ inches}.$$

If you go on to take a course in calculus, you will learn a technique for finding the exact value of x that produces a minimum area. In this case, that value is $x = 6\sqrt{2} \approx 8.485$.

Group Activity *Asymptotes of Graphs of Rational Functions*

Discuss whether or not it is possible for the graph of a rational function to cross its horizontal asymptote or its slant asymptote. Use the graphs of the following functions to investigate these questions. What can you conclude?

$$f(x) = \frac{x}{x^2 + 1} \qquad \text{and} \qquad g(x) = \frac{x^3}{x^2 + 1}$$

3.6 /// EXERCISES

In Exercises 1–4, use a graphing utility to graph $f(x) = 2/x$ and the function g in the same viewing rectangle. Describe the relationship between the two graphs.

1. $g(x) = f(x) + 1$ **2.** $g(x) = f(x - 1)$

3. $g(x) = -f(x)$ **4.** $g(x) = \frac{1}{2}f(x + 2)$

In Exercises 5–8, use a graphing utility to graph $f(x) = 2/x^2$ and the function g in the same viewing rectangle. Describe the relationship between the two graphs.

5. $g(x) = f(x) - 2$ **6.** $g(x) = -f(x)$

7. $g(x) = f(x - 2)$ **8.** $g(x) = \frac{1}{4}f(x)$

In Exercises 9–12, use a graphing utility to graph $f(x) = 4/x^3$ and the function g in the same viewing rectangle. Describe the relationship between the two graphs.

9. $g(x) = f(x + 2)$ **10.** $g(x) = f(x) + 1$

11. $g(x) = -f(x)$ **12.** $g(x) = \frac{1}{4}f(x)$

In Exercises 13–30, sketch the graph of the rational function. As sketching aids, check for intercepts, symmetry, vertical asymptotes, and horizontal asymptotes. Use a graphing utility to verify your graph.

13. $f(x) = \dfrac{1}{x + 2}$ **14.** $f(x) = \dfrac{1}{x - 3}$

15. $h(x) = \dfrac{-1}{x + 2}$ **16.** $g(x) = \dfrac{1}{3 - x}$

17. $C(x) = \dfrac{5 + 2x}{1 + x}$ **18.** $P(x) = \dfrac{1 - 3x}{1 - x}$

19. $g(x) = \dfrac{1}{x + 2} + 2$ **20.** $f(t) = \dfrac{1 - 2t}{t}$

21. $f(x) = \dfrac{x^2}{x^2 + 9}$ **22.** $f(x) = 2 - \dfrac{3}{x^2}$

23. $h(x) = \dfrac{x^2}{x^2 - 9}$ **24.** $g(x) = \dfrac{x}{x^2 - 9}$

25. $g(s) = \dfrac{s}{s^2 + 1}$ **26.** $f(x) = -\dfrac{1}{(x - 2)^2}$

27. $g(x) = \dfrac{4(x + 1)}{x(x - 4)}$ **28.** $h(x) = \dfrac{2}{x^2(x - 2)}$

29. $f(x) = \dfrac{3x}{x^2 - x - 2}$ **30.** $f(x) = \dfrac{2x}{x^2 + x - 2}$

In Exercises 31–40, use a graphing utility to obtain the graph of the function. Give the domain of the function and identify any vertical or horizontal asymptotes.

31. $f(x) = \dfrac{2 + x}{1 - x}$ **32.** $f(x) = \dfrac{3 - x}{2 - x}$

33. $f(t) = \dfrac{3t + 1}{t}$ **34.** $h(x) = \dfrac{1}{x - 3} + 1$

35. $h(t) = \dfrac{4}{t^2 + 1}$ **36.** $g(x) = -\dfrac{x}{(x - 2)^2}$

37. $f(t) = \dfrac{2t^2}{t^2 - 4}$ **38.** $f(x) = \dfrac{x + 4}{x^2 + x - 6}$

39. $f(x) = \dfrac{20x}{x^2 + 1} - \dfrac{1}{x}$

40. $f(x) = 5\left(\dfrac{1}{x - 4} - \dfrac{1}{x + 2}\right)$

Exploration **In Exercises 41 and 42, use a graphing utility to obtain the graph of the function. What do you observe about its asymptotes?**

41. $h(x) = \dfrac{6x}{\sqrt{x^2 + 1}}$ **42.** $g(x) = \dfrac{4|x - 2|}{x + 1}$

Exploration **In Exercises 43 and 44, use a graphing utility to obtain the graph of the function. What do you observe about its asymptotes?**

43. $f(x) = \dfrac{4(x - 1)^2}{x^2 - 4x + 5}$

44. $g(x) = \dfrac{3x^4 - 5x + 3}{x^4 + 1}$

Think About It In Exercises 45 and 46, use a graphing utility to obtain the graph of the function. Explain why there is no vertical asymptote when a superficial examination of the function may indicate that there should be one.

45. $h(x) = \dfrac{6 - 2x}{3 - x}$

46. $g(x) = \dfrac{x^2 + x - 2}{x - 1}$

47. *True or False?* If the graph of a rational function f has a vertical asymptote at $x = 5$, it is possible to sketch the graph without lifting your pencil from the paper. Explain.

48. *Essay* Write a paragraph discussing whether every rational function has a vertical asymptote.

In Exercises 49–56, sketch the graph of the rational function. As sketching aids, check for intercepts, symmetry, vertical asymptotes, and slant asymptotes. Use a graphing utility to verify your graph.

49. $f(x) = \dfrac{2x^2 + 1}{x}$

50. $f(x) = \dfrac{1 - x^2}{x}$

51. $g(x) = \dfrac{x^2 + 1}{x}$

52. $h(x) = \dfrac{x^2}{x - 1}$

53. $f(x) = \dfrac{x^3}{x^2 - 1}$

54. $g(x) = \dfrac{x^3}{2x^2 - 8}$

55. $f(x) = \dfrac{x^2 - x + 1}{x - 1}$

56. $f(x) = \dfrac{2x^2 - 5x + 5}{x - 2}$

Graphical Reasoning In Exercises 57–60, (a) use the graph to determine any x-intercepts of the rational function, and (b) set $y = 0$ and solve the resulting equation to confirm your result in part (a).

57. $y = \dfrac{x + 1}{x - 3}$

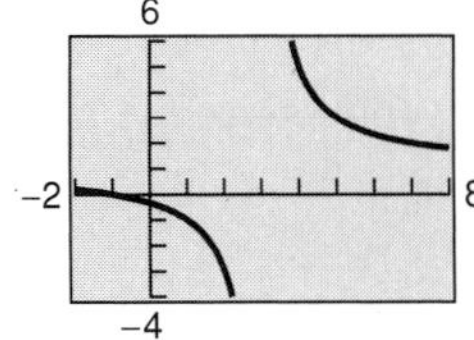

58. $y = \dfrac{2x}{x - 3}$

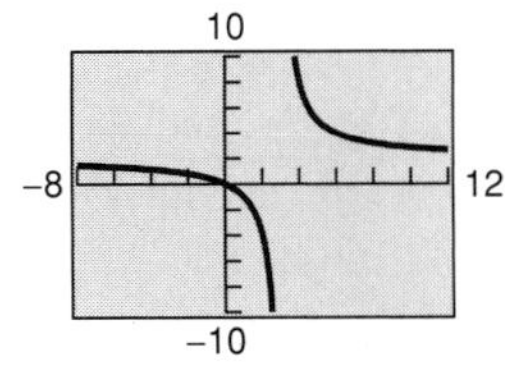

59. $y = \dfrac{1}{x} - x$

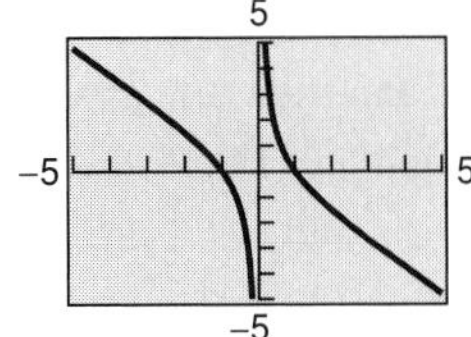

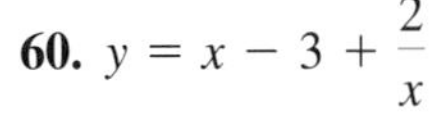

60. $y = x - 3 + \dfrac{2}{x}$

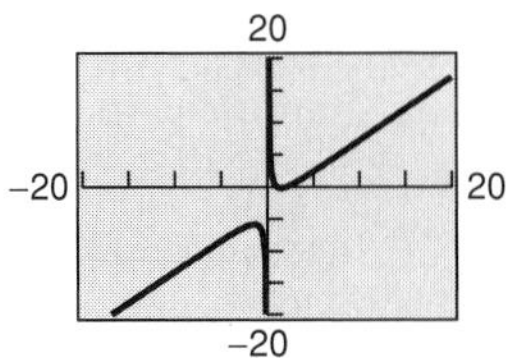

In Exercises 61–64, use a graphing utility to graph the rational function. Give the domain of the function and identify any asymptotes. Then zoom out far enough so that the graph appears as a line. Identify the line.

61. $y = \dfrac{2x^2 + x}{x + 1}$

62. $y = \dfrac{x^2 + 5x + 8}{x + 3}$

63. $y = \dfrac{1 + 3x^2 - x^3}{x^2}$

64. $y = \dfrac{12 - 2x - x^2}{2(4 + x)}$

Graphical Reasoning In Exercises 65–68, (a) use a graphing utility to graph the function and determine any x-intercepts, and (b) set $y = 0$ and solve the resulting equation to confirm your result in part (a).

65. $y = \dfrac{1}{x + 5} + \dfrac{4}{x}$

66. $y = 20\left(\dfrac{2}{x + 1} - \dfrac{3}{x}\right)$

67. $y = x - \dfrac{6}{x - 1}$

68. $y = x - \dfrac{9}{x}$

69. *Concentration of a Mixture* A 1000-liter tank contains 50 liters of a 25% brine solution. You add x liters of a 75% brine solution to the tank.

(a) Show that the concentration C of the final mixture is given by

$$C = \frac{3x + 50}{4(x + 50)}.$$

(b) Determine the domain of the function on the basis of the physical constraints of the problem.

(c) Use a graphing utility to graph the function. As the tank is filled, what happens to the rate at which the concentration of brine increases? What does the concentration of brine appear to approach?

70. *Dimensions of a Rectangle* A rectangular region of length x and width y has an area of 500 square meters.

(a) Express the width y as a function of x.

(b) Determine the domain of the function on the basis of the physical constraints of the problem.

(c) Sketch a graph of the function and determine the width of the rectangle if $x = 30$ meters.

In Exercises 71 and 72, use a graphing utility to obtain the graph of the function and locate any relative maximum or minimum points on the graph.

71. $f(x) = \dfrac{3(x+1)}{x^2+x+1}$ **72.** $C(x) = x + \dfrac{32}{x}$

73. *Page Design* A page that is x inches wide and y inches high contains 30 square inches of print. The margins at the top and bottom are 2 inches and the margins on each side are 1 inch (see figure).

(a) Show that the total area A of the page is given by

$$A = \frac{2x(2x+11)}{x-2}.$$

(b) Determine the domain of the function on the basis of the physical constraints of the problem.

(c) Use a graphing utility to graph the area function and approximate the page size such that the minimum amount of paper will be used.

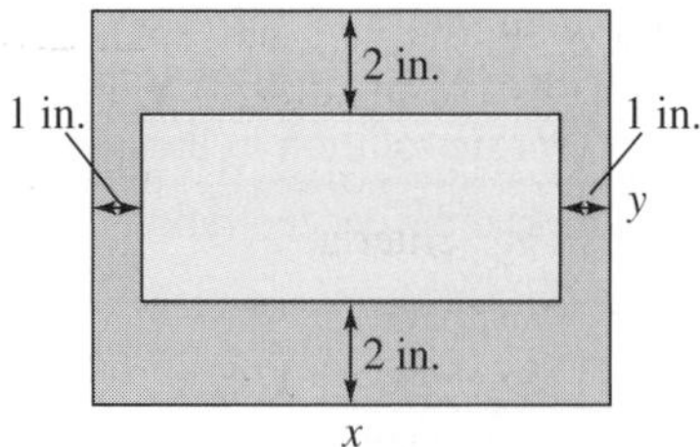

74. *Minimum Area* A right triangle is formed in the first quadrant by the x-axis, the y-axis, and a line segment through the point (3, 2) (see figure).

(a) Show that an equation of the line segment is

$$y = \frac{2(a-x)}{a-3}, \qquad 0 \le x \le a.$$

(b) Show that the area of the triangle is $A = \dfrac{a^2}{a-3}$.

(c) Use a graphing utility to graph the area function and estimate the value of a that yields a minimum area. Estimate the minimum area.

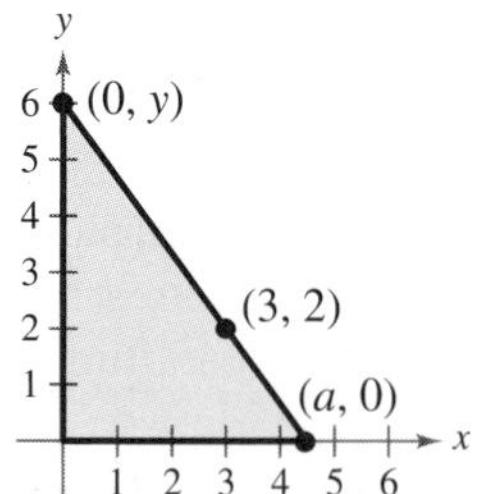

75. *Ordering and Transportation Cost* The ordering and transportation cost C for the components used in manufacturing a certain product is given by

$$C = 100\left(\frac{200}{x^2} + \frac{x}{x+30}\right), \qquad 1 \le x$$

where C is measured in thousands of dollars and x is the order size in hundreds. Use a graphing utility to obtain the graph of the cost function and, from the graph, estimate the order size that minimizes cost.

76. *Average Cost* The cost of producing x units of a product is $C = 0.2x^2 + 10x + 5$, and therefore the average cost per unit is

$$\overline{C} = \frac{C}{x} = \frac{0.2x^2 + 10x + 5}{x}, \qquad 0 < x.$$

Sketch the graph of the average cost function, and estimate the number of units that should be produced to minimize the average cost per unit.

77. *Medicine* The concentration of a certain chemical in the bloodstream t hours after injection into muscle tissue is given by

$$C = \frac{3t^2 + t}{t^3 + 50}, \qquad 0 \le t.$$

(a) Determine the horizontal asymptote of the function and interpret its meaning in the context of the problem.

(b) Use a graphing utility to graph the function and approximate the time when the bloodstream concentration is greatest.

78. *Numerical and Graphical Analysis* A driver averaged 50 miles per hour on the round trip between home and a city 100 miles away. The average speeds for going and returning were x and y miles per hour.

(a) Show that $y = \dfrac{25x}{x - 25}$.

(b) Determine the vertical and horizontal asymptotes of the function.

(c) Use a graphing utility to complete the table. What do you observe?

x	30	35	40	45	50	55	60
y							

(d) Use a graphing utility to graph the function.

(e) Is it possible to average 20 miles per hour in one direction and still average 50 miles per hour on the round trip? Explain.

79. *Comparing Models* The per capita consumption y (in pounds) of lowfat milk in the United States from 1983 through 1992 is given in the table. (Source: U.S. Department of Agriculture)

Year	1983	1984	1985	1986	1987
y	75.4	78.6	83.3	88.1	89.6

Year	1988	1989	1990	1991	1992
y	89.9	96.3	98.3	99.9	99.3

For each of the following, let t be the time in years, with $t = 3$ corresponding to 1983.

(a) A model for the data is given by $y = \dfrac{5000t}{45t + 69}$.

Use a graphing utility to plot the data points and graph the model in the same viewing rectangle.

(b) Use the regression capabilities of a graphing utility to fit a line to the data.

(c) Use the regression capabilities of a graphing utility to fit a parabola to the data.

(d) Which of the three models would you recommend as a predictor of future lowfat milk consumption? Explain your reasoning.

80. *Data Analysis* The numbers N (in thousands) of insured commercial banks in the United States for the years 1984 through 1993 are given in the table. (Source: U.S. Federal Deposit Insurance Corporation)

Year	1984	1985	1986	1987	1988
N	14.5	14.4	14.2	13.7	13.1

Year	1989	1990	1991	1992	1993
N	12.7	12.3	11.9	11.5	11.0

For each of the following, let t be the time in years, with $t = 4$ corresponding to 1984.

(a) Use the regression capabilities of a graphing utility to fit a line to the data. Use a graphing utility to plot the data points and graph the model in the same viewing rectangle.

(b) Fit a rational model to the data. Take the reciprocal of N to generate the points $(t, 1/N)$. Use the regression capabilities of a graphing utility to fit a line to this data. The resulting line has the form

$$\frac{1}{N} = at + b.$$

Solve for N. Use a graphing utility to plot the data points and graph the rational model in the same viewing rectangle.

(c) Use a graphing utility to create a table giving the predicted number of banks based on each model for each of the years in the given table. Which model do you prefer? Why?

81. *Think About It* Write a rational function satisfying the following criteria.

Vertical asymptote: $x = 2$
Slant asymptote: $y = x + 1$
Zero of the function: $x = -2$

82. *Think About It* Write a rational function satisfying the following criteria.

Vertical asymptote: $x = -1$
Horizontal asymptote: $y = 2$
Zero of the function: $x = 3$

Focus on Concepts

In this chapter you studied several concepts related to polynomial and rational functions. You can use the following questions to check your understanding of several of these basic concepts. The answers to these questions are given in the back of the book.

1. *Modeling Company Profits* The profits P (in millions of dollars) for a company are modeled by a quadratic function of the form

$P = at^2 + bt + c$

where t represents the year. If you were president of the company, which of the models would you prefer? Explain your reasoning.

(a) a is positive and $-b/(2a) \le t$.

(b) a is positive and $t \le -b/(2a)$.

(c) a is negative and $-b/(2a) \le t$.

(d) a is negative and $t \le -b/(2a)$.

2. *Graphs of Polynomial Functions* Describe a polynomial function that could represent each graph. (Indicate the degree of the function and the sign of its leading coefficient.)

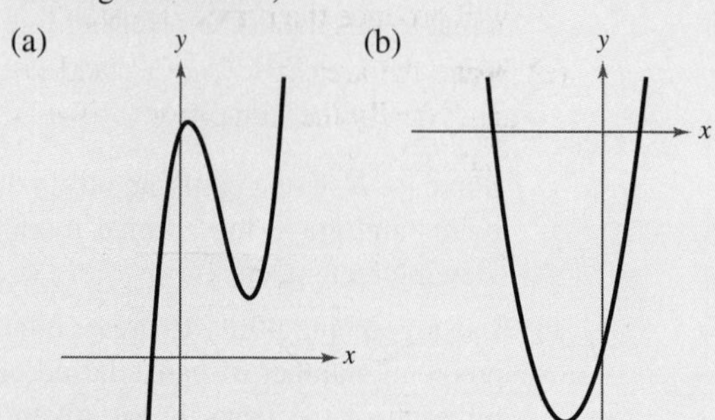

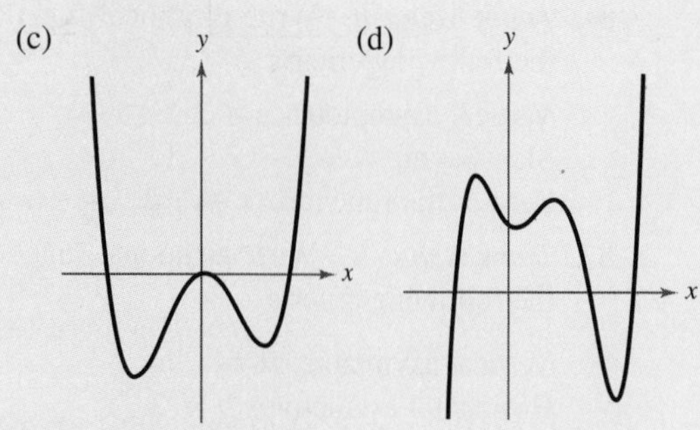

3. *Zeros of Polynomial Functions* The real line shows all of the real zeros of a polynomial function.

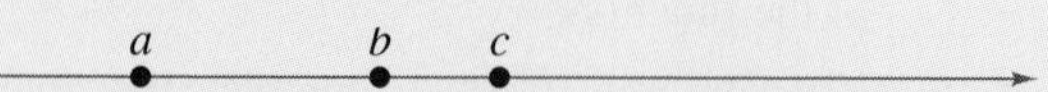

(a) Could the function be a second-degree polynomial function? If so, sketch its graph.

(b) Could the function be a third-degree polynomial function? If so, sketch its graph.

(c) Could the function be a fourth-degree polynomial function? If so, sketch its graph.

(d) Could the function be a fifth-degree polynomial function? If so, sketch its graph.

Graphical Reasoning In Exercises 4–7, write a function whose graph resembles the given graph. Explain your reasoning.

4.

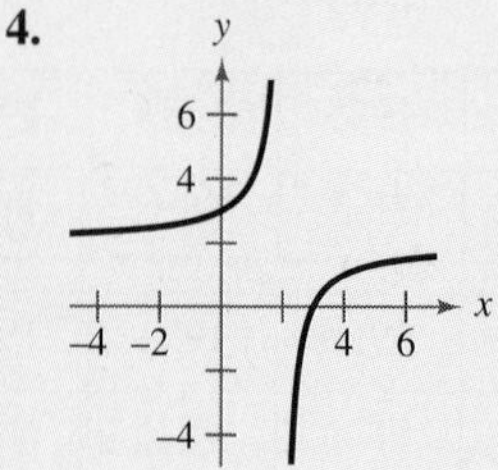

5.

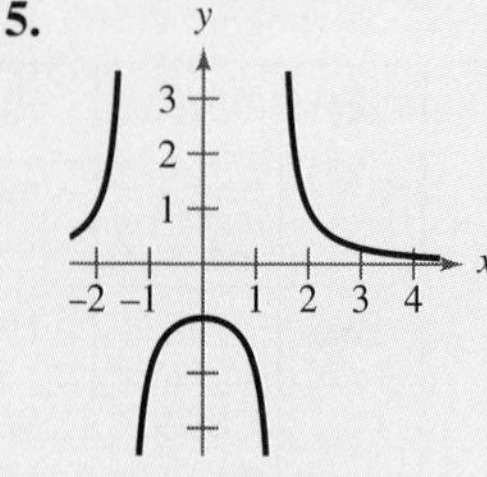

6.

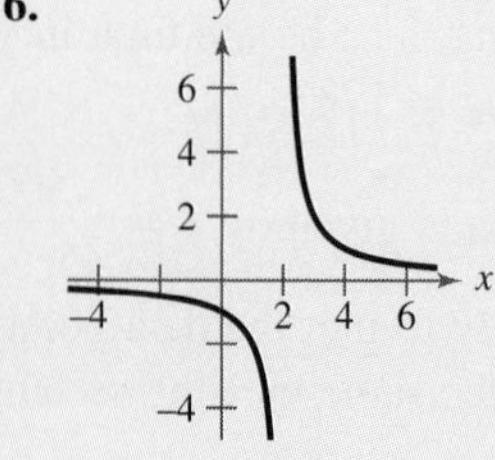

7.

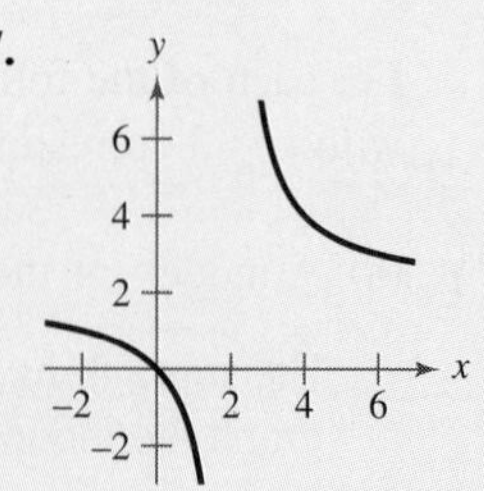

3 /// REVIEW EXERCISES

In Exercises 1–4, sketch the graph of the quadratic function. Identify the vertex and the intercepts.

1. $f(x) = \left(x + \frac{3}{2}\right)^2 + 1$

2. $f(x) = (x - 4)^2 - 4$

3. $f(x) = \dfrac{1}{3}(x^2 + 5x - 4)$

4. $f(x) = 3x^2 - 12x + 11$

In Exercises 5 and 6, find the quadratic function that has the indicated vertex and whose graph passes through the given point.

5. Vertex: $(1, -4)$; Point: $(2, -3)$

6. Vertex: $(2, 3)$; Point: $(-1, 6)$

Graphical Reasoning **In Exercises 7 and 8, use a graphing utility to graph each equation in the same viewing rectangle. Describe how each graph differs from the graph of $y = x^2$.**

7. (a) $y = 2x^2$ (b) $y = -2x^2$

(c) $y = x^2 + 2$ (d) $y = (x + 2)^2$

8. (a) $y = x^2 - 4$ (b) $y = 4 - x^2$

(c) $y = (x - 3)^2$ (d) $y = \frac{1}{2}x^2 - 1$

In Exercises 9–16, find the maximum or minimum value of the quadratic function.

9. $g(x) = x^2 - 2x$

10. $f(x) = x^2 + 8x + 10$

11. $f(x) = 6x - x^2$

12. $h(x) = 3 + 4x - x^2$

13. $f(t) = -2t^2 + 4t + 1$

14. $h(x) = 4x^2 + 4x + 13$

15. $h(x) = x^2 + 5x - 4$

16. $f(x) = 4x^2 + 4x + 5$

17. *Numerical, Graphical, and Analytical Analysis* A rectangle is inscribed in the region bounded by the x-axis, the y-axis, and the graph of $x + 2y - 8 = 0$ (see figure).

(a) Complete six rows of a table like the one below.

x	y	Area
1	$4 - \frac{1}{2}(1)$	$(1)\left[4 - \frac{1}{2}(1)\right] = \frac{7}{2}$
2	$4 - \frac{1}{2}(2)$	$(2)\left[4 - \frac{1}{2}(2)\right] = 6$

(b) Use a graphing utility to generate additional rows of the table. Use the table to estimate the dimensions that will produce the maximum area.

(c) Write the area A as a function of x. Determine the domain of the function in the context of the problem.

(d) Use a graphing utility to graph the area function. Use the graph to approximate the dimensions that will produce the maximum area.

(e) Write the area function in standard form to find analytically the dimensions that will produce the maximum area.

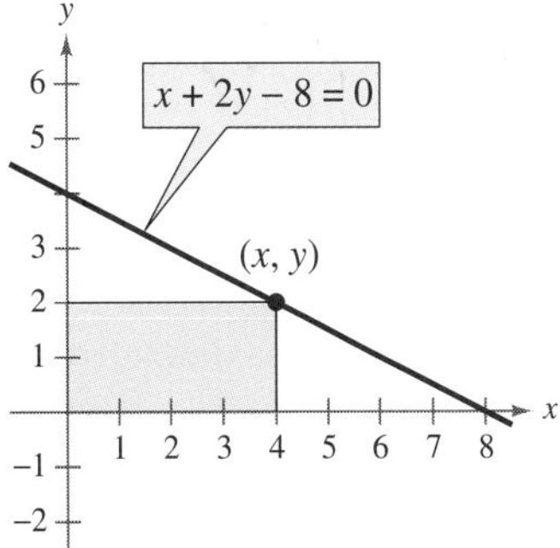

18. *Maximum Profit* Let x be the amount (in hundreds of dollars) a company spends on advertising, and let P be the profit, where

$$P = 230 + 20x - \tfrac{1}{2}x^2.$$

How much advertising will yield a maximum profit?

In Exercises 19–22, determine the right-hand and left-hand behavior of the graph of the polynomial function.

19. $f(x) = -x^2 + 6x + 9$

20. $f(x) = \frac{1}{2}x^3 + 2x$

21. $g(x) = \frac{3}{4}(x^4 + 3x^2 + 2)$

22. $h(x) = -x^5 - 7x^2 + 10x$

Graphical Analysis **In Exercises 23 and 24, use a graphing utility to graph the functions f and g in the same viewing rectangle. Zoom out far enough so that the right-hand and left-hand behavior of f and g appear identical.**

23. $f(x) = \frac{1}{2}x^3 - 2x + 1, \quad g(x) = \frac{1}{2}x^3$

24. $f(x) = -x^4 + 2x^3, \quad g(x) = -x^4$

In Exercises 25–30, use a graphing utility to graph the function.

25. $g(x) = x^4 - x^3 - 2x^2$

26. $h(x) = -2x^3 - x^2 + x$

27. $f(t) = t^3 - 3t$

28. $f(x) = -x^3 + 3x - 2$

29. $f(x) = x(x + 3)^2$

30. $f(t) = t^4 - 4t^2$

31. *Volume* A rectangular package can have a maximum combined length and girth (perimeter of a cross section) of 216 centimeters (see figure).

(a) Write the volume V as a function of x.

(b) Use a graphing utility to graph the volume function, and use the graph to estimate the dimensions of the package of maximum volume.

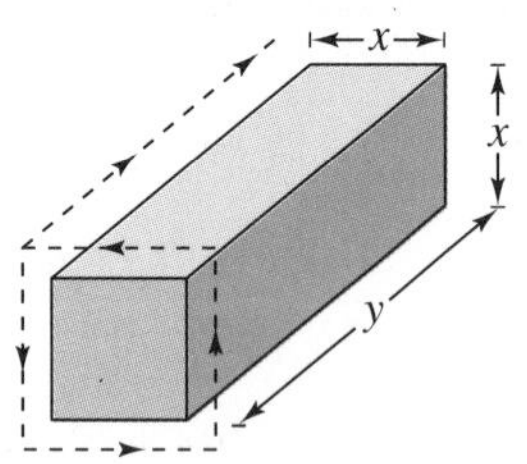

32. *Volume* Rework Exercise 31 for a cylindrical package. (The cross sections are circular.)

Graphical Analysis **In Exercises 33 and 34, use a graphing utility to graph the two equations in the same viewing rectangle. Use the graphs to verify that the expressions are equivalent. Verify the results analytically.**

33. $y_1 = \dfrac{x^2}{x - 2}, \quad y_2 = x + 2 + \dfrac{4}{x - 2}$

34. $y_1 = \dfrac{x^4 + 1}{x^2 + 2}, \quad y_2 = x^2 - 2 + \dfrac{5}{x^2 + 2}$

In Exercises 35–40, perform the division.

35. $\dfrac{24x^2 - x - 8}{3x - 2}$

36. $\dfrac{4x + 7}{3x - 2}$

37. $\dfrac{x^4 - 3x^2 + 2}{x^2 - 1}$

38. $\dfrac{3x^4}{x^2 - 1}$

39. $\dfrac{x^4 + x^3 - x^2 + 2x}{x^2 + 2x}$

40. $\dfrac{6x^4 + 10x^3 + 13x^2 - 5x + 2}{2x^2 - 1}$

In Exercises 41–44, divide by synthetic division.

41. $(0.25x^4 - 4x^3) \div (x - 2)$

42. $(2x^3 + 2x^2 - x + 2) \div \left(x - \frac{1}{2}\right)$

43. $(6x^4 - 4x^3 - 27x^2 + 18x) \div \left(x - \frac{2}{3}\right)$

44. $(0.1x^3 + 0.3x^2 - 0.5) \div (x - 5)$

In Exercises 45 and 46, use synthetic division to decide whether the x-values are zeros of the function.

45. $f(x) = 2x^3 + 3x^2 - 20x - 21$

(a) $x = 4$ (b) $x = -1$

(c) $x = -\frac{7}{2}$ (d) $x = 0$

46. $f(x) = 20x^4 + 9x^3 - 14x^2 - 3x$

(a) $x = -1$ (b) $x = \frac{3}{4}$

(c) $x = 0$ (d) $x = 1$

In Exercises 47 and 48, find a polynomial with integer coefficients that has the given zeros.

47. $-1, -1, \frac{1}{3}, -\frac{1}{2}$

48. $2, -3, 1 - 2i, 1 + 2i$

In Exercises 49–54, find all the zeros of the function.

49. $f(x) = 4x^3 - 11x^2 + 10x - 3$

50. $f(x) = 10x^3 + 21x^2 - x - 6$

51. $f(x) = 6x^3 - 5x^2 + 24x - 20$

52. $f(x) = x^3 - 1.3x^2 - 1.7x + 0.6$

53. $f(x) = 6x^4 - 25x^3 + 14x^2 + 27x - 18$

54. $f(x) = 5x^4 + 126x^2 + 25$

In Exercises 55–58, use a graphing utility to (a) graph the function, (b) determine the number of real zeros of the function, and (c) approximate the real zeros of the function to the nearest hundredth.

55. $f(x) = x^4 + 2x + 1$

56. $g(x) = x^3 - 3x^2 + 3x + 2$

57. $h(x) = x^3 - 6x^2 + 12x - 10$

58. $f(x) = x^5 + 2x^3 - 3x - 20$

59. *Data Analysis* Sales S (in billions of dollars) of recreational vehicles in the United States for the years 1980 through 1993 are listed in the table.

Year	1980	1981	1982	1983	1984
S	1.2	1.8	1.7	3.4	4.1

Year	1985	1986	1987	1988	1989
S	3.5	3.9	4.5	4.8	4.5

Year	1990	1991	1992	1993
S	4.1	3.6	4.4	4.8

A model for the data is

$$S = 1.209 + 0.290t + 0.176t^2 - 0.031t^3 + 0.0013t^4$$

where S is sales in billions of dollars and t is the time in years, with $t = 0$ corresponding to 1980. (Source: National Sporting Goods Association)

(a) Use a graphing utility to plot the data points and graph the model in the same viewing rectangle.

(b) The table shows that sales were down from 1989 through 1992. Give a possible explanation. Does the model show the downturn in sales?

(c) Use a graphing utility to approximate the magnitude of the decrease in sales during the slump described in part (b). Was the actual decrease more or less than indicated by the model?

(d) Use the model to estimate sales in 1995.

60. *Age of the Groom* The average age of the groom in a wedding for a given age of the bride can be approximated by the model

$$y = -0.00428x^2 + 1.442x - 3.136, \quad 20 \le x \le 55$$

where y is the age of the groom and x is the age of the bride. For what age of the bride is the average age of the groom 30? (Source: U.S. National Center for Health Statistics)

In Exercises 61–68, sketch the graph of the rational function. As sketching aids, check for intercepts, symmetry, vertical asymptotes, horizontal asymptotes, and slant asymptotes.

61. $f(x) = \dfrac{-5}{x^2}$

62. $h(x) = \dfrac{x - 3}{x - 2}$

63. $P(x) = \dfrac{x^2}{x^2 + 1}$

64. $f(x) = \dfrac{2x}{x^2 + 4}$

65. $f(x) = \dfrac{x}{x^2 + 1}$

66. $h(x) = \dfrac{4}{(x - 1)^2}$

67. $f(x) = \dfrac{2x^3}{x^2 + 1}$

68. $y = \dfrac{2x^2}{x^2 - 4}$

In Exercises 69–72, use a graphing utility to graph the function. Identify any vertical, horizontal, or slant asymptotes.

69. $s(x) = \dfrac{8x^2}{x^2 + 4}$

70. $y = \dfrac{5x}{x^2 - 4}$

71. $g(x) = \dfrac{x^2 + 1}{x + 1}$

72. $y = \dfrac{1}{x + 3} + 2$

Think About It **In Exercises 73 and 74, write a rational function f having the specified characteristics.**

73. Vertical asymptotes: $x = -3, x = 4$
Horizontal asymptote: $y = 2$

74. Vertical asymptote: $x = 5$
Slant asymptote: $y = 2x$

75. *Average Cost* A business has a cost of $C = 0.5x + 500$ for producing x units. The average cost per unit is

$$\overline{C} = \frac{C}{x} = \frac{0.5x + 500}{x}, \qquad 0 < x.$$

Determine the average cost per unit as x increases without bound. (Find the horizontal asymptote.)

76. *Seizure of Illegal Drugs* The cost in millions of dollars for the U.S. government to seize $p\%$ of a certain illegal drug as it enters the country is given by

$$C = \frac{528p}{100 - p}, \qquad 0 \le p \le 100.$$

(a) Find the cost of seizing 25%.

(b) Find the cost of seizing 50%.

(c) Find the cost of seizing 75%.

(d) According to this model, would it be possible to seize 100% of the drug?

77. *Population of Fish* The Parks and Wildlife Commission introduces 80,000 fish into a large human-made lake. The population of the fish in thousands is given by

$$N = \frac{20(4 + 3t)}{1 + 0.05t}, \qquad 0 \le t$$

where t is time in years.

(a) Use a graphing utility to graph the function.

(b) Find the populations when t is 5, 10, and 25.

(c) What is the limiting number of fish in the lake as time increases? Explain your reasoning.

78. *Numerical and Graphical Analysis* A right triangle is formed in the first quadrant by the x- and y-axes and a line through the point (2, 3).

(a) Draw a figure that illustrates the problem. Label the known and unknown quantities.

(b) Verify that the area of the triangle is given by

$$A = \frac{3x^2}{2(x - 2)}, \qquad x > 2.$$

(c) Use a graphing utility to generate a table giving the area for values of x. Starting the table with $x = 2.5$, increment x in steps of 0.5. Continue until you can approximate the dimensions of the triangle of minimum area.

(d) Use a graphing utility to graph the area function. Use the graph to approximate the dimensions of the triangle of minimum area.

(e) Determine the slant asymptote of the area function. Explain its meaning.

79. *Think About It* Use a graphing utility to graph the functions

$$f(x) = \frac{x^2 - 9}{x + 3} \quad \text{and} \quad g(x) = x - 3$$

in the same viewing rectangle. Does the graphing utility show the difference in the domains of the functions? Explain.

CHAPTER PROJECT *Graphs of Polynomial and Rational Functions*

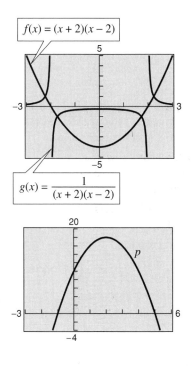

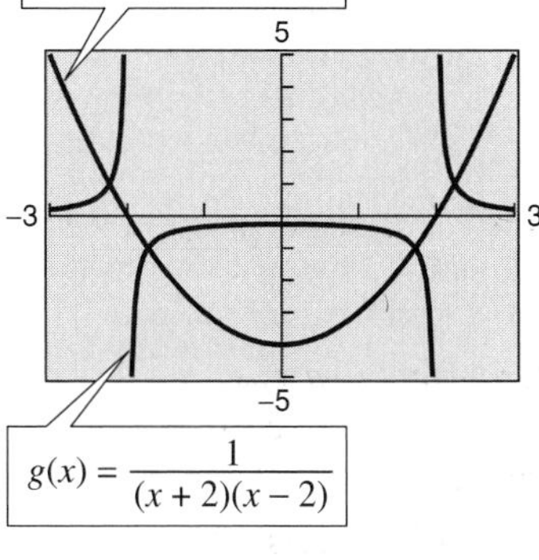

In this project, you will compare the graph of a polynomial function with the graph of its reciprocal function. Consider the polynomial function

$$f(x) = x^2 - 4 = (x + 2)(x - 2).$$

The reciprocal of this function is the rational function

$$g(x) = \frac{1}{f(x)} = \frac{1}{(x + 2)(x - 2)}.$$

The graphs of f and g are shown at the left.

(a) How are the zeros of f related to the vertical asymptotes of g?

(b) What is the minimum point of the graph of f? How is this point related to the graph of g? In general, how are the relative minimums and relative maximums of a polynomial function f related to the relative minimums and relative maximums of the reciprocal function $g(x) = 1/f(x)$?

(c) The reciprocal function g has a horizontal asymptote at $y = 0$. What does this tell you about the behavior of f for very large and very small values of x?

(d) The graph of a polynomial function P is shown at the left. Use the graph of P to sketch the graph of its reciprocal function $q(x) = 1/P(x)$.

Questions for Further Exploration

1. The polynomial function

$$f(x) = x^2 + 9$$

has no real zeros. What does this indicate about the graph of the reciprocal function $g(x) = 1/f(x)$?

2. Use a graphing utility to graph the function

$$f(x) = (x - 1)(x - 2)(x - 3)$$

and its reciprocal function $g(x) = 1/f(x)$ in the same viewing rectangle. Compare the features of the graphs of f and g.

3. You are given the graphs of two functions at the right. Each function is of the form $g(x) = 1/f(x)$, where $f(x)$ is a polynomial. Use each graph to sketch the graph of the corresponding polynomial function f.

Figure for 3

(a)

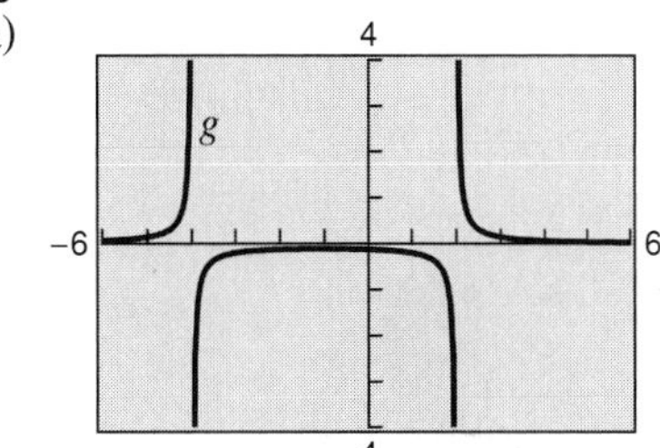

(b)

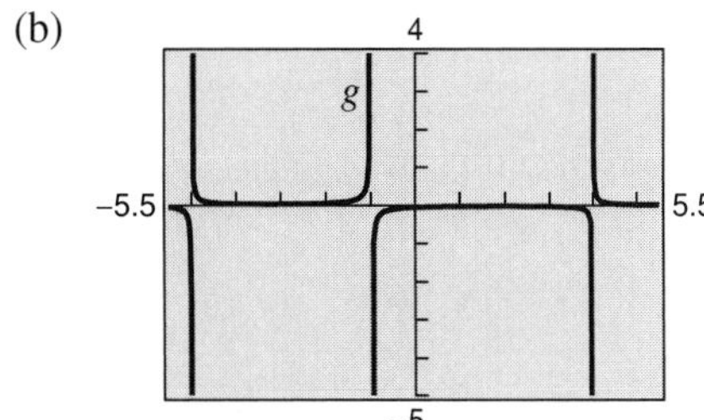

3 /// CHAPTER TEST

Take this test as you would take a test in class. After you are done, check your work against the answers given in the back of the book.

The *Interactive* CD-ROM provides answers to the Chapter Tests and Cumulative Tests. It also offers Chapter Pre-Tests (that test key skills and concepts covered in previous chapters) and Chapter Post-Tests, both of which have randomly generated exercises with diagnostic capabilities.

1. Describe how the graph of g differs from the graph of $f(x) = x^2$.
   (a) $g(x) = 2 - x^2$ (b) $g(x) = \left(x - \frac{3}{2}\right)^2$
2. Identify the vertex and intercepts of the graph of $y = x^2 + 4x + 3$.
3. Find an equation of the parabola shown at the right.
4. The path of a ball is given by $y = -\frac{1}{20}x^2 + 3x + 5$, where y is the height in feet and x is the horizontal distance in feet.
   (a) Find the maximum height of the ball.
   (b) Which term determines the height at which the ball was thrown? Does changing this term change the coordinates of the maximum height of the ball? Explain.
5. Divide by long division: $(3x^3 + 4x - 1) \div (x^2 + 1)$.
6. Divide by synthetic division: $(2x^4 - 5x^2 - 3) \div (x - 2)$.

Figure for 3

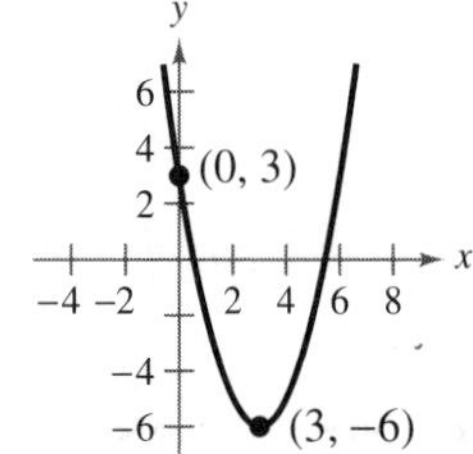

In Exercises 7 and 8, list all the possible rational zeros of the function. Use a graphing utility to graph the function and find all the rational zeros.

7. $g(t) = 2t^4 - 3t^3 + 16t - 24$
8. $h(x) = 3x^5 + 2x^4 - 3x - 2$

In Exercises 9 and 10, use the root-finding capabilities of a graphing utility to approximate the real zeros of the function accurate to three decimal places.

9. $f(x) = x^4 - x^3 - 1$
10. $f(x) = 3x^5 + 2x^4 - 12x - 8$

In Exercises 11 and 12, find a polynomial function with integer coefficients that has the given zeros.

11. $0, 3, 3 + i, 3 - i$
12. $1 + \sqrt{3}i, 1 - \sqrt{3}i, 2, 2$

In Exercises 13 and 14, use a graphing utility to graph the function.

13. $h(x) = \dfrac{4}{x^2} - 1$
14. $g(x) = \dfrac{x^2 + 2}{x - 1}$

15. Find a rational function with vertical asymptotes at $x = \pm 3$ and a horizontal asymptote at $y = 4$.

Exponential and Logarithmic Functions

Chapter 4

Automobiles are designed with crumple zones that allow the occupants to move short distances when the automobiles come to abrupt stops. The greater the distance moved, the fewer g's the crash victims experience. (One g is equal to the acceleration due to gravity.) In crash tests with vehicles moving at 90 kilometers per hour, analysts measured the numbers y of g's that were undergone during deceleration by crash dummies that were permitted to move distances of x meters during impact.

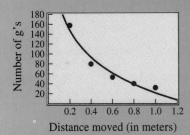

x	0.2	0.4	0.6	0.8	1.0
y	158	80	53	40	32

You can use a graphing utility to draw a scatter plot of the data and fit an appropriate model to the data. Using natural logarithmic regression capabilities, you can find one model to be $y = 21.37 - 78.58 \ln x$. The graph shows the data points along with the graph of the model. (See Exercise 94 on page 358.)

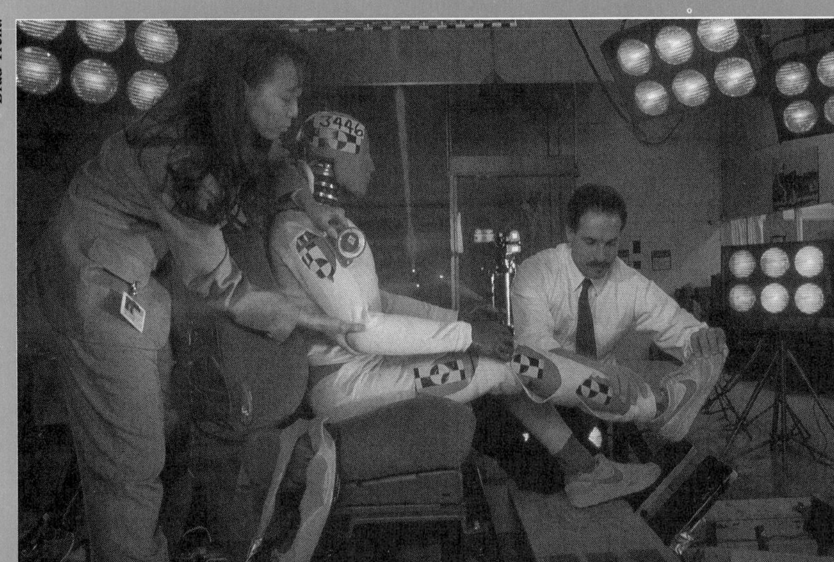

Brad Trent.

At a General Motors lab, engineer Bonnie Cheung and physicist Stephen Rouhana prepare a dummy for a simulated automobile crash. The laser helps position the dummy.

4.1 Exponential Functions and Their Graphs

Exponential Functions / Graphs of Exponential Functions / The Natural Base e / Compound Interest / Other Applications

Exponential Functions

So far, this text has dealt mainly with **algebraic functions,** which include polynomial functions and rational functions. In this chapter you will study two types of nonalgebraic functions—*exponential* functions and *logarithmic* functions. These functions are examples of **transcendental functions.**

Note The base $a = 1$ is excluded because it yields

$f(x) = 1^x = 1.$

This is a constant function, not an exponential function.

Definition of Exponential Function

The **exponential function f with base a** is denoted by

$$f(x) = a^x$$

where $a > 0$, $a \neq 1$, and x is any real number.

You already know how to evaluate a^x for integer and rational values of x. For example, you know that $4^3 = 64$ and $4^{1/2} = 2$. However, to evaluate 4^x for any real number x, you need to interpret forms with *irrational* exponents. For the purposes of this text, it is sufficient to think of

$$a^{\sqrt{2}} \quad (\text{where } \sqrt{2} \approx 1.414214)$$

as that value having the successively closer approximations

$$a^{1.4}, a^{1.41}, a^{1.414}, a^{1.4142}, a^{1.41421}, a^{1.414214}, \ldots .$$

Example 1 shows how to use a calculator to evaluate an exponential expression.

Library of Functions

The exponential function $f(x) = a^x$ is different from all the functions you have studied so far because the variable x is an *exponent.* The domain, like those of polynomial functions, is the set of all real numbers.

EXAMPLE 1 **Evaluating Exponential Expressions**

Use a calculator to evaluate each expression.

a. $2^{-3.1}$ **b.** $2^{-\pi}$

Solution

Number	*Graphing Calculator Keystrokes*	*Display*
a. $2^{-3.1}$	2 [∧] [(-)] 3.1 [ENTER]	0.1166291
b. $2^{-\pi}$	2 [∧] [(-)] π [ENTER]	0.1133147

Graphs of Exponential Functions

The graphs of all exponential functions have similar characteristics, as shown in Examples 2, 3, and 4.

Figure 4.1

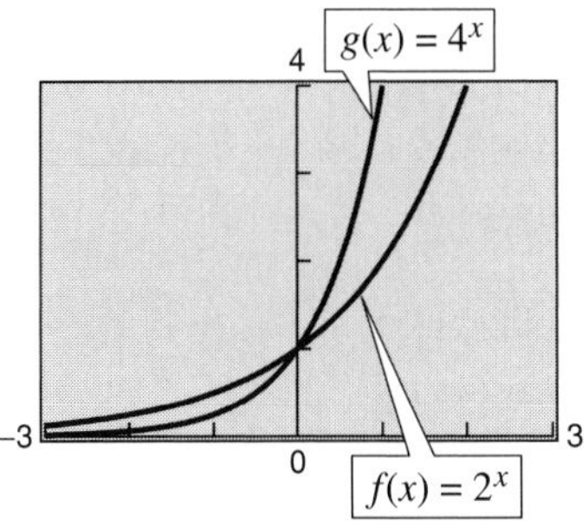

EXAMPLE 2 Graphs of $y = a^x$

In the same coordinate plane, sketch the graph of each function.

a. $f(x) = 2^x$

b. $g(x) = 4^x$

Solution

The table below lists some values for each function, and Figure 4.1 shows their graphs. Note that both graphs are increasing. Moreover, the graph of $g(x) = 4^x$ is increasing more rapidly than the graph of $f(x) = 2^x$.

x	-2	-1	0	1	2	3
2^x	$\frac{1}{4}$	$\frac{1}{2}$	1	2	4	8
4^x	$\frac{1}{16}$	$\frac{1}{4}$	1	4	16	64

Figure 4.2

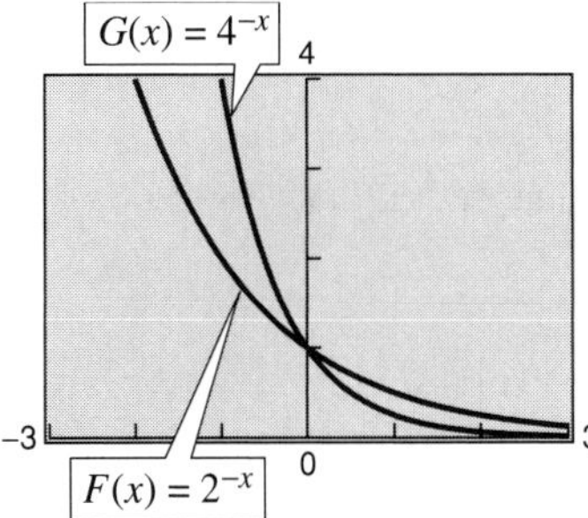

EXAMPLE 3 Graphs of $y = a^{-x}$

In the same coordinate plane, sketch the graph of each function.

a. $F(x) = 2^{-x}$

b. $G(x) = 4^{-x}$

Solution

The table below lists some values for each function, and Figure 4.2 shows their graphs. Note that both graphs are decreasing. Moreover, the graph of $G(x) = 4^{-x}$ is decreasing more rapidly than the graph of $F(x) = 2^{-x}$.

x	-3	-2	-1	0	1	2
2^{-x}	8	4	2	1	$\frac{1}{2}$	$\frac{1}{4}$
4^{-x}	64	16	4	1	$\frac{1}{4}$	$\frac{1}{16}$

Note The tables in Examples 2 and 3 were evaluated by hand. You could, of course, use a graphing utility to construct tables with even more values.

> **Study Tip**
>
> Be sure you see that the functions $y_1 = \left(\frac{1}{2}\right)^x$ and $y_2 = 2^{-x}$ are the same. In general, $\left(\frac{1}{a}\right)^x = a^{-x}$.

Comparing the functions in Examples 2 and 3, observe that

$$F(x) = 2^{-x} = f(-x) \quad \text{and} \quad G(x) = 4^{-x} = g(-x).$$

Consequently, the graph of F is a reflection (in the y-axis) of the graph of f. The graph of G and g have the same relationship. The graphs in Figures 4.1 and 4.2 are typical of the exponential functions a^x and a^{-x}. They have one y-intercept and one horizontal asymptote (the x-axis), and they are continuous. The basic characteristics of these exponential functions are summarized in Figure 4.3.

Graph of $y = a^x$

- Domain: $(-\infty, \infty)$
- Range: $(0, \infty)$
- Intercept: $(0, 1)$
- Increasing
- x-axis is a horizontal asymptote ($a^x \to 0$ as $x \to -\infty$)
- Continuous

Graph of $y = a^{-x}$

- Domain: $(-\infty, \infty)$
- Range: $(0, \infty)$
- Intercept: $(0, 1)$
- Decreasing
- x-axis is a horizontal asymptote ($a^{-x} \to 0$ as $x \to \infty$)
- Continuous

Figure 4.3

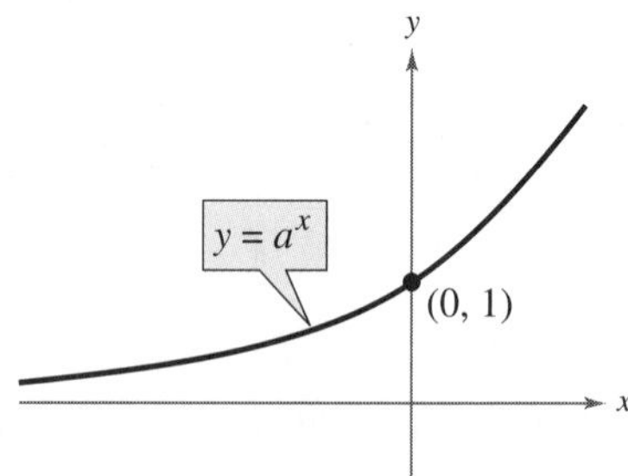

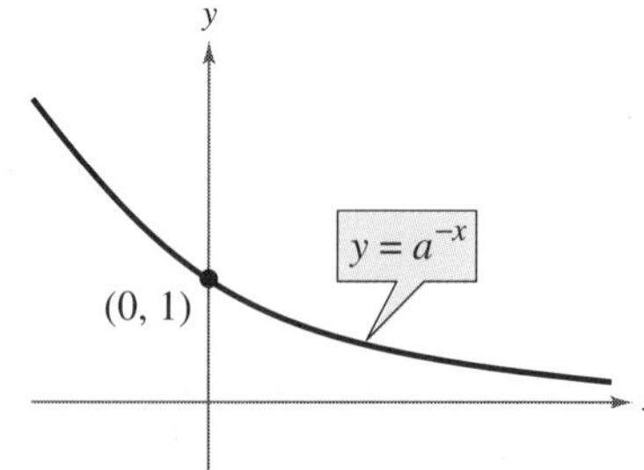

EXPLORATION

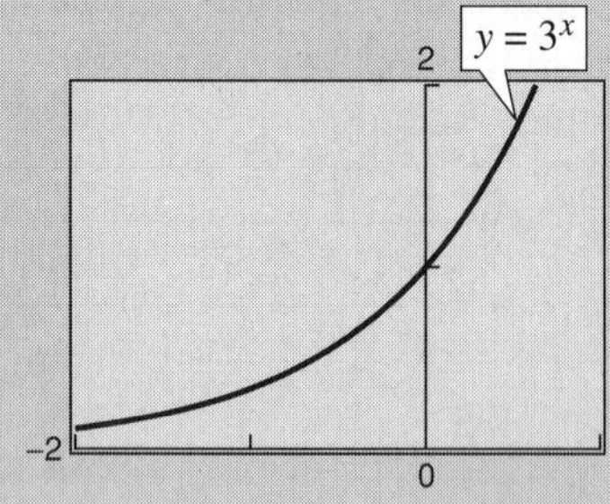

Use a graphing utility to graph $y = a^x$ with $a = 3, 5$, and 7 in the same viewing rectangle. (Use a viewing rectangle in which $-2 \le x \le 1$ and $0 \le y \le 2$.) For instance, the graph of $y = 3^x$ is shown at the left. How do the graphs compare with each other? Which graph is on the top in the interval $(-\infty, 0)$? Which is on the bottom? Which graph is on the top in the interval $(0, \infty)$? Which is on the bottom? Repeat this experiment with the graphs of $y = a^x$ for $a = \frac{1}{3}, \frac{1}{5}$, and $\frac{1}{7}$. (Use a viewing rectangle in which $-1 \le x \le 2$ and $0 \le y \le 2$.) What can you conclude about the relationship between the function's behavior and the value of a?

In the following example, notice how the graph of $y = a^x$ can be used to sketch the graphs of functions of the form

$$f(x) = b \pm a^{x+c}.$$

EXAMPLE 4 **Graphs of Exponential Functions**

Each of the following graphs is a transformation of the graph of $f(x) = 3^x$, as shown in Figure 4.4.

a. Because $g(x) = 3^{x+1} = f(x+1)$, the graph of g can be obtained by shifting the graph of f one unit to the left.

b. Because $h(x) = 3^x - 2 = f(x) - 2$, the graph of h can be obtained by shifting the graph of f down two units.

c. Because $k(x) = -3^x = -f(x)$, the graph of k can be obtained by reflecting the graph of f in the x-axis.

d. Because $j(x) = 3^{-x} = f(-x)$, the graph of j can be obtained by reflecting the graph of f in the y-axis.

Note In Figure 4.4, notice that the transformations in parts (a), (c), and (d) keep the x-axis as a horizontal asymptote, but the transformation in part (b) yields a new horizontal asymptote of $y = -2$. Also, be sure to note how the y-intercept is affected by each transformation.

Figure 4.4

(a)

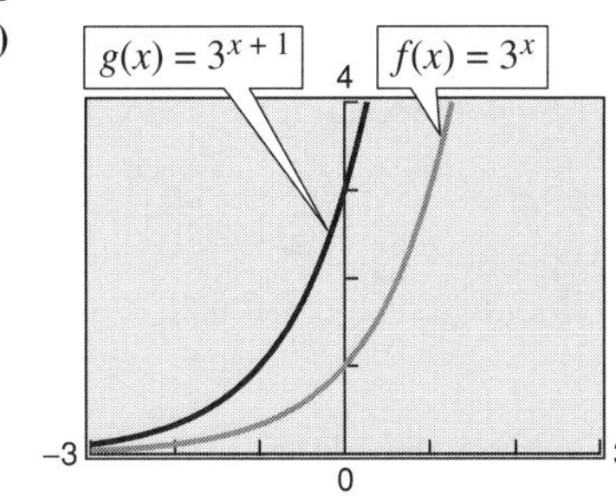

(b)

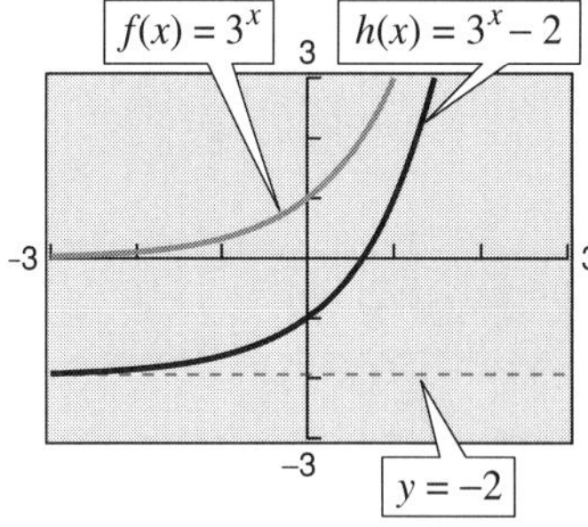

(c)

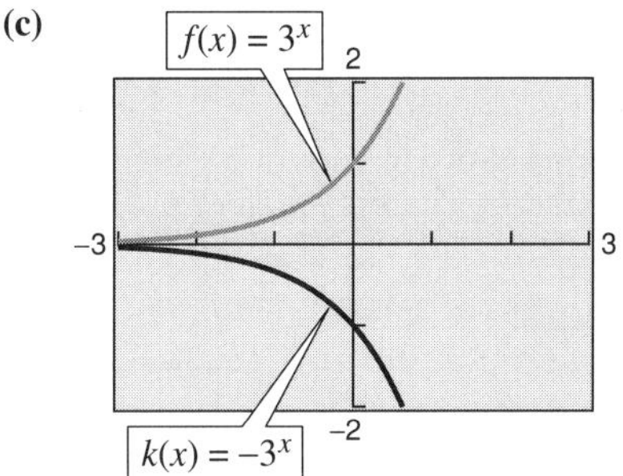

(d)

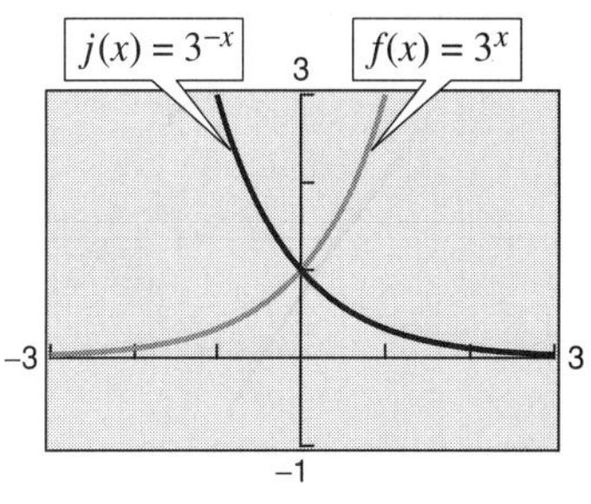

The following table generated on a *TI-82* or a *TI-83* shows some points of the graphs in Figure 4.4(a). The functions $f(x)$ and $g(x)$ are represented by Y_1 and Y_2. respectively. Explain how you can use the table to describe the transformation.

X	Y1	Y2
−3	0.11111	0.03704
−2	0.33333	0.11111
−1	1	0.33333
0	3	1
1	9	3
2	27	9
3	81	27
X = −3		

The Natural Base e

EXPLORATION

Use your graphing utility to graph the functions

$y_1 = 2^x$
$y_2 = e^x$
$y_3 = 3^x$

in the same viewing rectangle. From the relative positions of these graphs, make a guess as to the value of the real number e. Then try to find a number a such that the graphs of $y_2 = e^x$ and $y_4 = a^x$ are as close as possible.

For many applications, the convenient choice for a base is the irrational number

$$e \approx 2.71828 \ldots .$$

This number is called the **natural base.** The function $f(x) = e^x$ is the **natural exponential function.** Its graph is shown in Figure 4.5. Be sure you see that for the exponential function $f(x) = e^x$, e is the constant 2.71828 . . . , whereas x is the variable.

Figure 4.5

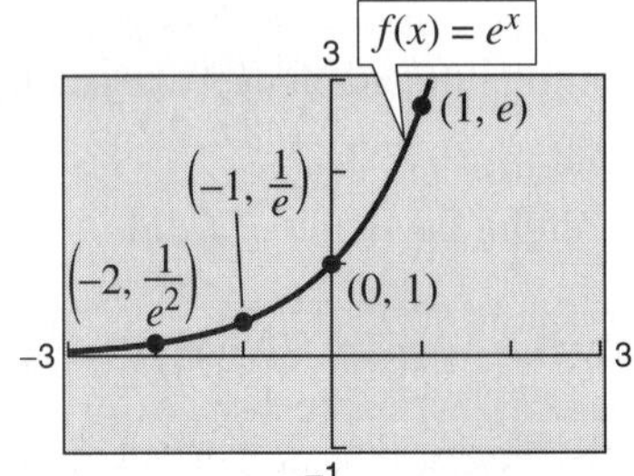

EXAMPLE 5 Approximation of the Number e

Evaluate the expression $[1 + (1/x)]^x$ for several large values of x to see that the values approach $e \approx 2.71828$ as x increases without bound.

Solution

Using a calculator, you can complete a table like that shown below. From this table, it seems reasonable to conclude that

$$\left(1 + \frac{1}{x}\right)^x \to e \quad \text{as} \quad x \to \infty.$$

You can further confirm this conclusion by using a graphing utility to graph

$$f(x) = \left(1 + \frac{1}{x}\right)^x$$

and $y = e$ in the same viewing rectangle, as shown in Figure 4.6. Note that as x increases, the graph of f gets closer and closer to the line given by $y = e$.

Figure 4.6

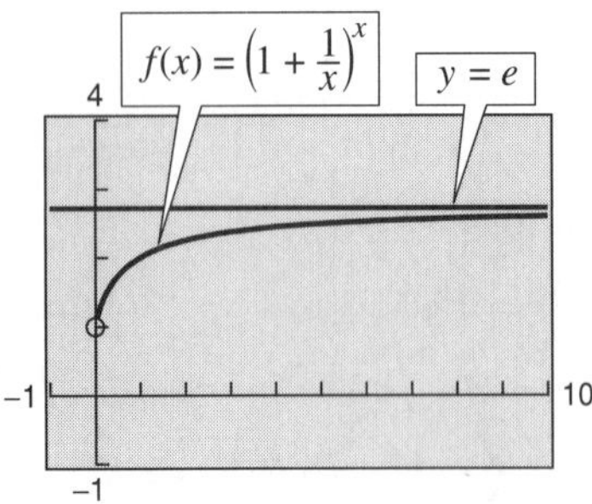

x	10	100	1000	10,000	100,000	1,000,000
$\left(1 + \frac{1}{x}\right)^x$	2.59374	2.70481	2.71692	2.71815	2.71827	2.71828

EXPLORATION

Use a graphing utility to graph $y = (1 + x)^{1/x}$. What is the domain of this function? Describe the behavior of the graph near $x = 0$. Is there a y-intercept? How does the behavior of the graph near $x = 0$ relate to the result of Example 5?

EXAMPLE 6 Evaluating the Natural Exponential Function

Use a calculator to evaluate each expression.

a. e^{-2} **b.** e^{-1} **c.** e^{1} **d.** e^{2}

Solution

Number	*Graphing Calculator Keystrokes*	*Display*
a. e^{-2}	e^x (-) 2 ENTER	0.1353353
b. e^{-1}	e^x (-) 1 ENTER	0.3678794
c. e^{1}	e^x 1 ENTER	2.7182818
d. e^{2}	e^x 2 ENTER	7.3890561

The *Interactive* CD-ROM shows every example with its solution; clicking on the *Try It!* button brings up similar problems. Guided Examples and Integrated Examples show step-by-step solutions to additional examples. Integrated Examples are related to several concepts in the section.

EXAMPLE 7 Graphing Natural Exponential Functions

Sketch the graph of each natural exponential function.

a. $f(x) = 2e^{0.24x}$

b. $g(x) = \frac{1}{2}e^{-0.58x}$

Solution

To sketch these two graphs, you can use a calculator to construct a table of values, as shown at the left. After constructing the table, plot the points and connect them with smooth curves, as shown in Figure 4.7. Note that the graph in part (a) is increasing, whereas the graph in part (b) is decreasing. Use a graphing calculator to verify these graphs.

x	$f(x)$	$g(x)$
−3	0.974	2.849
−2	1.238	1.595
−1	1.573	0.893
0	2.000	0.500
1	2.543	0.280
2	3.232	0.157
3	4.109	0.088

Figure 4.7

(a)

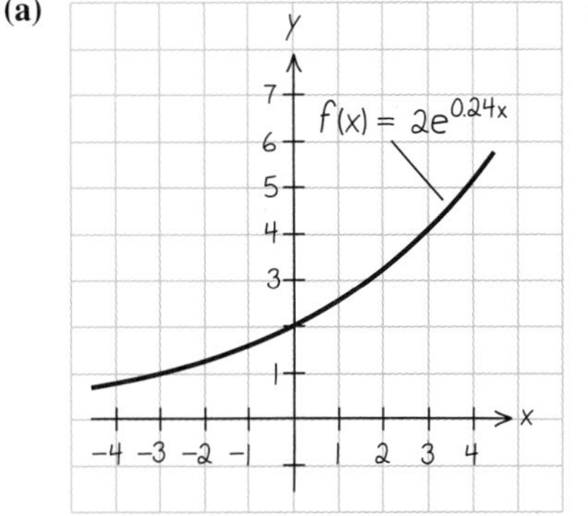

(b)

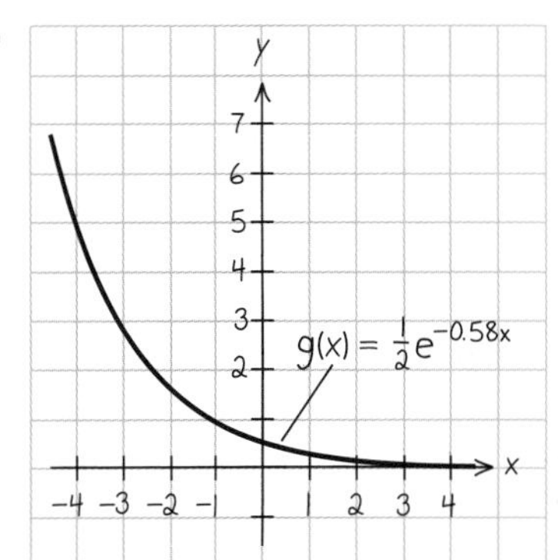

Compound Interest

One of the most familiar examples of exponential growth is that of an investment earning *continuously compounded interest.* Suppose a principal P is invested at an annual interest rate r, compounded once a year. If the interest is added to the principal at the end of the year, the balance is

$$\begin{aligned} P_1 &= P + Pr \\ &= P(1 + r). \end{aligned}$$

This pattern of multiplying the previous principal by $1 + r$ is then repeated each successive year, as shown in the table.

Time in Years	Balance After Each Compounding
0	$P = P$
1	$P_1 = P(1 + r)$
2	$P_2 = P_1(1 + r) = P(1 + r)(1 + r) = P(1 + r)^2$
3	$P_3 = P_2(1 + r) = P(1 + r)^2(1 + r) = P(1 + r)^3$
$\vdots$	$\vdots$
n	$P_n = P(1 + r)^n$

Study Tip

The interest rate r in the formula for compound interest should be written as a decimal. For example, an interest rate of 7% would be written $r = 0.07$.

To accommodate more frequent (quarterly, monthly, or daily) compounding of interest, let n be the number of compoundings per year and let t be the number of years. Then the rate per compounding is r/n, and the account balance after t years is

$$A = P\left(1 + \frac{r}{n}\right)^{nt}. \qquad \text{Amount with } n \text{ compoundings per year}$$

If you let the number of compoundings n increase without bound, you approach **continuous compounding.** In the formula for n compoundings per year, let $m = n/r$. This produces

$$A = P\left(1 + \frac{r}{n}\right)^{nt} = P\left(1 + \frac{1}{m}\right)^{mrt} = P\left[\left(1 + \frac{1}{m}\right)^{m}\right]^{rt}.$$

As m increases without bound, you know from Example 5 that $[1 + (1/m)]^m$ approaches e. Hence, for continuous compounding, it follows that

$$P\left[\left(1 + \frac{1}{m}\right)^{m}\right]^{rt} \rightarrow P[e]^{rt}$$

and you can write $A = Pe^{rt}$. This result is part of the reason that e is the "natural" choice for a base of an exponential function.

Formulas for Compound Interest

After t years, the balance A in an account with principal P and annual interest rate r (expressed as a decimal) is given by the following formulas.

1. For n compoundings per year: $A = P\left(1 + \dfrac{r}{n}\right)^{nt}$
2. For continuous compounding: $A = Pe^{rt}$

Real Life

EXAMPLE 8 **Finding the Balance for Compound Interest**

A sum of \$9000 is invested at an annual interest rate of 8.5%, compounded annually. Find the balance in the account after 3 years.

Figure 4.8

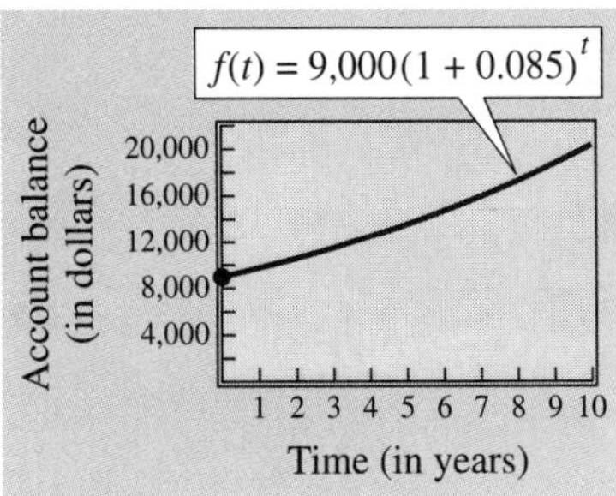

Solution

In this case, $P = 9000$, $r = 8.5\% = 0.085$, $n = 1$, and $t = 3$. Using the formula

$$A = P\left(1 + \frac{r}{n}\right)^{nt},$$

you have

$$A = 9000(1 + 0.085)^3 = 9000(1.085)^3 \approx \$11{,}495.60.$$

The graph of $f(t) = 9000(1 + 0.085)^t$ is shown in Figure 4.8.

Real Life

EXAMPLE 9 **Finding Compound Interest**

A total of \$12,000 is invested at an annual interest rate of 9%. Find the balance after 5 years if it is compounded

a. quarterly. **b.** continuously.

Solution

a. For quarterly compoundings, you have $n = 4$. Thus, in 5 years at 9%, the balance is

$$A = P\left(1 + \frac{r}{n}\right)^{nt} = 12{,}000\left(1 + \frac{0.09}{4}\right)^{4(5)} = \$18{,}726.11.$$

b. For continuous compounding, the balance is

$$A = Pe^{rt} = 12{,}000e^{0.09(5)} = \$18{,}819.75.$$

Note that continuous compounding yields \$93.64 more than quarterly compounding.

Other Applications

Figure 4.9

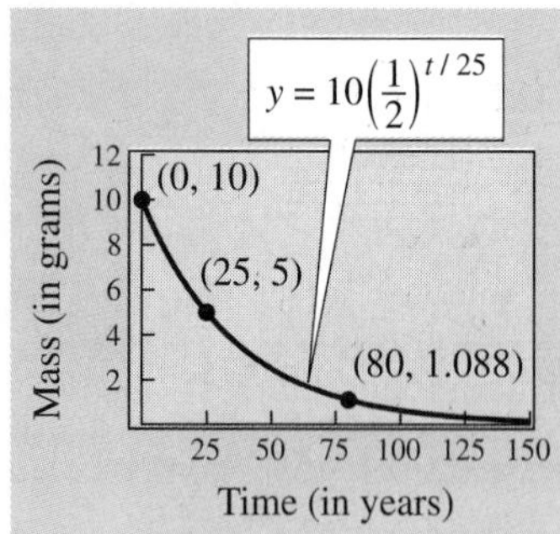

Real Life

EXAMPLE 10 Radioactive Decay

Let y represent the mass of a quantity of a radioactive element whose half-life is 25 years. After t years, the mass (in grams) is given by

$$y = 10\left(\frac{1}{2}\right)^{t/25}.$$

a. What is the initial mass (when $t = 0$)?

b. How much of the initial mass is present after 80 years?

Solution

a. When $t = 0$, the mass is

$$y = 10\left(\frac{1}{2}\right)^{0} = 10(1) = 10 \text{ grams.}$$

b. When $t = 80$, the mass is

$$y = 10\left(\frac{1}{2}\right)^{80/25} = 10(0.5)^{3.2} \approx 1.088 \text{ grams.}$$

The graph of this function is shown in Figure 4.9.

Real Life

EXAMPLE 11 Population Growth

The approximate number of fruit flies in an experimental population after t hours is given by

$$Q(t) = 20e^{0.03t}, \qquad t \geq 0.$$

a. Find the initial number of fruit flies in the population.

b. How large is the population of fruit flies after 72 hours?

c. Sketch the graph of Q.

Figure 4.10

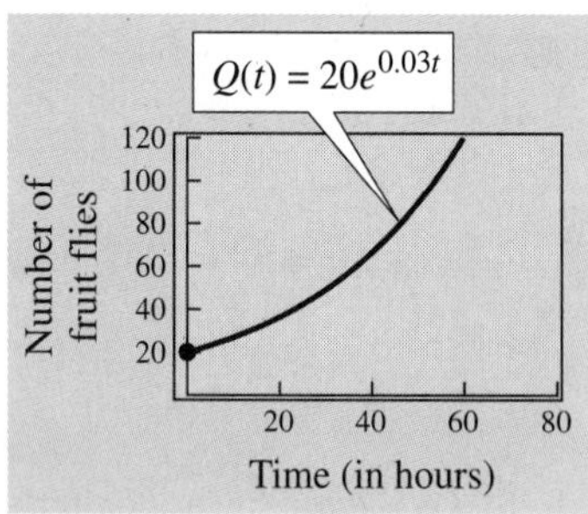

Solution

a. To find the initial population, evaluate $Q(t)$ at $t = 0$.

$$Q(0) = 20e^{0.03(0)} = 20e^{0} = 20(1) = 20 \text{ flies}$$

b. After 72 hours, the population size is

$$Q(72) = 20e^{0.03(72)} = 20e^{2.16} \approx 173 \text{ flies.}$$

c. The graph of Q is shown in Figure 4.10.

Group Activity *Exponential Growth*

Consider the following sequences of numbers.

Sequence 1: 2, 4, 6, 8, 10, 12, . . . , $2n$

Sequence 2: 2, 4, 8, 16, 32, 64, . . . , 2^n

The first sequence, $f(n) = 2n$, is an example of **linear growth.** The second sequence, $f(n) = 2^n$, is an example of **exponential growth.** Which of the following sequences represents linear growth and which represents exponential growth? Can you find a linear function and an exponential function that represent the two sequences? (*Hint:* Look for a *shift*, of one or more units, from a linear or exponential growth pattern.)

a. 2, 5, 8, 11, 14, . . . **b.** 4, 10, 28, 82, 244, . . .

Later you will see that sequences that represent linear growth are *arithmetic sequences* and sequences that represent exponential growth are *geometric sequences.*

The *Interactive* CD-ROM contains step-by-step solutions to all odd-numbered Section and Review Exercises. It also provides Tutorial Exercises, which link to Guided Examples for additional help.

4.1 /// EXERCISES

In Exercises 1–10, use a calculator to evaluate the expression. Round your result to three decimal places.

1. $(3.4)^{5.6}$ **2.** $5000(2^{-1.5})$

3. $(1.005)^{400}$ **4.** $8^{2\pi}$

5. $5^{-\pi}$ **6.** $\sqrt[3]{4395}$

7. $100^{\sqrt{2}}$ **8.** $e^{1/2}$

9. $e^{-3/4}$ **10.** $e^{3.2}$

Think About It **In Exercises 11–14, use properties of exponents to determine which functions (if any) are the same.**

11. $f(x) = 3^{x-2}$
$g(x) = 3^x - 9$
$h(x) = \frac{1}{9}(3x)$

12. $f(x) = 4^x + 12$
$g(x) = 2^{2x+6}$
$h(x) = 64(4^x)$

13. $f(x) = 16(4^{-x})$
$g(x) = \left(\frac{1}{4}\right)^{x-2}$
$h(x) = 16(2^{-2x})$

14. $f(x) = 5^{-x} + 3$
$g(x) = 5^{3-x}$
$h(x) = -5^{x-3}$

In Exercises 15–22, graph the exponential function *by hand*. Identify the following features of the graph.

(a) Asymptotes
(b) Intercepts
(c) Increasing or decreasing

15. $g(x) = 5^x$ **16.** $f(x) = \left(\frac{3}{2}\right)^x$

17. $f(x) = \left(\frac{1}{5}\right)^x = 5^{-x}$ **18.** $h(x) = \left(\frac{3}{2}\right)^{-x}$

19. $h(x) = 5^{x-2}$ **20.** $g(x) = \left(\frac{3}{2}\right)^{x+2}$

21. $g(x) = 5^{-x} - 3$ **22.** $f(x) = \left(\frac{3}{2}\right)^{-x} + 2$

In Exercises 23–30, match the exponential function with its graph. [The graphs are labeled (a), (b), (c), (d), (e), (f), (g), and (h).]

(a)
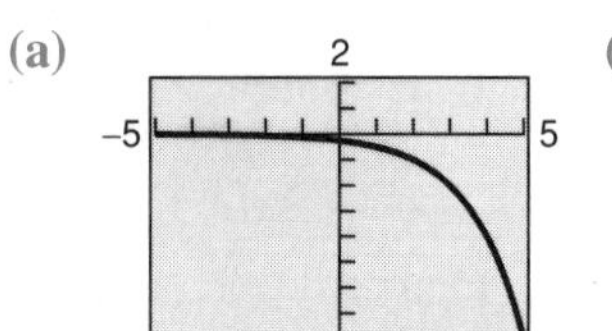

(b)
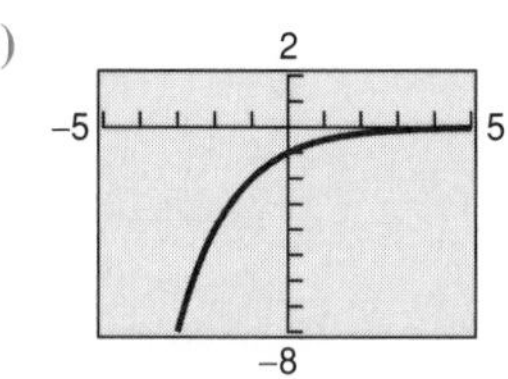

(c)
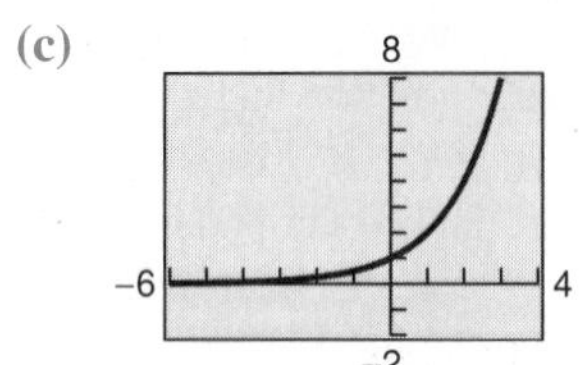

(d)
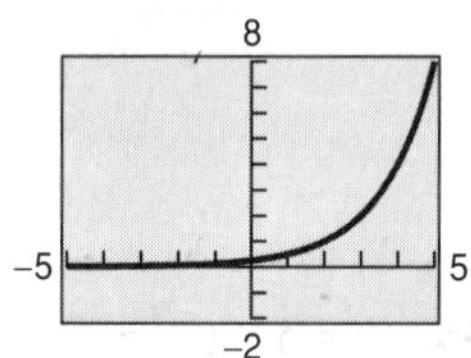

(e)
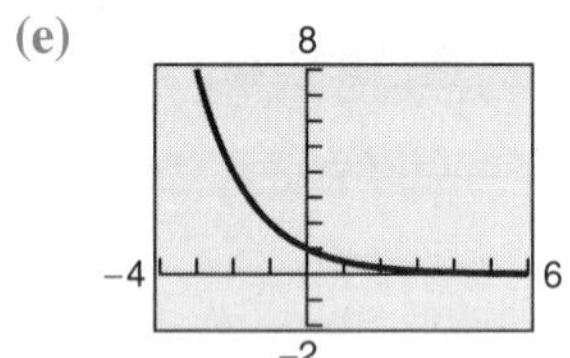

(f)
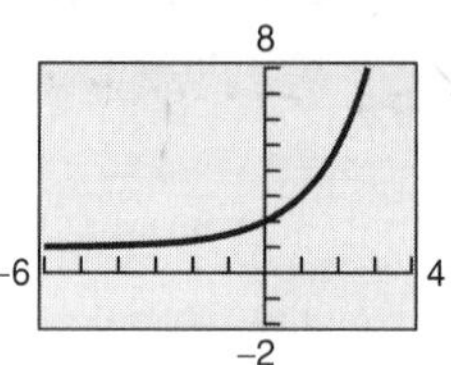

(g)
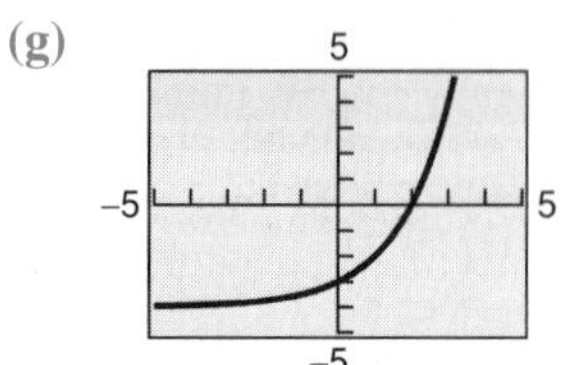

(h)
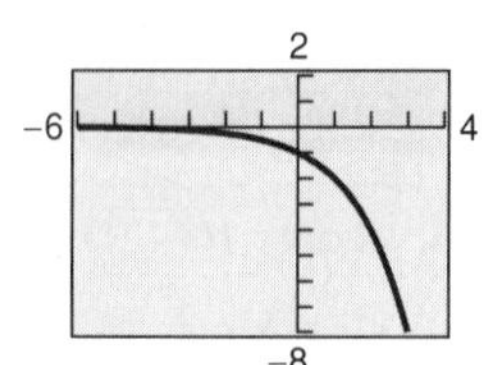

23. $f(x) = 2^x$

24. $f(x) = -2^x$

25. $f(x) = 2^{-x}$

26. $f(x) = -2^{-x}$

27. $f(x) = 2^x - 4$

28. $f(x) = 2^x + 1$

29. $f(x) = -2^{x-2}$

30. $f(x) = 2^{x-2}$

In Exercises 31–40, use a graphing utility to graph the exponential function. Identify any asymptotes of the graph.

31. $y = 2^{-x^2}$

32. $y = 3^{-|x|}$

33. $y = 3^{x-2} + 1$

34. $y = 4^{x+1} - 2$

35. $y = 1.08^{-5x}$

36. $y = 1.08^{5x}$

37. $s(t) = 2e^{0.12t}$

38. $s(t) = 3e^{-0.2t}$

39. $g(x) = 1 + e^{-x}$

40. $h(x) = e^{x-2}$

41. *Exploration* Consider the functions $f(x) = 3^x$ and $g(x) = 4^x$.

(a) Use a graphing utility to complete the table and use the table to estimate the solution of the inequality $4^x < 3^x$.

x	-1	-0.5	0	0.5	1
$f(x)$					
$g(x)$					

(b) Use a graphing utility to graph $f(x)$ and $g(x)$ in the same viewing rectangle. Use the graphs to solve the inequalities.

(i) $4^x < 3^x$ (ii) $4^x > 3^x$

42. *Exploration* Consider the functions $f(x) = \left(\frac{1}{2}\right)^x$ and $g(x) = \left(\frac{1}{4}\right)^x$.

(a) Use a graphing utility to complete the table and use the table to estimate the solution of the inequality $\left(\frac{1}{4}\right)^x < \left(\frac{1}{2}\right)^x$.

x	-1	-0.5	0	0.5	1
$f(x)$					
$g(x)$					

(b) Use a graphing utility to graph $f(x)$ and $g(x)$ in the same viewing rectangle. Use the graphs to solve the inequalities.

(i) $\left(\frac{1}{4}\right)^x < \left(\frac{1}{2}\right)^x$ (ii) $\left(\frac{1}{4}\right)^x > \left(\frac{1}{2}\right)^x$

43. Use the graph of $f(x) = 3^x$ to graph each of the functions. Identify the transformation.

(a) $g(x) = f(x - 2) = 3^{x-2}$

(b) $h(x) = -\frac{1}{2}f(x) = -\frac{1}{2}(3^x)$

(c) $q(x) = f(-x) + 3 = 3^{-x} + 3$

44. Use a graphing utility to graph each function. Use the graph to find any asymptotes of the function.

(a) $f(x) = \dfrac{8}{1 + e^{-0.5x}}$ (b) $g(x) = \dfrac{8}{1 + e^{-0.5/x}}$

45. Use a graphing utility to graph each function. Use the graph to find where the function is increasing and decreasing, and approximate any relative maximum or minimum values.

(a) $f(x) = x^2e^{-x}$ (b) $g(x) = x2^{3-x}$

46. *Comparing Functions* Use a graphing utility to graph $y_1 = e^x$ and each of the functions $y_2 = x^2$, $y_3 = x^3$, $y_4 = \sqrt{x}$, and $y_5 = |x|$. Which function increases at the fastest rate for "large" values of x?

47. *Conjecture* Use the result of Exercise 46 to make a conjecture about the rates of growth of $y_1 = e^x$ and $y = x^n$ where n is a natural number and x is "large."

48. *Essay* Use the results of Exercises 46 and 47 to describe what is implied when it is stated that a quantity is growing exponentially.

49. *Graphical Analysis* Use a graphing utility to graph

$$f(x) = \left(1 + \frac{0.5}{x}\right)^x \quad \text{and} \quad g(x) = e^{0.5}$$

in the same viewing rectangle. What is the relationship between f and g as x increases without bound?

50. *Conjecture* Use the result of Exercise 49 to make a conjecture about the value of

$$\left(1 + \frac{r}{x}\right)^x$$

as x increases without bound.

51. *Trust Fund* You deposit \$5000 in a trust fund that pays 7.5% interest, compounded continuously, and you specify that the balance of the fund will be given to the college from which you graduated after the money has earned interest for 50 years. How much will your college receive?

52. *Trust Fund* On the day of your grandchild's birth, you deposit \$25,000 in a trust fund that pays 8.75% interest, compounded continuously. Determine the balance in this account on your grandchild's 25th birthday.

Compound Interest **In Exercises 53–56, complete the table to determine the balance A for P dollars invested at rate r for t years and compounded n times per year.**

n	1	2	4	12	365	Continuous
A						

53. $P = \$2500$, $r = 12\%$, $t = 10$ years

54. $P = \$1000$, $r = 10\%$, $t = 10$ years

55. $P = \$2500$, $r = 12\%$, $t = 20$ years

56. $P = \$1000$, $r = 10\%$, $t = 40$ years

Compound Interest **In Exercises 57–60, complete the table to determine the amount of money P that should be invested at rate r to produce a final balance of \$100,000 in t years.**

t	1	10	20	30	40	50
P						

57. $r = 9\%$, compounded continuously

58. $r = 12\%$, compounded continuously

59. $r = 10\%$, compounded monthly

60. $r = 7\%$, compounded daily

61. *Demand Function* The demand equation for a certain product is given by

$$p = 5000\left(1 - \frac{4}{4 + e^{-0.002x}}\right)$$

where p is the price and x is the number of units.

(a) Use a graphing utility to graph the demand function for $x > 0$ and $p > 0$.

(b) Find the price p for a demand of $x = 500$ units.

(c) Use the graph in part (a) to approximate the highest price that will still yield a demand of at least 600 units.

62. *Graphical Reasoning* There are two options for investing \$500. The first earns 7% compounded annually and the second earns 7% simple interest. The figure shows the growth of each investment over a 30-year period.

(a) Identify the two types of investments in the figure. Explain your reasoning.

(b) Verify your answer in part (a) by finding the equations that model the investment growth and using a graphing utility to graph the models.

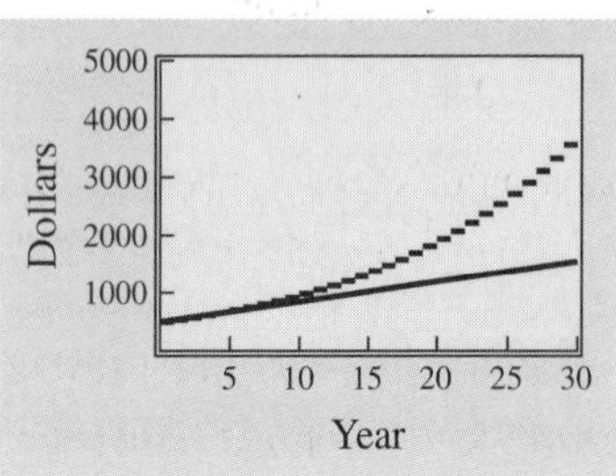

63. *Bacteria Growth* A certain type of bacteria increases according to the model

$$P(t) = 100e^{0.2197t}$$

where t is the time in hours. Find (a) $P(0)$, (b) $P(5)$, and (c) $P(10)$.

64. *Population Growth* The population of a town increases according to the model

$$P(t) = 2500e^{0.0293t}$$

where t is the time in years, with $t = 0$ corresponding to 1990. Use the model to estimate the population in (a) 2000 and (b) 2010.

65. *Radioactive Decay* Let Q represent the mass of a quantity of radium 226, whose half-life is 1620 years. The quantity of radium present after t years is given by

$$Q = 25\left(\tfrac{1}{2}\right)^{t/1620}.$$

(a) Determine the initial quantity (when $t = 0$).

(b) Determine the quantity present after 1000 years.

(c) Use a graphing utility to graph the function over the interval $t = 0$ to $t = 5000$.

66. *Data Analysis* To estimate the amount of defoliation caused by gypsy moths, a forester counts the number x of egg masses on $\frac{1}{40}$ of an acre (circle of radius 18.6 feet) during the fall. The percent of defoliation y the next spring is given in the table. (Source: USDA, Forest Service)

x	0	25	50	75	100
y	12	44	81	96	99

(a) Use a graphing utility to plot the data points.

(b) A model for the data is given by

$$y = \frac{300}{3 + 17e^{-0.065x}}.$$

Use a graphing utility to graph the model in the same viewing rectangle used in part (a). How well does the model fit the data?

(c) Use a graphing utility to create a table comparing the model with the sample data.

(d) Estimate the percent of defoliation if 36 egg masses are counted on $\frac{1}{40}$ acre.

(e) Use the graph to estimate the number of egg masses per $\frac{1}{40}$ acre if it is seen that approximately $\frac{2}{3}$ of the forest is defoliated the next spring.

67. *Data Analysis* A meteorologist measures the atmospheric pressure P (in kilograms per square meter) at altitude h (in kilometers). The data is shown in the table.

h	0	5	10	15	20
P	10,332	5583	2376	1240	517

(a) Use a graphing utility to plot the data points.

(b) A model for the data is given by

$$P = 10{,}958e^{-0.15h}.$$

Use a graphing utility to graph the model in the same viewing rectangle used in part (a). How well does the model fit the data?

(c) Use a graphing utility to create a table comparing the model with the sample data.

(d) Estimate the pressure at a height of 8 kilometers.

(e) Estimate the altitude at which the pressure is 2000 kilograms per square meter.

68. *Radioactive Decay* Let Q represent the mass of a quantity of carbon 14, whose half-life is 5730 years. The quantity present after t years is given by

$$Q = 10\left(\frac{1}{2}\right)^{t/5730}.$$

(a) Determine the initial quantity (when $t = 0$).

(b) Determine the quantity present after 2000 years.

(c) Sketch the graph of the function over the interval $t = 0$ to $t = 10{,}000$.

69. *Data Analysis* A cup of water at an initial temperature of 78°C is placed in a room at a constant temperature of 21°C. The temperature of the water is measured every 5 minutes for a period of $\frac{1}{2}$ hour. The results are recorded in the table, where t is the time in minutes and T is the temperature in degrees Celsius.

t	0	5	10	15
T	78.0°	66.0°	57.5°	51.2°

t	20	25	30
T	46.3°	42.5°	39.6°

(a) Use the regression capabilities of a graphing utility to fit a line to the data. Use the graphing utility to plot the data points and the regression line. Does the data appear linear? Explain.

(b) Use the regression capabilities of a graphing utility to fit a parabola to the data. Use the graphing utility to plot the data points and the regression parabola. Does the data appear quadratic? Even though the quadratic model appears to be a "good" fit, explain why it may not be a good model for predicting the temperature of the water when $t = 60$.

(c) The graph of the model should be asymptotic with the temperature of the room. Subtract the room temperature from each of the temperatures in the table. Use a graphing utility to fit an exponential model to the revised data. Add the room temperature to this regression model. Use a graphing utility to plot the original data points and the model.

(d) Explain why the procedure in part (c) was necessary for finding the exponential model.

70. *Inflation* If the annual rate of inflation averages 4% over the next 10 years, the approximate cost C of goods or services during any year in that decade will be given by

$$C(t) = P(1.04)^t$$

where t is the time in years and P is the present cost. If the price of an oil change for your car is presently \$23.95, estimate the price 10 years from now.

71. *Depreciation* After t years, the value of a car that costs \$20,000 is given by

$$V(t) = 20{,}000\left(\frac{3}{4}\right)^t.$$

Use a graphing utility to graph the function and determine the value of the car 2 years after it was purchased.

72. *True or False?* $e = 271{,}801/99{,}990$. Explain.

73. *Think About It* Without using a graphing utility, explain why you know that $2^{\sqrt{2}}$ is greater than 2, but less than 4.

74. *Think About It* Which functions are exponential?

(a) $3x$ (b) $3x^2$

(c) 3^x (d) 2^{-x}

75. *Finding a Pattern* Use a graphing utility to compare the graph of the function $y = e^x$ with the graphs of the following functions.

(a) $y_1 = 1 + \dfrac{x}{1!}$

(b) $y_2 = 1 + \dfrac{x}{1!} + \dfrac{x^2}{2!}$

(c) $y_3 = 1 + \dfrac{x}{1!} + \dfrac{x^2}{2!} + \dfrac{x^3}{3!}$

76. *Finding a Pattern* Identify the pattern of successive polynomials given in Exercise 75. Extend the pattern one more term and compare the graph of the resulting polynomial function with the graph of $y = e^x$. What do you think this pattern implies?

77. Given the exponential function $f(x) = a^x$, show that

(a) $f(u + v) = f(u) \cdot f(v)$ (b) $f(2x) = [f(x)]^2$.

4.2 Logarithmic Functions and Their Graphs

Logarithmic Functions / Graphs of Logarithmic Functions / The Natural Logarithmic Function / Application

Logarithmic Functions

In Section 1.7, you studied the concept of the inverse of a function. There, you learned that if a function has the property that no horizontal line intersects its graph more than once, the function must have an inverse. By looking back at the graphs of the exponential functions introduced in Section 4.1, you will see that every function of the form $f(x) = a^x$ passes the "Horizontal Line Test" and therefore must have an inverse. This inverse function is called the **logarithmic function with base *a*.**

Note The equations

$$y = \log_a x \quad \text{and} \quad x = a^y$$

are equivalent. The first equation is in logarithmic form and the second is in exponential form.

Definition of Logarithmic Function

For $x > 0$ and $0 < a \neq 1$,

$$y = \log_a x \quad \text{if and only if} \quad x = a^y.$$

The function given by

$$f(x) = \log_a x$$

is called the **logarithmic function with base *a*.**

Library of Functions

The logarithmic function is the inverse of the exponential function. Its domain is the set of positive real numbers and its range is the set of all real numbers. Because of the inverse properties of logarithms and exponents, the exponential equation $a^0 = 1$ implies that $\log_a 1 = 0$.

When evaluating logarithms, remember that *a logarithm is an exponent.* This means that $\log_a x$ is the exponent to which a must be raised to obtain x. For instance, $\log_2 8 = 3$ because 2 must be raised to the third power to get 8.

EXAMPLE 1 Evaluating Logarithms

a. $\log_2 32 = 5$ because $2^5 = 32$.

b. $\log_3 27 = 3$ because $3^3 = 27$.

c. $\log_4 2 = \frac{1}{2}$ because $4^{1/2} = \sqrt{4} = 2$.

d. $\log_{10} \frac{1}{100} = -2$ because $10^{-2} = \frac{1}{10^2} = \frac{1}{100}$.

e. $\log_3 1 = 0$ because $3^0 = 1$.

f. $\log_2 2 = 1$ because $2^1 = 2$.

The logarithmic function with base 10 is called the **common logarithmic function.** On most calculators, this function is denoted by [LOG].

EXAMPLE 2 Evaluating Logarithms on a Calculator

Use a calculator to evaluate each expression.

a. $\log_{10} 10$ **b.** $2\log_{10} 2.5$ **c.** $\log_{10}(-2)$

Solution

Number	*Graphing Calculator Keystrokes*	*Display*
a. $\log_{10} 10$	[LOG] 10 [ENTER]	1
b. $2\log_{10} 2.5$	2 [LOG] 2.5 [ENTER]	0.7958800
c. $\log_{10}(-2)$	[LOG] [(-)] 2 [ENTER]	ERROR

Note that the calculator displays an error message when you try to evaluate $\log_{10}(-2)$. The reason for this is that the domain of every logarithmic function is the set of *positive* real numbers.

Study Tip

Because $\log_a x$ is the inverse function of a^x, it follows that the domain of $\log_a x$ is the range of a^x, $(0, \infty)$. In other words, $\log_a x$ is defined only if x is positive.

Some graphing utilities such as the *TI-85*, do not give an error message for $\log_{10}(-2)$. Instead, the graphing utility will display a complex number. For the purpose of this text, however, we will continue to say that the domain of a logarithmic function is the set of positive *real* numbers.

The following properties follow directly from the definition of the logarithmic function with base a.

Properties of Logarithms

1. $\log_a 1 = 0$ because $a^0 = 1$.
2. $\log_a a = 1$ because $a^1 = a$.
3. $\log_a a^x = x$ because $a^x = a^x$.
4. If $\log_a x = \log_a y$, then $x = y$.

EXAMPLE 3 Using Properties of Logarithms

Solve each equation for x.

a. $\log_2 x = \log_2 3$

b. $\log_4 4 = x$

Solution

a. Using Property 4, you can conclude that $x = 3$.

b. Using Property 2, you can conclude that $x = 1$.

Graphs of Logarithmic Functions

To sketch the graph of $y = \log_a x$, you can use the fact that the graphs of inverse functions are reflections of each other in the line $y = x$.

Figure 4.11

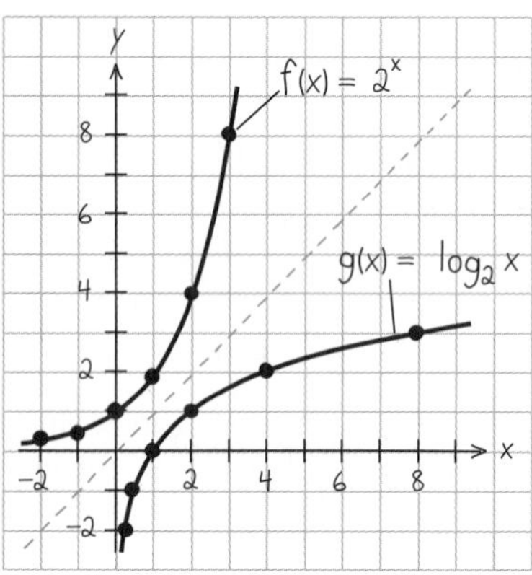

EXAMPLE 4 **Graphs of Exponential and Logarithmic Functions**

In the same coordinate plane, sketch the graph of each function.

a. $f(x) = 2^x$ **b.** $g(x) = \log_2 x$

Solution

a. For $f(x) = 2^x$, construct a table of values, as follows.

x	-2	-1	0	1	2	3
2^x	$\frac{1}{4}$	$\frac{1}{2}$	1	2	4	8

By plotting these points and connecting them with a smooth curve, you obtain the graph of $f(x)$ shown in Figure 4.11.

b. Because $g(x) = \log_2 x$ is the inverse of $f(x) = 2^x$, the graph of g is obtained by reflecting the graph of f in the line $y = x$, as shown in Figure 4.11.

Before you can confirm the result of Example 4 with a graphing utility, you need to know how to enter $\log_2 x$. You will learn how to do this using the *Change-of-Base Formula* discussed in Section 4.3.

Figure 4.12

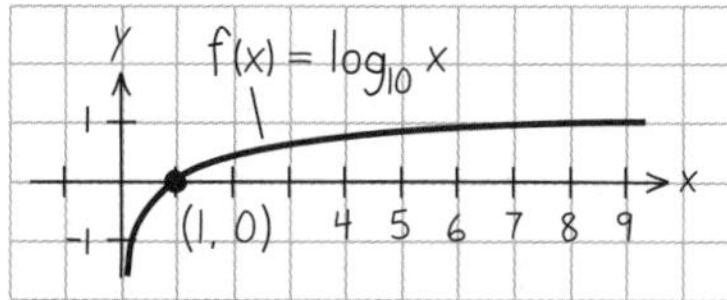

Note Compare Figure 4.12 with that obtained with a graphing utility. Note that the domain and range are $(0, \infty)$ and $(-\infty, \infty)$, respectively.

EXAMPLE 5 **Sketching the Graph of a Logarithmic Function**

Sketch the graph of the common logarithmic function $f(x) = \log_{10} x$.

Solution

Begin by constructing a table of values. Note that some of the values can be obtained without a calculator, whereas others require a calculator. Next, plot the points and connect them with a smooth curve, as shown in Figure 4.12.

	Without Calculator				With Calculator		
x	$\frac{1}{100}$	$\frac{1}{10}$	1	10	2	5	8
$\log_{10} x$	-2	-1	0	1	0.301	0.699	0.903

The nature of the graph in Figure 4.12 is typical of functions of the form $f(x) = \log_a x$, $a > 1$. They have one x-intercept and one vertical asymptote. Notice how slowly the graph rises for $x > 1$. In Figure 4.12, you would need to move out to $x = 1000$ before the graph rose to $y = 3$. The basic characteristics of logarithmic graphs are summarized in Figure 4.13.

Note In Figure 4.13, note that the vertical asymptote occurs at $x = 0$, where $\log_a x$ is *undefined.*

Figure 4.13

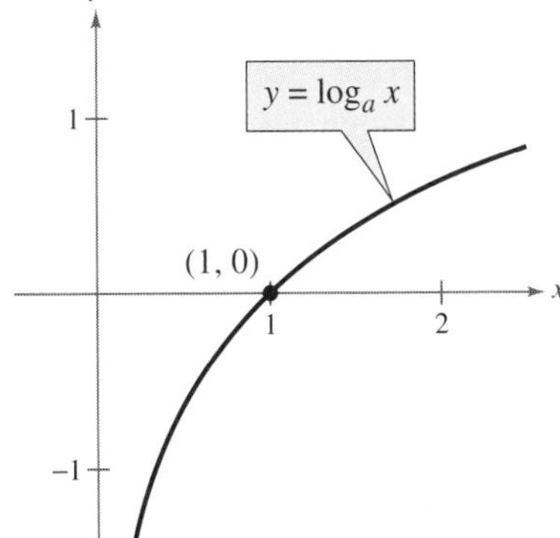

Graph of $y = \log_a x$, $a > 1$

- Domain: $(0, \infty)$
- Range: $(-\infty, \infty)$
- Intercept: $(1, 0)$
- Increasing
- y-axis is a vertical asymptote ($\log_a x \to -\infty$ as $x \to 0^+$)
- Continuous
- Reflection of graph of $y = a^x$ about the line $y = x$

The *Interactive* CD-ROM offers graphing utility emulators of the *TI-82* and *TI-83*, which can be used with the Examples, Explorations, Technology notes, and Exercises.

EXPLORATION

Use a graphing utility to graph $y = \log_{10} x$ and $y = 8$ in the same viewing rectangle. Do the graphs intersect? If so, find a viewing rectangle that shows the point of intersection. What is the point of intersection?

EXAMPLE 6 **Sketching the Graphs of Logarithmic Functions**

The graph of each of the following functions is similar to the graph of $f(x) = \log_{10} x$, as shown in Figure 4.14.

a. Because $g(x) = \log_{10}(x - 1) = f(x - 1)$, the graph of g can be obtained by shifting the graph of f one unit to the right.

b. Because $h(x) = 2 + \log_{10} x = 2 + f(x)$, the graph of h can be obtained by shifting the graph of f two units up.

Note In Figure 4.14(a), notice how the transformation affects the vertical asymptote.

Figure 4.14

(a)

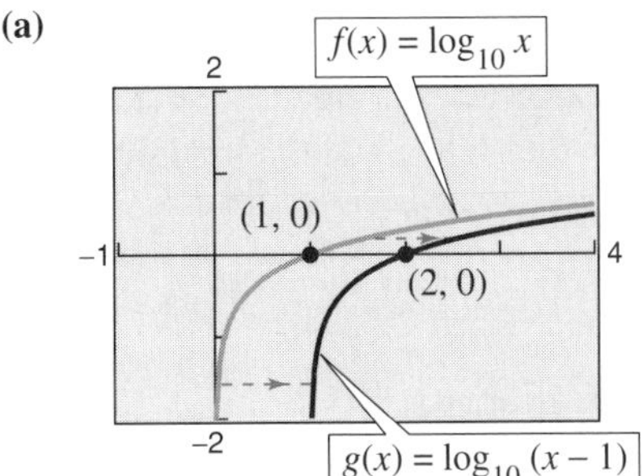

(b)

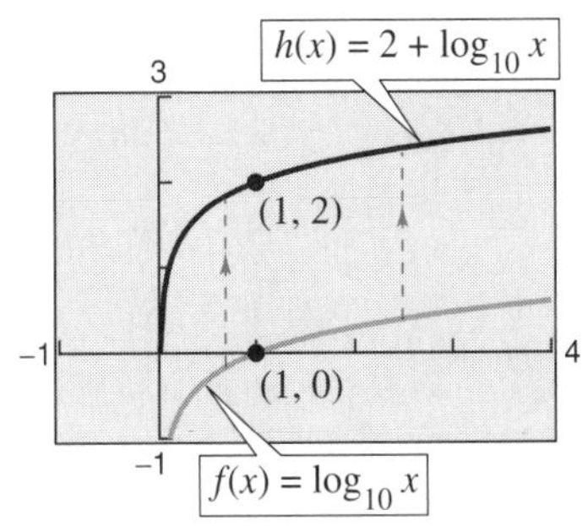

The Natural Logarithmic Function

EXPLORATION

Because the natural exponential function $y = e^x$ passes the Horizontal Line Test, it has an inverse. Use the Draw Inverse feature of your graphing utility to graph $y = e^x$ and its inverse. Compare this graph with that of the natural logarithmic function. What can you conclude?

The most widely used base for logarithmic functions is the number e, where

$$e \approx 2.718281828 \ldots .$$

The logarithmic function with base e is the **natural logarithmic function** and is denoted by the special symbol $\ln x$, read as "el en of x."

The Natural Logarithmic Function

The function defined by

$$f(x) = \log_e x = \ln x, \quad x > 0$$

is called the **natural logarithmic function.**

The four properties of logarithms listed on page 331 are also valid for natural logarithms.

Properties of Natural Logarithms

1. $\ln 1 = 0$ because $e^0 = 1$.
2. $\ln e = 1$ because $e^1 = e$.
3. $\ln e^x = x$ because $e^x = e^x$.
4. If $\ln x = \ln y$, then $x = y$.

Figure 4.15

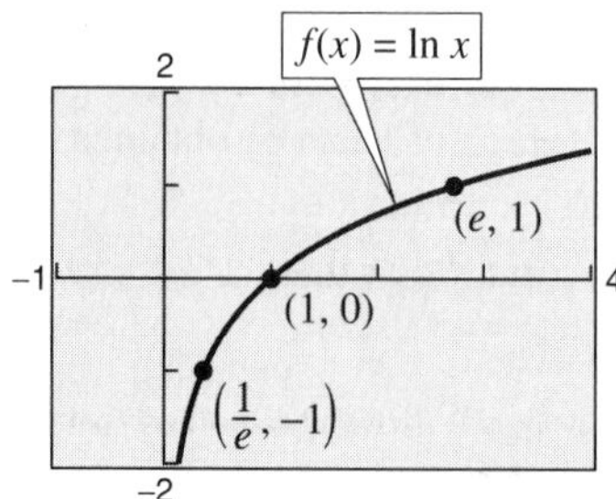

The graph of the natural logarithmic function is shown in Figure 4.15. Try using a graphing utility to confirm this graph. What is the domain of the natural logarithmic function?

EXAMPLE 7 **Using Properties of Natural Logarithms**

a. $\ln \dfrac{1}{e} = \ln e^{-1} = -1$ Property 3

b. $\ln e^2 = 2$ Property 3

c. $\ln e^0 = 0$ Property 1

d. $2 \ln e = 2(1) = 2$ Property 2

On most calculators, the natural logarithm is denoted by [LN], as illustrated in Example 8.

Use a graphing utility to graph f and h in Example 9. How can you use the graphs to verify the domains of the functions?

EXAMPLE 8 Evaluating the Natural Logarithmic Function

Use a calculator to evaluate each expression.

a. $\ln 2$ **b.** $\ln 0.3$ **c.** $\ln e^2$ **d.** $\ln(-1)$

Solution

	Number	*Graphing Calculator Keystrokes*	*Display*
a.	$\ln 2$	[LN] 2 [ENTER]	0.6931472
b.	$\ln 0.3$	[LN] .3 [ENTER]	−1.2039728
c.	$\ln e^2$	[LN] [e^x] 2 [ENTER]	2
d.	$\ln(-1)$	[LN] [(-)] 1 [ENTER]	ERROR

In Example 8, be sure you see that $\ln(-1)$ gives an error message on most calculators. This occurs because the domain of $\ln x$ is the set of *positive* real numbers (see Figure 4.15). Hence, $\ln(-1)$ is undefined.

EXAMPLE 9 Finding the Domains of Logarithmic Functions

Find the domain of each function.

a. $f(x) = \ln(x - 2)$ **b.** $g(x) = \ln(2 - x)$ **c.** $h(x) = \ln x^2$

Figure 4.16

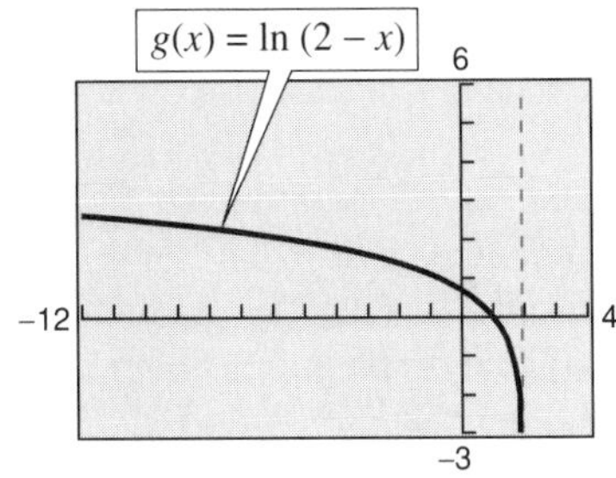

Solution

a. Because $\ln(x - 2)$ is defined only if $x - 2 > 0$, it follows that the domain of f is $(2, \infty)$.

b. Because $\ln(2 - x)$ is defined only if $2 - x > 0$, it follows that the domain of g is $(-\infty, 2)$. The graph of g is shown in Figure 4.16.

c. Because $\ln x^2$ is defined only if $x^2 > 0$, it follows that the domain of h is all real numbers except $x = 0$.

Note In Example 9, suppose you had been asked to analyze the function $h(x) = \ln|x - 2|$. How would the domain of this function compare with the domains of the functions given in parts (a) and (b) of the example?

Application

EXAMPLE 10 **Human Memory Model** *Real Life*

Students participating in a psychological experiment attended several lectures on a subject. At the end of the last lecture, and every month for the next year, the students were tested to see how much of the material they remembered. The average scores for the group were given by the *human memory model*

$$f(t) = 75 - 6\ln(t + 1), \quad 0 \le t \le 12$$

where t is the time in months.

a. What was the average score on the original $(t = 0)$ exam?

b. What was the average score at the end of $t = 2$ months?

c. What was the average score at the end of $t = 6$ months?

Solution

a. The original average score was

$$f(0) = 75 - 6\ln 1 = 75 - 6(0) = 75.$$

b. After 2 months, the average score was

$$f(2) = 75 - 6\ln 3 \approx 75 - 6(1.0986) \approx 68.4.$$

c. After 6 months, the average score was

$$f(6) = 75 - 6\ln 7 \approx 75 - 6(1.9459) \approx 63.3.$$

The graph of f in an appropriate viewing rectangle is shown in Figure 4.17.

EXPLORATION

Use a graphing utility to determine how many months it would take for the average score in Example 10 to decrease to 60. Explain your method of solving the problem. Describe another way that you can use a graphing utility to determine the answer.

Figure 4.17

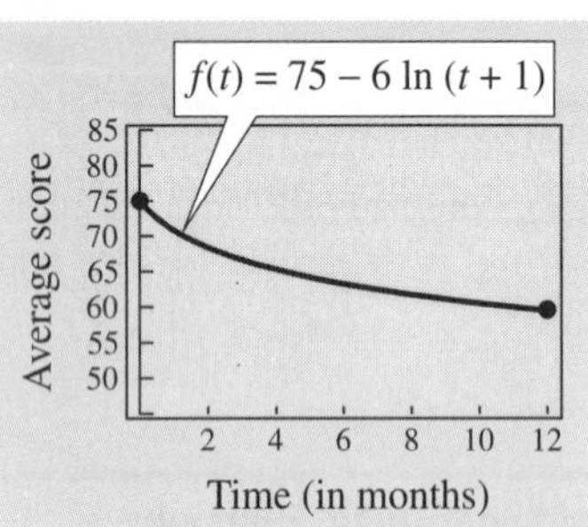

Group Activity

Transforming Logarithmic Functions

x	$f_1(x)$	$f_2(x)$
−3	Undefined	Undefined
−2.99999	−13.816	−17.816
−2.9999	−9.210	−13.210
−2	0	−4
0	1.099	−2.901
3	1.792	−2.208
6	2.197	−1.803

The table at the left gives selected points for two natural logarithmic functions of the form $f(x) = b \pm \ln(x + c)$. Study the table. What can you infer? Compare the given data with data for $f(x) = \ln x$. Try to find a natural logarithmic function that fits each set of data. Explain the method you used. (*Hint:* You might find it easier to find the value of c first.)

4.2 /// EXERCISES

In Exercises 1–8, write the logarithmic equation in exponential form. For example, the exponential form of $\log_5 25 = 2$ is $5^2 = 25$.

1. $\log_4 64 = 3$

2. $\log_3 81 = 4$

3. $\log_7 \frac{1}{49} = -2$

4. $\log_{10} \frac{1}{1000} = -3$

5. $\log_{32} 4 = \frac{2}{5}$

6. $\log_{16} 8 = \frac{3}{4}$

7. $\ln 1 = 0$

8. $\ln 4 = 1.386\ldots$

In Exercises 9–18, write the exponential equation in logarithmic form.

9. $5^3 = 125$

10. $8^2 = 64$

11. $81^{1/4} = 3$

12. $9^{3/2} = 27$

13. $6^{-2} = \frac{1}{36}$

14. $10^{-3} = 0.001$

15. $e^3 = 20.0855\ldots$

16. $e^0 = 1$

17. $e^x = 4$

18. $u^v = w$

In Exercises 19–30, evaluate the expression without using a calculator.

19. $\log_2 16$

20. $\log_2\left(\frac{1}{8}\right)$

21. $\log_{16} 4$

22. $\log_{27} 9$

23. $\log_7 1$

24. $\log_{10} 1000$

25. $\log_{10} 0.01$

26. $\log_{10} 10$

27. $\ln e^3$

28. $\ln 1$

29. $\log_a a^2$

30. $\log_a \dfrac{1}{a}$

In Exercises 31–40, use a calculator to evaluate the logarithm. Round to three decimal places.

31. $\log_{10} 345$

32. $\log_{10}\left(\frac{4}{5}\right)$

33. $\log_{10} 145$

34. $\log_{10} 12.5$

35. $\ln 18.42$

36. $\ln\sqrt{42}$

37. $\ln\left(1 + \sqrt{3}\right)$

38. $\ln\left(\sqrt{5} - 2\right)$

39. $\ln 0.32$

40. $\ln 0.75$

In Exercises 41–44, describe the relationship between the graphs of f and g. What is the relationship between the functions f and g?

41. $f(x) = 3^x$
$g(x) = \log_3 x$

42. $f(x) = 5^x$
$g(x) = \log_5 x$

43. $f(x) = e^x$
$g(x) = \ln x$

44. $f(x) = 10^x$
$g(x) = \log_{10} x$

In Exercises 45–50, use the graph of $y = \log_3 x$ to match the given function with its graph. [The graphs are labeled (a), (b), (c), (d), (e), and (f).]

(a)

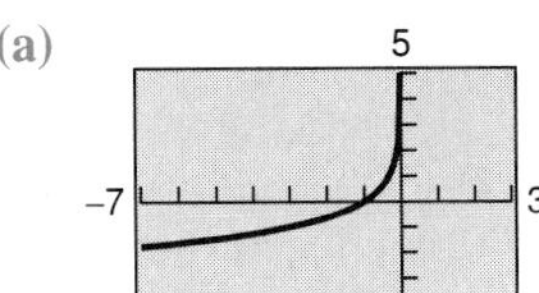

(b)

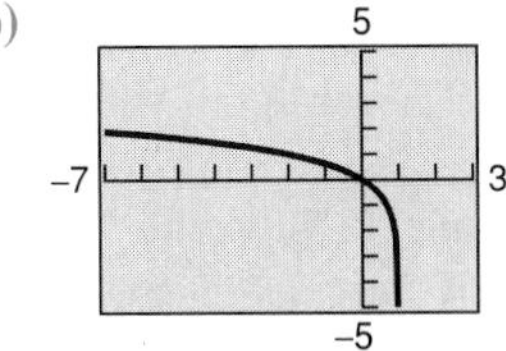

(c)

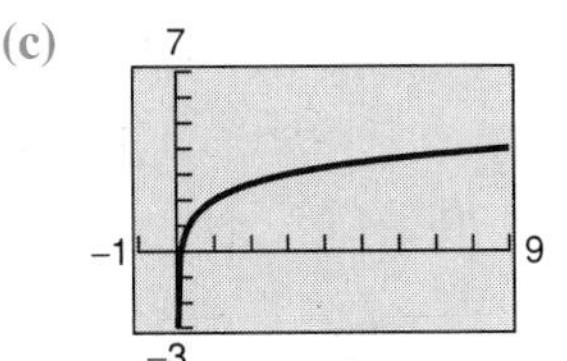

(d)

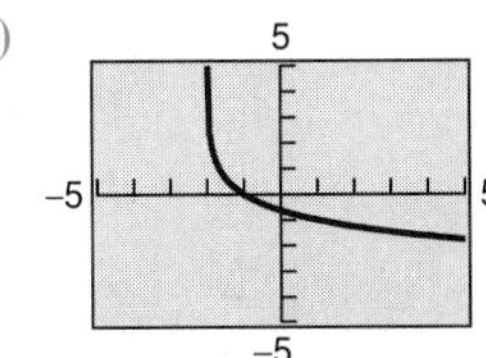

(e)

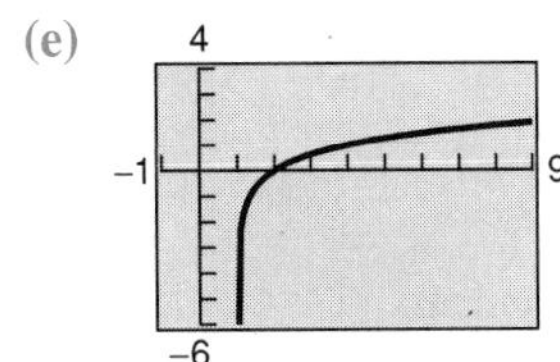

(f)

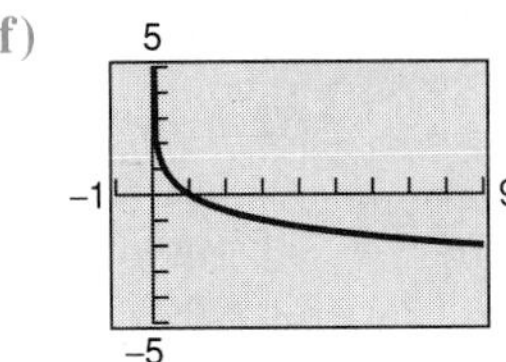

45. $f(x) = \log_3 x + 2$

46. $f(x) = -\log_3 x$

47. $f(x) = -\log_3(x + 2)$

48. $f(x) = \log_3(x - 1)$

49. $f(x) = \log_3(1 - x)$

50. $f(x) = -\log_3(-x)$

In Exercises 51–56, find the domain, vertical asymptote, and x-intercept of the logarithmic function, and sketch its graph by hand.

51. $f(x) = \log_4 x$

52. $g(x) = \log_6 x$

53. $h(x) = \log_4(x - 3)$

54. $f(x) = -\log_6(x + 2)$

55. $y = -\log_3 x + 2$

56. $y = \log_5(x - 1) + 4$

In Exercises 57–62, use a graphing utility to graph the logarithmic function. Determine the domain and identify any vertical asymptote and x-intercept.

57. $y = \log_{10}\left(\dfrac{x}{5}\right)$

58. $y = \log_{10}(-x)$

59. $f(x) = \ln(x - 2)$

60. $h(x) = \ln(x + 1)$

61. $g(x) = \ln(-x)$

62. $f(x) = \ln(3 - x)$

In Exercises 63–66, use a graphing utility to graph the function. What is the domain? Use the graph to determine the intervals in which the function is increasing and decreasing and approximate any relative maximum or minimum values of the function.

63. $f(x) = \dfrac{x}{2} - \ln\dfrac{x}{4}$

64. $g(x) = \dfrac{12 \ln x}{x}$

65. $h(x) = 4x \ln x$

66. $f(x) = \dfrac{x}{\ln x}$

67. *Population Growth* The population of a town will double in

$$t = \frac{10 \ln 2}{\ln 67 - \ln 50} \text{ years.}$$

Find t.

68. *Human Memory Model* Students in a mathematics class were given an exam and then tested monthly with an equivalent exam. The average score for the class was given by the human memory model

$$f(t) = 80 - 17 \log_{10}(t + 1), \quad 0 \le t \le 12$$

where t is the time in months.

(a) What was the average score on the original exam ($t = 0$)?

(b) What was the average score after 4 months?

(c) What was the average score after 10 months?

69. *Exploration* The table of values was obtained by evaluating a function. Determine which of the statements may be true and which must be false.

x	1	2	8
y	0	1	3

(a) y is an exponential function of x.

(b) y is a logarithmic function of x.

(c) x is an exponential function of y.

(d) y is a linear function of x.

70. *Exploration* Use a graphing utility to compare the graph of the function $y = \ln x$ with the graphs of the following functions.

(a) $y = x - 1$

(b) $y = (x - 1) - \frac{1}{2}(x - 1)^2$

(c) $y = (x - 1) - \frac{1}{2}(x - 1)^2 + \frac{1}{3}(x - 1)^3$

71. *Finding a Pattern* Identify the pattern of successive polynomials given in Exercise 70. Extend the pattern one more term and compare the graph of the resulting polynomial function with the graph of $y = \ln x$. What do you think the pattern implies?

72. *Graphical Analysis* Use a graphing utility to graph f and g in the same viewing rectangle and determine which is increasing at the greater rate for "large" values of x. What can you conclude about the rate of growth of the natural logarithmic function?

(a) $f(x) = \ln x, \quad g(x) = \sqrt{x}$

(b) $f(x) = \ln x, \quad g(x) = \sqrt[4]{x}$

73. *Investment Time* A principal P, invested at $9\frac{1}{2}\%$ and compounded continuously, increases to an amount K times the original principal after t years, where t is given by

$$t = (\ln K)/0.095.$$

(a) Complete the table and interpret your results.

K	1	2	4	6	8	10	12
t							

(b) Use a graphing utility to graph the function.

74. *Data Analysis* The table gives the temperatures T (°F) at which water boils at selected pressures p (pounds per square inch). (Source: Standard Handbook of Mechanical Engineers)

p	5	10	14.696 (1atm)
T	162.24°	193.21°	212.00°

p	20	30	40
T	227.96°	250.33°	267.25°

p	60	80	100
T	292.71°	312.03°	327.81°

A model that approximates this data is

$$T = 87.97 + 34.96 \ln x + 7.91\sqrt{x}.$$

(a) Use a graphing utility to plot the data points and graph the model in the same viewing rectangle. How well does the model fit the data?

(b) Use the graph to estimate the pressure required for the boiling point of water to exceed 300°F.

75. *Data Analysis* A meteorologist measures the atmospheric pressure P (in kilograms per square meter) at altitude h (in kilometers). The data is shown below.

h	0	5	10	15	20
P	10,332	5583	2376	1240	517

A model for the data is given by

$$P = 10{,}957e^{-0.15h}.$$

(a) Use a graphing utility to plot the data points and graph the model in the same viewing rectangle.

(b) Use a graphing utility to plot the points $(h, \ln P)$. Use the regression capabilities of the graphing utility to fit a regression line to the revised data points.

(c) The line in part (b) has the form $\ln P = ah + b$. Write the line in exponential form.

(d) Verify, graphically and analytically, that the result of part (c) is equivalent to the given exponential model for the data.

76. *World Population Growth* The time t in years for the world population to double if it is increasing at a continuous rate of r is given by

$$t = \frac{\ln 2}{r}.$$

Complete the table. What do your results imply?

r	0.005	0.010	0.015	0.020	0.025	0.030
t						

77. *Tractrix* A person walking along a dock (the y-axis) drags a boat by a 10-foot rope. The boat travels along a path known as a *tractrix*. The equation of this path is

$$y = 10 \ln\left(\frac{10 + \sqrt{100 - x^2}}{x}\right) - \sqrt{100 - x^2}.$$

(a) Use a graphing utility to obtain a graph of the function. What is the domain of the function?

(b) Identify any asymptotes of the graph.

(c) Determine the position of the person when the x-coordinate of the position of the boat is $x = 2$.

(d) Let $(0, p)$ be the position of the person. Determine p as a function of x, the x-coordinate of the position of the boat.

(e) Use a graphing utility to graph the function p. When does the position of the person change most for a small change in the position of the boat? Explain.

78. *Sound Intensity* The relationship between the number of decibels β and the intensity of a sound I in watts per square centimeter is given by

$$\beta = 10 \log_{10}\left(\frac{I}{10^{-16}}\right).$$

(a) Determine the number of decibels of a sound with an intensity of 10^{-4} watts per square centimeter.

(b) Determine the number of decibels of a sound with an intensity of 10^{-6} watts per square centimeter.

(c) The intensity of the sound in part (a) is 100 times as great as that in part (b). Is the number of decibels 100 times as great? Explain.

Ventilation Rates In Exercises 79 and 80, use the model

$$y = 80.4 - 11 \ln x, \qquad 100 \le x \le 1500$$

which approximates the minimum required ventilation rate in terms of the air space per child in a public school classroom. In the model, x is the air space per child in cubic feet and y is the ventilation rate in cubic feet per minute.

79. Use a graphing utility to graph the function and approximate the required ventilation rate if there is 300 cubic feet of air space per child.

80. A classroom is designed for 30 students. The air-conditioning system in the room has the capacity to move 450 cubic feet of air per minute.

(a) Determine the ventilation rate per child, assuming that the room is filled to capacity.

(b) Use the graph of Exercise 79 to estimate the air space required per child.

(c) Determine the minimum number of square feet of floor space required for the room if the ceiling height is 30 feet.

81. *Work* The work (in foot-pounds) done in compressing an initial volume of 9 cubic feet of air at a pressure of 15 pounds per square inch to a volume of 3 cubic feet is

$$W = 19{,}440(\ln 9 - \ln 3).$$

Find W.

82. (a) Use a graphing utility to complete the table for the function

$$f(x) = \frac{\ln x}{x}.$$

x	1	5	10	10^2	10^4	10^6
$f(x)$						

(b) Use the table to determine what $f(x)$ approaches as x increases without bound.

(c) Use a graphing utility to check the result of part (b).

Monthly Payment In Exercises 83–86, use the model

$$t = 10.042 \ln\left(\frac{x}{x - 1250}\right), \qquad 1250 < x$$

which approximates the length of a home mortgage of \$150,000 at 10% in terms of the monthly payment. In the model, t is the length of the mortgage in years and x is the monthly payment in dollars (see figure).

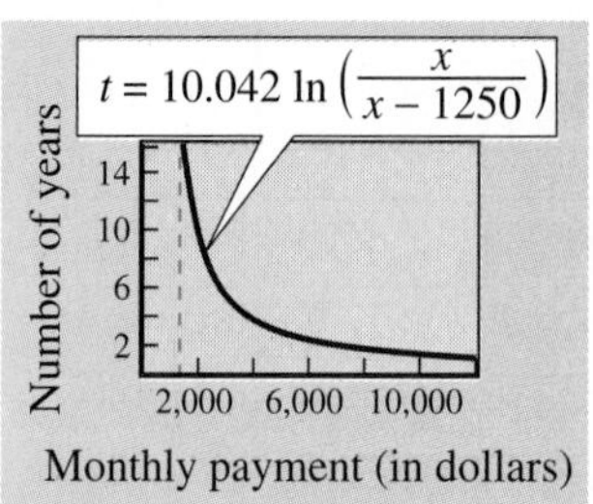

83. Use the model to approximate the length of the mortgage (for \$150,000 at 10%) if the monthly payment is \$1316.35.

84. Use the model to approximate the length of the mortgage (for \$150,000 at 10%) if the monthly payment is \$1982.26.

85. Approximate the total amount paid over the term of the mortgage with a monthly payment of \$1316.35. What amount of the total is interest costs?

86. Approximate the total amount paid over the term of the mortgage with a monthly payment of \$1982.26. What amount of the total is interest costs?

87. *Exploration* Answer the following for the function $f(x) = \log_{10} x$. Do not use a calculator.

(a) What is the domain of f?

(b) Find f^{-1}.

(c) If x is a real number between 1000 and 10,000, in what interval will $f(x)$ be found?

(d) Determine the interval in which x will be found if $f(x)$ is negative.

(e) If $f(x)$ is increased by one unit, x must have been increased by what factor?

(f) If $f(x_1) = 3n$ and $f(x_2) = n$, what is the ratio of x_1 to x_2?

4.3 Properties of Logarithms

Change of Base / Properties of Logarithms / Rewriting Logarithmic Expressions / Application

John Napier, a Scottish mathematician, developed logarithms as a way to simplify some of the tedious calculations of his day. Beginning in 1594, Napier worked about 20 years on the invention of logarithms. Napier was only partially successful in his quest to simplify tedious calculations. Nonetheless, the development of logarithms was a step forward and received immediate recognition.

Change of Base

Most calculators have only two types of log keys, one for common logarithms (base 10) and one for natural logarithms (base e). Although common logs and natural logs are the most frequently used, you may occasionally need to evaluate logarithms to other bases. To do this, you can use the following *change-of-base formula.*

Change-of-Base Formula

Let a, b, and x be positive real numbers such that $a \neq 1$ and $b \neq 1$. Then $\log_a x$ is given by

$$\log_a x = \frac{\log_b x}{\log_b a}.$$

One way to look at the change-of-base formula is that logarithms to base a are simply *constant multiples* of logarithms to base b. The constant multiplier is $1/(\log_b a)$.

Note Notice in Examples 1 and 2 that the result is the same whether common logarithms or natural logarithms are used in the change-of-base formula.

EXAMPLE 1 **Changing Bases Using Common Logarithms**

a. $\log_4 30 = \dfrac{\log_{10} 30}{\log_{10} 4} \approx \dfrac{1.47712}{0.60206} \approx 2.4534$

b. $\log_2 14 = \dfrac{\log_{10} 14}{\log_{10} 2} \approx \dfrac{1.14613}{0.30103} \approx 3.8074$

EXAMPLE 2 **Changing Bases Using Natural Logarithms**

a. $\log_4 30 = \dfrac{\ln 30}{\ln 4} \approx \dfrac{3.40120}{1.38629} \approx 2.4534$

b. $\log_2 14 = \dfrac{\ln 14}{\ln 2} \approx \dfrac{2.63906}{0.693147} \approx 3.8074$

Properties of Logarithms

You know from the previous section that the logarithmic function with base a is the *inverse* of the exponential function with base a. Thus, it makes sense that the properties of exponents should have corresponding properties involving logarithms. For instance, the exponential property $a^0 = 1$ has the corresponding logarithmic property $\log_a 1 = 0$.

Note There is no general property that can be used to rewrite $\log_a(u \pm v)$. Specifically, $\log_a(x + y)$ is not equal to $\log_a x + \log_a y$.

Properties of Logarithms

Let a be a positive number such that $a \neq 1$, and let n be a real number. If u and v are positive real numbers, the following properties are true.

1. $\log_a(uv) = \log_a u + \log_a v$	**1.** $\ln(uv) = \ln u + \ln v$
2. $\log_a \dfrac{u}{v} = \log_a u - \log_a v$	**2.** $\ln \dfrac{u}{v} = \ln u - \ln v$
3. $\log_a u^n = n \log_a u$	**3.** $\ln u^n = n \ln u$

Think About the Proof

To prove Property 1, let $x = \log_a u$ and $y = \log_a v$. The corresponding exponential forms of these two equations are $a^x = u$ and $a^y = v$. Multiplying u and v produces

$$uv = a^x a^y = a^{x+y}.$$

Can you see how to use this equation to complete the proof? The details of this proof are listed in the appendix.

EXAMPLE 3 Using Properties of Logarithms

Write the logarithm in terms of ln 2 and ln 3.

a. $\ln 6$ **b.** $\ln \dfrac{2}{27}$

Solution

a. $\ln 6 = \ln(2 \cdot 3)$ Rewrite 6 as $2 \cdot 3$.

$= \ln 2 + \ln 3$ Property 1

b. $\ln \dfrac{2}{27} = \ln 2 - \ln 27$ Property 2

$= \ln 2 - \ln 3^3$ Rewrite 27 as 3^3.

$= \ln 2 - 3 \ln 3$ Property 3

EXAMPLE 4 Using Properties of Logarithms

Use the properties of logarithms to verify that $-\ln \frac{1}{2} = \ln 2$.

Solution

$$-\ln \tfrac{1}{2} = -\ln(2^{-1}) = -(-1) \ln 2 = \ln 2$$

Try checking this result on your calculator.

Rewriting Logarithmic Expressions

The properties of logarithms are useful for rewriting logarithmic expressions in forms that simplify the operations of algebra. This is true because they convert complicated products, quotients, and exponential forms into simpler sums, differences, and products, respectively.

EXAMPLE 5 **Rewriting the Logarithm of a Product**

$$\begin{aligned}\log_{10} 5x^3y &= \log_{10} 5 + \log_{10} x^3y\\ &= \log_{10} 5 + \log_{10} x^3 + \log_{10} y\\ &= \log_{10} 5 + 3\log_{10} x + \log_{10} y\end{aligned}$$

EXAMPLE 6 **Rewriting the Logarithm of a Quotient**

$$\begin{aligned}\ln \frac{\sqrt{3x-5}}{7} &= \ln(3x-5)^{1/2} - \ln 7\\ &= \frac{1}{2}\ln(3x-5) - \ln 7\end{aligned}$$

EXPLORATION

Use a graphing utility to graph the functions

$$y = \ln x - \ln(x-3)$$

and

$$y = \ln \frac{x}{x-3}$$

in the same viewing rectangle. Does the graphing utility show the functions with the same domain? If so, should it? Explain your reasoning.

In Examples 5 and 6, the properties of logarithms were used to *expand* logarithmic expressions. In Examples 7 and 8, this procedure is reversed and the properties of logarithms are used to *condense* logarithmic expressions.

EXAMPLE 7 **Condensing a Logarithmic Expression**

$$\begin{aligned}\tfrac{1}{2}\log_{10} x + 3\log_{10}(x+1) &= \log_{10} x^{1/2} + \log_{10}(x+1)^3\\ &= \log_{10}\left[\sqrt{x}\cdot(x+1)^3\right]\end{aligned}$$

EXAMPLE 8 **Condensing a Logarithmic Expression**

$$\begin{aligned}2\ln(x+2) - \ln x &= \ln(x+2)^2 - \ln x\\ &= \ln \frac{(x+2)^2}{x}\end{aligned}$$

Application

Real Life

EXAMPLE 9 Finding a Mathematical Model

The table gives the mean distance from the sun x and the orbital period y of the six planets that are closest to the sun. In the table, the mean distance is given in terms of astronomical units (where the earth's mean distance is defined as 1.0), and the period is given in terms of years. Find an equation that expresses y as a function of x.

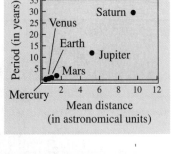

Planet	Mercury	Venus	Earth	Mars	Jupiter	Saturn
Period, y	0.241	0.615	1.0	1.881	11.861	29.457
Mean Distance, x	0.387	0.723	1.0	1.523	5.203	9.541

Figure 4.18

Solution

The points in the table are plotted in Figure 4.18. From this figure it is not clear how to find an equation that relates y and x. To solve this problem, take the natural log of each of the x- and y-values given in the table. This produces the following results.

Planet	Mercury	Venus	Earth	Mars	Jupiter	Saturn
$\ln y$	−1.423	−0.486	0.0	0.632	2.473	3.383
$\ln x$	−0.949	−0.324	0.0	0.421	1.649	2.256

Figure 4.19

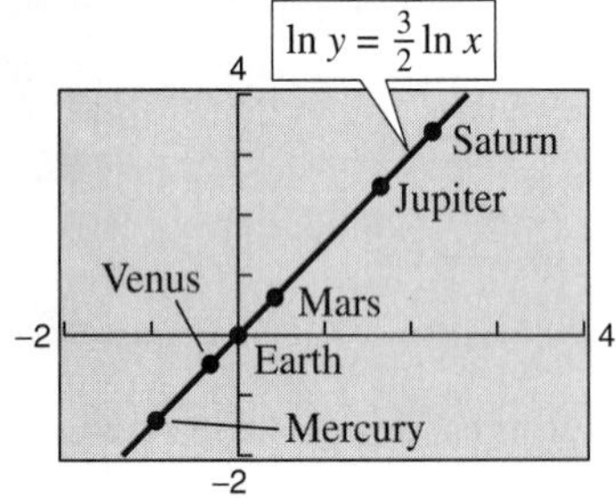

Now, by plotting the points in the second table, you can see that all six of the points appear to lie in a line (see Figure 4.19). You can use a graphical approach or the algebraic approach discussed in Section 2.6 to find that the slope of this line is $\frac{3}{2}$, and you can therefore conclude that $\ln y = \frac{3}{2} \ln x$. [Try to convert this to $y = f(x)$ form.]

Note This Group Activity is related to the one on page 22 in Section P.2. If you didn't work on that activity, try both activities now. Which approach do you prefer? Explain your reasoning.

Group Activity

Kepler's Law

The relationship described in Example 9 was first discovered by Johannes Kepler. Use properties of logarithms to rewrite the relationship so that y is expressed as a function of x. Then verify the values in the first table.

4.3 /// EXERCISES

In Exercises 1 and 2, use a graphing utility to graph the two functions in the same viewing rectangle. What do the graphs suggest? Explain your reasoning.

1. $f(x) = \log_{10} x$
$g(x) = \dfrac{\ln x}{\ln 10}$

2. $f(x) = \ln x$
$g(x) = \dfrac{\log_{10} x}{\log_{10} e}$

In Exercises 3–6, use the change-of-base formula to write the given logarithm as a multiple of a common logarithm. For instance, $\log_2 3 = (1/\log_{10} 2) \log_{10} 3$.

3. $\log_3 5$ **4.** $\log_4 10$
5. $\log_2 x$ **6.** $\ln 5$

In Exercises 7–10, use the change-of-base formula to write the given logarithm as a multiple of a natural logarithm. For instance, $\log_2 3 = (1/\ln 2) \ln 3$.

7. $\log_3 5$ **8.** $\log_4 10$
9. $\log_2 x$ **10.** $\log_{10} 5$

In Exercises 11–18, evaluate the logarithm using the change-of-base formula. Round your result to three decimal places.

11. $\log_3 7$ **12.** $\log_7 4$
13. $\log_{1/2} 4$ **14.** $\log_4(0.55)$
15. $\log_9(0.4)$ **16.** $\log_{20} 125$
17. $\log_{15} 1250$ **18.** $\log_{1/3}(0.015)$

In Exercises 19–38, use the properties of logarithms to write the expression as a sum, difference, and/or constant multiple of logarithms. (Assume all variables are positive.)

19. $\log_{10} 5x$ **20.** $\log_{10} 10z$
21. $\log_{10} \dfrac{5}{x}$ **22.** $\log_{10} \dfrac{y}{2}$
23. $\log_8 x^4$ **24.** $\log_6 z^{-3}$
25. $\ln \sqrt{z}$ **26.** $\ln \sqrt[3]{t}$
27. $\ln xyz$ **28.** $\ln \dfrac{xy}{z}$
29. $\ln \sqrt{a-1}, \quad a > 1$ **30.** $\ln\left(\dfrac{x^2-1}{x^3}\right), \quad x > 1$
31. $\ln z(z-1)^2, \quad z > 1$ **32.** $\ln \sqrt{\dfrac{x^2}{y^3}}$
33. $\ln \sqrt[3]{\dfrac{x}{y}}$ **34.** $\ln \dfrac{x}{\sqrt{x^2+1}}$
35. $\ln \dfrac{x^4\sqrt{y}}{z^5}$ **36.** $\ln \sqrt{x^2(x+2)}$
37. $\log_b \dfrac{x^2}{y^2z^3}$ **38.** $\log_b \dfrac{\sqrt{x}y^4}{z^4}$

Graphical Analysis In Exercises 39 and 40, use a graphing utility to graph the two equations in the same viewing rectangle. What do the graphs suggest? Explain your reasoning.

39. $y_1 = \ln[x^3(x+4)], \quad y_2 = 3\ln x + \ln(x+4)$

40. $y_1 = \ln\left(\dfrac{\sqrt{x}}{x-2}\right), \quad y_2 = \frac{1}{2}\ln x - \ln(x-2)$

In Exercises 41–60, write the expression as the logarithm of a single quantity.

41. $\ln x + \ln 2$ **42.** $\ln y + \ln z$
43. $\log_4 z - \log_4 y$ **44.** $\log_5 8 - \log_5 t$
45. $2\log_2(x+4)$ **46.** $-4\log_6 2x$
47. $\frac{1}{3}\log_3 5x$ **48.** $\frac{3}{2}\log_7(z-2)$
49. $\ln x - 3\ln(x+1)$ **50.** $2\ln 8 + 5\ln z$
51. $\ln(x-2) - \ln(x+2)$
52. $3\ln x + 2\ln y - 4\ln z$
53. $\ln x - 2[\ln(x+2) + \ln(x-2)]$
54. $4[\ln z + \ln(z+5)] - 2\ln(z-5)$

55. $\frac{1}{3}[2 \ln(x + 3) + \ln x - \ln(x^2 - 1)]$
56. $2[\ln x - \ln(x + 1) - \ln(x - 1)]$
57. $\frac{1}{3}[\ln y + 2 \ln(y + 4)] - \ln(y - 1)$
58. $\frac{1}{2}[\ln(x + 1) + 2 \ln(x - 1)] + 3 \ln x$
59. $2 \ln 3 - \frac{1}{2} \ln(x^2 + 1)$
60. $\frac{3}{2} \ln 5t^6 - \frac{3}{4} \ln t^4$

Graphical Analysis **In Exercises 61 and 62, use a graphing utility to graph the two equations in the same viewing rectangle. What do the graphs suggest? Verify your conclusion algebraically.**

61. $y_1 = 2[\ln 8 - \ln(x^2 + 1)]$, $y_2 = \ln\left[\dfrac{64}{(x^2 + 1)^2}\right]$

62. $y_1 = \ln x + \frac{1}{3} \ln(x + 1)$, $y_2 = \ln(x\sqrt[3]{x + 1})$

Think About It **In Exercises 63 and 64, use a graphing utility to graph the two equations in the same viewing rectangle. Are the expressions equivalent? Explain.**

63. $y_1 = \ln x^2$, $y_2 = 2 \ln x$
64. $y_1 = \frac{1}{4} \ln[x^4(x^2 + 1)]$, $y_2 = \ln x + \frac{1}{4} \ln(x^2 + 1)$

65. *Think About It* Use a graphing utility to graph

$$f(x) = \ln \frac{x}{2}, \quad g(x) = \frac{\ln x}{\ln 2}, \quad h(x) = \ln x - \ln 2$$

in the same viewing rectangle. Which two functions have identical graphs? Explain why.

66. *Exploration* Approximate the natural logarithms of as many integers as possible between 1 and 20 given that $\ln 2 \approx 0.6931$, $\ln 3 \approx 1.0986$, and $\ln 5 \approx 1.6094$. (Do not use a calculator.)

In Exercises 67–80, find the exact value of the logarithm without using a calculator. (If this is not possible, state the reason.)

67. $\log_3 9$
68. $\log_6 \sqrt[3]{6}$
69. $\log_4 16^{1.2}$
70. $\log_5 \left(\frac{1}{125}\right)$
71. $\log_3 (-9)$
72. $\log_2 (-16)$
73. $\log_5 75 - \log_5 3$
74. $\log_4 2 + \log_4 32$
75. $\ln e^2 - \ln e^5$
76. $3 \ln e^4$
77. $\log_{10} 0$
78. $\ln 1$
79. $\ln e^{4.5}$
80. $\ln \sqrt[4]{e^3}$

In Exercises 81–88, use the properties of logarithms to simplify the logarithmic expression.

81. $\log_4 8$
82. $\log_5\left(\frac{1}{15}\right)$
83. $\log_7 \sqrt{70}$
84. $\log_2(4^2 \cdot 3^4)$
85. $\log_5 \left(\frac{1}{250}\right)$
86. $\log_{10} \left(\frac{9}{300}\right)$
87. $\ln(5e^6)$
88. $\ln \dfrac{6}{e^2}$

89. *Sound Intensity* The relationship between the number of decibels β and the intensity of a sound I in watts per square centimeter is given by

$$\beta = 10 \log_{10}\left(\frac{I}{10^{-16}}\right).$$

Use the properties of logarithms to write the formula in simpler form, and determine the number of decibels of a sound with an intensity of 10^{-10} watts per square centimeter.

90. *Human Memory Model* Students participating in a psychological experiment attended several lectures. After the last lecture, and every month for the next year, the students were tested to see how much of the material they remembered. The average scores for the group were given by the memory model

$$f(t) = 90 - 15 \log_{10}(t + 1), \quad 0 \le t \le 12$$

where t is the time in months.

(a) Use a graphing utility to graph the function over the specified domain.

(b) What was the average score on the original exam $(t = 0)$?

(c) What was the average score after 6 months?

(d) What was the average score after 12 months?

(e) When would the average score have decreased to 75?

91. *Comparing Models* A cup of water at an initial temperature of 78°C is placed in a room at a constant temperature of 21°C. The temperature of the water is measured every 5 minutes for a period of $\frac{1}{2}$ hour. The results are recorded as ordered pairs of the form (t, T), where t is the time in minutes and T is the temperature in degrees Celsius.

(0, 78.0°), (5, 66.0°), (10, 57.5°), (15, 51.2°), (20, 46.3°), (25, 42.5°), (30, 39.6°)

(a) The graph of the model for the data should be asymptotic with the temperature of the room. Subtract the room temperature from each of the temperatures in the ordered pairs. Use a graphing utility to plot the data points (t, T) and $(t, T - 21)$.

(b) Use the regression capabilities of a graphing utility to fit an exponential model to the revised data. This model will be of the form

$$T - 21 = ab^x.$$

Solve for T and graph the model. Compare the result with the plot of the original data.

(c) Take the natural logarithms of the revised temperatures. Use a graphing utility to plot the points $(t, \ln(T - 21))$ and observe that the points appear linear. Use the regression capabilities of a graphing utility to fit a line to this data. The resulting line has the form

$$\ln(T - 21) = at + b.$$

Use the properties of logarithms to solve for T. Verify that the result is equivalent to the model in part (b).

(d) Fit a rational model to the data. Take the reciprocals of the y-coordinates of the revised data to generate the points

$$\left(t, \frac{1}{T - 21}\right).$$

Use a graphing utility to plot these points and observe that they appear linear. Use the regression capabilities of a graphing utility to fit a line to this data. The resulting line has the form

$$\frac{1}{T - 21} = at + b.$$

Solve for T, and use a graphing utility to graph the rational function and the original data points.

92. *Essay* Write a short paragraph explaining why the transformations of the data in Exercise 91 were necessary to obtain the models. Why did taking the logarithms of the temperatures lead to a linear scatter plot? Why did taking the reciprocals of the temperatures lead to a linear scatter plot?

In Exercises 93–96, use the change-of-base formula and a graphing utility to graph each function.

93. $f(x) = \log_2 x$

94. $f(x) = \log_7 x$

95. $g(x) = \log_3 x^{1/2}$

96. $f(t) = \log_5 \dfrac{t}{3}$

True or False? **In Exercises 97–102, determine whether the statement is true or false given that $f(x) = \ln x$. If the statement is false, state why or give an example showing that it is false.**

97. $f(0) = 0$

98. $f(ax) = f(a) + f(x), \quad a > 0, x > 0$

99. $f(x - 2) = f(x) - f(2), \quad x > 2$

100. $\sqrt{f(x)} = \frac{1}{2} f(x)$

101. If $f(u) = 2f(v)$, then $v = u^2$.

102. If $f(x) < 0$, then $0 < x < 1$.

103. Prove that $\log_b \dfrac{u}{v} = \log_b u - \log_b v$.

104. Prove that $\log_b u^n = n \log_b u$.

Review **Solve Exercises 105–108 as a review of the skills and problem-solving techniques you learned in previous sections. Simplify the expression.**

105. $\dfrac{24xy^{-2}}{16x^{-3}y}$

106. $\left(\dfrac{2x^2}{3y}\right)^{-3}$

107. $(18x^3y^4)^{-3}(18x^3y^4)^3$

108. $xy(x^{-1} + y^{-1})^{-1}$

4.4 Solving Exponential and Logarithmic Equations

Introduction / Solving Exponential Equations / Solving Logarithmic Equations / Approximating Solutions / Application

Introduction

So far in this chapter, you have studied the definitions, graphs, and properties of exponential and logarithmic functions. In this section, you will study procedures for *solving equations* involving exponential and logarithmic functions. As a simple example, consider the exponential equation $2^x = 32$. One property of exponential functions states that $a^x = a^y$ if and only if $x = y$. You can obtain the solution by rewriting the sample equation in the form $2^x = 2^5$, which implies that $x = 5$.

You can use your graphing utility to verify the inverse properties of logarithmic and exponential functions. For instance, try graphing

$$y_1 = e^{\ln x}$$
$$y_2 = \ln e^x.$$

Which of these graphs shows only part of the line $y = x$? Why?

Although this method works in some cases, it does not work for an equation as simple as $e^x = 7$. In such a case, solution procedures are based on the fact that the exponential and logarithmic functions are inverses of each other. The following comparisons show the **inverse properties** of exponential and logarithmic functions.

	Base a	*Base e*
1.	$\log_a a^x = x$	$\ln e^x = x$
2.	$a^{\log_a x} = x$	$e^{\ln x} = x$

To solve $e^x = 7$, you can take the natural logarithms of both sides to obtain

$$e^x = 7 \qquad \text{Original equation}$$
$$\ln e^x = \ln 7 \qquad \text{Take logarithms of both sides.}$$
$$x = \ln 7 \qquad \ln e^x = x \text{ because } e^x = e^x.$$

Here are some guidelines for solving exponential and logarithmic equations.

Solving Exponential and Logarithmic Equations

1. *To solve an exponential equation,* first isolate the exponential expression, then take the logarithms of both sides and solve for the variable.
2. *To solve a logarithmic equation,* rewrite the equation in exponential form and solve for the variable.

Solving Exponential Equations

EXAMPLE 1 **Solving an Exponential Equation**

$e^x = 72$	Original equation
$\ln e^x = \ln 72$	Take logarithms of both sides.
$x = \ln 72$	Inverse property of logs and exponents
$x \approx 4.277$	

The solution is $x = \ln 72$. Check this solution in the original equation.

EXAMPLE 2 **Solving an Exponential Equation**

$e^x + 5 = 60$	Original equation
$e^x = 55$	Combine like terms.
$\ln e^x = \ln 55$	Take logarithms of both sides.
$x = \ln 55$	Inverse property of logs and exponents
$x \approx 4.007$	

Figure 4.20

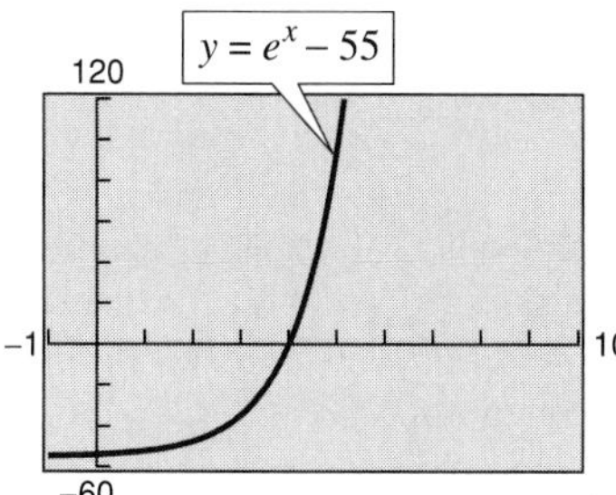

The solution is $x = \ln 55$. Check this solution in the original equation. The graph of $y = e^x - 55$ is shown in Figure 4.20. Use the root feature of your graphing utility to find the x-intercept and thus confirm the solution.

EXAMPLE 3 **Solving an Exponential Equation**

$4e^{2x} = 5$	Original equation
$e^{2x} = \dfrac{5}{4}$	Divide both sides by 4.
$\ln e^{2x} = \ln \dfrac{5}{4}$	Take logarithms of both sides.
$2x = \ln \dfrac{5}{4}$	Inverse property of logs and exponents
$x = \dfrac{1}{2} \ln \dfrac{5}{4}$	Solve for x.
$x \approx 0.112$	

Figure 4.21

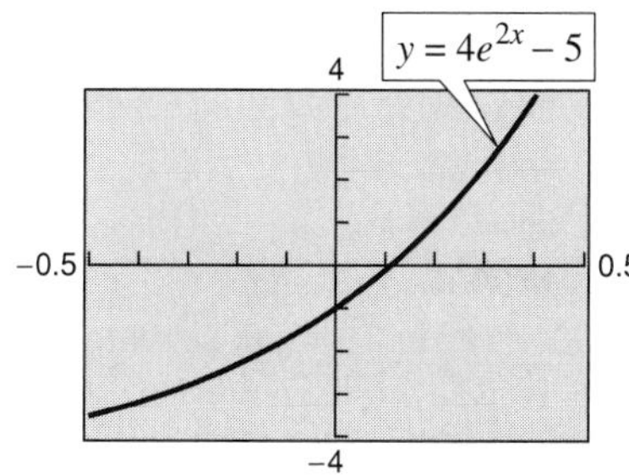

The solution is $x = \frac{1}{2} \ln \frac{5}{4}$. Check this solution in the original equation. The graph of $y = 4e^{2x} - 5$, shown in Figure 4.21, helps to confirm this solution.

When an equation involves two or more exponential expressions, you can still use a procedure similar to that demonstrated in the first three examples. However, the algebra is a bit more complicated and a graphical approach is often easier. Study the next example carefully.

EXAMPLE 4 Solving an Exponential Equation

Solve $e^{2x} - 3e^x + 2 = 0$.

Solution

Figure 4.22

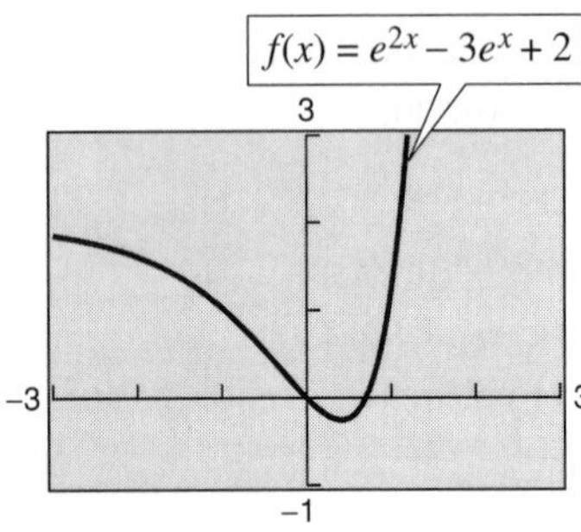

The graph of the function $f(x) = e^{2x} - 3e^x + 2$ in Figure 4.22 indicates that there are two solutions: one near $x = 0$ and one near $x = 0.7$. You can verify that $x = 0$ is indeed a solution by noting that $f(0) = e^{2(0)} - 3e^0 + 2 = 1 - 3 + 2 = 0$. Similarly, you can use the root feature to determine that $x \approx 0.693$. You can also solve this problem algebraically.

$$e^{2x} - 3e^x + 2 = 0 \quad \text{Original equation}$$

$$(e^x)^2 - 3e^x + 2 = 0 \quad \text{Quadratic form}$$

$$(e^x - 2)(e^x - 1) = 0 \quad \text{Factor.}$$

$$e^x - 2 = 0 \qquad e^x - 1 = 0 \quad \text{Set factors equal to zero.}$$

$$e^x = 2 \qquad e^x = 1$$

$$x = \ln 2 \qquad x = 0 \quad \text{Solutions}$$

The equation has two solutions, $x = \ln 2 \approx 0.693$ and $x = 0$, which confirm the graphical analysis.

EXAMPLE 5 A Base Other Than e

Solve $2^x = 10$.

Solution

$$2^x = 10 \quad \text{Original equation}$$

$$\ln 2^x = \ln 10 \quad \text{Take logarithms of both sides.}$$

$$x \ln 2 = \ln 10 \quad \text{Property of logarithms}$$

$$x = \frac{\ln 10}{\ln 2} \quad \text{Solve for } x.$$

Figure 4.23

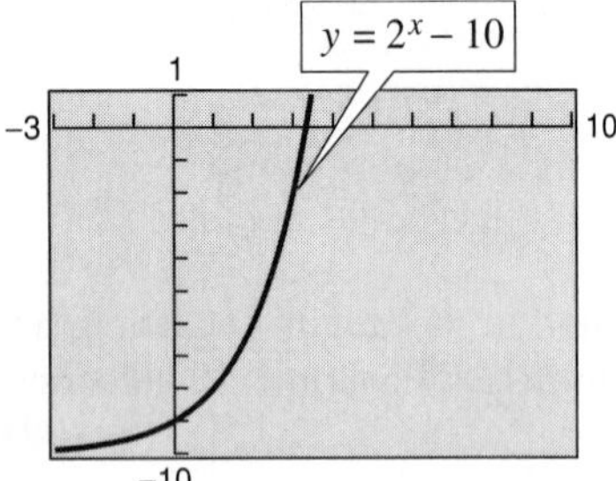

The equation has one solution: $x = \ln 10/\ln 2 \approx 3.32$. Check this solution in the original equation. The graph of $y = 2^x - 10$, shown in Figure 4.23, helps to confirm this solution. (*Note:* Using the change-of-base formula, you could write the solution as $x = \log_2 10$.)

Solving Logarithmic Equations

To solve a logarithmic equation such as $\ln x = 3$, you can write the equation in exponential form as follows.

$$\ln x = 3 \quad \text{Logarithmic form}$$
$$e^{\ln x} = e^3 \quad \text{Exponentiate both sides.}$$
$$x = e^3 \quad \text{Exponential form}$$

This procedure is called *exponentiating* both sides of an equation. It is applied after the logarithmic expression has been isolated.

EXAMPLE 6 **Solving a Logarithmic Equation**

$$5 + 2 \ln x = 4 \quad \text{Original equation}$$
$$2 \ln x = -1 \quad \text{Combine like terms.}$$
$$\ln x = -\frac{1}{2} \quad \text{Isolate the logarithmic expression.}$$
$$e^{\ln x} = e^{-1/2} \quad \text{Exponentiate both sides.}$$
$$x = e^{-1/2} \quad \text{Inverse property of exponents and logs}$$
$$x \approx 0.607$$

The equation has one solution: $x = e^{-1/2}$. Check this solution in the original equation. The graph of $y = 1 + 2 \ln x$, shown in Figure 4.24, helps to confirm this solution.

Figure 4.24

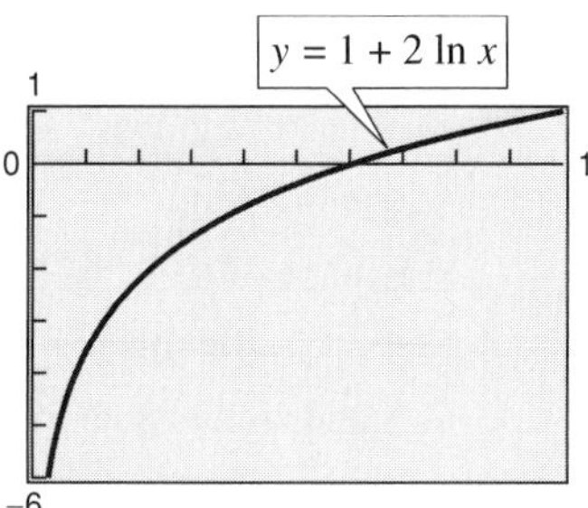

EXAMPLE 7 **Solving a Logarithmic Equation**

$$2 \ln 3x = 4 \quad \text{Original equation}$$
$$\ln 3x = 2 \quad \text{Isolate the logarithmic expression.}$$
$$e^{\ln 3x} = e^2 \quad \text{Exponentiate both sides.}$$
$$3x = e^2 \quad \text{Inverse property of exponents and logs}$$
$$x = \frac{1}{3}e^2 \quad \text{Solve for } x.$$
$$x \approx 2.463$$

The equation has one solution: $x = \frac{1}{3}e^2$. Check this solution in the original equation. The graph of $y = -4 + 2 \ln 3x$, shown in Figure 4.25, helps to confirm this solution.

Figure 4.25

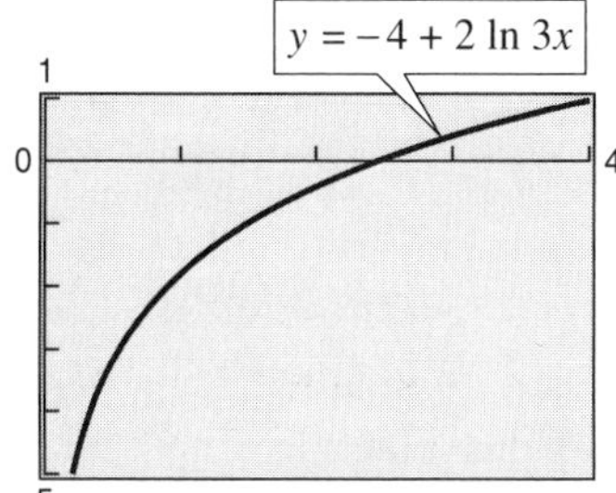

Complicated equations involving logarithmic expressions can be solved using a graphing utility, as demonstrated in Example 8.

Figure 4.26

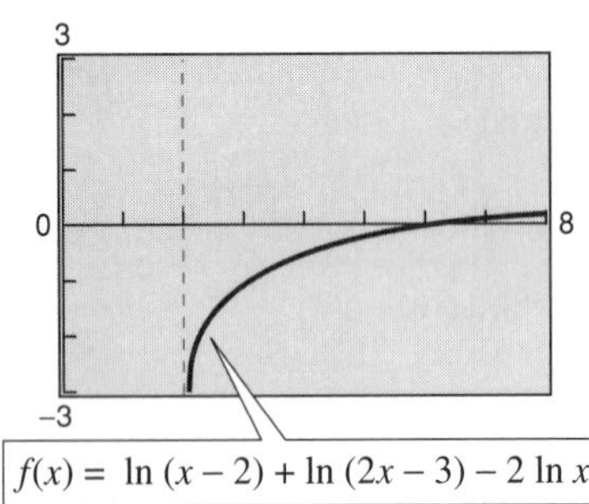

EXAMPLE 8 Solving a Logarithmic Equation

Solve for x in the equation $\ln(x-2) + \ln(2x-3) = 2\ln x$.

Solution

The graph of the function $f(x) = \ln(x-2) + \ln(2x-3) - 2\ln x$, shown in Figure 4.26, indicates that the only zero is $x = 6$. This can be verified by substituting $x = 6$ into the original equation. You could also solve this equation algebraically. Notice that in this case the technique produces an extraneous solution.

$$\ln(x-2) + \ln(2x-3) = 2\ln x \qquad \text{Original equation}$$

$$\ln[(x-2)(2x-3)] = \ln x^2 \qquad \text{Properties of logarithms}$$

$$\ln(2x^2 - 7x + 6) = \ln x^2$$

$$e^{\ln(2x^2-7x+6)} = e^{\ln x^2} \qquad \text{Exponentiate both sides.}$$

$$2x^2 - 7x + 6 = x^2 \qquad \text{Inverse property of exponents and logs}$$

$$x^2 - 7x + 6 = 0 \qquad \text{Quadratic form}$$

$$(x-6)(x-1) = 0 \qquad \text{Factor.}$$

$$x - 6 = 0 \Rightarrow x = 6 \qquad \text{Set 1st factor equal to 0.}$$

$$x - 1 = 0 \Rightarrow x = 1 \qquad \text{Set 2nd factor equal to 0.}$$

Finally, by checking these two "solutions" in the original equation, you can conclude that $x = 1$ is not valid. Can you see why? Thus, the only solution is $x = 6$.

EXPLORATION

There are many ways to solve the equation given in Example 8. One way would be to think of the solution as the intersection of two graphs (the graphs of the left and right sides of the equation). Try using a graphing utility to graph the functions

$$y = \ln(x-2) + \ln(2x-3)$$

and

$$y = 2\ln x$$

in the same viewing rectangle. Does the point of intersection agree with the solution found in the example?

EXAMPLE 9 **The Change-of-Base Formula**

Prove the change-of-base formula: $\log_a x = \dfrac{\log_b x}{\log_b a}$.

Solution

Begin by letting $y = \log_a x$ and writing the equivalent exponential form $a^y = x$. Now, taking the logarithms *with base b* of both sides produces the following.

$$\log_b a^y = \log_b x$$

$$y \log_b a = \log_b x$$

$$y = \frac{\log_b x}{\log_b a}$$

$$\log_a x = \frac{\log_b x}{\log_b a}$$

For solving exponential or logarithmic equations, the following properties are useful.

1. $x = y$ if and only if $\log_a x = \log_a y$.
2. $x = y$ if and only if $a^x = a^y$, $a > 0$, $a \neq 1$.

Can you see where these properties were used in the examples in this section?

Approximating Solutions

Equations that involve combinations of algebraic functions, exponential functions, and/or logarithmic functions can be very difficult to solve by algebraic procedures. Here again, you can take advantage of a graphing utility.

EXAMPLE 10 **Approximating the Solution of an Equation**

Approximate the solution of $\ln x = x^2 - 2$.

Solution

To begin, use a graphing utility to graph

$$y = -x^2 + 2 + \ln x$$

as shown in Figure 4.27. From this graph, you can see that the equation has two solutions. Next, using the root or zero feature, you can approximate the two solutions to be $x \approx 0.138$ and $x \approx 1.564$.

Figure 4.27

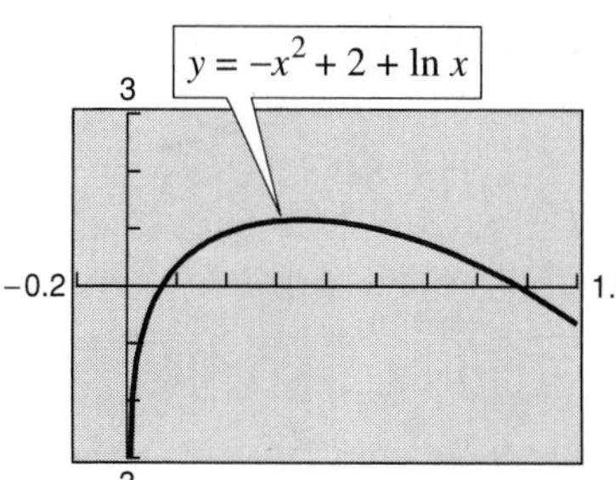

Application

EXAMPLE 11 Consumer Price Index for Sugar *Real Life*

Figure 4.28

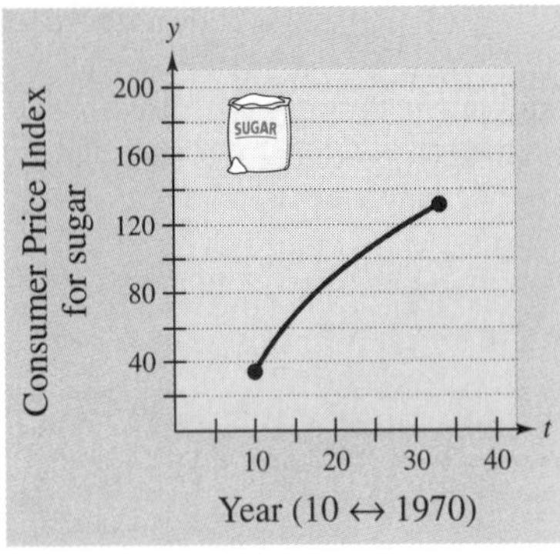

From 1970 to 1993, the Consumer Price Index (CPI) value y for a fixed amount of sugar for the year t can be modeled by the equation $y = -169.8 + 86.8 \ln t$, where $t = 10$ represents 1970 (see Figure 4.28). During which year did the price of sugar reach 4 times its 1970 price of 30.5 on the CPI? (Source: U.S. Bureau of Labor Statistics)

Solution

$-169.8 + 86.8 \ln t = y$	Given model
$-169.8 + 86.8 \ln t = 122$	Let $y = (4)(30.5) = 122$.
$86.8 \ln t = 291.8$	Add 169.8 to both sides.
$\ln t \approx 3.362$	Divide both sides by 86.8.
$e^{\ln t} \approx e^{3.362}$	Exponentiate both sides.
$t \approx 28.8$	Inverse property of exponents and logs

The solution is $t \approx 28.8$ years. Because $t = 10$ represents 1970, it follows that the price of sugar reached four times its 1970 price in 1988.

Group Activity — *Comparing Mathematical Models*

The table below gives the monthly amount of traffic y (in millions of packets) on the NSFNET, the backbone of the Internet, for January of each year x from 1988 through 1995, with $x = 8$ representing 1988. (Source: Network Information Center)

x	8	9	10	11	12	13	14	15
y	85	468	2466	5868	12,153	24,016	45,283	60,399

(a) Create a scatter plot of the data. Find a linear model for the data, and add its graph to your scatter plot. According to this model, when will monthly network traffic reach 150,000 million packets?

(b) Create a new table giving values for $\ln x$ and $\ln y$, and create a scatter plot of the data. Use the method illustrated in Example 9 in Section 4.3 to find a model for the data, and add its graph to your scatter plot. According to this model, when will monthly network traffic reach 150,000 million packets?

(c) Solve the model in part (b) for y, and add its graph to your scatter plot in part (a). Which model best fits the data? Which model will best predict future traffic levels? Explain your reasoning.

4.4 /// EXERCISES

In Exercises 1–6, determine whether the x-values are solutions of the equation.

1. $4^{2x-7} = 64$
(a) $x = 5$
(b) $x = 2$

2. $2^{3x+1} = 32$
(a) $x = -1$
(b) $x = 2$

3. $3e^{x+2} = 75$
(a) $x = -2 + e^{25}$
(b) $x = -2 + \ln 25$
(c) $x \approx 1.2189$

4. $5^{2x+3} = 812$
(a) $x = -1.5 + \log_5 \sqrt{812}$
(b) $x \approx 0.5813$
(c) $x = \frac{1}{2}\left(-3 + \dfrac{\ln 812}{\ln 5}\right)$

5. $\log_4(3x) = 3$
(a) $x \approx 20.3560$
(b) $x = -4$
(c) $x = \frac{64}{3}$

6. $\ln(x - 1) = 3.8$
(a) $x = 1 + e^{3.8}$
(b) $x \approx 45.7012$
(c) $x = 1 + \ln 3.8$

In Exercises 7–10, approximate the point of intersection of the graphs of f and g. Then solve the equation $f(x) = g(x)$ algebraically.

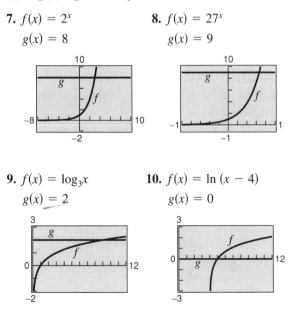

7. $f(x) = 2^x$
$g(x) = 8$

8. $f(x) = 27^x$
$g(x) = 9$

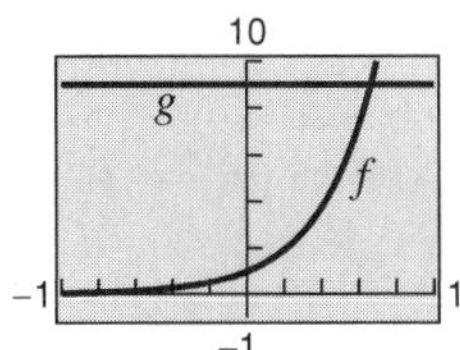

9. $f(x) = \log_3 x$
$g(x) = 2$

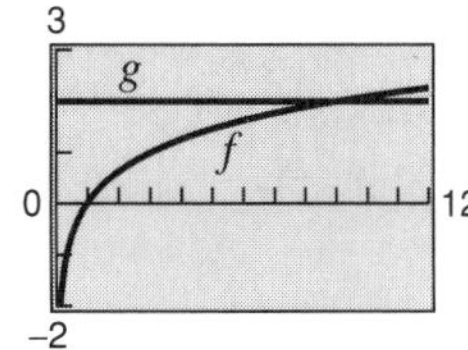

10. $f(x) = \ln(x - 4)$
$g(x) = 0$

In Exercises 11–20, solve for x.

11. $4^x = 16$
12. $3^x = 243$
13. $7^x = \frac{1}{49}$
14. $8^x = 4$
15. $\left(\frac{3}{4}\right)^x = \frac{27}{64}$
16. $3^{x-1} = 27$
17. $\log_4 x = 3$
18. $\log_5 5x = 2$
19. $\log_{10} x = -1$
20. $\ln(2x - 1) = 0$

In Exercises 21–26, simplify the expression.

21. $\ln e^{x^2}$
22. $\ln e^{2x-1}$
23. $e^{\ln(5x+2)}$
24. $-1 + \ln e^{2x}$
25. $e^{\ln x^2}$
26. $-8 + e^{\ln x^3}$

In Exercises 27–34, solve the exponential equation algebraically. Round the result to three decimal places.

27. $e^x = 10$
28. $4e^x = 91$
29. $7 - 2e^x = 5$
30. $-14 + 3e^x = 11$
31. $500e^{-x} = 300$
32. $1000e^{-4x} = 75$
33. $10^x = 42$
34. $10^x = 570$

In Exercises 35–38, let $f(x)$ represent the left side of the equation. Complete the table to obtain a rough estimate of the solution, and then use a graphing utility to graph both sides of the equation to obtain a better estimate of the solution (use three-decimal-place accuracy).

35. $e^{3x} = 12$

x	0.6	0.7	0.8	0.9	1.0
$f(x)$					

36. $e^{2x} = 50$

x	1.6	1.7	1.8	1.9	2.0
$f(x)$					

37. $20(100 - e^{x/2}) = 500$

x	5	6	7	8	9
$f(x)$					

38. $\dfrac{400}{1 + e^{-x}} = 350$

x	0	1	2	3	4
$f(x)$					

In Exercises 39–50, use the root feature of a graphing utility to solve the exponential equation accurate to three decimal places.

39. $e^{2x} - 4e^x - 5 = 0$

40. $e^{2x} - 5e^x + 6 = 0$

41. $3^{2x} = 80$

42. $6^{5x} = 3000$

43. $5^{-t/2} = 0.20$

44. $4^{-3t} = 0.10$

45. $2^{3-x} = 565$

46. $\dfrac{3000}{2 + e^{2x}} = 2$

47. $8(10^{3x}) = 12$

48. $3(5^{x-1}) = 21$

49. $\left(1 + \dfrac{0.10}{12}\right)^{12t} = 2$

50. $\left(1 + \dfrac{0.065}{365}\right)^{365t} = 4$

In Exercises 51–54, use a graphing utility to graph the function and approximate its zero accurate to three decimal places.

51. $g(x) = 6e^{1-x} - 25$

52. $f(x) = 3e^{3x/2} - 962$

53. $g(t) = e^{0.09t} - 3$

54. $h(t) = e^{0.125t} - 8$

In Exercises 55–60, solve the equation algebraically. Round the result to three decimal places.

55. $\ln x = -3$

56. $\ln x = 2$

57. $\ln\sqrt{x + 2} = 1$

58. $\ln(x + 1)^2 = 2$

59. $\log_{10}(x + 4) - \log_{10}x = \log_{10}(x + 2)$

60. $\log_4 x - \log_4(x - 1) = \frac{1}{2}$

In Exercises 61–64, let $f(x)$ represent the left side of the equation. Complete the table to obtain a rough estimate of the solution, and then use a graphing utility to graph both sides of the equation to obtain a better estimate of the solution (use three-decimal-place accuracy).

61. $\ln 2x = 2.4$

x	2	3	4	5	6
$f(x)$					

62. $3 \ln 5x = 10$

x	4	5	6	7	8
$f(x)$					

63. $6 \log_3(0.5x) = 11$

x	12	13	14	15	16
$f(x)$					

64. $5 \log_{10}(x - 2) = 11$

x	150	155	160	165	170
$f(x)$					

In Exercises 65–76, use the root or zero feature of a graphing utility to solve the logarithmic equation accurate to three decimal places.

65. $\log_{10}(z - 3) = 2$

66. $\log_{10}x^2 = 6$

67. $2 \ln x = 7$

68. $\ln 4x = 1$

69. $\ln x + \ln(x - 2) = 1$

70. $\ln x + \ln(x + 3) = 1$

71. $\log_3 x + \log_3(x^2 - 8) = \log_3 8x$

72. $\log_2 x + \log_2(x + 2) = \log_2(x + 6)$

73. $\ln(x + 5) = \ln(x - 1) - \ln(x + 1)$

74. $\ln(x + 1) - \ln(x - 2) = \ln x^2$

75. $\ln x + \ln(x^2 + 1) = 8$

76. $\log_{10} 8x - \log_{10}(1 + \sqrt{x}) = 2$

In Exercises 77–80, use a graphing utility to approximate the point of intersection of the graphs. Round the result to three decimal places.

77. $y_1 = 7$
$y_2 = 2^x$

78. $y_1 = 500$
$y_2 = 1500e^{-x/2}$

79. $y_1 = 3$
$y_2 = \ln x$

80. $y_1 = 10$
$y_2 = 4\ln(x - 2)$

Compound Interest **In Exercises 81 and 82, find the time required for a \$1000 investment to double at interest rate r, compounded continuously.**

81. $r = 0.085$

82. $r = 0.12$

83. *Think About It* Is the time required for the investments in Exercises 81 and 82 to quadruple twice as long as the time for them to double? Give a reason for your answer and verify your answer algebraically.

84. *Essay* Write a paragraph explaining whether or not the time required for an investment to double is dependent on the size of the investment.

Compound Interest **In Exercises 85 and 86, find the time required for a \$1000 investment to triple at interest rate r, compounded continuously.**

85. $r = 0.085$

86. $r = 0.12$

87. *Average Heights* The percents of American males and females between the ages of 18 and 24 who are at least x inches tall are given by

$$m(x) = \frac{100}{1 + e^{-0.6114(x-69.71)}} \quad \text{Males}$$

$$f(x) = \frac{100}{1 + e^{-0.66607(x-64.51)}} \quad \text{Females}$$

where m and f are the percents and x is the height in inches. (Source: U.S. National Center for Health Statistics)

(a) Use a graphing utility to graph the two functions in the same viewing rectangle.

(b) Use the graph to determine the horizontal asymptotes of the functions.

(c) What is the median height for each sex?

88. *Human Memory Model* In a group project in learning theory, a mathematical model for the proportion P of correct responses after n trials was found to be

$$P = \frac{0.83}{1 + e^{-0.2n}}.$$

(a) Use a graphing utility to graph the function.

(b) Use the graph to determine the horizontal asymptotes of the function. Interpret the meaning of the upper asymptote in the context of the problem.

(c) After how many trials will 60% of the responses be correct?

89. *Demand Function* The demand equation for a certain product is given by

$$p = 500 - 0.5(e^{0.004x}).$$

Find the demands x for prices of (a) $p = \$350$ and (b) $p = \$300$.

90. *Demand Function* The demand equation for a certain product is given by

$$p = 5000\left(1 - \frac{4}{4 + e^{-0.002x}}\right).$$

Find the demands x for prices of (a) $p = \$600$ and (b) $p = \$400$.

91. *Forest Yield* The yield V (in millions of cubic feet per acre) for a forest at age t years is given by

$$V = 6.7e^{-48.1/t}.$$

(a) Use a graphing utility to graph the function.

(b) Determine the horizontal asymptote of the function. Interpret its meaning in the context of the problem.

(c) Find the time necessary to obtain a yield of 1.3 million cubic feet.

92. *Trees per Acre* The number of trees per acre N of a certain species is approximated by the model

$$N = 68(10^{-0.04x}), \qquad 5 \le x \le 40$$

where x is the average diameter of the trees three feet above the ground. Use the model to approximate the average diameter of the trees in a test plot when $N = 21$.

93. *Data Analysis* An object at a temperature of 160°C is removed from a furnace and placed in a room at 20°C. The temperature T of the object is measured each hour h and recorded in the table.

h	0	1	2	3	4	5
T	160°	90°	56°	38°	29°	24°

A model for this data is

$T = 20[1 + 7(2^{-h})]$.

(a) Use a graphing utility to plot the data points and graph the model in the same viewing rectangle.

(b) Use the graph to identify the horizontal asymptote of the model and interpret the asymptote in the context of the problem.

(c) Approximate the time when the temperature of the object is 100°C.

94. *Chapter Opener* To protect the occupants of automobiles in accidents, the automobiles are designed with crumple zones. Most people will black out under a force of 5 g's (five times the force of gravity) if it is applied for 10 seconds or more, but for very short durations, humans have withstood as much as 40 g's. In crash tests with vehicles moving at 90 kilometers per hour, analysts measured the numbers y of g's that were undergone during deceleration by crash dummies that were permitted (by crumple zones and passive restraint systems) to move distances of x meters during impact. The data is shown in the table.

x	0.2	0.4	0.6	0.8	1
y	158	80	53	40	32

(a) Use the regression capabilities of a graphing utility to find a natural logarithmic model for the data. A second model for the data is $y = -3.00 + 11.88 \ln x + (36.94/x)$. Graph both models with the data points in the same viewing window. Which model do you think is a better choice? Comment on accuracy and simplicity.

(b) Use both models to estimate the distance traveled during impact if the passenger deceleration must not exceed 30 g's. Comment on the difference in distances traveled given by the two models and the practical implications of choosing a model when designing the crumple zones.

Review Solve Exercises 95–100 as a review of the skills and problem-solving techniques you learned in previous sections. Match the equation with its graph, and identify any intercepts. [The graphs are labeled (a), (b), (c), (d), (e), and (f).]

(a)
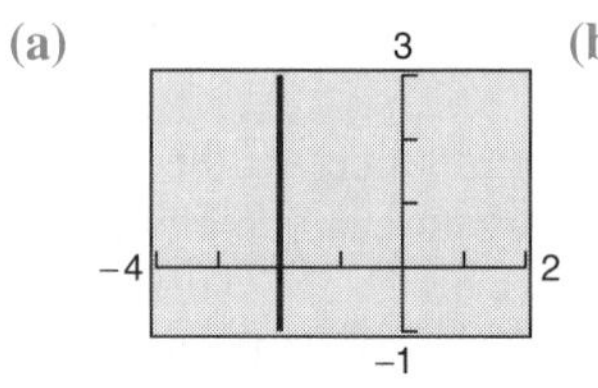

(b)
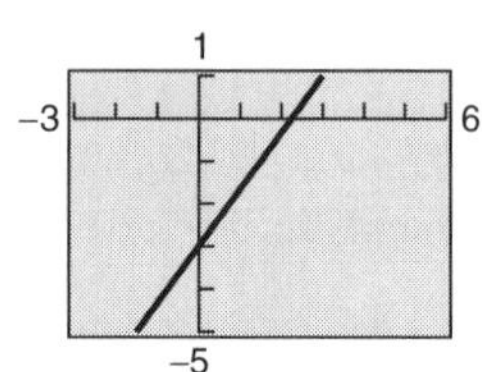

(c)
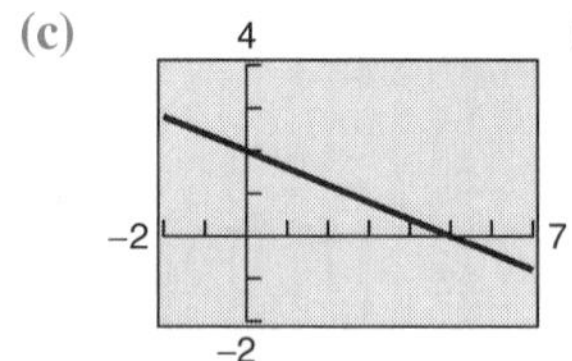

(d)
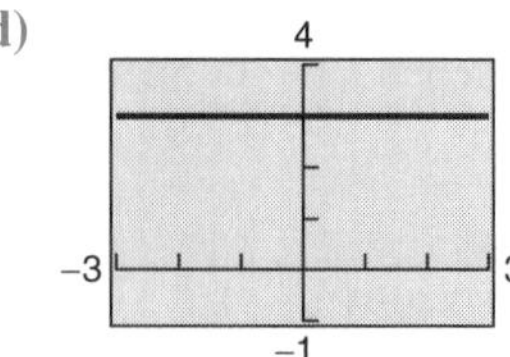

(e)
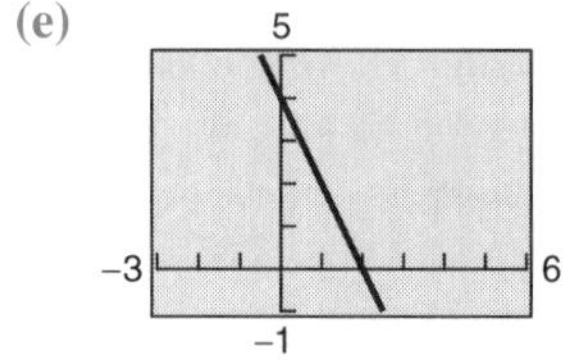

(f)
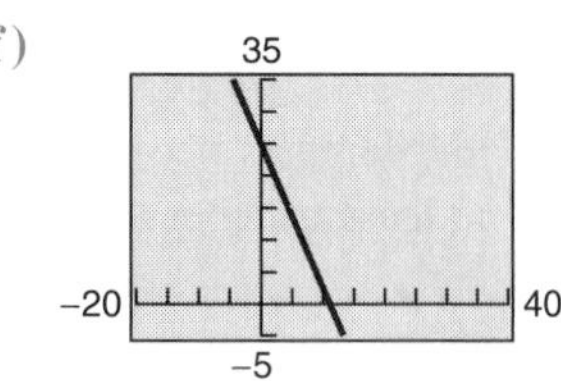

95. $4x - 3y - 9 = 0$

96. $2x + 5y - 10 = 0$

97. $y = 25 - 2.25x$

98. $\dfrac{x}{2} + \dfrac{y}{4} = 1$

99. $y - 3 = 0$

100. $x + 2 = 0$

4.5 Exponential and Logarithmic Models

Introduction / Exponential Growth and Decay / Gaussian Models / Logistics Growth Models / Logarithmic Models

Introduction

The five most common types of mathematical models involving exponential functions and logarithmic functions are as follows.

1. Exponential growth: $y = ae^{bx}, \quad b > 0$

2. Exponential decay: $y = ae^{-bx}, \quad b > 0$

3. Gaussian model: $y = ae^{-(x-b)^2/c}$

4. Logistics growth model: $y = \dfrac{a}{1 + be^{-(x-c)/d}}$

5. Logarithmic models: $y = a + b \ln x, \quad y = a + b \log_{10} x$

Note The compound interest problems studied in Section 4.1 are examples of exponential growth.

The basic shapes of these graphs are shown in Figure 4.29.

Figure 4.29

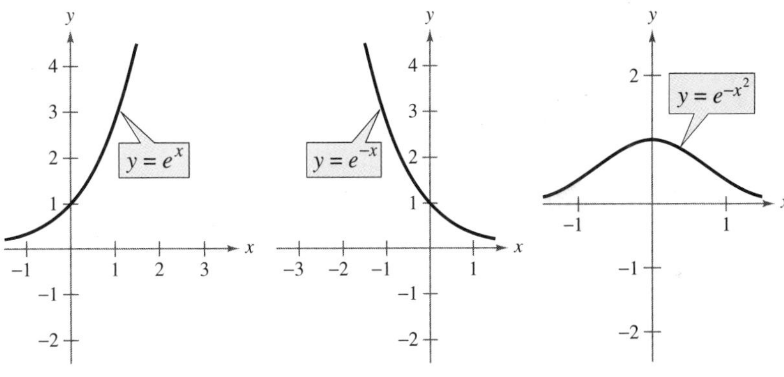

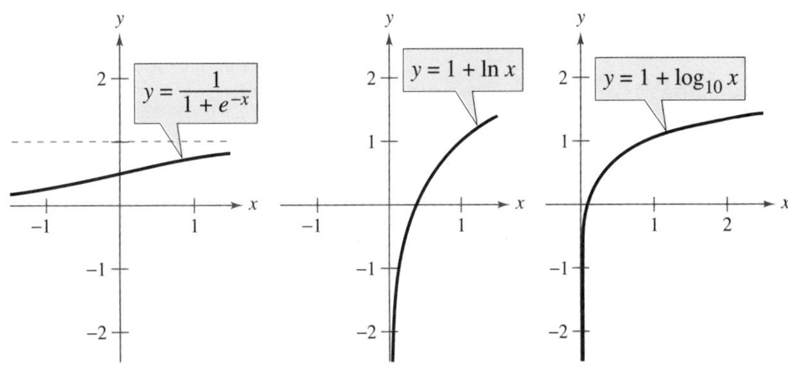

Study Tip

You can often gain quite a bit of insight into a situation modeled by an exponential or logarithmic function by identifying and interpreting the function's asymptotes. Use the graphs in Figure 4.29 to identify the asymptotes of each function.

Exponential Growth and Decay

Real Life

EXAMPLE 1 Population Increase

The world population (in millions) from 1980 through 1992 is shown in the table. The scatter plot of the data is shown in Figure 4.30(a). (Source: Statistical Office of the United Nations)

Year	1980	1985	1986	1987	1988	1989	1990	1991	1992
Population	4453	4850	4936	5024	5112	5202	5294	5384	5478

Figure 4.30

(a)

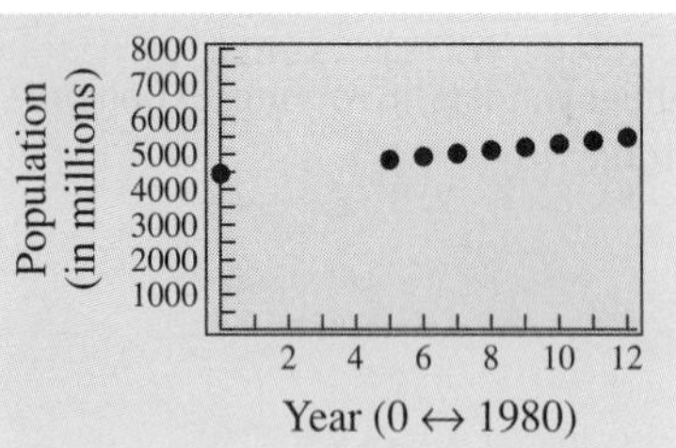

(b)

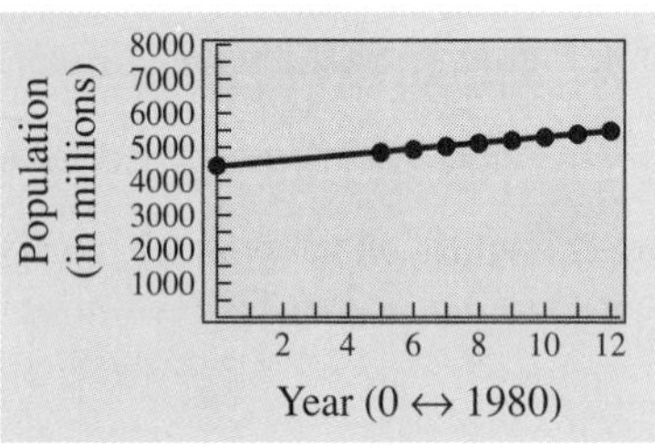

An exponential growth model that approximates this data is given by

$$P = 4451e^{0.017303t}, \quad 0 \le t \le 12$$

where P is the population (in millions) and $t = 0$ represents 1980. Compare the values given by the model with the estimates given by the United Nations. According to this model, when will the world population reach 6 billion?

Solution

The following table compares the two sets of population figures. The graph of the model is shown in Figure 4.30(b).

Year	1980	1985	1986	1987	1988	1989	1990	1991	1992
Population	4453	4850	4936	5024	5112	5202	5294	5384	5478
Model	4451	4853	4938	5024	5112	5201	5292	5384	5478

To find when the world population will reach 6 billion, let $P = 6000$ in the model and solve for t.

$$4451e^{0.017303t} = P \quad \text{Given model}$$

$$4451e^{0.017303t} = 6000 \quad \text{Let } P = 6000.$$

$$e^{0.017303t} \approx 1.348 \quad \text{Divide both sides by 4451.}$$

$$\ln e^{0.017303t} \approx \ln 1.348 \quad \text{Take logarithms of both sides.}$$

$$0.017303t \approx 0.2986 \quad \text{Inverse property of logs and exponents}$$

$$t \approx 17.26 \quad \text{Divide both sides by 0.017303.}$$

According to the model, the world population will reach 6 billion in 1997.

Note An exponential model increases (or decreases) by the same percent each year. What is the annual percent increase for the model in Example 1?

In Example 1, you were given the exponential growth model. But suppose this model had not been given; how could you have found such a model? One technique for doing this is demonstrated in Example 2.

Some graphing utilities have a curve-fitting feature that can be used to find models that represent data. For instance, to find the exponential growth model that approximates the data in the table in Example 1, you can use the following steps using a *TI-82* or a *TI-83*.

1. Press STAT ENTER.
2. Enter the data in L_1 and L_2.
3. Press STAT ▷.
4. Select *ExpReg:* and press ENTER ENTER.

Write the exponential growth model you obtain. How does it compare with the model in Example 1?

EXAMPLE 2 Finding an Exponential Growth Model

Find an exponential growth model whose graph passes through the points (0, 4453) and (7, 5024), as shown in Figure 4.31(a).

Solution

The general form of the model is

$$y = ae^{bx}.$$

From the fact that the graph passes through the point (0, 4453), you know that $y = 4453$ when $x = 0$. By substituting these values into the general form of the model, you have

$$4453 = ae^0$$
$$4453 = a.$$

In a similar way, from the fact that the graph passes through the point (7, 5024), you know that $y = 5024$ when $x = 7$. By substituting these values into the model, you have

$$5024 = 4453e^{7b} \quad \Rightarrow \quad b = \frac{1}{7}\ln\frac{5024}{4453} \approx 0.01724.$$

Thus, the exponential growth model is

$$y = 4453e^{0.01724x}.$$

The graph of the model is shown in Figure 4.31(b).

Figure 4.31

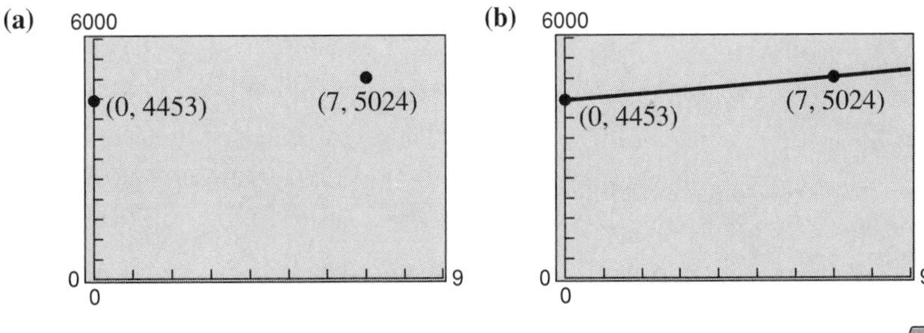

Figure 4.32

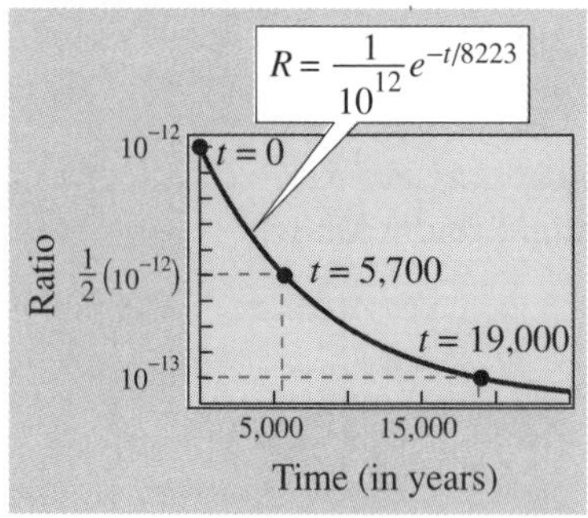

In living organic material, the ratio of the content of radioactive carbon (carbon 14) to the content of nonradioactive carbon (carbon 12) is about 1 to 10^{12}. When organic material dies, its carbon 12 content remains fixed, whereas its radioactive carbon 14 begins to decay with a half-life of about 5700 years. To estimate the age of dead organic material, scientists use the following formula, which denotes the ratio of carbon 14 to carbon 12 present at any time t (in years).

$$R = \frac{1}{10^{12}}e^{-t/8223}$$

The graph of R is shown in Figure 4.32. Note that R decreases as t increases.

Real Life

EXAMPLE 3 **Carbon Dating**

The ratio of carbon 14 to carbon 12 in a newly discovered fossil is

$$R = \frac{1}{10^{13}}.$$

Estimate the age of the fossil.

Solution

In the carbon dating model, substitute the given value of R to obtain the following.

$$\frac{1}{10^{12}}e^{-t/8223} = R \qquad \text{Given model}$$

$$\frac{e^{-t/8223}}{10^{12}} = \frac{1}{10^{13}} \qquad \text{Let } R = \frac{1}{10^{13}}.$$

$$e^{-t/8223} = \frac{1}{10} \qquad \text{Multiply both sides by } 10^{12}.$$

$$\ln e^{-t/8223} = \ln \frac{1}{10} \qquad \text{Take logarithms of both sides.}$$

$$-\frac{t}{8223} \approx -2.3026 \qquad \text{Inverse property of logs and exponents}$$

$$t \approx 18,934 \qquad \text{Solve for } t.$$

Thus, to the nearest thousand years, you can estimate the age of the fossil to be 19,000 years. How could you verify this answer with a graphing utility?

Note The carbon dating model in Example 3 assumed that the carbon 14/carbon 12 ratio was one part in 10,000,000,000,000. Suppose an error in measurement occurred and the actual ratio was only one part in 8,000,000,000,000. The fossil age corresponding to the actual ratio would then be approximately 17,000 years. Try checking this result.

Gaussian Models

As mentioned at the beginning of this section, Gaussian models are of the form

$$y = ae^{-(x-b)^2/c}.$$

This type of model is commonly used in probability and statistics to represent populations that are **normally distributed.** For *standard* normal distributions, the model takes the form

$$y = \frac{1}{\sigma\sqrt{2\pi}}e^{-x^2/2\sigma^2}$$

where σ is the standard deviation (σ is the lowercase Greek letter sigma). The graph of a Gaussian model is called a **bell-shaped curve.** Try assigning a value to σ and sketching a normal distribution curve with a graphing utility. Can you see why it is called a bell-shaped curve?

Real Life

EXAMPLE 4 SAT Scores

In 1993, the Scholastic Aptitude Test (SAT) scores for males roughly followed a normal distribution given by

$$y = 0.0026e^{-(x-500)^2/48{,}000}, \quad 200 \le x \le 800$$

where x is the SAT score for mathematics. Sketch the graph of this function. From the graph, estimate the average SAT score. (Source: College Board)

Solution

The graph of the function is given in Figure 4.33. From the graph, you can see that the average mathematics score for males in 1988 was 500.

Figure 4.33

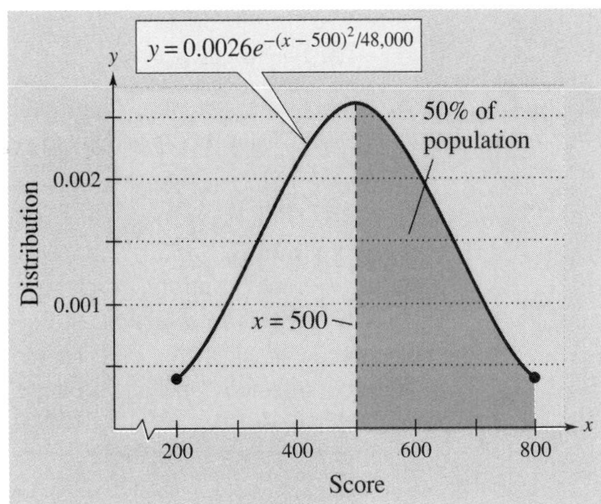

Logistics Growth Models

Figure 4.34 Logistics Curve

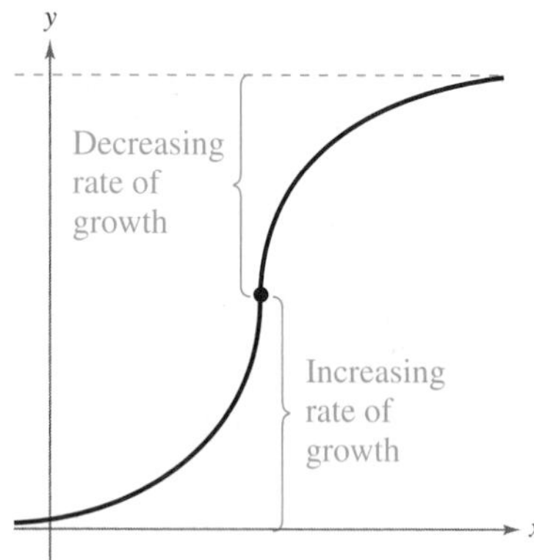

Some populations initially have rapid growth, followed by a declining rate of growth, as indicated by the graph in Figure 4.34. One model for describing this type of growth pattern is the **logistics curve** given by the function

$$y = \frac{a}{1 + be^{-(x-c)/d}}$$

where y is the population size and x is the time. An example is a bacteria culture allowed to grow initially under ideal conditions, followed by less favorable conditions that inhibit growth. A logistics growth curve is also called a **sigmoidal curve.**

Real Life

EXAMPLE 5 Spread of a Virus

On a college campus of 5000 students, one student returns from vacation with a contagious flu virus. The spread of the virus is modeled by

$$y = \frac{5000}{1 + 4999e^{-0.8t}}, \quad 0 \le t$$

where y is the total number infected after t days. The college will cancel classes when 40% or more of the students are ill.

a. How many students are infected after 5 days?

b. After how many days will the college cancel classes?

Solution

a. After 5 days, the number infected is

$$y = \frac{5000}{1 + 4999e^{-0.8(5)}} = \frac{5000}{1 + 4999e^{-4}} \approx 54.$$

b. In this case, the number infected must be $(0.40)(5000) = 2000$ to cancel classes. Therefore, you can solve for t in the following equation.

$$2000 = \frac{5000}{1 + 4999e^{-0.8t}}$$

Figure 4.35

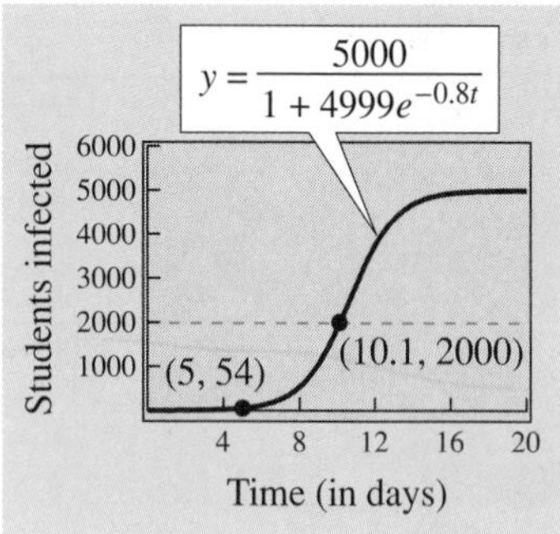

Using a graphing utility, graph $y = 2000$ and $y = 5000/(1 + 4999e^{-0.8t})$ and find the point of intersection. In Figure 4.35, you can see that the point of intersection occurs near $t \approx 10.1$. Hence, after 10 days, at least 40% of the students will be infected, and the college will cancel classes. Verify this answer algebraically.

Logarithmic Models

Real Life

EXAMPLE 6 **Magnitudes of Earthquakes**

On the Richter scale, the magnitude R of an earthquake of intensity I is given by

$$R = \log_{10} \frac{I}{I_0}$$

where $I_0 = 1$ is the minimum intensity used for comparison. Find the intensities per unit of area for the following earthquakes. (Intensity is a measure of the wave energy of the earthquake.)

a. Tokyo and Yokohama, Japan, in 1923, $R = 8.3$

b. Kobe, Japan, in 1995, $R = 7.2$

Solution

a. Because $I_0 = 1$ and $R = 8.3$, you have

$$8.3 = \log_{10} I$$
$$I = 10^{8.3} \approx 199{,}526{,}000.$$

b. For $R = 7.2$, you have $7.2 = \log_{10} I$, and $I = 10^{7.2} \approx 15{,}849{,}000$.

Note that an increase of 1.1 units on the Richter scale (from 7.2 to 8.3) represents an intensity change by a factor of

$$\frac{199{,}526{,}000}{15{,}849{,}000} \approx 13.$$

In other words, the earthquake in 1923 had a magnitude about 13 times greater than that of the 1995 quake.

Group Activity *Identifying Appropriate Models*

Decide for each data set which model from this section would provide the best fit. Discuss your reasoning with others in your group.

Data Set A: (18.4, 1.07), (19, 1.21), (20, 1.45), (22, 1.85), (23.5, 1.99), (25, 1.96), (26.3, 1.80), (27, 1.67), (29.7, 1.04), (31, 0.75)

Data Set B: (1.5, 11.03), (2, 12.47), (3.5, 15.26), (5, 17.05), (7.8, 19.27), (9, 19.99), (10.2, 20.61), (13.6, 22.05), (19.3, 23.80), (27, 25.48)

4.5 /// EXERCISES

In Exercises 1–6, match the function with its graph. [The graphs are labeled (a) through (f).]

(a)

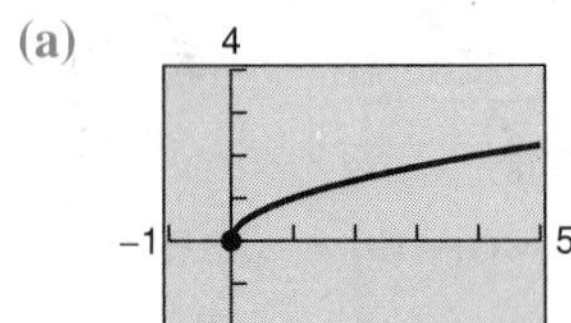

(b)

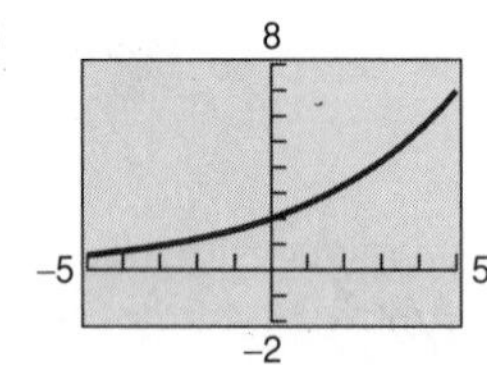

(c)

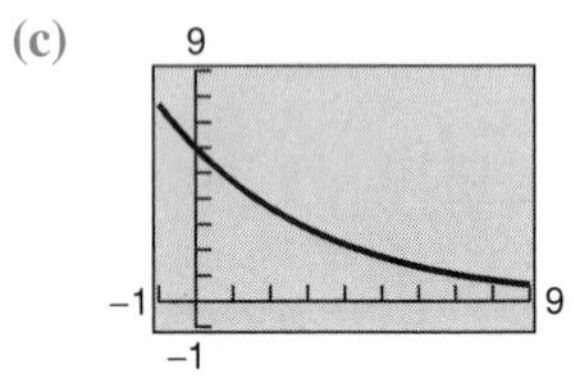

(d)

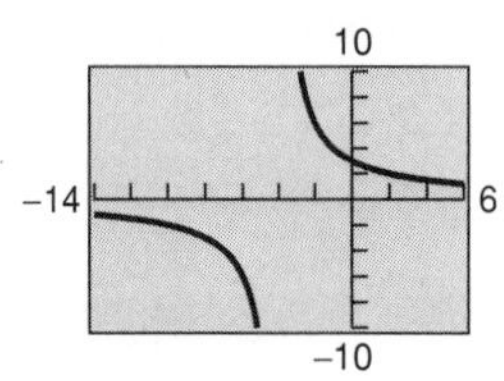

(e)

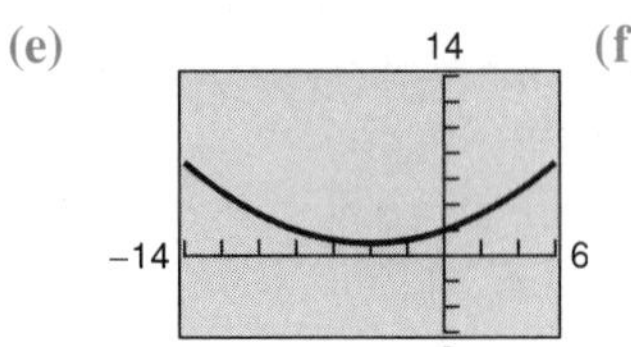

(f)

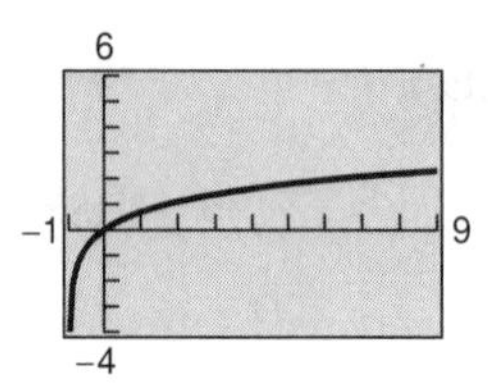

1. $y = 2e^{x/4}$

2. $y = 6e^{-x/4}$

3. $y = \frac{1}{16}(x^2 + 8x + 32)$

4. $y = \dfrac{12}{x + 4}$

5. $y = \ln(x + 1)$

6. $y = \sqrt{x}$

Compound Interest **In Exercises 7–14, complete the table for a savings account in which interest is compounded continuously.**

	Initial Investment	*Annual % Rate*	*Time to Double*	*Amount After 10 Years*
7.	\$1000	12%		
8.	\$20,000	$10\frac{1}{2}$%		
9.	\$750		$7\frac{3}{4}$ yr	
10.	\$10,000		5 yr	
11.	\$500			\$1292.85
12.	\$600			\$19,205.00
13.		4.5%		\$10,000.00
14.		8%		\$20,000.00

Compound Interest **In Exercises 15 and 16, determine the principal P that must be invested at rate r, compounded monthly, so that \$500,000 will be available for retirement in t years.**

15. $r = 7\frac{1}{2}\%, t = 20$

16. $r = 12\%, t = 40$

Compound Interest **In Exercises 17 and 18, determine the time necessary for \$1000 to double if it is invested at interest rate r compounded (a) annually, (b) monthly, (c) daily, and (d) continuously.**

17. $r = 11\%$

18. $r = 10\frac{1}{2}\%$

19. *Compound Interest* Complete the table for the time t necessary for P dollars to triple if interest is compounded continuously at rate r.

r	2%	4%	6%	8%	10%	12%
t						

20. *Modeling Data* Draw a scatter plot of the data in Exercise 19. Use the regression capabilities of a graphing utility to find a model for the data.

21. *Compound Interest* Complete the table for the time t necessary for P dollars to triple if interest is compounded annually at rate r.

r	2%	4%	6%	8%	10%	12%
t						

22. *Modeling Data* Draw a scatter plot of the data in Exercise 21. Use the regression capabilities of a graphing utility to find a model for the data.

23. *Comparing Investments* If \$1 is invested in an account over a 10-year period, the amount in the account, where t represents the time in years, is

$$A = 1 + 0.075[\![t]\!] \quad \text{or} \quad A = e^{0.07t}$$

depending on whether the account pays simple interest at $7\frac{1}{2}\%$ or continuous compound interest at 7%. Use a graphing utility to graph each function in the same viewing rectangle. Which grows at a faster rate?

24. *Comparing Investments* If \$1 is invested in an account over a 10-year period, the amount in the account, where t represents the time in years, is

$$A = 1 + 0.06[\![t]\!] \quad \text{or} \quad A = \left(1 + \frac{0.055}{365}\right)^{[\![365t]\!]}$$

depending on whether the account pays simple interest at 6% or compound interest at $5\frac{1}{2}\%$ compounded daily. Use a graphing utility to graph each function in the same viewing rectangle. Which grows at a faster rate?

In Exercises 25–28, complete the table for the given radioactive isotope.

	Isotope	Half-Life (years)	Initial Quantity	Amount After 1000 Years
25.	^{226}Ra	1620	10g	
26.	^{226}Ra	1620		1.5g
27.	^{14}C	5730	3g	
28.	^{230}Pu	24,360		0.4g

In Exercises 29–32, find the exponential model $y = ae^{bx}$ that fits the points given in the graph or table.

29.

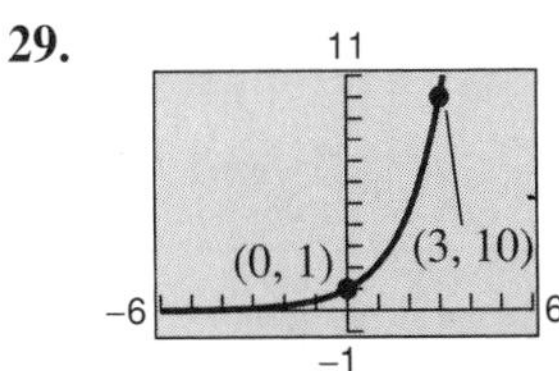

30.

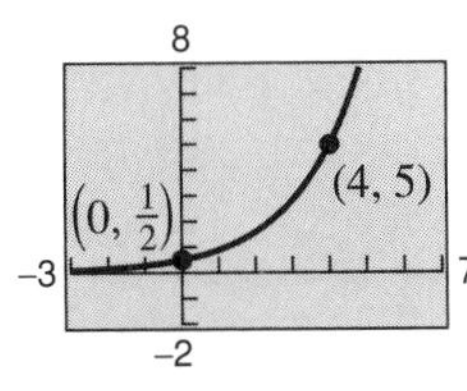

31.

x	y
0	1
3	$\frac{1}{4}$

32.

x	y
0	5
4	1

33. *Population* The population P of a city is given by

$$P = 105{,}300e^{0.015t}$$

where $t = 0$ represents 1990. According to this model, when will the population reach 150,000?

34. *Population* The population P of a city is given by

$$P = 240{,}360e^{0.012t}$$

where $t = 0$ represents 1990. According to this model, when will the population reach 275,000?

35. *Population* The population P of a city is given by

$$P = 2500e^{kt}$$

where $t = 0$ represents 1990. In 1945, the population was 1350. Find the value of k, and use this value to predict the population in the year 2010.

36. *Population* The population P of a city is given by

$$P = 140{,}500e^{kt}$$

where $t = 0$ represents 1990. In 1960, the population was 100,250. Find the value of k, and use this value to predict the population in the year 2000.

Population **Exercises 37–40 give the populations (in millions) of four cities in 1990 and the projected populations (in millions) for the year 2000. Find the exponential growth model $y = ae^{bt}$ for each population by letting $t = 0$ correspond to 1990. Use the model to predict the population of the city in 2010. (Source: U.S. Bureau of the Census, International Database)**

	City	1990	2000
37.	Dhaka, Bangladesh	4.22	6.49
38.	Houston, Texas	2.30	2.65
39.	Detroit, Michigan	3.00	2.74
40.	London, United Kingdom	9.17	8.57

41. *Think About It* In Exercises 37 and 38, you can see that the populations of Dhaka and Houston are growing at different rates. What constant in the equation $y = ae^{bt}$ is determined by these different growth rates? Discuss the relationship between the different growth rates and the magnitude of the constant.

42. *Think About It* In Exercises 37 and 39, you can see that one population is increasing while the other is decreasing. What constant in the equation $y = ae^{bt}$ reflects this difference? Explain.

43. *Bacteria Growth* The number of bacteria N in a culture is given by the model

$$N = 100e^{kt}$$

where t is the time in hours. If $N = 300$ when $t = 5$, estimate the time required for the population to double in size. Verify your estimate graphically.

44. *Bacteria Growth* The number of bacteria N in a culture is given by the model

$$N = 250e^{kt}$$

where t is the time in hours. If $N = 280$ when $t = 10$, estimate the time required for the population to double in size. Verify your estimate graphically.

45. *Radioactive Decay* The half-life of radioactive radium (radium 226) is 1620 years. What percent of a present amount of radioactive radium will remain after 100 years?

46. *Radioactive Decay* Carbon 14 dating assumes that the carbon dioxide on earth today has the same radioactive content as it did centuries ago. If this is true, the amount of carbon 14 absorbed by a tree that grew several centuries ago should be the same as the amount of carbon 14 absorbed by a tree growing today. A piece of ancient charcoal contains only 15% as much radioactive carbon as a piece of modern charcoal. How long ago was the tree burned to make the ancient charcoal if the half-life of carbon 14 is 5730 years?

47. *Depreciation* A computer that cost \$4600 new has a book value of \$3000 after 2 years. Find the value of the computer after 3 years by using the exponential model

$$y = ae^{bt}.$$

48. *Comparing Models* A car that cost \$22,000 new has a book value of \$13,000 after 2 years.

(a) Find the straight-line model $V = mt + b$.

(b) Find the exponential model $V = ae^{kt}$.

(c) Use a graphing utility to graph the two models in the same viewing rectangle. Which model depreciates faster in the first 2 years?

(d) Find the book values of the car after 1 year and after 3 years using each model.

(e) Interpret the slope of the straight-line model.

49. *Sales* The sales S (in thousands of units) of a new product after it has been on the market t years are given by

$$S(t) = 100(1 - e^{kt}).$$

Fifteen thousand units of the new product were sold the first year.

(a) Complete the model by solving for k.

(b) Use a graphing utility to graph the model.

(c) Use the graph to estimate the number of units sold after 5 years.

50. *Sales and Advertising* After discontinuing all advertising for a certain product in 1994, the manufacturer noted that sales began to drop according to the model

$$S = \frac{500,000}{1 + 0.6e^{kt}}$$

where S represents the number of units sold and $t = 0$ represents 1994. In 1996, the company sold 300,000 units.

(a) Complete the model by solving for k.

(b) Estimate sales in 1999.

51. *Sales and Advertising* The sales S (in thousands of units) of a product after x hundred dollars is spent on advertising is given by

$$S = 10(1 - e^{kx}).$$

When \$500 is spent on advertising, 2500 units are sold.

(a) Complete the model by solving for k.

(b) Estimate the number of units that will be sold if advertising expenditures are raised to \$700.

52. *Profits* Because of a slump in the economy, a company finds that its annual profits have dropped from $742,000 in 1994 to $632,000 in 1996. If the profit follows an exponential pattern of decline, what is the expected profit for 1997? (Let $t = 0$ represent 1994.)

53. *Learning Curve* The management at a factory has found that the maximum number of units a worker can produce in a day is 30. The learning curve for the number of units N produced per day after a new employee has worked t days is given by

$$N = 30(1 - e^{kt}).$$

After 20 days on the job, a new employee produces 19 units (see figure).

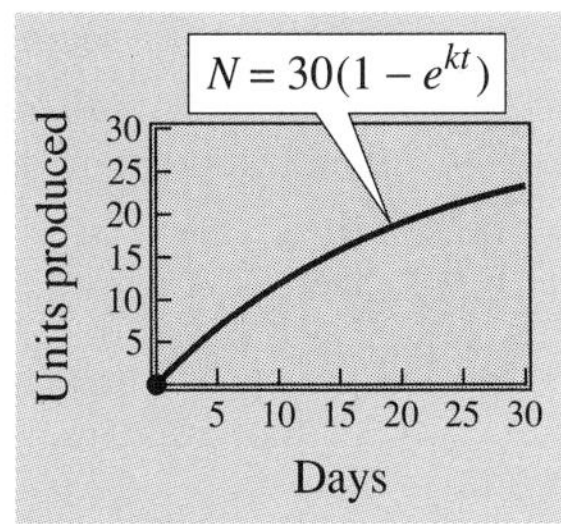

(a) Find the learning curve for this employee (first find the value of k).

(b) How many days should pass before this employee is producing 25 units per day?

(c) Is the employee's production increasing at a linear rate? Explain your reasoning.

54. *Endangered Species* A conservation organization releases 100 animals of an endangered species into a game preserve. The organization believes that the preserve has a carrying capacity of 1000 animals and that the growth of the herd will follow the logistics curve

$$p(t) = \frac{1000}{1 + 9e^{-0.1656t}}$$

where t is measured in months.

(a) Use a graphing utility to graph the function. Use the graph to determine the horizontal asymptotes and interpret the meaning of the larger asymptote in the context of the problem.

(b) Estimate the population after 5 months.

(c) When will the population reach 500?

Earthquake Magnitudes **In Exercises 55 and 56, use the Richter scale (see page 365) for measuring the magnitudes of earthquakes.**

55. Find the magnitude R of an earthquake of intensity I (let $I_0 = 1$).

(a) $I = 80{,}500{,}000$

(b) $I = 48{,}275{,}000$

56. Find the intensity I of an earthquake measuring R on the Richter scale (let $I_0 = 1$).

(a) Colombia in 1906, $R = 8.6$

(b) Los Angeles in 1971, $R = 6.7$

Intensity of Sound **In Exercises 57–60, use the following information. For Exercises 57 and 58, determine the level of sound (in decibels) for the given sound intensity.**

The level of sound β, in decibels, with an intensity I is given by $\beta(I) = 10 \log_{10} \frac{I}{I_0}$, where I_0 is an intensity of 10^{-16} watts per square centimeter, corresponding roughly to the faintest sound that can be heard by the human ear.

57. (a) $I = 10^{-14}$ watts per cm² (faint whisper)

(b) $I = 10^{-9}$ watts per cm² (busy street corner)

(c) $I = 10^{-6.5}$ watts per cm² (air hammer)

(d) $I = 10^{-4}$ watts per cm² (threshold of pain)

58. (a) $I = 10^{-13}$ watts per cm² (whisper)

(b) $I = 10^{-7.5}$ watts per cm² (jet 4 miles from takeoff)

(c) $I = 10^{-7}$ watts per cm² (diesel truck at 25 feet)

(d) $I = 10^{-4.5}$ watts per cm² (auto horn at 3 feet)

59. *Noise Level* As a result of the installation of noise suppression materials, the noise level in an auditorium was reduced from 93 to 80 decibels. Find the percent decrease in the intensity level of the noise due to the installation of these materials.

60. *Noise Level* As a result of the installation of a muffler, the level of noise emitted by an engine was reduced from 88 to 72 decibels. Find the percent decrease in the intensity level of the noise due to the installation of the muffler.

Acidity **In Exercises 61–66, use the acidity model given by pH $= -\log_{10}[H^+]$, where acidity (pH) is a measure of the hydrogen ion concentration $[H^+]$ (measured in moles of hydrogen per liter) of a solution.**

61. Find the pH if $[H^+] = 2.3 \times 10^{-5}$.

62. Find the pH if $[H^+] = 11.3 \times 10^{-6}$.

63. Compute $[H^+]$ for a solution in which pH $= 5.8$.

64. Compute $[H^+]$ for a solution in which pH $= 3.2$.

65. A certain fruit has a pH of 2.5 and an antacid tablet has a pH of 9.5. The hydrogen ion concentration of the fruit is how many times the concentration of the tablet?

66. If the pH of a solution is decreased by one unit, the hydrogen ion concentration is increased by what factor?

67. *Home Mortgage* A $120,000 home mortgage for 35 years at $9\frac{1}{2}\%$ has a monthly payment of $985.93. Part of the monthly payment goes for the interest charge on the unpaid balance, and the remainder of the payment is used to reduce the principal. The amount that goes for interest is given by

$$u = M - \left(M - \frac{Pr}{12}\right)\left(1 + \frac{r}{12}\right)^{12t}$$

and the amount that goes toward reduction of the principal is given by

$$v = \left(M - \frac{Pr}{12}\right)\left(1 + \frac{r}{12}\right)^{12t}.$$

In these formulas, P is the size of the mortgage, r is the interest rate, M is the monthly payment, and t is the time in years.

(a) Use a graphing utility to graph each function in the same viewing rectangle. (The viewing rectangle should show all 35 years of mortgage payments.)

(b) In the early years of the mortgage, the larger part of the monthly payment goes for what purpose? Approximate the time when the monthly payment is evenly divided between interest and principal reduction.

(c) Repeat parts (a) and (b) for a repayment period of 20 years ($M = \$1118.56$). What can you conclude?

68. *Home Mortgage* The total interest u paid on a home mortgage of P dollars at interest rate r for t years is given by

$$u = P\left[\frac{rt}{1 - \left(\dfrac{1}{1 + r/12}\right)^{12t}} - 1\right].$$

Consider a $120,000 home mortgage at $9\frac{1}{2}\%$.

(a) Use a graphing utility to graph the total interest function.

(b) Approximate the length of the mortgage when the total interest paid is the same as the size of the mortgage. Is it possible to pay twice as much in interest charges as the size of one's mortgage?

69. *Data Analysis* The time t (in seconds) required to attain a speed of s miles per hour from a standing start for a 1995 Dodge Avenger is given in the table. (Source: *Road & Track*, March 1995)

s	30	40	50	60	70	80	90
t	3.4	5.0	7.0	9.3	12.0	15.8	20.0

Two models for this data are

$$t_1 = 40.757 + 0.556s - 15.817 \ln s$$

and

$$t_2 = 1.2259 + 0.0023s^2.$$

(a) Use a graphing utility to fit a linear model t_3 and an exponential model t_4 to the data.

(b) Use a graphing utility to graph the data points and each of the four models.

(c) Use a graphing utility to create a table comparing the given data with the estimates obtained from each model.

(d) Use the results of part (c) to find the sum of the absolute values of the differences between the data and the estimated values given by each model. Based on the four sums, which model do you think best fits the data? Explain.

70. *Comparing Models* A 1990 Chevrolet Beretta with a six-cylinder engine, automatic transmission, and air conditioning had a retail price of \$11,500. A local dealership used the following guide for the approximate value of the car for the years 1990 through 1995. (Source: National Automobile Dealer's Association)

Year	1990	1991	1992
Value	\$11,500	\$9315	\$9200

Year	1993	1994	1995
Value	\$7935	\$7130	\$6095

Let V represent the value of the automobile in the year t, with $t = 0$ corresponding to 1990.

(a) Use the regression capabilities of a graphing utility to find linear and quadratic models for the data. Use the graphing utility to plot the data and graph the models in the same viewing rectangle.

(b) What does the slope represent in the linear model in part (a)?

(c) Do you think the quadratic model is realistic? Explain.

(d) Use the regression capabilities of a graphing utility to fit an exponential model to the data. Use the graphing utility to plot the data and graph the model in the same viewing rectangle. How well does the model fit the data?

(e) Fit a rational model to the data. Take the reciprocal of the depreciated values of the automobile to generate the points $\left(t, \frac{1}{V}\right)$. Use the regression capabilities of a graphing utility to fit a line to this data. The resulting line has the form

$$\frac{1}{V} = at + b.$$

Solve for V and use the graphing utility to graph the rational function and the original data points in the same viewing rectangle. How well does the model fit the data?

(f) Determine the horizontal asymptotes of the exponential and rational models. Interpret their meanings in the context of the problem.

71. *Essay* Use your school's library or some other reference source to write a paper describing John Napier's work with logarithms.

72. *Essay* Before the development of electronic calculators and graphing utilities, some computations were done on a slide rule. Use your school's library or some other reference source to write a paper describing the use of logarithmic scales on a slide rule.

73. *Estimating the Time of Death* At 8:30 A.M., a coroner was called to the home of a person who had died during the night. In order to estimate the time of death, the coroner took the person's temperature twice. At 9:00 A.M. the temperature was 85.7°, and at 9:30 A.M. the temperature was 82.8°. From these two temperatures the coroner was able to determine that the time elapsed since death and the body temperature were related by the formula

$$t = -2.5 \ln \frac{T - 70}{98.6 - 70}$$

with t representing the time in hours elapsed since death and T representing the temperature (in degrees Fahrenheit) of the person's body. Assume that the person had a normal body temperature of 98.6° at death and that the room temperature was a constant 70°. (This formula is derived from a general cooling principle called Newton's Law of Cooling.) Use the formula to estimate the time of death of the person.

Review **Solve Exercises 74–77 as a review of the skills and problem-solving techniques you learned in previous sections. Divide by synthetic division.**

74. $\dfrac{4x^3 + 4x^2 - 39x + 36}{x + 4}$

75. $\dfrac{8x^3 - 36x^2 + 54x - 27}{x - \frac{3}{2}}$

76. $(2x^3 - 8x^2 + 3x - 9) \div (x - 4)$

77. $(x^4 - 3x + 1) \div (x + 5)$

4.6 Exploring Data: Nonlinear Models

Classifying Scatter Plots / Fitting Nonlinear Models to Data / Application

Classifying Scatter Plots

In Section 2.6, you saw how to fit linear models to data. In real life, many relationships between two variables are nonlinear. A scatter plot can be used to give you an idea of which type of model will best fit a set of data.

EXAMPLE 1 Classifying Scatter Plots

Decide whether each set of data could be best modeled by an exponential model, $y = ab^x$, or a logarithmic model, $y = a + b \ln x$.

a. (0.9, 1.9), (1.3, 2.4), (1.3, 2.2), (1.4, 2.4), (1.6, 2.7), (1.8, 3.0), (2.1, 3.4), (2.1, 3.3), (2.5, 4.2), (2.9, 5.1), (3.2, 6.0), (3.3, 6.2), (3.6, 7.3), (4.0, 8.8), (4.2, 9.8), (4.3, 10.2)

b. (0.9, 3.2), (1.3, 4.0), (1.3, 3.8), (1.4, 4.2), (1.6, 4.5), (1.8, 4.8), (2.1, 5.1), (2.1, 5.0), (2.5, 5.5), (2.9, 5.8), (3.2, 6.0), (3.3, 6.2), (3.6, 6.2), (4.0, 6.6), (4.2, 6.7), (4.3, 6.8)

Study Tip

You can change an exponential model of the form $y = ab^x$ to one of the form $y = ae^{cx}$ by rewriting b in the form

$$b = e^{\ln b}.$$

For instance, $y = 3(2^x)$ can be written as

$$y = 3(2^x) = 3e^{(\ln 2)x} \approx 3e^{0.693x}.$$

Solution

Begin by entering the data into a graphing utility. You should obtain the scatter plots shown in Figure 4.36.

Figure 4.36

(a)
12
0
0
5

(b)

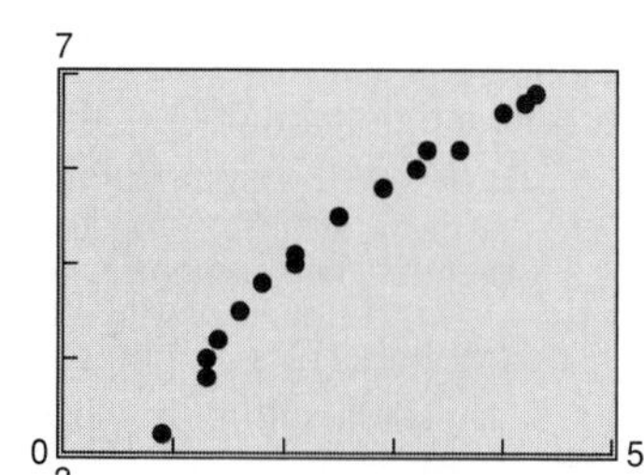

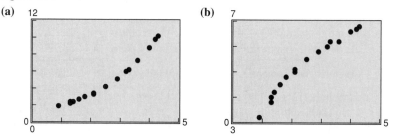

From the scatter plots, it appears that the data in part (a) can be modeled by an exponential function and the data in part (b) can be modeled by a logarithmic function.

Fitting Nonlinear Models to Data

Once you have used a scatter plot to determine the type of model to be fit to a set of data, there are several ways that you can actually find the model. Each method is best used with a computer or calculator, rather than with hand calculations.

EXAMPLE 2 Fitting a Model to Data

Fit the following data, from Example 1(a), with a quadratic model, an exponential model, and a power model. Which model do you think fits best?

(0.9, 1.9), (1.3, 2.4), (1.3, 2.2), (1.4, 2.4), (1.6, 2.7), (1.8, 3.0), (2.1, 3.4), (2.1, 3.3), (2.5, 4.2), (2.9, 5.1), (3.2, 6.0), (3.3, 6.2), (3.6, 7.3), (4.0, 8.8), (4.2, 9.8), (4.3, 10.2)

Solution

Begin by entering the data into a calculator or computer that has least squares regression programs. Then run the regression programs for quadratic, exponential, and power models.

Quadratic:	$y = ax^2 + bx + c$	$a = 0.589$	$b = -0.685$	$c = 2.188$
Exponential:	$y = ab^x$	$a = 1.203$	$b = 1.646$	
Power:	$y = ax^b$	$a = 1.667$	$b = 1.134$	

Note Deciding which model best fits a set of data is a question that is studied in detail in statistics. Basically, the model that fits best is the one whose sum of squared differences is the least. In Example 2, the sums of the squared differences are 0.124 for the quadratic model, 0.057 for the exponential model, and 5.98 for the power model.

To decide which model best fits the data, you can compare the y-values given by each model with the actual y-values. The model whose y-values are closest to the actual values is the one that fits best. In this case, the best-fitting model is the exponential model. The graphs of all three models are shown in Figure 4.37.

Figure 4.37

Quadratic

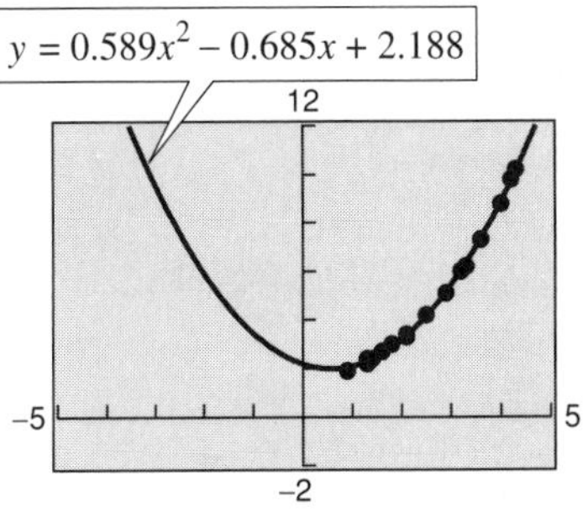

Exponential

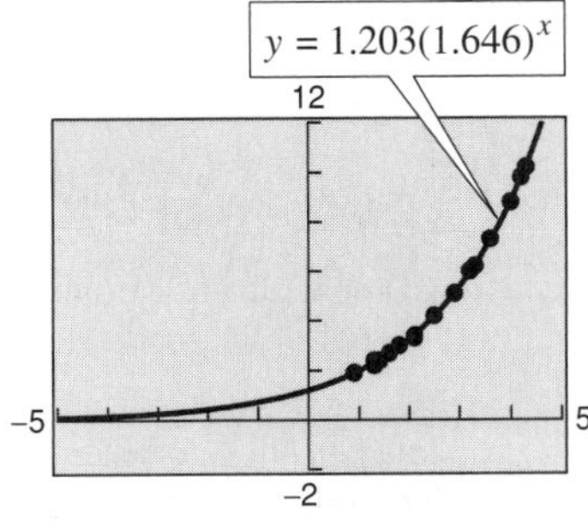

Power

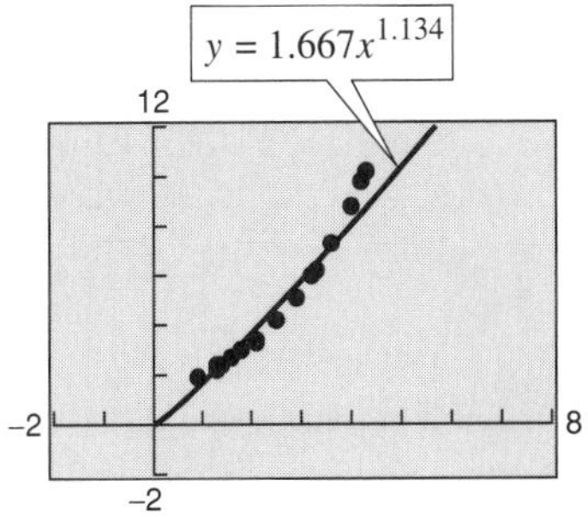

EXAMPLE 3 Fitting a Model to Data

Fit the following data, from Example 1(b), with a logarithmic model and a power model. Which model do you think fits best?

(0.9, 3.2), (1.3, 4.0), (1.3, 3.8), (1.4, 4.2), (1.6, 4.5), (1.8, 4.8), (2.1, 5.1), (2.1, 5.0), (2.5, 5.5), (2.9, 5.8), (3.2, 6.0), (3.3, 6.2), (3.6, 6.2), (4.0, 6.6), (4.2, 6.7), (4.3, 6.8)

Solution

Begin by entering the data into a calculator or computer that has least squares regression programs. Then run the regression programs for logarithmic and power models.

Logarithmic:	$y = a + b \ln x$	$a = 3.377$	$b = 2.302$
Power:	$y = ax^b$	$a = 3.522$	$b = 0.462$

To decide which model best fits the data, you can compare the y-values given by each model with the actual y-values. The model whose y-values are closest to the actual values is the one that fits best. In this case, both models have good fits, but the logarithmic model is slightly better. The graphs of both models are shown in Figure 4.38.

Note In Example 3, the sum of the squared differences for the logarithmic model is 0.085 and the sum of the squared differences for the power model is 0.220.

Figure 4.38

Logarithmic

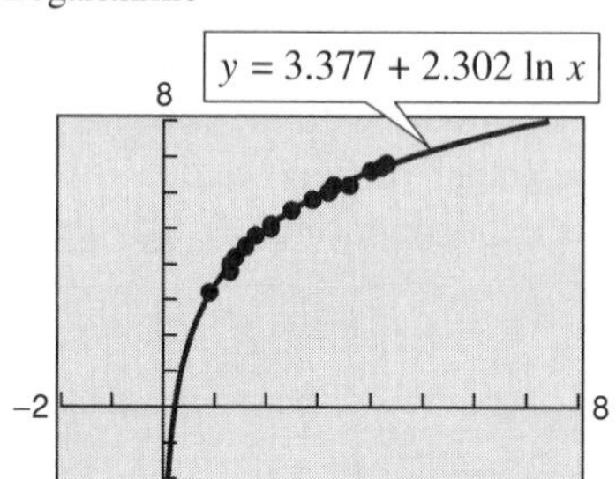

Power

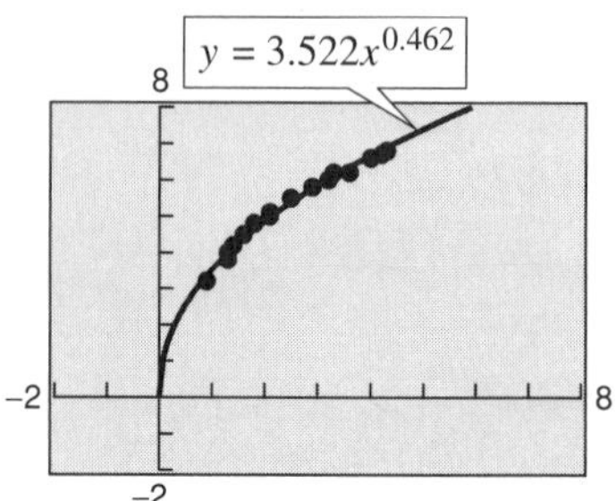

EXPLORATION

Try using a calculator or computer to fit the data in Example 3 with a quadratic model of the form

$$y = ax^2 + bx + c.$$

Do you think this model fits better than the logarithmic model in Example 3? Explain your reasoning.

Application

Real Life

EXAMPLE 4 Finding an Exponential Model

The total amounts A (in billions of dollars) spent on health care in the United States in the years 1970 through 1991, are shown below. Find a model for the data, and use the model to predict the amount spent in 1998. In the list of data points (t, A), t represents the year, with $t = 0$ corresponding to 1970. (Source: U.S. Health Care Financing Administration)

(0, 74.4), (1, 82.3), (2, 92.3), (3, 102.5), (4, 116.1), (5, 132.9), (6, 152.2), (7, 172.0), (8, 193.7), (9, 217.2), (10, 250.1), (11, 290.2), (12, 326.1), (13, 358.6), (14, 389.6), (15, 422.6), (16, 454.9), (17, 494.2), (18, 546.1), (19, 604.3), (20, 675.0), (21, 751.8)

Solution

Figure 4.39

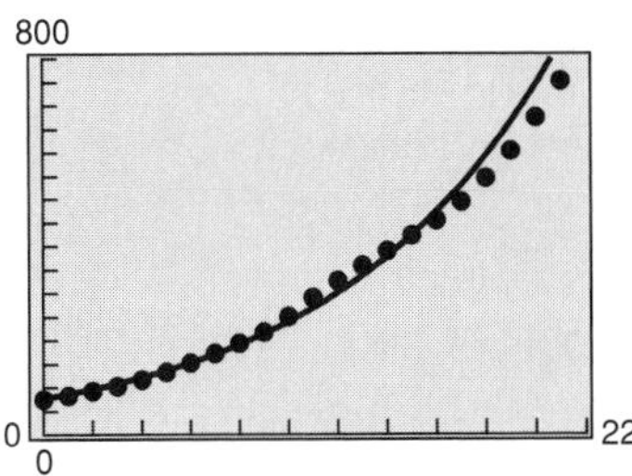

Begin by entering the data into a computer or calculator that has least squares regression programs. Then plot the data, as shown in Figure 4.39. From the scatter plot, it appears that an exponential model is a good fit. After running the exponential regression program, you should obtain

$$A = 77.27(1.12)^t \quad \text{or} \quad A = 77.27e^{0.1133t}.$$

(The correlation coefficient is $r = 0.997$, which implies that the model is a good fit to the data.) From the model, you can see that the amount spent on health care from 1970 through 1991 had an average annual increase of 12%. From this model, you can predict the 1998 amount to be

$$A = 77.27(1.12)^{28} \approx 1845.5 \text{ billion dollars}$$

which is more than twice the amount spent in 1991.

Group Activity

Fitting a Model to Data

The numbers y (in millions) of long-playing albums sold in the United States in the years 1975 through 1992 are listed below. The data is given as ordered pairs of the form (t, y), where t is the year, with $t = 5$ representing 1975. Create a scatter plot of the data. With others in your group, decide which type of model best fits this data. Then find the model.

(5, 257.0), (6, 273.0), (7, 344.0), (8, 341.3), (9, 318.3), (10, 322.8), (11, 295.2), (12, 243.9), (13, 209.6), (14, 204.6), (15, 167.0), (16, 125.2), (17, 107.0), (18, 72.4), (19, 34.6), (20, 11.7), (21, 4.8), (22, 2.3)

4.6 /// EXERCISES

In Exercises 1–8, determine if the scatter plot could best be modeled by a linear model, a quadratic model, an exponential model, a logarithmic model, a Gaussian model, or a logistics model.

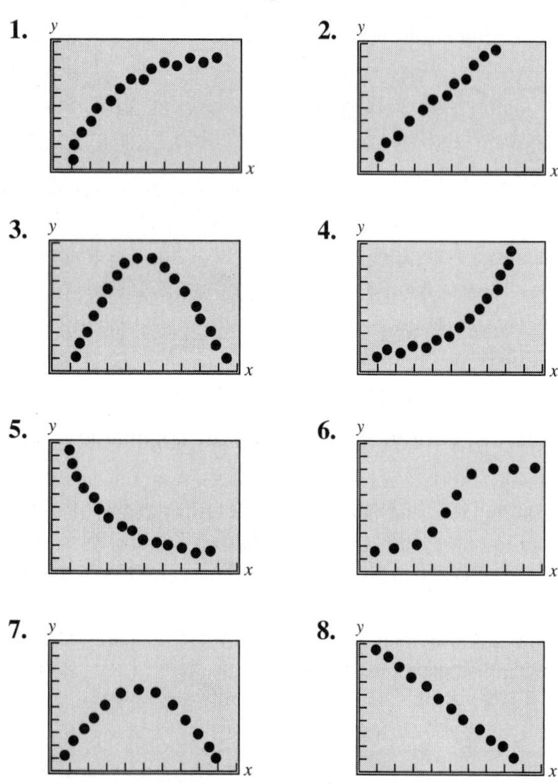

In Exercises 9–14, sketch a scatter plot of the data. Decide whether the data could best be modeled by a linear, an exponential, or a logarithmic model.

9. (1, 2.0), (1.5, 3.5), (2, 4.0), (4, 5.8), (6, 7.0), (8, 7.8)

10. (1, 5.8), (1.5, 6.0), (2, 6.5), (4, 7.6), (6, 8.9), (8, 10.0)

11. (1, 4.4), (1.5, 4.7), (2, 5.5), (4, 9.9), (6, 18.1), (8, 33.0)

12. (1, 11.0), (1.5, 9.6), (2, 8.2), (4, 4.5), (6, 2.5), (8, 1.4)

13. (1, 7.5), (1.5, 7.0), (2, 6.8), (4, 5.0), (6, 3.5), (8, 2.0)

14. (1, 5.0), (1.5, 6.0), (2, 6.4), (4, 7.8), (6, 8.6), (8, 9.0)

In Exercises 15–18, use a graphing utility to find an exponential model, $y = ab^x$, for the points. Use the graphing utility to obtain a scatter plot and the graph of the model.

15. (0, 4), (1, 5), (2, 6), (3, 8), (4, 12)

16. (0, 6.0), (2, 8.9), (4, 20.0), (6, 34.3), (8, 61.1), (10, 120.5)

17. (0, 10.0), (1, 6.1), (2, 4.2), (3, 3.8), (4, 3.6)

18. (−3, 120.2), (0, 80.5), (3, 64.8), (6, 58.2), (10, 55.0)

In Exercises 19–22, use a graphing utility to find a logarithmic model, $y = a + b \ln x$, for the points. Use the graphing utility to obtain a scatter plot and the graph of the model.

19. (1, 2.0), (2, 3.0), (3, 3.5), (4, 4.0), (5, 4.1), (6, 4.2), (7, 4.5)

20. (1, 8.5), (2, 11.4), (4, 12.8), (6, 13.6), (8, 14.2), (10, 14.6)

21. (1, 10), (2, 6), (3, 6), (4, 5), (5, 3), (6, 2)

22. (3, 14.6), (6, 11.0), (9, 9.0), (12, 7.6), (15, 6.5)

In Exercises 23–26, use a graphing utility to find a power model, $y = ax^b$, for the points. Use the graphing utility to obtain a scatter plot and the graph of the model.

23. (1, 2.0), (2, 3.4), (5, 6.7), (6, 7.3), (10, 12.0)

24. (0.5, 1.0), (2, 12.5), (4, 33.2), (6, 65.7), (8, 98.5), (10, 150.0)

25. (1, 10.0), (2, 4.0), (3, 0.7), (4, 0.1)

26. (2, 450), (4, 385), (6, 345), (8, 332), (10, 312)

27. *Breaking Strength* The breaking strength y (in tons) of a steel cable of diameter d (in inches) is given by the data in the table.

d	0.50	0.75	1.00	1.25	1.50	1.75
y	9.85	21.8	38.3	59.2	84.4	114.0

(a) Use the regression capabilities of a graphing utility to fit an appropriate model to the data.

(b) Use the graphing utility to plot the data and graph the model.

(c) Use the model to predict the breaking strength for a cable of diameter 2 inches.

28. *Foreign Travel* The numbers of United States citizens y (in millions) who traveled to foreign countries in the years 1984 through 1993 are given in the table, where $t = 4$ represents the year 1984. (Source: U.S. Bureau of Economic Analysis)

t	4	5	6	7	8
y	34.0	34.7	37.2	39.4	40.7

t	9	10	11	12	13
y	41.1	44.6	41.6	43.9	45.5

(a) Use the regression capabilities of a graphing utility to fit an appropriate model to the data.

(b) Use the graphing utility to plot the data and graph the model.

(c) Use the model to predict the number of foreign travelers in the year 2001.

29. *World Population* The world population y (in billions) for the years 1983 through 1994 is given in the table, where $x = 3$ corresponds to 1983. (Source: U.S. Bureau of the Census, International Database)

x	3	4	5	6	7	8
y	4.68	4.77	4.85	4.94	5.02	5.11

x	9	10	11	12	13	14
y	5.20	5.29	5.38	5.48	5.55	5.64

(a) Use the regression capabilities of a graphing utility to fit a linear model to the data.

(b) Use the regression capabilities of a graphing utility to fit an exponential model to the data.

(c) Population growth is often exponential. For the 12 years of data given, is the exponential model better than the linear model? Explain.

(d) Use each model to predict the population in the year 2001.

30. *Atmospheric Pressure* The atmospheric pressure decreases with increasing altitude. At sea level, the average air pressure is 1.033227 kilograms per square centimeter, and this pressure is called one atmosphere. Variations in weather conditions cause changes in the atmospheric pressure of up to ± 5 percent. The table gives the pressures p (in atmospheres) at given altitudes h (in kilometers).

h	0	5	10	15	20	25
p	1	0.55	0.25	0.12	0.06	0.02

(a) Use a graphing utility to attempt to find the logarithmic model $p = a + b \ln h$ for the data. Explain why the result is an error message.

(b) Use a graphing utility to find the logarithmic model $h = a + b \ln p$ for the data.

(c) Use a graphing utility to plot the data and graph the logarithmic model.

(d) Use the model to estimate the altitude at which the pressure is 0.75 atmosphere.

(e) Use the graph to estimate the pressure at an altitude of 13 kilometers.

31. *Lumber* The table gives the total domestic production x and the domestic consumption y of lumber in the United States from 1983 through 1990. The measurements are in millions of board feet. (Source: *Current Industrial Reports*)

x	34.6	37.1	36.4	42.0
y	48.7	52.7	53.5	57.4

x	44.9	44.6	43.6	43.9
y	61.8	58.8	58.8	54.5

(a) Use a graphing utility to find a power model for the data.

(b) Use a graphing utility to plot the data and graph the power model in the same viewing rectangle.

(c) Use the model to estimate consumption for a production level of 50 million board feet.

32. *National Health Expenditures* The table gives the national health expenditures y (in billions of dollars) for the years 1982 through 1991. The years are given by x, where $x = 2$ corresponds to 1982. (Source: U.S. Health Care Financing Administration)

x	2	3	4	5	6
y	326.1	358.6	389.6	422.6	454.9

x	7	8	9	10	11
y	494.2	546.1	604.3	675.0	751.8

(a) Use a graphing utility to find an exponential model for the data. Use the graphing utility to plot the data and graph the exponential model in the same viewing rectangle.

(b) Use a graphing utility to find a logarithmic model for the data. Use the graphing utility to plot the data and graph the logarithmic model in the same viewing rectangle.

(c) Use the graphs plotted in parts (a) and (b) to determine which model is better. If the rate of growth of health care costs could be slowed, which model may be better for the future? Explain.

(d) Use each model to predict health care costs in the year 2001.

33. *Frequency* Students in a physics lab measured the frequency f (in hertz) of a plucked wire with tension T (in newtons). The data is plotted on the graph.

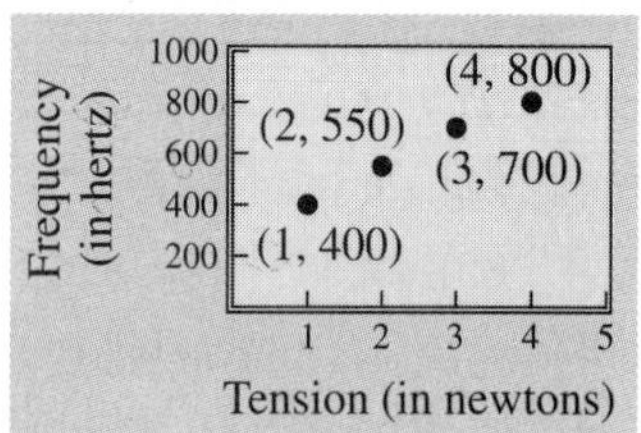

(a) Use a graphing utility to find a power model for the data. Use the graphing utility to plot the data and graph the model in the same viewing rectangle.

(b) Use the model to estimate the frequency at $T = 1.5$.

(c) Based on the limited number of observations, it appears as though the frequency is proportional to what common power of x?

34. *Wind Turbine* Tests on a wind turbine in a wind tunnel measured the power output P (in kilowatts) at various wind speeds S (in miles per hour). The test results are plotted on the graph.

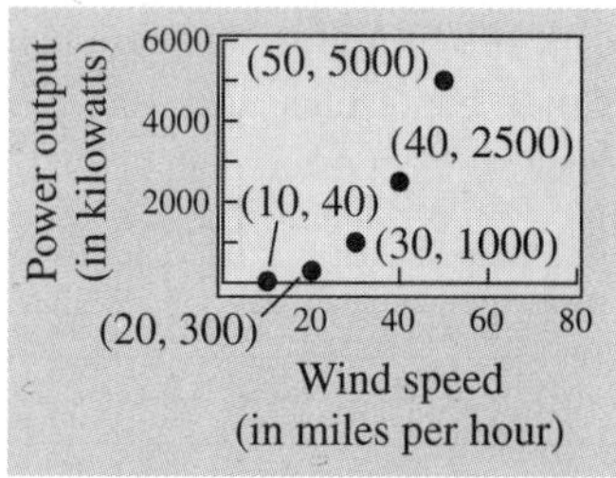

(a) Use a graphing utility to find a power model for the data. Use the graphing utility to plot the data and graph the model in the same viewing rectangle.

(b) Use the model to estimate the frequency at $S = 45$.

(c) Based on the limited number of observations, it appears as though the power output is proportional to what power of x?

35. *Comparing Models* The amounts y (in billions of dollars) donated to charity (by individuals, foundations, corporations, and charitable bequests) in the years 1983 through 1992 in the United States are given in the table, where $x = 3$ corresponds to 1983. (Source: AAFRC Trust for Philanthropy)

x	3	4	5	6	7
y	63.2	68.6	73.2	83.9	90.3

x	8	9	10	11	12
y	98.4	107.0	111.7	116.8	124.3

(a) Use the regression capabilities of a graphing utility to find the following models for the data.

$y_1 = ax + b$

$y_2 = a + b \ln x$

$y_3 = ab^x$

$y_4 = ax^b$

(b) Use the graphing utility to graph the data and each of the models. Use the graphs to select the model that you think best fits the data.

(c) For each of the models y_i $(i = 1, 2, 3, 4)$, complete the following table.

x	y	$y - y_i$	$(y - y_i)^2$
3	63.2		
4	68.6		
5	73.2		
6	83.9		
7	90.3		
8	98.4		
9	107.0		
10	111.7		
11	116.8		
12	124.3		

(d) For each model, find the sum of the entries in the last column of the table in part (c). Use the results to select the best model for the data.

(e) Explain what the sums in part (d) represent.

36. *Comparing Models* A V8 car engine is coupled to a dynamometer, and the horsepower y is measured at different engine speeds x (in thousands of revolutions per minute). The results are shown in the table.

x	1	2	3	4	5	6
y	40	85	140	200	225	245

(a) Use the regression capabilities of a graphing utility to find the following models for the data.

$y_1 = ax^3 + bx^2 + cx + d$

$y_2 = a + b \ln x$

$y_3 = ab^x$

$y_4 = ax^b$

(b) Use the graphing utility to graph the data and each of the models. Use the graphs to select the model that you think best fits the data.

(c) For each of the models y_i $(i = 1, 2, 3, 4)$, complete the following table.

x	y	$y - y_i$	$(y - y_i)^2$
1	40		
2	85		
3	140		
4	200		
5	225		
6	245		

(d) For each model, find the sum of the entries in the last column of the table in part (c). Use the results to select the best model for the data.

(e) Explain what the sums in part (d) represent.

Focus on Concepts

In this chapter, you studied several concepts that are related to exponential and logarithmic functions. You can use the following questions to check your understanding of several of these basic concepts. The answers to these questions are given in the back of the book.

1. *Comparing Graphs* The graphs of $y = e^{kt}$ are shown for $k = a, b, c$, and d. Use the graphs to order a, b, c, and d. Which of the four values are negative? Which are positive?

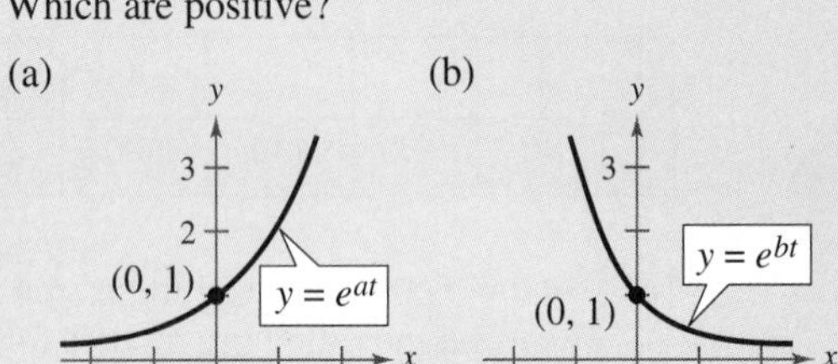

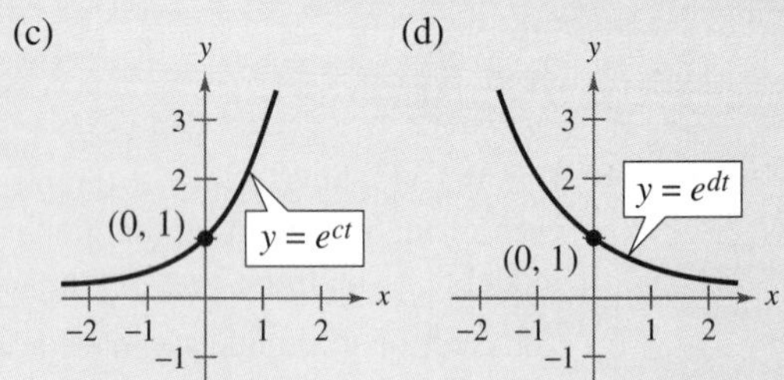

2. *True or False?* Rewrite each verbal statement as an equation. Then decide whether the statement is true or false. If it is false, give an example that shows it is false.

(a) The logarithm of the product of two numbers is equal to the sum of the logarithms of the numbers.

(b) The logarithm of the sum of two numbers is equal to the product of the logarithms of the numbers.

(c) The logarithm of the difference of two numbers is equal to the difference of the logarithms of the numbers.

(d) The logarithm of the quotient of two numbers is equal to the difference of the logarithms of the numbers.

3. *Investing Money* You are investing P dollars at an annual rate r, compounded continuously, for t years. Which of the following would be most advantageous? Explain your reasoning.

(a) Double the amount you invest.

(b) Double your interest rate.

(c) Double the number of years.

4. Identify the model as linear, logarithmic, exponential, logistic, or none of the above. Explain your reasoning.

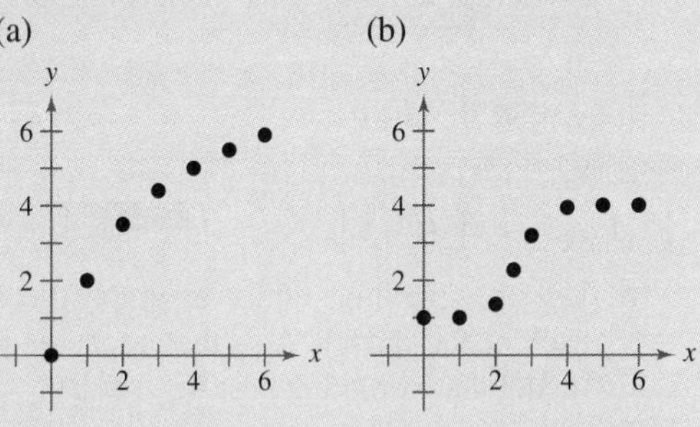

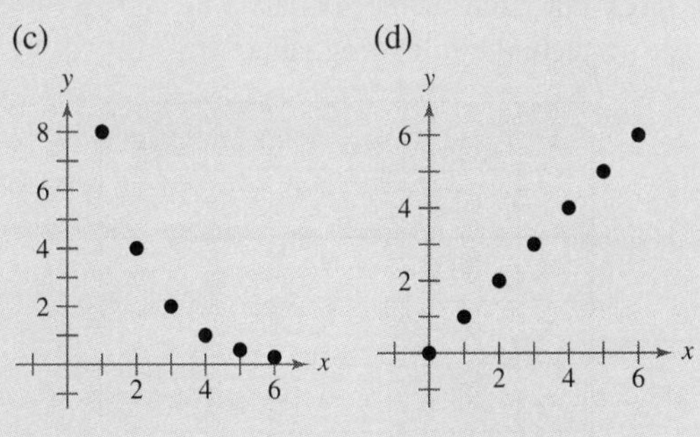

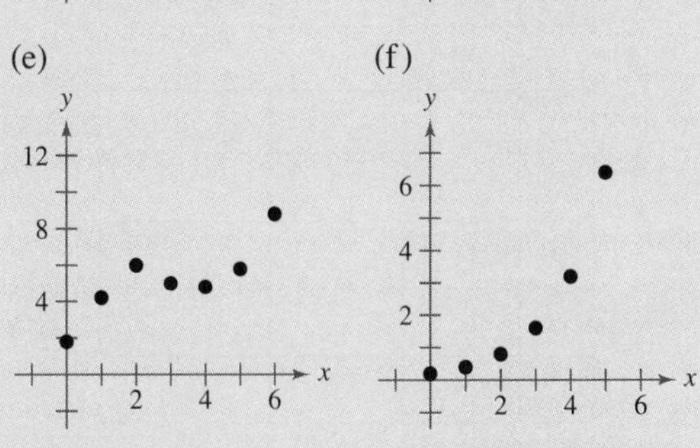

4 /// REVIEW EXERCISES

In Exercises 1–6, match the function with its graph. [The graphs are labeled (a), (b), (c), (d), (e), and (f).]

(a)

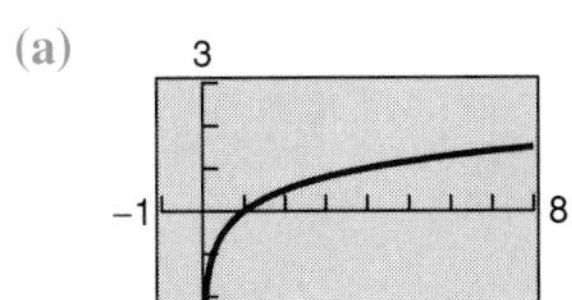

(b)

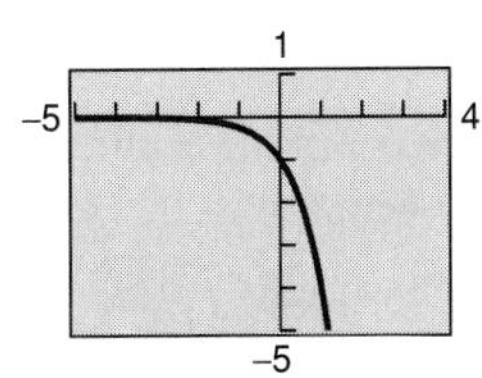

(c)

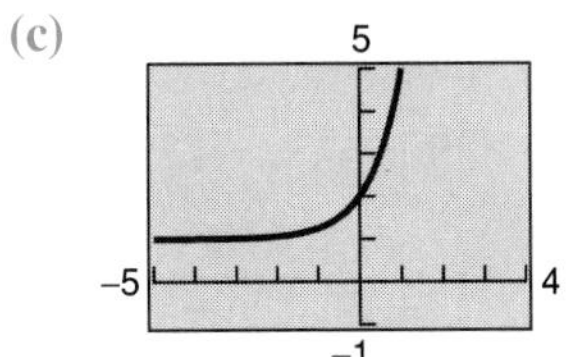

(d)

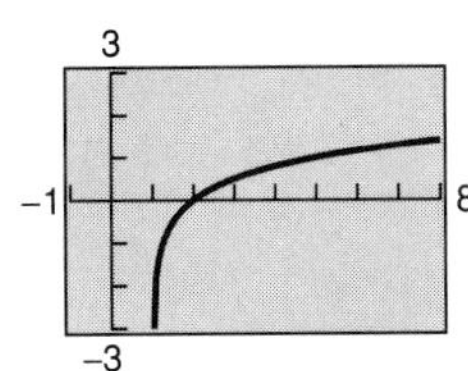

(e)

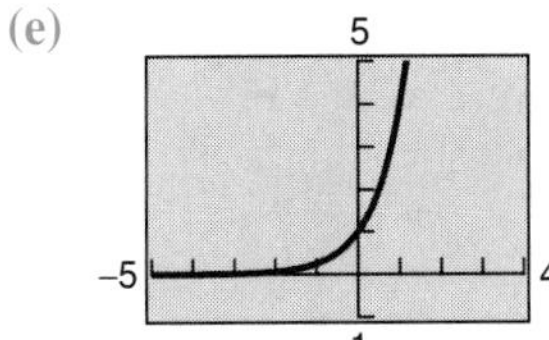

(f)

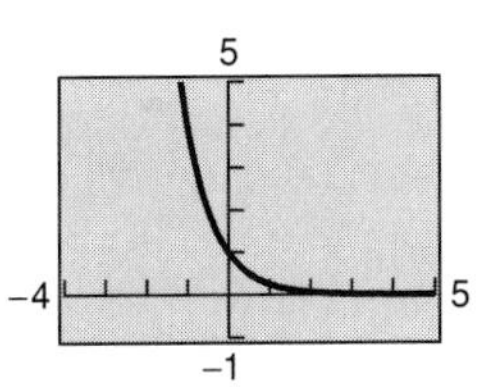

1. $f(x) = 4^x$

2. $f(x) = 4^{-x}$

3. $f(x) = -4^x$

4. $f(x) = 4^x + 1$

5. $f(x) = \log_4 x$

6. $f(x) = \log_4(x - 1)$

In Exercises 7–14, sketch the graph of the function. Use a graphing utility to verify your graph.

7. $f(x) = 6^x$

8. $f(x) = 0.3^x$

9. $g(x) = 6^{-x}$

10. $g(x) = 0.3^{-x}$

11. $h(x) = e^{-x/2}$

12. $h(x) = 2 - e^{-x/2}$

13. $f(x) = e^{x+2}$

14. $s(t) = 4e^{-2/t}, \quad t > 0$

In Exercises 15–18, use a graphing utility to graph the function. Identify any asymptotes.

15. $g(t) = 8 - 0.5e^{-t/4}$

16. $h(x) = 12(1 + e^{-x/2})$

17. $f(x) = \dfrac{10}{1 + 2^{-0.05x}}$

18. $g(x) = 200e^{4/x}$

In Exercises 19 and 20, complete the table to determine the balance A for P dollars invested at rate r for t years and compounded n times per year.

n	1	2	4	12	365	Continuous
A						

19. $P = \$3,500,\ r = 10.5\%,\ t = 10$ years

20. $P = \$2,000,\ r = 12\%,\ t = 30$ years

In Exercises 21 and 22, complete the table to determine the amount of money P that should be invested at rate r to produce a final balance of $200,000 in t years.

t	1	10	20	30	40	50
P						

21. $r = 8\%$, compounded continuously

22. $r = 10\%$, compounded monthly

23. *Depreciation* After t years, the value of a car that originally cost $14,000 is given by

$$V(t) = 14{,}000\left(\frac{3}{4}\right)^t.$$

(a) Use a graphing utility to graph the function.

(b) Find the value of the car 2 years after it was purchased.

(c) According to the model, when does the car depreciate most rapidly? Is this realistic? Explain.

24. *Trust Fund* On the day a child is born, a deposit of \$50,000 is made in a trust fund that pays 8.75% interest, compounded continuously. Determine the balance in the account after 35 years.

25. *Drug Decomposition* A solution of a certain drug contains 500 units per milliliter when prepared. It is analyzed after 40 days and is found to contain 300 units per milliliter. Assuming that the rate of decomposition is proportional to the amount present, the equation giving the amount A after t days is

$$A = 500e^{-0.013t}.$$

Use a graphing utility to graph the model. Approximate A graphically and analytically at $t = 60$.

26. *Waiting Times* The average time between incoming calls at a switchboard is 3 minutes. The probability of waiting less than t minutes for the next incoming call is approximated by the model

$$F(t) = 1 - e^{-t/3}.$$

If a call has just come in, find the probability that the next call will come within

(a) $\frac{1}{2}$ minute. (b) 2 minutes. (c) 5 minutes.

27. *Fuel Efficiency* A certain automobile gets 28 miles per gallon of gasoline for speeds up to 50 miles per hour. Over 50 miles per hour, the number of miles per gallon drops at the rate of 12% for each additional 10 miles per hour. If s is the speed and y is the number of miles per gallon, then

$$y = 28e^{0.6-0.012s}, \qquad s \geq 50.$$

Use the model to complete the table.

s	50	55	60	65	70
y	28				

28. *Inflation* If the inflation rate averages 4.5% over the next 10 years, the approximate cost C of goods or services t years from now is given by

$$C(t) = P(1.045)^t$$

where P is the present cost. If the price of a tire is presently \$69.95, estimate the price 10 years from now.

In Exercises 29–34, sketch the graph of the function. Use a graphing utility to verify your graph.

29. $g(x) = \log_2 x$ **30.** $g(x) = \log_5 x$

31. $f(x) = \ln x + 3$ **32.** $f(x) = \ln(x - 3)$

33. $h(x) = \ln(e^{x-1})$ **34.** $f(x) = \frac{1}{4} \ln x$

In Exercises 35 and 36, use a graphing utility to graph the function.

35. $y = \log_{10}(x^2 + 1)$ **36.** $y = \sqrt{x} \ln(x + 1)$

In Exercises 37 and 38, write the equation in logarithmic form.

37. $4^3 = 64$ **38.** $25^{3/2} = 125$

In Exercises 39–46, evaluate the expression *by hand*.

39. $\log_{10} 1000$ **40.** $\log_9 3$

41. $\log_3 \frac{1}{9}$ **42.** $\log_4 \frac{1}{16}$

43. $\ln e^7$ **44.** $\log_a \frac{1}{a}$

45. $\ln 1$ **46.** $\ln e^{-3}$

In Exercises 47–50, evaluate the logarithm using the change-of-base formula. Do each problem twice, once with common logarithms and once with natural logarithms. Round the result to three decimal places.

47. $\log_4 9$ **48.** $\log_{1/2} 5$

49. $\log_{12} 200$ **50.** $\log_3 0.28$

In Exercises 51–56, write the expression as a sum, difference, and/or multiple of logarithms.

51. $\log_5 5x^2$ **52.** $\log_7 \frac{\sqrt{x}}{4}$

53. $\log_{10} \frac{5\sqrt{y}}{x^2}$ **54.** $\ln \left|\frac{x - 1}{x + 1}\right|$

55. $\ln[(x^2 + 1)(x - 1)]$ **56.** $\ln \sqrt[5]{\frac{4x^2 - 1}{4x^2 + 1}}$

In Exercises 57–62, write the expression as the logarithm of a single quantity.

57. $\log_2 5 + \log_2 x$

58. $\log_6 y - 2 \log_6 z$

59. $\frac{1}{2} \ln|2x - 1| - 2 \ln|x + 1|$

60. $5 \ln|x - 2| - \ln|x + 2| - 3 \ln|x|$

61. $\ln 3 + \frac{1}{3} \ln(4 - x^2) - \ln x$

62. $3[\ln x - 2 \ln(x^2 + 1)] + 2 \ln 5$

True or False? **In Exercises 63–68, determine whether the statement is true or false.**

63. $\log_b b^{2x} = 2x$

64. $e^{x-1} = \dfrac{e^x}{e}$

65. The domain of the function $f(x) = \ln x$ is the set of all real numbers.

66. $\ln(x + y) = \ln x + \ln y$

67. $\ln(x + y) = \ln(x \cdot y)$

68. $\log\left(\dfrac{10}{x}\right) = 1 - \log x$

In Exercises 69–72, approximate the logarithm using the properties of logarithms and given that $\log_b 2 \approx 0.3562$, $\log_b 3 \approx 0.5646$, and $\log_b 5 \approx 0.8271$.

69. $\log_b 25$

70. $\log_b\left(\dfrac{25}{9}\right)$

71. $\log_b \sqrt{3}$

72. $\log_b 30$

73. *Climb Rate* The time t, in minutes, for a small plane to climb to an altitude of h feet is given by

$$t = 50 \log_{10} \frac{18{,}000}{18{,}000 - h}$$

where 18,000 feet is the plane's absolute ceiling.

(a) Determine the domain of the function appropriate for the context of the problem.

(b) Use a graphing utility to graph the time function and identify any asymptotes.

(c) As the plane approaches its absolute ceiling, what can be said about the time required to further increase its altitude?

(d) Find the time for the plane to climb to an altitude of 4000 feet.

74. *Snow Removal* The number of miles s of roads cleared of snow is approximated by the model

$$s = 25 - \frac{13 \ln(h/12)}{\ln 3}, \qquad 2 \le h \le 15$$

where h is the depth of the snow in inches. Use this model to find s when $h = 10$ inches.

In Exercises 75–80, solve the exponential equation. Round the solution to three decimal places.

75. $e^x = 12$

76. $e^{3x} = 25$

77. $3e^{-5x} = 132$

78. $14e^{3x+2} = 560$

79. $e^{2x} - 6x + 8 = 0$

80. $e^{2x} - 7e^x + 10 = 0$

In Exercises 81–86, solve the logarithmic equation. Round the solution to three decimal places.

81. $\ln 3x = 8.2$

82. $2 \ln 4x = 15$

83. $\ln x - \ln 3 = 2$

84. $\ln\sqrt{x + 1} = 2$

85. $\log(x - 1) = \log(x - 2) - \log(x + 2)$

86. $\log(1 - x) = -1$

In Exercises 87–90, use a graphing utility to solve the equation. Round the solution to two decimal places.

87. $2^{0.6x} - 3x = 0$

88. $25e^{-0.3x} = 12$

89. $2 \ln(x + 3) + 3x = 8$

90. $6 \log_{10}(x^2 + 1) - x = 0$

In Exercises 91–94, find the exponential function $y = Ae^{bx}$ that passes through the two points.

91. $(0, 2), (4, 3)$

92. $\left(0, \frac{1}{2}\right), (5, 5)$

93. $(0, 4), \left(5, \frac{1}{2}\right)$

94. $(0, 2), (5, 1)$

95. *Demand Function* The demand equation for a certain product is given by

$$p = 500 - 0.5e^{0.004x}.$$

Find the demands x for prices of (a) $p = \$450$ and (b) $p = \$400$.

96. *Typing Speed* In a typing class, the average number of words per minute typed after t weeks of lessons was found to be

$$N = \frac{157}{1 + 5.4e^{-0.12t}}.$$

Find the time necessary to type (a) 50 words per minute and (b) 75 words per minute.

97. *Compound Interest* A deposit of \$750 is made in a savings account for which the interest is compounded continuously. The balance will double in $7\frac{3}{4}$ years.

(a) What is the annual interest rate for this account?

(b) Find the balance in the account after 10 years.

(c) The *effective yield* of a savings plan is the percent increase in the balance after 1 year. Find the effective yield.

98. *Compound Interest* A deposit of \$10,000 is made in a savings account for which the interest is compounded continuously. The balance will double in 5 years.

(a) What is the annual interest rate for this account?

(b) Find the balance after 1 year.

(c) The *effective yield* of a savings plan is the percent increase in the balance after 1 year. Find the effective yield.

99. *Sound Intensity* The relationship between the number of decibels β and the intensity of a sound I in watts per centimeter squared is given by

$$\beta = 10 \log_{10}\left(\frac{I}{10^{-16}}\right).$$

Determine the intensity of a sound in watts per centimeter squared if the decibel level is 125.

100. *Earthquake Magnitudes* On the Richter scale, the magnitude R of an earthquake of intensity I is given by

$$R = \log_{10} \frac{I}{I_0}$$

where $I_0 = 1$ is the minimum intensity used for comparison. Find the intensity per unit of area for the following values of R.

(a) $R = 8.4$ (b) $R = 6.85$ (c) $R = 9.1$

101. *Exponential Regression* Use a graphing utility to find an exponential model $y = ab^x$ through the points (0, 250), (4, 135), (6, 92), and (10, 67). Sketch a scatter plot and the graph of the exponential model.

102. *Data Analysis* The data in the table gives the yield y (in milligrams) of a chemical reaction after t minutes.

t	1	2	3	4
y	1.5	7.4	10.2	13.4

t	5	6	7	8
y	15.8	16.3	18.2	18.3

(a) Use a graphing utility to fit the linear model $y = at + b$ to the data.

(b) Use a graphing utility to fit the logarithmic model $y = a + b \ln t$ to the data.

(c) Create a scatter plot for the data, and graph the linear and logarithmic models. Which is a better model for the data? Explain your reasoning.

103. *Sporting Goods Sales* The sales y (in billions of dollars) of sporting goods for the years 1984 through 1993 are given in the table, where $x = 4$ corresponds to 1984. (Source: National Sporting Goods Association)

x	4	5	6	7	8
y	26.4	27.4	30.6	33.9	42.1

x	9	10	11	12	13
y	45.2	44.1	42.9	42.4	44.6

(a) Use a graphing utility to find the power model $y = ax^b$ for the data. Use the graphing utility to plot the data and graph the model.

(b) Use the model to predict sales in the year 2001.

CHAPTER PROJECT *A Graphical Approach to Compound Interest*

In this project, you will use a graphing utility to compare savings plans. For instance, suppose you are depositing \$1000 in a savings account and are given the following options.

- 6.2% annual interest rate, compounded annually
- 6.1% annual interest rate, compounded quarterly
- 6.0% annual interest rate, compounded continuously

(a) For each option, write a function that gives the balance as a function of the time t (in years).

(b) Graph all three functions in the same viewing rectangle. Can you find a viewing rectangle that distinguishes among the graphs of the three functions? If so, describe the viewing rectangle.

(c) Find the balances for the three options after 25, 50, 75, and 100 years. Is the option that yields the greatest balance after 25 years the same option that yields the greatest balances after 50, 75, and 100 years? Explain.

(d) The **effective yield** of a savings plan is the percent increase in the balance after 1 year. Find the effective yields for the three options listed above. How can the effective yield be used to decide which option is best?

Questions for Further Exploration

1. You deposit \$25,000 in an account to accrue interest for 40 years. The account pays 8% compounded annually. Assume that the income tax on the earned interest is 30%. Which of the following plans produces a larger balance after all income tax is paid?

(a) *Deferred* The income tax on the interest that is earned is paid in one lump sum at the end of 40 years.

(b) *Not Deferred* The income tax on the interest that is earned each year is paid at the end of each year.

2. Which of the following would produce a larger balance? Explain.

(a) 8.02% annual interest rate, compounded monthly

(b) 8.03% annual interest rate, compounded daily

(c) 8% annual interest rate, compounded continuously

3. You deposit \$1000 in each of two savings accounts. The interest for the accounts is paid according to the two options described in Question 1. How long would it take for the balance in one of the accounts to exceed the balance in the other account by \$100? By \$100,000?

4. No income tax is due on the interest earned in some types of investments. You deposit \$25,000 into an account. Which of the following plans is better? Explain.

(a) *Tax-Free* The account pays 5%, compounded annually. There is no income tax on the earned interest.

(b) *Tax-Deferred* The account pays 7%, compounded annually. At maturity, the earned interest is taxable at a rate of 40%.

4 /// CHAPTER TEST

Take this test as you would take a test in class. After you are done, check your work against the answers given in the back of the book.

The *Interactive* CD-ROM provides answers to the Chapter Tests and Cumulative Tests. It also offers Chapter Pre-Tests (that test key skills and concepts covered in previous chapters) and Chapter Post-Tests, both of which have randomly generated exercises with diagnostic capabilities.

1. Sketch the graph of the function $f(x) = 2^{-x/3}$.
2. Use a graphing utility to graph f and determine its horizontal asymptotes.

$$f(x) = \frac{1000}{1 + 4e^{-0.2x}}$$

3. Determine the principal that will yield \$200,000 when invested at 8% compounded daily for 20 years.
4. Write the logarithmic equation $\log_4 64 = 3$ in exponential form.
5. Sketch the graph of the function $g(x) = \log_3(x - 2)$.
6. Use the properties of logarithms to expand $\ln\left(\frac{6x^2}{\sqrt{x^2 + 1}}\right)$.

In Exercises 7–9, solve the equation. Round to three decimal places.

7. $e^{x/2} = 450$
8. $\left(1 + \frac{0.06}{4}\right)^{4t} = 3$
9. $5 \ln(x + 4) = 22$

10. A truck that costs \$28,000 new has a depreciated value of \$20,000 after 1 year. Find the value of the truck when it is 3 years old by using the exponential model $y = ae^{bx}$.
11. The population of a certain species t years after it is introduced into a new habitat is given by $p(t) = 1200/(1 + 3e^{-t/5})$.
    (a) Determine the initial size of the population.
    (b) Determine the population after 5 years.
    (c) After how many years will the population be 800?
12. By observation, identify the equation that corresponds to the graph shown in the figure. Explain your reasoning.

    (a) $y = 6e^{-x^2/2}$ (b) $y = \dfrac{6}{1 + e^{-x/2}}$ (c) $y = 6(1 - e^{-x^2/2})$

Figure for 12

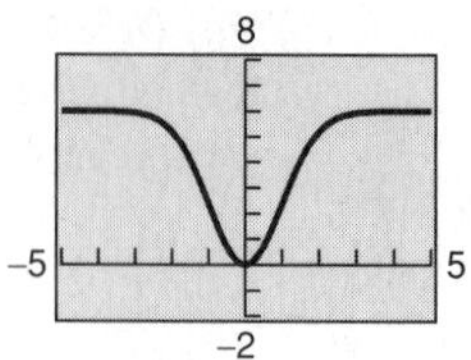

13. The numbers of cellular telephone subscribers y (in millions) for the years 1990 through 1993 are given by (0, 5.3), (1, 7.6), (2, 11.0), and (3, 16.0), where x is the time in years, with $x = 0$ corresponding to 1990. Use a graphing utility to fit an exponential model to the data. Sketch a scatter plot of the data, and graph the model in the same viewing rectangle. (Source: Cellular Telecommunications Industry Association)

Chapter 5

Systems of Equations and Inequalities

In 1975, pickups, sport utilities, and vans accounted for only 21% of all light vehicles sold in the United States. By 1995, they accounted for almost 40%.

These three types of "trucks" all come in both full-size models and compact models. From 1990 to 1994, the sales (in thousands) of compact pickups P and compact sport utilities S were as follows.

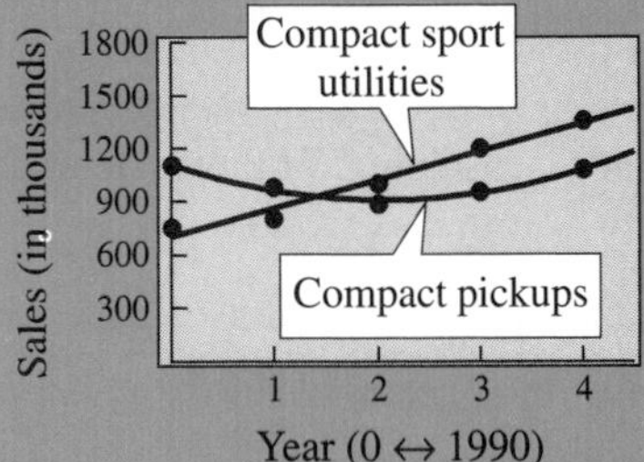

Year	P	S
1990	1102.9	756.3
1991	980.7	807.9
1992	887.8	998.4
1993	954.1	1203.2
1994	1080.8	1359.1

These sales can be modeled by the following quadratic and linear models.

$$P = 47t^2 - 195t + 1109$$

$$S = 160t + 705$$

You can use a graphing utility to graph these two models in the same viewing rectangle. By using the zoom and trace features, you can verify that the curves intersect at approximately (1.4, 928). This indicates that compact sport utilities began outselling compact pickups in 1992. (See Exercises 77 and 78 on page 422.)

Alan Levenson/Tony Stone Images.

In 1994, in the full-size category, pickups outsold sport utilities. But in the compact category, for the third year in a row, sport utilities outsold pickups.

5.1 Solving Systems of Equations

The Method of Substitution / Graphical Approach to Finding Solutions / Applications

The Method of Substitution

EXPLORATION

Use a graphing utility to graph

$$y_1 = -2x + 5$$
$$y_2 = 1.5x - 2$$

in the same viewing rectangle. Find the coordinates of the point of intersection of these two lines. Do you obtain the same coordinates as are obtained algebraically in the discussion at the right?

Up to this point in the text, most problems have involved either a function of one variable or a single equation in two variables. However, many problems in science, business, and engineering involve two or more equations in two or more variables. To solve such problems, you need to find solutions of **systems of equations.** Here is an example of a system of two equations in x and y.

$$2x + y = 5 \qquad \text{Equation 1}$$
$$3x - 2y = 4 \qquad \text{Equation 2}$$

A **solution** of this system is an ordered pair that satisfies each equation in the system. Finding the set of all such solutions is called **solving the system of equations.** For instance, the ordered pair (2, 1) is a solution of this system. To check this, you can substitute 2 for x and 1 for y in *each* equation.

$$2x + y = 5 \qquad \text{Equation 1}$$
$$2(2) + 1 \stackrel{?}{=} 5 \qquad \text{Substitute 2 for } x \text{ and 1 for } y.$$
$$4 + 1 = 5 \qquad \text{Solution checks in Equation 1.} \checkmark$$
$$3x - 2y = 4 \qquad \text{Equation 2}$$
$$3(2) - 2(1) \stackrel{?}{=} 4 \qquad \text{Substitute 2 for } x \text{ and 1 for } y.$$
$$6 - 2 = 4 \qquad \text{Solution checks in Equation 2.} \checkmark$$

In this chapter you will study three ways to solve equations, beginning with the **method of substitution.**

The Method of Substitution

1. *Solve* one of the equations for one variable in terms of the other.
2. *Substitute* the expression found in Step 1 into the other equation to obtain an equation in one variable.
3. *Solve* the equation obtained in Step 2.
4. *Back-substitute* the solution in Step 3 into the expression obtained in Step 1 to find the value of the other variable.
5. *Check* that the solution satisfies *each* of the original equations.

EXPLORATION

Use a graphing utility to graph $y = -x + 4$ and $y = x - 2$ in the same viewing rectangle. Use the trace feature to find the coordinates of the point of intersection. Are the coordinates the same as the solution found in Example 1? Explain.

On the *TI-82* or *TI-83*, set the calculator to dot mode and use the following keystrokes.

ZOOM (zoom)(4:ZDecimal)
ENTER

What changes does this make in the coordinates of the point?

EXAMPLE 1 Solving a System of Equations

Solve the system of equations.

$x + y = 4$ Equation 1

$x - y = 2$ Equation 2

Solution

Begin by solving for y in Equation 1.

$y = 4 - x$ Solve for y in Equation 1.

Next, substitute this expression for y into Equation 2 and solve the resulting single-variable equation for x.

$x - y = 2$ Equation 2

$x - (4 - x) = 2$ Substitute $4 - x$ for y.

$x - 4 + x = 2$ Simplify.

$2x = 6$ Combine like terms.

$x = 3$ Solve for x.

Finally, you can solve for y by *back-substituting* $x = 3$ into the equation $y = 4 - x$, to obtain

$y = 4 - x$ Revised Equation 1

$y = 4 - 3$ Substitute 3 for x.

$y = 1.$ Solve for y.

The solution is the ordered pair (3, 1). You can check this as follows.

Check

$x + y = 4$ Equation 1

$3 + 1 \stackrel{?}{=} 4$ Substitute for x and y. 

$4 = 4$ Solution checks in Equation 1.

$x - y = 2$ Equation 2

$3 - 1 \stackrel{?}{=} 2$ Substitute for x and y. 

$2 = 2$ Solution checks in Equation 2.

Study Tip

Because many steps are required to solve a system of equations, it is very easy to make errors in arithmetic. Thus, we *strongly* suggest that you always *check your solution by substituting it into each equation in the original system.*

Note The term *back-substitution* implies that you work *backwards.* First you solve for one of the variables, and then you substitute that value *back* into one of the equations in the system to find the value of the other variable.

The *Interactive* CD-ROM offers graphing utility emulators of the *TI-82* and *TI-83*, which can be used with the Examples, Explorations, Technology notes, and Exercises.

One way to check the answers you obtain in this section is to use a graphing utility. For instance, enter the two equations in Example 2

$$y_1 = 12{,}000 - x$$

$$y_2 = \frac{1180 - 0.09x}{0.11}$$

and find an appropriate viewing rectangle that shows where the lines intersect. Then use the zoom and trace features to find their point of intersection. Does this point agree with the solution obtained at the right?

Real Life

EXAMPLE 2 **Solving a System by Substitution**

A total of $12,000 is invested in two funds paying 9% and 11% simple interest. The yearly interest is $1180. How much is invested at each rate?

Solution

Verbal Model: [9% fund] + [11% fund] = [Total investment]

[9% interest] + [11% interest] = [Total interest]

Labels:

Amount in 9% fund $= x$	(dollars)
Interest for 9% fund $= 0.09x$	(dollars)
Amount in 11% fund $= y$	(dollars)
Interest for 11% fund $= 0.11y$	(dollars)
Total investment $= \$12{,}000$	(dollars)
Total interest $= \$1180$	(dollars)

System:

$$x + y = 12{,}000 \qquad \text{Equation 1}$$

$$0.09x + 0.11y = 1180 \qquad \text{Equation 2}$$

To begin, it is convenient to multiply both sides of Equation 2 by 100 to obtain $9x + 11y = 118{,}000$. This eliminates the need to work with decimals.

$$9x + 11y = 118{,}000 \qquad \text{Revised Equation 2}$$

To solve this system, you can solve for x in Equation 1.

$$x = 12{,}000 - y \qquad \text{Revised Equation 1}$$

Next, substitute this expression for x into Revised Equation 2 and solve the resulting equation for y.

$$9x + 11y = 118{,}000 \qquad \text{Revised Equation 2}$$

$$9(12{,}000 - y) + 11y = 118{,}000 \qquad \text{Substitute } 12{,}000 - y \text{ for } x.$$

$$108{,}000 - 9y + 11y = 118{,}000 \qquad \text{Distributive Property}$$

$$2y = 10{,}000 \qquad \text{Combine like terms.}$$

$$y = 5000 \qquad \text{Amount in 11\% fund.}$$

Finally, back-substitute the value $y = 5000$ to solve for x.

$$x = 12{,}000 - y \qquad \text{Revised Equation 1}$$

$$x = 12{,}000 - 5000 \qquad \text{Substitute 5000 for } y.$$

$$x = 7000 \qquad \text{Amount in 9\% fund}$$

The solution is $(7000, 5000)$. Check this in the original problem.

The equations in Examples 1 and 2 are linear. Substitution can also be used to solve systems in which one or both of the equations are nonlinear.

EXAMPLE 3 **Substitution: Two-Solution Case**

Solve the system of equations.

$x^2 - x - y = 1$ Equation 1

$-x + y = -1$ Equation 2

Solution

Begin by solving for y in Equation 2 to obtain $y = x - 1$. Next, substitute this expression for y into Equation 1 and solve for x.

$x^2 - x - y = 1$ Equation 1

$x^2 - x - (x - 1) = 1$ Substitute $x - 1$ for y.

$x^2 - 2x + 1 = 1$ Simplify.

$x^2 - 2x = 0$ Standard form

$x(x - 2) = 0$ Factor.

$x = 0, 2$ Solve for x.

Back-substituting these values of x to solve for the corresponding values of y produces the solutions $(0, -1)$ and $(2, 1)$. Check these in the original system.

EXPLORATION

Use a graphing utility to graph the two equations in Example 3

$y_1 = x^2 - x - 1$

$y_2 = x - 1$

in the same viewing rectangle. How many solutions do you think this system has?

Repeat this experiment for the equations in Example 4. How many solutions does this system have? Explain your reasoning.

EXAMPLE 4 **Substitution: No-Solution Case**

Solve the system of equations.

$-x + y = 4$ Equation 1

$x^2 + y = 3$ Equation 2

Solution

Begin by solving for y in Equation 1 to obtain $y = x + 4$. Next, substitute this expression for y into Equation 2 and solve for x.

$x^2 + y = 3$ Equation 2

$x^2 + (x + 4) = 3$ Substitute $x + 4$ for y.

$x^2 + x + 1 = 0$ Simplify.

$$x = \frac{-1 \pm \sqrt{1^2 - 4(1)(1)}}{2}$$ Quadratic Formula

Because this yields two complex values, the equation $x^2 + x + 1 = 0$ has no *real* solution. Hence, this system has no *real* solution.

The *Interactive* CD-ROM shows every example with its solution; clicking on the *Try It!* button brings up similar problems. Guided Examples and Integrated Examples show step-by-step solutions to additional examples. Integrated Examples are related to several concepts in the section.

Graphical Approach to Finding Solutions

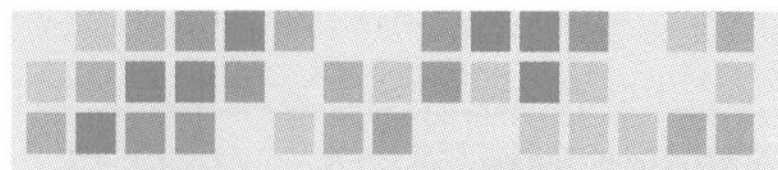

Use a graphing utility to graph the equations in Figure 5.1(a) in the same viewing rectangle. To find the point of intersection of the graphs, use the following keystrokes on a *TI-82* or *TI-83*.

CALC (5:intersect) ENTER

Cursor to the first graph and press ENTER. Then cursor to the second graph and press ENTER ENTER.

From Examples 2, 3, and 4, you can see that a system of two equations in two unknowns can have exactly one solution, more than one solution, or no solution. In practice, you can gain insight about the location and number of solutions of a system of equations by graphing each of the equations in the same coordinate plane. The solutions of the system correspond to the **points of intersection** of the graphs. For instance, in Figure 5.1(a), the two equations graph as two lines with a *single point* of intersection. The two equations in Example 3 graph as a parabola and a line with *two points* of intersection, as shown in Figure 5.1(b). The two equations in Example 4 graph as a line and a parabola that happen to have *no points* of intersection, as shown in Figure 5.1(c).

Figure 5.1

(a) One Intersection Point **(b)** Two Intersection Points **(c)** No Intersection Points

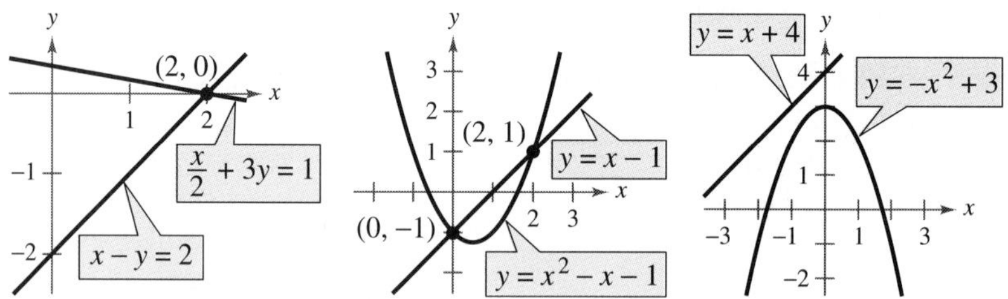

Note Example 5 shows you the value of a graphical approach to solving systems of equations in two variables. Notice what would happen if you tried only the substitution method in Example 5. It would be difficult to solve this equation for x using standard algebraic techniques.

EXAMPLE 5 **Solving a System of Equations**

Solve the system of equations.

$$y = \ln x \quad \text{Equation 1}$$
$$x + y = 1 \quad \text{Equation 2}$$

Solution

From the graphs of these equations, shown in Figure 5.2, it is clear that there is only one point of intersection. Also, it appears that (1, 0) is the solution point. You can confirm this by substituting in *both* equations.

Figure 5.2

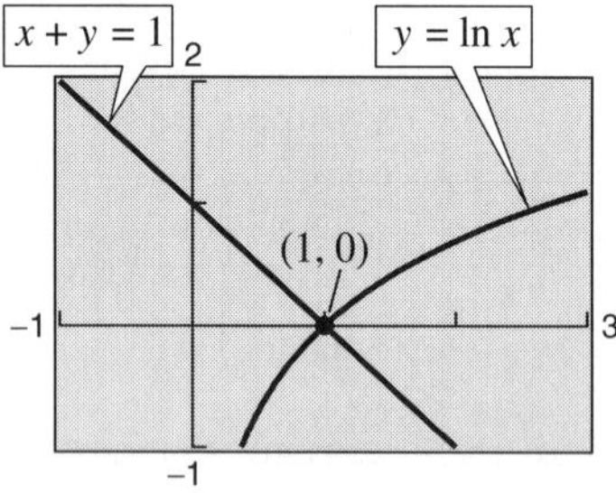

Check

$$0 = \ln 1 \quad \text{Equation 1 checks. ✓}$$
$$1 + 0 = 1 \quad \text{Equation 2 checks. ✓}$$

Applications

The total cost C of producing x units of a product typically has two components: the initial cost and the cost per unit. When enough units have been sold so that the total revenue R equals the total cost, the sales are said to have reached the **break-even point.** You will find that the break-even point corresponds to the point of intersection of the cost and revenue curves.

Real Life

EXAMPLE 6 **An Application: Break-Even Analysis**

A small business invests \$10,000 in equipment to produce a product. Each unit of the product costs \$0.65 to produce and is sold for \$1.20. How many items must be sold before the business breaks even?

Solution

The total cost of producing x units is

$$\text{Total cost} = \text{Cost per unit} \cdot \text{Number of units} + \text{Initial cost}$$

$$C = 0.65x + 10{,}000. \qquad \text{Equation 1}$$

The revenue obtained by selling x units is

$$\text{Total revenue} = \text{Price per unit} \cdot \text{Number of units}$$

$$R = 1.2x. \qquad \text{Equation 2}$$

Because the break-even point occurs when $R = C$, you have

$$1.2x = 0.65x + 10{,}000 \qquad \text{Equate } R \text{ and } C.$$

$$0.55x = 10{,}000 \qquad \text{Subtract } 0.65x \text{ from both sides.}$$

$$x = \frac{10{,}000}{0.55} \qquad \text{Divide both sides by 0.55.}$$

$$x \approx 18{,}182 \text{ units.} \qquad \text{Use a calculator.}$$

Figure 5.3

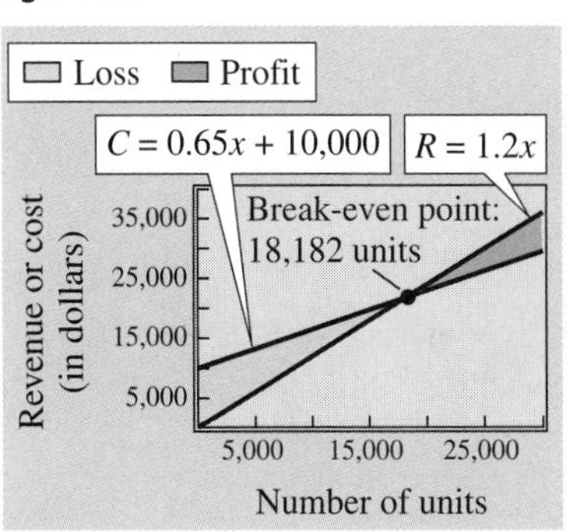

Note in Figure 5.3 that sales less than the break-even point correspond to an overall loss, whereas sales greater than the break-even point correspond to a profit. Verify the break-even point using the intersection feature of a graphing utility.

Note Another way to view the solution in Example 6 is to consider the profit function $P = R - C$. The break-even point occurs when the profit is 0, which is the same as saying that $R = C$.

Real Life

EXAMPLE 7 State Populations

From 1985 to 1993, the population of Arizona was increasing at a faster rate than the population of South Carolina. Two models that approximate the populations P are

$$P = 2785.8 + 88.8t \quad \text{Arizona}$$

$$P = 3079.3 + 42.9t \quad \text{South Carolina}$$

where $t = 5$ represents 1985 (see Figure 5.4). According to these two models, when would you expect the population of Arizona to have exceeded the population of South Carolina? (Source: U.S. Bureau of the Census)

Figure 5.4

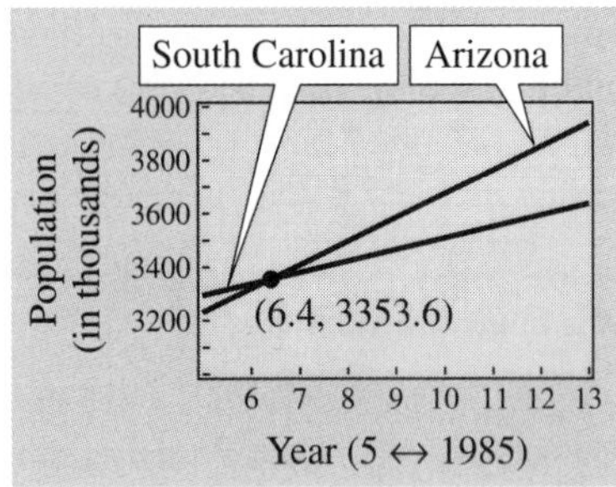

Solution

Because the first equation has already been solved for P in terms of t, you can substitute this value into the second equation and solve for t, as follows.

$$2785.8 + 88.8t = 3079.3 + 42.9t$$

$$88.8t - 42.9t = 3079.3 - 2785.8$$

$$45.9t = 293.5$$

$$t \approx 6.4$$

Thus, from the given models, you would expect that the population of Arizona exceeded the population of South Carolina sometime during 1986.

Group Activity — *Points of Intersection*

In this section, you learned that the graphs of two equations can intersect at zero, one, or more points. Use a graphing utility to graph each of the following systems. Decide if the system has no solution, one solution, two solutions, or more than two solutions.

a. $-0.5x^2 + 0.25x + y = -1.5$
$1.75x + y = -2.625$

b. $-0.5x^2 + 0.25x + y = -1.5$
$-1.25x + y = -4.5$

c. $-0.5x^2 + 0.25x + y = -1.5$
$3.5x + y = 2.25$

Create three other systems of equations, one with no solution, one with one solution, and one with two solutions. Trade your systems with a group member. Solve and compare your results.

5.1 /// EXERCISES

In Exercises 1–10, solve the system by the method of substitution. Check your solution graphically.

1. $2x + y = 6$
$-x + y = 0$

2. $x - y = -4$
$x + 2y = 5$

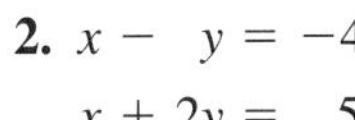

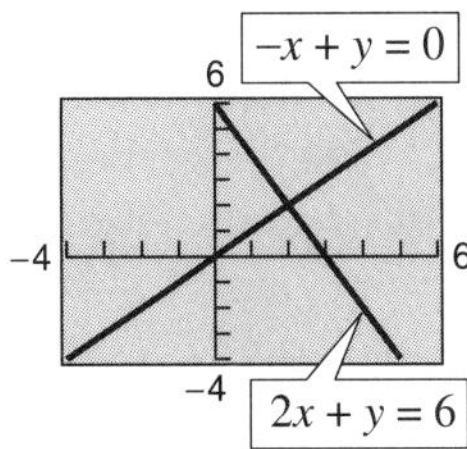

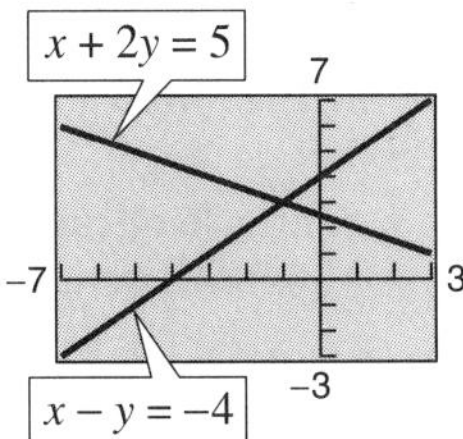

3. $x - y = -4$
$x^2 - y = -2$

4. $3x + y = 2$
$x^3 + y = 0$

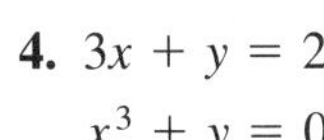

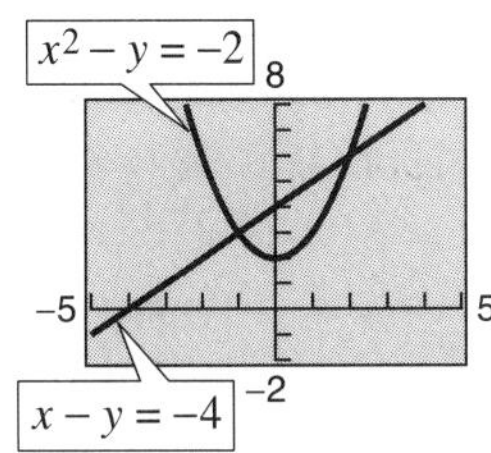

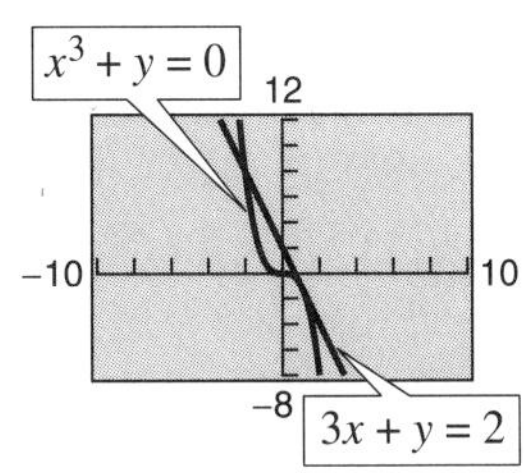

5. $x - 3y = 15$
$x^2 + y^2 = 25$

6. $x + y = 0$
$x^3 - 5x - y = 0$

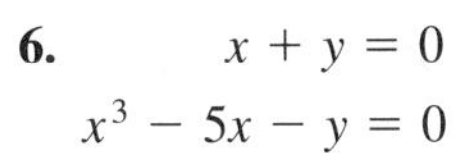

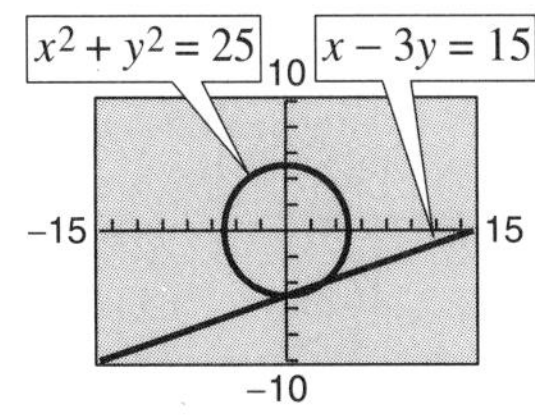

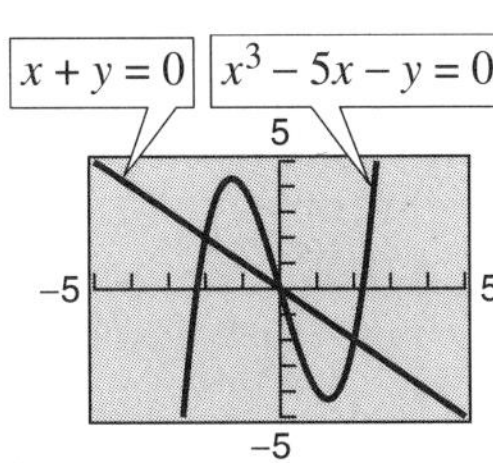

7. $x^2 + y = 0$
$x^2 - 4x - y = 0$

8. $y = -2x^2 + 2$
$y = 2(x^4 - 2x^2 + 1)$

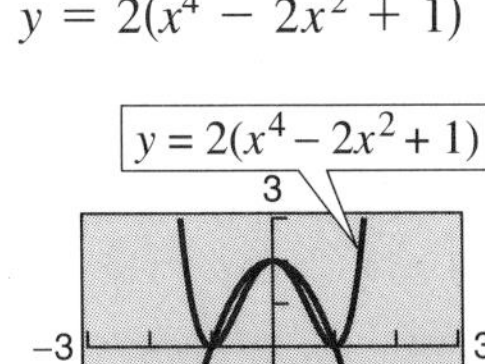

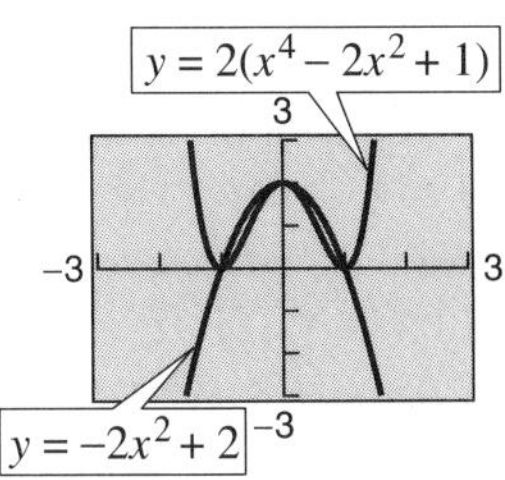

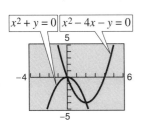

9. $x - 6y = -8$
$x^2 - 4y^3 = 0$

10. $y = x^3 - 3x^2 + 4$
$y = -2x + 4$

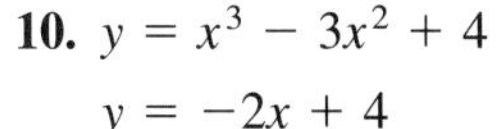

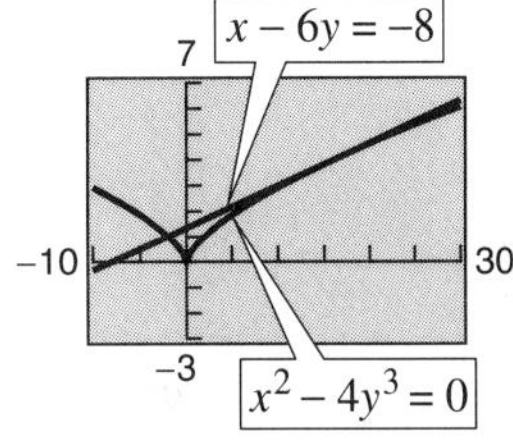

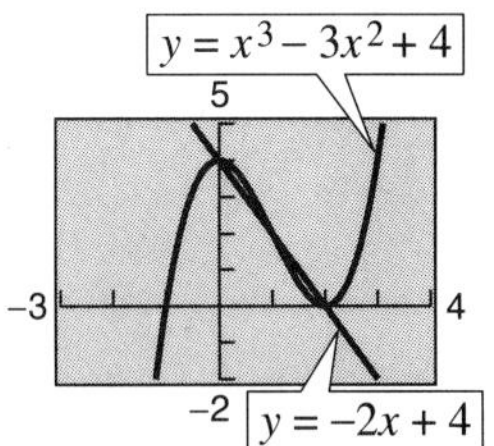

In Exercises 11–22, solve the system by the method of substitution. Use a graphing utility to verify your result.

11. $x - y = 0$
$5x - 3y = 10$

12. $x + 2y = 1$
$5x - 4y = -23$

13. $2x - y + 2 = 0$
$4x + y - 5 = 0$

14. $6x - 3y - 4 = 0$
$x + 2y - 4 = 0$

15. $1.5x + 0.8y = 2.3$
$0.3x - 0.2y = 0.1$

16. $30x - 40y - 33 = 0$
$10x + 20y - 21 = 0$

17. $\frac{1}{5}x + \frac{1}{2}y = 8$
$x + y = 20$

18. $\frac{1}{2}x + \frac{3}{4}y = 10$
$\frac{3}{4}x - y = 4$

19. $2x - y = 4$
$-4x + 2y = -12$

20. $-\frac{2}{3}x + y = -2$
$2x - 3y = 6$

21. $x - y = 0$
$2x + y = 0$

22. $x - 2y = 0$
$3x - y = 0$

In Exercises 23–28, solve the system of equations graphically.

23. $-x + 2y = 2$
$3x + y = 15$

24. $x + y = 0$
$3x - 2y = 10$

25. $x - 3y = -2$
$5x + 3y = 17$

26. $-x + 2y = 1$
$x - y = 2$

27. $x + y = 4$
$x^2 + y^2 - 4x = 0$

28. $x - y + 3 = 0$
$x^2 - 4x + 7 = y$

In Exercises 29–40, use a graphing utility to approximate all points of intersection of the graph of the system of equations.

29. $7x + 8y = 24$
$x - 8y = 8$

30. $x - y = 0$
$5x - 2y = 6$

31. $2x - y + 3 = 0$
$x^2 + y^2 - 4x = 0$

32. $3x - 2y = 0$
$x^2 + y^2 = 4$

33. $x^2 + y^2 = 8$
$y = x^2$

34. $x^2 + y^2 = 25$
$(x - 8)^2 + y^2 = 41$

35. $y = e^x$
$x - y + 1 = 0$

36. $x + 2y = 8$
$y = \log_2 x$

37. $y = \sqrt{x}$
$y = x$

38. $x - y = 3$
$x - y^2 = 1$

39. $x^2 + y^2 = 169$
$x^2 - 8y = 104$

40. $x^2 + y^2 = 4$
$2x^2 - y = 2$

In Exercises 41–52, solve the system graphically or algebraically. Explain why you chose the method you used.

41. $y = 2x$
$y = x^2 + 1$

42. $x + y = 4$
$x^2 + y = 2$

43. $3x - 7y + 6 = 0$
$x^2 - y^2 = 4$

44. $x^2 + y^2 = 25$
$2x + y = 10$

45. $x - 2y = 4$
$x^2 - y = 0$

46. $y = (x + 1)^3$
$y = \sqrt{x - 1}$

47. $y - e^{-x} = 1$
$y - \ln x = 3$

48. $y = x^3 - 2x^2 + x - 1$
$y = -x^2 + 3x - 1$

49. $y = x^4 - 2x^2 + 1$
$y = 1 - x^2$

50. $x^2 + y = 4$
$e^x - y = 0$

51. $xy - 1 = 0$
$2x - 4y + 7 = 0$

52. $x - 2y = 1$
$y = \sqrt{x - 1}$

The *Interactive* CD-ROM contains step-by-step solutions to all odd-numbered Section and Review Exercises. It also provides Tutorial Exercises, which link to Guided Examples for additional help.

53. *Think About It* When solving a system of equations by substitution, how do you recognize that the system has no solution?

54. *Essay* Write a brief paragraph describing any advantages of substitution over the graphical method of solving a system of equations.

Break-Even Analysis **In Exercises 55–58, use a graphing utility to graph the cost and revenue functions in the same viewing rectangle. Find the sales x necessary to break even ($R = C$) and the corresponding revenue R obtained by selling x units. (Round to the nearest whole unit.)**

	Cost	*Revenue*
55.	$C = 8650x + 250{,}000$	$R = 9950x$
56.	$C = 5.5\sqrt{x} + 10{,}000$	$R = 3.29x$
57.	$C = 2.65x + 350{,}000$	$R = 4.15x$
58.	$C = 0.08x + 50{,}000$	$R = 0.25x$

59. *Break-Even Point* A small business invests \$16,000 to produce an item that will sell for \$5.95. Each unit can be produced for \$3.45.

(a) Write cost and revenue functions for x units produced and sold.

(b) Use a graphing utility to graph the cost and revenue functions. Use the graph to approximate the number of units that must be sold to break even.

(c) Verify the result of part (b) algebraically.

60. *Break-Even Point* A small business has an initial investment of \$5000. The unit cost of the product is \$21.60, and the selling price is \$34.10. How many units must be sold to break even?

61. *Investment Portfolio* A total of \$20,000 is invested in two funds paying 6.5% and 8.5% simple interest. The 6.5% investment has a lower risk. The investor wants a yearly interest income of \$1600 from the two investments. What is the most that can be invested at 6.5% to meet this requirement?

62. *Investment Portfolio* A total of $25,000 is invested in two funds paying 6% and 8.5% simple interest. The 6% investment has a lower risk. The investor wants a yearly interest income of $2000 from the two investments.

(a) Write a system of equations in which one equation represents the total amount invested and the other equation represents the $2000 required in interest. Let x and y represent the amounts invested at 6% and 8.5%, respectively.

(b) Use a graphing utility to graph the two equations. As the amount invested at 6% increases, how does the amount invested at 8.5% change and how does the amount of interest change? Explain.

(c) What is the most that can be invested at 6% to meet the requirement of $2000 per year in interest?

63. *Choice of Two Jobs* You are offered two different jobs selling dental supplies.

- One company offers a straight commission of 6% of sales.
- The other company offers a salary of $250 per week plus 3% of sales.

How much would you have to sell in a week in order to make the straight commission offer better?

64. *Choice of Two Jobs* You are offered two different jobs selling college textbooks.

- One company offers an annual salary of $20,000 plus a year-end bonus of 1% of your total sales.
- The other company offers an annual salary of $15,000 plus a year-end bonus of 2% of your total sales.

Determine the annual sales that make the second offer better.

65. *Market Equilibrium* The supply and demand curves for a business dealing with wheat are given by

Supply: $p = 1.45 + 0.00014x^2$

Demand: $p = (2.388 - 0.007x)^2$

where p is the price in dollars per bushel and x is the quantity in bushels per day. Use a graphing utility to graph the supply and demand equations and find the market equilibrium. (The *market equilibrium* is the point of intersection of the graphs for $x > 0$.)

66. *Log Volume* You are offered two different rules for estimating the number of board feet in a log that is 16 feet long. One is the *Doyle Log Rule* and is modeled by

$$V = (D - 4)^2, \qquad 5 \le D \le 40$$

and the other is the *Scribner Log Rule* and is modeled by

$$V = 0.79D^2 - 2D - 4, \qquad 5 \le D \le 40$$

where D is the diameter of the log and V is its volume in board feet.

(a) Use a graphing utility to graph the log rules in the same viewing rectangle.

(b) For what diameter do the two rules agree?

(c) If you were selling large logs, which rule would you use? Explain your reasoning.

Geometry **In Exercises 67–70, find the dimensions of the rectangle meeting the specified conditions.**

	Perimeter	*Condition*
67.	30 meters	The length is 3 meters greater than the width.
68.	280 centimeters	The width is 20 centimeters less than the length.
69.	42 inches	The width is three-fourths the length.
70.	210 feet	The length is one and one-half times the width.

71. *Geometry* What are the dimensions of a rectangular tract of land if its perimeter is 40 miles and its area is 96 square miles?

72. *Geometry* What are the dimensions of an isosceles right triangle with a 2-inch hypotenuse and an area of 1 square inch?

73. *Exploration* Find an equation of a line whose graph intersects the graph of the parabola $y = x^2$ at the following numbers of points. (There is more than one correct answer.)

(a) Two points

(b) One point

(c) No points

74. *Hyperbolic Mirror* In a hyperbolic mirror, light rays directed to one focus will be reflected to the other focus. The mirror in the figure has the equation

$$\frac{x^2}{25} - \frac{y^2}{36} = 1.$$

At which point on the mirror will light from the point (0, 10) reflect to the focus?

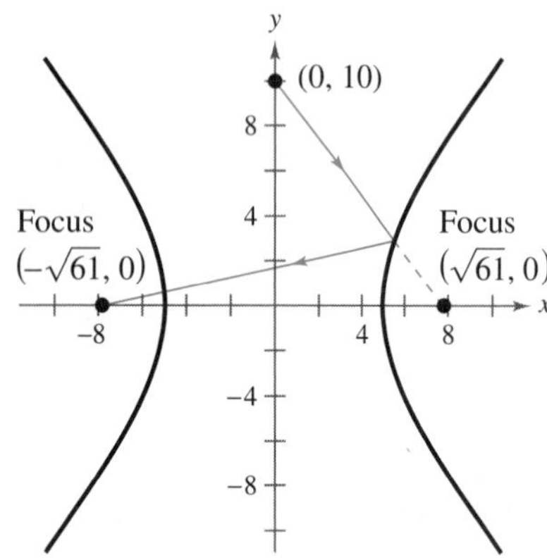

75. *Data Analysis* The table gives the numbers y of millions of short tons of newsprint produced in the years 1990 through 1993 in the United States. (Source: American Paper Institute)

Year	1990	1991	1992	1993
y	6.6	6.8	7.1	7.1

Let t represent the time in years, with $t = 0$ corresponding to 1990.

(a) Use the regression capabilities of a graphing utility to fit a linear model and a quadratic model to the data.

(b) Use the graphing utility to graph the data and the two models in the same viewing rectangle.

(c) Approximate the points of intersection of the graphs of the models.

(d) Use the models to predict newsprint production in 1994. Which model do you think gives the more accurate prediction? Explain.

76. *Conjecture*

(a) Use a graphing utility to graph the system of equations

$$y = b^x$$
$$y = x^b$$

for $b = 2$ and $b = 4$.

(b) For a fixed value of $b > 1$, make a conjecture about the number of points of intersection of the graphs in part (a).

Review **Solve Exercises 77–82 as a review of the skills and problem-solving techniques you learned in previous sections. Find the general form of the equation of the line through the two points.**

77. $(-2, 7), (5, 5)$

78. $(3.5, 4), (10, 6)$

79. $(6, 3), (10, 3)$

80. $(4, -2), (4, 5)$

81. $\left(\frac{3}{5}, 0\right), (4, 6)$

82. $\left(-\frac{7}{3}, 8\right), \left(\frac{5}{2}, \frac{1}{2}\right)$

5.2 Systems of Linear Equations in Two Variables

The Method of Elimination / Graphical Interpretation of Solutions / Applications

The Method of Elimination

In Section 5.1, you studied two methods for solving a system of equations: substitution and graphing. Now you will study the **method of elimination.** The key step in this method is to obtain, for one of the variables, coefficients that differ only in sign so that *adding* the equations eliminates the variable.

$$3x + 5y = 7 \qquad \text{Equation 1}$$
$$-3x - 2y = -1 \qquad \text{Equation 2}$$
$$3y = 6 \qquad \text{Add equations.}$$

Note that by adding the two equations, you eliminate the variable x and obtain a single equation in y. Solving this equation for y produces $y = 2$, which you can then back-substitute into one of the original equations to solve for x.

EXAMPLE 1 The Method of Elimination

Solve the system of linear equations.

$$3x + 2y = 4 \qquad \text{Equation 1}$$
$$5x - 2y = 8 \qquad \text{Equation 2}$$

Solution

You can eliminate y by adding the two equations.

$$3x + 2y = 4 \qquad \text{Equation 1}$$
$$5x - 2y = 8 \qquad \text{Equation 2}$$
$$8x = 12 \qquad \text{Add equations.}$$

Therefore, $x = \frac{3}{2}$. By back-substituting, you can solve for y.

$$3x + 2y = 4 \qquad \text{Equation 1}$$
$$3\left(\tfrac{3}{2}\right) + 2y = 4 \qquad \text{Substitute } \tfrac{3}{2} \text{ for } x.$$
$$y = -\tfrac{1}{4} \qquad \text{Solve for } y.$$

The solution is $\left(\frac{3}{2}, -\frac{1}{4}\right)$. You can check the solution *algebraically* by substituting into the original system, or *graphically,* as shown in Figure 5.5.

Figure 5.5

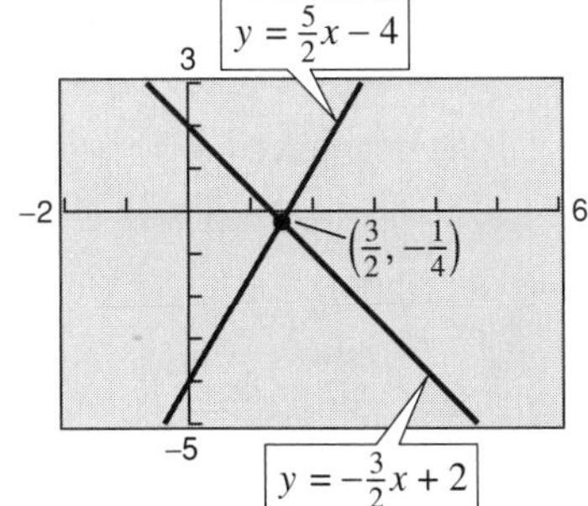

Note Try using the method of substitution to solve the system given in Example 1. Which method do you think is easier? Many people find that the method of elimination is more efficient.

EXAMPLE 2 **The Method of Elimination**

Solve the system of linear equations.

$$2x - 3y = -7 \quad \text{Equation 1}$$
$$3x + y = -5 \quad \text{Equation 2}$$

Solution

For this system, you can obtain coefficients that differ only in sign by multiplying Equation 2 by 3.

$$2x - 3y = -7 \quad \Rightarrow \quad 2x - 3y = -7 \quad \text{Equation 1}$$
$$3x + y = -5 \quad \Rightarrow \quad 9x + 3y = -15 \quad \text{Multiply Equation 2 by 3.}$$
$$11x = -22 \quad \text{Add equations.}$$

Thus, you can see that $x = -2$. By back-substituting this value of x into Equation 1, you can solve for y, as follows.

$$2x - 3y = -7 \quad \text{Equation 1}$$
$$2(-2) - 3y = -7 \quad \text{Substitute } -2 \text{ for } x.$$
$$-3y = -3 \quad \text{Collect like terms.}$$
$$y = 1 \quad \text{Solve for } y.$$

The solution is $(-2, 1)$. You can check the solution algebraically, as follows, or graphically, as shown in Figure 5.6.

Check

$$2(-2) - 3(1) \stackrel{?}{=} -7 \quad \text{Substitute into Equation 1.}$$
$$-4 - 3 = -7 \quad \text{Equation 1 checks. } \checkmark$$
$$3(-2) + 1 \stackrel{?}{=} -5 \quad \text{Substitute into Equation 2.}$$
$$-6 + 1 = -5 \quad \text{Equation 2 checks. } \checkmark$$

Study Tip

To obtain coefficients (for one of the variables) that differ only in sign, you often need to multiply one or both of the equations by suitably chosen constants.

Figure 5.6

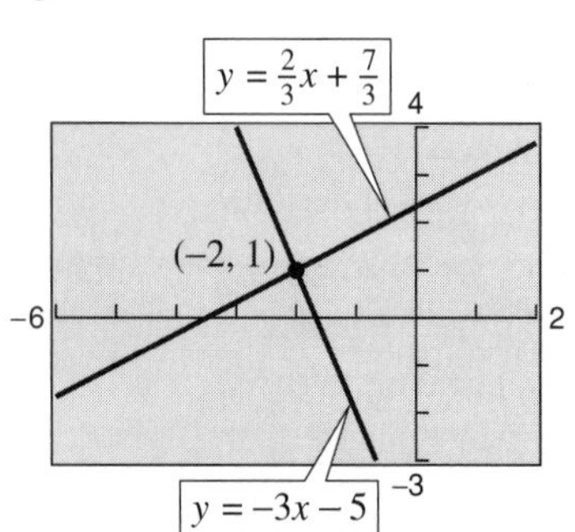

In Example 2, the two systems of linear equations

$$\begin{aligned} 2x - 3y &= -7 \\ 3x + y &= -5 \end{aligned} \quad \text{and} \quad \begin{aligned} 2x - 3y &= -7 \\ 9x + 3y &= -15 \end{aligned}$$

are called **equivalent** because they have precisely the same solution set. The operations that can be performed on a system of linear equations to produce an equivalent system are (1) interchanging any two equations, (2) multiplying an equation by a nonzero constant, and (3) adding a multiple of one equation to any other equation in the system.

EXPLORATION

Use a graphing utility to graph each system of equations.

a. $y = 5x + 1$
$y - x = -5$

b. $3y = 4x - 1$
$-8x + 2 = -6y$

c. $2y = -x + 3$
$-4 = y + \frac{1}{2}x$

Determine the number of solutions each system has. Explain your reasoning.

The Method of Elimination

To use the **method of elimination** to solve a system of two linear equations in x and y, use the following steps.

1. Obtain coefficients for x (or y) that differ only in sign by multiplying all terms of one or both equations by suitably chosen constants.
2. Add the equations to eliminate one variable, and solve the resulting equation.
3. Back-substitute the value obtained in Step 2 into either of the original equations and solve for the other variable.
4. Check your solution in both of the original equations.

EXAMPLE 3 **The Method of Elimination**

Solve the system of linear equations.

$$5x + 3y = 9 \quad \text{Equation 1}$$
$$2x - 4y = 14 \quad \text{Equation 2}$$

Solution

You can obtain coefficients that differ only in sign by multiplying Equation 1 by 4 and multiplying Equation 2 by 3.

$$5x + 3y = 9 \implies 20x + 12y = 36 \quad \text{Multiply Equation 1 by 4.}$$
$$2x - 4y = 14 \implies 6x - 12y = 42 \quad \text{Multiply Equation 2 by 3.}$$
$$26x = 78 \quad \text{Add equations.}$$

Figure 5.7

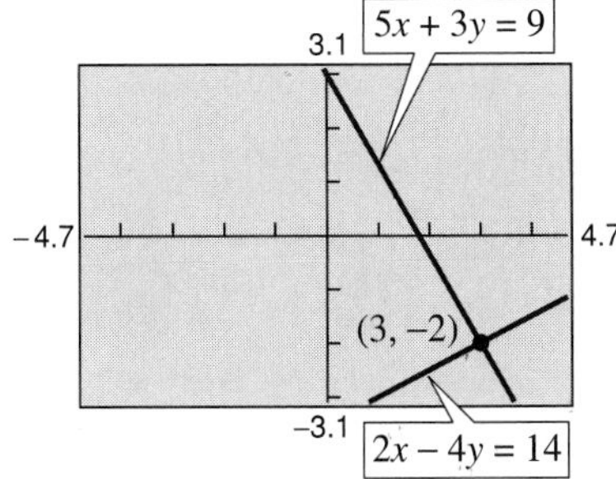

From this equation, you can see that $x = 3$. By back-substituting this value of x into Equation 2, you can solve for y, as follows.

$$2x - 4y = 14 \quad \text{Equation 2}$$
$$2(3) - 4y = 14 \quad \text{Substitute 3 for } x.$$
$$-4y = 8 \quad \text{Collect like terms.}$$
$$y = -2 \quad \text{Solve for } y.$$

The solution is $(3, -2)$. You can check the solution algebraically by substituting into the original system, or graphically, as shown in Figure 5.7.

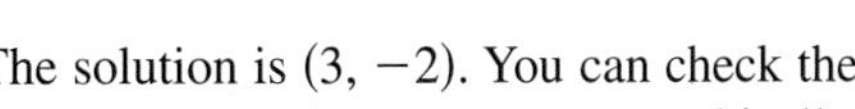

Graphical Interpretation of Solutions

It is possible for a *general* system of equations to have exactly one solution, two or more solutions, or no solution. If a system of *linear* equations has two different solutions, it must have an *infinite* number of solutions. To see why this is true, consider the following graphical interpretations of a system of two linear equations in two variables. (Remember that the graph of a linear equation in two variables is a straight line.)

Study Tip

Keep in mind that the terminology and methods discussed in this and the following section apply only to systems of *linear* equations.

Graphical Interpretation of Solutions

For a system of two linear equations in two variables, the number of solutions is given by one of the following.

Number of Solutions	*Graphical Interpretation*
1. Exactly one solution	The two lines intersect at one point.
2. Infinitely many solutions	The two lines are identical.
3. No solution	The two lines are parallel.

The graphical interpretations presented above are further illustrated in Figure 5.8. In the second graph, note that the two lines coincide. This case is illustrated in Example 5. In the third graph, note that the two lines are parallel. This case is illustrated in Example 4.

Figure 5.8

Consistent	Consistent	Inconsistent
Two lines that intersect	Two lines that coincide	Two parallel lines
One point of intersection	Infinitely many points	No point of intersection

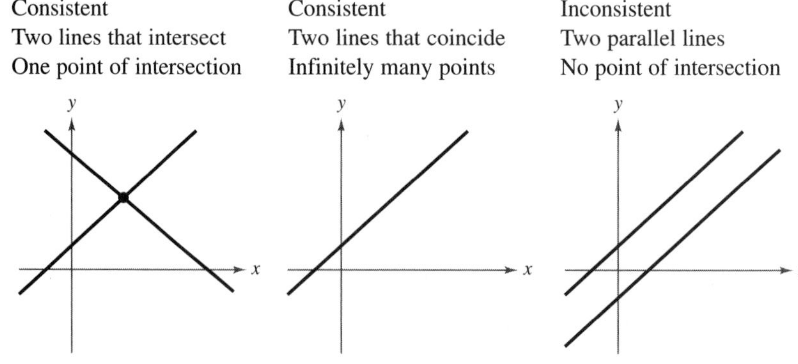

A system of linear equations is called **consistent** if it has at least one solution, and it is called **inconsistent** if it has no solution. In Examples 4 and 5, note how you can use the method of elimination to determine that a system of linear equations has no solution or infinitely many solutions.

Figure 5.9

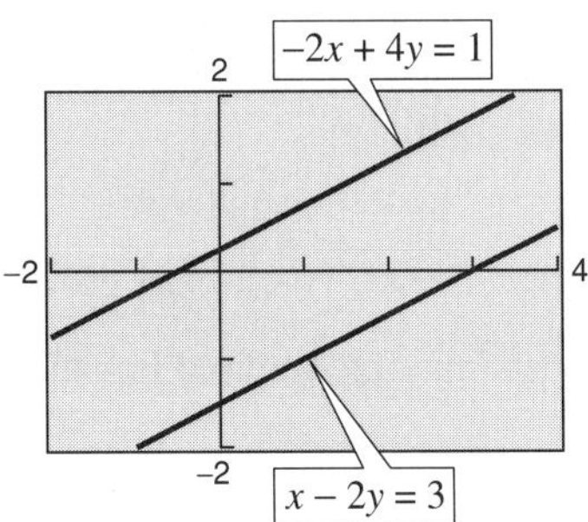

EXAMPLE 4 The Method of Elimination: No-Solution Case

Solve the system of linear equations.

$$x - 2y = 3 \quad \text{Equation 1}$$
$$-2x + 4y = 1 \quad \text{Equation 2}$$

Solution

To obtain coefficients that differ only in sign, multiply Equation 1 by 2.

$$x - 2y = 3 \quad \Longrightarrow \quad 2x - 4y = 6 \quad \text{Multiply Equation 1 by 2.}$$
$$-2x + 4y = 1 \quad \Longrightarrow \quad -2x + 4y = 1 \quad \text{Equation 2}$$
$$0 = 7 \quad \text{False statement}$$

Because there are no values of x and y for which $0 = 7$, you can conclude that the system is inconsistent and has no solution. The lines corresponding to the two equations in this system are shown in Figure 5.9. Note that the two lines are parallel, and therefore have no point of intersection.

In Example 4, note that the occurrence of a false statement, such as $0 = 7$, indicates that the system has no solution. In the next example, note that the occurrence of a statement that is true for all values of the variables—in this case, $0 = 0$—indicates that the system has infinitely many solutions.

Note In Example 5, notice that you could have reached the conclusion that the equations are equivalent by multiplying Equation 1 by 2 to obtain Equation 2.

EXAMPLE 5 The Method of Elimination: Many-Solutions Case

Solve the system of linear equations.

$$2x - y = 1 \quad \text{Equation 1}$$
$$4x - 2y = 2 \quad \text{Equation 2}$$

Solution

To obtain coefficients that differ only in sign, multiply Equation 2 by $-\frac{1}{2}$.

$$2x - y = 1 \quad \Longrightarrow \quad 2x - y = 1 \quad \text{Equation 1}$$
$$4x - 2y = 2 \quad \Longrightarrow \quad -2x + y = -1 \quad \text{Multiply Equation 2 by } -\tfrac{1}{2}.$$
$$0 = 0 \quad \text{Add equations.}$$

Because the two equations turn out to be equivalent (have the same solution set), you can conclude that the system has infinitely many solutions. The solution set consists of all points (x, y) lying on the line

$$2x - y = 1. \quad \text{See Figure 5.10.}$$

Figure 5.10

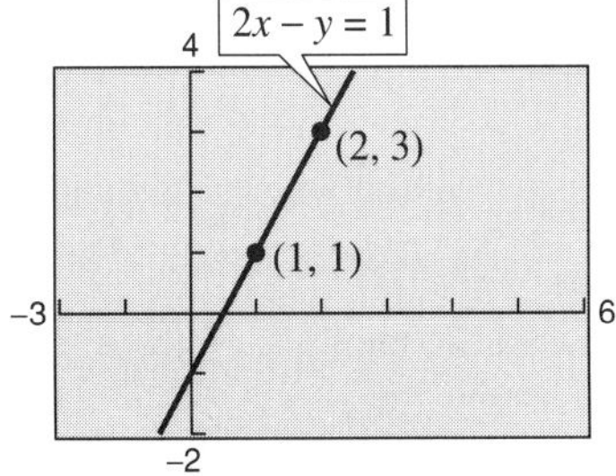

The general solution of the linear system

$$ax + by = c$$
$$dx + ey = f$$

is $x = (ce - bf)/(ae - db)$ and $y = (af - cd)/(ae - db)$. If $ae - db = 0$, the system does not have a unique solution. A *TI-82* or *TI-83* program for solving such a system is given below. Programs for other graphing calculator models may be found in the appendix. Try using this program to check the solution of the system in Example 6.

```
PROGRAM:SOLVE
:Disp "AX+BY=C"
:Prompt A
:Prompt B
:Prompt C
:Disp "DX+EY=F"
:Prompt D
:Prompt E
:Prompt F
:If AE-DB=0
:Then
:Disp "NO UNIQUE
SOLUTION"
:Else
:(CE-BF)/(AE-DB)→X
:(AF-CD)/(AE-DB)→Y
:Disp X
:Disp Y
:End
```

Example 6 illustrates a strategy for solving a system of linear equations that has decimal coefficients.

EXAMPLE 6 A Linear System Having Decimal Coefficients

Solve the system of linear equations.

$$0.02x - 0.05y = -0.38 \quad \text{Equation 1}$$
$$0.03x + 0.04y = 1.04 \quad \text{Equation 2}$$

Solution

Because the coefficients are two-place decimals, begin by multiplying each equation by 100 to produce a system with integer coefficients.

$$2x - 5y = -38 \quad \text{Revised Equation 1}$$
$$3x + 4y = 104 \quad \text{Revised Equation 2}$$

Now, to obtain coefficients that differ only in sign, multiply Revised Equation 1 by 3 and multiply Revised Equation 2 by -2.

$$2x - 5y = -38 \Rightarrow 6x - 15y = -114 \quad \text{Multiply Revised Equation 1 by 3.}$$
$$3x + 4y = 104 \Rightarrow -6x - 8y = -208 \quad \text{Multiply Revised Equation 2 by -2.}$$
$$-23y = -322 \quad \text{Add equations.}$$

Thus, you can conclude that

$$y = \frac{-322}{-23} = 14.$$

Back-substituting this value into Revised Equation 2 produces the following.

$$3x + 4y = 104 \quad \text{Revised Equation 2}$$
$$3x + 4(14) = 104 \quad \text{Substitute 14 for } y.$$
$$3x = 48 \quad \text{Collect like terms.}$$
$$x = 16 \quad \text{Solve for } x.$$

The solution is (16, 14). Check this in the original system, as follows.

Check

$$0.02(16) - 0.05(14) \stackrel{?}{=} -0.38 \quad \text{Substitute into Equation 1.}$$
$$0.32 - 0.70 = -0.38 \quad \text{Equation 1 checks. ✓}$$

$$0.03(16) + 0.04(14) \stackrel{?}{=} 1.04 \quad \text{Substitute into Equation 2.}$$
$$0.48 + 0.56 = 1.04 \quad \text{Equation 2 checks. ✓}$$

Applications

At this point, you may be asking the question "How can I tell which application problems can be solved using a system of linear equations?" The answer comes from the following considerations.

1. Does the problem involve more than one unknown quantity?
2. Are there two (or more) equations or conditions to be satisfied?

If one or both of these conditions occur, the appropriate mathematical model for the problem may be a system of linear equations. Example 7 shows how to construct such a model.

Real Life

EXAMPLE 7 **An Application of a Linear System**

An airplane flying into a headwind travels the 2000-mile flying distance between two cities in 4 hours and 24 minutes. On the return flight, the same distance is traveled in 4 hours. Find the airspeed of the plane and the speed of the wind, assuming that both remain constant.

Solution

The two unknown quantities are the speeds of the wind and the plane. If r_1 is the speed of the plane and r_2 is the speed of the wind, then

$r_1 - r_2 =$ speed of the plane *against* the wind

$r_1 + r_2 =$ speed of the plane *with* the wind

as shown in Figure 5.11. Using the formula

Distance $=$ (rate)(time)

for these two speeds, you obtain the following equations.

$$2000 = (r_1 - r_2)\left(4 + \frac{24}{60}\right)$$

$$2000 = (r_1 + r_2)(4)$$

Figure 5.11

Original flight

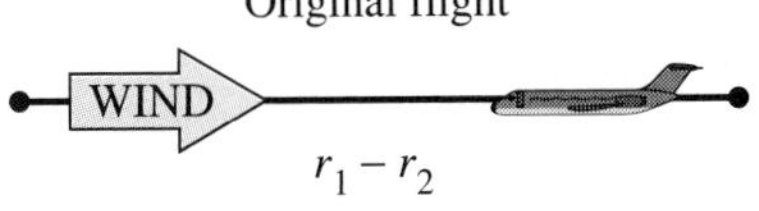

$r_1 - r_2$

Return flight

WIND

$r_1 + r_2$

These two equations simplify as follows.

$$5000 = 11r_1 - 11r_2 \qquad \text{Equation 1}$$

$$500 = r_1 + r_2 \qquad \text{Equation 2}$$

By elimination, the solution is

$$r_1 = \frac{5250}{11} \approx 477.27 \text{ miles per hour} \qquad \text{Speed of plane}$$

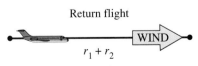

$$r_2 = \frac{250}{11} \approx 22.73 \text{ miles per hour.} \qquad \text{Speed of wind}$$

Real Life

EXAMPLE 8 **Finding the Point of Equilibrium**

The demand and supply functions for a certain type of calculator are given by

$$p = 150 - 0.00001x \qquad \text{Equation 1: demand equation}$$
$$p = 60 + 0.00002x \qquad \text{Equation 2: supply equation}$$

Figure 5.12

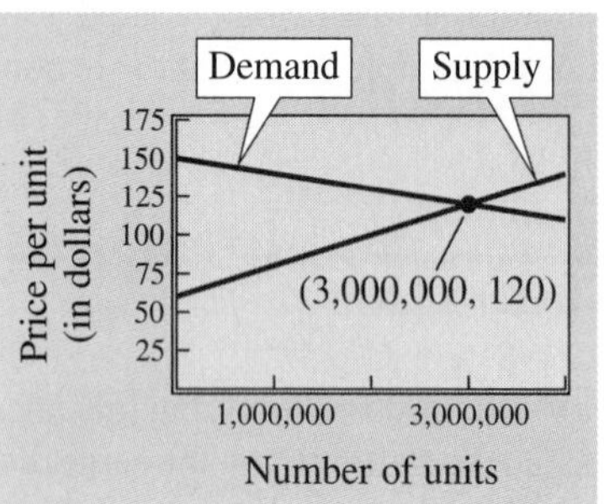

where p is the price in dollars and x is the number of units. Find the point of equilibrium for this market. The point of equilibrium is defined by the price p and number of units x that satisfy both the demand and supply equations.

Solution

Begin by substituting the value of p given in Equation 2 into Equation 1.

$$p = 150 - 0.00001x \qquad \text{Equation 1}$$
$$60 + 0.00002x = 150 - 0.00001x \qquad \text{Substitute } 60 + 0.00002x \text{ for } p.$$
$$0.00003x = 90 \qquad \text{Collect like terms.}$$
$$x = 3{,}000{,}000 \qquad \text{Solve for } x.$$

Thus, the point of equilibrium occurs when the demand and supply are each 3 million units. (See Figure 5.12.) The price that corresponds to this x-value is obtained by back-substituting $x = 3{,}000{,}000$ into either of the original equations. For instance, back-substituting into Equation 1 produces

$$p = 150 - 0.00001(3{,}000{,}000) = 150 - 30 = \$120.$$

Try back-substituting $x = 3{,}000{,}000$ into Equation 2 to see that you obtain the same price.

Note In a free market, the demands for many products are related to the prices of the products. As prices decrease, consumer demand increases and the amounts that producers are able or willing to supply decrease.

Group Activity

Creating Consistent and Inconsistent Systems

For each of the following systems, find the value of k such that the system has infinitely many solutions.

a. $4x + 3y = -8$
$\quad x + ky = -2$

b. $3x - 12y = 9$
$\quad x - 4y = k$

Is it possible to find values of k such that the systems have no solutions? Explain why or why not for each system. Is it possible to find values of k such that the systems have unique solutions? Explain why or why not for each system. Summarize the differences between these two systems of equations.

5.2 /// EXERCISES

In Exercises 1–10, solve the system of equations by the method of elimination. Label each graphed line with its equation.

1. $2x + y = 5$
$x - y = 1$

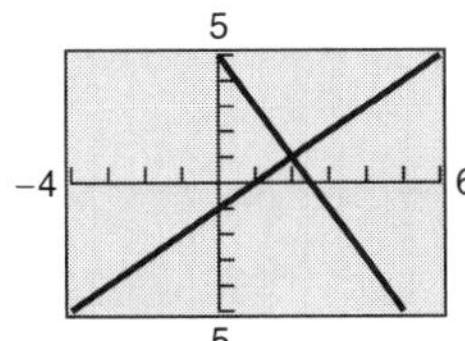

2. $x + 3y = 1$
$-x + 2y = 4$

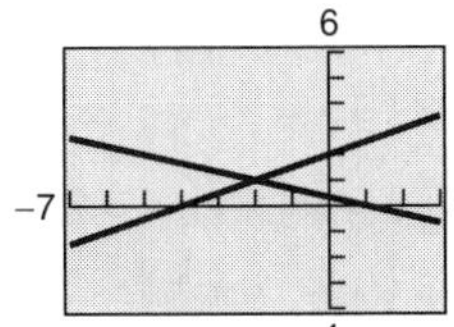

3. $x + y = 0$
$3x + 2y = 1$

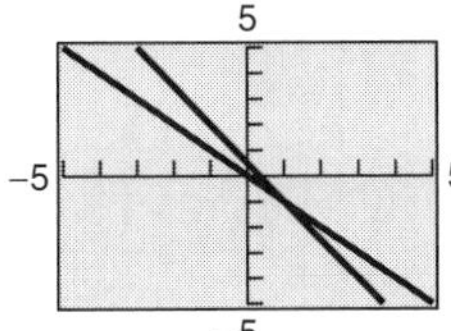

4. $2x - y = 3$
$4x + 3y = 21$

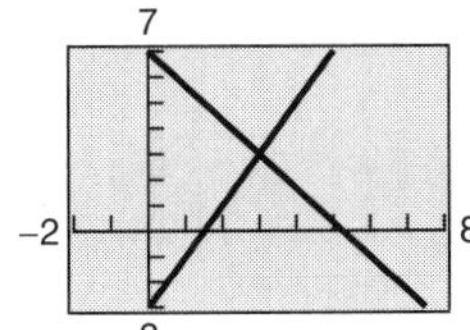

5. $x - y = 2$
$-2x + 2y = 5$

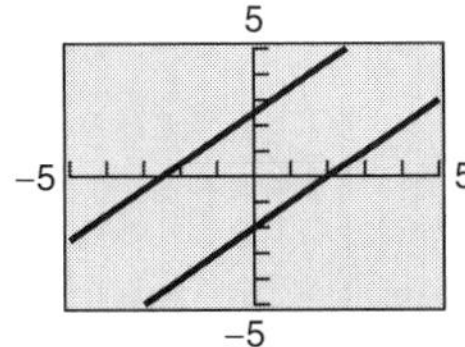

6. $3x + 2y = 3$
$6x + 4y = 14$

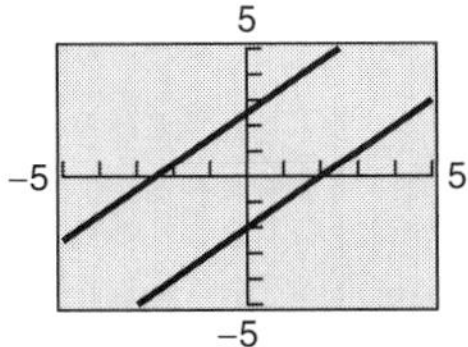

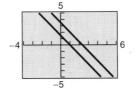

7. $3x - 2y = 5$
$-6x + 4y = -10$

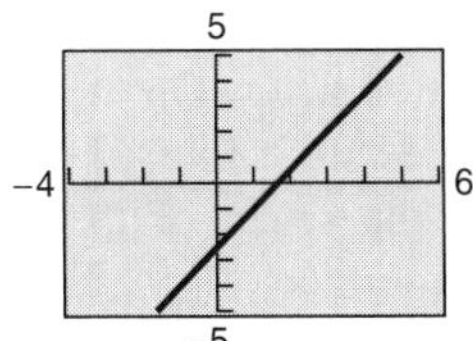

8. $x - 2y = 4$
$6x + 2y = 10$

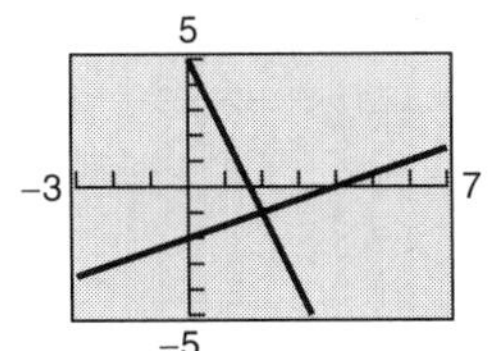

9. $9x + 3y = 1$
$3x - 6y = 5$

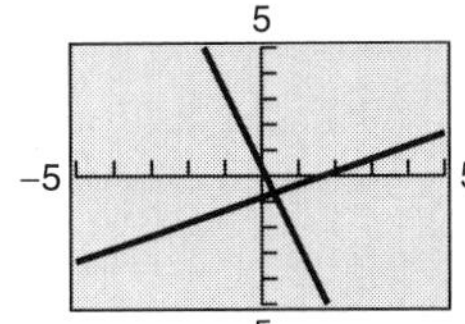

10. $5x + 3y = -18$
$2x - 6y = 1$

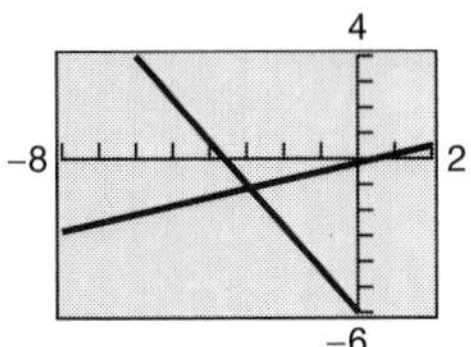

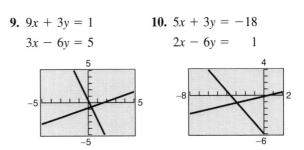

In Exercises 11–20, solve the system by the method of elimination and check any solution algebraically.

11. $x + 2y = 4$
$x - 2y = 1$

12. $3x - 5y = 2$
$2x + 5y = 13$

13. $2x + 3y = 18$
$5x - y = 11$

14. $x + 7y = 12$
$3x - 5y = 10$

15. $3x + 2y = 10$
$2x + 5y = 3$

16. $8r + 16s = 20$
$16r + 50s = 55$

17. $2u + v = 120$
$u + 2v = 120$

18. $5u + 6v = 24$
$3u + 5v = 18$

19. $6r - 5s = 3$
$10r - 12s = 5$

20. $1.8x + 1.2y = 4$
$9x + 6y = 3$

In Exercises 21–30, solve the system by the method of elimination and verify any solution with a graphing utility.

21. $\dfrac{x}{4} + \dfrac{y}{6} = 1$
$x - y = 3$

22. $\frac{2}{3}x + \frac{1}{6}y = \frac{2}{3}$
$4x + y = 4$

23. $\dfrac{x+3}{4} + \dfrac{y-1}{3} = 1$
$2x - y = 12$

24. $\dfrac{x-1}{2} + \dfrac{y+2}{3} = 4$
$x - 2y = 5$

25. $2.5x - 3y = 1.5$
$10x - 12y = 6$

26. $0.02x - 0.05y = -0.19$
$0.03x + 0.04y = 0.52$

27. $0.05x - 0.03y = 0.21$
$0.07x + 0.02y = 0.16$

28. $0.2x - 0.5y = -27.8$
$0.3x + 0.4y = 68.7$

29. $4b + 3m = 3$
$3b + 11m = 13$

30. $3b + 3m = 7$
$3b + 5m = 3$

In Exercises 31–34, use a graphing utility to graph the lines in the system. Use the graphs to determine if the system is consistent or inconsistent. If the system is consistent, determine the number of solutions.

31. $\frac{1}{5}x - \frac{1}{3}y = 1$
$-3x + 5y = 9$

32. $2x + y = 5$
$x - 2y = -1$

33. $2x - 5y = 0$
$x - y = 3$

34. $4x - 6y = 7$
$2x - 3y = 3.5$

In Exercises 35–38, use a graphing utility to graph the two equations. Use the graphs to approximate the solution of the system.

35. $8x + 9y = 42$
$6x - y = 16$

36. $\frac{3}{2}x - \frac{1}{5}y = 8$
$-2x + 3y = 3$

37. $4y = -8$
$7x - 2y = 25$

38. $0.5x + 2.2y = 9$
$6x + 0.4y = -22$

In Exercises 39–42, use any method to solve the system.

39. $3x - 5y = 7$
$2x + y = 9$

40. $-x + 3y = 17$
$4x + 3y = 7$

41. $y = 2x - 5$
$y = 5x - 11$

42. $7x + 3y = 16$
$y = x + 2$

Exploration **In Exercises 43 and 44, find a system of linear equations that has the given solution. (There is more than one correct answer.)**

43. $\left(3, \frac{5}{2}\right)$

44. $(8, -2)$

Think About It **In Exercises 45 and 46, the graphs of the two equations appear to be parallel. Yet, when the system is solved algebraically, it is found that the system does have a solution. Find the solution and explain why it does not appear on the portion of the graph that is shown.**

45. $100y - x = 200$
$99y - x = -198$

46. $21x - 20y = 0$
$13x - 12y = 120$

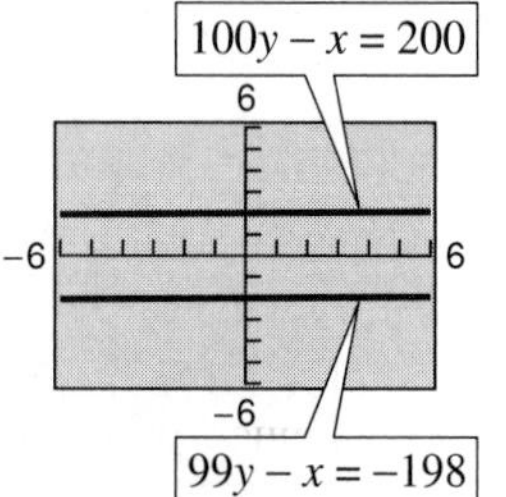

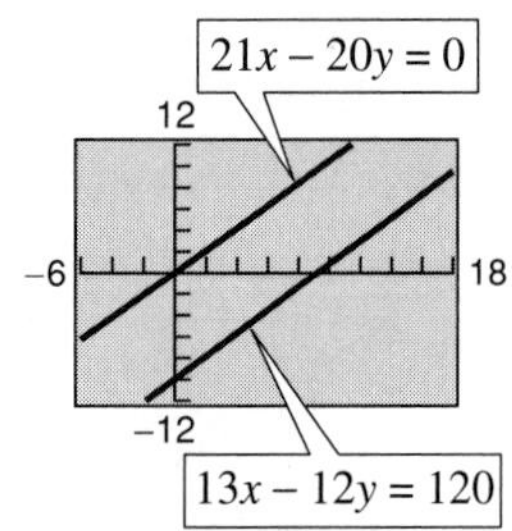

47. *Essay* Briefly explain whether or not it is possible for a consistent system of linear equations to have exactly two solutions.

48. *Think About It* Give examples of (a) a system of linear equations that has no solution and (b) a system that has an infinite number of solutions.

Exploration **In Exercises 49 and 50, find the value of *k* such that the system of equations is inconsistent.**

49. $4x - 8y = -3$
$2x + ky = 16$

50. $15x + 3y = 6$
$-10x + ky = 9$

51. *Airplane Speed* An airplane flying into a headwind travels the 1800-mile flying distance between two cities in 3 hours and 36 minutes. On the return flight, the distance is traveled in 3 hours. Find the airspeed of the plane and the speed of the wind, assuming that both remain constant.

52. *Airplane Speed* Two planes start from the same airport and fly in opposite directions. The second plane starts $\frac{1}{2}$ hour after the first plane, but its speed is 80 kilometers per hour faster. Find the groundspeed of each plane if 2 hours after the first plane departs the planes are 3200 kilometers apart.

53. *Acid Mixture* Ten liters of a 30% acid solution is obtained by mixing a 20% solution with a 50% solution.

(a) Write a system of equations in which one equation represents the amount of final mixture required and the other represents the amount of acid in the final mixture. Let x and y represent the amounts of 20% and 50% solutions, respectively.

(b) Use a graphing utility to graph the two equations in part (a). As the amount of the 20% solution increases, how does the amount of the 50% solution change?

(c) How much of each solution is required to obtain the specified concentration in the final mixture?

54. *Fuel Mixture* Five hundred gallons of 89 octane gasoline is obtained by mixing 87 octane gasoline with 92 octane gasoline. How much of each type must be used?

55. *Investment Portfolio* A total of \$12,000 is invested in two corporate bonds that pay 10.5% and 12% simple interest. The investor wants an annual interest income of \$1350 from the investments. What is the most that can be invested in the 10.5% bond?

56. *Investment Portfolio* A total of \$32,000 is invested in two municipal bonds that pay 5.75% and 6.25% simple interest. The investor wants an annual interest income of \$1900 from the investments. What is the most that can be invested in the 5.75% bond?

57. *Ticket Sales* Five hundred tickets were sold for a certain performance of a play. The tickets for adults and children sold for \$7.50 and \$4.00, respectively, and the total receipts for the performance were \$3312.50. How many of each kind of ticket were sold?

58. *Shoe Sales* On Saturday night, the manager of a shoe store evaluates the receipts of the previous week's sales. Two hundred and forty pairs of two different styles of tennis shoes were sold. One style sold for \$66.95 and the other sold for \$84.95. The total receipts were \$17,652. The cash register that was supposed to record the number of each type of shoe sold malfunctioned. Can you recover the information? If so, how many shoes of each type were sold?

59. *Driving Distances* In a trip of 300 kilometers, two people drive. One person drives three times as far as the other. Find the distance that each person drives.

60. *Truck Scheduling* A contractor hires two trucking companies to haul 1600 tons of crushed stone for a highway construction project. The contracts state that one company is to haul four times as much as the other. Find the amount hauled by each company.

Fitting a Line to Data **In Exercises 61–68, find the least squares regression line $y = ax + b$ for the points**

$$(x_1, y_1), (x_2, y_2), \ldots, (x_n, y_n).$$

To find the line, solve the following system for a and b.

$$nb + \left(\sum_{i=1}^{n} x_i\right)a = \sum_{i=1}^{n} y_i$$

$$\left(\sum_{i=1}^{n} x_i\right)b + \left(\sum_{i=1}^{n} x_i^2\right)a = \sum_{i=1}^{n} x_i y_i$$

Then use the linear regression capabilities of a graphing utility to confirm the result.

61. $5b + 10a = 20.2$
$10b + 30a = 50.1$

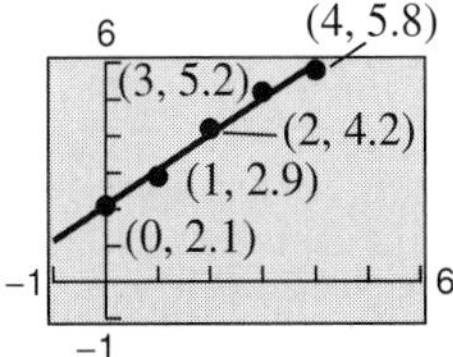

62. $5b + 10a = 11.7$
$10b + 30a = 25.6$

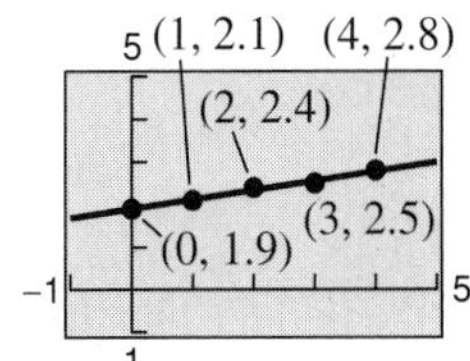

63. $7b + 21a = 35.1$
$21b + 91a = 114.2$

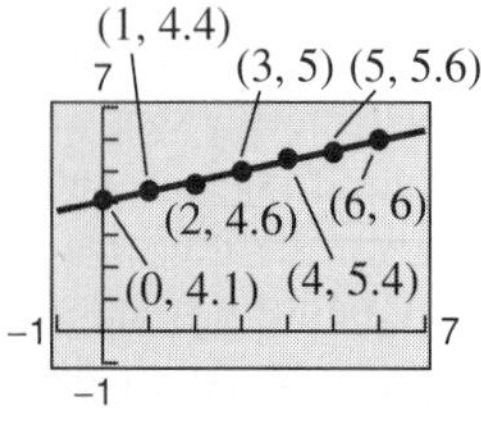

64. $6b + 15a = 23.6$
$15b + 55a = 48.8$

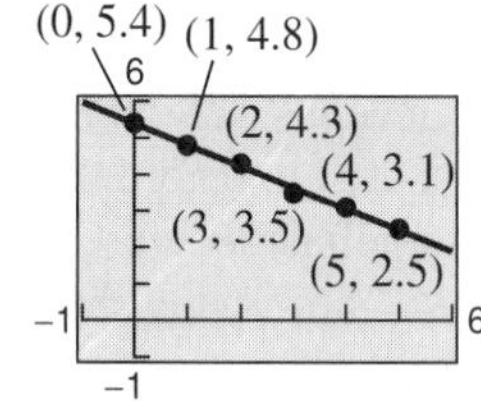

65. $(-2, 0), (0, 1), (2, 3)$

66. $(-3, 0), (-1, 1), (1, 1), (3, 2)$

67. $(0, 4), (1, 3), (1, 1), (2, 0)$

68. $(1, 0), (2, 0), (3, 0), (3, 1), (4, 1), (4, 2), (5, 2), (6, 2)$

69. *Analyzing Data* A farmer used four test plots to determine the relationship between wheat yield in bushels per acre and the amount of fertilizer in hundreds of pounds per acre. The results are given in the table.

Fertilizer, x	1.0	1.5	2.0	2.5
Yield, y	32	41	48	53

Use the technique demonstrated in Exercises 61–68 to find the line that best fits the data. Then use the line to estimate the yield for a fertilizer application of 160 pounds per acre.

70. *Analyzing Data* A store manager wants to know the demand for a certain product as a function of the price. The daily sales for the different prices of the product are given in the table.

Price, x	\$1.00	\$1.25	\$1.50
Demand, y	450	375	330

Use the technique demonstrated in Exercises 61–68 to find the line that best fits the data. Then use the line to predict the demand when the price is \$1.40.

Supply and Demand **In Exercises 71–74, find the point of equilibrium of the demand and supply equations.**

	Demand	*Supply*
71.	$p = 50 - 0.5x$	$p = 0.125x$
72.	$p = 100 - 0.05x$	$p = 25 + 0.1x$
73.	$p = 140 - 0.00002x$	$p = 80 + 0.00001x$
74.	$p = 400 - 0.0002x$	$p = 225 + 0.0005x$

Advanced Applications **In Exercises 75–78, solve the system of equations for u and v. While solving for these variables, consider the transcendental functions as constants. (Systems of this type are found in a course in differential equations.)**

75. $u \sin x + v \cos x = 0$
$u \cos x - v \sin x = \sec x$

76. $u \cos 2x + v \sin 2x = 0$
$u(-2 \sin 2x) + v(2 \cos 2x) = \csc 2x$

77. $ue^x + vxe^x = 0$
$ue^x + v(x + 1)e^x = e^x \ln x$

78. $ue^{2x} + vxe^{2x} = 0$
$u(2e^{2x}) + v(2x + 1)e^{2x} = \dfrac{e^{2x}}{x}$

Review **Solve Exercises 79–82 as a review of the skills and problem-solving techniques you learned in previous sections. Determine the domain of the function.**

79. $f(x) = x^2 - 2x$

80. $g(x) = \sqrt[3]{x - 2}$

81. $h(x) = \sqrt{25 - x^2}$

82. $s(t) = \dfrac{4}{9 - t^2}$

5.3 Multivariable Linear Systems

Row-Echelon Form and Back-Substitution / Gaussian Elimination / Nonsquare Systems / Graphical Interpretation of Three-Variable Systems / Applications

Row-Echelon Form and Back-Substitution

The method of elimination can be applied to a system of linear equations in more than two variables. When elimination is used to solve a system of linear equations, the goal is to rewrite the system in a form to which back-substitution can be applied. To see how this works, consider the following two systems of linear equations.

$$\begin{aligned} x - 2y + 3z &= 9 \\ -x + 3y &= -4 \\ 2x - 5y + 5z &= 17 \end{aligned} \qquad\qquad \begin{aligned} x - 2y + 3z &= 9 \\ y + 3z &= 5 \\ z &= 2 \end{aligned}$$

The system on the right is said to be in **row-echelon form,** which means that it has a "stair-step" pattern with leading coefficients of 1. After comparing the two systems, it should be clear that it is easier to solve the system on the right.

One of the most influential Chinese mathematics books was the *Chui-chang suan-shu* or *Nine Chapters on the Mathematical Art* (written in approximately 250 B.C.). Chapter Eight of the Nine Chapters contained solutions of systems of linear equations using positive and negative numbers. One such system was

$$\begin{aligned} 3x + 2y + 3z &= 39 \\ 2x + 3y + 3z &= 34 \\ 2x + 2y + 3z &= 26 \end{aligned}$$

This system was solved using column operations on a matrix. Matrices (plural for matrix) will be discussed in the next chapter.

EXAMPLE 1 Using Back-Substitution

Solve the system of linear equations.

$$\begin{aligned} x - 2y + 3z &= 9 && \text{Equation 1} \\ y + 3z &= 5 && \text{Equation 2} \\ z &= 2 && \text{Equation 3} \end{aligned}$$

Solution

From Equation 3, you know the value of z. To solve for y, substitute $z = 2$ into Equation 2 to obtain

$$\begin{aligned} y + 3(2) &= 5 && \text{Substitute 2 for } z. \\ y &= -1. && \text{Solve for } y. \end{aligned}$$

Finally, substitute $y = -1$ and $z = 2$ into Equation 1 to obtain

$$\begin{aligned} x - 2(-1) + 3(2) &= 9 && \text{Substitute } -1 \text{ for } y \text{ and 2 for } z. \\ x &= 1. && \text{Solve for } x. \end{aligned}$$

The solution is $x = 1$, $y = -1$, and $z = 2$, which can be written as the **ordered triple** $(1, -1, 2)$. Check this in the original system of equations.

Gaussian Elimination

Two systems of equations are **equivalent** if they have the same solution set. To solve a system that is not in row-echelon form, first convert it to an *equivalent* system that *is* in row-echelon form by using one or more of the elementary row operations shown at the left. This process is called **Gaussian elimination,** after the German mathematician Carl Friedrich Gauss (1777–1855).

Note

Elementary Row Operations

1. Interchange two equations.
2. Multiply one of the equations by a nonzero constant.
3. Add a multiple of one equation to another equation.

EXAMPLE 2 Using Elimination to Solve a System

Solve the system of linear equations.

$$\begin{aligned} x - 2y + 3z &= 9 && \text{Equation 1} \\ -x + 3y &= -4 && \text{Equation 2} \\ 2x - 5y + 5z &= 17 && \text{Equation 3} \end{aligned}$$

Solution

There are many ways to begin, but we suggest working from the upper left corner, saving the x in the upper left position and eliminating the other x's from the first column.

$$\begin{aligned} x - 2y + 3z &= 9 \\ y + 3z &= 5 \\ 2x - 5y + 5z &= 17 \end{aligned}$$

Adding the first equation to the second equation produces a new second equation.

$$\begin{aligned} x - 2y + 3z &= 9 \\ y + 3z &= 5 \\ -y - z &= -1 \end{aligned}$$

Adding -2 times the first equation to the third equation produces a new third equation.

Now that all but the first x have been eliminated from the first column, go to work on the second column. (You need to eliminate y from the third equation.)

$$\begin{aligned} x - 2y + 3z &= 9 \\ y + 3z &= 5 \\ 2z &= 4 \end{aligned}$$

Adding the second equation to the third equation produces a new third equation.

Finally, you need a coefficient of 1 for z in the third equation.

$$\begin{aligned} x - 2y + 3z &= 9 \\ y + 3z &= 5 \\ z &= 2 \end{aligned}$$

Multiplying the third equation by $\frac{1}{2}$ produces a new third equation.

This is the same system that was solved in Example 1, and, as in that example, you can conclude that the solution is $x = 1$, $y = -1$, and $z = 2$. The solution can also be written as the ordered triple $(1, -1, 2)$.

Study Tip

In Example 2, you can check the solution by substituting $x = 1$, $y = -1$, and $z = 2$ into each original equation, as follows.

Equation 1:

$1 - 2(-1) + 3(2) = 9$ ✓

Equation 2:

$-(1) + 3(-1) = -4$ ✓

Equation 3:

$2(1) - 5(-1) + 5(2) = 17$ ✓

The next example involves an inconsistent system—one that has no solution. The key to recognizing an inconsistent system is that at some stage in the elimination process, you obtain an absurdity such as $0 = -2$.

EXAMPLE 3 **An Inconsistent System**

Solve the system of linear equations.

$$\begin{aligned} x - 3y + z &= 1 && \text{Equation 1} \\ 2x - y - 2z &= 2 && \text{Equation 2} \\ x + 2y - 3z &= -1 && \text{Equation 3} \end{aligned}$$

Solution

$$\begin{aligned} x - 3y + z &= 1 \\ 5y - 4z &= 0 \\ x + 2y - 3z &= -1 \end{aligned}$$

Adding -2 times the first equation to the second equation produces a new second equation.

$$\begin{aligned} x - 3y + z &= 1 \\ 5y - 4z &= 0 \\ 5y - 4z &= -2 \end{aligned}$$

Adding -1 times the first equation to the third equation produces a new third equation.

$$\begin{aligned} x - 3y + z &= 1 \\ 5y - 4z &= 0 \\ 0 &= -2 \end{aligned}$$

Adding -1 times the second equation to the third equation produces a new third equation.

Because the third "equation" is impossible, you can conclude that this system is inconsistent and therefore has no solution. Moreover, because this system is equivalent to the original system, you can conclude that the original system also has no solution.

As with a system of linear equations in two variables, the number of solutions of a system of linear equations in more than two variables must fall into one of three categories.

The Number of Solutions of a Linear System

For a system of linear equations, exactly one of the following is true.

1. There is exactly one solution.
2. There are infinitely many solutions.
3. There is no solution.

EXAMPLE 4 **A System with Infinitely Many Solutions**

Solve the system of linear equations.

$$\begin{aligned} x + y - 3z &= -1 \quad &\text{Equation 1}\\ y - z &= 0 \quad &\text{Equation 2}\\ -x + 2y &= 1 \quad &\text{Equation 3}\end{aligned}$$

Solution

$$\begin{aligned} x + y - 3z &= -1\\ y - z &= 0\\ 3y - 3z &= 0\end{aligned}$$

Adding the first equation to the third equation produces a new third equation.

$$\begin{aligned} x + y - 3z &= -1\\ y - z &= 0\\ 0 &= 0\end{aligned}$$

Adding -3 times the second equation to the third equation produces a new third equation.

This means that Equation 3 depends on Equations 1 and 2 in the sense that it gives us no additional information about the variables. Thus, the original system is equivalent to the system

$$\begin{aligned} x + y - 3z &= -1\\ y - z &= 0.\end{aligned}$$

In the last equation, solve for y in terms of z to obtain $y = z$. Back-substituting for y into the previous equation produces $x = 2z - 1$. Finally, letting $z = a$, the solutions to the given system are all of the form

$$x = 2a - 1, \qquad y = a, \qquad \text{and} \qquad z = a$$

where a is a real number. Thus, every ordered triple of the form

$$(2a - 1, a, a), \qquad a \text{ is a real number}$$

is a solution of the system.

Study Tip

There are an infinite number of solutions to Example 4, but they all are of a specific form. By selecting, for example, a-values of 0, 1, and 3, you can verify that $(-1, 0, 0)$, $(1, 1, 1)$, and $(5, 3, 3)$ are specific solutions. It is incorrect to say simply that the solution to Example 4 is "infinite." You must also specify the form of the solutions.

In Example 4, there are other ways to write the same infinite set of solutions. For instance, the solutions could have been written as

$$\left(b, \tfrac{1}{2}(b + 1), \tfrac{1}{2}(b + 1)\right), \qquad b \text{ is a real number.}$$

Try convincing yourself of this by substituting $a = 0$, $a = 1$, $a = 2$, and $a = 3$ into the solution listed in Example 4. Then substitute $b = -1$, $b = 1$, $b = 3$, and $b = 5$ into the solution listed above. The resulting sets of ordered triples should be the same. Thus, when comparing descriptions of an infinite solution set, keep in mind that there is more than one way to describe the set.

Nonsquare Systems

So far, each system of linear equations has been **square,** which means that the number of equations is equal to the number of variables. In a **nonsquare** system, the number of equations differs from the number of variables. A system of linear equations cannot have a unique solution unless there are at least as many equations as there are variables in the system.

EXAMPLE 5 A System with Fewer Equations than Variables

Solve the system of linear equations.

$$x - 2y + z = 2 \qquad \text{Equation 1}$$

$$2x - y - z = 1 \qquad \text{Equation 2}$$

Solution

Begin by rewriting the system in row-echelon form, as follows.

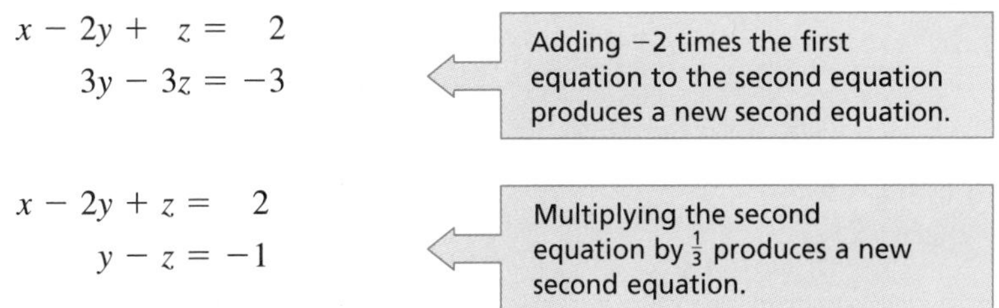

$$\begin{aligned} x - 2y + z &= 2 \\ 3y - 3z &= -3 \end{aligned}$$

Adding −2 times the first equation to the second equation produces a new second equation.

$$\begin{aligned} x - 2y + z &= 2 \\ y - z &= -1 \end{aligned}$$

Multiplying the second equation by $\frac{1}{3}$ produces a new second equation.

Solving for y in terms of z, you get $y = z - 1$, and back-substitution into Equation 1 yields

$$\begin{aligned} x - 2(z - 1) + z &= 2 \\ x - 2z + 2 + z &= 2 \\ x &= z. \end{aligned}$$

Finally, by letting $z = a$, you have the solution

$$x = a, \qquad y = a - 1, \qquad \text{and} \qquad z = a$$

where a is a real number. Thus, every ordered triple of the form

$$(a, a - 1, a), \qquad a \text{ is a real number}$$

is a solution of the system.

Note In Example 5, try choosing some values of a to obtain different solutions of the system, such as (1, 0, 1), (2, 1, 2), and (3, 2, 3). Then check each of the solutions in the original system.

Figure 5.13

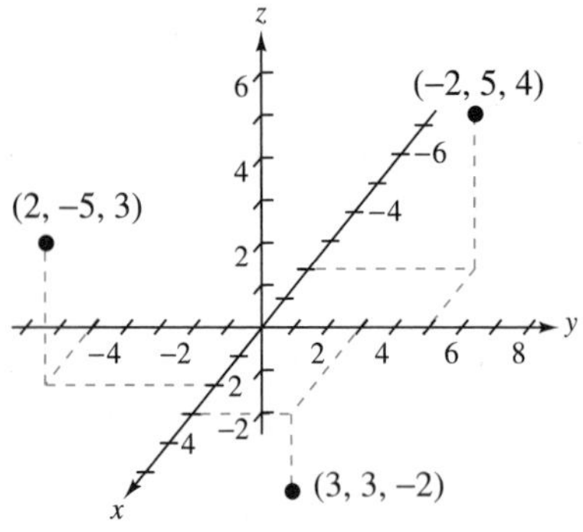

Graphical Interpretation of Three-Variable Systems

Solutions of equations with three variables can be pictured with a **three-dimensional coordinate system.** To construct such a system, begin with the xy-coordinate plane in a horizontal position. Then draw the z-axis as a vertical line through the origin.

Every ordered triple (x, y, z) corresponds to a point on the three-dimensional coordinate system. For instance, the points corresponding to

$$(-2, 5, 4), \quad (2, -5, 3), \quad \text{and} \quad (3, 3, -2)$$

are shown in Figure 5.13.

The **graph** of an equation in three variables consists of all points (x, y, z) that are solutions of the equation. The graph of a linear equation in three variables is a *plane.* Sketching graphs on a three-dimensional coordinate system is difficult because the sketch itself is only two-dimensional.

Figure 5.14

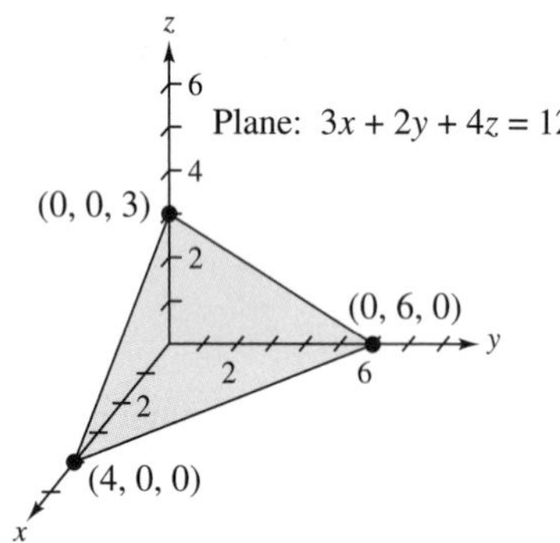

One technique for sketching a plane is to find the three points at which the plane intersects the axes. For instance, the plane given by

$$3x + 2y + 4z = 12$$

intersects the x-axis at the point $(4, 0, 0)$, the y-axis at the point $(0, 6, 0)$, and the z-axis at the point $(0, 0, 3)$.

By plotting these three points, connecting them with line segments, and shading the resulting triangular region, you can picture a portion of the graph, as shown in Figure 5.14.

The graph of a system of three linear equations in three variables consists of *three* planes. When these planes intersect in a single point, the system has exactly one solution. When the three planes have no point in common, the system has no solution. When the three planes contain a common line, the system has infinitely many solutions. (See Figure 5.15.)

Figure 5.15

Exactly One Solution

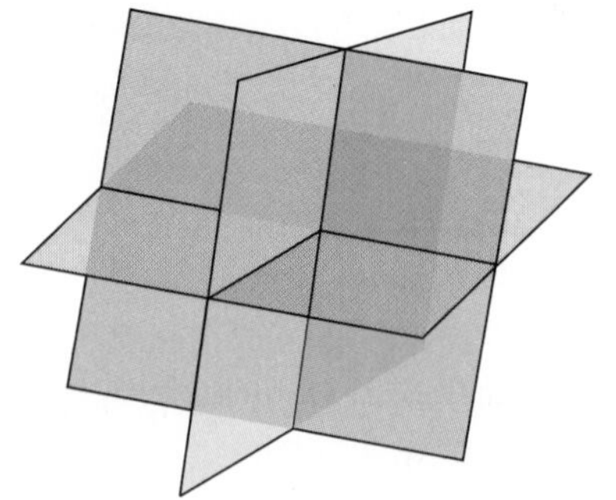

No Solution

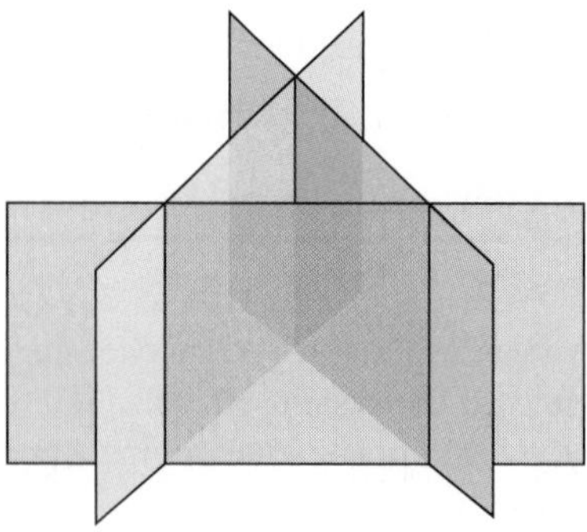

Infinitely Many Solutions

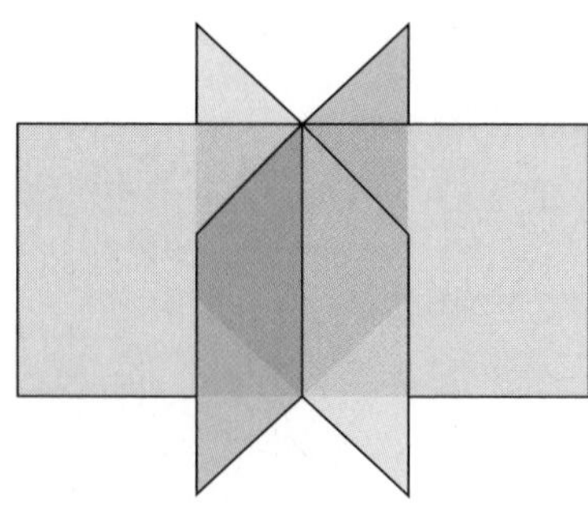

> **Study Tip**
>
> When you use a system of linear equations to solve an application, it is wise to interpret your solution in the context of the problem to see if it makes sense. For instance, in Example 6 the solution results in a position equation of $s = -16t^2 + 48t + 20$, which implies that the object was thrown upward at a velocity of 48 feet per second from a height of 20 feet. The object undergoes a constant downward acceleration of 32 feet per second squared. (Physics will tell you that this is the value of the acceleration due to gravity.)

Applications

Real Life

EXAMPLE 6 Vertical Motion

The height at time t of an object that is moving in a (vertical) line with constant acceleration a is given by the **position equation**

$$s = \frac{1}{2}at^2 + v_0t + s_0.$$

The height s is measured in feet, t is measured in seconds, v_0 is the initial velocity (at $t = 0$) in feet per second, and s_0 is the initial height. Find the values of a, v_0, and s_0, if $s = 52$ at $t = 1$, $s = 52$ at $t = 2$, and $s = 20$ at $t = 3$.

Solution

You can obtain three linear equations in a, v_0, and s_0 as follows.

When $t = 1$: $\frac{1}{2}a(1)^2 + v_0(1) + s_0 = 52$ ⇨ $a + 2v_0 + 2s_0 = 104$

When $t = 2$: $\frac{1}{2}a(2)^2 + v_0(2) + s_0 = 52$ ⇨ $2a + 2v_0 + s_0 = 52$

When $t = 3$: $\frac{1}{2}a(3)^2 + v_0(3) + s_0 = 20$ ⇨ $9a + 6v_0 + 2s_0 = 40$

Solving this system yields $a = -32$, $v_0 = 48$, and $s_0 = 20$.

EXAMPLE 7 Data Analysis: Curve-Fitting

Find a quadratic equation, $y = ax^2 + bx + c$, whose graph passes through the points $(-1, 3)$, $(1, 1)$, and $(2, 6)$.

Solution

Because the graph of $y = ax^2 + bx + c$ passes through the points $(-1, 3)$, $(1, 1)$, and $(2, 6)$, you can write the following.

When $x = -1, y = 3$: $a(-1)^2 + b(-1) + c = 3$

When $x = 1, y = 1$: $a(1)^2 + b(1) + c = 1$

When $x = 2, y = 6$: $a(2)^2 + b(2) + c = 6$

This produces the following system of linear equations.

$a - b + c = 3$ Equation 1

$a + b + c = 1$ Equation 2

$4a + 2b + c = 6$ Equation 3

The solution of this system is $a = 2$, $b = -1$, and $c = 0$. Thus, the equation of the parabola is $y = 2x^2 - x$, as shown in Figure 5.16.

Figure 5.16

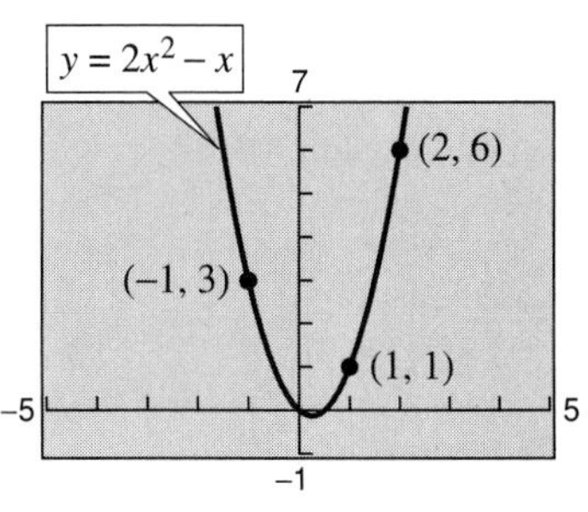

Group Activity

Mathematical Modeling

x	1	3	5
y	59	49	59

Suppose you work for an outerwear manufacturer, and the marketing department is concerned about sales trends in Georgia. Your manager has asked you to investigate climatic data in hopes of explaining sales patterns and gives you the table at the left, which gives the average monthly temperature y in degrees Fahrenheit for Savannah, Georgia, for the month x, where $x = 1$ corresponds to November.

Construct a scatter plot of the data. Decide which type of mathematical model might be appropriate for the data and use the methods you have learned thus far to fit an appropriate model. Your manager would like to know the average monthly temperatures for December ($x = 2$) and February ($x = 4$). Explain to your manager how you found your model, what it represents, and how it may be used to find the December and February average temperatures. Investigate the usefulness of this model for the rest of the year. Would you recommend using the model to predict average monthly temperatures for the whole year or just part of the year? Explain your reasoning. (Source: National Climatic Data Center)

5.3 /// EXERCISES

In Exercises 1–6, use back-substitution to solve the system of linear equations.

1. $\begin{aligned} 2x - y + 5z &= 24 \\ y + 2z &= 6 \\ z &= 4 \end{aligned}$

2. $\begin{aligned} 4x - 3y - 2z &= 21 \\ 6y - 5z &= -8 \\ z &= -2 \end{aligned}$

3. $\begin{aligned} 2x + y - 3z &= 10 \\ y &= 2 \\ y - z &= 4 \end{aligned}$

4. $\begin{aligned} x &= 8 \\ 2x + 3y &= 10 \\ x - y + 2z &= 22 \end{aligned}$

5. $\begin{aligned} 4x - 2y + z &= 8 \\ 2z &= 4 \\ -y + z &= 4 \end{aligned}$

6. $\begin{aligned} 5x - 8z &= 22 \\ 3y - 5z &= 10 \\ z &= -4 \end{aligned}$

In Exercises 7 and 8, perform the row operation and write the equivalent system.

7. Add Equation 1 to Equation 2.

$$\begin{aligned} x - 2y + 3z &= 5 && \text{Equation 1} \\ -x + 3y - 5z &= 4 && \text{Equation 2} \\ 2x - 3z &= 0 && \text{Equation 3} \end{aligned}$$

What did this operation accomplish?

8. Add -2 times Equation 1 to Equation 3.

$$\begin{aligned} x - 2y + 3z &= 5 && \text{Equation 1} \\ -x + 3y - 5z &= 4 && \text{Equation 2} \\ 2x - 3z &= 0 && \text{Equation 3} \end{aligned}$$

What did this operation accomplish?

In Exercises 9–34, solve the system of linear equations and check any solution algebraically.

9. $\begin{aligned} x + y + z &= 6 \\ 2x - y + z &= 3 \\ 3x \phantom{{}-y} - z &= 0 \end{aligned}$

10. $\begin{aligned} x + y + z &= 2 \\ -x + 3y + 2z &= 8 \\ 4x + y \phantom{{}+2z} &= 4 \end{aligned}$

11. $\begin{aligned} 2x \phantom{{}+3y} + 2z &= 2 \\ 5x + 3y \phantom{{}-4z} &= 4 \\ 3y - 4z &= 4 \end{aligned}$

12. $\begin{aligned} 4x + y - 3z &= 11 \\ 2x - 3y + 2z &= 9 \\ x + y + z &= -3 \end{aligned}$

13. $\begin{aligned} 6y + 4z &= -12 \\ 3x + 3y \phantom{{}+4z} &= 9 \\ 2x \phantom{{}+3y} - 3z &= 10 \end{aligned}$

14. $\begin{aligned} 2x + 4y + z &= -4 \\ 2x - 4y + 6z &= 13 \\ 4x - 2y + z &= 6 \end{aligned}$

15. $\begin{aligned} 3x - 2y + 4z &= 1 \\ x + y - 2z &= 3 \\ 2x - 3y + 6z &= 8 \end{aligned}$

16. $\begin{aligned} 5x - 3y + 2z &= 3 \\ 2x + 4y - z &= 7 \\ x - 11y + 4z &= 3 \end{aligned}$

17. $\begin{aligned} 3x + 3y + 5z &= 1 \\ 3x + 5y + 9z &= 0 \\ 5x + 9y + 17z &= 0 \end{aligned}$

18. $\begin{aligned} 2x + y + 3z &= 1 \\ 2x + 6y + 8z &= 3 \\ 6x + 8y + 18z &= 5 \end{aligned}$

19. $\begin{aligned} x + 2y - 7z &= -4 \\ 2x + y + z &= 13 \\ 3x + 9y - 36z &= -33 \end{aligned}$

20. $\begin{aligned} 2x + y - 3z &= 4 \\ 4x \phantom{{}+3y} + 2z &= 10 \\ -2x + 3y - 13z &= -8 \end{aligned}$

21. $\begin{aligned} 3x - 3y + 6z &= 6 \\ x + 2y - z &= 5 \\ 5x - 8y + 13z &= 7 \end{aligned}$

22. $\begin{aligned} x \phantom{{}-2y} + 4z &= 13 \\ 4x - 2y + z &= 7 \\ 2x - 2y - 7z &= -19 \end{aligned}$

23. $\begin{aligned} x - 2y + 5z &= 2 \\ 4x \phantom{{}-2y} - z &= 0 \end{aligned}$

24. $\begin{aligned} x - 3y + 2z &= 18 \\ 5x - 13y + 12z &= 80 \end{aligned}$

25. $\begin{aligned} 2x - 3y + z &= -2 \\ -4x + 9y \phantom{{}+z} &= 7 \end{aligned}$

26. $\begin{aligned} 2x + 3y + 3z &= 7 \\ 4x + 18y + 15z &= 44 \end{aligned}$

27. $\begin{aligned} x \phantom{{}+2y-z} + 3w &= 4 \\ 2y - z - w &= 0 \\ 3y \phantom{{}-z} - 2w &= 1 \\ 2x - y + 4z \phantom{{}-w} &= 5 \end{aligned}$

28. $\begin{aligned} x + y + z + w &= 6 \\ 2x + 3y \phantom{{}+z} - w &= 0 \\ -3x + 4y + z + 2w &= 4 \\ x + 2y - z + w &= 0 \end{aligned}$

29. $\begin{aligned} x \phantom{{}+y} + 4z &= 1 \\ x + y + 10z &= 10 \\ 2x - y + 2z &= -5 \end{aligned}$

30. $\begin{aligned} 3x - 2y - 6z &= -4 \\ -3x + 2y + 6z &= 1 \\ x - y - 5z &= -3 \end{aligned}$

31. $\begin{aligned} 2x + 3y \phantom{{}-z} &= 0 \\ 4x + 3y - z &= 0 \\ 8x + 3y + 3z &= 0 \end{aligned}$

32. $\begin{aligned} 4x + 3y + 17z &= 0 \\ 5x + 4y + 22z &= 0 \\ 4x + 2y + 19z &= 0 \end{aligned}$

33. $\begin{aligned} 12x + 5y + z &= 0 \\ 23x + 4y - z &= 0 \end{aligned}$

34. $\begin{aligned} 5x + 5y - z &= 0 \\ 10x + 5y + 2z &= 0 \\ 5x + 15y - 9z &= 0 \end{aligned}$

35. *Think About It* Are the two systems of equations equivalent? Give reasons for your answer.

$\begin{aligned} x + 3y - z &= 6 \\ 2x - y + 2z &= 1 \\ 3x + 2y - z &= 2 \end{aligned} \qquad \begin{aligned} x + 3y - z &= 6 \\ -7y + 4z &= 1 \\ -7y - 4z &= -16 \end{aligned}$

36. *Think About It* When using Gaussian elimination to solve a system of linear equations, how can you recognize that it has no solution? Give an example that illustrates your answer.

Exploration **In Exercises 37 and 38, find a system of linear equations that has the given solution. (The answer is not unique.)**

37. $(4, -1, 2)$

38. $\left(-\frac{3}{2}, 4, -7\right)$

In Exercises 39–42, find the equation of the parabola

$$y = ax^2 + bx + c$$

that passes through the given points. To verify your result, use a graphing utility to plot the points and graph the parabola.

39. $(0, 0), (2, -2), (4, 0)$

40. $(0, 3), (1, 4), (2, 3)$

41. $(2, 0), (3, -1), (4, 0)$

42. $(1, 3), (2, 2), (3, -3)$

In Exercises 43–46, find the equation of the circle

$$x^2 + y^2 + Dx + Ey + F = 0$$

that passes through the given points. To verify your result, use a graphing utility to plot the points and graph the circle.

43. $(0, 0), (2, 2), (4, 0)$

44. $(0, 0), (0, 6), (3, 3)$

45. $(-3, -1), (2, 4), (-6, 8)$

46. $(0, 0), (0, -2), (3, 0)$

Vertical Motion **In Exercises 47–50, find the position equation $s = \frac{1}{2}at^2 + v_0t + s_0$ for an object at the given heights moving vertically at the specified times.**

47. At $t = 1$ second, $s = 128$ feet
At $t = 2$ seconds, $s = 80$ feet
At $t = 3$ seconds, $s = 0$ feet

48. At $t = 1$ second, $s = 48$ feet
At $t = 2$ seconds, $s = 64$ feet
At $t = 3$ seconds, $s = 48$ feet

49. At $t = 1$ second, $s = 452$ feet
At $t = 2$ seconds, $s = 260$ feet
At $t = 3$ seconds, $s = 116$ feet

50. At $t = 1$ second, $s = 132$ feet
At $t = 2$ seconds, $s = 100$ feet
At $t = 3$ seconds, $s = 36$ feet

51. *Investments* An inheritance of \$16,000 was divided among three investments yielding a total of \$990 in interest per year. The interest rates were 5%, 6%, and 7%. Find the amount in each investment if the 5% and 6% investments were \$3000 and \$2000 less than the 7% investment, respectively.

52. *Investments* A total of \$1520 a year is received in interest from three investments. The interest rates for the three investments are 5%, 7%, and 8%. The 5% investment is half of the 7% investment, and the 7% investment is \$1500 less than the 8% investment. Find the amount in each investment.

53. *Borrowing* A small corporation borrowed \$775,000 to expand its product line. Some of the money was borrowed at 8%, some at 9%, and some at 10%. How much was borrowed at each rate if the annual interest was \$67,000 and the amount borrowed at 8% was four times the amount borrowed at 10%?

54. *Borrowing* A small corporation borrowed \$800,000 to expand its product line. Some of the money was borrowed at 8%, some at 9%, and some at 10%. How much was borrowed at each rate if the annual interest was \$67,000 and the amount borrowed at 8% was five times the amount borrowed at 10%?

Investment Portfolio **In Exercises 55 and 56, consider an investor with a portfolio totaling \$500,000 that is invested in certificates of deposit, municipal bonds, blue-chip stocks, and growth or speculative stocks. How much is put in each type of investment?**

55. The certificates of deposit pay 10% annually, and the municipal bonds pay 8% annually. Over a 5-year period, the investor expects the blue-chip stocks to return 12% annually and the growth stocks to return 13% annually. The investor wants a combined annual return of 10% and also wants to have only one-fourth of the portfolio invested in stocks.

56. The certificates of deposit pay 9% annually, and the municipal bonds pay 5% annually. Over a 5-year period, the investor expects the blue-chip stocks to return 12% annually and the growth stocks to return 14% annually. The investor wants a combined annual return of 10% and also wants to have only one-fourth of the portfolio invested in stocks.

57. *Crop Spraying* A mixture of 12 liters of chemical A, 16 liters of chemical B, and 26 liters of chemical C is required to kill a certain destructive crop insect. Commercial spray X contains 1, 2, and 2 parts, respectively, of these chemicals. Commercial spray Y contains only chemical C. Commercial spray Z contains only chemicals A and B in equal amounts. How much of each type of commercial spray is needed to get the desired mixture?

58. *Chemistry* A chemist needs 10 liters of a 25% acid solution. The solution is to be mixed from three solutions whose concentrations are 10%, 20%, and 50%. How many liters of each solution should the chemist use to satisfy the following?

(a) Use as little as possible of the 50% solution.

(b) Use as much as possible of the 50% solution.

(c) Use 2 liters of the 50% solution.

59. *Truck Scheduling* A small company that manufactures products A and B has an order for 15 units of product A and 16 units of product B. The company has trucks of three different sizes that can haul the products in the amounts shown in the table.

Truck	Large	Medium	Small
Product A	6	4	0
Product B	3	4	3

How many trucks of each size are needed to deliver the order? Give *two* possible solutions.

60. *Electrical Networks* When Kirchhoff's Laws are applied to the electrical network in the figure, the currents I_1, I_2, and I_3 are the solution of the system

$$\begin{aligned} I_1 - I_2 + I_3 &= 0 \\ 3I_1 + 2I_2 &= 7 \\ 2I_2 + 4I_3 &= 8. \end{aligned}$$

Find the currents.

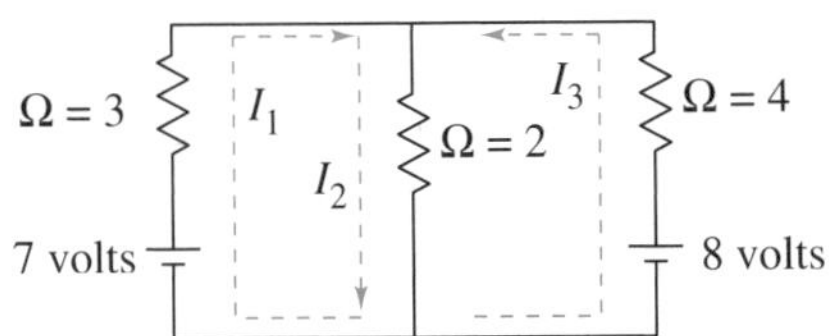

61. *Pulley System* A system of pulleys is loaded with 128-pound and 32-pound weights (see figure). The tensions t_1 and t_2 in the ropes and the acceleration a of the 32-pound weight are found by solving the system

$$\begin{aligned} t_1 - 2t_2 &= 0 \\ t_1 - 2a &= 128 \\ t_2 + a &= 32 \end{aligned}$$

where t_1 and t_2 are measured in pounds and a is in feet per second squared. Solve the system.

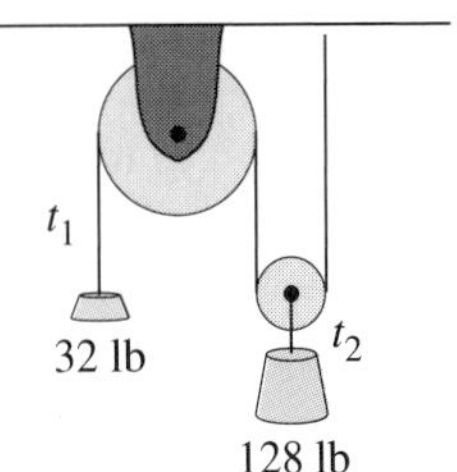

62. *Pulley System* If the 32-pound weight is replaced with a 64-pound weight in the pulley system described in Exercise 61, the new pulley system is modeled by the following system of equations.

$$\begin{aligned} t_1 - 2t_2 &= 0 \\ t_1 - 2a &= 128 \\ t_2 + 2a &= 64 \end{aligned}$$

Solve the system, and use your answer for the acceleration to describe what (if anything) is happening in the system.

Three-Dimensional Graphics **In Exercises 63–66, sketch the plane represented by the linear equation. Then list four points that lie in the plane.**

63. $2x + 3y + 4z = 12$

64. $x + y + z = 6$

65. $2x + y + z = 4$

66. $x + 2y + 2z = 6$

Fitting a Parabola **In Exercises 67–70, find the least squares regression parabola $y = ax^2 + bx + c$ for the points $(x_1, y_1), (x_2, y_2), \ldots, (x_n, y_n)$. To find the parabola, solve the following system of linear equations for a, b, and c. Then use the least squares regression capabilities of a graphing utility to confirm the result.**

$$nc + \left(\sum_{i=1}^{n} x_i\right)b + \left(\sum_{i=1}^{n} x_i^2\right)a = \sum_{i=1}^{n} y_i$$

$$\left(\sum_{i=1}^{n} x_i\right)c + \left(\sum_{i=1}^{n} x_i^2\right)b + \left(\sum_{i=1}^{n} x_i^3\right)a = \sum_{i=1}^{n} x_i y_i$$

$$\left(\sum_{i=1}^{n} x_i^2\right)c + \left(\sum_{i=1}^{n} x_i^3\right)b + \left(\sum_{i=1}^{n} x_i^4\right)a = \sum_{i=1}^{n} x_i^2 y_i$$

67. **68.**

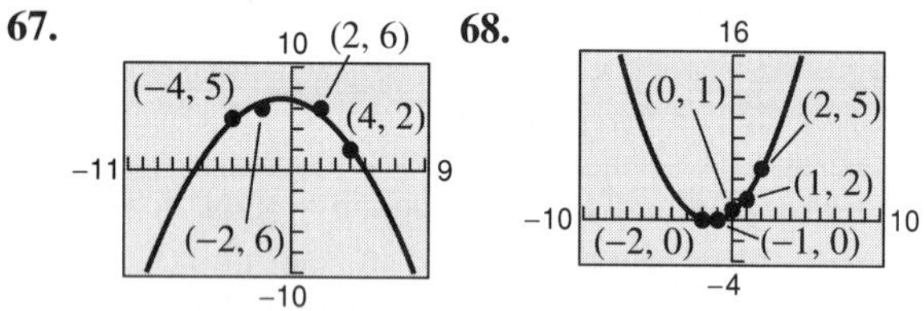

69. **70.**

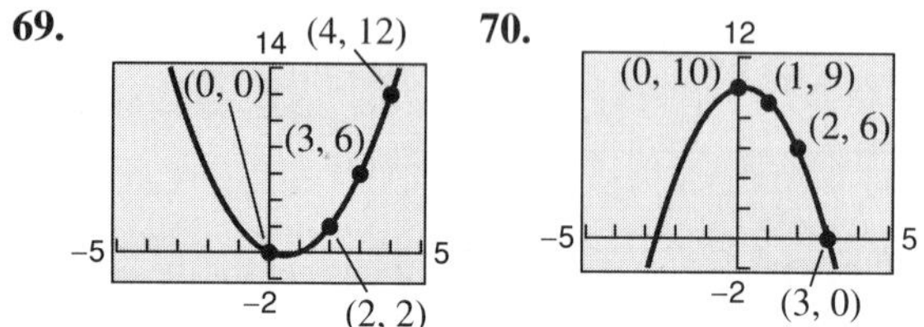

71. *Analyzing Data* In testing of a new braking system on an automobile, the speeds in miles per hour and the stopping distances in feet were recorded as follows.

Speed, x	20	30	40	50	60
Stopping Distance, y	25	55	105	188	300

(a) Find the least squares regression parabola for the data. Then use a graphing utility to graph the parabola and the data on the same set of axes.

(b) Use the model to estimate the stopping distance if the speed is 70 miles per hour.

72. *Analyzing Data* A wildlife management team studied the reproduction rates of deer in five tracts of a wildlife preserve. Each tract contained 5 acres. For each tract, the number of females and the percent of females that had offspring the following year were recorded. The results are given in the table.

Number, x	80	100	120	140	160
Percent, y	80	75	68	55	30

(a) Find the least squares regression parabola for the data.

(b) Use a graphing utility to graph the parabola and the data in the same viewing rectangle.

(c) Use the model to predict the percent of females that would have offspring if $x = 170$.

Advanced Applications **In Exercises 73–76, find values of x, y, and λ that satisfy the system. These systems arise in certain optimization problems in calculus; λ is called a Lagrange multiplier.**

73. $y + \lambda = 0$, $x + \lambda = 0$, $x + y - 10 = 0$

74. $2x + \lambda = 0$, $2y + \lambda = 0$, $x + y - 4 = 0$

75. $2x - 2x\lambda = 0$, $-2y + \lambda = 0$, $y - x^2 = 0$

76. $2 + 2y + 2\lambda = 0$, $2x + 1 + \lambda = 0$, $2x + y - 100 = 0$

77. *Chapter Opener* Interpret the slope of the model for sales of compact sport utilities on page 387.

78. *Chapter Opener* If the models on page 387 are assumed to be accurate in forecasting future sales, will there be a time when compact pickup sales again exceed compact utility sales? If so, when?

Review **Solve Exercises 79–82 as a review of the skills and problem-solving techniques you learned in previous sections.**

79. What is $7\frac{1}{2}\%$ of 85?

80. 225 is what percent of 150?

81. 0.5% of what number is 400?

82. 48% of what number is 132?

5.4 Partial Fractions

Introduction / Partial Fraction Decomposition

Introduction

In this section, you will learn to write a rational expression as the sum of two or more simpler rational expressions. For example, the rational expression $(x + 7)/(x^2 - x - 6)$ can be written as the sum of two fractions with linear denominators. That is,

$$\frac{x + 7}{x^2 - x - 6} = \frac{2}{x - 3} + \frac{-1}{x + 2}.$$

Each fraction on the right side of the equation is a **partial fraction,** and together they make up the **partial fraction decomposition** of the left side.

Note One of the most important applications of partial fractions is in calculus. If you go on to take a course in calculus, you will learn how partial fractions can be used in a calculus operation called antidifferentiation.

Decomposition of $N(x)/D(x)$ into Partial Fractions

1. *Divide if improper:* If $N(x)/D(x)$ is an improper fraction, divide the denominator into the numerator to obtain

$$\frac{N(x)}{D(x)} = (\text{polynomial}) + \frac{N_1(x)}{D(x)}$$

and apply Steps 2, 3, and 4 (below) to the proper rational expression $N_1(x)/D(x)$.

2. *Factor denominator:* Completely factor the denominator into factors of the form

$$(px + q)^m \qquad \text{and} \qquad (ax^2 + bx + c)^n$$

where $(ax^2 + bx + c)$ is irreducible.

3. *Linear factors:* For *each* factor of the form $(px + q)^m$, the partial fraction decomposition must include the following sum of m fractions.

$$\frac{A_1}{(px + q)} + \frac{A_2}{(px + q)^2} + \cdots + \frac{A_m}{(px + q)^m}$$

4. *Quadratic factors:* For *each* factor of the form $(ax^2 + bx + c)^n$, the partial fraction decomposition must include the following sum of n fractions.

$$\frac{B_1x + C_1}{ax^2 + bx + c} + \frac{B_2x + C_2}{(ax^2 + bx + c)^2} + \cdots + \frac{B_nx + C_n}{(ax^2 + bx + c)^n}$$

Partial Fraction Decomposition

Algebraic techniques for determining the constants in the numerators of partial fractions are demonstrated in the examples that follow. Note that the techniques vary slightly, depending on the type of factors in the denominator: linear or quadratic, distinct or repeated.

You can graphically check the decomposition found in Example 1. To do this, graph

$$y_1 = \frac{x+7}{x^2 - x - 6}$$

and

$$y_2 = \frac{2}{x-3} - \frac{1}{x+2}$$

in the same viewing rectangle. Their graphs should be identical, as shown below.

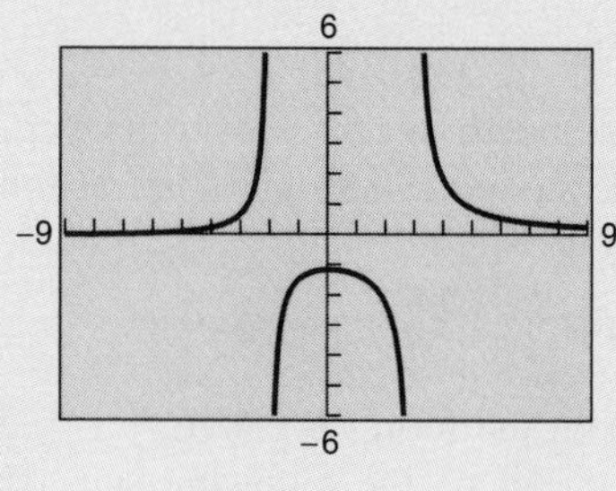

EXAMPLE 1 **Distinct Linear Factors**

Write the partial fraction decomposition for

$$\frac{x+7}{x^2 - x - 6}.$$

Solution

Because $x^2 - x - 6 = (x-3)(x+2)$, you should include one partial fraction with a constant numerator for each linear factor of the denominator and write

$$\frac{x+7}{x^2 - x - 6} = \frac{A}{x-3} + \frac{B}{x+2}.$$

Multiplying both sides of this equation by the least common denominator, $(x-3)(x+2)$, leads to the **basic equation**

$$x + 7 = A(x+2) + B(x-3). \qquad \text{Basic equation}$$

Because this equation is true for all x, you can substitute any *convenient* values of x that will help you determine the constants A and B. Values of x that are especially convenient are ones that make the factors $(x+2)$ and $(x-3)$ equal to zero. For instance, let $x = -2$. Then

$$-2 + 7 = A(-2+2) + B(-2-3) \qquad \text{Substitute } -2 \text{ for } x.$$
$$5 = -5B$$
$$-1 = B.$$

To solve for A, let $x = 3$ and obtain

$$3 + 7 = A(3+2) + B(3-3) \qquad \text{Substitute 3 for } x.$$
$$10 = 5A$$
$$2 = A.$$

Therefore, the partial fraction decomposition is

$$\frac{x+7}{x^2 - x - 6} = \frac{2}{x-3} - \frac{1}{x+2}$$

as indicated at the beginning of this section. Check this result by combining the two partial fractions on the right side of the equation.

The next example shows how to find the partial fraction decomposition for a rational function whose denominator has a repeated linear factor.

EXPLORATION

Partial fraction decomposition is practical only for rational functions whose denominators factor "nicely." For example, the factorization of the denominator $x^2 - x - 5$ is

$$\left(x - \frac{1 - \sqrt{21}}{2}\right)\left(x - \frac{1 + \sqrt{21}}{2}\right).$$

Write the basic equation and try to complete the decomposition for

$$\frac{x + 7}{x^2 - x - 5}.$$

What problems do you encounter?

EXAMPLE 2 Repeated Linear Factors

Write the partial fraction decomposition for

$$\frac{5x^2 + 20x + 6}{x^3 + 2x^2 + x}.$$

Solution

Because the denominator factors as

$$x^3 + 2x^2 + x = x(x^2 + 2x + 1) = x(x + 1)^2$$

you should include one partial fraction with a constant numerator for each power of x and $(x + 1)$ and write

$$\frac{5x^2 + 20x + 6}{x(x + 1)^2} = \frac{A}{x} + \frac{B}{x + 1} + \frac{C}{(x + 1)^2}.$$

Multiplying by the LCD, $x(x + 1)^2$, leads to the basic equation

$$\begin{aligned} 5x^2 + 20x + 6 &= A(x + 1)^2 + Bx(x + 1) + Cx. && \text{Basic equation} \\ &= Ax^2 + 2Ax + A + Bx^2 + Bx + Cx \\ &= (A + B)x^2 + (2A + B + C)x + A. && \text{Polynomial form} \end{aligned}$$

By equating coefficients of like terms on opposite sides of the equation, you obtain the following system of linear equations.

$$\begin{aligned} A + B \qquad &= 5 \\ 2A + B + C &= 20 \\ A \qquad\qquad &= 6 \end{aligned}$$

Substituting 6 for A in the first equation produces

$$\begin{aligned} 6 + B &= 5 \\ B &= -1. \end{aligned}$$

Substituting 6 for A and -1 for B in the second equation produces

$$\begin{aligned} 2(6) + (-1) + C &= 20 \\ C &= 9. \end{aligned}$$

Therefore, the partial fraction decomposition is

$$\frac{5x^2 + 20x + 6}{x(x + 1)^2} = \frac{6}{x} - \frac{1}{x + 1} + \frac{9}{(x + 1)^2}.$$

The method of partial fractions was introduced by John Bernoulli (1667–1748), a Swiss mathematician who was instrumental in the early development of calculus. John Bernoulli was a professor at the University of Basel and taught many outstanding students, the most famous of whom was Leonhard Euler.

EXAMPLE 3 **Distinct Linear and Quadratic Factors**

Write the partial fraction decomposition for

$$\frac{x^2 + 4x + 4}{x^3 - x^2 + 2x - 2}.$$

Solution

Because the denominator factors as

$$x^3 - x^2 + 2x - 2 = (x - 1)(x^2 + 2)$$

you should include one partial fraction with a constant numerator and one partial fraction with a linear numerator and write

$$\frac{x^2 + 4x + 4}{x^3 - x^2 + 2x - 2} = \frac{A}{x - 1} + \frac{Bx + C}{x^2 + 2}.$$

Multiplying by the LCD, $(x - 1)(x^2 + 2)$, yields the basic equation

$$\begin{aligned} x^2 + 4x + 4 &= A(x^2 + 2) + (Bx + C)(x - 1) && \text{Basic equation} \\ &= Ax^2 + 2A + Bx^2 - Bx + Cx - C \\ &= (A + B)x^2 + (-B + C)x + 2A - C. && \text{Polynomial form} \end{aligned}$$

By equating coefficients of like terms on opposite sides of the equation, you obtain the following system of linear equations.

$$\begin{aligned} A + B &= 1 \\ -B + C &= 4 \\ 2A - C &= 4 \end{aligned}$$

Using the techniques for solving multivariable linear systems that you learned in Section 5.3, you can solve this system to find that $A = 3$, $B = -2$, and $C = 2$. Therefore, the partial fraction decomposition is

$$\frac{x^2 + 4x + 4}{x^3 - x^2 + 2x - 2} = \frac{3}{x - 1} + \frac{-2x + 2}{x^2 + 2}.$$

Note You can check a partial fraction decomposition algebraically by substituting x-values into each side of the equation. For instance, when $x = 0$, you have

$$\frac{0^2 + 4(0) + 4}{0^3 - 0^2 + 2(0) - 2} \stackrel{?}{=} \frac{3}{0 - 1} + \frac{-2(0) + 2}{0^2 + 2}$$

$$-\frac{4}{2} = -\frac{3}{1} + \frac{2}{2}. \quad ✓$$

The next example shows how to find the partial fraction decomposition for a rational function whose denominator has a repeated quadratic factor.

EXAMPLE 4 **Repeated Quadratic Factors**

Write the partial fraction decomposition for

$$\frac{8x^3 + 13x}{(x^2 + 2)^2}.$$

Solution

Include one partial fraction with a linear numerator for each power of $(x^2 + 2)$, and write

$$\frac{8x^3 + 13x}{(x^2 + 2)^2} = \frac{Ax + B}{x^2 + 2} + \frac{Cx + D}{(x^2 + 2)^2}.$$

Multiplying by the LCD, $(x^2 + 2)^2$, yields the basic equation

$$\begin{aligned} 8x^3 + 13x &= (Ax + B)(x^2 + 2) + Cx + D && \text{Basic equation} \\ &= Ax^3 + 2Ax + Bx^2 + 2B + Cx + D \\ &= Ax^3 + Bx^2 + (2A + C)x + (2B + D). && \text{Polynomial form} \end{aligned}$$

Equating coefficients of like terms,

$$8x^3 + 0x^2 + 13x + 0 = Ax^3 + Bx^2 + (2A + C)x + (2B + D)$$

produces the following system of linear equations.

$$\begin{aligned} A \qquad\qquad\qquad &= 8 \\ B \qquad\qquad &= 0 \\ 2A + \qquad C \qquad &= 13 \\ 2B + \qquad D &= 0 \end{aligned}$$

Substituting 8 for A in the third equation produces

$$C = -3$$

and substituting 0 for B in the fourth equation produces

$$D = 0.$$

Therefore, the partial fraction decomposition is

$$\frac{8x^3 + 13x}{(x^2 + 2)^2} = \frac{8x}{x^2 + 2} + \frac{-3x}{(x^2 + 2)^2}.$$

Study Tip

Keep in mind that for *improper* rational expressions such as

$$\frac{N(x)}{D(x)} = \frac{2x^3 + x^2 - 7x + 7}{x^2 + x - 2}$$

you must first divide before applying partial fraction decomposition, such as

$$\frac{N(x)}{D(x)} = 2x - 1 + \frac{-2x + 5}{x^2 + x - 2}.$$

Group Activity *Error Analysis*

Suppose you are tutoring a student in algebra. In trying to find a partial fraction decomposition, your student writes the following.

$$\frac{x^2+1}{x(x-1)} = \frac{A}{x} + \frac{B}{x-1}$$

$$x^2 + 1 = A(x-1) + Bx \qquad \text{Basic equation}$$

$$x^2 + 1 = (A+B)x - A$$

Your student then forms the following system of linear equations.

$$A + B = 0$$

$$-A = 1$$

Solve the system and check the partial fraction decomposition it yields. Has your student worked the problem correctly? If not, what went wrong?

5.4 /// EXERCISES

In Exercises 1–6, write the form of the partial fraction decomposition of the rational expression. Do not solve for the constants.

1. $\dfrac{7}{x^2 - 14x}$
2. $\dfrac{x-2}{x^2+4x+3}$
3. $\dfrac{12}{x^3 - 10x^2}$
4. $\dfrac{4x^2+3}{(x-5)^3}$
5. $\dfrac{2x-3}{x^3+10x}$
6. $\dfrac{x-1}{x(x^2+1)^2}$

In Exercises 7–30, write the partial fraction decomposition for the rational expression. Check your result algebraically.

7. $\dfrac{1}{x^2-1}$
8. $\dfrac{1}{4x^2-9}$
9. $\dfrac{1}{x^2+x}$
10. $\dfrac{3}{x^2-3x}$
11. $\dfrac{1}{2x^2+x}$
12. $\dfrac{5}{x^2+x-6}$
13. $\dfrac{3}{x^2+x-2}$
14. $\dfrac{x+1}{x^2+4x+3}$
15. $\dfrac{x^2+12x+12}{x^3-4x}$
16. $\dfrac{x+2}{x(x-4)}$
17. $\dfrac{4x^2+2x-1}{x^2(x+1)}$
18. $\dfrac{2x-3}{(x-1)^2}$
19. $\dfrac{3x}{(x-3)^2}$
20. $\dfrac{6x^2+1}{x^2(x-1)^3}$
21. $\dfrac{x^2-1}{x(x^2+1)}$
22. $\dfrac{x}{(x-1)(x^2+x+1)}$
23. $\dfrac{x^2}{x^4-2x^2-8}$
24. $\dfrac{2x^2+x+8}{(x^2+4)^2}$
25. $\dfrac{x}{16x^4-1}$
26. $\dfrac{x^2-4x+7}{(x+1)(x^2-2x+3)}$

27. $\dfrac{x^2 + 5}{(x + 1)(x^2 - 2x + 3)}$

28. $\dfrac{x + 1}{x^3 + x}$

29. $\dfrac{x^4}{(x - 1)^3}$

30. $\dfrac{x^2 - x}{x^2 + x + 1}$

In Exercises 31–38, write the partial fraction decomposition for the rational expression. Use a graphing utility to check your result graphically.

31. $\dfrac{5 - x}{2x^2 + x - 1}$

32. $\dfrac{3x^2 - 7x - 2}{x^3 - x}$

33. $\dfrac{x - 1}{x^3 + x^2}$

34. $\dfrac{4x^2 - 1}{2x(x + 1)^2}$

35. $\dfrac{x^2 + x + 2}{(x^2 + 2)^2}$

36. $\dfrac{x^3}{(x + 2)^2(x - 2)^2}$

37. $\dfrac{2x^3 - 4x^2 - 15x + 5}{x^2 - 2x - 8}$

38. $\dfrac{x^3 - x + 3}{x^2 + x - 2}$

In Exercises 39–42, write the partial fraction decomposition for the rational expression. Check your result algebraically. Then assign a value to the constant a and check the result graphically.

39. $\dfrac{1}{a^2 - x^2}$

40. $\dfrac{1}{x(x + a)}$

41. $\dfrac{1}{y(a - y)}$

42. $\dfrac{1}{(x + 1)(a - x)}$

Graphical Analysis In Exercises 43–46, write the partial fraction decomposition for the rational function. Identify the graph of the rational function and the graphs of each term of its decomposition. State any relationship between the vertical asymptotes of the rational function and the vertical asymptotes of the terms of the decomposition.

43. $y = \dfrac{x - 12}{x(x - 4)}$

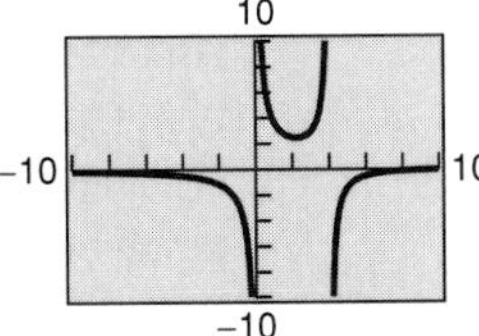

44. $y = \dfrac{2(x + 1)^2}{x(x^2 + 1)}$

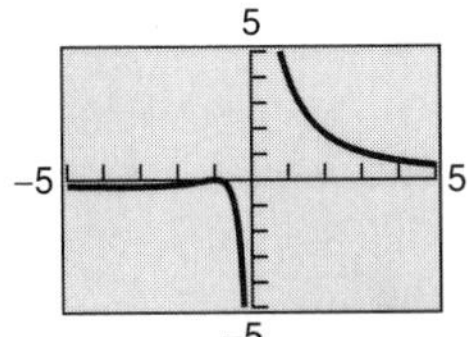

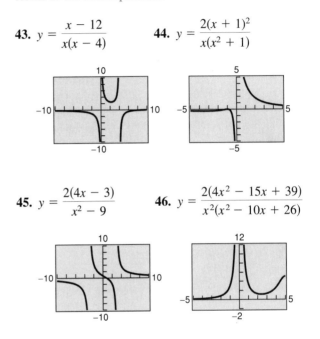

45. $y = \dfrac{2(4x - 3)}{x^2 - 9}$

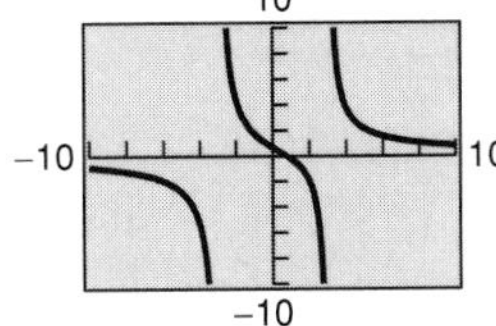

46. 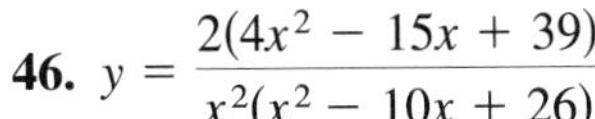
$y = \dfrac{2(4x^2 - 15x + 39)}{x^2(x^2 - 10x + 26)}$

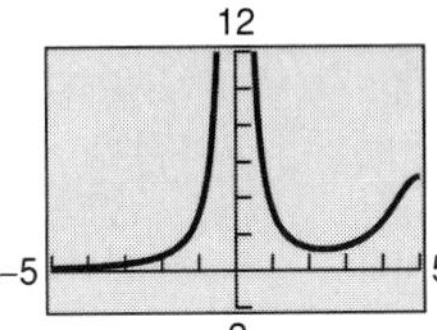

47. *Exhaust Temperatures* The magnitude of the range of exhaust temperatures in degrees Fahrenheit in an experimental diesel engine is approximated by

$$R = \frac{2000(4 - 3x)}{(11 - 7x)(7 - 4x)}, \qquad 0 \le x \le 1$$

where x is the relative load.

(a) Write the partial fraction decomposition for the rational function.

(b) The decomposition in part (a) is the difference of two fractions. The absolute values of the terms give the expected maximum and minimum temperatures of the exhaust gases. Use a graphing utility to graph each term.

5.5 Systems of Inequalities

The Graph of an Inequality / Systems of Inequalities / Applications

The Graph of an Inequality

The following statements are inequalities in two variables.

$$3x - 2y < 6 \qquad \text{and} \qquad 2x^2 + 3y^2 \geq 6$$

An ordered pair (a, b) is a **solution of an inequality** in x and y if the inequality is true when a and b are substituted for x and y, respectively. The **graph** of an inequality is the collection of all solutions of the inequality. To sketch the graph of an inequality, begin by sketching the graph of the *corresponding equation.* The graph of the equation will normally separate the plane into two or more regions. In each such region, one of the following must be true.

1. *All* points in the region are solutions of the inequality.
2. *No* point in the region is a solution of the inequality.

Thus, you can determine whether the points in an entire region satisfy the inequality by simply testing *one* point in the region.

Sketching the Graph of an Inequality in Two Variables

1. Replace the inequality sign with an equal sign, and sketch the graph of the resulting equation. (Use a dashed line for $<$ or $>$ and a solid line for $\leq$ or $\geq$.)
2. Test one point in each of the regions formed by the graph in Step 1. If the point satisfies the inequality, shade the entire region to denote that every point in the region satisfies the inequality.

Figure 5.17

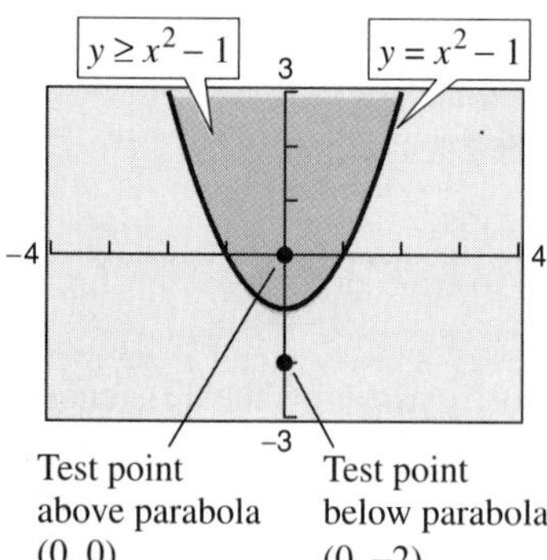

EXAMPLE 1 Sketching the Graph of an Inequality

Sketch the graph of $y \geq x^2 - 1$.

Solution

The graph of the corresponding *equation* $y = x^2 - 1$ is a parabola, as shown in Figure 5.17. By testing the point $(0, 0)$ *above* the parabola and the point $(0, -2)$ *below* the parabola, you can see that the points that satisfy the inequality are those lying above (or on) the parabola.

The inequality given in Example 1 is a nonlinear inequality in two variables. Most of the following examples involve **linear inequalities** such as $ax + by < c$. The graph of a linear inequality is a half-plane lying on one side of the line $ax + by = c$.

EXAMPLE 2 Sketching the Graphs of Linear Inequalities

Sketch the graph of each linear inequality.

a. $x > -2$ **b.** $y \leq 3$

Solution

a. The graph of the corresponding equation $x = -2$ is a vertical line. The points that satisfy the inequality $x > -2$ are those lying to the right of this line, as shown in Figure 5.18.

b. The graph of the corresponding equation $y = 3$ is a horizontal line. The points that satisfy the inequality $y \leq 3$ are those lying below (or on) this line, as shown in Figure 5.19.

Figure 5.18

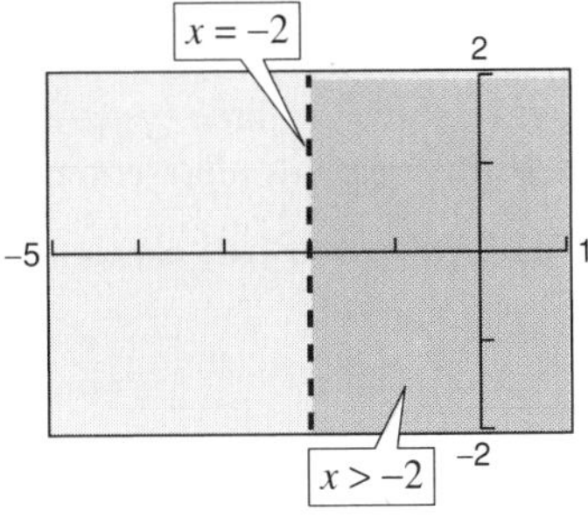

Figure 5.19

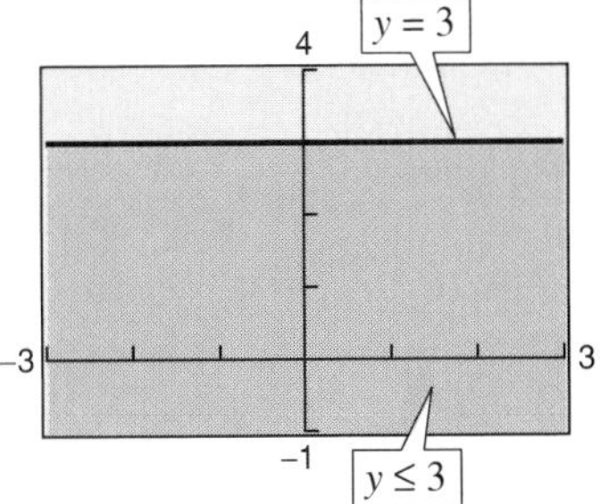

Study Tip

To graph a linear inequality, it can help to write the inequality in slope-intercept form. For instance, by writing $x - y < 2$ in Example 3 in the form

$$y > x - 2$$

you can see that the solution points lie *above* the line $x - y = 2$ (or $y = x - 2$), as shown in Figure 5.20.

Figure 5.20

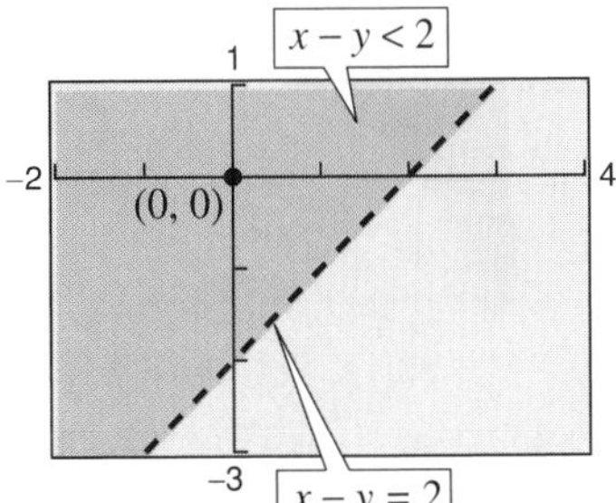

EXAMPLE 3 Sketching the Graph of a Linear Inequality

Sketch the graph of $x - y < 2$.

Solution

The graph of the corresponding equation $x - y = 2$ is a line, as shown in Figure 5.20. Because the origin (0, 0) satisfies the inequality, the graph consists of the half-plane lying above the line. (Try checking a point below the line. Regardless of which point you choose, you will see that it does not satisfy the inequality.)

Systems of Inequalities

Many practical problems in business, science, and engineering involve systems of linear inequalities. A **solution** of a system of inequalities in x and y is a point (x, y) that satisfies each inequality in the system.

To sketch the graph of a system of inequalities in two variables, first sketch the graph of each individual inequality (on the same coordinate system) and then find the region that is *common* to every graph in the system. For systems of *linear* inequalities, it is helpful to find the vertices of the solution region.

EXAMPLE 4 Solving a System of Inequalities

Sketch the graph (and label the vertices) of the solution set of the system.

$$\begin{aligned} x - y &< 2 \\ x &> -2 \\ y &\le 3 \end{aligned}$$

Solution

The graphs of these inequalities are shown in Figures 5.18 to 5.20. The triangular region common to all three graphs can be found by superimposing the graphs on the same coordinate system, as shown in Figure 5.21. To find the vertices of the region, solve the three systems of corresponding equations obtained by taking *pairs* of equations representing the boundaries of the individual regions.

Vertex A: $(-2, -4)$	*Vertex B:* $(5, 3)$	*Vertex C:* $(-2, 3)$
$x - y = 2$	$x - y = 2$	$x = -2$
$x = -2$	$y = 3$	$y = 3$

Figure 5.21

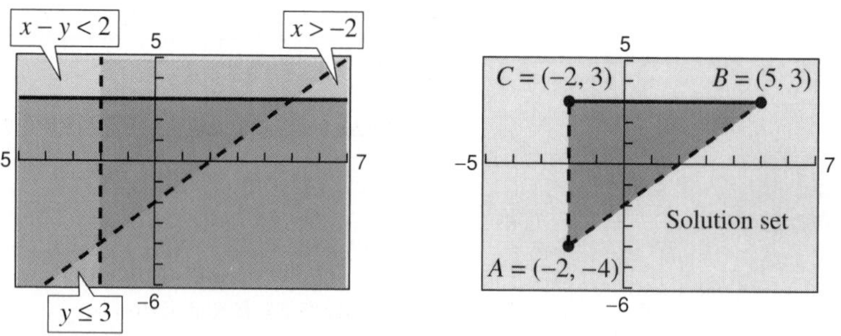

For the triangular region shown in Figure 5.21, each point of intersection of a pair of boundary lines corresponds to a vertex. With more complicated regions, two border lines can sometimes intersect at a point that is not a vertex of the region, as shown in Figure 5.22. To keep track of which points of intersection are actually vertices of the region, we suggest that you sketch the region and refer to your sketch as you find each point of intersection.

Figure 5.22

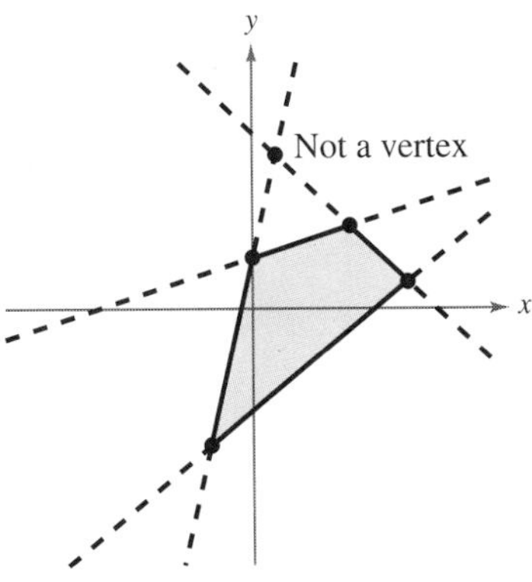

EXAMPLE 5 **Solving a System of Inequalities**

Sketch the region containing all points that satisfy the system.

$$x^2 - y \leq 1$$
$$-x + y \leq 1$$

Solution

As shown in Figure 5.23, the points that satisfy the inequality $x^2 - y \leq 1$ are the points lying above (or on) the parabola given by

$$y = x^2 - 1. \quad \text{Parabola}$$

The points that satisfy the inequality $-x + y \leq 1$ are the points lying below (or on) the line given by

$$y = x + 1. \quad \text{Line}$$

To find the points of intersection of the parabola and the line, solve the system of corresponding equations.

$$x^2 - y = 1$$
$$-x + y = 1$$

Using the method of substitution, you can find the solutions to be $(-1, 0)$ and $(2, 3)$, as shown in Figure 5.23.

Figure 5.23

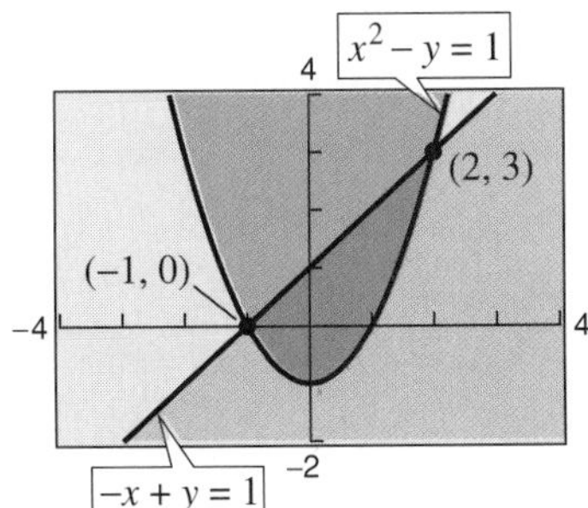

When solving a system of inequalities, you should be aware that the system might have no solution. For instance, the system

$$x + y > 3$$
$$x + y < -1$$

has no solution points, because the quantity $(x + y)$ cannot be both less than -1 and greater than 3, as shown in Figure 5.24.

Figure 5.24 No Solution

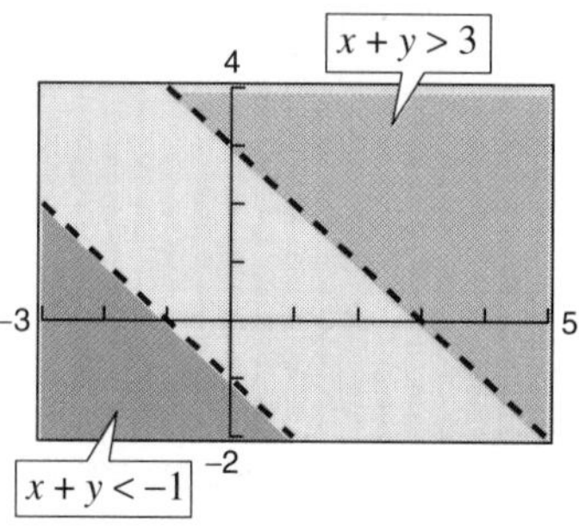

Figure 5.25 Unbounded Region

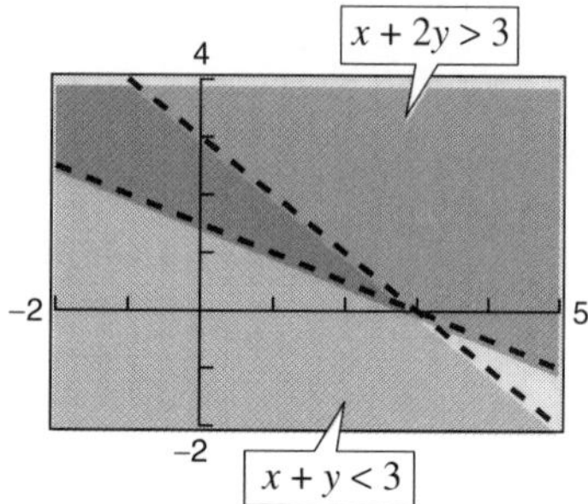

Another possibility is that the solution set of a system of inequalities can be unbounded. For instance, the solution set of

$$x + \quad y < 3$$
$$x + 2y > 3$$

forms an *infinite wedge,* as shown in Figure 5.25.

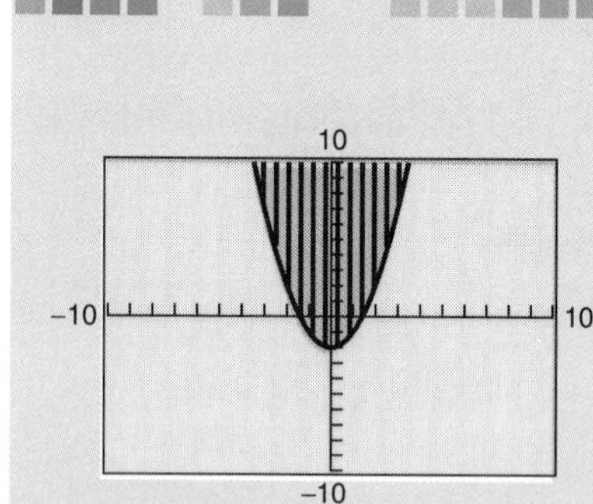

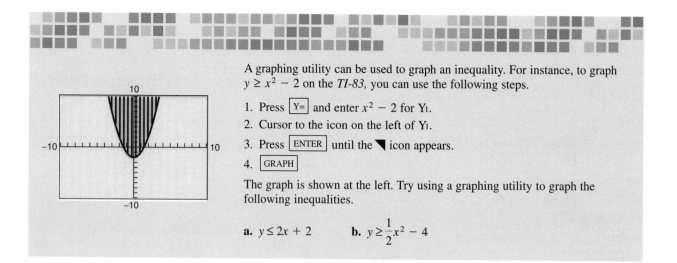

A graphing utility can be used to graph an inequality. For instance, to graph $y \geq x^2 - 2$ on the *TI-83,* you can use the following steps.

1. Press [Y=] and enter $x^2 - 2$ for Y1.
2. Cursor to the icon on the left of Y1.
3. Press [ENTER] until the ◥ icon appears.
4. [GRAPH]

The graph is shown at the left. Try using a graphing utility to graph the following inequalities.

a. $y \leq 2x + 2$ **b.** $y \geq \frac{1}{2}x^2 - 4$

Figure 5.26

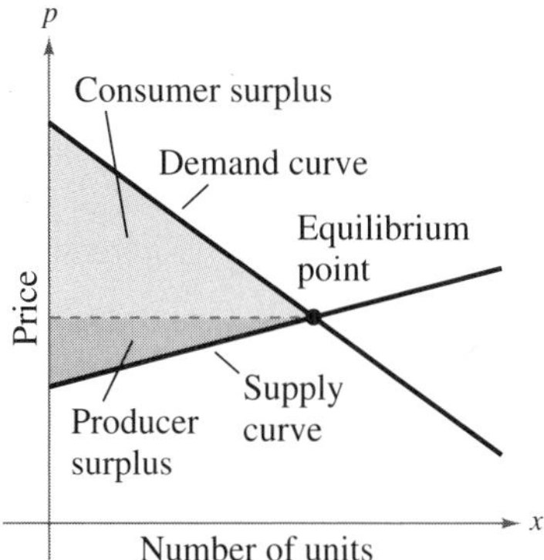

Applications

Example 8 in Section 5.2 discussed the *point of equilibrium* for a demand function and a supply function. The next example discusses two related concepts that economists call **consumer surplus** and **producer surplus.** As shown in Figure 5.26, the consumer surplus is defined as the area of the region that lies *below* the demand curve, *above* the horizontal line passing through the equilibrium point, and to the right of the p-axis. Similarly, the producer surplus is defined as the area of the region that lies *above* the supply curve, *below* the horizontal line passing through the equilibrium point, and to the right of the p-axis. The consumer surplus is a measure of the amount that consumers would have been willing to pay *above what they actually paid,* whereas the producer surplus is a measure of the amount that producers would have been willing to receive *below what they actually received.*

Real Life

EXAMPLE 6 Consumer Surplus and Producer Surplus

The demand and supply functions for a certain type of calculator are given by

$$p = 150 - 0.00001x \qquad \text{Demand equation}$$
$$p = 60 + 0.00002x \qquad \text{Supply equation}$$

where p is the price in dollars and x represents the number of units. Find the consumer surplus and producer surplus for these two equations.

Solution

Begin by finding the point of equilibrium by solving the equation

$$60 + 0.00002x = 150 - 0.00001x.$$

In Example 8 in Section 5.2, you saw that the solution is $x = 3{,}000{,}000$, which corresponds to an equilibrium price of $p = \$120$. Thus, the consumer surplus and producer surplus are the areas of the following triangular regions.

Consumer Surplus	*Producer Surplus*
$p \le 150 - 0.00001x$	$p \ge 60 + 0.00002x$
$p \ge 120$	$p \le 120$
$x \ge 0$	$x \ge 0$

Figure 5.27

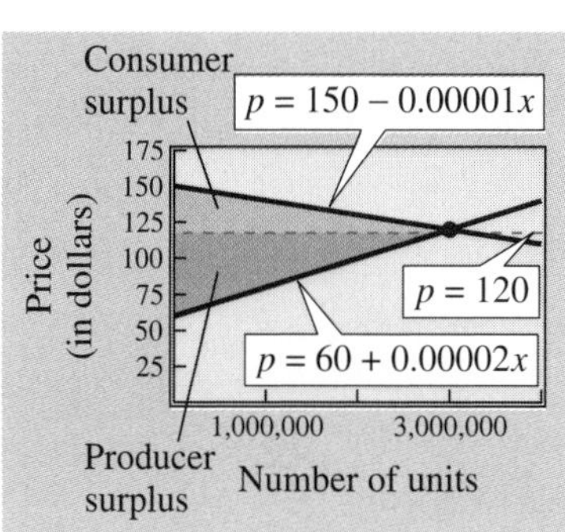

In Figure 5.27, you can see that the consumer and producer surpluses are

Consumer surplus $= \frac{1}{2}(\text{base})(\text{height}) = \frac{1}{2}(3{,}000{,}000)(30) = \$45{,}000{,}000$

Producer surplus $= \frac{1}{2}(\text{base})(\text{height}) = \frac{1}{2}(3{,}000{,}000)(60) = \$90{,}000{,}000.$

Real Life

EXAMPLE 7 **Nutrition**

The liquid portion of a diet is to provide at least 300 calories, 36 units of vitamin A, and 90 units of vitamin C daily. A cup of dietary drink X provides 60 calories, 12 units of vitamin A, and 10 units of vitamin C. A cup of dietary drink Y provides 60 calories, 6 units of vitamin A, and 30 units of vitamin C. Set up a system of linear inequalities that describes the minimum daily requirements for calories and vitamins.

Solution

Begin by letting x and y represent the following.

$x =$ number of cups of dietary drink X

$y =$ number of cups of dietary drink Y

To meet the minimum daily requirements, the following inequalities must be satisfied.

Figure 5.28

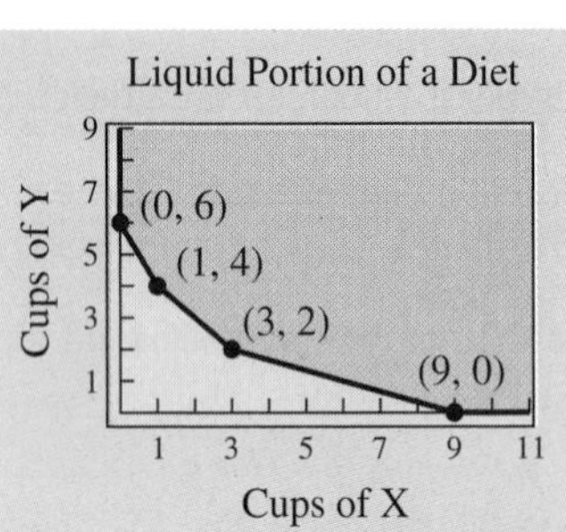

$$\begin{aligned} &\textit{For Calories:} && 60x + 60y \ge 300 \\ &\textit{For Vitamin A:} && 12x + 6y \ge 36 \\ &\textit{For Vitamin C:} && 10x + 30y \ge 90 \\ & && x \ge 0 \\ & && y \ge 0 \end{aligned}$$

The last two inequalities are included because x and y cannot be negative. The graph of this system of inequalities is shown in Figure 5.28. (More is said about this application in Example 7 in Section 5.6.)

Group Activity

Graphing Systems of Inequalities

The Technology feature on page 434 describes how to sketch the graph of a single inequality with a graphing utility. With the user's guide for your graphing utility, find how to graph a *system* of inequalities. Then use the graphing utility to graph the following systems.

a. $y \le 4 - x^2$

$y \ge 2x - 3$

b. $y \le 4 - x^2$

$y \ge x^2 - 4$

5.5 /// EXERCISES

In Exercises 1–8, match the inequality with its graph. [The graphs are labeled (a), (b), (c), (d), (e), (f), (g), and (h).]

(a)

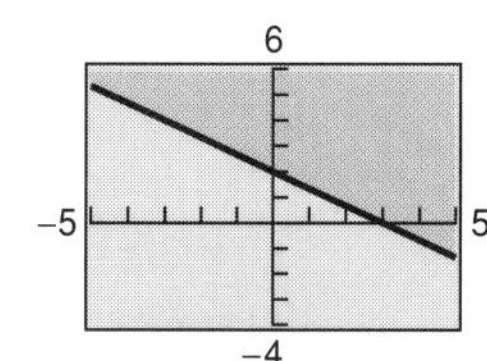

(b)

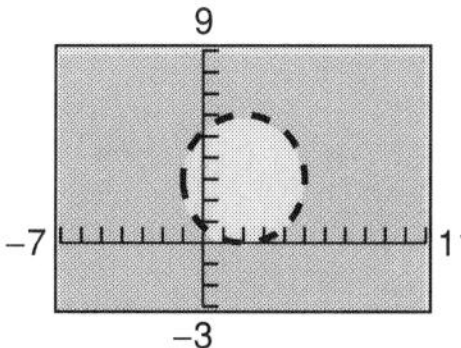

(c)

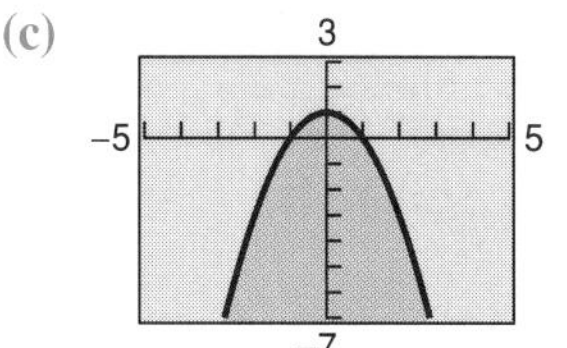

(d)

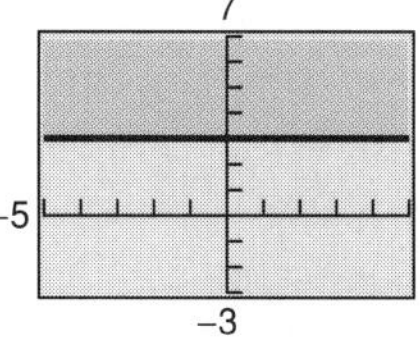

(e)

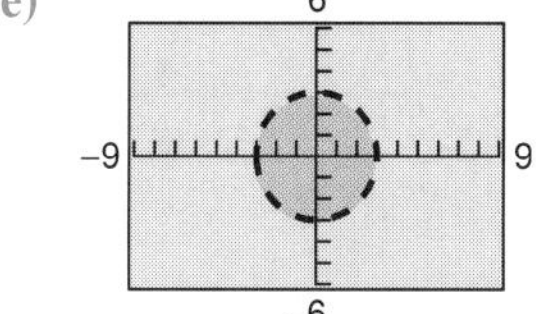

(f)

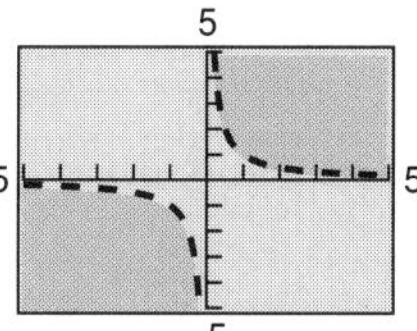

(g)

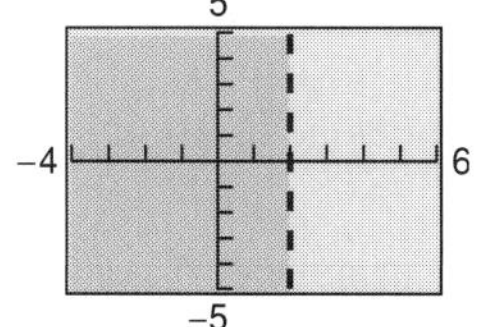

(h)

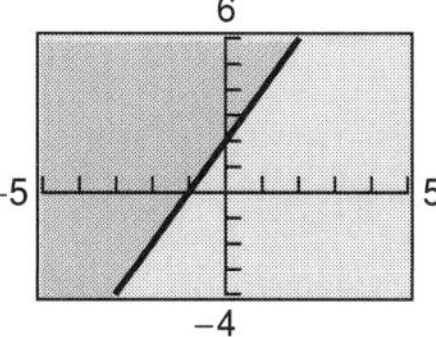

1. $x < 2$

2. $y \geq 3$

3. $2x + 3y \geq 6$

4. $2x - y \leq -2$

5. $x^2 + y^2 < 9$

6. $(x - 2)^2 + (y - 3)^2 > 9$

7. $xy > 1$

8. $y \leq 1 - x^2$

In Exercises 9–18, sketch the graph of the inequality.

9. $x \geq 2$

10. $x \leq 4$

11. $y \geq -1$

12. $y \leq 3$

13. $y < 2 - x$

14. $y > 2x - 4$

15. $2y - x \geq 4$

16. $5x + 3y \geq -15$

17. $y^2 - x < 0$

18. $(x + 1)^2 + y^2 < 9$

In Exercises 19–24, use a graphing utility to graph the inequality. Shade the region representing the solution.

19. $y(1 + x^2) \leq 1$

20. $y < \ln x$

21. $y \geq \frac{2}{3}x - 1$

22. $y \leq 6 - \frac{3}{2}x$

23. $x^2 + 5y - 10 \leq 0$

24. $2x^2 - y - 3 > 0$

In Exercises 25–30, write an inequality for the shaded region shown in the figure.

25.

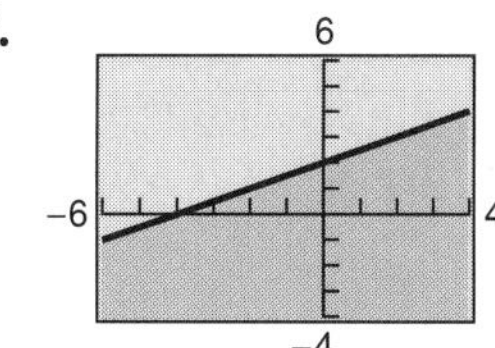

26.

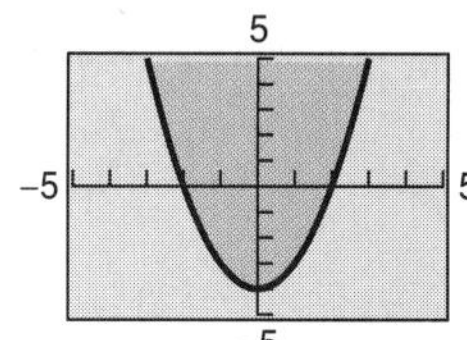

27.

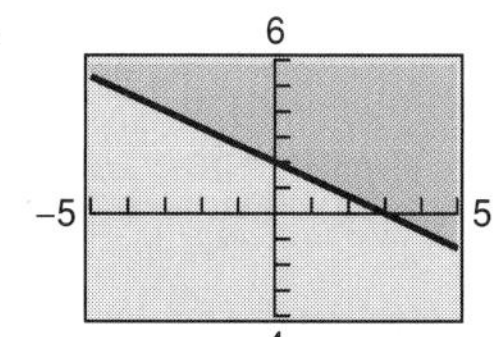

28.

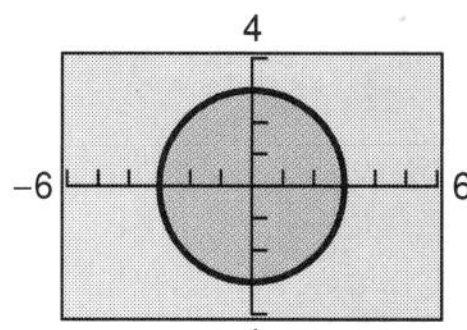

29.

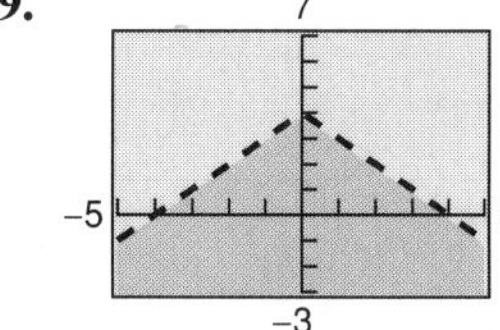

30.

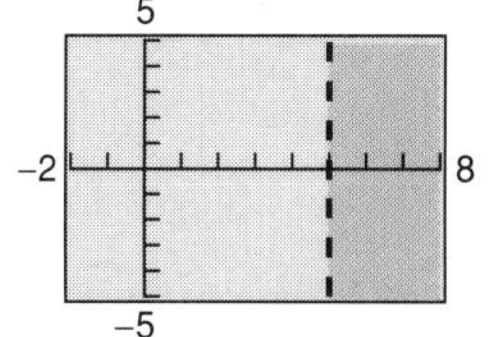

In Exercises 31–44, sketch the graph of the solution of the system of inequalities. Verify your sketch by using a graphing utility to shade the region representing the solution of the system.

31. $\begin{aligned} x + y &\le 1 \\ -x + y &\le 1 \\ y &\ge 0 \end{aligned}$

32. $\begin{aligned} 3x + 2y &< 6 \\ x \quad &> 0 \\ y &> 0 \end{aligned}$

33. $\begin{aligned} x + y &\le 5 \\ x \quad &\ge 2 \\ y &\ge 0 \end{aligned}$

34. $\begin{aligned} 2x^2 + y &\ge 2 \\ x \quad &\le 2 \\ y &\le 1 \end{aligned}$

35. $\begin{aligned} -3x + 2y &< 6 \\ x - 4y &> -2 \\ 2x + y &< 3 \end{aligned}$

36. $\begin{aligned} x - 7y &> -36 \\ 5x + 2y &> 5 \\ 6x - 5y &> 6 \end{aligned}$

37. $\begin{aligned} 2x + y &> 2 \\ 6x + 3y &< 2 \end{aligned}$

38. $\begin{aligned} x - 2y &< -6 \\ 5x - 3y &> -9 \end{aligned}$

39. $\begin{aligned} x \quad &\ge 1 \\ x - 2y &\le 3 \\ 3x + 2y &\ge 9 \\ x + y &\le 6 \end{aligned}$

40. $\begin{aligned} x - y^2 &> 0 \\ x - y &< 2 \end{aligned}$

41. $\begin{aligned} x^2 + y^2 &\le 9 \\ x^2 + y^2 &\ge 1 \end{aligned}$

42. $\begin{aligned} x^2 + y^2 &\le 25 \\ 4x - 3y &\le 0 \end{aligned}$

43. $\begin{aligned} x &> y^2 \\ x &< y + 2 \end{aligned}$

44. $\begin{aligned} x &< 2y - y^2 \\ 0 &< x + y \end{aligned}$

In Exercises 45–50, use a graphing utility to graph the inequalities. Shade the region representing the solution of the system.

45. $\begin{aligned} y &\le \sqrt{3x} + 1 \\ y &\ge x^2 + 1 \end{aligned}$

46. $\begin{aligned} y &< -x^2 + 2x + 3 \\ y &> x^2 - 4x + 3 \end{aligned}$

47. $\begin{aligned} y &< x^3 - 2x + 1 \\ y &> -2x \\ x &\le 1 \end{aligned}$

48. $\begin{aligned} y &\ge x^4 - 2x^2 + 1 \\ y &\le 1 - x^2 \end{aligned}$

49. $\begin{aligned} x^2y &\ge 1 \\ 0 < x &\le 4 \\ y &\le 4 \end{aligned}$

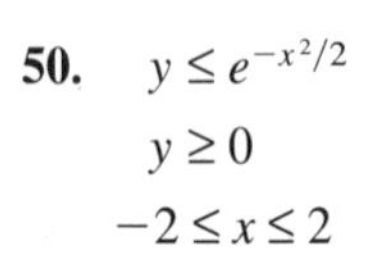

50. $\begin{aligned} y &\le e^{-x^2/2} \\ y &\ge 0 \\ -2 \le x &\le 2 \end{aligned}$

In Exercises 51–60, derive a set of inequalities to describe the region.

51.

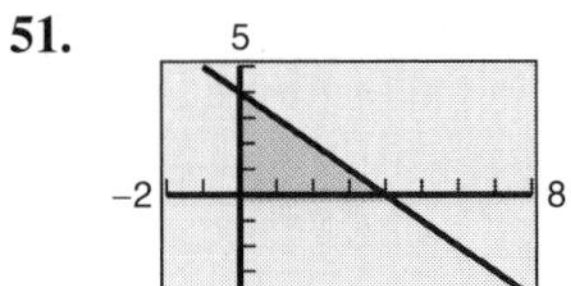

52.

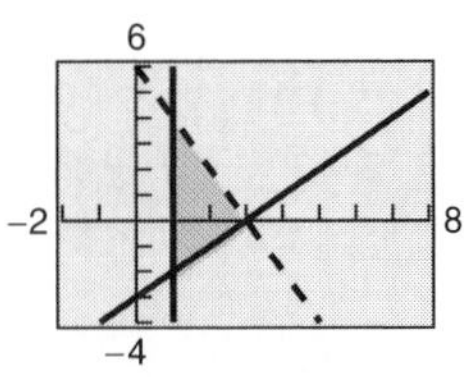

53.

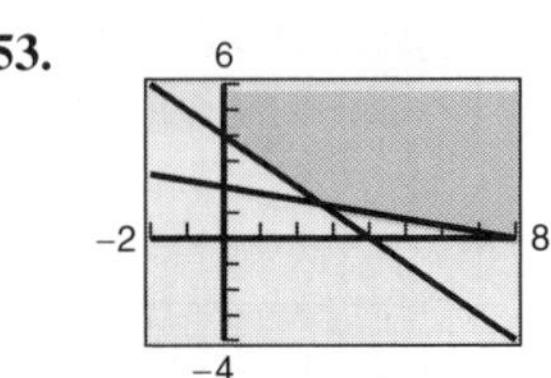

54.

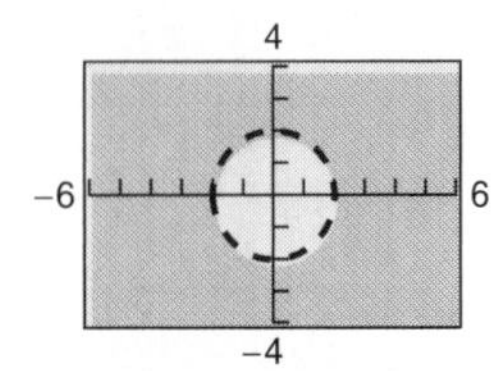

55.

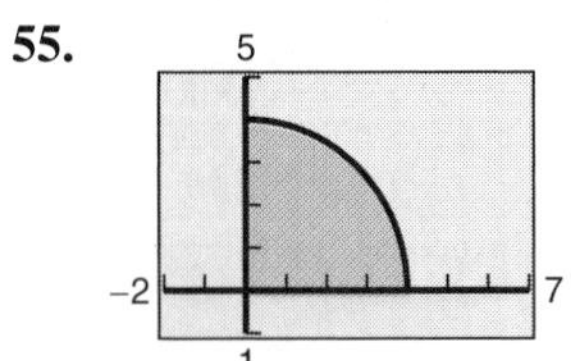

56.

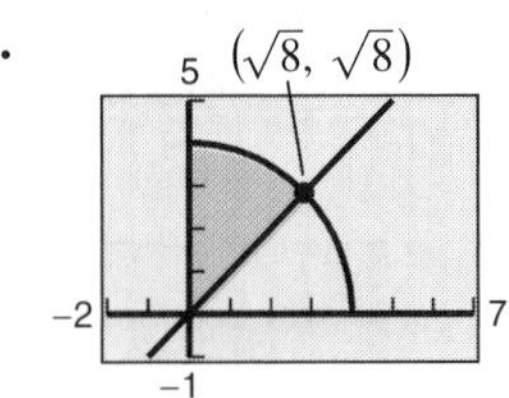

57. Rectangular region with vertices at (2, 1), (5, 1), (5, 7), and (2, 7)

58. Parallelogrammatic region with vertices at (0, 0), (4, 0), (1, 4), and (5, 4)

59. Triangular region with vertices at (0, 0), (5, 0), and (2, 3)

60. Triangular region with vertices at (−1, 0), (1, 0), and (0, 1)

In Exercises 61–67, (a) find a system of inequalities that models the problem, and (b) use a graphing utility to shade the region representing the solution of the system.

61. *Investment* A person plans to invest $20,000 in two different interest-bearing accounts. Each account is to contain at least $5000, and one account should have at least twice the amount that is in the other account.

62. *Concert Ticket Sales* Two types of tickets are to be sold for a concert. One type costs \$15 per ticket and the other costs \$25 per ticket. The promoter of the concert must sell at least 15,000 tickets, including at least 8000 of the \$15 tickets and at least 4000 of the \$25 tickets, and the gross receipts must total at least \$275,000 in order for the concert to be held.

63. *Furniture Production* A furniture company can sell all the tables and chairs it produces. Each table requires 1 hour of assembly and $1\frac{1}{3}$ hours of finishing. Each chair requires $1\frac{1}{2}$ hours of assembly and $1\frac{1}{2}$ hours of finishing. The company's assembly center is available 12 hours per day, and its finishing center is available 15 hours per day.

64. *Computer Inventory* A store sells two models of a certain brand of computer. Because of the demand, it is necessary to stock twice as many units of model A as units of model B. The costs to the store for the two models are \$800 and \$1200, respectively. The management does not want more than \$20,000 in computer inventory at any one time, and it wants at least four model A computers and two model B computers in inventory at all times.

65. *Diet Supplement* A dietitian is asked to design a special diet supplement using two different foods. Each ounce of food X contains 20 units of calcium, 15 units of iron, and 10 units of vitamin B. Each ounce of food Y contains 10 units of calcium, 10 units of iron, and 20 units of vitamin B. The minimum daily requirements in the diet are 280 units of calcium, 160 units of iron, and 180 units of vitamin B.

66. *Diet Supplement* A dietitian is asked to design a special diet supplement using two different foods. Each ounce of food X contains 20 units of calcium, 15 units of iron, and 10 units of vitamin B. Each ounce of food Y contains 10 units of calcium, 10 units of iron, and 20 units of vitamin B. The minimum daily requirements in the diet are 300 units of calcium, 150 units of iron, and 200 units of vitamin B.

67. *Physical Fitness Facility* You plan an indoor running track with an exercise floor inside the track (see figure). The track must be at least 125 meters long, and the exercise floor must have an area of at least 500 square meters.

Figure for 67

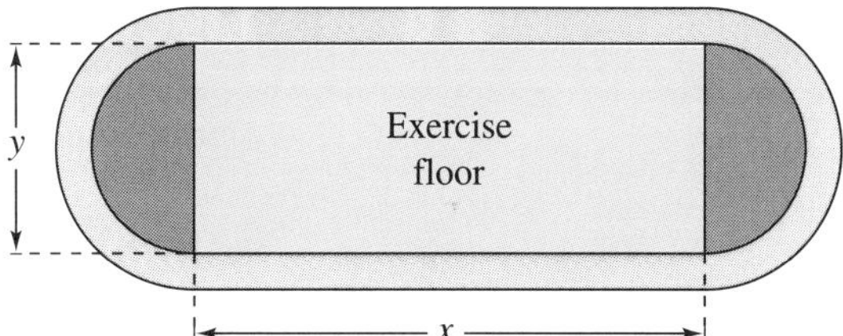

68. *Graphical Reasoning* Two concentric circles have radii of x and y meters, where $y > x$ (see figure). The area between the boundaries of the circles must be at least 10 square meters.
   (a) Find an inequality describing the constraints on the circles.
   (b) Use a graphing utility to graph the inequality in part (a). Graph the line $y = x$ in the same viewing rectangle.
   (c) Identify the graph of the line in relation to the boundary of the inequality. Explain its meaning in the context of the problem.

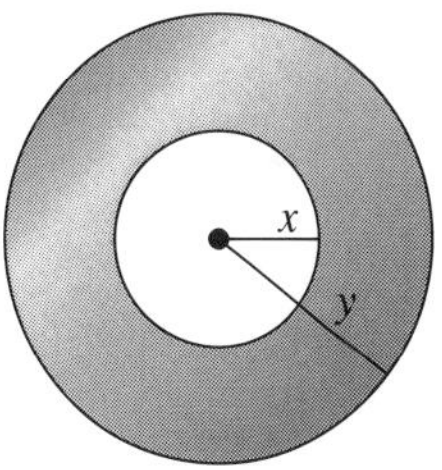

Consumer and Producer Surpluses **In Exercises 69–72, use a graphing utility to shade the regions representing the consumer surplus and producer surplus for the supply and demand equations.**

	Demand	*Supply*
69.	$p = 50 - 0.5x$	$p = 0.125x$
70.	$p = 60 - x$	$p = 10 + \frac{7}{3}x$
71.	$p = 300 - x$	$p = 100 + x$
72.	$p = 140 - 0.00002x$	$p = 80 + 0.00001x$

73. *Think About It* After graphing the boundary of an inequality in x and y, how do you decide on which side of the boundary the solution set of the inequality lies?

5.6 Linear Programming

Linear Programming: A Graphical Approach / Applications

Linear Programming: A Graphical Approach

Many applications in business and economics involve a process called **optimization,** in which you are asked to find the minimum or maximum value of a quantity. In this section you will study an optimization strategy called **linear programming.**

A two-dimensional linear programming problem consists of a linear **objective function** and a system of linear inequalities called **constraints.** The objective function gives the quantity that is to be maximized (or minimized), and the constraints determine the set of **feasible solutions.** For example, suppose you are asked to maximize the value of

$$z = ax + by \qquad \text{Objective function}$$

subject to a set of constraints that determines the region in Figure 5.29. Because every point in the region satisfies each constraint, it is not clear how you should go about finding the point that yields a maximum value of z. Fortunately, it can be shown that if there is an optimal solution, it must occur at one of the vertices. This means that *you can find the maximum value by testing z at each of the vertices.*

Figure 5.29

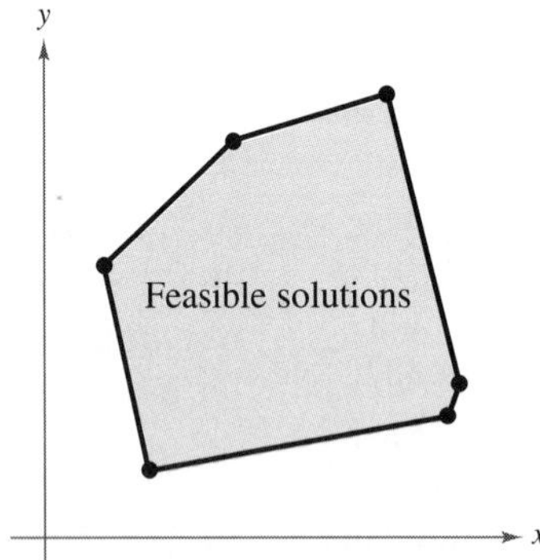

Optimal Solution of a Linear Programming Problem

If a linear programming problem has a solution, it must occur at a vertex of the set of feasible solutions. If there is more than one solution, at least one of them must occur at such a vertex. In either case, the value of the objective function is unique.

EXAMPLE 1 **Solving a Linear Programming Problem**

Find the maximum value of

$$z = 3x + 2y \qquad \text{Objective function}$$

subject to the following constraints.

$$\left.\begin{aligned} x &\geq 0 \\ y &\geq 0 \\ x + 2y &\leq 4 \\ x - y &\leq 1 \end{aligned}\right\} \quad \text{Constraints}$$

Figure 5.30

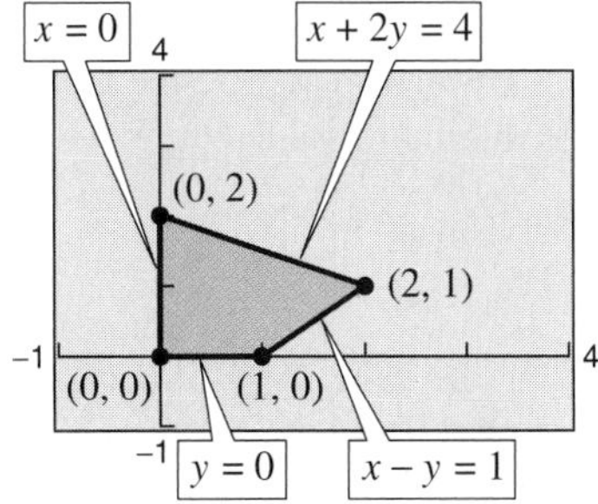

Solution

The constraints form the region shown in Figure 5.30. At the four vertices of this region, the objective function has the following values.

At (0, 0): $z = 3(0) + 2(0) = 0$

At (1, 0): $z = 3(1) + 2(0) = 3$

At (2, 1): $z = 3(2) + 2(1) = 8$ Maximum value of z

At (0, 2): $z = 3(0) + 2(2) = 4$

Thus, the maximum value of z is 8, and this occurs when $x = 2$ and $y = 1$.

Note In Example 1, try testing some of the *interior* points in the region. You will see that the corresponding values of z are less than 8. Here are some examples.

At (1, 1): $z = 3(1) + 2(1) = 5$

At $\left(1, \frac{1}{2}\right)$: $z = 3(1) + 2\left(\frac{1}{2}\right) = 4$

At $\left(\frac{1}{2}, \frac{3}{2}\right)$: $z = 3\left(\frac{1}{2}\right) + 2\left(\frac{3}{2}\right) = \frac{9}{2}$

To see why the maximum value of the objective function in Example 1 must occur at a vertex, consider writing the objective function in the form

$$y = -\frac{3}{2}x + \frac{z}{2} \qquad \text{Family of lines}$$

Figure 5.31

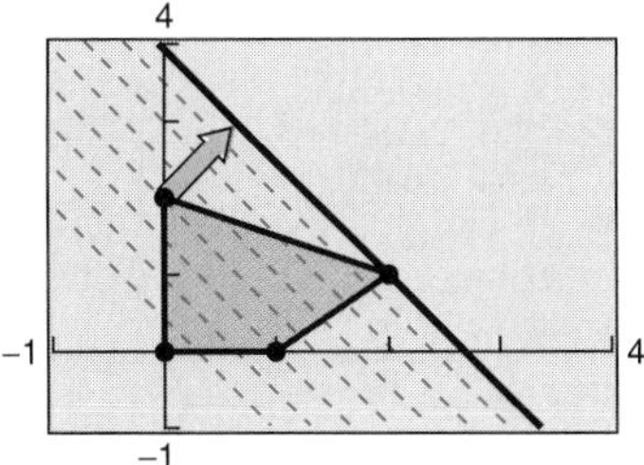

where $z/2$ is the y-intercept of the objective function. This equation represents a family of lines, each of slope $-\frac{3}{2}$. Of these infinitely many lines, you want the one that has the largest z-value while still intersecting the region determined by the constraints. In other words, of all the lines with a slope of $-\frac{3}{2}$, you want the one that has the largest y-intercept *and* intersects the given region, as shown in Figure 5.31. It should be clear that such a line will pass through one (or more) of the vertices of the region.

Solving a Linear Programming Problem

To solve a linear programming problem involving two variables by the graphical method, use the following steps.

1. Sketch the region corresponding to the system of constraints. (The points inside or on the boundary of the region are called *feasible solutions*.)
2. Find the vertices of the region.
3. Test the objective function at each of the vertices and select the values of the variables that optimize the objective function. For a bounded region, both a minimum and a maximum value will exist. (For an unbounded region, *if* an optimal solution exists, it will occur at a vertex.)

Study Tip

Remember that a vertex of a region can be found using a system of linear equations. The system will consist of the equations of the lines passing through the vertex.

These guidelines will work whether the objective function is to be maximized or minimized. For instance, the same test used in Example 1 to find the maximum value of z can be used to conclude that the minimum value of z is 0 and that this value occurs at the vertex $(0, 0)$.

EXAMPLE 2 Solving a Linear Programming Problem

Find the maximum value of

$$z = 4x + 6y \qquad \text{Objective function}$$

where $x \geq 0$ and $y \geq 0$, subject to the following constraints.

$$\left.\begin{aligned} -x + y &\leq 11 \\ x + y &\leq 27 \\ 2x + 5y &\leq 90 \end{aligned}\right\} \quad \text{Constraints}$$

Figure 5.32

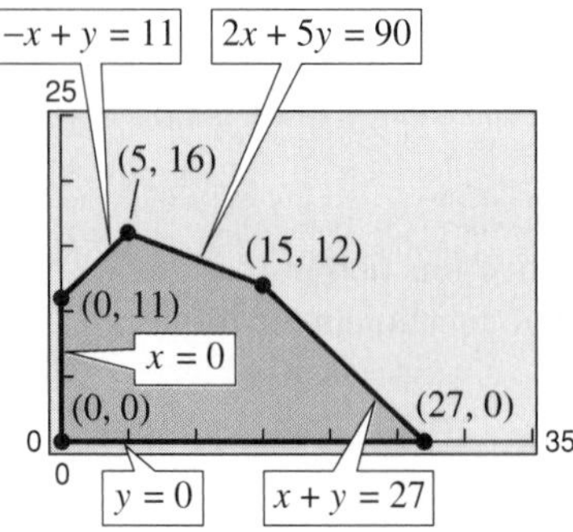

Solution

The region bounded by the constraints is shown in Figure 5.32. By testing the objective function at each vertex, you obtain the following.

$$\begin{aligned} &\text{At } (0, 0): && z = 4(0) + 6(0) = 0 \\ &\text{At } (0, 11): && z = 4(0) + 6(11) = 66 \\ &\text{At } (5, 16): && z = 4(5) + 6(16) = 116 \\ &\text{At } (15, 12): && z = 4(15) + 6(12) = 132 \qquad \text{Maximum value of } z \\ &\text{At } (27, 0): && z = 4(27) + 6(0) = 108 \end{aligned}$$

Thus, the maximum value of z is 132, and this occurs when $x = 15$ and $y = 12$.

The next example shows that the same basic procedure can be used to solve a problem in which the objective function is to be *minimized*.

EXAMPLE 3 **Minimizing an Objective Function**

Find the minimum value of

$$z = 5x + 7y \qquad \text{Objective function}$$

where $x \geq 0$ and $y \geq 0$, subject to the following constraints.

$$\left.\begin{aligned} 2x + 3y &\geq 6 \\ 3x - y &\leq 15 \\ -x + y &\leq 4 \\ 2x + 5y &\leq 27 \end{aligned}\right\} \quad \text{Constraints}$$

Solution

The region bounded by the constraints is shown in Figure 5.33. By testing the objective function at each vertex, you obtain the following.

Figure 5.33

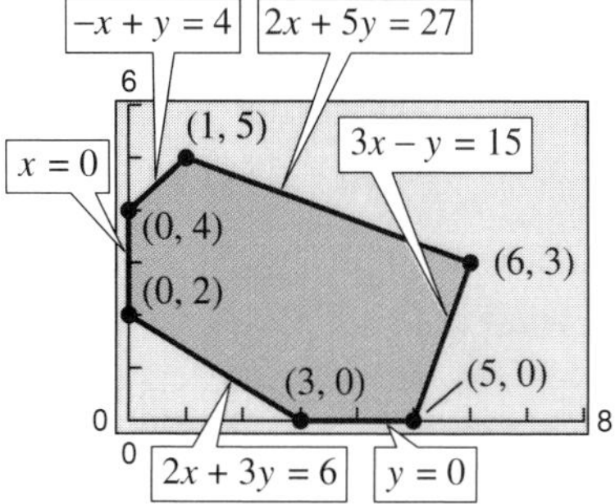

At (0, 2): $z = 5(0) + 7(2) = 14$ Minimum value of z

At (0, 4): $z = 5(0) + 7(4) = 28$

At (1, 5): $z = 5(1) + 7(5) = 40$

At (6, 3): $z = 5(6) + 7(3) = 51$

At (5, 0): $z = 5(5) + 7(0) = 25$

At (3, 0): $z = 5(3) + 7(0) = 15$

Thus, the minimum value of z is 14, and this occurs when $x = 0$ and $y = 2$.

EXAMPLE 4 **Maximizing an Objective Function**

Find the maximum value of

$$z = 5x + 7y \qquad \text{Objective function}$$

where $x \geq 0$ and $y \geq 0$, subject to the following constraints.

$$\left.\begin{aligned} 2x + 3y &\geq 6 \\ 3x - y &\leq 15 \\ -x + y &\leq 4 \\ 2x + 5y &\leq 27 \end{aligned}\right\} \quad \text{Constraints}$$

Solution

This linear programming problem is identical to that given in Example 3 above, except that the objective function is *maximized* instead of minimized. Using the values of z at the vertices shown above, you can conclude that the maximum value of z is 51, and that this value occurs when $x = 6$ and $y = 3$.

It is possible for the maximum (or minimum) value in a linear programming problem to occur at *two* different vertices. For instance, at the vertices of the region shown in Figure 5.34, the objective function

$$z = 2x + 2y \qquad \text{Objective function}$$

has the following values.

At (0, 0): $z = 2(0) + 2(0) = 0$

At (0, 4): $z = 2(0) + 2(4) = 8$

At (2, 4): $z = 2(2) + 2(4) = 12$ Maximum value of z

At (5, 1): $z = 2(5) + 2(1) = 12$ Maximum value of z

At (5, 0): $z = 2(5) + 2(0) = 10$

Figure 5.34

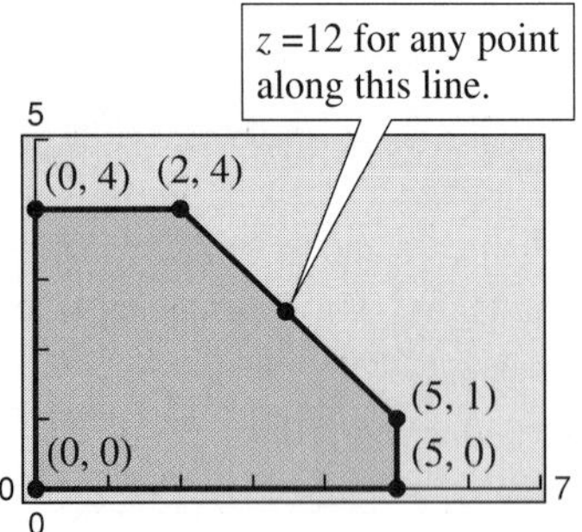

In this case, you can conclude that the objective function has a maximum value (of 12) not only at the vertices (2, 4) and (5, 1), but also at *any point on the line segment connecting these two vertices,* as shown in Figure 5.34. Note that the objective function

$$y = -x + \tfrac{1}{2}z$$

has the same slope as the line through the vertices (2, 4) and (5, 1).

Some linear programming problems have no optimal solution. This can occur if the region determined by the constraints is *unbounded.* Example 5 illustrates such a problem.

EXAMPLE 5 **An Unbounded Region**

Find the maximum value of

$$z = 4x + 2y \qquad \text{Objective function}$$

where $x \geq 0$ and $y \geq 0$, subject to the following constraints.

$$\left.\begin{aligned} x + 2y &\geq 4 \\ 3x + y &\geq 7 \\ -x + 2y &\leq 7 \end{aligned}\right\} \text{ Constraints}$$

Solution

The region determined by the constraints is shown in Figure 5.35. For this unbounded region, there is no maximum value of z. To see this, note that the point $(x, 0)$ lies in the region for all values of $x \geq 4$. By choosing large values of x, you can obtain values of

$$z = 4(x) + 2(0) = 4x$$

that are as large as you want. Thus, there is no maximum value of z. For this problem, there *is* a minimum value of $z = 10$, which occurs at the vertex (2, 1).

Figure 5.35

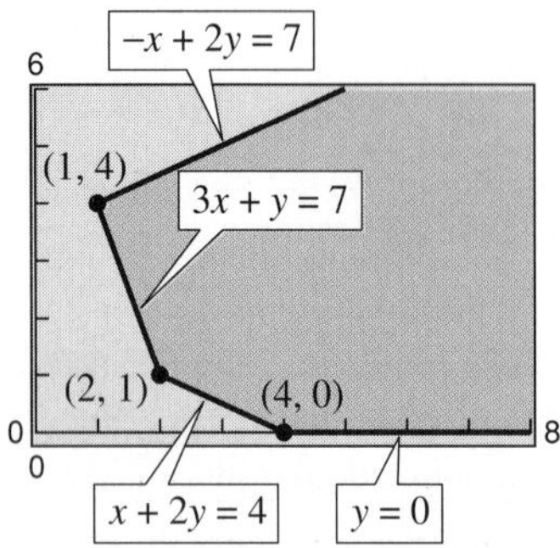

Applications

Example 6 shows how linear programming can be used to find the maximum profit in a business application.

EXAMPLE 6 **Maximum Profit** *Real Life*

A manufacturer wants to maximize the profit for two products. Product I yields a profit of \$1.50 per unit, and product II yields a profit of \$2.00 per unit. Market tests and available resources have indicated the following constraints.

1. The combined production level should not exceed 1200 units per month.
2. The demand for product II is no more than half the demand for product I.
3. The production level of product I is less than or equal to 600 units plus three times the production level of product II.

Solution

If you let x be the number of units of product I and y be the number of units of product II, the objective function (for the combined profit) is given by

$$P = 1.5x + 2y. \qquad \text{Objective function}$$

The three constraints translate into the following linear inequalities.

1. $x + y \le 1200 \quad \Rightarrow \quad x + y \le 1200$
2. $y \le \frac{1}{2}x \quad \Rightarrow \quad -x + 2y \le 0$
3. $x \le 3y + 600 \quad \Rightarrow \quad x - 3y \le 600$

Because neither x nor y can be negative, you also have the two additional constraints of $x \ge 0$ and $y \ge 0$. Figure 5.36 shows the region determined by the constraints. To find the maximum profit, test the value of P at the vertices of the region.

Figure 5.36

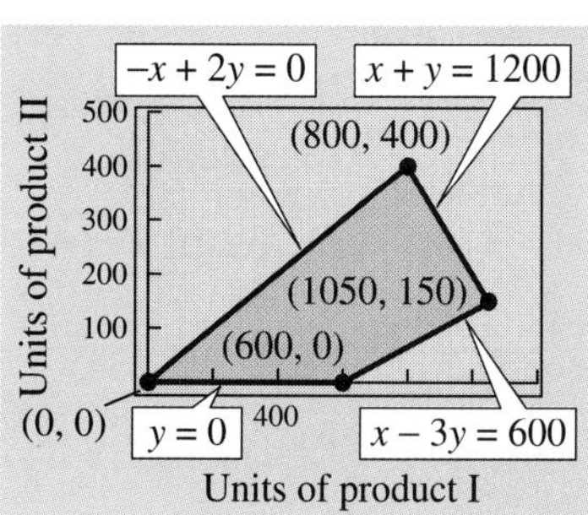

At (0, 0):	$P = 1.5(0) + 2(0)$	$= 0$	
At (800, 400):	$P = 1.5(800) + 2(400)$	$= 2000$	Maximum profit
At (1050, 150):	$P = 1.5(1050) + 2(150)$	$= 1875$	
At (600, 0):	$P = 1.5(600) + 2(0)$	$= 900$	

Thus, the maximum profit is \$2000, and it occurs when the monthly production consists of 800 units of product I and 400 units of product II.

Note In Example 6, suppose the manufacturer improved the production of product I so that it yielded a profit of \$2.50 per unit. How would this affect the number of units the manufacturer should sell to obtain a maximum profit?

Real Life

EXAMPLE 7 Minimum Cost

The liquid portion of a diet is to provide at least 300 calories, 36 units of vitamin A, and 90 units of vitamin C daily. A cup of dietary drink X costs \$0.12 and provides 60 calories, 12 units of vitamin A, and 10 units of vitamin C. A cup of dietary drink Y costs \$0.15 and provides 60 calories, 6 units of vitamin A, and 30 units of vitamin C. How many cups of each drink should be consumed each day to minimize the cost and still meet the daily requirements?

Solution

As in Example 7 on page 436, let x be the number of cups of dietary drink X and let y be the number of cups of dietary drink Y.

Figure 5.37

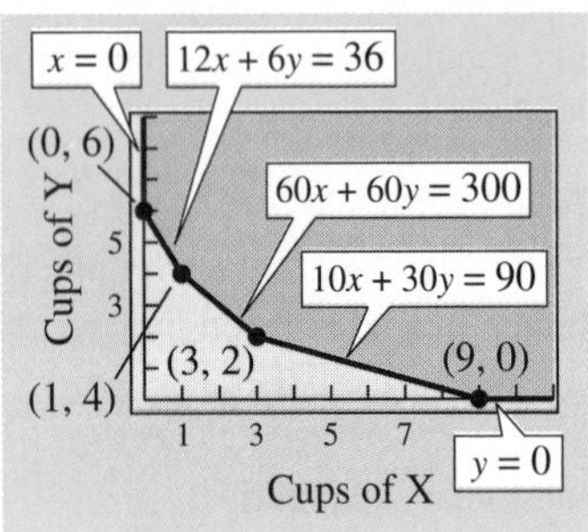

$$\left.\begin{array}{lrl} \textit{For Calories:} & 60x + 60y & \geq 300 \\ \textit{For Vitamin A:} & 12x + 6y & \geq 36 \\ \textit{For Vitamin C:} & 10x + 30y & \geq 90 \\ & x & \geq 0 \\ & y & \geq 0 \end{array}\right\} \quad \text{Constraints}$$

The cost C is given by $C = 0.12x + 0.15y$. Objective function

The graph of the region determined by the constraints is shown in Figure 5.37. To determine the minimum cost, test C at each vertex of the region.

$$\begin{array}{ll} \text{At } (0, 6): & C = 0.12(0) + 0.15(6) = 0.90 \\ \text{At } (1, 4): & C = 0.12(1) + 0.15(4) = 0.72 \\ \text{At } (3, 2): & C = 0.12(3) + 0.15(2) = 0.66 \quad \text{Minimum value of } C \\ \text{At } (9, 0): & C = 0.12(9) + 0.15(0) = 1.08 \end{array}$$

Thus, the minimum cost is \$0.66 per day, and this occurs when three cups of drink X and two cups of drink Y are consumed each day.

Group Activity

Creating a Linear Programming Problem

Sketch the region determined by the constraints $x \geq 0$, $y \geq 0$, $x + 2y \leq 10$, and $x + y \leq 7$. Find, if possible, an objective function of the form $z = ax + by$ that has a maximum at the indicated vertex of the region.

a. Maximum at (0, 5) **b.** Maximum at (4, 3)

c. Maximum at (7, 0) **d.** Maximum at (0, 0)

5.6 /// EXERCISES

In Exercises 1–12, find the minimum and maximum values of the objective function, subject to the indicated constraints. (For each exercise, the graph of the region determined by the constraints is provided.)

1. Objective function:

$z = 4x + 5y$

Constraints:

$x \geq 0$

$y \geq 0$

$x + y \leq 6$

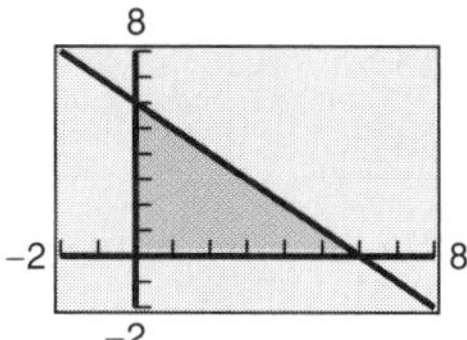

2. Objective function:

$z = 2x + 8y$

Constraints:

$x \geq 0$

$y \geq 0$

$2x + y \leq 4$

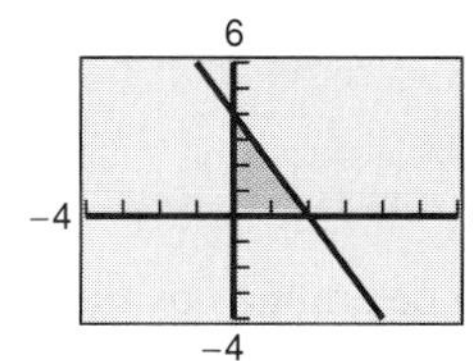

3. Objective function:

$z = 10x + 6y$

Constraints:
(See Exercise 1.)

4. Objective function:

$z = 7x + 3y$

Constraints:
(See Exercise 2.)

5. Objective function:

$z = 3x + 2y$

Constraints:

$x \geq 0$

$y \geq 0$

$x + 3y \leq 15$

$4x + y \leq 16$

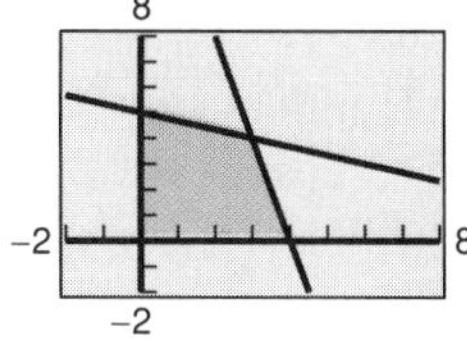

6. Objective function:

$z = 4x + 3y$

Constraints:

$x \geq 0$

$2x + 3y \geq 6$

$3x - 2y \leq 9$

$x + 5y \leq 20$

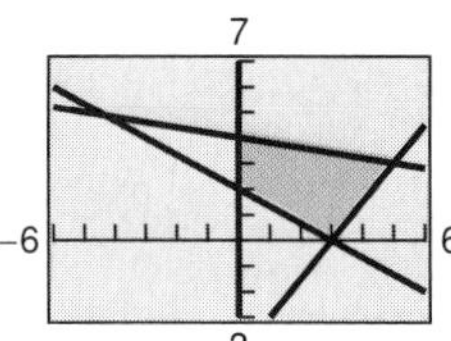

7. Objective function:

$z = 5x + 0.5y$

Constraints:
(See Exercise 5.)

8. Objective function:

$z = x + 6y$

Constraints:
(See Exercise 6.)

9. Objective function:

$z = 10x + 7y$

Constraints:

$0 \leq x \leq 60$

$0 \leq y \leq 45$

$5x + 6y \leq 420$

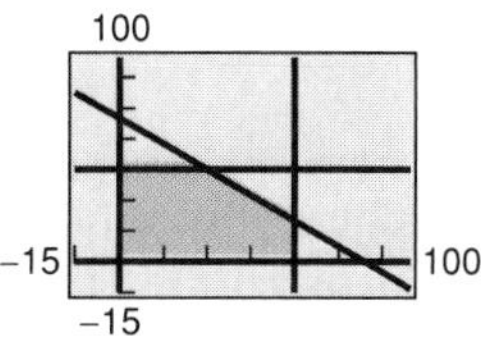

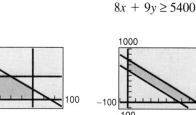

10. Objective function:

$z = 50x + 35y$

Constraints:

$x \geq 0$

$y \geq 0$

$8x + 9y \leq 7200$

$8x + 9y \geq 5400$

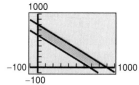

11. Objective function:

$z = 25x + 30y$

Constraints:
(See Exercise 9.)

12. Objective function:

$z = 16x + 18y$

Constraints:
(See Exercise 10.)

In Exercises 13–20, sketch the constraint region. Then find the minimum and maximum values of the objective function, subject to the constraints.

13. Objective function:

$z = 6x + 10y$

Constraints:

$x \geq 0$

$y \geq 0$

$2x + 5y \leq 10$

14. Objective function:

$z = 7x + 8y$

Constraints:

$x \geq 0$

$y \geq 0$

$x + \frac{1}{2}y \leq 4$

15. Objective function:

$z = 9x + 24y$

Constraints:
(See Exercise 13.)

16. Objective function:

$z = 7x + 2y$

Constraints:
(See Exercise 14.)

17. Objective function:

$z = 4x + 5y$

Constraints:

$$\begin{aligned} x &\geq 0 \\ y &\geq 0 \\ x + y &\geq 8 \\ 3x + 5y &\geq 30 \end{aligned}$$

18. Objective function:

$z = 4x + 5y$

Constraints:

$$\begin{aligned} x &\geq 0 \\ y &\geq 0 \\ 2x + 2y &\leq 10 \\ x + 2y &\leq 6 \end{aligned}$$

19. Objective function:

$z = 2x + 7y$

Constraints:
(See Exercise 17.)

20. Objective function:

$z = 2x - y$

Constraints:
(See Exercise 18.)

In Exercises 21–26, use a graphing utility to sketch the region determined by the constraints. Then find the minimum and maximum values of the objective function, subject to the constraints.

21. Objective function:

$z = 4x + y$

Constraints:

$$\begin{aligned} x &\geq 0 \\ y &\geq 0 \\ x + 2y &\leq 40 \\ 2x + 3y &\geq 72 \end{aligned}$$

22. Objective function:

$z = x$

Constraints:

$$\begin{aligned} x &\geq 0 \\ y &\geq 0 \\ 2x + 3y &\leq 60 \\ 2x + y &\leq 28 \\ 4x + y &\leq 48 \end{aligned}$$

23. Objective function:

$z = x + 4y$

Constraints:
(See Exercise 21.)

24. Objective function:

$z = y$

Constraints:
(See Exercise 22.)

25. Objective function:

$z = 2x + 3y$

Constraints:
(See Exercise 21.)

26. Objective function:

$z = 3x + 2y$

Constraints:
(See Exercise 22.)

Exploration **In Exercises 27–30, (a) use a graphing utility to graph the region bounded by the following constraints.**

$$\begin{aligned} 3x + y &\leq 15 \\ 4x + 3y &\leq 30 \\ x &\geq 0 \\ y &\geq 0 \end{aligned}$$

(b) Graph the objective function for the given maximum value of z in the same viewing rectangle as the graph of the constraints. (c) Use the graph to determine the feasible point or points that yield the maximum. Explain how you arrived at your answer.

	Objective Function	*Maximum*
27.	$z = 2x + y$	$z = 12$
28.	$z = 5x + y$	$z = 25$
29.	$z = x + y$	$z = 10$
30.	$z = 3x + y$	$z = 15$

Exploration **In Exercises 31–34, (a) use a graphing utility to graph the region bounded by the following constraints.**

$$\begin{aligned} x + 4y &\leq 20 \\ x + y &\leq 8 \\ 3x + 2y &\leq 21 \\ x &\geq 0 \\ y &\geq 0 \end{aligned}$$

(b) Graph the objective function for the given maximum value of z in the same viewing rectangle as the graph of the constraints. (c) Use the graph to determine the feasible point or points that yield the maximum. Explain how you arrived at your answer.

	Objective Function	*Maximum*
31.	$z = x + 5y$	$z = 25$
32.	$z = 2x + 4y$	$z = 24$
33.	$z = 4x + 5y$	$z = 36$
34.	$z = 4x + y$	$z = 28$

Think About It **In Exercises 35–38, find an objective function that has a maximum or minimum value at the indicated vertex of the constraint region shown below. (There are many correct answers.)**

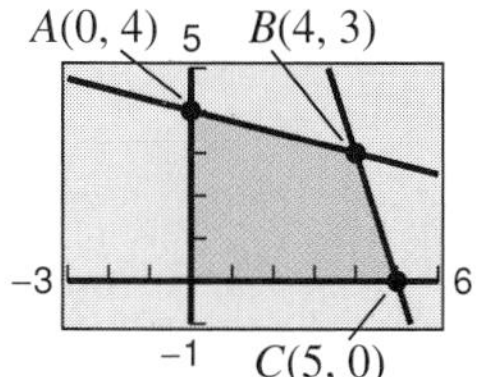

35. The maximum occurs at vertex A.

36. The maximum occurs at vertex B.

37. The maximum occurs at vertex C.

38. The minimum occurs at vertex C.

39. *Maximum Profit* A merchant plans to sell two models of compact disc players at costs of \$250 and \$400. The \$250 model yields a profit of \$45, and the \$400 model yields a profit of \$50. The merchant estimates that the total monthly demand will not exceed 250 units. The merchant does not want to invest more than \$70,000 in inventory for these products. Find the number of units of each model that should be stocked in order to maximize profit.

40. *Maximum Profit* A fruit grower has 150 acres of land available to raise two crops, A and B. It takes 1 day to trim an acre of crop A and 2 days to trim an acre of crop B, and there are 240 days per year available for trimming. It takes 0.3 day to pick an acre of crop A and 0.1 day to pick an acre of crop B, and there are 30 days available for picking. The profits are \$140 per acre for crop A and \$235 per acre for crop B. Find the number of acres of each fruit that should be planted to maximize profit.

41. *Minimum Cost* Two gasolines, type A (\$1.13 per gallon) and type B (\$1.28 per gallon), have octane ratings of 80 and 92, respectively. Determine the blend of minimum cost with an octane rating of at least 90. (*Hint:* Let x be the fraction of each gallon that is type A and let y be the fraction that is type B.)

42. *Maximum Revenue* An accounting firm has 900 hours of staff time and 100 hours of reviewing time available each week. The firm charges \$2000 for an audit and \$300 for a tax return. Each audit requires 100 hours of staff time and 10 hours of review time. Each tax return requires 12.5 hours of staff time and 2.5 hours of review time. What numbers of audits and tax returns will yield the maximum revenue?

43. *Maximum Revenue* The accounting firm in Exercise 42 lowers its charge for an audit to \$1000. What numbers of audits and tax returns will yield the maximum revenue?

44. *Maximum Profit* A manufacturer produces two models of bicycles. The amounts of time (in hours) required for assembling, painting, and packaging the two models are as follows.

	Model A	Model B
Assembling	2	2.5
Painting	4	1
Packaging	1	0.75

The total amounts of time available for assembling, painting, and packaging are 4000, 4800, and 1500 hours, respectively. The profits per unit are \$45 (model A) and \$50 (model B). How many of each model should be produced to maximize profit?

45. *Maximum Profit* A manufacturer produces two models of bicycles. The amounts of time (in hours) required for assembling, painting, and packaging the two models are as follows.

	Model A	Model B
Assembling	2.5	3
Painting	2	1
Packaging	0.75	1.25

The total amounts of time available for assembling, painting, and packaging are 4000, 2500, and 1500 hours, respectively. The profits per unit are \$50 (model A) and \$52 (model B). How many of each model should be produced to maximize profit?

46. *Minimum Cost* A farming cooperative mixes two brands of cattle feed. Brand X costs \$25 per bag and contains 2 units of nutritional element A, 2 units of element B, and 2 units of element C. Brand Y costs \$20 per bag and contains 1 unit of nutritional element A, 9 units of element B, and 3 units of element C. Find the number of bags of each brand that should be mixed to produce a mixture having a minimum cost per bag. The minimum requirements for nutrients A, B, and C are 12 units, 36 units, and 24 units, respectively.

In Exercises 47–52, the linear programming problem has an unusual characteristic. Sketch a graph of the solution region for the problem and describe the unusual characteristic. The objective function is to be maximized in each case.

47. Objective function:

$z = 2.5x + y$

Constraints:

$x \geq 0$

$y \geq 0$

$3x + 5y \leq 15$

$5x + 2y \leq 10$

48. Objective function:

$z = x + y$

Constraints:

$x \geq 0$

$y \geq 0$

$-x + y \leq 1$

$-x + 2y \leq 4$

49. Objective function:

$z = -x + 2y$

Constraints:

$x \geq 0$

$y \geq 0$

$x \leq 10$

$x + y \leq 7$

50. Objective function:

$z = x + y$

Constraints:

$x \geq 0$

$y \geq 0$

$-x + y \leq 0$

$-3x + y \geq 3$

51. Objective function:

$z = 3x + 4y$

Constraints:

$x \geq 0$

$y \geq 0$

$x + y \leq 1$

$2x + y \leq 4$

52. Objective function:

$z = x + 2y$

Constraints:

$x \geq 0$

$y \geq 0$

$x + 2y \leq 4$

$2x + y \leq 4$

In Exercises 53 and 54, determine values of t such that the objective function has a maximum value at the indicated vertex.

53. Objective function:

$z = 3x + ty$

Constraints:

$x \geq 0$

$y \geq 0$

$x + 3y \leq 15$

$4x + y \leq 16$

(a) $(0, 5)$

(b) $(3, 4)$

54. Objective function:

$z = 3x + ty$

Constraints:

$x \geq 0$

$y \geq 0$

$x + 2y \geq 4$

$x - y \leq 1$

(a) $(2, 1)$

(b) $(0, 2)$

Review **Solve Exercises 55–58 as a review of the skills and problem-solving techniques you learned in previous sections. Simplify the compound fraction.**

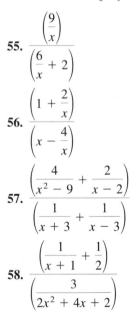

55. $\dfrac{\left(\dfrac{9}{x}\right)}{\left(\dfrac{6}{x} + 2\right)}$

56. $\dfrac{\left(1 + \dfrac{2}{x}\right)}{\left(x - \dfrac{4}{x}\right)}$

57. $\dfrac{\left(\dfrac{4}{x^2 - 9} + \dfrac{2}{x - 2}\right)}{\left(\dfrac{1}{x + 3} + \dfrac{1}{x - 3}\right)}$

58. $\dfrac{\left(\dfrac{1}{x + 1} + \dfrac{1}{2}\right)}{\left(\dfrac{3}{2x^2 + 4x + 2}\right)}$

Focus on Concepts

In this chapter, you studied several concepts that are required for solving systems of equations and inequalities. You can use the following questions to check your understanding of several of these basic concepts. The answers to these questions are given in the back of the book.

1. What is meant by a solution of a system of equations in two variables?

2. When solving a system of equations by substitution, how do you recognize that the system has no solution?

3. When solving a system of equations by elimination, how do you recognize that the system has no solution?

4. Describe any advantages of the algebraic method over the graphical method of solving a system of equations.

5. A system of two equations in two unknowns is solved and has a finite number of solutions. Determine the maximum number of solutions of the system satisfying each of the following conditions.

(a) Both equations are linear.

(b) One equation is linear and the other is quadratic.

(c) Both equations are quadratic.

6. Explain what is meant by an inconsistent system of linear equations.

7. How can you tell graphically that a system of linear equations in two variables has no solution? Give an example.

8. Describe the operations on a system of linear equations that produce equivalent systems of equations.

9. Are the following two systems of equations equivalent? Give reasons for your answer.

$$\begin{aligned} x + 3y - z &= 6 \\ 2x - y + 2z &= 1 \\ 3x + 2y - z &= 2 \end{aligned} \qquad \begin{aligned} x + 3y - z &= 6 \\ -7y + 4z &= 1 \\ -7y - 4z &= -16 \end{aligned}$$

10. One of the following systems is inconsistent and the other has one solution. How can you determine which is which by observation?

$$\begin{aligned} 3x - 5y &= 3 \\ -12x + 20y &= 8 \end{aligned} \qquad \begin{aligned} 3x - 5y &= 3 \\ 9x - 20y &= 6 \end{aligned}$$

In Exercises 11–14, match the system of inequalities with the graph of its solution. [The graphs are labeled (a), (b), (c), and (d).]

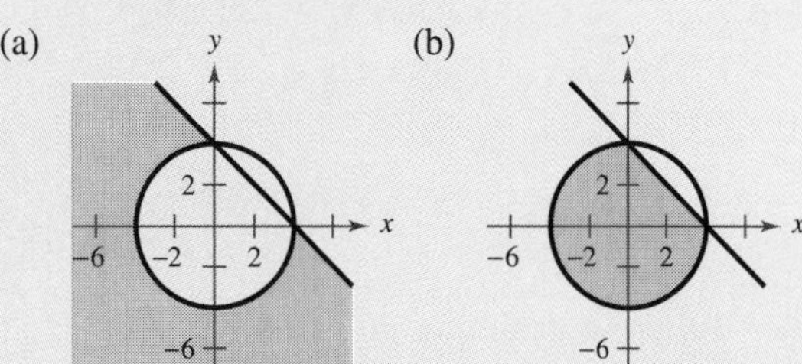

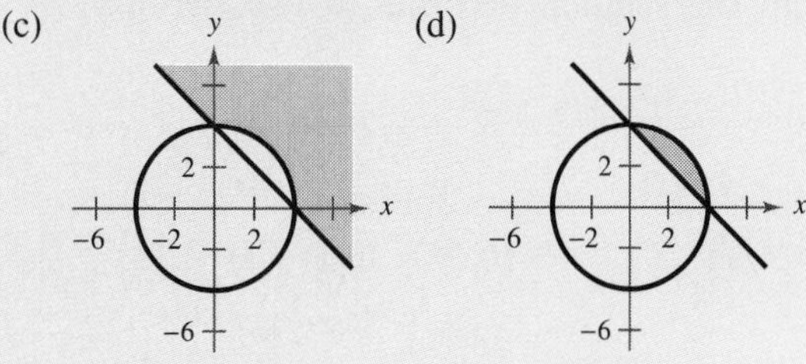

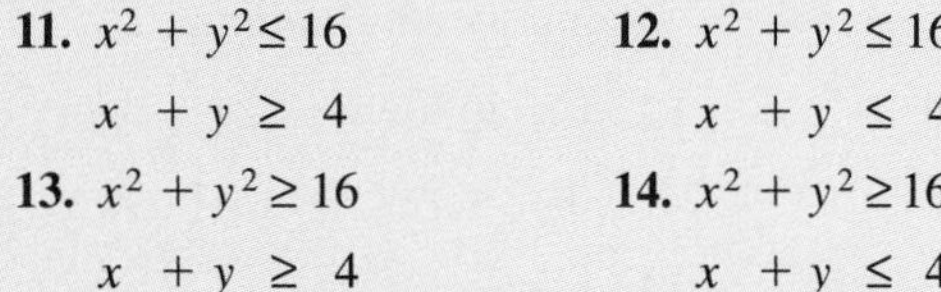

11. $x^2 + y^2 \le 16$
$x + y \ge 4$

12. $x^2 + y^2 \le 16$
$x + y \le 4$

13. $x^2 + y^2 \ge 16$
$x + y \ge 4$

14. $x^2 + y^2 \ge 16$
$x + y \le 4$

15. The graph of the solution of the inequality $x + 2y < 6$ is given in the figure. Describe how the solution set would change for each of the following.

(a) $x + 2y \le 6$ (b) $x + 2y > 6$

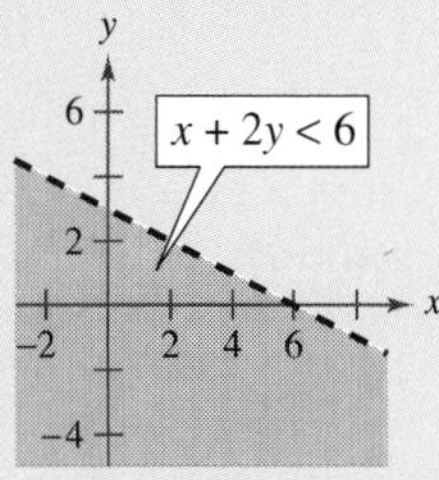

5 /// REVIEW EXERCISES

In Exercises 1–6, solve the system by the method of substitution.

1. $x + y = 2$
$x - y = 0$

2. $2x = 3(y - 1)$
$y = x$

3. $x^2 - y^2 = 9$
$x - y = 1$

4. $x^2 + y^2 = 169$
$3x + 2y = 39$

5. $y = 2x^2$
$y = x^4 - 2x^2$

6. $x = y + 3$
$x = y^2 + 1$

In Exercises 7–10, use a graphing utility to solve the system of equations. If you cannot identify the exact solution, find the solution accurate to two decimal places.

7. $y^2 - 2y + x = 0$
$x + y = 0$

8. $y = 2x^2 - 4x + 1$
$y = x^2 - 4x + 3$

9. $y = 2(6 - x)$
$y = 2^{x-2}$

10. $y = \ln(x - 1) - 3$
$y = 4 - \frac{1}{2}x$

In Exercises 11–18, solve the system by elimination.

11. $2x - y = 2$
$6x + 8y = 39$

12. $40x + 30y = 24$
$20x - 50y = -14$

13. $0.2x + 0.3y = 0.14$
$0.4x + 0.5y = 0.20$

14. $12x + 42y = -17$
$30x - 18y = 19$

15. $3x - 2y = 0$
$3x + 2(y + 5) = 10$

16. $7x + 12y = 63$
$2x + 3y = 15$

17. $1.25x - 2y = 3.5$
$5x - 8y = 14$

18. $1.5x + 2.5y = 8.5$
$6x + 10y = 24$

In Exercises 19 and 20, find a system of linear equations having the given solution. (There is more than one correct answer.)

19. $\left(\frac{4}{3}, 3\right)$

20. $(-6, 8)$

21. *Break-Even Point* You set up a business and make an initial investment of \$10,000. The unit cost of the product is \$2.85 and the selling price is \$4.95. How many units must you sell to break even?

22. *Choice of Two Jobs* You are offered two sales jobs. One company offers an annual salary of \$22,500 plus a year-end bonus of 1.5% of your total sales. The other company offers a salary of \$20,000 plus a year-end bonus of 2% of total sales. What amount of sales will make the second offer better? Explain.

23. *Flying Speeds* Two planes leave Pittsburgh and Philadelphia at the same time, each going to the other city. One plane flies 25 miles per hour faster than the other. Find the ground speed of each plane if the cities are 275 miles apart and the planes pass one another after 40 minutes of flying time.

24. *Dimensions of a Rectangle* The perimeter of a rectangle is 480 meters and its length is 150% of its width. Find the dimensions of the rectangle.

Supply and Demand **In Exercises 25 and 26, find the point of equilibrium.**

	Demand Function	*Supply Function*
25.	$p = 37 - 0.0002x$	$p = 22 + 0.00001x$
26.	$p = 120 - 0.0001x$	$p = 45 + 0.0002x$

In Exercises 27–30, solve the system of equations. Use a graphing utility to verify your solution.

27.
$x + 3y - z = 13$
$2x - 5z = 23$
$4x - y - 2z = 14$

28.
$x + 2y + 6z = 4$
$-3x + 2y - z = -4$
$4x + 2z = 16$

29.
$x - 2y + z = -6$
$2x - 3y = -7$
$-x + 3y - 3z = 11$

30.
$2x + 6z = -9$
$3x - 2y + 11z = -16$
$3x - y + 7z = -11$

Exploration In Exercises 31 and 32, find a system of linear equations having the given solution. (There is more than one correct answer.)

31. $(4, -1, 3)$ **32.** $\left(5, \frac{3}{2}, 2\right)$

In Exercises 33 and 34, find the equation of the parabola $y = ax^2 + bx + c$ that passes through the given points. Use a graphing utility to verify your result.

33.

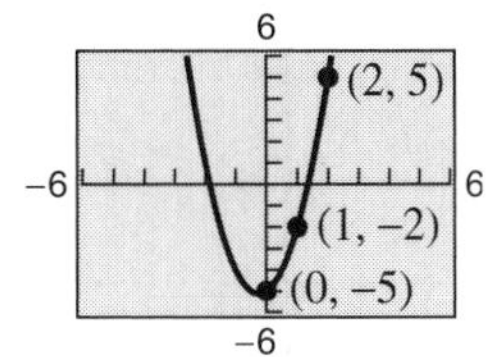

34.

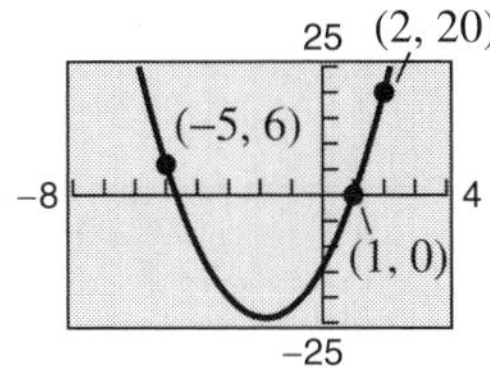

35. *Investments* An inheritance of \$20,000 was divided among three investments yielding \$1780 in interest per year. The interest rates for the three investments were 7%, 9%, and 11%. Find the amount placed in each investment if the second and third were \$3000 and \$1000 less than the first, respectively.

36. *Data Analysis* Let x and y represent the median ages at first marriage for women and men, respectively. For 7 selected years since 1970, these median ages are given by the following ordered pairs. (Source: U.S. Center for Health Statistics)

(20.6, 22.5), (20.8, 22.7), (21.8, 23.6), (23.0, 24.8), (23.3, 25.1), (23.6, 25.3), (23.7, 25.5)

(a) Find the least squares regression line $y = ax + b$ for the data by solving the following system of linear equations.

$$\begin{aligned} 7b + 156.8a &= 169.5 \\ 156.8b + 3522.78a &= 3806.8 \end{aligned}$$

(b) Use a graphing utility to plot the data and graph the regression line in the same viewing rectangle.

(c) Is the line a good model for the data? Explain.

(d) What information is given by the slope of the regression line? Explain.

In Exercises 37–42, write the partial fraction decomposition for the rational expression.

37. $\dfrac{4 - x}{x^2 + 6x + 8}$ **38.** $\dfrac{-x}{x^2 + 3x + 2}$

39. $\dfrac{x^2}{x^2 + 2x - 15}$ **40.** $\dfrac{9}{x^2 - 9}$

41. $\dfrac{x^2 + 2x}{x^3 - x^2 + x - 1}$ **42.** $\dfrac{3x^3 + 4x}{(x^2 + 1)^2}$

In Exercises 43–50, sketch a graph of the solution set of the system of inequalities. Use a graphing utility to verify your result.

43. $\begin{aligned} x + 2y &\le 160 \\ 3x + y &\le 180 \\ x &\ge 0 \\ y &\ge 0 \end{aligned}$ **44.** $\begin{aligned} 2x + 3y &\le 24 \\ 2x + y &\le 16 \\ x &\ge 0 \\ y &\ge 0 \end{aligned}$

45. $\begin{aligned} 3x + 2y &\ge 24 \\ x + 2y &\ge 12 \\ 2 \le x &\le 15 \\ y &\le 15 \end{aligned}$ **46.** $\begin{aligned} 2x + y &\ge 16 \\ x + 3y &\ge 18 \\ 0 \le x &\le 25 \\ 0 \le y &\le 25 \end{aligned}$

47. $\begin{aligned} y &< x + 1 \\ y &> x^2 - 1 \end{aligned}$ **48.** $\begin{aligned} y &\le 6 - 2x - x^2 \\ y &\ge x + 6 \end{aligned}$

49. $\begin{aligned} 2x - 3y &\ge 0 \\ 2x - y &\le 8 \\ y &\ge 0 \end{aligned}$ **50.** $\begin{aligned} x^2 + y^2 &\le 9 \\ (x - 3)^2 + y^2 &\le 9 \end{aligned}$

In Exercises 51 and 52, derive a set of inequalities to describe the region.

51. *Parallelogram:* Vertices at (1, 5), (3, 1), (6, 10), (8, 6)

52. *Triangle:* Vertices at (1, 2), (6, 7), (8, 1)

In Exercises 53 and 54, find a system of inequalities that models the description. Use a graphing utility to graph and shade the solution of the system.

53. *Fruit Distribution* A Pennsylvania fruit grower has 1500 bushels of apples that are to be divided between markets in Harrisburg and Philadelphia. These two markets need at least 400 bushels and 600 bushels, respectively.

54. *Inventory Costs* A warehouse operator has 24,000 square feet of floor space in which to store two products. Each unit of product I requires 20 square feet of floor space and costs \$12 per day to store. Each unit of product II requires 30 square feet of floor space and costs \$8 per day to store. The total storage cost per day cannot exceed \$12,400.

In Exercises 55 and 56, use a graphing utility to shade the regions representing the consumer surplus and producer surplus for the supply and demand equations.

	Demand	*Supply*
55.	$p = 160 - 0.0001x$	$p = 70 + 0.0002x$
56.	$p = 130 - 0.0002x$	$p = 30 + 0.0003x$

In Exercises 57–60, use a graphing utility to sketch the region determined by the constraints. Then find the minimum or maximum values of the objective function, subject to the constraints.

57. Maximize:

$z = 3x + 4y$

Constraints:

$$x \geq 0$$
$$y \geq 0$$
$$2x + 5y \leq 50$$
$$4x + y \leq 28$$

58. Minimize:

$z = 10x + 7y$

Constraints:

$$x \geq 0$$
$$y \geq 0$$
$$2x + y \geq 100$$
$$x + y \geq 75$$

59. Minimize:

$z = 1.75x + 2.25y$

Constraints:

$$x \geq 0$$
$$y \geq 0$$
$$2x + y \geq 25$$
$$3x + 2y \geq 45$$

60. Maximize:

$z = 50x + 70y$

Constraints:

$$x \geq 0$$
$$y \geq 0$$
$$x + 2y \leq 1500$$
$$5x + 2y \leq 3500$$

61. *Maximum Revenue* A student is working part time as a cosmetologist to pay college expenses. The student may work no more than 24 hours per week. Haircuts cost \$17 and require an average of 20 minutes, and permanents cost \$60 and require an average of 1 hour and 10 minutes. What combination of haircuts and/or permanents will yield a maximum revenue?

62. *Maximum Profit* A manufacturer produces products A and B, yielding profits of \$18 and \$24, respectively. Each product must go through three processes that require the amounts of time per unit shown in the table.

Process	Hours for Product A	Hours for Product B	Hours Available per Day
I	4	2	24
II	1	2	9
III	1	1	8

Find the daily production levels for the two products that will maximize profit.

63. *Minimum Cost* A pet supply company mixes two brands of dry dog food. Brand X costs \$15 per bag and contains eight units of nutritional element A, one unit of nutritional element B, and two units of nutritional element C. Brand Y costs \$30 per bag and contains two units of nutritional element A, one unit of nutritional element B, and seven units of nutritional element C. Each bag of mixed dog food must contain at least 16 units, 5 units, and 20 units of nutritional elements A, B, and C, respectively. Find the numbers of bags of brands X and Y that should be mixed to produce a mixture meeting the minimum nutritional requirements and having a minimum cost per bag.

64. *Minimum Cost* Two gasolines, type A and type B, have octane ratings of 80 and 92, respectively. Type A costs \$1.25 per gallon and type B costs \$1.55 per gallon. Determine the blend of minimum cost with an octane rating of at least 88. (*Hint:* Let x be the fraction of each gallon that is type A and let y be the fraction that is type B.)

65. *Exploration* Find k_1 and k_2 such that the following system of equations has an infinite number of solutions.

$$3x - 5y = 8$$
$$2x + k_1 y = k_2$$

CHAPTER PROJECT *Fitting Models to Data*

In this project, you will find and use models relating to newspaper circulation in the United States. (Source: Editor and Publisher Company)

(a) The numbers (in millions) of morning, evening, and Sunday newspapers sold each day in the United States in the years 1981 through 1992 are shown in the first table below. Use the linear regression capabilities of a graphing utility to fit least squares regression models to the three sets of data. Then use the models to project the numbers of morning, evening, and Sunday newspapers that will be sold each day in 1998. In the table, $t = 1$ represents 1981.

(b) The numbers of newspaper companies in the United States from 1981 through 1992 are shown in the second table below. Find a linear model for each set of data.

(c) From 1981 through 1992, the circulation of morning newspapers increased. However, because the number of morning newspaper companies also increased, the competition for morning newspaper readers became keener. Did the average circulation per morning newspaper company increase or decrease? Explain.

(d) If you had the opportunity to invest in a company that published only one type of newspaper (morning, evening, or Sunday), which would you choose? Explain your reasoning.

Daily Newspaper Circulation (in millions)

Year, t	1	2	3	4	5	6	7	8	9	10	11	12
Morning	30.6	33.2	33.8	35.4	36.4	37.4	39.1	40.4	40.7	41.3	41.5	42.4
Evening	30.9	29.3	28.8	27.7	26.4	25.1	23.7	22.2	21.8	21.0	19.2	17.8
Sunday	55.2	56.3	56.7	57.5	58.8	58.9	60.1	61.5	62.0	62.6	62.1	62.2

Numbers of Newspaper Companies

Year, t	1	2	3	4	5	6	7	8	9	10	11	12
Morning	408	434	446	458	482	499	511	529	530	559	571	596
Evening	1352	1310	1284	1257	1220	1188	1166	1141	1125	1084	1042	996
Sunday	755	768	772	783	798	802	820	840	847	863	875	891

3–5 /// CUMULATIVE TEST

Take this test as you would take a test in class. After you are done, check your work against the answers given in the back of the book.

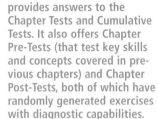

The *Interactive* CD-ROM provides answers to the Chapter Tests and Cumulative Tests. It also offers Chapter Pre-Tests (that test key skills and concepts covered in previous chapters) and Chapter Post-Tests, both of which have randomly generated exercises with diagnostic capabilities.

In Exercises 1–6, use a graphing utility to graph the function.

1. $h(x) = -(x^2 + 4x)$ **2.** $f(t) = \frac{1}{4}t(t - 2)^2$ **3.** $g(s) = \dfrac{2s}{s - 3}$

4. $g(s) = \dfrac{2s^2}{s - 3}$ **5.** $f(x) = 6(2^{-x})$ **6.** $g(x) = \log_3 x$

7. Let x be the amount (in hundreds of dollars) that a company spends on advertising, and let P be the profit (in thousands of dollars), where $P = 230 + 20x - \frac{1}{2}x^2$. How much advertising will maximize the profit?

8. Find all the zeros of $f(x) = x^3 + 2x^2 + 4x + 8$.

9. Approximate the real zero of $g(x) = x^3 + 3x^2 - 6$ to the nearest hundredth.

10. Write $2\ln x - \frac{1}{2}\ln(x + 5)$ as a logarithm of a single quantity.

11. You deposit \$2500 in an account earning 7.5% interest, compounded continuously. Find the balance after 25 years.

In Exercises 12 and 13, solve the equation algebraically. Verify graphically.

12. $6e^{2x} = 72$ **13.** $\log_2 x + \log_2 5 = 6$

In Exercises 14 and 15, solve the system by the specified method.

14. Substitution: $\begin{aligned} 2x - y^2 &= 0 \\ x - y &= 4 \end{aligned}$ **15.** Graphical: $\begin{aligned} y &= \log_3 x \\ y &= -\tfrac{1}{3}x + 2 \end{aligned}$

Figure for 17

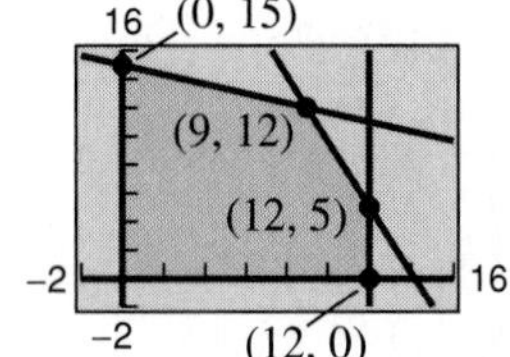

16. Find the equation of the parabola $y = ax^2 + bx + c$ passing through the points $(0, 6)$, $(-2, 2)$, and $\left(3, \frac{9}{2}\right)$.

17. Derive a set of inequalities to describe the region shown in the figure.

18. A merchant plans to sell two models of compact disc players. One model sells for \$275 and yields a profit of \$55, and the other model sells for \$400 and yields a profit of \$75. The merchant estimates that the total monthly demand will not exceed 300 units. The merchant does not want to invest more than \$100,000 in inventory for these products. Find the number of units of each model that should be stocked in order to maximize profit.

Matrices and Determinants

Chapter 6

The times (in minutes) for the winning men's and women's 1000-meter speed skating events at the winter Olympics are shown below. (In 1994, the winter Olympics occurred only 2 years after the previous winter Olympics.)

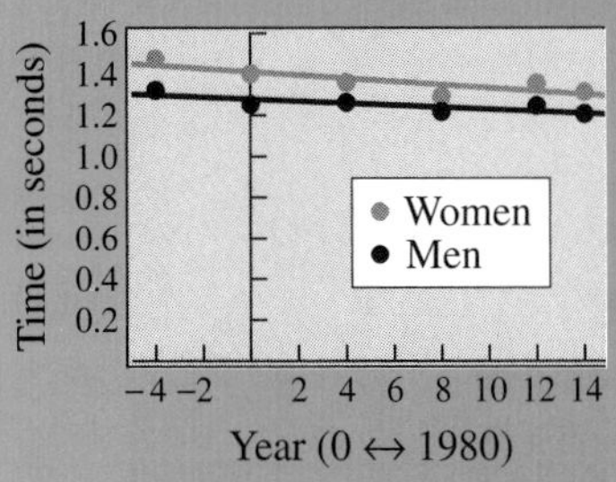

Year	Men	Women
1976	1.322	1.474
1980	1.253	1.402
1984	1.263	1.360
1988	1.217	1.294
1992	1.248	1.358
1994	1.207	1.312

By using a graphing utility, you can determine that the best-fitting linear models are

$s = 1.279 - 0.0049t$ Men

$s = 1.411 - 0.0078t$ Women

where s is the time (in minutes) and t is the year, with $t = 0$ representing 1980. According to these two models, the women's times are decreasing a little more rapidly than the men's times. (See Exercises 81 and 82 on page 472.)

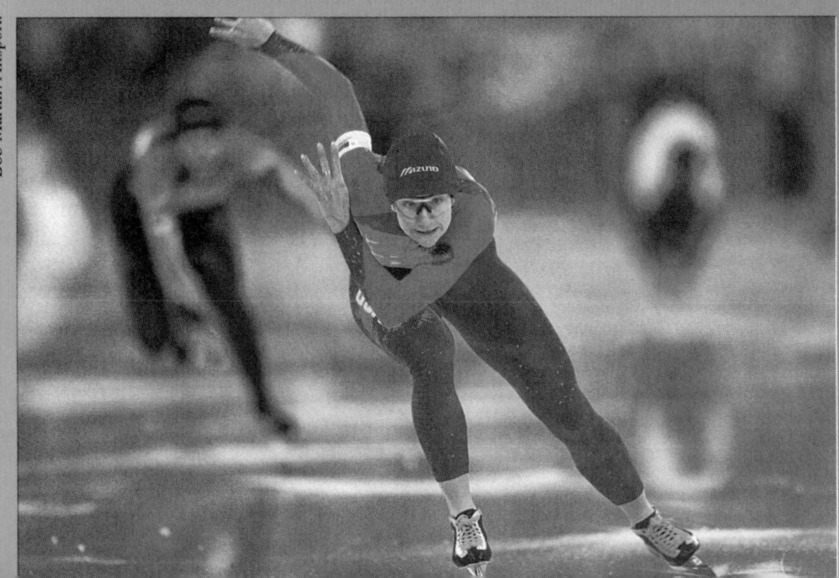

Bob Martin/Allsport.

Bonnie Blair won the 1000-meter women's speed skating event in the 1992 and 1994 winter Olympics. These were the first times this event was ever won by an American.

6.1 Matrices and Systems of Equations

Matrices / Elementary Row Operations / Gaussian Elimination with Back-Substitution / Gauss-Jordan Elimination

Matrices

In this section you will study a streamlined technique for solving systems of linear equations. This technique involves the use of a rectangular array of real numbers called a **matrix.**

Note The plural of matrix is *matrices.*

Definition of Matrix

If m and n are positive integers, an $m \times n$ (read "m by n") **matrix** is a rectangular array

$$\underbrace{\begin{bmatrix} a_{11} & a_{12} & a_{13} & \cdots & a_{1n} \\ a_{21} & a_{22} & a_{23} & \cdots & a_{2n} \\ a_{31} & a_{32} & a_{33} & \cdots & a_{3n} \\ \vdots & \vdots & \vdots & & \vdots \\ a_{m1} & a_{m2} & a_{m3} & \cdots & a_{mn} \end{bmatrix}}_{n \text{ columns}} \Bigg\} \; m \text{ rows}$$

in which each **entry** a_{ij} of the matrix is a real number. An $m \times n$ matrix has m **rows** (horizontal lines) and n **columns** (vertical lines).

The entry in the ith row and jth column is denoted by the *double subscript* notation a_{ij}. A matrix having m rows and n columns is said to be of **order** $m \times n$. If $m = n$, the matrix is **square** of order n. For a square matrix, the entries $a_{11}, a_{22}, a_{33}, \ldots$ are the **main diagonal** entries.

EXAMPLE 1 **Examples of Matrices**

a. *Order:* 1×1

$$[2]$$

b. *Order:* 1×4

$$\begin{bmatrix} 1 & -3 & 0 & \frac{1}{2} \end{bmatrix}$$

c. *Order:* 2×2

$$\begin{bmatrix} 0 & 0 \\ 0 & 0 \end{bmatrix}$$

d. *Order:* 3×2

$$\begin{bmatrix} 5 & 0 \\ 2 & -2 \\ -7 & 4 \end{bmatrix}$$

Note A matrix that has only one row is called a **row matrix,** and a matrix that has only one column is called a **column matrix.**

A matrix derived from a system of linear equations (each written in standard form with the constant term on the right) is the **augmented matrix** of the system. Moreover, the matrix derived from the coefficients of the system (but not including the constant terms) is the **coefficient matrix** of the system.

System	*Augmented Matrix*	*Coefficient Matrix*
$\begin{aligned} x - 4y + 3z &= 5 \\ -x + 3y - z &= -3 \\ 2x \quad\quad - 4z &= 6 \end{aligned}$	$\begin{bmatrix} 1 & -4 & 3 & \vdots & 5 \\ -1 & 3 & -1 & \vdots & -3 \\ 2 & 0 & -4 & \vdots & 6 \end{bmatrix}$	$\begin{bmatrix} 1 & -4 & 3 \\ -1 & 3 & -1 \\ 2 & 0 & -4 \end{bmatrix}$

Note Note the use of 0 for the missing y-variable in the third equation, and also note the fourth column (of constant terms) in the augmented matrix.

When forming either the coefficient matrix or the augmented matrix of a system, you should begin by vertically aligning the variables in the equations and using 0's for the missing variables.

Given System	*Line Up Variables.*	*Form Augmented Matrix.*
$\begin{aligned} x + 3y &= 9 \\ -y + 4z &= -2 \\ x - 5z &= 0 \end{aligned}$	$\begin{aligned} x + 3y \quad\quad &= 9 \\ -y + 4z &= -2 \\ x \quad\quad - 5z &= 0 \end{aligned}$	$\begin{bmatrix} 1 & 3 & 0 & \vdots & 9 \\ 0 & -1 & 4 & \vdots & -2 \\ 1 & 0 & -5 & \vdots & 0 \end{bmatrix}$

Elementary Row Operations

In Section 5.3, you studied three operations that can be used on a system of linear equations to produce an equivalent system.

1. Interchange two equations.
2. Multiply an equation by a nonzero constant.
3. Add a multiple of an equation to another equation.

In matrix terminology, these three operations correspond to **elementary row operations.** An elementary row operation on an augmented matrix of a given system of linear equations produces a new augmented matrix corresponding to a new (but equivalent) system of linear equations. Two matrices are **row-equivalent** if one can be obtained from the other by a sequence of elementary row operations.

Elementary Row Operations

1. Interchange two rows.
2. Multiply a row by a nonzero constant.
3. Add a multiple of a row to another row.

Although elementary row operations are simple to perform, they involve a lot of arithmetic. Because it is easy to make a mistake, we suggest that you get in the habit of noting the elementary row operations performed in each step so that you can go back and check your work.

The *Interactive* CD-ROM shows every example with its solution; clicking on the *Try It!* button brings up similar problems. Guided Examples and Integrated Examples show step-by-step solutions to additional examples. Integrated Examples are related to several concepts in the section.

EXAMPLE 2 Elementary Row Operations

a. Interchange the first and second rows.

Original Matrix *New Row-Equivalent Matrix*

$$\begin{bmatrix} 0 & 1 & 3 & 4 \\ -1 & 2 & 0 & 3 \\ 2 & -3 & 4 & 1 \end{bmatrix} \qquad \begin{matrix} R_2 \\ R_1 \\ \end{matrix} \begin{bmatrix} -1 & 2 & 0 & 3 \\ 0 & 1 & 3 & 4 \\ 2 & -3 & 4 & 1 \end{bmatrix}$$

b. Multiply the first row by $\frac{1}{2}$.

Original Matrix *New Row-Equivalent Matrix*

$$\begin{bmatrix} 2 & -4 & 6 & -2 \\ 1 & 3 & -3 & 0 \\ 5 & -2 & 1 & 2 \end{bmatrix} \qquad \begin{matrix} \frac{1}{2}R_1 \rightarrow \\ \\ \end{matrix} \begin{bmatrix} 1 & -2 & 3 & -1 \\ 1 & 3 & -3 & 0 \\ 5 & -2 & 1 & 2 \end{bmatrix}$$

c. Add -2 times the first row to the third row.

Original Matrix *New Row-Equivalent Matrix*

$$\begin{bmatrix} 1 & 2 & -4 & 3 \\ 0 & 3 & -2 & -1 \\ 2 & 1 & 5 & -2 \end{bmatrix} \qquad \begin{matrix} \\ \\ -2R_1 + R_3 \rightarrow \end{matrix} \begin{bmatrix} 1 & 2 & -4 & 3 \\ 0 & 3 & -2 & -1 \\ 0 & -3 & 13 & -8 \end{bmatrix}$$

Note that the elementary row operation is written beside the row that is *changed.*

The *Interactive* CD-ROM offers graphing utility emulators of the *TI-82* and *TI-83*, which can be used with the Examples, Explorations, Technology notes, and Exercises.

Most graphing utilities can perform elementary row operations on matrices. For instance, on a *TI-82* or *TI-83*, you can perform the elementary row operation shown in Example 2(c) as follows.

1. Use the matrix edit feature to enter the matrix as [A].
2. Choose the "* row + (" feature in the matrix math menu.

   * row + (−2, [A], 1, 3) ENTER

The new row-equivalent matrix will be displayed. To do a sequence of row operations, use ANS in place of [A] in each operation. If you want to save this new matrix, you must do this with separate steps.

In Example 2 of Section 5.3, you used Gaussian elimination with back-substitution to solve a system of linear equations. The next example demonstrates the matrix version of Gaussian elimination. The two methods are essentially the same. The basic difference is that with matrices you do not need to keep writing the variables.

EXAMPLE 3 **Using Elementary Row Operations**

Linear System

$$\begin{aligned} x - 2y + 3z &= 9 \\ -x + 3y &= -4 \\ 2x - 5y + 5z &= 17 \end{aligned}$$

Associated Augmented Matrix

$$\begin{bmatrix} 1 & -2 & 3 & \vdots & 9 \\ -1 & 3 & 0 & \vdots & -4 \\ 2 & -5 & 5 & \vdots & 17 \end{bmatrix}$$

Add the first equation to the second equation.

$$\begin{aligned} x - 2y + 3z &= 9 \\ y + 3z &= 5 \\ 2x - 5y + 5z &= 17 \end{aligned}$$

Add the first row to the second row ($R_1 + R_2$).

$$\begin{matrix} \\ R_1 + R_2 \rightarrow \\ \\ \end{matrix} \begin{bmatrix} 1 & -2 & 3 & \vdots & 9 \\ 0 & 1 & 3 & \vdots & 5 \\ 2 & -5 & 5 & \vdots & 17 \end{bmatrix}$$

Add -2 times the first equation to the third equation.

$$\begin{aligned} x - 2y + 3z &= 9 \\ y + 3z &= 5 \\ -y - z &= -1 \end{aligned}$$

Add -2 times the first row to the third row ($-2R_1 + R_3$).

$$\begin{matrix} \\ \\ -2R_1 + R_3 \rightarrow \end{matrix} \begin{bmatrix} 1 & -2 & 3 & \vdots & 9 \\ 0 & 1 & 3 & \vdots & 5 \\ 0 & -1 & -1 & \vdots & -1 \end{bmatrix}$$

Add the second equation to the third equation.

$$\begin{aligned} x - 2y + 3z &= 9 \\ y + 3z &= 5 \\ 2z &= 4 \end{aligned}$$

Add the second row to the third row ($R_2 + R_3$).

$$\begin{matrix} \\ \\ R_2 + R_3 \rightarrow \end{matrix} \begin{bmatrix} 1 & -2 & 3 & \vdots & 9 \\ 0 & 1 & 3 & \vdots & 5 \\ 0 & 0 & 2 & \vdots & 4 \end{bmatrix}$$

Multiply the third equation by $\frac{1}{2}$.

$$\begin{aligned} x - 2y + 3z &= 9 \\ y + 3z &= 5 \\ z &= 2 \end{aligned}$$

Multiply the third row by $\frac{1}{2}$.

$$\begin{matrix} \\ \\ \frac{1}{2}R_3 \rightarrow \end{matrix} \begin{bmatrix} 1 & -2 & 3 & \vdots & 9 \\ 0 & 1 & 3 & \vdots & 5 \\ 0 & 0 & 1 & \vdots & 2 \end{bmatrix}$$

At this point, you can use back-substitution to find that the solution is $x = 1$, $y = -1$, and $z = 2$, as was done in Example 2 of Section 5.3.

Note Remember that you can check a solution by substituting the values of x, y, and z into each equation in the original system.

The last matrix in Example 3 is said to be in **row-echelon form.** The term *echelon* refers to the stair-step pattern formed by the nonzero elements of the matrix. To be in this form, a matrix must have the following properties.

Some graphing utilities, such as the *TI-85, TI-92,* and *HP-48G,* can automatically transform a matrix to row-echelon form and reduced row-echelon form. Read your user's manual to see if your calculator has this capability. If so, use it to verify the results in this section.

Row-Echelon Form and Reduced Row-Echelon Form

A matrix in **row-echelon form** has the following properties.

1. All rows consisting entirely of zeros occur at the bottom of the matrix.
2. For each row that does not consist entirely of zeros, the first nonzero entry is 1 (called a **leading 1**).
3. For two successive (nonzero) rows, the leading 1 in the higher row is farther to the left than the leading 1 in the lower row.

A matrix in *row-echelon form* is in **reduced row-echelon form** if every column that has a leading 1 has zeros in every position above and below its leading 1.

EXAMPLE 4 **Row-Echelon Form**

The following matrices are in row-echelon form.

a. $\begin{bmatrix} 1 & 2 & -1 & 4 \\ 0 & 1 & 0 & 3 \\ 0 & 0 & 1 & -2 \end{bmatrix}$ **b.** $\begin{bmatrix} 0 & 1 & 0 & 5 \\ 0 & 0 & 1 & 3 \\ 0 & 0 & 0 & 0 \end{bmatrix}$

c. $\begin{bmatrix} 1 & -5 & 2 & -1 & 3 \\ 0 & 0 & 1 & 3 & -2 \\ 0 & 0 & 0 & 1 & 4 \\ 0 & 0 & 0 & 0 & 1 \end{bmatrix}$ **d.** $\begin{bmatrix} 1 & 0 & 0 & -1 \\ 0 & 1 & 0 & 2 \\ 0 & 0 & 1 & 3 \\ 0 & 0 & 0 & 0 \end{bmatrix}$

The matrices in (b) and (d) also happen to be in *reduced* row-echelon form. The following matrices are not in row-echelon form.

e. $\begin{bmatrix} 1 & 2 & -3 & 4 \\ 0 & 2 & 1 & -1 \\ 0 & 0 & 1 & -3 \end{bmatrix}$ **f.** $\begin{bmatrix} 1 & 2 & -1 & 2 \\ 0 & 0 & 0 & 0 \\ 0 & 1 & 2 & -4 \end{bmatrix}$

Every matrix has a row-equivalent matrix that is in row-echelon form. For instance, in Example 4, you can change the matrix in part (e) to row-echelon form by multiplying its second row by $\frac{1}{2}$. What elementary row operation could you perform on the matrix in part (f) so that it would be in row-echelon form?

Gaussian Elimination with Back-Substitution

EXAMPLE 5 **Gaussian Elimination with Back-Substitution**

Solve the system.

$$\begin{aligned} y + z - 2w &= -3 \\ x + 2y - z \quad\quad &= 2 \\ 2x + 4y + z - 3w &= -2 \\ x - 4y - 7z - w &= -19 \end{aligned}$$

Study Tip

Gaussian elimination with back-substitution works well for solving systems of linear equations by hand or with a computer. For this algorithm, the order in which the elementary row operations are performed is important. We suggest operating from *left to right by columns,* using elementary row operations to obtain zeros in all entries directly below the leading 1's.

Solution

$$\begin{matrix} R_2 \\ R_1 \\ \\ \\ \end{matrix} \begin{bmatrix} 1 & 2 & -1 & 0 & \vdots & 2 \\ 0 & 1 & 1 & -2 & \vdots & -3 \\ 2 & 4 & 1 & -3 & \vdots & -2 \\ 1 & -4 & -7 & -1 & \vdots & -19 \end{bmatrix}$$

First column has leading 1 in upper left corner.

$$\begin{matrix} \\ \\ -2R_1 + R_3 \rightarrow \\ -R_1 + R_4 \rightarrow \end{matrix} \begin{bmatrix} 1 & 2 & -1 & 0 & \vdots & 2 \\ 0 & 1 & 1 & -2 & \vdots & -3 \\ 0 & 0 & 3 & -3 & \vdots & -6 \\ 0 & -6 & -6 & -1 & \vdots & -21 \end{bmatrix}$$

First column has zeros below its leading 1.

$$\begin{matrix} \\ \\ \\ 6R_2 + R_4 \rightarrow \end{matrix} \begin{bmatrix} 1 & 2 & -1 & 0 & \vdots & 2 \\ 0 & 1 & 1 & -2 & \vdots & -3 \\ 0 & 0 & 3 & -3 & \vdots & -6 \\ 0 & 0 & 0 & -13 & \vdots & -39 \end{bmatrix}$$

Second column has zeros below its leading 1.

$$\begin{matrix} \\ \\ \frac{1}{3}R_3 \rightarrow \\ \\ \end{matrix} \begin{bmatrix} 1 & 2 & -1 & 0 & \vdots & 2 \\ 0 & 1 & 1 & -2 & \vdots & -3 \\ 0 & 0 & 1 & -1 & \vdots & -2 \\ 0 & 0 & 0 & -13 & \vdots & -39 \end{bmatrix}$$

Third column has zeros below its leading 1.

$$\begin{matrix} \\ \\ \\ -\frac{1}{13}R_4 \rightarrow \end{matrix} \begin{bmatrix} 1 & 2 & -1 & 0 & \vdots & 2 \\ 0 & 1 & 1 & -2 & \vdots & -3 \\ 0 & 0 & 1 & -1 & \vdots & -2 \\ 0 & 0 & 0 & 1 & \vdots & 3 \end{bmatrix}$$

Fourth column has a leading 1.

The matrix is now in row-echelon form, and the corresponding system is

$$\begin{aligned} x + 2y - z \quad\quad &= 2 \\ y + z - 2w &= -3 \\ z - w &= -2 \\ w &= 3. \end{aligned}$$

Using back-substitution, you can determine that the solution is $x = -1$, $y = 2$, $z = 1$, and $w = 3$. Check this in the original system of equations.

Gaussian Elimination with Back-Substitution

1. Write the augmented matrix of the system of linear equations.
2. Use elementary row operations to rewrite the augmented matrix in row-echelon form.
3. Write the system of linear equations corresponding to the matrix in row-echelon form, and use back-substitution to find the solution.

When solving a system of linear equations, remember that it is possible for the system to have no solution. If, in the elimination process, you obtain a row with zeros except for the last entry, it is unnecessary to continue the elimination process. You can simply conclude that the system is inconsistent.

EXAMPLE 6 **A System with No Solution**

Solve the system.

$$\begin{aligned} x - y + 2z &= 4 \\ x + z &= 6 \\ 2x - 3y + 5z &= 4 \\ 3x + 2y - z &= 1 \end{aligned}$$

Solution

$$\begin{bmatrix} 1 & -1 & 2 & \vdots & 4 \\ 1 & 0 & 1 & \vdots & 6 \\ 2 & -3 & 5 & \vdots & 4 \\ 3 & 2 & -1 & \vdots & 1 \end{bmatrix} \quad \begin{matrix} \\ -R_1 + R_2 \rightarrow \\ -2R_1 + R_3 \rightarrow \\ -3R_1 + R_4 \rightarrow \end{matrix} \begin{bmatrix} 1 & -1 & 2 & \vdots & 4 \\ 0 & 1 & -1 & \vdots & 2 \\ 0 & -1 & 1 & \vdots & -4 \\ 0 & 5 & -7 & \vdots & -11 \end{bmatrix}$$

$$\begin{matrix} \\ \\ R_2 + R_3 \rightarrow \\ \\ \end{matrix} \begin{bmatrix} 1 & -1 & 2 & \vdots & 4 \\ 0 & 1 & -1 & \vdots & 2 \\ 0 & 0 & 0 & \vdots & -2 \\ 0 & 5 & -7 & \vdots & -11 \end{bmatrix}$$

Note that the third row of this matrix consists of zeros except for the last entry. This means that the original system of linear equations is *inconsistent.* You can see why this is true by converting back to a system of linear equations.

$$\begin{aligned} x - y + 2z &= 4 \\ y - z &= 2 \\ 0 &= -2 \\ 5y - 7z &= -11 \end{aligned}$$

Because the third equation is not possible, the system has no solution.

Gauss-Jordan Elimination

With Gaussian elimination, elementary row operations are applied to a matrix to obtain a (row-equivalent) row-echelon form. A second method of elimination, called **Gauss-Jordan elimination,** after Carl Friedrich Gauss (1777–1855) and Wilhelm Jordan (1842–1899), continues the reduction process until a *reduced* row-echelon form is obtained. This procedure is demonstrated in the following example.

For a demonstration of a graphical approach to Gauss-Jordan elimination on a 2 × 3 matrix, see the graphing calculator program for this section in the appendix.

EXAMPLE 7 Gauss-Jordan Elimination

Use Gauss-Jordan elimination to solve the system.

$$\begin{aligned} x - 2y + 3z &= 9 \\ -x + 3y \qquad &= -4 \\ 2x - 5y + 5z &= 17 \end{aligned}$$

Solution

In Example 3, Gaussian elimination was used to obtain the row-echelon form

$$\begin{bmatrix} 1 & -2 & 3 & \vdots & 9 \\ 0 & 1 & 3 & \vdots & 5 \\ 0 & 0 & 1 & \vdots & 2 \end{bmatrix}.$$

Now, rather than using back-substitution, apply additional elementary row operations until you obtain a matrix in *reduced* row-echelon form. To do this, you must produce zeros above each of the leading 1's, as follows.

$$\begin{matrix} 2R_2 + R_1 \rightarrow \\ \\ \\ \end{matrix} \begin{bmatrix} 1 & 0 & 9 & \vdots & 19 \\ 0 & 1 & 3 & \vdots & 5 \\ 0 & 0 & 1 & \vdots & 2 \end{bmatrix}$$

Second column has zeros above its leading 1.

$$\begin{matrix} -9R_3 + R_1 \rightarrow \\ -3R_3 + R_2 \rightarrow \\ \\ \end{matrix} \begin{bmatrix} 1 & 0 & 0 & \vdots & 1 \\ 0 & 1 & 0 & \vdots & -1 \\ 0 & 0 & 1 & \vdots & 2 \end{bmatrix}$$

Third column has zeros above its leading 1.

Now, converting back to a system of linear equations, you have

$$\begin{aligned} x \qquad &= 1 \\ y \quad &= -1 \\ z &= 2. \end{aligned}$$

The beauty of Gauss-Jordan elimination is that, from the reduced row-echelon form, you can simply read the solution.

Note Which technique do you prefer: Gaussian elimination or Gauss-Jordan elimination?

The elimination procedures described in this section employ an algorithmic approach that is easily adapted to computer use. However, the procedure makes no effort to avoid fractional coefficients. For instance, if the system given in Example 7 had been listed as

$$\begin{aligned} 2x - 5y + 5z &= 17 \\ x - 2y + 3z &= 9 \\ -x + 3y \quad\quad &= -4 \end{aligned}$$

the procedure would have required multiplication of the first row by $\frac{1}{2}$, which would have introduced fractions in the first row. For hand computations, fractions can sometimes be avoided by judiciously choosing the order in which the elementary row operations are applied.

EXAMPLE 8 **A System with an Infinite Number of Solutions**

Solve the system.

$$\begin{aligned} 2x + 4y - 2z &= 0 \\ 3x + 5y \quad\quad &= 1 \end{aligned}$$

Solution

$$\begin{bmatrix} 2 & 4 & -2 & \vdots & 0 \\ 3 & 5 & 0 & \vdots & 1 \end{bmatrix} \quad \tfrac{1}{2}R_1 \rightarrow \begin{bmatrix} 1 & 2 & -1 & \vdots & 0 \\ 3 & 5 & 0 & \vdots & 1 \end{bmatrix}$$

$$-3R_1 + R_2 \rightarrow \begin{bmatrix} 1 & 2 & -1 & \vdots & 0 \\ 0 & -1 & 3 & \vdots & 1 \end{bmatrix}$$

$$-R_2 \rightarrow \begin{bmatrix} 1 & 2 & -1 & \vdots & 0 \\ 0 & 1 & -3 & \vdots & -1 \end{bmatrix}$$

$$-2R_2 + R_1 \rightarrow \begin{bmatrix} 1 & 0 & 5 & \vdots & 2 \\ 0 & 1 & -3 & \vdots & -1 \end{bmatrix}$$

The corresponding system of equations is

$$\begin{aligned} x \quad + 5z &= 2 \\ y - 3z &= -1. \end{aligned}$$

Solving for x and y in terms of z, you have $x = -5z + 2$ and $y = 3z - 1$. Then, letting $z = a$, the solution set has the form

$$(-5a + 2, 3a - 1, a)$$

where a is a real number. Try substituting values for a to obtain a few solutions. Then check each solution in the original system of equations.

Note You have seen that the row-echelon form of a given matrix *is not* unique; however, the *reduced* row-echelon form of a given matrix *is* unique. Try applying Gauss-Jordan elimination to the row-echelon matrix given at the right to see that you obtain the same reduced row-echelon form as in Example 7.

It is worth noting that the row-echelon form of a matrix is not unique. That is, two different sequences of elementary row operations may yield different row-echelon forms. For instance, the following sequence of elementary row operations on the matrix in Example 3 produces a slightly different row-echelon form.

$$\begin{bmatrix} 1 & -2 & 3 & \vdots & 9 \\ -1 & 3 & 0 & \vdots & -4 \\ 2 & -5 & 5 & \vdots & 17 \end{bmatrix} \qquad \begin{matrix} R_2 \\ R_1 \end{matrix} \begin{bmatrix} -1 & 3 & 0 & \vdots & -4 \\ 1 & -2 & 3 & \vdots & 9 \\ 2 & -5 & 5 & \vdots & 17 \end{bmatrix}$$

$$-R_1 \rightarrow \begin{bmatrix} 1 & -3 & 0 & \vdots & 4 \\ 1 & -2 & 3 & \vdots & 9 \\ 2 & -5 & 5 & \vdots & 17 \end{bmatrix}$$

$$\begin{matrix} \\ -R_1 + R_2 \rightarrow \\ -2R_1 + R_3 \rightarrow \end{matrix} \begin{bmatrix} 1 & -3 & 0 & \vdots & 4 \\ 0 & 1 & 3 & \vdots & 5 \\ 0 & 1 & 5 & \vdots & 9 \end{bmatrix}$$

$$\begin{matrix} \\ \\ -R_2 + R_3 \rightarrow \end{matrix} \begin{bmatrix} 1 & -3 & 0 & \vdots & 4 \\ 0 & 1 & 3 & \vdots & 5 \\ 0 & 0 & 2 & \vdots & 4 \end{bmatrix}$$

$$\begin{matrix} \\ \\ \frac{1}{2}R_3 \rightarrow \end{matrix} \begin{bmatrix} 1 & -3 & 0 & \vdots & 4 \\ 0 & 1 & 3 & \vdots & 5 \\ 0 & 0 & 1 & \vdots & 2 \end{bmatrix}$$

The corresponding system of linear equations is

$$\begin{aligned} x - 3y &= 4 \\ y + 3z &= 5 \\ z &= 2. \end{aligned}$$

Try using back-substitution on this system to see that you obtain the same solution that was obtained in Example 3.

Group Activity *Error Analysis*

One of your classmates has submitted the following steps for a solution of a system by Gauss-Jordan elimination. Find the error(s) in the solution and discuss how to explain the error(s) to your classmate.

$$\begin{bmatrix} 1 & 1 & \vdots & 4 \\ 2 & 3 & \vdots & 5 \end{bmatrix} \quad -2R_1 + R_2 \rightarrow \begin{bmatrix} 1 & 1 & \vdots & 4 \\ 0 & 1 & \vdots & 5 \end{bmatrix}$$

$$-1R_2 + R_1 \rightarrow \begin{bmatrix} 1 & 0 & \vdots & 4 \\ 0 & 1 & \vdots & 5 \end{bmatrix}$$

6.1 /// EXERCISES

In Exercises 1–6, determine the order of the matrix.

1. $\begin{bmatrix} 4 & -2 \\ 7 & 0 \\ 0 & 8 \end{bmatrix}$

2. $\begin{bmatrix} 5 & -3 & 8 & 7 \end{bmatrix}$

3. $\begin{bmatrix} 2 \\ 36 \\ 3 \end{bmatrix}$

4. $\begin{bmatrix} -3 & 7 & 15 & 0 \\ 0 & 0 & 3 & 3 \\ 1 & 1 & 6 & 7 \end{bmatrix}$

5. $\begin{bmatrix} 33 & 45 \\ -9 & 20 \end{bmatrix}$

6. $\begin{bmatrix} 4 \end{bmatrix}$

In Exercises 7–10, form the augmented matrix for the system of linear equations.

7. $\begin{aligned} 4x - 3y &= -5 \\ -x + 3y &= 12 \end{aligned}$

8. $\begin{aligned} 7x + 4y &= 22 \\ 5x - 9y &= 15 \end{aligned}$

9. $\begin{aligned} x + 10y - 2z &= 2 \\ 5x - 3y + 4z &= 0 \\ 2x + y \quad\quad &= 6 \end{aligned}$

10. $\begin{aligned} 7x - 5y + z &= 13 \\ 19x \quad\quad - 8z &= 10 \end{aligned}$

In Exercises 11–14, write the system of linear equations represented by the augmented matrix. (Use variables x, y, z, and w.)

11. $\left[\begin{array}{cc:c} 1 & 2 & 7 \\ 2 & -3 & 4 \end{array}\right]$

12. $\left[\begin{array}{cc:c} 7 & -5 & 0 \\ 8 & 3 & -2 \end{array}\right]$

13. $\left[\begin{array}{ccc:c} 2 & 0 & 5 & -12 \\ 0 & 1 & -2 & 7 \\ 6 & 3 & 0 & 2 \end{array}\right]$

14. $\left[\begin{array}{cccc:c} 9 & 12 & 3 & 0 & 0 \\ -2 & 18 & 5 & 2 & 10 \\ 1 & 7 & -8 & 0 & -4 \end{array}\right]$

In Exercises 15–18, determine whether the matrix is in row-echelon form. If it is, determine if it is also in reduced row-echelon form.

15. $\begin{bmatrix} 1 & 0 & 0 & 0 \\ 0 & 1 & 1 & 5 \\ 0 & 0 & 0 & 0 \end{bmatrix}$

16. $\begin{bmatrix} 1 & 3 & 0 & 0 \\ 0 & 0 & 1 & 8 \\ 0 & 0 & 0 & 0 \end{bmatrix}$

17. $\begin{bmatrix} 2 & 0 & 4 & 0 \\ 0 & -1 & 3 & 6 \\ 0 & 0 & 1 & 5 \end{bmatrix}$

18. $\begin{bmatrix} 1 & 0 & 2 & 1 \\ 0 & 1 & -3 & 10 \\ 0 & 0 & 1 & 0 \end{bmatrix}$

In Exercises 19–22, fill in the blanks using elementary row operations to form a row-equivalent matrix.

19. $\begin{bmatrix} 1 & 4 & 3 \\ 2 & 10 & 5 \end{bmatrix}$

$\begin{bmatrix} 1 & 4 & 3 \\ 0 & \square & -1 \end{bmatrix}$

20. $\begin{bmatrix} 3 & 6 & 8 \\ 4 & -3 & 6 \end{bmatrix}$

$\begin{bmatrix} 1 & \square & \frac{8}{3} \\ 4 & -3 & 6 \end{bmatrix}$

21. $\begin{bmatrix} 1 & 1 & 4 & -1 \\ 3 & 8 & 10 & 3 \\ -2 & 1 & 12 & 6 \end{bmatrix}$

$\begin{bmatrix} 1 & 1 & 4 & -1 \\ 0 & 5 & \square & \square \\ 0 & 3 & \square & \square \end{bmatrix}$

$\begin{bmatrix} 1 & 1 & 4 & -1 \\ 0 & 1 & -\frac{2}{5} & \frac{6}{5} \\ 0 & 3 & \square & \square \end{bmatrix}$

22. $\begin{bmatrix} 2 & 4 & 8 & 3 \\ 1 & -1 & -3 & 2 \\ 2 & 6 & 4 & 9 \end{bmatrix}$

$\begin{bmatrix} 1 & \square & \square & \square \\ 1 & -1 & -3 & 2 \\ 2 & 6 & 4 & 9 \end{bmatrix}$

$\begin{bmatrix} 1 & 2 & 4 & \frac{3}{2} \\ 0 & \square & -7 & \frac{1}{2} \\ 0 & 2 & \square & \square \end{bmatrix}$

23. Perform the *sequence* of row operations on the matrix. What did the operations accomplish?

$\begin{bmatrix} 1 & 2 & 3 \\ 2 & -1 & -4 \\ 3 & 1 & -1 \end{bmatrix}$

(a) Add -2 times Row 1 to Row 2.

(b) Add -3 times Row 1 to Row 3.

(c) Add -1 times Row 2 to Row 3.

(d) Multiply Row 2 by $-\frac{1}{5}$.

(e) Add -2 times Row 2 to Row 1.

24. Perform the *sequence* of row operations on the matrix. What did the operations accomplish?

$$\begin{bmatrix} 7 & 1 \\ 0 & 2 \\ -3 & 4 \\ 4 & 1 \end{bmatrix}$$

(a) Add Row 3 to Row 4.

(b) Interchange Rows 1 and 4.

(c) Add 3 times Row 1 to Row 3.

(d) Add −7 times Row 1 to Row 4.

(e) Multiply Row 2 by $\frac{1}{2}$.

(f) Add the appropriate multiples of Row 2 to Rows 1, 3, and 4.

In Exercises 25–28, write the matrix in row-echelon form. Remember that the row-echelon form of a matrix is not unique.

25. $\begin{bmatrix} 1 & 1 & 0 & 5 \\ -2 & -1 & 2 & -10 \\ 3 & 6 & 7 & 14 \end{bmatrix}$

26. $\begin{bmatrix} 1 & 2 & -1 & 3 \\ 3 & 7 & -5 & 14 \\ -2 & -1 & -3 & 8 \end{bmatrix}$

27. $\begin{bmatrix} 1 & -1 & -1 & 1 \\ 5 & -4 & 1 & 8 \\ -6 & 8 & 18 & 0 \end{bmatrix}$

28. $\begin{bmatrix} 1 & -3 & 0 & -7 \\ -3 & 10 & 1 & 23 \\ 4 & -10 & 2 & -24 \end{bmatrix}$

In Exercises 29–32, use the matrix capabilities of a graphing utility to write the matrix in *reduced* row-echelon form.

29. $\begin{bmatrix} 3 & 3 & 3 \\ -1 & 0 & -4 \\ 2 & 4 & -2 \end{bmatrix}$

30. $\begin{bmatrix} 1 & 3 & 2 \\ 5 & 15 & 9 \\ 2 & 6 & 10 \end{bmatrix}$

31. $\begin{bmatrix} 1 & 2 & 3 & -5 \\ 1 & 2 & 4 & -9 \\ -2 & -4 & -4 & 3 \\ 4 & 8 & 11 & -14 \end{bmatrix}$

32. $\begin{bmatrix} 1 & -3 \\ -1 & 8 \\ 0 & 4 \\ -2 & 10 \end{bmatrix}$

In Exercises 33–36, write the system of linear equations represented by the augmented matrix. Then use back-substitution to find the solution. (Use variables *x*, *y*, and *z*.)

33. $\left[\begin{array}{cc:c} 1 & -2 & 4 \\ 0 & 1 & -3 \end{array}\right]$

34. $\left[\begin{array}{cc:c} 1 & 5 & 0 \\ 0 & 1 & -1 \end{array}\right]$

35. $\left[\begin{array}{ccc:c} 1 & -1 & 2 & 4 \\ 0 & 1 & -1 & 2 \\ 0 & 0 & 1 & -2 \end{array}\right]$

36. $\left[\begin{array}{ccc:c} 1 & 2 & -2 & -1 \\ 0 & 1 & 1 & 9 \\ 0 & 0 & 1 & -3 \end{array}\right]$

In Exercises 37–40, an augmented matrix that represents a system of linear equations (in variables *x*, *y*, and *z*) has been reduced using Gauss-Jordan elimination. Write the solution represented by the augmented matrix.

37. $\left[\begin{array}{cc:c} 1 & 0 & 7 \\ 0 & 1 & -5 \end{array}\right]$

38. $\left[\begin{array}{cc:c} 1 & 0 & -2 \\ 0 & 1 & 4 \end{array}\right]$

39. $\left[\begin{array}{ccc:c} 1 & 0 & 0 & -4 \\ 0 & 1 & 0 & -8 \\ 0 & 0 & 1 & 2 \end{array}\right]$

40. $\left[\begin{array}{ccc:c} 1 & 0 & 0 & 3 \\ 0 & 1 & 0 & -1 \\ 0 & 0 & 1 & 0 \end{array}\right]$

In Exercises 41–56, solve the system of equations. Use Gaussian elimination with back-substitution or Gauss-Jordan elimination.

41. $\begin{aligned} x + 2y &= 7 \\ 2x + y &= 8 \end{aligned}$

42. $\begin{aligned} 2x + 6y &= 16 \\ 2x + 3y &= 7 \end{aligned}$

43. $\begin{aligned} -3x + 5y &= -22 \\ 3x + 4y &= 4 \\ 4x - 8y &= 32 \end{aligned}$

44. $\begin{aligned} x + 2y &= 0 \\ x + y &= 6 \\ 3x - 2y &= 8 \end{aligned}$

45. $\begin{aligned} 8x - 4y &= 7 \\ 5x + 2y &= 1 \end{aligned}$

46. $\begin{aligned} 2x - y &= -0.1 \\ 3x + 2y &= 1.6 \end{aligned}$

47. $\begin{aligned} -x + 2y &= 1.5 \\ 2x - 4y &= 3 \end{aligned}$

48. $\begin{aligned} x - 3y &= 5 \\ -2x + 6y &= -10 \end{aligned}$

49. $\begin{aligned} x \qquad - 3z &= -2 \\ 3x + y - 2z &= 5 \\ 2x + 2y + z &= 4 \end{aligned}$

50. $\begin{aligned} 2x - y + 3z &= 24 \\ 2y - z &= 14 \\ 7x - 5y \qquad &= 6 \end{aligned}$

51. $\begin{aligned} x + y - 5z &= 3 \\ x \qquad - 2z &= 1 \\ 2x - y - z &= 0 \end{aligned}$

52. $\begin{aligned} 2x \qquad + 3z &= 3 \\ 4x - 3y + 7z &= 5 \\ 8x - 9y + 15z &= 9 \end{aligned}$

53. $\begin{aligned} x + 2y + z &= 8 \\ 3x + 7y + 6z &= 26 \end{aligned}$

54. $\begin{aligned} 4x + 12y - 7z - 20w &= 22 \\ 3x + 9y - 5z - 28w &= 30 \end{aligned}$

55. $\begin{aligned} x + 2y &= 0 \\ -x - y &= 0 \end{aligned}$

56. $\begin{aligned} x + 2y &= 0 \\ 2x + 4y &= 0 \end{aligned}$

In Exercises 57–62, use the matrix capabilities of a graphing utility to reduce the augmented matrix and solve the system of equations.

57. $\begin{aligned} 3x + 3y + 12z &= 6 \\ x + y + 4z &= 2 \\ 2x + 5y + 20z &= 10 \\ -x + 2y + 8z &= 4 \end{aligned}$

58. $\begin{aligned} 2x + 10y + 2z &= 6 \\ x + 5y + 2z &= 6 \\ x + 5y + z &= 3 \\ -3x - 15y - 3z &= -9 \end{aligned}$

59. $\begin{aligned} 2x + y - z + 2w &= -6 \\ 3x + 4y \qquad + w &= 1 \\ x + 5y + 2z + 6w &= -3 \\ 5x + 2y - z - w &= 3 \end{aligned}$

60. $\begin{aligned} x + 2y + 2z + 4w &= 11 \\ 3x + 6y + 5z + 12w &= 30 \end{aligned}$

61. $\begin{aligned} x + y + z &= 0 \\ 2x + 3y + z &= 0 \\ 3x + 5y + z &= 0 \end{aligned}$

62. $\begin{aligned} x + 2y + z + 3w &= 0 \\ x - y \qquad + w &= 0 \\ y - z + 2w &= 0 \end{aligned}$

63. *Think About It* The augmented matrix represents a system of linear equations (in variables x, y, and z) that has been reduced using Gauss-Jordan elimination. Write a system of equations with nonzero coefficients that is represented by the reduced matrix. (The answer is not unique.)

$$\left[\begin{array}{ccc:c} 1 & 0 & 3 & -2 \\ 0 & 1 & 4 & 1 \\ 0 & 0 & 0 & 0 \end{array}\right]$$

64. *Think About It*

(a) Describe the row-echelon form of an augmented matrix that corresponds to a system of linear equations that is inconsistent.

(b) Describe the row-echelon form of an augmented matrix that corresponds to a system of linear equations that has an infinite number of solutions.

65. *Borrowing Money* A small corporation borrowed \$1,500,000 to expand its product line. Some of the money was borrowed at 8%, some at 9%, and some at 12%. How much was borrowed at each rate if the annual interest was \$133,000 and the amount borrowed at 8% was 4 times the amount borrowed at 12%?

66. *Borrowing Money* A small corporation borrowed \$500,000 to expand its product line. Some of the money was borrowed at 9%, some at 10%, and some at 12%. How much was borrowed at each rate if the annual interest was \$52,000 and the amount borrowed at 10% was $2\frac{1}{2}$ times the amount borrowed at 9%?

67. *Partial Fractions* Write the partial fraction decomposition for $(4x^2)/[(x + 1)^2(x - 1)]$.

68. *Electrical Network* The currents in an electrical network are given by the solution of the system

$$\begin{aligned} I_1 - I_2 + I_3 &= 0 \\ 2I_1 + 2I_2 &= 7 \\ 2I_2 + 4I_3 &= 8 \end{aligned}$$

where I_1, I_2, and I_3 are measured in amperes. Solve the system of equations.

In Exercises 69–74, find the specified equation that passes through the points. Use a graphing utility to verify your result.

69. Parabola:

$y = ax^2 + bx + c$

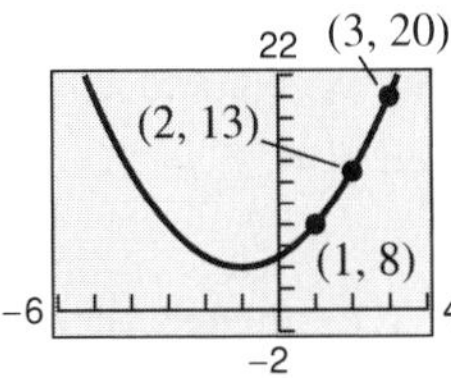

70. Parabola:

$y = ax^2 + bx + c$

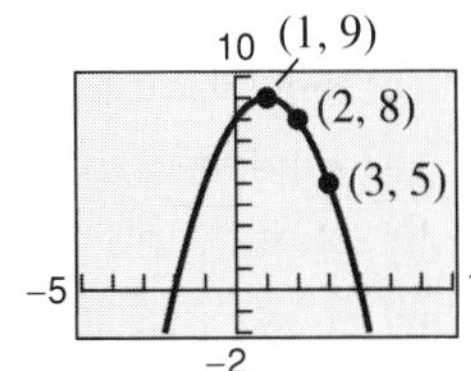

71. Cubic:

$y = ax^3 + bx^2 + cx + d$

(−2, 2) 5 (2, 2)

−6 4

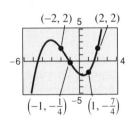

$\left(-1, -\frac{1}{4}\right)$ −5 $\left(1, -\frac{7}{4}\right)$

72. Cubic:

$y = ax^3 + bx^2 + cx + d$

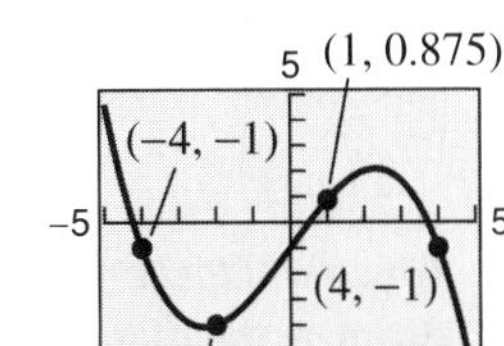

73. Quartic:

$y = ax^4 + \cdots + dx + e$

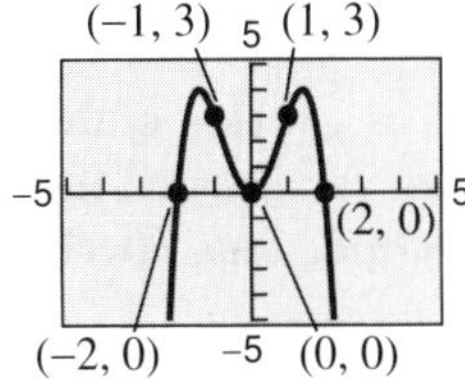

74. Quartic:

$y = ax^4 + \cdots + dx + e$

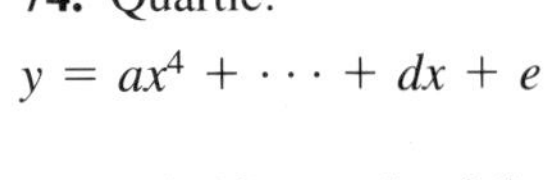

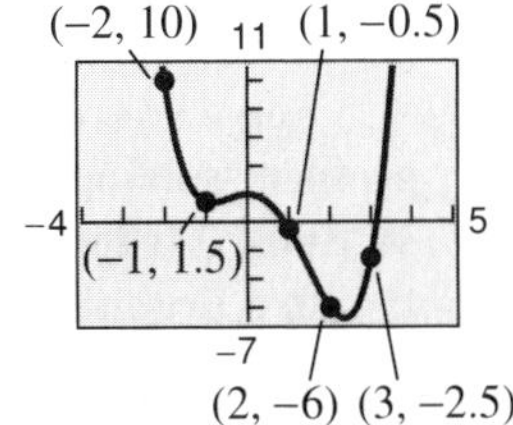

75. *Reading a Graph* The bar graph gives the value y, in millions of dollars, for new orders of civil jet transport aircraft built by U.S. companies in the years 1990 through 1992. (Source: Aerospace Industries Association of America)

(a) Find the equation of the parabola that passes through the points. Let $t = 0$ represent 1990.

(b) Use a graphing utility to graph the parabola.

(c) Use the equation in part (a) to estimate y in 1993.

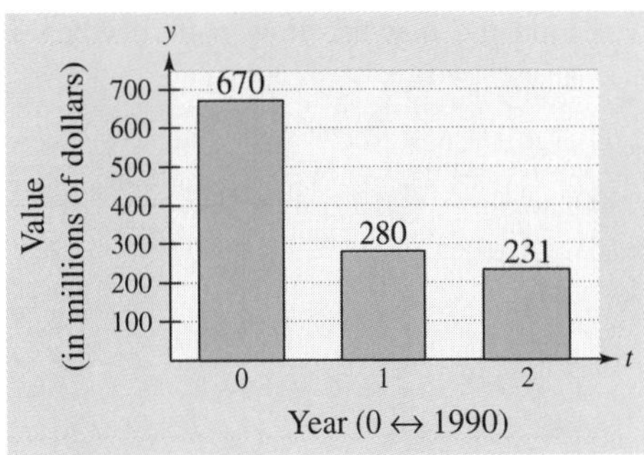

76. *Mathematical Modeling* After the path of a ball thrown by a baseball player is videotaped, it is analyzed on a television set with a grid covering the screen. The tape is paused three times, and the position of the ball is measured each time. The coordinates are approximately (0, 5.0), (15, 9.6), and (30, 12.4). (The x-coordinate measures the horizontal distance from the player in feet, and the y-coordinate is the height of the ball in feet.)

(a) Find the equation of the parabola $y = ax^2 + bx + c$ that passes through the three points.

(b) Use a graphing utility to graph the parabola. Approximate the maximum height of the ball and the point at which the ball strikes the ground.

(c) Find analytically the maximum height of the ball and the point at which it strikes the ground.

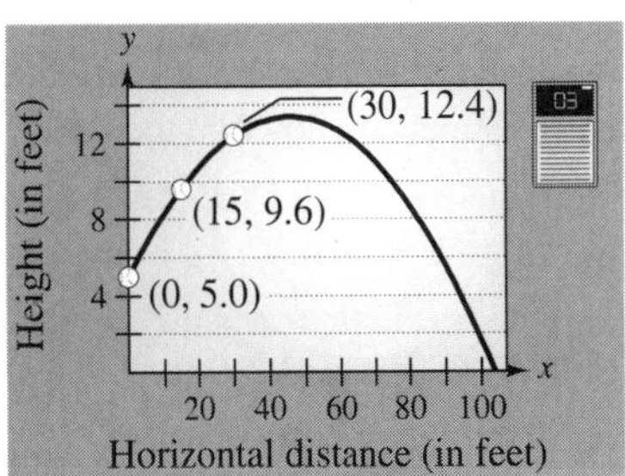

Network Analysis **In Exercises 77–80, answer the questions about the specified network. (In a network it is assumed that the total flow into each junction is equal to the total flow out of the junction.)**

77. Water is flowing through a network of pipes (in thousands of cubic meters per hour). (See figure.)

(a) Solve this system for the water flow represented by x_i, $i = 1, 2, 3, 4, 5, 6, 7$.

(b) Find the network flow pattern when $x_6 = x_7 = 0$.

(c) Find the network flow pattern when $x_5 = 1000$ and $x_6 = 0$.

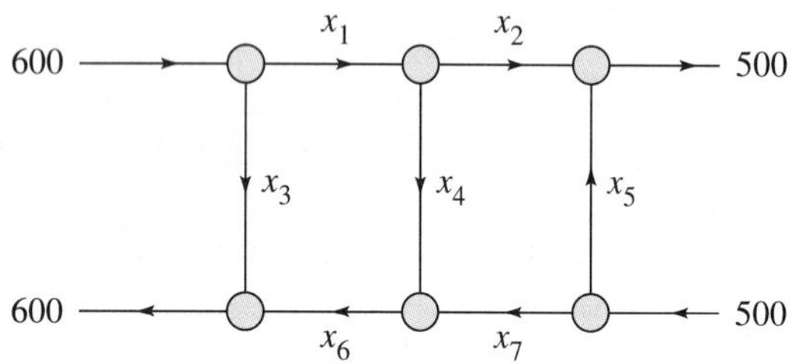

78. The flow of traffic (in vehicles per hour) through a network of streets is shown in the figure.

(a) Solve this system for the traffic flow represented by x_i, $i = 1, 2, 3, 4, 5$.

(b) Find the traffic flow when $x_2 = 200$ and $x_3 = 50$.

(c) Find the traffic flow when $x_2 = 150$ and $x_3 = 0$.

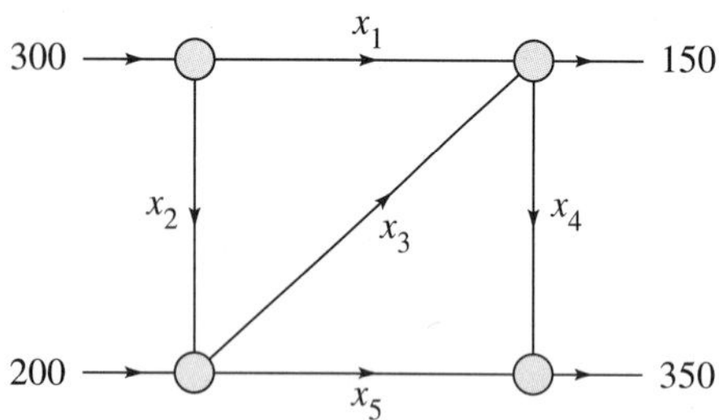

79. The flow of traffic (in vehicles per hour) through a network of streets is shown in the figure.

(a) Solve this system for the traffic flow represented by x_i, $i = 1, 2, 3, 4$.

(b) Find the traffic flow when $x_4 = 0$.

(c) Find the traffic flow when $x_4 = 100$.

Figure for 79

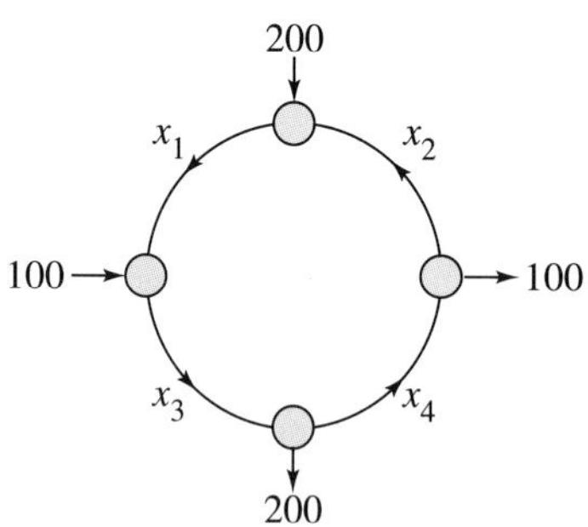

80. The flow of traffic (in vehicles per hour) through a network of streets is shown in the figure.

(a) Solve this system for the traffic flow represented by x_i, $i = 1, 2, 3, 4, 5$.

(b) Find the traffic flow when $x_3 = 0$ and $x_5 = 100$.

(c) Find the traffic flow when $x_3 = x_5 = 100$.

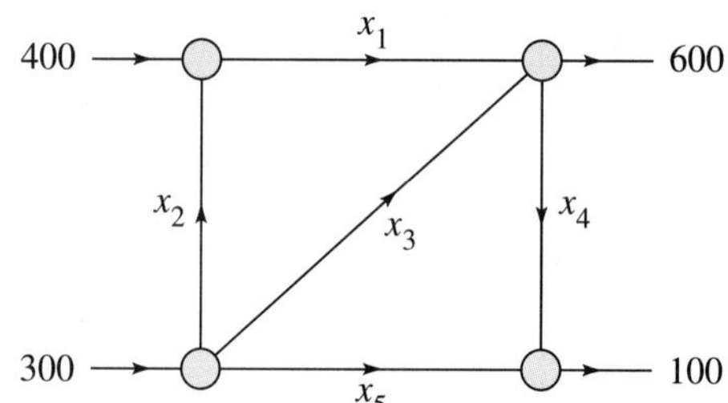

81. *Chapter Opener* Use the models on page 457 to estimate the men's and women's winning times in the 1000-meter speed skating events in the year 2002.

82. *Chapter Opener* If the models on page 457 continue to represent the winning times in the 1000-meter speed skating events, in which winter Olympics will the women's time be less than the men's time?

Review **Solve Exercises 83–86 as a review of the skills and problem-solving techniques you learned in previous sections. Graph the function, and check each graph with a graphing utility.**

83. $f(x) = 2^{x-1}$

84. $g(x) = 3^{-x/2}$

85. $h(x) = \log_2(x - 1)$

86. $f(x) = 3 + \ln x$

6.2 Operations with Matrices

Equality of Matrices / Matrix Addition and Scalar Multiplication / Matrix Multiplication / Applications

Equality of Matrices

In Section 6.1, you used matrices to solve systems of linear equations. Matrices, however, can do much more than this. There is a rich mathematical theory of matrices, and its applications are numerous. This section and the next two introduce some fundamentals of matrix theory. It is standard mathematical convention to represent matrices in any of the following three ways.

1. A matrix can be denoted by an uppercase letter such as A, B, or C.
2. A matrix can be denoted by a representative element enclosed in brackets, such as $[a_{ij}]$, $[b_{ij}]$, or $[c_{ij}]$.
3. A matrix can be denoted by a rectangular array of numbers such as

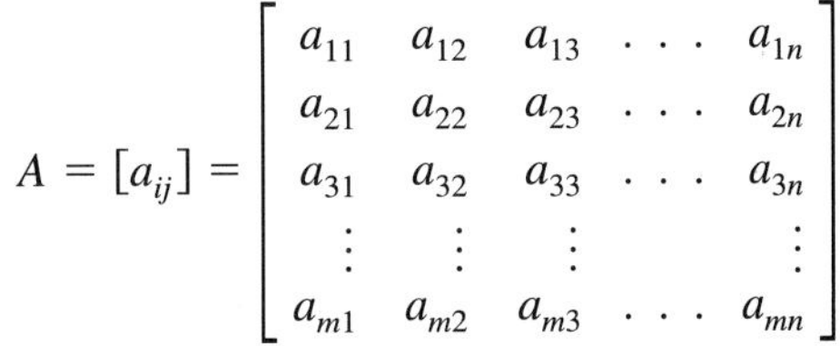

$$A = [a_{ij}] = \begin{bmatrix} a_{11} & a_{12} & a_{13} & \cdots & a_{1n} \\ a_{21} & a_{22} & a_{23} & \cdots & a_{2n} \\ a_{31} & a_{32} & a_{33} & \cdots & a_{3n} \\ \vdots & \vdots & \vdots & & \vdots \\ a_{m1} & a_{m2} & a_{m3} & \cdots & a_{mn} \end{bmatrix}.$$

Two matrices $A = [a_{ij}]$ and $B = [b_{ij}]$ are **equal** if they have the same order $(m \times n)$ and $a_{ij} = b_{ij}$ for $1 \le i \le m$ and $1 \le j \le n$. In other words, two matrices are equal if their corresponding entries are equal.

A British mathematician, Arthur Cayley, invented matrices around 1858. Cayley was a Cambridge University graduate and a lawyer by profession. His ground-breaking work on matrices was begun as he studied the theory of transformations. Cayley also was instrumental in the development of determinants. Cayley and two American mathematicians, Benjamin Peirce (1809–1880) and his son Charles S. Peirce (1839–1914), are credited with developing "matrix algebra."

EXAMPLE 1 Equality of Matrices

Solve for a_{11}, a_{12}, a_{21}, and a_{22} in the following matrix equation.

$$\begin{bmatrix} a_{11} & a_{12} \\ a_{21} & a_{22} \end{bmatrix} = \begin{bmatrix} 2 & -1 \\ -3 & 0 \end{bmatrix}$$

Solution

Because two matrices are equal only if their corresponding entries are equal, you can conclude that

$$a_{11} = 2, \quad a_{12} = -1, \quad a_{21} = -3, \quad \text{and} \quad a_{22} = 0.$$

Matrix Addition and Scalar Multiplication

You can **add** two matrices (of the same order) by adding their corresponding entries.

Definition of Matrix Addition

If $A = [a_{ij}]$ and $B = [b_{ij}]$ are matrices of order $m \times n$, their **sum** is the $m \times n$ matrix given by

$$A + B = [a_{ij} + b_{ij}].$$

The sum of two matrices of different orders is undefined.

Most graphing utilities can perform matrix addition and scalar multiplication. If you have such a graphing utility, duplicate the matrix operations in Examples 2 and 3. Try adding two matrices of different orders such as

$$A = \begin{bmatrix} 1 & 2 \\ 3 & 4 \end{bmatrix} \quad \text{and}$$

$$B = \begin{bmatrix} 5 \\ 6 \end{bmatrix}.$$

What error message does your utility display?

EXAMPLE 2 **Addition of Matrices**

a. $\begin{bmatrix} -1 & 2 \\ 0 & 1 \end{bmatrix} + \begin{bmatrix} 1 & 3 \\ -1 & 2 \end{bmatrix} = \begin{bmatrix} -1+1 & 2+3 \\ 0-1 & 1+2 \end{bmatrix} = \begin{bmatrix} 0 & 5 \\ -1 & 3 \end{bmatrix}$

b. $\begin{bmatrix} 0 & 1 & -2 \\ 1 & 2 & 3 \end{bmatrix} + \begin{bmatrix} 0 & 0 & 0 \\ 0 & 0 & 0 \end{bmatrix} = \begin{bmatrix} 0 & 1 & -2 \\ 1 & 2 & 3 \end{bmatrix}$

c. $\begin{bmatrix} 1 \\ -3 \\ -2 \end{bmatrix} + \begin{bmatrix} -1 \\ 3 \\ 2 \end{bmatrix} = \begin{bmatrix} 0 \\ 0 \\ 0 \end{bmatrix}$

d. The sum of

$$A = \begin{bmatrix} 2 & 1 & 0 \\ 4 & 0 & -1 \\ 3 & -2 & 2 \end{bmatrix} \quad \text{and} \quad B = \begin{bmatrix} 0 & 1 \\ -1 & 3 \\ 2 & 4 \end{bmatrix}$$

is undefined.

In work with matrices, numbers are usually referred to as **scalars.** In this text, scalars will always be real numbers. You can multiply a matrix A by a scalar c by multiplying each entry in A by c.

Definition of Scalar Multiplication

If $A = [a_{ij}]$ is an $m \times n$ matrix and c is a scalar, the **scalar multiple** of A by c is the $m \times n$ matrix given by

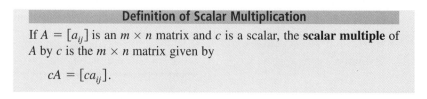

$$cA = [ca_{ij}].$$

The symbol $-A$ represents the scalar product $(-1)A$. Moreover, if A and B are of the same order, $A - B$ represents the sum of A and $(-1)B$. That is,

$$A - B = A + (-1)B.$$ Subtraction of matrices

EXAMPLE 3 **Scalar Multiplication and Matrix Subtraction**

For the following matrices, find (a) $3A$, (b) $-B$, and (c) $3A - B$.

$$A = \begin{bmatrix} 2 & 2 & 4 \\ -3 & 0 & -1 \\ 2 & 1 & 2 \end{bmatrix} \quad \text{and} \quad B = \begin{bmatrix} 2 & 0 & 0 \\ 1 & -4 & 3 \\ -1 & 3 & 2 \end{bmatrix}$$

Solution

a. $$3A = 3\begin{bmatrix} 2 & 2 & 4 \\ -3 & 0 & -1 \\ 2 & 1 & 2 \end{bmatrix}$$ Scalar multiplication

$$= \begin{bmatrix} 3(2) & 3(2) & 3(4) \\ 3(-3) & 3(0) & 3(-1) \\ 3(2) & 3(1) & 3(2) \end{bmatrix}$$ Multiply each entry by 3.

$$= \begin{bmatrix} 6 & 6 & 12 \\ -9 & 0 & -3 \\ 6 & 3 & 6 \end{bmatrix}$$ Simplify.

b. $$-B = (-1)\begin{bmatrix} 2 & 0 & 0 \\ 1 & -4 & 3 \\ -1 & 3 & 2 \end{bmatrix}$$ Definition of negation

$$= \begin{bmatrix} -2 & 0 & 0 \\ -1 & 4 & -3 \\ 1 & -3 & -2 \end{bmatrix}$$ Multiply each entry by -1.

c. $$3A - B = \begin{bmatrix} 6 & 6 & 12 \\ -9 & 0 & -3 \\ 6 & 3 & 6 \end{bmatrix} - \begin{bmatrix} 2 & 0 & 0 \\ 1 & -4 & 3 \\ -1 & 3 & 2 \end{bmatrix}$$ Matrix subtraction

$$= \begin{bmatrix} 4 & 6 & 12 \\ -10 & 4 & -6 \\ 7 & 0 & 4 \end{bmatrix}$$ Subtract corresponding entries.

EXPLORATION

Select two 3×2 matrices A and B. Enter them into your graphing utility and calculate $A + B$ and $B + A$. What do you observe?

Now select a real number c and calculate $c(A + B)$ and $cA + cB$. What do you observe?

It is often convenient to rewrite the scalar multiple cA by factoring c out of every entry in the matrix. For instance, in the following example, the scalar $\frac{1}{2}$ has been factored out of the matrix.

$$\begin{bmatrix} \frac{1}{2} & -\frac{3}{2} \\ \frac{5}{2} & \frac{1}{2} \end{bmatrix} = \frac{1}{2}\begin{bmatrix} 1 & -3 \\ 5 & 1 \end{bmatrix}$$

The properties of matrix addition and scalar multiplication are similar to those of addition and multiplication of real numbers.

Properties of Matrix Addition and Scalar Multiplication

Let A, B, and C be $m \times n$ matrices and let c and d be scalars.

1. $A + B = B + A$	Commutative Property of Matrix Addition
2. $A + (B + C) = (A + B) + C$	Associative Property of Matrix Addition
3. $(cd)A = c(dA)$	Associative Property of Scalar Multiplication
4. $1A = A$	Scalar Identity
5. $c(A + B) = cA + cB$	Distributive Property
6. $(c + d)A = cA + dA$	Distributive Property

Note that the Associative Property of Matrix Addition allows you to write expressions such as $A + B + C$ without ambiguity because the same sum occurs no matter how the matrices are grouped. In other words, you obtain the same sum whether you group $A + B + C$ as $(A + B) + C$ or as $A + (B + C)$. This same reasoning applies to sums of four or more matrices.

EXAMPLE 4 **Addition of More Than Two Matrices**

By adding corresponding entries, you obtain the following sum of four matrices.

$$\begin{bmatrix} 1 \\ 2 \\ -3 \end{bmatrix} + \begin{bmatrix} -1 \\ -1 \\ 2 \end{bmatrix} + \begin{bmatrix} 0 \\ 1 \\ 4 \end{bmatrix} + \begin{bmatrix} 2 \\ -3 \\ -2 \end{bmatrix} = \begin{bmatrix} 2 \\ -1 \\ 1 \end{bmatrix}$$

Most graphing utilities can add and subtract matrices and multiply matrices by scalars. For instance, on a *TI-82* or *TI-83*, you can find the sum of

$$A = \begin{bmatrix} 2 & -3 \\ -1 & 0 \end{bmatrix} \quad \text{and} \quad B = \begin{bmatrix} -1 & 4 \\ 2 & -5 \end{bmatrix}$$

by entering the matrices and then using the following keystrokes.

[*A*] [+] [*B*] [ENTER]

One important property of addition of real numbers is that the number 0 is the additive identity. That is, $c + 0 = c$ for any real number c. For matrices, a similar property holds. That is, if A is an $m \times n$ matrix and O is the $m \times n$ **zero matrix** consisting entirely of zeros, then $A + O = A$.

In other words, O is the **additive identity** for the set of all $m \times n$ matrices. For example, the following matrices are the additive identities for the set of all 2×3 and 2×2 matrices.

$$O = \underbrace{\begin{bmatrix} 0 & 0 & 0 \\ 0 & 0 & 0 \end{bmatrix}}_{\text{Zero } 2 \times 3 \text{ matrix}} \quad \text{and} \quad O = \underbrace{\begin{bmatrix} 0 & 0 \\ 0 & 0 \end{bmatrix}}_{\text{Zero } 2 \times 2 \text{ matrix}}$$

Note The algebra of real numbers and the algebra of matrices also have important differences, which will be discussed later.

The algebra of real numbers and the algebra of matrices have many similarities. For example, compare the following solutions.

Real Numbers (*Solve for x.*)	$m \times n$ *Matrices* (*Solve for X.*)
$x + a = b$	$X + A = B$
$x + a + (-a) = b + (-a)$	$X + A + (-A) = B + (-A)$
$x + 0 = b - a$	$X + O = B - A$
$x = b - a$	$X = B - A$

EXAMPLE 5 Solving a Matrix Equation

Solve for X in the equation $3X + A = B$, where

$$A = \begin{bmatrix} 1 & -2 \\ 0 & 3 \end{bmatrix} \quad \text{and} \quad B = \begin{bmatrix} -3 & 4 \\ 2 & 1 \end{bmatrix}.$$

Solution

Begin by solving the equation for X to obtain

$$3X = B - A \quad \Rightarrow \quad X = \frac{1}{3}(B - A).$$

Now, using the matrices A and B, you have

$$\begin{aligned} X &= \frac{1}{3}\left(\begin{bmatrix} -3 & 4 \\ 2 & 1 \end{bmatrix} - \begin{bmatrix} 1 & -2 \\ 0 & 3 \end{bmatrix}\right) \\ &= \frac{1}{3}\begin{bmatrix} -4 & 6 \\ 2 & -2 \end{bmatrix} \\ &= \begin{bmatrix} -\frac{4}{3} & 2 \\ \frac{2}{3} & -\frac{2}{3} \end{bmatrix}. \end{aligned}$$

Matrix Multiplication

Note The definition of matrix multiplication indicates a *row-by-column* multiplication, where the entry in the ith row and jth column of the product AB is obtained by multiplying the entries in the ith row of A by the corresponding entries in the jth column of B and then adding the results. Example 6 illustrates this process.

The third basic matrix operation is **matrix multiplication.** At first glance, the following definition may seem unusual. You will see later, however, that this definition of the product of two matrices has many practical applications.

Definition of Matrix Multiplication

If $A = [a_{ij}]$ is an $m \times n$ matrix and $B = [b_{ij}]$ is an $n \times p$ matrix, the **product** AB is an $m \times p$ matrix

$$AB = [c_{ij}]$$

where $c_{ij} = a_{i1}b_{1j} + a_{i2}b_{2j} + a_{i3}b_{3j} + \cdot \cdot \cdot + a_{in}b_{nj}$.

Some graphing utilities, such as the *TI-82* and *TI-83*, are able to add, subtract, and multiply matrices. If you have such a graphing utility, enter the matrices

$$A = \begin{bmatrix} 1 & 2 & 3 \\ 2 & -5 & 1 \end{bmatrix} \text{ and}$$

$$B = \begin{bmatrix} -3 & 2 & 1 \\ 4 & -2 & 0 \\ 1 & 2 & 3 \end{bmatrix}$$

and use the following keystrokes to find the product of the matrices.

[A] [×] [B] [ENTER]

You should get:

$$\begin{bmatrix} 8 & 4 & 10 \\ -25 & 16 & 5 \end{bmatrix}.$$

EXAMPLE 6 **Finding the Product of Two Matrices**

Find the product AB where

$$A = \begin{bmatrix} -1 & 3 \\ 4 & -2 \\ 5 & 0 \end{bmatrix} \quad \text{and} \quad B = \begin{bmatrix} -3 & 2 \\ -4 & 1 \end{bmatrix}.$$

Solution

First, note that the product AB is defined because the number of columns of A is equal to the number of rows of B. Moreover, the product AB has order 3×2, and is of the form

$$\begin{bmatrix} -1 & 3 \\ 4 & -2 \\ 5 & 0 \end{bmatrix}\begin{bmatrix} -3 & 2 \\ -4 & 1 \end{bmatrix} = \begin{bmatrix} c_{11} & c_{12} \\ c_{21} & c_{22} \\ c_{31} & c_{32} \end{bmatrix}.$$

To find the entries of the product, multiply each row of A by each column of B, as follows. Use a graphing utility to check this result.

$$AB = \begin{bmatrix} -1 & 3 \\ 4 & -2 \\ 5 & 0 \end{bmatrix}\begin{bmatrix} -3 & 2 \\ -4 & 1 \end{bmatrix}$$

$$= \begin{bmatrix} (-1)(-3) + (3)(-4) & (-1)(2) + (3)(1) \\ (4)(-3) + (-2)(-4) & (4)(2) + (-2)(1) \\ (5)(-3) + (0)(-4) & (5)(2) + (0)(1) \end{bmatrix}$$

$$= \begin{bmatrix} -9 & 1 \\ -4 & 6 \\ -15 & 10 \end{bmatrix}$$

Be sure you understand that for the product of two matrices to be defined, the number of columns of the first matrix must equal the number of rows of the second matrix. That is, the middle two indices must be the same and the outside two indices give the order of the product, as shown in the following diagram.

$$\begin{array}{ccccc} A & & B & = & AB \\ m \times n & & n \times p & & m \times p \end{array}$$

equal (middle indices n and n)

order of AB (outside indices m and p)

EXPLORATION

Use a graphing utility to multiply the matrices

$$A = \begin{bmatrix} 1 & 2 \\ 3 & 4 \end{bmatrix} \quad \text{and}$$

$$B = \begin{bmatrix} 0 & 1 \\ 2 & 3 \end{bmatrix}.$$

Do you obtain the same result for the product AB as for the product BA? What does this tell you about matrix multiplication and commutativity?

EXAMPLE 7 Matrix Multiplication

a. $\begin{bmatrix} 1 & 0 & 3 \\ 2 & -1 & -2 \end{bmatrix} \begin{bmatrix} -2 & 4 & 2 \\ 1 & 0 & 0 \\ -1 & 1 & -1 \end{bmatrix} = \begin{bmatrix} -5 & 7 & -1 \\ -3 & 6 & 6 \end{bmatrix}$

$2 \times 3 \qquad 3 \times 3 \qquad 2 \times 3$

b. $\begin{bmatrix} 3 & 4 \\ -2 & 5 \end{bmatrix} \begin{bmatrix} 1 & 0 \\ 0 & 1 \end{bmatrix} = \begin{bmatrix} 3 & 4 \\ -2 & 5 \end{bmatrix}$

$2 \times 2 \qquad 2 \times 2 \qquad 2 \times 2$

c. $\begin{bmatrix} 1 & 2 \\ 1 & 1 \end{bmatrix} \begin{bmatrix} -1 & 2 \\ 1 & -1 \end{bmatrix} = \begin{bmatrix} 1 & 0 \\ 0 & 1 \end{bmatrix}$

$2 \times 2 \qquad 2 \times 2 \qquad 2 \times 2$

d. $\begin{bmatrix} 1 & -2 & -3 \end{bmatrix} \begin{bmatrix} 2 \\ -1 \\ 1 \end{bmatrix} = \begin{bmatrix} 1 \end{bmatrix}$

$1 \times 3 \qquad 3 \times 1 \qquad 1 \times 1$

e. $\begin{bmatrix} 2 \\ -1 \\ 1 \end{bmatrix} \begin{bmatrix} 1 & -2 & -3 \end{bmatrix} = \begin{bmatrix} 2 & -4 & -6 \\ -1 & 2 & 3 \\ 1 & -2 & -3 \end{bmatrix}$

$3 \times 1 \qquad 1 \times 3 \qquad 3 \times 3$

Note In parts (d) and (e) of Example 7, note that the two products are different. Matrix multiplication is not, in general, commutative. That is, for most matrices, $AB \neq BA$.

f. The product AB for the following matrices is not defined.

$$A = \begin{bmatrix} -2 & 1 \\ 1 & -3 \\ 1 & 4 \end{bmatrix} \quad \text{and} \quad B = \begin{bmatrix} -2 & 3 & 1 & 4 \\ 0 & 1 & -1 & 2 \\ 2 & -1 & 0 & 1 \end{bmatrix}$$

$3 \times 2 \qquad 3 \times 4$

The general pattern for matrix multiplication is as follows. To obtain the entry in the ith row and the jth column of the product AB, use the ith row of A and the jth column of B.

$$\begin{bmatrix} a_{11} & a_{12} & a_{13} & \cdots & a_{1n} \\ a_{21} & a_{22} & a_{23} & \cdots & a_{2n} \\ a_{31} & a_{32} & a_{33} & \cdots & a_{3n} \\ \vdots & \vdots & \vdots & & \vdots \\ a_{i1} & a_{i2} & a_{i3} & \cdots & a_{in} \\ \vdots & \vdots & \vdots & & \vdots \\ a_{m1} & a_{m2} & a_{m3} & \cdots & a_{mn} \end{bmatrix} \begin{bmatrix} b_{11} & b_{12} & \cdots & b_{1j} & \cdots & b_{1p} \\ b_{21} & b_{22} & \cdots & b_{2j} & \cdots & b_{2p} \\ b_{31} & b_{32} & \cdots & b_{3j} & \cdots & b_{3p} \\ \vdots & \vdots & & \vdots & & \vdots \\ b_{n1} & b_{n2} & \cdots & b_{nj} & \cdots & b_{np} \end{bmatrix} = \begin{bmatrix} c_{11} & c_{12} & \cdots & c_{1j} & \cdots & c_{1p} \\ c_{21} & c_{22} & \cdots & c_{2j} & \cdots & c_{2p} \\ \vdots & \vdots & & \vdots & & \vdots \\ c_{i1} & c_{i2} & \cdots & c_{ij} & \cdots & c_{ip} \\ \vdots & \vdots & & \vdots & & \vdots \\ c_{m1} & c_{m2} & \cdots & c_{mj} & \cdots & c_{mp} \end{bmatrix}$$

$$a_{i1}b_{1j} + a_{i2}b_{2j} + a_{i3}b_{3j} + \cdots + a_{in}b_{nj} = c_{ij}$$

Properties of Matrix Multiplication

Let A, B, and C be matrices and let c be a scalar.

1. $A(BC) = (AB)C$ — Associative Property of Matrix Multiplication
2. $A(B + C) = AB + AC$ — Distributive Property
3. $(A + B)C = AC + BC$ — Distributive Property
4. $c(AB) = (cA)B = A(cB)$ — Associative Property of Scalar Multiplication

The $n \times n$ matrix that consists of 1's on its main diagonal and 0's elsewhere is called the **identity matrix of order n** and is denoted by

$$I_n = \begin{bmatrix} 1 & 0 & 0 & \cdots & 0 \\ 0 & 1 & 0 & \cdots & 0 \\ 0 & 0 & 1 & \cdots & 0 \\ \vdots & \vdots & \vdots & & \vdots \\ 0 & 0 & 0 & \cdots & 1 \end{bmatrix}. \qquad \text{Identity matrix}$$

Note that an identity matrix must be *square*. When the order is understood to be n, you can denote I_n simply by I. If A is an $n \times n$ matrix, the identity matrix has the property that $AI_n = A$ and $I_nA = A$. For example,

$$\begin{bmatrix} 3 & -2 & 5 \\ 1 & 0 & 4 \\ -1 & 2 & -3 \end{bmatrix} \begin{bmatrix} 1 & 0 & 0 \\ 0 & 1 & 0 \\ 0 & 0 & 1 \end{bmatrix} = \begin{bmatrix} 3 & -2 & 5 \\ 1 & 0 & 4 \\ -1 & 2 & -3 \end{bmatrix}$$

and

$$\begin{bmatrix} 1 & 0 & 0 \\ 0 & 1 & 0 \\ 0 & 0 & 1 \end{bmatrix} \begin{bmatrix} 3 & -2 & 5 \\ 1 & 0 & 4 \\ -1 & 2 & -3 \end{bmatrix} = \begin{bmatrix} 3 & -2 & 5 \\ 1 & 0 & 4 \\ -1 & 2 & -3 \end{bmatrix}.$$

Applications

One application of matrix multiplication is representation of a system of linear equations. Note how the system

$$\begin{aligned} a_{11}x_1 + a_{12}x_2 + a_{13}x_3 &= b_1 \\ a_{21}x_1 + a_{22}x_2 + a_{23}x_3 &= b_2 \\ a_{31}x_1 + a_{32}x_2 + a_{33}x_3 &= b_3 \end{aligned}$$

can be written as the matrix equation $AX = B$, where A is the *coefficient matrix* of the system, and X and B are column matrices.

Note The column matrix B is also called a *constant* matrix. Its entries are the constant terms in the system of equations.

$$\underset{A}{\begin{bmatrix} a_{11} & a_{12} & a_{13} \\ a_{21} & a_{22} & a_{23} \\ a_{31} & a_{32} & a_{33} \end{bmatrix}} \times \underset{X}{\begin{bmatrix} x_1 \\ x_2 \\ x_3 \end{bmatrix}} = \underset{B}{\begin{bmatrix} b_1 \\ b_2 \\ b_3 \end{bmatrix}}$$

EXAMPLE 8 Solving a System of Linear Equations

Solve the matrix equation $AX = B$ for X, where

Coefficient matrix Column matrix

$$A = \begin{bmatrix} 1 & -2 & 1 \\ 0 & 1 & 2 \\ 2 & 3 & -2 \end{bmatrix} \quad \text{and} \quad B = \begin{bmatrix} -4 \\ 4 \\ 2 \end{bmatrix}.$$

Solution

As a system of linear equations, $AX = B$ is as follows.

$$\begin{aligned} x_1 - 2x_2 + x_3 &= -4 \\ x_2 + 2x_3 &= 4 \\ 2x_1 + 3x_2 - 2x_3 &= 2 \end{aligned}$$

Using Gauss-Jordan elimination on the augmented matrix of this system, you obtain the following reduced row-echelon matrix.

$$\left[\begin{array}{ccc|c} 1 & 0 & 0 & -1 \\ 0 & 1 & 0 & 2 \\ 0 & 0 & 1 & 1 \end{array}\right]$$

Thus, the solution of the system of linear equations is $x_1 = -1$, $x_2 = 2$, and $x_3 = 1$, and the solution of the matrix equation is

$$X = \begin{bmatrix} x_1 \\ x_2 \\ x_3 \end{bmatrix} = \begin{bmatrix} -1 \\ 2 \\ 1 \end{bmatrix}.$$

Use a graphing utility to verify that $AX = B$.

Real Life

EXAMPLE 9 **Softball Team Expenses**

Two softball teams submit equipment lists to their sponsors.

	Women's Team	*Men's Team*
Bats	12	15
Balls	45	38
Gloves	15	17

Each bat costs \$48, each ball costs \$4, and each glove costs \$42. Use matrices to find the total cost of equipment for each team.

Solution

The equipment lists and the costs per item can be written in matrix form as

$$E = \begin{bmatrix} 12 & 15 \\ 45 & 38 \\ 15 & 17 \end{bmatrix} \quad \text{and} \quad C = [48 \quad 4 \quad 42].$$

The total cost of equipment for each team is given by the product

$$CE = [48 \quad 4 \quad 42] \begin{bmatrix} 12 & 15 \\ 45 & 38 \\ 15 & 17 \end{bmatrix} = [1386 \;\; 1586].$$

Thus, the total cost of equipment for the women's team is \$1386, and the total cost of equipment for the men's team is \$1586.

Group Activity

Problem Posing

Write a matrix multiplication application problem that uses the matrix

$$A = \begin{bmatrix} 20 & 42 & 33 \\ 17 & 30 & 50 \end{bmatrix}.$$

Exchange problems with another student in your class. Form the matrices that represent the problem, and solve the problem. Interpret your solution in the context of the problem. Check with the creator of the problem to see if you are correct. Discuss other ways to represent and/or approach the problem.

6.2 /// EXERCISES

In Exercises 1–4, find x and y.

1. $\begin{bmatrix} x & -2 \\ 7 & y \end{bmatrix} = \begin{bmatrix} -4 & -2 \\ 7 & 22 \end{bmatrix}$

2. $\begin{bmatrix} -5 & x \\ y & 8 \end{bmatrix} = \begin{bmatrix} -5 & 13 \\ 12 & 8 \end{bmatrix}$

3. $\begin{bmatrix} 16 & 4 & 5 & 4 \\ -3 & 13 & 15 & 6 \\ 0 & 2 & 4 & 0 \end{bmatrix} = \begin{bmatrix} 16 & 4 & 2x+1 & 4 \\ -3 & 13 & 15 & 3x \\ 0 & 2 & 3y-5 & 0 \end{bmatrix}$

4. $\begin{bmatrix} x+2 & 8 & -3 \\ 1 & 2y & 2x \\ 7 & -2 & y+2 \end{bmatrix} = \begin{bmatrix} 2x+6 & 8 & -3 \\ 1 & 18 & -8 \\ 7 & -2 & 11 \end{bmatrix}$

In Exercises 5–10, find (a) $A + B$, (b) $A - B$, (c) $3A$, and (d) $3A - 2B$.

5. $A = \begin{bmatrix} 1 & -1 \\ 2 & -1 \end{bmatrix}$, $B = \begin{bmatrix} 2 & -1 \\ -1 & 8 \end{bmatrix}$

6. $A = \begin{bmatrix} 1 & 2 \\ 2 & 1 \end{bmatrix}$, $B = \begin{bmatrix} -3 & -2 \\ 4 & 2 \end{bmatrix}$

7. $A = \begin{bmatrix} 6 & -1 \\ 2 & 4 \\ -3 & 5 \end{bmatrix}$, $B = \begin{bmatrix} 1 & 4 \\ -1 & 5 \\ 1 & 10 \end{bmatrix}$

8. $A = \begin{bmatrix} 2 & 1 & 1 \\ -1 & -1 & 4 \end{bmatrix}$, $B = \begin{bmatrix} 2 & -3 & 4 \\ -3 & 1 & -2 \end{bmatrix}$

9. $A = \begin{bmatrix} 2 & 2 & -1 & 0 & 1 \\ 1 & 1 & -2 & 0 & -1 \end{bmatrix}$,

$B = \begin{bmatrix} 1 & 1 & -1 & 1 & 0 \\ -3 & 4 & 9 & -6 & -7 \end{bmatrix}$

10. $A = \begin{bmatrix} 3 \\ 2 \\ -1 \end{bmatrix}$, $B = \begin{bmatrix} -4 \\ 6 \\ 2 \end{bmatrix}$

In Exercises 11–14, solve for X given

$$A = \begin{bmatrix} -2 & -1 \\ 1 & 0 \\ 3 & -4 \end{bmatrix} \text{ and } B = \begin{bmatrix} 0 & 3 \\ 2 & 0 \\ -4 & -1 \end{bmatrix}.$$

11. $X = 3A - 2B$

12. $2X = 2A - B$

13. $2X + 3A = B$

14. $2A + 4B = -2X$

In Exercises 15–20, find (a) AB, (b) BA, and, if possible, (c) A^2. (*Note:* $A^2 = AA$.)

15. $A = \begin{bmatrix} 1 & 2 \\ 4 & 2 \end{bmatrix}$, $B = \begin{bmatrix} 2 & -1 \\ -1 & 8 \end{bmatrix}$

16. $A = \begin{bmatrix} 2 & -1 \\ 1 & 4 \end{bmatrix}$, $B = \begin{bmatrix} 0 & 0 \\ 3 & -3 \end{bmatrix}$

17. $A = \begin{bmatrix} 3 & -1 \\ 1 & 3 \end{bmatrix}$, $B = \begin{bmatrix} 1 & -3 \\ 3 & 1 \end{bmatrix}$

18. $A = \begin{bmatrix} 1 & -1 \\ 1 & 1 \end{bmatrix}$, $B = \begin{bmatrix} 1 & 3 \\ -3 & 1 \end{bmatrix}$

19. $A = \begin{bmatrix} 1 & -1 & 7 \\ 2 & -1 & 8 \\ 3 & 1 & -1 \end{bmatrix}$, $B = \begin{bmatrix} 1 & 1 & 2 \\ 2 & 1 & 1 \\ 1 & -3 & 2 \end{bmatrix}$

20. $A = \begin{bmatrix} 3 & 2 & 1 \end{bmatrix}$, $B = \begin{bmatrix} 2 \\ 3 \\ 0 \end{bmatrix}$

In Exercises 21–28, find AB, if possible.

21. $A = \begin{bmatrix} 2 & 1 \\ -3 & 4 \\ 1 & 6 \end{bmatrix}$, $B = \begin{bmatrix} 0 & -1 & 0 \\ 4 & 0 & 2 \\ 8 & -1 & 7 \end{bmatrix}$

22. $A = \begin{bmatrix} 0 & -1 & 0 \\ 4 & 0 & 2 \\ 8 & -1 & 7 \end{bmatrix}$, $B = \begin{bmatrix} 2 & 1 \\ -3 & 4 \\ 1 & 6 \end{bmatrix}$

23. $A = \begin{bmatrix} -1 & 3 \\ 4 & -5 \\ 0 & 2 \end{bmatrix}$, $B = \begin{bmatrix} 1 & 2 \\ 0 & 7 \end{bmatrix}$

24. $A = \begin{bmatrix} 1 & 0 & 0 \\ 0 & 4 & 0 \\ 0 & 0 & -2 \end{bmatrix}$, $B = \begin{bmatrix} 3 & 0 & 0 \\ 0 & -1 & 0 \\ 0 & 0 & 5 \end{bmatrix}$

25. $A = \begin{bmatrix} 5 & 0 & 0 \\ 0 & -8 & 0 \\ 0 & 0 & 7 \end{bmatrix}$, $B = \begin{bmatrix} \frac{1}{5} & 0 & 0 \\ 0 & -\frac{1}{8} & 0 \\ 0 & 0 & \frac{1}{2} \end{bmatrix}$

26. $A = \begin{bmatrix} 10 \\ 12 \end{bmatrix}$, $B = \begin{bmatrix} 6 & -2 & 1 & 6 \end{bmatrix}$

27. $A = \begin{bmatrix} 0 & 0 & 5 \\ 0 & 0 & -3 \\ 0 & 0 & 4 \end{bmatrix}$, $B = \begin{bmatrix} 6 & -11 & 4 \\ 8 & 16 & 4 \\ 0 & 0 & 0 \end{bmatrix}$

28. $A = \begin{bmatrix} 1 & 0 & 3 & -2 \\ 6 & 13 & 8 & -17 \end{bmatrix}$, $B = \begin{bmatrix} 1 & 6 \\ 4 & 2 \end{bmatrix}$

In Exercises 29–34, use the matrix capabilities of a graphing utility to find AB.

29. $A = \begin{bmatrix} 5 & 6 & -3 \\ -2 & 5 & 1 \\ 10 & -5 & 5 \end{bmatrix}$, $B = \begin{bmatrix} 1 & -1 & 2 \\ 8 & 1 & 4 \\ 4 & -2 & 9 \end{bmatrix}$

30. $A = \begin{bmatrix} 11 & -12 & 4 \\ 14 & 10 & 12 \\ 6 & -2 & 9 \end{bmatrix}$, $B = \begin{bmatrix} 12 & 10 \\ -5 & 12 \\ 15 & 16 \end{bmatrix}$

31. $A = \begin{bmatrix} -3 & 8 & -6 & 8 \\ -12 & 15 & 9 & 6 \\ 5 & -1 & 1 & 5 \end{bmatrix}$,

$B = \begin{bmatrix} 3 & 1 & 6 \\ 24 & 15 & 14 \\ 16 & 10 & 21 \\ 8 & -4 & 10 \end{bmatrix}$

32. $A = \begin{bmatrix} -2 & 4 & 8 \\ 21 & 5 & 6 \\ 13 & 2 & 6 \end{bmatrix}$,

$B = \begin{bmatrix} 2 & 0 \\ -7 & 15 \\ 32 & 14 \\ 0.5 & 1.6 \end{bmatrix}$

33. $A = \begin{bmatrix} 9 & 10 & -38 & 18 \\ 100 & -50 & 250 & 75 \end{bmatrix}$,

$B = \begin{bmatrix} 52 & -85 & 27 & 45 \\ 40 & -35 & 60 & 82 \end{bmatrix}$

34. $A = \begin{bmatrix} 15 & -18 \\ -4 & 12 \\ -8 & 22 \end{bmatrix}$,

$B = \begin{bmatrix} -7 & 22 & 1 \\ 8 & 16 & 24 \end{bmatrix}$

In Exercises 35–38, find matrices A, X, and B such that the system of linear equations can be written as the matrix equation $AX = B$. Solve the system of equations. Use a graphing utility to check your result.

35. $\begin{aligned} -x + y &= 4 \\ -2x + y &= 0 \end{aligned}$

36. $\begin{aligned} x - 2y + 3z &= 9 \\ -x + 3y - z &= -6 \\ 2x - 5y + 5z &= 17 \end{aligned}$

37. $\begin{aligned} 2x + 3y &= 5 \\ x + 4y &= 10 \end{aligned}$

38. $\begin{aligned} x + y - 3z &= -1 \\ -x + 2y &= 1 \\ -y + z &= 0 \end{aligned}$

In Exercises 39–42, use the matrix capabilities of a graphing utility to find

$$f(A) = a_0 I_n + a_1 A + a_2 A^2 + \cdots + a_n A^n.$$

39. $f(x) = x^2 - 5x + 2$, $A = \begin{bmatrix} 2 & 0 \\ 4 & 5 \end{bmatrix}$

40. $f(x) = x^2 - 7x + 6$, $A = \begin{bmatrix} 5 & 4 \\ 1 & 2 \end{bmatrix}$

41. $f(x) = x^3 - 10x^2 + 31x - 30$, $A = \begin{bmatrix} 3 & 1 & 4 \\ 0 & 2 & 6 \\ 0 & 0 & 5 \end{bmatrix}$

42. $f(x) = x^2 - 10x + 24$, $A = \begin{bmatrix} 8 & -4 \\ 2 & 2 \end{bmatrix}$

43. *Think About It* If a, b, and c are real numbers such that $c \neq 0$ and $ac = bc$, then $a = b$. However, if A, B, and C are nonzero matrices such that $AC = BC$, then A is *not necessarily* equal to B. Illustrate this using the following matrices.

$$A = \begin{bmatrix} 0 & 1 \\ 0 & 1 \end{bmatrix}, \quad B = \begin{bmatrix} 1 & 0 \\ 1 & 0 \end{bmatrix}, \quad C = \begin{bmatrix} 2 & 3 \\ 2 & 3 \end{bmatrix}$$

44. *Think About It* If a and b are real numbers such that $ab = 0$, then $a = 0$ or $b = 0$. However, if A and B are matrices such that $AB = 0$, it is *not necessarily* true that $A = 0$ or $B = 0$. Illustrate this using the following matrices.

$$A = \begin{bmatrix} 3 & 3 \\ 4 & 4 \end{bmatrix}, \quad B = \begin{bmatrix} 1 & -1 \\ -1 & 1 \end{bmatrix}$$

Think About It **In Exercises 45–54, use matrices A and B each of order 2×3, C of order 3×2, and D of order 2×2. Determine whether the matrices are of proper order to perform the operation(s). If so, give the order of the answer.**

45. $A + 2C$

46. $B - 3C$

47. AB

48. BC

49. $BC - D$

50. $CB - D$

51. $(CA)D$

52. $(BC)D$

53. $D(A - 3B)$

54. $(BC - D)A$

55. *Factory Production* A certain corporation has three factories, each of which manufactures two products. The number of units of product i produced at factory j in one day is represented by a_{ij} in the matrix

$$A = \begin{bmatrix} 60 & 40 & 20 \\ 30 & 90 & 60 \end{bmatrix}.$$

Find the production levels if production is increased by 20%. (*Hint:* Because an increase of 20% corresponds to 100% + 20%, multiply the given matrix by 1.2.)

56. *Factory Production* A certain corporation has four factories, each of which manufactures two products. The number of units of product i produced at factory j in one day is represented by a_{ij} in the matrix

$$A = \begin{bmatrix} 100 & 90 & 70 & 30 \\ 40 & 20 & 60 & 60 \end{bmatrix}.$$

Find the production levels if production is increased by 10%.

57. *Crop Production* A fruit grower raises two crops, which are shipped to three outlets. The number of units of crop i that are shipped to outlet j is represented by a_{ij} in the matrix

$$A = \begin{bmatrix} 100 & 75 & 75 \\ 125 & 150 & 100 \end{bmatrix}.$$

The profit per unit is represented by the matrix

$$B = [\$3.75 \quad \$7.00].$$

Find the product BA, and state what each entry of the product represents.

58. *Revenue* A manufacturer produces three models of a product, which are shipped to two warehouses. The number of units of model i that are shipped to warehouse j is represented by a_{ij} in the matrix

$$A = \begin{bmatrix} 5{,}000 & 4{,}000 \\ 6{,}000 & 10{,}000 \\ 8{,}000 & 5{,}000 \end{bmatrix}.$$

The price per unit is represented by the matrix

$$B = [\$20.50 \quad \$26.50 \quad \$29.50].$$

Compute BA and interpret the result.

Exploration **In Exercises 59 and 60, let $i = \sqrt{-1}$.**

59. Consider the matrix

$$A = \begin{bmatrix} i & 0 \\ 0 & i \end{bmatrix}.$$

Find A^2, A^3, and A^4. Identify any similarities with i^2, i^3, and i^4.

60. Find and identify A^2 for the matrix

$$A = \begin{bmatrix} 0 & -i \\ i & 0 \end{bmatrix}.$$

61. *Inventory Levels* A company sells five models of computers through three retail outlets. The inventories are given by S.

$$S = \overset{\text{Model}}{\overset{\begin{matrix} A & B & C & D & E \end{matrix}}{\begin{bmatrix} 3 & 2 & 2 & 3 & 0 \\ 0 & 2 & 3 & 4 & 3 \\ 4 & 2 & 1 & 3 & 2 \end{bmatrix}}} \begin{matrix} 1 \\ 2 \\ 3 \end{matrix} \Big\} \text{ Outlet}$$

The wholesale and retail prices are given by T.

$$T = \overset{\text{Price}}{\overset{\begin{matrix} \text{Wholesale} & \text{Retail} \end{matrix}}{\begin{bmatrix} \$840 & \$1100 \\ \$1200 & \$1350 \\ \$1450 & \$1650 \\ \$2650 & \$3000 \\ \$3050 & \$3200 \end{bmatrix}}} \begin{matrix} A \\ B \\ C \\ D \\ E \end{matrix} \Big\} \text{ Model}$$

Compute ST and interpret the result.

62. *Labor/Wage Requirements* A company that manufactures boats has the following labor-hour and wage requirements.

Labor per Boat

$$S = \begin{bmatrix} 1.0 \text{ hr} & 0.5 \text{ hr} & 0.2 \text{ hr} \\ 1.6 \text{ hr} & 1.0 \text{ hr} & 0.2 \text{ hr} \\ 2.5 \text{ hr} & 2.0 \text{ hr} & 0.4 \text{ hr} \end{bmatrix}$$

Department: columns Cutting, Assembly, Packaging; Boat size: rows Small, Medium, Large

Wages per Hour

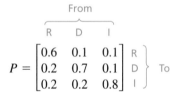

$$T = \begin{bmatrix} \$12 & \$10 \\ \$9 & \$8 \\ \$6 & \$5 \end{bmatrix}$$

Plant: columns A, B; Department: rows Cutting, Assembly, Packaging

Compute ST and interpret the result.

63. *Voting Preference* The matrix

$$P = \begin{bmatrix} 0.6 & 0.1 & 0.1 \\ 0.2 & 0.7 & 0.1 \\ 0.2 & 0.2 & 0.8 \end{bmatrix}$$

From: columns R, D, I; To: rows R, D, I

is called a stochastic matrix. Each entry $p_{ij}(i \neq j)$ represents the proportion of the voting population that changes from party i to party j, and p_{ii} represents the proportion that remains loyal to the party from one election to the next. Compute and interpret P^2.

64. *Voting Preference* Use a graphing utility to find P^3, P^4, P^5, P^6, P^7, and P^8 for the matrix given in Exercise 63. Can you detect a pattern as P is raised to higher powers?

65. *Exploration* Let A and B be unequal diagonal matrices of the same order. (A *diagonal matrix* is a square matrix in which each entry not on the main diagonal is zero.) Determine the products AB for several pairs of such matrices. Make a conjecture about a quick rule for such products.

66. *Exploration* Consider matrices of the form

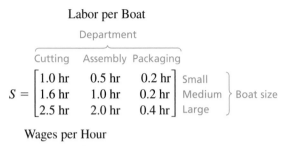

$$A = \begin{bmatrix} 0 & a_{12} & a_{13} & a_{14} & \cdots & a_{1n} \\ 0 & 0 & a_{23} & a_{24} & \cdots & a_{2n} \\ 0 & 0 & 0 & a_{34} & \cdots & a_{3n} \\ \vdots & \vdots & \vdots & \vdots & \cdots & \vdots \\ 0 & 0 & 0 & 0 & \cdots & a_{(n-1)n} \\ 0 & 0 & 0 & 0 & \cdots & 0 \end{bmatrix}.$$

(a) Write a 2×2 matrix and a 3×3 matrix in the form of A.

(b) Use a graphing utility to raise each of the matrices to higher powers. Describe the result.

(c) Use the result of part (b) to make a conjecture about powers of A if A is a 4×4 matrix. Use a graphing utility to test your conjecture.

(d) Use the results of parts (b) and (c) to make a conjecture about powers of A if A is an $n \times n$ matrix.

Review **Solve Exercises 67–70 as a review of the skills and problem-solving techniques you learned in previous sections. Simplify the logarithmic expression.**

67. $\log_4 \sqrt{32}$

68. $\log_5(5^3 \cdot 4)$

69. $\log_2 \dfrac{12}{5}$

70. $\ln \dfrac{100}{e^2}$

6.3 The Inverse of a Square Matrix

The Inverse of a Matrix / Finding Inverse Matrices / The Inverse of a 2 × 2 Matrix / Systems of Linear Equations

The Inverse of a Matrix

This section further develops the algebra of matrices. To begin, consider the real number equation $ax = b$. To solve this equation for x, multiply both sides of the equation by a^{-1} (provided that $a \neq 0$).

$$\begin{aligned} ax &= b \\ (a^{-1}a)x &= a^{-1}b \\ (1)x &= a^{-1}b \\ x &= a^{-1}b \end{aligned}$$

The number a^{-1} is called the *multiplicative inverse of a* because $a^{-1}a = 1$. The definition of the multiplicative inverse of a matrix is similar.

Note The symbol A^{-1} is read "A inverse."

Definition of the Inverse of a Square Matrix

Let A be an $n \times n$ matrix. If there exists a matrix A^{-1} such that

$$AA^{-1} = I_n = A^{-1}A$$

A^{-1} is called the **inverse** of A.

Note Recall that it is not always true that $AB = BA$, even if both products are defined. However, if A and B are both square matrices and $AB = I_n$, it can be shown that $BA = I_n$. Hence, in Example 1, you need only to check that $AB = I_2$.

EXAMPLE 1 The Inverse of a Matrix

Show that B is the inverse of A, where

$$A = \begin{bmatrix} -1 & 2 \\ -1 & 1 \end{bmatrix} \quad \text{and} \quad B = \begin{bmatrix} 1 & -2 \\ 1 & -1 \end{bmatrix}.$$

Solution

To show that B is the inverse of A, show that $AB = I = BA$, as follows.

$$AB = \begin{bmatrix} -1 & 2 \\ -1 & 1 \end{bmatrix}\begin{bmatrix} 1 & -2 \\ 1 & -1 \end{bmatrix} = \begin{bmatrix} -1+2 & 2-2 \\ -1+1 & 2-1 \end{bmatrix} = \begin{bmatrix} 1 & 0 \\ 0 & 1 \end{bmatrix}$$

$$BA = \begin{bmatrix} 1 & -2 \\ 1 & -1 \end{bmatrix}\begin{bmatrix} -1 & 2 \\ -1 & 1 \end{bmatrix} = \begin{bmatrix} -1+2 & 2-2 \\ -1+1 & 2-1 \end{bmatrix} = \begin{bmatrix} 1 & 0 \\ 0 & 1 \end{bmatrix}$$

If a matrix A has an inverse, A is called **invertible** (or **nonsingular**); otherwise, A is called **singular.** A nonsquare matrix cannot have an inverse. To see this, note that if A is of order $m \times n$ and B is of order $n \times m$ (where $m \neq n$), the products AB and BA are of different orders and therefore cannot be equal to each other. Not all square matrices possess inverses (see the matrix at the bottom of page 490). If, however, a matrix does have an inverse, that inverse is unique. The following example shows how to use systems of equations to find the inverse of a matrix.

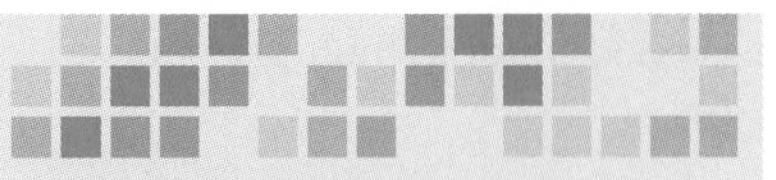

Most graphing utilities have the capability of finding the inverse of a square matrix. For instance, to find the inverse of the matrix

$$A = \begin{bmatrix} 2 & -3 & 1 \\ -1 & 2 & -1 \\ -2 & 0 & 1 \end{bmatrix}$$

on a *TI-82* or *TI-83*, enter the matrix. Then use the following keystrokes

[A] $\boxed{x^{-1}}$ $\boxed{\text{ENTER}}$

After you find A^{-1}, store it as [B] and use the graphing utility to find [A] × [B] and [B] × [A]. What can you conclude?

EXAMPLE 2 **Finding the Inverse of a Matrix**

Find the inverse of

$$A = \begin{bmatrix} 1 & 4 \\ -1 & -3 \end{bmatrix}.$$

Solution

To find the inverse of A, try to solve the matrix equation $AX = I$ for X.

$$\overset{A}{\begin{bmatrix} 1 & 4 \\ -1 & -3 \end{bmatrix}} \overset{X}{\begin{bmatrix} x_{11} & x_{12} \\ x_{21} & x_{22} \end{bmatrix}} = \overset{I}{\begin{bmatrix} 1 & 0 \\ 0 & 1 \end{bmatrix}}$$

$$\begin{bmatrix} x_{11} + 4x_{21} & x_{12} + 4x_{22} \\ -x_{11} - 3x_{21} & -x_{12} - 3x_{22} \end{bmatrix} = \begin{bmatrix} 1 & 0 \\ 0 & 1 \end{bmatrix}$$

Equating corresponding entries, you obtain the following two systems of linear equations.

$$\begin{aligned} x_{11} + 4x_{21} &= 1 \\ -x_{11} - 3x_{21} &= 0 \end{aligned} \qquad \begin{aligned} x_{12} + 4x_{22} &= 0 \\ -x_{12} - 3x_{22} &= 1 \end{aligned}$$

From the first system you can determine that $x_{11} = -3$ and $x_{21} = 1$, and from the second system you can determine that $x_{12} = -4$ and $x_{22} = 1$. Therefore, the inverse of A is

$$X = A^{-1} = \begin{bmatrix} -3 & -4 \\ 1 & 1 \end{bmatrix}.$$

You can use matrix multiplication to check this result.

Check

$$AA^{-1} = \begin{bmatrix} 1 & 4 \\ -1 & -3 \end{bmatrix}\begin{bmatrix} -3 & -4 \\ 1 & 1 \end{bmatrix} = \begin{bmatrix} 1 & 0 \\ 0 & 1 \end{bmatrix} \quad ✓$$

$$A^{-1}A = \begin{bmatrix} -3 & -4 \\ 1 & 1 \end{bmatrix}\begin{bmatrix} 1 & 4 \\ -1 & -3 \end{bmatrix} = \begin{bmatrix} 1 & 0 \\ 0 & 1 \end{bmatrix} \quad ✓$$

Finding Inverse Matrices

In Example 2, note that the two systems of linear equations have the *same coefficient matrix A*. Rather than solve the two systems represented by

$$\begin{bmatrix} 1 & 4 & \vdots & 1 \\ -1 & -3 & \vdots & 0 \end{bmatrix} \text{ and } \begin{bmatrix} 1 & 4 & \vdots & 0 \\ -1 & -3 & \vdots & 1 \end{bmatrix}$$

separately, you can solve them *simultaneously* by **adjoining** the identity matrix to the coefficient matrix to obtain

$$\overbrace{\begin{bmatrix} 1 & 4 \\ -1 & -3 \end{bmatrix}}^{A} \vdots \overbrace{\begin{bmatrix} 1 & 0 \\ 0 & 1 \end{bmatrix}}^{I}.$$

Then, applying Gauss-Jordan elimination to this matrix, you can solve *both* systems with a single elimination process, as follows.

$$\begin{bmatrix} 1 & 4 & \vdots & 1 & 0 \\ -1 & -3 & \vdots & 0 & 1 \end{bmatrix}$$

$$R_1 + R_2 \rightarrow \begin{bmatrix} 1 & 4 & \vdots & 1 & 0 \\ 0 & 1 & \vdots & 1 & 1 \end{bmatrix}$$

$$-4R_2 + R_1 \rightarrow \begin{bmatrix} 1 & 0 & \vdots & -3 & -4 \\ 0 & 1 & \vdots & 1 & 1 \end{bmatrix}$$

Thus, from the "doubly augmented" matrix $[A \ : \ I]$, you obtained the matrix $[I \ : \ A^{-1}]$.

$$\overbrace{\begin{bmatrix} 1 & 4 \\ -1 & -3 \end{bmatrix}}^{A} \vdots \overbrace{\begin{bmatrix} 1 & 0 \\ 0 & 1 \end{bmatrix}}^{I} \Longrightarrow \overbrace{\begin{bmatrix} 1 & 0 \\ 0 & 1 \end{bmatrix}}^{I} \vdots \overbrace{\begin{bmatrix} -3 & -4 \\ 1 & 1 \end{bmatrix}}^{A^{-1}}$$

This procedure (or algorithm) works for any square matrix that has an inverse.

> **EXPLORATION**
>
> Select two 2×2 matrices A and B that have inverses. Enter them into your graphing utility and calculate $(AB)^{-1}$. Then calculate $B^{-1}A^{-1}$ and $A^{-1}B^{-1}$. Make a conjecture about the inverse of a product of two invertible matrices.

Finding an Inverse Matrix

Let A be a square matrix of order n.

1. Write the $n \times 2n$ matrix that consists of the given matrix A on the left and the $n \times n$ identity matrix I on the right to obtain $[A \ \vdots \ I]$. Note that we separate the matrices A and I by a dotted line. We call this process **adjoining** the matrices A and I.
2. If possible, row reduce A to I using elementary row operations on the *entire* matrix $[A \ \vdots \ I]$. The result will be the matrix $[I \ \vdots \ A^{-1}]$. If this is not possible, A is not invertible.
3. Check your work by multiplying to see that $AA^{-1} = I = A^{-1}A$.

Verify the computations in Example 3 with your graphing utility. Enter the 3×6 matrix $[A \;\vdots\; I]$ and row reduce it to the matrix $[I \;\vdots\; A^{-1}]$, as follows.

$$\begin{bmatrix} 1 & -1 & 0 & 1 & 0 & 0 \\ 1 & 0 & -1 & 0 & 1 & 0 \\ 6 & -2 & -3 & 0 & 0 & 1 \end{bmatrix}$$

$$\downarrow$$

$$\begin{bmatrix} 1 & 0 & 0 & -2 & -3 & 1 \\ 0 & 1 & 0 & -3 & -3 & 1 \\ 0 & 0 & 1 & -2 & -4 & 1 \end{bmatrix}$$

What happens if you try this method to find the inverse of

$$A = \begin{bmatrix} 1 & 2 & 0 \\ 3 & -1 & 2 \\ -2 & 3 & -2 \end{bmatrix}?$$

EXAMPLE 3 **Finding the Inverse of a Matrix**

Find the inverse of $A = \begin{bmatrix} 1 & -1 & 0 \\ 1 & 0 & -1 \\ 6 & -2 & -3 \end{bmatrix}$.

Solution

Begin by adjoining the identity matrix to A to form the matrix

$$[A \;\vdots\; I] = \left[\begin{array}{ccc|ccc} 1 & -1 & 0 & 1 & 0 & 0 \\ 1 & 0 & -1 & 0 & 1 & 0 \\ 6 & -2 & -3 & 0 & 0 & 1 \end{array}\right].$$

Using elementary row operations to obtain the form $[I \;\vdots\; A^{-1}]$ results in

$$\left[\begin{array}{ccc|ccc} 1 & 0 & 0 & -2 & -3 & 1 \\ 0 & 1 & 0 & -3 & -3 & 1 \\ 0 & 0 & 1 & -2 & -4 & 1 \end{array}\right].$$

Therefore, the matrix A is invertible and its inverse is

$$A^{-1} = \begin{bmatrix} -2 & -3 & 1 \\ -3 & -3 & 1 \\ -2 & -4 & 1 \end{bmatrix}.$$

Try using a graphing utility to confirm this result by multiplying A by A^{-1} to obtain I.

The process shown in Example 3 applies to any $n \times n$ matrix A. If A has an inverse, this process will find it. If A does not have an inverse, the process will tell us so. For instance, the following matrix has no inverse.

$$A = \begin{bmatrix} 1 & 2 & 0 \\ 3 & -1 & 2 \\ -2 & 3 & -2 \end{bmatrix}$$

To confirm that matrix A above has no inverse, begin by adjoining the identity matrix to A to form

$$[A \;\vdots\; I] = \left[\begin{array}{ccc|ccc} 1 & 2 & 0 & 1 & 0 & 0 \\ 3 & -1 & 2 & 0 & 1 & 0 \\ -2 & 3 & -2 & 0 & 0 & 1 \end{array}\right].$$

Then use elementary row operations to obtain

$$\left[\begin{array}{ccc|ccc} 1 & 2 & 0 & 1 & 0 & 0 \\ 0 & -7 & 2 & -3 & 1 & 0 \\ 0 & 0 & 0 & -1 & 1 & 1 \end{array}\right].$$

At this point in the elimination process you can see that it is impossible to obtain the identity matrix I on the left. Therefore, A is not invertible.

The Inverse of a 2 × 2 Matrix

Using Gauss-Jordan elimination to find the inverse of a matrix works well (even as a computer technique) for matrices of order 3 × 3 or greater. For 2 × 2 matrices, however, many people prefer to use a formula for the inverse rather than Gauss-Jordan elimination. This simple formula, which works *only* for 2 × 2 matrices, is explained as follows. If A is a 2 × 2 matrix given by

$$A = \begin{bmatrix} a & b \\ c & d \end{bmatrix}$$

then A is invertible if and only if $ad - bc \neq 0$. If $ad - bc \neq 0$, the inverse is given by

$$A^{-1} = \frac{1}{ad - bc}\begin{bmatrix} d & -b \\ -c & a \end{bmatrix}.$$

Try verifying this inverse by multiplication.

EXPLORATION

Use a graphing utility with matrix operations to find the inverse of the matrix

$$A = \begin{bmatrix} 1 & -3 \\ -2 & 6 \end{bmatrix}.$$

What message appears on the screen? Why does the graphing utility display this message?

Note The denominator $ad - bc$ is called the **determinant** of the 2 × 2 matrix A. You will study determinants in the next section.

EXAMPLE 4 Finding the Inverse of a 2 × 2 Matrix

If possible, find the inverse of the matrix.

a. $A = \begin{bmatrix} 3 & -1 \\ -2 & 2 \end{bmatrix}$

b. $B = \begin{bmatrix} 3 & -1 \\ -6 & 2 \end{bmatrix}$

Solution

a. For the matrix A, apply the formula for the inverse of a 2 × 2 matrix to obtain

$$ad - bc = (3)(2) - (-1)(-2) = 4.$$

Because this quantity is not zero, the inverse is formed by interchanging the entries on the main diagonal, changing the signs of the other two entries, and multiplying by the scalar $\frac{1}{4}$, as follows.

$$A^{-1} = \frac{1}{4}\begin{bmatrix} 2 & 1 \\ 2 & 3 \end{bmatrix} = \begin{bmatrix} \frac{1}{2} & \frac{1}{4} \\ \frac{1}{2} & \frac{3}{4} \end{bmatrix}$$

b. For the matrix B, you have

$$ad - bc = (3)(2) - (-1)(-6) = 0$$

which means that B is not invertible.

The formula $X = A^{-1}B$ is used on most graphing utilities to solve linear systems that have invertible coefficient matrices. That is, you enter the $n \times n$ coefficient matrix [A] and the $n \times 1$ column matrix [B]. The solution X is given by $[A]^{-1}[B]$.

Systems of Linear Equations

You know that a system of linear equations can have exactly one solution, infinitely many solutions, or no solution. If the coefficient matrix A of a *square* system (a system that has the same number of equations as variables) is invertible, the system has a unique solution, which is given as follows.

A System of Equations with a Unique Solution

If A is an invertible matrix, the system of linear equations represented by $AX = B$ has a unique solution given by

$$X = A^{-1}B.$$

Note Use Gauss-Jordan elimination or a graphing utility to verify A^{-1} for the system of equations in Example 5.

EXAMPLE 5 **Solving a System of Equations Using an Inverse**

Use an inverse matrix to solve the system.

$$\begin{aligned} 2x + 3y + z &= -1 \\ 3x + 3y + z &= 1 \\ 2x + 4y + z &= -2 \end{aligned}$$

Solution

$$X = A^{-1}B = \begin{bmatrix} -1 & 1 & 0 \\ -1 & 0 & 1 \\ 6 & -2 & -3 \end{bmatrix} \begin{bmatrix} -1 \\ 1 \\ -2 \end{bmatrix} = \begin{bmatrix} 2 \\ -1 \\ -2 \end{bmatrix}$$

Thus, the solution is $x = 2$, $y = -1$, and $z = -2$.

Group Activity

Finding an Inverse Matrix

Use a graphing utility to decide which of the following matrices is (are) invertible.

a. $A = \begin{bmatrix} -3 & 2 \\ 7 & 4 \end{bmatrix}$

b. $B = \begin{bmatrix} 1 & -4 & 2 \\ 2 & -9 & 5 \\ 1 & -5 & 4 \end{bmatrix}$

c. $C = \begin{bmatrix} -4 & 6 & -4 \\ 2 & 4 & 0 \\ 6 & -2 & 4 \end{bmatrix}$

6.3 /// EXERCISES

In Exercises 1–8, show that B is the inverse of A.

1. $A = \begin{bmatrix} 2 & 1 \\ 5 & 3 \end{bmatrix}$, $B = \begin{bmatrix} 3 & -1 \\ -5 & 2 \end{bmatrix}$

2. $A = \begin{bmatrix} 1 & -1 \\ -1 & 2 \end{bmatrix}$, $B = \begin{bmatrix} 2 & 1 \\ 1 & 1 \end{bmatrix}$

3. $A = \begin{bmatrix} 1 & 2 \\ 3 & 4 \end{bmatrix}$, $B = \begin{bmatrix} -2 & 1 \\ \frac{3}{2} & -\frac{1}{2} \end{bmatrix}$

4. $A = \begin{bmatrix} 1 & -1 \\ 2 & 3 \end{bmatrix}$, $B = \begin{bmatrix} \frac{3}{5} & \frac{1}{5} \\ -\frac{2}{5} & \frac{1}{5} \end{bmatrix}$

5. $A = \begin{bmatrix} -2 & 2 & 3 \\ 1 & -1 & 0 \\ 0 & 1 & 4 \end{bmatrix}$, $B = \dfrac{1}{3}\begin{bmatrix} -4 & -5 & 3 \\ -4 & -8 & 3 \\ 1 & 2 & 0 \end{bmatrix}$

6. $A = \begin{bmatrix} 2 & -17 & 11 \\ -1 & 11 & -7 \\ 0 & 3 & -2 \end{bmatrix}$, $B = \begin{bmatrix} 1 & 1 & 2 \\ 2 & 4 & -3 \\ 3 & 6 & -5 \end{bmatrix}$

7. $A = \begin{bmatrix} 2 & 0 & 1 & 1 \\ 3 & 0 & 0 & 1 \\ -1 & 1 & -2 & 1 \\ 4 & -1 & 1 & 0 \end{bmatrix}$,

$B = \begin{bmatrix} -1 & 2 & -1 & -1 \\ -4 & 9 & -5 & -6 \\ 0 & 1 & -1 & -1 \\ 3 & -5 & 3 & 3 \end{bmatrix}$

8. $A = \begin{bmatrix} -1 & 1 & 0 & -1 \\ 1 & -1 & 1 & 0 \\ -1 & 1 & 2 & 0 \\ 0 & -1 & 1 & 1 \end{bmatrix}$,

$B = \dfrac{1}{3}\begin{bmatrix} -3 & 1 & 1 & -3 \\ -3 & -1 & 2 & -3 \\ 0 & 1 & 1 & 0 \\ -3 & -2 & 1 & 0 \end{bmatrix}$

In Exercises 9–24, find the inverse of the matrix (if it exists).

9. $\begin{bmatrix} 2 & 0 \\ 0 & 3 \end{bmatrix}$

10. $\begin{bmatrix} 1 & 2 \\ 3 & 7 \end{bmatrix}$

11. $\begin{bmatrix} 1 & -2 \\ 2 & -3 \end{bmatrix}$

12. $\begin{bmatrix} -7 & 33 \\ 4 & -19 \end{bmatrix}$

13. $\begin{bmatrix} -1 & 1 \\ -2 & 1 \end{bmatrix}$

14. $\begin{bmatrix} 11 & 1 \\ -1 & 0 \end{bmatrix}$

15. $\begin{bmatrix} 2 & 4 \\ 4 & 8 \end{bmatrix}$

16. $\begin{bmatrix} 2 & 3 \\ 1 & 4 \end{bmatrix}$

17. $\begin{bmatrix} 2 & 7 & 1 \\ -3 & -9 & 2 \end{bmatrix}$

18. $\begin{bmatrix} -2 & 5 \\ 6 & -15 \\ 0 & 1 \end{bmatrix}$

19. $\begin{bmatrix} 1 & 1 & 1 \\ 3 & 5 & 4 \\ 3 & 6 & 5 \end{bmatrix}$

20. $\begin{bmatrix} 1 & 2 & 2 \\ 3 & 7 & 9 \\ -1 & -4 & -7 \end{bmatrix}$

21. $\begin{bmatrix} 1 & 0 & 0 \\ 3 & 4 & 0 \\ 2 & 5 & 5 \end{bmatrix}$

22. $\begin{bmatrix} 1 & 0 & 0 \\ 3 & 0 & 0 \\ 2 & 5 & 5 \end{bmatrix}$

23. $\begin{bmatrix} -8 & 0 & 0 & 0 \\ 0 & 1 & 0 & 0 \\ 0 & 0 & 4 & 0 \\ 0 & 0 & 0 & -5 \end{bmatrix}$

24. $\begin{bmatrix} 1 & 3 & -2 & 0 \\ 0 & 2 & 4 & 6 \\ 0 & 0 & -2 & 1 \\ 0 & 0 & 0 & 5 \end{bmatrix}$

In Exercises 25–34, use the matrix capabilities of a graphing utility to find the inverse of the matrix (if it exists).

25. $\begin{bmatrix} 1 & 2 & -1 \\ 3 & 7 & -10 \\ -5 & -7 & -15 \end{bmatrix}$

26. $\begin{bmatrix} 10 & 5 & -7 \\ -5 & 1 & 4 \\ 3 & 2 & -2 \end{bmatrix}$

27. $\begin{bmatrix} 1 & 1 & 2 \\ 3 & 1 & 0 \\ -2 & 0 & 3 \end{bmatrix}$

28. $\begin{bmatrix} 3 & 2 & 2 \\ 2 & 2 & 2 \\ -4 & 4 & 3 \end{bmatrix}$

29. $\begin{bmatrix} 0.1 & 0.2 & 0.3 \\ -0.3 & 0.2 & 0.2 \\ 0.5 & 0.4 & 0.4 \end{bmatrix}$

30. $\begin{bmatrix} 2 & 0 & 0 \\ 0 & 3 & 0 \\ 0 & 0 & 5 \end{bmatrix}$

31. $\begin{bmatrix} 1 & 0 & 3 & 0 \\ 0 & 2 & 0 & 4 \\ 1 & 0 & 3 & 0 \\ 0 & 2 & 0 & 4 \end{bmatrix}$

32. $\begin{bmatrix} -1 & 0 & 1 & 0 \\ 0 & 2 & 0 & -1 \\ 2 & 0 & -1 & 0 \\ 0 & -1 & 0 & 1 \end{bmatrix}$

33. $\begin{bmatrix} 1 & -2 & -1 & -2 \\ 3 & -5 & -2 & -3 \\ 2 & -5 & -2 & -5 \\ -1 & 4 & 4 & 11 \end{bmatrix}$

34. $\begin{bmatrix} 4 & 8 & -7 & 14 \\ 2 & 5 & -4 & 6 \\ 0 & 2 & 1 & -7 \\ 3 & 6 & -5 & 10 \end{bmatrix}$

35. If A is a 2×2 matrix given by

$$A = \begin{bmatrix} a & b \\ c & d \end{bmatrix}$$

then A is invertible if and only if $ad - bc \neq 0$. If $ad - bc \neq 0$, verify that the inverse is given by

$$A^{-1} = \frac{1}{ad - bc} \begin{bmatrix} d & -b \\ -c & a \end{bmatrix}.$$

36. Use the result of Exercise 35 to find the inverse of each matrix.

(a) $\begin{bmatrix} 5 & -2 \\ 2 & 3 \end{bmatrix}$

(b) $\begin{bmatrix} 7 & 12 \\ -8 & -5 \end{bmatrix}$

In Exercises 37–40, use an inverse matrix to solve the system of linear equations. (Use the inverse matrix found in Exercise 11.)

37. $\begin{aligned} x - 2y &= 5 \\ 2x - 3y &= 10 \end{aligned}$

38. $\begin{aligned} x - 2y &= 0 \\ 2x - 3y &= 3 \end{aligned}$

39. $\begin{aligned} x - 2y &= 4 \\ 2x - 3y &= 2 \end{aligned}$

40. $\begin{aligned} x - 2y &= 1 \\ 2x - 3y &= -2 \end{aligned}$

In Exercises 41 and 42, use an inverse matrix to solve the system of linear equations. (Use the inverse matrix found in Exercise 19.)

41. $\begin{aligned} x + y + z &= 0 \\ 3x + 5y + 4z &= 5 \\ 3x + 6y + 5z &= 2 \end{aligned}$

42. $\begin{aligned} x + y + z &= -1 \\ 3x + 5y + 4z &= 2 \\ 3x + 6y + 5z &= 0 \end{aligned}$

In Exercises 43 and 44, use an inverse matrix and the matrix capabilities of a graphing utility to solve the system of linear equations. (Use the inverse matrix found in Exercise 33.)

43. $\begin{aligned} x_1 - 2x_2 - x_3 - 2x_4 &= 0 \\ 3x_1 - 5x_2 - 2x_3 - 3x_4 &= 1 \\ 2x_1 - 5x_2 - 2x_3 - 5x_4 &= -1 \\ -x_1 + 4x_2 + 4x_3 + 11x_4 &= 2 \end{aligned}$

44. $\begin{aligned} x_1 - 2x_2 - x_3 - 2x_4 &= 1 \\ 3x_1 - 5x_2 - 2x_3 - 3x_4 &= -2 \\ 2x_1 - 5x_2 - 2x_3 - 5x_4 &= 0 \\ -x_1 + 4x_2 + 4x_3 + 11x_4 &= -3 \end{aligned}$

In Exercises 45–52, use an inverse matrix to solve (if possible) the system of linear equations.

45. $\begin{aligned} 3x + 4y &= -2 \\ 5x + 3y &= 4 \end{aligned}$

46. $\begin{aligned} 18x + 12y &= 13 \\ 30x + 24y &= 23 \end{aligned}$

47. $\begin{aligned} -0.4x + 0.8y &= 1.6 \\ 2x - 4y &= 5 \end{aligned}$

48. $\begin{aligned} 13x - 6y &= 17 \\ 26x - 12y &= 8 \end{aligned}$

49. $\begin{aligned} 3x + 6y &= 6 \\ 6x + 14y &= 11 \end{aligned}$

50. $\begin{aligned} 3x + 2y &= 1 \\ 2x + 10y &= 6 \end{aligned}$

51. $\begin{aligned} 4x - y + z &= -5 \\ 2x + 2y + 3z &= 10 \\ 5x - 2y + 6z &= 1 \end{aligned}$

52. $\begin{aligned} 4x - 2y + 3z &= -2 \\ 2x + 2y + 5z &= 16 \\ 8x - 5y - 2z &= 4 \end{aligned}$

In Exercises 53–56, use the matrix capabilities of a graphing utility to solve (if possible) the system of linear equations.

53. $\begin{aligned} 5x - 3y + 2z &= 2 \\ 2x + 2y - 3z &= 3 \\ -x + 7y - 8z &= 4 \end{aligned}$

54. $\begin{aligned} 2x + 3y + 5z &= 4 \\ 3x + 5y + 9z &= 7 \\ 5x + 9y + 17z &= 13 \end{aligned}$

55. $\begin{aligned} 7x - 3y \qquad + 2w &= 41 \\ -2x + y \qquad - w &= -13 \\ 4x \qquad + z - 2w &= 12 \\ -x + y \qquad - w &= -8 \end{aligned}$

56. $\begin{aligned} 2x + 5y \qquad + w &= 11 \\ x + 4y + 2z - 2w &= -7 \\ 2x - 2y + 5z + w &= 3 \\ x \qquad - 3w &= -1 \end{aligned}$

Bond Investments **In Exercises 57–60, consider a person who invests in AAA-rated bonds, A-rated bonds, and B-rated bonds. The average yields are 6.5% on AAA-bonds, 7% on A-bonds, and 9% on B-bonds. The person invests twice as much in B-bonds as in A-bonds. Let x, y, and z represent the amounts invested in AAA-, A-, and B-bonds, respectively.**

$$\begin{aligned} x + y + z &= \text{(total investment)} \\ 0.065x + 0.07y + 0.09z &= \text{(annual return)} \\ 2y - z &= 0 \end{aligned}$$

Use the inverse of the coefficient matrix of this system to find the amount invested in each type of bond.

57. Total investment = \$25,000
Annual return = \$1900

58. Total investment = \$45,000
Annual return = \$3750

59. Total investment = \$12,000
Annual return = \$835

60. Total investment = \$500,000
Annual return = \$38,000

61. *Essay* Write a brief paragraph explaining the advantage of using an inverse matrix to solve the systems of linear equations in Exercises 37–44.

62. *True or False?* Multiplication of an invertible matrix and its inverse is commutative. Give an example to demonstrate your answer.

Circuit Analysis **In Exercises 63 and 64, consider the circuit in the figure. The currents I_1, I_2, and I_3, in amperes, are given by the solution of the system of linear equations**

$$\begin{aligned} 2I_1 \qquad + 4I_3 &= E_1 \\ I_2 + 4I_3 &= E_2 \\ I_1 + I_2 - I_3 &= 0 \end{aligned}$$

where E_1 and E_2 are voltages. Use the inverse of the coefficient matrix of this system to find the unknown currents for the given voltages.

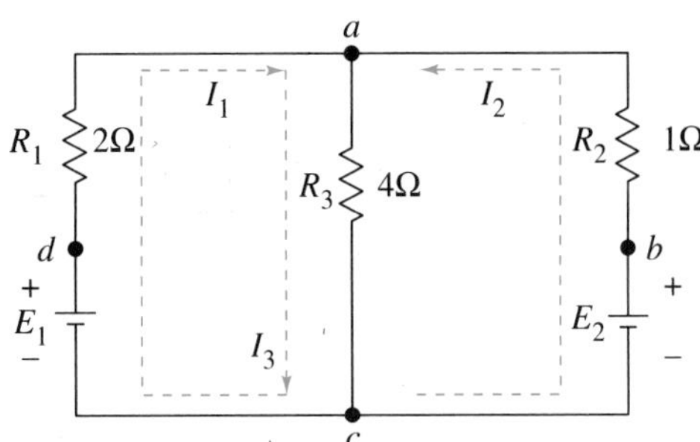

63. $E_1 = 14$ V, $E_2 = 28$ V

64. $E_1 = 10$ V, $E_2 = 10$ V

65. *Exploration* Consider the matrices of the form

$$A = \begin{bmatrix} a_{11} & 0 & 0 & 0 & \cdots & 0 \\ 0 & a_{22} & 0 & 0 & \cdots & 0 \\ 0 & 0 & a_{33} & 0 & \cdots & 0 \\ \vdots & \vdots & \vdots & \vdots & \cdots & \vdots \\ 0 & 0 & 0 & 0 & \cdots & a_{nn} \end{bmatrix}.$$

(a) Write a 2×2 matrix and a 3×3 matrix in the form of A. Find the inverse of each.

(b) Use the result of part (a) to make a conjecture about the inverse of a matrix of the form of A.

6.4 The Determinant of a Square Matrix

The Determinant of a 2 × 2 Matrix / Minors and Cofactors / The Determinant of a Square Matrix / Triangular Matrices

The Determinant of a 2 × 2 Matrix

Every *square* matrix can be associated with a real number called its **determinant.** Determinants have many uses, and several will be discussed in this and the next section. Historically, the use of determinants arose from special number patterns that occur when systems of linear equations are solved. For instance, the system

$$a_1x + b_1y = c_1$$
$$a_2x + b_2y = c_2$$

has a solution given by

$$x = \frac{c_1b_2 - c_2b_1}{a_1b_2 - a_2b_1} \quad \text{and} \quad y = \frac{a_1c_2 - a_2c_1}{a_1b_2 - a_2b_1}$$

provided that $a_1b_2 - a_2b_1 \neq 0$. Note that the denominator of each fraction is the same. This denominator is called the **determinant** of the coefficient matrix of the system.

Coefficient Matrix	*Determinant*
$A = \begin{bmatrix} a_1 & b_1 \\ a_2 & b_2 \end{bmatrix}$	$\det(A) = a_1b_2 - a_2b_1$

The determinant of the matrix A can also be denoted by vertical bars on both sides of the matrix, as indicated in the following definition.

Definition of the Determinant of a 2 × 2 Matrix

The **determinant** of the matrix

$$A = \begin{bmatrix} a_1 & b_1 \\ a_2 & b_2 \end{bmatrix}$$

is given by

$$\det(A) = |A| = \begin{vmatrix} a_1 & b_1 \\ a_2 & b_2 \end{vmatrix} = a_1b_2 - a_2b_1.$$

Note In this text, $\det(A)$ and $|A|$ are used interchangeably to represent the determinant of A. Although vertical bars are also used to denote the absolute value of a real number, the context will show which use is intended.

A convenient method for remembering the formula for the determinant of a 2×2 matrix is shown in the following diagram.

$$\det(A) = \begin{vmatrix} a_1 & b_1 \\ a_2 & b_2 \end{vmatrix} = a_1b_2 - a_2b_1$$

Note that the determinant is given by the difference of the products of the two diagonals of the matrix.

EXAMPLE 1 **The Determinant of a 2×2 Matrix**

Find the determinant of each matrix.

a. $A = \begin{bmatrix} 2 & -3 \\ 1 & 2 \end{bmatrix}$ **b.** $B = \begin{bmatrix} 2 & 1 \\ 4 & 2 \end{bmatrix}$ **c.** $C = \begin{bmatrix} 0 & \frac{3}{2} \\ 2 & 4 \end{bmatrix}$

Solution

Note Notice in Example 1 that the determinant of a matrix can be positive, zero, or negative.

a. $\det(A) = \begin{vmatrix} 2 & -3 \\ 1 & 2 \end{vmatrix} = 2(2) - 1(-3) = 4 + 3 = 7$

b. $\det(B) = \begin{vmatrix} 2 & 1 \\ 4 & 2 \end{vmatrix} = 2(2) - 4(1) = 4 - 4 = 0$

c. $\det(C) = \begin{vmatrix} 0 & \frac{3}{2} \\ 2 & 4 \end{vmatrix} = 0(4) - 2(\frac{3}{2}) = 0 - 3 = -3$

The determinant of a matrix of order 1×1 is defined simply as the entry of the matrix. For instance, if $A = [-2]$, $\det(A) = -2$.

Most graphing utilities can evaluate the determinant of a matrix. For instance, on a *TI-82* or *TI-83,* you can evaluate the determinant of

$$A = \begin{bmatrix} 2 & -3 \\ 1 & 2 \end{bmatrix}$$

by entering the matrix as [A] and then choosing the "det" feature in the matrix math menu.

det [A] ENTER

The result should be 7, as in Example 1(a). Try evaluating determinants of other matrices. What happens when you try to evaluate the determinant of a nonsquare matrix?

Minors and Cofactors

To define the determinant of a square matrix of order 3×3 or higher, it is convenient to introduce the concepts of **minors** and **cofactors.**

Minors and Cofactors of a Square Matrix

If A is a square matrix, the **minor** M_{ij} of the entry a_{ij} is the determinant of the matrix obtained by deleting the ith row and jth column of A. The **cofactor** C_{ij} of the entry a_{ij} is given by

$$C_{ij} = (-1)^{i+j}M_{ij}.$$

Sign Pattern for Cofactors

$$\begin{bmatrix} + & - & + \\ - & + & - \\ + & - & + \end{bmatrix}$$

3×3 matrix

$$\begin{bmatrix} + & - & + & - \\ - & + & - & + \\ + & - & + & - \\ - & + & - & + \end{bmatrix}$$

4×4 matrix

$$\begin{bmatrix} + & - & + & - & + & \cdots \\ - & + & - & + & - & \cdots \\ + & - & + & - & + & \cdots \\ - & + & - & + & - & \cdots \\ + & - & + & - & + & \cdots \\ \vdots & \vdots & \vdots & \vdots & \vdots & \end{bmatrix}$$

$n \times n$ matrix

Note In the sign pattern for cofactors above, notice that *odd* positions (where $i + j$ is odd) have negative signs and *even* positions (where $i + j$ is even) have positive signs.

EXAMPLE 2 **Finding the Minors and Cofactors of a Matrix**

Find all the minors and cofactors of

$$A = \begin{bmatrix} 0 & 2 & 1 \\ 3 & -1 & 2 \\ 4 & 0 & 1 \end{bmatrix}.$$

Solution

To find the minor M_{11}, delete the first row and first column of A and evaluate the determinant of the resulting matrix.

$$\begin{bmatrix} 0 & 2 & 1 \\ 3 & -1 & 2 \\ 4 & 0 & 1 \end{bmatrix}, \quad M_{11} = \begin{vmatrix} -1 & 2 \\ 0 & 1 \end{vmatrix} = -1(1) - 0(2) = -1$$

Similarly, to find M_{12}, delete the first row and second column.

$$\begin{bmatrix} 0 & 2 & 1 \\ 3 & -1 & 2 \\ 4 & 0 & 1 \end{bmatrix}, \quad M_{12} = \begin{vmatrix} 3 & 2 \\ 4 & 1 \end{vmatrix} = 3(1) - 4(2) = -5$$

Continuing this pattern, you obtain the following minors.

$$\begin{array}{lll} M_{11} = -1 & M_{12} = -5 & M_{13} = 4 \\ M_{21} = 2 & M_{22} = -4 & M_{23} = -8 \\ M_{31} = 5 & M_{32} = -3 & M_{33} = -6 \end{array}$$

Now, to find the cofactors, combine these minors with the checkerboard pattern of signs shown at the left (for a 3×3 matrix) to obtain the following.

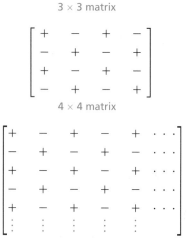

$$\begin{array}{lll} C_{11} = -1 & C_{12} = 5 & C_{13} = 4 \\ C_{21} = -2 & C_{22} = -4 & C_{23} = 8 \\ C_{31} = 5 & C_{32} = 3 & C_{33} = -6 \end{array}$$

The Determinant of a Square Matrix

The following definition is called **inductive** because it uses determinants of matrices of order $n - 1$ to define the determinant of a matrix of order n.

Note Try checking that for a 2×2 matrix this definition yields

$$|A| = a_{11}a_{22} - a_{21}a_{12}$$

as previously defined.

Determinant of a Square Matrix

If A is a square matrix (of order 2×2 or greater), the determinant of A is the sum of the entries in any row (or column) of A multiplied by their respective cofactors. For instance, expanding along the first row yields

$$|A| = a_{11}C_{11} + a_{12}C_{12} + \cdots + a_{1n}C_{1n}.$$

Applying this definition to find a determinant is called **expanding by cofactors.**

EXAMPLE 3 The Determinant of a Matrix of Order 3×3

Find the determinant of

$$A = \begin{bmatrix} 0 & 2 & 1 \\ 3 & -1 & 2 \\ 4 & 0 & 1 \end{bmatrix}.$$

Solution

Note that this is the same matrix that was given in Example 2. There you found the cofactors of the entries in the first row to be

$$C_{11} = -1, \quad C_{12} = 5, \quad \text{and} \quad C_{13} = 4.$$

Therefore, by the definition of the determinant of a square matrix, you have

$$\begin{aligned} |A| &= a_{11}C_{11} + a_{12}C_{12} + a_{13}C_{13} && \text{First-row expansion} \\ &= 0(-1) + 2(5) + 1(4) \\ &= 14. \end{aligned}$$

In Example 3, the determinant was found by expanding by the cofactors in the first row. You could have used any row or column. For instance, you could have expanded along the second row to obtain

$$\begin{aligned} |A| &= a_{21}C_{21} + a_{22}C_{22} + a_{23}C_{23} && \text{Second-row expansion} \\ &= 3(-2) + (-1)(-4) + 2(8) \\ &= 14. \end{aligned}$$

When expanding by cofactors, you do not need to find cofactors of zero entries, because zero times its cofactor is zero.

$$a_{ij}C_{ij} = (0)C_{ij} = 0$$

Thus, the row (or column) containing the most zeros is usually the best choice for expansion by cofactors. This is demonstrated in the next example.

EXAMPLE 4 **The Determinant of a Matrix of Order 4 × 4**

Find the determinant of

$$A = \begin{bmatrix} 1 & -2 & 3 & 0 \\ -1 & 1 & 0 & 2 \\ 0 & 2 & 0 & 3 \\ 3 & 4 & 0 & 2 \end{bmatrix}.$$

Solution

After inspecting this matrix, you can see that three of the entries in the third column are zeros. Thus, you can eliminate some of the work in the expansion by using the third column.

$$|A| = 3(C_{13}) + 0(C_{23}) + 0(C_{33}) + 0(C_{43})$$

Because C_{23}, C_{33}, and C_{43} have zero coefficients, you need only find the cofactor C_{13}. To do this, delete the first row and third column of A and evaluate the determinant of the resulting matrix.

$$C_{13} = (-1)^{1+3}\begin{vmatrix} -1 & 1 & 2 \\ 0 & 2 & 3 \\ 3 & 4 & 2 \end{vmatrix}$$ Delete 1st row and 3rd column.

$$= \begin{vmatrix} -1 & 1 & 2 \\ 0 & 2 & 3 \\ 3 & 4 & 2 \end{vmatrix}$$ Simplify.

Expanding by cofactors in the second row yields the following.

$$C_{13} = 0(-1)^3\begin{vmatrix} 1 & 2 \\ 4 & 2 \end{vmatrix} + 2(-1)^4\begin{vmatrix} -1 & 2 \\ 3 & 2 \end{vmatrix} + 3(-1)^5\begin{vmatrix} -1 & 1 \\ 3 & 4 \end{vmatrix}$$

$$= 0 + 2(1)(-8) + 3(-1)(-7)$$

$$= 5$$

Thus, you obtain

$$|A| = 3C_{13} = 3(5) = 15.$$

Study Tip

Although most graphing utilities can calculate the determinant of a square matrix, it is also important to know how to calculate them *by hand*.

Note Try using a graphing utility to confirm the result of Example 4.

Triangular Matrices

Evaluating determinants of matrices of order 4 or higher can be tedious. There is, however, an important exception: the determinant of a **triangular** matrix. A square matrix is **upper triangular** if it has all zero entries below its main diagonal and **lower triangular** if it has all zero entries above its main diagonal. A matrix that is both upper and lower triangular is called **diagonal.** That is, a diagonal matrix is one in which all entries above and below the main diagonal are zero.

EXPLORATION

The formula for the determinant of a triangular matrix (discussed at the right) is only one of many properties of matrices. You can use a computer or calculator to discover other properties. For instance, how is $|cA|$ related to $|A|$? How are $|A|$ and $|B|$ related to $|AB|$?

Upper Triangular Matrix *Lower Triangular Matrix*

$$\begin{bmatrix} a_{11} & a_{12} & a_{13} & \cdots & a_{1n} \\ 0 & a_{22} & a_{23} & \cdots & a_{2n} \\ 0 & 0 & a_{33} & \cdots & a_{3n} \\ \vdots & \vdots & \vdots & & \vdots \\ 0 & 0 & 0 & \cdots & a_{nn} \end{bmatrix} \qquad \begin{bmatrix} a_{11} & 0 & 0 & \cdots & 0 \\ a_{21} & a_{22} & 0 & \cdots & 0 \\ a_{31} & a_{32} & a_{33} & \cdots & 0 \\ \vdots & \vdots & \vdots & & \vdots \\ a_{n1} & a_{n2} & a_{n3} & \cdots & a_{nn} \end{bmatrix}$$

To find the determinant of a triangular matrix of any order, simply form the product of the entries on the main diagonal.

EXAMPLE 5 The Determinant of a Triangular Matrix

a. $\begin{vmatrix} 2 & 0 & 0 & 0 \\ 4 & -2 & 0 & 0 \\ -5 & 6 & 1 & 0 \\ 1 & 5 & 3 & 3 \end{vmatrix} = (2)(-2)(1)(3) = -12$

b. $\begin{vmatrix} -1 & 0 & 0 & 0 & 0 \\ 0 & 3 & 0 & 0 & 0 \\ 0 & 0 & 2 & 0 & 0 \\ 0 & 0 & 0 & 4 & 0 \\ 0 & 0 & 0 & 0 & -2 \end{vmatrix} = (-1)(3)(2)(4)(-2) = 48$

Group Activity

The Determinant of a Triangular Matrix

Write an argument that explains why the determinant of a 3×3 triangular matrix is the product of its main-diagonal entries.

$$\begin{bmatrix} a_{11} & a_{12} & a_{13} \\ 0 & a_{22} & a_{23} \\ 0 & 0 & a_{33} \end{bmatrix} = a_{11}a_{22}a_{33}$$

6.4 /// EXERCISES

In Exercises 1–16, find the determinant of the matrix.

1. $[5]$

2. $[-8]$

3. $\begin{bmatrix} 2 & 1 \\ 3 & 4 \end{bmatrix}$

4. $\begin{bmatrix} -3 & 1 \\ 5 & 2 \end{bmatrix}$

5. $\begin{bmatrix} 5 & 2 \\ -6 & 3 \end{bmatrix}$

6. $\begin{bmatrix} 2 & -2 \\ 4 & 3 \end{bmatrix}$

7. $\begin{bmatrix} -7 & 6 \\ \frac{1}{2} & 3 \end{bmatrix}$

8. $\begin{bmatrix} 4 & -3 \\ 0 & 0 \end{bmatrix}$

9. $\begin{bmatrix} 2 & 6 \\ 0 & 3 \end{bmatrix}$

10. $\begin{bmatrix} 2 & -3 \\ -6 & 9 \end{bmatrix}$

11. $\begin{bmatrix} 2 & -1 & 0 \\ 4 & 2 & 1 \\ 4 & 2 & 1 \end{bmatrix}$

12. $\begin{bmatrix} -2 & 2 & 3 \\ 1 & -1 & 0 \\ 0 & 1 & 4 \end{bmatrix}$

13. $\begin{bmatrix} 6 & 3 & -7 \\ 0 & 0 & 0 \\ 4 & -6 & 3 \end{bmatrix}$

14. $\begin{bmatrix} 1 & 1 & 2 \\ 3 & 1 & 0 \\ -2 & 0 & 3 \end{bmatrix}$

15. $\begin{bmatrix} -1 & 2 & -5 \\ 0 & 3 & 4 \\ 0 & 0 & 3 \end{bmatrix}$

16. $\begin{bmatrix} 1 & 0 & 0 \\ -4 & -1 & 0 \\ 5 & 1 & 5 \end{bmatrix}$

In Exercises 17–20, use the matrix capabilities of a graphing utility to find the determinant of the matrix.

17. $\begin{bmatrix} 0.3 & 0.2 & 0.2 \\ 0.2 & 0.2 & 0.2 \\ -0.4 & 0.4 & 0.3 \end{bmatrix}$

18. $\begin{bmatrix} 0.1 & 0.2 & 0.3 \\ -0.3 & 0.2 & 0.2 \\ 0.5 & 0.4 & 0.4 \end{bmatrix}$

19. $\begin{bmatrix} 1 & 4 & -2 \\ 3 & 6 & -6 \\ -2 & 1 & 4 \end{bmatrix}$

20. $\begin{bmatrix} 2 & 3 & 1 \\ 0 & 5 & -2 \\ 0 & 0 & -2 \end{bmatrix}$

In Exercises 21–24, find all (a) minors and (b) cofactors of the matrix.

21. $\begin{bmatrix} 3 & 4 \\ 2 & -5 \end{bmatrix}$

22. $\begin{bmatrix} 11 & 0 \\ -3 & 2 \end{bmatrix}$

23. $\begin{bmatrix} 3 & -2 & 8 \\ 3 & 2 & -6 \\ -1 & 3 & 6 \end{bmatrix}$

24. $\begin{bmatrix} -2 & 9 & 4 \\ 7 & -6 & 0 \\ 6 & 7 & -6 \end{bmatrix}$

In Exercises 25–30, find the determinant of the matrix by the method of expansion by cofactors. Expand using the indicated row or column.

25. $\begin{bmatrix} -3 & 2 & 1 \\ 4 & 5 & 6 \\ 2 & -3 & 1 \end{bmatrix}$ (a) Row 1 (b) Column 2

26. $\begin{bmatrix} -3 & 4 & 2 \\ 6 & 3 & 1 \\ 4 & -7 & -8 \end{bmatrix}$ (a) Row 2 (b) Column 3

27. $\begin{bmatrix} 5 & 0 & -3 \\ 0 & 12 & 4 \\ 1 & 6 & 3 \end{bmatrix}$ (a) Row 2 (b) Column 2

28. $\begin{bmatrix} 10 & -5 & 5 \\ 30 & 0 & 10 \\ 0 & 10 & 1 \end{bmatrix}$ (a) Row 3 (b) Column 1

29. $\begin{bmatrix} 6 & 0 & -3 & 5 \\ 4 & 13 & 6 & -8 \\ -1 & 0 & 7 & 4 \\ 8 & 6 & 0 & 2 \end{bmatrix}$ (a) Row 2 (b) Column 2

30. $\begin{bmatrix} 10 & 8 & 3 & -7 \\ 4 & 0 & 5 & -6 \\ 0 & 3 & 2 & 7 \\ 1 & 0 & -3 & 2 \end{bmatrix}$ (a) Row 3 (b) Column 1

In Exercises 31–40, find the determinant of the matrix. Expand by cofactors on the row or column that appears to make the computations easiest.

31. $\begin{bmatrix} 1 & 4 & -2 \\ 3 & 2 & 0 \\ -1 & 4 & 3 \end{bmatrix}$

32. $\begin{bmatrix} 2 & -1 & 3 \\ 1 & 4 & 4 \\ 1 & 0 & 2 \end{bmatrix}$

33. $\begin{bmatrix} 2 & 4 & 6 \\ 0 & 3 & 1 \\ 0 & 0 & -5 \end{bmatrix}$

34. $\begin{bmatrix} -3 & 0 & 0 \\ 7 & 11 & 0 \\ 1 & 2 & 2 \end{bmatrix}$

35. $\begin{bmatrix} 2 & 6 & 6 & 2 \\ 2 & 7 & 3 & 6 \\ 1 & 5 & 0 & 1 \\ 3 & 7 & 0 & 7 \end{bmatrix}$ **36.** $\begin{bmatrix} 3 & 6 & -5 & 4 \\ -2 & 0 & 6 & 0 \\ 1 & 1 & 2 & 2 \\ 0 & 3 & -1 & -1 \end{bmatrix}$

37. $\begin{bmatrix} 5 & 3 & 0 & 6 \\ 4 & 6 & 4 & 12 \\ 0 & 2 & -3 & 4 \\ 0 & 1 & -2 & 2 \end{bmatrix}$ **38.** $\begin{bmatrix} 1 & 4 & 3 & 2 \\ -5 & 6 & 2 & 1 \\ 0 & 0 & 0 & 0 \\ 3 & -2 & 1 & 5 \end{bmatrix}$

39. $\begin{bmatrix} 3 & 2 & 4 & -1 & 5 \\ -2 & 0 & 1 & 3 & 2 \\ 1 & 0 & 0 & 4 & 0 \\ 6 & 0 & 2 & -1 & 0 \\ 3 & 0 & 5 & 1 & 0 \end{bmatrix}$

40. $\begin{bmatrix} 5 & 2 & 0 & 0 & -2 \\ 0 & 1 & 4 & 3 & 2 \\ 0 & 0 & 2 & 6 & 3 \\ 0 & 0 & 3 & 4 & 1 \\ 0 & 0 & 0 & 0 & 2 \end{bmatrix}$

In Exercises 41–48, use the matrix capabilities of a graphing utility to evaluate the determinant.

41. $\begin{vmatrix} 3 & 8 & -7 \\ 0 & -5 & 4 \\ 8 & 1 & 6 \end{vmatrix}$ **42.** $\begin{vmatrix} 5 & -8 & 0 \\ 9 & 7 & 4 \\ -8 & 7 & 1 \end{vmatrix}$

43. $\begin{vmatrix} 7 & 0 & -14 \\ -2 & 5 & 4 \\ -6 & 2 & 12 \end{vmatrix}$ **44.** $\begin{vmatrix} 3 & 0 & 0 \\ -2 & 5 & 0 \\ 12 & 5 & 7 \end{vmatrix}$

45. $\begin{vmatrix} 1 & -1 & 8 & 4 \\ 2 & 6 & 0 & -4 \\ 2 & 0 & 2 & 6 \\ 0 & 2 & 8 & 0 \end{vmatrix}$ **46.** $\begin{vmatrix} 0 & -3 & 8 & 2 \\ 8 & 1 & -1 & 6 \\ -4 & 6 & 0 & 9 \\ -7 & 0 & 0 & 14 \end{vmatrix}$

47. $\begin{vmatrix} 3 & -2 & 4 & 3 & 1 \\ -1 & 0 & 2 & 1 & 0 \\ 5 & -1 & 0 & 3 & 2 \\ 4 & 7 & -8 & 0 & 0 \\ 1 & 2 & 3 & 0 & 2 \end{vmatrix}$

48. $\begin{vmatrix} -2 & 0 & 0 & 0 & 0 \\ 0 & 3 & 0 & 0 & 0 \\ 0 & 0 & -1 & 0 & 0 \\ 0 & 0 & 0 & 2 & 0 \\ 0 & 0 & 0 & 0 & -4 \end{vmatrix}$

In Exercises 49–52, evaluate the determinants to verify the equation.

49. $\begin{vmatrix} w & x \\ y & z \end{vmatrix} = -\begin{vmatrix} y & z \\ w & x \end{vmatrix}$

50. $\begin{vmatrix} w & cx \\ y & cz \end{vmatrix} = c\begin{vmatrix} w & x \\ y & z \end{vmatrix}$

51. $\begin{vmatrix} w & x \\ y & z \end{vmatrix} = \begin{vmatrix} w & x + cw \\ y & z + cy \end{vmatrix}$

52. $\begin{vmatrix} w & x \\ cw & cx \end{vmatrix} = 0$

In Exercises 53 and 54, evaluate the determinant to verify the equation.

53. $\begin{vmatrix} 1 & x & x^2 \\ 1 & y & y^2 \\ 1 & z & z^2 \end{vmatrix} = (y - x)(z - x)(z - y)$

54. $\begin{vmatrix} a + b & a & a \\ a & a + b & a \\ a & a & a + b \end{vmatrix} = b^2(3a + b)$

In Exercises 55 and 56, solve for x.

55. $\begin{vmatrix} x - 1 & 2 \\ 3 & x - 2 \end{vmatrix} = 0$

56. $\begin{vmatrix} x - 2 & -1 \\ -3 & x \end{vmatrix} = 0$

In Exercises 57–62, evaluate the determinant, where the entries are functions. Determinants of this type occur in calculus.

57. $\begin{vmatrix} 4u & -1 \\ -1 & 2v \end{vmatrix}$ **58.** $\begin{vmatrix} 3x^2 & -3y^2 \\ 1 & 1 \end{vmatrix}$

59. $\begin{vmatrix} e^{2x} & e^{3x} \\ 2e^{2x} & 3e^{3x} \end{vmatrix}$ **60.** $\begin{vmatrix} e^{-x} & xe^{-x} \\ -e^{-x} & (1 - x)e^{-x} \end{vmatrix}$

61. $\begin{vmatrix} x & \ln x \\ 1 & 1/x \end{vmatrix}$ **62.** $\begin{vmatrix} x & x \ln x \\ 1 & 1 + \ln x \end{vmatrix}$

In Exercises 63–66, find (a) $|A|$, (b) $|B|$, (c) AB, and (d) $|AB|$.

63. $A = \begin{bmatrix} -1 & 0 \\ 0 & 3 \end{bmatrix}, \quad B = \begin{bmatrix} 2 & 0 \\ 0 & -1 \end{bmatrix}$

64. $A = \begin{bmatrix} -2 & 1 \\ 4 & -2 \end{bmatrix}, \quad B = \begin{bmatrix} 1 & 2 \\ 0 & -1 \end{bmatrix}$

65. $A = \begin{bmatrix} -1 & 2 & 1 \\ 1 & 0 & 1 \\ 0 & 1 & 0 \end{bmatrix}, \quad B = \begin{bmatrix} -1 & 0 & 0 \\ 0 & 2 & 0 \\ 0 & 0 & 3 \end{bmatrix}$

66. $A = \begin{bmatrix} 2 & 0 & 1 \\ 1 & -1 & 2 \\ 3 & 1 & 0 \end{bmatrix}, \quad B = \begin{bmatrix} 2 & -1 & 4 \\ 0 & 1 & 3 \\ 3 & -2 & 1 \end{bmatrix}$

67. *Exploration* Find square matrices A and B to demonstrate that

$$|A + B| \neq |A| + |B|.$$

68. *Exploration* Consider square matrices in which the entries are consecutive integers. An example of such a matrix is

$$\begin{bmatrix} 4 & 5 & 6 \\ 7 & 8 & 9 \\ 10 & 11 & 12 \end{bmatrix}.$$

(a) Use a graphing utility to evaluate four determinants of this type. Make a conjecture based on the results.

(b) Verify your conjecture.

69. *Essay* Write a brief paragraph explaining the difference between a square matrix and its determinant.

70. *Think About It* If A is a matrix of order 3×3 such that $|A| = 5$, is it possible to find $|2A|$? Explain.

In Exercises 71–73, a property of determinants is given. State how the property has been applied to the given determinants and use a graphing utility to verify the results.

71. If A and B are square matrices and B is obtained from A by interchanging two rows of A or interchanging two columns of A, then $|B| = -|A|$.

(a) $\begin{vmatrix} 1 & 3 & 4 \\ -7 & 2 & -5 \\ 6 & 1 & 2 \end{vmatrix} = -\begin{vmatrix} 1 & 4 & 3 \\ -7 & -5 & 2 \\ 6 & 2 & 1 \end{vmatrix}$

(b) $\begin{vmatrix} 1 & 3 & 4 \\ -2 & 2 & 0 \\ 1 & 6 & 2 \end{vmatrix} = -\begin{vmatrix} 1 & 6 & 2 \\ -2 & 2 & 0 \\ 1 & 3 & 4 \end{vmatrix}$

72. If A and B are square matrices and B is obtained from A by adding a multiple of a row of A to another row of A or by adding a multiple of a column of A to another column of A, then $|B| = |A|$.

(a) $\begin{vmatrix} 1 & -3 \\ 5 & 2 \end{vmatrix} = \begin{vmatrix} 1 & -3 \\ 0 & 17 \end{vmatrix}$

(b) $\begin{vmatrix} 5 & 4 & 2 \\ 2 & -3 & 4 \\ 7 & 6 & 3 \end{vmatrix} = \begin{vmatrix} 1 & 10 & -6 \\ 2 & -3 & 4 \\ 7 & 6 & 3 \end{vmatrix}$

73. If A and B are square matrices and B is obtained from A by multiplying a row of A by a nonzero constant c or multiplying a column of A by a nonzero constant c, then $|B| = c|A|$.

(a) $\begin{vmatrix} 5 & 10 & 15 \\ 2 & -3 & 4 \\ 2 & -7 & 1 \end{vmatrix} = 5\begin{vmatrix} 1 & 2 & 3 \\ 2 & -3 & 4 \\ 2 & -7 & 1 \end{vmatrix}$

(b) $\begin{vmatrix} 1 & 8 & -3 \\ 3 & -12 & 6 \\ 7 & 4 & 9 \end{vmatrix} = 12\begin{vmatrix} 1 & 2 & -1 \\ 3 & -3 & 2 \\ 7 & 1 & 3 \end{vmatrix}$

6.5 Applications of Matrices and Determinants

Area of a Triangle / Lines in the Plane / Cramer's Rule / Cryptography

Area of a Triangle

In this section, you will study some additional applications of matrices and determinants. The first involves a formula for finding the area of a triangle whose vertices are given by three points on a rectangular coordinate system.

Area of a Triangle

The area of a triangle with vertices (x_1, y_1), (x_2, y_2), and (x_3, y_3) is given by

$$\text{Area} = \pm\frac{1}{2}\begin{vmatrix} x_1 & y_1 & 1 \\ x_2 & y_2 & 1 \\ x_3 & y_3 & 1 \end{vmatrix}$$

where the symbol $(\pm)$ indicates that the appropriate sign should be chosen to yield a positive area.

EXAMPLE 1 **Finding the Area of a Triangle**

Find the area of a triangle whose vertices are $(1, 0)$, $(2, 2)$, and $(4, 3)$, as shown in Figure 6.1.

Figure 6.1

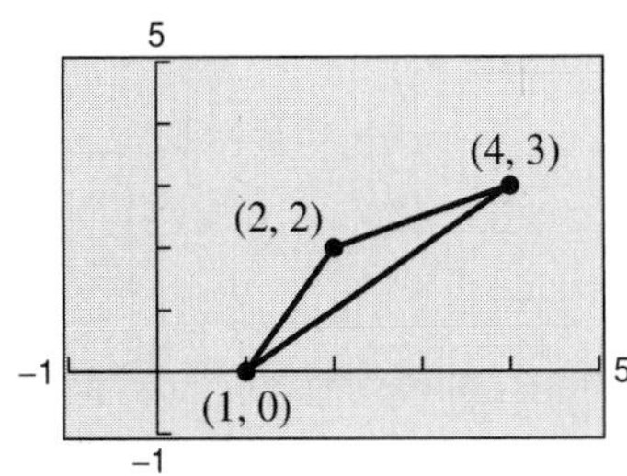

Solution

Let $(x_1, y_1) = (1, 0)$, $(x_2, y_2) = (2, 2)$, and $(x_3, y_3) = (4, 3)$. Then, to find the area of a triangle, evaluate the determinant

$$\begin{vmatrix} x_1 & y_1 & 1 \\ x_2 & y_2 & 1 \\ x_3 & y_3 & 1 \end{vmatrix} = \begin{vmatrix} 1 & 0 & 1 \\ 2 & 2 & 1 \\ 4 & 3 & 1 \end{vmatrix}$$

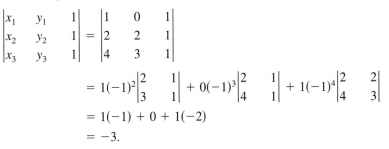

$$= 1(-1)^2\begin{vmatrix} 2 & 1 \\ 3 & 1 \end{vmatrix} + 0(-1)^3\begin{vmatrix} 2 & 1 \\ 4 & 1 \end{vmatrix} + 1(-1)^4\begin{vmatrix} 2 & 2 \\ 4 & 3 \end{vmatrix}$$

$$= 1(-1) + 0 + 1(-2)$$

$$= -3.$$

Using this value, you can conclude that the area of the triangle is

$$\text{Area} = -\frac{1}{2}\begin{vmatrix} 1 & 0 & 1 \\ 2 & 2 & 1 \\ 4 & 3 & 1 \end{vmatrix} = -\frac{1}{2}(-3) = \frac{3}{2}.$$

EXAMPLE 2 Finding the Area of a Triangle

Find the area of the triangle whose vertices are $\left(\frac{1}{3}, -\frac{2}{7}\right)$, $\left(6\frac{1}{4}, 4\frac{1}{2}\right)$, and $\left(8, -5\frac{1}{5}\right)$, as shown in Figure 6.2.

Figure 6.2

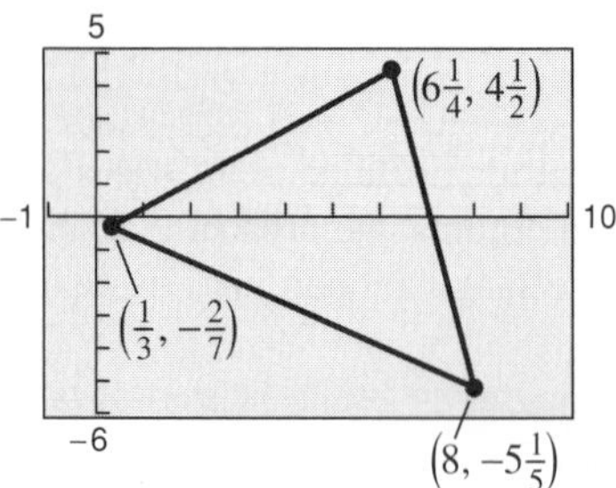

Solution

Let $(x_1, y_1) = \left(\frac{1}{3}, -\frac{2}{7}\right)$, $(x_2, y_2) = \left(6\frac{1}{4}, 4\frac{1}{2}\right)$, and $(x_3, y_3) = \left(8, -5\frac{1}{5}\right)$. Then, to find the area of the triangle, evaluate the determinant

$$\begin{vmatrix} x_1 & y_1 & 1 \\ x_2 & y_2 & 1 \\ x_3 & y_3 & 1 \end{vmatrix} = \begin{vmatrix} \frac{1}{3} & -\frac{2}{7} & 1 \\ 6\frac{1}{4} & 4\frac{1}{2} & 1 \\ 8 & -5\frac{1}{5} & 1 \end{vmatrix}.$$

Using the matrix capabilities of a graphing utility, you find the value of the determinant to be $-65.7\overline{6}$.

Now you can use this value to conclude that the area of the triangle is

$$\text{Area} = -\tfrac{1}{2}(-65.7\overline{6}) = 32.88\overline{3}.$$

Lines in the Plane

Figure 6.3

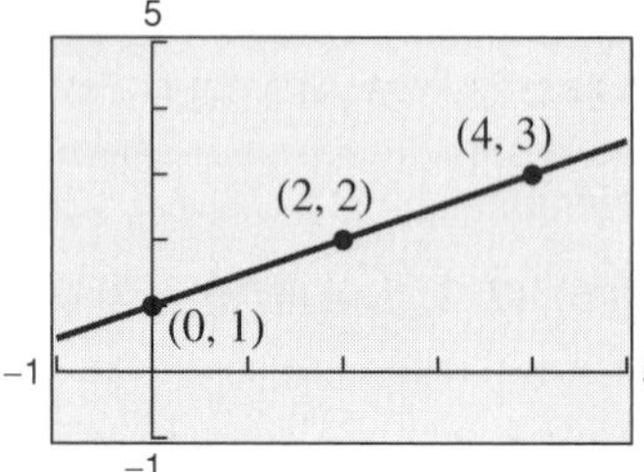

Suppose the three points in Example 1 had been on the same line. What would have happened had the area formula been applied to three such points? The answer is that the determinant would have been zero. Consider, for instance, the three collinear points (0, 1), (2, 2), and (4, 3), as shown in Figure 6.3. The area of the "triangle" that has these three points as vertices is

$$\frac{1}{2}\begin{vmatrix} 0 & 1 & 1 \\ 2 & 2 & 1 \\ 4 & 3 & 1 \end{vmatrix} = \frac{1}{2}\left[0(-1)^2\begin{vmatrix} 2 & 1 \\ 3 & 1 \end{vmatrix} + 1(-1)^3\begin{vmatrix} 2 & 1 \\ 4 & 1 \end{vmatrix} + 1(-1)^4\begin{vmatrix} 2 & 2 \\ 4 & 3 \end{vmatrix}\right]$$

$$= \frac{1}{2}[0 + (-1)(-2) + 1(-2)] = 0.$$

This result is generalized as follows.

Test for Collinear Points

Three points (x_1, y_1), (x_2, y_2), and (x_3, y_3) are collinear (lie on the same line) if and only if

$$\begin{vmatrix} x_1 & y_1 & 1 \\ x_2 & y_2 & 1 \\ x_3 & y_3 & 1 \end{vmatrix} = 0.$$

Figure 6.4

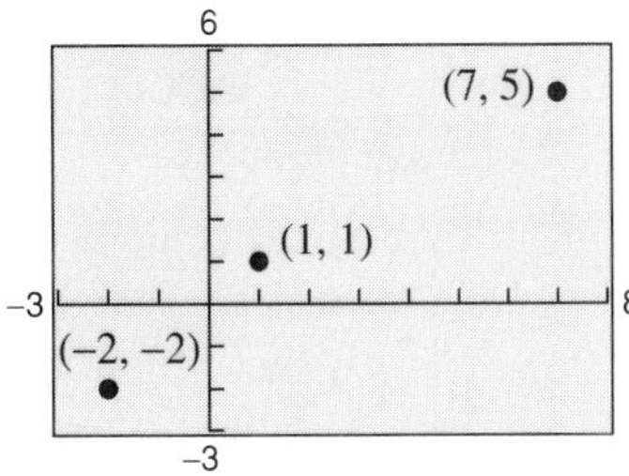

EXAMPLE 3 Testing for Collinear Points

Determine whether the points $(-2, -2)$, $(1, 1)$, and $(7, 5)$ lie on the same line. (See Figure 6.4.)

Solution

Letting $(x_1, y_1) = (-2, -2)$, $(x_2, y_2) = (1, 1)$, and $(x_3, y_3) = (7, 5)$, you have

$$\begin{vmatrix} x_1 & y_1 & 1 \\ x_2 & y_2 & 1 \\ x_3 & y_3 & 1 \end{vmatrix} = \begin{vmatrix} -2 & -2 & 1 \\ 1 & 1 & 1 \\ 7 & 5 & 1 \end{vmatrix}$$

$$= -2(-1)^2\begin{vmatrix} 1 & 1 \\ 5 & 1 \end{vmatrix} + (-2)(-1)^3\begin{vmatrix} 1 & 1 \\ 7 & 1 \end{vmatrix} + 1(-1)^4\begin{vmatrix} 1 & 1 \\ 7 & 5 \end{vmatrix}$$

$$= -2(-4) + 2(-6) + 1(-2)$$

$$= -6.$$

Because the value of this determinant *is not* zero, you can conclude that the three points do not lie on the same line.

You can use the following steps on a *TI-82* or *TI-83* graphing calculator to check whether three points are collinear.

1. Plot the points by entering Pt-On(x, y) for each point. [Pt-On(can be found in the DRAW POINTS menu.]
2. Draw a line from the two farthest points by entering Line (x_1, y_1, x_2, y_2).

Use the steps above to check the results of Example 3. Explain how the graph shows that the points are not collinear. Why is Step 2 important in determining if points are collinear?

The test for collinear points can be adapted to another use. That is, if you are given two points on a rectangular coordinate system, you can find an equation of the line passing through the two points, as follows.

Two-Point Form of the Equation of a Line

An equation of the line passing through the distinct points (x_1, y_1) and (x_2, y_2) is given by

$$\begin{vmatrix} x & y & 1 \\ x_1 & y_1 & 1 \\ x_2 & y_2 & 1 \end{vmatrix} = 0.$$

Note that this method of finding the equation of a line works for all lines, including horizontal and vertical lines. For instance, the equation of the vertical line through $(2, 0)$ and $(2, 2)$ is

$$\begin{vmatrix} x & y & 1 \\ 2 & 0 & 1 \\ 2 & 2 & 1 \end{vmatrix} = 0$$

$$-2x + 4 = 0$$

$$x = 2.$$

Figure 6.5

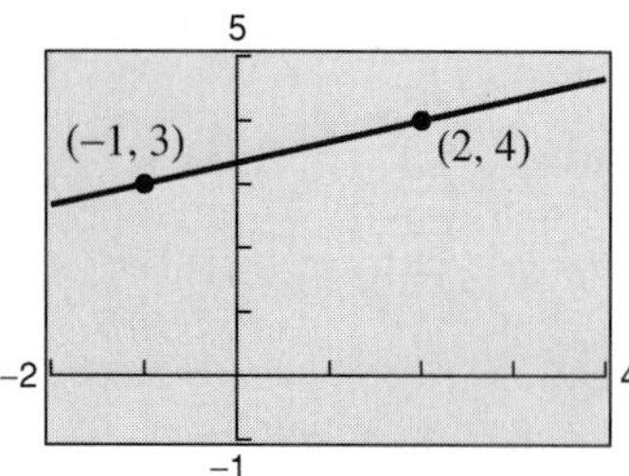

EXAMPLE 4 **Finding an Equation of a Line**

Find an equation of the line passing through the two points (2, 4) and (−1, 3), as shown in Figure 6.5.

Solution

Applying the determinant formula for the equation of a line produces

$$\begin{vmatrix} x & y & 1 \\ 2 & 4 & 1 \\ -1 & 3 & 1 \end{vmatrix} = 0.$$

To evaluate this determinant, you can expand by cofactors along the first row to obtain the following.

$$\begin{aligned} x(-1)^2\begin{vmatrix} 4 & 1 \\ 3 & 1 \end{vmatrix} + y(-1)^3\begin{vmatrix} 2 & 1 \\ -1 & 1 \end{vmatrix} + 1(-1)^4\begin{vmatrix} 2 & 4 \\ -1 & 3 \end{vmatrix} &= x - 3y + 10 \\ &= 0 \end{aligned}$$

Therefore, an equation of the line is

$$x - 3y + 10 = 0.$$

Note There are a variety of ways to check that the equation of the line in Example 4 is correct. You can check it algebraically using the techniques you learned in Section 1.2, or you can check it graphically by plotting the points and graphing the line in the same viewing rectangle.

Cramer's Rule

So far, you have studied three methods for solving a system of linear equations: substitution, elimination (with equations), and elimination (with matrices). You will now study one more method, **Cramer's Rule,** named after Gabriel Cramer (1704–1752). This rule uses determinants to write the solution of a system of linear equations. To see how Cramer's Rule works, take another look at the solution described at the beginning of Section 6.4. There, it was pointed out that the system

$$\begin{aligned} a_1x + b_1y &= c_1 \\ a_2x + b_2y &= c_2 \end{aligned}$$

has a solution given by

$$x = \frac{c_1b_2 - c_2b_1}{a_1b_2 - a_2b_1} \quad \text{and} \quad y = \frac{a_1c_2 - a_2c_1}{a_1b_2 - a_2b_1}$$

provided that $a_1b_2 - a_2b_1 \neq 0$. Each numerator and denominator in this solution can be expressed as a determinant, as follows.

$$x = \frac{c_1b_2 - c_2b_1}{a_1b_2 - a_2b_1} = \frac{\begin{vmatrix} c_1 & b_1 \\ c_2 & b_2 \end{vmatrix}}{\begin{vmatrix} a_1 & b_1 \\ a_2 & b_2 \end{vmatrix}}, \qquad y = \frac{a_1c_2 - a_2c_1}{a_1b_2 - a_2b_1} = \frac{\begin{vmatrix} a_1 & c_1 \\ a_2 & c_2 \end{vmatrix}}{\begin{vmatrix} a_1 & b_1 \\ a_2 & b_2 \end{vmatrix}}$$

Relative to the original system, the denominator for x and y is simply the determinant of the *coefficient* matrix of the system. This determinant is denoted by D. The numerators for x and y are denoted by D_x and D_y, respectively. They are formed by using the column of constants as replacements for the coefficients of x and y, as follows.

Coefficient Matrix	D	D_x	D_y
$\begin{bmatrix} a_1 & b_1 \\ a_2 & b_2 \end{bmatrix}$	$\begin{vmatrix} a_1 & b_1 \\ a_2 & b_2 \end{vmatrix}$	$\begin{vmatrix} c_1 & b_1 \\ c_2 & b_2 \end{vmatrix}$	$\begin{vmatrix} a_1 & c_1 \\ a_2 & c_2 \end{vmatrix}$

EXAMPLE 5 **Using Cramer's Rule for a 2 × 2 System**

Use Cramer's Rule to solve the following system of linear equations.

$$\begin{aligned} 4x - 2y &= 10 \\ 3x - 5y &= 11 \end{aligned}$$

Solution

To begin, find the determinant of the coefficient matrix.

$$D = \begin{vmatrix} 4 & -2 \\ 3 & -5 \end{vmatrix} = -20 - (-6) = -14$$

Because this determinant is not zero, you can apply Cramer's Rule to find the solution, as follows.

$$x = \frac{D_x}{D} = \frac{\begin{vmatrix} 10 & -2 \\ 11 & -5 \end{vmatrix}}{-14} = \frac{(-50) - (-22)}{-14} = \frac{-28}{-14} = 2$$

$$y = \frac{D_y}{D} = \frac{\begin{vmatrix} 4 & 10 \\ 3 & 11 \end{vmatrix}}{-14} = \frac{44 - 30}{-14} = \frac{14}{-14} = -1$$

Therefore, the solution is $x = 2$ and $y = -1$. Check this in the original system.

Cramer's Rule generalizes easily to systems of n equations in n variables. The value of each variable is given as the quotient of two determinants. The denominator is the determinant of the coefficient matrix, and the numerator is the determinant of the matrix formed by replacing the column corresponding to the variable (being solved for) with the column representing the constants. For instance, the solution for x_3 in the system

$$\begin{aligned} a_{11}x_1 + a_{12}x_2 + a_{13}x_3 &= b_1 \\ a_{21}x_1 + a_{22}x_2 + a_{23}x_3 &= b_2 \\ a_{31}x_1 + a_{32}x_2 + a_{33}x_3 &= b_3 \end{aligned}$$

is given by

$$x_3 = \frac{|A_3|}{|A|} = \frac{\begin{vmatrix} a_{11} & a_{12} & b_1 \\ a_{21} & a_{22} & b_2 \\ a_{31} & a_{32} & b_3 \end{vmatrix}}{\begin{vmatrix} a_{11} & a_{12} & a_{13} \\ a_{21} & a_{22} & a_{23} \\ a_{31} & a_{32} & a_{33} \end{vmatrix}}.$$

Cramer's Rule

If a system of n linear equations in n variables has a coefficient matrix A with a *nonzero* determinant $|A|$, the solution is given by

$$x_1 = \frac{|A_1|}{|A|}, \quad x_2 = \frac{|A_2|}{|A|}, \quad . . . , \quad x_n = \frac{|A_n|}{|A|}$$

where the ith column of A_i is the column of constants in the system of equations. If the coefficient matrix is zero, the system has either no solution or infinitely many solutions.

EXAMPLE 6 **Using Cramer's Rule for a 3 × 3 System**

Use Cramer's Rule, if possible, to solve the following system of linear equations.

$$\begin{aligned} -x \phantom{{}-y} + z &= 4 \\ 2x - y + z &= -3 \\ y - 3z &= 1 \end{aligned}$$

Solution

Using the matrix capabilities of a graphing utility to evaluate the determinant of the coefficient matrix A, you find that Cramer's Rule cannot be applied because $|A| = 0$.

Cryptography

A **cryptogram** is a message written according to a secret code. (The Greek word *kryptos* means "hidden.") Matrix multiplication can be used to **encode** and **decode** messages. To begin, you need to assign a number to each letter in the alphabet (with 0 assigned to a blank space), as follows.

0 = _	9 = I	18 = R
1 = A	10 = J	19 = S
2 = B	11 = K	20 = T
3 = C	12 = L	21 = U
4 = D	13 = M	22 = V
5 = E	14 = N	23 = W
6 = F	15 = O	24 = X
7 = G	16 = P	25 = Y
8 = H	17 = Q	26 = Z

Then the message is converted to numbers and partitioned into **uncoded row matrices,** each having n entries, as demonstrated in Example 7.

EXAMPLE 7 **Forming Uncoded Row Matrices**

Write the uncoded row matrices of order 1×3 for the message

MEET ME MONDAY.

Solution

Partitioning the message (including blank spaces, but ignoring punctuation) into groups of three produces the following uncoded row matrices.

$$[13 \quad 5 \quad 5] \quad [20 \quad 0 \quad 13] \quad [5 \quad 0 \quad 13] \quad [15 \quad 14 \quad 4] \quad [1 \quad 25 \quad 0]$$

M E E T _ M E _ M O N D A Y _

Note that a blank space is used to fill out the last uncoded row matrix.

To **encode** a message, choose an $n \times n$ invertible matrix A and multiply the uncoded row matrices by A (on the right) to obtain **coded row matrices.** Here is an example.

Uncoded Matrix *Encoding Matrix A* *Coded Matrix*

$$[13 \quad 5 \quad 5] \begin{bmatrix} 1 & -2 & 2 \\ -1 & 1 & 3 \\ 1 & -1 & -4 \end{bmatrix} = [13 \quad -26 \quad 21]$$

This technique is further illustrated in Example 8.

An efficient method for encoding the message at the right with your graphing utility is to enter A as a 3×3 matrix. Let B be the 5×3 matrix whose rows are the uncoded row matrices,

$$B = \begin{bmatrix} 13 & 5 & 5 \\ 20 & 0 & 13 \\ 5 & 0 & 13 \\ 15 & 14 & 4 \\ 1 & 25 & 0 \end{bmatrix}.$$

The product BA gives the coded row matrices.

EXAMPLE 8 **Encoding a Message**

Use the following matrix to encode the message MEET ME MONDAY.

$$A = \begin{bmatrix} 1 & -2 & 2 \\ -1 & 1 & 3 \\ 1 & -1 & -4 \end{bmatrix}$$

Solution

The coded row matrices are obtained by multiplying each of the uncoded row matrices found in Example 7 by the matrix A, as follows.

Uncoded Matrix *Encoding Matrix A* *Coded Matrix*

$$[13 \quad 5 \quad 5]\begin{bmatrix} 1 & -2 & 2 \\ -1 & 1 & 3 \\ 1 & -1 & -4 \end{bmatrix} = [13 \quad -26 \quad 21]$$

$$[20 \quad 0 \quad 13]\begin{bmatrix} 1 & -2 & 2 \\ -1 & 1 & 3 \\ 1 & -1 & -4 \end{bmatrix} = [33 \quad -53 \quad -12]$$

$$[5 \quad 0 \quad 13]\begin{bmatrix} 1 & -2 & 2 \\ -1 & 1 & 3 \\ 1 & -1 & -4 \end{bmatrix} = [18 \quad -23 \quad -42]$$

$$[15 \quad 14 \quad 4]\begin{bmatrix} 1 & -2 & 2 \\ -1 & 1 & 3 \\ 1 & -1 & -4 \end{bmatrix} = [5 \quad -20 \quad 56]$$

$$[1 \quad 25 \quad 0]\begin{bmatrix} 1 & -2 & 2 \\ -1 & 1 & 3 \\ 1 & -1 & -4 \end{bmatrix} = [-24 \quad 23 \quad 77]$$

Thus, the sequence of coded row matrices is

$$[13 \ -26 \ \ 21][33 \ -53 -12][18 \ -23 -42][5 \ -20 \ \ 56][-24 \ \ 23 \ \ 77].$$

Finally, removing the matrix notation produces the following cryptogram.

$$13 \ -26 \ \ 21 \ \ 33 \ -53 -12 \ \ 18 \ -23 -42 \ \ 5 \ -20 \ \ 56 \ \ -24 \ \ 23 \ \ 77$$

For those who do not know the matrix A, decoding the cryptogram found in Example 8 is difficult. But for an authorized receiver who knows the matrix A, decoding is simple. The receiver need only multiply the coded row matrices by A^{-1} (on the right) to retrieve the uncoded row matrices. Here is an example.

$$\underbrace{[13 \ -26 \quad 21]}_{\text{Coded}}A^{-1} = \underbrace{[13 \quad 5 \quad 5]}_{\text{Uncoded}}$$

EXAMPLE 9 **Decoding a Message**

Use the inverse of the matrix $A = \begin{bmatrix} 1 & -2 & 2 \\ -1 & 1 & 3 \\ 1 & -1 & -4 \end{bmatrix}$ to decode the cryptogram

13 −26 21 33 −53 −12 18 −23 −42 5 −20 56 −24 23 77.

Solution

Partition the message into groups of three to form the coded row matrices. Then, multiply each coded row matrix by A^{-1} (on the right).

Coded Matrix *Decoding Matrix* A^{-1} *Decoded Matrix*

$$[13 \quad -26 \quad 21]\begin{bmatrix} -1 & -10 & -8 \\ -1 & -6 & -5 \\ 0 & -1 & -1 \end{bmatrix} = [13 \quad 5 \quad 5]$$

$$[33 \quad -53 \quad -12]\begin{bmatrix} -1 & -10 & -8 \\ -1 & -6 & -5 \\ 0 & -1 & -1 \end{bmatrix} = [20 \quad 0 \quad 13]$$

$$[18 \quad -23 \quad -42]\begin{bmatrix} -1 & -10 & -8 \\ -1 & -6 & -5 \\ 0 & -1 & -1 \end{bmatrix} = [5 \quad 0 \quad 13]$$

$$[5 \quad -20 \quad 56]\begin{bmatrix} -1 & -10 & -8 \\ -1 & -6 & -5 \\ 0 & -1 & -1 \end{bmatrix} = [15 \quad 14 \quad 4]$$

$$[-24 \quad 23 \quad 77]\begin{bmatrix} -1 & -10 & -8 \\ -1 & -6 & -5 \\ 0 & -1 & -1 \end{bmatrix} = [1 \quad 25 \quad 0]$$

Thus, the message is as follows.

[13 5 5] [20 0 13] [5 0 13] [15 14 4] [1 25 0]
M E E T M E M O N D A Y

Group Activity *Cryptography*

Create your own numeric code for the alphabet (such as on page 511), and use it to convert a message of your own into numbers. Create an invertible $n \times n$ matrix A to encode your message. Exchange your numeric code, encoded message, and matrix A with another group. Find the necessary decoding matrix and decode the message you received.

6.5 /// EXERCISES

In Exercises 1–10, use a determinant to find the area of the triangle with the given vertices.

1. 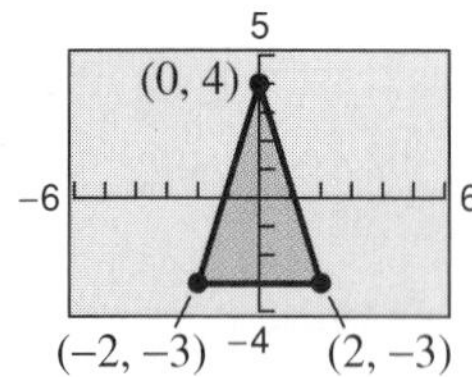

2. 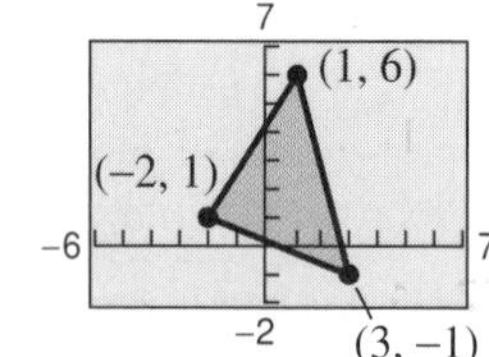

3. (0, 0), (1, 5), (3, 1)

4. (0, 0), (4, 5), (5, −2)

5. $\left(0, \frac{1}{2}\right), \left(\frac{5}{2}, 0\right), (4, 3)$

6. (0, 4), (2, 3), (5, 0)

7. (4, 5), (6, 1), (7, 9)

8. (0, −2), (−1, 4), (3, 5)

9. (−3, 5), (2, 6), (3, −5)

10. (−2, 4), (1, 5), (3, −2)

In Exercises 11 and 12, find a value of x such that the triangle has an area of 4.

11. $(-5, 1), (0, 2), (-2, x)$

12. $(-4, 2), (-3, 5), (-1, x)$

In Exercises 13–16, use Cramer's Rule to solve (if possible) the system of equations.

13. $\begin{aligned} 3x + 4y &= -2 \\ 5x + 3y &= 4 \end{aligned}$

14. $\begin{aligned} -0.4x + 0.8y &= 1.6 \\ 0.2x + 0.3y &= 2.2 \end{aligned}$

15. $\begin{aligned} 4x - y + z &= -5 \\ 2x + 2y + 3z &= 10 \\ 5x - 2y + 6z &= 1 \end{aligned}$

16. $\begin{aligned} 4x - 2y + 3z &= -2 \\ 2x + 2y + 5z &= 16 \\ 8x - 5y - 2z &= 4 \end{aligned}$

In Exercises 17 and 18, use a graphing utility and Cramer's Rule to solve (if possible) the system of equations.

17. $\begin{aligned} 3x + 3y + 5z &= 1 \\ 3x + 5y + 9z &= 2 \\ 5x + 9y + 17z &= 4 \end{aligned}$

18. $\begin{aligned} 2x + 3y + 5z &= 4 \\ 3x + 5y + 9z &= 7 \\ 5x + 9y + 17z &= 13 \end{aligned}$

19. *Area of a Region* A large region of forest has been infected with gypsy moths. The region is roughly triangular, as shown in the figure. From the northernmost vertex A of the region, the distances to the other vertices are 25 miles south and 10 miles east (for vertex B), and 20 miles south and 28 miles east (for vertex C). Use a graphing utility to approximate the number of square miles in this region.

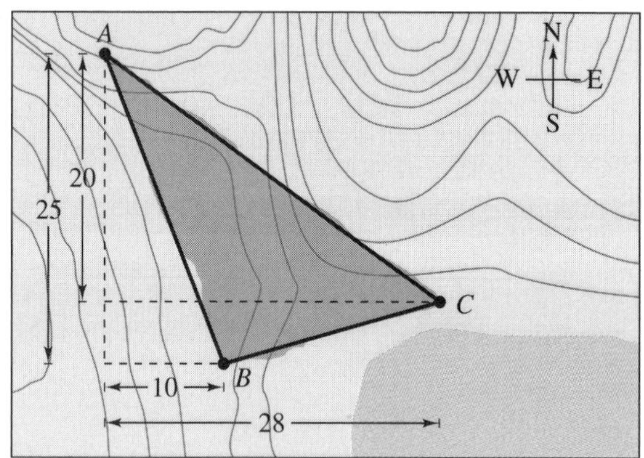

20. *Area of a Region* You own a triangular tract of land, as shown in the figure. To estimate the number of square feet in the tract, you start at one vertex, walk 65 feet east and 50 feet north to the second vertex, and then walk 85 feet west and 30 feet north to the third vertex. Use a graphing utility to determine how many square feet there are in the tract of land.

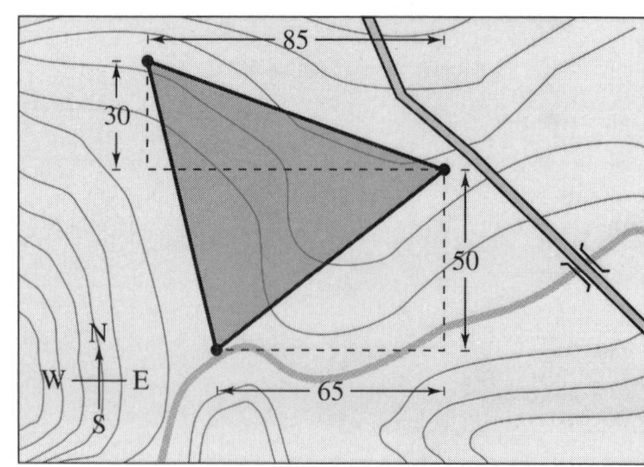

In Exercises 21–26, use the determinant feature of a graphing utility to decide if the points are collinear.

21. $(3, -1), (0, -3), (12, 5)$

22. $(-3, -5), (6, 1), (10, 2)$

23. $\left(2, -\frac{1}{2}\right), (-4, 4), (6, -3)$

24. $(0, 1), (4, -2), (-8, 7)$

25. $(0, 2), (1, 2.4), (-1, 1.6)$

26. $(2, 3), (3, 3.5), (-1, 2)$

In Exercises 27–32, use a determinant to find an equation of the line through the points.

27. $(0, 0), (5, 3)$

28. $(0, 0), (-2, 2)$

29. $(-4, 3), (2, 1)$

30. $(10, 7), (-2, -7)$

31. $\left(-\frac{1}{2}, 3\right), \left(\frac{5}{2}, 1\right)$

32. $\left(\frac{2}{3}, 4\right), (6, 12)$

In Exercises 33 and 34, find x such that the points are collinear.

33. $(2, -5), (4, x), (5, -2)$

34. $(-6, 2), (-5, x), (-3, 5)$

In Exercises 35 and 36, find the uncoded 1×3 row matrices for the message. Then encode the message using the matrix.

Message — *Matrix*

35. TROUBLE IN RIVER CITY — $\begin{bmatrix} 1 & -1 & 0 \\ 1 & 0 & -1 \\ -6 & 2 & 3 \end{bmatrix}$

36. PLEASE SEND MONEY — $\begin{bmatrix} 4 & 2 & 1 \\ -3 & -3 & -1 \\ 3 & 2 & 1 \end{bmatrix}$

In Exercises 37–40, write a cryptogram for the message using the matrix

$$A = \begin{bmatrix} 1 & 2 & 2 \\ 3 & 7 & 9 \\ -1 & -4 & -7 \end{bmatrix}.$$

37. LANDING SUCCESSFUL

38. BEAM ME UP SCOTTY

39. HAPPY BIRTHDAY

40. OPERATION OVERLORD

In Exercises 41 and 42, use A^{-1} to decode the cryptogram.

41. $A = \begin{bmatrix} 1 & 2 \\ 3 & 5 \end{bmatrix}$

11, 21, 64, 112, 25, 50, 29, 53, 23, 46, 40, 75, 55, 92

42. $A = \begin{bmatrix} 1 & -1 & 0 \\ 1 & 0 & -1 \\ -6 & 2 & 3 \end{bmatrix}$

9, −1, −9, 38, −19, −19, 28, −9, −19, −80, 25, 41, −64, 21, 31, 9, −5, −4

In Exercises 43 and 44, decode the cryptogram by using the inverse of the matrix

$$A = \begin{bmatrix} 1 & 2 & 2 \\ 3 & 7 & 9 \\ -1 & -4 & -7 \end{bmatrix}.$$

43. 20, 17, −15, −12, −56, −104, 1, −25, −65, 62, 143, 181

44. 13, −9, −59, 61, 112, 106, −17, −73, −131, 11, 24, 29, 65, 144, 172

45. The following cryptogram was encoded with a 2×2 matrix.

8, 21, −15, −10, −13, −13, 5, 10, 5, 25,
5, 19, −1, 6, 20, 40, −18, −18, 1, 16

The last word of the message is _RON. What is the message?

46. The following cryptogram was encoded with a 2×2 matrix.

5, 2, 25, 11, −2, −7, −15, −15, 32,
14, −8, −13, 38, 19, −19, −19, 37, 16

The last word of the message is _SUE. What is the message?

Focus on Concepts

In this chapter, you studied several concepts that are required in the study of matrices and determinants and their applications. You can use the following questions to check your understanding of several of these basic concepts. The answers to these questions are given in the back of the book.

1. Describe the three elementary row operations that can be performed on an augmented matrix.

2. What is the relationship between the three elementary row operations on an augmented matrix and the operations that lead to equivalent systems of equations?

3. In your own words, describe the difference between a matrix in row-echelon form and a matrix in reduced row-echelon form.

In Exercises 4–7, the row-echelon form of an augmented matrix that corresponds to a system of linear equations is given. Use the matrix to determine whether the system is consistent or inconsistent, and if it is consistent, determine the number of solutions.

4. $\begin{bmatrix} 1 & 2 & 3 & \vdots & 9 \\ 0 & 1 & -2 & \vdots & 2 \\ 0 & 0 & 0 & \vdots & 0 \end{bmatrix}$

5. $\begin{bmatrix} 1 & 2 & 3 & \vdots & 9 \\ 0 & 1 & -2 & \vdots & 2 \\ 0 & 0 & 0 & \vdots & 8 \end{bmatrix}$

6. $\begin{bmatrix} 1 & 2 & 3 & \vdots & 9 \\ 0 & 1 & -2 & \vdots & 2 \\ 0 & 0 & 1 & \vdots & -3 \end{bmatrix}$

7. $\begin{bmatrix} 1 & 2 & 3 & 10 & 6 & \vdots & 0 \\ 0 & 1 & -5 & -2 & 0 & \vdots & 5 \\ 0 & 0 & 1 & 12 & 0 & \vdots & -2 \\ 0 & 0 & 0 & 1 & 1 & \vdots & 0 \end{bmatrix}$

In Exercises 8–10, determine if the matrix operations (a) $A + 3B$ and (b) AB can be performed. If not, state why.

8. $A = \begin{bmatrix} 2 & -2 \\ 3 & 5 \end{bmatrix}, \quad B = \begin{bmatrix} -3 & 10 \\ 12 & 8 \end{bmatrix}$

9. $A = \begin{bmatrix} 5 & 4 \\ -7 & 2 \\ 11 & 2 \end{bmatrix}, \quad B = \begin{bmatrix} 4 & 12 \\ 20 & 40 \\ 15 & 30 \end{bmatrix}$

10. $A = \begin{bmatrix} 5 & 4 \\ -7 & 2 \\ 11 & 2 \end{bmatrix}, \quad B = \begin{bmatrix} 4 & 12 \\ 20 & 40 \end{bmatrix}$

11. Under what conditions does a matrix have an inverse?

12. Explain the difference between a square matrix and its determinant.

13. Is it possible to find the determinant of a 4×5 matrix? Explain.

14. What is meant by the cofactor of an entry of a matrix? How is it used to find the determinant of the matrix?

15. Three people were asked to solve a system of equations using an augmented matrix. Each person reduced the matrix to row-echelon form. The reduced matrices were

$$\begin{bmatrix} 1 & 2 & \vdots & 3 \\ 0 & 1 & \vdots & 1 \end{bmatrix}$$

$$\begin{bmatrix} 1 & 0 & \vdots & 1 \\ 0 & 1 & \vdots & 1 \end{bmatrix}$$

and

$$\begin{bmatrix} 1 & 2 & \vdots & 3 \\ 0 & 0 & \vdots & 0 \end{bmatrix}.$$

Could all three be right? Explain.

6 /// REVIEW EXERCISES

In Exercises 1 and 2, form the augmented matrix for the system of linear equations.

1. $\begin{aligned} 3x - 10y &= 15 \\ 5x + 4y &= 22 \end{aligned}$

2. $\begin{aligned} 8x - 7y + 4z &= 12 \\ 3x - 5y + 2z &= 20 \\ 5x + 3y - 3z &= 26 \end{aligned}$

In Exercises 3 and 4, write the system of linear equations represented by the augmented matrix. (Use variables x, y, z, and w.)

3. $\left[\begin{array}{ccc|c} 5 & 1 & 7 & -9 \\ 4 & 2 & 0 & 10 \\ 9 & 4 & 2 & 3 \end{array}\right]$

4. $\left[\begin{array}{cccc|c} 13 & 16 & 7 & 3 & 2 \\ 1 & 21 & 8 & 5 & 12 \\ 4 & 10 & -4 & 3 & -1 \end{array}\right]$

In Exercises 5 and 6, write the matrix in *reduced* row-echelon form.

5. $\begin{bmatrix} 0 & 1 & 1 \\ 1 & 2 & 3 \\ 2 & 2 & 2 \end{bmatrix}$

6. $\begin{bmatrix} 1 & 1 & 1 & 0 \\ 1 & 1 & 0 & 1 \\ 1 & 0 & 1 & 1 \\ 0 & 1 & 1 & 1 \end{bmatrix}$

In Exercises 7–10, use the matrix capabilities of a graphing utility to write the matrix in *reduced* row-echelon form.

7. $\begin{bmatrix} 3 & -2 & 1 & 0 \\ 4 & -3 & 0 & 1 \end{bmatrix}$

8. $\begin{bmatrix} 1 & 1 & 2 & 1 & 0 & 0 \\ -1 & 0 & 3 & 0 & 1 & 0 \\ 1 & 2 & 8 & 0 & 0 & 1 \end{bmatrix}$

9. $\begin{bmatrix} 1 & 3 & 4 \\ 0 & 1 & 1 \\ 2 & 4 & 6 \end{bmatrix}$

10. $\begin{bmatrix} 4 & 8 & 16 \\ 3 & -1 & 2 \\ -2 & 10 & 12 \end{bmatrix}$

In Exercises 11–22, use matrices and elementary row operations to solve (if possible) the system of equations.

11. $\begin{aligned} 5x + 4y &= 2 \\ -x + y &= -22 \end{aligned}$

12. $\begin{aligned} 2x - 5y &= 2 \\ 3x - 7y &= 1 \end{aligned}$

13. $\begin{aligned} 2x + y &= 0.3 \\ 3x - y &= -1.3 \end{aligned}$

14. $\begin{aligned} 0.2x - 0.1y &= 0.07 \\ 0.4x - 0.5y &= -0.01 \end{aligned}$

15. $\begin{aligned} 2x + y + 2z &= 4 \\ 2x + 2y &= 5 \\ 2x - y + 6z &= 2 \end{aligned}$

16. $\begin{aligned} 2x + 3y + z &= 10 \\ 2x - 3y - 3z &= 22 \\ 4x - 2y + 3z &= -2 \end{aligned}$

17. $\begin{aligned} 4x + 4y + 4z &= 5 \\ 4x - 2y - 8z &= 1 \\ 5x + 3y + 8z &= 6 \end{aligned}$

18. $\begin{aligned} 2x + 3y + 3z &= 3 \\ 6x + 6y + 12z &= 13 \\ 12x + 9y - z &= 2 \end{aligned}$

19. $\begin{aligned} -x + y + 2z &= 1 \\ 2x + 3y + z &= -2 \\ 5x + 4y + 2z &= 4 \end{aligned}$

20. $\begin{aligned} 3x + 21y - 29z &= -1 \\ 2x + 15y - 21z &= 0 \end{aligned}$

21. $\begin{aligned} x + 2y + 6z &= 1 \\ 2x + 5y + 15z &= 4 \\ 3x + y + 3z &= -6 \end{aligned}$

22. $\begin{aligned} x + 2y + w &= 3 \\ -3y + 3z &= 0 \\ 4x + 4y + z + 2w &= 0 \\ 2x + z &= 3 \end{aligned}$

23. *Think About It* Describe the row-echelon form of an augmented matrix that corresponds to a system of linear equations that has a unique solution.

24. *Partial Fractions* Write the partial fraction decomposition for the rational expression

$$\frac{x+9}{(x+1)(x+2)^2} = \frac{A}{x+1} + \frac{B}{x+2} + \frac{C}{(x+2)^2}.$$

In Exercises 25–28, use the matrix capabilities of a graphing utility to reduce the augmented matrix and solve the system of equations.

25. $\begin{aligned} x - 3y &= -2 \\ x + y &= 2 \end{aligned}$

26. $\begin{aligned} -x + 3y &= 5 \\ 4x - y &= 2 \end{aligned}$

27. $\begin{aligned} x + 2y - z &= 7 \\ -y - z &= 4 \\ 4x \qquad - z &= 16 \end{aligned}$

28. $\begin{aligned} 3x \qquad + 6z &= 0 \\ -2x + y \qquad &= 5 \\ y + 2z &= 3 \end{aligned}$

In Exercises 29–36, perform the matrix operations. If it is not possible, explain why.

29. $\begin{bmatrix} 2 & 1 & 0 \\ 0 & 5 & -4 \end{bmatrix} - 3\begin{bmatrix} 5 & 3 & -6 \\ 0 & -2 & 5 \end{bmatrix}$

30. $-2\begin{bmatrix} 1 & 2 \\ 5 & -4 \\ 6 & 0 \end{bmatrix} + 8\begin{bmatrix} 7 & 1 \\ 1 & 2 \\ 1 & 4 \end{bmatrix}$

31. $\begin{bmatrix} 1 & 2 \\ 5 & -4 \\ 6 & 0 \end{bmatrix}\begin{bmatrix} 6 & -2 & 8 \\ 4 & 0 & 0 \end{bmatrix}$

32. $\begin{bmatrix} 1 & 5 & 6 \\ 2 & -4 & 0 \end{bmatrix}\begin{bmatrix} 6 & -2 & 8 \\ 4 & 0 & 0 \end{bmatrix}$

33. $\begin{bmatrix} 1 & 5 & 6 \\ 2 & -4 & 0 \end{bmatrix}\begin{bmatrix} 6 & 4 \\ -2 & 0 \\ 8 & 0 \end{bmatrix}$

34. $\begin{bmatrix} 4 \\ 6 \end{bmatrix}\begin{bmatrix} 6 & -2 \end{bmatrix}$

35. $\begin{bmatrix} 1 & 3 & 2 \\ 0 & 2 & -4 \\ 0 & 0 & 3 \end{bmatrix}\begin{bmatrix} 4 & -3 & 2 \\ 0 & 3 & -1 \\ 0 & 0 & 2 \end{bmatrix}$

36. $\begin{bmatrix} 2 & 1 \\ 6 & 0 \end{bmatrix}\left(\begin{bmatrix} 4 & 2 \\ -3 & 1 \end{bmatrix} + \begin{bmatrix} -2 & 4 \\ 0 & 4 \end{bmatrix}\right)$

In Exercises 37–40, use a graphing utility to perform the matrix operations.

37. $3\begin{bmatrix} 8 & -2 & 5 \\ 1 & 3 & -1 \end{bmatrix} + 6\begin{bmatrix} 4 & -2 & -3 \\ 2 & 7 & 6 \end{bmatrix}$

38. $-5\begin{bmatrix} 2 & 0 \\ 7 & -2 \\ 8 & 2 \end{bmatrix} + 4\begin{bmatrix} 4 & -2 \\ 6 & 11 \\ -1 & 3 \end{bmatrix}$

39. $\begin{bmatrix} 4 & 1 \\ 11 & -7 \\ 12 & 3 \end{bmatrix}\begin{bmatrix} 3 & -5 & 6 \\ 2 & -2 & -2 \end{bmatrix}$

40. $\begin{bmatrix} -2 & 3 & 10 \\ 4 & -2 & 2 \end{bmatrix}\begin{bmatrix} 1 & 1 \\ -5 & 2 \\ 3 & 2 \end{bmatrix}$

In Exercises 41–44, solve for X given

$$A = \begin{bmatrix} -4 & 0 \\ 1 & -5 \\ -3 & 2 \end{bmatrix} \quad \text{and} \quad B = \begin{bmatrix} 1 & 2 \\ -2 & 1 \\ 4 & 4 \end{bmatrix}.$$

41. $X = 3A - 2B$

42. $6X = 4A + 3B$

43. $3X + 2A = B$

44. $2A - 5B = 3X$

45. Write the system of linear equations represented by the matrix equation

$$\begin{bmatrix} 5 & 4 \\ -1 & 1 \end{bmatrix}\begin{bmatrix} x \\ y \end{bmatrix} = \begin{bmatrix} 2 \\ -22 \end{bmatrix}.$$

46. Write the matrix equation $AX = B$ for the following system of linear equations.

$$\begin{aligned} 2x + 3y + z &= 10 \\ 2x - 3y - 3z &= 22 \\ 4x - 2y + 3z &= -2 \end{aligned}$$

In Exercises 47–50, use a graphing utility to find the inverse of the matrix (if it exists).

47. $\begin{bmatrix} 2 & 6 \\ 3 & -6 \end{bmatrix}$

48. $\begin{bmatrix} 3 & -10 \\ 4 & 2 \end{bmatrix}$

49. $\begin{bmatrix} 2 & 0 & 3 \\ -1 & 1 & 1 \\ 2 & -2 & 1 \end{bmatrix}$

50. $\begin{bmatrix} 1 & 4 & 6 \\ 2 & -3 & 1 \\ -1 & 18 & 16 \end{bmatrix}$

In Exercises 51–58, evaluate the determinant. Use a graphing utility to verify your result.

51. $\begin{vmatrix} 50 & -30 \\ 10 & 5 \end{vmatrix}$

52. $\begin{vmatrix} 8 & 5 \\ 2 & -4 \end{vmatrix}$

53. $\begin{vmatrix} 10 & 8 \\ -6 & -4 \end{vmatrix}$

54. $\begin{vmatrix} x & x^2 \\ 1 & 2x \end{vmatrix}$

55. $\begin{vmatrix} 1 & 0 & -2 \\ 0 & 1 & 0 \\ -2 & 0 & 1 \end{vmatrix}$

56. $\begin{vmatrix} 0 & 3 & 1 \\ 5 & -2 & 1 \\ 1 & 6 & 1 \end{vmatrix}$

57. $\begin{vmatrix} 3 & 0 & -4 & 0 \\ 0 & 8 & 1 & 2 \\ 6 & 1 & 8 & 2 \\ 0 & 3 & -4 & 1 \end{vmatrix}$

58. $\begin{vmatrix} -5 & 6 & 0 & 0 \\ 0 & 1 & -1 & 2 \\ -3 & 4 & -5 & 1 \\ 1 & 6 & 0 & 3 \end{vmatrix}$

In Exercises 59–66, use a graphing utility to solve (if possible) the system of linear equations using the inverse of the coefficient matrix.

59. $\begin{aligned} x + 2y &= -1 \\ 3x + 4y &= -5 \end{aligned}$

60. $\begin{aligned} x + 3y &= 23 \\ -6x + 2y &= -18 \end{aligned}$

61. $\begin{aligned} -3x - 3y - 4z &= 2 \\ y + z &= -1 \\ 4x + 3y + 4z &= -1 \end{aligned}$

62. $\begin{aligned} x - 3y - 2z &= 8 \\ -2x + 7y + 3z &= -19 \\ x - y - 3z &= 3 \end{aligned}$

63. $\begin{aligned} x + 3y + 2z &= 2 \\ -2x - 5y - z &= 10 \\ 2x + 4y &= -12 \end{aligned}$

64. $\begin{aligned} 2x + 4y &= -12 \\ 3x + 4y - 2z &= -14 \\ -x + y + 2z &= -6 \end{aligned}$

65. $\begin{aligned} -x + y + z &= 6 \\ 4x - 3y + z &= 20 \\ 2x - y + 3z &= 8 \end{aligned}$

66. $\begin{aligned} 2x + 3y - 4z &= 1 \\ x - y + 2z &= -4 \\ 3x + 7y - 10z &= 0 \end{aligned}$

In Exercises 67–70, use Cramer's Rule to solve (if possible) the system of equations.

67. $\begin{aligned} x + 2y &= 5 \\ -x + y &= 1 \end{aligned}$

68. $\begin{aligned} 2x - y &= -10 \\ 3x + 2y &= -1 \end{aligned}$

69. $\begin{aligned} 20x + 8y &= 11 \\ 12x - 24y &= 21 \end{aligned}$

70. $\begin{aligned} 13x - 6y &= 17 \\ 26x - 12y &= 8 \end{aligned}$

In Exercises 71–74, use a graphing utility and Cramer's Rule to solve (if possible) the system of equations.

71. $\begin{aligned} 3x + 6y &= 5 \\ 6x + 14y &= 11 \end{aligned}$

72. $\begin{aligned} -0.4x + 0.8y &= 1.6 \\ 0.2x + 0.3y &= 2.2 \end{aligned}$

73. $\begin{aligned} 5x - 3y + 2z &= 2 \\ 2x + 2y - 3z &= 3 \\ x - 7y + 8z &= -4 \end{aligned}$

74. $\begin{aligned} 14x - 21y - 7z &= 10 \\ -4x + 2y - 2z &= 4 \\ 56x - 21y + 7z &= 5 \end{aligned}$

75. *Mixture Problem* A florist wants to arrange a dozen flowers consisting of two varieties: carnations and roses. Carnations cost \$0.75 each and roses cost \$1.50 each. How many of each should the florist use so that the arrangement will cost \$12.00?

76. *Mixture Problem* One hundred liters of a 60% acid solution is obtained by mixing a 75% solution with a 50% solution. How many liters of each must be used to obtain the desired mixture?

77. *Fitting a Parabola to Three Points* Find an equation of the parabola $y = ax^2 + bx + c$ that passes through the points $(-1, 2)$, $(0, 3)$, and $(1, 6)$.

78. *Break-Even Point* A small business invests \$25,000 in equipment to produce a product. Each unit of the product costs \$3.75 to produce and is sold for \$5.25. How many items must be sold before the business breaks even?

79. *Data Analysis* The median prices y (in thousands of dollars) of one-family houses sold in the United States in the years 1981 through 1993 are shown in the figure. The least squares regression line $y = a + bt$ for this data is found by solving the system

$$13a + 91b = 1107$$
$$91a + 819b = 8404.7$$

where $t = 1$ represents 1981. (Source: National Association of Realtors)

(a) Use a graphing utility to solve this system.

(b) Use a graphing utility to graph the regression line.

(c) Interpret the meaning of the slope of the regression line in the context of the problem.

(d) Use the regression line to estimate the median price of homes in 1995.

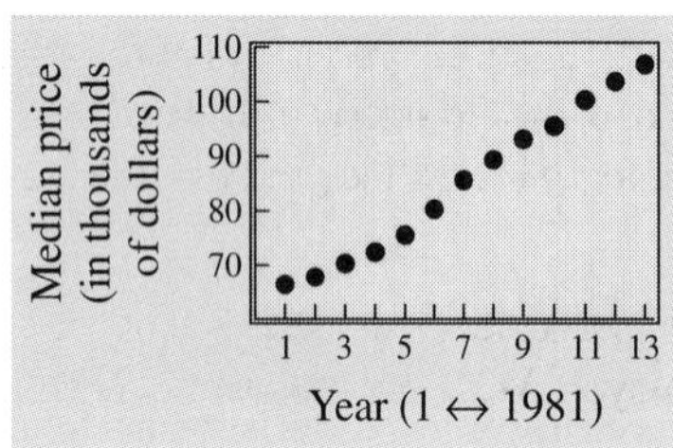

80. Solve the equation $\begin{vmatrix} 2-\lambda & 5 \\ 3 & -8-\lambda \end{vmatrix} = 0.$

In Exercises 81–84, use a determinant to find the area of the triangle with the given vertices.

81. (1, 0), (5, 0), (5, 8)

82. (−4, 0), (4, 0), (0, 6)

83. (1, 2), (4, −5), (3, 2)

84. $\left(\frac{3}{2}, 1\right), \left(4, -\frac{1}{2}\right), (4, 2)$

In Exercises 85–88, use a determinant to find an equation of the line through the given points.

85. (−4, 0), (4, 4)

86. (2, 5), (6, −1)

87. $\left(-\frac{5}{2}, 3\right), \left(\frac{7}{2}, 1\right)$

88. (−0.8, 0.2), (0.7, 3.2)

89. Verify that

$$\begin{vmatrix} a_{11} & a_{12} & a_{13} \\ a_{21} & a_{22} & a_{23} \\ a_{31}+c_1 & a_{32}+c_2 & a_{33}+c_3 \end{vmatrix} = \begin{vmatrix} a_{11} & a_{12} & a_{13} \\ a_{21} & a_{22} & a_{23} \\ a_{31} & a_{32} & a_{33} \end{vmatrix} + \begin{vmatrix} a_{11} & a_{12} & a_{13} \\ a_{21} & a_{22} & a_{23} \\ c_1 & c_2 & c_3 \end{vmatrix}.$$

90. *Circuit Analysis* Consider the circuit in the figure. The currents I_1, I_2, and I_3 in amperes are given by the solution of the system of linear equations. Use the inverse of the coefficient matrix of this system to find the unknown currents.

$$I_1 + I_2 + I_3 = 0$$
$$4I_1 - 10I_2 = 12$$
$$10I_2 - 2I_3 = -6$$

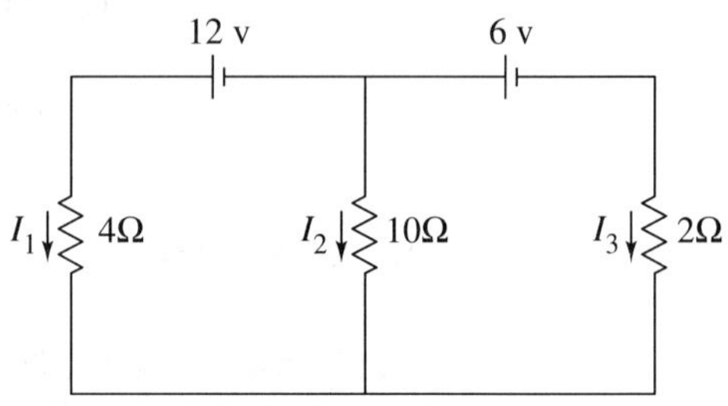

91. *Think About It* If A is a 3×3 matrix and $|A| = 2$, what is the value of $|4A|$? Give the reason for your answer.

CHAPTER PROJECT *Row Operations and Graphing*

In this project, you will investigate the graphical interpretation of elementary row operations.

(a) Solve the following systems by hand using Gauss-Jordan elimination.

$$\begin{aligned} 2x - 4y &= 9 \\ x + 5y &= 15 \end{aligned} \qquad \begin{aligned} 6x + 2y &= -19 \\ 3x - y &= -5 \end{aligned}$$

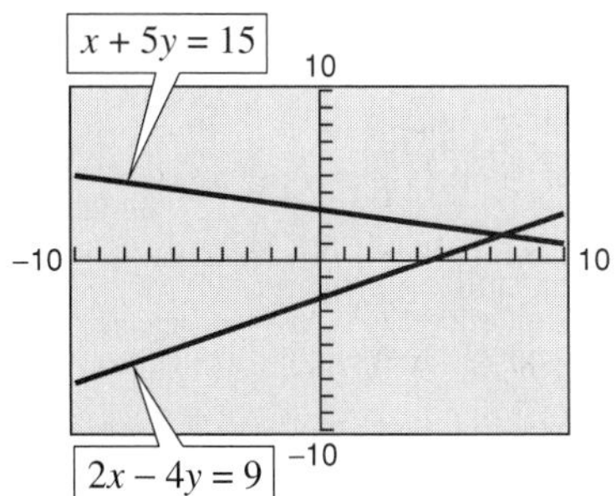

Graph of the system
$2x - 4y = 9$
$x + 5y = 15$

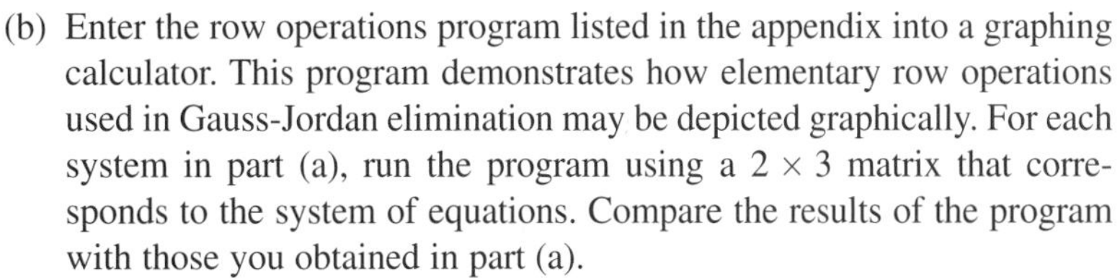
(b) Enter the row operations program listed in the appendix into a graphing calculator. This program demonstrates how elementary row operations used in Gauss-Jordan elimination may be depicted graphically. For each system in part (a), run the program using a 2×3 matrix that corresponds to the system of equations. Compare the results of the program with those you obtained in part (a).

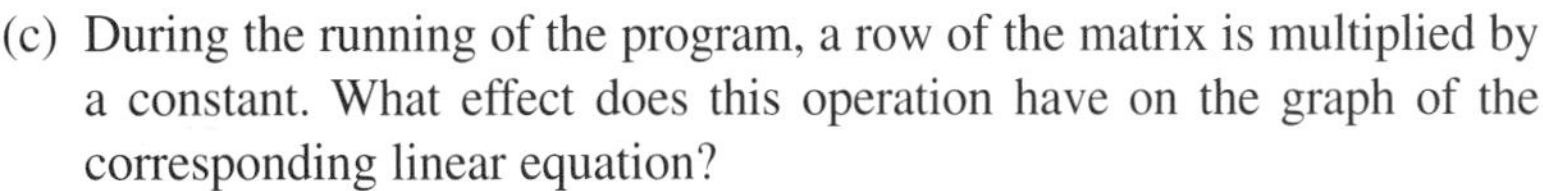
(c) During the running of the program, a row of the matrix is multiplied by a constant. What effect does this operation have on the graph of the corresponding linear equation?

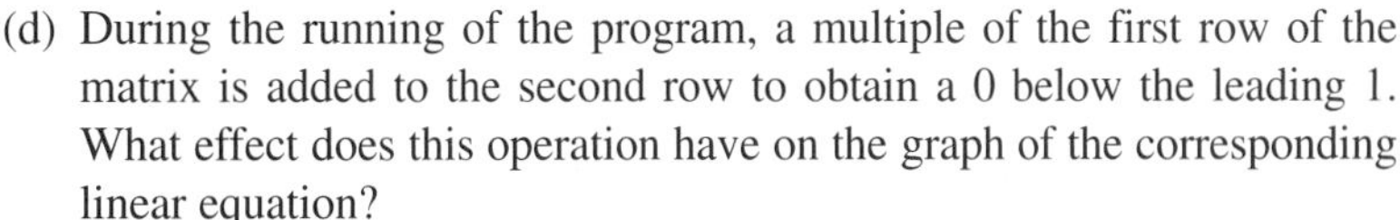
(d) During the running of the program, a multiple of the first row of the matrix is added to the second row to obtain a 0 below the leading 1. What effect does this operation have on the graph of the corresponding linear equation?

(e) Each time the 2×3 matrix is transformed, the graph of the corresponding linear equations is displayed. What do you notice about the point of intersection each time?

Questions for Further Exploration

1. Is finding a point of intersection using the program more or less accurate than finding the point of intersection using the zoom and trace features? Explain your reasoning and give an example.

2. Run the program to find the solution to the following linear system.

$$\begin{aligned} 2x &= -15 \\ 3x + 5y &= 3 \end{aligned}$$

Why is only one line drawn in all but the last screen? Verify the program's solution by hand.

3. Run the program using the following linear system.

$$\begin{aligned} 2y &= -3 \\ -2x + y &= 3 \end{aligned}$$

Describe what happens and why.

4. A system of equations with three variables has a corresponding 3×4 augmented matrix. Write a program that will transform a 3×4 matrix into reduced row-echelon form. At the end of the program display the final matrix.

6 /// CHAPTER TEST

Take this test as you would take a test in class. After you are done, check your work against the answers given in the back of the book.

The *Interactive* CD-ROM provides answers to the Chapter Tests and Cumulative Tests. It also offers Chapter Pre-Tests (that test key skills and concepts covered in previous chapters) and Chapter Post-Tests, both of which have randomly generated exercises with diagnostic capabilities.

In Exercises 1 and 2, write the matrix in reduced row-echelon form. Use a graphing utility to verify your result.

1. $\begin{bmatrix} 1 & -1 & 5 \\ 6 & 2 & 3 \\ 5 & 3 & -3 \end{bmatrix}$ **2.** $\begin{bmatrix} 1 & 0 & -1 & 2 \\ -1 & 1 & 1 & -3 \\ 1 & 1 & -1 & 1 \\ 3 & 2 & -3 & 4 \end{bmatrix}$

3. Use the matrix capabilities of a graphing utility to reduce the augmented matrix and solve the system of equations.

$$\begin{aligned} 4x + 3y - 2z &= 14 \\ -x - y + 2z &= -5 \\ 3x + y - 4z &= 8 \end{aligned}$$

4. Find the equation of the parabola $y = ax^2 + bx + c$ that passes through the points in the figure. Use a graphing utility to verify your result.

Figure for 4

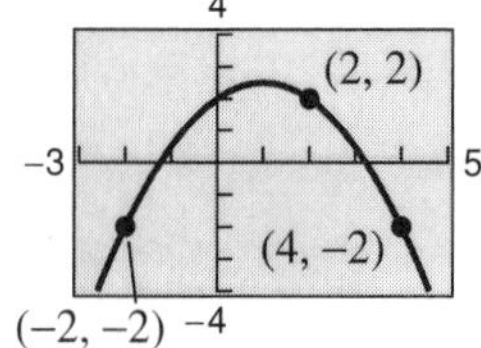

5. Find (a) $A - B$, (b) $3A$, and (c) $3A - 2B$.

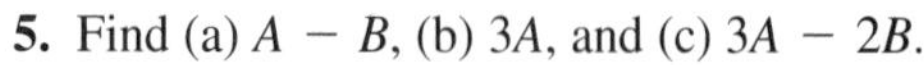

$$A = \begin{bmatrix} 5 & 4 & 4 \\ -4 & -4 & 0 \end{bmatrix}, \quad B = \begin{bmatrix} 4 & -1 & 6 \\ -4 & 0 & -3 \end{bmatrix}$$

6. Find AB, if possible.

$$A = \begin{bmatrix} 2 & -2 & 6 \\ 3 & -1 & 7 \\ 2 & 0 & -2 \end{bmatrix}, \quad B = \begin{bmatrix} 4 & 4 \\ 3 & 2 \\ 1 & -2 \end{bmatrix}$$

7. Find A^{-1} for $A = \begin{bmatrix} -6 & 4 \\ 10 & -5 \end{bmatrix}$.

8. Use the result of Exercise 7 to solve the system.

$$\begin{aligned} -6x + 4y &= 10 \\ 10x - 5y &= 20 \end{aligned}$$

9. Evaluate the determinant of the matrix

$$\begin{bmatrix} 4 & 0 & 3 \\ 1 & -8 & 2 \\ 3 & 2 & 2 \end{bmatrix}.$$

10. Use a determinant to find the area of the triangle in the figure.

Figure for 10

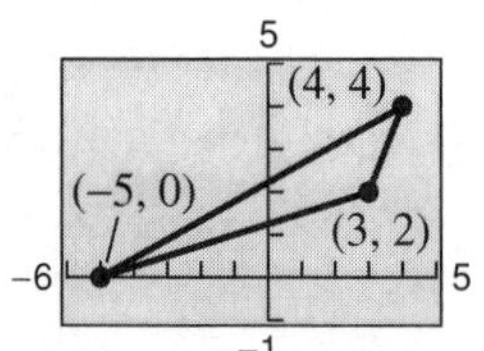

Sequences and Probability

Chapter 7

Burney Le Boeuf, shown in the inset photo, is a marine biologist at the University of California at Santa Cruz. He has spent many years studying elephant seals.

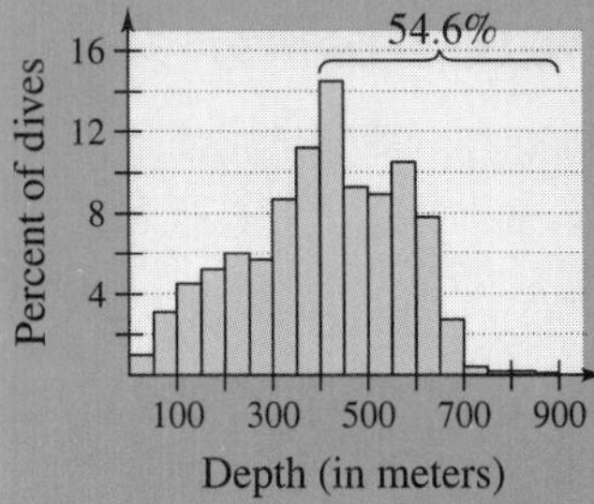

Of all mammals, reptiles, and birds, elephant seals are the champions of deep sea dives. One bull elephant seal was measured at a depth of nearly 1 mile! "Our theories were blown clear out of the water," says Le Boeuf. (Source: Smithsonian Magazine, September, 1995, "Elephant Seals, the Champion Divers of the Deep.")

The frequency distribution on the left shows the recorded depths of over 10,000 dives by female elephant seals. From this data, you can make predictions about the depth of a particular dive. For instance, because 54.6% of all dives were in depths greater than 400 meters, you can conclude that the probability that a particular dive will exceed 400 meters is 0.546. (See Exercises 59 and 60 on page 596.)

Frans Lanting—Minden Pictures.

Researchers from the University of California at Santa Cruz are trying to block a tranquilized bull elephant seal from entering the water. They will then attach instruments to the bull to measure the depths and durations of its dives.

7.1 Sequences and Summation Notation

Sequences / Factorial Notation / Summation Notation / Application

Sequences

In mathematics, the word *sequence* is used in much the same way as in ordinary English. Saying that a collection is listed *in sequence* means that it is ordered so that it has a first member, a second member, a third member, and so on.

Mathematically, you can think of a sequence as a *function* whose domain is the set of positive integers. Rather than using function notation, however, sequences are usually written using subscript notation, as indicated in the following definition.

Note On occasion, it is convenient to begin subscripting a sequence with 0 instead of 1 so that the terms of the sequence become

$$a_0, a_1, a_2, a_3, \ldots .$$

Definition of Sequence

An **infinite sequence** is a function whose domain is the set of positive integers. The function values

$$a_1, a_2, a_3, a_4, \ldots, a_n, \ldots$$

are the **terms** of the sequence. If the domain of the function consists of the first n positive integers only, the sequence is a **finite sequence.**

To graph a sequence using a graphing utility, set the utility to "seq" and "dot" modes and enter the sequence into U*n*. The graph of the sequence in Example 1(a) is shown below.

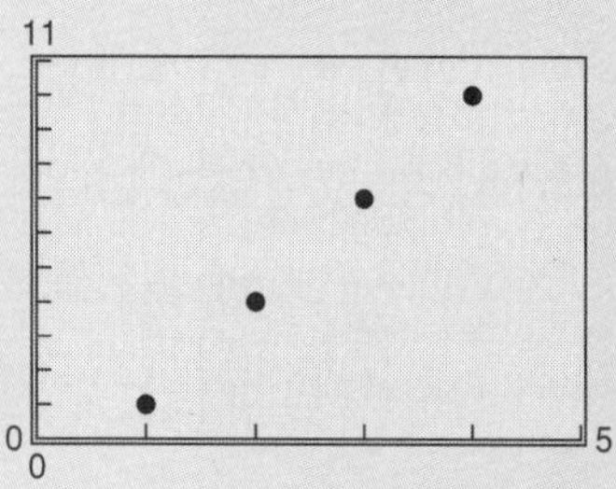

Graph the sequence in Example 1(b) and use the trace key to identify its terms.

EXAMPLE 1 **Finding the Terms of a Sequence**

a. The first four terms of the sequence given by $a_n = 3n - 2$ are

$a_1 = 3(1) - 2 = 1$ 1st term

$a_2 = 3(2) - 2 = 4$ 2nd term

$a_3 = 3(3) - 2 = 7$ 3rd term

$a_4 = 3(4) - 2 = 10.$ 4th term

b. The first four terms of the sequence given by $a_n = 3 + (-1)^n$ are

$a_1 = 3 + (-1)^1 = 3 - 1 = 2$ 1st term

$a_2 = 3 + (-1)^2 = 3 + 1 = 4$ 2nd term

$a_3 = 3 + (-1)^3 = 3 - 1 = 2$ 3rd term

$a_4 = 3 + (-1)^4 = 3 + 1 = 4.$ 4th term

EXPLORATION

Use the table feature of a graphing utility to create a table showing the terms of the sequences

$$a_n = \frac{(-1)^{n+1}}{2n-1}$$

and

$$a_n = \frac{(-1)^n}{2n-1}.$$

Are the terms the same? If not, how do they differ?

EXAMPLE 2 Finding the Terms of a Sequence

The first four terms of the sequence given by $a_n = \dfrac{(-1)^n}{2n-1}$ are

$$a_1 = \frac{(-1)^1}{2(1)-1} = \frac{-1}{2-1} = -1 \qquad \text{1st term}$$

$$a_2 = \frac{(-1)^2}{2(2)-1} = \frac{1}{4-1} = \frac{1}{3} \qquad \text{2nd term}$$

$$a_3 = \frac{(-1)^3}{2(3)-1} = \frac{-1}{6-1} = -\frac{1}{5} \qquad \text{3rd term}$$

$$a_4 = \frac{(-1)^4}{2(4)-1} = \frac{1}{8-1} = \frac{1}{7}. \qquad \text{4th term}$$

Simply listing the first few terms is not sufficient to define a unique sequence—the nth term *must be given.* To see this, consider the following sequences, both of which have the same first three terms.

$$\frac{1}{2}, \frac{1}{4}, \frac{1}{8}, \frac{1}{16}, \ldots, \frac{1}{2^n}, \ldots$$

$$\frac{1}{2}, \frac{1}{4}, \frac{1}{8}, \frac{1}{15}, \ldots, \frac{6}{(n+1)(n^2-n+6)}, \ldots$$

EXAMPLE 3 Finding the nth Term of a Sequence

Write an expression for the apparent nth term (a_n) of each sequence.

a. 1, 3, 5, 7, . . . **b.** 2, 5, 10, 17, . . .

Solution

a. *n:* 1 2 3 4 . . . n

Terms: 1 3 5 7 . . . a_n

Apparent Pattern: Each term is 1 less than twice n, which implies that

$$a_n = 2n - 1.$$

b. *n:* 1 2 3 4 . . . n

Terms: 2 5 10 17 . . . a_n

Apparent Pattern: Each term is 1 more than the square of n, which implies that

$$a_n = n^2 + 1.$$

Note Some sequences are defined **recursively.** To define a sequence recursively, you need to be given one or more of the first few terms. All other terms of the sequence are then defined using previous terms. A well-known example is the Fibonacci Sequence shown in Example 4.

The *Interactive* CD-ROM shows every example with its solution; clicking on the *Try It!* button brings up similar problems. Guided Examples and Integrated Examples show step-by-step solutions to additional examples. Integrated Examples are related to several concepts in the section.

EXAMPLE 4 **A Sequence That Is Defined Recursively**

The Fibonacci Sequence is defined recursively, as follows.

$$a_0 = 1,\ a_1 = 1,\ a_k = a_{k-2} + a_{k-1}, \qquad \text{where } k \geq 2$$

Write the first six terms of this sequence.

Solution

$a_0 = 1$ — 0th term is given.

$a_1 = 1$ — 1st term is given.

$a_2 = a_0 + a_1 = 1 + 1 = 2$ — Use recursive formula.

$a_3 = a_1 + a_2 = 1 + 2 = 3$ — Use recursive formula.

$a_4 = a_2 + a_3 = 2 + 3 = 5$ — Use recursive formula.

$a_5 = a_3 + a_4 = 3 + 5 = 8$ — Use recursive formula.

Factorial Notation

Some very important sequences in mathematics involve terms that are defined with special types of products called **factorials.**

Definition of Factorial

If n is a positive integer, ***n* factorial** is defined by

$$n! = 1 \cdot 2 \cdot 3 \cdot 4 \cdot \cdot \cdot (n-1) \cdot n.$$

As a special case, zero factorial is defined as $0! = 1$.

Here are some values of $n!$ for the first several nonnegative integers. Notice that $0! = 1$ by definition.

$0! = 1$

$1! = 1$

$2! = 1 \cdot 2 = 2$

$3! = 1 \cdot 2 \cdot 3 = 6$

$4! = 1 \cdot 2 \cdot 3 \cdot 4 = 24$

$5! = 1 \cdot 2 \cdot 3 \cdot 4 \cdot 5 = 120$

The value of n does not have to be very large before the value of $n!$ becomes huge. For instance, $10! = 3{,}628{,}800$.

Factorials follow the same conventions for order of operations as do exponents. For instance,

$$2n! = 2(n!) = 2(1 \cdot 2 \cdot 3 \cdot 4 \cdots n)$$

whereas $(2n)! = 1 \cdot 2 \cdot 3 \cdot 4 \cdots 2n$.

EXAMPLE 5 **Finding the Terms of a Sequence Involving Factorials**

List the first five terms of the sequence given by $a_n = 2^n/n!$. Begin with $n = 0$.

Solution

$$a_0 = \frac{2^0}{0!} = \frac{1}{1} = 1 \qquad \text{0th term}$$

$$a_1 = \frac{2^1}{1!} = \frac{2}{1} = 2 \qquad \text{1st term}$$

$$a_2 = \frac{2^2}{2!} = \frac{4}{2} = 2 \qquad \text{2nd term}$$

$$a_3 = \frac{2^3}{3!} = \frac{8}{6} = \frac{4}{3} \qquad \text{3rd term}$$

$$a_4 = \frac{2^4}{4!} = \frac{16}{24} = \frac{2}{3} \qquad \text{4th term}$$

Most graphing utilities have the capability to compute $n!$. Use your utility to compare $3 \cdot 5!$ and $(3 \cdot 5)!$. How large a value of $n!$ will your graphing utility allow you to compute?

When working with fractions involving factorials, you will often find that the fractions can be reduced.

EXAMPLE 6 **Evaluating Factorial Expressions**

Evaluate each factorial expression.

a. $\dfrac{8!}{2! \cdot 6!}$ **b.** $\dfrac{2! \cdot 6!}{3! \cdot 5!}$ **c.** $\dfrac{n!}{(n-1)!}$

Solution

a. $$\frac{8!}{2! \cdot 6!} = \frac{\cancel{1 \cdot 2 \cdot 3 \cdot 4 \cdot 5 \cdot 6} \cdot 7 \cdot 8}{1 \cdot 2 \cdot \cancel{1 \cdot 2 \cdot 3 \cdot 4 \cdot 5 \cdot 6}} = \frac{7 \cdot 8}{2} = 28$$

b. $$\frac{2! \cdot 6!}{3! \cdot 5!} = \frac{\cancel{1 \cdot 2} \cdot \cancel{1 \cdot 2 \cdot 3 \cdot 4 \cdot 5} \cdot 6}{\cancel{1 \cdot 2} \cdot 3 \cdot \cancel{1 \cdot 2 \cdot 3 \cdot 4 \cdot 5}} = \frac{6}{3} = 2$$

c. $$\frac{n!}{(n-1)!} = \frac{\cancel{1 \cdot 2 \cdot 3 \cdots (n-1)} \cdot n}{\cancel{1 \cdot 2 \cdot 3 \cdots (n-1)}} = n$$

Summation Notation

There is a convenient notation for the sum of the terms of a finite sequence. It is called **summation notation** or **sigma notation** because it involves the use of the uppercase Greek letter sigma, written as Σ.

Definition of Summation Notation

The sum of the first n terms of a sequence is represented by

$$\sum_{i=1}^{n} a_i = a_1 + a_2 + a_3 + a_4 + \cdots + a_n$$

where i is called the **index of summation,** n is the **upper limit of summation,** and 1 is the **lower limit of summation.**

EXAMPLE 7 **Sigma Notation for Sums**

a. $\displaystyle\sum_{i=1}^{5} 3i = 3(1) + 3(2) + 3(3) + 3(4) + 3(5)$

$= 3(1 + 2 + 3 + 4 + 5)$

$= 3(15) = 45$

b. $\displaystyle\sum_{k=3}^{6} (1 + k^2) = (1 + 3^2) + (1 + 4^2) + (1 + 5^2) + (1 + 6^2)$

$= 10 + 17 + 26 + 37 = 90$

c. $\displaystyle\sum_{i=0}^{8} \frac{1}{i!} = \frac{1}{0!} + \frac{1}{1!} + \frac{1}{2!} + \frac{1}{3!} + \frac{1}{4!} + \frac{1}{5!} + \frac{1}{6!} + \frac{1}{7!} + \frac{1}{8!}$

$\displaystyle = 1 + 1 + \frac{1}{2} + \frac{1}{6} + \frac{1}{24} + \frac{1}{120} + \frac{1}{720} + \frac{1}{5040} + \frac{1}{40{,}320}$

≈ 2.71828

For this summation, note that the sum is very close to the irrational number $e \approx 2.718281828$. It can be shown that as more terms of the sequence whose nth term is $1/n!$ are added, the sum becomes closer and closer to e.

Note In Example 7, note that the lower limit of a summation does not have to be 1. Also note that the index of summation does not have to be the letter i. For instance, in part (b), the letter k is the index of summation.

Think About the Proof

Use the Distributive Property for real numbers to prove Property 1 of the Properties of Sums. The details of the proof are given in the appendix.

Properties of Sums

1. $\sum_{i=1}^{n} ca_i = c\sum_{i=1}^{n} a_i$, c is any constant

2. $\sum_{i=1}^{n} (a_i + b_i) = \sum_{i=1}^{n} a_i + \sum_{i=1}^{n} b_i$

3. $\sum_{i=1}^{n} (a_i - b_i) = \sum_{i=1}^{n} a_i - \sum_{i=1}^{n} b_i$

Variations in the upper and lower limits of summation can produce quite different-looking summation notations for *the same sum.* For example, consider the following two sums.

$$\sum_{i=1}^{5} 3(2^i) = 3\sum_{i=1}^{5} 2^i = 3(2^1 + 2^2 + 2^3 + 2^4 + 2^5)$$

$$\sum_{i=0}^{4} 3(2^{i+1}) = 3\sum_{i=0}^{4} 2^{i+1} = 3(2^1 + 2^2 + 2^3 + 2^4 + 2^5)$$

Notice that both sums have identical terms.

Figure 7.1 The total area is 2.

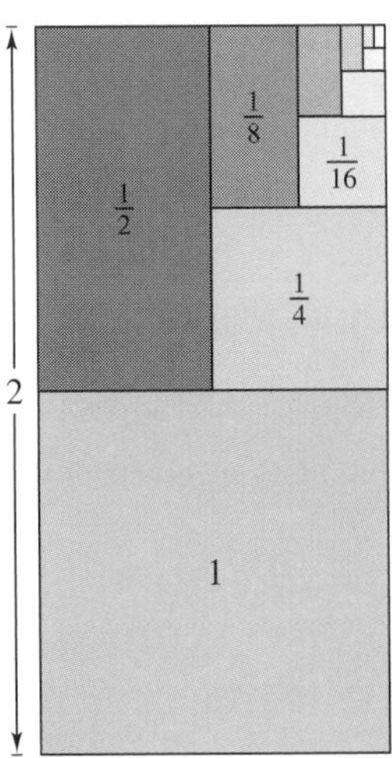

The sum of the terms of any *finite* sequence must be a finite number. An important discovery in mathematics was that for some special types of *infinite* sequences, the sum of *all* the terms is a finite number. For instance, it can be shown that the sum of all of the terms of the sequence whose nth term is $1/2^n$ (with n beginning at 0) is

$$\sum_{n=0}^{\infty} \frac{1}{2^n} = 1 + \frac{1}{2} + \frac{1}{4} + \frac{1}{8} + \frac{1}{16} + \frac{1}{32} + \cdots = 2.$$

See Figure 7.1.

Note The summation of the terms of a sequence is a **series.** The summation

$$\sum_{i=1}^{\infty} a_i = a_1 + a_2 + a_3 + \cdots$$

is an **infinite series.** Infinite series have important uses in calculus.

EXAMPLE 8 **Finding the Sum of an Infinite Sequence**

$$\begin{aligned}\sum_{n=1}^{\infty} \frac{3}{10^n} &= \frac{3}{10^1} + \frac{3}{10^2} + \frac{3}{10^3} + \frac{3}{10^4} + \frac{3}{10^5} + \cdots \\ &= 0.3 + 0.03 + 0.003 + 0.0003 + 0.00003 + \cdots \\ &= 0.33333\cdots \\ &= \frac{1}{3}\end{aligned}$$

Application

Sequences have many applications in business and science. One is illustrated in Example 9.

EXAMPLE 9 **Population of the United States** *Real Life*

From 1950 to 1993, the resident population of the United States can be approximated by the model

$$a_n = \sqrt{23{,}107 + 874.7n + 2.74n^2}, \qquad n = 0, 1, \ldots, 43$$

where a_n is the population in millions and n represents the calendar year, with $n = 0$ corresponding to 1950. Find the last five terms of this finite sequence. (Source: U.S. Bureau of Census)

Solution

The last five terms of this finite sequence are as follows.

$$a_{39} = \sqrt{23{,}107 + 874.7(39) + 2.74(39)^2} \approx 247.8 \qquad \text{1989 population}$$

$$a_{40} = \sqrt{23{,}107 + 874.7(40) + 2.74(40)^2} \approx 250.0 \qquad \text{1990 population}$$

$$a_{41} = \sqrt{23{,}107 + 874.7(41) + 2.74(41)^2} \approx 252.1 \qquad \text{1991 population}$$

$$a_{42} = \sqrt{23{,}107 + 874.7(42) + 2.74(42)^2} \approx 254.3 \qquad \text{1992 population}$$

$$a_{43} = \sqrt{23{,}107 + 874.7(43) + 2.74(43)^2} \approx 256.5 \qquad \text{1993 population}$$

The bar graph in Figure 7.2 graphically represents the populations given by this sequence for the entire 44-year period from 1950 to 1993.

Figure 7.2

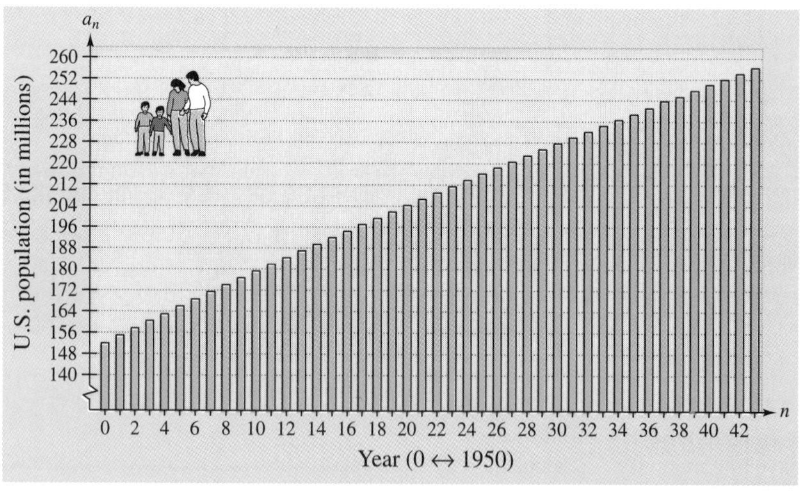

Group Activity

A Summation Program

A graphing calculator can be programmed to calculate the sum of the first n terms of a sequence. For instance, the program below lists the program steps for the *TI-82* and *TI-83*. See the appendix for programs for other graphing calculator models.

`PROGRAM:SUM`	Title of program
`:Prompt M`	M is the lower limit of summation.
`:Prompt N`	N is the upper limit of summation.
`:0 → S`	Initialize S for use as stored sum.
`:For (X, M, N)`	Loop from lower limit to upper limit.
`:S + Y1 → S`	Evaluate function stored in Y1 and add to sum.
`:Disp S`	Display partial sum.
`:End`	End For loop.

To use this program to find the sum of a sequence, store the nth term of the sequence as Y1 and run the program. For instance, to find the sum

$$\sum_{i=1}^{10} 2^i$$

you should press [Y=], enter 2^X, and then use the keystroke sequence

[PRGM], cursor to program SUM, [ENTER] [ENTER]
1 [ENTER] 10 [ENTER].

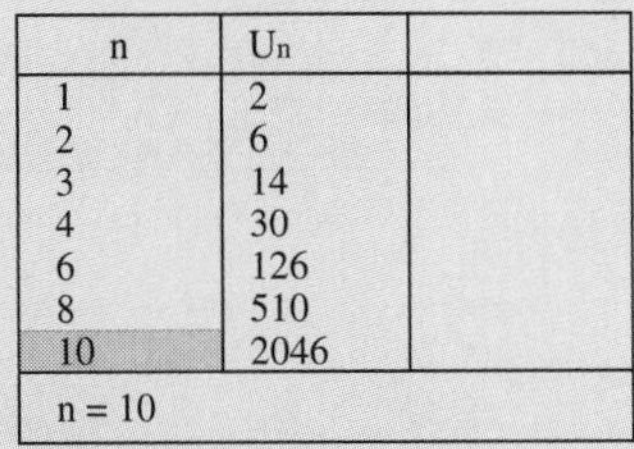

n	Un	
1	2	
2	6	
3	14	
4	30	
6	126	
8	510	
10	2046	
n = 10		

The result should be 2046. (Note that the calculator displays 10 numbers—the last number displayed is the final sum.) To pause the program after each partial sum, insert :Pause after :Disp S. Then press [ENTER] to display the next partial sum. Try using this program to find the following sums.

a. $\sum_{i=1}^{15}(i-2)^2$ **b.** $\sum_{i=1}^{20}3^{(i-1)}$ **c.** $\sum_{i=1}^{10}\sqrt{e^i}$

Another way to calculate the sum of a sequence using the *TI-82* or *TI-83* is to combine the **sum** and **seq(** commands under the LIST menu. The partial sums can be displayed in a table using seq mode by choosing the "ask" feature in the TblSet menu, as shown at the left. Try using this method to find the sums in (a), (b), and (c).

The *Interactive* CD-ROM contains step-by-step solutions to all odd-numbered Section and Review Exercises. It also provides Tutorial Exercises, which link to Guided Examples for additional help.

7.1 /// EXERCISES

In Exercises 1–22, write the first five terms of the sequence. (Assume n begins with 1.)

1. $a_n = 2n + 1$
2. $a_n = 4n - 3$
3. $a_n = 2^n$
4. $a_n = \left(\frac{1}{2}\right)^n$
5. $a_n = (-2)^n$
6. $a_n = \left(-\frac{1}{2}\right)^n$
7. $a_n = \dfrac{n+1}{n}$
8. $a_n = \dfrac{n}{n+1}$
9. $a_n = \dfrac{6n}{3n^2 - 1}$
10. $a_n = \dfrac{3n^2 - n + 4}{2n^2 + 1}$
11. $a_n = \dfrac{1 + (-1)^n}{n}$
12. $a_n = 1 + (-1)^n$
13. $a_n = 3 - \dfrac{1}{2^n}$
14. $a_n = \dfrac{3^n}{4^n}$
15. $a_n = \dfrac{1}{n^{3/2}}$
16. $a_n = \dfrac{10}{n^{2/3}}$
17. $a_n = \dfrac{3^n}{n!}$
18. $a_n = \dfrac{n!}{n}$
19. $a_n = \dfrac{(-1)^n}{n^2}$
20. $a_n = (-1)^n\left(\dfrac{n}{n+1}\right)$
21. $a_n = \frac{2}{3}$
22. $a_n = n(n-1)(n-2)$

In Exercises 23 and 24, find the indicated term of the sequence.

23. $a_n = (-1)^n(3n - 2)$

$a_{25} = \blacksquare$

24. $a_n = \dfrac{2^n}{n!}$

$a_{10} = \blacksquare$

In Exercises 25–28, write the first five terms of the sequence defined recursively.

25. $a_1 = 28, \quad a_{k+1} = a_k - 4$
26. $a_1 = 15, \quad a_{k+1} = a_k + 3$
27. $a_1 = 3, \quad a_{k+1} = 2(a_k - 1)$
28. $a_1 = 32, \quad a_{k+1} = \frac{1}{2}a_k$

In Exercises 29–34, use a graphing utility to graph the first 10 terms of the sequence.

29. $a_n = \dfrac{2}{3}n$
30. $a_n = 2 - \dfrac{4}{n}$
31. $a_n = 16(-0.5)^{n-1}$
32. $a_n = 8(0.75)^{n-1}$
33. $a_n = \dfrac{2n}{n+1}$
34. $a_n = \dfrac{3n^2}{n^2+1}$

In Exercises 35–38, match the sequence with its graph. [The graphs are labeled (a), (b), (c), and (d).]

(a)
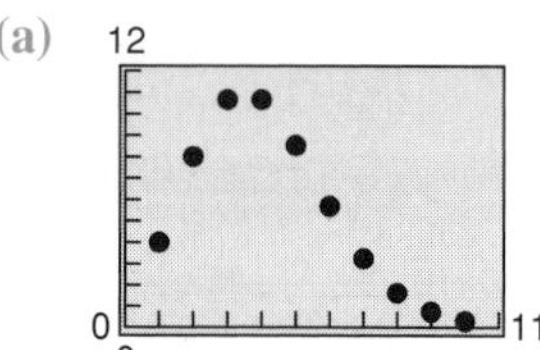

(b)
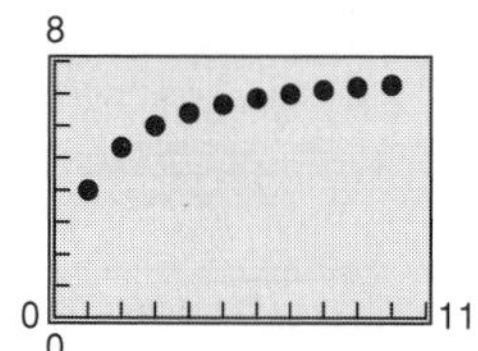

(c)
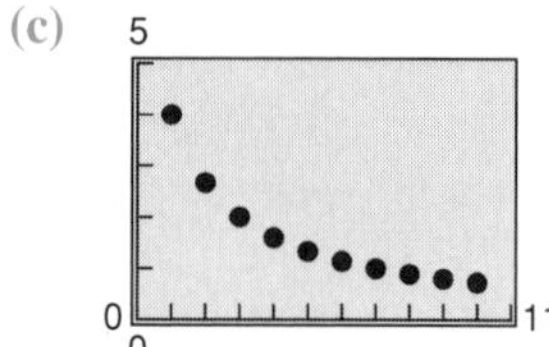

(d)
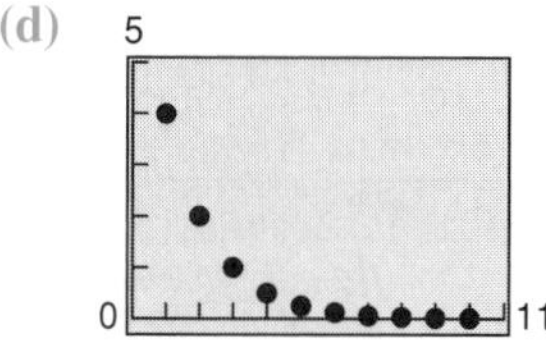

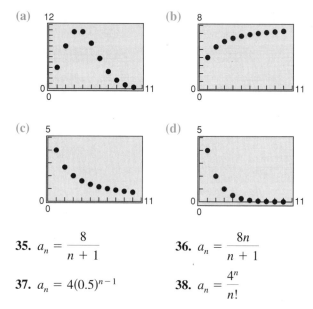

35. $a_n = \dfrac{8}{n+1}$
36. $a_n = \dfrac{8n}{n+1}$
37. $a_n = 4(0.5)^{n-1}$
38. $a_n = \dfrac{4^n}{n!}$

In Exercises 39–46, simplify the ratio of factorials.

39. $\dfrac{4!}{6!}$
40. $\dfrac{4!}{7!}$
41. $\dfrac{10!}{8!}$
42. $\dfrac{25!}{23!}$
43. $\dfrac{(n+1)!}{n!}$
44. $\dfrac{(n+2)!}{n!}$
45. $\dfrac{(2n-1)!}{(2n+1)!}$
46. $\dfrac{(2n+2)!}{(2n)!}$

In Exercises 47–60, write an expression for the *most apparent* nth term of the sequence. (Assume n begins with 1.)

47. $1, 4, 7, 10, 13, \ldots$

48. $3, 7, 11, 15, 19, \ldots$

49. $0, 3, 8, 15, 24, \ldots$

50. $1, \frac{1}{4}, \frac{1}{9}, \frac{1}{16}, \frac{1}{25}, \ldots$

51. $\frac{2}{3}, \frac{3}{4}, \frac{4}{5}, \frac{5}{6}, \frac{6}{7}, \ldots$

52. $\frac{2}{1}, \frac{3}{3}, \frac{4}{5}, \frac{5}{7}, \frac{6}{9}, \ldots$

53. $\frac{1}{2}, \frac{-1}{4}, \frac{1}{8}, \frac{-1}{16}, \ldots$

54. $\frac{1}{3}, \frac{2}{9}, \frac{4}{27}, \frac{8}{81}, \ldots$

55. $1 + \frac{1}{1}, 1 + \frac{1}{2}, 1 + \frac{1}{3}, 1 + \frac{1}{4}, 1 + \frac{1}{5}, \ldots$

56. $1 + \frac{1}{2}, 1 + \frac{3}{4}, 1 + \frac{7}{8}, 1 + \frac{15}{16}, 1 + \frac{31}{32}, \ldots$

57. $1, \frac{1}{2}, \frac{1}{6}, \frac{1}{24}, \frac{1}{120}, \ldots$

58. $2, -4, 6, -8, 10, \ldots$

59. $1, -1, 1, -1, 1, \ldots$

60. $1, 2, \frac{2^2}{2}, \frac{2^3}{6}, \frac{2^4}{24}, \frac{2^5}{120}, \ldots$

In Exercises 61–64, write the first five terms of the sequence defined recursively. Use the pattern to write the nth term of the sequence as a function of n. (Assume n begins with 1.)

61. $a_1 = 6, \quad a_{k+1} = a_k + 2$

62. $a_1 = 25, \quad a_{k+1} = a_k - 5$

63. $a_1 = 81, \quad a_{k+1} = \frac{1}{3}a_k$

64. $a_1 = 14, \quad a_{k+1} = -2a_k$

In Exercises 65–76, find the sum.

65. $\sum_{i=1}^{5}(2i + 1)$

66. $\sum_{i=1}^{6}(3i - 1)$

67. $\sum_{k=1}^{4} 10$

68. $\sum_{k=1}^{5} 6$

69. $\sum_{i=0}^{4} i^2$

70. $\sum_{i=0}^{5} 3i^2$

71. $\sum_{k=0}^{3} \frac{1}{k^2 + 1}$

72. $\sum_{j=3}^{5} \frac{1}{j}$

73. $\sum_{i=1}^{4}[(i - 1)^2 + (i + 1)^3]$

74. $\sum_{k=2}^{5}(k + 1)(k - 3)$

75. $\sum_{i=1}^{4} 2^i$

76. $\sum_{j=0}^{4}(-2)^j$

In Exercises 77–80, use a calculator to find the sum.

77. $\sum_{j=1}^{6}(24 - 3j)$

78. $\sum_{j=1}^{10} \frac{3}{j + 1}$

79. $\sum_{k=0}^{4} \frac{(-1)^k}{k + 1}$

80. $\sum_{k=0}^{4} \frac{(-1)^k}{k!}$

In Exercises 81–90, use sigma notation to write the sum.

81. $\frac{1}{3(1)} + \frac{1}{3(2)} + \frac{1}{3(3)} + \cdots + \frac{1}{3(9)}$

82. $\frac{5}{1 + 1} + \frac{5}{1 + 2} + \frac{5}{1 + 3} + \cdots + \frac{5}{1 + 15}$

83. $\left[2\left(\frac{1}{8}\right) + 3\right] + \left[2\left(\frac{2}{8}\right) + 3\right] + \cdots + \left[2\left(\frac{8}{8}\right) + 3\right]$

84. $\left[1 - \left(\frac{1}{6}\right)^2\right] + \left[1 - \left(\frac{2}{6}\right)^2\right] + \cdots + \left[1 - \left(\frac{6}{6}\right)^2\right]$

85. $3 - 9 + 27 - 81 + 243 - 729$

86. $1 - \frac{1}{2} + \frac{1}{4} - \frac{1}{8} + \cdots - \frac{1}{128}$

87. $\frac{1}{1^2} - \frac{1}{2^2} + \frac{1}{3^2} - \frac{1}{4^2} + \cdots - \frac{1}{20^2}$

88. $\frac{1}{1 \cdot 3} + \frac{1}{2 \cdot 4} + \frac{1}{3 \cdot 5} + \cdots + \frac{1}{10 \cdot 12}$

89. $\frac{1}{4} + \frac{3}{8} + \frac{7}{16} + \frac{15}{32} + \frac{31}{64}$

90. $\frac{1}{2} + \frac{2}{4} + \frac{6}{8} + \frac{24}{16} + \frac{120}{32} + \frac{720}{64}$

91. *Compound Interest* A deposit of \$5000 is made in an account that earns 8% interest compounded quarterly. The balance in the account after n quarters is given by

$$A_n = 5000\left(1 + \frac{0.08}{4}\right)^n, \qquad n = 1, 2, 3, \ldots .$$

(a) Compute the first eight terms of this sequence.

(b) Find the balance in this account after 10 years by computing the 40th term of the sequence.

92. *Compound Interest* A deposit of \$100 is made *each month* in an account that earns 12% interest compounded monthly. The balance in the account after n months is given by

$$A_n = 100(101)[(1.01)^n - 1], \qquad n = 1, 2, 3, \ldots .$$

(a) Compute the first six terms of this sequence.

(b) Find the balance in this account after 5 years by computing the 60th term of the sequence.

(c) Find the balance in this account after 20 years by computing the 240th term of the sequence.

93. *Per Capita Hospital Care* The average per capita annual cost for hospital care from 1981 to 1991 is approximated by the model

$$a_n = 510.13 + 16.37n + 3.23n^2, \; n = 1, \ldots, 11$$

where a_n is the per capita cost in dollars and n is the year, with $n = 1$ corresponding to 1981. Find the terms of this finite sequence and use a graphing utility to construct a bar graph that represents the sequence. (Source: U.S. Health Care Financing Administration)

94. *Federal Debt* It took more than 200 years for the United States to accumulate a \$1 trillion debt. Then it took just 8 years to get to a \$3 trillion debt. The federal debt during the decade of the 1980s is approximated by the model

$$a_n = 0.1\sqrt{82 + 9n^2}, \qquad n = 0, \ldots, 10$$

where a_n is the debt in trillions and n is the year, with $n = 0$ corresponding to 1980. Find the terms of this finite sequence and use a graphing utility to construct a bar graph that represents the sequence. (Source: Treasury Department)

95. *Corporate Income* The net incomes a_n (in millions of dollars) of Wal-Mart for the years 1985 through 1994 are shown in the figure. These incomes can be approximated by the model

$$a_n = 129.9 + 0.9n^3, \qquad n = 5, \ldots, 14$$

where $n = 5$ represents 1985. Use this model to approximate the total net income from 1985 through 1994. Compare this sum with the result of adding the incomes shown in the figure. (Source: Wal-Mart)

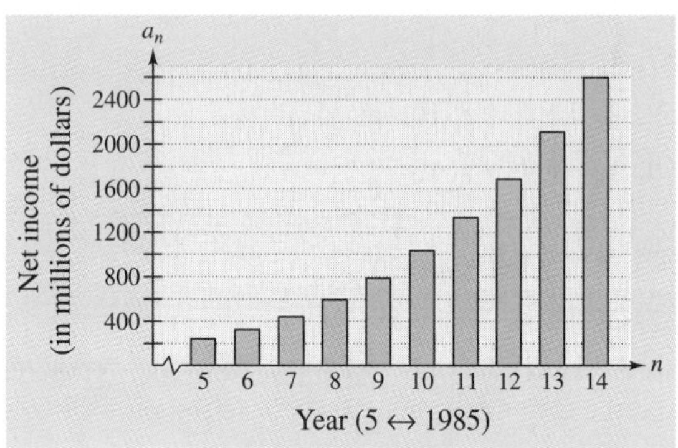

96. *Corporate Dividends* From 1985 through 1994, the dividends a_n declared per share of common stock of Procter & Gamble Company can be approximated by the model

$$a_n = 1.39 + 0.18n - 1.02 \ln n, \quad n = 5, \ldots, 14$$

where $n = 5$ represents 1985. Use this model to approximate the total dividends per share of common stock from 1985 through 1994. (Source: Procter & Gamble Company)

Fibonacci Sequence **In Exercises 97 and 98, use the Fibonacci Sequence. (See Example 4.)**

97. Write the first 12 terms of the Fibonacci sequence a_n and the first 10 terms of the sequence given by

$$b_n = \frac{a_{n+1}}{a_n}, \qquad n > 1.$$

98. Using the definition for b_n in Exercise 97, show that b_n can be defined recursively by

$$b_n = 1 + \frac{1}{b_{n-1}}.$$

7.2 Arithmetic Sequences

Arithmetic Sequences / The Sum of an Arithmetic Sequence / Applications

Arithmetic Sequences

A sequence whose consecutive terms have a common difference is called an **arithmetic sequence.**

EXPLORATION

Consider the following sequences.

$1, 4, 7, 10, 13, \ldots, 3n - 2, \ldots$

$-5, 1, 7, 13, 19, \ldots, 6n - 11, \ldots$

$\frac{5}{2}, \frac{3}{2}, \frac{1}{2}, -\frac{1}{2}, \ldots, \frac{7}{2} - n, \ldots$

What relationship do you observe between successive terms of these sequences? Sequences with this property are called **arithmetic sequences.**

Definition of Arithmetic Sequence

A sequence is **arithmetic** if the differences between consecutive terms are the same. Thus, the sequence

$$a_1, a_2, a_3, a_4, \ldots, a_n, \ldots$$

is arithmetic if there is a number d such that

$$a_2 - a_1 = d, \quad a_3 - a_2 = d, \quad a_4 - a_3 = d$$

and so on. The number d is the **common difference** of the arithmetic sequence.

You can generate the arithmetic sequence in Example 1(a) with a *TI-82* or *TI-83* using the following steps.

7 [ENTER]

4 [+] [ANS]

Now press [ENTER] repeatedly to generate the terms of the sequence.

The *Interactive* CD-ROM offers graphing utility emulators of the *TI-82* and *TI-83*, which can be used with the Examples, Explorations, Technology notes, and Exercises.

EXAMPLE 1 Examples of Arithmetic Sequences

a. The sequence whose nth term is $4n + 3$ is arithmetic. For this sequence, the common difference between consecutive terms is 4.

$$7, 11, 15, 19, \ldots, 4n + 3, \ldots$$

$11 - 7 = 4$

b. The sequence whose nth term is $7 - 5n$ is arithmetic. For this sequence, the common difference between consecutive terms is -5.

$$2, -3, -8, -13, \ldots, 7 - 5n, \ldots$$

$-3 - 2 = -5$

c. The sequence whose nth term is $\frac{1}{4}(n + 3)$ is arithmetic. For this sequence, the common difference between consecutive terms is $\frac{1}{4}$.

$$1, \frac{5}{4}, \frac{3}{2}, \frac{7}{4}, \ldots, \frac{n+3}{4}, \ldots$$

$\frac{5}{4} - 1 = \frac{1}{4}$

Note An arithmetic sequence $a_n = dn + c$ can be thought of as "counting by d's" after a shift of c units from d. For instance, the sequence

$$2, 6, 10, 14, 18, \ldots$$

has a common difference of 4, so you are counting by 4's after a shift of 2 units below 4. Thus, the nth term is $4n - 2$. Similarly, the nth term of the sequence

$$6, 11, 16, 21, \ldots$$

is $5n + 1$ because you are counting by 5's after a shift of 1 unit above 5.

In Example 1, notice that each of the arithmetic sequences has an nth term that is of the form $dn + c$, where the common difference of the sequence is d. This result is summarized as follows.

The nth Term of an Arithmetic Sequence

The nth term of an arithmetic sequence has the form

$$a_n = dn + c$$

where d is the common difference between consecutive terms of the sequence and $c = a_1 - d$.

EXAMPLE 2 Finding the nth Term of an Arithmetic Sequence

Find a formula for the nth term of the arithmetic sequence whose common difference is 3 and whose first term is 2.

Solution

Because the sequence is arithmetic, you know that the formula for the nth term is of the form $a_n = dn + c$. Moreover, because the common difference is $d = 3$, the formula must have the form

$$a_n = 3n + c.$$

Because $a_1 = 2$, it follows that

$$c = a_1 - d = 2 - 3 = -1.$$

Thus, the formula for the nth term is

$$a_n = 3n - 1.$$

The sequence therefore has the following form.

$$2, 5, 8, 11, 14, \ldots, 3n - 1, \ldots$$

Another way to find a formula for the nth term of the sequence in Example 2 is to begin by writing the terms of the sequence.

a_1	a_2	a_3	a_4	a_5	a_6	a_7	. . .
2	2 + 3	5 + 3	8 + 3	11 + 3	14 + 3	17 + 3	. . .
2	5	8	11	14	17	20	. . .

From these terms, you can reason that the nth term is of the form

$$a_n = dn + c = 3n - 1.$$

EXAMPLE 3 **Finding the *n*th Term of an Arithmetic Sequence**

The fourth term of an arithmetic sequence is 20, and the 13th term is 65. Write the first several terms of this sequence.

Solution

The fourth and 13th terms of the sequence are related by

$$a_{13} = a_4 + 9d.$$

Using $a_4 = 20$ and $a_{13} = 65$, you can conclude that $d = 5$, which implies that the sequence is as follows.

a_1	a_2	a_3	a_4	a_5	a_6	a_7	a_8	a_9	a_{10}	a_{11}	
5,	10,	15,	20,	25,	30,	35,	40,	45,	50,	55,	. . .

Study Tip

If you substitute $a_1 - d$ for c in the formula $a_n = dn + c$, the nth term of an arithmetic sequence has the alternative recursive formula

$$a_n = a_1 + (n - 1)d.$$

Use this formula to solve Example 4 and Example 9.

If you know the nth term of an arithmetic sequence *and* you know the common difference of the sequence, you can find the $(n + 1)$th term by using the *recursive formula*

$$a_{n+1} = a_n + d.$$

With this formula, you can find any term of an arithmetic sequence, *provided* that you know the previous term. For instance, if you know the first term, you can find the second term. Then, knowing the second term, you can find the third term, and so on.

EXAMPLE 4 **Using a Recursive Formula**

Find the ninth term of the arithmetic sequence that begins with 2 and 9.

Solution

For this sequence, the common difference is $d = 9 - 2 = 7$. There are two ways to find the ninth term. One way is simply to write out the first nine terms (by repeatedly adding 7).

2, 9, 16, 23, 30, 37, 44, 51, 58

Another way to find the ninth term is first to find a formula for the nth term. Because the first term is 2, it follows that

$$c = a_1 - d = 2 - 7 = -5.$$

Therefore, a formula for the nth term is $a_n = 7n - 5$, which implies that the ninth term is $a_9 = 7(9) - 5 = 58$.

The Sum of an Arithmetic Sequence

There is a simple formula for the *sum* of a finite arithmetic sequence.

Note Be sure you see that this formula works only for *arithmetic* sequences.

The Sum of a Finite Arithmetic Sequence

The sum of a finite arithmetic sequence with n terms is given by

$$S_n = \frac{n}{2}(a_1 + a_n).$$

EXAMPLE 5 **Finding the Sum of an Arithmetic Sequence**

Find the sum: $1 + 3 + 5 + 7 + 9 + 11 + 13 + 15 + 17 + 19$.

Solution

To begin, notice that the sequence is arithmetic (with a common difference of 2). Moreover, the sequence has 10 terms. Thus, the sum of the sequence is

$$\begin{aligned} S_n &= 1 + 3 + 5 + 7 + 9 + 11 + 13 + 15 + 17 + 19 \\ &= \frac{n}{2}(a_1 + a_n) \\ &= \frac{10}{2}(1 + 19) && n = 10 \\ &= 5(20) \\ &= 100. \end{aligned}$$

Think About the Proof

To prove the formula for the sum of a finite arithmetic sequence, consider writing the sum in two different ways. One way is to write the sum as

$$\begin{aligned} S_n = a_1 &+ (a_1 + d) \\ &+ (a_1 + 2d) + \cdots \\ &+ [a_1 + (n - 1)d]. \end{aligned}$$

In the second way, you repeatedly subtract d from the nth term to obtain

$$\begin{aligned} S_n = a_n &+ (a_n - d) \\ &+ (a_n - 2d) + \cdots \\ &+ [a_n - (n - 1)d]. \end{aligned}$$

Can you discover a way to combine these two versions of S_n to obtain the following formula?

$$S_n = \frac{n}{2}(a_1 + a_n)$$

The details of this proof are shown in the appendix.

EXAMPLE 6 **Finding the Sum of an Arithmetic Sequence**

Find the sum of the integers from 1 to 100.

Solution

The integers from 1 to 100 form an arithmetic sequence that has 100 terms. Thus, you can use the formula for the sum of an arithmetic sequence, as follows.

$$\begin{aligned} S_n &= 1 + 2 + 3 + 4 + 5 + 6 + \cdots + 99 + 100 \\ &= \frac{n}{2}(a_1 + a_n) \\ &= \frac{100}{2}(1 + 100) && n = 100 \\ &= 50(101) \\ &= 5050 \end{aligned}$$

EXAMPLE 7 Finding the Sum of an Arithmetic Sequence

Find the sum of the first 150 terms of the arithmetic sequence

$$5, 16, 27, 38, 49, \ldots .$$

Solution

For this arithmetic sequence, you have $a_1 = 5$ and $d = 16 - 5 = 11$. Thus, $c = a_1 - d = 5 - 11 = -6$, and the nth term is

$$a_n = 11n - 6.$$

Therefore, $a_{150} = 11(150) - 6 = 1644$, and the sum of the first 150 terms is

$$\begin{aligned} S_n &= \frac{n}{2}(a_1 + a_n) \\ &= \frac{150}{2}(5 + 1644) \\ &= 75(1649) \\ &= 123{,}675. \end{aligned}$$

Note The sum of the first n terms of an infinite sequence is called the ***n*th partial sum.**

Applications

EXAMPLE 8 Seating Capacity *Real Life*

An auditorium has 20 rows of seats. There are 20 seats in the first row, 21 seats in the second row, 22 seats in the third row, and so on. (See Figure 7.3.) How many seats are there in all 20 rows?

Solution

The numbers of seats in the 20 rows form an arithmetic sequence in which the common difference is $d = 1$. Because $c = a_1 - d = 20 - 1 = 19$, you can determine that the formula for the nth term of the sequence is $a_n = n + 19$. Therefore, the 20th term in the sequence is $a_{20} = 20 + 19 = 39$, and the total number of seats is

$$\begin{aligned} S_n &= 20 + 21 + 22 + \cdots + 39 \\ &= \frac{n}{2}(a_1 + a_{20}) \\ &= \frac{20}{2}(20 + 39) \\ &= 10(59) \\ &= 590. \end{aligned}$$

Figure 7.3

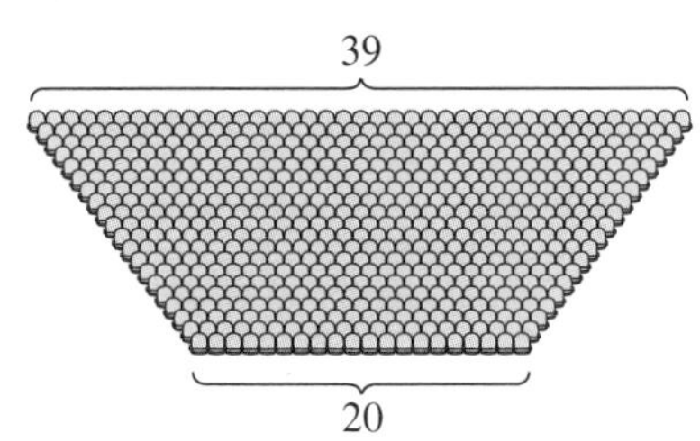

Real Life

EXAMPLE 9 **Total Sales**

A small business sells \$10,000 worth of products during its first year. The owner of the business has set a goal of increasing annual sales by \$7500 each year for 9 years. Assuming that this goal is met, find the total sales during the first 10 years this business is in operation.

Solution

The annual sales form an arithmetic sequence in which $a_1 = 10{,}000$ and $d = 7500$. Thus, $c = a_1 - d = 10{,}000 - 7500 = 2500$, and the nth term of the sequence is

$$a_n = 7500n + 2500.$$

This implies that the 10th term of the sequence is $a_{10} = 77{,}500$. The sum of the first 10 terms of the sequence is

$$\begin{aligned} S_n &= \frac{n}{2}(a_1 + a_{10}) \\ &= \frac{10}{2}(10{,}000 + 77{,}500) \\ &= 5(87{,}500) \\ &= 437{,}500. \end{aligned}$$

Thus, the total sales for the first 10 years is \$437,500.

Group Activity

Numerical Relationships

Decide whether it is possible to fill in the blanks in each of the following such that the resulting sequence is arithmetic. If so, find a recursive formula for the sequence.

a. −7, __, __, __, __, __, 11

b. 17, __, __, __, __, __, __, __, __, 71

c. 2, 6, __, __, 162

d. 4, 7.5, __, __, __, __, __, __, __, __, 39

e. 8, 12, __, __, __, 60.75

7.2 /// EXERCISES

In Exercises 1–10, determine whether the sequence is arithmetic. If it is, find the common difference.

1. $10, 8, 6, 4, 2, \ldots$
2. $4, 7, 10, 13, 16, \ldots$
3. $1, 2, 4, 8, 16, \ldots$
4. $3, \frac{5}{2}, 2, \frac{3}{2}, 1, \ldots$
5. $\frac{9}{4}, 2, \frac{7}{4}, \frac{3}{2}, \frac{5}{4}, \ldots$
6. $\frac{1}{3}, \frac{2}{3}, \frac{4}{3}, \frac{8}{3}, \frac{16}{3}, \ldots$
7. $-12, -8, -4, 0, 4, \ldots$
8. $\ln 1, \ln 2, \ln 3, \ln 4, \ln 5, \ldots$
9. $5.3, 5.7, 6.1, 6.5, 6.9, \ldots$
10. $1^2, 2^2, 3^2, 4^2, 5^2, \ldots$

In Exercises 11–18, write the first five terms of the sequence. Determine whether the sequence is arithmetic, and if it is, find the common difference.

11. $a_n = 5 + 3n$
12. $a_n = (2^n)n$
13. $a_n = \dfrac{1}{n + 1}$
14. $a_n = 1 + (n - 1)4$
15. $a_n = 100 - 3n$
16. $a_n = 2^{n-1}$
17. $a_n = 3 + \dfrac{(-1)^n 2}{n}$
18. $a_n = (-1)^n$

In Exercises 19–24, write the first five terms of the arithmetic sequence. Find the common difference and write the *n*th term of the sequence as a function of *n*.

19. $a_1 = 15, \quad a_{k+1} = a_k + 4$
20. $a_1 = 2, \quad a_{k+1} = a_k + \frac{2}{3}$
21. $a_1 = 200, \quad a_{k+1} = a_k - 10$
22. $a_1 = 72, \quad a_{k+1} = a_k - 6$
23. $a_1 = \frac{3}{2}, \quad a_{k+1} = a_k - \frac{1}{4}$
24. $a_1 = 0.375, \quad a_{k+1} = a_k + 0.25$

In Exercises 25–32, write the first five terms of the arithmetic sequence.

25. $a_1 = 5, d = 6$
26. $a_1 = 5, d = -\frac{3}{4}$
27. $a_1 = -2.6, d = -0.4$
28. $a_1 = 16.5, d = 0.25$
29. $a_1 = 2, a_{12} = 46$
30. $a_4 = 16, a_{10} = 46$
31. $a_8 = 26, a_{12} = 42$
32. $a_3 = 19, a_{15} = -1.7$

In Exercises 33–44, find a formula for a_n for the arithmetic sequence.

33. $a_1 = 1, d = 3$
34. $a_1 = 15, d = 4$
35. $a_1 = 100, d = -8$
36. $a_1 = 0, d = -\frac{2}{3}$
37. $a_1 = x, d = 2x$
38. $a_1 = -y, d = 5y$
39. $4, \frac{3}{2}, -1, -\frac{7}{2}, \ldots$
40. $10, 5, 0, -5, -10, \ldots$
41. $a_1 = 5, a_4 = 15$
42. $a_1 = -4, a_5 = 16$
43. $a_3 = 94, a_6 = 85$
44. $a_5 = 190, a_{10} = 115$

In Exercises 45–48, match the sequence with its graph. [The graphs are labeled (a), (b), (c), and (d).]

(a)
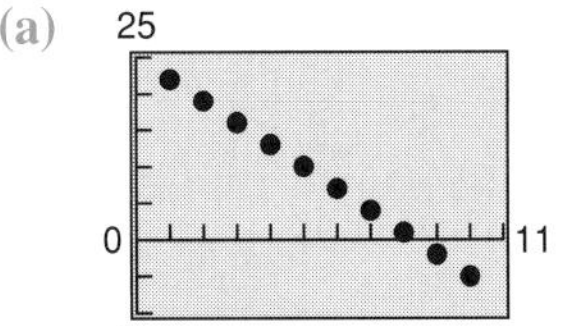

(b)
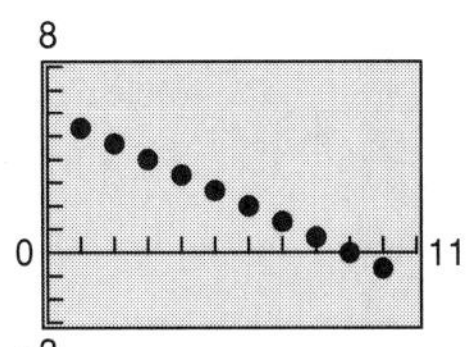

(c)
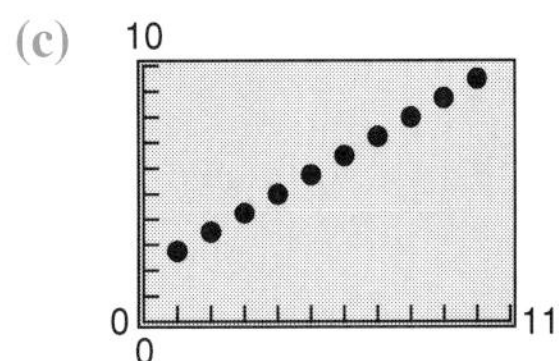

(d)
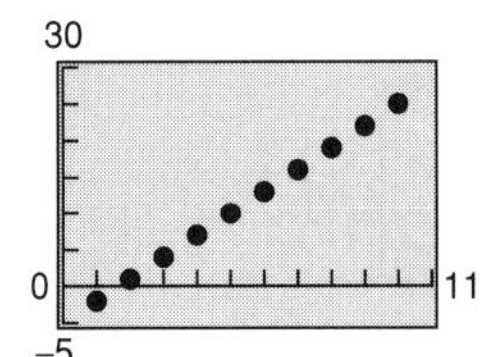

45. $a_n = -\frac{2}{3}n + 6$
46. $a_n = 3n - 5$
47. $a_n = 2 + \frac{3}{4}n$
48. $a_n = 25 - 3n$

In Exercises 49–52, use a graphing utility to graph the first 10 terms of the sequence.

49. $a_n = 15 - \frac{3}{2}n$
50. $a_n = -5 + 2n$
51. $a_n = 0.2n + 3$
52. $a_n = -0.3n + 8$

53. *Think About It* Describe the geometric pattern of the graph of an arithmetic sequence. Explain.

54. *Think About It* Explain how to use the first two terms of an arithmetic sequence to find the nth term.

In Exercises 55–62, find the nth partial sum of the arithmetic sequence.

55. 8, 20, 32, 44, . . . , $n = 10$
56. 2, 8, 14, 20, . . . , $n = 25$
57. $-6, -2, 2, 6, \ldots, \quad n = 50$
58. 0.5, 0.9, 1.3, 1.7, . . . , $n = 10$
59. 40, 37, 34, 31, . . . , $n = 10$
60. 1.50, 1.45, 1.40, 1.35, . . . , $n = 20$
61. $a_1 = 100, \quad a_{25} = 220, \quad n = 25$
62. $a_1 = 15, \quad a_{100} = 307, \quad n = 100$

In Exercises 63–70, find the sum.

63. $\sum_{n=1}^{50} n$
64. $\sum_{n=1}^{100} 2n$
65. $\sum_{n=1}^{100} 5n$
66. $\sum_{n=51}^{100} 7n$
67. $\sum_{n=11}^{30} n - \sum_{n=1}^{10} n$
68. $\sum_{n=51}^{100} n - \sum_{n=1}^{50} n$
69. $\sum_{n=1}^{500} (n + 3)$
70. $\sum_{n=1}^{250} (1000 - n)$

In Exercises 71–76, use a calculator to find the sum.

71. $\sum_{n=1}^{20} (2n + 5)$
72. $\sum_{n=1}^{100} \frac{n + 4}{2}$
73. $\sum_{n=0}^{50} (1000 - 5n)$
74. $\sum_{n=0}^{100} \frac{8 - 3n}{16}$
75. $\sum_{i=1}^{60} \left(250 - \frac{8}{3}i\right)$
76. $\sum_{j=1}^{200} (4.5 + 0.025j)$

77. Find the sum of the first 100 positive odd integers.

78. Find the sum of the integers from -10 to 50.

Job Offer **In Exercises 79 and 80, consider a job offer with the given starting salary and guaranteed salary increase for the first 5 years of employment.**

(a) Determine the person's salary during the sixth year of employment.

(b) Determine the person's total compensation from the company through 6 full years of employment.

	Starting Salary	*Annual Raise*
79.	\$32,500	\$1500
80.	\$36,800	\$1750

81. *Seating Capacity* Determine the seating capacity of an auditorium with 30 rows of seats if there are 20 seats in the first row, 24 seats in the second row, 28 seats in the third row, and so on.

82. *Seating Capacity* Determine the seating capacity of an auditorium with 36 rows of seats if there are 15 seats in the first row, 18 seats in the second row, 21 seats in the third row, and so on.

83. *Brick Pattern* A brick patio has the approximate shape of a trapezoid (see figure). The patio has 18 rows of bricks. The first row has 14 bricks and the 18th row has 31 bricks. How many bricks are in the patio?

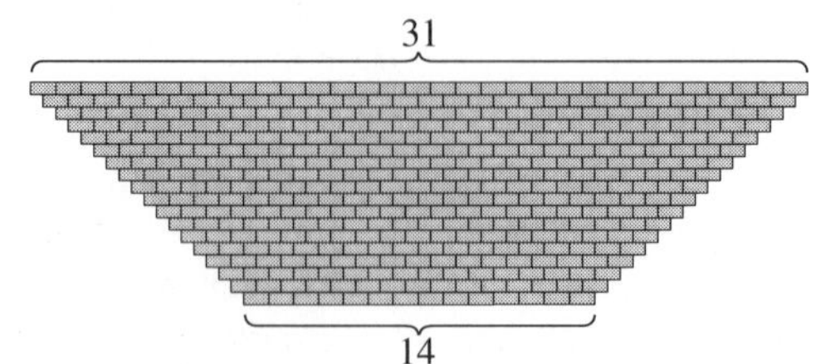

84. *Number of Logs* Logs are stacked in a pile as shown in the figure. The top row has 15 logs and the bottom row has 21 logs. How many logs are in the stack?

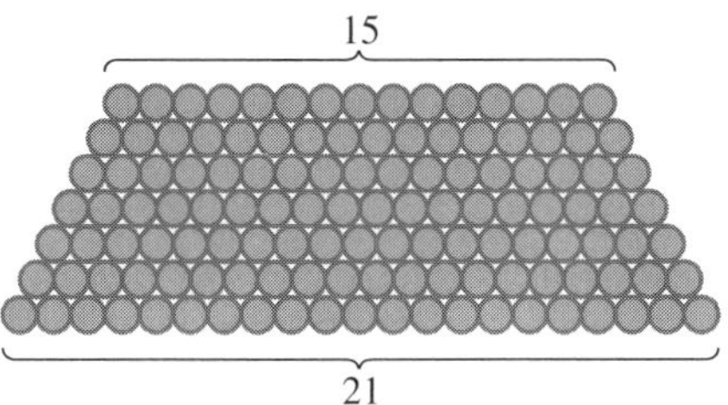

85. *Auditorium Seating* Each row in a small auditorium has two more seats than the preceding row (see figure). Find the seating capacity of the auditorium if the front row seats 25 people and there are 15 rows of seats.

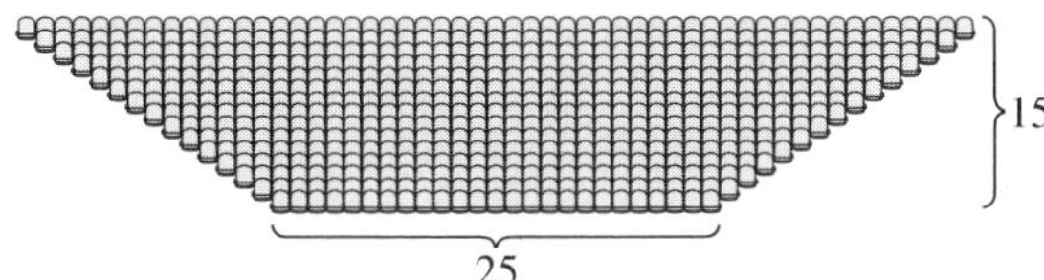

86. *Baling Hay* In the first two trips around a field baling hay, a farmer makes 93 bales and 89 bales, respectively (see figure). Because each trip is shorter than the preceding trip, the farmer estimates that the same pattern will continue. Estimate the total number of bales made if there are another six trips around the field.

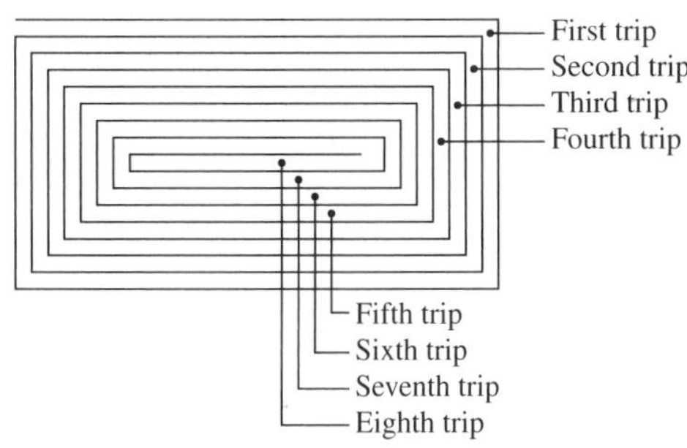

87. *Grandfather Clock* Each hour, a grandfather clock strikes the number of times corresponding to the hour of the day. How many times does the clock strike in a day?

88. *Falling Object* An object (with negligible air resistance) is dropped from an airplane. During the first second of fall, the object falls 4.9 meters; during the second second, it falls 14.7 meters; during the third second, it falls 24.5 meters; during the fourth second, it falls 34.3 meters. If this arithmetic pattern continues, how many meters will the object have fallen after 10 seconds?

89. *Pattern Recognition*

(a) Compute the following five sums of positive odd integers.

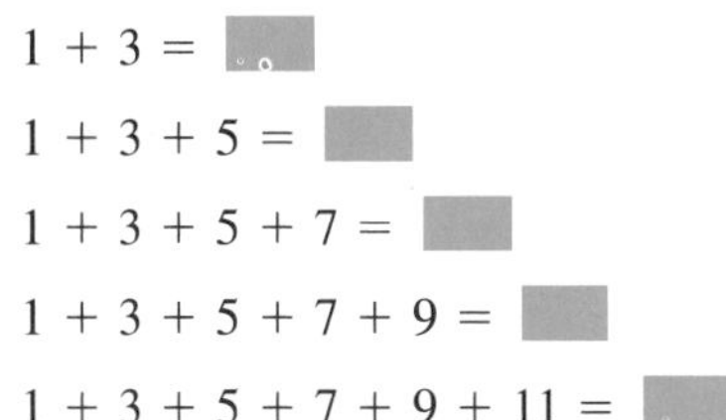

(b) Use the sums in part (a) to make a conjecture about the sums of positive odd integers. Check your conjecture for the sum

$1 + 3 + 5 + 7 + 9 + 11 + 13 = \square.$

(c) Verify your conjecture analytically.

90. *Think About It* In each of the following, an operation is performed on each term of an arithmetic sequence. Determine if the resulting sequence is arithmetic and, if so, state the common difference.

(a) A constant C is added to each term.

(b) Each term is multiplied by a nonzero constant C.

(c) Each term is squared.

91. *Think About It* The sum of the first 20 terms of an arithmetic sequence with a common difference of 3 is 650. Find the first term.

92. *Think About It* The sum of the first n terms of an arithmetic sequence with first term a_1 and common difference d is S_n. Determine the sum if the first term is increased by 5. Explain.

7.3 Geometric Sequences

Geometric Sequences / The Sum of a Geometric Sequence / Application

EXPLORATION

Consider the following three sequences.

$$1, 3, 9, 27, 81, 243, \ldots, \frac{3^n}{3}, \ldots$$

$$-2, 4, -8, 16, -32, 64, \ldots, (-2)^n, \ldots$$

$$\frac{1}{2}, \frac{1}{4}, \frac{1}{8}, \frac{1}{16}, \ldots, \left(\frac{1}{2}\right)^n, \ldots$$

What relationship do you observe between successive terms of these sequences? Sequences with this property are called **geometric sequences.**

Geometric Sequences

In Section 7.2, you learned that a sequence whose consecutive terms have a common *difference* is an arithmetic sequence. In this section, you will study another important type of sequence called a **geometric sequence.** Consecutive terms of a geometric sequence have a common *ratio.*

Definition of Geometric Sequence

A sequence is **geometric** if the ratios of consecutive terms are the same.

$$\frac{a_2}{a_1} = r, \quad \frac{a_3}{a_2} = r, \quad \frac{a_4}{a_3} = r, \quad \ldots, \qquad r \neq 0$$

The number r is the **common ratio** of the sequence.

EXAMPLE 1 **Examples of Geometric Sequences**

a. The sequence whose nth term is 2^n is geometric. For this sequence, the common ratio between consecutive terms is 2.

$$2, 4, 8, 16, \ldots, 2^n, \ldots$$

$$\tfrac{4}{2} = 2$$

b. The sequence whose nth term is $4(3^n)$ is geometric. For this sequence, the common ratio between consecutive terms is 3.

$$12, 36, 108, 324, \ldots, 4(3^n), \ldots$$

$$\tfrac{36}{12} = 3$$

c. The sequence whose nth term is $\left(-\frac{1}{3}\right)^n$ is geometric. For this sequence, the common ratio between consecutive terms is $-\frac{1}{3}$.

$$-\frac{1}{3}, \frac{1}{9}, -\frac{1}{27}, \frac{1}{81}, \ldots, \left(-\frac{1}{3}\right)^n, \ldots$$

$$\frac{1/9}{-1/3} = -\frac{1}{3}$$

In Example 1, notice that each of the geometric sequences has an nth term that is of the form ar^n, where the common ratio of the sequence is r.

The nth Term of a Geometric Sequence

The nth term of a geometric sequence has the form

$$a_n = a_1 r^{n-1}$$

where r is the common ratio of consecutive terms of the sequence. Thus, every geometric sequence can be written in the following form.

$$\begin{array}{ccccccc} a_1, & a_2, & a_3, & a_4, & a_5, \ldots, & a_n, & \ldots \\ \downarrow & \downarrow & \downarrow & \downarrow & \downarrow \ldots & \downarrow & \ldots \\ a_1, & a_1 r, & a_1 r^2, & a_1 r^3, & a_1 r^4, \ldots, & a_1 r^{n-1}, & \ldots \end{array}$$

Note If you know the nth term of a geometric sequence, you can find the $(n+1)$th term by multiplying by r. That is, $a_{n+1} = ra_n$.

You can generate the geometric sequence in Example 2 with a graphing utility using the following steps.

3 [ENTER]

2 [×] [ANS]

Now press [ENTER] repeatedly to generate the terms of the sequence.

EXAMPLE 2 **Finding the Terms of a Geometric Sequence**

Write the first five terms of the geometric sequence whose first term is $a_1 = 3$ and whose common ratio is $r = 2$.

Solution

Starting with 3, repeatedly multiply by 2 to obtain the following.

$a_1 = 3$ — 1st term

$a_2 = 3(2^1) = 6$ — 2nd term

$a_3 = 3(2^2) = 12$ — 3rd term

$a_4 = 3(2^3) = 24$ — 4th term

$a_5 = 3(2^4) = 48$ — 5th term

EXAMPLE 3 **Finding a Term of a Geometric Sequence**

Find the 15th term of the geometric sequence whose first term is 20 and whose common ratio is 1.05.

Solution

$a_{15} = a_1 r^{n-1}$ — Formula for geometric sequence

$= 20(1.05)^{15-1}$ — Substitute for a_1, r, and n.

≈ 39.599 — Use a calculator.

EXAMPLE 4 **Finding a Term of a Geometric Sequence**

Find the 12th term of the geometric sequence

5, 15, 45, ...

Solution

The common ratio of this sequence is $r = 15/5 = 3$. Because the first term is $a_1 = 5$, you can determine the 12th term ($n = 12$) to be

$$\begin{aligned} a_{12} &= a_1 r^{n-1} && \text{Formula for geometric sequence} \\ &= 5(3)^{12-1} && \text{Substitute for } a_1, r, \text{ and } n. \\ &= 5(177{,}147) && \text{Use a calculator.} \\ &= 885{,}735. && \text{Simplify.} \end{aligned}$$

If you know any two terms of a geometric sequence, you can use that information to find a formula for the nth term of the sequence.

Study Tip

Remember that r is the common ratio of consecutive terms of a geometric sequence. So, in Example 5,

$$\begin{aligned} a_{10} &= a_1 r^9 \\ &= a_1 \cdot r \cdot r \cdot r \cdot r^6 \\ &= a_1 \cdot \frac{a_2}{a_1} \cdot \frac{a_3}{a_2} \cdot \frac{a_4}{a_3} \cdot r^6 \\ &= a_4 r^6. \end{aligned}$$

EXAMPLE 5 **Finding a Term of a Geometric Sequence**

The fourth term of a geometric sequence is 125, and the 10th term is 125/64. Find the 14th term. (Assume that the terms of the sequence are positive.)

Solution

The 10th term is related to the fourth term by the equation

$$a_{10} = a_4 r^6. \qquad \text{Multiply 4th term by } r^{10-4}.$$

Because $a_{10} = 125/64$ and $a_4 = 125$, you can solve for r as follows.

$$\frac{125}{64} = 125r^6$$

$$\frac{1}{64} = r^6$$

$$\frac{1}{2} = r$$

You can obtain the 14th term by multiplying the 10th term by r^4.

$$\begin{aligned} a_{14} &= a_{10} r^4 \\ &= \frac{125}{64}\left(\frac{1}{2}\right)^4 = \frac{125}{1024} \end{aligned}$$

The Sum of a Geometric Sequence

The formula for the sum of a *finite* geometric sequence is as follows.

The Sum of a Finite Geometric Sequence

The sum of the geometric sequence

$$a_1, a_1r, a_1r^2, a_1r^3, a_1r^4, \ldots, a_1r^{n-1}$$

with common ratio $r \neq 1$ is given by

$$S_n = a_1\left(\frac{1 - r^n}{1 - r}\right).$$

Think About the Proof

To prove the formula for the sum of a finite geometric sequence, consider the following two equations.

$$S_n = a_1 + a_1r + a_1r^2 + \cdots + a_1r^{n-2} + a_1r^{n-1}$$

$$rS_n = a_1r + a_1r^2 + a_1r^3 + \cdots + a_1r^{n-1} + a_1r^n$$

Can you discover a way to simplify the difference of these two equations to obtain the formula for the sum of a geometric sequence? The details of the proof are given in the appendix.

EXAMPLE 6 **Finding the Sum of a Finite Geometric Sequence**

Find the sum $\sum_{n=1}^{12} 4(0.3)^n$.

Solution

By writing out a few terms, you have

$$\sum_{n=1}^{12} 4(0.3)^n = 4(0.3) + 4(0.3)^2 + 4(0.3)^3 + \cdots + 4(0.3)^{12}.$$

Now, because $a_1 = 4(0.3)$, $r = 0.3$, and $n = 12$, you can apply the formula for the sum of a finite geometric sequence to obtain

$$\sum_{n=1}^{12} 4(0.3)^n = a_1\left(\frac{1 - r^n}{1 - r}\right)$$

$$= 4(0.3)\left[\frac{1 - (0.3)^{12}}{1 - 0.3}\right]$$

$$\approx 1.714.$$

Using a *TI-82* or *TI-83* graphing calculator, you can calculate the sum of the sequence in Example 6 as

sum(seq(4(.3)∧N, N, 1, 12, 1)) = 1.7142848.

Calculate the sum beginning at $n = 0$. You should obtain a sum of 5.7142848.

When using the formula for the sum of a geometric sequence, be careful to check that the index begins at $i = 1$. If the index begins at $i = 0$, you must adjust the formula for the nth partial sum. For instance, if the index in Example 6 had begun with $n = 0$, the sum would have been

$$\sum_{n=0}^{12} 4(0.3)^n = 4 + \sum_{n=1}^{12} 4(0.3)^n \approx 4 + 1.714 = 5.714.$$

Study Tip

Notice that the formula for the sum of an infinite geometric sequence requires that $|r| < 1$. What happens if $r = 1$ or $r = -1$? Give examples of infinite geometric sequences for which $|r| > 1$ and convince yourself that they do not have finite sums.

The formula for the sum of a *finite* geometric sequence can, depending on the value of r, be extended to produce a formula for the sum of an *infinite* geometric sequence. Specifically, if the common ratio r has the property that $|r| < 1$, it can be shown that r^n becomes arbitrarily close to zero as n increases without bound. Consequently,

$$a_1\left(\frac{1-r^n}{1-r}\right) \longrightarrow a_1\left(\frac{1-0}{1-r}\right) \quad \text{as} \quad n \longrightarrow \infty.$$

This result is summarized as follows.

The Sum of an Infinite Geometric Sequence

If $|r| < 1$, the infinite geometric sequence

$$a_1, a_1r, a_1r^2, a_1r^3, \ldots, a_1r^{n-1}, \ldots$$

has the sum

$$S = \frac{a_1}{1-r}.$$

Use a graphing utility to create a table showing the partial sums of the sequence in Example 7(a) by entering

Un = sum(seq(4(.6)∧(N−1), N, 1, n, 1)).

How can you determine the upper limit of the sequence from the table?

EXAMPLE 7 **Finding the Sum of an Infinite Geometric Sequence**

Find the sum of each infinite geometric sequence.

a. $\displaystyle\sum_{n=1}^{\infty} 4(0.6)^{n-1}$ **b.** $\displaystyle\sum_{n=1}^{\infty} 3(0.1)^{n-1}$

Solution

a. $$\sum_{n=1}^{\infty} 4(0.6)^{n-1} = 4 + 4(0.6) + 4(0.6)^2 + 4(0.6)^3 + \cdots + 4(0.6)^{n-1} + \cdots$$

$$= \frac{4}{1-(0.6)} \qquad \frac{a_1}{1-r}$$

$$= 10$$

b. $$\sum_{n=1}^{\infty} 3(0.1)^{n-1} = 3 + 3(0.1) + 3(0.1)^2 + 3(0.1)^3 + \cdots + 3(0.1)^{n-1} + \cdots$$

$$= \frac{3}{1-(0.1)} \qquad \frac{a_1}{1-r}$$

$$= \frac{10}{3}$$

$$\approx 3.33$$

Application

Note The type of investment account in Example 8 is an *annuity*.

Real Life

EXAMPLE 8 **Compound Interest**

A deposit of \$50 is made on the first day of each month in a savings account that pays 6% compounded monthly. What is the balance at the end of 2 years?

Solution

The first deposit will gain interest for 24 months, and its balance will be

$$A_{24} = 50\left(1 + \frac{0.06}{12}\right)^{24} = 50(1.005)^{24}.$$

The second deposit will gain interest for 23 months, and its balance will be

$$A_{23} = 50\left(1 + \frac{0.06}{12}\right)^{23} = 50(1.005)^{23}.$$

The last deposit will gain interest for only 1 month, and its balance will be

$$A_1 = 50\left(1 + \frac{0.06}{12}\right)^{1} = 50(1.005).$$

The total balance in the account will be the sum of the balances of the 24 deposits. Using the formula for the sum of a finite geometric sequence, with $A_1 = 50(1.005)$ and $r = 1.005$, you have

$$S_n = 50(1.005)\left[\frac{1 - (1.005)^{24}}{1 - 1.005}\right] = \$1277.96.$$

Group Activity *An Experiment*

You will need a piece of string or yarn, a pair of scissors, and a tape measure. Measure out any length of string at least 5 feet long. Double over the string and cut it in half. Take one of the resulting halves, double it over, and cut it in half. Continue this process until you are no longer able to cut a length of string in half. How many cuts were you able to make? Construct a sequence of the resulting string lengths after each cut, starting with the original length of the string. Find a formula for the *n*th term of this sequence. How many cuts could you theoretically make? Discuss why you were not able to make that many cuts.

7.3 /// EXERCISES

In Exercises 1–10, determine whether the sequence is geometric. If it is, find the common ratio.

1. $5, 15, 45, 135, \ldots$

2. $3, 12, 48, 192, \ldots$

3. $3, 12, 21, 30, \ldots$

4. $1, -2, 4, -8, \ldots$

5. $1, -\frac{1}{2}, \frac{1}{4}, -\frac{1}{8}, \ldots$

6. $5, 1, 0.2, 0.04, \ldots$

7. $\frac{1}{2}, \frac{2}{3}, \frac{3}{4}, \frac{4}{5}, \ldots$

8. $9, -6, 4, -\frac{8}{3}, \ldots$

9. $1, \frac{1}{2}, \frac{1}{3}, \frac{1}{4}, \ldots$

10. $\frac{1}{5}, \frac{2}{7}, \frac{3}{9}, \frac{4}{11}, \ldots$

In Exercises 11–20, write the first five terms of the geometric sequence.

11. $a_1 = 2, r = 3$

12. $a_1 = 6, r = 2$

13. $a_1 = 1, r = \frac{1}{2}$

14. $a_1 = 1, r = \frac{1}{3}$

15. $a_1 = 5, r = -\frac{1}{10}$

16. $a_1 = 6, r = -\frac{1}{4}$

17. $a_1 = 1, r = e$

18. $a_1 = 2, r = \sqrt{3}$

19. $a_1 = 3, r = \dfrac{x}{2}$

20. $a_1 = 5, r = 2x$

In Exercises 21–26, write the first five terms of the geometric sequence. Determine the common ratio and write the *n*th term of the sequence as a function of *n*.

21. $a_1 = 64, \quad a_{k+1} = \frac{1}{2}a_k$

22. $a_1 = 81, \quad a_{k+1} = \frac{1}{3}a_k$

23. $a_1 = 4, \quad a_{k+1} = 3a_k$

24. $a_1 = 5, \quad a_{k+1} = -2a_k$

25. $a_1 = 6, \quad a_{k+1} = -\frac{3}{2}a_k$

26. $a_1 = 36, \quad a_{k+1} = -\frac{2}{3}a_k$

In Exercises 27–38, find the *n*th term of the geometric sequence.

27. $a_1 = 4, r = \frac{1}{2}, n = 10$

28. $a_1 = 5, r = \frac{3}{2}, n = 8$

29. $a_1 = 6, r = -\frac{1}{3}, n = 12$

30. $a_1 = 8, r = \sqrt{5}, n = 9$

31. $a_1 = 100, r = e^x, n = 9$

32. $a_1 = 1, r = -\dfrac{x}{3}, n = 7$

33. $a_1 = 500, r = 1.02, n = 40$

34. $a_1 = 1000, r = 1.005, n = 60$

35. $a_1 = 16, a_4 = \frac{27}{4}, n = 3$

36. $a_2 = 3, a_5 = \frac{3}{64}, n = 1$

37. $a_2 = -18, a_5 = \frac{2}{3}, n = 6$

38. $a_3 = \frac{16}{3}, a_5 = \frac{64}{27}, n = 7$

In Exercises 39–42, match the sequence with its graph. [The graphs are labeled (a), (b), (c), and (d).]

(a)

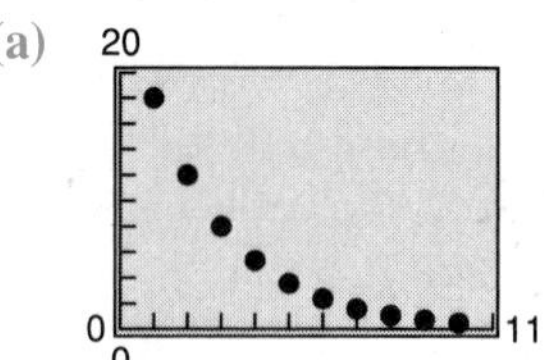

(b)

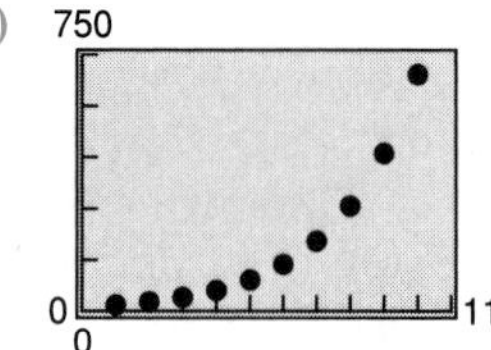

(c)

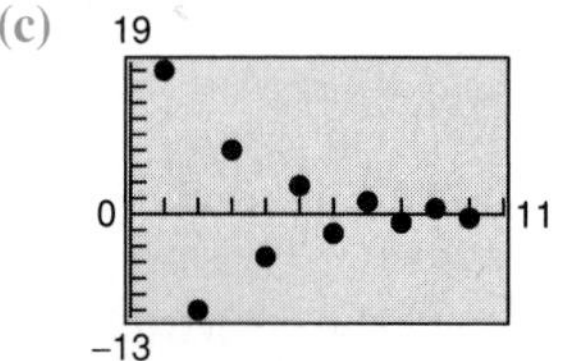

(d)

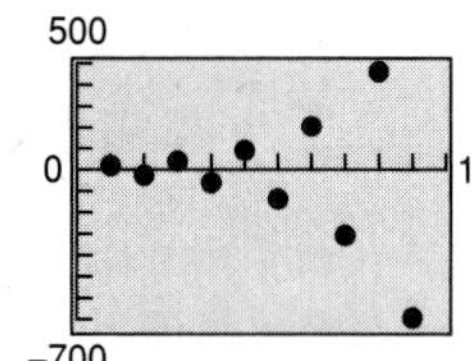

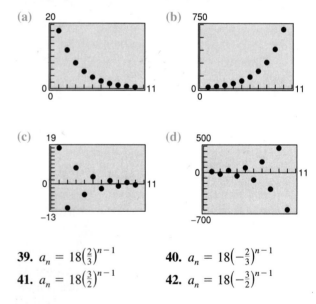

39. $a_n = 18\left(\frac{2}{3}\right)^{n-1}$

40. $a_n = 18\left(-\frac{2}{3}\right)^{n-1}$

41. $a_n = 18\left(\frac{3}{2}\right)^{n-1}$

42. $a_n = 18\left(-\frac{3}{2}\right)^{n-1}$

In Exercises 43–46, use a graphing utility to graph the first 10 terms of the sequence.

43. $a_n = 12(-0.75)^{n-1}$

44. $a_n = 12(-0.4)^{n-1}$

45. $a_n = 2(1.3)^{n-1}$

46. $a_n = 2(-1.4)^{n-1}$

47. *Essay* Write a brief paragraph explaining why the terms of a geometric sequence decrease in magnitude when $-1 < r < 1$.

48. *Essay* Write a brief paragraph explaining how to use the first two terms of a geometric sequence to find the nth term.

49. *Compound Interest* A principal of \$1000 is invested at 10% interest. Find the amount after 10 years if the interest is compounded (a) annually, (b) semiannually, (c) quarterly, (d) monthly, and (e) daily.

50. *Compound Interest* A principal of \$2500 is invested at 12% interest. Find the amount after 20 years if the interest is compounded (a) annually, (b) semiannually, (c) quarterly, (d) monthly, and (e) daily.

51. *Depreciation* A company buys a machine for \$135,000 and it depreciates at a rate of 30% per year. (In other words, at the end of each year the depreciated value is 70% of what it was at the beginning of the year.) Find the depreciated value of the machine after 5 full years.

52. *Population Growth* A city of 250,000 people is growing at a rate of 1.3% per year. Estimate the population of the city 30 years from now.

In Exercises 53 and 54, find the first four terms of the sequence of partial sums of the geometric sequence. In a sequence of partial sums, the term S_n is the sum of the first n terms of the sequence. For instance, S_2 is the sum of the first two terms.

53. $8, -4, 2, -1, \frac{1}{2}, \ldots$

54. $8, 12, 18, 27, \frac{81}{2}, \ldots$

In Exercises 55–64, find the sum.

55. $\sum_{n=1}^{9} 2^{n-1}$

56. $\sum_{n=1}^{9} (-2)^{n-1}$

57. $\sum_{i=1}^{7} 64\left(-\frac{1}{2}\right)^{i-1}$

58. $\sum_{i=1}^{6} 32\left(\frac{1}{4}\right)^{i-1}$

59. $\sum_{n=0}^{20} 3\left(\frac{3}{2}\right)^{n}$

60. $\sum_{n=0}^{15} 2\left(\frac{4}{3}\right)^{n}$

61. $\sum_{i=1}^{10} 8\left(-\frac{1}{4}\right)^{i-1}$

62. $\sum_{i=1}^{10} 5\left(-\frac{1}{3}\right)^{i-1}$

63. $\sum_{n=0}^{5} 300(1.06)^{n}$

64. $\sum_{n=0}^{6} 500(1.04)^{n}$

In Exercises 65 and 66, use summation notation to express the sum.

65. $5 + 15 + 45 + \cdots + 3645$

66. $2 - \frac{1}{2} + \frac{1}{8} - \cdots + \frac{1}{2048}$

67. *Annuities* A deposit of \$100 is made at the beginning of each month in an account that pays 10% interest, compounded monthly. The balance A in the account at the end of 5 years is given by

$$A = 100\left(1 + \frac{0.10}{12}\right)^{1} + \cdots + 100\left(1 + \frac{0.10}{12}\right)^{60}.$$

Find A.

68. *Annuities* A deposit of \$50 is made at the beginning of each month in an account that pays 12% interest, compounded monthly. The balance A in the account at the end of 5 years is given by

$$A = 50\left(1 + \frac{0.12}{12}\right)^{1} + \cdots + 50\left(1 + \frac{0.12}{12}\right)^{60}.$$

Find A.

69. *Annuities* A deposit of P dollars is made at the beginning of each month in an account earning an annual interest rate r, compounded monthly. The balance A after t years is

$$A = P\left(1 + \frac{r}{12}\right) + P\left(1 + \frac{r}{12}\right)^{2} + \cdots + P\left(1 + \frac{r}{12}\right)^{12t}.$$

Show that the balance is given by

$$A = P\left[\left(1 + \frac{r}{12}\right)^{12t} - 1\right]\left(1 + \frac{12}{r}\right).$$

70. *Annuities* A deposit of P dollars is made at the beginning of each month in an account earning an annual interest rate r, compounded continuously. The balance A after t years is

$$A = Pe^{r/12} + Pe^{2r/12} + \cdots + Pe^{12tr/12}.$$

Show that the balance is given by

$$A = \frac{Pe^{r/12}(e^{rt} - 1)}{e^{r/12} - 1}.$$

Annuities In Exercises 71–74, consider making monthly deposits of P dollars in a savings account earning an annual interest rate r. Use the results of Exercises 69 and 70 to find the balance A after t years if the interest is compounded (a) monthly and (b) continuously.

71. $P = \$50,\ r = 7\%,\ t = 20$ years

72. $P = \$75,\ r = 9\%,\ t = 25$ years

73. $P = \$100,\ r = 10\%,\ t = 40$ years

74. $P = \$20,\ r = 6\%,\ t = 50$ years

75. *Annuities* Consider an initial deposit of P dollars in an account earning an annual interest rate r, compounded monthly. At the end of each month, a withdrawal of W dollars will occur and the account will be depleted in t years. The amount of the initial deposit required is given by

$$P = W\left(1 + \frac{r}{12}\right)^{-1} + W\left(1 + \frac{r}{12}\right)^{-2} + \cdots + W\left(1 + \frac{r}{12}\right)^{-12t}.$$

Show that the initial deposit is given by

$$P = W\left(\frac{12}{r}\right)\left[1 - \left(1 + \frac{r}{12}\right)^{-12t}\right].$$

76. *Annuities* Determine the amount required in an individual retirement account for an individual who retires at age 65 and wants an income of \$2000 from the account each month for 20 years. Use the result of Exercise 75, and assume that the account earns 9% compounded monthly.

77. *Geometry* The sides of a square are 16 inches in length. A new square is formed by connecting the midpoints of the sides of the original square, and two of the triangles are shaded (see figure). If this process is repeated five more times, determine the total area of the shaded region.

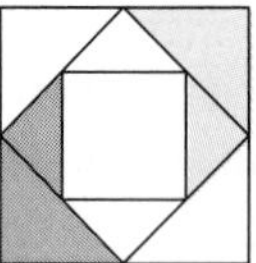

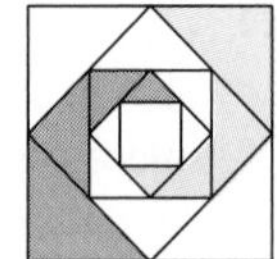

78. *Corporate Revenue* The annual revenues a_n (in billions of dollars) for the Coca-Cola Company for 1985 through 1994 can be approximated by the model

$$a_n = 3.49e^{0.108n}, \quad n = 5, 6, 7, \ldots, 14$$

where $n = 5$ represents 1985. Use this model and the formula for the sum of a geometric sequence to approximate the total revenue earned during this 10-year period. (Source: The Coca-Cola Company)

79. *Think About It* Suppose you go to work for a company that pays \$0.01 the first day, \$0.02 the second day, \$0.04 the third day, and so on. If the daily wage keeps doubling, what will your total income be for working (a) 29 days? (b) 30 days? (c) 31 days?

80. *Salary* A company is offering a job with a salary of \$30,000 for the first year. Suppose that during the next 39 years, there is a 5% raise each year. Determine the total compensation over the 40-year period.

In Exercises 81–88, find the sum.

81. $\displaystyle\sum_{n=0}^{\infty} \left(\tfrac{1}{2}\right)^n$

82. $\displaystyle\sum_{n=0}^{\infty} 2\left(\tfrac{2}{3}\right)^n$

83. $\displaystyle\sum_{n=0}^{\infty} \left(-\tfrac{1}{2}\right)^n$

84. $\displaystyle\sum_{n=0}^{\infty} 2\left(-\tfrac{2}{3}\right)^n$

85. $\displaystyle\sum_{n=0}^{\infty} 4\left(\tfrac{1}{4}\right)^n$

86. $\displaystyle\sum_{n=0}^{\infty} \left(\tfrac{1}{10}\right)^n$

87. $8 + 6 + \frac{9}{2} + \frac{27}{8} + \cdots$

88. $3 - 1 + \frac{1}{3} - \frac{1}{9} + \cdots$

In Exercises 89–92, find the rational number representation of the repeating decimal.

89. $0.\overline{36}$ **90.** $0.\overline{297}$

91. $0.3\overline{18}$ **92.** $1.3\overline{8}$

Graphical Reasoning **In Exercises 93 and 94, use a graphing utility to graph the function. Identify the horizontal asymptote of the graph and determine its relationship to the sum.**

93. $f(x) = 6\left[\dfrac{1 - (0.5)^x}{1 - (0.5)}\right], \quad \displaystyle\sum_{n=0}^{\infty} 6\left(\frac{1}{2}\right)^n$

94. $f(x) = 2\left[\dfrac{1 - (0.8)^x}{1 - (0.8)}\right], \quad \displaystyle\sum_{n=0}^{\infty} 2\left(\frac{4}{5}\right)^n$

95. *Distance* A ball is dropped from a height of 16 feet. Each time it drops h feet, it rebounds $0.81h$ feet.

(a) Find the total distance traveled by the ball.

(b) The ball takes the following time for each fall.

$$\begin{aligned}
s_1 &= -16t^2 + 16, & s_1 &= 0 \text{ if } t = 1 \\
s_2 &= -16t^2 + 16(0.81), & s_2 &= 0 \text{ if } t = 0.9 \\
s_3 &= -16t^2 + 16(0.81)^2, & s_3 &= 0 \text{ if } t = (0.9)^2 \\
s_4 &= -16t^2 + 16(0.81)^3, & s_4 &= 0 \text{ if } t = (0.9)^3 \\
&\vdots & &\vdots \\
s_n &= -16t^2 + 16(0.81)^{n-1}, & s_n &= 0 \text{ if } t = (0.9)^{n-1}
\end{aligned}$$

Beginning with s_2, the ball takes the same amount of time to bounce up as it does to fall, and thus the total time elapsed before it comes to rest is

$$t = 1 + 2\sum_{n=1}^{\infty} (0.9)^n.$$

Find this total.

Review **Solve Exercises 96–99 as a review of the skills and problem-solving techniques you learned in previous sections.**

96. The ratio of cement to sand in a 90-pound bag of dry mix is 1 to 4. Find the number of pounds of sand in the bag.

97. A truck traveled at an average speed of 50 miles per hour on a 200-mile trip. On the return trip, the average speed was 42 miles per hour. Find the average speed for the round trip.

98. Find two consecutive positive even integers whose product is 624.

99. Suppose your friend can mow a lawn in 4 hours and you can mow it in 6 hours. How long will it take both of you to mow the lawn?

7.4 Mathematical Induction

Introduction / Sums of Powers of Integers / Pattern Recognition / Finite Differences

Introduction

In this section you will study a form of mathematical proof called **mathematical induction.** It is important that you clearly see the logical need for it, so let's take a closer look at a problem discussed on page 538.

$$S_1 = 1 = 1^2$$
$$S_2 = 1 + 3 = 2^2$$
$$S_3 = 1 + 3 + 5 = 3^2$$
$$S_4 = 1 + 3 + 5 + 7 = 4^2$$
$$S_5 = 1 + 3 + 5 + 7 + 9 = 5^2$$

Judging from the pattern formed by these first five sums, it appears that the sum of the first n integers is

$$S_n = 1 + 3 + 5 + 7 + 9 + \cdots + (2n - 1) = n^2.$$

Although this particular formula *is* valid, it is important for you to see that recognizing a pattern and then simply *jumping to the conclusion* that the pattern must be true for all values of n is *not* a logically valid method of proof. There are many examples in which a pattern appears to be developing for small values of n but then fails at some point. One of the most famous cases of this was the conjecture by the French mathematician Pierre de Fermat (1601–1665), who speculated that all numbers of the form

$$F_n = 2^{2^n} + 1, \quad n = 0, 1, 2, \ldots$$

are prime. For $n = 0, 1, 2, 3$, and 4, the conjecture is true.

$$F_0 = 3, F_1 = 5, F_2 = 17, F_3 = 257, F_4 = 65{,}537$$

The size of the next Fermat number ($F_5 = 4{,}294{,}967{,}297$) is so great that it was difficult for Fermat to determine whether or not it was prime. However, another well-known mathematician, Leonhard Euler (1707–1783), later found a factorization

$$F_5 = 4{,}294{,}967{,}297 = 641(6{,}700{,}417)$$

which proved that F_5 is not prime, and therefore Fermat's conjecture was false.

Just because a rule, pattern, or formula seems to work for several values of n, you cannot simply decide that it is valid for *all* values of n without going through a *legitimate proof.*

Note It is important to recognize that both parts of the Principle of Mathematical Induction are necessary.

The Principle of Mathematical Induction

Let P_n be a statement involving the positive integer n. If

1. P_1 is true, and
2. the truth of P_k implies the truth of P_{k+1}, for every positive k,

then P_n must be true for all positive integers n.

To apply the Principle of Mathematical Induction, you need to be able to determine the statement P_{k+1} for a given statement P_k.

EXAMPLE 1 A Preliminary Example

Find P_{k+1} for the following.

a. $P_k : S_k = \dfrac{k^2(k+1)^2}{4}$

b. $P_k : S_k = 1 + 5 + 9 + \cdots + [4(k-1) - 3] + (4k - 3)$

c. $P_k : 3^k \geq 2k + 1$

Solution

a. $P_{k+1} : S_{k+1} = \dfrac{(k+1)^2(k+1+1)^2}{4}$ Replace k by $k + 1$.

$= \dfrac{(k+1)^2(k+2)^2}{4}.$ Simplify.

b. $P_{k+1} : S_{k+1} = 1 + 5 + 9 + \cdots + \{4[(k+1) - 1] - 3\} + [4(k+1) - 3]$

$= 1 + 5 + 9 + \cdots + (4k - 3) + (4k + 1).$

c. $P_{k+1} : 3^{k+1} \geq 2(k+1) + 1$

$3^{k+1} \geq 2k + 3.$

Figure 7.4

A well-known illustration used to explain why the Principle of Mathematical Induction works is the unending line of dominoes represented by Figure 7.4. If the line actually contains infinitely many dominoes, it is clear that you could not knock down the entire line by knocking down only *one domino* at a time. However, suppose it were true that each domino would knock down the next one as it fell. Then you could knock them all down simply by pushing the first one and starting a chain reaction. Mathematical induction works in the same way. If the truth of P_k implies the truth of P_{k+1} and if P_1 is true, the chain reaction proceeds as follows: P_1 implies P_2, P_2 implies P_3, P_3 implies P_4, and so on.

EXAMPLE 2 Using Mathematical Induction

Use mathematical induction to prove the following formula.

$$S_n = 1 + 3 + 5 + 7 + \cdots + (2n - 1)$$
$$= n^2$$

Solution

Mathematical induction consists of two distinct parts. First, you must show that the formula is true when $n = 1$.

1. When $n = 1$, the formula is valid, because

$$S_1 = 1 = 1^2.$$

The second part of mathematical induction has two steps. The first step is to assume that the formula is valid for *some* integer k. The second step is to use this assumption to prove that the formula is valid for the next integer, $k + 1$.

2. Assuming that the formula

$$S_k = 1 + 3 + 5 + 7 + \cdots + (2k - 1)$$
$$= k^2$$

is true, you must show that the formula $S_{k+1} = (k + 1)^2$ is true.

$$\begin{aligned} S_{k+1} &= 1 + 3 + 5 + 7 + \cdots + (2k - 1) + [2(k + 1) - 1] \\ &= [1 + 3 + 5 + 7 + \cdots + (2k - 1)] + (2k + 2 - 1) \\ &= S_k + (2k + 1) && \text{Group terms to form } S_k. \\ &= k^2 + 2k + 1 && \text{Replace } S_k \text{ by } k^2. \\ &= (k + 1)^2 \end{aligned}$$

Combining the results of parts (1) and (2), you can conclude by mathematical induction that the formula is valid for *all* positive integer values of n.

Note When using mathematical induction to prove a *summation* formula (such as the one in Example 2), it is helpful to think of S_{k+1} as $S_{k+1} = S_k + a_{k+1}$, where a_{k+1} is the $(k + 1)$ term of the original sum.

It occasionally happens that a statement involving natural numbers is not true for the first $k - 1$ positive integers but is true for all values of $n \geq k$. In these instances, you use a slight variation of the Principle of Mathematical Induction in which you verify P_k rather than P_1. This variation is called the **extended principle of mathematical induction.** To see the validity of this, note from Figure 7.4 that all but the first $k - 1$ dominoes can be knocked down by knocking over the kth domino. This suggests that you can prove a statement P_n to be true for $n \geq k$ by showing that P_k is true and that P_k implies P_{k+1}. In Exercises 35–38 in this section, you are asked to apply this extension of mathematical induction.

EXAMPLE 3 **Using Mathematical Induction**

Use mathematical induction to prove the following formula.

$$S_n = 1^2 + 2^2 + 3^2 + 4^2 + \cdots + n^2 = \frac{n(n+1)(2n+1)}{6}$$

Solution

1. When $n = 1$, the formula is valid, because

$$S_1 = 1^2 = \frac{1(2)(3)}{6}.$$

2. Assuming that

$$S_k = 1^2 + 2^2 + 3^2 + 4^2 + \cdots + k^2 = \frac{k(k+1)(2k+1)}{6}$$

you must show that

$$S_{k+1} = \frac{(k+1)(k+2)(2k+3)}{6}.$$

To do this, write the following.

$$\begin{aligned} S_{k+1} &= S_k + a_{k+1} \\ &= (1^2 + 2^2 + 3^2 + 4^2 + \cdots + k^2) + (k+1)^2 \\ &= \frac{k(k+1)(2k+1)}{6} + (k+1)^2 \\ &= \frac{k(k+1)(2k+1) + 6(k+1)^2}{6} \\ &= \frac{(k+1)[k(2k+1) + 6(k+1)]}{6} \\ &= \frac{(k+1)(2k^2 + 7k + 6)}{6} \\ &= \frac{(k+1)(k+2)(2k+3)}{6} \end{aligned}$$

Combining the results of parts (1) and (2), you can conclude by mathematical induction that the formula is valid for *all* $n \geq 1$.

Note When proving a formula with mathematical induction, the only statement that you *need* to verify is P_1. As a check, however, it is a good idea to try verifying some of the other statements. For instance, in Example 3, try verifying S_2 and S_3.

Sums of Powers of Integers

The formula in Example 3 is one of a collection of useful summation formulas. We summarize this and other formulas dealing with the sums of various powers of the first n positive integers, as follows.

Sums of Powers of Integers

1. $1 + 2 + 3 + 4 + \cdots + n = \dfrac{n(n+1)}{2}$

2. $1^2 + 2^2 + 3^2 + 4^2 + \cdots + n^2 = \dfrac{n(n+1)(2n+1)}{6}$

3. $1^3 + 2^3 + 3^3 + 4^3 + \cdots + n^3 = \dfrac{n^2(n+1)^2}{4}$

4. $1^4 + 2^4 + 3^4 + 4^4 + \cdots + n^4 = \dfrac{n(n+1)(2n+1)(3n^2+3n-1)}{30}$

5. $1^5 + 2^5 + 3^5 + 4^5 + \cdots + n^5 = \dfrac{n^2(n+1)^2(2n^2+2n-1)}{12}$

Note Each of these formulas for sums can be proven by mathematical induction. (See Exercises 11–13, 15, 16.)

EXAMPLE 4 **Finding a Sum of Powers of Integers**

Find $\sum_{n=1}^{7} n^3 = 1^3 + 2^3 + 3^3 + 4^3 + 5^3 + 6^3 + 7^3$.

Solution

Using the formula for the sum of the cubes of the first n positive integers, you obtain the following.

$$\begin{aligned}\sum_{n=1}^{7} n^3 &= 1^3 + 2^3 + 3^3 + 4^3 + 5^3 + 6^3 + 7^3 \\ &= \frac{7^2(7+1)^2}{4} \\ &= \frac{49(64)}{4} \\ &= 784\end{aligned}$$

Check this sum by adding the numbers 1, 8, 27, 64, 125, 216, and 343.

EXAMPLE 5 **Proving an Inequality by Mathematical Induction**

Prove that $n < 2^n$ for all positive integers n.

Solution

1. For $n = 1$, the formula is true, because

$$1 < 2^1.$$

2. Assuming that

$$k < 2^k$$

you need to show that $k + 1 < 2^{k+1}$. For $n = k$, you have

$$2^{k+1} = 2(2^k) > 2(k) = 2k. \qquad \text{By assumption}$$

Because $2k = k + k > k + 1$ for all $k > 1$, it follows that

$$2^{k+1} > 2k > k + 1$$

or

$$k + 1 < 2^{k+1}.$$

Therefore, $n < 2^n$ for all integers $n \geq 1$.

Pattern Recognition

Although choosing a formula on the basis of a few observations *does not* guarantee the validity of a formula, pattern recognition *is* important. Once you have a pattern or formula that you think works, you can try using mathematical induction to prove your formula.

Note Some common patterns to look for when you must find the nth term of a sequence are:

Linear	$an \pm b$
Quadratic	$n^2 \pm b$
Cubic	$n^3 \pm b$
Exponential	$2^n \pm b,\ 3^n \pm b$
Factorial	$n!,\ (2n)!,\ (2n-1)!$

Finding a Formula for the *n*th Term of a Sequence

To find a formula for the nth term of a sequence, consider the following guidelines.

1. Calculate the first several terms of the sequence. It is often a good idea to write the terms in both simplified and factored forms.
2. Try to find a recognizable pattern for the terms and write a formula for the nth term of the sequence. This is your *hypothesis* or *conjecture.* You might try computing one or two more terms in the sequence to test your hypothesis.
3. Use mathematical induction to prove your hypothesis.

EXAMPLE 6 **Finding a Formula for a Finite Sum**

Find a formula for the following finite sum.

$$\frac{1}{1 \cdot 2} + \frac{1}{2 \cdot 3} + \frac{1}{3 \cdot 4} + \frac{1}{4 \cdot 5} + \cdots + \frac{1}{n(n+1)}$$

Solution

Begin by writing out the first few sums.

$$S_1 = \frac{1}{1 \cdot 2} = \frac{1}{2} = \frac{1}{1+1}$$

$$S_2 = \frac{1}{1 \cdot 2} + \frac{1}{2 \cdot 3} = \frac{4}{6} = \frac{2}{3} = \frac{2}{2+1}$$

$$S_3 = \frac{1}{1 \cdot 2} + \frac{1}{2 \cdot 3} + \frac{1}{3 \cdot 4} = \frac{9}{12} = \frac{3}{4} = \frac{3}{3+1}$$

$$S_4 = \frac{1}{1 \cdot 2} + \frac{1}{2 \cdot 3} + \frac{1}{3 \cdot 4} + \frac{1}{4 \cdot 5} = \frac{48}{60} = \frac{4}{5} = \frac{4}{4+1}$$

From this sequence, it appears that the formula for the kth sum is

$$\begin{aligned} S_k &= \frac{1}{1 \cdot 2} + \frac{1}{2 \cdot 3} + \frac{1}{3 \cdot 4} + \frac{1}{4 \cdot 5} + \cdots + \frac{1}{k(k+1)} \\ &= \frac{k}{k+1}. \end{aligned}$$

To prove the validity of this hypothesis, use mathematical induction, as follows. Note that you have already verified the formula for $n = 1$, so you can begin by assuming that the formula is valid for $n = k$ and trying to show that it is valid for $n = k + 1$.

$$\begin{aligned} S_{k+1} &= \left[\frac{1}{1 \cdot 2} + \frac{1}{2 \cdot 3} + \frac{1}{3 \cdot 4} + \frac{1}{4 \cdot 5} + \cdots + \frac{1}{k(k+1)}\right] + \frac{1}{(k+1)(k+2)} \\ &= \frac{k}{k+1} + \frac{1}{(k+1)(k+2)} \\ &= \frac{k(k+2)+1}{(k+1)(k+2)} \\ &= \frac{k^2+2k+1}{(k+1)(k+2)} \\ &= \frac{(k+1)^2}{(k+1)(k+2)} \\ &= \frac{k+1}{k+2} \end{aligned}$$

Thus, the hypothesis is valid.

Study Tip

To show that the formula in Example 6 is valid for $n = k + 1$, you have to verify that

$$S_{k+1} = \frac{k+1}{(k+1)+1} = \frac{k+1}{k+2}.$$

Finite Differences

The **first differences** of a sequence are found by subtracting consecutive terms. The **second differences** are found by subtracting consecutive first differences. The first and second differences of the sequence 3, 5, 8, 12, 17, 23, . . . are as follows.

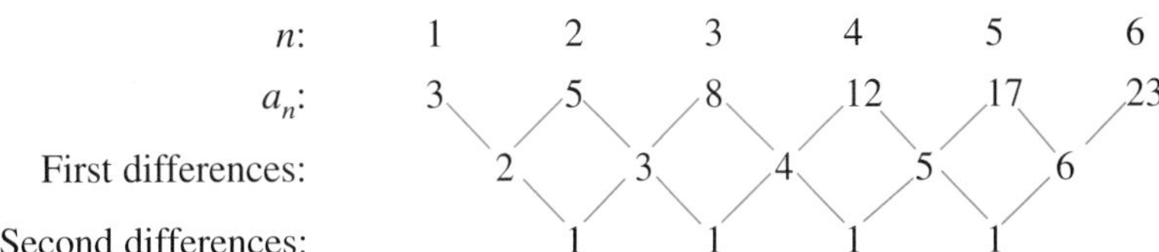

For this sequence, the second differences are all the same. When this happens, the sequence has a perfect quadratic model. If the first differences are all the same, the sequence has a linear model—that is, it is arithmetic.

EXAMPLE 7 Finding a Quadratic Model

Find the quadratic model for the sequence

3, 5, 8, 12, 17, 23, ...

Solution

You know the model has the form

$$a_n = an^2 + bn + c.$$

By substituting 1, 2, and 3 for n, you can obtain a system of three linear equations in three variables.

$a_1 = a(1)^2 + b(1) + c = 3$ Substitute 1 for n.

$a_2 = a(2)^2 + b(2) + c = 5$ Substitute 2 for n.

$a_3 = a(3)^2 + b(3) + c = 8$ Substitute 3 for n.

You now have a system of three equations in a, b, and c.

$a + b + c = 3$ Equation 1

$4a + 2b + c = 5$ Equation 2

$9a + 3b + c = 8$ Equation 3

Using the techniques discussed in Chapter 5, you can find the solution to be $a = \frac{1}{2}$, $b = \frac{1}{2}$, and $c = 2$. Thus, the quadratic model is

$$a_n = \frac{1}{2}n^2 + \frac{1}{2}n + 2.$$

Try checking the values of a_1, a_2, and a_3.

Group Activity

The Sum of the Angles of a Regular Polygon

A *regular n*-sided polygon is a polygon that has n equal sides and n equal angles. For instance, an equilateral triangle is a regular three-sided polygon. Each angle of an equilateral triangle measures 60°, and the sum of all three angles is 180°. Similarly, the sum of the four angles of a regular four-sided polygon (a square) is 360°. The figure below shows the sums of the angles of the first eight regular polygons.

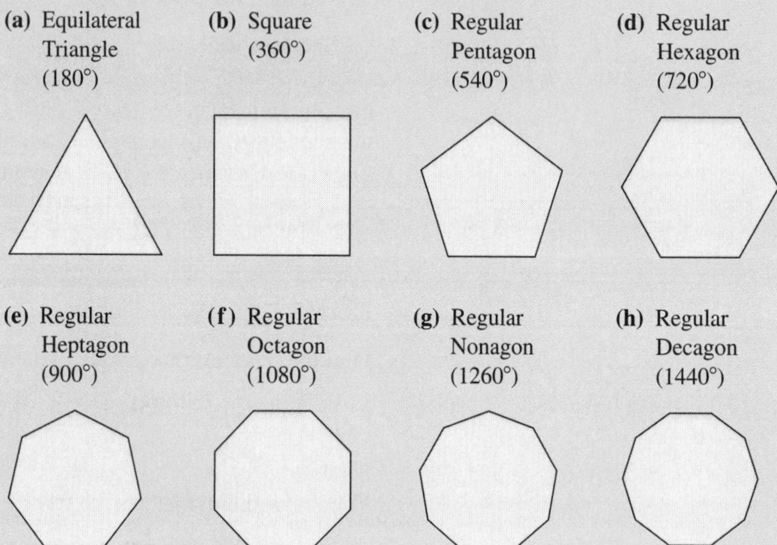

a. Use the data given in the figure to complete the first row of the table. Then complete the second and third rows of the table by finding the first and second differences.

n	3		4		5		6		7		8		9		10
a_n	180		360		?		?		?		?		?		?
First Differences		?		?		?		?		?		?		?	
Second Differences			?		?		?		?		?		?		

b. From the result of the table, does the data fit a linear model, a quadratic model, or neither? Explain your reasoning.

c. Find a model that fits the data. Then use the result to find a model that gives the number of degrees in each angle of a regular n-sided polygon.

7.4 /// EXERCISES

In Exercises 1–4, find P_{k+1} for the given P_k.

1. $P_k = \dfrac{5}{k(k+1)}$

2. $P_k = \dfrac{1}{(k+1)(k+3)}$

3. $P_k = \dfrac{k^2(k+1)^2}{4}$

4. $P_k = \dfrac{k}{2}(3k-1)$

In Exercises 5–18, use mathematical induction to prove the formula for every positive integer n.

5. $2 + 4 + 6 + 8 + \cdots + 2n = n(n+1)$

6. $3 + 7 + 11 + 15 + \cdots + (4n-1) = n(2n+1)$

7. $2 + 7 + 12 + 17 + \cdots + (5n-3) = \dfrac{n}{2}(5n-1)$

8. $1 + 4 + 7 + 10 + \cdots + (3n-2) = \dfrac{n}{2}(3n-1)$

9. $1 + 2 + 2^2 + 2^3 + \cdots + 2^{n-1} = 2^n - 1$

10. $2(1 + 3 + 3^2 + 3^3 + \cdots + 3^{n-1}) = 3^n - 1$

11. $1 + 2 + 3 + 4 + \cdots + n = \dfrac{n(n+1)}{2}$

12. $1^2 + 2^2 + 3^2 + 4^2 + \cdots + n^2 = \dfrac{n(n+1)(2n+1)}{6}$

13. $1^3 + 2^3 + 3^3 + 4^3 + \cdots + n^3 = \dfrac{n^2(n+1)^2}{4}$

14. $\left(1 + \dfrac{1}{1}\right)\left(1 + \dfrac{1}{2}\right)\left(1 + \dfrac{1}{3}\right) \cdots \left(1 + \dfrac{1}{n}\right) = n + 1$

15. $\displaystyle\sum_{i=1}^{5} i^5 = \frac{n^2(n+1)^2(2n^2+2n-1)}{12}$

16. $\displaystyle\sum_{i=1}^{n} i^4 = \frac{n(n+1)(2n+1)(3n^2+3n-1)}{30}$

17. $\displaystyle\sum_{i=1}^{n} i(i+1) = \frac{n(n+1)(n+2)}{3}$

18. $\displaystyle\sum_{i=1}^{n} \frac{1}{(2i-1)(2i+1)} = \frac{n}{2n+1}$

In Exercises 19–28, find the sum using the formulas for the sums of powers of integers.

19. $\displaystyle\sum_{n=1}^{20} n$

20. $\displaystyle\sum_{n=1}^{50} n$

21. $\displaystyle\sum_{n=1}^{6} n^2$

22. $\displaystyle\sum_{n=1}^{10} n^3$

23. $\displaystyle\sum_{n=1}^{5} n^4$

24. $\displaystyle\sum_{n=1}^{8} n^5$

25. $\displaystyle\sum_{n=1}^{6} (n^2 - n)$

26. $\displaystyle\sum_{n=1}^{10} (n^3 - n^2)$

27. $\displaystyle\sum_{i=1}^{6} (6i - 8i^3)$

28. $\displaystyle\sum_{j=1}^{4} \left(2 + \tfrac{5}{2}j - \tfrac{3}{2}j^2\right)$

In Exercises 29–34, find a formula for the sum of the first n terms of the sequence.

29. $1, 5, 9, 13, \ldots$

30. $25, 22, 19, 16, \ldots$

31. $1, \frac{9}{10}, \frac{81}{100}, \frac{729}{1000}, \ldots$

32. $3, -\frac{9}{2}, \frac{27}{4}, -\frac{81}{8}, \ldots$

33. $\dfrac{1}{4}, \dfrac{1}{12}, \dfrac{1}{24}, \dfrac{1}{40}, \ldots, \dfrac{1}{2n(n-1)}, \ldots$

34. $\dfrac{1}{2 \cdot 3}, \dfrac{1}{3 \cdot 4}, \dfrac{1}{4 \cdot 5}, \dfrac{1}{5 \cdot 6}, \ldots, \dfrac{1}{(n+1)(n+2)}, \ldots$

In Exercises 35–38, prove the inequality for the indicated integer values of n.

35. $n! > 2^n, \quad n \geq 4$

36. $\left(\frac{4}{3}\right)^n > n, \quad n \geq 7$

37. $\dfrac{1}{\sqrt{1}} + \dfrac{1}{\sqrt{2}} + \dfrac{1}{\sqrt{3}} + \cdots + \dfrac{1}{\sqrt{n}} > \sqrt{n}, \quad n \geq 2$

38. $\left(\dfrac{x}{y}\right)^{n+1} < \left(\dfrac{x}{y}\right)^n, \quad n \geq 1$ and $0 < x < y$

In Exercises 39–48, use mathematical induction to prove the given property for all positive integers n.

39. $(ab)^n = a^n b^n$

40. $\left(\frac{a}{b}\right)^n = \frac{a^n}{b^n}$

41. If $x_1 \neq 0,\ x_2 \neq 0, \ldots, x_n \neq 0$, then

$(x_1 x_2 x_3 \cdot \cdot \cdot x_n)^{-1} = x_1^{-1} x_2^{-1} x_3^{-1} \cdot \cdot \cdot x_n^{-1}.$

42. If $x_1 > 0,\ x_2 > 0, \ldots, x_n > 0$, then

$\ln(x_1 x_2 x_3 \cdot \cdot \cdot x_n) = \ln x_1 + \ln x_2 + \ln x_3 + \cdot \cdot \cdot + \ln x_n.$

43. Generalized Distributive Law:

$x(y_1 + y_2 + \cdot \cdot \cdot + y_n) = xy_1 + xy_2 + \cdot \cdot \cdot + xy_n$

44. $(a + bi)^n$ and $(a - bi)^n$ are complex conjugates for all $n \geq 1$.

45. *Trigonometry* $\sin(x + n\pi) = (-1)^n \sin x$

46. *Trigonometry* $\tan(x + n\pi) = \tan x$

47. A factor of $(n^3 + 3n^2 + 2n)$ is 3.

48. A factor of $(2^{2n-1} + 3^{2n-1})$ is 5.

49. *Essay* In your own words, explain what is meant by a proof by mathematical induction.

50. *Think About It* What conclusion can be drawn from the given information about the sequence of statements P_n?

(a) P_3 is true and P_k implies P_{k+1}.

(b) $P_1, P_2, P_3, \ldots, P_{50}$ are all true.

(c) P_1, P_2, and P_3 are all true, but the truth of P_k does not imply that P_{k+1} is true.

(d) P_2 is true and P_{2k} implies P_{2k+2}.

In Exercises 51–54, write the first five terms of the sequence.

51. $a_0 = 1$
$a_n = a_{n-1} + 2$

52. $a_0 = 10$
$a_n = 4a_{n-1}$

53. $a_0 = 4$
$a_1 = 2$
$a_n = a_{n-1} - a_{n-2}$

54. $a_0 = 0$
$a_1 = 2$
$a_n = a_{n-1} + 2a_{n-2}$

In Exercises 55–64, write the first five terms of the sequence where $a_1 = f(1)$. Then calculate the first and second differences of the sequence. Does the sequence have a linear model, a quadratic model, or neither?

55. $f(1) = 0$
$a_n = a_{n-1} + 3$

56. $f(1) = 2$
$a_n = n - a_{n-1}$

57. $f(1) = 3$
$a_n = a_{n-1} - n$

58. $f(2) = -3$
$a_n = -2a_{n-1}$

59. $a_0 = 0$
$a_n = a_{n-1} + n$

60. $a_0 = 2$
$a_n = (a_{n-1})^2$

61. $f(1) = 2$
$a_n = a_{n-1} + 2$

62. $f(1) = 0$
$a_n = a_{n-1} + 2n$

63. $a_0 = 1$
$a_n = a_{n-1} + n^2$

64. $a_0 = 0$
$a_n = a_{n-1} - 1$

In Exercises 65–68, find a quadratic model for the sequence with the indicated terms.

65. $a_0 = 3,\ a_1 = 3,\ a_4 = 15$

66. $a_0 = 7,\ a_1 = 6,\ a_3 = 10$

67. $a_0 = -3,\ a_2 = 1,\ a_4 = 9$

68. $a_0 = 3,\ a_2 = 0,\ a_6 = 36$

Review **Solve Exercises 69–72 as a review of the skills and problem-solving techniques you learned in previous sections. Solve the system of equations.**

69.
$$\begin{aligned} y &= x^2 \\ -3x + 2y &= 2 \end{aligned}$$

70.
$$\begin{aligned} x - y^3 &= 0 \\ x - 2y^2 &= 0 \end{aligned}$$

71.
$$\begin{aligned} x - y \quad\ \ &= -1 \\ x + 2y - 2z &= 3 \\ 3x - y + 2z &= 3 \end{aligned}$$

72.
$$\begin{aligned} 2x + y - 2z &= 1 \\ x \qquad - z &= 1 \\ 3x + 3y + z &= 12 \end{aligned}$$

7.5 The Binomial Theorem

Binomial Coefficients / Pascal's Triangle / Binomial Expansions

Binomial Coefficients

Recall that a **binomial** is a polynomial that has two terms. In this section, you will study a formula that provides a quick method of raising a binomial to a power. To begin, let's look at the expansion of $(x + y)^n$ for several values of n.

$$(x + y)^0 = 1$$
$$(x + y)^1 = x + y$$
$$(x + y)^2 = x^2 + 2xy + y^2$$
$$(x + y)^3 = x^3 + 3x^2y + 3xy^2 + y^3$$
$$(x + y)^4 = x^4 + 4x^3y + 6x^2y^2 + 4xy^3 + y^4$$
$$(x + y)^5 = x^5 + 5x^4y + 10x^3y^2 + 10x^2y^3 + 5xy^4 + y^5$$

There are several observations you can make about these expansions.

1. In each expansion, there are $n + 1$ terms.
2. In each expansion, x and y have symmetric roles. The powers of x decrease by 1 in successive terms, whereas the powers of y increase by 1.
3. The sum of the powers of each term is n. For instance, in the expansion of $(x + y)^5$, the sum of the powers of each term is 5.

$$(x + y)^5 = x^5 + \overbrace{5x^4y^1}^{4+1=5} + \overbrace{10x^3y^2}^{3+2=5} + 10x^2y^3 + 5x^1y^4 + y^5$$

4. The coefficients increase and then decrease in a symmetric pattern.

Think About the Proof

Use mathematical induction to prove the Binomial Theorem. The details of the proof are given in the appendix.

The coefficients of a binomial expansion are called **binomial coefficients.** To find them, you can use the following theorem.

The Binomial Theorem

In the expansion of $(x + y)^n$

$$(x + y)^n = x^n + nx^{n-1}y + \cdots + {}_nC_r\, x^{n-r}y^r + \cdots + nxy^{n-1} + y^n$$

the coefficient of $x^{n-r}y^r$ is given by

$${}_nC_r = \frac{n!}{(n-r)!r!}.$$

Note The symbol $\binom{n}{r}$ is often used in place of ${}_nC_r$ to denote binomial coefficients.

EXAMPLE 1 **Finding Binomial Coefficients**

Find the binomial coefficients.

a. ${}_8C_2$ **b.** ${}_{10}C_3$ **c.** ${}_7C_0$ **d.** ${}_8C_8$

Solution

a. $${}_8C_2 = \frac{8!}{6! \cdot 2!} = \frac{(8 \cdot 7) \cdot \cancel{6!}}{\cancel{6!} \cdot 2!} = \frac{8 \cdot 7}{2 \cdot 1} = 28$$

b. $${}_{10}C_3 = \frac{10!}{7! \cdot 3!} = \frac{(10 \cdot 9 \cdot 8) \cdot \cancel{7!}}{\cancel{7!} \cdot 3!} = \frac{10 \cdot 9 \cdot 8}{3 \cdot 2 \cdot 1} = 120$$

c. $${}_7C_0 = \frac{\cancel{7!}}{\cancel{7!} \cdot 0!} = 1$$

d. $${}_8C_8 = \frac{\cancel{8!}}{0! \cdot \cancel{8!}} = 1$$

Note When $r \neq 0$ and $r \neq n$, as in parts (a) and (b) above, there is a simple pattern for evaluating binomial coefficients.

$${}_8C_2 = \frac{\overbrace{8 \cdot 7}^{\text{2 factors}}}{\underbrace{2 \cdot 1}_{\text{2 factorial}}} \quad \text{and} \quad {}_{10}C_3 = \frac{\overbrace{10 \cdot 9 \cdot 8}^{\text{3 factors}}}{\underbrace{3 \cdot 2 \cdot 1}_{\text{3 factorial}}}$$

EXAMPLE 2 **Finding Binomial Coefficients**

Find the binomial coefficients.

a. ${}_7C_3$ **b.** ${}_7C_4$ **c.** ${}_{12}C_1$ **d.** ${}_{12}C_{11}$

Solution

a. $${}_7C_3 = \frac{7 \cdot \cancel{6} \cdot 5}{\cancel{3} \cdot \cancel{2} \cdot 1} = 35$$

b. $${}_7C_4 = \frac{7 \cdot \cancel{6} \cdot 5 \cdot \cancel{4}}{\cancel{4} \cdot \cancel{3} \cdot \cancel{2} \cdot 1} = 35$$

c. $${}_{12}C_1 = \frac{12}{1} = 12$$

d. $${}_{12}C_{11} = \frac{12!}{1! \cdot 11!} = \frac{(12) \cdot \cancel{11!}}{1! \cdot \cancel{11!}} = \frac{12}{1} = 12$$

Note It is not a coincidence that the results in parts (a) and (b) of Example 2 are the same and that the results in parts (c) and (d) are the same. In general, it is true that

$${}_nC_r = {}_nC_{n-r}.$$

This shows the symmetric property of binomial coefficients that was identified earlier.

Pascal's Triangle

There is a convenient way to remember a pattern for binomial coefficients. By arranging the coefficients in a triangular pattern, you obtain the following array, which is called **Pascal's Triangle.** This triangle is named after the famous French mathematician Blaise Pascal (1623–1662).

$$
\begin{array}{ccccccccccccccc}
&&&&&&&1&&&&&&&\\
&&&&&&1&&1&&&&&&\\
&&&&&1&&2&&1&&&&&\\
&&&&1&&3&&3&&1&&&&\\
&&&1&&4&&6&&4&&1&&&\\
&&1&&5&&10&&10&&5&&1&&\\
&1&&6&&15&&20&&15&&6&&1&\\
1&&7&&21&&35&&35&&21&&7&&1
\end{array}
$$

Note The top row in Pascal's Triangle is called the *zero row* because it corresponds to the binomial expansion

$$(x + y)^0 = 1.$$

Similarly, the next row is called the *first row* because it corresponds to the binomial expansion

$$(x + y)^1 = 1(x) + 1(y).$$

In general, the *nth row* in Pascal's Triangle gives the coefficients of $(x + y)^n$.

The first and last number in each row of Pascal's Triangle is 1. Every other number in each row is formed by adding the two numbers immediately above the number. Pascal noticed that numbers in this triangle are precisely the same numbers that are the coefficients of binomial expansions, as follows.

$$
\begin{aligned}
(x + y)^0 &= 1\\
(x + y)^1 &= 1x + 1y\\
(x + y)^2 &= 1x^2 + 2xy + 1y^2\\
(x + y)^3 &= 1x^3 + 3x^2y + 3xy^2 + 1y^3\\
(x + y)^4 &= 1x^4 + 4x^3y + 6x^2y^2 + 4xy^3 + 1y^4\\
(x + y)^5 &= 1x^5 + 5x^4y + 10x^3y^2 + 10x^2y^3 + 5xy^4 + 1y^5\\
(x + y)^6 &= 1x^6 + 6x^5y + 15x^4y^2 + 20x^3y^3 + 15x^2y^4 + 6xy^5 + 1y^6\\
(x + y)^7 &= 1x^7 + 7x^6y + 21x^5y^2 + 35x^4y^3 + 35x^3y^4 + 21x^2y^5 + 7xy^6 + 1y^7
\end{aligned}
$$

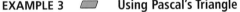

EXAMPLE 3 **Using Pascal's Triangle**

Use the seventh row of Pascal's Triangle to find the binomial coefficients.

$${}_8C_0, {}_8C_1, {}_8C_2, {}_8C_3, {}_8C_4, {}_8C_5, {}_8C_6, {}_8C_7, {}_8C_8$$

Solution

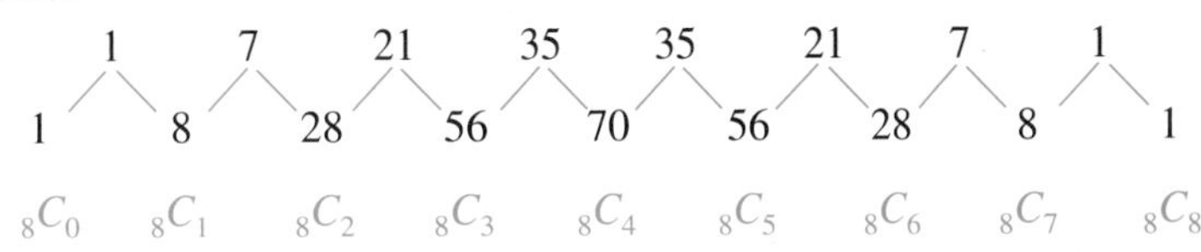

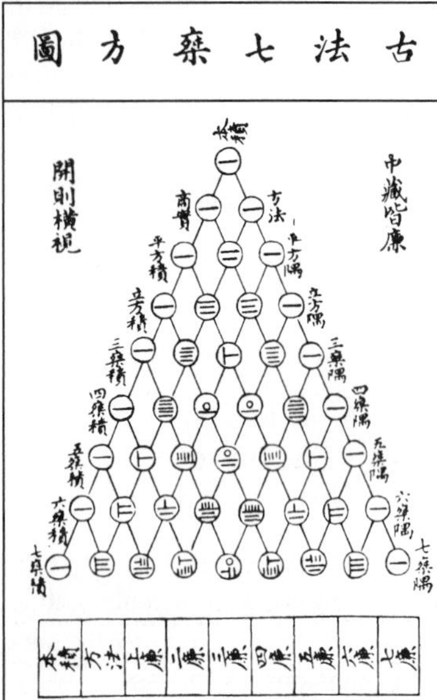

"Pascal's" Triangle and forms of the Binomial Theorem were known in Eastern cultures prior to the Western "discovery" of the theorem. A Chinese text *Precious Mirror* contains a triangle of binomial expansions through the eighth power.

Binomial Expansions

As mentioned at the beginning of this section, when you write out the coefficients for a binomial that is raised to a power, you are **expanding a binomial.** The formulas for binomial coefficients give you an easy way to expand binomials, as demonstrated in the next three examples.

EXAMPLE 4 Expanding a Binomial

Write the expansion for the expression

$$(x + 1)^3.$$

Solution

The binomial coefficients from the third row of Pascal's Triangle are

1, 3, 3, 1.

Therefore, the expansion is as follows.

$$\begin{aligned}(x + 1)^3 &= (1)x^3 + (3)x^2(1) + (3)x(1^2) + (1)(1^3)\\ &= x^3 + 3x^2 + 3x + 1\end{aligned}$$

To expand binomials representing *differences*, rather than sums, you alternate signs. Here are two examples.

$$(x - 1)^3 = x^3 - 3x^2 + 3x - 1$$
$$(x - 1)^4 = x^4 - 4x^3 + 6x^2 - 4x + 1$$

EXAMPLE 5 Expanding a Binomial

Write the expansion for the expression

$$(x - 3)^4.$$

Solution

The binomial coefficients from the fourth row of Pascal's Triangle are

1, 4, 6, 4, 1.

Therefore, the expansion is as follows.

$$\begin{aligned}(x - 3)^4 &= (1)x^4 - (4)x^3(3) + (6)x^2(3^2) - (4)x(3^3) + (1)(3^4)\\ &= x^4 - 12x^3 + 54x^2 - 108x + 81\end{aligned}$$

EXAMPLE 6 **Expanding a Binomial**

Write the expansion for $(x - 2y)^4$.

Solution

Use the fourth row of Pascal's Triangle, as follows.

$$\begin{aligned}(x - 2y)^4 &= (1)x^4 - (4)x^3(2y) + (6)x^2(2y)^2 - (4)x(2y)^3 + (1)(2y)^4 \\ &= x^4 - 8x^3y + 24x^2y^2 - 32xy^3 + 16y^4\end{aligned}$$

EXAMPLE 7 **Finding a Term in a Binomial Expansion**

Find the sixth term of $(a + 2b)^8$.

Solution

From the Binomial Theorem, you can see that the $r + 1$ term is ${}_nC_r\, x^{n-r}\, y^r$. So in this case, $6 = r + 1$ means that $r = 5$. Because $n = 8$, $x = a$, and $y = 2b$, the sixth term in the binomial expansion is

$$\begin{aligned}{}_8C_5 a^{8-5}(2b)^5 &= 56 \cdot a^3 \cdot (2b)^5 \\ &= 56(2^5)a^3b^5 \\ &= 1792a^3b^5\end{aligned}$$

Group Activity *Error Analysis*

Suppose you are a math instructor and receive the following solutions from one of your students on a quiz. Find the error(s) in each solution, and discuss ways that your student could avoid the error(s) in the future.

a. Find the second term in the expansion of

$(2x - 3y)^5$.

$5(2x)^4(3y)^2 = 720x^4y^2$

b. Find the fourth term in the expansion of

$\left(\frac{1}{2}x + 7y\right)^6$.

${}_6C_4\left(\frac{1}{2}x\right)^2(7y)^4 = 9003.75x^2y^4$

7.5 /// EXERCISES

In Exercises 1–10, evaluate ${}_nC_r$.

1. ${}_5C_3$
2. ${}_8C_6$
3. ${}_{12}C_0$
4. ${}_{20}C_{20}$
5. ${}_{20}C_{15}$
6. ${}_{12}C_5$
7. ${}_{100}C_{98}$
8. ${}_{10}C_4$
9. ${}_{100}C_2$
10. ${}_{10}C_6$

11. *Essay* In your own words, explain how to form the rows of Pascal's Triangle.
12. Form the first nine rows of Pascal's Triangle.

In Exercises 13–16, evaluate using Pascal's Triangle.

13. ${}_7C_4$
14. ${}_6C_3$
15. ${}_8C_5$
16. ${}_8C_7$

In Exercises 17–36, use the Binomial Theorem to expand and simplify the expression.

17. $(x + 1)^4$
18. $(x + 1)^6$
19. $(a + 2)^3$
20. $(a + 3)^4$
21. $(y - 2)^4$
22. $(y - 2)^5$
23. $(x + y)^5$
24. $(x + y)^6$
25. $(r + 3s)^6$
26. $(x + 2y)^4$
27. $(x - y)^5$
28. $(2x - y)^5$
29. $(1 - 2x)^3$
30. $(5 - 3y)^3$
31. $(x^2 + 5)^4$
32. $(x^2 + y^2)^6$
33. $\left(\frac{1}{x} + y\right)^5$
34. $\left(\frac{1}{x} + 2y\right)^6$
35. $2(x - 3)^4 + 5(x - 3)^2$
36. $3(x + 1)^5 - 4(x + 1)^3$

In Exercises 37–40, expand the binomial using Pascal's Triangle to determine the coefficients.

37. $(2t - s)^5$
38. $(x + 2y)^5$
39. $(3 - 2z)^4$
40. $(3y + 2)^5$

In Exercises 41–48, find the coefficient a of the given term in the expansion of the binomial.

	Binomial	*Term*
41.	$(x + 3)^{12}$	ax^5
42.	$(x^2 + 3)^{12}$	ax^8
43.	$(x - 2y)^{10}$	ax^8y^2
44.	$(4x - y)^{10}$	ax^2y^8
45.	$(3x - 2y)^9$	ax^4y^5
46.	$(2x - 3y)^8$	ax^6y^2
47.	$(x^2 + y)^{10}$	ax^8y^6
48.	$(z^2 - 1)^{12}$	az^6

49. *Think About It* How many terms are in the expansion of $(x + y)^n$?
50. *Think About It* How do the expansions of $(x + y)^n$ and $(x - y)^n$ differ?

In Exercises 51–54, use the Binomial Theorem to expand and simplify the expression.

51. $(\sqrt{x} + 3)^4$
52. $(2\sqrt{t} - 1)^3$
53. $(x^{2/3} - y^{1/3})^3$
54. $(u^{3/5} + 2)^5$

In Exercises 55–58, expand the binomial in the difference quotient and simplify.

$$\frac{f(x + h) - f(x)}{h}$$

55. $f(x) = x^3$
56. $f(x) = x^4$
57. $f(x) = \sqrt{x}$
58. $f(x) = \frac{1}{x}$

In Exercises 59–64, use the Binomial Theorem to expand the complex number. Simplify your result.

59. $(1 + i)^4$

60. $(2 - i)^5$

61. $(2 - 3i)^6$

62. $\left(5 + \sqrt{-9}\right)^3$

63. $\left(-\frac{1}{2} + \frac{\sqrt{3}}{2}i\right)^3$

64. $\left(5 - \sqrt{3}i\right)^4$

Probability **In Exercises 65–68, consider n independent trials of an experiment where each trial has two possible outcomes called success and failure. The probability of a success on each trial is p and the probability of a failure is $q = 1 - p$. In this context the term ${}_nC_k\, p^k q^{n-k}$ in the expansion of $(p + q)^n$ gives the probability of k successes in the n trials of the experiment.**

65. A fair coin is tossed seven times. To find the probability of obtaining 4 heads, evaluate the term

${}_7C_4\left(\frac{1}{2}\right)^4\left(\frac{1}{2}\right)^3$

in the expansion $\left(\frac{1}{2} + \frac{1}{2}\right)^7$.

66. The probability of a baseball player getting a hit on any given time at bat is $\frac{1}{4}$. To find the probability that the player gets 3 hits during the next 10 times at bat, evaluate the term

${}_{10}C_3\left(\frac{1}{4}\right)^3\left(\frac{3}{4}\right)^7$

in the expansion $\left(\frac{1}{4} + \frac{3}{4}\right)^{10}$.

67. The probability of a sales representative making a sale with any one customer is $\frac{1}{3}$. The sales representative makes 8 contacts a day. To find the probability of making 4 sales, evaluate the term

${}_8C_4\left(\frac{1}{3}\right)^4\left(\frac{2}{3}\right)^4$

in the expansion $\left(\frac{1}{3} + \frac{2}{3}\right)^8$.

68. To find the probability that the sales representative in Exercise 67 makes 4 sales if the probability of a sale with any one customer is $\frac{1}{2}$, evaluate the term

${}_8C_4\left(\frac{1}{2}\right)^4\left(\frac{1}{2}\right)^4$

in the expansion $\left(\frac{1}{2} + \frac{1}{2}\right)^8$.

Approximation **In Exercises 69–72, use the Binomial Theorem to approximate the given quantity accurate to three decimal places. For example, in Exercise 69, use the expansion**

$(1.02)^8 = (1 + 0.02)^8 = 1 + 8(0.02) + 28(0.02)^2 + \cdots.$

69. $(1.02)^8$

70. $(2.005)^{10}$

71. $(2.99)^{12}$

72. $(1.98)^9$

Graphical Reasoning **In Exercises 73–76, use a graphing utility to obtain the graph of f and g in the same viewing rectangle. What is the relationship between the two graphs? Use the Binomial Theorem to write the polynomial function g in standard form.**

73. $f(x) = x^3 - 4x, \quad g(x) = f(x + 4)$

74. $f(x) = -x^4 + 4x^2 - 1, \quad g(x) = f(x - 3)$

75. $f(x) = -x^2 + 3x + 2, \quad g(x) = f(x - 2)$

76. $f(x) = 2x^2 - 4x + 1, \quad g(x) = f(x + 3)$

In Exercises 77–80, prove the given property for all integers r and n where $0 \le r \le n$.

77. ${}_nC_r = {}_nC_{n-r}$

78. ${}_nC_0 - {}_nC_1 + {}_nC_2 - \cdots \pm {}_nC_n = 0$

79. ${}_{n+1}C_r = {}_nC_r + {}_nC_{r-1}$

80. The sum of the numbers in the nth row of Pascal's Triangle is 2^n.

Exploration **In Exercises 81 and 82, use a graphing utility to evaluate and determine which two are equal.**

81. (a) ${}_{12}C_5$ (b) $({}_6C_5)^2$
(c) ${}_{11}C_5 + {}_{11}C_4$ (d) ${}_6C_5 + {}_6C_5$

82. (a) ${}_{25}C_6$ (b) $2({}_{25}C_2 + {}_{25}C_4)$
(c) $\sum_{k=0}^{5}[({}_{10}C_k)({}_8C_{5-k})]$ (d) ${}_{18}C_5$

Graphical Reasoning **In Exercises 83 and 84, use a graphing utility to obtain the graphs of the functions in the given order and in the same viewing rectangle. Compare the graphs. Which two functions have identical graphs and why?**

83. (a) $f(x) = (1 - x)^3$

(b) $g(x) = 1 - 3x$

(c) $h(x) = 1 - 3x + 3x^2$

(d) $p(x) = 1 - 3x + 3x^2 - x^3$

84. (a) $f(x) = \left(1 - \frac{1}{2}x\right)^4$

(b) $g(x) = 1 - 2x + \frac{3}{2}x^2$

(c) $h(x) = 1 - 2x + \frac{3}{2}x^2 - \frac{1}{2}x^3$

(d) $p(x) = 1 - 2x + \frac{3}{2}x^2 - \frac{1}{2}x^3 + \frac{1}{16}x^4$

85. *Life Insurance* The average amount of life insurance per household $f(t)$ (in thousands of dollars) from 1970 through 1992 can be approximated by the model

$$f(t) = 0.1506t^2 + 0.7361t + 21.1374, \qquad 0 \le t \le 22$$

where $t = 0$ represents 1970 (see figure). You want to adjust this model so that $t = 0$ corresponds to 1980 rather than 1970. To do this, you shift the graph of f 10 units *to the left* and obtain

$$g(t) = f(t + 10).$$

(Source: American Council of Life Insurance)

(a) Write $g(t)$ in standard form.

(b) Use a graphing utility to graph f and g in the same viewing rectangle.

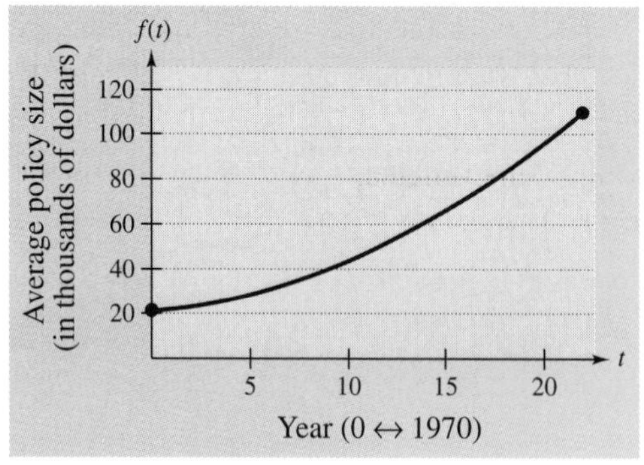

86. *Health Maintenance Organizations* The number of people $f(t)$ (in millions) enrolled in health maintenance organizations in the United States from 1976 through 1992 can be approximated by the model

$$f(t) = 0.1043t^2 + 0.7100t + 4.6852, \qquad 0 \le t \le 16$$

where $t = 0$ represents 1976 (see figure). You want to adjust this model so that $t = 0$ corresponds to 1980 rather than 1976. To do this, you shift the graph of f four units *to the left* and obtain

$$g(t) = f(t + 4).$$

(Source: Group Health Insurance Association of America)

(a) Write $g(t)$ in standard form.

(b) Use a graphing utility to graph f and g in the same viewing rectangle.

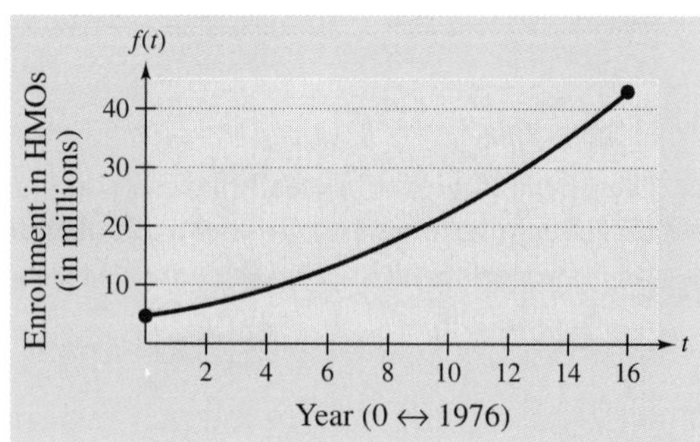

Review **Solve Exercises 87–90 as a review of the skills and problem-solving techniques you learned in previous sections. Describe the relationship between the graphs of f and g.**

87. $g(x) = f(x) + 8$

88. $g(x) = f(x - 3)$

89. $g(x) = f(-x)$

90. $g(x) = -f(x)$

7.6 Counting Principles

Simple Counting Problems / Counting Principles / Permutations / Combinations

Simple Counting Problems

The last two sections of this chapter present a brief introduction to some of the basic counting principles and their application to probability. In the next section, you will see that much of probability has to do with counting the number of ways an event can occur.

Real Life

EXAMPLE 1 Selecting Pairs of Numbers at Random

Eight pieces of paper are numbered from 1 to 8 and placed in a box. One piece of paper is drawn from the box, its number is written down, and the piece of paper is replaced in the box. Then, a piece of paper is again drawn from the box, and its number is written down. Finally, the two numbers are added together. How many different ways can a total of 12 be obtained?

Solution

To solve this problem, count the different ways that a total of 12 can be obtained using two numbers from 1 to 8.

First number	4	5	6	7	8
Second number	8	7	6	5	4

From this list, you can see that a total of 12 can occur in five different ways.

Real Life

EXAMPLE 2 Selecting Pairs of Numbers at Random

Note The difference between the counting problems in Examples 1 and 2 can be distinguished by saying that the random selection in Example 1 occurs **with replacement,** whereas the random selection in Example 2 occurs **without replacement,** which eliminates the possibility of choosing two 6's.

Eight pieces of paper are numbered from 1 to 8 and placed in a box. Two pieces of paper are drawn from the box, and the numbers on the paper are written down and totaled. How many different ways can a total of 12 be obtained?

Solution

To solve this problem, count the different ways that a total of 12 can be obtained *using two different numbers* from 1 to 8.

First number	4	5	7	8
Second number	8	7	5	4

Thus, a total of 12 can be obtained in four different ways.

Counting Principles

Examples 1 and 2 describe simple counting problems in which you can *list* each possible way that an event can occur. When it is possible, this is always the best way to solve a counting problem. However, some events can occur in so many different ways that it is not feasible to write out the entire list. In such cases, you must rely on formulas and counting principles. The most important of these is the **Fundamental Counting Principle.**

Note The Fundamental Counting Principle can be extended to three or more events. For instance, the number of ways that three events E_1, E_2, and E_3 can occur is $m_1 \cdot m_2 \cdot m_3$.

Fundamental Counting Principle

Let E_1 and E_2 be two events. The first event E_1 can occur in m_1 different ways. After E_1 has occurred, E_2 can occur in m_2 different ways. The number of ways that the two events can occur is $m_1 \cdot m_2$.

EXAMPLE 3 Using the Fundamental Counting Principle

Real Life

How many different pairs of letters from the English alphabet are possible?

Solution

This experiment has two events. The first event is the choice of the first letter, and the second event is the choice of the second letter. Because the English alphabet contains 26 letters, it follows that the number of letter pairs is $26 \cdot 26 = 676$.

EXAMPLE 4 Using the Fundamental Counting Principle

Real Life

Telephone numbers in the United States have 10 digits. The first three are the *area code* and the next seven are the *local telephone number.* How many different telephone numbers are possible within each area code? (Note that a local telephone number cannot begin with 0 or 1.)

Solution

Because the first digit cannot be 0 or 1, there are only eight choices for the first digit. For each of the other six digits, there are 10 choices.

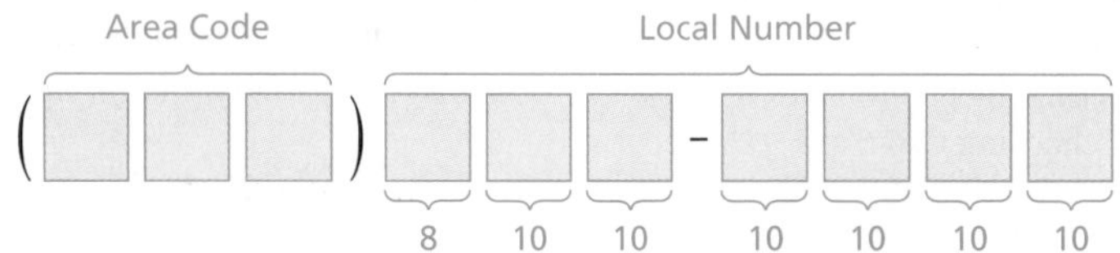

Thus, the number of local telephone numbers that are possible within each area code is $8 \cdot 10 \cdot 10 \cdot 10 \cdot 10 \cdot 10 \cdot 10 = 8{,}000{,}000$.

Permutations

One important application of the Fundamental Counting Principle is in determining the number of ways that n elements can be arranged (in order). An ordering of n elements is called a **permutation** of the elements.

Definition of Permutation

A **permutation** of n different elements is an ordering of the elements such that one element is first, one is second, one is third, and so on.

EXAMPLE 5 Finding the Number of Permutations of n Elements

How many permutations are possible for the letters A, B, C, D, E, and F?

Solution

Consider the following reasoning.

First position:	Any of the *six* letters.
Second position:	Any of the remaining *five* letters.
Third position:	Any of the remaining *four* letters.
Fourth position:	Any of the remaining *three* letters.
Fifth position:	Either of the remaining *two* letters.
Sixth position:	The *one* remaining letter.

Thus, the number of choices for the six positions are as follows.

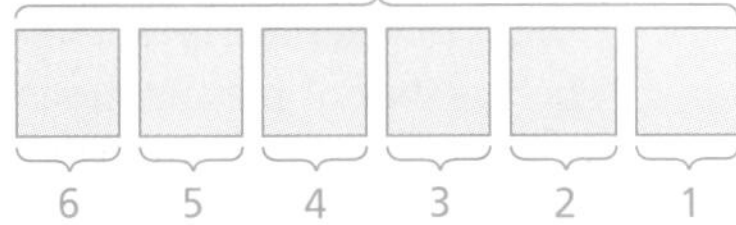

The total number of permutations of the six letters is $6! = 720$.

Number of Permutations of n Elements

The number of permutations of n elements is given by

$$n \cdot (n-1) \cdots 4 \cdot 3 \cdot 2 \cdot 1 = n!.$$

In other words, there are $n!$ different ways that n elements can be ordered.

Occasionally, you are interested in ordering a *subset* of a collection of elements rather than the entire collection. For example, you might want to choose (and order) r elements out of a collection of n elements. Such an ordering is called a **permutation of n elements taken r at a time.**

Real Life

EXAMPLE 6 **Counting Horse Race Finishes**

Eight horses are running in a race. In how many different ways can these horses come in first, second, and third? (Assume that there are no ties.)

Solution

Here are the different possibilities.

Win (first position):	*Eight* choices
Place (second position):	*Seven* choices
Show (third position):	*Six* choices

Using the Fundamental Counting Principle, multiply these three numbers together to obtain the following.

Different orders of horses

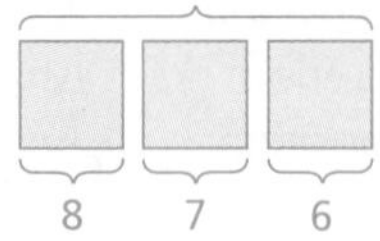

Thus, there are $8 \cdot 7 \cdot 6 = 336$ different orders.

Permutations of n Elements Taken r at a Time

The number of permutations of n elements taken r at a time is

$$_nP_r = \frac{n!}{(n-r)!} = n(n-1)(n-2)\cdot\cdot\cdot(n-r+1).$$

Using this formula, you can rework Example 6 to find that the number of permutations of eight horses taken three at a time is

$$\begin{aligned} _8P_3 &= \frac{8!}{5!} \\ &= \frac{8 \cdot 7 \cdot 6 \cdot \cancel{5!}}{\cancel{5!}} \\ &= 336 \end{aligned}$$

which is the same answer obtained in Example 6.

Remember that for permutations, order is important. Thus, if you are looking at the possible permutations of the letters A, B, C, and D taken three at a time, the permutations (A, B, D) and (B, A, D) would be different because the *order* of the elements is different.

Suppose, however, that you are asked to find the possible permutations of the letters A, A, B, and C. The total number of permutations of the four letters would be ${}_4P_4 = 4!$. However, not all of these arrangements would be *distinguishable* because there are two A's in the list. To find the number of distinguishable permutations, you can use the following formula.

Distinguishable Permutations

Suppose a set of n objects has n_1 of one kind of object, n_2 of a second kind, n_3 of a third kind, and so on, with $n = n_1 + n_2 + n_3 + \cdots + n_k$. Then the number of **distinguishable permutations** of the n objects is

$$\frac{n!}{n_1! \cdot n_2! \cdot n_3! \cdots n_k!}.$$

EXAMPLE 7 Distinguishable Permutations

In how many distinguishable ways can the letters in BANANA be written?

Solution

This word has six letters, of which three are A's, two are N's, and one is a B. Thus, the number of distinguishable ways the letters can be written is

$$\frac{6!}{3! \cdot 2! \cdot 1!} = \frac{6 \cdot 5 \cdot 4 \cdot \cancel{3!}}{\cancel{3!} \cdot 2!} = 60.$$

The 60 different "words" are as follows.

AAABNN	AAANBN	AAANNB	AABANN	AABNAN	AABNNA
AANABN	AANANB	AANBAN	AANBNA	AANNAB	AANNBA
ABAANN	ABANAN	ABANNA	ABNAAN	ABNANA	ABNNAA
ANAABN	ANAANB	ANABAN	ANABNA	ANANAB	ANANBA
ANBAAN	ANBANA	ANBNAA	ANNAAB	ANNABA	ANNBAA
BAAANN	BAANAN	BAANNA	BANAAN	BANANA	BANNAA
BNAAAN	BNAANA	BNANAA	BNNAAA	NAAABN	NAAANB
NAABAN	NAABNA	NAANAB	NAANBA	NABAAN	NABANA
NABNAA	NANAAB	NANABA	NANBAA	NBAAAN	NBAANA
NBANAA	NBNAAA	NNAAAB	NNAABA	NNABAA	NNBAAA

Combinations

When one counts the number of possible permutations of a set of elements, order is important. As a final topic in this section, we look at a method of selecting subsets of a larger set in which order *is not* important. Such subsets are called **combinations of n elements taken r at a time.** For instance, the combinations

$$\{A, B, C\} \quad \text{and} \quad \{B, A, C\}$$

are equivalent because both sets contain the same three elements, and the order in which the elements are listed is not important. Hence, you would count only one of the two sets. A common example of how a combination occurs is a card game in which the player is free to reorder the cards after they have been dealt.

EXAMPLE 8 Combinations of n Elements Taken r at a Time

In how many different ways can three letters be chosen from the letters A, B, C, D, and E? (The order of the three letters is not important.)

Solution

The following subsets represent the different combinations of three letters that can be chosen from five letters.

$$\{A, B, C\} \qquad \{A, B, D\}$$
$$\{A, B, E\} \qquad \{A, C, D\}$$
$$\{A, C, E\} \qquad \{A, D, E\}$$
$$\{B, C, D\} \qquad \{B, C, E\}$$
$$\{B, D, E\} \qquad \{C, D, E\}$$

From this list, you can conclude that there are 10 different ways that three letters can be chosen from five letters.

Most graphing utilities have keys that will evaluate the formulas for the number of permutations or combinations of n elements taken r at a time. For instance, on a *TI-82* or *TI-83*, you can evaluate ${}_8C_5$ as follows.

8 [MATH] (PRB) (3 : nCr) 5 [ENTER]

The display should be 56. You can evaluate ${}_8P_5$ (the number of permutations of eight elements taken five at a time) in a similar way.

Combinations of n Elements Taken r at a Time

The number of combinations of n elements taken r at a time is

$$_nC_r = \frac{n!}{(n-r)!r!}.$$

Note that the formula for $_nC_r$ is the same one given for binomial coefficients. To see how this formula is used, let's solve the counting problem in Example 8. In that problem, you are asked to find the number of combinations of five elements taken three at a time. Thus, $n = 5$, $r = 3$, and the number of combinations is

$$_5C_3 = \frac{5!}{2!3!} = \frac{5 \cdot \overset{2}{\cancel{4}} \cdot \cancel{3!}}{\cancel{2} \cdot 1 \cdot \cancel{3!}} = 10$$

which is the same answer obtained in Example 8.

Real Life

EXAMPLE 9 **Counting Card Hands**

A standard poker hand consists of five cards dealt from a deck of 52. How many different poker hands are possible? (After the cards are dealt, the player may reorder them, and therefore order is not important.)

Solution

You can find the number of different poker hands by using the formula for the number of combinations of 52 elements taken five at a time, as follows.

$$_{52}C_5 = \frac{52!}{47!5!} = \frac{52 \cdot 51 \cdot 50 \cdot 49 \cdot 48 \cdot 47!}{5 \cdot 4 \cdot 3 \cdot 2 \cdot 1 \cdot 47!} = 2{,}598{,}960$$

Group Activity

Problem Posing

According to NASA, each space shuttle astronaut consumes an average of 3000 calories per day. An evening meal normally consists of a main dish, a vegetable dish, and two different desserts. The space shuttle food list contains 10 items classified as main dishes, eight vegetable dishes, and 13 desserts. How many different evening meal menus are possible? Create two other problems that could be asked about the evening meal menus, and solve them. (Source: NASA)

7.6 /// EXERCISES

Random Selection In Exercises 1–8, determine the number of ways a computer can randomly generate one or more such integers from 1 through 12.

1. An odd integer
2. An even integer
3. A prime integer
4. An integer that is greater than 9
5. An integer that is divisible by 4
6. An integer that is divisible by 7
7. Two integers whose sum is 8
8. Two *distinct* integers whose sum is 8

9. *Entertainment Systems* A customer can choose one of two amplifiers, one of four disc players, and one of six speaker models for an entertainment system. Determine the number of possible system configurations.
10. *Computer Systems* A customer in a computer store can choose one of three monitors, one of two keyboards, and one of four computers. If all the choices are compatible, determine the number of possible system configurations.
11. *Job Applicants* A college needs two additional faculty members: a chemist and a statistician. In how many ways can these positions be filled if there are three applicants for the chemistry position and four applicants for the statistics position?
12. *Course Schedule* A college student is preparing a course schedule for the next semester. The student may select one of two mathematics courses, one of three science courses, and one of five courses from the social sciences and humanities. How many schedules are possible?
13. *True-False Exam* In how many ways can a six-question true-false exam be answered? (Assume that no questions are omitted.)
14. *True-False Exam* In how many ways can a 10-question true-false exam be answered? (Assume that no questions are omitted.)
15. *Toboggan Ride* Four people are lining up for a ride on a toboggan, but only two of the four are willing to take the first position. With that constraint, in how many ways can the four people be seated on the toboggan?
16. *Aircraft Boarding* Ten people are boarding an aircraft. Four have tickets for first class and board before those in the economy class. In how many ways can the 10 people board the aircraft?
17. *License Plate Numbers* In a certain state the automobile license plates consist of two letters followed by a four-digit number. How many distinct license plate numbers can be formed?
18. *License Plate Numbers* In a certain state the automobile license plates consist of two letters followed by a four-digit number. To avoid confusion between "O" and "zero" and "I" and "one," the letters "O" and "I" are not used. How many distinct license plate numbers can be formed?
19. *Three-Digit Numbers* How many three-digit numbers can be formed under the following conditions?
    (a) The leading digit cannot be zero.
    (b) The leading digit cannot be zero and no repetition of digits is allowed.
    (c) The leading digit cannot be zero and the number must be a multiple of 5.
    (d) The number is at least 400.
20. *Four-Digit Numbers* How many four-digit numbers can be formed under the following conditions?
    (a) The leading digit cannot be zero.
    (b) The leading digit cannot be zero and no repetition of digits is allowed.
    (c) The leading digit cannot be zero and the number must be less than 5000.
    (d) The leading digit cannot be zero and the number must be even.

21. *Combination Lock* A combination lock will open when the right choice of three numbers (from 1 to 40, inclusive) is selected. How many different lock combinations are possible?

22. *Combination Lock* A combination lock will open when the right choice of three numbers (from 1 to 50, inclusive) is selected. How many different lock combinations are possible?

23. *Concert Seats* Three couples have reserved seats in a given row for a concert. In how many different ways can they be seated if

(a) there are no seating restrictions?

(b) the two members of each couple wish to sit together?

24. *Single File* In how many orders can three girls and two boys walk through a doorway single file if

(a) there are no restrictions?

(b) the girls walk through before the boys?

In Exercises 25–30, evaluate ${}_nP_r$.

25. ${}_4P_4$ **26.** ${}_5P_5$

27. ${}_8P_3$ **28.** ${}_{20}P_2$

29. ${}_5P_4$ **30.** ${}_7P_4$

In Exercises 31 and 32, solve for n.

31. $14 \cdot {}_nP_3 = {}_{n+2}P_4$ **32.** ${}_nP_5 = 18 \cdot {}_{n-2}P_4$

In Exercises 33–38, evaluate using a calculator.

33. ${}_{20}P_5$ **34.** ${}_{100}P_5$

35. ${}_{100}P_3$ **36.** ${}_{10}P_8$

37. ${}_{20}C_5$ **38.** ${}_{10}C_7$

39. *Think About It* Can your calculator evaluate ${}_{100}P_{80}$? If not, explain why.

40. *Essay* Explain in words the meaning of ${}_nP_r$.

41. Write all permutations of the letters A, B, C, and D.

42. Write all the permutations of the letters A, B, C, and D if the letters B and C must remain between the letters A and D.

43. *Posing for a Photograph* In how many ways can five children line up in a row?

44. *Riding in a Car* In how many ways can six people sit in a six-passenger car?

45. *Choosing Officers* From a pool of 12 candidates, the offices of president, vice-president, secretary, and treasurer will be filled. In how many different ways can the offices be filled?

46. *Assembly Line Production* Four processes are involved in assembling a certain product, and they can be performed in any order. The management wants to test each order to determine which is the least time-consuming. How many different orders will have to be tested?

In Exercises 47–50, find the number of distinguishable permutations of the group of letters.

47. A, A, G, E, E, E, M

48. B, B, B, T, T, T, T, T

49. A, L, G, E, B, R, A

50. M, I, S, S, I, S, S, I, P, P, I

51. Write all the possible selections of two letters that can be formed from the letters A, B, C, D, E, and F. (The order of the two letters is not important.)

52. Write all the possible selections of three letters that can be formed from the letters A, B, C, D, E, and F. (The order of the three letters is not important.)

53. *Forming an Experimental Group* In order to conduct a certain experiment, four students are randomly selected from a class of 20. How many different groups of four students are possible?

54. *Test Questions* You can answer any 10 questions from a total of 12 questions on an exam. In how many different ways can you select the questions?

55. *Lottery Choices* There are 40 numbers in a particular state lottery. In how many ways can a player select six of the numbers?

56. *Lottery Choices* There are 50 numbers in a particular state lottery. In how many ways can a player select six of the numbers?

57. *Number of Subsets* How many subsets of four elements can be formed from a set of 100 elements?

58. *Number of Subsets* How many subsets of five elements can be formed from a set of 80 elements?

59. *Geometry* Three points that are not on a line determine three lines. How many lines are determined by seven points, no three of which are on a line?

60. *Defective Units* A shipment of 12 microwave ovens contains three defective units. In how many ways can a vending company purchase four of these units and receive (a) all good units, (b) two good units, and (c) at least two good units?

61. *Job Applicants* An employer interviews eight people for four openings in the company. Three of the eight people are women. If all eight are qualified, in how many ways can the employer fill the four positions if (a) the selection is random and (b) exactly two women are selected?

62. *Poker Hand* Five cards are selected from an ordinary deck of 52 playing cards. In how many ways can you get a full house? (A full house consists of three of one kind and two of another. For example, A-A-A-5-5 and K-K-K-10-10 are full houses.)

63. *Forming a Committee* Four people are to be selected at random from a group of four couples. In how many ways can this be done, given the following conditions?

(a) There are no restrictions.

(b) The group must have at least one couple.

(c) Each couple must be represented in the group.

64. *Interpersonal Relationships* The complexity of the interpersonal relationships increases dramatically as the size of a group increases. Determine the number of different two-person relationships in a group of people of size (a) 3, (b) 8, (c) 12, and (d) 20.

In Exercises 65–68, find the number of diagonals of the polygon. (A line segment connecting any two nonadjacent vertices is called a *diagonal* of the polygon.)

65. Pentagon

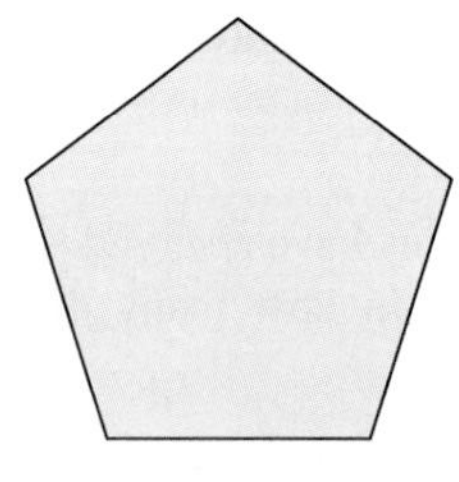

66. Hexagon

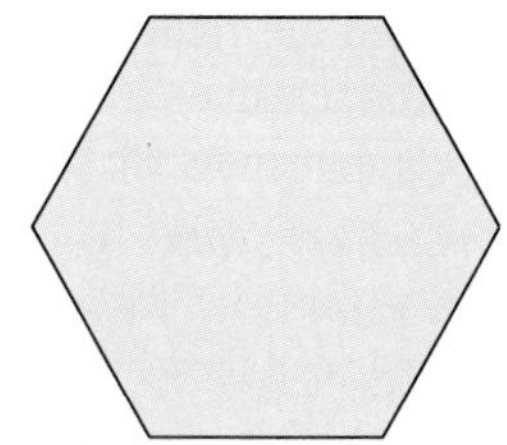

67. Octagon

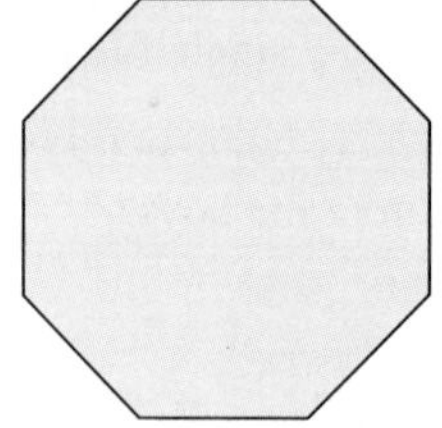

68. Decagon

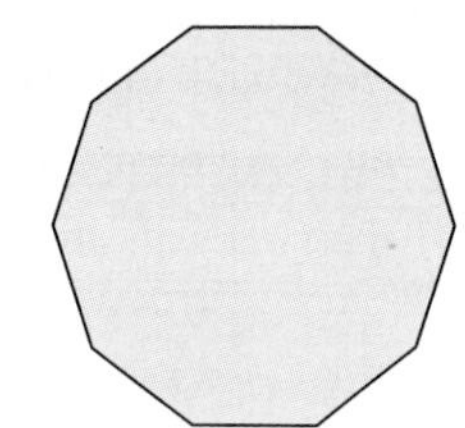

In Exercises 69–72, prove the identity.

69. ${}_nP_{n-1} = {}_nP_n$

70. ${}_nC_n = {}_nC_0$

71. ${}_nC_{n-1} = {}_nC_1$

72. ${}_nC_r = \dfrac{{}_nP_r}{r!}$

Review **Solve Exercises 73–76 as a review of the skills and problem-solving techniques you learned in previous sections. Solve the equation. (Round approximate answers to two decimal places.)**

73. $\sqrt{x-3} = x - 6$

74. $\dfrac{4}{t} + \dfrac{3}{2t} = 1$

75. $\log_2(x - 3) = 5$

76. $e^{x/3} = 16$

7.7 Probability

The Probability of an Event / Mutually Exclusive Events / Independent Events / The Complement of an Event

Blaise Pascal (1623–1662) was a French mathematician and scientist. He, along with Pierre de Fermat, laid the foundation for the mathematical theory of probability.

The Probability of an Event

Any happening whose result is uncertain is called an **experiment.** The possible results of the experiment are **outcomes,** the set of all possible outcomes of the experiment is the **sample space** of the experiment, and any subcollection of a sample space is an **event.**

For instance, when a six-sided die is tossed, the sample space can be represented by the numbers from 1 through 6. For this experiment, each of the outcomes is *equally likely.*

To describe sample spaces in such a way that each outcome is equally likely, you must sometimes distinguish between various outcomes in ways that appear artificial. Example 1 illustrates such a situation.

EXAMPLE 1 **Finding the Sample Space** *Real Life*

Find the sample space for each of the following.

a. One coin is tossed. **b.** Two coins are tossed. **c.** Three coins are tossed.

Solution

a. Because the coin will land either heads up (denoted by H) or tails up (denoted by T), the sample space is $S = \{H, T\}$.

b. Because either coin can land heads up or tails up, the possible outcomes are as follows.

HH = heads up on both coins

HT = heads up on first coin and tails up on second coin

TH = tails up on first coin and heads up on second coin

TT = tails up on both coins

Thus, the sample space is $S = \{HH, HT, TH, TT\}$. Note that this list distinguishes between the two cases HT and TH, even though these two outcomes appear to be similar.

c. Following the notation of part (b), the sample space is

$$S = \{HHH, HHT, HTH, HTT, THH, THT, TTH, TTT\}.$$

To calculate the probability of an event, count the number of outcomes in the event and in the sample space. The *number of outcomes* in event E is denoted by $n(E)$ and the number of outcomes in the sample space S is denoted by $n(S)$. The probability that event E will occur is given by $n(E)/n(S)$.

EXPLORATION

Toss two coins 40 times and write down the number of heads that occur on each toss (0, 1, or 2). How many times did two heads occur? How many times would you expect two heads to occur if you did the experiment 1000 times?

The Probability of an Event

If an event E has $n(E)$ equally likely outcomes and its sample space S has $n(S)$ equally likely outcomes, the **probability** of event E is

$$P(E) = \frac{n(E)}{n(S)}.$$

Because the number of outcomes in an event must be less than or equal to the number of outcomes in the sample space, the probability of an event must be a number from 0 to 1, inclusive. That is,

$$0 \leq P(E) \leq 1.$$

If $P(E) = 0$, event E *cannot occur*, and E is called an **impossible event.** If $P(E) = 1$, event E *must occur*, and E is called a **certain event.**

Real Life

EXAMPLE 2 Finding the Probability of an Event

a. Two coins are tossed. What is the probability that both land heads up?

b. A card is drawn from a standard deck of playing cards. What is the probability that it is an ace?

Solution

a. Following the procedure in Example 1(b), let

$$E = \{HH\}$$

and

$$S = \{HH, HT, TH, TT\}.$$

The probability of getting two heads is

$$P(E) = \frac{n(E)}{n(S)} = \frac{1}{4}.$$

b. Because there are 52 cards in a standard deck of playing cards and there are four aces (one in each suit), the probability of drawing an ace is

$$P(E) = \frac{n(E)}{n(S)} = \frac{4}{52} = \frac{1}{13}.$$

Figure 7.5

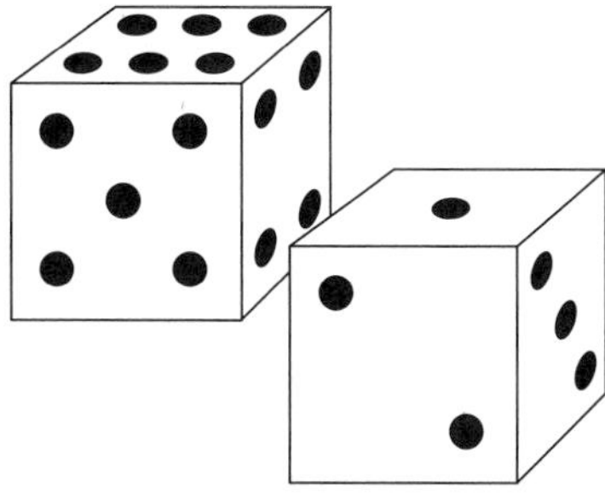

Real Life

EXAMPLE 3 **Finding the Probability of an Event**

Two six-sided dice are tossed. What is the probability that the total of the two dice is 7? (See Figure 7.5.)

Solution

Because there are six possible outcomes on each die, you can use the Fundamental Counting Principle to conclude that there are $6 \cdot 6$ or 36 different outcomes when two dice are tossed. To find the probability of rolling a total of 7, you must first count the number of ways this can occur.

First Die	1	2	3	4	5	6
Second Die	6	5	4	3	2	1

Thus, a total of 7 can be rolled in six ways, which means that the probability of rolling a 7 is

$$P(E) = \frac{n(E)}{n(S)} = \frac{6}{36} = \frac{1}{6}.$$

You could have written out each sample space in Examples 2 and 3 and simply counted the outcomes in the desired events. For larger sample spaces, however, you should use the counting principles discussed in Section 7.6.

Real Life

EXAMPLE 4 **Finding the Probability of an Event**

Twelve-sided dice, as shown in Figure 7.6, can be constructed (in the shape of regular dodecahedrons) so that each of the numbers from 1 to 6 appears twice on each die. Prove that these dice can be used in any game requiring ordinary six-sided dice without changing the probabilities of different outcomes.

Figure 7.6

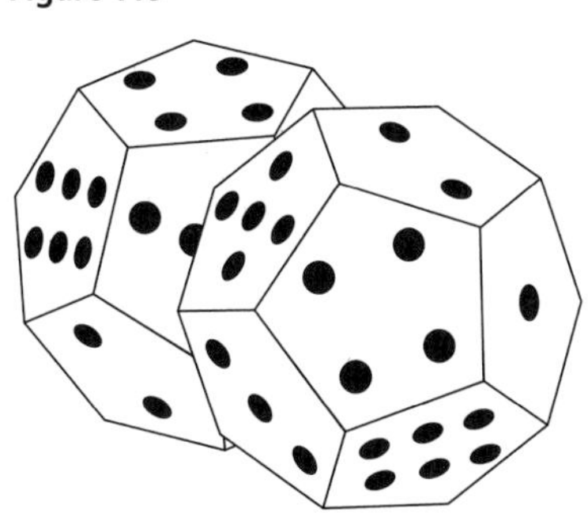

Solution

For an ordinary six-sided die, each of the numbers 1, 2, 3, 4, 5, and 6 occurs only once, so the probability of any particular number coming up is

$$P(E) = \frac{n(E)}{n(S)} = \frac{1}{6}.$$

For one of the 12-sided dice, each number occurs twice, so the probability of any particular number coming up is

$$P(E) = \frac{n(E)}{n(S)} = \frac{2}{12} = \frac{1}{6}.$$

Real Life

EXAMPLE 5 **The Probability of Winning a Lottery**

In a state lottery, a player chooses six different numbers from 1 to 40. If these six numbers match the six numbers drawn by the lottery commission, the player wins (or shares) the top prize. What is the probability of winning the top prize?

Solution

To find the number of elements in the sample space, use the formula for the number of combinations of 40 elements taken six at a time.

$$n(S) = {}_{40}C_6 = \frac{40 \cdot 39 \cdot 38 \cdot 37 \cdot 36 \cdot 35}{6 \cdot 5 \cdot 4 \cdot 3 \cdot 2 \cdot 1} = 3{,}838{,}380$$

If a person buys only one ticket, the probability of winning is

$$P(E) = \frac{n(E)}{n(S)} = \frac{1}{3{,}838{,}380}.$$

Real Life

EXAMPLE 6 **Random Selection**

The numbers of colleges and universities in the United States in 1992 are shown in Figure 7.7. One institution is selected at random. What is the probability that the institution is in one of the three southern regions? (Source: U.S. National Center for Education Statistics)

Solution

From the figure, the total number of colleges and universities is 3628. Because there are $600 + 272 + 289 = 1161$ colleges and universities in the three southern regions, the probability that the institution is in one of these regions is

$$P(E) = \frac{n(E)}{n(S)} = \frac{1161}{3628} \approx 0.320.$$

Figure 7.7

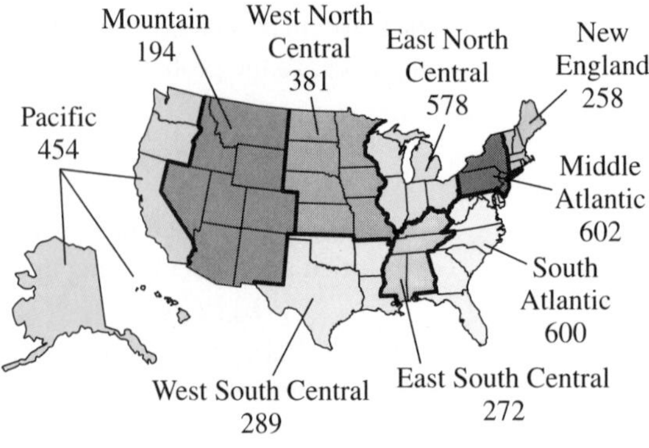

Mutually Exclusive Events

Two events A and B (from the same sample space) are **mutually exclusive** if A and B have no outcomes in common. In the terminology of sets, the **intersection of A and B** is the empty set and

$$P(A \cap B) = 0.$$

For instance, if two dice are tossed, the event A of rolling a total of 6 and the event B of rolling a total of 9 are mutually exclusive. To find the probability that one or the other of two mutually exclusive events will occur, you can *add* their individual probabilities.

Probability of the Union of Two Events

If A and B are events in the same sample space, the probability of A *or* B occurring is given by

$$P(A \cup B) = P(A) + P(B) - P(A \cap B).$$

If A and B are mutually exclusive, then

$$P(A \cup B) = P(A) + P(B).$$

Real Life

EXAMPLE 7 The Probability of a Union

One card is selected from a standard deck of 52 playing cards. What is the probability that the card is either a heart or a face card?

Solution

Because the deck has 13 hearts, the probability of selecting a heart (event A) is $P(A) = \frac{13}{52}$. Similarly, because the deck has 12 face cards, the probability of selecting a face card (event B) is $P(B) = \frac{12}{52}$. Because three of the cards are hearts and face cards (see Figure 7.8), it follows that $P(A \cap B) = \frac{3}{52}$. Finally, applying the formula for the probability of the union of two events, you can conclude that the probability of selecting a heart or a face card is

$$\begin{aligned} P(A \cup B) &= P(A) + P(B) - P(A \cap B) \\ &= \frac{13}{52} + \frac{12}{52} - \frac{3}{52} \\ &= \frac{22}{52} \\ &\approx 0.423. \end{aligned}$$

Figure 7.8

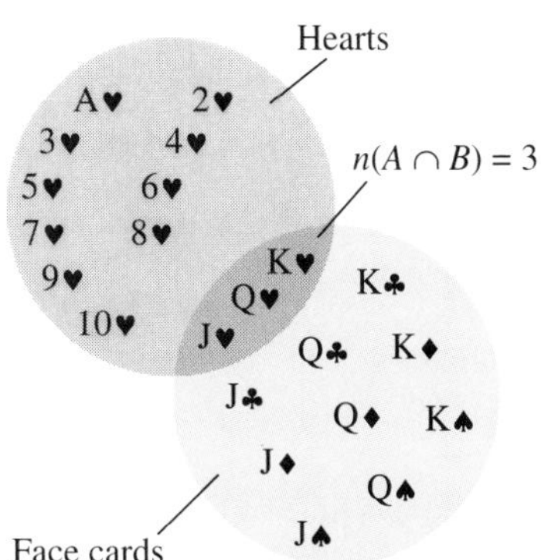

Real Life

EXAMPLE 8 Probability of Mutually Exclusive Events

The personnel department of a company has compiled data on employees' numbers of years of service. The results are shown in the table.

Years of Service	Number of Employees
0–4	157
5–9	89
10–14	74
15–19	63
20–24	42
25–29	38
30–34	37
35–39	21
40–44	8

If an employee is chosen at random, what is the probability that the employee has had nine or fewer years of service?

Solution

To begin, add the number of employees and find that the total is 529. Next, let event A represent choosing an employee with 0 to 4 years of service and let event B represent choosing an employee with 5 to 9 years of service. Then

$$P(A) = \frac{157}{529} \quad \text{and} \quad P(B) = \frac{89}{529}.$$

Because A and B have no outcomes in common, you can conclude that these two events are mutually exclusive and that

$$\begin{aligned} P(A \cup B) = P(A) + P(B) &= \frac{157}{529} + \frac{89}{529} \\ &= \frac{246}{529} \\ &\approx 0.465. \end{aligned}$$

Thus, the probability of choosing an employee who has nine or fewer years of service is about 0.465.

Independent Events

Two events are **independent** if the occurrence of one has no effect on the occurrence of the other. To find the probability that two independent events will occur, *multiply* the probabilities of each. For instance, rolling a total of 12 with two six-sided dice has no effect on the outcome of future rolls of the dice.

Probability of Independent Events

If A and B are independent events, the probability that both A and B will occur is

$$P(A \text{ and } B) = P(A) \cdot P(B).$$

Real Life

EXAMPLE 9 Probability of Independent Events

A random number generator on a computer selects three integers from 1 to 20. What is the probability that all three numbers are less than or equal to 5?

Solution

The probability of selecting a number from 1 to 5 is

$$P(A) = \frac{5}{20} = \frac{1}{4}.$$

Thus, the probability that all three numbers are less than or equal to 5 is

$$P(A) \cdot P(A) \cdot P(A) = \left(\frac{1}{4}\right)\left(\frac{1}{4}\right)\left(\frac{1}{4}\right)$$
$$= \frac{1}{64}.$$

Real Life

EXAMPLE 10 Probability of Independent Events

In 1992, 56% of the population of the United States were 30 years old or older. Suppose that in a survey, 10 people were chosen at random from the population. What is the probability that all 10 were 30 years old or older? (Source: U.S. Bureau of the Census)

Solution

Let A represent choosing a person who was 30 years old or older. Because the probability of choosing a person who was 30 years old or older was 0.56, you can conclude that the probability that all 10 people were 30 years old or older is

$$[P(A)]^{10} = (0.56)^{10} \approx 0.0030.$$

EXPLORATION

You are in a class with 22 other people. What is the probability that the birthdays of at least two of the 23 people fall on the same day of the year?

The complement of the probability that at least two people have the same birthday is the probability that all 23 birthdays are different. So, first find the probability that all 23 people have different birthdays and then find the complement.

Now, determine the probability that in a room with 50 people at least two people have the same birthday.

The Complement of an Event

The **complement of an event** A is the collection of all outcomes in the sample space that are *not* in A. The complement of event A is denoted by A'. Because $P(A \text{ or } A') = 1$ and because A and A' are mutually exclusive, it follows that $P(A) + P(A') = 1$. Therefore, the probability of A' is given by

$$P(A') = 1 - P(A).$$

For instance, if the probability of *winning* a certain game is

$$P(A) = \frac{1}{4}$$

the probability of *losing* the game is

$$P(A') = 1 - \frac{1}{4} = \frac{3}{4}.$$

Probability of a Complement

Let A be an event and let A' be its complement. If the probability of A is $P(A)$, the probability of the complement is

$$P(A') = 1 - P(A).$$

EXAMPLE 11 Finding the Probability of a Complement

Real Life

A manufacturer has determined that a certain machine averages one faulty unit for every 1000 it produces. What is the probability that an order of 200 units will have one or more faulty units?

Solution

To solve this problem as stated, you would need to find the probabilities of having exactly one faulty unit, exactly two faulty units, exactly three faulty units, and so on. However, using complements, you can simply find the probability that all units are perfect and then subtract this value from 1. Because the probability that any given unit is perfect is 999/1000, the probability that all 200 units are perfect is

$$P(A) = \left(\frac{999}{1000}\right)^{200} \approx 0.8186.$$

Therefore, the probability that at least one unit is faulty is

$$P(A') = 1 - P(A) \approx 0.1814.$$

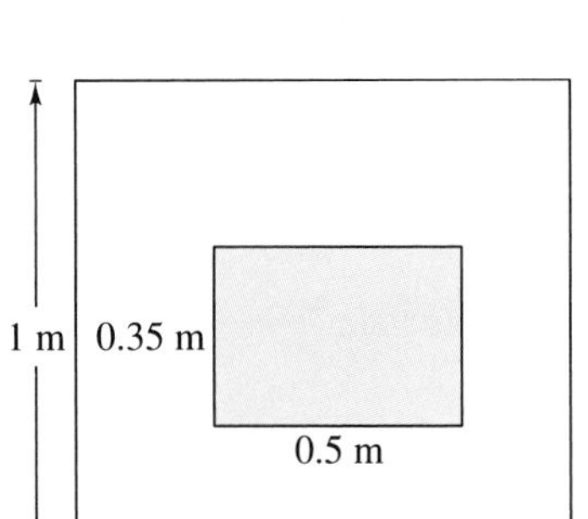

Group Activity — An Experiment in Geometric Probability

In this section you have been finding probabilities from a *theoretical* point of view. Another way to find probabilities is from an *experimental* point of view. For instance, suppose you want to find the probability of hitting the shaded portion of the rectangular target shown in the figure when a dart is thrown.

a. What is the theoretical probability that a thrown dart hits the shaded portion? Explain your reasoning.

b. The following *TI-82/TI-83* program can be used to simulate the outcome of a dart throw. (Programs for other calculator models may be found in the appendix.) Discuss how the program works. What does it do?

```
PROGRAM:DARTS
:0→K
:0→W
:Input "HOW MANY THROWS?",N
:Lbl 1
:K+1→K
:rand→X:rand→Y
:If X<.25 or X>.75
:Goto 2
:If Y<.3 or Y>.65
:Goto 2
:W+1→W
:Lbl 2
:If K=N
:Then
:Disp "WINS=",W
:Disp "LOSSES=",(N–W)
:Stop
:End
:Goto 1
```

c. Use the program to simulate 30 throws of a dart. What proportion of the throws landed inside the shaded area? This proportion is an experimental probability. How does it compare with the theoretical probability you calculated in part (a)?

d. Combine your results with those of the rest of your class. What proportion of the pooled throws landed in the shaded region? How does this experimental probability compare with the theoretical probability and the experimental probability from part (c)?

e. Can you make a conjecture about the relationship between the difference in the experimental and theoretical probabilities and the number of trials used to find the experimental probability? Explain how you might be able to get an experimental probability that more closely approximates the theoretical probability.

f. Discuss the types of situations in which using simulations to estimate probabilities would be useful.

7.7 /// EXERCISES

In Exercises 1–6, determine the sample space for the given experiment.

1. A coin and a six-sided die are tossed.
2. A six-sided die is tossed twice and the sum of the points is recorded.
3. A taste tester has to rank three varieties of yogurt, A, B, and C, according to preference.
4. Two marbles are selected from a sack containing two red marbles, two blue marbles, and one black marble. The color of each marble is recorded.
5. Two county supervisors are selected from five supervisors, A, B, C, D, and E, to study a recycling plan.
6. A sales representative makes presentations about a product in three homes per day. In each home there may be a sale (denote by S) or there may be no sale (denote by F).

Heads or Tails **In Exercises 7–10, find the probability in the experiment of tossing a coin three times. Use the sample space $S = \{HHH, HHT, HTH, HTT, THH, THT, TTH, TTT\}$.**

7. The probability of getting exactly one tail
8. The probability of getting a head on the first toss
9. The probability of getting at least one head
10. The probability of getting at least two heads

Drawing a Card **In Exercises 11–14, find the probability in the experiment of selecting one card from a standard deck of 52 playing cards.**

11. The card is a face card.
12. The card is not a face card.
13. The card is a red face card.
14. The card is a 6 or lower. (Aces are low.)

Tossing a Die **In Exercises 15–20, find the probability in the experiment of tossing a six-sided die twice.**

15. The sum is 4.
16. The sum is at least 7.
17. The sum is less than 11.
18. The sum is 2, 3, or 12.
19. The sum is odd and no more than 7.
20. The sum is odd or prime.

Drawing Marbles **In Exercises 21–24, find the probability in the experiment of drawing two marbles (without replacement) from a bag containing one green, two yellow, and three red marbles.**

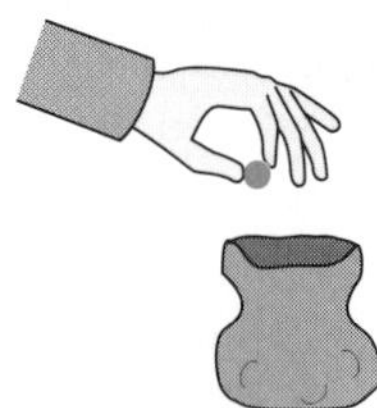

21. Both marbles are red.
22. Both marbles are yellow.
23. Neither marble is yellow.
24. The marbles are of different colors.

In Exercises 25 and 26, you are given the probability that an event *will* happen. Find the probability that the event *will not* happen.

25. $P(E) = 0.7$
26. $P(E) = 0.36$

In Exercises 27 and 28, you are given the probability that an event *will not* happen. Find the probability that the event *will* happen.

27. $P(E') = 0.15$
28. $P(E') = 0.84$

29. *Graphical Reasoning* At the end of 1994 there were approximately 2.5 million minimum-wage workers in the United States. The figure gives the age profile of these workers. (Source: U.S. Bureau of Labor Statistics)

(a) Estimate the number of minimum-wage workers in the age category 16–19.

(b) A person is selected at random from the population of minimum-wage workers. What is the probability that the person is in the 25–34 age group?

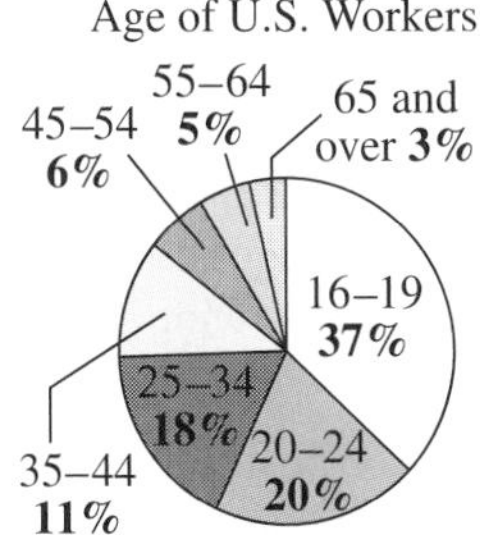

30. *Graphical Reasoning* In 1991 there were approximately 101 million workers in the civilian labor force in the United States. The figure gives the educational attainments of these workers. (Source: U.S. Bureau of Labor Statistics)

(a) Estimate the number of workers whose highest educational attainment was a high school education.

(b) A person is selected at random from the civilian work force. What is the probability that the person has 1 to 3 years of college education?

(c) A person is selected at random from the civilian work force. What is the probability that the person has more than a high school education?

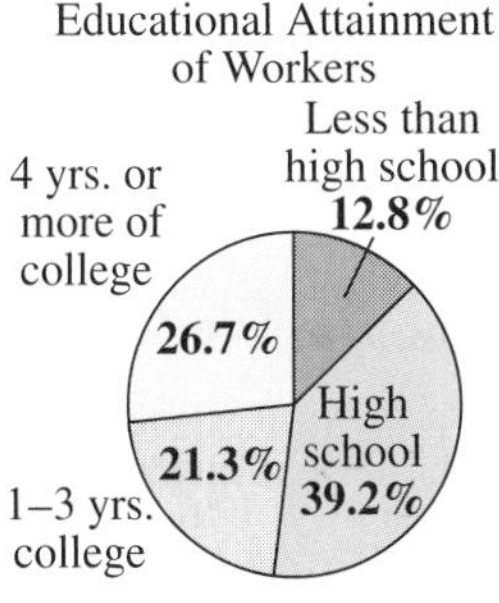

31. *Data Analysis* A study of the effectiveness of a flu vaccine was conducted with a sample of 500 people. Some in the study were given no vaccine, some were given one injection, and some were given two injections. The results of the study are given below.

	No Vaccine	One Injection	Two Injections	Total
Flu	7	2	13	22
No flu	149	52	277	478
Total	156	54	290	500

A person is selected at random from the sample. Find the specified probability.

(a) The person had two injections.

(b) The person did not get the flu.

(c) The person got the flu and had one injection.

32. *Data Analysis* A sample of 100 college students were interviewed to determine their political party affiliations and whether they favored a balanced-budget amendment to the constitution. The results of the study are given in the table.

	Favor	Not favor	Unsure	Total
Democrat	23	25	7	55
Republican	32	9	4	45
Total	55	34	11	100

A person is selected at random from the sample. Find the probability that the described person is selected.

(a) A person who doesn't favor the amendment

(b) A Republican

(c) A Democrat who favors the amendment

33. *Alumni Association* A college sends a survey to selected members of the class of 1995. Of the 1254 people who graduated that year, 672 are women, and of those, 124 went on to graduate school. Of the 582 male graduates, 198 went on to graduate school. If an alumni member is selected at random, what is the probability that the person is (a) female, (b) male, and (c) female and did not attend graduate school?

34. *Post–High School Education* In a high school graduating class of 72 students, 28 are on the honor roll. Of these, 18 are going on to college, and of the other 44 students, 12 are going on to college. If a student is selected at random from the class, what is the probability that the person chosen is (a) going to college, (b) not going to college, and (c) on the honor roll, but not going to college?

35. *Winning an Election* Taylor, Moore, and Jenkins are candidates for public office. It is estimated that Moore and Jenkins have about the same probability of winning, and Taylor is believed to be twice as likely to win as either of the others. Find the probability of each candidate winning the election.

36. *Winning an Election* Three people have been nominated for president of a class. From a poll, it is estimated that the first has a 37% chance of winning and the second has a 44% chance of winning. What is the probability that the third candidate will win?

In Exercises 37–48, the sample spaces are large and you should use the counting principles discussed in Section 7.6.

37. *Preparing for a Test* A class is given a list of 20 study problems from which 10 will be part of an upcoming exam. If a given student knows how to solve 15 of the problems, find the probability that the student will be able to answer (a) all 10 questions on the exam, (b) exactly eight questions on the exam, and (c) at least nine questions on the exam.

38. *Preparing for a Test* A class is given a list of eight study problems from which five will be part of an upcoming exam. If a given student knows how to solve six of the problems, find the probability that the student will be able to answer (a) all five questions on the exam, (b) exactly four questions on the exam, and (c) at least four questions on the exam.

39. *Letter Mix-Up* Four letters and envelopes are addressed to four different people. If the letters are randomly inserted into the envelopes, what is the probability that (a) exactly one is inserted in the correct envelope and (b) at least one is inserted in the correct envelope?

40. *Payroll Mix-Up* Five paychecks and envelopes are addressed to five different people. If the paychecks are randomly inserted into the envelopes, what is the probability that (a) exactly one is inserted in the correct envelope and (b) at least one is inserted in the correct envelope?

41. *Game Show* On a game show you are given five digits to arrange in the proper order to form the price of a car. If you are correct, you win the car. What is the probability of winning, given the following conditions?

(a) You guess the position of each digit.

(b) You know the first digit and guess the others.

42. *Game Show* On a game show you are given four digits to arrange in the proper order to form the price of a car. If you are correct, you win the car. What is the probability of winning, given the following conditions?

(a) You guess the position of each digit.

(b) You know the first digit and guess the others.

43. *Drawing Cards from a Deck* Two cards are selected at random from an ordinary deck of 52 playing cards. Find the probability that two aces are selected, given the following conditions.

(a) The cards are drawn in sequence, with the first card being replaced and the deck reshuffled prior to the second drawing.

(b) The two cards are drawn consecutively, without replacement.

44. *Poker Hand* Five cards are drawn from an ordinary deck of 52 playing cards. What is the probability of getting a full house?

45. *Defective Units* A shipment of 12 microwave ovens contains three defective units. A vending company has ordered four of these units, and because all are packaged identically, the selection will be random.

(a) What is the probability that all four units are good?

(b) What is the probability that exactly two units are good?

(c) What is the probability that at least two units are good?

46. *Defective Units* A shipment of 20 compact disc players contains four defective units. A retail outlet has ordered five of these units.

(a) What is the probability that all five units are good?

(b) What is the probability that exactly four units are good?

(c) What is the probability that at least one unit is good?

47. *Random Number Generator* Two integers (from 1 through 30) are chosen by a random number generator. What is the probability that (a) the numbers are both even, (b) one number is even and one is odd, (c) both numbers are less than 10, and (d) the same number is chosen twice?

48. *Random Number Generator* Two integers (from 1 through 40) are chosen by a random number generator. What is the probability that (a) the numbers are both even, (b) one number is even and one is odd, (c) both numbers are less than 30, and (d) the same number is chosen twice?

49. *Backup System* A space vehicle has an independent backup system for one of its communication networks. The probability that either system will function satisfactorily for the duration of a flight is 0.985. What is the probability that during a given flight (a) both systems function satisfactorily, (b) at least one system functions satisfactorily, and (c) both systems fail?

50. *Backup Vehicle* A fire company keeps two rescue vehicles to serve the community. Because of the demand on the company's time and the chance of mechanical failure, the probability that a specific vehicle is available when needed is 90%. If the availability of one vehicle is *independent* of the other, find the probability that (a) both vehicles are available at a given time, (b) neither vehicle is available at a given time, and (c) at least one vehicle is available at a given time.

51. *Making a Sale* A sales representative makes sales at approximately one-fourth of all calls. If, on a given day, the representative contacts five potential clients, what is the probability that a sale will be made with (a) all five contacts, (b) none of the contacts, and (c) at least one contact?

52. *A Boy or a Girl?* Assume that the probability of the birth of a child of a particular sex is 50%. In a family with four children, what is the probability that (a) all the children are boys, (b) all the children are the same sex, and (c) there is at least one boy?

53. *Is That Cash or Charge?* Suppose that the methods used by shoppers to pay for merchandise are as shown in the pie graph. If two shoppers are chosen at random, what is the probability that both shoppers paid for their purchases only in cash?

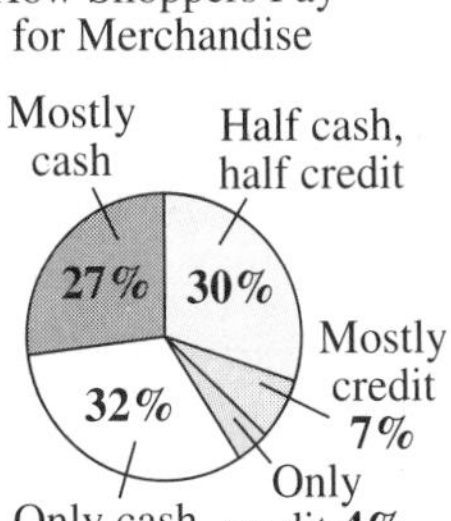

54. *Flexible Work Hours* In a survey, people were asked if they would prefer to work flexible hours—even if it meant slower career advancement—so they could spend more time with their families. The results of the survey are shown in the figure below. Suppose that three people from the survey were chosen at random. What is the probability that all three people would prefer flexible work hours?

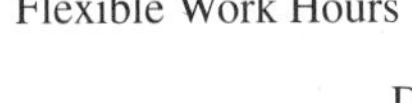

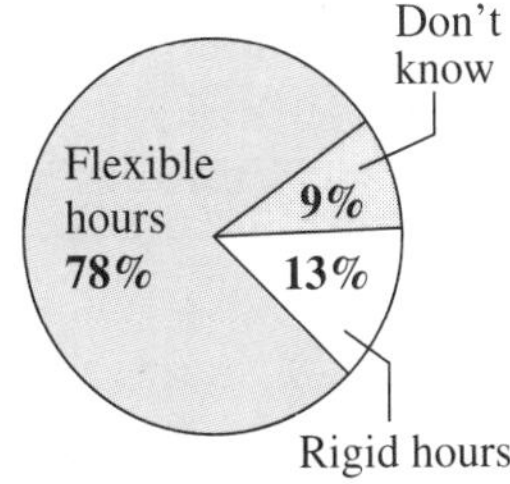

55. *Geometry* You and a friend agree to meet at your favorite fast-food restaurant between 5:00 and 6:00 P.M. The one who arrives first will wait 15 minutes for the other, after which the first person will leave (see figure). What is the probability that the two of you will actually meet, assuming that your arrival times are random within the hour?

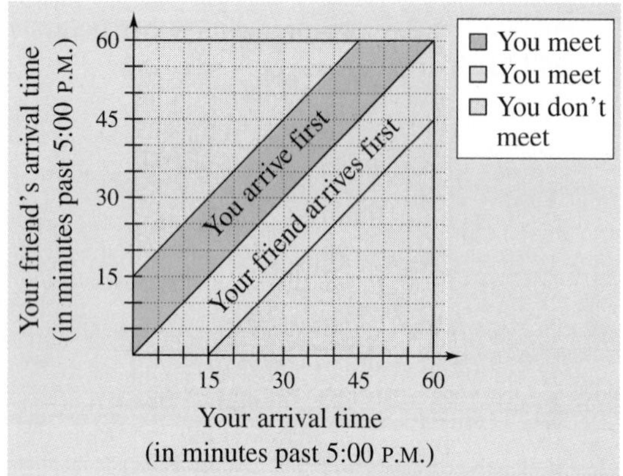

56. *Estimating π* A coin of diameter d is dropped onto a paper that contains a grid of squares d units on a side (see figure).

(a) Find the probability that the coin covers a vertex of one of the squares on the grid.

(b) Perform the experiment 100 times and use the results to approximate π.

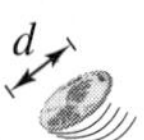

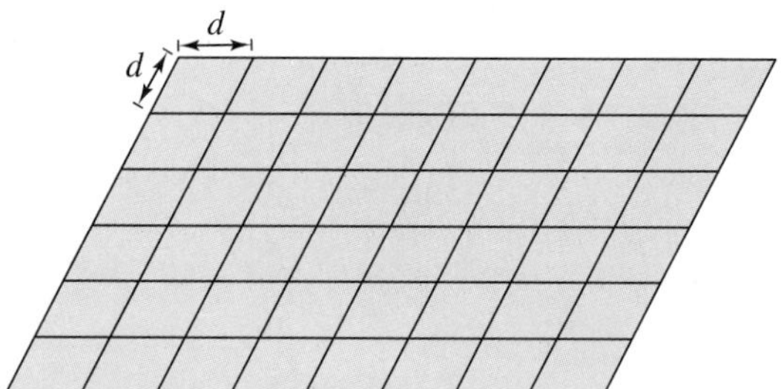

57. *Pattern Recognition and Exploration* Consider a group of n people.

(a) Explain why the following gives the probability that the n people have distinct birthdays.

$$n = 2: \quad \frac{365}{365} \cdot \frac{364}{365} = \frac{365 \cdot 364}{365^2}$$

$$n = 3: \quad \frac{365}{365} \cdot \frac{364}{365} \cdot \frac{363}{365} = \frac{365 \cdot 364 \cdot 363}{365^3}$$

(b) Use the pattern in part (a) to write an expression for the probability that four people ($n = 4$) have distinct birthdays.

(c) Let P_n be the probability that the n people have distinct birthdays. Verify that this probability can be obtained recursively by

$$P_1 = 1 \quad \text{and} \quad P_n = \frac{365 - (n - 1)}{365} P_{n-1}.$$

(d) Explain why $Q_n = 1 - P_n$ gives the probability that at least two people in a group of n people have the same birthday.

(e) Use the results of parts (c) and (d) to complete the table.

n	10	15	20	23	30	40	50
P_n							
Q_n							

(f) How many people must be in a group before the probability of at least two of them having the same birthday is greater than $\frac{1}{2}$? Explain.

58. *Think About It* The weather forecast indicates that the probability of rain is 40%. Explain what this means?

59. *Chapter Opener* If the probability that a dive of an elephant seal will exceed 400 meters is 0.546, determine the probability that a dive will be 400 meters or less.

60. *Chapter Opener* Use the histogram on page 523 to estimate the probability that a dive will be between 300 and 500 meters.

Focus on Concepts

In this chapter, you studied several concepts that are required in the study of sequences, counting principles, and probability. You can use the following questions to check your understanding of several of these basic concepts. The answers to these questions are given in the back of the book.

1. An infinite sequence is a function. What is the domain of the function?

2. How do the two sequences differ?

   (a) $a_n = \dfrac{(-1)^n}{n}$ (b) $a_n = \dfrac{(-1)^{n+1}}{n}$

In Exercises 3–6, decide whether the statement is true. Explain your reasoning.

3. $\dfrac{(n+2)!}{n!} = (n+2)(n+1)$

4. $\displaystyle\sum_{i=1}^{5}(i^3 + 2i) = \sum_{i=1}^{5} i^3 + \sum_{i=1}^{5} 2i$

5. $\displaystyle\sum_{k=1}^{8} 3k = 3\sum_{k=1}^{8} k$

6. $\displaystyle\sum_{j=1}^{6} 2^j = \sum_{j=3}^{8} 2^{j-2}$

7. In your own words, explain what makes a sequence (a) arithmetic and (b) geometric.

8. The graphs of two sequences are shown below. Identify each sequence as arithmetic or geometric. Explain your reasoning.

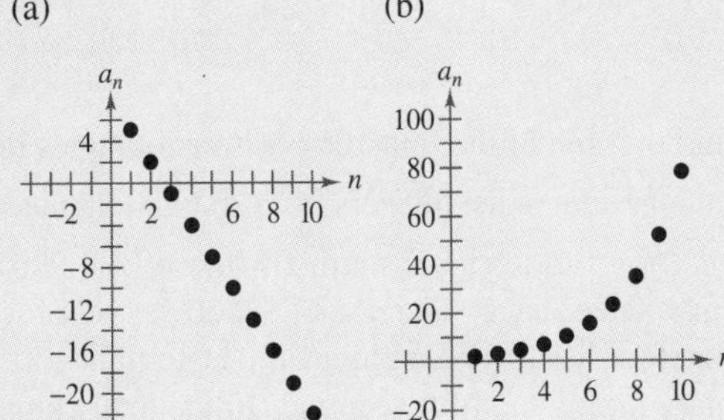

9. Explain what a recursion formula is.

10. Explain why the terms of a geometric sequence decrease when $0 < r < 1$.

In Exercises 11–14, match the sequence or sum of a sequence with its graph without doing any calculations. Explain your reasoning. [The graphs are labeled (a), (b), (c), and (d).]

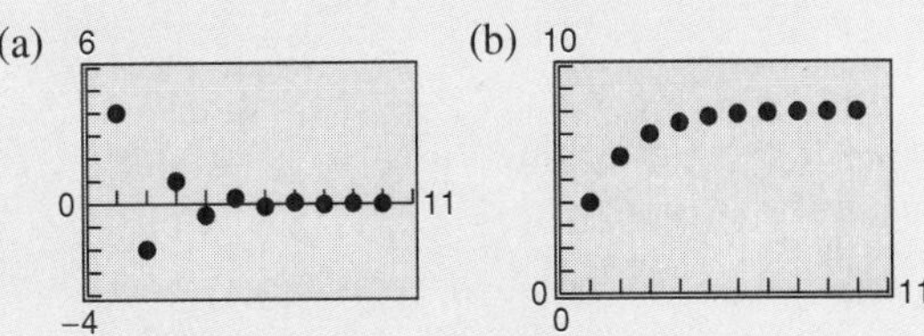

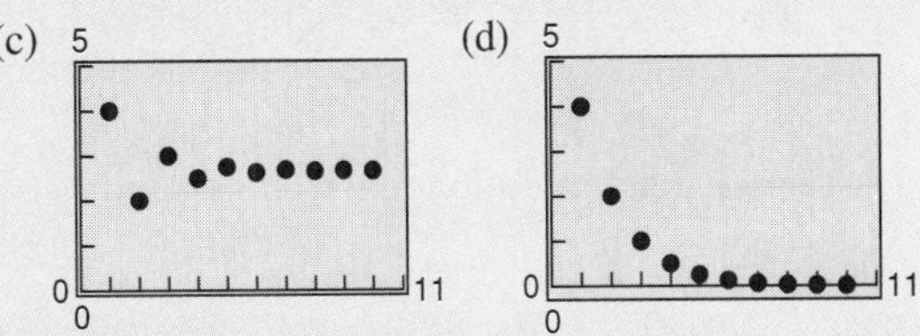

11. $a_n = 4\left(\frac{1}{2}\right)^{n-1}$

12. $a_n = 4\left(-\frac{1}{2}\right)^{n-1}$

13. $a_n = \displaystyle\sum_{k=1}^{n} 4\left(\frac{1}{2}\right)^{k-1}$

14. $a_n = \displaystyle\sum_{k=1}^{n} 4\left(-\frac{1}{2}\right)^{k-1}$

15. How do the expansions of $(x+y)^n$ and $(x-y)^n$ differ?

16. What is the relationship between ${}_nC_r$ and ${}_nC_{n-r}$?

17. Without calculating the numbers, determine which of the following is greater. Explain.

    (a) The combination of 10 elements taken 6 at a time

    (b) The permutation of 10 elements taken 6 at a time

18. The probability of an event must be a real number in what interval? Is the interval open or closed?

19. The probability of an event is $\frac{2}{3}$. What is the probability that the event does not occur? Explain.

20. The weather forecast indicates that the probability of rain is 60%. Explain what this means.

7 /// REVIEW EXERCISES

In Exercises 1–4, write the first five terms of the sequence. (Assume n begins with 1.)

1. $a_n = 2 + \frac{6}{n}$

2. $a_n = \frac{5n}{2n - 1}$

3. $a_n = \frac{72}{n!}$

4. $a_n = n(n - 1)$

In Exercises 5–8, use a graphing utility to graph the first 10 terms of the sequence.

5. $a_n = \frac{3}{2}n$

6. $a_n = 4(0.4)^{n-1}$

7. $a_n = \frac{3n}{n + 2}$

8. $a_n = 5 - \frac{3}{n}$

In Exercises 9–12, use sigma notation to write the sum.

9. $\frac{1}{2(1)} + \frac{1}{2(2)} + \frac{1}{2(3)} + \cdots + \frac{1}{2(20)}$

10. $2(1^2) + 2(2^2) + 2(3^2) + \cdots + 2(9^2)$

11. $\frac{1}{2} + \frac{2}{3} + \frac{3}{4} + \cdots + \frac{9}{10}$

12. $1 - \frac{1}{3} + \frac{1}{9} - \frac{1}{27} + \cdots$

In Exercises 13–20, find the sum.

13. $\sum_{i=1}^{6} 5$

14. $\sum_{k=2}^{5} 4k$

15. $\sum_{j=1}^{4} \frac{6}{j^2}$

16. $\sum_{i=1}^{8} \frac{i}{i + 1}$

17. $\sum_{k=1}^{10} 2k^3$

18. $\sum_{j=0}^{4} (j^2 + 1)$

19. $\sum_{n=0}^{10} (n^2 + 3)$

20. $\sum_{n=1}^{100} \left(\frac{1}{n} - \frac{1}{n + 1}\right)$

In Exercises 21–24, write the first five terms of the arithmetic sequence.

21. $a_1 = 3,\ d = 4$

22. $a_1 = 8,\ d = -2$

23. $a_4 = 10,\ a_{10} = 28$

24. $a_2 = 14,\ a_6 = 22$

In Exercises 25–28, write the first five terms of the arithmetic sequence defined recursively. Determine the common difference and write the nth term of the sequence as a function of n.

25. $a_1 = 35 \qquad a_{k+1} = a_k - 3$

26. $a_1 = 15 \qquad a_{k+1} = a_k + \frac{5}{2}$

27. $a_1 = 9 \qquad a_{k+1} = a_k + 7$

28. $a_1 = 100 \qquad a_{k+1} = a_k - 5$

In Exercises 29 and 30, write an expression for the nth term of the arithmetic sequence and find the sum of the first 20 terms of the sequence.

29. $a_1 = 100,\ d = -3$

30. $a_1 = 10,\ a_3 = 28$

In Exercises 31–34, find the sum.

31. $\sum_{j=1}^{10} (2j - 3)$

32. $\sum_{j=1}^{8} (20 - 3j)$

33. $\sum_{k=1}^{11} \left(\frac{2}{3}k + 4\right)$

34. $\sum_{k=1}^{25} \left(\frac{3k + 1}{4}\right)$

35. Find the sum of the first 100 positive multiples of 5.

36. Find the sum of the integers from 20 to 80 (inclusive).

37. *Job Offer* A job has a starting salary of \$34,000 and a guaranteed salary increase of \$2250 per year for the first four years of employment. Determine (a) the salary during the fifth year and (b) the total compensation through 5 full years of employment.

38. *Baling Hay* In the first two trips baling hay around a field, a farmer makes 123 bales and 112 bales, respectively. Because each trip is shorter than the preceding trip, the farmer estimates that the same pattern will continue. Estimate the total number of bales made if there are another six trips around the field.

In Exercises 39–42, write the first five terms of the geometric sequence.

39. $a_1 = 4,\ r = -\frac{1}{4}$

40. $a_1 = 2,\ r = 2$

41. $a_1 = 9,\ a_3 = 4$

42. $a_1 = 2,\ a_3 = 12$

In Exercises 43–46, write the first five terms of the geometric sequence defined recursively. Determine the common ratio and write the nth term of the sequence as a function of n.

43. $a_1 = 120 \quad a_{k+1} = \frac{1}{3}a_k$

44. $a_1 = 200 \quad a_{k+1} = 0.1a_k$

45. $a_1 = 25 \quad a_{k+1} = -\frac{3}{5}a_k$

46. $a_1 = 18 \quad a_{k+1} = \frac{5}{3}a_k$

In Exercises 47 and 48, write an expression for the nth term of the geometric sequence and find the sum of the first 20 terms of the sequence.

47. $a_1 = 16,\ a_2 = -8$

48. $a_1 = 100,\ r = 1.05$

In Exercises 49–54, find the sum.

49. $\sum_{i=1}^{7} 2^{i-1}$

50. $\sum_{i=1}^{5} 3^{i-1}$

51. $\sum_{i=1}^{\infty} \left(\frac{7}{8}\right)^{i-1}$

52. $\sum_{i=1}^{\infty} \left(\frac{1}{3}\right)^{i-1}$

53. $\sum_{k=1}^{\infty} 4\left(\frac{2}{3}\right)^{k-1}$

54. $\sum_{k=1}^{\infty} 1.3\left(\frac{1}{10}\right)^{k-1}$

In Exercises 55 and 56, use a graphing utility to find the sum.

55. $\sum_{i=1}^{10} 10\left(\frac{3}{5}\right)^{i-1}$

56. $\sum_{i=1}^{25} 100(1.06)^{i-1}$

57. *Depreciation* A company buys a machine for \$120,000. During the next 5 years, the machine will depreciate at a rate of 30% per year. (That is, at the end of each year, the depreciated value is 70% of what it was at the beginning of the year.)

(a) Find the formula for the nth term of a geometric sequence that gives the value of the machine t full years after it was purchased.

(b) Find the depreciated value of the machine at the end of 5 full years.

58. *Total Compensation* A job pays a salary of \$32,000 the first year. During the next 39 years, there is a 5.5% raise each year. What is the total salary over the 40-year period?

59. *Compound Interest* A deposit of \$200 is made at the beginning of each month for 2 years in an account that pays 6%, compounded monthly. What is the balance in the account at the end of the 2 years?

60. *Compound Interest* A deposit of \$100 is made at the beginning of each month for 10 years in an account that pays 6.5%, compounded monthly. What is the balance in the account at the end of the 10 years?

In Exercises 61–64, use mathematical induction to prove the formula for every positive integer n.

61. $1 + 4 + \cdots + (3n - 2) = \frac{n}{2}(3n - 1)$

62. $1 + \frac{3}{2} + 2 + \frac{5}{2} + \cdots + \frac{1}{2}(n + 1) = \frac{n}{4}(n + 3)$

63. $\sum_{i=0}^{n-1} ar^i = \frac{a(1 - r^n)}{1 - r}$

64. $\sum_{k=0}^{n-1} (a + kd) = \frac{n}{2}[2a + (n - 1)d]$

In Exercises 65–68, evaluate ${}_nC_r$ or ${}_nP_r$. Use the $\boxed{{}_nC_r}$ or $\boxed{{}_nP_r}$ feature of a graphing utility to verify your answer.

65. ${}_6C_4$

66. ${}_{10}C_7$

67. ${}_8P_5$

68. ${}_{12}P_3$

In Exercises 69–74, use the Binomial Theorem to expand the binomial. Simplify your answer. (Remember that $i = \sqrt{-1}$.)

69. $\left(\dfrac{x}{2} + y\right)^4$

70. $(a - 3b)^5$

71. $\left(\dfrac{2}{x} - 3x\right)^6$

72. $(3x + y^2)^7$

73. $(5 + 2i)^4$

74. $(4 - 5i)^3$

75. *Amateur Radio* A Novice Amateur Radio license consists of two letters, one digit, and then three more letters. How many different licenses can be issued if no restrictions are placed on the letters or digits?

76. *Morse Code* In Morse code, each character is transmitted using a sequence of dits and dahs. How many different characters can be formed by a sequence of three dits and dahs? (These can be repeated. For example, dit-dit-dit represents the letter *s*.)

77. *Matching Socks* A man has five pairs of socks (no two pairs are the same color). If he randomly selects two socks from a drawer, what is the probability that he gets a matched pair?

78. *Bookshelf Order* A child returns a five-volume set of books to a bookshelf. The child is not able to read, and hence cannot distinguish one volume from another. What is the probability that the books are shelved in the correct order?

79. *Roll of the Dice* Are the chances of rolling a 3 with one die the same as rolling a total of 6 with two dice? If not, which has the higher probability?

80. *Roll of the Dice* A six-sided die is rolled six times. What is the probability that each side appears exactly once?

81. *Tossing a Coin* Find the probability of obtaining at least one tail when a coin is tossed five times.

82. *Parental Independence* Suppose that in a survey, senior citizens were asked if they would live with their children when they reached the point of not being able to live alone. The results are shown in the figure. If three senior citizens who could not live alone are randomly selected, what is the probability that all three are *not* living with their children?

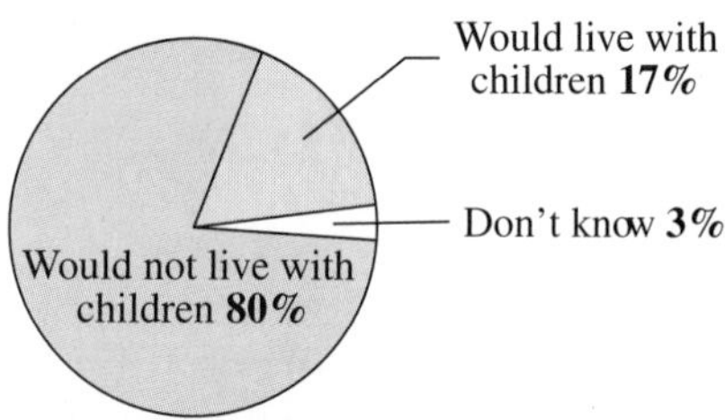

83. *Card Game* Five cards are drawn from an ordinary deck of 52 playing cards. Find the probability of getting two pairs. (For example, the hand could be A-A-5-5-Q or 4-4-7-7-K.)

84. *Data Analysis* A sample of college students, faculty members, and administrators were asked whether they favored a proposed increase in the annual activity fee to enhance student life on campus. The results of the study are given in the table.

	Students	Faculty	Admin.	Total
Favor	237	37	18	292
Oppose	163	38	7	208
Total	400	75	25	500

A person is selected at random from the sample. Find the specified probability.

(a) The person is not in favor of the proposal.

(b) The person is a student.

(c) The person is a faculty member and is in favor of the proposal.

CHAPTER PROJECT *Exploring Difference Quotients*

In this project, you will explore difference quotients and see how they can be used to find average rates.

Suppose you are driving from Atlanta to Miami. The trip is about 700 miles and takes you 12 hours. Your average speed is

$$\text{Average speed} = \frac{\text{distance}}{\text{time}} = \frac{700}{12} \approx 58.3 \text{ miles per hour.}$$

This concept can be generalized as follows. Let f be a function defined on the interval $[a, b]$. The **average rate of change of f** from a to b is given by

$$\text{Average rate of change} = \frac{f(b) - f(a)}{b - a}.$$

This expression is called a **difference quotient.**

(a) Calculate the average rate of change of $f(x) = 3x + 4$ on the interval $[2, 6]$. Select any other interval and show that you obtain the same average rate of change.

(b) Calculate the average rates of change of $f(x) = x^2$ on the intervals $[1, 3]$ and $[4, 6]$. Are they equal? Explain.

(c) During a 1-hour trip, your average speed is 50 miles per hour. Discuss the relationship between this speed and the speeds shown on your speedometer.

Questions for Further Exploration

1. Let $f(x) = x^2 - 2$. Calculate the average rates of change of f on the intervals $[1, 3]$, $[1, 2]$, $[1, 1.5]$, and $[1, 1.1]$.

(a) Find an expression for the average rate of change of f on the interval $[1, 1 + h]$, where h is any real number.

(b) Create a table that shows the average rate of change of f for several values of h. Choose values of h that get closer and closer to 0. What value does the average rate of change of f approach as h approaches 0?

(c) The answer to part (b) is the slope of the tangent line to the graph of f at the point $(1, f(1))$. Graph this line and f in the same viewing rectangle. Describe the behavior of the graphs near the point $(1, f(1))$.

2. The data in the table gives the costs of first-class postage in the United States for selected years from 1968 to 1995, where $t = 0$ corresponds to 1900. (Source: U.S. Postal Service)

t	68	71	74	75	78
Postage	\$0.06	\$0.08	\$0.10	\$0.13	\$0.15

t	81	85	88	91	95
Postage	\$0.20	\$0.22	\$0.25	\$0.29	\$0.32

(a) Find the average rate of change of the cost of postage between each two adjacent time intervals.

(b) When was the average rate of change largest? smallest?

(c) Are there intervals over which the average rate of change was zero? What does this mean?

7 /// CHAPTER TEST

Take this test as you would take a test in class. After you are done, check your work against the answers given in the back of the book.

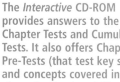

The *Interactive* CD-ROM provides answers to the Chapter Tests and Cumulative Tests. It also offers Chapter Pre-Tests (that test key skills and concepts covered in previous chapters) and Chapter Post-Tests, both of which have randomly generated exercises with diagnostic capabilities.

In Exercises 1 and 2, write the first five terms of the sequence.

1. $a_n = \left(-\frac{2}{3}\right)^{n-1}$. (Begin with $n = 1$.)

2. $a_1 = 12$ and $a_{k+1} = a_k + 4$

In Exercises 3 and 4, find a formula for the nth term of the sequence.

3. Arithmetic: $a_1 = 5000$, $d = -100$

4. Geometric: $a_1 = 4$, $a_{k+1} = \frac{1}{2}a_k$

5. Use sigma notation to write: $\dfrac{2}{3(1) + 1} + \dfrac{2}{3(2) + 1} + \cdot \cdot \cdot + \dfrac{2}{3(12) + 1}$.

6. Find the sum of the first 50 positive multiples of 3.

7. Find the common ratio: $2, -3, \frac{9}{2}, -\frac{27}{4}, \frac{81}{8}, \ldots$

8. Find the sum of the first eight terms of the geometric sequence in which $a_1 = 3$ and $a_2 = 6$.

9. Find the balance in an increasing annuity in which a principal of \$50 is deposited at the beginning of each month for 25 years. Assume that the amount in the fund is compounded monthly at 8%.

10. Evaluate: ${}_{20}C_3$.

11. Find the coefficient of the term x^3y^5 in the expansion of $(x + y)^8$.

12. How many distinct license plates can be issued with one letter followed by a three-digit number?

13. Four students are randomly selected from a class of 25 to answer questions from a reading assignment. In how many ways can the four be selected?

14. A card is drawn from a standard deck of 52 playing cards. Find the probability that it is a red face card.

15. Suppose that two spark plugs require replacement in a four-cylinder engine. If the mechanic randomly removes two plugs, find the probability that they are the two defective plugs.

16. On a game show you are given five digits to arrange in the proper order to represent the price of a car. You know the first digit, but must guess the remaining four. What is your probability of winning?

Conics and Parametric Equations

Chapter 8

On July 16, 1994, the comet Shoemaker-Levy 9 collided with Jupiter, the largest planet in our solar system. The dark spots in the photo of Jupiter in the inset show the results of the collision. The impact was filmed from the space probe Galileo, on its way to Jupiter.

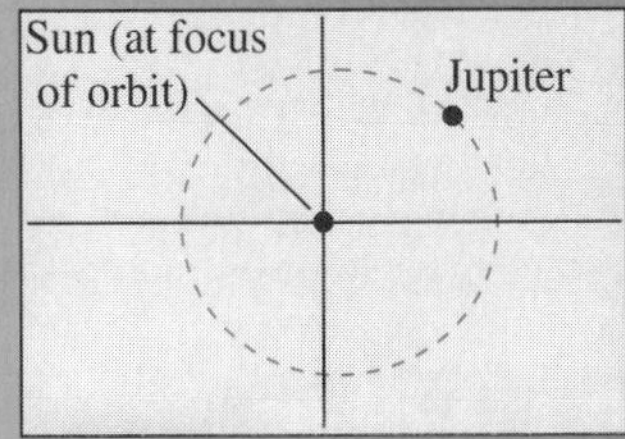

All the planets in our solar system travel in elliptical paths about the sun. Comets, on the other hand, can have elliptical, parabolic, or hyperbolic paths.

Since the solar system's formation, thousands of comets have collided with the planets and their moons. For evidence of this, all you need to do is look at the moon's surface through a telescope. (See Exercises 91 and 92 on page 626.)

Alan Levenson; (inset) Courtesy of the Space Telescope Science Institute/NASA.

Astronomers David Levy, Carolyn Shoemaker, and Eugene Shoemaker (from left to right) are shown with the 18-inch Schmidt telescope they used to discover the comet that is now named after them.

8.1 Conics

Introduction / Parabolas / Ellipses / Hyperbolas

Introduction

Apollonius, a prominent mathematician of the third century B.C., was the first to realize that the three types of conics—parabola, ellipse, and hyperbola—could be generated by intersecting a plane with a double-napped cone.

Conic sections were discovered during the classical Greek period, 600 to 300 B.C. This early Greek study was largely concerned with the geometrical properties of conics. It was not until the early 17th century that the broad applicability of conics became apparent and played a prominent role in the early development of calculus.

Figure 8.1

Circle Ellipse Parabola Hyperbola

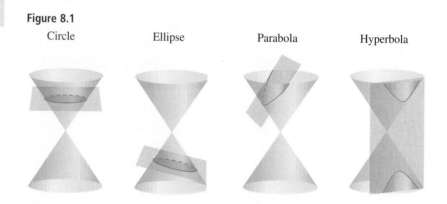

A **conic section** (or simply **conic**) is the intersection of a plane and a double-napped cone. Notice in Figure 8.1 that in the formation of the four basic conics, the intersecting plane does not pass through the vertex of the cone. When the plane does pass through the vertex, the resulting figure is a **degenerate conic,** as shown in Figure 8.2.

Figure 8.2

Point

Line

Two intersecting lines

There are several ways to approach the study of conics. You could begin by defining conics in terms of the intersections of planes and cones, as the Greeks did, or you could define them algebraically, in terms of the general second-degree equation

$$Ax^2 + Bxy + Cy^2 + Dx + Ey + F = 0.$$

However, we will use a third approach, in which each of the conics is defined as a *locus* (collection) of points satisfying a certain geometric property. For example, in Section P.5, you saw how the definition of a circle as *the collection of all points* (x, y) *that are equidistant from a fixed point* (h, k) led easily to the standard equation of a circle

$$(x - h)^2 + (y - k)^2 = r^2. \qquad \text{Equation of circle}$$

Parabolas

In Section 3.1, you learned that the graph of the quadratic function $f(x) = ax^2 + bx + c$ is a parabola that opens upward or downward. The following definition of a parabola is more general in the sense that it is independent of the orientation of the parabola.

Figure 8.3

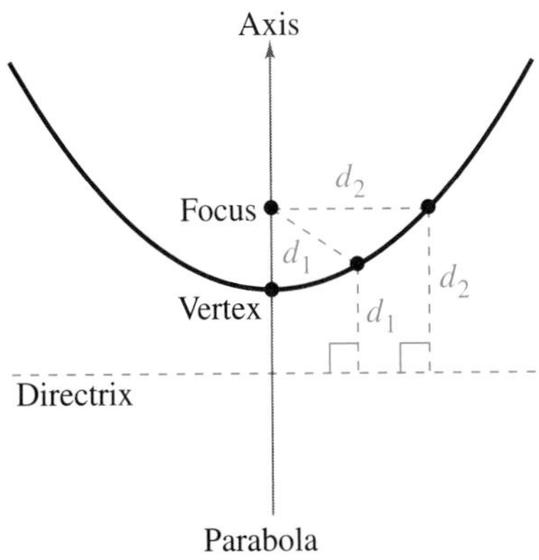

Parabola

Definition of Parabola

A **parabola** is the set of all points (x, y) that are equidistant from a fixed line, the **directrix,** and a fixed point, the **focus,** not on the line. See Figure 8.3.

The midpoint between the focus and the directrix is the **vertex,** and the line passing through the focus and the vertex is the **axis** of the parabola.

Standard Equation of a Parabola (Vertex at Origin)

The **standard form of the equation of a parabola** with vertex at $(0, 0)$ and directrix $y = -p$ is

$$x^2 = 4py, \qquad p \neq 0. \qquad \text{Vertical axis}$$

For directrix $x = -p$, the equation is

$$y^2 = 4px, \qquad p \neq 0. \qquad \text{Horizontal axis}$$

The focus is on the axis p units (directed distance) from the vertex.

Think About the Proof

The proofs of the two cases are similar. To prove the case with a vertical axis, use the diagram in Figure 8.4(a). Begin by assuming that the directrix, $y = -p$, is parallel to the x-axis. In Figure 8.4(a), assume that $p > 0$. Because p is the *directed* distance from the vertex to the focus, the focus must lie above the vertex. How does this information allow you to complete the proof? The details of the proof are given in the appendix.

Notice that a parabola can have a vertical or a horizontal axis. Examples of each are shown in Figure 8.4.

Figure 8.4

(a) Parabola with Vertical Axis **(b)** Parabola with Horizontal Axis

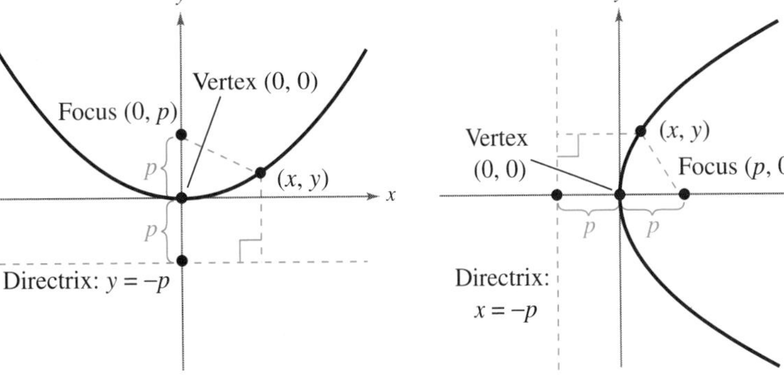

EXAMPLE 1 **Finding the Focus of a Parabola**

Find the focus of the parabola whose equation is $y = -2x^2$.

Solution

Because the squared term in the equation involves x, you know that the axis is vertical, and the equation is of the form $x^2 = 4py$. You can write the given equation in this form as follows.

$$x^2 = -\frac{1}{2}y$$

$$x^2 = 4\left(-\frac{1}{8}\right)y \qquad \text{Standard form}$$

Thus, $p = -\frac{1}{8}$. Because p is negative, the parabola opens downward (see Figure 8.5), and the focus of the parabola is $(0, p) = \left(0, -\frac{1}{8}\right)$.

Figure 8.5

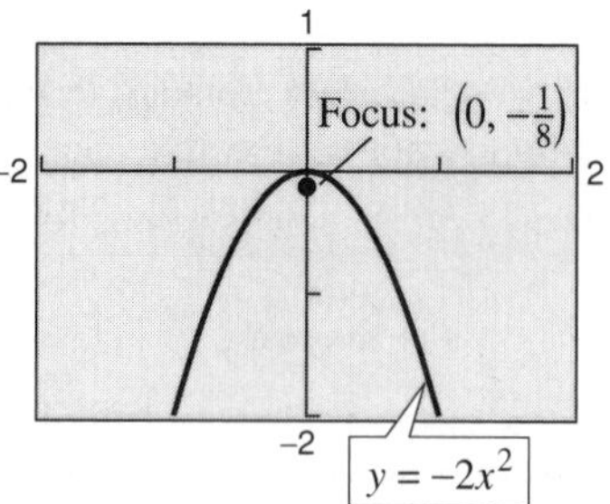

EXAMPLE 2 **A Parabola with a Horizontal Axis**

Write the standard form of the equation of the parabola with vertex at the origin and focus at (2, 0).

Solution

The axis of the parabola is horizontal, passing through (0, 0) and (2, 0), as shown in Figure 8.6. Thus, the standard form is $y^2 = 4px$. Because the focus is $p = 2$ units from the vertex, the equation is

$$y^2 = 4(2)x$$

$$y^2 = 8x.$$

The equation $y^2 = 8x$ does not define y as a function of x. Thus, to use a graphing utility to graph $y^2 = 8x$, you need to break the graph into two equations

$$y = 2\sqrt{2x} \quad \text{and} \quad y = -2\sqrt{2x}$$

each of which is a function of x.

Figure 8.6

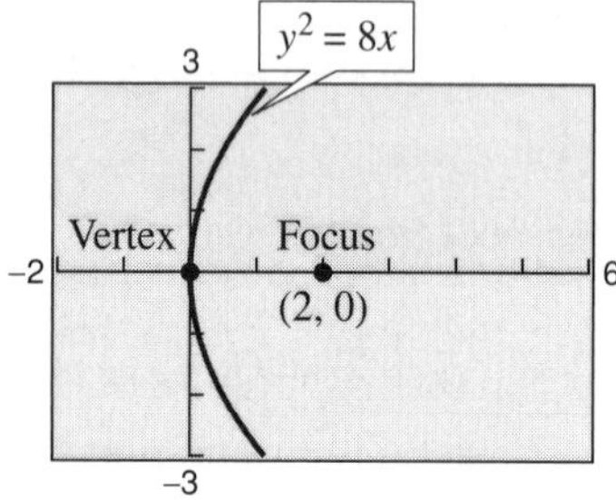

Parabolas occur in a wide variety of applications. For instance, a parabolic reflector can be formed by revolving a parabola about its axis. The resulting surface has the property that all incoming rays parallel to the axis are reflected through the focus of the parabola—this is the principle behind the construction of the parabolic mirrors used in reflecting telescopes. Conversely, the light rays emanating from the focus of a parabolic reflector used in a flashlight are all parallel to one another, as shown in Figure 8.7.

Figure 8.7

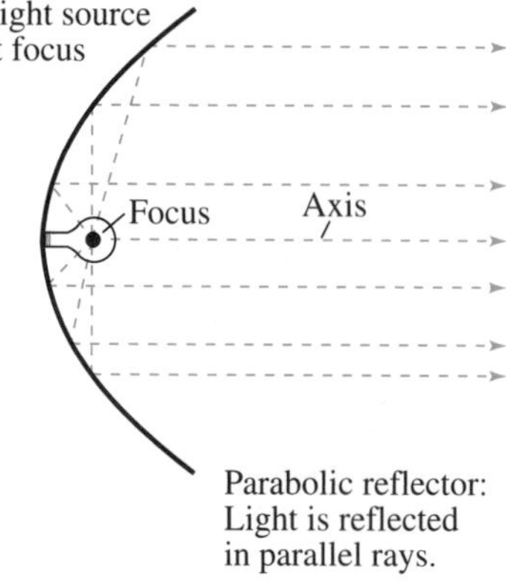

Parabolic reflector: Light is reflected in parallel rays.

Figure 8.8

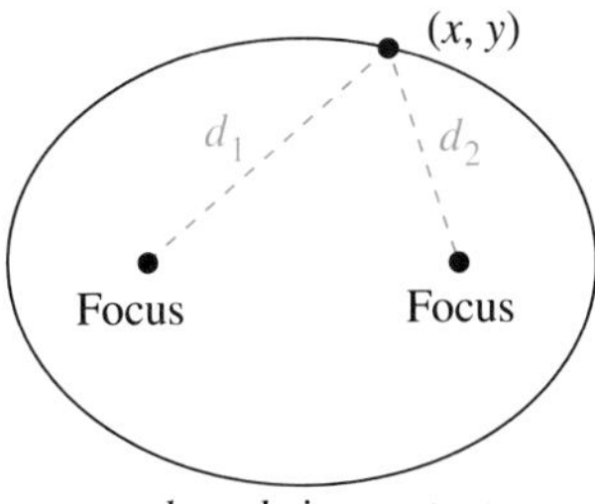

$d_1 + d_2$ is constant.

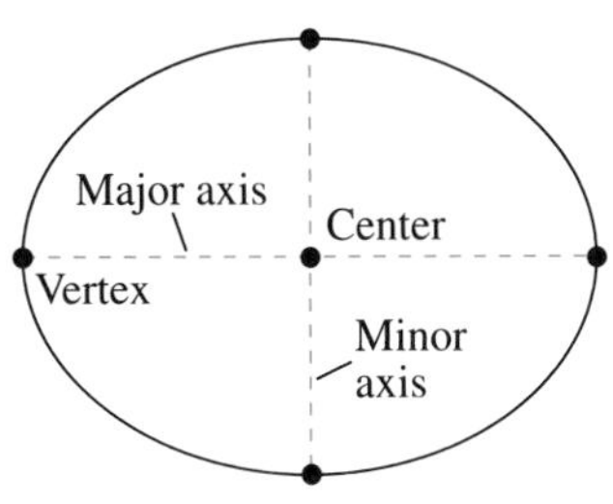

A computer animation of this concept appears in the *Interactive* CD-ROM.

Figure 8.9

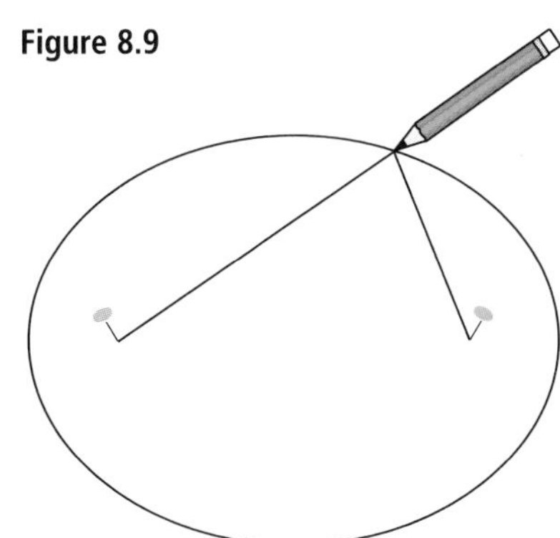

Ellipses

Definition of Ellipse

An **ellipse** is the set of all points (x, y) the sum of whose distances from two distinct fixed points **(foci)** is constant. See Figure 8.8.

The line through the foci intersects the ellipse at two points **(vertices).** The chord joining the vertices is the **major axis,** and its midpoint is the **center** of the ellipse. The chord perpendicular to the major axis at the center is the **minor axis** of the ellipse.

You can visualize the definition of an ellipse by imagining two thumbtacks placed at the foci, as shown in Figure 8.9. If the ends of a fixed length of string are fastened to the thumbtacks and the string is drawn taut with a pencil, the path traced by the pencil will be an ellipse.

The standard form of the equation of an ellipse takes one of two forms, depending on whether the major axis is horizontal or vertical.

Standard Equation of an Ellipse (Center at Origin)

The **standard form of the equation of an ellipse** with center at the origin and major and minor axes of lengths $2a$ and $2b$ (where $0 < b < a$), is

$$\frac{x^2}{a^2} + \frac{y^2}{b^2} = 1 \quad \text{or} \quad \frac{x^2}{b^2} + \frac{y^2}{a^2} = 1.$$

The vertices and foci lie on the major axis, a and c units, respectively, from the center, as shown in Figure 8.10. Moreover, a, b, and c are related by the equation $c^2 = a^2 - b^2$.

Figure 8.10

(a) Major axis is horizontal. Minor axis is vertical.

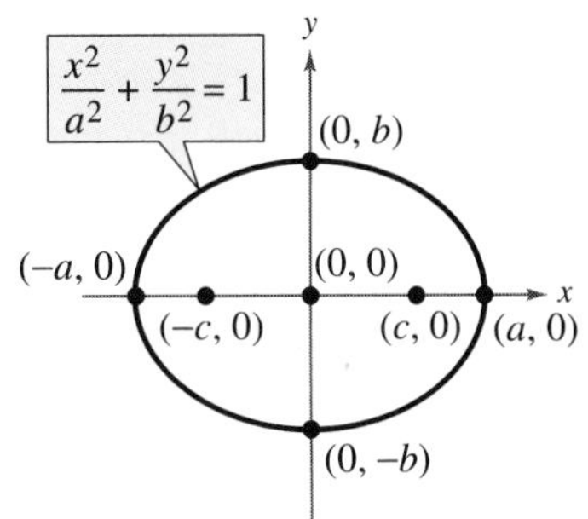

(b) Major axis is vertical. Minor axis is horizontal.

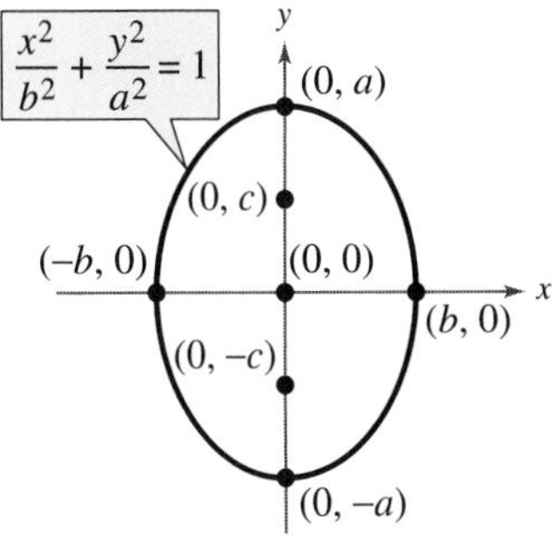

EXAMPLE 3 Finding the Standard Equation of an Ellipse

Find the standard form of the equation of the ellipse that has a major axis of length 6 and foci at $(-2, 0)$ and $(2, 0)$, as shown in Figure 8.11.

Figure 8.11

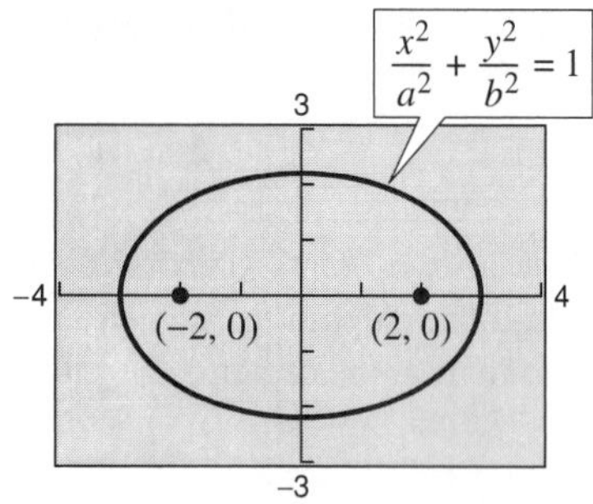

Solution

Because the foci occur at $(-2, 0)$ and $(2, 0)$, the center of the ellipse is $(0, 0)$, and the major axis is horizontal. Thus, the ellipse has an equation of the form

$$\frac{x^2}{a^2} + \frac{y^2}{b^2} = 1. \quad \text{Standard form}$$

Because the length of the major axis is 6,

$$2a = 6 \quad \text{Length of major axis}$$

which implies that $a = 3$. Moreover, the distance from the center to either focus is $c = 2$. Finally,

$$b^2 = a^2 - c^2 = 3^2 - 2^2 = 9 - 4 = 5$$

which yields the equation

$$\frac{x^2}{9} + \frac{y^2}{5} = 1.$$

The *Interactive* CD-ROM shows every example with its solution; clicking on the *Try It!* button brings up similar problems. Guided Examples and Integrated Examples show step-by-step solutions to additional examples. Integrated Examples are related to several concepts in the section.

EXAMPLE 4 Sketching an Ellipse

Sketch the ellipse given by $4x^2 + y^2 = 36$, and identify the vertices.

Solution

$$4x^2 + y^2 = 36 \quad \text{Original equation}$$

$$\frac{4x^2}{36} + \frac{y^2}{36} = \frac{36}{36}$$

$$\frac{x^2}{3^2} + \frac{y^2}{6^2} = 1 \quad \text{Standard form}$$

Because the denominator of the y^2-term is larger than the denominator of the x^2-term, you can conclude that the major axis is vertical. Moreover, because $a = 6$, the vertices are $(0, -6)$ and $(0, 6)$. Finally, because $b = 3$, the endpoints of the minor axis are $(-3, 0)$ and $(3, 0)$, as shown in Figure 8.12. Note that you can sketch the ellipse by locating the endpoints of the two axes. Because 3^2 is the denominator of the x^2-term, move three units to the *right and left* of the center to locate the endpoints of the horizontal axis. Similarly, because 6^2 is the denominator of the y^2-term, move six units *up and down* from the center to locate the endpoints of the vertical axis.

Note You can verify the graph in Example 4 with a graphing utility by entering $y_1 = \sqrt{36 - 4x^2}$ and $y_2 = -\sqrt{36 - 4x^2}$ and using a square viewing rectangle.

Figure 8.12

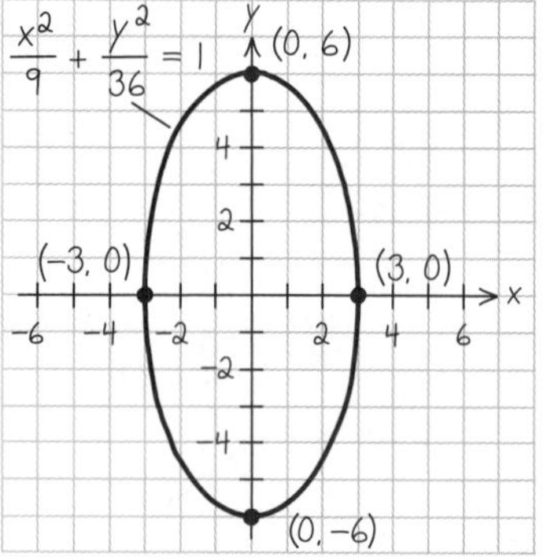

Hyperbolas

The definition of a **hyperbola** is similar to that of an ellipse. The difference is that, for an ellipse, the *sum* of the distances between the foci and a point on the ellipse is constant, whereas for a hyperbola it is the *difference* of these distances that is constant.

Figure 8.13

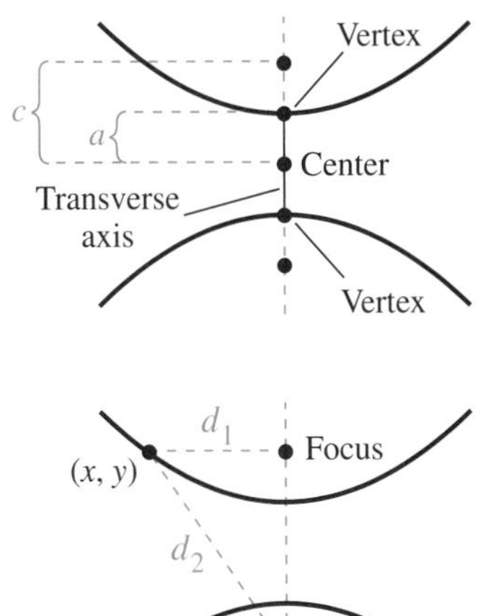

$d_2 - d_1$ is constant.

Definition of Hyperbola

A **hyperbola** is the set of all points (x, y) the difference of whose distances from two distinct fixed points **(foci)** is constant. See Figure 8.13.

The graph of a hyperbola has two disconnected parts **(branches).** The line through the two foci intersects the hyperbola at two points **(vertices).** The line segment connecting the vertices is the **transverse axis,** and the midpoint of the transverse axis is the **center** of the hyperbola.

Standard Equation of a Hyperbola (Center at Origin)

The **standard form of the equation of a hyperbola** with center at the origin (where $a \neq 0$ and $b \neq 0$) is

$$\frac{x^2}{a^2} - \frac{y^2}{b^2} = 1 \quad \text{or} \quad \frac{y^2}{a^2} - \frac{x^2}{b^2} = 1.$$

The vertices and foci are a and c units from the center, respectively, and $b^2 = c^2 - a^2$. See Figure 8.14.

Figure 8.14

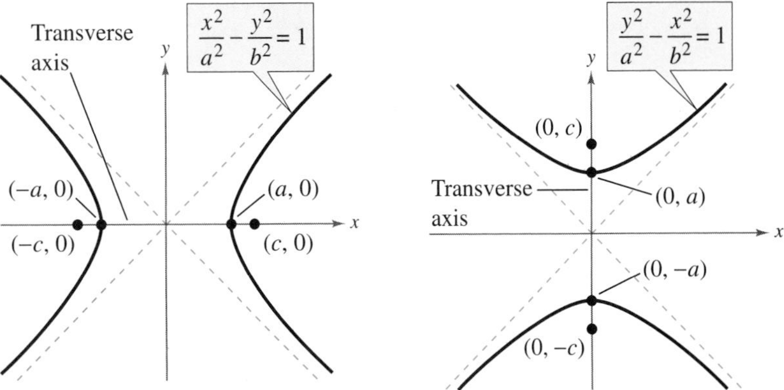

EXAMPLE 5 Finding the Standard Equation of a Hyperbola

Find the standard form of the equation of the hyperbola with foci at $(-3, 0)$ and $(3, 0)$ and vertices at $(-2, 0)$ and $(2, 0)$, as shown in Figure 8.15.

Figure 8.15

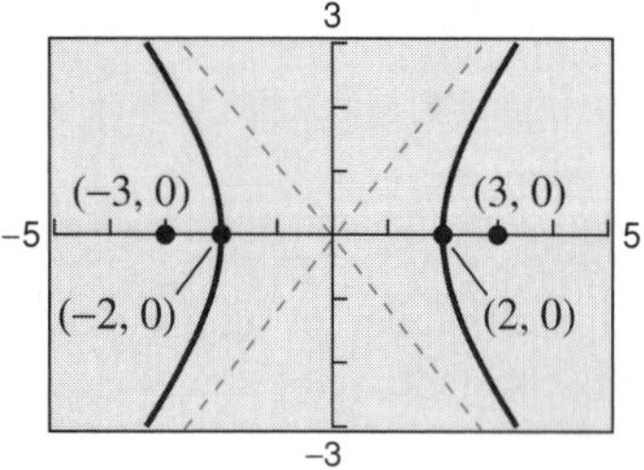

Solution

It can be seen that $c = 3$ because the foci are three units from the center. Moreover, $a = 2$ because the vertices are two units from the center. Thus, it follows that

$$b^2 = c^2 - a^2 = 3^2 - 2^2 = 9 - 4 = 5.$$

Because the transverse axis is horizontal, the standard form of the equation is

$$\frac{x^2}{a^2} - \frac{y^2}{b^2} = 1.$$

Finally, substitute $a^2 = 4$ and $b^2 = 5$ to obtain

$$\frac{x^2}{4} - \frac{y^2}{5} = 1. \qquad \text{Standard form}$$

To graph the hyperbola in Example 5, first solve for y^2 to obtain $y^2 = 5(x^2 - 4)/4$. Then enter the positive and negative square roots of the right-hand side of the equation to obtain the following.

$$y_1 = \sqrt{5(x^2 - 4)}/2$$
$$y_2 = -\sqrt{5(x^2 - 4)}/2$$

An important aid in sketching the graph of a hyperbola is the determination of its **asymptotes,** as shown in Figure 8.16. Each hyperbola has two asymptotes that intersect at the center of the hyperbola. Furthermore, the asymptotes pass through the corners of a rectangle of dimensions $2a$ by $2b$. The line segment of length $2b$, joining $(0, b)$ and $(0, -b)$ [or $(-b, 0)$ and $(b, 0)$], is the **conjugate axis** of the hyperbola.

Figure 8.16

Transverse axis is horizontal.

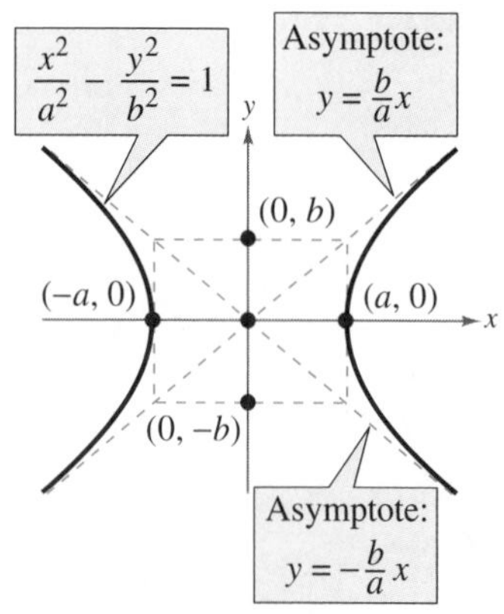

Transverse axis is vertical.

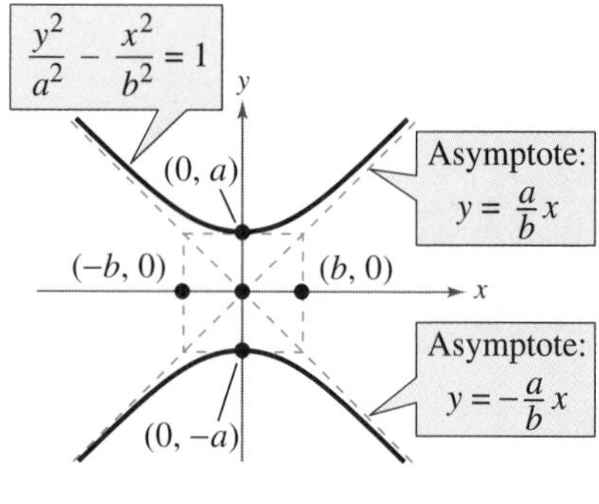

Asymptotes of a Hyperbola (Center at Origin)

The **asymptotes of a hyperbola** with center at (0, 0) are

$$y = \frac{b}{a}x \quad \text{and} \quad y = -\frac{b}{a}x \qquad \text{Transverse axis is horizontal.}$$

or

$$y = \frac{a}{b}x \quad \text{and} \quad y = -\frac{a}{b}x. \qquad \text{Transverse axis is vertical.}$$

EXPLORATION

Use a graphing utility to graph the hyperbola in Example 6. Does your graph look like that shown in Figure 8.17(b)? If not, what must you do to get both the upper and lower portions of the hyperbola? Explain your reasoning.

EXAMPLE 6 **Sketching the Graph of a Hyperbola**

Sketch the graph of the hyperbola whose equation is $4x^2 - y^2 = 16$.

Solution

$$4x^2 - y^2 = 16 \qquad \text{Original equation}$$

$$\frac{4x^2}{16} - \frac{y^2}{16} = \frac{16}{16}$$

$$\frac{x^2}{2^2} - \frac{y^2}{4^2} = 1 \qquad \text{Standard form}$$

Because the x^2-term is positive, you can conclude that the transverse axis is horizontal and the vertices occur at $(-2, 0)$ and $(2, 0)$. Moreover, the endpoints of the conjugate axis occur at $(0, -4)$ and $(0, 4)$, and you can sketch the rectangle shown in Figure 8.17(a). Finally, by drawing the asymptotes through the corners of this rectangle, you can complete the sketch shown in Figure 8.17(b).

Figure 8.17

(a)

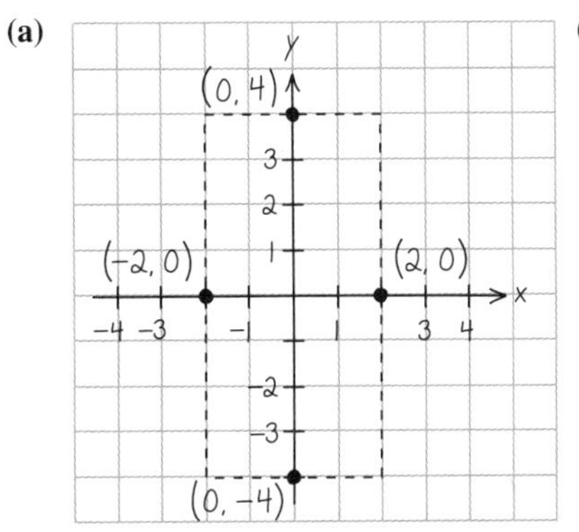

(b)

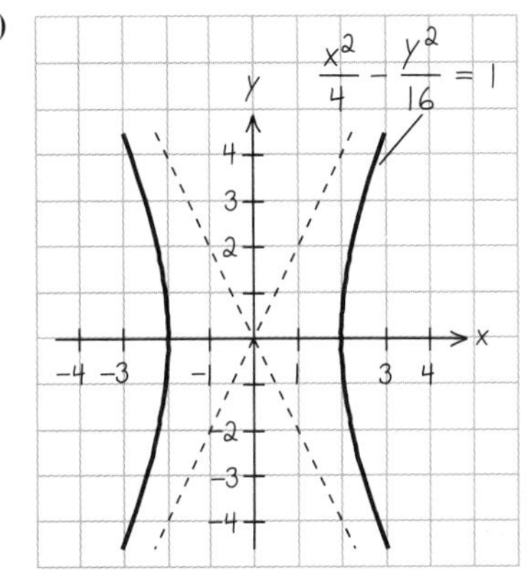

EXAMPLE 7 **Finding the Standard Equation of a Hyperbola**

Find the standard form of the equation of the hyperbola having vertices at $(0, -3)$ and $(0, 3)$ and with asymptotes $y = -2x$ and $y = 2x$, as shown in Figure 8.18.

Figure 8.18

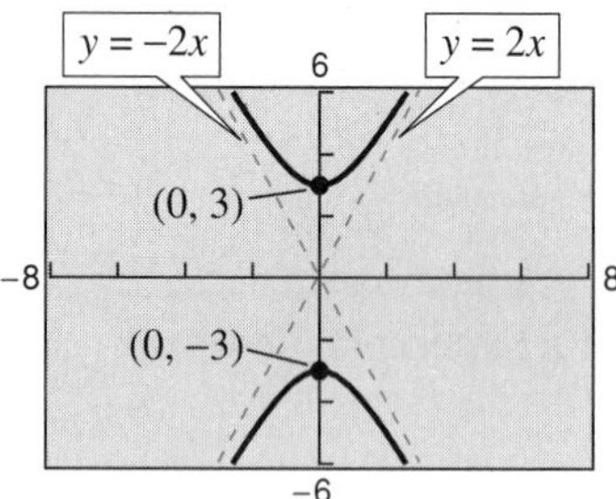

Solution

Because the transverse axis is vertical, the asymptotes are of the form

$$y = \frac{a}{b}x \quad \text{and} \quad y = -\frac{a}{b}x.$$

Thus,

$$\frac{a}{b} = 2$$

and because $a = 3$, you can determine that $b = \frac{3}{2}$. Finally, you can conclude that the hyperbola has the following equation.

$$\frac{y^2}{3^2} - \frac{x^2}{(3/2)^2} = 1 \qquad \text{Standard form}$$

Group Activity

Hyperbolas in Applications

At the beginning of this section, we mentioned that each type of conic section can be formed by the intersection of a plane and a double-napped cone. The figure below shows three examples of how such intersections can occur in physical situations.

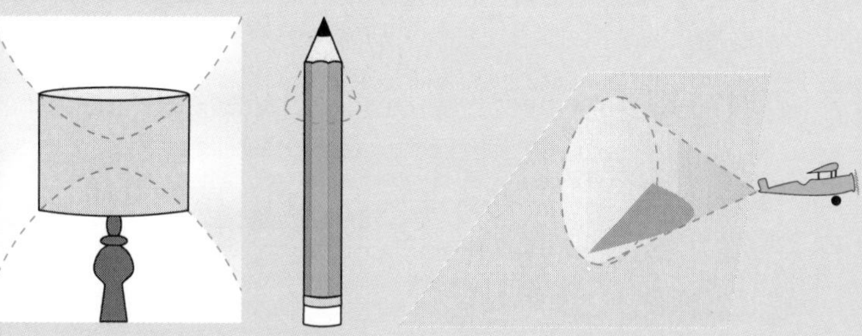

Identify the cone and hyperbola (or portion of a hyperbola) in each of the three situations. Can you think of other examples of physical situations in which hyperbolas are formed?

8.1 /// EXERCISES

In Exercises 1–8, match the equation with its graph. [The graphs are labeled (a), (b), (c), (d), (e), (f), (g), and (h).]

(a)
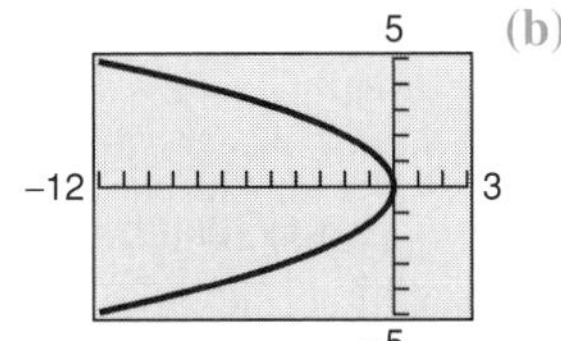

(b)
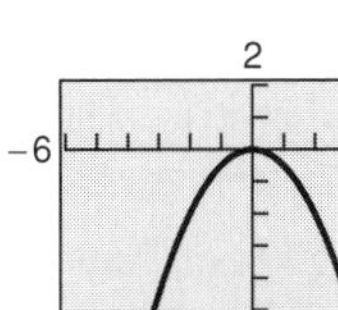

(c)
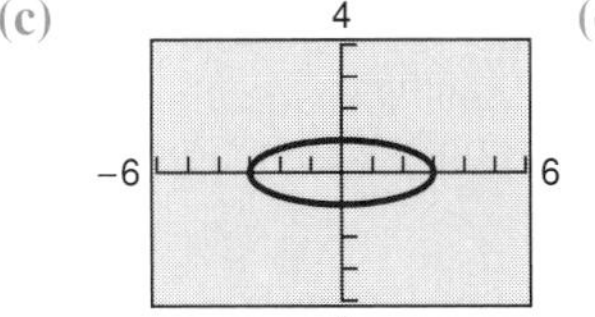

(d)
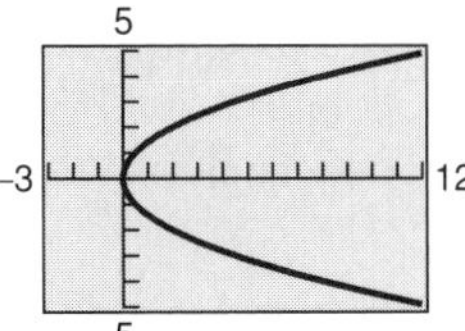

(e)
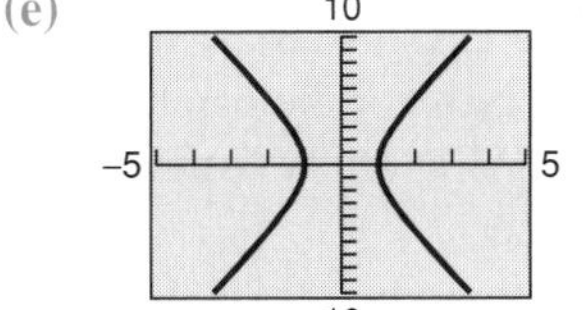

(f)
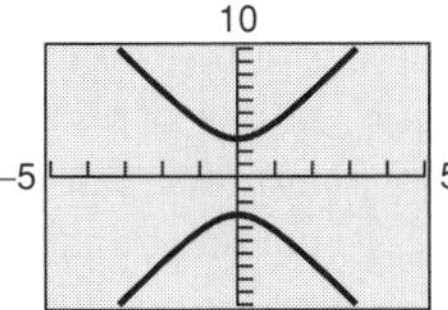

(g)
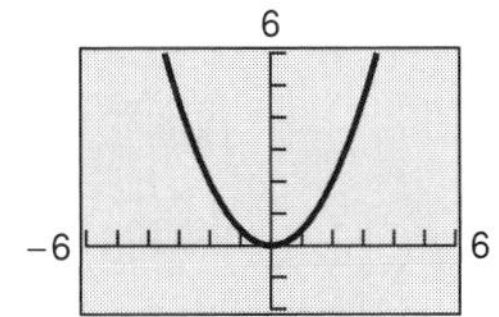

(h)
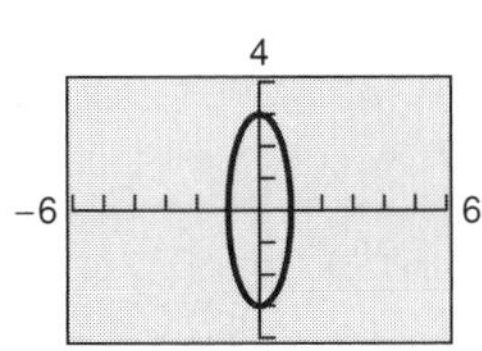

1. $x^2 = 2y$

2. $x^2 = -2y$

3. $y^2 = 2x$

4. $y^2 = -2x$

5. $\dfrac{x^2}{1} + \dfrac{y^2}{9} = 1$

6. $\dfrac{x^2}{9} + \dfrac{y^2}{1} = 1$

7. $\dfrac{x^2}{1} - \dfrac{y^2}{9} = 1$

8. $\dfrac{y^2}{9} - \dfrac{x^2}{1} = 1$

The *Interactive* CD-ROM contains step-by-step solutions to all odd-numbered Section and Review Exercises. It also provides Tutorial Exercises, which link to Guided Examples for additional help.

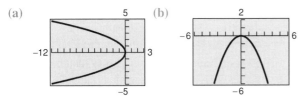

In Exercises 9–14, find the vertex and focus of the parabola and sketch its graph.

9. $y = \frac{1}{2}x^2$

10. $y = 2x^2$

11. $y^2 = -6x$

12. $y^2 = 3x$

13. $x^2 + 8y = 0$

14. $x + y^2 = 0$

In Exercises 15 and 16, use a graphing utility to graph the parabola and its tangent line. Identify the point of tangency.

	Parabola	*Tangent Line*
15.	$y^2 - 8x = 0$	$x - y + 2 = 0$
16.	$x^2 + 12y = 0$	$x + y - 3 = 0$

In Exercises 17–26, find an equation of the parabola with vertex at the origin.

17. Focus: $\left(0, -\frac{3}{2}\right)$

18. Focus: $(2, 0)$

19. Focus: $(-2, 0)$

20. Focus: $(0, -2)$

21. Directrix: $y = -1$

22. Directrix: $x = 3$

23. Directrix: $y = 2$

24. Directrix: $x = -2$

25. Horizontal axis and passes through the point $(4, 6)$

26. Vertical axis and passes through the point $(-2, -2)$

In Exercises 27 and 28, find an equation of the parabola and determine the coordinates of the focus.

27.
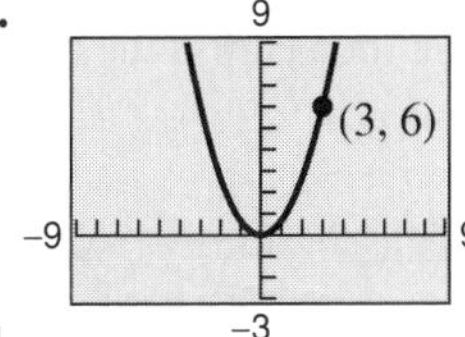

28.
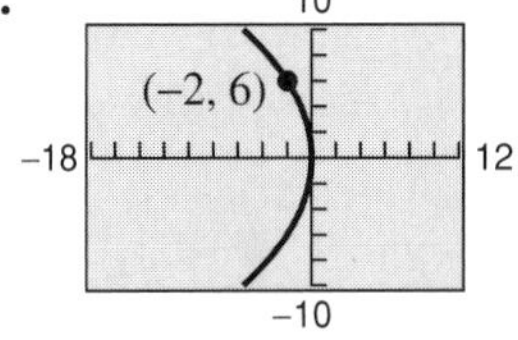

29. *Satellite Antenna* Write an equation for a cross section of the parabolic television dish antenna shown in the figure.

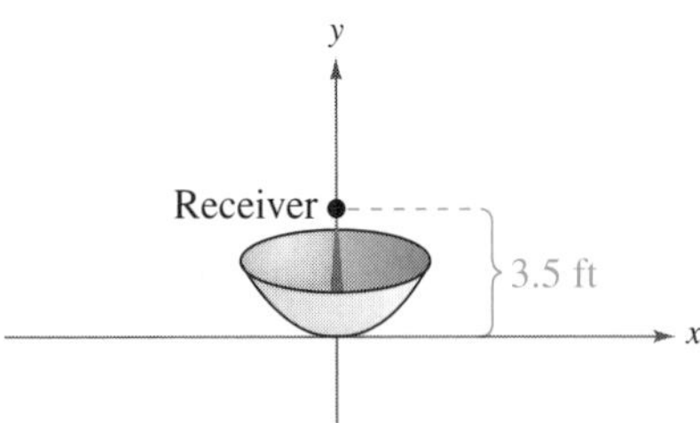

30. *Suspension Bridge* Each cable of a suspension bridge is suspended (in the shape of a parabola) between two towers that are 120 meters apart and rise 20 meters above the roadway. The cables touch the roadway midway between the towers.

(a) Draw a sketch of the bridge. Locate the origin of a rectangular coordinate system at the center of the roadway. Label the coordinates of the known points.

(b) Write an equation that models the cables.

(c) Complete the table by finding the heights of the suspension cables over the roadway at distances of x meters from the center of the bridge.

x	0	20	40	60
y				

31. *Beam Deflection* A simply supported beam is 64 feet long and has a load at the center (see figure). The deflection of the beam at its center is 1 inch. The shape of the deflected beam is parabolic.

(a) Find an equation of the parabola. (Assume that the origin is at the center of the beam.)

(b) How far from the center of the beam is the deflection $\frac{1}{2}$ inch?

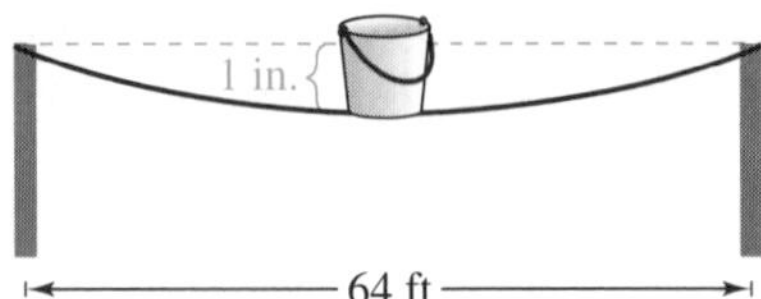

32. *Exploration* Consider the equation $x^2 = 4py$.

(a) Use a graphing utility to graph the parabolas for $p = 1, p = 2, p = 3,$ and $p = 4$. Describe the effect on the graph when p increases.

(b) Locate the focus of each parabola in part (a).

(c) For each parabola in part (a), find the length of the chord passing through the focus parallel to the directrix. How can the length of this chord be determined directly from $x^2 = 4py$?

(d) Explain how the result of part (c) can be used as a sketching aid when graphing parabolas.

33. *True or False?* It is possible for a parabola to intersect its directrix. Explain your reasoning.

34. *Exploration* Let (x_1, y_1) be the coordinates of a point on the parabola $x^2 = 4py$. The equation of the line tangent to the parabola at the point is

$$y - y_1 = \frac{x_1}{2p}(x - x_1).$$

(a) What is the slope of the tangent line?

(b) Find an equation of the tangent line at the endpoints of each of the chords for the parabolas in Exercise 32. Use a graphing utility to graph the parabola and tangent lines.

In Exercises 35–40, find the center and vertices of the ellipse and sketch its graph.

35. $\frac{x^2}{25} + \frac{y^2}{16} = 1$

36. $\frac{x^2}{144} + \frac{y^2}{169} = 1$

37. $\frac{x^2}{16} + \frac{y^2}{25} = 1$

38. $\frac{x^2}{169} + \frac{y^2}{144} = 1$

39. $\frac{x^2}{9} + \frac{y^2}{5} = 1$

40. $\frac{x^2}{28} + \frac{y^2}{64} = 1$

In Exercises 41 and 42, use a graphing utility to graph the ellipse. (*Hint:* Use two equations.)

41. $5x^2 + 3y^2 = 15$

42. $x^2 + 4y^2 = 4$

Think About It **In Exercises 43 and 44, which part of the ellipse $4x^2 + 9y^2 = 36$ is represented by the equation?**

43. $x = -\frac{3}{2}\sqrt{4 - y^2}$

44. $y = \frac{2}{3}\sqrt{9 - x^2}$

In Exercises 45–52, find an equation of the ellipse with center at the origin.

45.

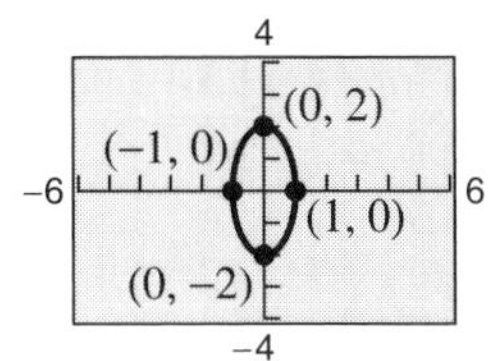

46.

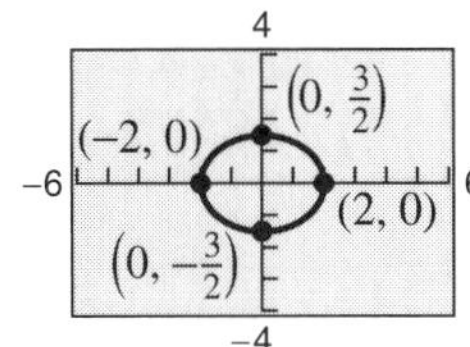

47. Vertices: $(\pm 5, 0)$; Foci: $(\pm 2, 0)$

48. Vertices: $(0, \pm 8)$; Foci: $(0, \pm 4)$

49. Foci: $(\pm 5, 0)$; Major axis of length 12

50. Foci: $(\pm 2, 0)$; Major axis of length 8

51. Vertices: $(0, \pm 5)$; Passes through the point $(4, 2)$

52. Major axis vertical; Passes through the points $(0, 4)$ and $(2, 0)$

53. *Fireplace Arch* A fireplace arch is to be constructed in the shape of a semiellipse. The opening is to have a height of 2 feet at the center and a width of 6 feet along the base (see figure). The contractor draws the outline of the ellipse by the method discussed on page 607. Give the required positions of the tacks and the length of the string.

Figure for 53

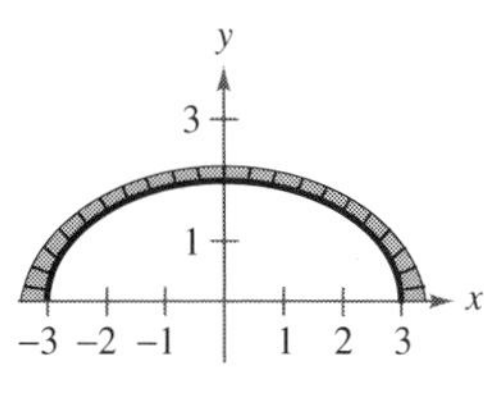

Figure for 54

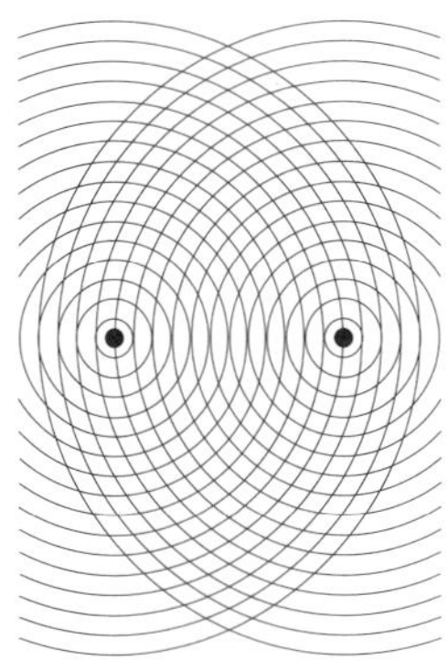

54. *Geometry* Sketch a graph of the ellipse that consists of all points (x, y) such that the sum of the distances between (x, y) and two fixed points is 16 units and the foci are located at the centers of the two sets of concentric circles in the figure.

55. *Think About It* On page 607 (see Figure 8.9) it is noted that an ellipse can be drawn using two thumbtacks, a string of fixed length (greater than the distance between the two tacks), and a pencil.

(a) What is the length of the string in terms of a?

(b) Explain why the path is an ellipse.

56. *Mountain Tunnel* A semielliptical arch over a tunnel for a road through a mountain has a major axis of 100 feet and a height at the center of 30 feet.

(a) Create a sketch to solve the problem. Draw the tunnel on a rectangular coordinate system with the center of the road entering the tunnel at the origin. Identify the coordinates of the known points.

(b) Find an equation of the semielliptical tunnel.

(c) Determine the height of the arch 5 feet from the edge of the tunnel.

57. *Think About It* Is the graph of $x^2 + 4y^4 = 4$ an ellipse? Explain.

58. *Think About It* The graph of $x^2 + y^2 = 0$ is a degenerate conic. Sketch the graph of this equation.

59. *Exploration* Consider the ellipse

$$\frac{x^2}{a^2} + \frac{y^2}{b^2} = 1, \qquad a + b = 20.$$

(a) The area of the ellipse is given by $A = \pi ab$. Write the area of the ellipse as a function of a.

(b) Find the equation of an ellipse with an area of 264 square centimeters.

(c) Complete the table using your equation from part (a), and make a conjecture about the shape of the ellipse with maximum area.

a	8	9	10	11	12	13
A						

(d) Use a graphing utility to graph the area function and use the graph to make a conjecture about the shape of the ellipse that yields the maximum area.

60. *Geometry* The area of the ellipse in the figure is twice the area of the circle. What is the length of the major axis? (The area of an ellipse is $A = \pi ab$.)

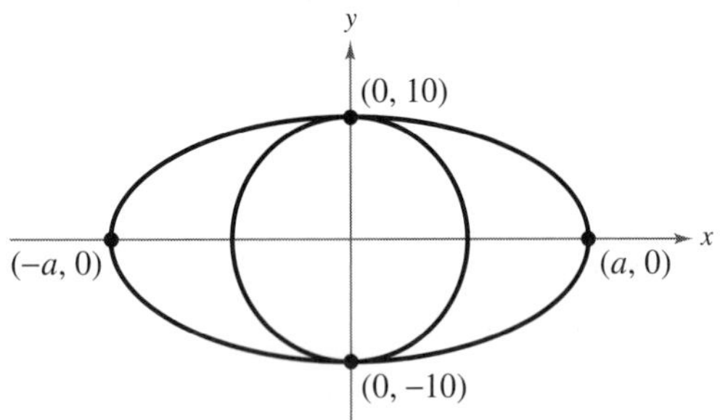

61. *Essay* Write a paragraph discussing the change in the shape and orientation of the graph of the ellipse

$$\frac{x^2}{a^2} + \frac{y^2}{16} = 1$$

as a increases from 1 to 8.

62. *Geometry* A line segment through a focus of an ellipse and with endpoints on the ellipse and perpendicular to the major axis is called a **latus rectum** of the ellipse. Therefore, an ellipse has two latera recta. Knowing the length of the latera recta is helpful in sketching an ellipse because it yields other points on the curve (see figure). Show that the length of each latus rectum is $2b^2/a$.

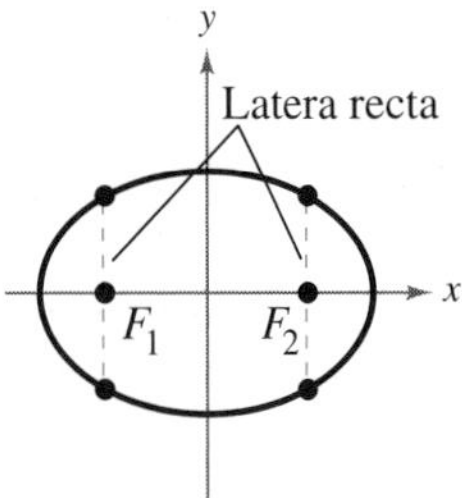

In Exercises 63–66, sketch the graph of the ellipse, making use of the latera recta (see Exercise 62).

63. $\dfrac{x^2}{4} + \dfrac{y^2}{1} = 1$

64. $\dfrac{x^2}{9} + \dfrac{y^2}{16} = 1$

65. $9x^2 + 4y^2 = 36$

66. $5x^2 + 3y^2 = 15$

In Exercises 67–72, find the center, vertices, and foci of the hyperbola and sketch its graph, using asymptotes as an aid.

67. $x^2 - y^2 = 1$

68. $\dfrac{x^2}{9} - \dfrac{y^2}{16} = 1$

69. $\dfrac{y^2}{1} - \dfrac{x^2}{4} = 1$

70. $\dfrac{y^2}{9} - \dfrac{x^2}{1} = 1$

71. $\dfrac{y^2}{25} - \dfrac{x^2}{144} = 1$

72. $\dfrac{x^2}{36} - \dfrac{y^2}{4} = 1$

In Exercises 73 and 74, use a graphing utility to graph the hyperbola and its asymptotes.

73. $2x^2 - 3y^2 = 6$

74. $3y^2 - 5x^2 = 15$

Think About It **In Exercises 75 and 76, state which part of the graph of the hyperbola $4x^2 - 9y^2 = 36$ is represented by the given equation.**

75. $y = -\frac{2}{3}\sqrt{x^2 - 9}$

76. $x = \frac{3}{2}\sqrt{y^2 + 4}$

In Exercises 77–84, find an equation of the specified hyperbola with center at the origin.

77. Vertices: $(0, \pm 2)$; Foci: $(0, \pm 4)$

78. Vertices: $(\pm 3, 0)$; Foci: $(\pm 5, 0)$

79. Vertices: $(\pm 1, 0)$; Asymptotes: $y = \pm 3x$

80. Vertices: $(0, \pm 3)$; Asymptotes: $y = \pm 3x$

81. Foci: $(0, \pm 8)$; Asymptotes: $y = \pm 4x$

82. Foci: $(\pm 10, 0)$; Asymptotes: $y = \pm \frac{3}{4}x$

83.

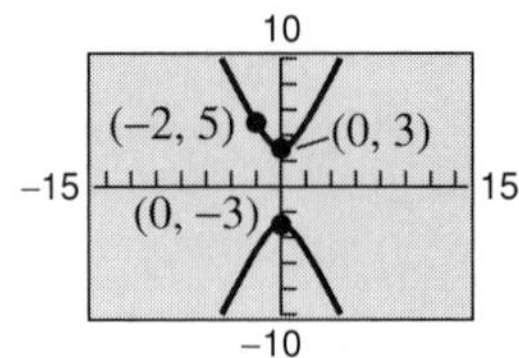

84.

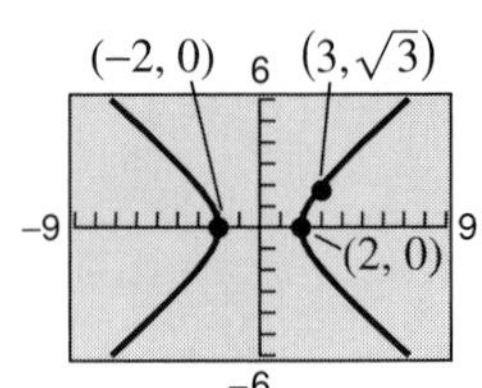

85. *Hyperbolic Mirror* A hyperbolic mirror (used in some telescopes) has the property that a light ray directed at the focus will be reflected to the other focus (see figure). The focus of a hyperbolic mirror has coordinates (24, 0). Find the vertex of the mirror if its mount has coordinates (24, 24).

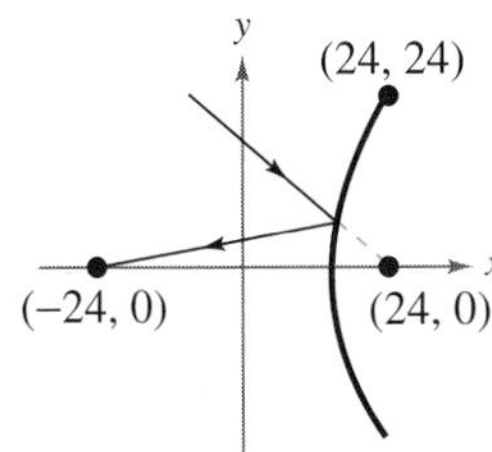

86. *LORAN* Long distance radio navigation for aircraft and ships uses synchronized pulses transmitted by widely separated transmitting stations. These pulses travel at the speed of light (186,000 miles per second). The difference in the times of arrival of these pulses at an aircraft or ship is constant on a hyperbola having the transmitting stations as foci. Assume that two stations, 300 miles apart, are positioned on the rectangular coordinate system at points with coordinates $(-150, 0)$ and $(150, 0)$ and that a ship is traveling on a path with coordinates $(x, 75)$ (see figure). Find the x-coordinate of the position of the ship when the time difference between the pulses from the transmitting stations is 1000 microseconds (0.001 second).

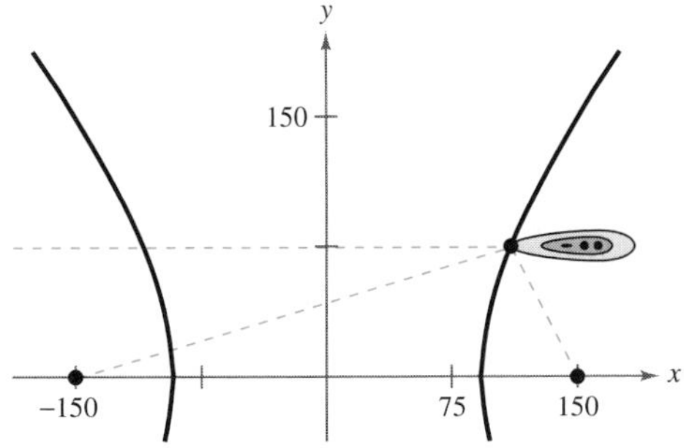

87. Use the definition of an ellipse to derive the standard form of the equation of an ellipse.

88. Use the definition of a hyperbola to derive the standard form of the equation of a hyperbola.

Review **Solve Exercises 89–94 as a review of the skills and problem-solving techniques you learned in previous sections.**

89. Factor each expression completely.

(a) $12x^2 + 7x - 10$

(b) $25x^3 - 60x^2 + 36x$

(c) $12z^4 + 17z^3 + 5z^2$

(d) $x^3 + 3x^2 - 4x - 12$

90. Find all solutions of each equation.

(a) $5x^2 + 8 = 0$

(b) $x^2 - 6x + 4 = 0$

(c) $4x^2 + 4x - 11 = 0$

(d) $x^4 - 18x^2 + 18 = 0$

91. Find a polynomial with integer coefficients that has the given zeros.

(a) 0, 3, 4 (b) $-6, 1$

(c) $-3, 1 + \sqrt{2}, 1 - \sqrt{2}$

(d) $3, 2 + i, 2 - i$

92. Find all the zeros of $f(x) = 2x^3 - 3x^2 + 50x - 75$ if one of the zeros is $x = \frac{3}{2}$.

93. List the possible rational zeros of the function $g(x) = 6x^4 + 7x^3 - 29x^2 - 28x + 20$.

94. Use a graphing utility to graph the function $h(x) = 2x^4 + x^3 - 19x^2 - 9x + 9$. Use the graph and the Rational Zero Test to find the zeros of h.

8.2 Translations of Conics

Vertical and Horizontal Shifts of Conics / Writing Equations of Conics in Standard Form

Vertical and Horizontal Shifts of Conics

In Section 8.1 you looked at conic sections whose graphs were in *standard position.* In this section you will study the equations of conic sections that have been shifted vertically or horizontally in the plane.

Standard Forms of Equations of Conics

Circle: Center $= (h, k)$, Radius $= r$

$$(x - h)^2 + (y - k)^2 = r^2$$

Ellipse: Center $= (h, k)$

Major axis length $= 2a$

Minor axis length $= 2b$

Hyperbola: Center $= (h, k)$

Transverse axis length $= 2a$

Conjugate axis length $= 2b$

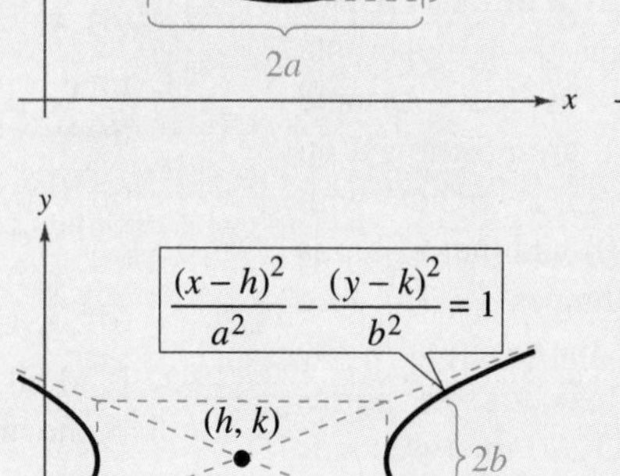

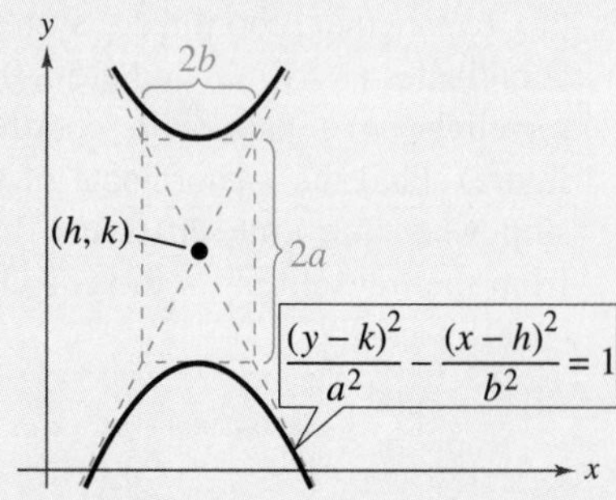

Parabola: Vertex $= (h, k)$

Directed distance from vertex to focus $= p$

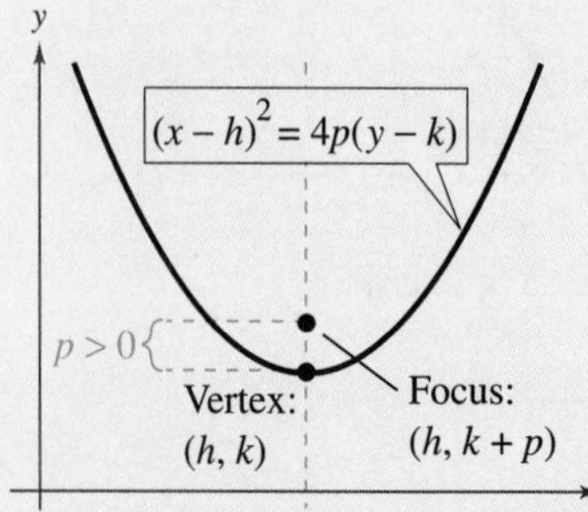

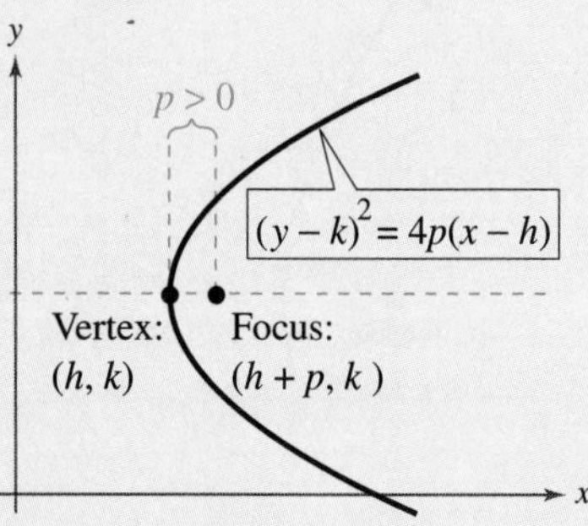

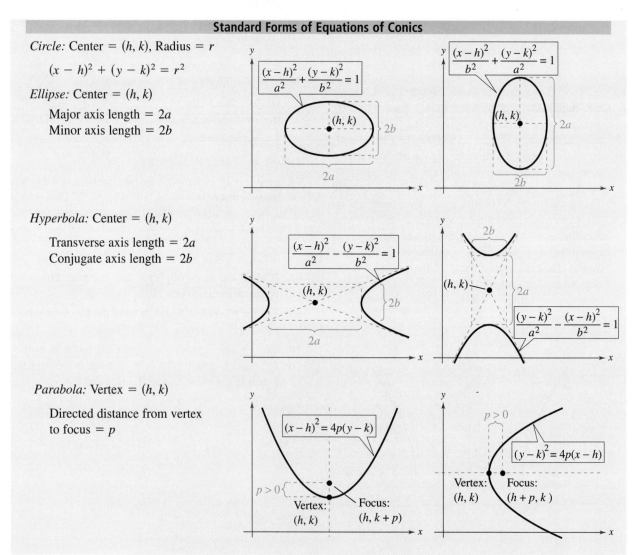

EXAMPLE 1 **Equations of Conic Sections**

a. The graph of

$$(x - 1)^2 + (y + 2)^2 = 3^2$$

is a circle whose center is the point $(1, -2)$ and whose radius is 3, as shown in Figure 8.19(a).

b. The graph of

$$\frac{(x - 2)^2}{3^2} + \frac{(y - 1)^2}{2^2} = 1$$

is an ellipse whose center is the point $(2, 1)$. The major axis of the ellipse is horizontal and of length $2(3) = 6$, and the minor axis of the ellipse is vertical and of length $2(2) = 4$, as shown in Figure 8.19(b).

c. The graph of

$$\frac{(x - 3)^2}{1^2} - \frac{(y - 2)^2}{3^2} = 1$$

is a hyperbola whose center is the point $(3, 2)$. The transverse axis is horizontal and of length $2(1) = 2$, and the conjugate axis is vertical and of length $2(3) = 6$, as shown in Figure 8.19(c).

d. The graph of

$$(x - 2)^2 = 4(-1)(y - 3)$$

is a parabola whose vertex is the point $(2, 3)$. The axis of the parabola is vertical. The focus is one unit above or below the vertex and, because $p = -1$, it follows that the focus lies *below* the vertex, as shown in Figure 8.19(d).

Figure 8.19

(a)

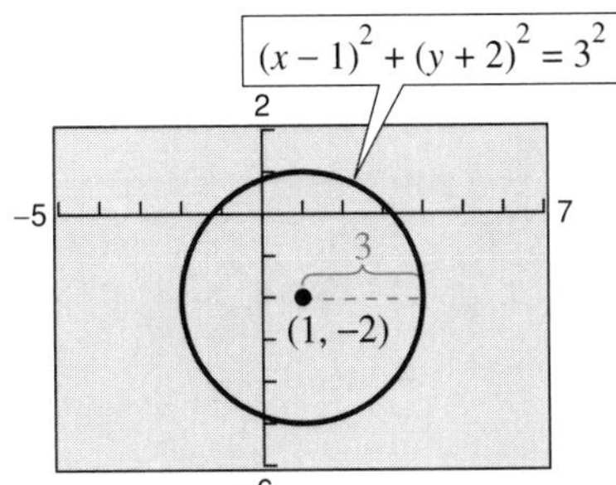

(b)

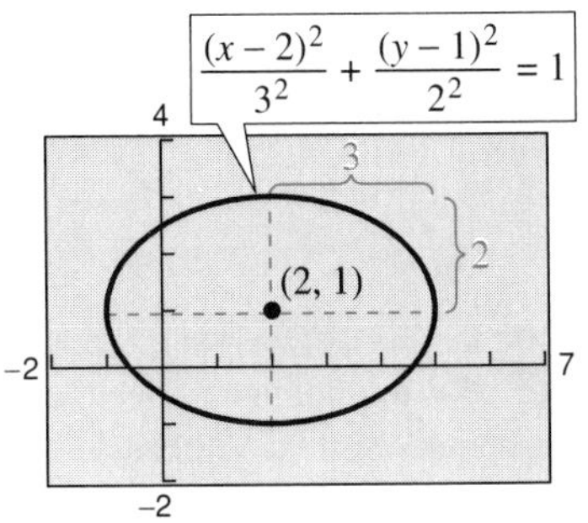

(c)

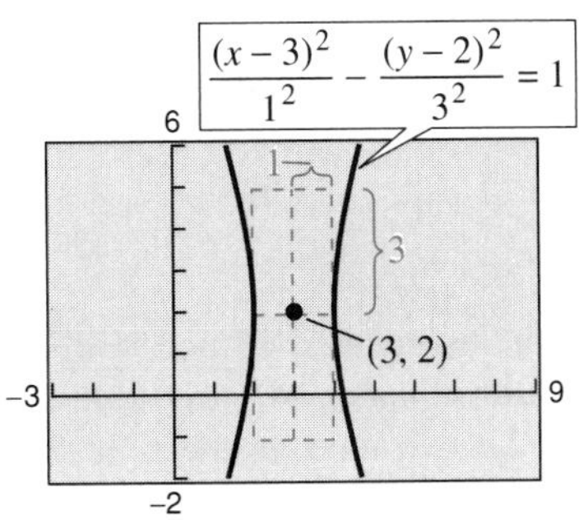

(d)

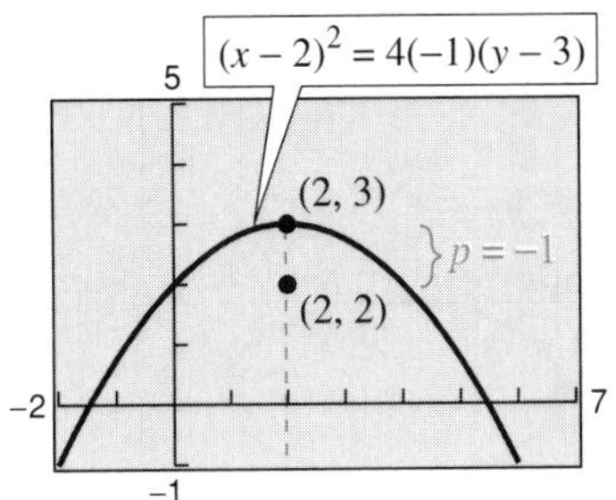

Writing Equations of Conics in Standard Form

EXAMPLE 2 **Finding the Standard Form of a Parabola**

Find the vertex and focus of the parabola $x^2 - 2x + 4y - 3 = 0$.

Solution

$$x^2 - 2x + 4y - 3 = 0 \quad \text{Original equation}$$
$$x^2 - 2x = -4y + 3 \quad \text{Group terms.}$$
$$x^2 - 2x + 1 = -4y + 3 + 1 \quad \text{Add 1 to both sides.}$$
$$(x - 1)^2 = -4y + 4 \quad \text{Completed square form}$$
$$(x - 1)^2 = 4(-1)(y - 1) \quad (x - h)^2 = 4p(y - k)$$

From this standard form, it follows that $h = 1$, $k = 1$, and $p = -1$. Because the axis is vertical and p is negative, the parabola opens downward. The vertex is $(h, k) = (1, 1)$ and the focus is $(h, k + p) = (1, 0)$. The graph is shown in Figure 8.20. You can check this with a graphing utility by graphing $y = \frac{1}{4}(-x^2 + 2x + 3)$.

Figure 8.20

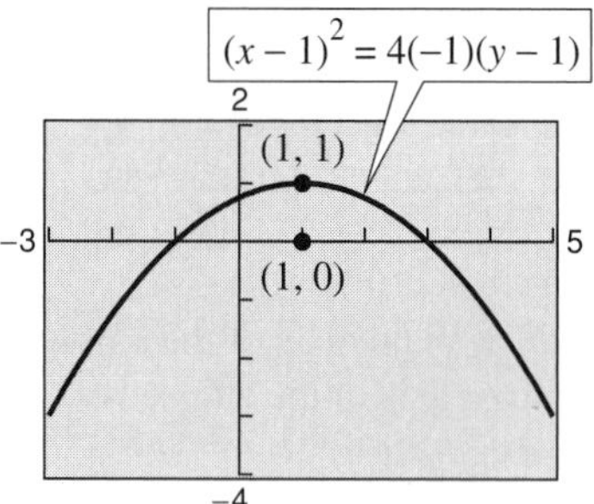

Note Note in Example 2 that p is the *directed distance* from the vertex to the focus. Because the axis of the parabola is vertical and $p = -1$, the focus is one unit *below* the vertex, and the parabola opens downward.

EXAMPLE 3 **Sketching an Ellipse**

Sketch the graph of the ellipse $x^2 + 4y^2 + 6x - 8y + 9 = 0$.

Solution

$$x^2 + 4y^2 + 6x - 8y + 9 = 0 \quad \text{Original equation}$$
$$(x^2 + 6x + \quad) + (4y^2 - 8y + \quad) = -9 \quad \text{Group terms.}$$
$$(x^2 + 6x + \quad) + 4(y^2 - 2y + \quad) = -9 \quad \text{Factor 4 out of } y\text{-terms.}$$
$$(x^2 + 6x + 9) + 4(y^2 - 2y + 1) = -9 + 9 + 4(1) \quad \text{Add 9 and 4 to both sides.}$$
$$(x + 3)^2 + 4(y - 1)^2 = 4 \quad \text{Completed square form}$$
$$\frac{(x + 3)^2}{4} + \frac{(y - 1)^2}{1} = 1 \quad \frac{(x - h)^2}{a^2} + \frac{(y - k)^2}{b^2} = 1$$

From this standard form, it follows that the center is $(h, k) = (-3, 1)$. Because the denominator of the x-term is $4 = a^2 = 2^2$, the endpoints of the major axis lie two units to the right and left of the center. Similarly, because the denominator of the y-term is $1 = b^2 = 1^2$, the endpoints of the minor axis lie one unit up and down from the center. The ellipse is shown in Figure 8.21.

Figure 8.21

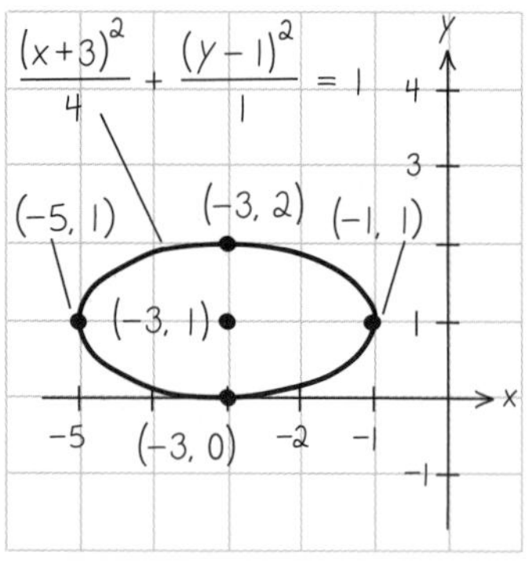

EXAMPLE 4 **Sketching a Hyperbola**

Sketch the graph of the hyperbola given by the equation

$$y^2 - 4x^2 + 4y + 24x - 41 = 0.$$

Figure 8.22

(a)

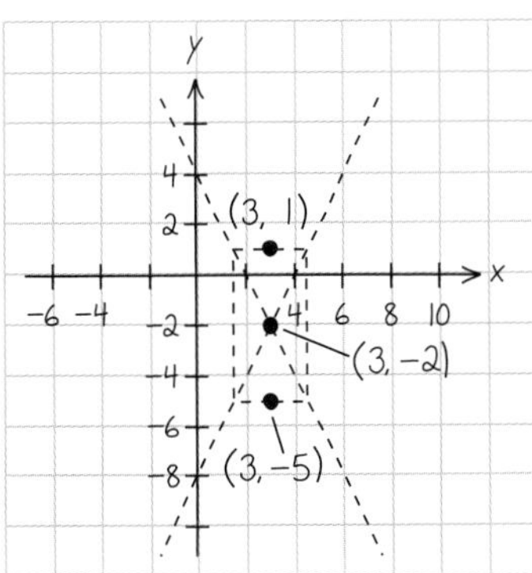

(b)

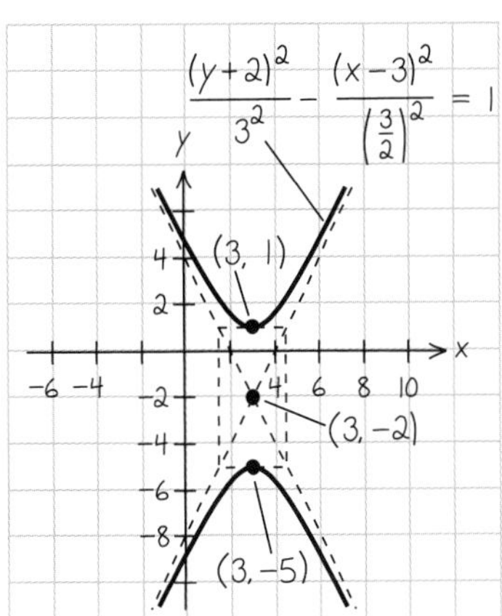

Solution

$$y^2 - 4x^2 + 4y + 24x - 41 = 0 \quad \text{Original equation}$$

$$(y^2 + 4y + \quad) - (4x^2 - 24x + \quad) = 41 \quad \text{Group terms.}$$

$$(y^2 + 4y + \quad) - 4(x^2 - 6x + \quad) = 41 \quad \text{Factor 4 out of } x\text{-terms.}$$

$$(y^2 + 4y + 4) - 4(x^2 - 6x + 9) = 41 + 4 - 4(9) \quad \text{Add 4, subtract 36.}$$

$$(y + 2)^2 - 4(x - 3)^2 = 9 \quad \text{Completed square form}$$

$$\frac{(y + 2)^2}{9} - \frac{4(x - 3)^2}{9} = 1 \quad \text{Divide both sides by 9.}$$

$$\frac{(y + 2)^2}{9} - \frac{(x - 3)^2}{9/4} = 1 \quad \text{Change 4 to } \frac{1}{1/4}.$$

$$\frac{(y + 2)^2}{3^2} - \frac{(x - 3)^2}{(3/2)^2} = 1 \quad \frac{(y-k)^2}{a^2} - \frac{(x-h)^2}{b^2} = 1$$

From this standard form, it follows that the transverse axis is vertical and the center lies at $(h, k) = (3, -2)$. Because the denominator of the y-term is $a^2 = 3^2$, you know that the vertices occur three units above and below the center.

$$(3, 1) \quad \text{and} \quad (3, -5) \qquad \text{Vertices}$$

To sketch the hyperbola, draw a rectangle whose top and bottom pass through the vertices. Because the denominator of the x-term is $b^2 = (3/2)^2$, locate the sides of the rectangle $\frac{3}{2}$ units to the right and left of the center, as shown in Figure 8.22(a). Finally, sketch the asymptotes by drawing lines through the opposite corners of the rectangle. Using these asymptotes, you can complete the graph of the hyperbola, as shown in Figure 8.22(b).

To find the foci in Example 4, first find c.

$$c^2 = a^2 + b^2 = 9 + \frac{9}{4} = \frac{45}{4} \quad \Longrightarrow \quad c = \frac{3\sqrt{5}}{2}$$

Because the transverse axis is vertical, the foci lie c units above and below the center.

$$\left(3, -2 + \tfrac{3}{2}\sqrt{5}\right) \quad \text{and} \quad \left(3, -2 - \tfrac{3}{2}\sqrt{5}\right) \qquad \text{Foci}$$

Figure 8.23

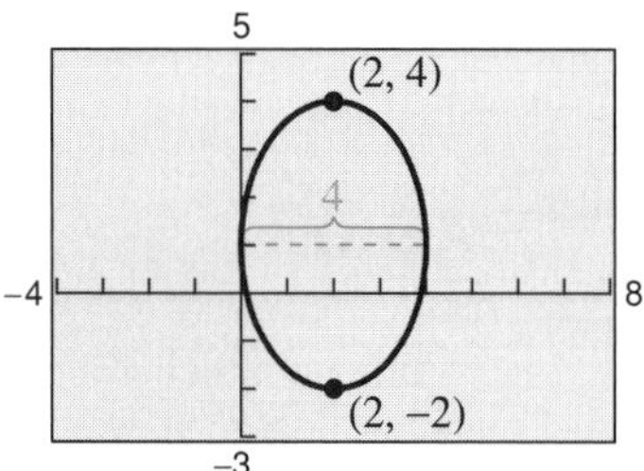

Figure 8.24

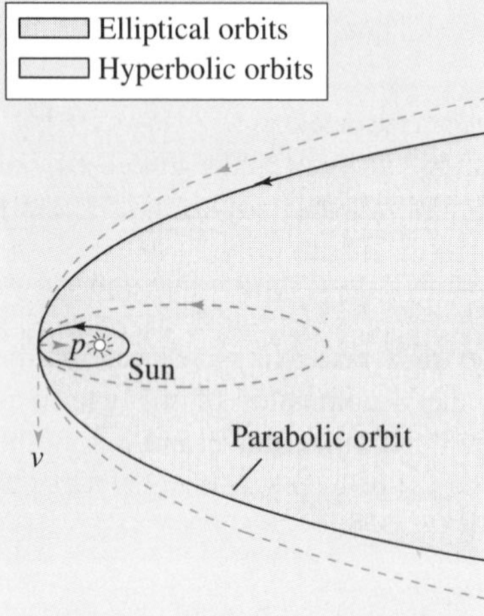

EXAMPLE 5 Writing the Equation of an Ellipse

Write the standard form of the equation of the ellipse whose vertices are $(2, -2)$ and $(2, 4)$. The length of the minor axis of the ellipse is 4, as shown in Figure 8.23.

Solution

The center of the ellipse lies at the midpoint of its vertices. Thus, the center is

$$(h, k) = (2, 1). \qquad \text{Center}$$

Because the vertices lie on a vertical line and are six units apart, it follows that the major axis is vertical and has a length of $2a = 6$. Thus, $a = 3$. Moreover, because the minor axis has a length of 4, it follows that $2b = 4$, which implies that $b = 2$. Therefore, the standard form of the ellipse is as follows.

$$\frac{(x-h)^2}{b^2} + \frac{(y-k)^2}{a^2} = 1 \qquad \text{Major axis is vertical.}$$

$$\frac{(x-2)^2}{2^2} + \frac{(y-1)^2}{3^2} = 1 \qquad \text{Standard form}$$

An interesting application of conic sections involves the orbits of comets in our solar system. Of the 610 comets identified prior to 1970, 245 have elliptical orbits, 295 have parabolic orbits, and 70 have hyperbolic orbits. For example, Halley's comet has an elliptical orbit, and reappearance of this comet can be predicted every 76 years. The center of the sun is a focus of each of these orbits, and each orbit has a vertex at the point where the comet is closest to the sun, as shown in Figure 8.24.

Group Activity

Identifying Equations of Conics

Make up an equation for one of the conics you studied in this section, and graph it with a graphing utility. Exchange printouts or graphing utility screen displays with another student. Try to reconstruct the equation of the conic that is represented by the graph you received. Compare results with the student who used your graph.

8.2 /// EXERCISES

In Exercises 1–8, find the vertex, focus, and directrix of the parabola, and sketch its graph without the aid of a graphing utility.

1. $(x - 1)^2 + 8(y + 2) = 0$
2. $(x + 3) + (y - 2)^2 = 0$
3. $\left(y + \frac{1}{2}\right)^2 = 2(x - 5)$
4. $\left(x + \frac{1}{2}\right)^2 = 4(y - 3)$
5. $y = \frac{1}{4}(x^2 - 2x + 5)$
6. $4x - y^2 - 2y - 33 = 0$
7. $y^2 + 6y + 8x + 25 = 0$
8. $y^2 - 4y - 4x = 0$

In Exercises 9–12, find the vertex, focus, and directrix of the parabola, and use a graphing utility to obtain its graph.

9. $y = -\frac{1}{6}(x^2 + 4x - 2)$
10. $x^2 - 2x + 8y + 9 = 0$
11. $y^2 + x + y = 0$
12. $y^2 - 4x - 4 = 0$

In Exercises 13–22, find an equation of the parabola.

13. 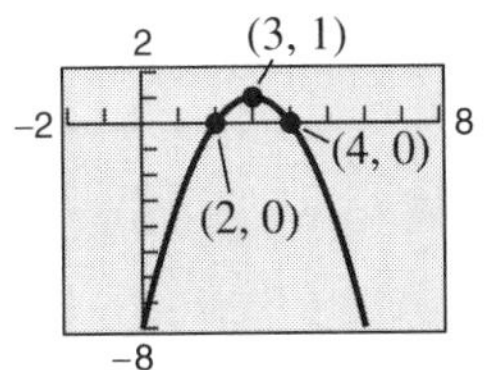

14. 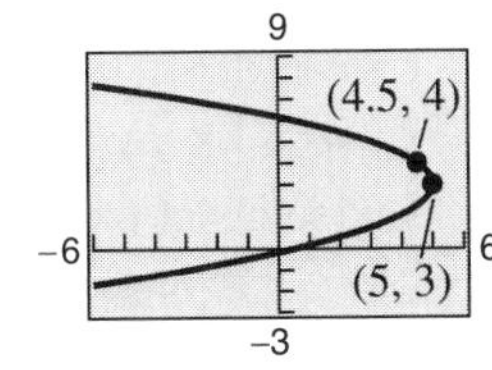

15. 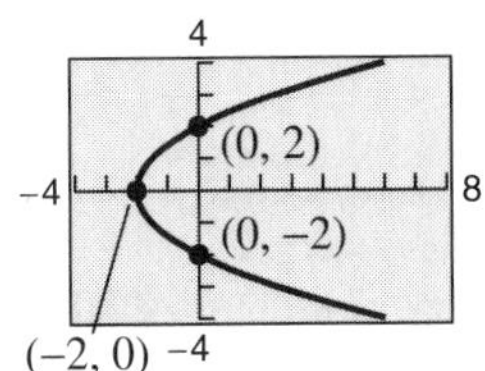

16. 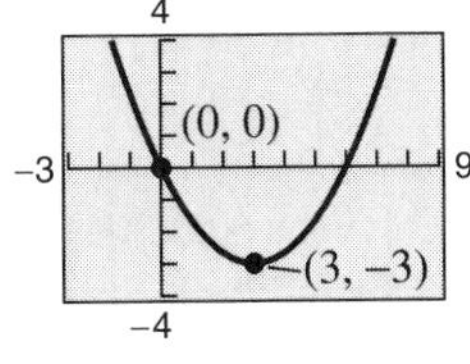

17. Vertex: $(3, 2)$; Focus: $(1, 2)$
18. Vertex: $(-1, 2)$; Focus: $(-1, 0)$
19. Vertex: $(0, 4)$; Directrix: $y = 2$
20. Vertex: $(-2, 1)$; Directrix: $x = 1$
21. Focus: $(2, 2)$; Directrix: $x = -2$
22. Focus: $(0, 0)$; Directrix: $y = 4$

Think About It In Exercises 23 and 24, change the equation so that its graph matches the description.

23. $(y - 3)^2 = 6(x + 1)$; Upper half of parabola
24. $(y + 1)^2 = 2(x - 2)$; Lower half of parabola

25. *Satellite Orbit* An earth satellite in a 100-mile-high circular orbit around the earth has a velocity of approximately 17,500 miles per hour. If this velocity is multiplied by $\sqrt{2}$, the satellite will have the minimum velocity necessary to escape the earth's gravity and it will follow a parabolic path with the center of the earth as the focus (see figure).
    (a) Find the escape velocity of the satellite.
    (b) Find an equation of its path (assume the radius of the earth is 4000 miles).

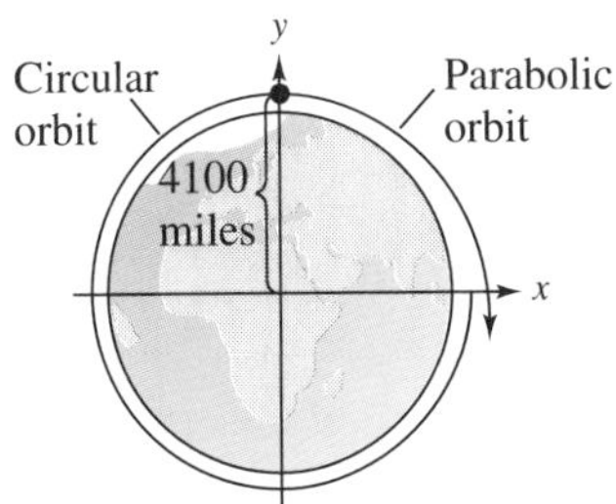

26. *Projectile Motion* A bomber is flying at an altitude of 30,000 feet and a speed of 540 miles per hour (792 feet per second). How many feet will a bomb dropped from the plane travel horizontally before it hits the target if its path is modeled by

$$y = 30{,}000 - \frac{x^2}{39{,}204}?$$

27. *Path of a Projectile* The path of a softball is given by the equation

$y = -0.08x^2 + x + 4.$

The coordinates x and y are measured in feet, with $x = 0$ corresponding to the position from which the ball was thrown.

(a) Use a graphing utility to graph the trajectory of the softball.

(b) Move the cursor along the path to approximate the highest point and the range of the trajectory.

28. *Revenue* The revenue R generated by the sale of x units is given by $R = 375x - \frac{3}{2}x^2$.

(a) Use a graphing utility to graph the function.

(b) Use the trace feature of the graphing utility to approximate *graphically* the sales that will maximize the revenue.

(c) Use a table to approximate *numerically* the sales that will maximize the revenue.

(d) Find the coordinates of the vertex to find *analytically* the sales that will maximize the revenue.

(e) Compare the results of parts (b) to (d). What did you learn by using all three approaches?

In Exercises 29 and 30, find the center, foci, and vertices of the ellipse, and graph the ellipse with the aid of a graphing utility.

29. $12x^2 + 20y^2 - 12x + 40y - 37 = 0$

30. $36x^2 + 9y^2 + 48x - 36y + 43 = 0$

In Exercises 31–36, find the center, foci, and vertices of the ellipse, and sketch its graph.

31. $\dfrac{(x-1)^2}{9} + \dfrac{(y-5)^2}{25} = 1$

32. $(x+2)^2 + \dfrac{(y+4)^2}{1/4} = 1$

33. $9x^2 + 4y^2 + 36x - 24y + 36 = 0$

34. $9x^2 + 4y^2 - 36x + 8y + 31 = 0$

35. $16x^2 + 25y^2 - 32x + 50y + 16 = 0$

36. $9x^2 + 25y^2 - 36x - 50y + 61 = 0$

In Exercises 37–48, find an equation for the ellipse.

37.

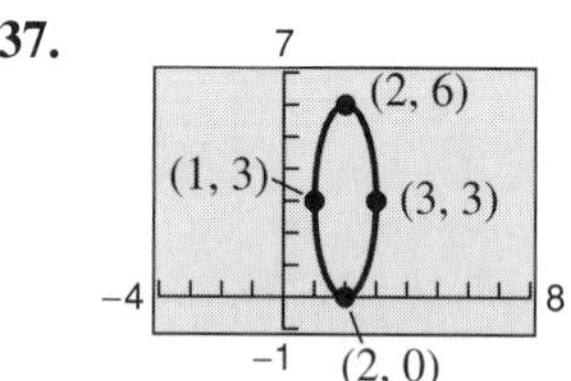

38.

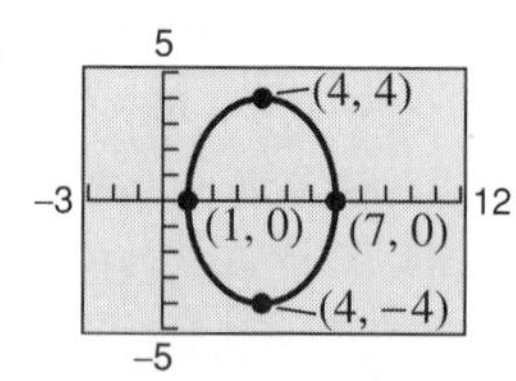

39.

40.

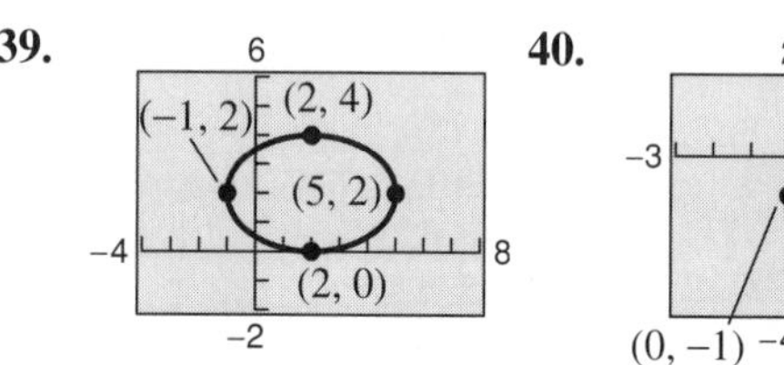

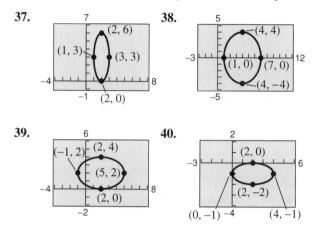

41. Vertices: (0, 2), (4, 2); Minor axis of length 2

42. Foci: (0, 0), (4, 0); Major axis of length 8

43. Foci: (0, 0), (0, 8); Major axis of length 16

44. Center: (2, −1); Vertex: $\left(2, \frac{1}{2}\right)$;
Minor axis of length 2

45. Vertices: (3, 1), (3, 9); Minor axis of length 6

46. Center: (3, 2); $a = 3c$; Foci: (1, 2), (5, 2)

47. Center: (0, 4); $a = 2c$; Vertices: (−4, 4), (4, 4)

48. Vertices: (5, 0), (5, 12);
Endpoints of the minor axis: (0, 6), (10, 6)

Think About It **In Exercises 49 and 50, change the equation so that its graph matches the description.**

49. $\dfrac{(x-3)^2}{9} + \dfrac{y^2}{4} = 1$; Right half of ellipse

50. $\dfrac{(x+1)^2}{16} + \dfrac{(y-2)^2}{25} = 1$; Bottom half of ellipse

In Exercises 51–56, e is called the eccentricity of the ellipse and is defined by $e = c/a$. It measures the flatness of the ellipse.

51. Find an equation of the ellipse with vertices $(\pm 5, 0)$ and eccentricity $e = 3/5$.

52. Find an equation of the ellipse with vertices $(0, \pm 8)$ and eccentricity $e = 1/2$.

53. *Planetary Motion* The planet Pluto moves in an elliptical orbit with the sun at one of the foci (see figure). The length of half of the major axis is 3.666×10^9 miles and the eccentricity is 0.248. Find the smallest distance (*perihelion*) and the greatest distance (*aphelion*) of Pluto from the sun.

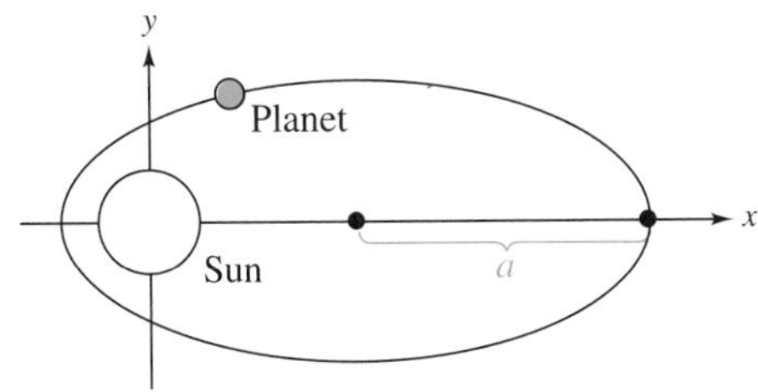

54. *Orbit of Saturn* Saturn moves in an elliptical orbit with the sun at one of the foci. The smallest distance and the greatest distance of the planet from the sun are 1.3495×10^9 and 1.5045×10^9 kilometers, respectively. Find the eccentricity of the orbit.

55. *Exploration* Consider the ellipse given by

$$\frac{x^2}{a^2} + \frac{y^2}{b^2} = 1.$$

(a) Show that the equation of the ellipse can be written as

$$\frac{(x-h)^2}{a^2} + \frac{(y-k)^2}{a^2(1-e^2)} = 1.$$

(b) Use a graphing utility to graph the ellipse

$$\frac{(x-2)^2}{4} + \frac{(y-3)^2}{4(1-e^2)} = 1$$

for $e = 0.95, 0.75, 0.5, 0.25$, and 0.

(c) Make a conjecture about the change in the shape of the ellipse as e approaches 0.

56. *Satellite Orbit* The first artificial satellite to orbit the earth was Sputnik I (launched by the former Soviet Union in 1957). Its highest point above the earth's surface was 938 kilometers, and its lowest point was 212 kilometers (see figure). The center of the earth is the focus of the elliptical orbit and the radius of the earth is 6378 kilometers. Find the eccentricity of the orbit.

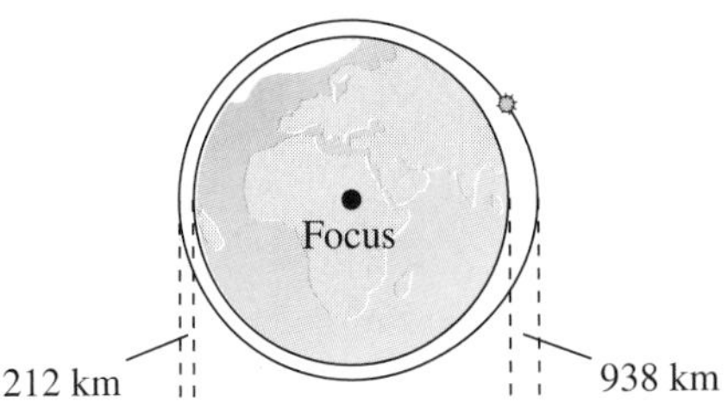

In Exercises 57–64, find the center, vertices, and foci of the hyperbola and sketch its graph. Sketch the asymptotes as an aid in obtaining the graph of the hyperbola.

57. $\dfrac{(x-1)^2}{4} - \dfrac{(y+2)^2}{1} = 1$

58. $\dfrac{(x+1)^2}{144} - \dfrac{(y-4)^2}{25} = 1$

59. $(y+6)^2 - (x-2)^2 = 1$

60. $\dfrac{(y-1)^2}{1/4} - \dfrac{(x+3)^2}{1/9} = 1$

61. $9x^2 - y^2 - 36x - 6y + 18 = 0$

62. $x^2 - 9y^2 + 36y - 72 = 0$

63. $x^2 - 9y^2 + 2x - 54y - 80 = 0$

64. $16y^2 - x^2 + 2x + 64y + 63 = 0$

In Exercises 65 and 66, find the center, vertices, and foci of the hyperbola and graph the hyperbola and its asymptotes with the aid of a graphing utility.

65. $9y^2 - x^2 + 2x + 54y + 62 = 0$

66. $9x^2 - y^2 + 54x + 10y + 55 = 0$

In Exercises 67–78, find an equation for the hyperbola.

67.

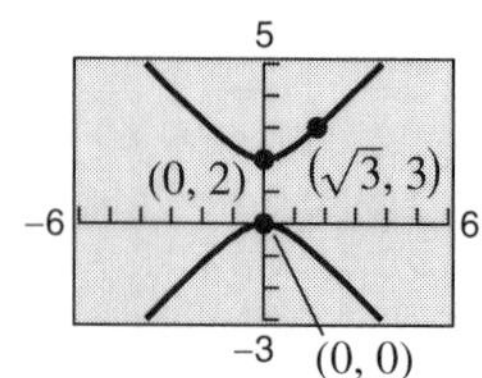

68.

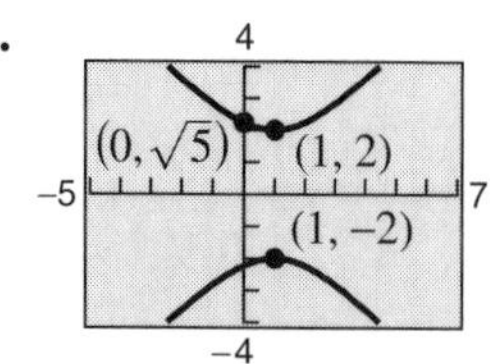

69.

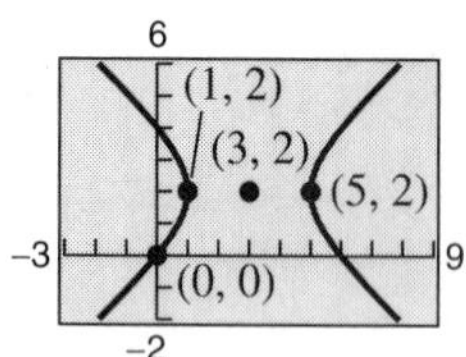

70.

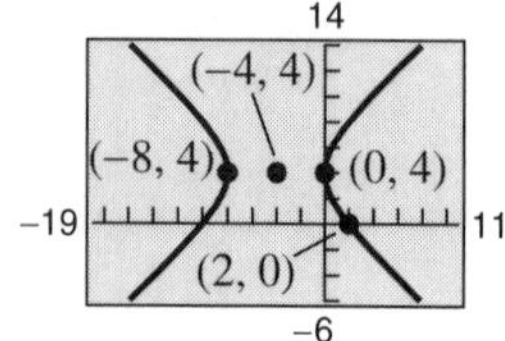

71. Vertices: (2, 0), (6, 0); Foci: (0, 0), (8, 0)

72. Vertices: (2, 3), (2, −3); Foci: (2, 5), (2, −5)

73. Vertices: (4, 1), (4, 9); Foci: (4, 0), (4, 10)

74. Vertices: (−2, 1), (2, 1); Foci: (−3, 1), (3, 1)

75. Vertices: (2, 3), (2, −3);
Passes through the point (0, 5)

76. Vertices: (−2, 1), (2, 1);
Passes through the point (4, 3)

77. Vertices: (0, 2), (6, 2);
Asymptotes: $y = \frac{2}{3}x$, $y = 4 - \frac{2}{3}x$

78. Vertices: (3, 0), (3, 4);
Asymptotes: $y = \frac{2}{3}x$, $y = 4 - \frac{2}{3}x$

Think About It **In Exercises 79 and 80, describe the part of the hyperbola**

$$\frac{(x-3)^2}{4} - \frac{(y-1)^2}{9} = 1$$

given by the equation.

79. $x = 3 - \frac{2}{3}\sqrt{9 + (y-1)^2}$

80. $y = 1 + \frac{3}{2}\sqrt{(x-3)^2 - 4}$

In Exercises 81–88, classify the graph of the equation as a circle, a parabola, an ellipse, or a hyperbola.

81. $x^2 + y^2 - 6x + 4y + 9 = 0$

82. $x^2 + 4y^2 - 6x + 16y + 21 = 0$

83. $4x^2 - y^2 - 4x - 3 = 0$

84. $y^2 - 4y - 4x = 0$

85. $4x^2 + 3y^2 + 8x - 24y + 51 = 0$

86. $4y^2 - 2x^2 - 4y - 8x - 15 = 0$

87. $25x^2 - 10x - 200y - 119 = 0$

88. $4x^2 + 4y^2 - 16y + 15 = 0$

In Exercises 89 and 90, use a graphing utility to graph the conics in the same viewing rectangle and approximate the coordinates of any points of intersection.

89. $x^2 + 9y^2 = 9, \quad y = x^2 - 4$

90. $x^2 + y^2 = 25, \quad x^2 - y^2 = 1$

91. *Chapter Opener* Halley's comet has an elliptical orbit with the sun at one focus. The eccentricity of the orbit is approximately 0.97. The length of the major axis of the orbit is approximately 36.23 astronomical units. (An astronomical unit is about 93 million miles.) Find an equation for the orbit. Place the center of the orbit at the origin, and place the major axis on the x-axis.

92. *Chapter Opener* The comet Encke has an elliptical orbit with the sun at one focus. Encke ranges from 0.34 to 4.08 astronomical units from the sun. Find an equation of the orbit. Place the center of the orbit at the origin and place the major axis on the x-axis.

8.3 Parametric Equations

Plane Curves / Sketching a Plane Curve / Eliminating the Parameter / Finding Parametric Equations for a Graph

Plane Curves

Up to this point, you have been representing a graph by a single equation involving the *two* variables x and y. In this section, you will study situations in which it is useful to introduce a *third* variable to represent a curve in the plane.

To see the usefulness of this procedure, consider the path followed by an object that is propelled into the air at an angle of 45°. If the initial velocity of the object is 48 feet per second, it can be shown that the object follows the parabolic path given by

$$y = -\frac{x^2}{72} + x, \qquad \text{Rectangular equation}$$

as shown in Figure 8.25. However, this equation does not tell the whole story. Although it does tell us *where* the object has been, it doesn't tell us *when* the object was at a given point (x, y) on the path. To determine this time, you can introduce a third variable t, which is called a **parameter.** It is possible to write both x and y as functions of t to obtain the **parametric equations**

$$x = 24\sqrt{2}t \qquad \text{Parametric equation for } x$$

and

$$y = -16t^2 + 24\sqrt{2}t. \qquad \text{Parametric equation for } y$$

From this set of equations you can determine that at time $t = 0$, the object is at the point $(0, 0)$. Similarly, at time $t = 1$, the object is at the point $(24\sqrt{2}, 24\sqrt{2} - 16)$, and so on.

For this particular motion problem, x and y are continuous functions of t, and the resulting path is a **plane curve.** (Recall that a *continuous function* is one whose graph can be traced without lifting the pencil from the paper.)

Figure 8.25

Curvilinear Motion: two variables for position, one variable for time

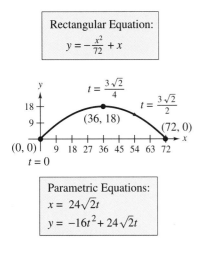

Rectangular Equation:
$y = -\frac{x^2}{72} + x$

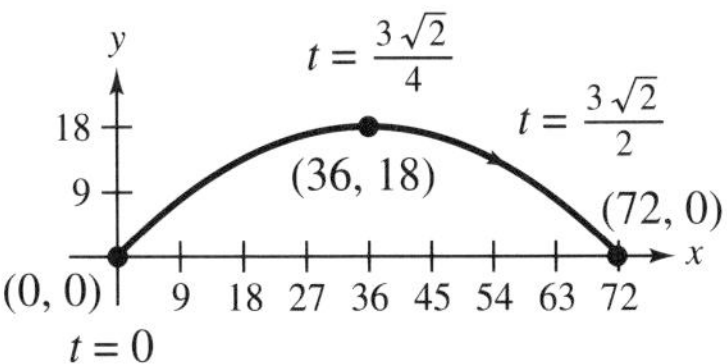

Parametric Equations:
$x = 24\sqrt{2}t$
$y = -16t^2 + 24\sqrt{2}t$

Library of Functions

Parametric equations consist of a pair of functions $x = f(t)$ and $y = g(t)$, each of which is a function of the parameter t. These equations define a plane curve, which might not be the graph of a function, as in Example 1. Most graphing utilities have a parametric mode.

Definition of a Plane Curve

If f and g are continuous functions of t on an interval I, the set of ordered pairs $(f(t), g(t))$ is a **plane curve** C. The equations

$$x = f(t) \qquad \text{and} \qquad y = g(t)$$

are **parametric equations** for C, and t is the **parameter.**

Sketching a Plane Curve

One way to sketch a curve represented by a pair of parametric equations is to plot points in the xy-plane. Each set of coordinates (x, y) is determined from a value chosen for the parameter t. By plotting the resulting points in the order of *increasing* values of t, you trace the curve in a specific direction. This is called the **orientation** of the curve.

EXAMPLE 1 Sketching a Curve

Sketch the curve given by the parametric equations

$$x = t^2 - 4 \quad \text{and} \quad y = \frac{t}{2}, \qquad -2 \le t \le 3.$$

Describe the orientation of the curve.

Solution

Using values of t in the given interval, the parametric equations yield the points (x, y) shown in the table.

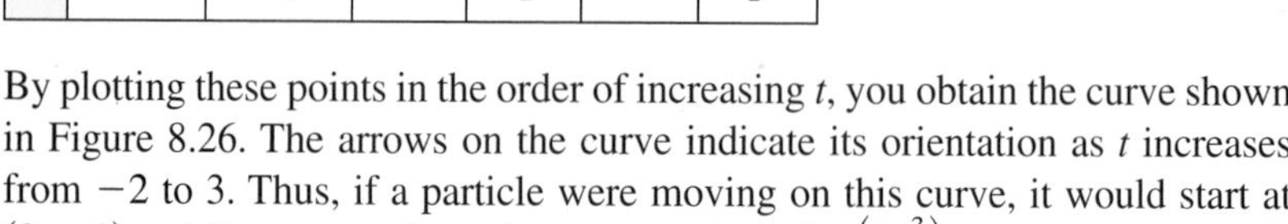

t	-2	-1	0	1	2	3
x	0	-3	-4	-3	0	5
y	-1	$-\frac{1}{2}$	0	$\frac{1}{2}$	1	$\frac{3}{2}$

By plotting these points in the order of increasing t, you obtain the curve shown in Figure 8.26. The arrows on the curve indicate its orientation as t increases from -2 to 3. Thus, if a particle were moving on this curve, it would start at $(0, -1)$ and then move along the curve to the point $\left(5, \frac{3}{2}\right)$.

Figure 8.26

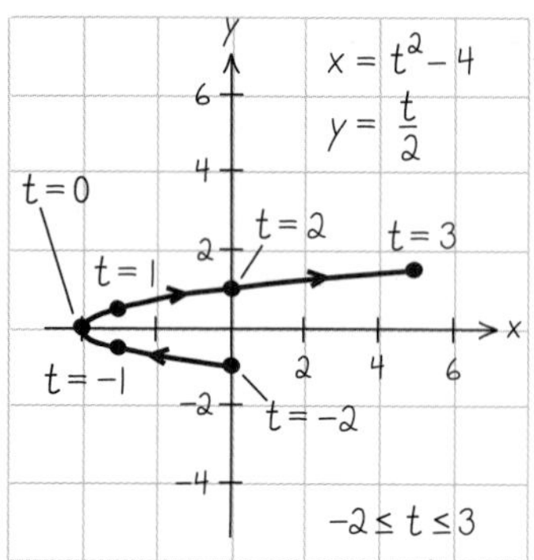

Figure 8.27

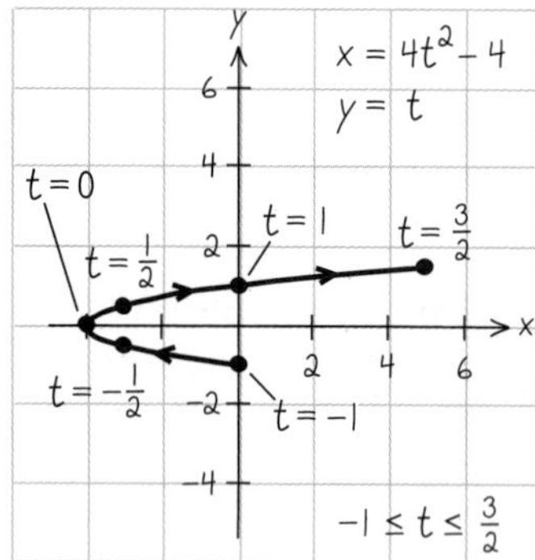

The graph shown in Figure 8.26 does not define y as a function of x. This points out one benefit of parametric equations—they can be used to represent graphs that are more general than graphs of functions.

Two different sets of parametric equations can have the same graph. For example, the set of parametric equations

$$x = 4t^2 - 4 \quad \text{and} \quad y = t, \qquad -1 \le t \le \frac{3}{2}$$

has the same graph as the set given in Example 1. However, by comparing the values of t in Figures 8.26 and 8.27, you can see that this second graph is traced out more *rapidly* (considering t as time) than the first graph. Thus, in applications, different parametric representations can be used to represent various *speeds* at which objects travel along a given path.

EXPLORATION

Use a graphing utility, set in parametric mode, to graph the curve given by

$$\text{X}_{1T} = \text{T} \text{ and } \text{Y}_{1T} = 1 - \text{T}^2.$$

Set the viewing rectangle so that $-4 \le x \le 4$ and $-12 \le y \le 2$. Now, use the graphing utility to graph the curve with various settings for t. Use the following.

(a) $0 \le t \le 3$

(b) $-3 \le t \le 0$

(c) $-3 \le t \le 3$

Compare the curves given by the different t settings. Repeat this experiment using $\text{X}_{1T} = -\text{T}$. How does this change the results?

The *Interactive* CD-ROM offers graphing utility emulators of the *TI-82* and *TI-83*, which can be used with the Examples, Explorations, Technology notes, and Exercises.

Another way to display a curve represented by a pair of parametric equations is to use a graphing utility.

EXAMPLE 2 Using a Graphing Utility in Parametric Mode

Use a graphing utility to graph the curves represented by the parametric equations. For which curve is y a function of x? (Use $-4 \le t \le 4$.)

a. $x = t^2$, $y = t^3$ **b.** $x = t$, $y = t^3$ **c.** $x = t^2$, $y = t$

Solution

Begin by setting the graphing utility to parametric mode. When choosing a viewing rectangle, you must set not only minimum and maximum values of x and y but also minimum and maximum values of t.

a. Enter the parametric equations for x and y.

$$\text{X}_{1T} = \text{T}^2, \qquad \text{Y}_{1T} = \text{T}^3$$

The curve is shown in Figure 8.28(a). From the graph, you can see that y *is not* a function of x.

b. Enter the parametric equations for x and y.

$$\text{X}_{1T} = \text{T}, \qquad \text{Y}_{1T} = \text{T}^3$$

The curve is shown in Figure 8.28(b). From the graph, you can see that y *is* a function of x.

c. Enter the parametric equations for x and y.

$$\text{X}_{1T} = \text{T}^2, \qquad \text{Y}_{1T} = \text{T}$$

The curve is shown in Figure 8.28(c). From the graph, you can see that y *is not* a function of x.

Figure 8.28

(a)

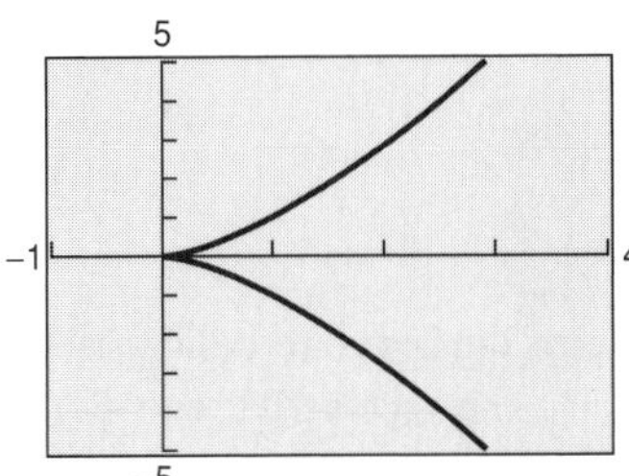

(b)

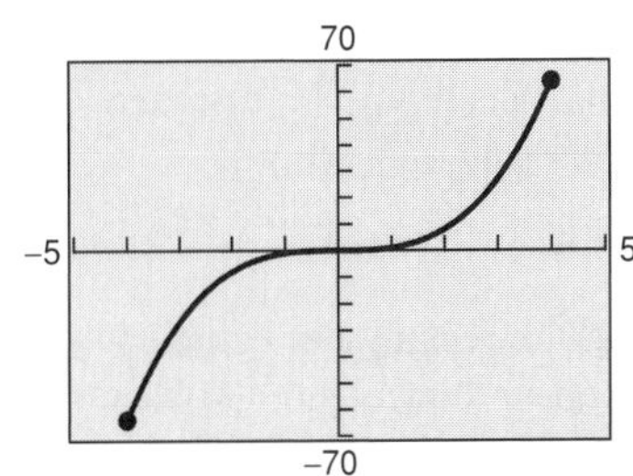

(c)

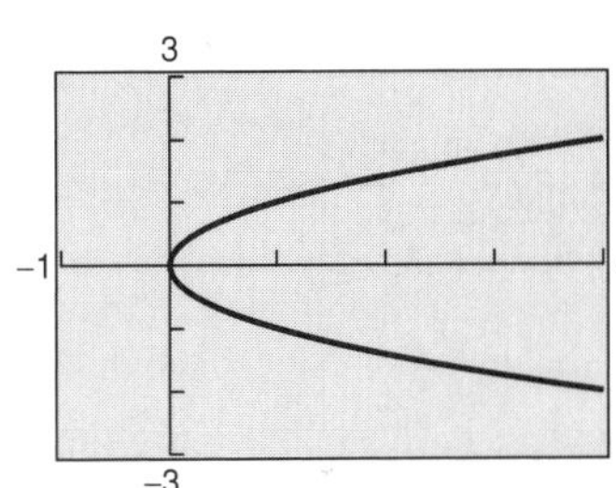

Eliminating the Parameter

Note It is important to realize that eliminating the parameter is primarily an aid to identifying the curve. If the parametric equations represent the path of a moving object, the graph alone is not sufficient to describe the object's motion. You still need the parametric equations to determine the *position, direction,* and *speed* at a given time.

Many curves that are represented by sets of parametric equations have graphs that can also be represented by rectangular equations (in x and y). The process of finding the rectangular equation is called **eliminating the parameter.**

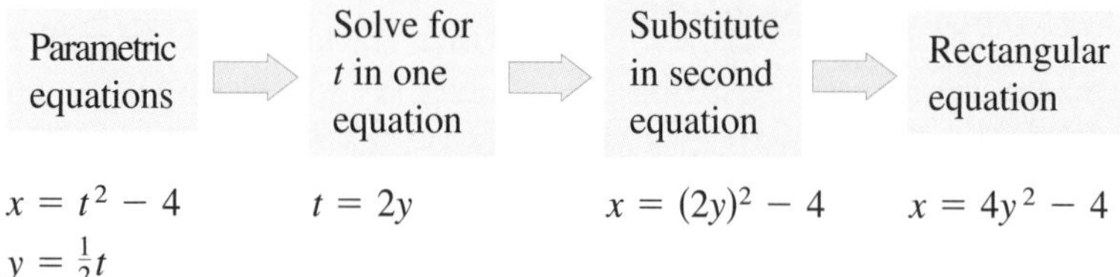

$x = t^2 - 4$ $\quad$ $t = 2y$ $\quad$ $x = (2y)^2 - 4$ $\quad$ $x = 4y^2 - 4$

$y = \frac{1}{2}t$

After eliminating the parameter, you can recognize that the curve is a parabola with a horizontal axis and vertex at $(-4, 0)$.

Converting equations from parametric to rectangular form can change the ranges of x and y. In such cases, you should restrict x and y in the rectangular equation so that its graph matches the graph of the parametric equations.

EXAMPLE 3 **Eliminating the Parameter**

Identify the curve represented by the equations

$$x = \frac{1}{\sqrt{t+1}} \quad \text{and} \quad y = \frac{t}{t+1}.$$

Solution

Solving for t in the equation for x produces

$$x = \frac{1}{\sqrt{t+1}} \quad \text{or} \quad \frac{1}{x^2} = t + 1$$

which implies that $t = (1/x^2) - 1$. Substituting in the equation for y, you obtain

$$\begin{aligned} y &= \frac{t}{t+1} \\ &= \frac{(1/x^2) - 1}{(1/x^2) - 1 + 1} \\ &= 1 - x^2. \end{aligned}$$

From the rectangular equation, you can recognize the curve to be a parabola that opens downward and has its vertex at $(0, 1)$. The rectangular equation is defined for all values of x. From the parametric equation for x, however, you can see that the curve is defined only when $-1 < t$. Thus, you should restrict the domain of x to positive values, as shown in Figure 8.29.

Figure 8.29

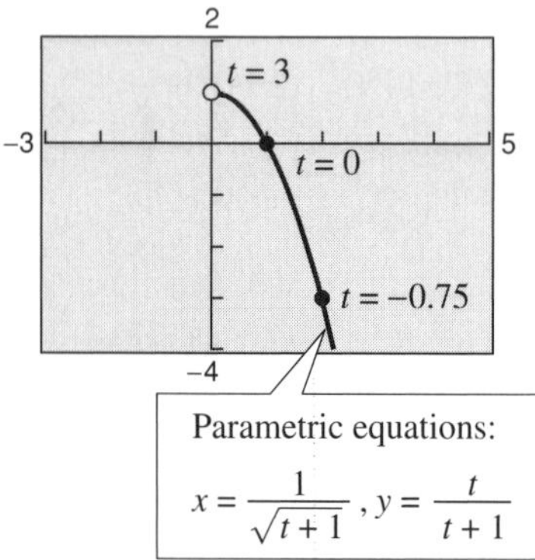

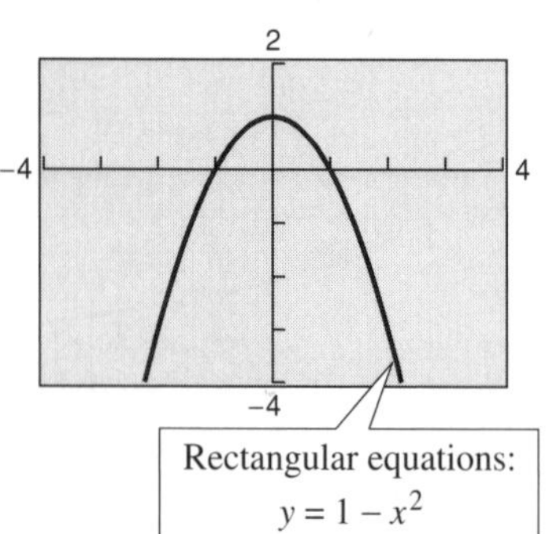

Figure 8.30

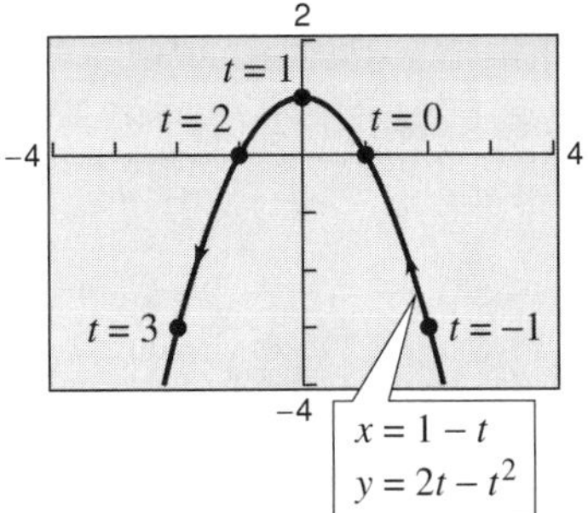

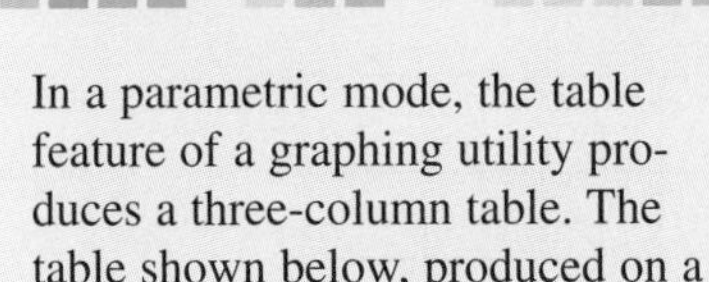

In a parametric mode, the table feature of a graphing utility produces a three-column table. The table shown below, produced on a *TI-83*, represents the graph in Figure 8.30.

T	XIT	YIT
−2	3	−8
−1	2	−3
0	1	0
1	0	1
2	−1	0
3	−2	−3
4	−3	−8
T=−2		

Finding Parametric Equations for a Graph

How can you determine a set of parametric equations for a given graph or a given physical description? From the discussion following Example 1, you know that such a representation is not unique. This is further demonstrated in Example 4.

EXAMPLE 4 Finding Parametric Equations for a Given Graph

Find a set of parametric equations to represent the graph of $y = 1 - x^2$ using the following parameters.

a. $t = x$ **b.** $t = 1 - x$

Solution

a. Letting $t = x$, you obtain the following parametric equations.

$x = t$ Parametric equation for x

$y = 1 - x^2$ Given rectangular equation

$= 1 - t^2$ Parametric equation for y

b. Letting $t = 1 - x$, you obtain the following parametric equations.

$x = 1 - t$ Parametric equation for x

$y = 1 - (1 - t)^2$ Substitute $1 - t$ for x.

$= 2t - t^2$ Parametric equation for y

In Figure 8.30, note how the resulting curve is oriented by the increasing values of t. For part (a), the curve would have the opposite orientation.

Group Activity

Changing the Orientation of a Curve

The orientation of a curve refers to the direction in which the curve is traced as the values of the parameter increase. For instance, as t increases, how is the folium of Descartes given by

$$x = \frac{3t}{1 + t^3} \quad \text{and} \quad y = \frac{3t^2}{1 + t^3}, \qquad -10 \le t \le 10$$

traced out? Find a parametric representation for which the curve is traced out in the opposite direction.

8.3 /// EXERCISES

In Exercises 1–8, match the equation with its graph. [The graphs are labeled (a), (b), (c), (d), (e), (f), (g), and (h).]

(a)

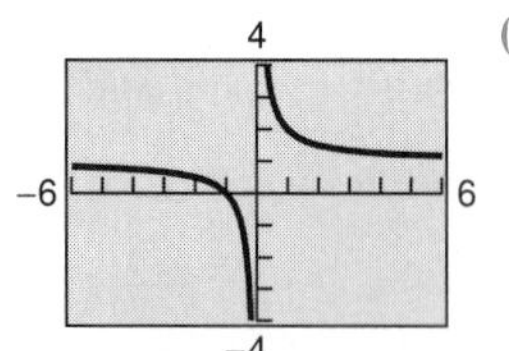

(b)

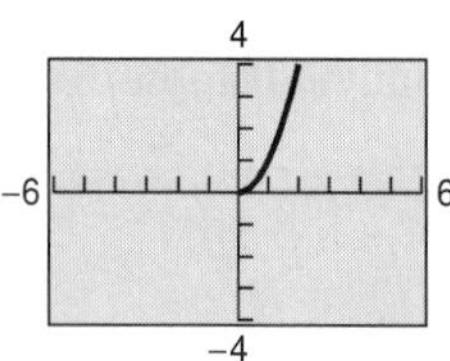

(c)

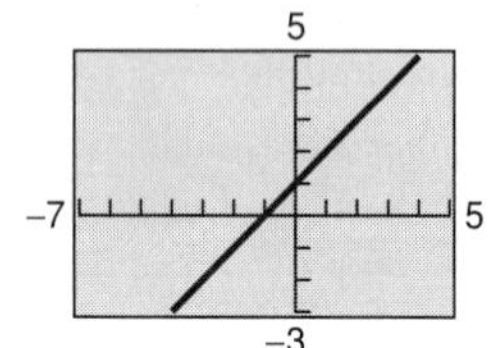

(d)

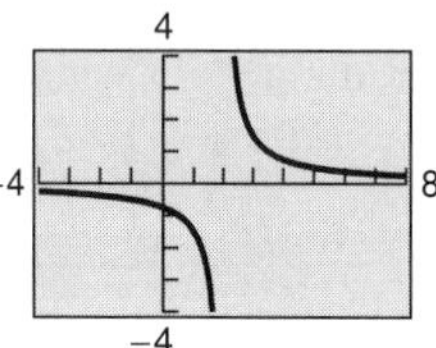

(e)

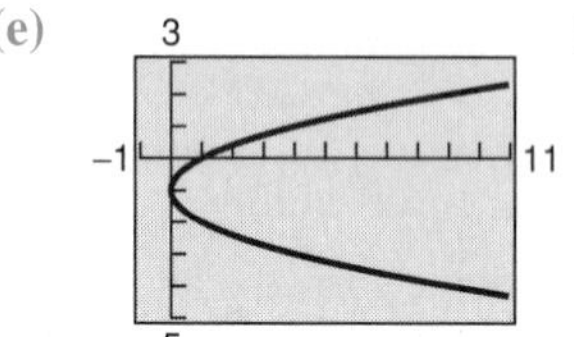

(f)

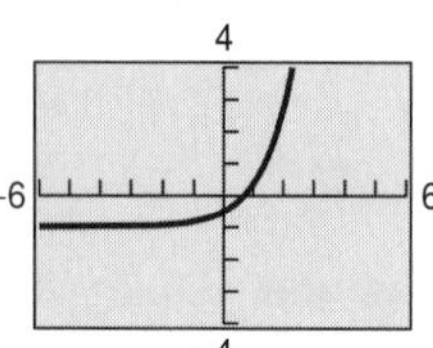

(g)

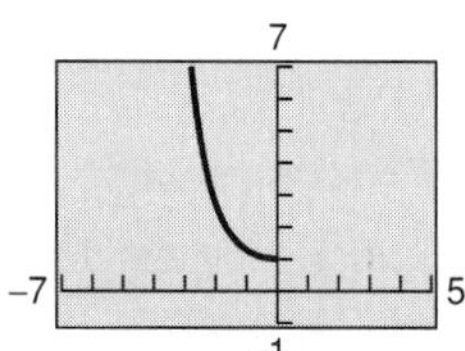

(h)

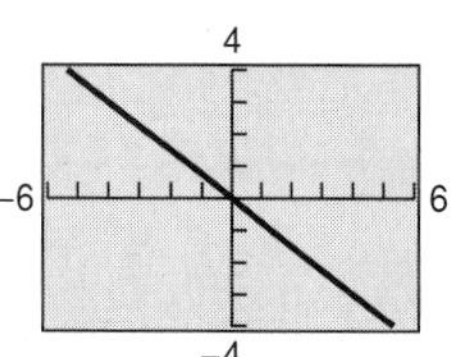

1. $x = t$
$y = t + 1$

2. $x = t$
$y = -\frac{3}{4}t$

3. $x = \sqrt{t}$
$y = t$

4. $x = t^2$
$y = t - 1$

5. $x = \dfrac{1}{t}$
$y = t + 1$

6. $x = \dfrac{1}{2}t$
$y = \dfrac{3}{t - 4}$

7. $x = \ln t$
$y = \frac{1}{2}t - 1$

8. $x = -2\sqrt{t}$
$y = e^t$

9. Consider the parametric equations

$x = \sqrt{t}$ and $y = 1 - t$.

(a) Complete the table.

t	0	1	2	3	4
x					
y					

(b) Plot the points (x, y) generated in part (a), and sketch a graph of the parametric equations.

(c) Use a graphing utility to graph the curve represented by the parametric equations.

(d) Find the rectangular equation by eliminating the parameter. Sketch its graph. How do the graphs differ from those in parts (b) and (c)?

10. Consider the parametric equations

$x = 2/t$ and $y = t - 1$.

(a) Complete the table.

t	-2	-1	1	2	3
x					
y					

(b) Plot the points (x, y) generated in part (a), and sketch a graph of the parametric equations.

(c) Use a graphing utility to graph the curve represented by the parametric equations.

(d) Find the rectangular equation by eliminating the parameter. Sketch its graph.

In Exercises 11–26, sketch the curve represented by the parametric equations (indicate the direction of the curve). Use a graphing utility to confirm your result. Then eliminate the parameter and write the corresponding rectangular equation whose graph represents the curve.

11. $x = t$, $y = -2t$

12. $x = t$, $y = \frac{1}{2}t$

13. $x = 3t - 1$, $y = 2t + 1$

14. $x = 3 - 2t$, $y = 2 + 3t$

15. $x = \frac{1}{4}t$, $y = t^2$

16. $x = t$, $y = t^3$

17. $x = t + 1$, $y = t^2$

18. $x = \sqrt{t}$, $y = 1 - t$

19. $x = t^3$, $y = \dfrac{1}{2}t^2$

20. $x = t - 1$, $y = \dfrac{t}{t - 1}$

21. $x = 2t$, $y = |t - 2|$

22. $x = |t - 1|$, $y = t + 2$

23. $x = e^{-t}$, $y = e^{3t}$

24. $x = e^{2t}$, $y = e^t$

25. $x = t^3$, $y = 3 \ln t$

26. $x = \ln 2t$, $y = t^2$

In Exercises 27–30, use a graphing utility to graph the curve represented by the parametric equations.

27. $x = 12t$, $y = -8t^2 + 32t$

28. $x = 3t/(1 + t^3)$, $y = 3t^2/(1 + t^3)$

29. $x = t/2$, $y = \ln(t^2 + 1)$

30. $x = 10 - 0.01e^t$, $y = 0.2t^2$

31. *True or False?* The two sets of parametric equations $x = t$, $y = t^2 + 1$ and $x = 3t$, $y = 9t^2 + 1$ correspond to the same rectangular equation.

32. *True or False?* The graph of the parametric equations $x = t^2$ and $y = t^2$ is the line $y = x$.

In Exercises 33–36, determine how the plane curves differ from each other.

33. (a) $x = t$, $y = 2t + 1$ (b) $x = 1/t$, $y = (2/t) + 1$
(c) $x = e^{-t}$, $y = 2e^{-t} + 1$ (d) $x = e^t$, $y = 2e^t + 1$

34. (a) $x = t$, $y = t^2 - 1$ (b) $x = t^2$, $y = t^4 - 1$
(c) $x = \dfrac{1}{t}$, $y = \dfrac{1}{t^2} - 1$ (d) $x = e^t$, $y = e^{2t} - 1$

35. (a) $x = 2\sqrt{t}$, $y = 4 - \sqrt{t}$ (b) $x = 2\sqrt[3]{t}$, $y = 4 - \sqrt[3]{t}$
(c) $x = 2(t + 1)$, $y = 3 - t$ (d) $x = -2t^2$, $y = 4 + t^2$

36. (a) $x = t$, $y = t$ (b) $x = t^2$, $y = t^2$
(c) $x = -t$, $y = -t$ (d) $x = t^3$, $y = t^3$

37. Graph the parametric equations $x = \sqrt[3]{t}$ and $y = t - 1$. Describe how the graph changes for each of the following.
(a) $0 \le t \le 1$ (b) $0 \le t \le 27$
(c) $-8 \le t \le 27$ (d) $-27 \le t \le 27$

38. Eliminate the parameter and obtain the standard form of the line through (x_1, y_1) and (x_2, y_2) if
$x = x_1 + t(x_2 - x_1)$
$y = y_1 + t(y_2 - y_1)$.

In Exercises 39–42, use the result of Exercise 38 to find a set of parametric equations for the line through the points.

39. $(0, 0), (5, -2)$

40. $(1, 4), (5, -2)$

41. $(-2, 3), (3, 10)$

42. $(-1, -4), (15, 20)$

43. Find a set of parametric equations and the interval for t for the graph of the line from $(3, -1)$ to $(3, 5)$. Use a graphing utility to verify your result.

44. Find a set of parametric equations and the interval for t for the graph of the line from $(-3, 4)$ to $(6, 4)$. Use a graphing utility to verify your result.

45. Change the parametric equations for the line segment described in Exercise 43 so that the orientation of the graph will be reversed. Use a graphing utility to verify your result.

46. Change the parametric equations for the line segment described in Exercise 44 so that the orientation of the graph will be reversed. Use a graphing utility to verify your result.

In Exercises 47–50, find two different sets of parametric equations for the given rectangular equation.

47. $y = 3x - 2$

48. $y = 1/x$

49. $y = x^3$

50. $y = x^2$

Projectile Motion **A projectile is launched at a height h feet above the ground at an angle of 45° with the horizontal. If the initial velocity is v_0 feet per second, the path of the projectile is modeled by the parametric equations**

$$x = \left(\frac{v_0\sqrt{2}}{2}\right)t \quad \text{and} \quad y = h + \left(\frac{v_0\sqrt{2}}{2}\right)t - 16t^2.$$

In Exercises 51 and 52, use a graphing utility to graph the paths of projectiles launched from ground level at the specified values of h and v_0. For each case, use the graph to approximate the maximum height and the range of the projectile.

51. (a) $h = 0$, $v_0 = 88$ ft/sec
(b) $h = 0$, $v_0 = 132$ ft/sec
(c) $h = 30$, $v_0 = 88$ ft/sec
(d) $h = 30$, $v_0 = 132$ ft/sec

52. (a) $h = 0$, $v_0 = 60$ ft/sec
(b) $h = 0$, $v_0 = 100$ ft/sec
(c) $h = 75$, $v_0 = 60$ ft/sec
(d) $h = 75°$, $v_0 = 100$ ft/sec

53. *Baseball* The center-field fence in a ballpark is 10 feet high and 400 feet from home plate. The ball is hit at a point 3 feet above the ground and leaves the bat at a speed of 150 feet per second (see figure).
(a) If the ball leaves the bat at an angle of 15° with the horizontal, the parametric equations for its path are $x = 145t$ and $y = 3 + 39t - 16t^2$. Use a graphing utility to sketch the path of the ball. Is the hit a home run?
(b) If the ball leaves the bat at an angle of 23° with the horizontal, the parametric equations for its path are $x = 138t$ and $y = 3 + 59t - 16t^2$. Use a graphing utility to sketch the path of the ball. Is the hit a home run?

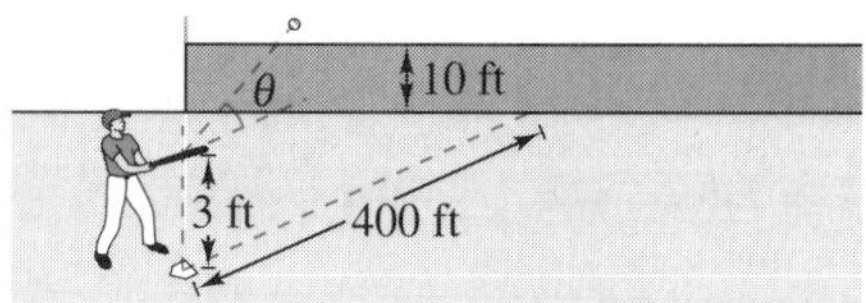

54. *Football* The quarterback of a football team releases a pass at a height of 7 feet above the playing field, and the football is caught at a height of 4 feet and 30 yards directly downfield. The pass is released at an angle of 35° with the horizontal. The parametric equations for the path of the football are

$$x = 0.82v_0t \quad \text{and} \quad y = 7 + 0.57v_0t - 16t^2$$

where v_0 is the speed of the football when it is released.
(a) Find the speed of the football when it is released and write a set of parametric equations for the path of the ball.
(b) Use a graphing utility to graph the path of the ball and approximate its maximum height.
(c) Find the time the receiver has to position himself after the quarterback releases the ball.

Focus on Concepts

In this chapter, you studied several concepts that are required in the study of conics and parametric equations. You can use the following questions to check your understanding of several of these basic concepts. The answers to these questions are given in the back of the book.

In Exercises 1 and 2, an equation and four variations are given. In your own words, describe how the graph of each variation differs from the graph of the first equation.

1. $y^2 = 8x$

(a) $(y - 2)^2 = 8x$ (b) $y^2 = 8(x + 1)$

(c) $y^2 = -8x$ (d) $y^2 = 4x$

2. $\dfrac{x^2}{4} + \dfrac{y^2}{9} = 1$

(a) $\dfrac{x^2}{9} + \dfrac{y^2}{4} = 1$ (b) $\dfrac{x^2}{4} + \dfrac{y^2}{4} = 1$

(c) $\dfrac{x^2}{4} + \dfrac{y^2}{25} = 1$ (d) $\dfrac{(x - 3)^2}{4} + \dfrac{y^2}{9} = 1$

3. Explain how the central rectangle of a hyperbola can be used to sketch its asymptotes.

4. Consider an ellipse with a horizontal major axis of length 10 units. The number b in the standard form of the equation of the ellipse must be less than what number? Explain the change in the shape of the ellipse as b approaches this number.

5. The graph of the parametric equations $x = t^3$ and $y = t - 1$ is shown below. Would the graph change for the equations $x = (-t)^3$ and $y = -t - 1$? If so, how would it change?

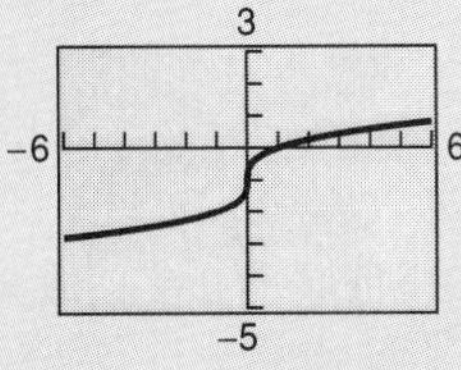

6. *True or False?* There is only one set of parametric equations that represent the line $y = 3 - 2x$. Explain.

In Exercises 7–10, match the set of parametric equations with the corresponding graph. For each case the interval for t is $-1 \le t \le 1$. [The graphs are labeled (a), (b), (c), and (d).]

(a)

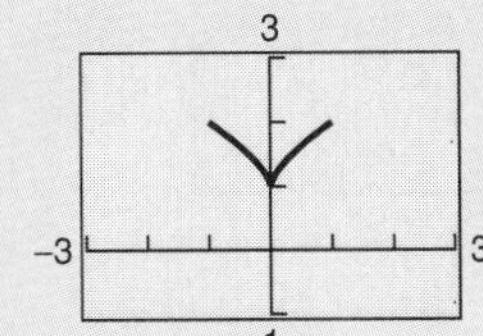

(b)

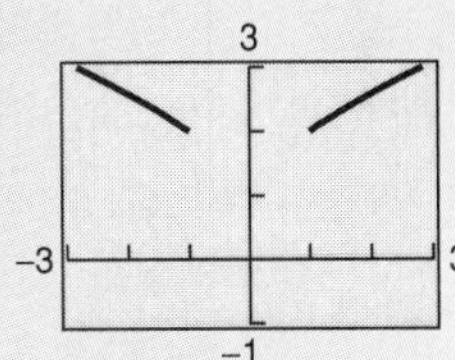

(c)

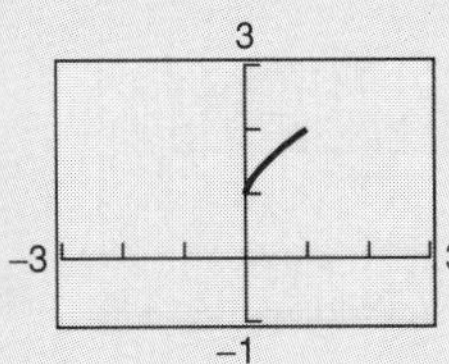

(d)

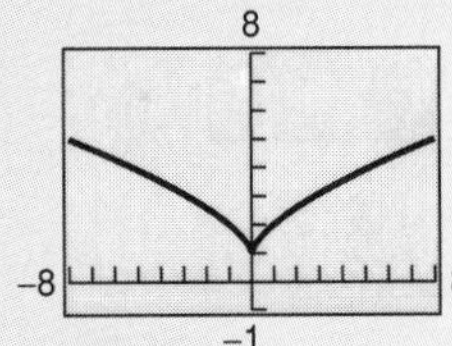

7. $x = t^3$
$y = t^2 + 1$

8. $x = t^6$
$y = t^4 + 1$

9. $x = (2t)^3$
$y = (2t)^2 + 1$

10. $x = 1/t^3$
$y = (1/t^2) + 1$

8 /// REVIEW EXERCISES

In Exercises 1–6, match the equation with the correct graph. [The graphs are labeled (a), (b), (c), (d), (e), and (f).]

(a)

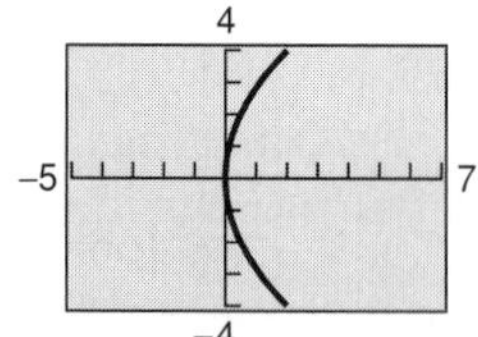

(b)

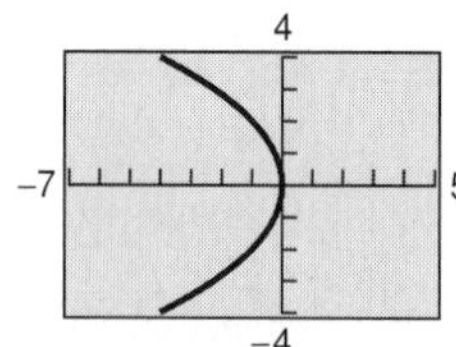

(c)

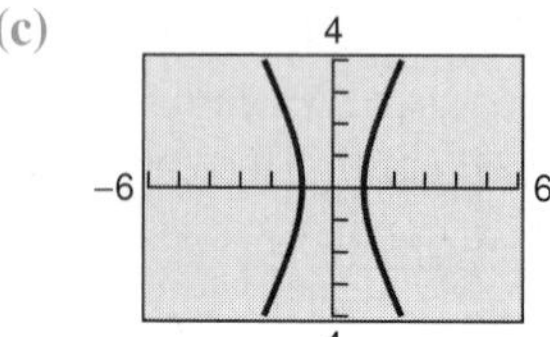

(d)

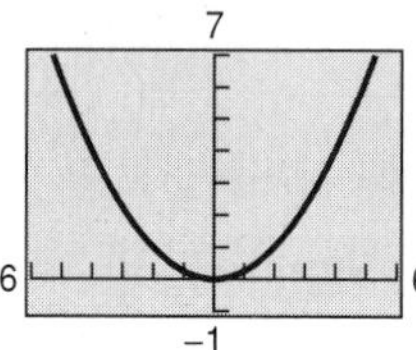

(e)

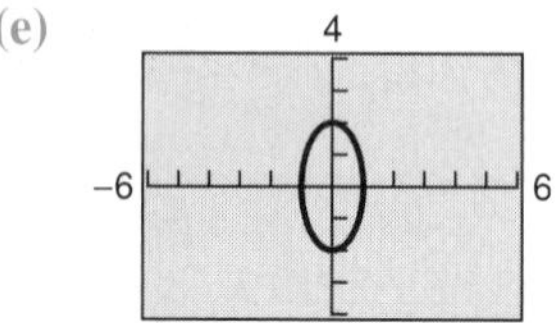

(f)

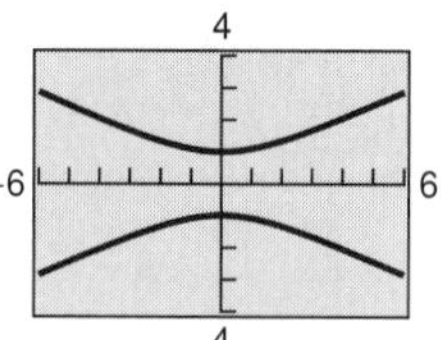

1. $4x^2 + y^2 = 4$
2. $x^2 = 4y$
3. $4x^2 - y^2 = 4$
4. $y^2 = -4x$
5. $x^2 - 5y^2 = -5$
6. $y^2 - 8x = 0$

In Exercises 7–14, identify the conic and sketch its graph. Solve for y and use a graphing utility to verify your answer.

7. $4x - y^2 = 0$
8. $8y + x^2 = 0$
9. $4x^2 + y^2 = 16$
10. $2x^2 + 6y^2 = 18$
11. $5y^2 - 4x^2 = 20$
12. $x^2 - y^2 = \frac{9}{4}$
13. $x^2 + y^2 - 2x - 4y + 5 = 0$
14. $4x^2 + y^2 - 16x + 15 = 0$

In Exercises 15–22, solve for y and use a graphing utility to graph the resulting equations. Identify the conic.

15. $x^2 - 6x + 2y + 9 = 0$
16. $y^2 - 12y - 8x + 20 = 0$
17. $x^2 + 9y^2 + 10x - 18y + 25 = 0$
18. $16x^2 + 16y^2 - 16x + 24y - 3 = 0$
19. $4x^2 - 4y^2 - 4x + 8y - 11 = 0$
20. $x^2 - 9y^2 + 10x + 18y + 7 = 0$
21. $x^2 - 10xy + y^2 + 1 = 0$
22. $40x^2 + 36xy + 25y^2 - 52 = 0$

In Exercises 23–28, find an equation of the parabola.

23.

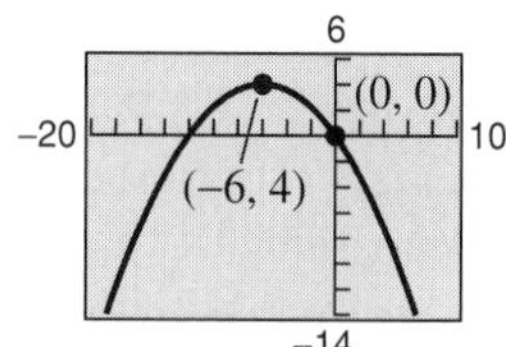

24.

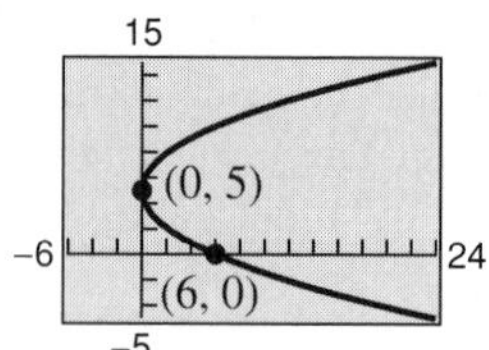

25. Vertex: (4, 2); Focus: (4, 0)
26. Vertex: (2, 0); Focus: (0, 0)
27. Vertex: (0, 2); Horizontal axis; Passes through (−1, 0)
28. Vertex: (2, 2); Directrix: $y = 0$

In Exercises 29–34, find an equation of the ellipse.

29.

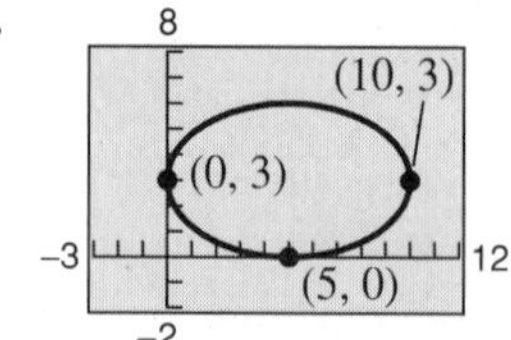

30.

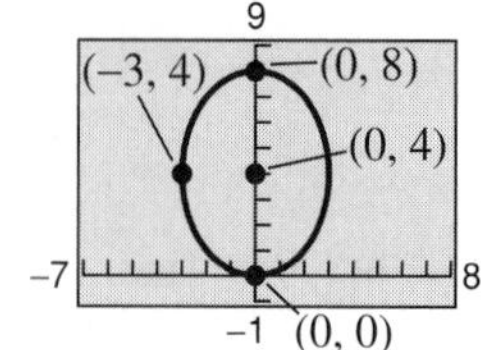

31. Vertices: $(-3, 0), (7, 0)$; Foci: $(0, 0), (4, 0)$

32. Vertices: $(2, 0), (2, 4)$; Foci: $(2, 1), (2, 3)$

33. Vertices: $(0, \pm 6)$; Passes through $(2, 2)$

34. Vertices: $(0, 1), (4, 1)$; Passes through $(2, 0)$

In Exercises 35–40, find an equation of the hyperbola.

35.

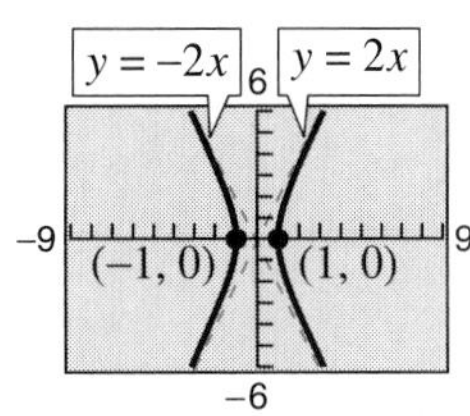

36.

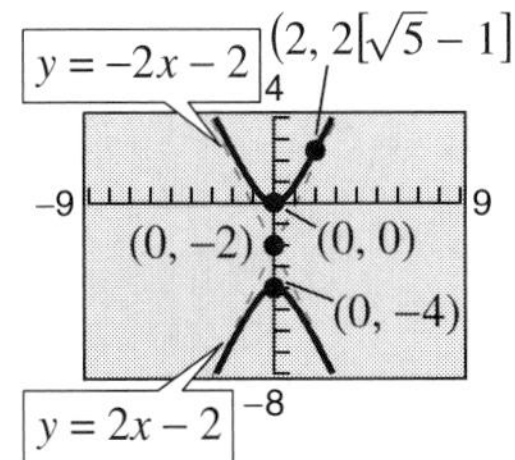

37. Vertices: $(0, \pm 1)$; Foci: $(0, \pm 3)$

38. Vertices: $(2, 2), (-2, 2)$; Foci: $(4, 2), (-4, 2)$

39. Foci: $(0, 0), (8, 0)$;

Asymptotes: $y = \pm 2(x - 4)$

40. Foci: $(3, \pm 2)$;

Asymptotes: $y = \pm 2(x - 3)$

41. *Satellite Antenna* A cross section of a large parabolic antenna (see figure) is given by

$$y = \frac{x^2}{200}, \qquad -100 \le x \le 100.$$

The receiving and transmitting equipment is positioned at the focus. Find the coordinates of the focus.

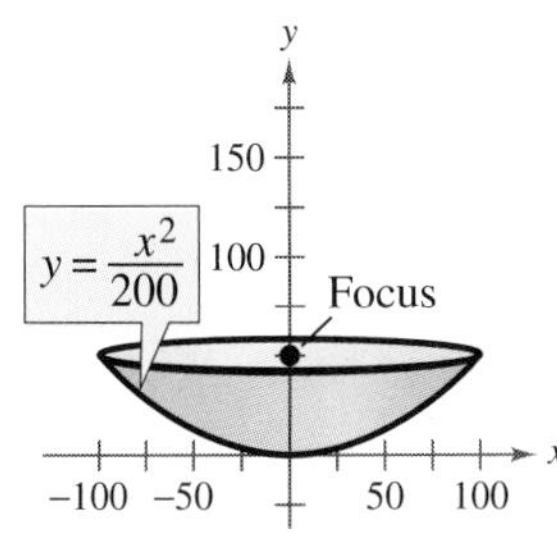

42. *Parabolic Archway* A parabolic archway is 12 meters high at the vertex. At a height of 10 meters the width of the archway is 8 meters (see figure). How wide is the archway at ground level?

Figure for 42

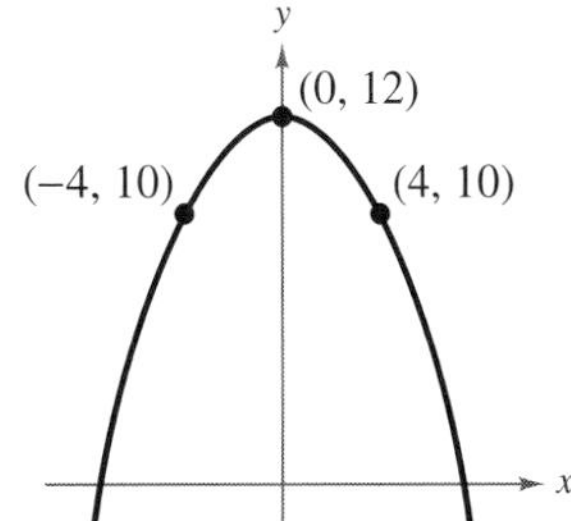

43. *Architecture* A church window is bounded on top by a parabola and below by the arc of a circle (see figure).

(a) Find equations for the parabola and circle.

(b) Use a graphing utility to complete the table giving the vertical distance d between the circle and the parabola for the given value of x.

x	0	1	2	3	4
d					

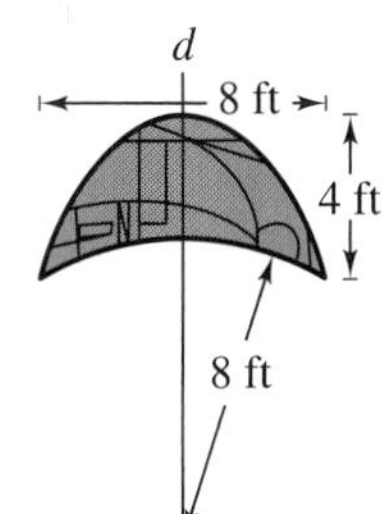

44. *CompuServe* CompuServe is a part of H & R Block, Inc. The pretax earnings (in millions of dollars) for the years 1990 through 1994 are given in the table. (Source: *1994 H & R Block, Inc. Annual Report*)

Year	1990	1991	1992	1993	1994
Earnings	40.3	48.6	55.4	74.0	102.3

(a) Use the regression capabilities of a graphing utility to find a quadratic model $y = at^2 + bt + c$, where y is the earnings in millions of dollars and t is the time in years, with $t = 0$ corresponding to 1990.

(b) Use a graphing utility to create a scatter plot of the data and graph the model.

(c) Use the model to estimate the earnings for 1995.

45. *Semielliptical Archway* A semielliptical archway is to be formed over the entrance to an estate. The arch is to be set on pillars that are 10 feet apart and is to have a height (atop the pillars) of 4 feet (see figure). Where should the foci be placed in order to sketch the semielliptical arch?

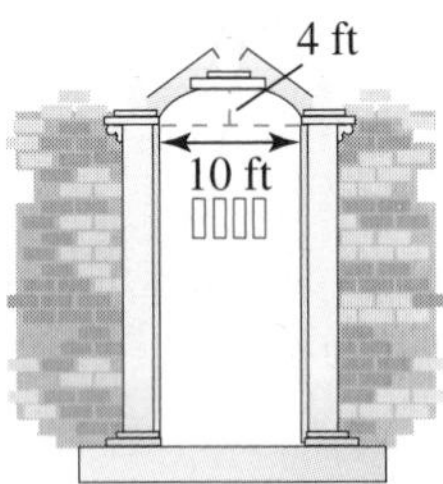

46. *Heating and Plumbing* Find the diameter d of the largest water pipe that can be placed behind a ventilation duct of diameter 60 centimeters as shown in the figure.

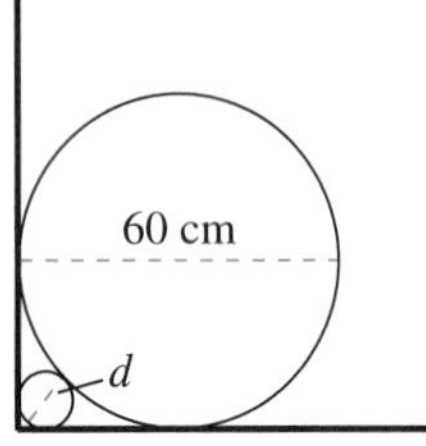

47. *Locating an Explosion* Two of your friends live 4 miles apart on the same "east-west" street, and you live halfway between them. You are talking on a three-way phone call when you hear an explosion. Six seconds after you hear the explosion your friend to the east hears it, and eight seconds after you hear the explosion your friend to the west hears it. Find equations of two hyperbolas that would locate the explosion. (Sound travels at a rate of 1100 feet per second.)

48. *True or False?* The graph of $(x^2/4) - y^4 = 1$ is an equation of a hyperbola. Explain.

The tangent line at the point (x_0, y_0) on the conic

$$\frac{x^2}{a^2} \pm \frac{y^2}{b^2} = 1 \quad \text{is given by} \quad \frac{x_0 x}{a^2} \pm \frac{y_0 y}{b^2} = 1.$$

In Exercises 49–52, (a) find an equation of the tangent line to the conic at the specified point, and (b) verify your result by using a graphing utility to graph the conic and the tangent line.

	Conic	*Point*
49.	$\frac{x^2}{100} + \frac{y^2}{25} = 1$	$(-8, 3)$
50.	$x^2 + 7y^2 = 16$	$(3, 1)$
51.	$\frac{x^2}{9} - y^2 = 1$	$(6, \sqrt{3})$
52.	$\frac{x^2}{4} - \frac{y^2}{2} = 1$	$(6, 4)$

In Exercises 53–64, sketch the curve represented by the parametric equations and write the corresponding rectangular equation by eliminating the parameter. Verify your result with a graphing utility.

53. $x = \sqrt[3]{t}$, $y = t$

54. $x = t$, $y = \sqrt[3]{t}$

55. $x = \frac{1}{t}$, $y = t$

56. $x = t$, $y = \frac{1}{t}$

57. $x = 2t$, $y = 4t$

58. $x = t^2$, $y = \sqrt{t}$

59. $x = 1 + 4t$, $y = 2 - 3t$

60. $x = t + 4$, $y = t^2$

61. $x = \frac{1}{t}$, $y = t^2$

62. $x = \frac{1}{t}$, $y = 2t + 3$

63. $x = 3$, $y = t$

64. $x = t$, $y = 2$

In Exercises 65–68, find a set of parametric equations for the line through the points.

65. $(3, 5), (8, 5)$

66. $(2, -1), (2, 4)$

67. $(-1, 6), (10, 0)$

68. $(0, 0), \left(\frac{5}{2}, 6\right)$

CHAPTER PROJECT *Parametric Equations and Inverse Functions*

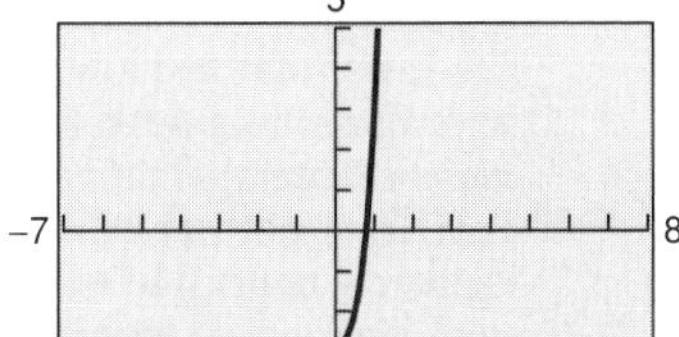

In this project, you will use the parametric mode of a graphing utility to sketch the inverse of a function.

(a) Consider the function given by $f(x) = x^5 + 4x^3 + x - 3$. You can graph this function in parametric mode using the following.

X1T = T

Y1T = T ∧ 5 + 4T ∧ 3 + T − 3

In the graph shown at the left, notice that the function passes the Horizontal Line Test and therefore has an inverse function. To graph the inverse function, you can interchange the roles of x and y in the given parametric equations, as follows.

X2T = T ∧ 5 + 4T ∧ 3 + T − 3

Y2T = T

Sketch the graph of f and f^{-1} in the same viewing rectangle.

(b) Compare the graphs of f and f^{-1}.

Questions for Further Exploration

1. Consider the function $f(x) = \sqrt{1 - x}$.

(a) Use the parametric mode of a graphing utility to sketch the graph of f^{-1}.

(b) Solve for $f^{-1}(x)$ algebraically and compare its graph with the graph found in part (a). Are the graphs the same?

2. Use the parametric mode of a graphing utility to sketch the graph of the inverse of $f(x) = e^x$. What is the name of this inverse function?

3. Use the parametric mode of a graphing utility to sketch the graph of

$$f(x) = \frac{1}{x^2 - 9}.$$

Why doesn't this function have an inverse? Explain how to restrict the domain of f to define a new function that does have an inverse. Verify your result graphically.

4. In your own words, explain why the graph of an invertible function and its inverse are mirror images of each other about the line $y = x$.

5. The function $F = \frac{9}{5}C + 32$ expresses the relationship between degrees Celsius C and degrees Fahrenheit F.

(a) Use a graphing utility to graph F and its inverse in the same viewing rectangle.

(b) Find the point at which the two graphs intersect.

(c) What is the significance of the point found in part (b)?

6. (a) Do all linear functions have inverses? If not, give an example of one that doesn't.

(b) Do any quadratic functions have inverses? If so, give an example of one that does.

(c) Do all cubic functions have inverses? If not, give an example of one that doesn't.

6–8 /// CUMULATIVE TEST

Take this test as you would take a test in class. After you are done, check your work against the answers given in the back of the book.

The *Interactive* CD-ROM provides answers to the Chapter Tests and Cumulative Tests. It also offers Chapter Pre-Tests (that test key skills and concepts covered in previous chapters) and Chapter Post-Tests, both of which have randomly generated exercises with diagnostic capabilities.

1. Find $2A - B$: $A = \begin{bmatrix} 6 & 2 & -3 \\ -1 & 4 & 5 \end{bmatrix}$, $\quad B = \begin{bmatrix} 1 & -1 & 1 \\ 4 & 5 & 10 \end{bmatrix}$
2. Find AB: $A = \begin{bmatrix} 4 & -3 \\ 2 & 1 \\ 5 & 0 \end{bmatrix}$, $\quad B = \begin{bmatrix} 3 & -2 \\ 1 & -3 \end{bmatrix}$
3. Find the inverse (if it exists): $\begin{bmatrix} 1 & 2 & -1 \\ 3 & 7 & -10 \\ -5 & -7 & -15 \end{bmatrix}$
4. Find the area of the triangle with vertices (0, 0), (6, 2), and (8, 10).
5. Find the sum of the first 20 terms of the arithmetic sequence 8, 12, 16, 20, ...
6. Find the sum: $\sum_{i=0}^{\infty} 3\left(\frac{1}{2}\right)^i$
7. Use mathematical induction to prove the formula $3 + 7 + 11 + 15 + \cdots + (4n - 1) = n(2n + 1)$.
8. Use the Binomial Theorem to expand and simplify $(z - 3)^4$.
9. The salary for the first year of a job is \$32,000. During the next 9 years there is a 5% raise each year. Determine the total compensation over the 10-year period.
10. A personnel manager has 10 applicants to fill three different positions. In how many ways can this be done, assuming that all the applicants are qualified for any of the three positions?
11. On a game show, the digits 3, 4, and 5 must be arranged in the proper order to form the price of an appliance. If they are arranged correctly, the contestant wins the appliance. What is the probability of winning if the contestant knows that the price is at least \$400?
12. Sketch the graph of $\frac{(x - 2)^2}{4} + \frac{(y + 1)^2}{9} = 1$.
13. Find an equation of the hyperbola with foci (0, 0) and (0, 4), and asymptotes $y = \pm\frac{1}{2}x + 2$.
14. Find a set of parametric equations of the line passing through the points (2, −3) and (6, 4). (The answer is not unique.)
15. Use a graphing utility to graph the curve represented by the parametric equations $x = 2t + 1$ and $y = t^2$.

Appendix A Think About the Proof

Section 3.3, Page 271

The Remainder Theorem

If a polynomial $f(x)$ is divided by $x - k$, the remainder is

$$r = f(k).$$

Proof /// From the Division Algorithm, you have

$$f(x) = (x - k)q(x) + r(x)$$

and because either $r(x) = 0$ or the degree of $r(x)$ is less than the degree of $x - k$, you know that $r(x)$ must be a constant. That is, $r(x) = r$. Now, by evaluating $f(x)$ at $x = k$, you have

$$f(k) = (k - k)q(k) + r = (0)q(k) + r = r.$$ ///

Section 3.3, Page 271

The Factor Theorem

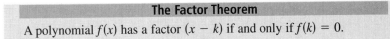

A polynomial $f(x)$ has a factor $(x - k)$ if and only if $f(k) = 0$.

Proof /// Using the Division Algorithm with the factor $(x - k)$, you have

$$f(x) = (x - k)q(x) + r(x).$$

By the Remainder Theorem, $r(x) = r = f(k)$, and you have

$$f(x) = (x - k)q(x) + f(k)$$

where $q(x)$ is a polynomial of lesser degree than $f(x)$. If $f(k) = 0$, then

$$f(x) = (x - k)q(x)$$

and you see that $(x - k)$ is a factor of $f(x)$. Conversely, if $(x - k)$ is a factor of $f(x)$, division of $f(x)$ by $(x - k)$ yields a remainder of 0. Hence, by the Remainder Theorem, you have $f(k) = 0$. ///

Section 3.4, Pages 281 and 282

Linear Factorization Theorem

If $f(x)$ is a polynomial of degree n

$$f(x) = a_n x^n + a_{n-1}x^{n-1} + \cdot \cdot \cdot + a_1x + a_0$$

where $n > 0$, f has precisely n linear factors

$$f(x) = a_n(x - c_1)(x - c_2) \cdot \cdot \cdot (x - c_n)$$

where $c_1, c_2, . . . , c_n$ are complex numbers and a_n is the leading coefficient of $f(x)$.

Proof /// Using the Fundamental Theorem, you know that f must have at least one zero, c_1. Consequently, $(x - c_1)$ is a factor of $f(x)$, and you have

$$f(x) = (x - c_1)f_1(x).$$

If the degree of $f_1(x)$ is greater than zero, you again apply the Fundamental Theorem to conclude that f_1 must have a zero c_2, which implies that

$$f(x) = (x - c_1)(x - c_2)f_2(x).$$

It is clear that the degree of $f_1(x)$ is $n - 1$, that the degree of $f_2(x)$ is $n - 2$, and that you can repeatedly apply the Fundamental Theorem n times until you obtain

$$f(x) = a_n(x - c_1)(x - c_2) \cdot \cdot \cdot (x - c_n)$$

where a_n is the leading coefficient of the polynomial $f(x)$. ///

Section 3.4, Page 285

Factors of a Polynomial

Every polynomial of degree $n > 0$ with real coefficients can be written as the product of linear and quadratic factors with real coefficients, where the quadratic factors have no real zeros.

Proof /// To begin, you use the Linear Factorization Theorem to conclude that $f(x)$ can be *completely* factored in the form

$$f(x) = d(x - c_1)(x - c_2)(x - c_3) \cdot \cdot \cdot (x - c_n).$$

If each c_i is real, there is nothing more to prove. If any c_i is complex ($c_i = a + bi$, $b \neq 0$), then, because the coefficients of $f(x)$ are real, you know that the conjugate $c_j = a - bi$ is also a zero. By multiplying the corresponding factors, you obtain

$$\begin{aligned}(x - c_i)(x - c_j) &= [x - (a + bi)][x - (a - bi)] \\ &= x^2 - 2ax + (a^2 + b^2)\end{aligned}$$

where each coefficient is real. ///

Section 4.3, Page 342

Properties of Logarithms

Let a be a positive number such that $a \neq 1$, and let n be a real number. If u and v are positive real numbers, the following properties are true.

1. $\log_a(uv) = \log_a u + \log_a v$ **1.** $\ln(uv) = \ln u + \ln v$

2. $\log_a \dfrac{u}{v} = \log_a u - \log_a v$ **2.** $\ln \dfrac{u}{v} = \ln u - \ln v$

3. $\log_a u^n = n \log_a u$ **3.** $\ln u^n = n \ln u$

Proof /// To prove Property 1, let

$$x = \log_a u \quad \text{and} \quad y = \log_a v.$$

The corresponding exponential forms of these two equations are

$$a^x = u \quad \text{and} \quad a^y = v.$$

Multiplying u and v produces $uv = a^x a^y = a^{x+y}$. The corresponding logarithmic form of $uv = a^{x+y}$ is $\log_a(uv) = x + y$. Hence, $\log_a(uv) = \log_a u + \log_a v$. ///

Section 7.1, Page 529

Properties of Sums

1. $\sum_{i=1}^{n} ca_i = c\sum_{i=1}^{n} a_i$, c is any constant

2. $\sum_{i=1}^{n} (a_i + b_i) = \sum_{i=1}^{n} a_i + \sum_{i=1}^{n} b_i$

3. $\sum_{i=1}^{n} (a_i - b_i) = \sum_{i=1}^{n} a_i - \sum_{i=1}^{n} b_i$

Proof /// Each of these properties follows directly from the Associative Property of Addition, the Commutative Property of Addition, and the Distributive Property of multiplication over addition. For example, note the use of the Distributive Property in the proof of Property 1.

$$\sum_{i=1}^{n} ca_i = ca_1 + ca_2 + ca_3 + \cdot \cdot \cdot + ca_n$$

$$= c(a_1 + a_2 + a_3 + \cdot \cdot \cdot + a_n) = c\sum_{i=1}^{n} a_i$$

///

Section 7.2, Page 538

The Sum of a Finite Arithmetic Sequence

The sum of a finite arithmetic sequence with n terms is given by

$$S_n = \frac{n}{2}(a_1 + a_n).$$

Proof /// Begin by generating the terms of the arithmetic sequence in two ways. In the first way, repeatedly add d to the first term to obtain

$$\begin{aligned} S_n &= a_1 + a_2 + a_3 + \cdots + a_{n-2} + a_{n-1} + a_n \\ &= a_1 + [a_1 + d] + [a_1 + 2d] + \cdots + [a_1 + (n-1)d]. \end{aligned}$$

In the second way, repeatedly subtract d from the nth term to obtain

$$\begin{aligned} S_n &= a_n + a_{n-1} + a_{n-2} + \cdots + a_3 + a_2 + a_1 \\ &= a_n + [a_n - d] + [a_n - 2d] + \cdots + [a_n - (n-1)d]. \end{aligned}$$

If you add these two versions of S_n, the multiples of d cancel and you obtain

$$\begin{aligned} 2S_n &= \overbrace{(a_1 + a_n) + (a_1 + a_n) + (a_1 + a_n) + \cdots + (a_1 + a_n)}^{n \text{ terms}} \\ &= n(a_1 + a_n). \end{aligned}$$

Thus, you have

$$S_n = \frac{n}{2}(a_1 + a_n).$$ ///

Section 7.3, Page 547

The Sum of a Finite Geometric Sequence

The sum of the geometric sequence

$$a_1,\ a_1r,\ a_1r^2,\ a_1r^3,\ a_1r^4,\ .\ .\ .\ ,\ a_1r^{n-1}$$

with common ratio $r \neq 1$ is given by

$$S_n = a_1\left(\frac{1 - r^n}{1 - r}\right).$$

Proof /// Begin by writing out the nth partial sum.

$$S_n = a_1 + a_1r + a_1r^2 + \cdot\cdot\cdot + a_1r^{n-2} + a_1r^{n-1}$$

Multiplication by r yields

$$rS_n = a_1r + a_1r^2 + a_1r^3 + \cdot\cdot\cdot + a_1r^{n-1} + a_1r^n.$$

Subtracting the second equation from the first yields

$$S_n - rS_n = a_1 - a_1r^n.$$

Therefore, $S_n(1 - r) = a_1(1 - r^n)$, and, because $r \neq 1$, you have

$$S_n = a_1\left(\frac{1 - r^n}{1 - r}\right).$$

///

Section 7.5, Page 565

The Binomial Theorem

In the expansion of $(x + y)^n$

$$(x + y)^n = x^n + nx^{n-1}y + \cdots + {}_nC_r\, x^{n-r}y^r + \cdots + nxy^{n-1} + y^n$$

the coefficient of $x^{n-r}y^r$ is given by

$${}_nC_r = \frac{n!}{(n-r)!r!}.$$

Proof /// The Binomial Theorem can be proved quite nicely using mathematical induction. The steps are straightforward but look a little messy, so we will present only an outline of the proof.

1. If $n = 1$, you have

$$(x + y)^1 = x^1 + y^1 = {}_1C_0x + {}_1C_1y$$

and the formula is valid.

2. Assuming that the formula is true for $n = k$, the coefficient of $x^{k-r}y^r$ is given by

$${}_kC_r = \frac{k!}{(k-r)!r!} = \frac{k(k-1)(k-2)\cdots(k-r+1)}{r!}.$$

To show that the formula is true for $n = k + 1$, look at the coefficient of $x^{k+1-r}y^r$ in the expansion of

$$(x + y)^{k+1} = (x + y)^k(x + y).$$

From the right-hand side, you can determine that the term involving $x^{k+1-r}y^r$ is the sum of two products.

$$\begin{aligned}
&({}_kC_rx^{k-r}y^r)(x) + ({}_kC_{r-1}x^{k+1-r}y^{r-1})(y)\\
&\quad = \left[\frac{k!}{(k-r)!r!} + \frac{k!}{(k-r+1)!(r-1)!}\right]x^{k+1-r}y^r\\
&\quad = \left[\frac{(k+1-r)k!}{(k+1-r)!r!} + \frac{k!r}{(k+1-r)!r!}\right]x^{k+1-r}y^r\\
&\quad = \left[\frac{k!(k+1-r+r)}{(k+1-r)!r!}\right]x^{k+1-r}y^r\\
&\quad = \left[\frac{(k+1)!}{(k+1-r)!r!}\right]x^{k+1-r}y^r\\
&\quad = {}_{k+1}C_rx^{k+1-r}y^r
\end{aligned}$$

Thus, by mathematical induction, the Binomial Theorem is valid for all positive integers n. ///

Section 8.1, Page 605

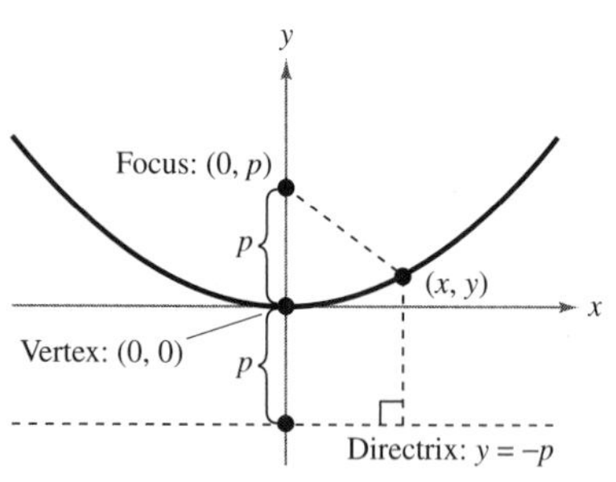

Standard Equation of a Parabola (Vertex at Origin)

The **standard form of the equation of a parabola** with vertex at $(0, 0)$ and directrix $y = -p$ is

$x^2 = 4py, \quad p \neq 0.$ Vertical axis

For directrix $x = -p$, the equation is

$y^2 = 4px, \quad p \neq 0.$ Horizontal axis

The focus is on the axis p units (directed distance) from the vertex.

Proof /// Because the two cases are similar, a proof will be given for the first case only. Suppose the directrix $(y = -p)$ is parallel to the x-axis. In the figure, you assume that $p > 0$, and because p is the directed distance from the vertex to the focus, the focus must lie above the vertex. Because the point (x, y) is equidistant from $(0, p)$ and $y = -p$, you can apply the Distance Formula to obtain

$$\sqrt{(x - 0)^2 + (y - p)^2} = y + p$$
$$x^2 + (y - p)^2 = (y + p)^2$$
$$x^2 + y^2 - 2py + p^2 = y^2 + 2py + p^2$$
$$x^2 = 4py.$$

///

Appendix B Programs

Evaluating an Algebraic Expression (Section P.3)

This program, shown in the Technology note on page 28, can be used to evaluate an algebraic expression in one variable at several values of the variable.

Note: On the *TI-83* and the *TI-82*, the "Lbl" and "Goto" commands may be entered through the "CTL" menu accessed by pressing the PRGM key. The "Disp" and "Input" commands may be entered through the "I/O" menu accessed by pressing the PRGM key. On the *TI-83*, the symbol "Y_1" may be entered through the "Y-VARS" and "Function" menus accessed by pressing the VARS key. On the *TI-82,* the symbol "Y_1" may be entered through the "Function" menu accessed by pressing the Y-VARS key. Keystroke sequences required for similar commands on other calculators will vary. Consult the user manual for your calculator.

TI-80

```
PROGRAM:EVALUAT
:LBL A
:INPUT "ENTER X",X
:DISP Y1
:GOTO A
```

To use this program, enter an expression in Y1. Expressions may also be evaluated directly on the *TI-80*'s home screen.

TI-81

```
Prgm1:EVALUATE
:Lbl 1
:Disp "ENTER X"
:Input X
:Disp Y1
:Goto 1
```

To use this program, enter an expression in Y_1. Expressions may also be evaluated directly on the *TI-81*'s home screen.

TI-83
TI-82

```
PROGRAM:EVALUATE
:Lbl A
:Input "ENTER X",X
:Disp Y1
:Goto A
```

To use this program, enter an expression in Y_1. Expressions may also be evaluated directly on the *TI-83*'s or *TI-82*'s home screen.

TI-85

```
PROGRAM:EVALUATE
:Lbl A
:Input "Enter x",x
:Disp y1
:Goto A
```

To use this program, enter an expression in y1.

TI-92

```
evaluate( )
Prgm
Lbl one
Input "enter x",x
Disp y1(x)
Goto one
EndPrgm
```

To use this program, enter an expression in y1. Expressions may also be evaluated on the *TI-92*'s home screen.

Casio fx-7700G

```
EVALUATE
Lbl 1
"X="?→X
"F(X)=":f1◢
Goto 1
```

To use this program, enter an expression in f1.

Casio fx-7700GE
Casio fx-9700GE
Casio CFX-9800G

```
EVALUATE↵
Lbl 1↵
"X="?→X↵
"F(X)=":f1◢
Goto 1
```

To use this program, enter an expression in f1.

Sharp EL-9200C
Sharp EL-9300C

```
evaluate
———————REAL
Goto top
Label equation
Y=f(X)
Return
Label top
Input X
Gosub equation
Print Y
Goto top
End
```

To use this program, replace f(X) with your expression in X.

HP-38G

Use the Solve aplet to evaluate an expression.

1. Press [LIB]. Highlight the Solve aplet. Press {{START}}.
2. Set your expression equal to y, enter the equation (y = *your expression*) in E1 and press {{OK}}. The equation should be checked.
3. Press [NUM].
4. Highlight the x-variable field. Enter a value for x and press {{OK}}.
5. Highlight the y-variable field and press {{SOLVE}}. The value of the expression will appear in the y-variable field.
6. Repeat steps 4 and 5 to evaluate the expression for other values of x.

Reflections and Shifts Program (Section 1.5)

This program, referenced in the Group Activity on page 134, will sketch a graph of the function $y = R(x + H)^2 + V$, where $R = \pm 1$, H is an integer between -6 and 6, and V is an integer between -3 and 3. This program gives you practice working with reflections, horizontal shifts, and vertical shifts.

Note: On the *TI-83* and the *TI-82*, the "int" and "rand" commands may be entered through the "NUM" and "PRB" menus, respectively, accessed by pressing the MATH key. The "=" and "<" symbols may be entered through the "TEST" menu accessed by pressing the TEST key. Other commands, such as "If," "Then," "Else," and "End," may be entered through the "CTL" menu accessed by pressing the PRGM key. The commands "Xmin," "Xmax," "Xscl," "Ymin," "Ymax," and "Yscl" may be entered through the "Window" menu accessed by pressing the VARS key. The commands "DispGraph" and "Pause" may be entered through the "I/O" and "CTL" menus, respectively, accessed by pressing the PRGM key. For additional keystroke instructions, see previous programs in this appendix. Keystroke sequences for similar commands on other calculators will vary. Consult the user manual for your calculator.

TI-80

```
PROGRAM:PARABOL
:-6+INT (12RAND)→H
:-3+INT (6RAND)→V
:RAND→R
:IF R <.5
:THEN
:-1→R
:ELSE
:1→R
:END
:"R(X+H)^2+V"→Y1
:-9→XMIN
:9→XMAX
:1→XSCL
:-6→YMIN
:6→YMAX
:1→YSCL
:DISPGRAPH
:PAUSE
:DISP "Y=R(X+H)^2+V^2"
:DISP "R=",R
:DISP "H=",H
:DISP "V=",V
:PAUSE
```

Press ENTER after viewing the graph to display the values of the integers.

TI-81

```
Prgm2:PARABOLA
:Rand→H
:-6+Int (12H)→H
:Rand→V
:-3+Int (6V)→V
:Rand→R
:If R <.5
:-1→R
:If R >.49
:1→R
:"R(X+H)^2+V"→Y1
:-9→Xmin
:9→Xmax
:1→Xscl
:-6→Ymin
:6→Ymax
:1→Yscl
:DispGraph
:Pause
:Disp "Y=R(X+H)^2+V"
:Disp "R="
:Disp R
:Disp "H="
:Disp H
:Disp "V="
:Disp V
:End
```

Press ENTER after viewing the graph to display the values of the integers.

TI-83
TI-82

```
PROGRAM:PARABOLA
:-6+int (12rand)→H
:-3+int (6rand)→V
:rand→R
:If R < .5
:Then
:-1→R
:Else
:1→R
:End
:"R(X+H)²+V"→Y1
:-9→Xmin
:9→Xmax
:1→Xscl
:-6→Ymin
:6→Ymax
:1→Yscl
:DispGraph
:Pause
:Disp "Y=R(X+H)²+V"
:Disp "R=",R
:Disp "H=",H
:Disp "V=",V
:Pause
```

Press ENTER after viewing the graph to display the values of the integers.

TI-85

```
PROGRAM:PARABOLA
:rand→H
:-6+int (12H)→H
:rand→V
:-3+int (6V)→V
:rand→R
:If R < .5
:-1→R
:If R > .49
:1→R
:y1=R(x+H)²+V
:-9→xMin
:9→xMax
:1→xScl
:-6→yMin
:6→yMax
:1→yScl
:DispG
:Pause
:Disp "Y=R(X+H)²+V"
:Disp "R=",R
:Disp "H=",H
:Disp "V=",V
:Pause
```

Press ENTER after viewing the graph to display the values of the integers.

TI-92

```
Parabola( )
Prgm
ClrHome
ClrIO
setMode("Split Screen",
   "Left-Right")
setMode("Split 1 App","Home")
setMode("Split 2 App","Graph")
-6+int (12rand( ))→h
-3+int (6rand( ))→v
rand( )→r
If r < .5 Then
   -1→r
      Else
   1→r
EndIf
r*(x+h)^2+v→y1(x)
-9→xmin
9→xmax
1→xscl
-6→ymin
6→ymax
1→yscl
DispG
Disp "y1(x)=r(x+h)^2+v"
Output 20,1, "r=":Output 20,11,r
Output 40,1, "h=":Output 40,11,h
Output 60,1, "v=":Output 60,11,v
Pause
setMode("Split Screen","Full")
EndPrgm
```

Casio fx-7700G

```
PARABOLA
-6+INT (12Ran#)→H
-3+INT (6Ran#)→V
-1→R:Ran#<0.5 ⇒1→R
Range -9,9,1,-6,6,1
Graph Y=R(X+H)²+V ◢
"Y=R(X+H)²+V"
"R=":R ◢
"H=":H ◢
"V=":V
```

Press [EXE] after viewing the graph to display the values of the integers.

Casio fx-7700GE
Casio fx-9700GE
Casio CFX-9800G

```
PARABOLA↵
-6+Int (12Ran#)→H↵
-3+Int (6Ran#)→V↵
Ran#→R↵
R< .5⇒-1→R↵
R≥ .5⇒1→R↵
Range -9,9,1,-6,6,1↵
Graph Y=R(X+H)²+V ◢
"Y=R(X+H)²+V"↵
"R=":R ◢
"H=":H ◢
"V=":V
```

Press [EXE] after viewing the graph to display the values of the integers.

Sharp EL-9200C
Sharp EL-9300C

```
parabola
————————REAL
h=int (random*12) -6
v=int (random*6) -3
s=(random*2) -1
r=s/abs s
Range -9,9,1,-6,6,1
Graph r(X+h)²+v
Wait
Print "y=r(X+h)²+v
Print r
Print h
Print v
End
```

Press [ENTER] after viewing the graph to display the values of the integers.

HP-38G

PARABOLA

```
PARABOLA PROGRAM
-6+INT(12RANDOM)▶H:
-3+INT(6RANDOM)▶V:
RANDOM ▶R:
IF R>.5
   THEN -1▶R:
   ELSE 1▶R:
END:
'R*(X+H)²+V'▶F1(X):
CHECK 1:

PARANS PROGRAM
ERASE:
DISP 2;"Y=R(X+H)²+V":
DISP 3;"R="R:
DISP 4;"H="H:
DISP 5;"V="V:
FREEZE:

PARABOLA.SV PROGRAM
SETVIEWS "RUN
PARABOLA";PARABOLA;1;
"ANSWER";PARANS;1;
"  ";PARABOLA.SV;0:
```

1. Press [LIB]. Highlight the Function aplet. Press {{SAVE}}. Enter the name PARABOLA for the new aplet and press {{OK}}.
2. Press ■ [SETUP-PLOT] and set XRNG: from −12 to 12, YRNG: from −6 to 6, and XTICK: and YTICK: to 1.
3. Enter the 3 programs PARABOLA, PARANS, PARABOLA.SV.
4. Run the program PARABOLA.SV.
5. Enter the PARABOLA aplet.
6. Press ■ [VIEWS]. Highlight RUN PARABOLA and press {{OK}}.
7. After viewing the graph press ■ [VIEWS]. Highlight ANSWER and press {{OK}} to see the values of the integers.
8. Press {{OK}} to return to the graph.
9. Repeat steps 6, 7, and 8 for a new parabola.

Graph Reflection Program (Section 1.7)

This program, shown in the Technology note on page 153, will graph a function f and its reflection in the line $y = x$.

Note: On the *TI-83* and the *TI-82*, the "While" command may be entered through the "CTL" menu accessed by pressing the PRGM key. The "Pt-On(" command may be entered through the "POINTS" menu accessed by pressing the DRAW key. For additional keystroke instructions, see previous programs in this appendix. Keystroke sequences required for similar commands on other calculators will vary. Consult the user manual for your calculator.

TI-80

```
PROGRAM:REFLECT
:47XMIN/63→YMIN
:47XMAX/63→YMAX
:XSCL→YSCL
:"X"→Y2
:DISPGRAPH
:(XMAX−XMIN)/62→I
:XMIN→X
:LBL A
:PT-ON(Y1,X)
:X+I→X
:If X>XMAX
:STOP
:GOTO A
```

To use this program, enter the function in Y1 and set a viewing rectangle.

TI-81

```
Prgm3:REFLECT
:2Xmin/3→Ymin
:2Xmax/3→Ymax
:Xscl→Yscl
:"X"→Y2
:DispGraph
:(Xmax−Xmin)/95→I
:Xmin→X
:Lbl 1
:Pt-On(Y1,X)
:X+I→X
:If X>Xmax
:End
:Goto 1
```

To use this program, enter the function in Y_1 and set a viewing rectangle.

TI-83
TI-82

```
PROGRAM:REFLECT
:63Xmin/95→Ymin
:63Xmax/95→Ymax
:Xscl→Yscl
:"X"→Y2
:DispGraph
:(Xmax−Xmin)/94→I
:Xmin→X
:While X≤Xmax
:Pt-On(Y1,X)
:X+I→X
:End
```

To use this program, enter the function in Y_1 and set a viewing rectangle.

TI-85

```
PROGRAM:REFLECT
:63*xMin/127→yMin
:63*xMax/127→yMax
:xScl→yScl
:y2=x
:DispG
:(xMax−xMin)/126→I
:xMin→x
:Lbl A
:PtOn(y1,x)
:x+I→x
:If x>xMax
:Stop
:Goto A
```

To use this program, enter the function in y1 and set a viewing rectangle.

TI-92

```
Prgm
103xmin/239→ymin
103xmax/239→ymax
xscl→yscl
x→y2(x)
DispG
(xmax−xmin)/238→n
xmin→x
While x<xmax
   PtOn y1(x),x
   x+n→x
EndWhile
EndPrgm
```

To use this program, enter a function in y1 and set an appropriate viewing window.

Casio fx-7700G

```
REFLECTION
"GRAPH -A TO A"
"A="?→A
Range -A,A,1,-2A÷3,2A÷3,1
Graph Y=f1
-A→B
Lbl 1
B→X
Plot f1,B
B+A÷32→B
B≤A⇒Goto1 :Graph Y=X
```

To use this program, enter the function in f_1.

Casio fx-7700GE

```
REFLECTION
"GRAPH -A TO A"↵
"A="?→A↵
Range -A,A,1,-2A÷3,2A÷3,1↵
Graph Y=f1↵
-A→B↵
Lbl 1↵
B→X↵
Plot f1,B↵
B+A÷32→B↵
B≤A⇒Goto1:Graph Y=X
```

To use this program, enter the function in f_1.

Casio fx-9700GE

```
REFLECTION↵
63Xmin÷127→A↵
63Xmax÷127→B↵
Xscl→C↵
Range , , , A, B, C↵
(Xmax−Xmin)÷126→I↵
Xmax→M↵
Xmin→D↵
Graph Y=f1↵
Lbl 1↵
D→X↵
Plot f1,D↵
D+I→D↵
D≤M⇒Goto 1:Graph Y=X
```

To use this program, enter a function in f_1 and set a viewing rectangle.

Casio CFX-9800G

```
REFLECTION↵
63Xmin÷95→A↵
63Xmax÷95→B↵
Xscl→C↵
Range , , , A, B, C↵
(Xmax−Xmin)÷94→I↵
Xmax→M↵
Xmin→D↵
Graph Y=f1↵
```

Casio CFX-9800G

(Continued)

```
Lbl 1↵
D→X↵
Plot f1,D↵
D+I→D↵
D≤M⇒Goto 1:Graph Y=X
```

To use this program, enter a function in f_1 and set a viewing rectangle.

Sharp EL-9200C
Sharp EL-9300C

```
reflection
————REAL
Goto top
Label equation
Y=f(X)
Return
Label rng
xmin=-10
xmax=10
xstp=(xmax−xmin)/10
ymin=2xmin/3
ymax=2xmax/3
ystp=xstp
Range xmin,xmax,xstp,ymin,
   ymax,ystp
Return
Label top
Gosub rng
Graph X
step=(xmax−xmin)/(94*2)
X=xmin
Label 1
Gosub equation
Plot X,Y
Plot Y, X
X=X+step
If X<=xmax Goto 1
End
```

To use this program, replace f(X) with your expression in X.

Quadratic Formula Program (Section 2.4)

This program, shown in the Technology note on page 201, will display the solutions to quadratic equations or the words "No Real Solution." To use the program, write the quadratic equation in standard form and enter the values of *a*, *b*, and *c*.

Note: On the *TI-82*, the "Prompt" command may be entered through the "I/O" menu accessed by pressing the PRGM key. For additional keystroke instructions, see previous programs in this appendix. Keystroke sequences for similar commands on other calculators will vary. Consult the user's manual for your calculator.

TI-80

```
PROGRAM:QUADRAT
:Disp "AX²+BX+C = 0"
:Input "ENTER A", A
:Input "ENTER B", B
:Input "ENTER C", C
:B²−4AC→D
:If D≥0
:Then
:(-B+√D)/(2A)→M
:Disp M
:(-B−√D)/(2A)→N
:Disp N
:Else
:Disp "NO REAL SOLUTION"
:End
```

TI-81

```
Prgm4: QUADRAT
:Disp "ENTER A"
:Input A
:Disp "ENTER B"
:Input B
:Disp "ENTER C"
:Input C
:B²−4AC→D
:If D<0
:Goto 1
:((-B+√D)/(2A))→M
:Disp M
:((-B−√D)/(2A))→N
:Disp N
:End
:Lbl 1
:Disp "NO REAL"
:Disp "SOLUTION"
:End
```

TI-82
TI-83

```
PROGRAM:QUADRAT
:Disp "AX²+BX+C=0"
:Prompt A
:Prompt B
:Prompt C
:B²−4AC→D
:If D≥0
:Then
:(-B+√D)/(2A)→M
:Disp M
:(-B−√D)/(2A)→N
:Disp N
:Else
:Disp "NO REAL SOLUTION"
:End
```

TI-85

```
PROGRAM:QUADRAT
:Disp "AX²+BX+C=0"
:Input "ENTER A", A
:Input "ENTER B", B
:Input "ENTER C", C
:B²−4*A*C→D
:(-B+√D)/(2A)→M
:Disp M
:(-B−√D)/(2A)→N
:Disp N
```

This program gives both real and complex answers. Solutions to quadratic equations are also available directly by using the *TI-85* POLY function.

TI-92

```
:quadrat( )
:Prgm
:setMode("Complex Format",
   "RECTANGULAR")
:Disp "AX^2+BX+C=0"
:Input "Enter A.",a
:Input "Enter B.",b
:Input "Enter C.",c
:b^2−4*a*c→d
:(-b+√(d))/(2*a)→m
:(-b−√(d))/(2*a)→n
:Disp m
:Disp n
:setMode("Complex Format",
   "REAL")
:EndPrgm
```

This program gives both real and complex answers.

Casio fx-7700G

```
QUADRATIC
"AX²+BX+C=0"
"A="?→A
"B="?→B
"C="?→C
B²−4AC→D
D<0 ⇒ Goto 1
"X=":(-B+√D)÷(2A) ◢
"OR X=":(-B−√D)÷(2A)
Goto 2
Lbl 1
"NO REAL SOLUTION"
Lbl 2
```

Casio fx-7700GE

```
QUADRATIC ↲
"AX²+BX+C=0" ↲
"A="?→A ↲
"B="?→B ↲
"C="?→C ↲
B²−4AC→D ↲
D<0 ⇒ Goto 1 ↲
(-B+√D)÷(2A) ◢
(-B−√D)÷(2A) ↲
Goto 2
Lbl 1
"NO REAL SOLUTION"
Lbl 2
```

Solutions to quadratic equations are also available directly from the Casio calculator's EQUATION MENU.

Casio fx-9700GE
Casio CFX-9800G

```
QUADRATIC ↲
"AX²+BX+C=0" ↲
"A="?→A ↲
"B="?→B ↲
"C="?→C ↲
B²−4AC→D ↲
(-B+√D)÷(2A) ◢
(-B−√D)÷(2A) ↲
```

Both real and complex answers are given. Solutions to quadratic equations are also available directly from the Casio calculator's EQUATION MENU.

Sharp EL-9200C
Sharp EL-9300C

```
Quadrat
————COMPLEX
Print "AX²+BX+C=0"
Input A
Input B
Input C
D=B²−4AC
M =(-B+√D)/(2A)
N =(-B−√D)/(2A)
Print M
Print N
End
```

This program is written in the program's complex mode, so both real and complex answers are given.

HP-38G

```
QUADRAT PROGRAM
INPUT A;"AX²+BX+C=0";
   "ENTER A";"";1:
INPUT B;"AX²+BX+C=0";
   "ENTER B";"";1:
INPUT C;"AX²+BX+C=0";
   "ENTER C";"";1:
B²−4AC▶D:
(-B+√D)/(2A)▶Z1:
(-B+√D)/(2A)▶Z2:
DISP 3;Z1:
DISP 5;Z2:
FREEZE
```

This program displays the answer in complex form (x, y) where x is the real part and y is the imaginary part.

Systems of Linear Equations (Section 5.2)

This program, shown in the Technology note on page 404, will display the solution of a system of two linear equations in two variables of the form

$$ax + by = c$$
$$dx + ey = f$$

if a unique solution exists.

Note: For help with *TI-83* or *TI-82* keystrokes, see previous programs in this appendix.

TI-80

```
PROGRAM:SOLVE
:DISP "AX+BY=C"
:INPUT "ENTER A",A
:INPUT "ENTER B",B
:INPUT "ENTER C",C
:DISP "DX+EY=F"
:INPUT "ENTER D",D
:INPUT "ENTER E",E
:INPUT "ENTER F",F
:IF AE−DB=0
:THEN
:DISP "NO UNIQUE"
:DISP "SOLUTION"
:ELSE
:(CE−BF)/(AE−DB)→X
:(AF−CD)/(AE−DB)→Y
:DISP X
:DISP Y
:END
```

TI-81

```
Prgm4:SOLVE
:Disp "AX+BY=C"
:Input A
:Input B
:Input C
:Disp "DX+EY=F"
:Input D
:Input E
:Input F
:If AE−DB=0
:Goto 1
:(CE−BF)/(AE−DB)→X
:(AF−CD)/(AE−DB)→Y
:Disp X
:Disp Y
:End
:Lbl 1
:Disp "NO UNIQUE SOLUTION"
:End
```

TI-83
TI-82

```
PROGRAM:SOLVE
:Disp "AX+BY=C"
:Prompt A
:Prompt B
:Prompt C
:Disp "DX+EY=F"
:Prompt D
:Prompt E
:Prompt F
:If AE−DB=0
:Then
:Disp "NO UNIQUE"
:Disp "SOLUTION"
:Else
:(CE−BF)/(AE−DB)→X
:(AF−CD)/(AE−DB)→Y
:Disp X
:Disp Y
:End
```

TI-85

```
PROGRAM:SOLVE
:Disp "AX+BY=C"
:Input "ENTER A",A
:Input "ENTER B",B
:Input "ENTER C",C
:Disp "DX+EY=F"
:Input "ENTER D",D
:Input "ENTER E",E
:Input "ENTER F",F
:If A*E−D*B==0
:Goto A
:(C*E−B*F)/(A*E−D*B)→X
:(A*F−C*D)/(A*E−D*B)→Y
:Disp X
:Disp Y
:Stop
:Lbl A
:Disp "NO UNIQUE SOLUTION"
```

TI-92

```
Solve( )
Prgm
ClrI0
Disp "Ax+By=C"
Input "Enter A.",a
Input "Enter B.",b
Input "Enter C.",c
ClrI0
Disp "Dx+Ey=F"
Input "Enter D.",d
Input "Enter E.",e
Input "Enter F.",f
If a*e−d*b=0 Then
    Disp "No unique solution"
  Else
    (c*e−b*f)/(a*e−d*b)→x
    (a*f−c*d)/(a*e−d*b)→y
    Disp x
    Disp y
EndIf
EndPrgm
```

Casio fx-7700G

```
SOLVE
"AX+BY=C"
"A="?→A
"B="?→B
"C="?→C
"DX+EY=F"
"D="?→D
"E="?→E
"F="?→F
AE−DB=0⇒Goto 1
"X=":(CE−BF)÷(AE−DB)◢
"Y=":(AF−CD)÷(AE−DB)
Goto 2
Lbl 1
"NO UNIQUE SOLUTION"
Lbl 2
```

Casio fx-7700GE
Casio fx-9700GE
Casio CFX-9800G

```
SOLVE↵
"AX+BY=C"↵
"A":?→A↵
"B":?→B↵
"C":?→C↵
"DX+EY=F"↵
"D":?→D↵
"E":?→E↵
"F":?→F↵
AE−DB=0⇒Goto 1↵
"X=":(CE−BF)÷(AE−DB) ◢
"Y=":(AF−CD)÷(AE−DB)↵
Goto 2↵
Lbl 1↵
"NO UNIQUE SOLUTION"↵
Lbl 2
```

Solutions to systems of linear equations are also available directly from the Casio calculator's EQUATION MENU.

Sharp EL-9200C
Sharp EL-9300C

```
solve
————REAL
Print "AX+BY=C"
Input A
Input B
Input C
Print "DX+EY=F"
Input D
Input E
Input F
If A*E−D*B=0 Goto 1
X=(C*E−B*F)/(A*E−D*B)
Y=(A*F−C*D)/(A*E−D*B)
Print X
Print Y
End
Label 1
Print "no unique solution"
End
```

Equations must be entered in the form: AX + BY = C; DX + EY = F. Uppercase letters are used so that the values can be accessed in the calculation mode of the calculator.

HP-38G

SOLVE

```
SOLVE PROGRAM
INPUT A;"AX+BY=C";
    "ENTER A";" ";1:
INPUT B;"AX+BY=C";
    "ENTER B";" ";1:
INPUT C;"AX+BY=C";
    "ENTER C";" ";1:
INPUT D;"DX+EY=F";
    "ENTER D";" ";1:
INPUT E;"DX+EY=F";
    "ENTER E";" ";1:
INPUT F;"DX+EY=F";
    "ENTER F";" ";1:
ERASE:
IF AE−DB==0
THEN DISP 3; "NO UNIQUE
    SOLUTION":
ELSE RUN "SOLVE.SOLN":
END:
FREEZE:
```

```
SOLVE.SOLN PROGRAM
(CE−BF)/(AE−DB)▶X:
(AF−CD)/(AE−DB)▶Y:
DISP 3;"X="X:
DISP 5;"Y="Y:
```

1. Input the 2 programs SOLVE and SOLVE.SOLN.
2. Run the SOLVE program.

Visualizing Row Operations (Section 6.1)

This program, referenced in the Technology note on page 465, demonstrates how elementary matrix row operations used in Gauss-Jordan elimination may be interpreted graphically. It asks the user to enter a 2×3 matrix that corresponds to a system of two linear equations. (The matrix entries should not be equivalent to either vertical or horizontal lines. This demonstration is also most effective if the y-intercepts of the lines are between -10 and 10.)

While the demonstration is running, you should notice that each elementary row operation creates an equivalent system. This equivalence is reinforced graphically because while the equations of the lines change with each elementary row operation, the point of intersection remains the same. You may want to run this program a second time to notice the relationship between the row operations and the graphs of the lines of the system.

TI-81

```
Prgm6:ROWOPS
:Disp "ENTER A"
:Disp "2 BY 3 MATRIX"
:Disp "A B C"
:Disp "D E F"
:Input A
:Input B
:Input C
:Input D
:Input E
:Input F
:A→[A](1,1)
:B→[A](1,2)
:C→[A](1,3)
:D→[A](2,1)
:E→[A](2,2)
:F→[A](2,3)
:ClrHome
:Disp "ORIGINAL MATRIX"
:Disp [A]
:Pause
:"B⁻¹(C−AX)"→Y2
:"E⁻¹(F−DX)"→Y1
:-10→Xmin
:10→Xmax
:1→Xscl
:-10→Ymin
:10→Ymax
:1→Yscl
:DispGraph
:Pause
:ClrHome
:Disp "OBTAIN LEADING"
:Disp "1 IN ROW 1"
:*row(A⁻¹,[A],1)→[A]
:Disp [A]
:Pause
:ClrDraw
:"(A/B)(C/A−X)"→Y2
:DispGraph
:Pause
:ClrHome
:Disp "OBTAIN 0 BELOW"
:Disp "LEADING 1 IN"
:Disp "COLUMN 1"
:*row+(-D,[A],1,2)→[A]
:Disp[A]
:Pause
:ClrDraw
:"(E−(BD/A))⁻¹(F−(DC/A))"→Y1
:DispGraph
:Pause
:ClrHome
:[A](2,2)→G
:If G=0
:Goto 1
:*row(G⁻¹,[A],2)→[A]
:Disp "OBTAIN LEADING"
```

(Continued on next page)

```
:Disp "1 IN ROW 2"
:Disp [A]
:Pause
:ClrDraw
:DispGraph
:Pause
:ClrHome
:Disp "OBTAIN 0 ABOVE"
:Disp "LEADING 1 IN"
:Disp "COLUMN 2"
:[A](1,2)→H
:*row+(-H,[A],2,1)→[A]
:Disp [A]
:Pause
:ClrDraw
:Y2−Off
:Line([A](1,3),-10,[A](1,3),10)
:DispGraph
:Pause
:ClrHome
:Disp "THE POINT OF"
:Disp "INTERSECTION IS"
:Disp "X="
:Disp [A](1,3)
:Disp "Y="
:Disp [A](2,3)
:End
:Lbl 1
:If [A](2,3)=0
:Disp "INFINITELY MANY"
:Disp "SOLUTIONS"
:If [A](2,3) ≠ 0
:Disp "INCONSISTENT"
:Disp "SYSTEM"
:End
```

To use this program, dimension matrix [A] as a 2×3 matrix. Press ENTER after each screen display to continue the program.

TI-83
TI-82

```
PROGRAM: ROWOPS
:Disp "ENTER A"
:Disp "2 BY 3 MATRIX:"
:Disp "A B C"
:Disp "D E F"
:Prompt A,B,C
:Prompt D,E,F
:A→[A](1,1):B→[A](1,2)
:C→[A](1,3):D→[A](2,1)
:E→[A](2,2):F→[A](2,3)
:ClrHome
:Disp "ORIGINAL MATRIX:"
:Pause [A]
:"B⁻¹(C−AX)"→Y2
:"E⁻¹(F−DX)"→Y1
:ZStandard:Pause:ClrHome
:Disp "OBTAIN LEADING"
:Disp "1 IN ROW 1"
:*row(A⁻¹,[A],1)→[A]
:Pause [A]:ClrDraw
:"(A/B)(C/A−X)"→Y2
:DispGraph:Pause:ClrHome
:Disp "OBTAIN 0 BELOW"
:Disp "LEADING 1 IN"
:Disp "COLUMN 1"
:*row+(-D,[A],1,2)→[A]
:Pause [A]:ClrDraw
:"(E−(BD/A))⁻¹(F−(DC/A))"→Y1
:DispGraph:Pause:ClrHome
:[A](2,2)→G
:If G=0
:Goto 1
:*row(G⁻¹,[A],2)→[A]
:Disp "OBTAIN LEADING"
:Disp "1 IN ROW 2"
:Pause [A]:ClrDraw
:DispGraph:Pause:ClrHome
:Disp "OBTAIN 0 ABOVE"
:Disp "LEADING 1 IN"
:Disp "COLUMN 2"
:[A](1,2)→H
:*row+(-H,[A],2,1)→[A]
:Pause [A]:ClrDraw:FnOff 2
:Vertical -(B/A)(E−(BD/A))⁻¹
  (F−DC/A)+C/A
:DispGraph:Pause:ClrHome
:Disp "THE POINT OF"
:Disp "INTERSECTION IS"
:Disp "X=",[A](1,3),"Y=",[A](2,3)
:Stop
:Lbl 1
If [A](2,3)=0
:Then
:Disp "INFINITELY MANY"
:Disp "SOLUTIONS"
:Else
:Disp "INCONSISTENT"
:Disp "SYSTEM"
:End
```

To use this program, dimension matrix [A] as a 2×3 matrix. Press ENTER after each screen display to continue the program.

TI-85

```
PROGRAM:ROWOPS
:Disp "enter a"
:Disp "2 by 3 matrix:"
:Disp "A B C"
:Disp "D E F"
:Prompt A,B,C
:Prompt D,E,F
:A→TEMP(1,1):B→TEMP(1,2)
:C→TEMP(1,3):D→TEMP(2,1)
:E→TEMP(2,2):F→TEMP(2,3)
:ClLCD
:Disp "original matrix:"
:Disp TEMP
:Pause
:y2=B⁻¹(C−A*x)
:y1=E⁻¹(F−D*x)
:ZStd:Pause:ClLCD
:Disp "obtain leading"
:Disp "1 in row 1"
:multR(A⁻¹,TEMP,1)→TEMP
:DispTEMP:Pause
:"(A/B)(C/A−X)"→y
:ClDrw:DispG:Pause:ClLCD
:Disp "obtain 0 below"
:Disp "leading 1 in"
:Disp "column 1"
:mRAdd(-D,TEMP,1,2)→TEMP
:Disp TEMP:Pause
:If TEMP(2,2)==0
:Goto A
:y1=(E−(B*D/A))⁻¹(F−(D*C/A))
:ClDrw:DispG:Pause:ClLCD
:TEMP(2,2)→G
:multR(G⁻¹,TEMP,2)→TEMP
:Disp "obtain leading"
:Disp "1 in row 2"
:Disp TEMP:Pause
:ClDrw:DispG:Pause:ClLCD
:Disp "obtain 0 above"
:Disp "leading 1 in"
:Disp "column 2"
:TEMP(1,2)→H
:mRAdd(-H,TEMP,2,1)→TEMP
:Disp TEMP
:Pause:FnOff 2:ClDrw
:Vert -(B/A)(E−(B*D/A))⁻¹
   (F−D*C/A)+C/A
:DispG:Pause:ClLCD
:Disp "the point of"
:Disp "intersection is"
:Disp "X=",TEMP(1,3),
   "Y=",TEMP(2,3)
:Stop
:Lbl A
:If TEMP(2,3)==0
:Then
:Disp "infinitely many"
:Disp "solutions"
:Else
:Disp "inconsistent"
:Disp "system"
:End
```

To use this program, dimension matrix TEMP as a 2×3 matrix. Press ENTER after each screen display to continue the program.

TI-92

```
rowops( )
Prgm
ClrIO
ClrHome
setMode("Split Screen","Left-
Right")
setMode("Split 1 App","Home")
setMode("Split 2 App","Graph")
Disp "ENTER A"
Disp "2 BY 3 MATRIX:"
Disp "A B C"
Disp "D E F"
Prompt a,b,c
Prompt d,e,f
[[a,b,c][d,e,f]]→mat1
ClrIO
b^(-1)*(c-a*x)→y2(x)
e^(-1)*(f-d*x)→y1(x)
ZoomStd
Disp "ORIGINAL MATRIX:"
Pause mat1
ClrIO
a/b*(c/a-x)→y2(x)
Disp "OBTAIN LEADING"
Disp "1 IN ROW 1"
mRow(a^(-1),mat1,1)→mat1
Pause mat1
ClrIO
(e-b*d/a)^(-1)*(f-d*c/a)→y1(x)
DispG
Disp "OBTAIN 0 BELOW"
Disp "LEADING 1 IN"
Disp "COLUMN 1"
mRowAdd(-d,mat1,1,2)→mat1
Pause mat1
ClrIO
mat1[2,2]→g
If g=0
Goto a1
Disp "OBTAIN LEADING"
Disp "1 IN ROW 2"
mRow(g^(-1),mat1,2)→mat1
Pause mat1
ClrIO
mat1[1,2]→h
FnOff 2
LineVert -b/a*(e-b*d/a)^(-1)*
   (f-d*c/a)+c/a
Disp "OBTAIN 0 ABOVE"
Disp "LEADING 1 IN"
Disp "COLUMN 2"
mRowAdd(-h,mat1,2,1)→mat1
Pause mat1
ClrIO
Disp "THE POINT OF"
Disp "INTERSECTION IS"
Disp "X=",mat1[1,3],"Y=",
   mat1[2,3]
Goto A2
Lbl a1
If mat1[2,3]=0 Then
Disp "INFINITELY MANY"
Disp "SOLUTIONS"
Else
Disp "INCONSISTENT"
Disp "SYSTEM"
EndIf
Lbl A2
Pause
setMode("Split Screen","Full")
EndPrgm
```

Press ENTER after each screen display to continue the program.

Casio fx-7700GE
Casio fx-9700GE
Casio CFX-7800G

```
ROWOPS ↵
"ENTER A"↵
"2 BY 3 MATRIX:"↵
"A B C"↵
"D E F"↵
"A="?→A:"B="?→B:
"C="?→C:"D="?→D:
"E="?→E:"F="?→F:↵
[[A,B,C][D,E,F]]→Mat A↵
Cls↵
"ORIGINAL MATRIX:" ◢
Mat A ◢
Range -10,10,1,-10,10,1↵
Graph Y=B⁻¹(C−AX)↵
Graph Y=E⁻¹(F−DX) ◢
Cls↵
"OBTAIN LEADING"↵
"1 IN ROW 1"◢
*Row A⁻¹,A,1↵
Mat A◢
Graph Y=(A÷B)(C÷A−X)↵
Graph Y=E⁻¹(F−DX) ◢
Cls↵
"OBTAIN 0 BELOW"↵
"LEADING 1 IN"↵
"COLUMN 1"◢
*Row+ -D,A,1,2↵
Mat A ◢
Graph Y=(A÷B)(C÷A−X)↵
GraphY=(E−(BD÷A))⁻¹
    (F−(DC÷A)) ◢
Cls↵
Mat A[2,2]→G↵
G=0 ⇒ Goto 1↵
*Row G⁻¹,A,2↵
"OBTAIN LEADING"↵
"1 IN ROW 2" ◢
Mat A◢
Graph Y=(A÷B)(C÷A−X)↵
Graph Y=(E−(BD÷A))⁻¹
    (F−(DC÷A)) ◢
Cls↵
"OBTAIN 0 ABOVE"↵
"LEADING 1 IN"↵
"COLUMN 2" ◢
Mat A[1,2]→H↵
*Row+ -H,A,2,1↵
Mat A ◢
Mat A[1,3]→J↵
Mat A[2,3]→K↵
Graph Y=K↵
Plot J,-10:Plot J,10:Line
"THE POINT OF"↵
"INTERSECTION IS"↵
"X=":J ◢
"Y=":K◢
Goto 3↵
Lbl 1↵
Mat A[2,3]=0 ⇒ Goto 2↵
"INCONSISTENT"↵
"SYSTEM"↵
Goto 3↵
Lbl 2↵
"INFINITELY MANY"↵
"SOLUTIONS"↵
Lbl 3
```

To use this program, dimension Mat A as a 2×3 matrix. Press EXE after each screen display to continue the program.

Sum Program (Section 7.1, page 531)

TI-80

```
PROGRAM:SUM
:INPUT "ENTER M", M
:INPUT "ENTER N", N
:0→S
:FOR(X,M,N)
:S+Y1→S
:DISP S
:END
```

To use this program, first store the nth term of the sequence in Y1 (in terms of X).

TI-81

```
Prgm5:SUM
:Disp "ENTER M"
:Input M
:Disp "ENTER N"
:Input N
:0→S
:Lbl 1
:M→X
:S+Y1→S
:Disp S
:IS>(M,N)
:Goto 1
:End
```

To use this program, first store the nth term of the sequence in Y1 (in terms of X).

TI-83
TI-82

```
PROGRAM:SUM
:Prompt M
:Prompt N
:0→S
:For(X,M,N)
:S+Y1→S
:Disp S
:End
```

To use this program, first store the nth term of the sequence in Y1 (in terms of X).

TI-85

```
PROGRAM:SUMS
:Prompt M
:Prompt N
:0→S
:For(x,M,N)
:S+y1→S
:Disp S
:End
```

To use this program, first store the nth term of the sequence in y1 (in terms of x).

TI-92

```
Summatn( )
Prgm
Input "Enter lower limit.",m
Input "Enter upper limit.",n
Σ(y1(x),x,m,n)→s
Disp "The partial sum is",s
EndPrgm
```

Sharp EL-9200C
Sharp EL-9300C

```
sum
————REAL
Goto top
Label sumit
s=s+(nth term)
Print s
m=m+1
Goto next
Label top
Input m
Input n
s=0
Label next
If m<=n Goto sumit
End
```

To use this program, first replace (nth term) with the nth term of the sequence in terms of m. For example, s=s+2^m, where 2^n is the nth term of the sequence.

Casio fx-7700G

```
SUM
"M="?→M
"N="?→N
0→S
Lbl 1
M→X
S+f1→S
"S=":S ◢
M+1→M
M≤N⇒1 Goto 1
```

To use this program, enter the nth term of the sequence into f_1 (in terms of X).

Casio fx-7700GE
Casio fx-9700GE
Casio CFX-9800G

```
SUM↵
"M="?→M↵
"N="?→N↵
0→S↵
Lbl 1↵
M→X↵
S+f1→S↵
"S=":S ◢
M+1→M↵
M≤N⇒1 Goto 1
```

To use this program, enter the nth term of the sequence into f_1 (in terms of X).

HP-38

```
SUM
SUM PROGRAM
INPUT M;"LOWER BOUND";
   "ENTER M";"";1:
INPUT N;"UPPER BOUND";
   "ENTER N";"";1:
0▶S:
ERASE:
SELECT "Function":
FOR I=M TO N
STEP 1;
RUN "SUM.STEP":
END:

SUM.STEP PROGRAM
S+F1(I)▶S:
DISP 4;"      "S:
FREEZE:
```

1. Input the 2 programs SUM and SUM.STEP.
2. Store the nth term of the sequence in the F1 function (in terms of x) in the Function aplet. Be sure that F1 is checked.
3. Run the SUM program.

Darts Program (Section 7.7, page 591)

TI-80

```
PROGRAM:DARTS
:CLRHOME
:0→K
:0→W
:INPUT "HOW MANY
   THROWS?",N
:LBL 1
:K+1→K
:RAND→X:RAND→Y
:IF X<.25:GOTO 2
:IF X>.75:GOTO 2
:IF X<.3:GOTO 2
:IF X>.65:GOTO 2
:W+1→W
:LBL 2
:IF K=N
:THEN
:DISP "WINS=",W
:DISP "LOSSES=",(N−W)
:STOP
:END
:GOTO 1
```

TI-81

```
Prgm6:DARTS
:ClrHome
:0→K
:0→W
:Disp "HOW MANY
   THROWS?"
:Input N
:Lbl 1
:K+1→K
:Rand→X
:Rand→Y
:If X<.25
:Goto 2
:If X>.75
:Goto 2
:If Y<.3
:Goto 2
:If X>.65
:Goto 2
:W+1→W
:Lbl 2
:If K≠N
:Goto 1
:DISP "WINS="
:Disp W
:DISP "LOSSES="
:N−W→L
:Disp L
:End
```

TI-83
TI-82

```
PROGRAM;DARTS
:ClrHome
:0→K
:0→W
:Input "HOW MANY
   THROWS?",N
:Lbl 1
:K+1→K
:rand→X:rand→Y
:IF X<.25 or X>.75
:Goto 2
:IF Y<.3 or Y>.65
:Goto 2
:W+1→W
:Lbl 2
:If K=N
:Then
:Disp "WINS=",W
:Disp "LOSSES=",(N−W)
:Stop
:End
:Goto 1
```

TI-85

```
PROGRAM:DARTS
:CILCD
:0→K
:0→W
:Input "HOW MANY
   THROWS?",N
:Lbl A
:K+1→K
:rand→X:rand→Y
:If X<.25
:Goto B
:If X>.75
:Goto B
:If Y<.3
:Goto B
:If Y>.65
:Goto B
:W+1→W
:Lbl B
:If K≠N
:Goto A
:DISP "WINS=",W
:DISP "LOSSES=",(N−W)
:Stop
```

TI-92

```
darts( )
Prgm
ClrIO
0→k
0→w
Input "How many throws?",n
Lbl a
k+1→k
rand( )→x
rand( )→y
If x<.25 or x>.75
Goto b
If y<.3 or y>.65
Goto b
w+1→w
Lbl b
If k=n Then
   Disp "Wins=",w
   Disp "Losses=",(n−w)
   Stop
EndIf
Goto a
EndPrgm
```

Casio fx-7700G

```
DARTS
Cls
0→K
0→W
"HOW MANY THROWS?"
?→N
Lbl 1
K+1→K
Ran#→X
Ran#→Y
X<.25⇒Goto 2
X>.75⇒Goto 2
Y<.3⇒Goto 2
Y>.65⇒Goto 2
W+1→W
Lbl 2
K≠N⇒Goto 1
"WINS="
W ◢
"LOSSES="
N−W
```

Casio fx-7700GE
Casio fx-9700GE
Casio CFX-9800G

```
DARTS↵
Cls↵
0→K↵
0→W↵
"HOW MANY THROWS?"↵
?→N↵
Lbl 1↵
K+1→K↵
Ran#→X↵
Ran#→Y↵
X<.25⇒Goto 2↵
X>.75⇒Goto 2↵
Y<.3⇒Goto 2↵
Y>.65⇒Goto 2↵
W+1→W↵
Lbl 2↵
K≠N⇒Goto 1↵
"WINS="↵
W ◢
"LOSSES="↵
N−W↵
```

Sharp EL-9200C
Sharp EL-9300C

```
darts
————————REAL
ClrT
k=0
w=0
Print "how many throws?"
Input n
Label a
k=k+1
x=random
y=random
If x<.25 Goto b
If x>.75 Goto b
If y<.3 Goto b
If y>.65 Goto b
w=w+1
If k≠n Goto a
Print "wins="
Print w
l=n−w
Print "losses="
Print 1
```

HP-38

DARTS

```
DARTS PROGRAM
0▶L:
INPUT N;"";"THROWS?";
"HOW MANY THROWS?";1:
FOR I=1 TO N
   STEP 1;
   RUN "DART.THROW":
END:
ERASE:
DISP 3;"WINS="N−L:
DISP 5;"LOSSES="L:
FREEZE:
```

```
DARTS.THROW PROGRAM
RANDOM▶X:
RANDOM▶Y:
CASE
IF X<.25 OR X>.75
THEN L+1▶L:
IF Y<.3 OR Y>.65
THEN L+1▶L:
END:
```

1. Input the 2 programs DARTS and DARTS.THROW.
2. Run the DARTS program.

CHAPTER P

Section P.1 *(page 10)*

1. (a) 5, 1 (b) $-9, 5, 0, 1$ (c) $-9, -\frac{7}{2}, 5, \frac{2}{3}, 0, 1$
(d) $\sqrt{2}$

3. (a) None (b) -13 (c) $2.01, 0.666\ldots, -13$
(d) $0.010110111\ldots$

5. (a) $\frac{6}{3}$ (b) $\frac{6}{3}$ (c) $-\frac{1}{3}, \frac{6}{3}, -7.5$ (d) $-\pi, \frac{1}{2}\sqrt{2}$

7. 0.625 **9.** $0.\overline{123}$ **11.** $-1 < 2.5$

13. $\frac{3}{2} < 7$

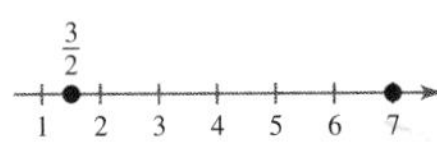

15. $-4 > -8$

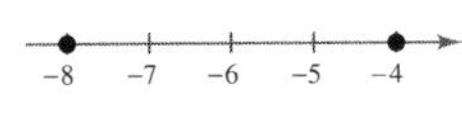

17. $\frac{5}{6} > \frac{2}{3}$

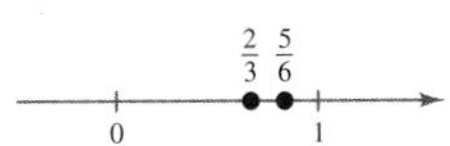

19. $x \le 5$ is the set of all real numbers less than or equal to 5. Unbounded

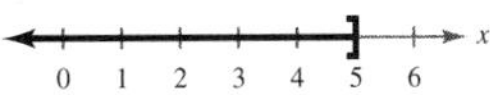

21. $x < 0$ is the set of all negative real numbers. Unbounded

23. $x \ge 4$ is the set of all real numbers greater than or equal to 4. Unbounded

25. $-2 < x < 2$ is the set of all real numbers greater than -2 and less than 2. Bounded

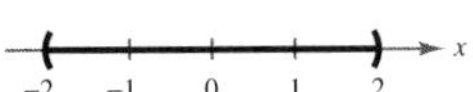

27. $-1 \le x < 0$ is the set of all negative real numbers greater than or equal to -1. Bounded

29. $\frac{127}{90}, \frac{584}{413}, \frac{7071}{5000}, \sqrt{2}, \frac{47}{33}$ **31.** $x < 0$ **33.** $y \ge 0$

35. $A \ge 30$ **37.** 10 **39.** $\pi - 3 \approx 0.1416$

41. -1 **43.** -9 **45.** 3.75 **47.** $|-3| > -|-3|$

49. $-5 = -|5|$ **51.** $-|-2| = -|2|$ **53.** 4

55. $\frac{5}{2}$ **57.** 51 **59.** $\frac{128}{75}$ **61.** $|x - 5| \le 3$

63. $|7 - 18| = 11$ miles **65.** $|y| \ge 6$

67. $|\$113{,}356 - \$112{,}700| = \$656 > \$500.00$
$0.05(\$112{,}700) = \$5635$
Because the actual expenses differ from the budget by more than \$500.00, there is failure to meet the "budget variance test."

69. $|\$37{,}335 - \$37{,}640| = \$305 < \$500$
$0.05(\$37{,}640) = \$1882$
Because the difference between the actual expenses and the budget is less than \$500 and less than 5% of the budgeted amount, there is compliance with the "budget variance test."

71. $|77.8 - 92.2| = 14.4$
There was a deficit of \$14.4 billion.

73. $|1031.3 - 1252.7| = 221.4$
There was a deficit of \$221.4 billion.

75. (a) No. If one is negative while the other is positive, they are unequal.
(b) $|u + v| \le |u| + |v|$

77. $7x, 4$ **79.** $4x^3, x, -5$ **81.** (a) -10 (b) -6

83. (a) 14 (b) 2

85. (a) Division by 0 is undefined. (b) 0

87. Commutative (addition)

89. Inverse (multiplication)

91. Distributive Property

93. Identity (multiplication)

95. Associative and Commutative (multiplication)

97. 0 **99.** Division by 0 is undefined. **101.** 6 **103.** $\frac{1}{2}$

105. $\frac{3}{8}$ **107.** $\frac{3}{10}$ **109.** 48 **111.** -2.57 **113.** 1.56

115.

n	1	0.5	0.01	0.0001	0.000001
$5/n$	5	10	500	50,000	5,000,000

117.

n	1	10	100	10,000	100,000
$5/n$	5	0.5	0.05	0.0005	0.00005

Section P.2 *(page 23)*

1. (a) 48 (b) 81 **3.** (a) 729 (b) -9
5. (a) 243 (b) $-\frac{3}{4}$ **7.** -1600 **9.** 2.125
11. -24 **13.** 5 **15.** (a) $-125z^3$ (b) $5x^6$
17. (a) $\dfrac{7}{x}$ (b) $\dfrac{4}{3}(x+y)^2$ **19.** (a) 1 (b) $\dfrac{1}{4x^4}$
21. (a) $\dfrac{10}{x}$ (b) $\dfrac{125x^9}{y^{12}}$ **23.** (a) 3^{3n} (b) $\dfrac{b^5}{a^5}$
25. $9^{1/2} = 3$ **27.** $\sqrt[5]{32} = 2$ **29.** $\sqrt{196} = 14$
31. $(-216)^{1/3} = -6$ **33.** $81^{3/4} = 27$
35. 3 **37.** 3 **39.** -125 **41.** $\frac{1}{8}$ **43.** -4
45. -7.225 **47.** 14.499 **49.** (a) $2\sqrt{2}$ (b) $2\sqrt[3]{3}$
51. (a) $6x\sqrt{2x}$ (b) $\dfrac{18}{z\sqrt{z}}$
53. (a) $2x\sqrt[3]{2x^2}$ (b) $\dfrac{5|x|\sqrt{3}}{y^2}$ **55.** 625 **57.** $\dfrac{2}{x}$
59. $\dfrac{1}{x^3}$, $x > 0$ **61.** (a) $\dfrac{\sqrt{3}}{3}$ (b) $4\sqrt[3]{4}$
63. (a) $\dfrac{x(5+\sqrt{3})}{11}$ (b) $3(\sqrt{6}-\sqrt{5})$ **65.** $\dfrac{2}{\sqrt{2}}$
67. $\dfrac{2}{3(\sqrt{5}-\sqrt{3})}$
69. (a) $3^{1/2} = \sqrt{3}$ (b) $(x+1)^{2/3} = \sqrt[3]{(x+1)^2}$
71. (a) $2\sqrt[4]{2}$ (b) $\sqrt[8]{2x}$ **73.** (a) $34\sqrt{2}$ (b) $22\sqrt{2}$
75. (a) $2\sqrt{x}$ (b) $4\sqrt{y}$ **77.** $\sqrt{5} + \sqrt{3} > \sqrt{5+3}$
79. $5 > \sqrt{3^2 + 2^2}$ **81.** 5.75×10^7 **83.** 8.99×10^{-5}
85. 524,000,000 **87.** 0.00000000048
89. (a) 954.448 (b) 3.077×10^{10}
91. (a) 67,082.039 (b) 39.791
93. When any positive integer is squared, the units digit is 0, 1, 4, 5, 6, or 9. Therefore, $\sqrt{5233}$ is not an integer.
95. $\dfrac{\pi}{2} \approx 1.57$ seconds **97.** 0.280 **99.** $\dfrac{25}{3}$ minutes

Section P.3 *(page 34)*

1. $-2x - 10$ **3.** $3x^3 - 2x + 2$
5. $8x^3 + 29x^2 + 11$ **7.** $3x^3 - 6x^2 + 3x$
9. $-15z^2 + 5z$ **11.** $-4x^4 + 4x$ **13.** $x^2 + 7x + 12$
15. $6x^2 - 7x - 5$ **17.** $4x^2 - 20xy + 25y^2$
19. $x^2 + 2xy + y^2 - 6x - 6y + 9$ **21.** $x^2 - 100$
23. $x^2 - 4y^2$ **25.** $m^2 - n^2 - 6m + 9$
27. $4r^4 - 25$ **29.** $x^3 + 3x^2 + 3x + 1$
31. $8x^3 - 12x^2y + 6xy^2 - y^3$ **33.** $2x^2 + 2x$
35. $u^4 - 16$ **37.** No; $(x^2 + 1) + (-x^2 + 3) = 4$
39. (a) $500r^2 + 1000r + 500$

(b)

r	$5\frac{1}{2}\%$	7%	8%
$500(1+r)^2$	\$556.51	\$572.45	\$583.20

r	$8\frac{1}{2}\%$	9%
$500(1+r)^2$	\$588.61	\$594.05

(c) Amount increases with increasing r.

41. $V = x(15 - 2x)\left(\dfrac{45 - 3x}{2}\right)$
$= \frac{3}{2}x(x - 15)(2x - 15)$

x (cm)	3	5	7
V (cu cm)	486	375	84

43. (a) $T = 0.14x^2 - 3.33x + 58.40$

(b)

x (mi/hr)	30	40	55
T (ft)	84.50	149.20	298.75

(c) Stopping distance increases at an accelerating rate as speed increases.

45. $(x+1)(x+4) = x(x+4) + 1(x+4)$
Distributive Property

47. $3(x + 2)$ **49.** $2x(x^2 - 3)$ **51.** $(x + 6)(x - 6)$
53. $(4y + 3)(4y - 3)$ **55.** $(x + 1)(x - 3)$
57. $(x - 2)^2$ **59.** $(2t + 1)^2$ **61.** $(x + 2)(x - 1)$

63. $(s-3)(s-2)$ **65.** $-(y+5)(y-4)$

67. $(3x-2)(x-1)$ **69.** $(5x+1)(x+5)$

71. $(x-2)(x^2+2x+4)$ **73.** $(y+4)(y^2-4y+16)$

75. $(x-1)(x^2+2)$ **77.** $(2x-1)(x^2-3)$

79. $x(x+3)(x-3)$ **81.** $x^2(x-4)$

83. $(x-1)^2$ **85.** $(1-2x)^2$ **87.** $-2x(x+1)(x-2)$

89. $(9x+1)(x+1)$ **91.** $(3x+1)(x^2+5)$

93. $x(x-4)(x^2+1)$ **95.** $-z(z+10)$

97. $(x+1)^2(x-1)^2$ **99.** $2(t-2)(t^2+2t+4)$

101. $(2x-1)(6x-1)$ **103.** $-(x+1)(x-3)(x+9)$

105. $7(x^2+1)(3x^2-1)$ **107.** $-2x(x-5)^3(x+5)$

109. $-(x^2+1)^4\left(\dfrac{x^2}{2}+1\right)$ **111.** (b) **113.** (a)

115.

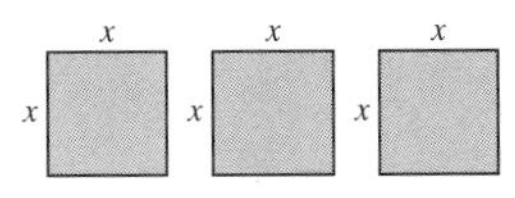

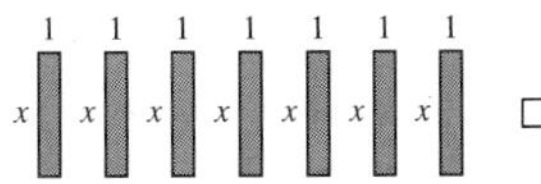

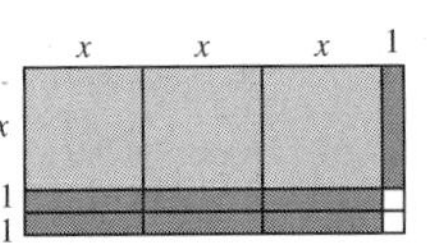

117.

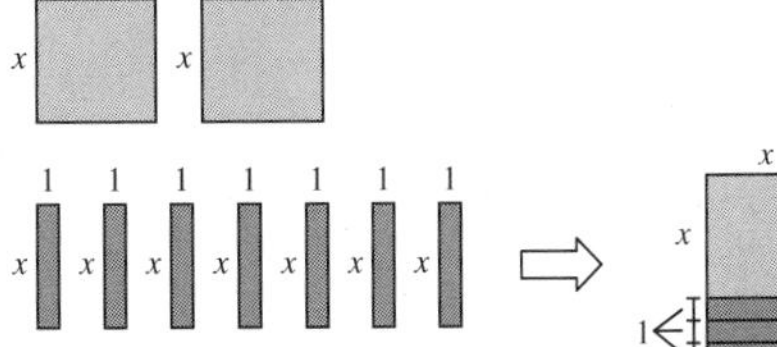

119. $4\pi(r+1)$ **121.** $4(6-x)(6+x)$

123. $-14,\ 14,\ -2,\ 2$ **125.** $2,\ -12$

127. $9x^2-9x-54=9(x^2-x-6)$

$$=9(x+2)(x-3)$$

129. (a) $\pi h(R-r)(R+r)$

(b) $2\pi\left[\left(\dfrac{R+r}{2}\right)(R-r)\right]h$

Section P.4 *(page 46)*

1. All real numbers **3.** All nonnegative real numbers

5. All real numbers x such that $x \neq 2$

7. All real numbers x such that $x \neq 0$ and $x \neq 4$

9. All real numbers x such that $x \geq -1$

11. $3x,\quad x \neq 0$ **13.** $\dfrac{3x}{2},\quad x \neq 0$ **15.** $\dfrac{3y}{y+1},\quad x \neq 0$

17. $-\dfrac{1}{2},\quad x \neq 5$ **19.** $\dfrac{x(x+3)}{x-2},\quad x \neq -2$

21. $\dfrac{y-4}{y+6},\quad y \neq 3$ **23.** $-(x^2+1)$ **25.** $z-2$

27.

x	0	1	2	3	4	5	6
$\dfrac{x^2-2x-3}{x-3}$	1	2	3	Undef.	5	6	7
$x+1$	1	2	3	4	5	6	7

The expressions are equivalent except at $x = 3$.

29. Only common factors of the numerator and denominator can be canceled. In this case, factors of terms were incorrectly canceled.

31. $\dfrac{\pi}{4}$ **33.** $\dfrac{1}{5(x-2)},\quad x \neq 1$

35. $\dfrac{r+1}{r},\quad r \neq 1$ **37.** $\dfrac{t-3}{(t+3)(t-2)},\quad t \neq -2$

39. $\dfrac{3}{2},\quad x \neq -y$ **41.** $x(x+1),\quad x \neq -1, 0$

43. $\dfrac{x+5}{x-1}$ **45.** $\dfrac{6x+13}{x+3}$ **47.** $-\dfrac{2}{x-2}$

49. $-\dfrac{x^2+3}{(x+1)(x-2)(x-3)}$ **51.** $\dfrac{2-x}{x^2+1}$

53. $-\dfrac{1}{(x^2+1)^5}$ **55.** $\dfrac{1}{2},\quad x \neq 2$

57. $-\dfrac{2x+h}{x^2(x+h)^2},\quad h \neq 0$ **59.** $\dfrac{2x-1}{2x},\quad x > 0$

61. $-\dfrac{1}{t^2\sqrt{t^2+1}}$ **63.** $\dfrac{1}{\sqrt{x+2}+\sqrt{x}}$

65. (a) $\frac{1}{16}$ minute (b) $\frac{x}{16}$ minutes

(c) $\frac{60}{16} = \frac{15}{4}$ minutes

67. $\frac{11x}{30}$ **69.** (a) 12.65% (b) $\frac{288(MN - P)}{N(MN + 12P)}$

71. (a)

t	0	1	2	3	4	5
T	75	63.3	55.9	51.3	48.3	46.4

(b)

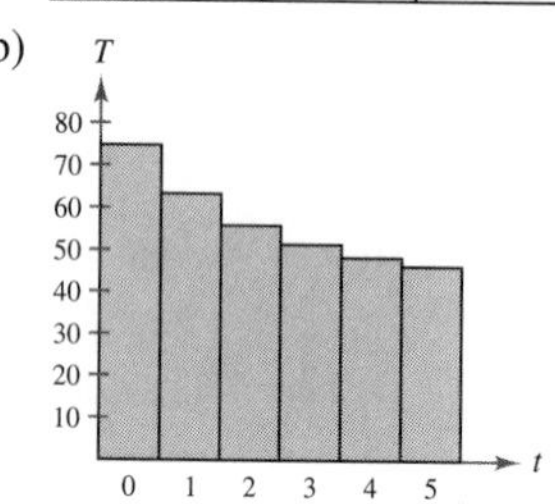

73. $\frac{x}{2(2x + 1)}$

Section P.5 *(page 55)*

1.

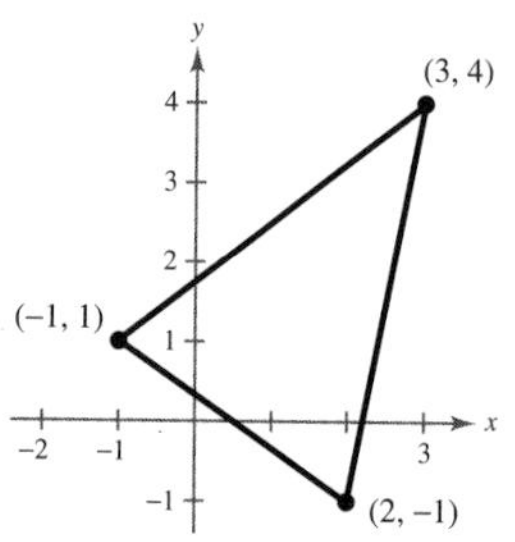

3.

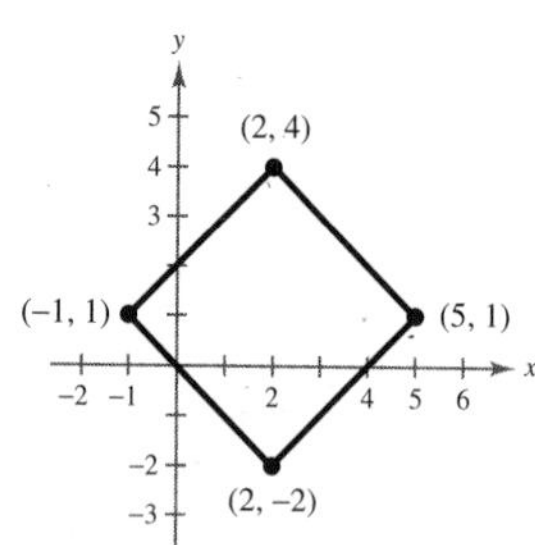

5. A: (2, 6); B: (−6, −2); C: (4, −4); D: (−3, 2)

7. A: (0, 5); B: (−3, −6); C: (1, −4.5); D: (−4, 2)

9. (−3, 4) **11.** (−5, −5)

13. Point on x-axis: $y = 0$; Point on y-axis: $x = 0$

15. Quadrant IV **17.** Quadrant II

19. Quadrant III or IV **21.** Quadrant III

23. Quadrants I and III

25.

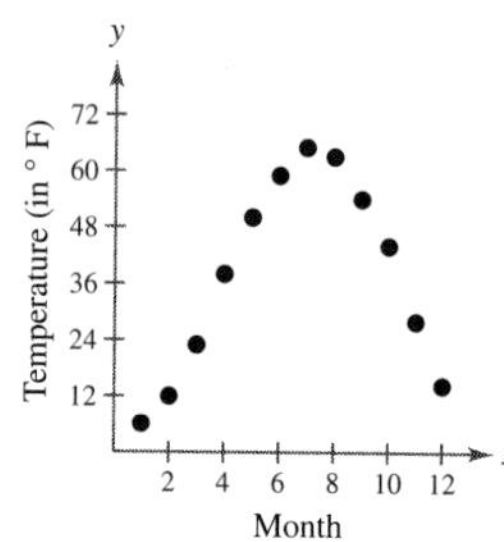

27. (0, 1), (4, 2), (1, 4)

29.

x	−2	−1	$-\frac{1}{2}$	0	$\frac{1}{2}$	1	2
y	3	$\frac{5}{2}$	$\frac{9}{4}$	2	$\frac{7}{4}$	$\frac{3}{2}$	1

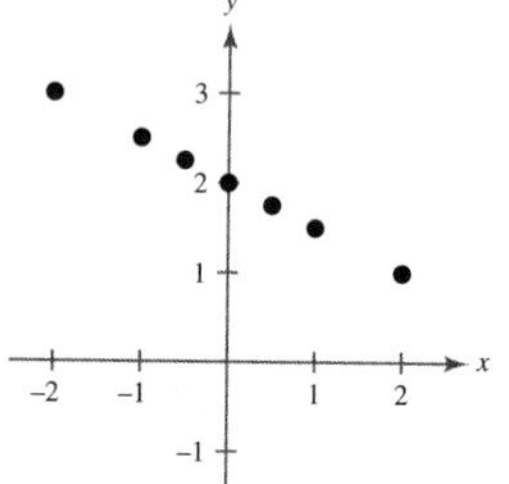

31. 1900% **33.** 1990: \$13.70 per 100 pounds

35. 1970s **37.** 65 **39.** $4^2 + 3^2 = 5^2$

41. $10^2 + 3^2 = (\sqrt{109})^2$ **43.** 8 **45.** 5

47. (a)

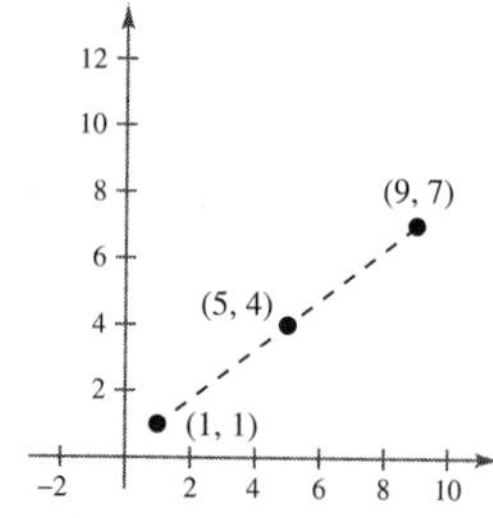

(b) 10

(c) (5, 4)

49. (a)

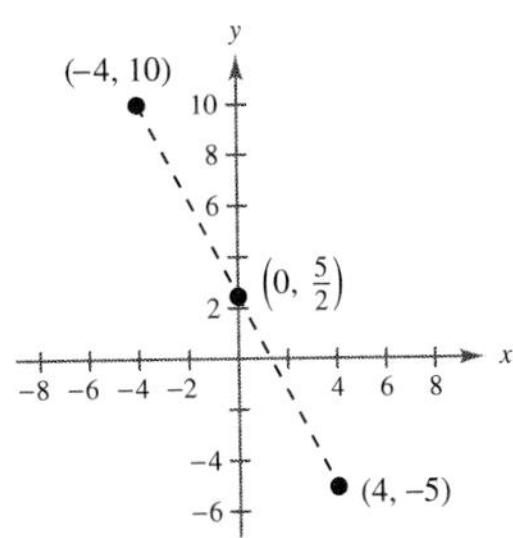

(b) 17
(c) $\left(0, \frac{5}{2}\right)$

51. (a)

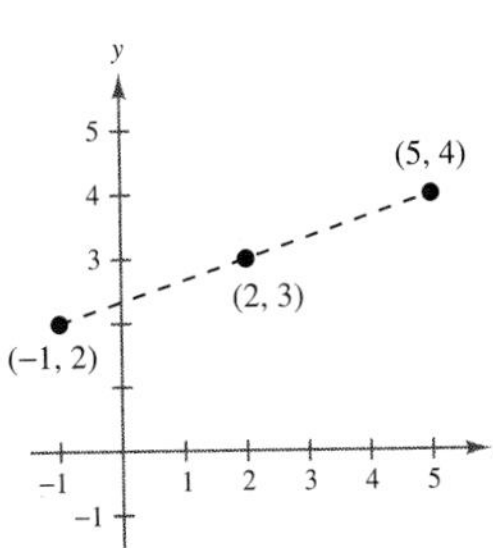

(b) $2\sqrt{10}$
(c) $(2, 3)$

53. (a)

(b) $\dfrac{\sqrt{82}}{3}$
(c) $\left(-1, \frac{7}{6}\right)$

55. (a)

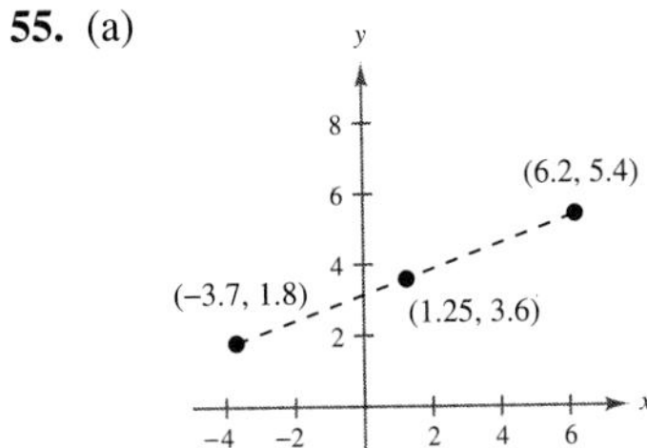

(b) $\sqrt{110.97}$
(c) $(1.25, 3.6)$

57. (a)

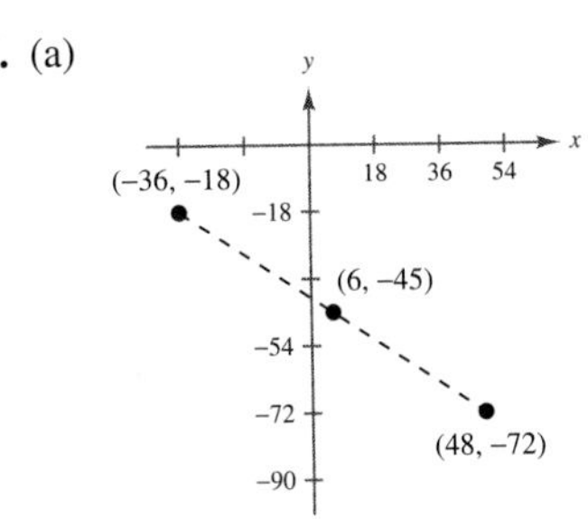

(b) $6\sqrt{277}$
(c) $(6, -45)$

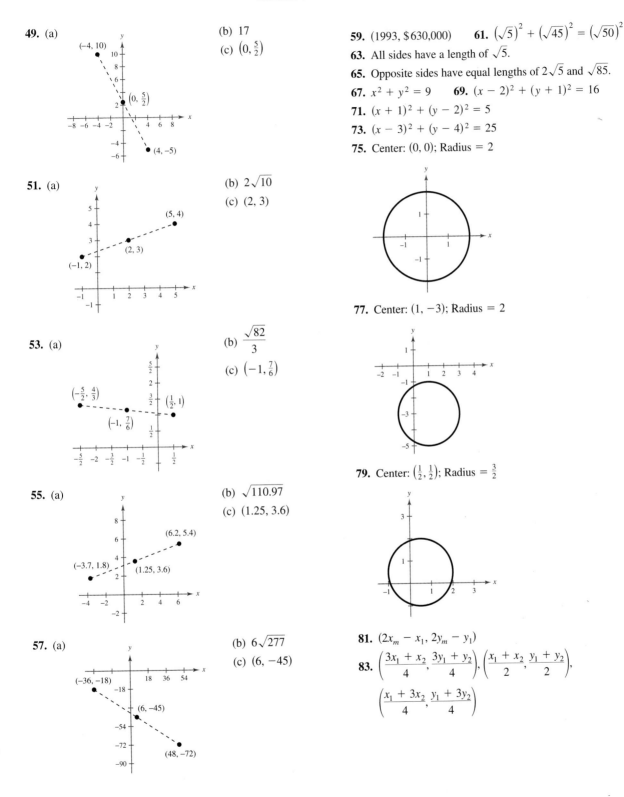

59. (1993, \$630,000) **61.** $\left(\sqrt{5}\right)^2 + \left(\sqrt{45}\right)^2 = \left(\sqrt{50}\right)^2$

63. All sides have a length of $\sqrt{5}$.

65. Opposite sides have equal lengths of $2\sqrt{5}$ and $\sqrt{85}$.

67. $x^2 + y^2 = 9$ **69.** $(x - 2)^2 + (y + 1)^2 = 16$

71. $(x + 1)^2 + (y - 2)^2 = 5$

73. $(x - 3)^2 + (y - 4)^2 = 25$

75. Center: $(0, 0)$; Radius $= 2$

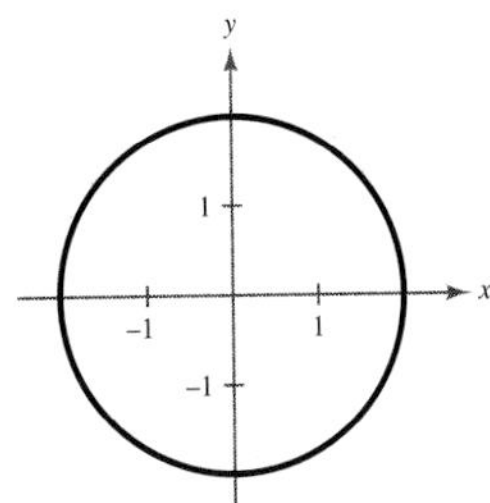

77. Center: $(1, -3)$; Radius $= 2$

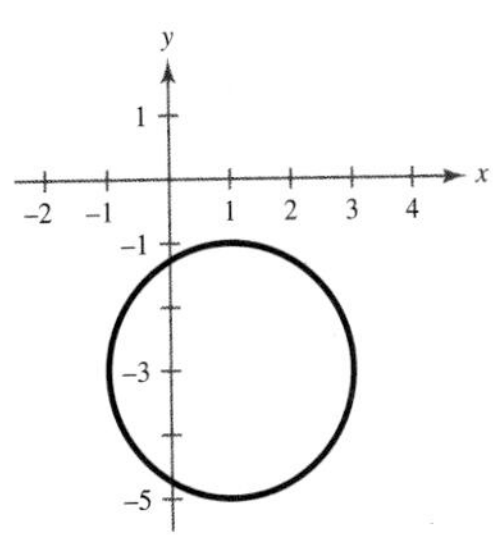

79. Center: $\left(\frac{1}{2}, \frac{1}{2}\right)$; Radius $= \frac{3}{2}$

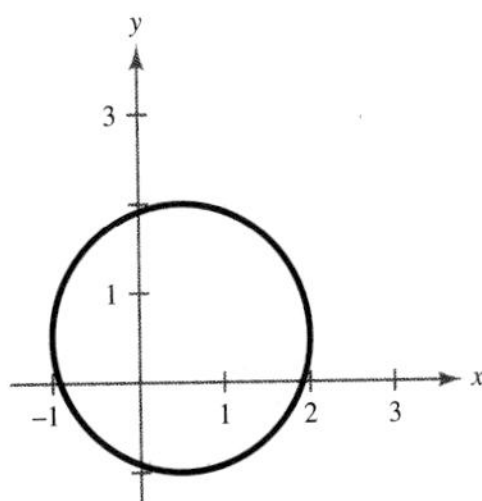

81. $(2x_m - x_1, 2y_m - y_1)$

83. $\left(\dfrac{3x_1 + x_2}{4}, \dfrac{3y_1 + y_2}{4}\right), \left(\dfrac{x_1 + x_2}{2}, \dfrac{y_1 + y_2}{2}\right),$
$\left(\dfrac{x_1 + 3x_2}{4}, \dfrac{y_1 + 3y_2}{4}\right)$

85. $5\sqrt{74} \approx 43$ yards

87.

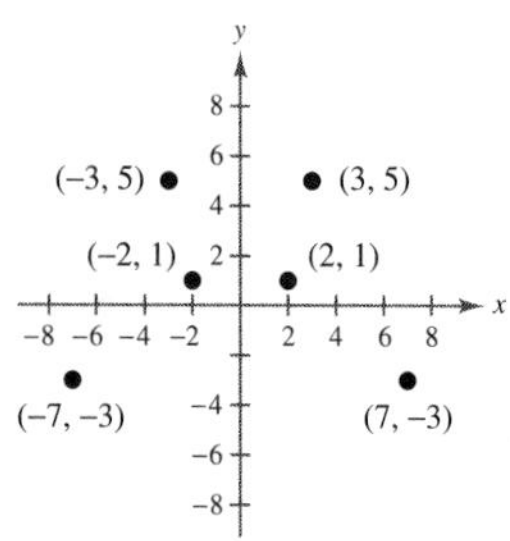

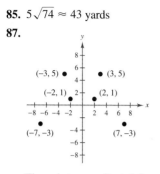

The points are reflected through the y-axis.

89. (a) 7 (b) Answers will vary.

Section P.6 *(page 65)*

1. (a) \$1.109 (b) [\$0.999, \$1.189]

3. 15

```
          ×
          ×     × × ×
        × × ×   × × × ×
        × × ×   × × × ×
×   ×   × × ×   × × × × ×   ×
+-+-+-+-+-+-+-+-+-+-+-+-+-+-+
10  12  14  16  18  20  22  24
```

5.

```
                      ×       ×
          ×     ×     ×       ×       ×
          ×   × ×     ×       ×       ×                     ×
×         ×   × ×     × × × × ×   × × × ×   ×           ×   ×
+-+-+-+-+-+-+-+-+-+-+-+-+-+-+-+-+-+-+-+-+-+-+-+-+-+-+-+-+-+-+
70  72  74  76  78  80  82  84  86  88  90  92  94  96  98  100
```

81 and 85

7.

Stems	Leaves
7	0 5 5 5 7 7 8 8 8
8	1 1 1 1 2 3 4 5 5 5 5 7 8 9 9 9
9	0 2 8
10	0 0

9.

Stems	Leaves
6	63 76 92
7	62 64
8	04 18 34 34 41 48 50 63 64 65 85 98
9	03 13 15 16 19 19 32 40 43 64 64 74 74
10	03 24 24 29 45 49 56 61 96 98
11	23 24 31 38 55 97
12	52 65
13	36 51
18	00

11. 500%

13.

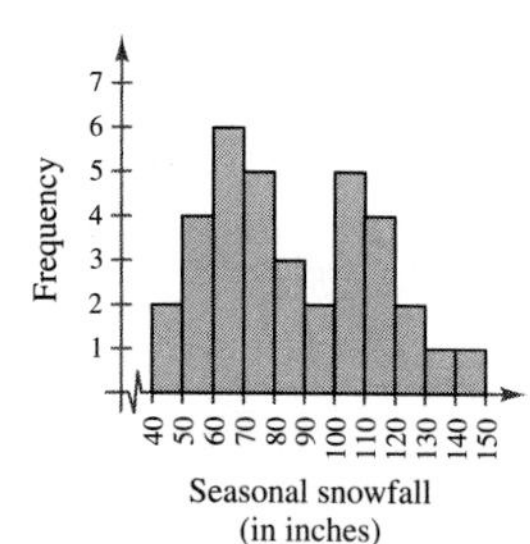

15.

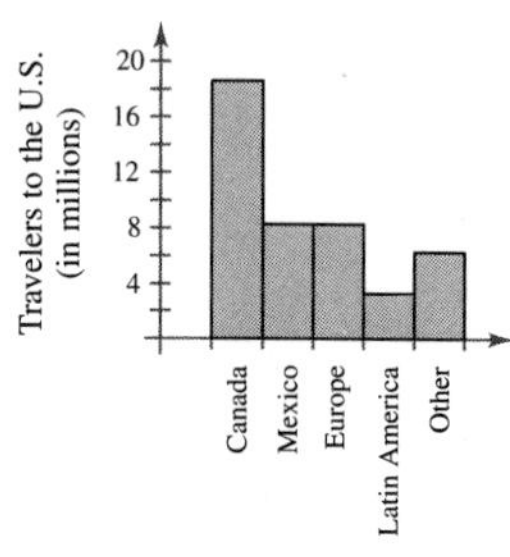

17. (a) 55.6%

(b) No. The trend limits the amount of funds available for capital improvements in industry.

19.

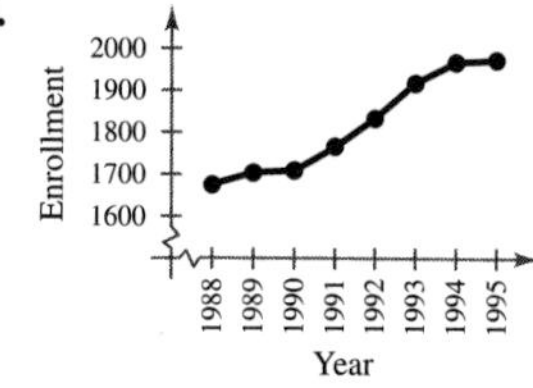

21.

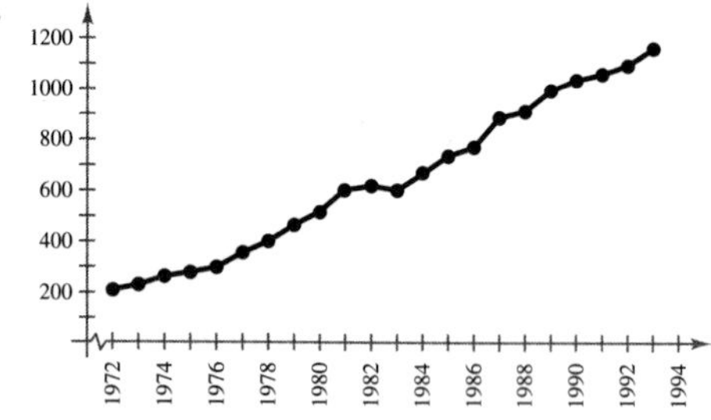

FOCUS ON CONCEPTS (page 68)

1.

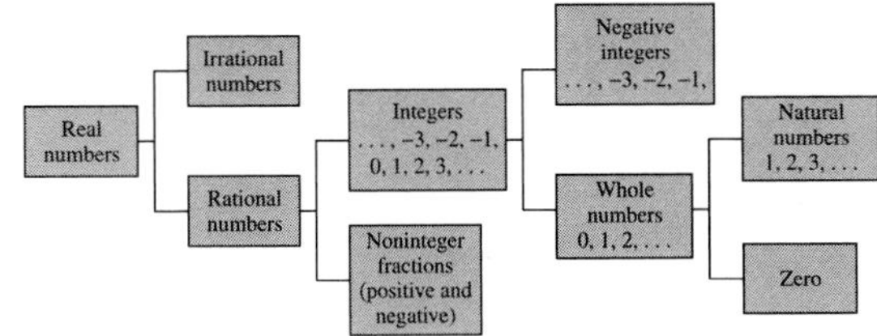

2. (a) $-A < 0$ (b) $-C > 0$
(c) $B - A < 0$ (d) $A - C > 0$

3. If $a < 0$, then $|a| = -a$. For example, let $a = -7$. Then $|-7| = -(-7) = 7$.

4. (a) The base for the exponent -1 is $3x$.
Therefore, $(3x)^{-1} = \dfrac{1}{3x}$.
(b) When multiplying, add exponents. $y^3 \cdot y^2 = y^5$
(c) Multiply the exponents to obtain $(a^2b^3)^4 = a^8b^{12}$.
(d) The square of a binomial contains a cross-product term. $(a + b)^2 = a^2 + 2ab + b^2$
(e) If $x < 0$, then $\sqrt{4x^2} > 0$ but $2x < 0$. $\sqrt{4x^2} = 2|x|$
(f) To add radicals, the index and the radicand must be the same.

5. A number written in scientific notation has the form $c \times 10^n$, where $1 \le c < 10$ and n is an integer. In scientific notation, the number 52.7×10^5 is 5.27×10^6.

6. (a) Yes. $(x^4 + 2x - 2) + (x^3 + 2) = x^4 + 3x$
(b) No. When adding third- and fourth-degree polynomials, the fourth-degree term of the fourth-degree polynomial will be in the sum.
(c) No. The sum will be fourth degree. The product of the two polynomials would be seventh degree.

7. The polynomial is written as a product.

8. Invert the divisor before multiplying.

9. (b) **10.** (c) **11.** (d) **12.** (a)

Review Exercises (page 69)

1. (a) 11 (b) 11, −14 (c) 11, −14, $-\frac{8}{9}$, $\frac{5}{2}$, 0.4
(d) $\sqrt{6}$

3. (a) $0.8\overline{3}$ (b) 0.875

5. The set consists of all real numbers less than or equal to 7.

7. $|x - 7| \ge 4$ **9.** $|y + 30| < 5$ **11.** −11

13. Associative Property of Addition

15. Commutative Property of Multiplication

17. (a) $-8z^3$ (b) $3a^3b^2$ **19.** (a) $\dfrac{3u^5}{v^4}$ (b) m^{-2}

21. 3.0296×10^{10} **23.** 483,300,000

25. (a) 11,414.125 (b) 18,380.160

27. $16^{1/2} = 4$ **29.** $2x^2$ **31.** $2 + \sqrt{3}$ **33.** $2\sqrt{2}$

35. $192\sqrt{2}$ **37.** −1 **39.** $16x^4$ **41.** 5

43. $-3x^2 - 7x + 1$ **45.** $4x^2 - 12x + 9$

47. $x^5 - 2x^4 + x^3 - x^2 + 2x - 1$

49. $y^6 + y^4 - y^3 - y$

51. $(x + 3)(x + 5) = 5(x + 3) + x(x + 3)$
Distributive Property

53. $x(x + 1)(x - 1)$ **55.** $(x + 10)(2x + 1)$

57. $(x - 1)(x^2 + 2)$

59. (a)

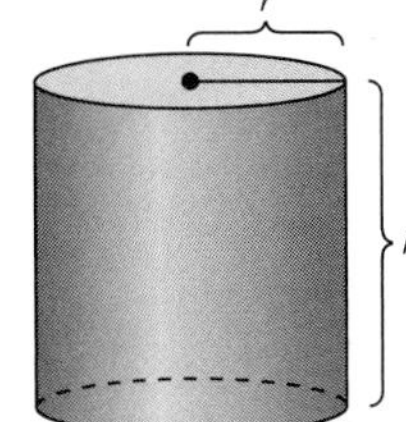

(b) $S = 2\pi r(r + h)$

61. $\dfrac{1}{x^2}$, $x \ne \pm 2$ **63.** $\dfrac{1}{5}x(5x - 6)$, $x \ne 0, -\dfrac{3}{2}$

65. $\dfrac{x^3 - x + 3}{(x - 1)(x + 2)}$ **67.** $\dfrac{x + 1}{x(x^2 + 1)}$

69. $-\dfrac{1}{xy(x + y)}$, $x \ne y$

71.

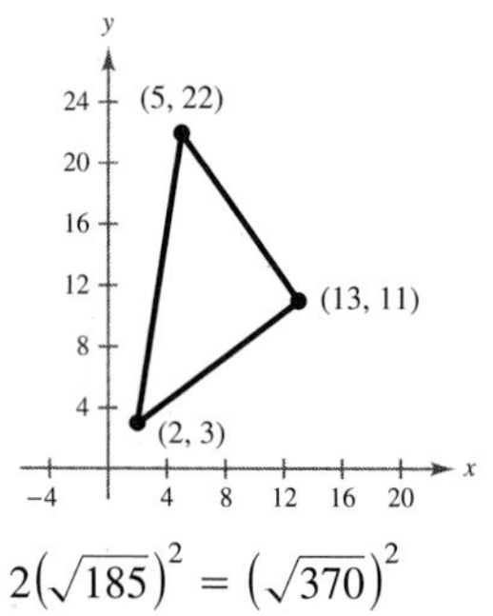

$2(\sqrt{185})^2 = (\sqrt{370})^2$

73. Quadrant IV

75. (a)

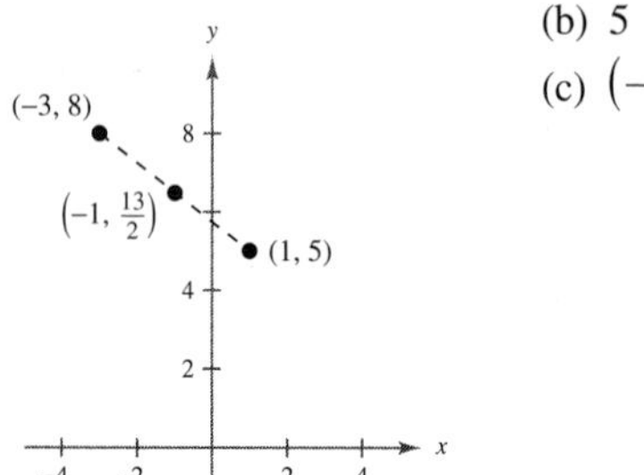

(b) 5 (c) $\left(-1, \frac{13}{2}\right)$

77.

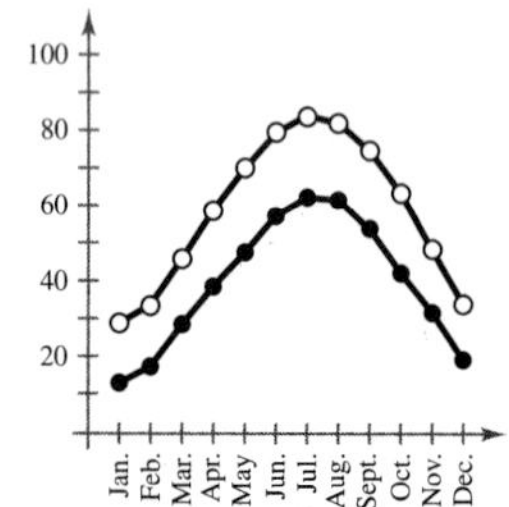

Chapter Test (page 73)

1. $-\frac{10}{3} > -|-4|$ **2.** 9.15 **3.** (a) -18 (b) $\frac{4}{27}$

4. (a) $-\frac{27}{125}$ (b) $\frac{8}{729}$ **5.** (a) 25 (b) 6

6. (a) 1.8×10^5 (b) 2.7×10^{13}

7. (a) $12z^8$ (b) $(u - 2)^{-7}$ **8.** (a) $\dfrac{3x^2}{y^2}$ (b) $\dfrac{2}{v}\sqrt[3]{\dfrac{2}{v^2}}$

9. (a) $15z\sqrt{2z}$ (b) $-10\sqrt{y}$ **10.** $2x^2 - 3x - 5$

11. $x^2 - 5$ **12.** $8, \quad x \neq 3$ **13.** $\dfrac{x - 1}{2x}, \quad x \neq \pm 1$

14. (a) $x^2(2x + 1)(x - 2)$ (b) $(x - 2)(x + 2)^2$

15. (a) $4\sqrt[3]{4}$ (b) $-3(1 + \sqrt{3})$

16.

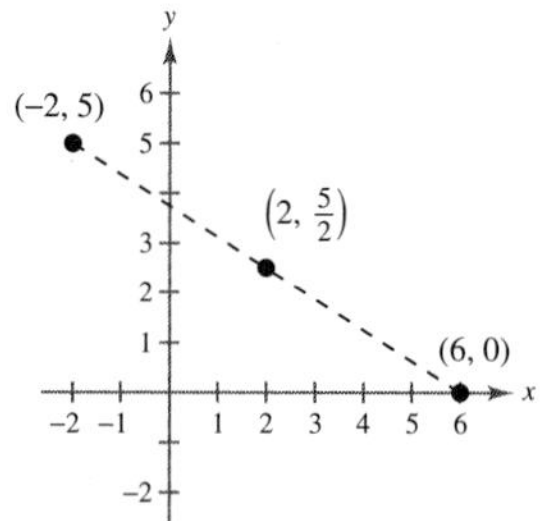

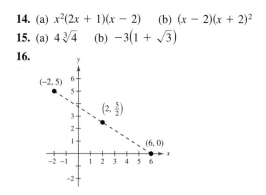

Midpoint: $\left(2, \frac{5}{2}\right)$; Distance: $\sqrt{89}$

17.

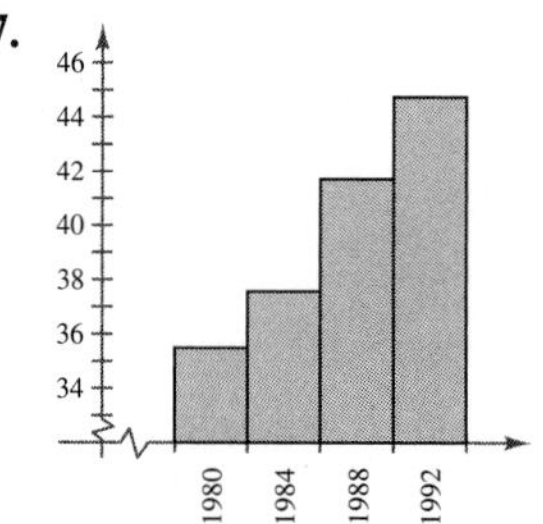

CHAPTER 1

Section 1.1 (page 84)

1. (a) Yes (b) Yes **3.** (a) No (b) Yes

5. (a) No (b) Yes **7.** (a) Yes (b) Yes

9.

x	-1	0	1	$\frac{3}{2}$	2
y	5	3	1	0	-1

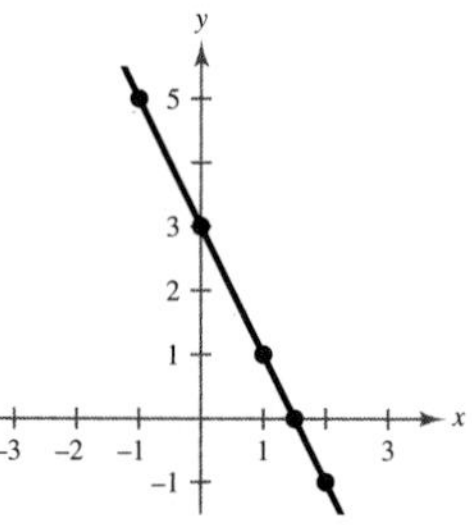

11.

x	-1	0	1	2	3
y	3	0	-1	0	3

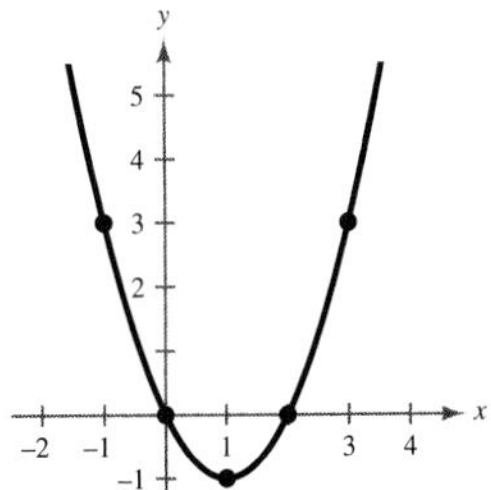

13.

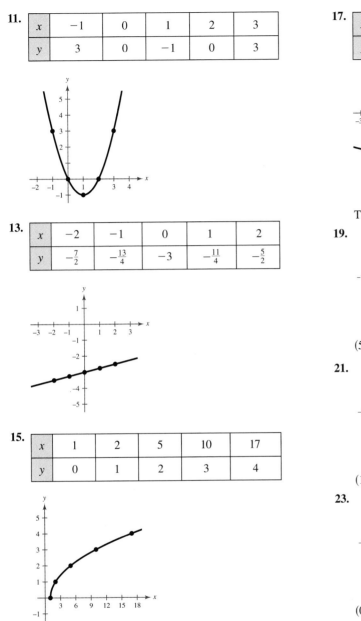

x	-2	-1	0	1	2
y	$-\frac{7}{2}$	$-\frac{13}{4}$	-3	$-\frac{11}{4}$	$-\frac{5}{2}$

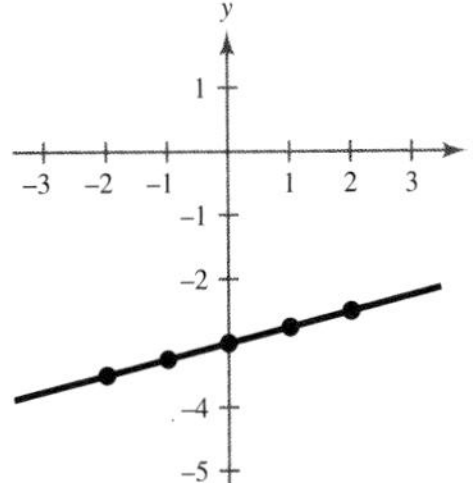

15.

x	1	2	5	10	17
y	0	1	2	3	4

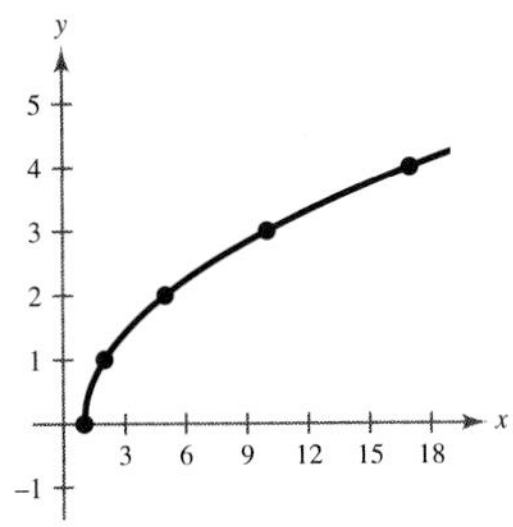

17.

x	-2	-1	0	1	2
y	$-\frac{5}{2}$	$-\frac{11}{4}$	-3	$-\frac{13}{4}$	$-\frac{7}{2}$

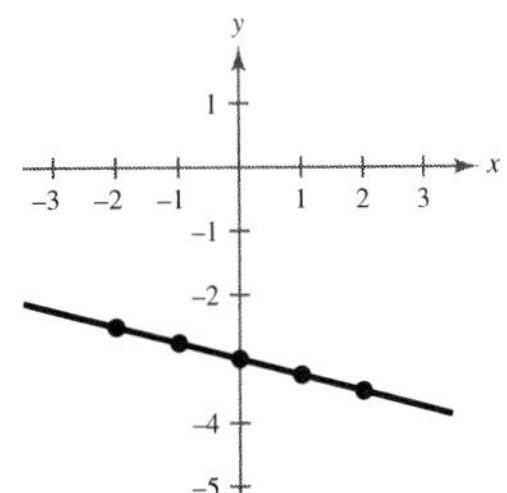

This line slopes downward to the right rather than upward.

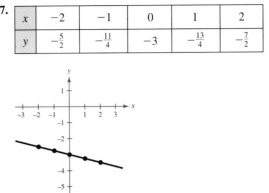

19.

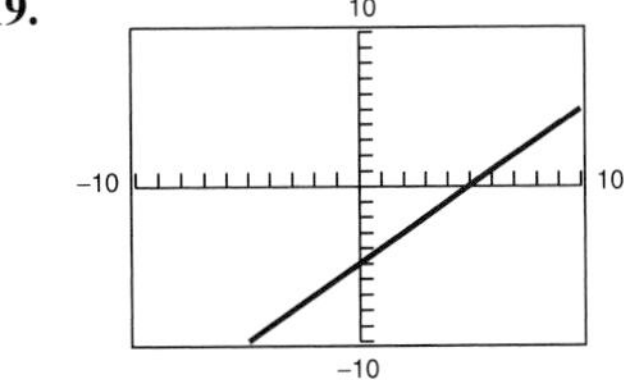

$(5, 0), (0, -5)$

21.

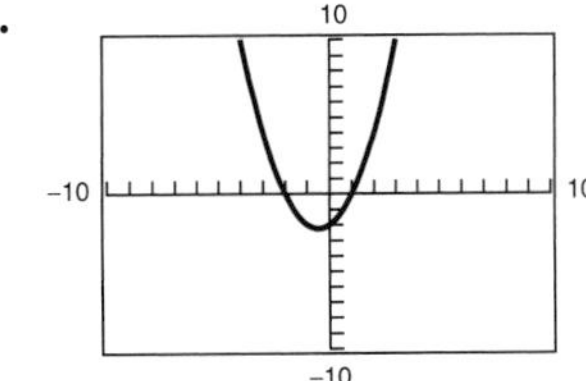

$(1, 0), (-2, 0), (0, -2)$

23.

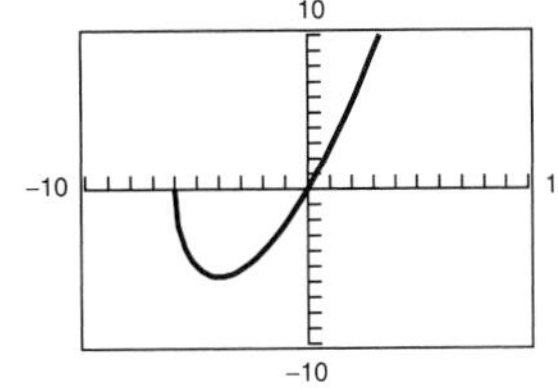

$(0, 0), (-6, 0)$

25.

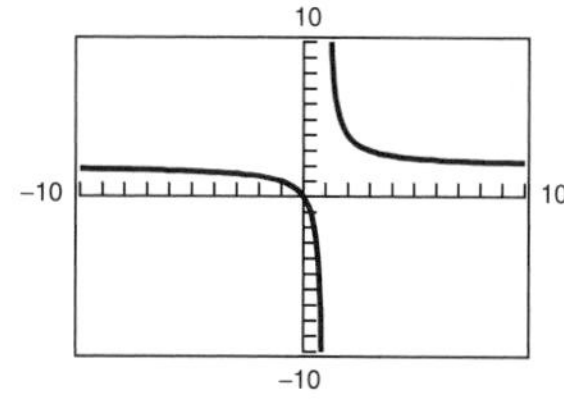

(0, 0)

27. (d) **29.** (f) **31.** (a)

33.

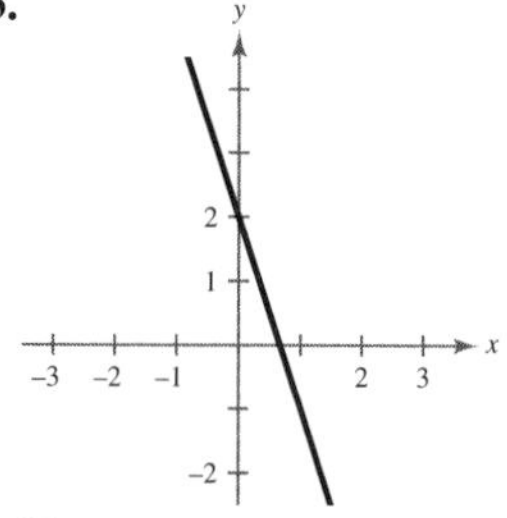

No symmetry

35.

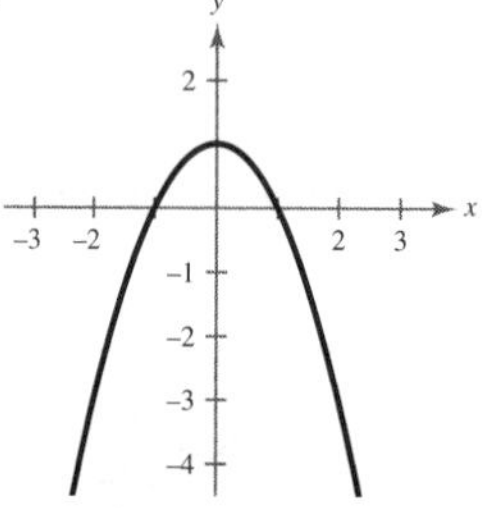

Symmetry: y-axis

37.

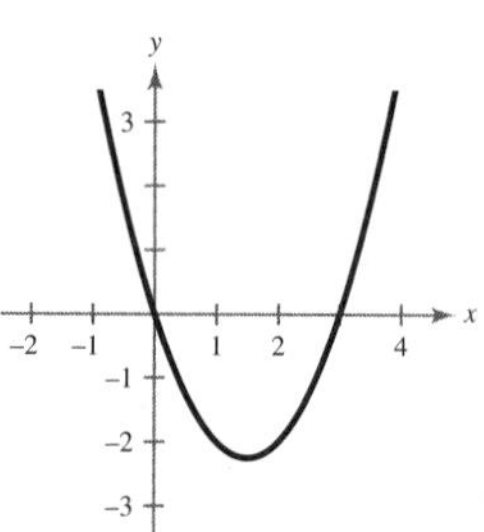

No symmetry

39.

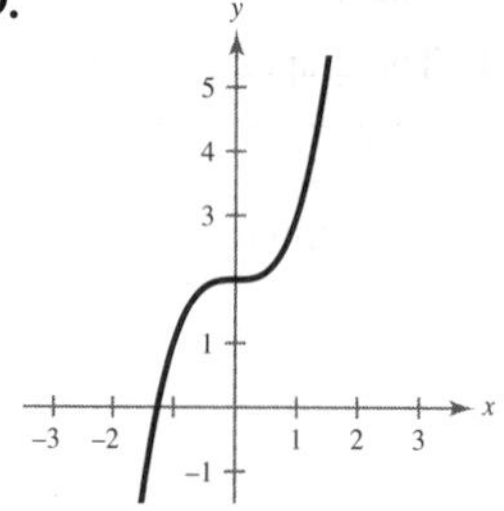

No symmetry

41.

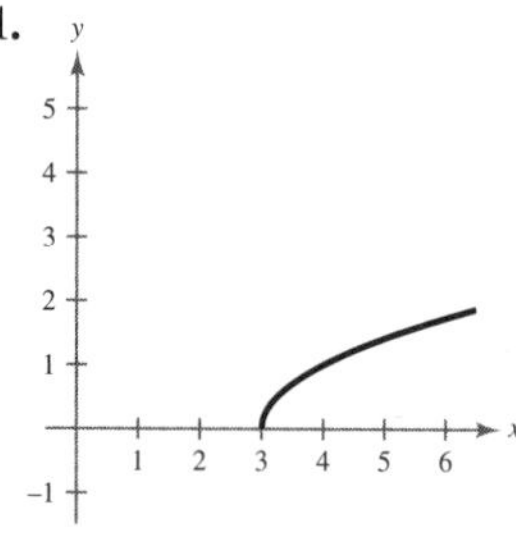

No symmetry

43.

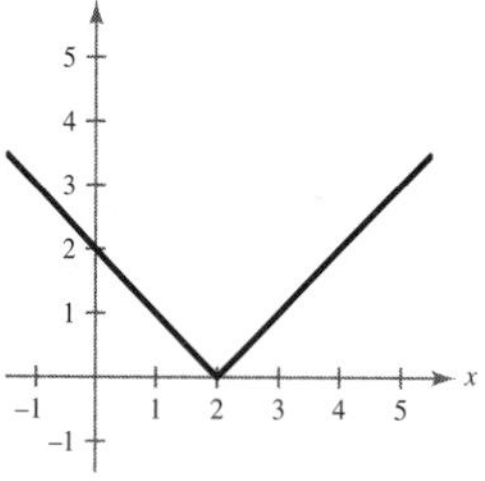

No symmetry

45.

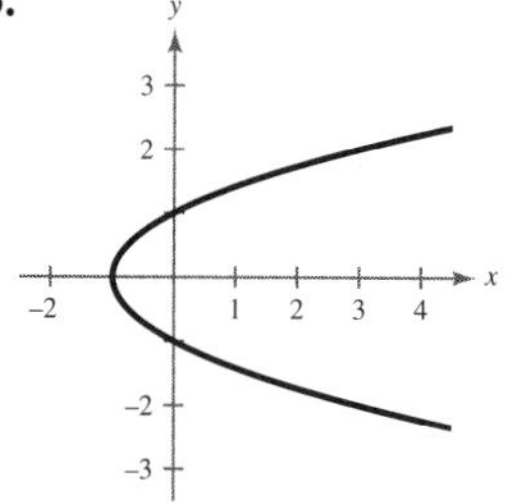

Symmetry: x-axis

47.

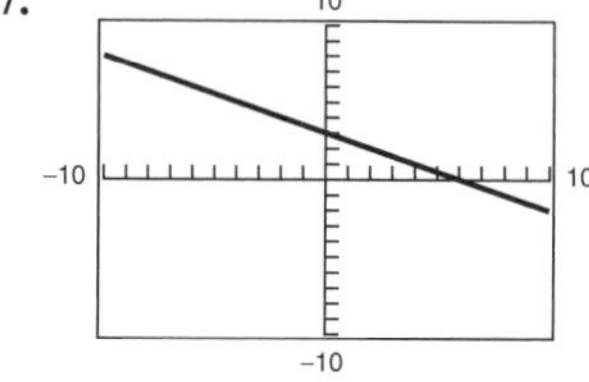

Intercepts: (6, 0), (0, 3)

49.

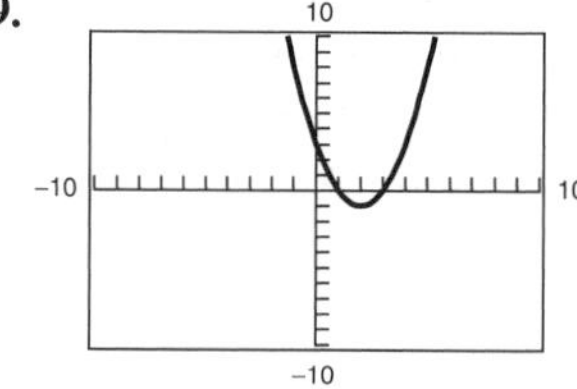

Intercepts: (3, 0), (1, 0), (0, 3)

51.

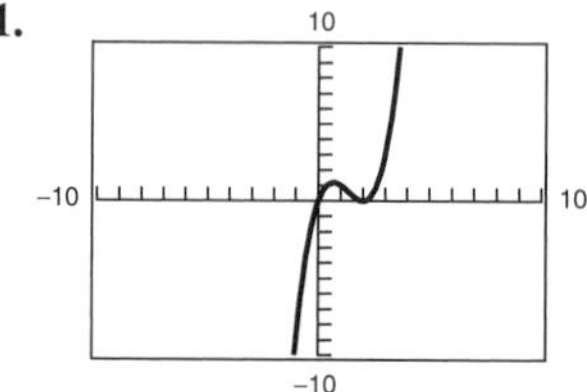

Intercepts: (0, 0), (2, 0)

53.

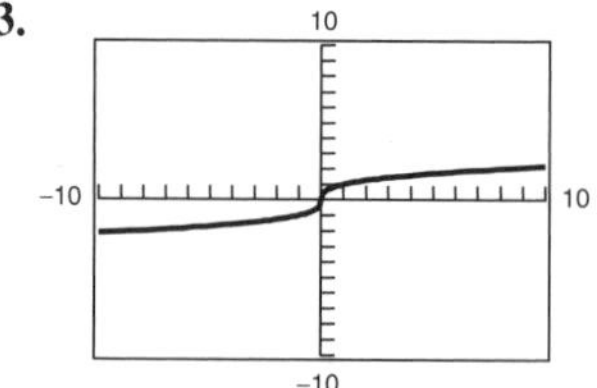

Intercept: (0, 0)

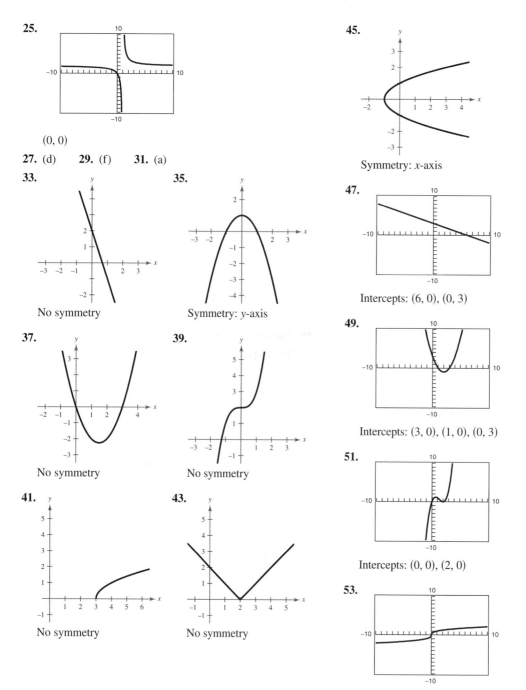

55.

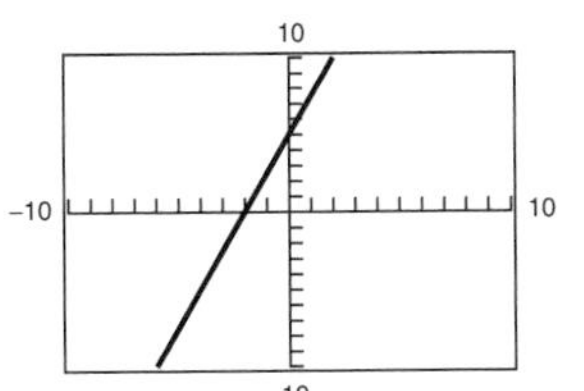

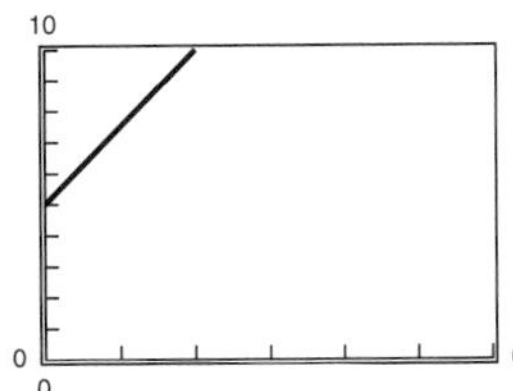

The standard setting gives a more complete graph.

57.

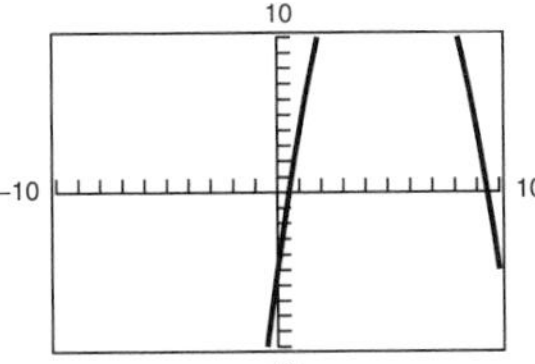

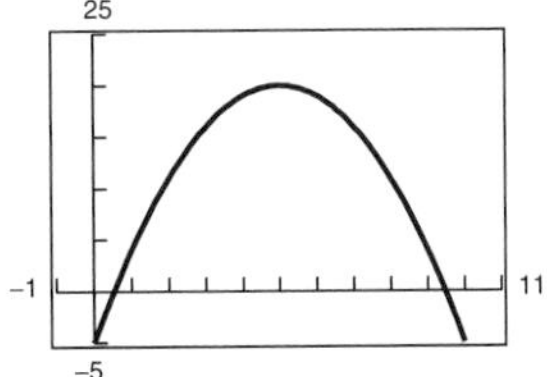

The specified setting gives a more complete graph.

59.

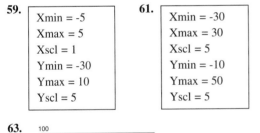

Xmin = -5
Xmax = 5
Xscl = 1
Ymin = -30
Ymax = 10
Yscl = 5

61.

Xmin = -30
Xmax = 30
Xscl = 5
Ymin = -10
Ymax = 50
Yscl = 5

63.

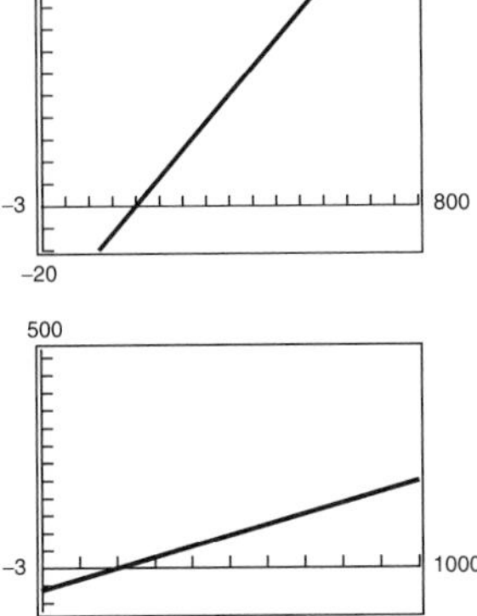

Xmin = -3
Xmax = 800
Xscl = 50
Ymin = -20
Ymax = 100
Yscl = 10

65. $y_1 = \sqrt{64 - x^2}$
$y_2 = -\sqrt{64 - x^2}$

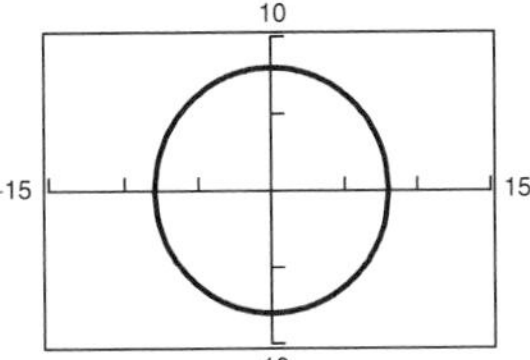

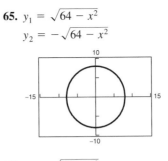

67. $y_1 = \sqrt{49 - x^2}$
$y_2 = -\sqrt{49 - x^2}$

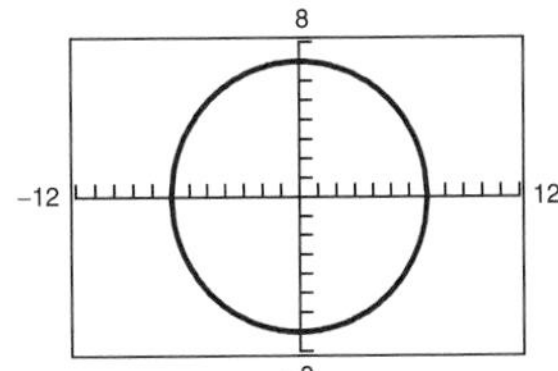

69. The graphs are identical. Distributive Property

71. The graphs are identical. Associative Property

73. (a)

Xmin = -1
Xmax = 9
Xscl = 1
Ymin = 60000
Ymax = 230000
Yscl = 10000

(b)

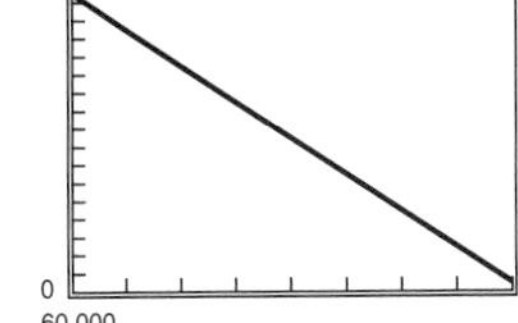

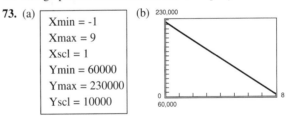

75. Change the viewing rectangle.

77. (a)

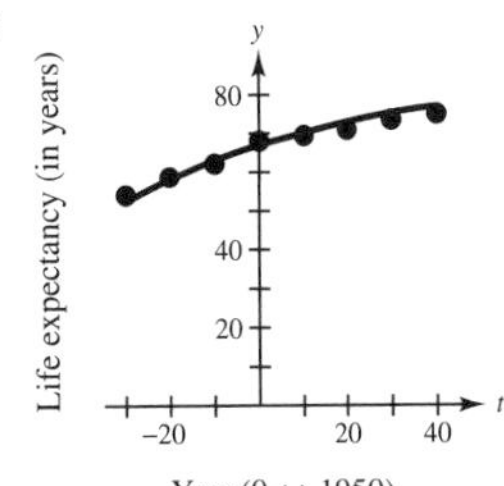

(b) 77.7 (c) 78.0

79.

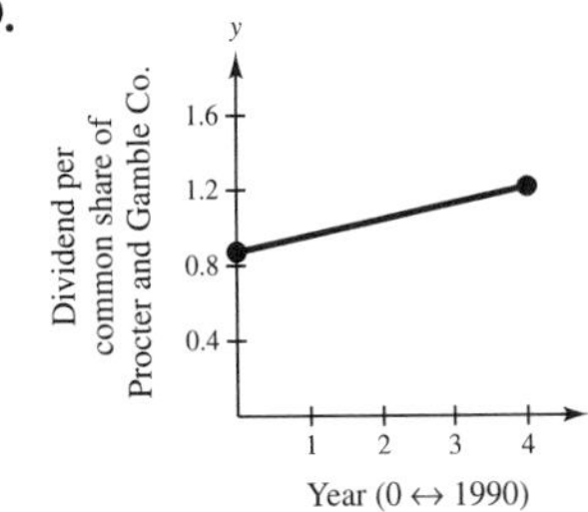

81.

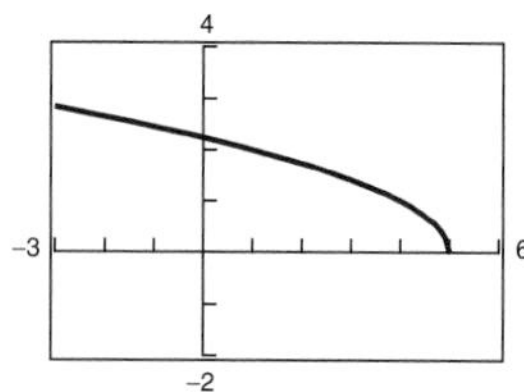

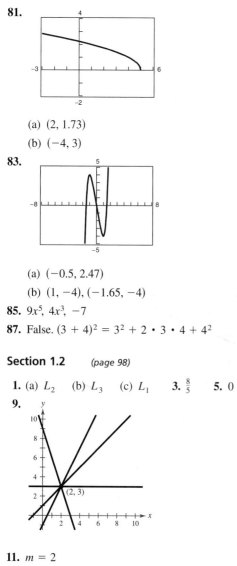

(a) (2, 1.73)

(b) (−4, 3)

83.

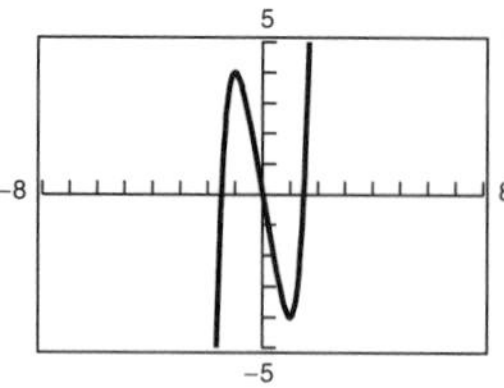

(a) (−0.5, 2.47)

(b) (1, −4), (−1.65, −4)

85. $9x^5$, $4x^3$, −7

87. False. $(3 + 4)^2 = 3^2 + 2 \cdot 3 \cdot 4 + 4^2$

Section 1.2 *(page 98)*

1. (a) L_2 (b) L_3 (c) L_1 **3.** $\frac{8}{5}$ **5.** 0 **7.** −4

9.

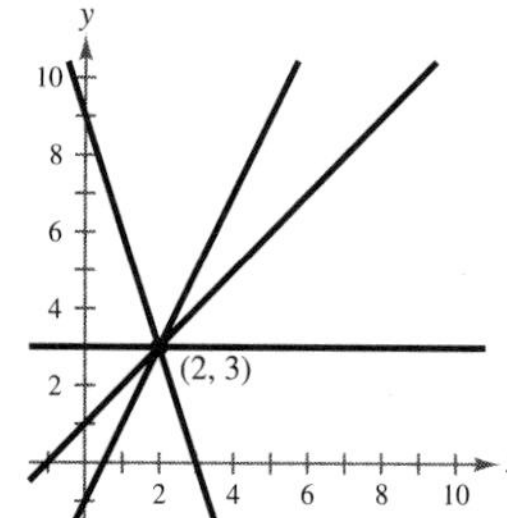

11. $m = 2$

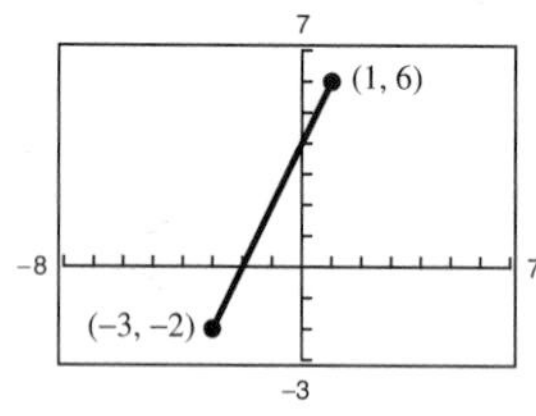

13. m is undefined.

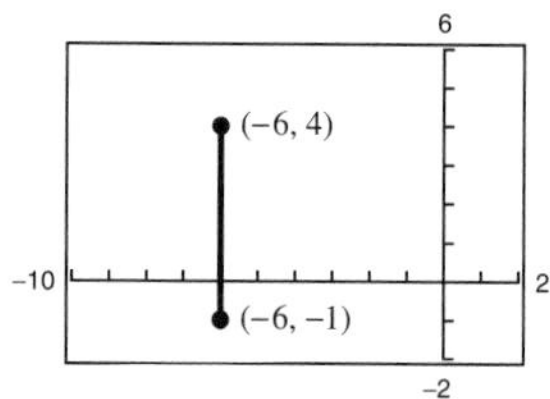

15. $m = \frac{4}{3}$

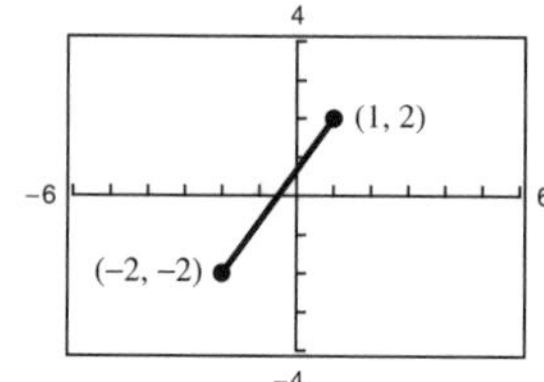

17. (0, 1), (3, 1), (−1, 1) **19.** (6, −5), (7, −4), (8, −3)

21. (−8, 0), (−8, 2), (−8, 3) **23.** Perpendicular

25. Parallel **27.** Yes. The rate of change remains the same on a line.

29. (a) Sales increasing 135 units per year. (b) No change in sales. (c) Sales decreasing 40 units per year.

31. (a)

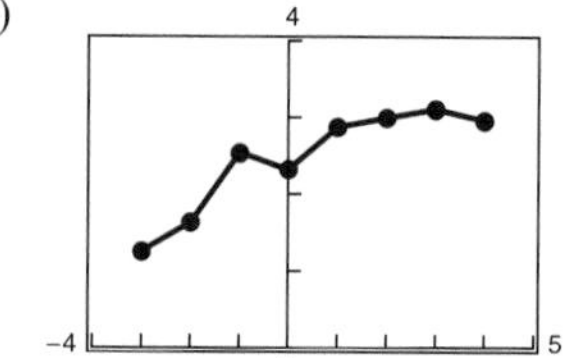

(b) Decreased most rapidly: 1989

Increased most rapidly: 1988

33.

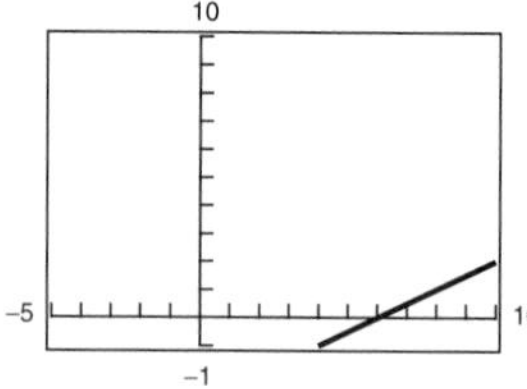

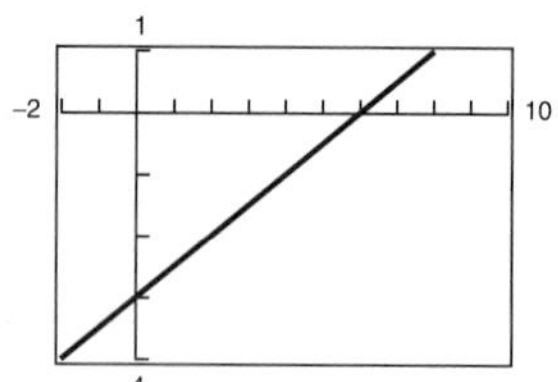

The second setting gives a more complete graph.

35. $m = 5$; Intercept: $(0, 3)$

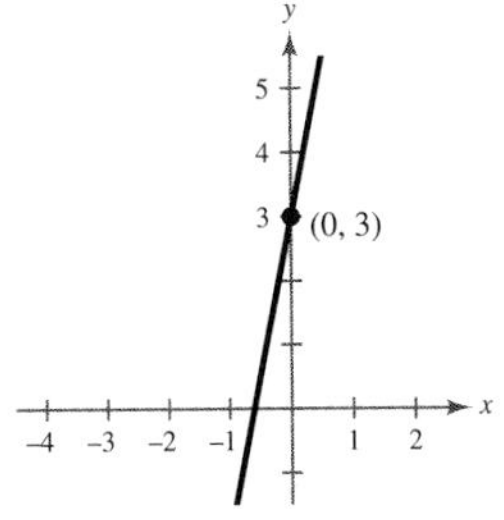

37. m is undefined.
There is no y-intercept.

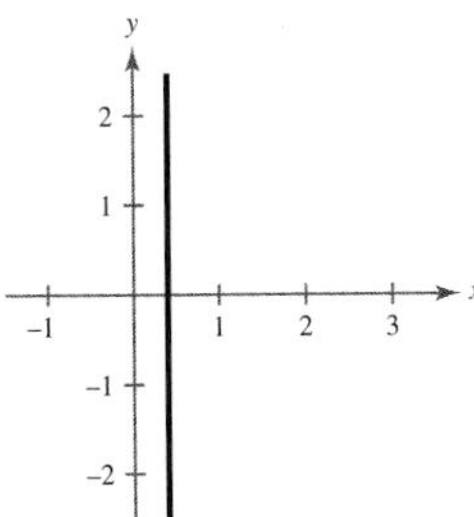

39. $m = -\frac{7}{6}$; Intercept: $(0, 5)$

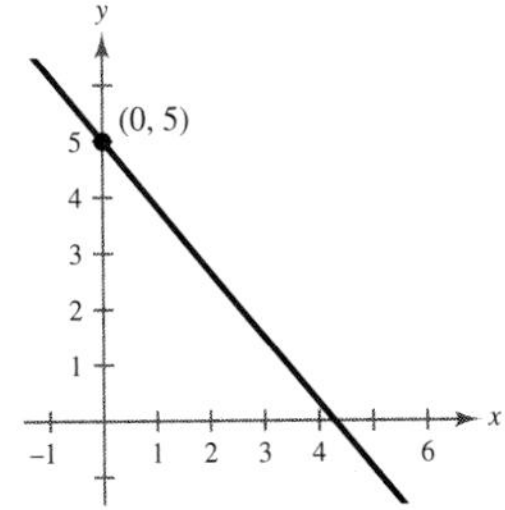

41. $3x + 5y - 10 = 0$

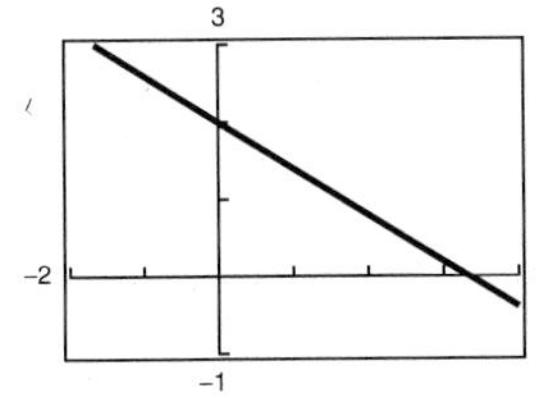

43. $x + 2y - 3 = 0$

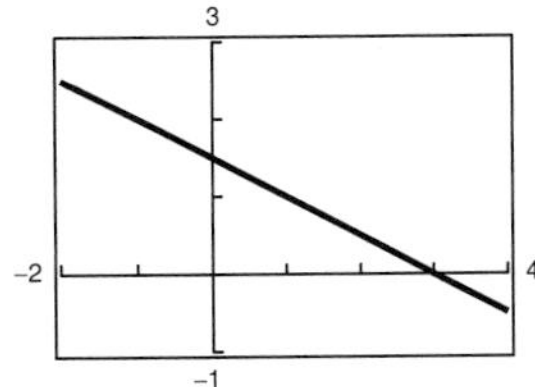

45. $x + 8 = 0$

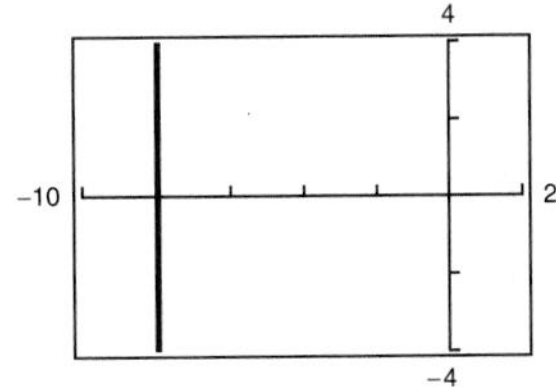

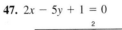

47. $2x - 5y + 1 = 0$

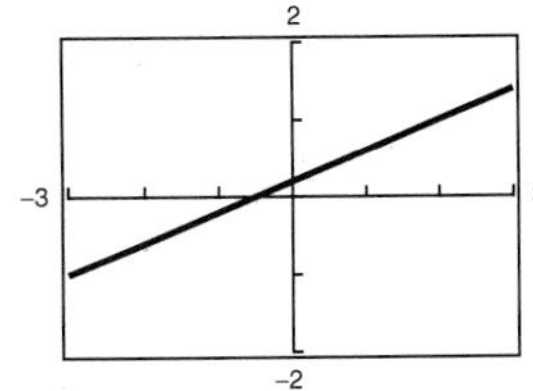

49. $16{,}666\frac{2}{3}$ feet ≈ 3.16 miles

51. $3x - y - 2 = 0$

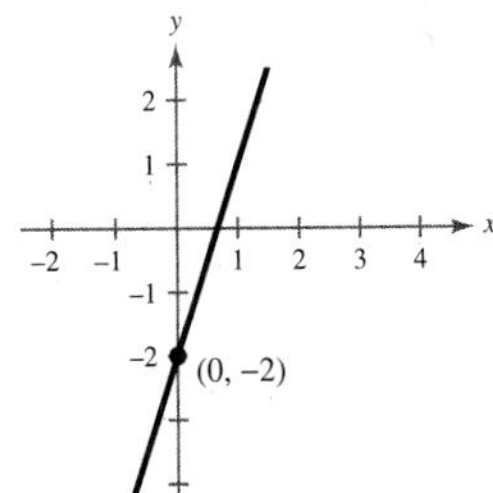

53. $2x + y = 0$

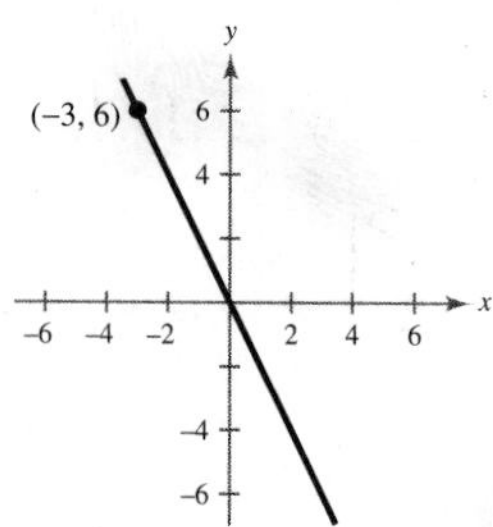

55. $x + 3y - 4 = 0$

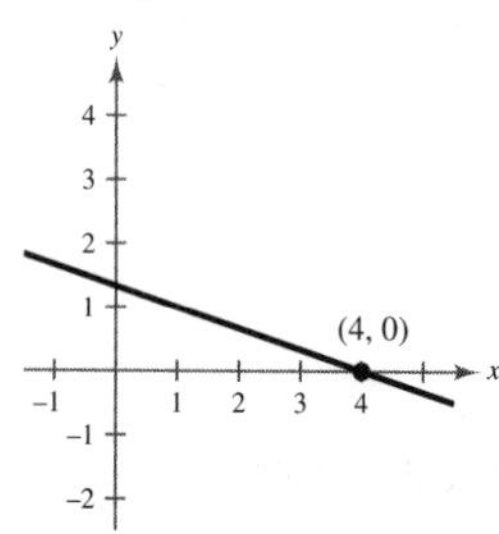

57. $x - 6 = 0$

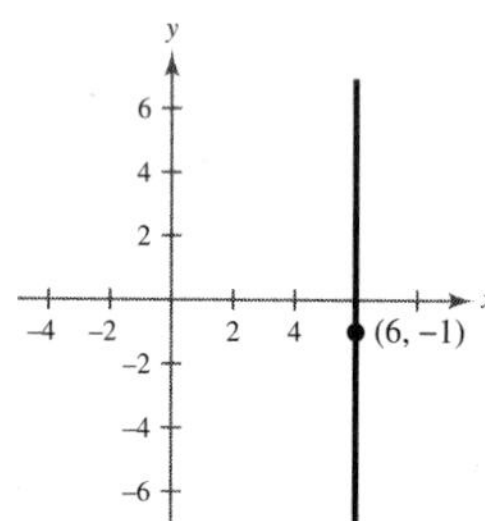

59. $8x - 6y - 17 = 0$

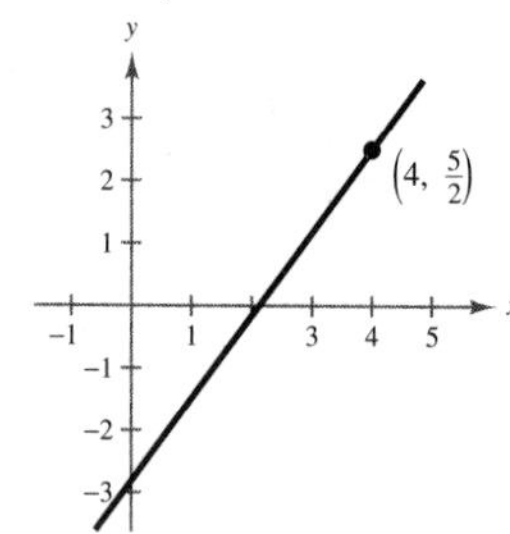

61.

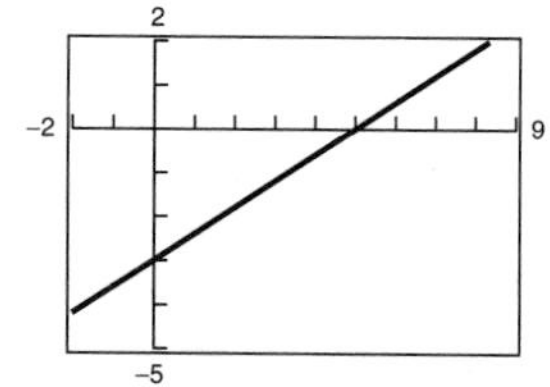

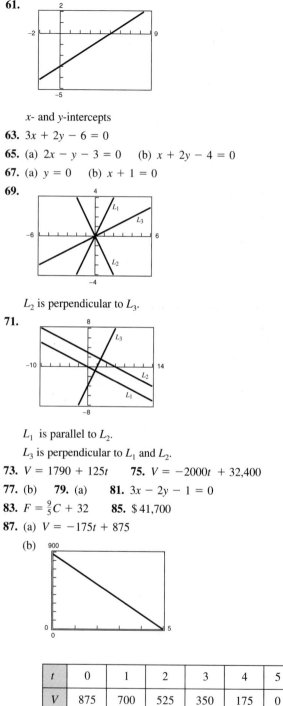

x- and y-intercepts

63. $3x + 2y - 6 = 0$

65. (a) $2x - y - 3 = 0$ (b) $x + 2y - 4 = 0$

67. (a) $y = 0$ (b) $x + 1 = 0$

69.

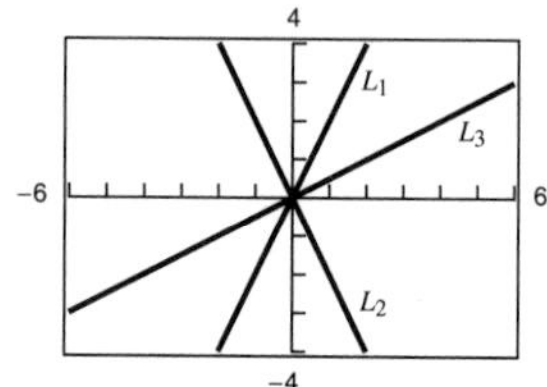

L_2 is perpendicular to L_3.

71.

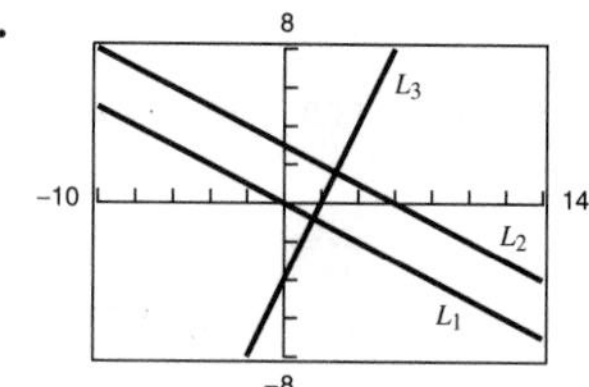

L_1 is parallel to L_2.

L_3 is perpendicular to L_1 and L_2.

73. $V = 1790 + 125t$ **75.** $V = -2000t + 32{,}400$

77. (b) **79.** (a) **81.** $3x - 2y - 1 = 0$

83. $F = \frac{9}{5}C + 32$ **85.** \$41,700

87. (a) $V = -175t + 875$

(b)

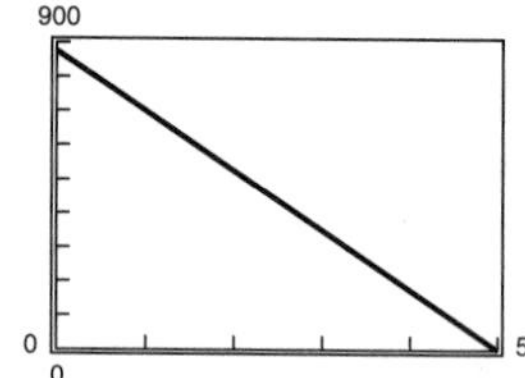

t	0	1	2	3	4	5
V	875	700	525	350	175	0

89. $S = 0.85L$

91. (a) $C = 16.75t + 36{,}500$ (b) $R = 27t$
(c) $P = 10.25t - 36{,}500$ (d) $t \approx 3561$ hours

93. (a) $y = 71.08t - 10.29$
(b)

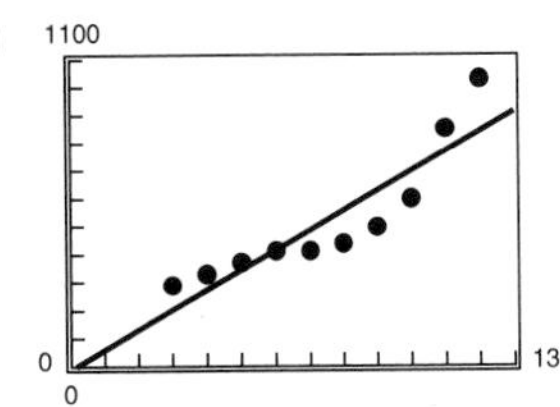

(c) \$1,411,000
(d) Average increase per year

95. (a) $x = 50 - \dfrac{p - 580}{15}$ (b) 45 (c) 49

Section 1.3 *(page 111)*

1. Yes **3.** No **5.** Yes **7.** No

9. (a) Function
(b) Not a function, because the element 1 in A corresponds to two elements, -2 and 1, in B
(c) Function

11. Each is a function. To each year there corresponds one and only one circulation.

13. Not a function **15.** Function **17.** Function

19. Not a function **21.** Function

23. (a) $\dfrac{1}{5}$ (b) 1 (c) $\dfrac{1}{4x + 1}$ (d) $\dfrac{1}{x + c + 1}$

25. (a) -1 (b) -9 (c) $2x - 5$

27. (a) 0 (b) -0.75 (c) $x^2 + 2x$

29. (a) 1 (b) 2.5 (c) $3 - 2|x|$

31. (a) $-\dfrac{1}{9}$ (b) Undefined (c) $\dfrac{1}{y^2 + 6y}$

33. (a) 1 (b) -1 (c) $\dfrac{|x - 1|}{x - 1}$

35. (a) -1 (b) 2 (c) 6

37.

x	-2	-1	0	1	2
$f(x)$	1	-2	-3	-2	1

39.

t	-5	-4	-3	-2	-1
$h(t)$	1	$\frac{1}{2}$	0	$\frac{1}{2}$	1

41.

x	-2	-1	0	1	2
$f(x)$	5	$\frac{9}{2}$	4	1	0

43. 5 **45.** ± 3 **47.** $2, -1$ **49.** 3, 0

51. All real numbers x **53.** All real numbers except $t = 0$

55. $y \ge 10$ **57.** $-1 \le x \le 1$

59. All real numbers except $x = 0, -2$

61. $(-2, 4), (-1, 1), (0, 0), (1, 1), (2, 4)$

63. $(-2, 0), (-1, 1), (0, \sqrt{2}), (1, \sqrt{3}), (2, 2)$

65. The domain is the set of inputs of the function, and the range is the set of outputs.

67. $g(x) = -2x^2$ **69.** $r(x) = \dfrac{32}{x}$ **71.** $3 + h$

73. $3x^2 + 3xc + c^2$ **75.** 3 **77.** $A = \dfrac{C^2}{4\pi}$

79. (a)

Height, x	*Length and Width*	*Volume, V*
1	$24 - 2(1)$	$1[24 - 2(1)]^2 = 484$
2	$24 - 2(2)$	$2[24 - 2(2)]^2 = 800$
3	$24 - 2(3)$	$3[24 - 2(3)]^2 = 972$
4	$24 - 2(4)$	$4[24 - 2(4)]^2 = 1024$
5	$24 - 2(5)$	$5[24 - 2(5)]^2 = 980$
6	$24 - 2(6)$	$6[24 - 2(6)]^2 = 864$

Maximum when $x = 4$

(b)

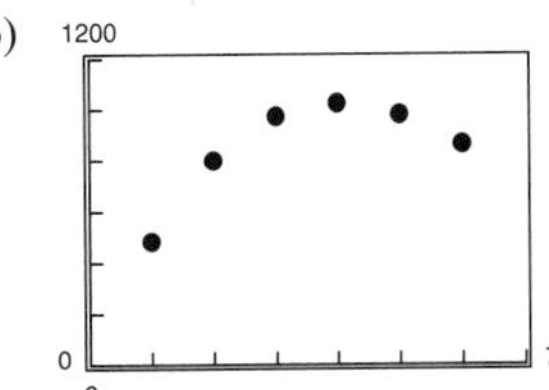

V is a function of x. $V = x(24 - 2x)^2, \quad 0 < x < 12$

81. $A = \dfrac{x^2}{2(x-2)}, \quad x > 2$

83. $V = x^2y = x^2(108 - 4x) = 108x^2 - 4x^3, \quad 0 < x < 27$

85. (a) $C = 12.30x + 98{,}000$ (b) $R = 17.98x$

(c) $P = 5.68x - 98{,}000$

87. (a) 28

(b) -17. Average decrease per year in the population.

(c)

t	1988	1989	1990	1991
N	9.0	19.8	43.9	53.6

t	1992	1993	1994	1995
N	30.9	16.7	10.2	6.9

Section 1.4 *(page 124)*

1. Domain: $(-\infty, -1], [1, \infty)$

Range: $[0, \infty)$

3. Domain: $[-4, 4]$

Range: $[0, 4]$

5.

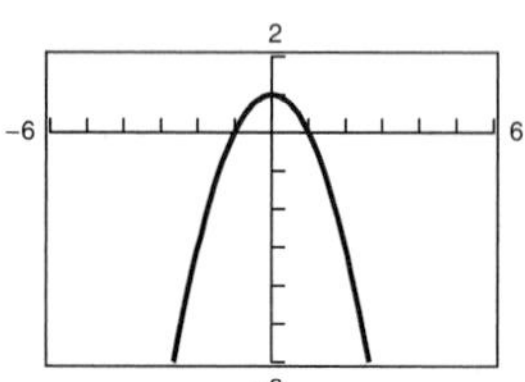

Domain: All real numbers

Range: $(-\infty, 1]$

7.

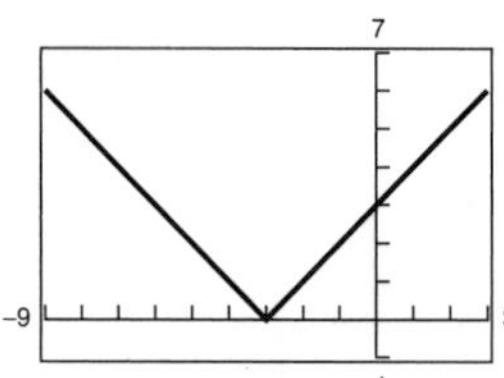

Domain: All real numbers

Range: $[0, \infty)$

9. Function. Graph the given function over the window shown in the figure.

11. Not a function. Solve for y and graph the resulting two functions.

13. Function. Solve for y and graph the resulting function.

15. Second setting **17.** First setting

19. (a) Increasing on $(-\infty, \infty)$ (b) Odd function

21. (a) Increasing on $(-\infty, 0), (2, \infty)$

Decreasing on $(0, 2)$

(b) Neither even nor odd

23. Yes. To each value of y there corresponds one and only one value of x.

25. (a)

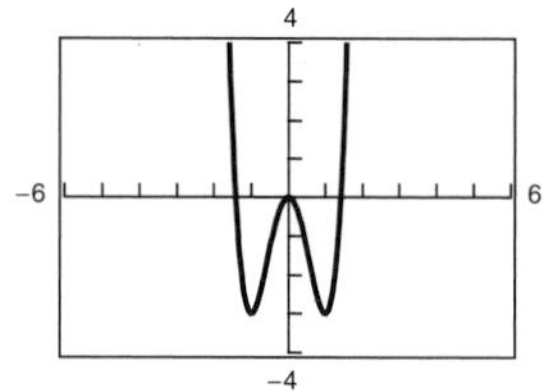

(b) Increasing on $(-1, 0), (1, \infty)$

Decreasing on $(-\infty, -1), (0, 1)$

(c) Even function

27. (a)

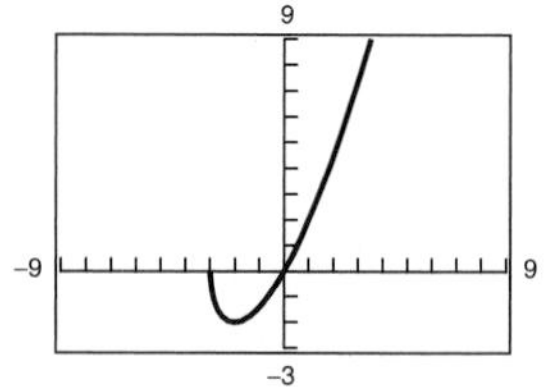

(b) Increasing on $(-2, \infty)$

Decreasing on $(-3, -2)$

(c) Neither even nor odd

29. Even function **31.** Odd function

33. Neither even nor odd **35.** (a) $\left(\frac{3}{2}, 4\right)$ (b) $\left(\frac{3}{2}, -4\right)$

37. Even function

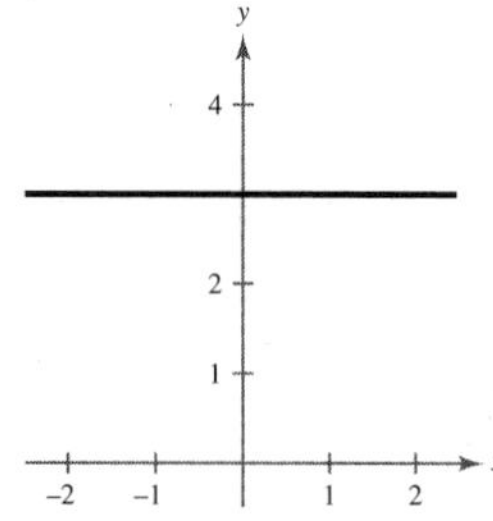

39. Neither even nor odd

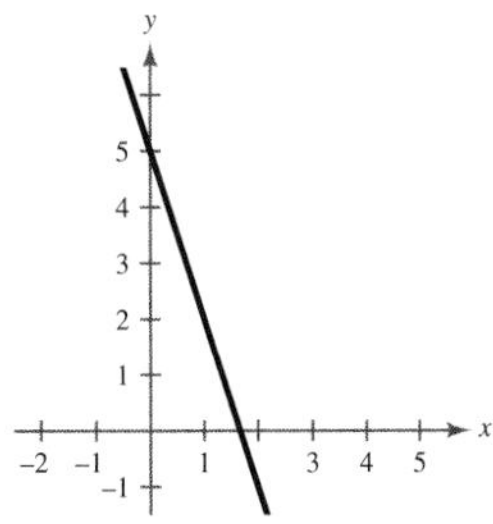

41. Neither even nor odd

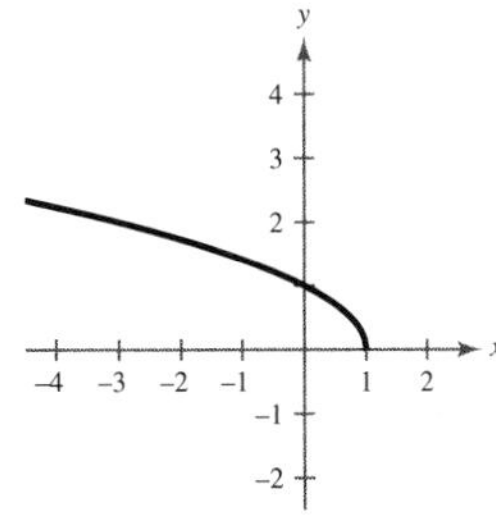

43. Neither even nor odd

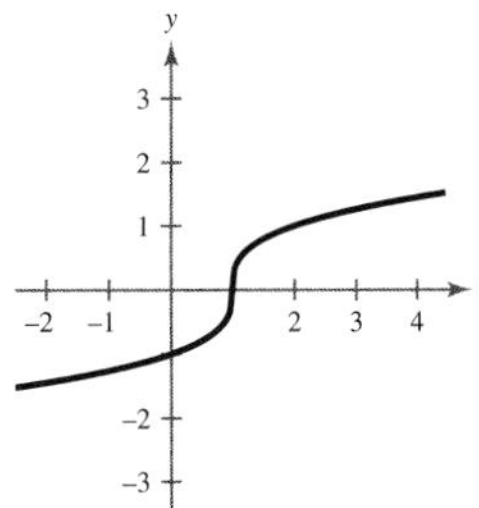

45. Neither even nor odd

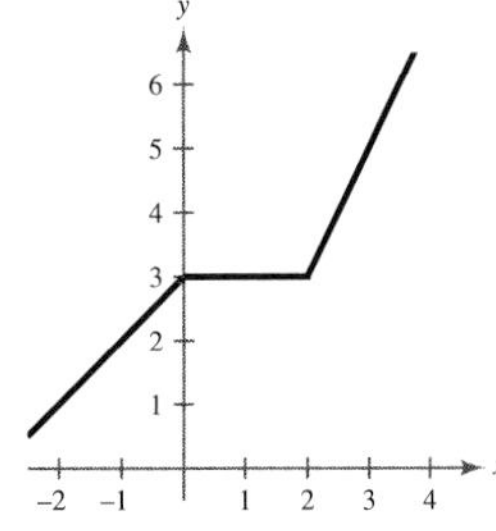

47.

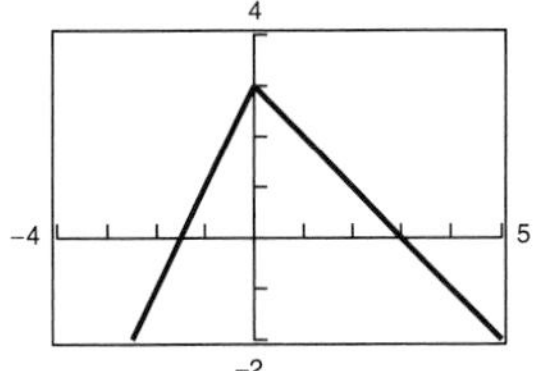

49.

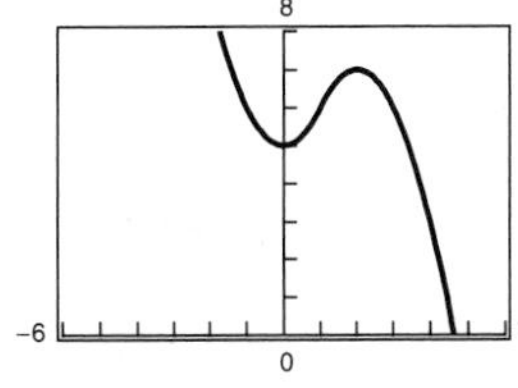

51.

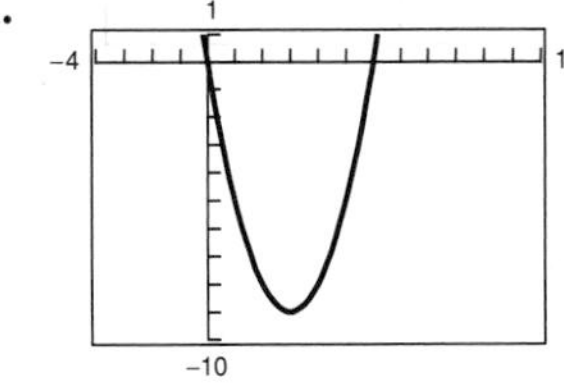

Relative minimum: $(3, -9)$

53.

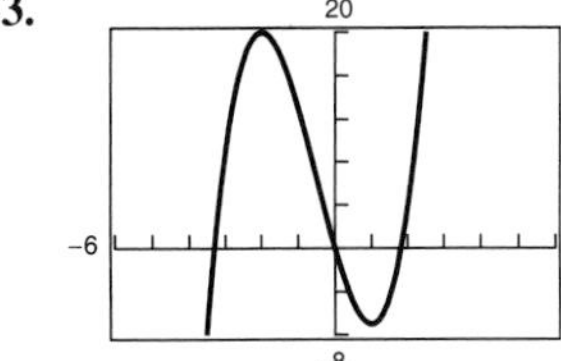

Relative minimum: $(1, -7)$

Relative maximum: $(-2, 20)$

55.

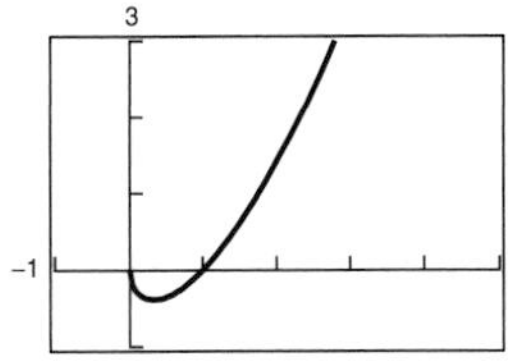

Minimum: $(0.33, -0.38)$

57. (a) Answers will vary.

(b)

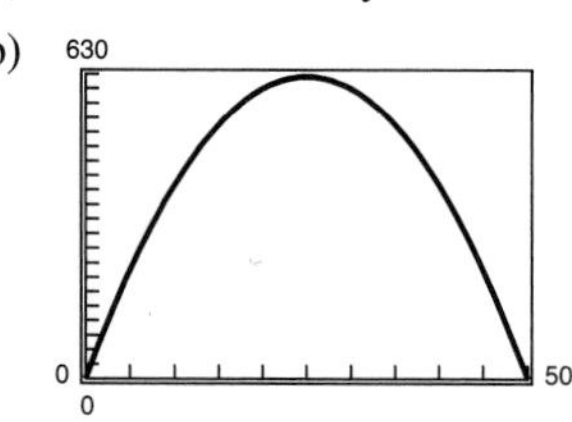

(c) 25×25 meters; 625 square meters

59.

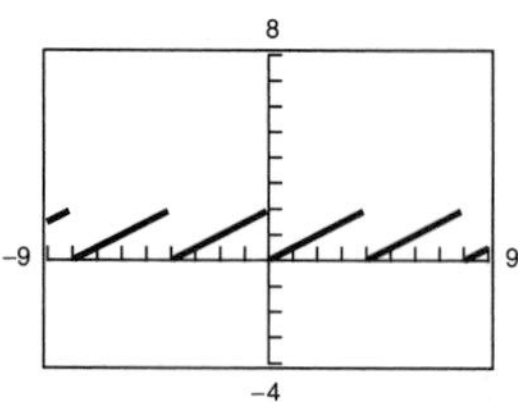

Domain: $(-\infty, \infty)$

Range: $[0, 2)$

Sawtooth pattern

61. (a) C_2 is the appropriate model because the cost does not increase until after the next minute of conversation has started.

(b) $7.85

63. $(-\infty, 4]$

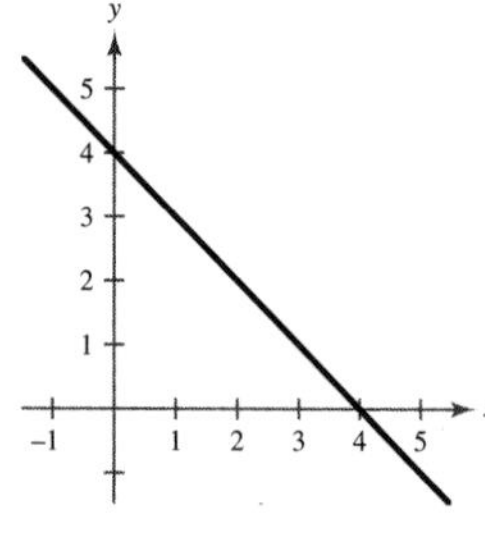

65. $(-\infty, -3]$, $[3, \infty)$

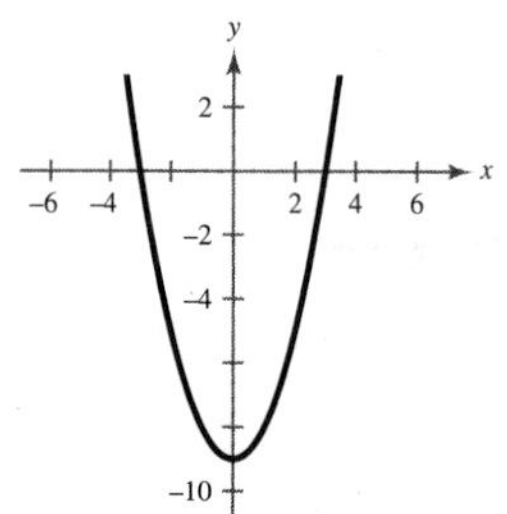

67. $[-1, 1]$

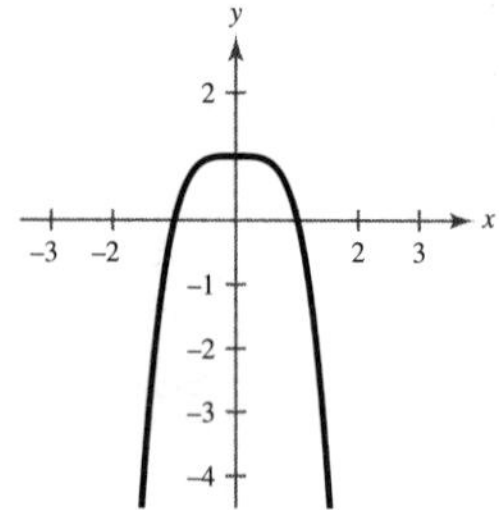

69. $(-\infty, \infty)$

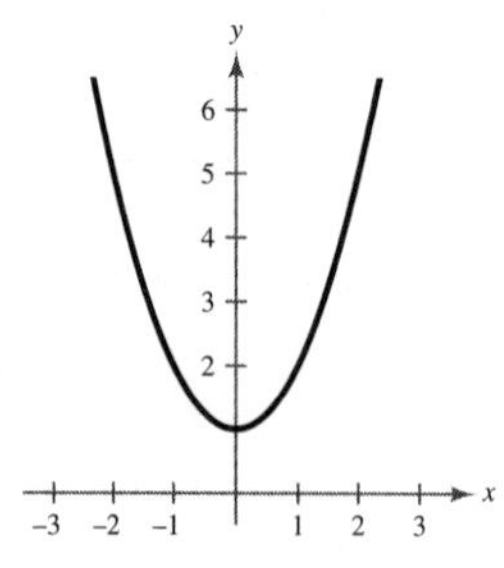

71. $h = -x^2 + 4x - 3$ **73.** $h = 2x - x^2$

75. $L = \frac{1}{2}y^2$

77. (a) $y = -87.49 + 16.28t - 4.82t^2 - 1.17t^3$

Domain: $-4 \le t \le 3$

(b)

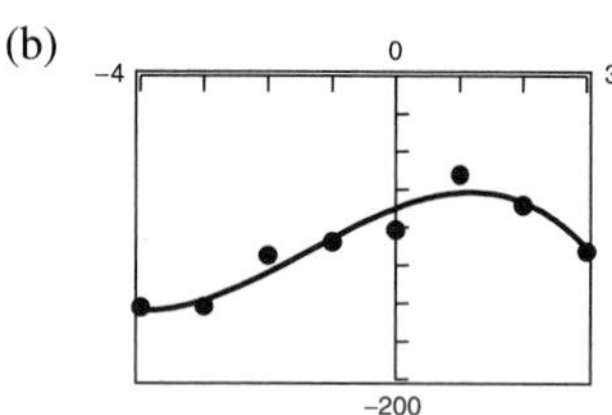

(c) 1992, 1990

(d) Because the balance would continue to decrease

Section 1.5 *(page 135)*

1.

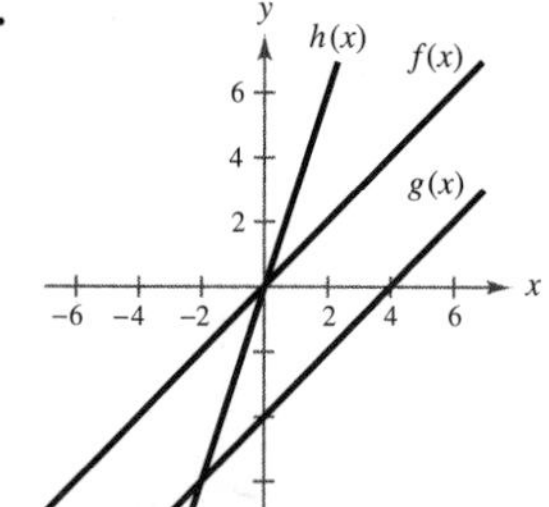

3.

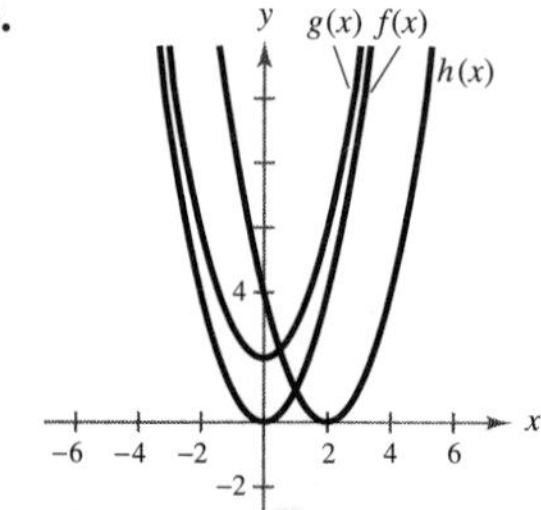

5.

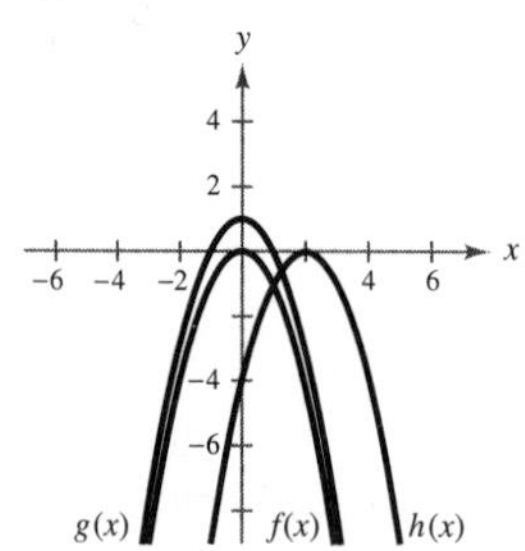

7.

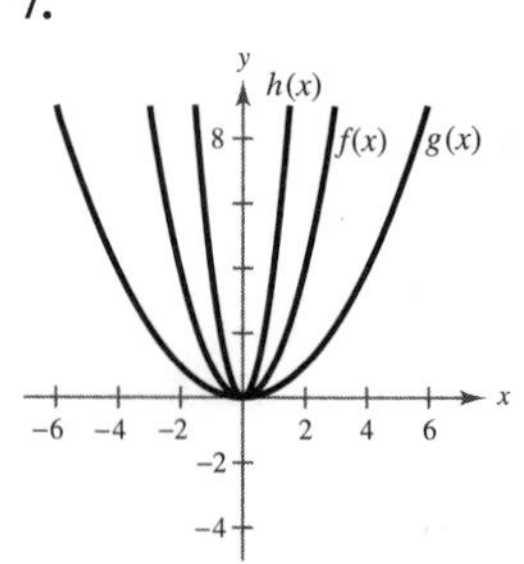

9.

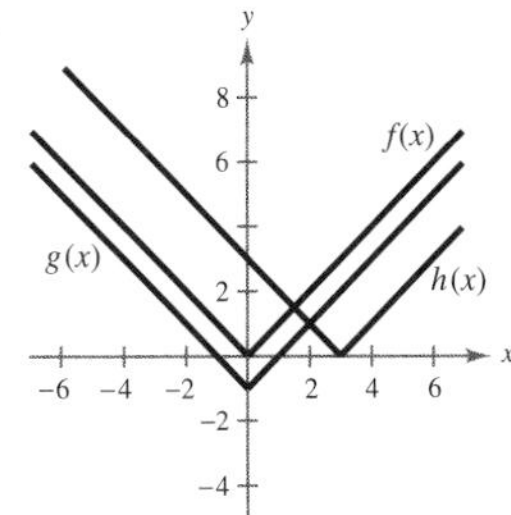

11. (a)

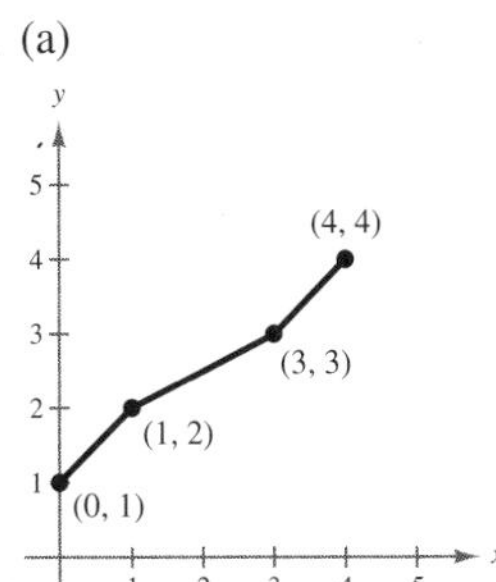

(b)

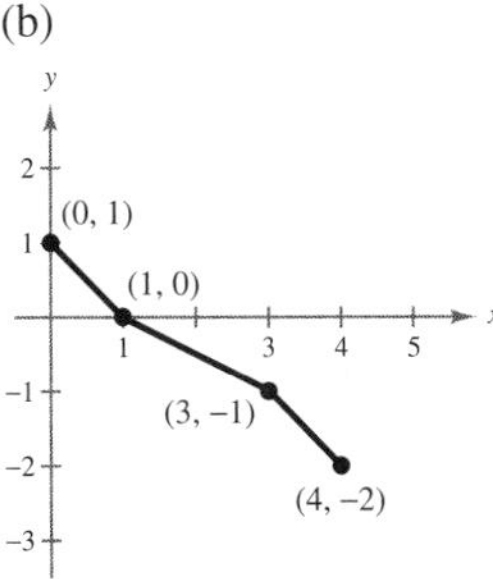

(c)

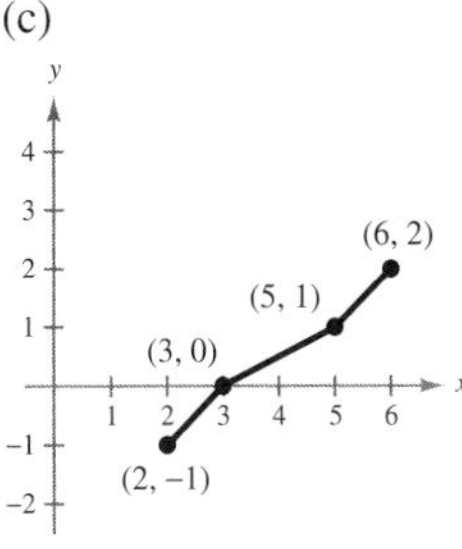

(d)

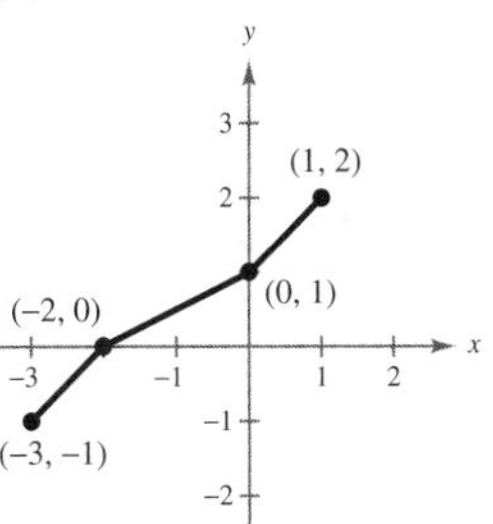

(e)

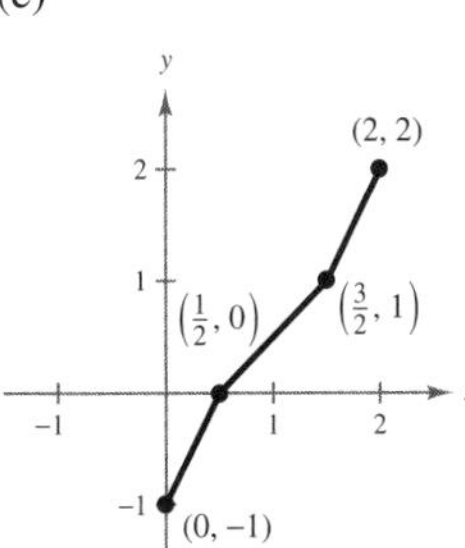

(f)

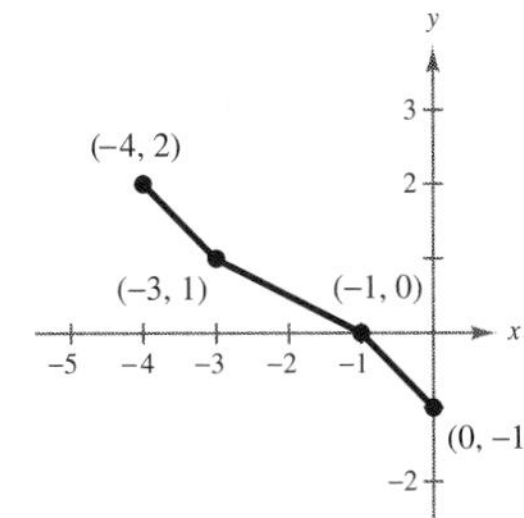

13. Horizontal shift of $y = x^3$
$y = (x - 2)^3$

15. Reflection in the x-axis of $y = x^2$
$y = -x^2$

17. Reflection in the x-axis and a vertical shift of $y = \sqrt{x}$
$y = 1 - \sqrt{x}$

19. Vertical shift of $y = x^2$
$y = x^2 - 1$

21. Reflection in the x-axis of $y = x^3$ followed by a vertical shift
$y = 1 - x^3$

23. Vertical shift of $y = x$
$y = x + 3$

25. Vertical shift **27.** Horizontal shift

29. Vertical stretch **31.** Horizontal shift

33. Reflection in the x-axis **35.** Vertical shrink

37.

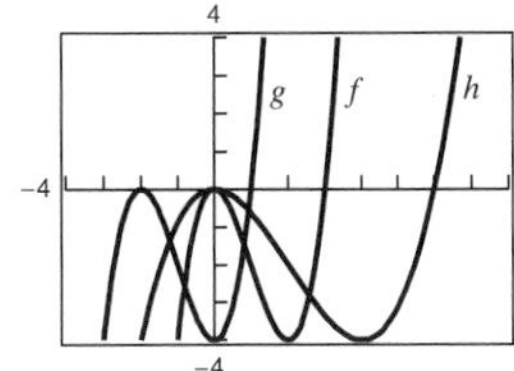

g is a horizontal shift and h is a horizontal stretch.

39.

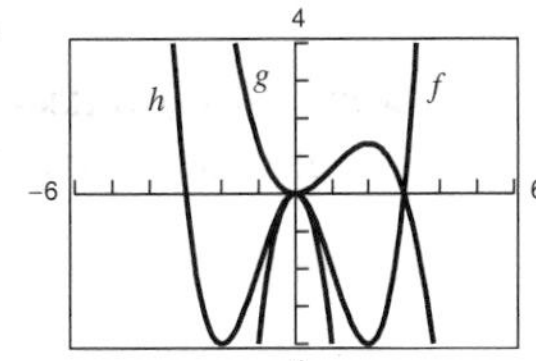

g is a vertical shrink and a reflection in the x-axis and h is a reflection in the y-axis.

41. Reflection in the x-axis followed by a vertical shift

43. A horizontal shift followed by a vertical shrink

45. A horizontal stretch followed by a vertical shift

47. $y = -(x^3 - 3x^2) + 1$

49. (a)

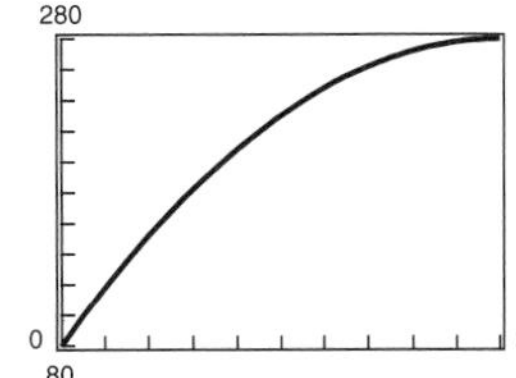

(b) $P(x) = -2420 + 20x - 0.5x^2$; vertical shift

(c) $P(x) = 80 + \frac{1}{5}x - \frac{x^2}{20{,}000}$; horizontal stretch

51. (a)

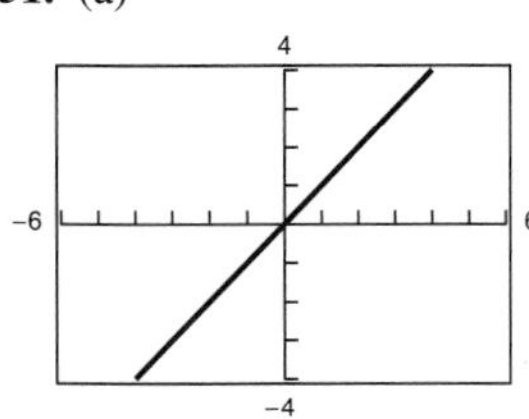

(b)

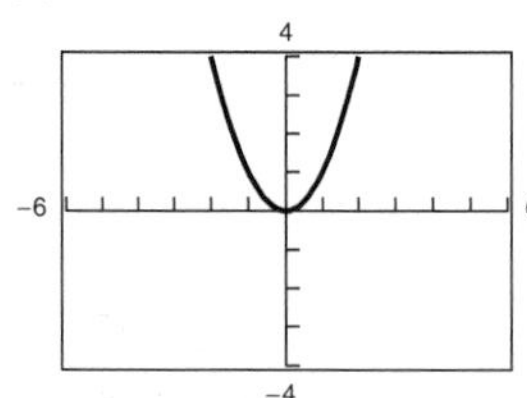

(c)

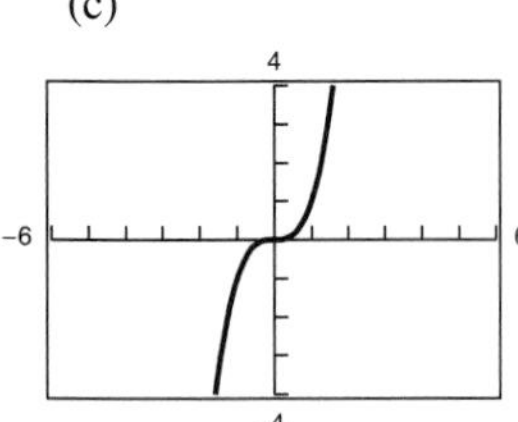

(d)

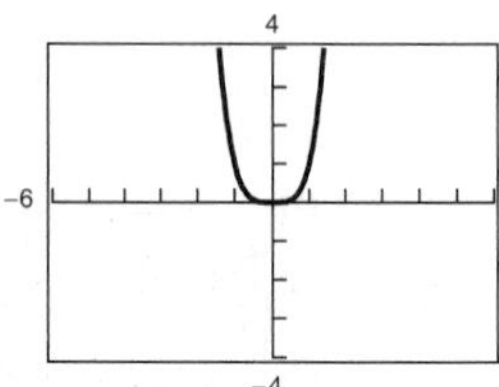

(e)

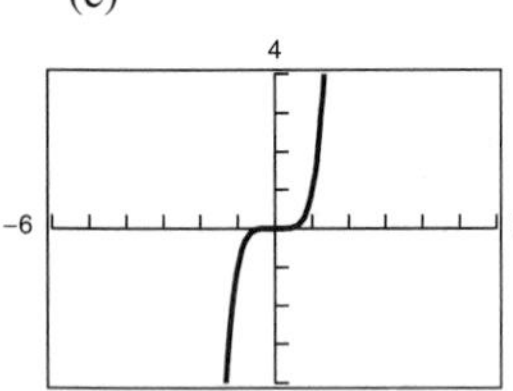

(f)

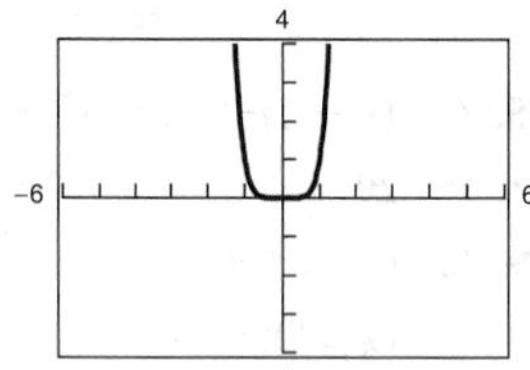

All the graphs pass through the origin. The graphs of the odd powers of x are symmetric to the origin and the graphs of the even powers are symmetric to the y-axis. As the powers increase, the graphs become flatter in the interval $-1 < x < 1$.

53.

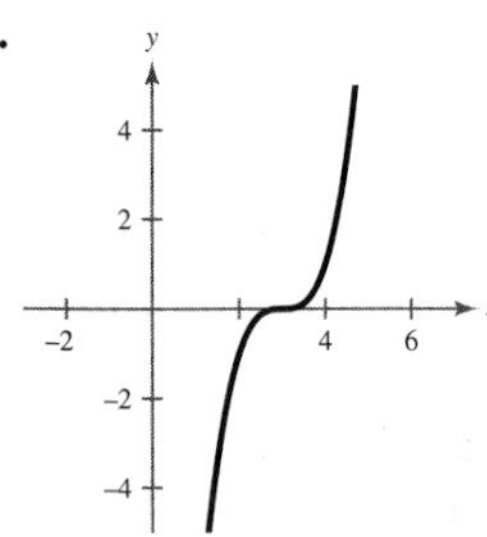

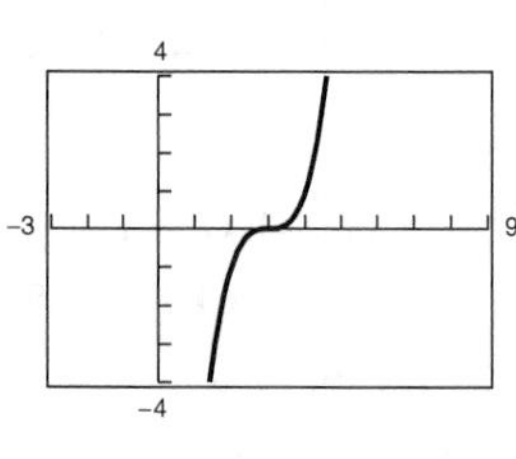

55.

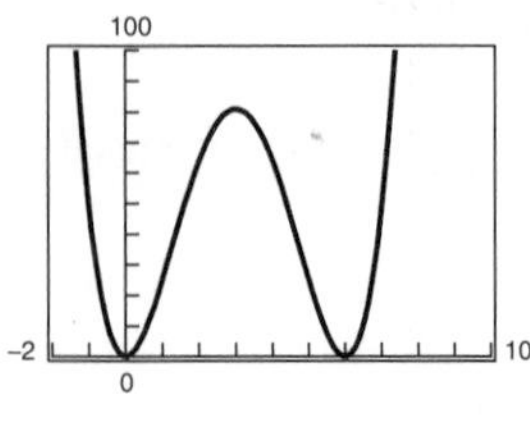

57.

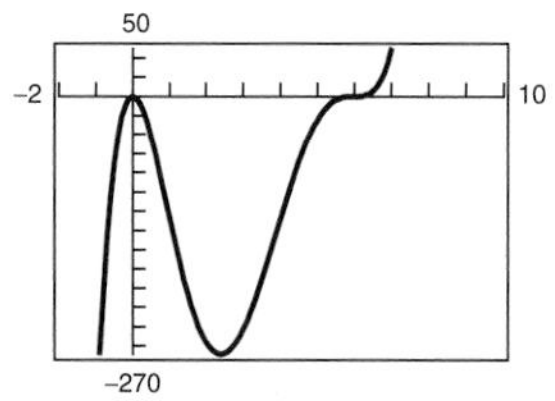

59. (a) To each time t there corresponds one and only one temperature T.

(b) 60°, 72°

(c) All the temperature changes would be 1 hour later.

(d) The temperature would be decreased by 1 degree.

61.

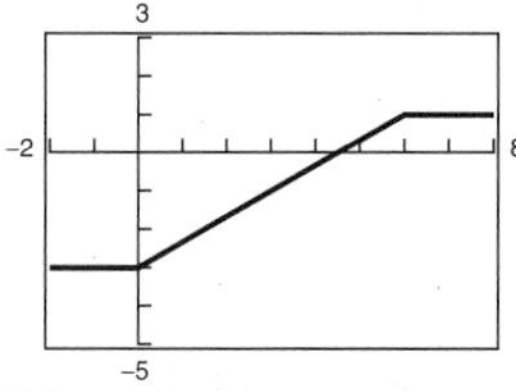

63.

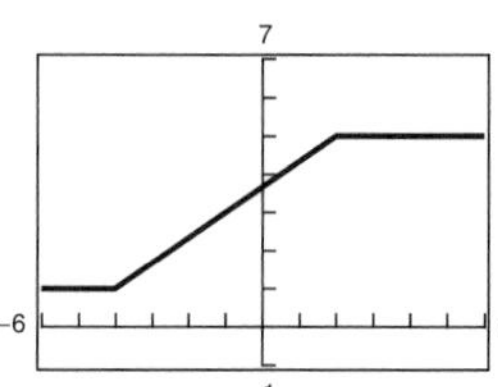

65.

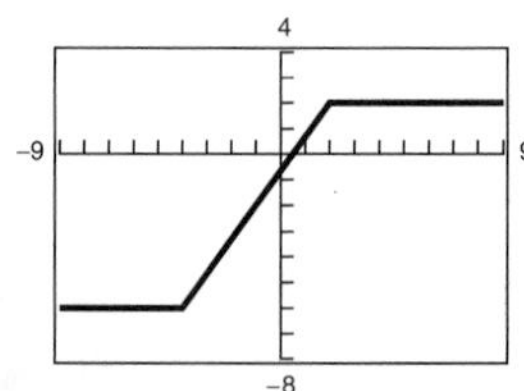

Section 1.6 *(page 145)*

1.

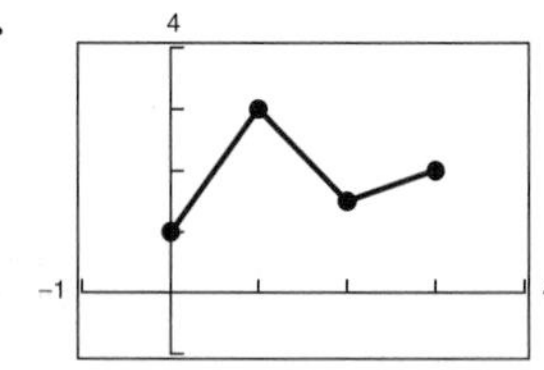

3.

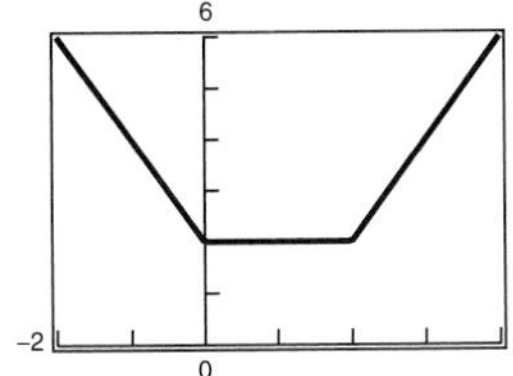

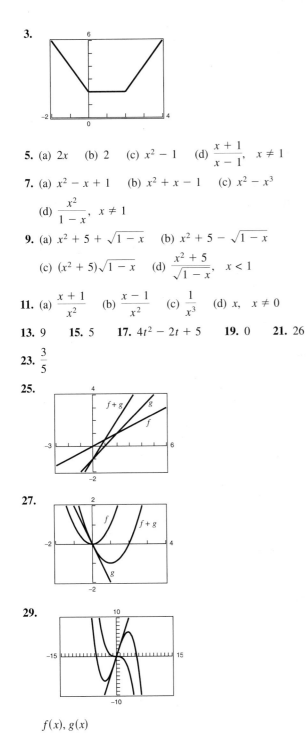

5. (a) $2x$ (b) 2 (c) $x^2 - 1$ (d) $\dfrac{x+1}{x-1}$, $x \neq 1$

7. (a) $x^2 - x + 1$ (b) $x^2 + x - 1$ (c) $x^2 - x^3$

(d) $\dfrac{x^2}{1-x}$, $x \neq 1$

9. (a) $x^2 + 5 + \sqrt{1-x}$ (b) $x^2 + 5 - \sqrt{1-x}$

(c) $(x^2 + 5)\sqrt{1-x}$ (d) $\dfrac{x^2+5}{\sqrt{1-x}}$, $x < 1$

11. (a) $\dfrac{x+1}{x^2}$ (b) $\dfrac{x-1}{x^2}$ (c) $\dfrac{1}{x^3}$ (d) x, $x \neq 0$

13. 9 **15.** 5 **17.** $4t^2 - 2t + 5$ **19.** 0 **21.** 26

23. $\dfrac{3}{5}$

25.

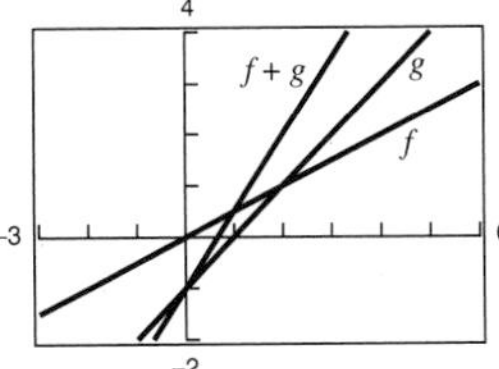

27.

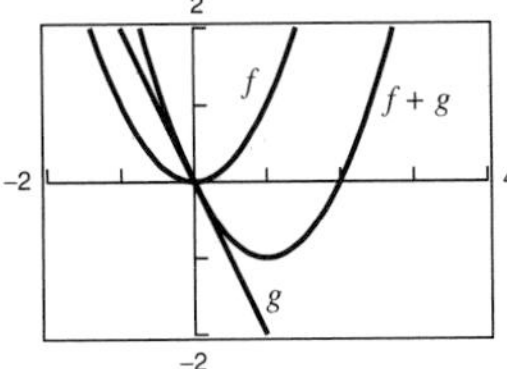

29.

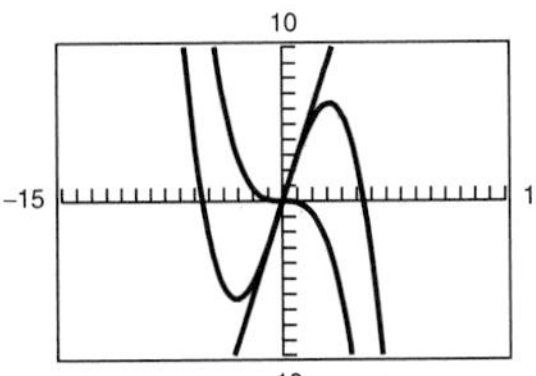

$f(x)$, $g(x)$

31. (a) $T = \frac{3}{4}x + \frac{1}{15}x^2$

(b)

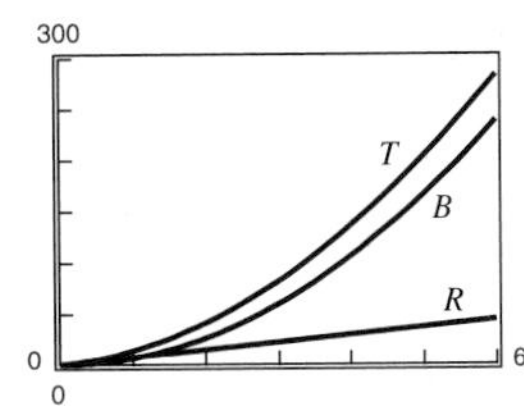

(c) B

33.

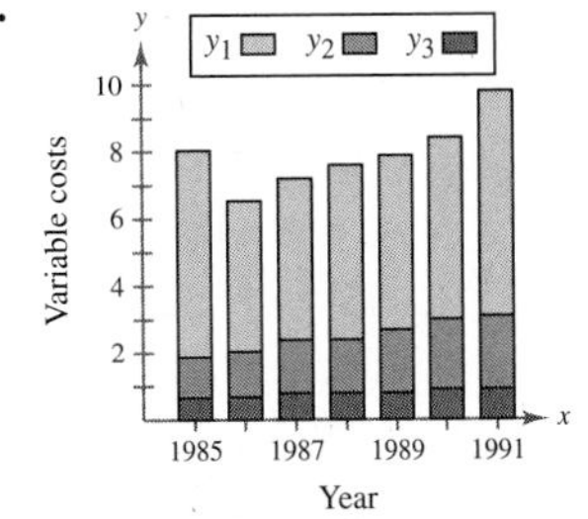

35. (a) $(x-1)^2$ (b) $x^2 - 1$ (c) x^4

37. (a) $20 - 3x$ (b) $-3x$ (c) $9x + 20$

39. (a) $(f \circ g)(x) = \sqrt{x^2 + 4}$

$(g \circ f)(x) = x + 4$, $x \geq -4$

Not equal

(b)

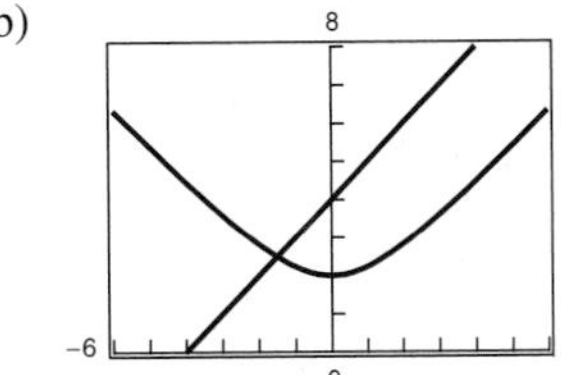

41. (a) $(f \circ g)(x) = x - \frac{8}{3}$

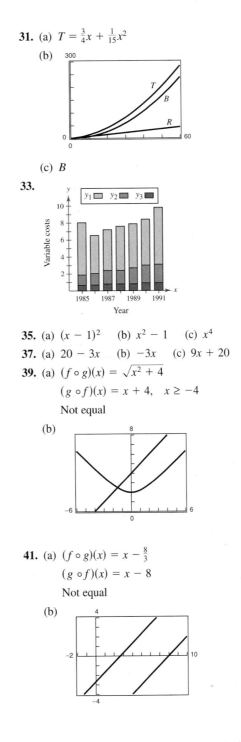

$(g \circ f)(x) = x - 8$

Not equal

(b)

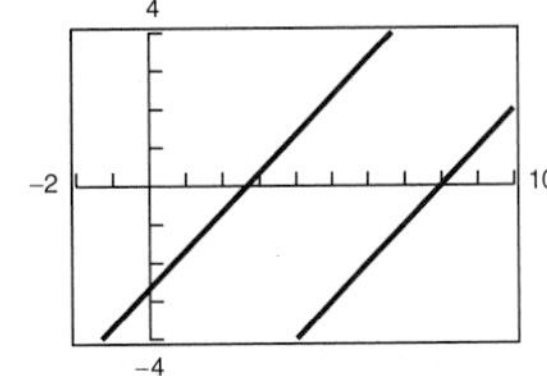

43. (a) $(f \circ g)(x) = x^4$
$(g \circ f)(x) = x^4$
Equal

(b)

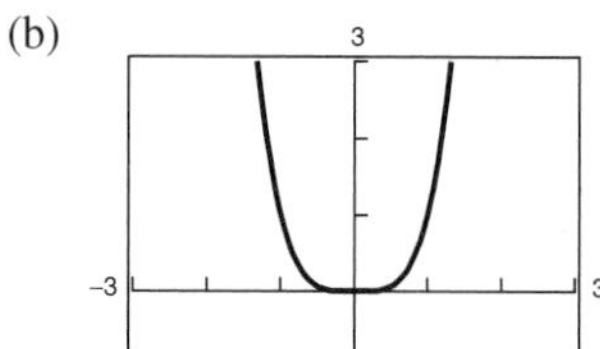

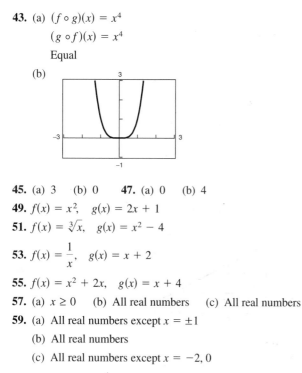

45. (a) 3 (b) 0 **47.** (a) 0 (b) 4

49. $f(x) = x^2$, $g(x) = 2x + 1$

51. $f(x) = \sqrt[3]{x}$, $g(x) = x^2 - 4$

53. $f(x) = \dfrac{1}{x}$, $g(x) = x + 2$

55. $f(x) = x^2 + 2x$, $g(x) = x + 4$

57. (a) $x \geq 0$ (b) All real numbers (c) All real numbers

59. (a) All real numbers except $x = \pm 1$
(b) All real numbers
(c) All real numbers except $x = -2, 0$

61. 3 **63.** $\dfrac{-4}{x(x + h)}$

65. $(A \circ r)(t) = 0.36\pi t^2$
$A \circ r$ represents the area of the circle at time t.

67. (a) $(C \circ x)(t) = 3000t + 750$
$C \circ x$ represents the cost after t production hours.

(b)

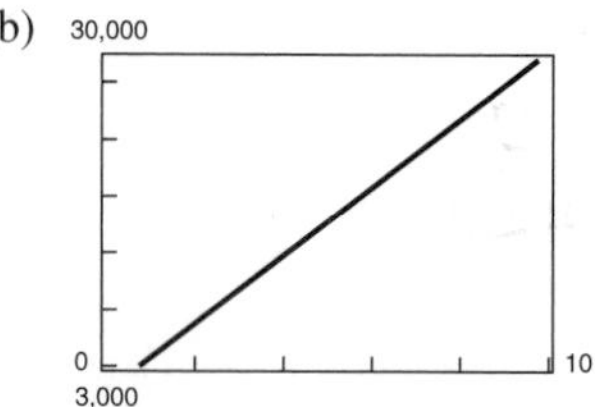

4.75 hours

69. $g(f(x))$ represents 3 percent of an amount over \$500,000.

71. Answers will vary. **73.** Answers will vary.

75. (a) $f(x) = (x^2 + 1) + (-2x)$

(b) $f(x) = \dfrac{-1}{(x + 1)(x - 1)} + \dfrac{x}{(x + 1)(x - 1)}$

77.

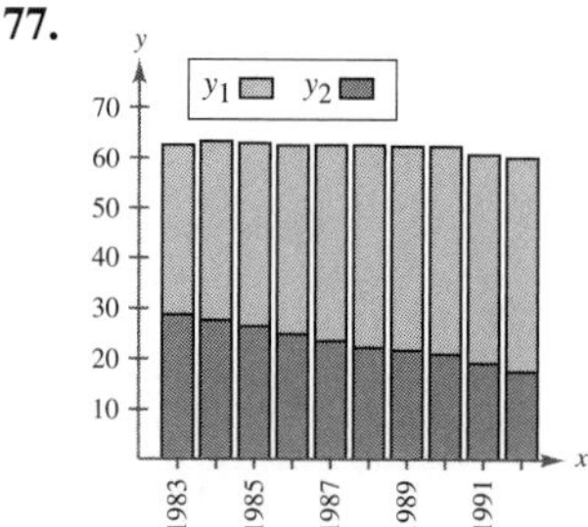

79.

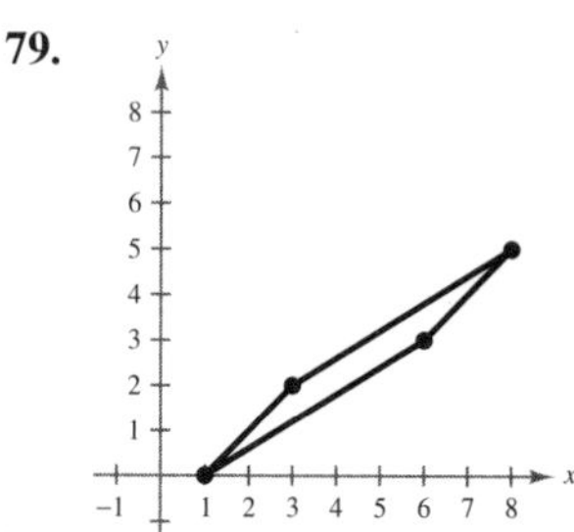

81. Quadrants I and III

Section 1.7 *(page 158)*

1. (c) **3.** (a) **5.** $f^{-1}(x) = \frac{1}{8}x$ **7.** $f^{-1}(x) = x - 10$

9. $f^{-1}(x) = x^3$

11. (a) $f(g(x)) = f\left(\dfrac{x}{2}\right) = 2\left(\dfrac{x}{2}\right) = x$

$g(f(x)) = g(2x) = \dfrac{(2x)}{2} = x$

(b)

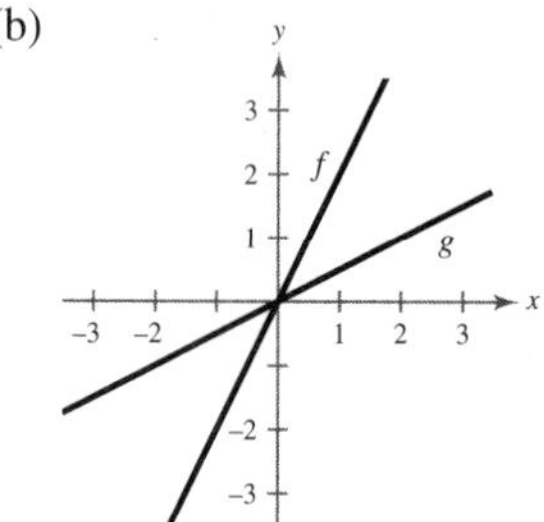

13. (a) $f(g(x)) = f\left(\frac{x-1}{5}\right) = 5\left(\frac{x-1}{5}\right) + 1 = x$

$g(f(x)) = g(5x+1) = \frac{(5x+1)-1}{5} = x$

(b)

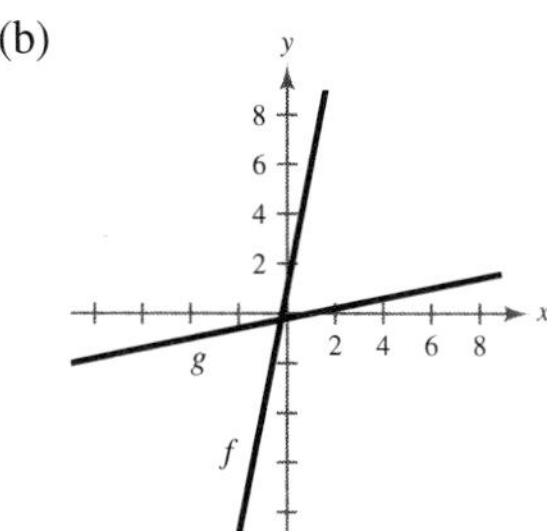

15. $f(g(x)) = f(\sqrt[3]{x}) = (\sqrt[3]{x})^3 = x$

$g(f(x)) = g(x^3) = \sqrt[3]{x^3} = x$

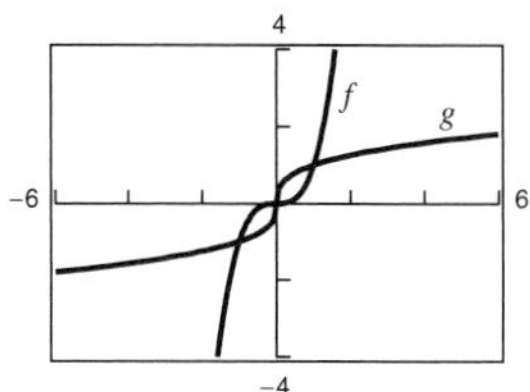

Reflections in the line $y = x$

17. $f(g(x)) = f(x^2+4), \quad x \ge 0$

$= \sqrt{(x^2+4) - 4} = x$

$g(f(x)) = g(\sqrt{x-4})$

$= (\sqrt{x-4})^2 + 4 = x$

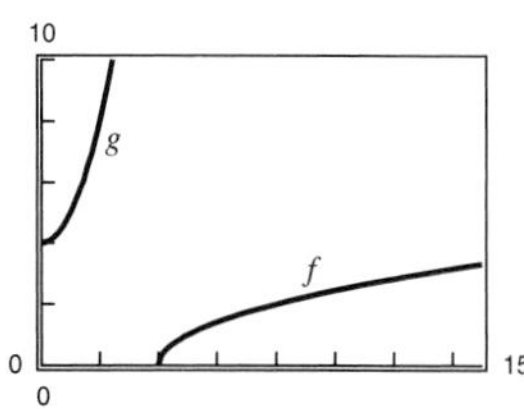

Reflections in the line $y = x$

19. $f(g(x)) = f(\sqrt[3]{1-x})$

$= 1 - (\sqrt[3]{1-x})^3 = x$

$g(f(x)) = g(1-x^3)$

$= \sqrt[3]{1-(1-x^3)} = x$

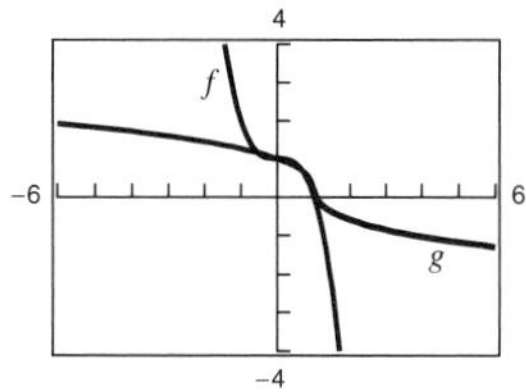

Reflections in the line $y = x$

21.

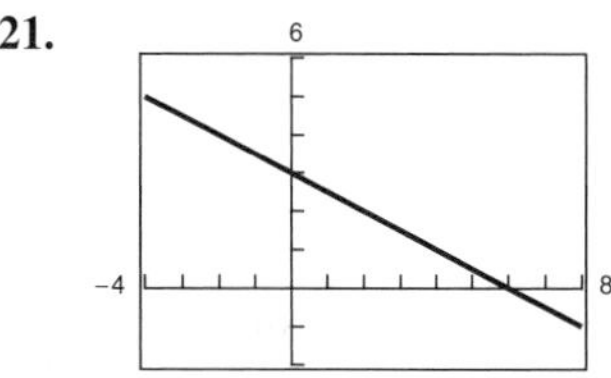

Yes

23.

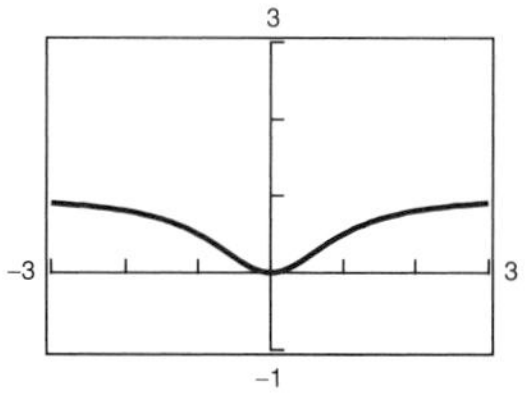

No

25.

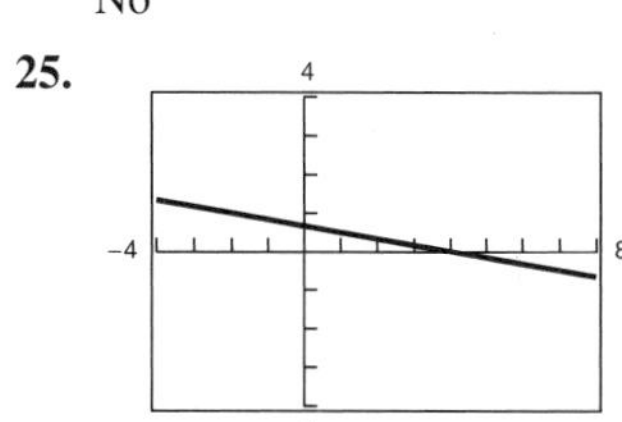

Yes

27.

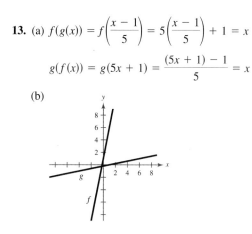

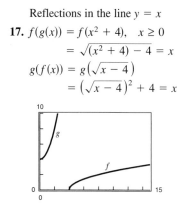

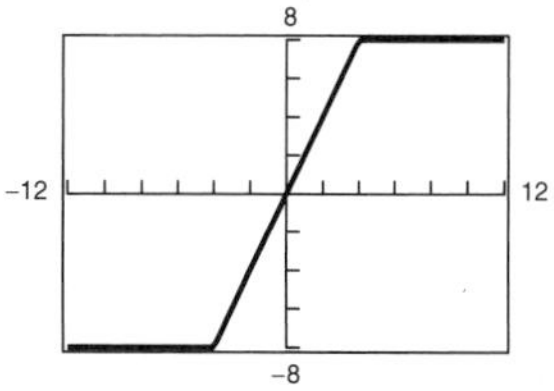

No

29.

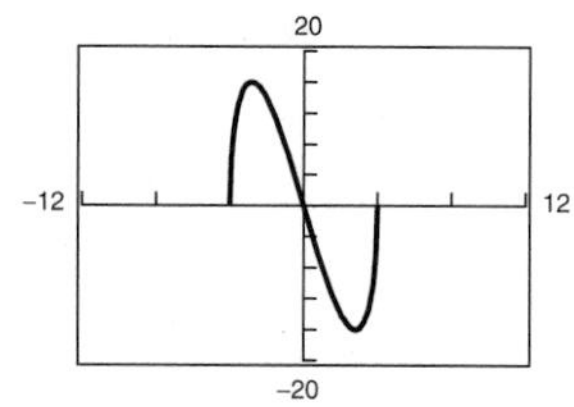

No

31. $f^{-1}(x) = \dfrac{x+3}{2}$

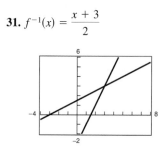

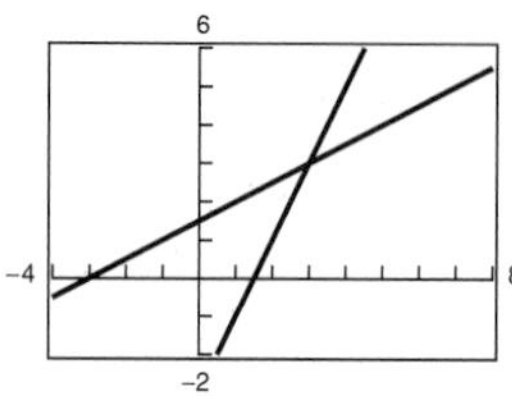

Reflections in the line $y = x$

33. $f^{-1}(x) = \sqrt[5]{x}$

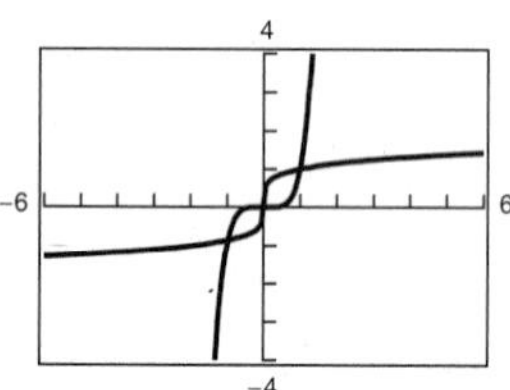

Reflections in the line $y = x$

35. $f^{-1}(x) = x^2, \quad x \ge 0$

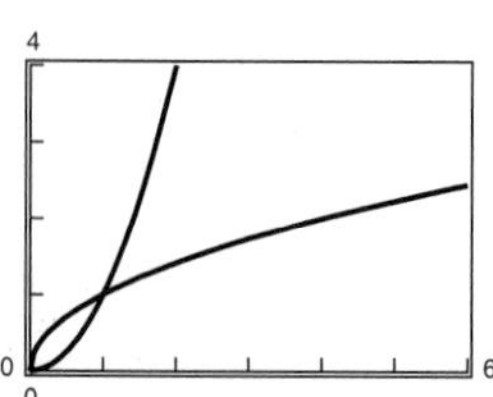

Reflections in the line $y = x$

37. $f^{-1}(x) = \sqrt{4 - x^2}, \quad 0 \le x \le 2$

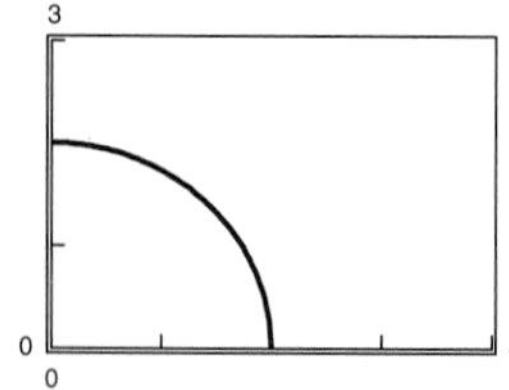

Reflections in the line $y = x$

39. $f^{-1}(x) = x^3 + 1$

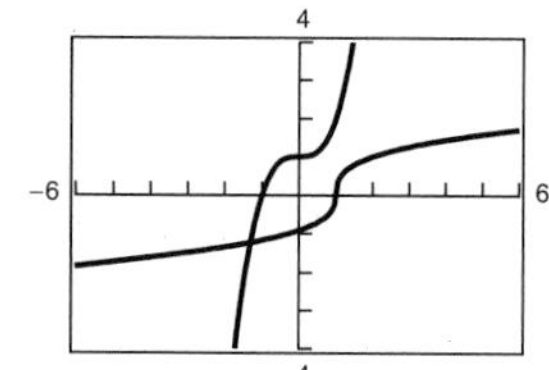

Reflections in the line $y = x$

41. No inverse **43.** $g^{-1}(x) = 8x$ **45.** No inverse

47. $f^{-1}(x) = \sqrt{x} - 3, \quad x \ge 0$ **49.** No inverse

51. $f^{-1}(x) = \dfrac{x^2 - 3}{2}, \quad x \ge 0$ **53.** No inverse

55. $f^{-1}(x) = -\sqrt{25 - x}, \quad x \le 25$

57. $y = \sqrt{x} + 2, \quad x \ge 0$ **59.** $y = x - 2, \quad x \ge 0$

61.

x	-4	-2	2	3
$f^{-1}(x)$	-2	-1	1	3

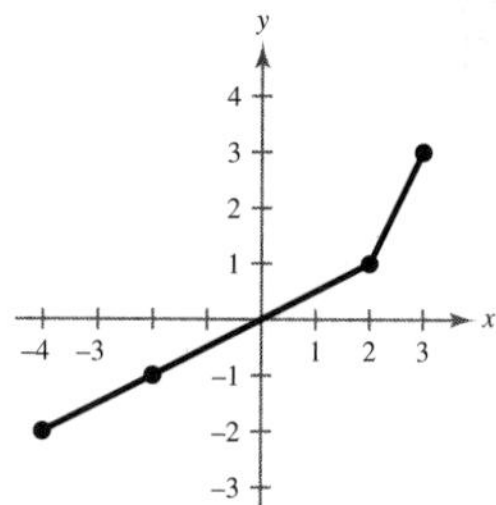

63. (a) and (b)

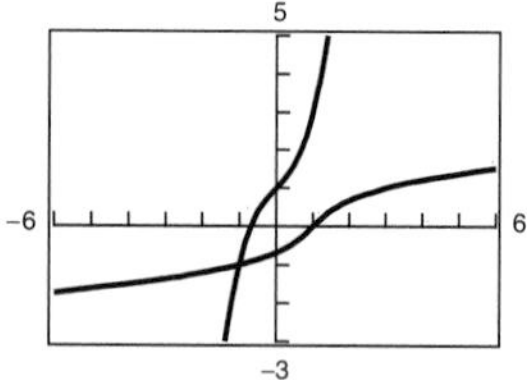

(c) Inverse function because it satisfies the Vertical Line Test

65. (a) and (b)

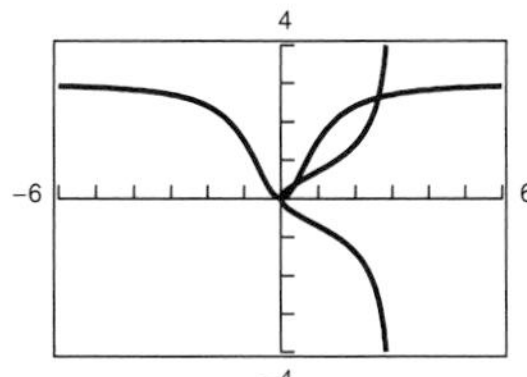

(c) Not an inverse function because it does not satisfy the Vertical Line Test

67. False, $f(x) = x^2$ **69.** True **71.** 32 **73.** 600

75. $2\sqrt[3]{x+3}$ **77.** $\dfrac{x+1}{2}$ **79.** $\dfrac{x+1}{2}$

81. Answers will vary.

83. (a) $y = \dfrac{x-8}{0.75}$

y = number of units produced

x = hourly wage

(b)

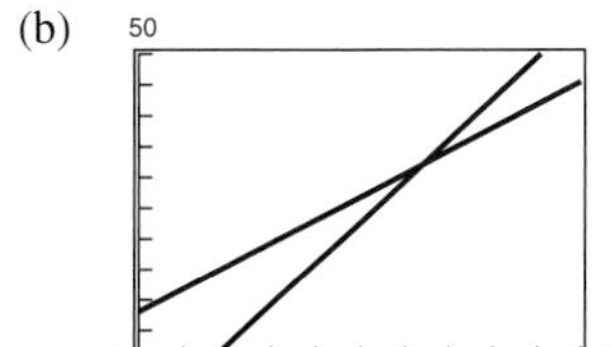

(c) $15.50

(d) 19

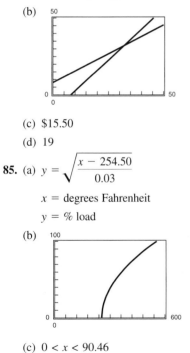

85. (a) $y = \sqrt{\dfrac{x - 254.50}{0.03}}$

x = degrees Fahrenheit

y = % load

(b)

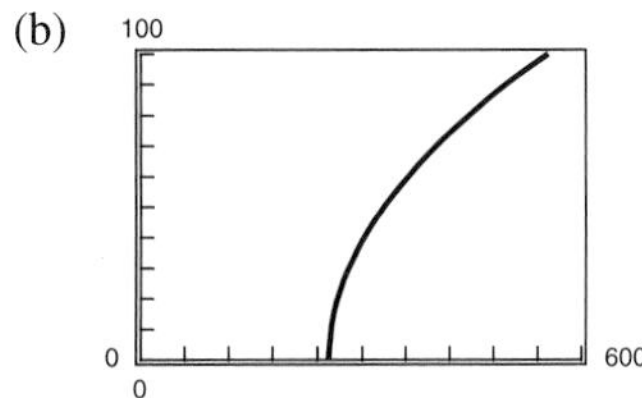

(c) $0 < x < 90.46$

FOCUS ON CONCEPTS *(page 161)*

1. No. The slope cannot be determined without knowing the scale on the y-axis. The slopes could be the same.

2. −4. The slope with the greatest magnitude corresponds to the steepest line.

3. V-intercept measures initial cost; Slope measures annual depreciation

4. No. The element 3 in the domain corresponds to two elements in the range.

5. (a)

Xmin = -15
Xmax = 6
Xscl = 3
Ymin = -18
Ymax = 6
Yscl = 3

(b)

Xmin = -24
Xmax = 36
Xscl = 6
Ymin = -54
Ymax = 12
Yscl = 6

6. (a) Even function. The graph is a reflection in the x-axis.

(b) Even function. The graph is a reflection in the y-axis.

(c) Even function. The graph is a vertical translation of f.

(d) Neither even nor odd. The graph is a horizontal translation of f.

7. (a) $g(t) = \frac{3}{4}f(t)$ (b) $g(t) = f(t) + 10{,}000$

(c) $g(t) = f(t-2)$

Review Exercises *(page 162)*

1.

x	-2	0	2	3	4
y	3	2	1	$\frac{1}{2}$	0

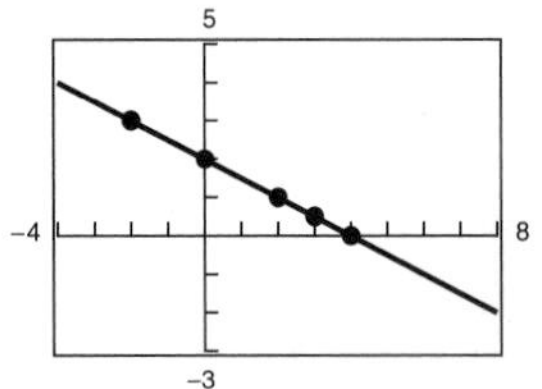

3.

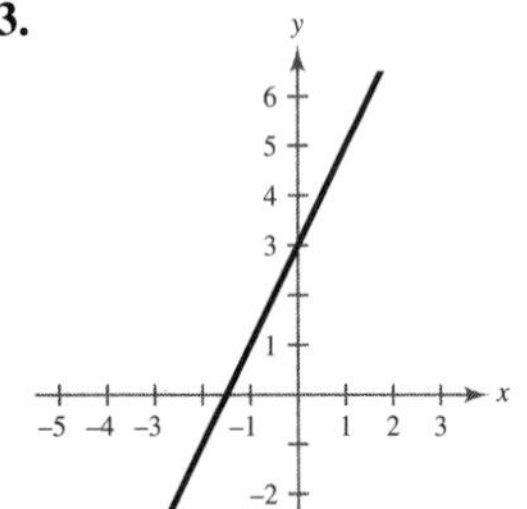

5.

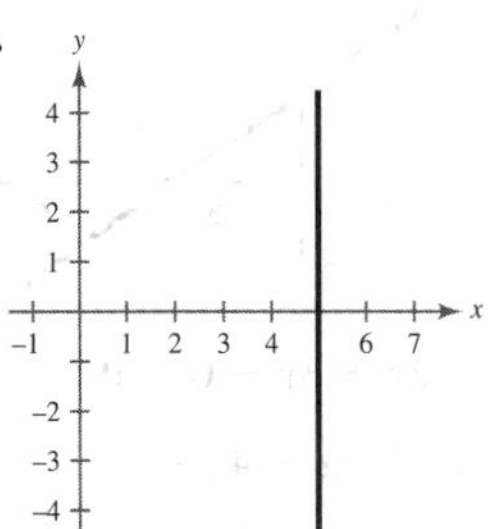

7.

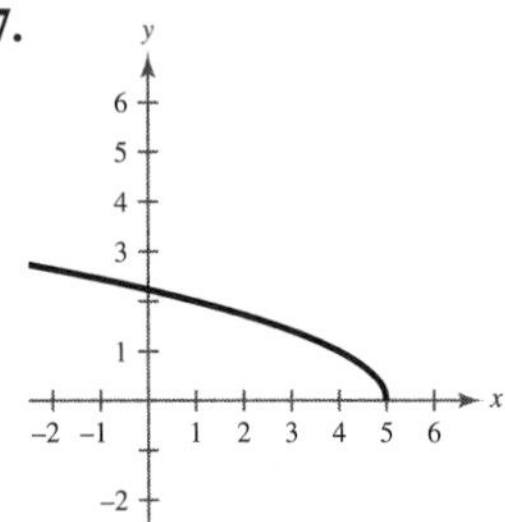

9.

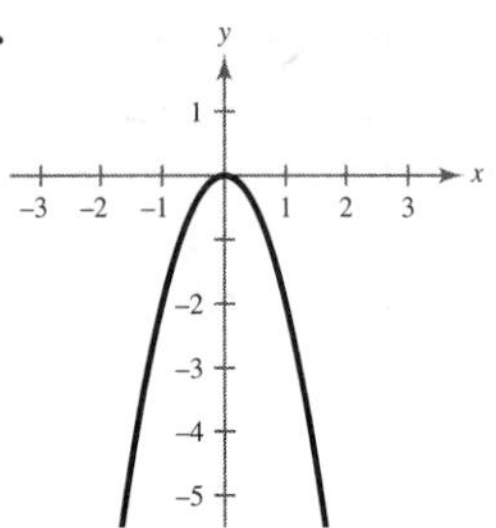

11.

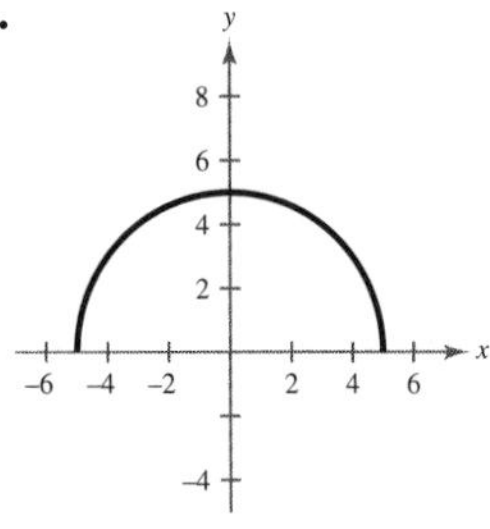

13.

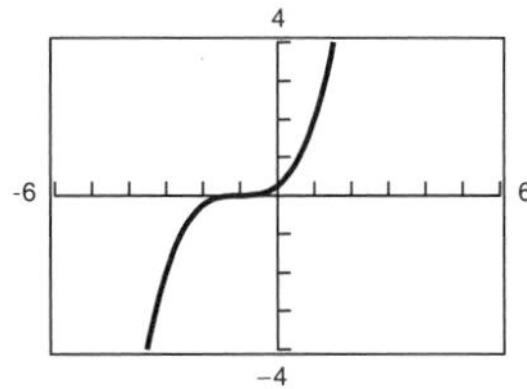

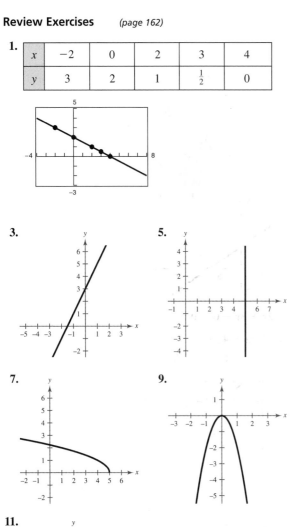

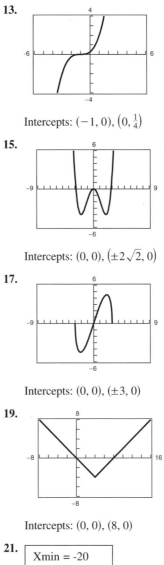

Intercepts: $(-1, 0)$, $\left(0, \frac{1}{4}\right)$

15.

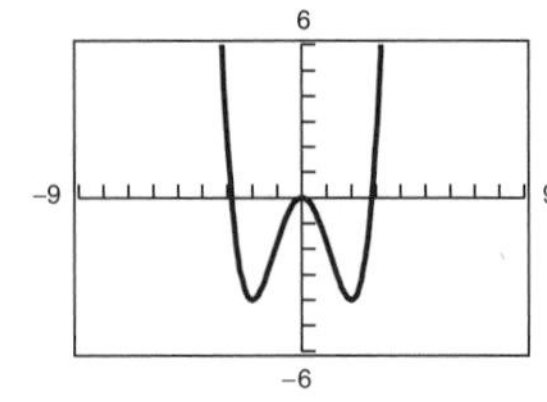

Intercepts: $(0, 0)$, $\left(\pm 2\sqrt{2}, 0\right)$

17.

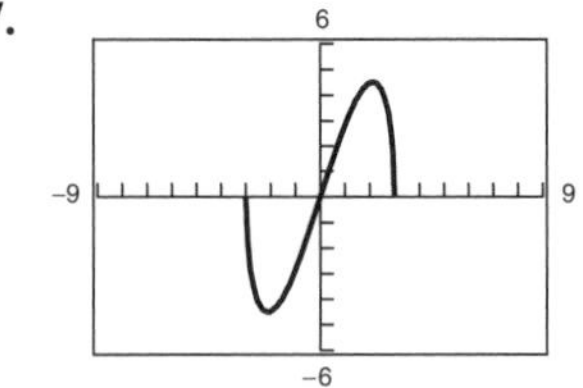

Intercepts: $(0, 0)$, $(\pm 3, 0)$

19.

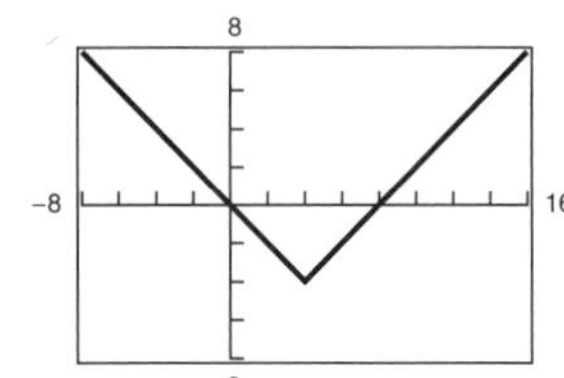

Intercepts: $(0, 0)$, $(8, 0)$

21.

Xmin = -20
Xmax = 50
Xscl = 10
Ymin = -2
Ymax = 1
Yscl = 0.5

23. (a)

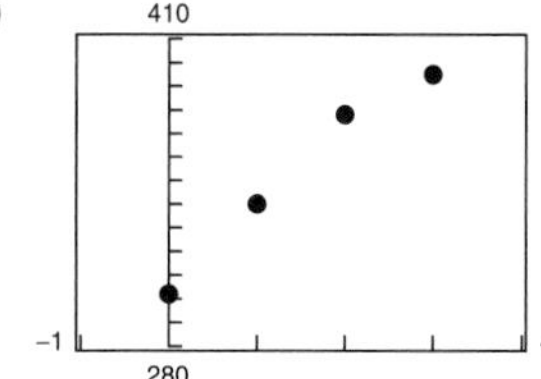

(b) $y = 31.7t + 306.2$

(c)

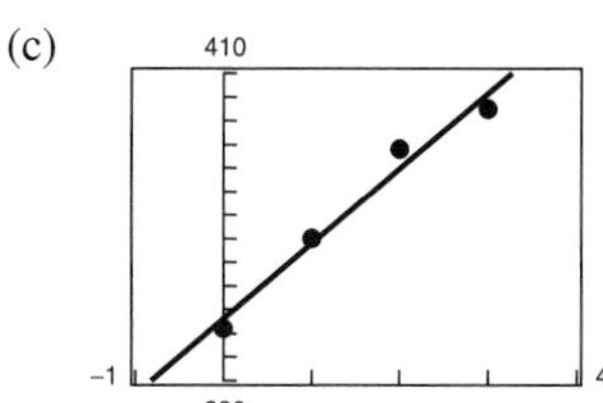

(d) 560, 623

25.

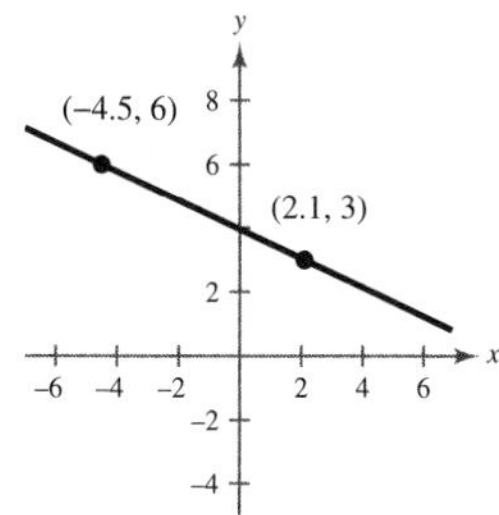

$m = -\frac{5}{11}$

27.

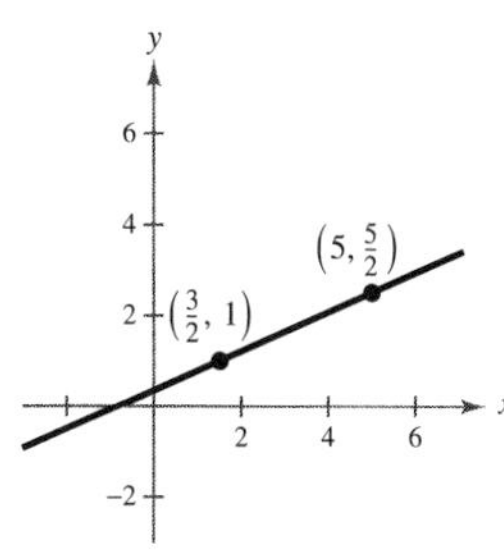

$m = \frac{3}{7}$

29. $t = \frac{7}{3}$ **31.** $t = 3$ **33.** $(6, 0), (10, 1), (-2, -2)$

35. $(-5, -7), (-4, -9), (-3, -11)$ **37.** $x = 0$

39. $5x - 12y + 2 = 0$ **41.** $2x - 7y + 2 = 0$

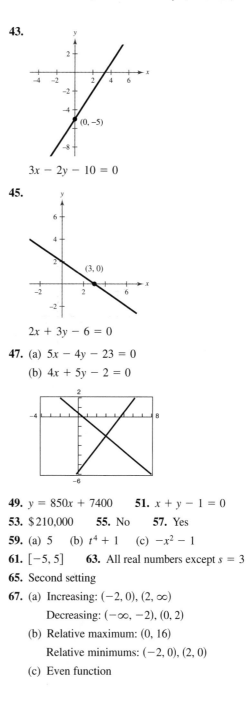

43.

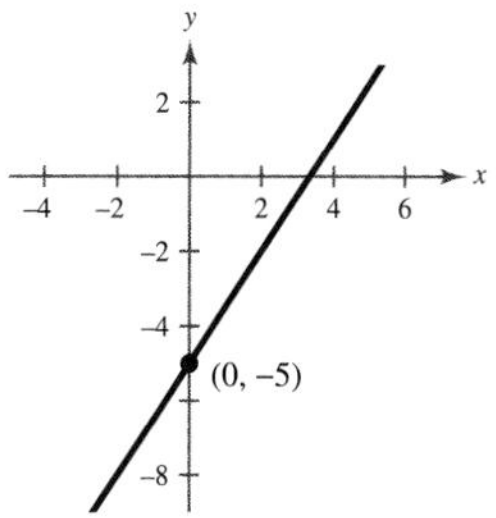

$3x - 2y - 10 = 0$

45.

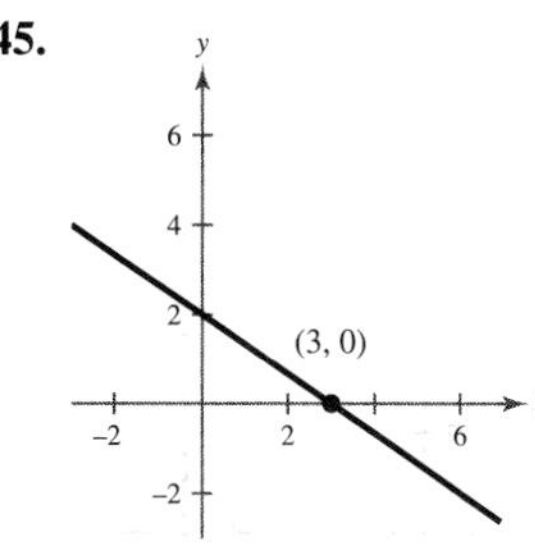

$2x + 3y - 6 = 0$

47. (a) $5x - 4y - 23 = 0$

(b) $4x + 5y - 2 = 0$

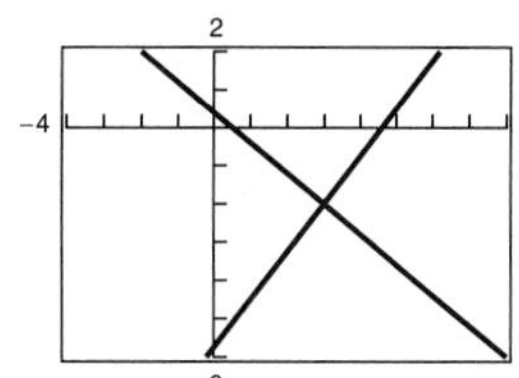

49. $y = 850x + 7400$ **51.** $x + y - 1 = 0$

53. $210,000 **55.** No **57.** Yes

59. (a) 5 (b) $t^4 + 1$ (c) $-x^2 - 1$

61. $[-5, 5]$ **63.** All real numbers except $s = 3$

65. Second setting

67. (a) Increasing: $(-2, 0), (2, \infty)$

Decreasing: $(-\infty, -2), (0, 2)$

(b) Relative maximum: $(0, 16)$

Relative minimums: $(-2, 0), (2, 0)$

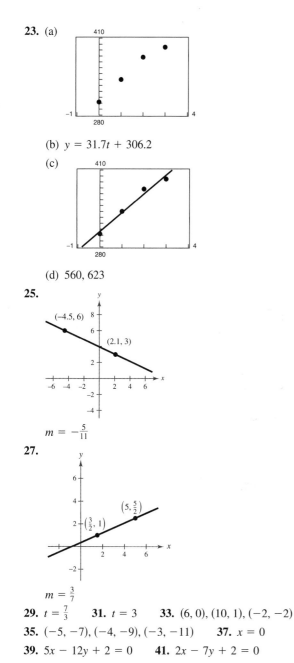

(c) Even function

69. (a) $f^{-1}(x) = 2x + 6$

(b)

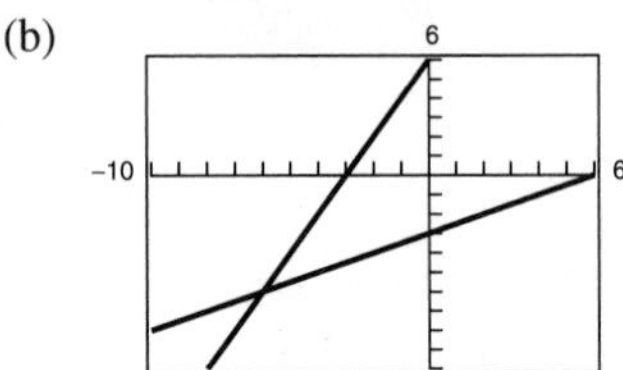

(c) $f^{-1}(f(x)) = f^{-1}\left(\frac{1}{2}x - 3\right)$

$= 2\left(\frac{1}{2}x - 3\right) + 6$

$= x$

$f(f^{-1}(x)) = f(2x + 6)$

$= \frac{1}{2}(2x + 6) - 3$

$= x$

71. (a) $f^{-1}(x) = x^2 - 1, \quad x \ge 0$

(b)

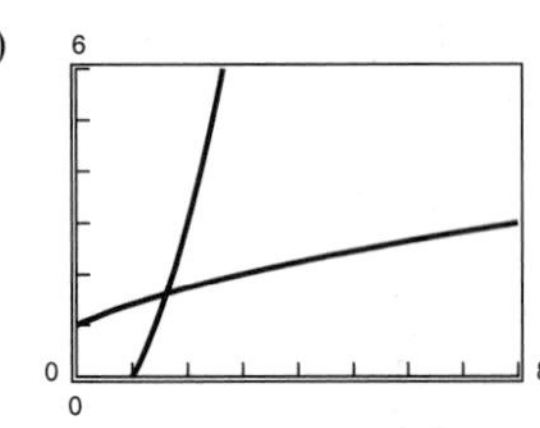

(c) $f^{-1}(f(x)) = f^{-1}(\sqrt{x + 1})$

$= (x + 1) - 1$

$= x$

$f(f^{-1}(x)) = f(x^2 - 1), \quad x \ge 0$

$= \sqrt{x^2 - 1 + 1}$

$= x$

73. $x \ge 4, \quad f^{-1}(x) = \sqrt{\frac{x}{2}} + 4$

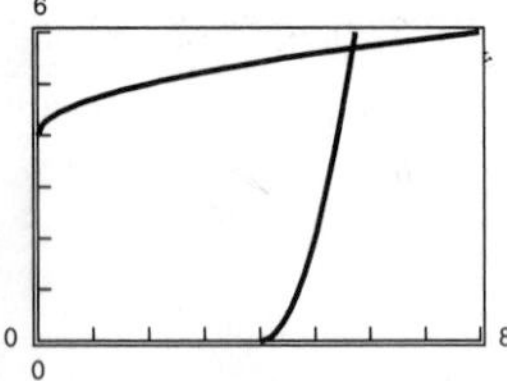

75. −7 **77.** 5 **79.** 23 **81.** 9

83. (a) $A = x(12 - x)$

(b) Domain: $[0, 12]$

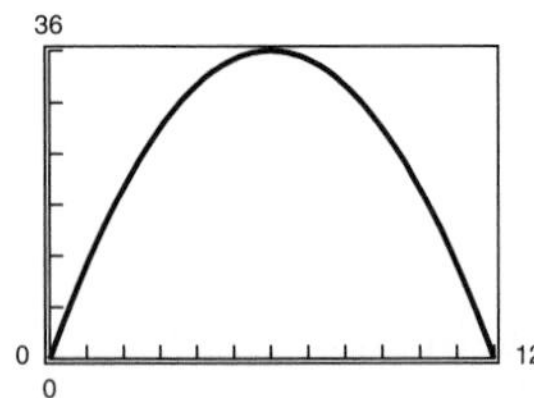

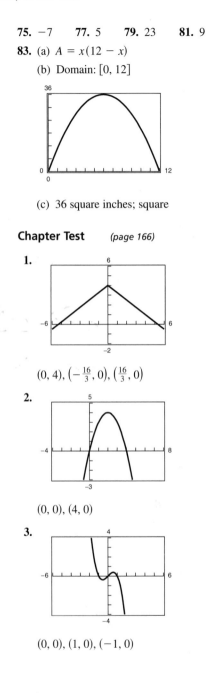

(c) 36 square inches; square

Chapter Test *(page 166)*

1.

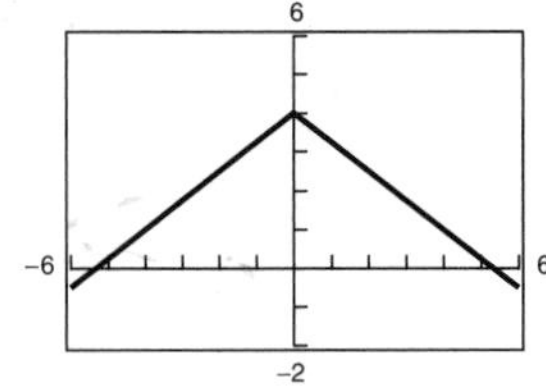

$(0, 4), \left(-\frac{16}{3}, 0\right), \left(\frac{16}{3}, 0\right)$

2.

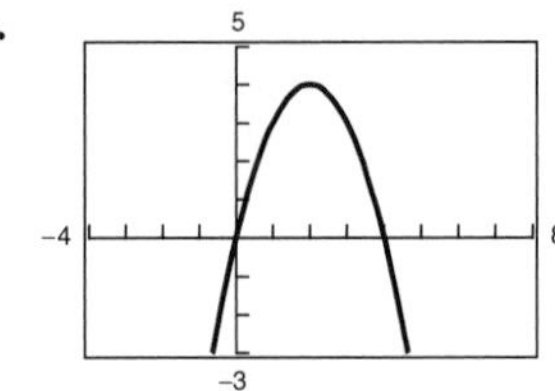

$(0, 0), (4, 0)$

3.

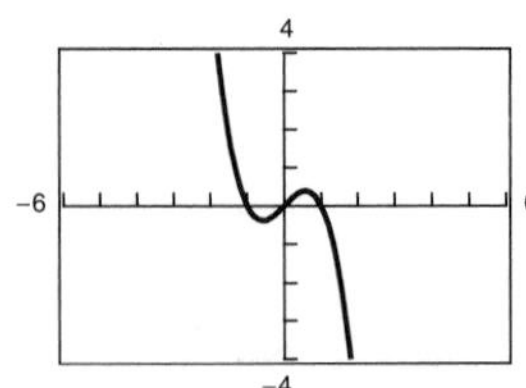

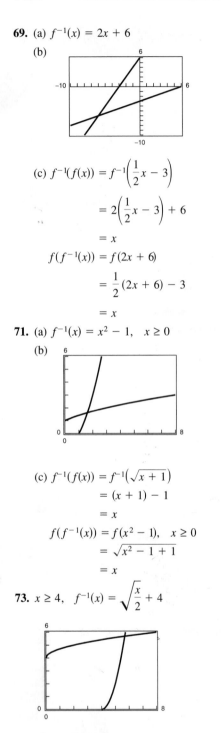

$(0, 0), (1, 0), (-1, 0)$

4.

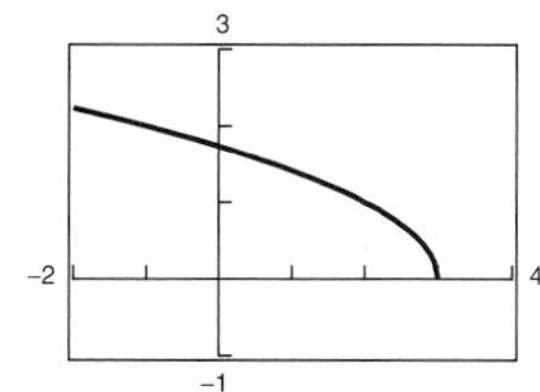

$(0, \sqrt{3}), (3, 0)$

5. $(1, -4), (4, \frac{1}{2}), (5, 2)$ **6.** $(0, -\frac{9}{2}), (3, 0)$

7. $2x - 5y + 20 = 0$

8. No. To some x there corresponds more than one value of y.

9. 7 **10.** $10 - \sqrt{6 - t}$ **11.** $\dfrac{\sqrt{3 - x} - 1}{2 - x}$

12. $(-\infty, 3]$ **13.** $C = 5.60x + 24{,}000$
$P = 3.60x - 24{,}000$

14.

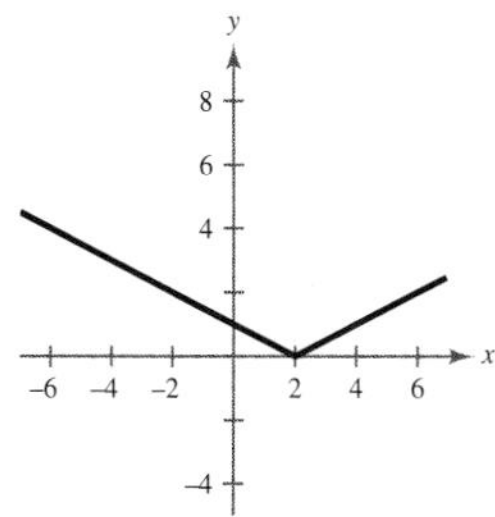

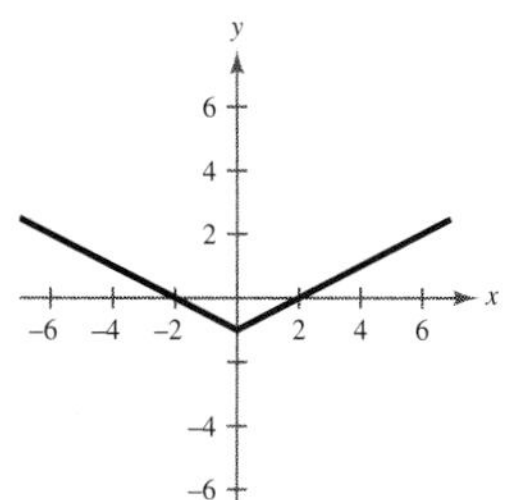

15. Increasing: $(-2, 0), (2, \infty)$
Decreasing: $(-\infty, -2), (0, 2)$

16. Increasing: $(-2, 2)$
Constant: $(-\infty, -2), (2, \infty)$

17. (a) $x^2 - \sqrt{2 - x}$, $(-\infty, 2]$
(b) $\dfrac{x^2}{\sqrt{2 - x}}$, $(-\infty, 2)$
(c) $2 - x$, $(-\infty, 2]$
(d) $2 - x^2$, $[0, \infty)$

18. Its graph must satisfy the Horizontal Line Test. Reflections in the line $y = x$.

CHAPTER 2

Section 2.1 *(page 176)*

1. (a) No (b) No (c) Yes (d) No

3. (a) Yes (b) No (c) No (d) No

5. (a) No (b) No (c) No (d) Yes

7. Identity **9.** Conditional **11.** Identity

13. Conditional

15. The equations have the same solutions and the one is derived from the other by the steps for generating equivalent equations given in this section.
$2x = 5,\ 2x + 3 = 8$

17. Given equation
Addition Property of Equality
Additive Inverse Property
Multiplication Property of Equality
Multiplicative Inverse Property

19. $x = 5$ **21.** $s = 6$ **23.** $x = \frac{7}{3}$ **25.** 3 **27.** 9

29. -26 **31.** -4 **33.** $-\frac{6}{5}$ **35.** 9 **37.** 10

39. 4 **41.** 5 **43.** No solution **45.** $\frac{11}{6}$ **47.** $\frac{5}{3}$

49. No solution **51.** 0 **53.** All real numbers

55. $6x + 15 = x$ **57.** 61.2 inches

59. $x = 10$ centimeters **61.** 72% increase

63. 43% decrease **65.** 3 hours

67. (a) 3.8 hours, 3.2 hours (b) 1.1 hours (c) 25.6 miles

69. $66\frac{2}{3}$ kilometers per hour **71.** 1.29 seconds

73. (a)

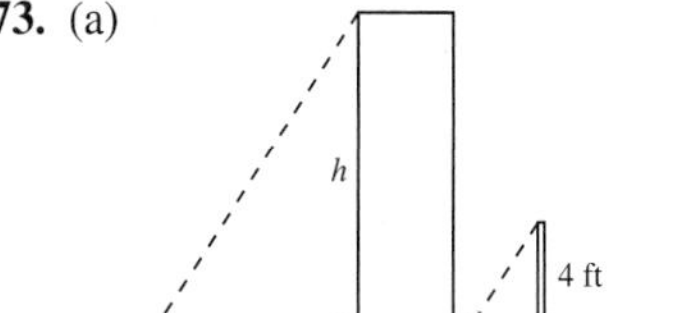

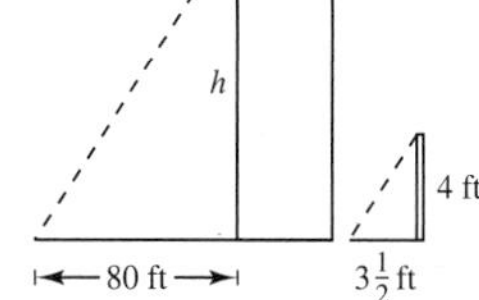

(b) 91.4 feet

75. $4000 **77.** 11.4% **79.** 50 pounds of each kind

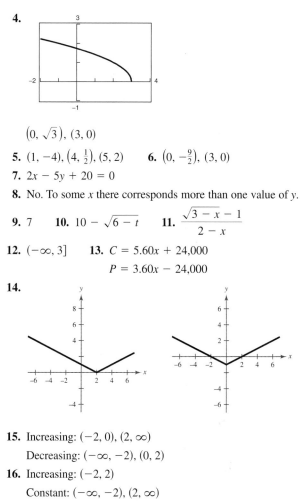

81. $x = 6$ **83.** $\dfrac{2A}{b}$ **85.** $\dfrac{2A - ah}{h}$

Section 2.2 *(page 187)*

1. $(5, 0), (0, -5)$ **3.** $(-2, 0), (1, 0), (0, -2)$

5. $(-2, 0), (0, 0)$ **7.** $(-1, 0), (5, 0), (0, -1)$

9. $(1, 0), \left(0, \frac{1}{2}\right)$

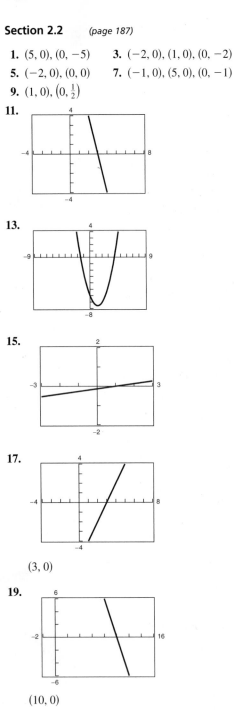

11.

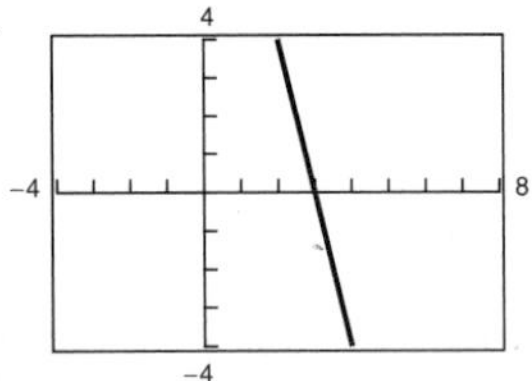

13.

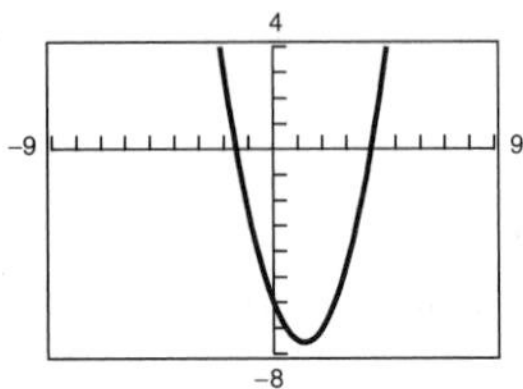

15.

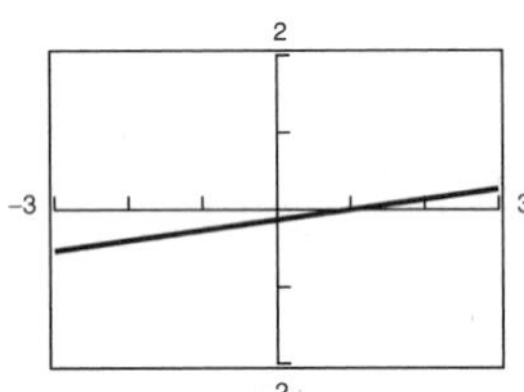

17.

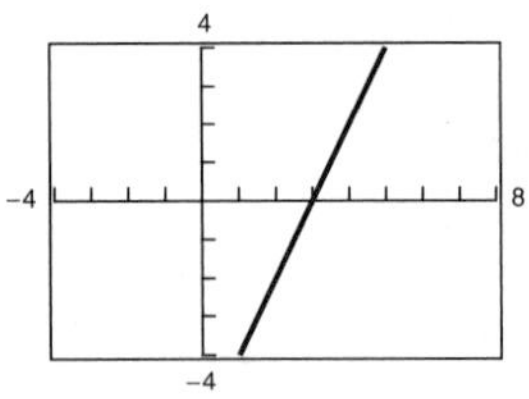

$(3, 0)$

19.

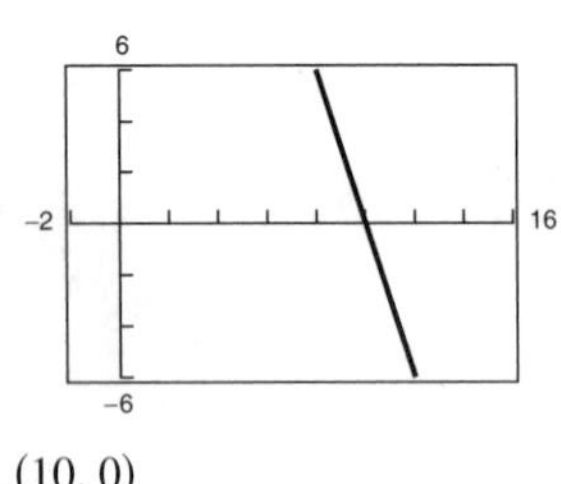

$(10, 0)$

21. $f(x) = 25(x - 3) - 12(x + 2) + 10 = 0$

23. $f(x) = \dfrac{2x}{3} + \dfrac{1}{x} - 10 = 0$

25. $f(x) = \dfrac{3}{x + 2} - \dfrac{4}{x - 2} - 5 = 0$

27. $\frac{15}{4}$ **29.** 6 **31.** $-\frac{5}{8}$ **33.** 2.172, 7.828

35. -1.379 **37.** $0.5, -3, 3$ **39.** $-0.717, 2.107$

41. -1.333 **43.** $-1, 7$

45. (a)

x	-1	0	1	2	3	4
$3.2x - 5.8$	-9	-5.8	-2.6	0.6	3.8	7

(b) $1 < x < 2$

(c)

x	1.5	1.6	1.7	1.8	1.9	2
$3.2x - 5.8$	-1	-0.68	-0.36	-0.04	0.28	0.6

(d) $1.8 < x < 1.9$

To improve accuracy, evaluate the expression in this interval and determine where the sign changes.

(e) $x = 1.8125$

47. $(1, 1)$ **49.** $(2, 2)$ **51.** $(-1, 3), (2, 6)$ **53.** $(4, 1)$

55. $(1.449, 1.899), (-3.449, -7.899)$

57. $(-2, 8), (1.333, 8)$ **59.** $(0, 0), (-2, 8), (2, 8)$

61. (a) 6.46 (b) $\frac{1.73}{0.27} \approx 6.41$

63. (a) $T(x) = \dfrac{x}{63} + \dfrac{280 - x}{54}$

(b) $0 \le x \le 280$

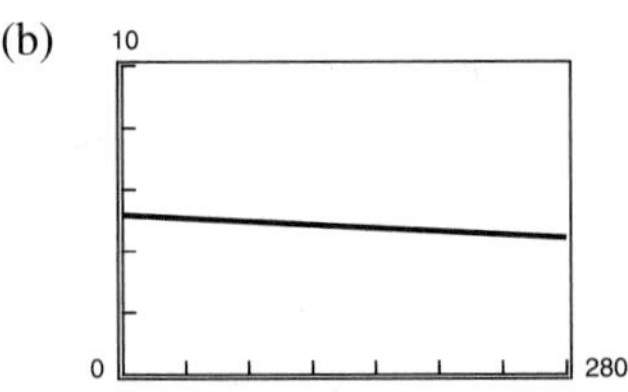

(c) 164.5 miles

65. (a) $C = 0.33(55 - x) + x$

(b) $0 \le x \le 55$

(c) 22.2

67. (a) $A(x) = 12x$

(b)

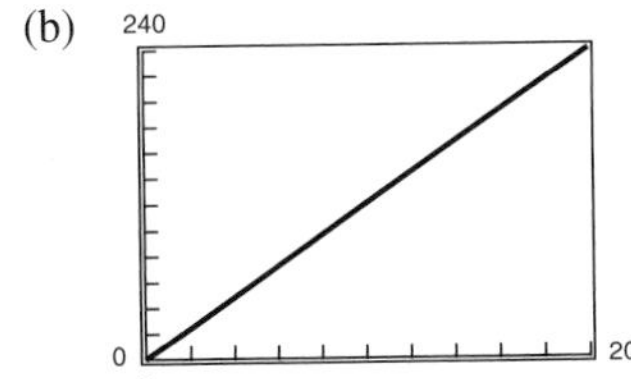

(c) 16.7

69. (a) 5

(b) One zero: $k = 25$

Three zeros: $k = 15$

(c) No. Because the range is $(-\infty, \infty)$, the function crosses the x-axis at least once.

71. $T = 10{,}000 + \frac{1}{2}x$ **73.** \$7600

75. 1990. To answer the question algebraically, solve the equation $0.43t + 30.86 = 31$. To answer the question graphically, draw the horizontal line $y = 31$ and note where it intersects the histogram.

Section 2.3 *(page 198)*

1. $a = -10,\ b = 6$ **3.** $a = 6,\ b = 5$

5. $4 + 3i$ **7.** $2 - 3\sqrt{3}i$ **9.** $5\sqrt{3}i$

11. $-1 - 6i$ **13.** 8 **15.** $0.3i$ **17.** $11 - i$

19. 4 **21.** $3 - 3\sqrt{2}i$ **23.** $-14 + 20i$

25. $\frac{1}{6} + \frac{7}{6}i$ **27.** $-2\sqrt{3}$ **29.** -10 **31.** $5 + i$

33. $12 + 30i$ **35.** 24 **37.** $-9 + 40i$ **39.** -10

41. $\sqrt{-6}\sqrt{-6} = \sqrt{6}i\sqrt{6}i = 6i^2 = -6$

43. 34 **45.** 9 **47.** 400 **49.** 8 **51.** $-6i$

53. $\frac{16}{41} + \frac{20}{41}i$ **55.** $\frac{3}{5} + \frac{4}{5}i$ **57.** $-7 - 6i$

59. $-\frac{9}{1681} + \frac{40}{1681}i$ **61.** $-\frac{1}{2} - \frac{5}{2}i$ **63.** $\frac{62}{949} + \frac{297}{949}i$

65. $i, -1, -i, 1, i, -1, -i, 1, i, -1, -i, 1, i, -1, -i, 1$

67. $-1 + 6i$ **69.** $-5i$ **71.** $-375\sqrt{3}i$

73. i **75.** $4 + 3i$ **77.** $6i$

79.

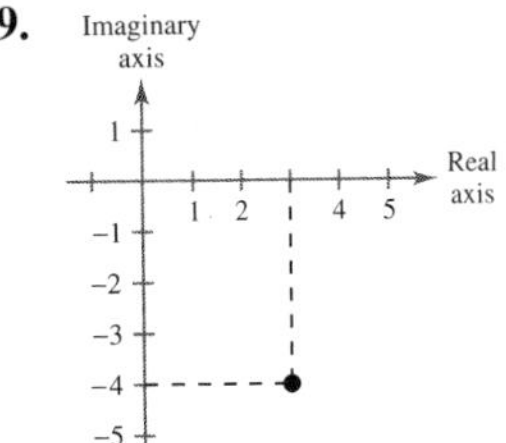

81.

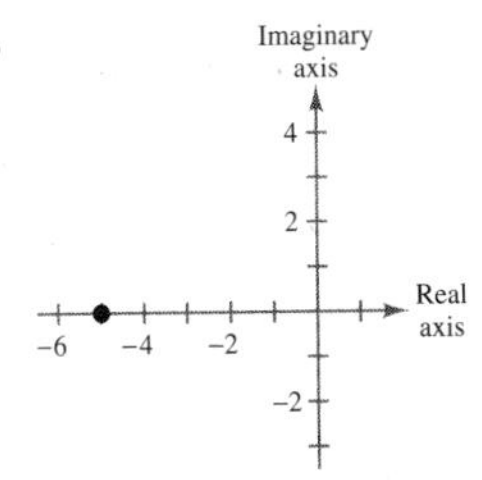

83. 0, 0, 0, 0, 0, 0

85. $0.5i, -0.25 + 0.5i, -0.1875 + 0.25i, -0.0273 + 0.4063i, -0.1643 + 0.4778i, -0.2013 + 0.3430i$

87. 1, 2, 5, 26, 677, 458,330

89. 8, 8, 8 **91.–93.** Answers will vary.

95. $x^3 + x^2 + 2x - 6$ **97.** $3x^2 + 11\frac{1}{2}x - 2$

99. $x^2 + y^2 + 2xy + 6x + 6y + 9$

101. $r = \sqrt{\dfrac{\alpha m_1 m_2}{F}}$ **103.** 88.9 kilometers per hour

Section 2.4 *(page 212)*

1. $2x^2 + 5x - 3 = 0$ **3.** $3x^2 - 60x - 10 = 0$

5. $0, -\frac{1}{2}$ **7.** $4, -2$ **9.** $3, -\frac{1}{2}$ **11.** $2, -6$

13. $\pm\sqrt{7}$; ± 2.65 **15.** $12 \pm 3\sqrt{2}$; 16.24, 7.76

17. $-3 \pm \sqrt{7}$ **19.** $1 \pm \dfrac{\sqrt{6}}{3}$

21.

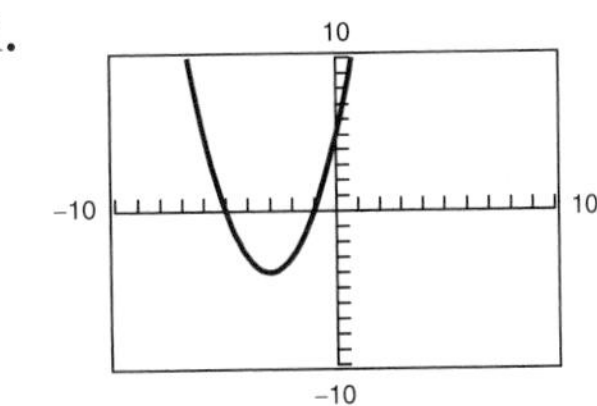

$(-1, 0), (-5, 0)$

23.

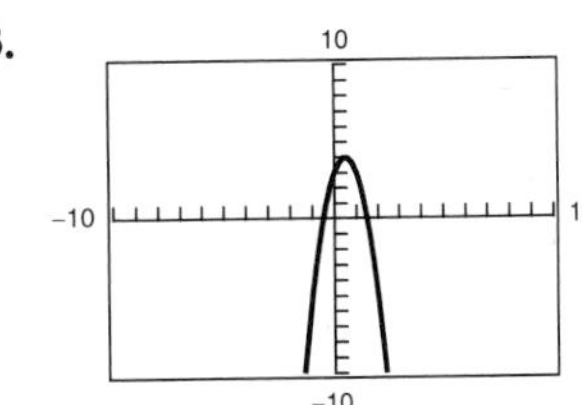

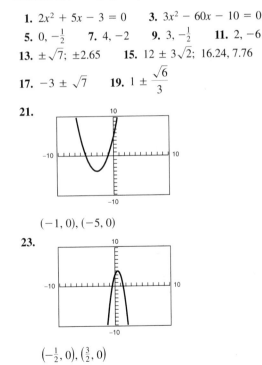

$\left(-\frac{1}{2}, 0\right), \left(\frac{3}{2}, 0\right)$

25.

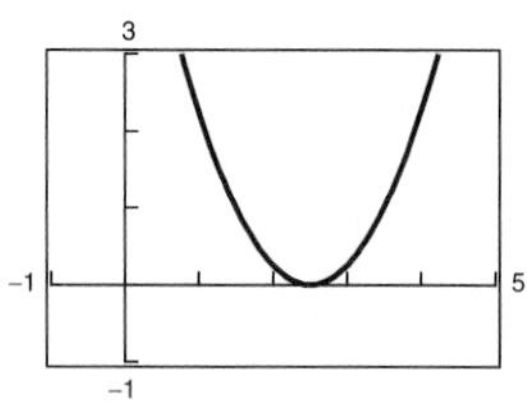

$\left(\frac{5}{2}, 0\right)$

27.

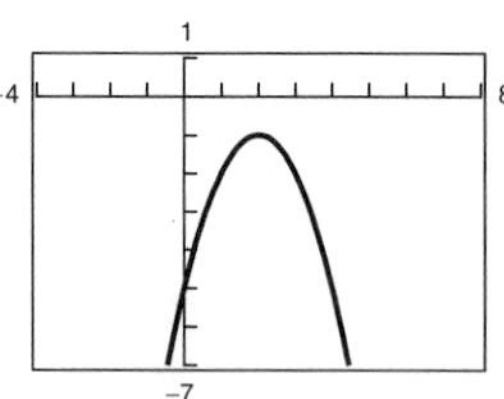

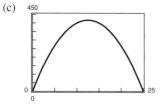

No x-intercepts; $2 \pm i$

29. No real solutions **31.** Two real solutions

33. $1 \pm \sqrt{3}$ **35.** $-4 \pm 2\sqrt{5}$ **37.** $\frac{2}{7}$ **39.** $1 \pm i$

41. $-\frac{3}{2}, -\frac{5}{2}$ **43.** $1 \pm \sqrt{2}$ **45.** $6, -12$

47. False. The product must equal zero to use the Zero-Factor Property.

49. $x^2 - 2x - 24 = 0$

51. (a) $A(x) = \frac{8}{3}x(25 - x)$

(b)

x	y	Area
2	$\frac{92}{3}$	$\frac{368}{3} \approx 123$
4	28	224
6	$\frac{76}{3}$	304
8	$\frac{68}{3}$	$\frac{1088}{3} \approx 363$
10	20	400
12	$\frac{52}{3}$	416
14	$\frac{44}{3}$	$\frac{1232}{3} \approx 411$

Approximate dimensions for maximum area: $24 \times \frac{52}{3}$

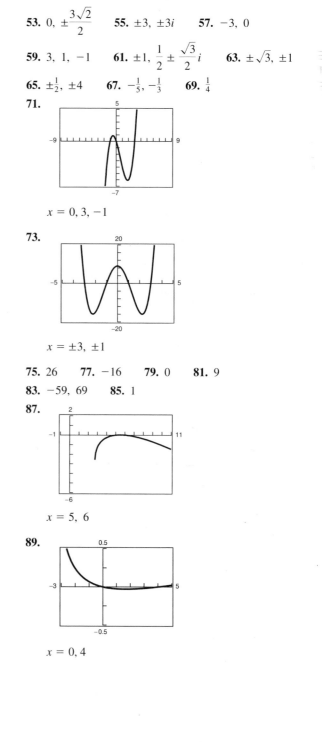

(c) 450, 0, 0, 25

Approximate dimensions for maximum area: $25 \times \frac{50}{3}$

(d) and (e) $15 \times 23\frac{1}{3}$ or 35×10

53. $0, \pm\frac{3\sqrt{2}}{2}$ **55.** $\pm 3, \pm 3i$ **57.** $-3, 0$

59. $3, 1, -1$ **61.** $\pm 1, \frac{1}{2} \pm \frac{\sqrt{3}}{2}i$ **63.** $\pm\sqrt{3}, \pm 1$

65. $\pm\frac{1}{2}, \pm 4$ **67.** $-\frac{1}{5}, -\frac{1}{3}$ **69.** $\frac{1}{4}$

71.

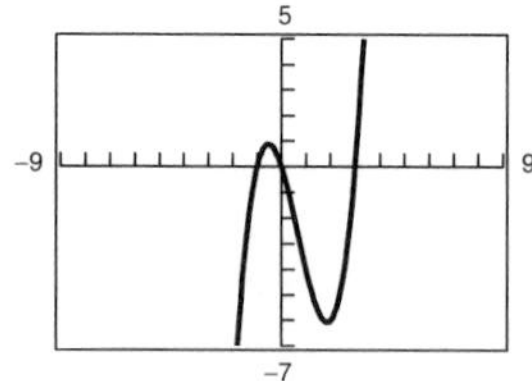

$x = 0, 3, -1$

73.

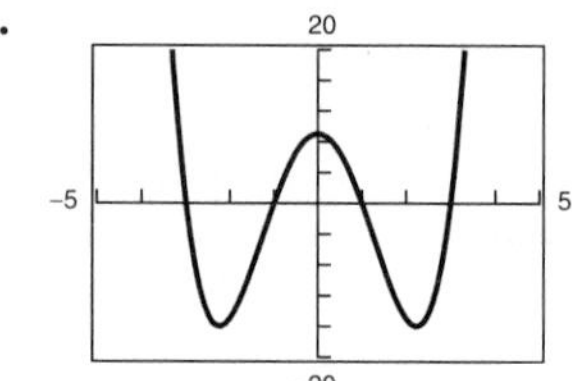

$x = \pm 3, \pm 1$

75. 26 **77.** -16 **79.** 0 **81.** 9

83. $-59, 69$ **85.** 1

87.

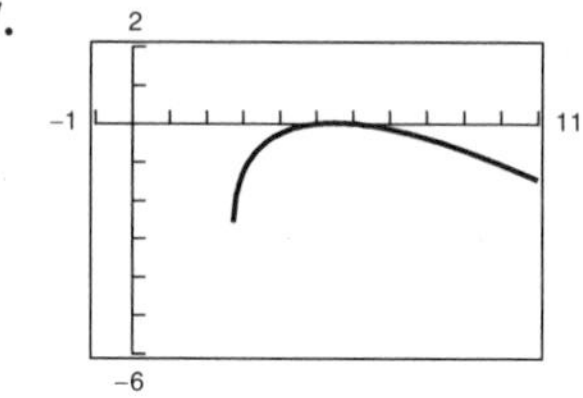

$x = 5, 6$

89.

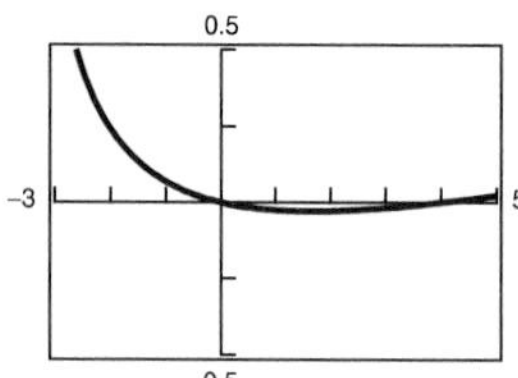

$x = 0, 4$

91. 4, -5 **93.** $\dfrac{-3 \pm \sqrt{21}}{6}$ **95.** 2, $-\dfrac{3}{2}$

97. 1, -3 **99.** 3, -2 **101.** $\sqrt{3}$, -3

103.

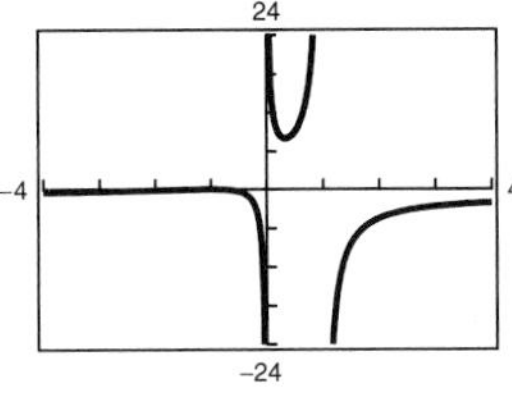

$x = -1$

105.

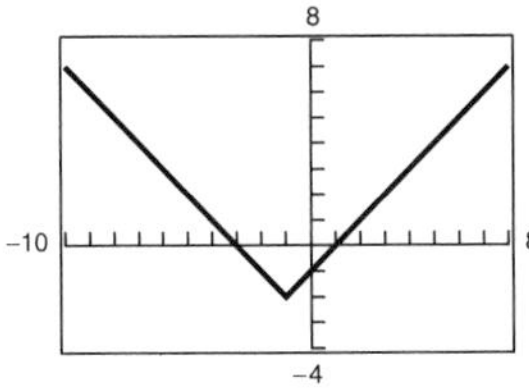

$x = 1, -3$

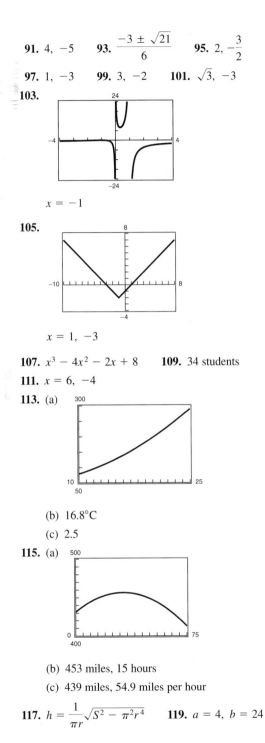

107. $x^3 - 4x^2 - 2x + 8$ **109.** 34 students

111. $x = 6, -4$

113. (a)

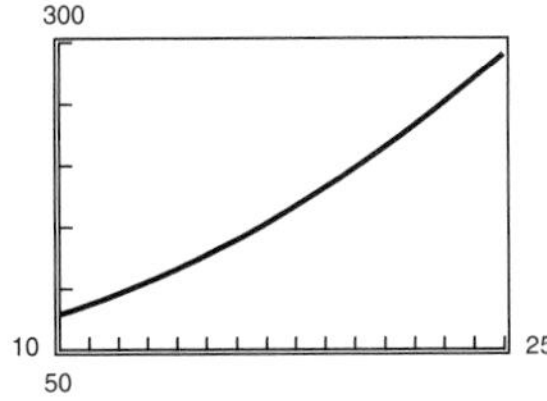

(b) 16.8°C

(c) 2.5

115. (a)

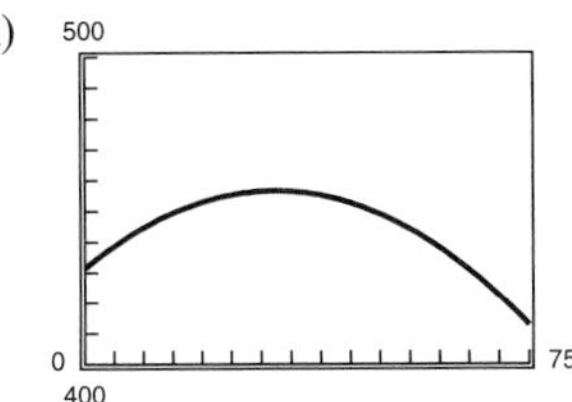

(b) 453 miles, 15 hours

(c) 439 miles, 54.9 miles per hour

117. $h = \dfrac{1}{\pi r}\sqrt{S^2 - \pi^2 r^4}$ **119.** $a = 4,\ b = 24$

Section 2.5 *(page 225)*

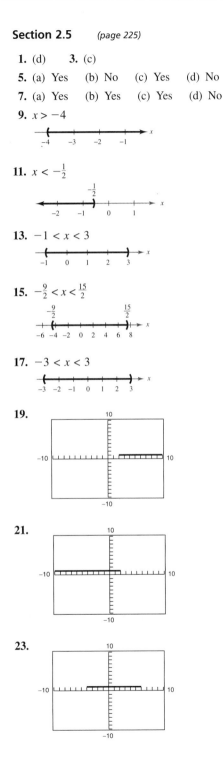

1. (d) **3.** (c)

5. (a) Yes (b) No (c) Yes (d) No

7. (a) Yes (b) Yes (c) Yes (d) No

9. $x > -4$

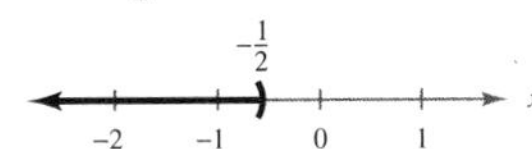

11. $x < -\frac{1}{2}$

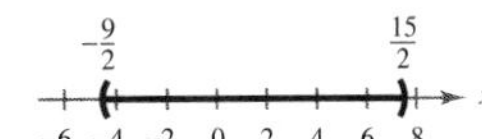

13. $-1 < x < 3$

15. $-\frac{9}{2} < x < \frac{15}{2}$

17. $-3 < x < 3$

19.

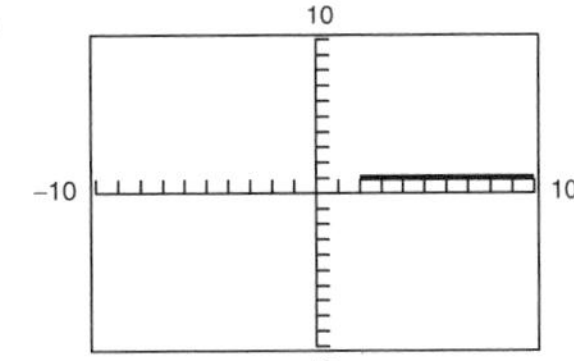

21.

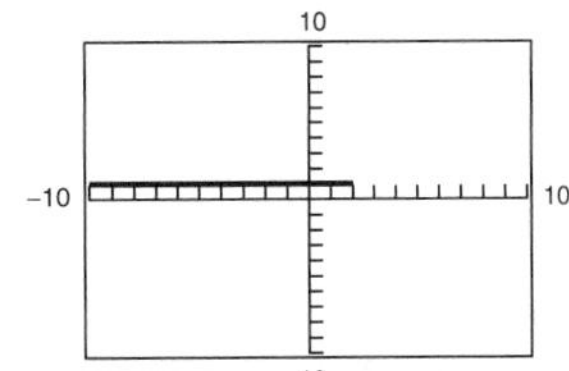

23.

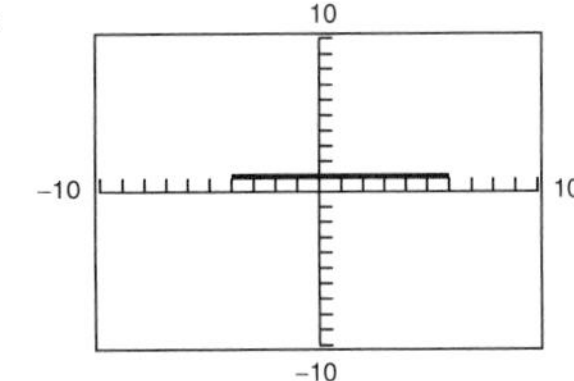

25.

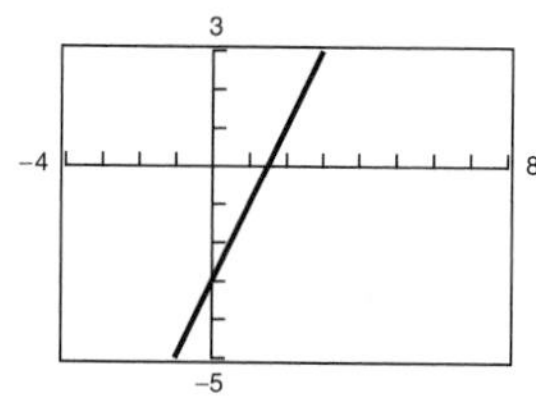

(a) $x \geq 2$

(b) $x \leq \frac{3}{2}$

27.

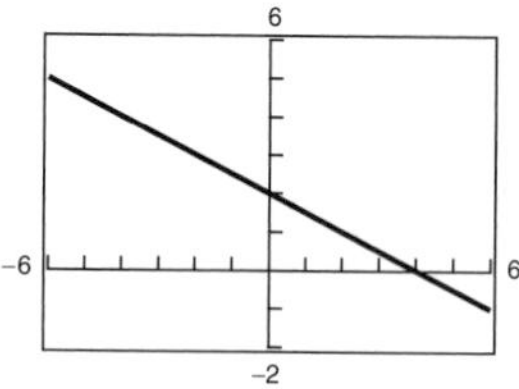

(a) $-2 \leq x \leq 4$

(b) $x \leq 4$

29. $[5, \infty)$

31. $x < -6,\ x > 6$

33. $16 \leq x \leq 24$

35. $x \leq 16,\ x \geq 24$

37. $x \leq -7,\ x \geq 13$

39. No solution

41.

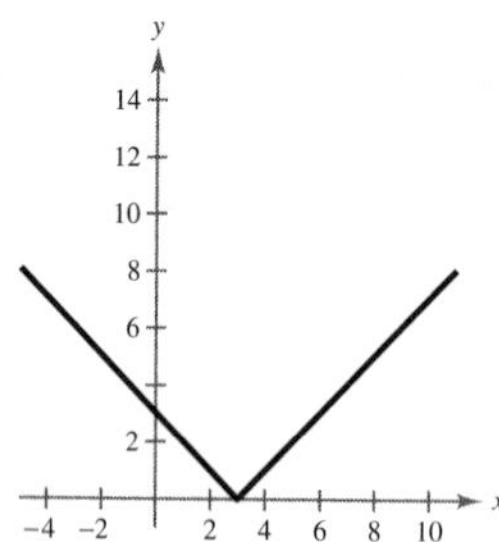

(a) $1 \leq x \leq 5$

(b) $x \leq -1,\ x \geq 7$

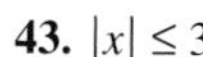

43. $|x| \leq 3$ **45.** $|x - 7| \geq 3$ **47.** $|x - 12| \leq 10$

49. (a)

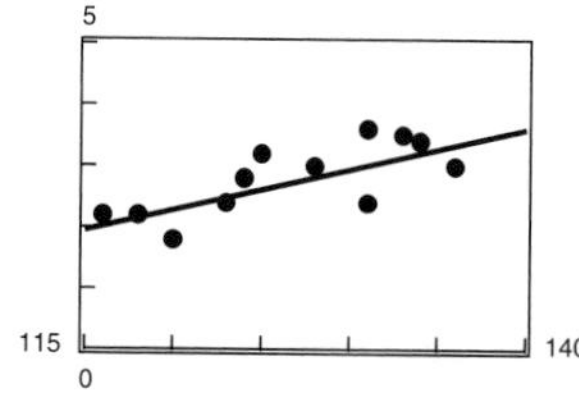

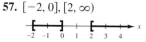

Linear

(b) $y = 0.067x - 5.638$

(c) $x \geq 129$

(d) IQ scores are not a particularly good predictor of GPA. Other factors, such as study habits, class attendance, and attitude, influence college performance.

51. $(-7, 3)$

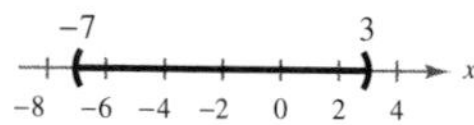

53. $(-\infty, -5], [1, \infty)$

55. $(-3, 2)$

57. $[-2, 0], [2, \infty)$

59.

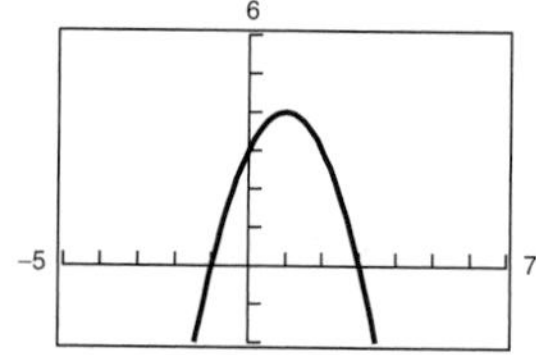

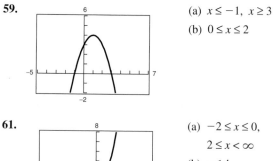

(a) $x \leq -1,\ x \geq 3$

(b) $0 \leq x \leq 2$

61.

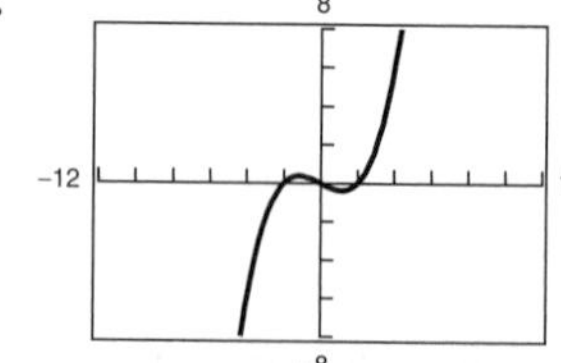

(a) $-2 \leq x \leq 0,$

$2 \leq x < \infty$

(b) $x \leq 4$

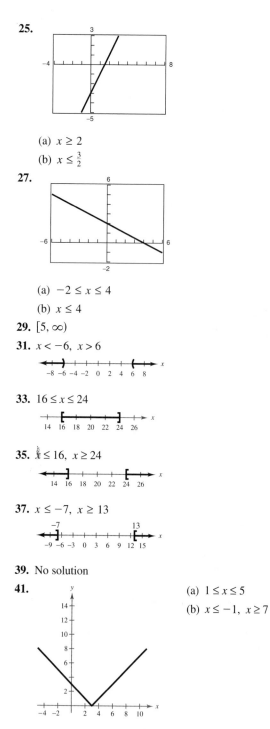

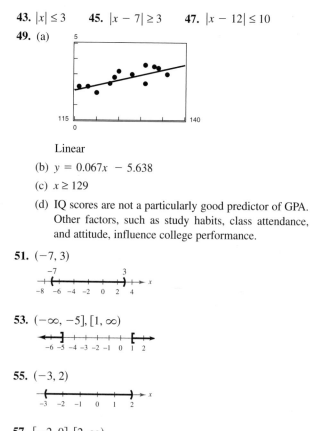

63. $(-\infty, -1)$, $(0, 1)$

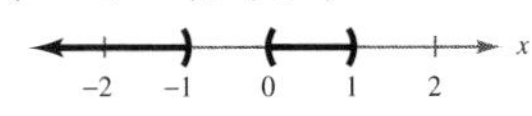

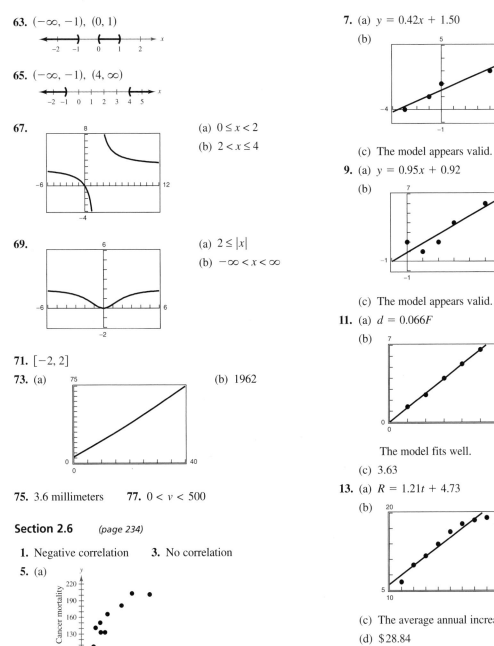

65. $(-\infty, -1)$, $(4, \infty)$

67.

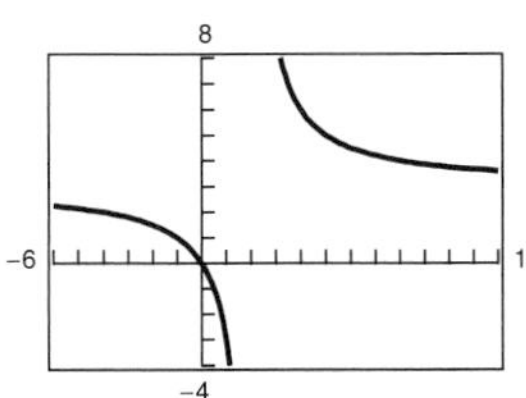

(a) $0 \le x < 2$

(b) $2 < x \le 4$

69.

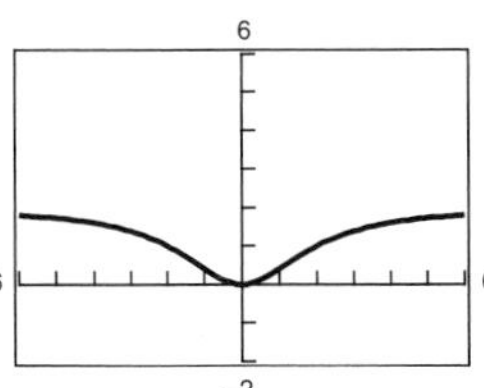

(a) $2 \le |x|$

(b) $-\infty < x < \infty$

71. $[-2, 2]$

73. (a)

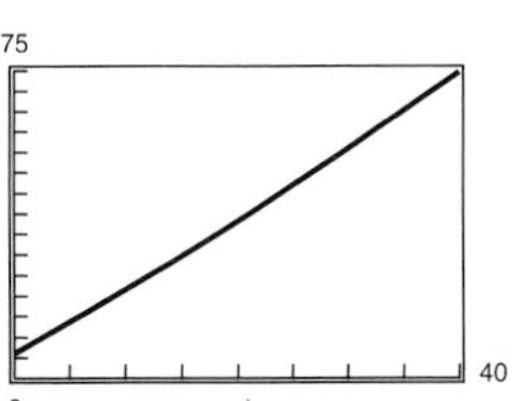

(b) 1962

75. 3.6 millimeters **77.** $0 < v < 500$

Section 2.6 *(page 234)*

1. Negative correlation **3.** No correlation

5. (a)

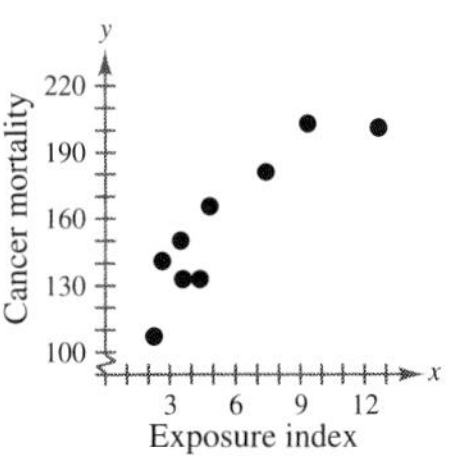

(b) Yes. The cancer mortality increases linearly with increased exposure to the carcinogenic substance.

7. (a) $y = 0.42x + 1.50$

(b)

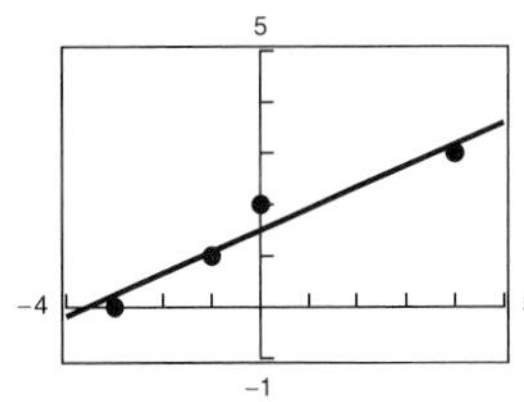

(c) The model appears valid.

9. (a) $y = 0.95x + 0.92$

(b)

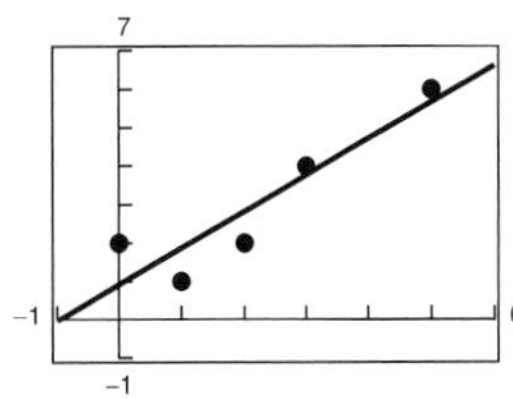

(c) The model appears valid.

11. (a) $d = 0.066F$

(b)

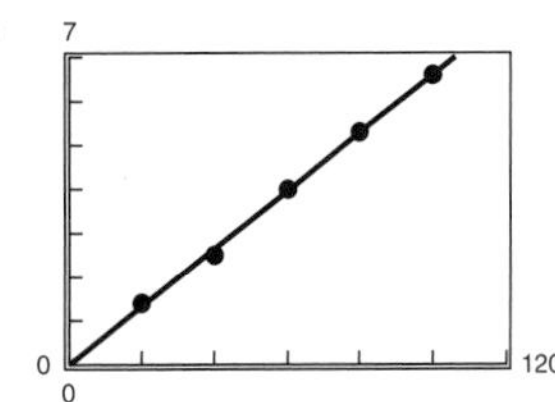

The model fits well.

(c) 3.63

13. (a) $R = 1.21t + 4.73$

(b)

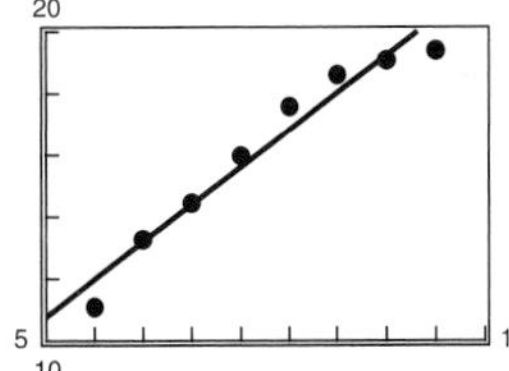

(c) The average annual increase in the monthly basic rate

(d) $28.84

15. (a) $y = 47.77x + 103.77$

(b) 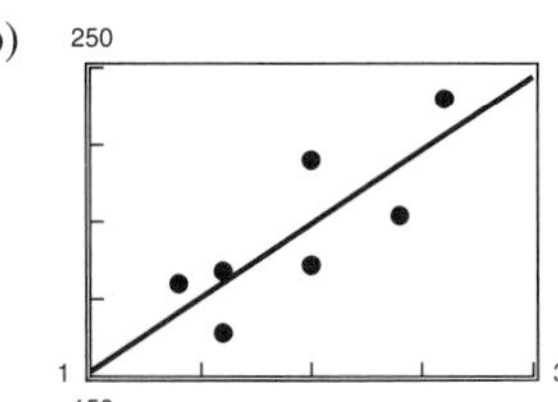

(c) The average increase in sales for increases in advertising expenditures

(d) $175,000

17. (a) $y = -0.64x + 1321.56$

(b) 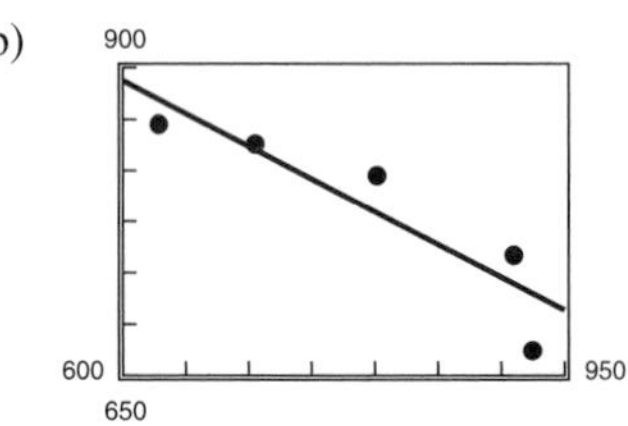

(c) The negative slope indicates that the amount of mortgage debt held by commercial banks is increasing as the amount held by savings institutions decreases.

FOCUS ON CONCEPTS (page 237)

1. An identity is true for all values of the variable and a conditional equation is true for only some values of the variable.

2. Equivalent equations have the same solution(s). An equation can be transformed into an equivalent equation by the following steps.

(a) Remove symbols of grouping, combine like terms, or reduce fractions.

(b) Add (or subtract) the same quantity to (from) both sides of the equation.

(c) Multiply (or divide) both sides of the equation by the same nonzero quantity.

(d) Interchange the two sides of the equation.

3. (a) Negative (b) Positive

4. (a) Neither (b) Both (c) Quadratic (d) Neither

5. Dividing by x does not yield an equivalent equation. $x = 0$ is also a solution.

6. False **7.** They are the same.

8. False. $(3 + 2i) + (5 - 2i) = 8$

9. For any real number x, $x^2 + 5 \geq 5$.

10. (b)

11. (a) $x = a$, $x = b$

(b) 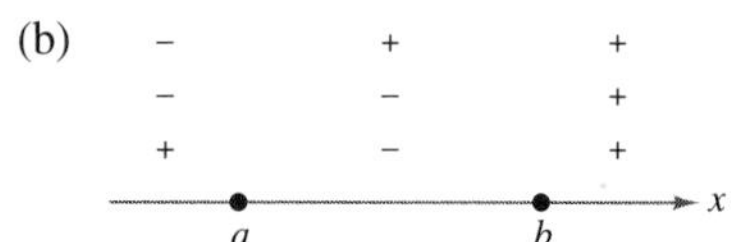

(c) The real zeros of the polynomial

12. False. Implies that the model fits the data well and that the slope of the model is negative

13. True

Review Exercises (page 238)

1. Identity **3.** (a) No (b) Yes (c) Yes (d) No

5. 20 **7.** $-\frac{1}{2}$ **9.** $\frac{1}{5}$ **11.** 0, 2 **13.** $-4 \pm 3\sqrt{2}$

15. $6 \pm \sqrt{6}$ **17.** $0, \frac{12}{5}$ **19.** 2, 6 **21.** 5 **23.** $\frac{25}{4}$

25. No solution **27.** $-124, 126$ **29.** $-2 \pm \frac{\sqrt{95}}{5}, -4$

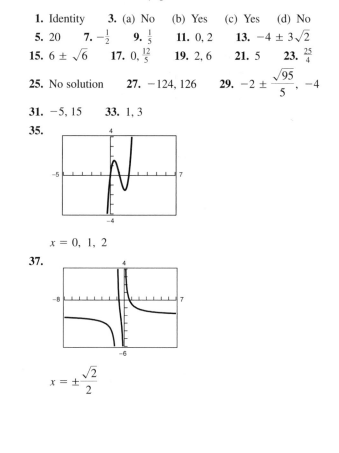

31. $-5, 15$ **33.** 1, 3

35. 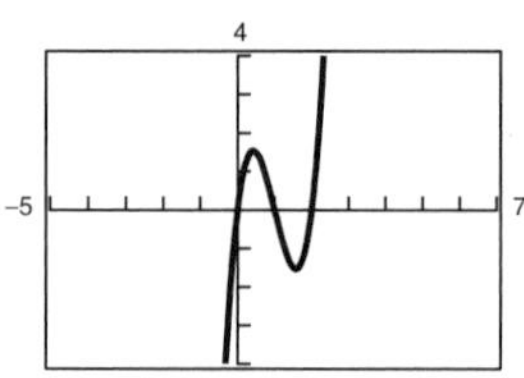

$x = 0, 1, 2$

37. 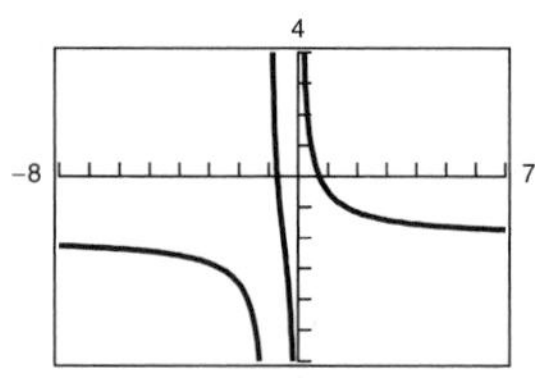

$x = \pm\frac{\sqrt{2}}{2}$

39. 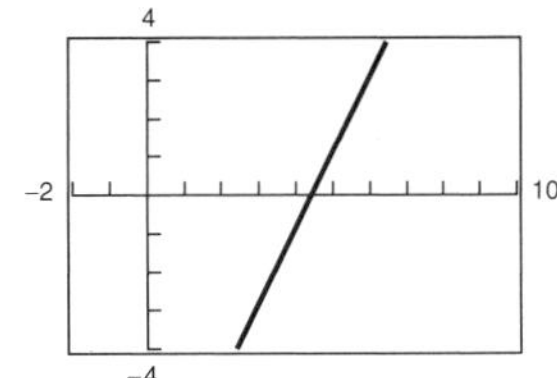

$x = \frac{40}{9}$

41. $r = \sqrt{\frac{3V}{\pi h}}$ **43.** $p = \frac{k}{3\pi r^2 L}$ **45.** $C = 4$

47. $3 + 7i$ **49.** $40 + 65i$ **51.** $-4 - 46i$

53. $1 - 6i$ **55.** $\frac{4}{3}i$ **57.** $x = \pm\sqrt{\frac{1}{3}}\,i$

59. $x = 0,\ 2 \pm i$ **61.** $\left(-\frac{5}{3}, \infty\right)$ **63.** $(-\infty, 3), (5, \infty)$

65. $(1, 3)$ **67.** $(-\infty, 0], [3, \infty)$ **69.** $x \le \frac{120}{7}$

71. $x > 4$ **73.** $[5, \infty)$

75. September: \$325,000; October: \$364,000

77. $2\frac{6}{7}$ liters **79.** 4 **81.** 56 miles per hour

83. (c), (b), (d), (a)

85. (a) $y = -1.20x + 64.27$

(b) 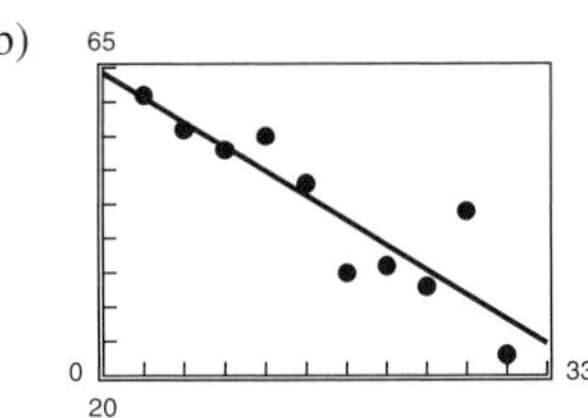

(c) Eliminate the point (27, 44); $y = -1.43x + 66.44$

Cumulative Test for Chapters P–2 *(page 242)*

1. $\frac{4x^3}{15y^5}$ **2.** $2x^2y\sqrt{6y}$ **3.** $5x - 6$

4. $x^3 - x^2 - 5x + 6$ **5.** $\frac{s - 1}{(s + 1)(s + 3)}$

6. $(3 + x)(7 - x)$ **7.** $x(1 + x)(1 - 6x)$

8. $2(3 - 2x)(9 + 6x + 4x^2)$

9. 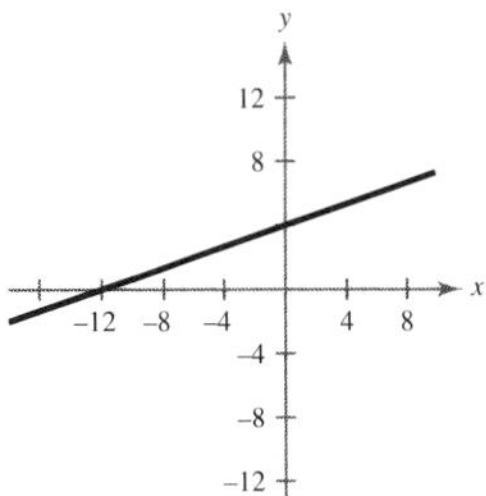

10. 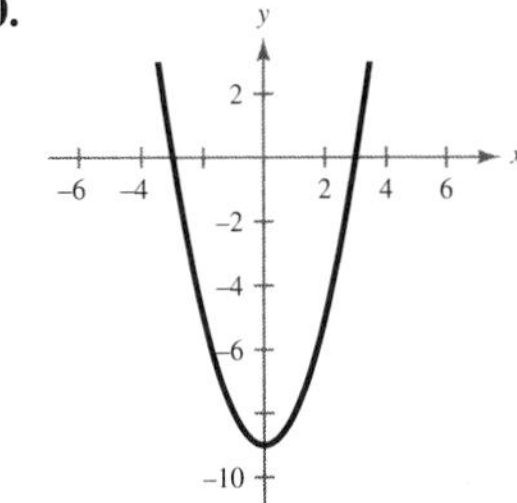

11. 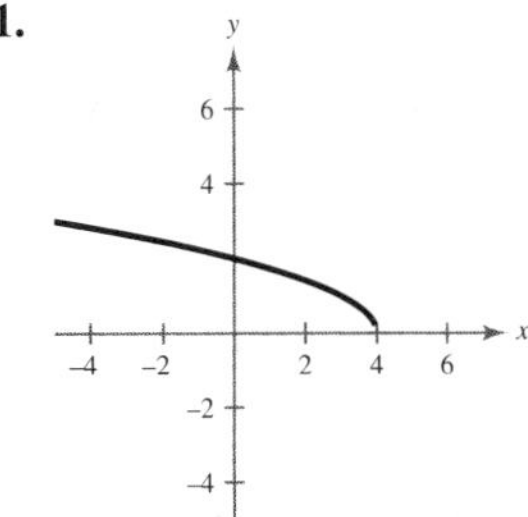

12. $y = 2x + 2$

13. No. To some x there correspond two values of y.

14. $f(6) = \frac{3}{2}$

$f(2)$ is undefined.

$f(s + 2) = \frac{s + 2}{s}$

15. (a) Vertical shrink

(b) Vertical shift two units upward

(c) Horizontal shift two units to the left

16. $h^{-1}(x) = \frac{1}{5}(x + 2)$ **17.** 7 **18.** 5

19. $-1 \pm \frac{\sqrt{3}}{3}$ **20.** 6 **21.** $y = 0.0568x - 1.5808,\ 3.4$

CHAPTER 3

Section 3.1 (page 251)

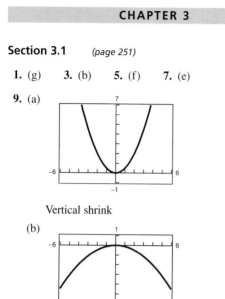

1. (g) **3.** (b) **5.** (f) **7.** (e)

9. (a)

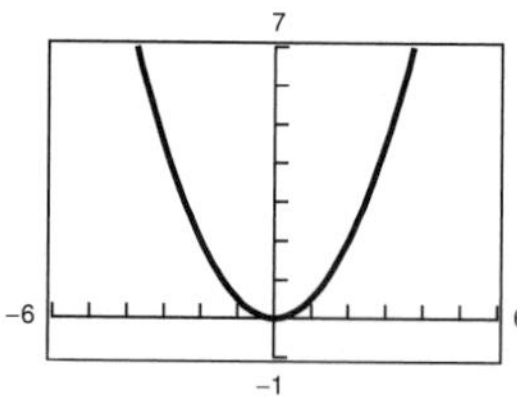

Vertical shrink

(b)

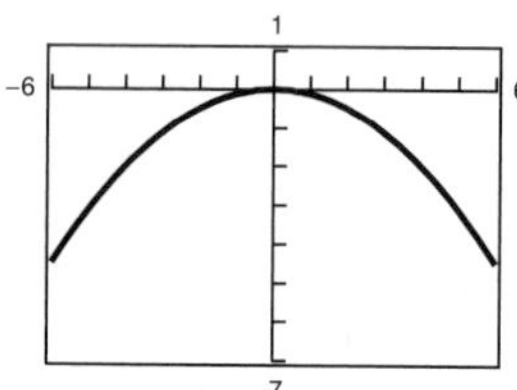

Vertical shrink and reflection in the x-axis

(c)

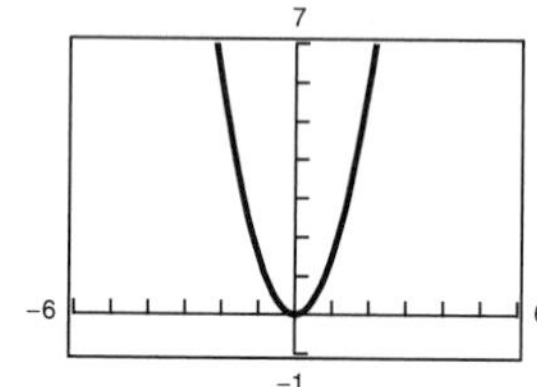

Vertical stretch

(d)

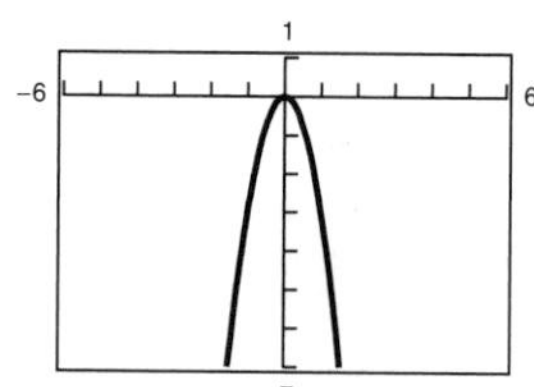

Vertical stretch and reflection in the x-axis

11. (a)

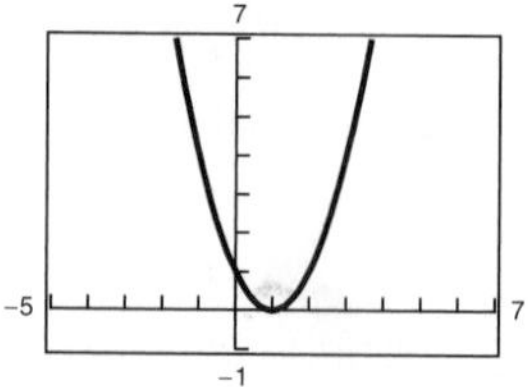

Horizontal translation

(b)

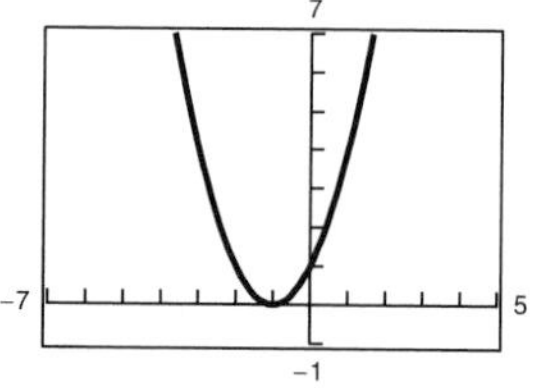

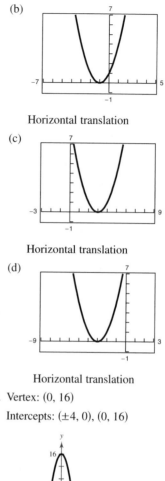

Horizontal translation

(c)

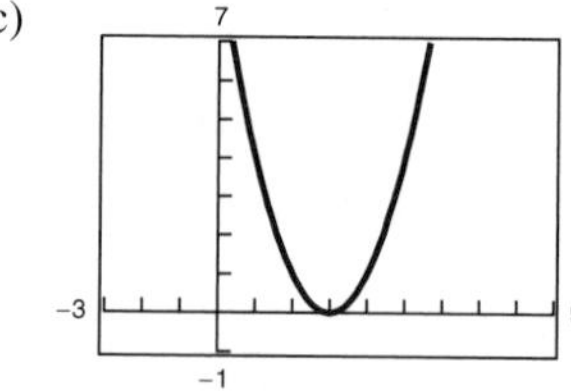

Horizontal translation

(d)

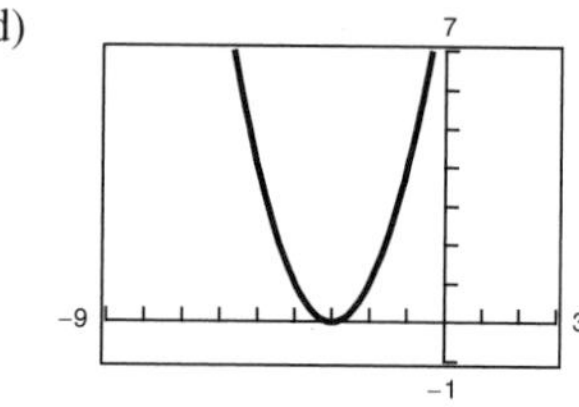

Horizontal translation

13. Vertex: (0, 16)

Intercepts: (±4, 0), (0, 16)

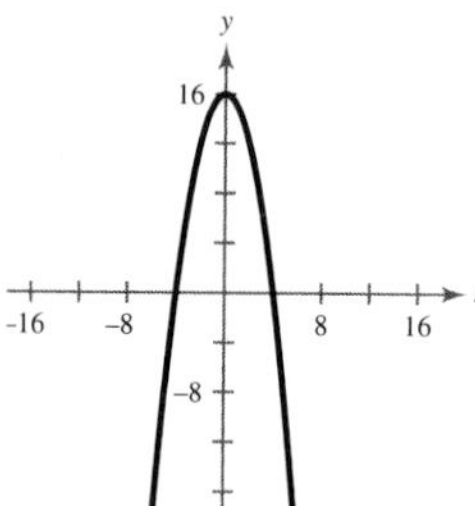

15. Vertex: (4, 0)

Intercepts: (4, 0), (0, 16)

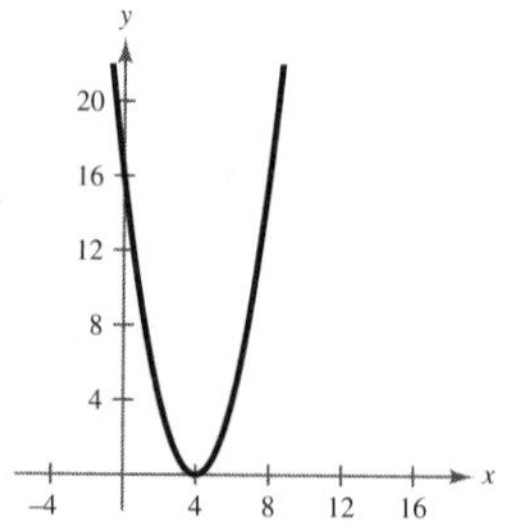

17. Vertex: $\left(\frac{1}{2}, 1\right)$
Intercept: $\left(0, \frac{5}{4}\right)$

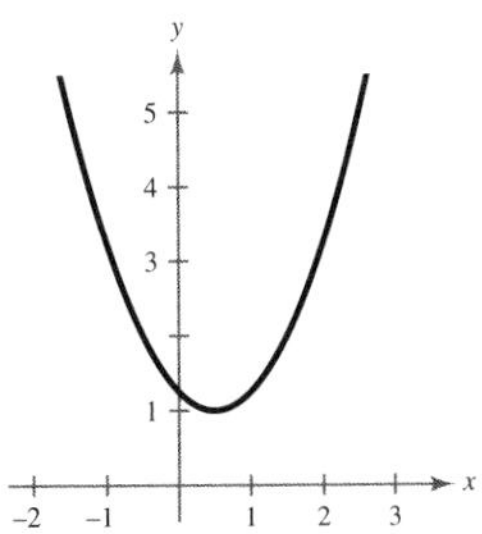

19. Vertex: $(1, 6)$
Intercepts: $\left(1 \pm \sqrt{6}, 0\right)$, $(0, 5)$

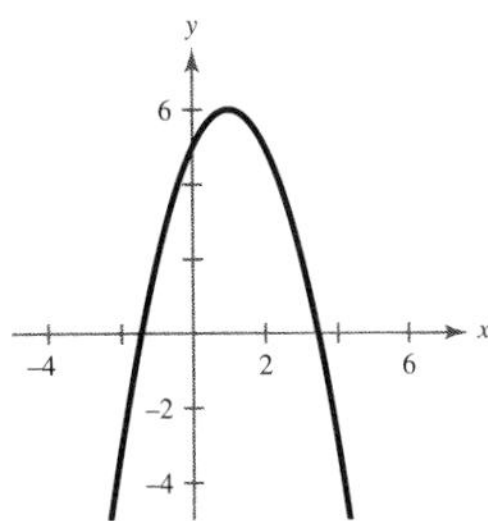

21. Vertex: $\left(\frac{1}{2}, 20\right)$
Intercept: $(0, 21)$

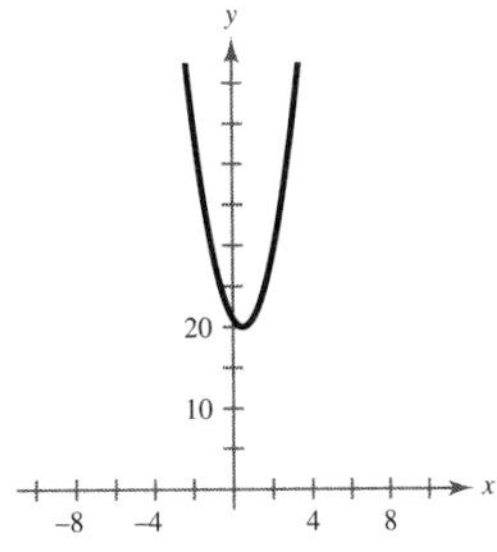

23. Vertex: $(-1, 4)$
Intercepts: $(1, 0)$, $(-3, 0)$, $(0, 3)$

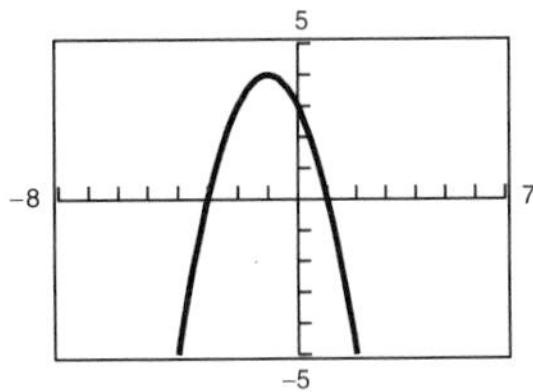

25. Vertex: $(4, -1)$
Intercepts: $\left(4 \pm \frac{1}{2}\sqrt{2}, 0\right)$, $(0, 31)$

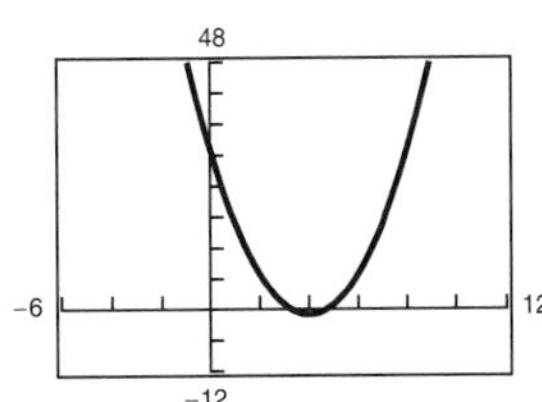

27. $y = (x - 1)^2$ **29.** $y = -2(x + 2)^2 + 2$
31. $f(x) = (x + 2)^2 + 5$ **33.** $f(x) = -\frac{1}{2}(x - 3)^2 + 4$
35. $f(x) = \frac{3}{4}(x - 5)^2 + 12$ **37.** $(\pm 4, 0)$
39. $(5, 0)$, $(-1, 0)$

41.

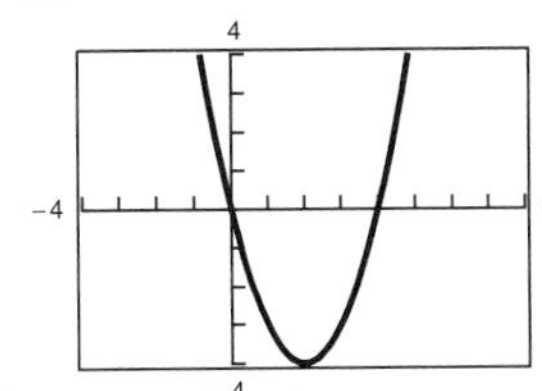

$(0, 0)$, $(4, 0)$

43.

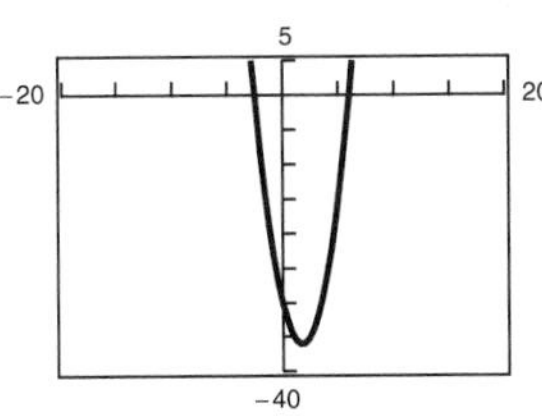

$\left(-\frac{5}{2}, 0\right)$, $(6, 0)$

45. $f(x) = x^2 - 2x - 3$
$g(x) = -x^2 + 2x + 3$

47. $f(x) = 2x^2 + 7x + 3$
$g(x) = -2x^2 - 7x - 3$

49. 55, 55

51. (a) $A = x(50 - x), 0 < x < 50$
(b)

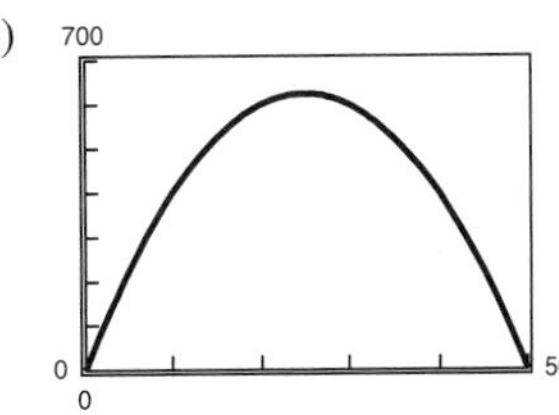

(c) 25 feet × 25 feet

53. 4500 units

55. (a)

x	y	*Area*
2	$\frac{1}{3}[200 - 4(2)]$	$(2)(2)\left(\frac{1}{3}\right)[200 - 4(2)] = 256$
4	$\frac{1}{3}[200 - 4(4)]$	$(2)(4)\left(\frac{1}{3}\right)[200 - 4(4)] \approx 491$
6	$\frac{1}{3}[200 - 4(6)]$	$(2)(6)\left(\frac{1}{3}\right)[200 - 4(6)] = 704$
8	$\frac{1}{3}[200 - 4(8)]$	$(2)(8)\left(\frac{1}{3}\right)[200 - 4(8)] = 896$
10	$\frac{1}{3}[200 - 4(10)]$	$(2)(10)\left(\frac{1}{3}\right)[200 - 4(10)] \approx 1067$
12	$\frac{1}{3}[200 - 4(12)]$	$(2)(12)\left(\frac{1}{3}\right)[200 - 4(12)] = 1216$

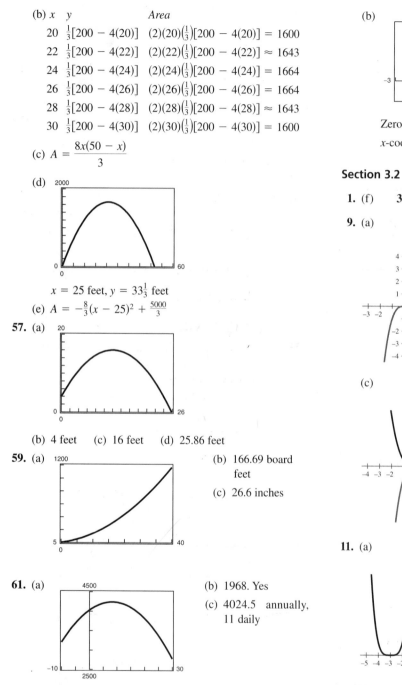

(b)

x	y	Area
20	$\frac{1}{3}[200 - 4(20)]$	$(2)(20)(\frac{1}{3})[200 - 4(20)] = 1600$
22	$\frac{1}{3}[200 - 4(22)]$	$(2)(22)(\frac{1}{3})[200 - 4(22)] \approx 1643$
24	$\frac{1}{3}[200 - 4(24)]$	$(2)(24)(\frac{1}{3})[200 - 4(24)] = 1664$
26	$\frac{1}{3}[200 - 4(26)]$	$(2)(26)(\frac{1}{3})[200 - 4(26)] = 1664$
28	$\frac{1}{3}[200 - 4(28)]$	$(2)(28)(\frac{1}{3})[200 - 4(28)] \approx 1643$
30	$\frac{1}{3}[200 - 4(30)]$	$(2)(30)(\frac{1}{3})[200 - 4(30)] = 1600$

(c) $A = \dfrac{8x(50 - x)}{3}$

(d)

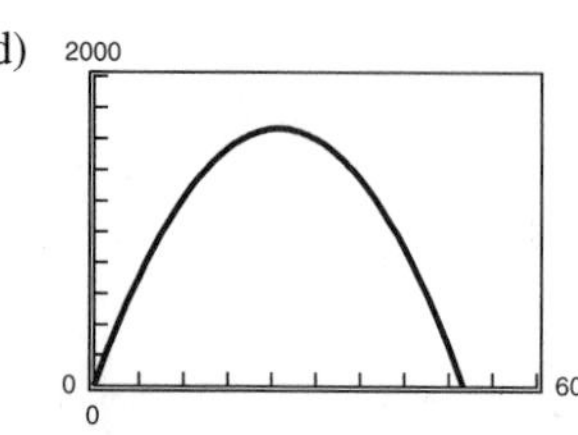

$x = 25$ feet, $y = 33\frac{1}{3}$ feet

(e) $A = -\frac{8}{3}(x - 25)^2 + \frac{5000}{3}$

57. (a)

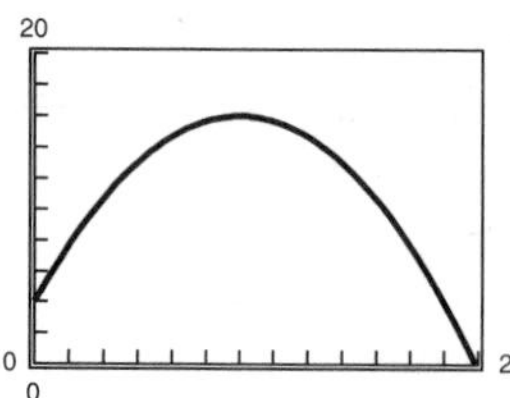

(b) 4 feet (c) 16 feet (d) 25.86 feet

59. (a)

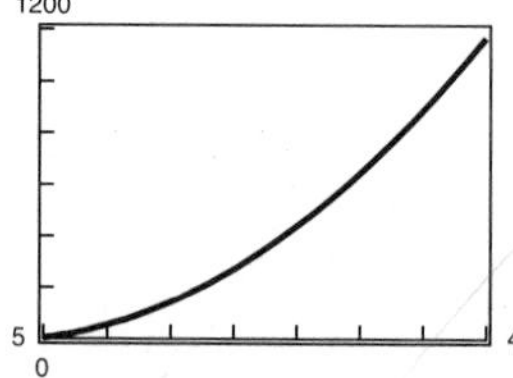

(b) 166.69 board feet

(c) 26.6 inches

61. (a)

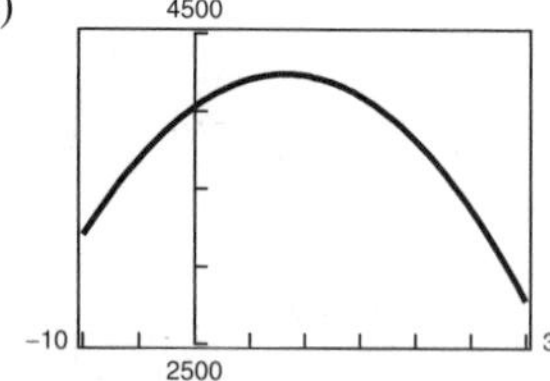

(b) 1968. Yes

(c) 4024.5 annually, 11 daily

63. (a) Answers will vary.

(b)

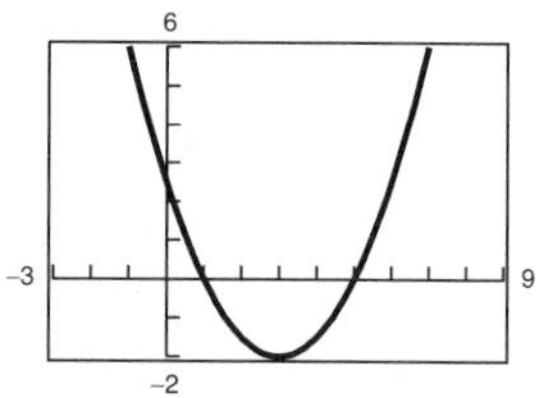

Zeros: $x = 1, 5$

x-coordinate of the vertex: 3

Section 3.2 *(page 263)*

1. (f) **3.** (c) **5.** (e) **7.** (g)

9. (a)

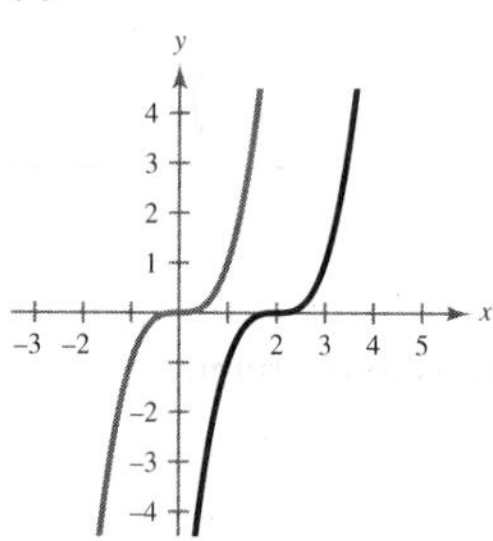

(b)

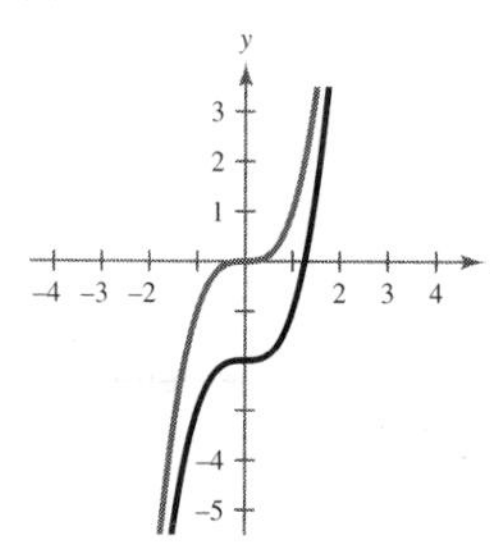

(c)

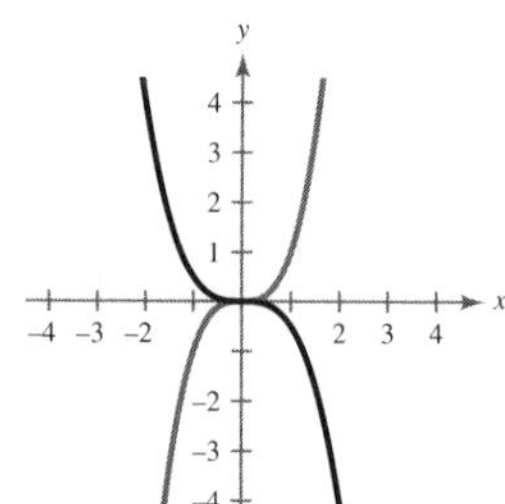

(d)

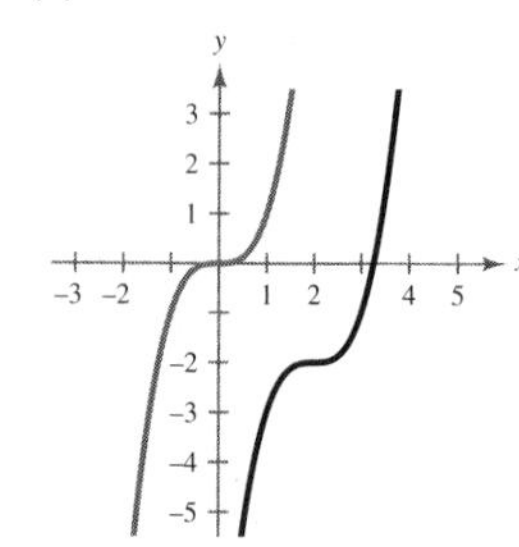

11. (a)

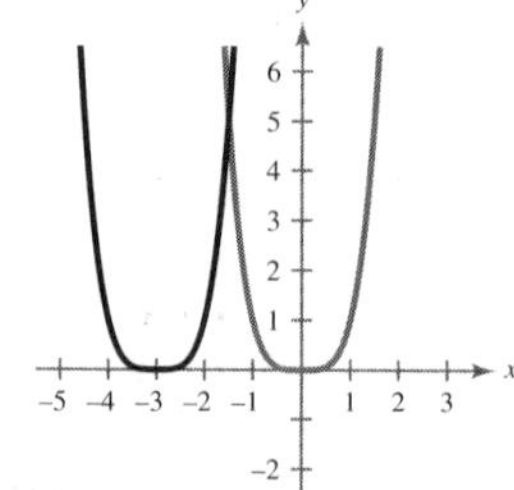

(b)

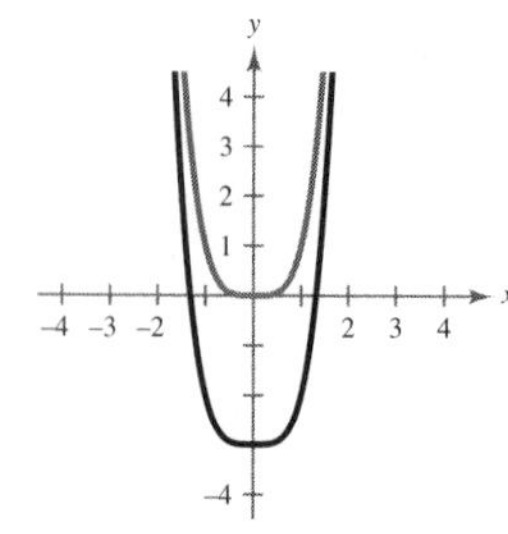

(c)

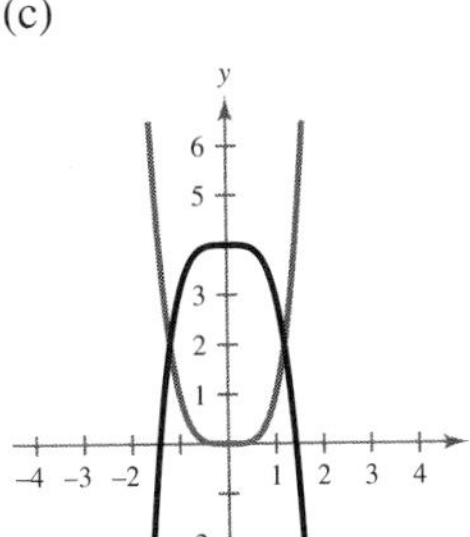

(d)

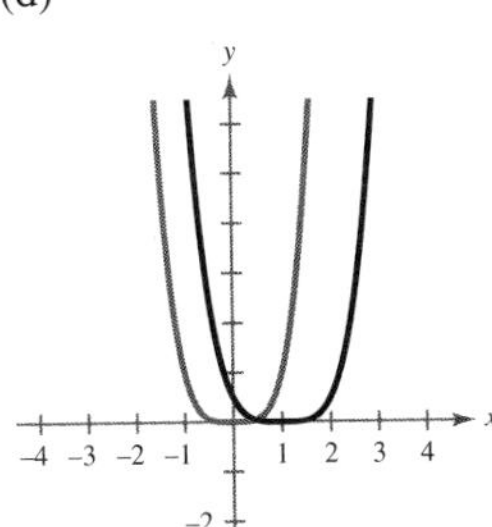

13.

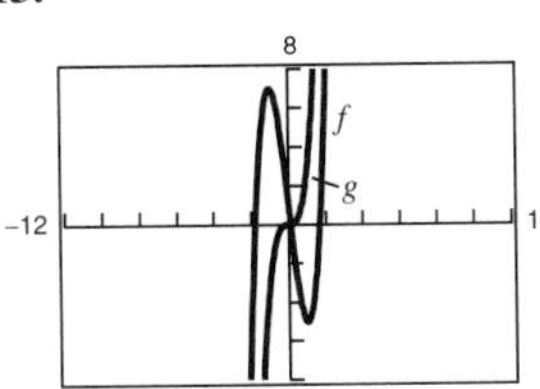

15.

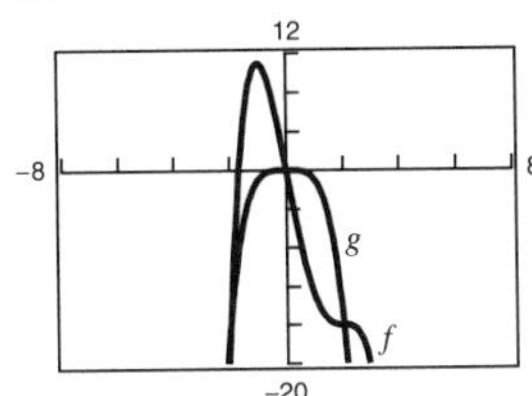

17. Rises to the left. Rises to the right. **19.** Falls to the left. Falls to the right.

21. Falls to the left. Rises to the right. **23.** Rises to the left. Falls to the right.

25. Falls to the left. Falls to the right. **27.** ± 5 **29.** 3

31. $1, -2$ **33.** 2, 0

35.

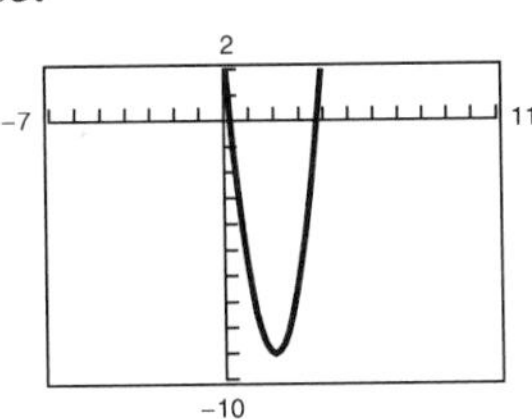

$(0.268, 0), (3.732, 0)$

37.

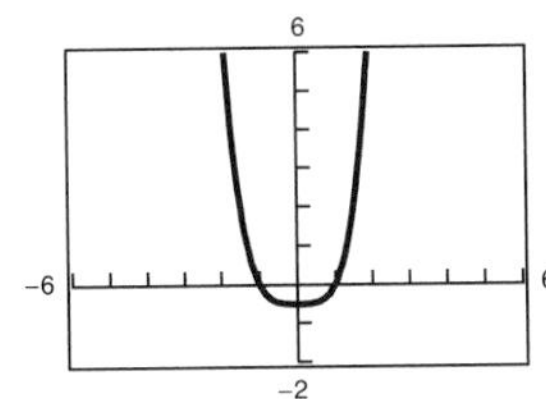

$(-1, 0), (1, 0)$

39.

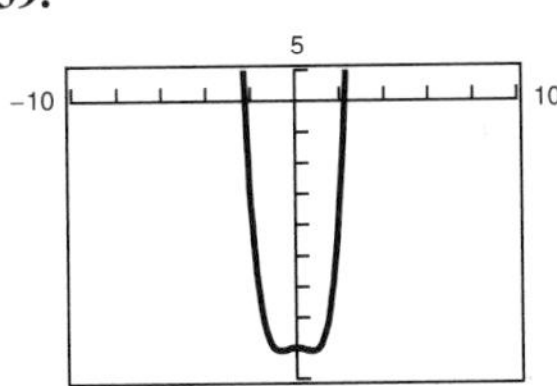

$(-2.236, 0), (2.236, 0)$

41.

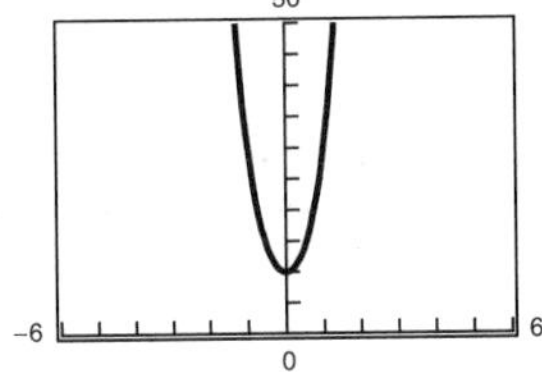

No x-intercepts

43.

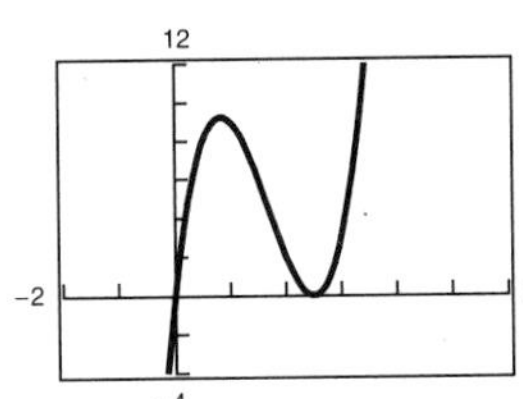

$(0, 0), \left(\frac{5}{2}, 0\right)$

45.

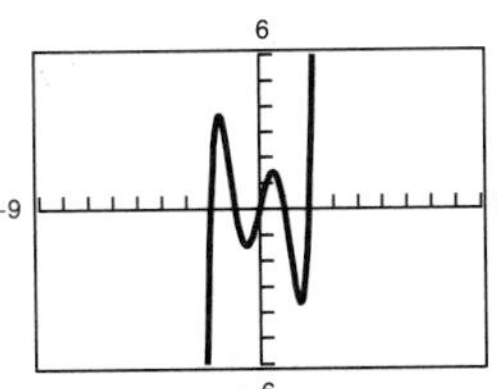

$(0, 0), (\pm 1, 0), (\pm 2, 0)$

47. $f(x) = x^2 - 10x$ **49.** $f(x) = x^2 + 4x - 12$

51. $f(x) = x^3 + 5x^2 + 6x$

53. $f(x) = x^4 - 4x^3 - 9x^2 + 36x$ **55.** $f(x) = x^2 - 2x - 2$

57. (a)

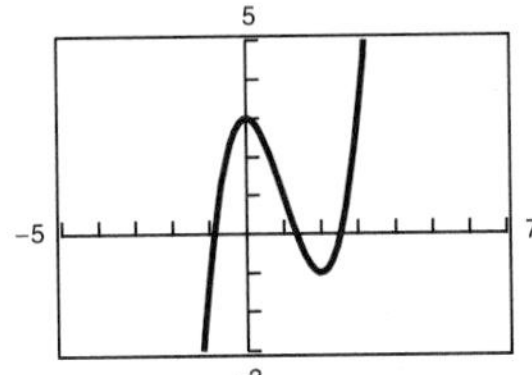

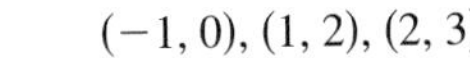
$(-1, 0), (1, 2), (2, 3)$

(b) $-0.879, 1.347, 2.532$

59. (a)

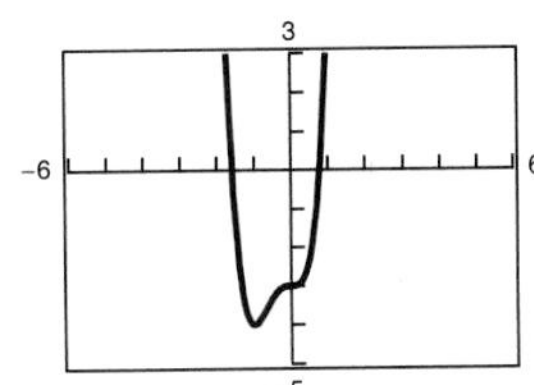

$(-2, -1), (0, 1)$

(b) $-1.585, 0.779$

61.

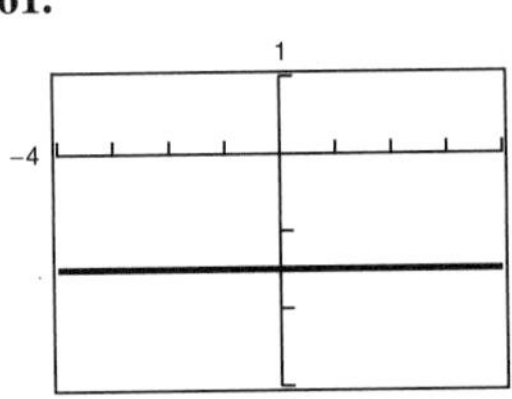

Xmin = -4
Xmax = 4
Xscl = 1
Ymin = -3
Ymax = 1
Yscl = 1

63.

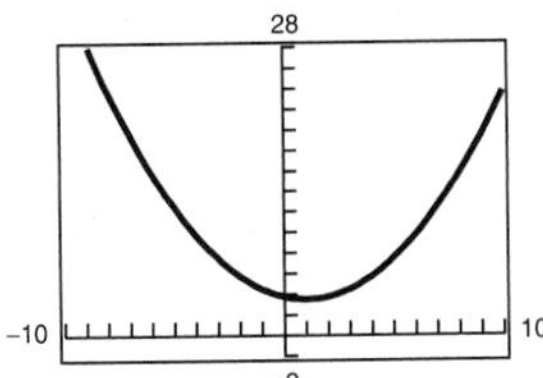

Xmin = -10
Xmax = 10
Xscl = 1
Ymin = -2
Ymax = 28
Yscl = 2

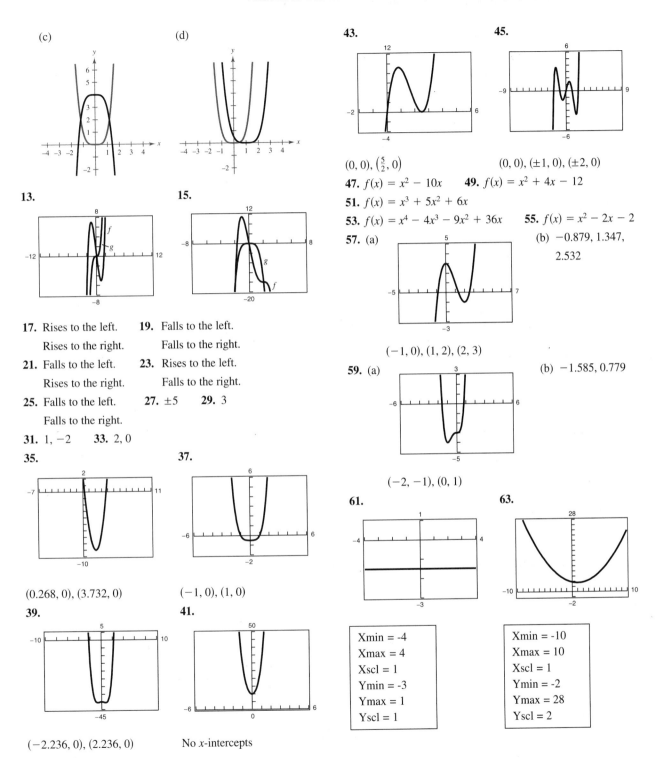

65.

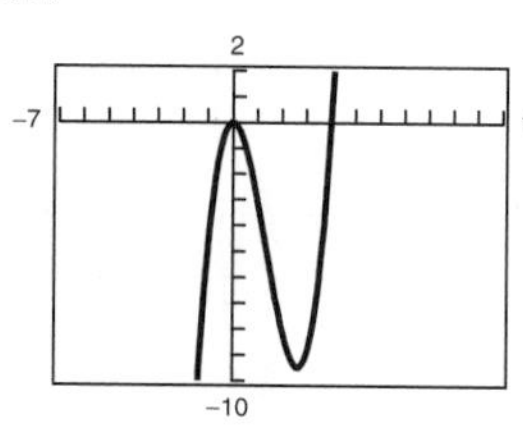

Two x-intercepts

67.

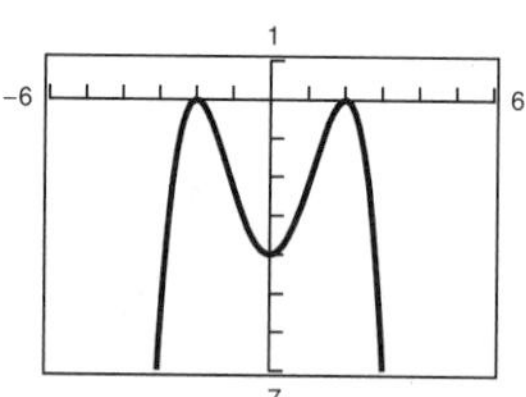

y-axis symmetry
Two x-intercepts

69.

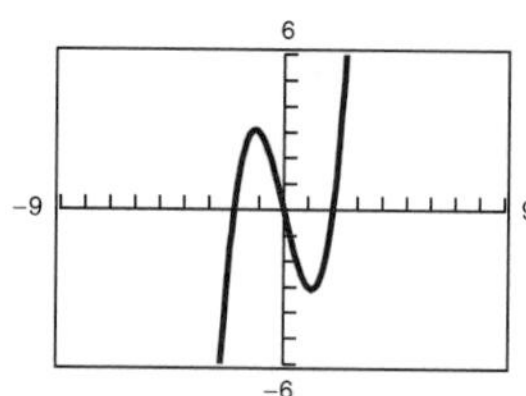

Symmetric to the origin
Three x-intercepts

71.

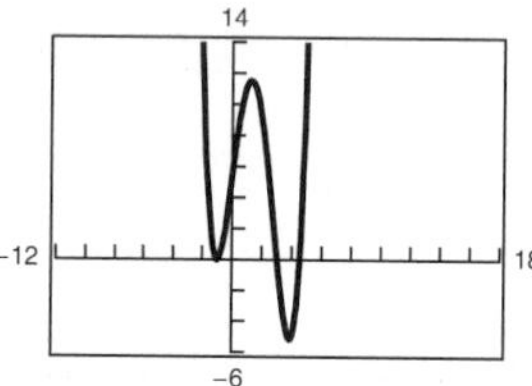

Three x-intercepts

73.

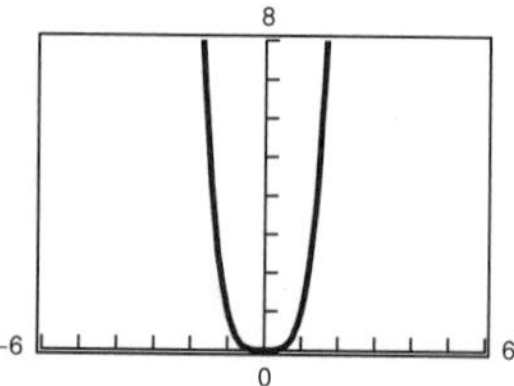

(a) Vertical shift of two units; even
(b) Horizontal shift of two units; neither even nor odd
(c) Reflection in the y-axis; even
(d) Reflection in the x-axis; even
(e) Horizontal stretch; even
(f) Vertical shrink; even
(g) $g(x) = x^3$; odd
(h) $g(x) = x^{16}$; even

75. (a) Answers will vary. (b) $0 < x < 6$

(c)

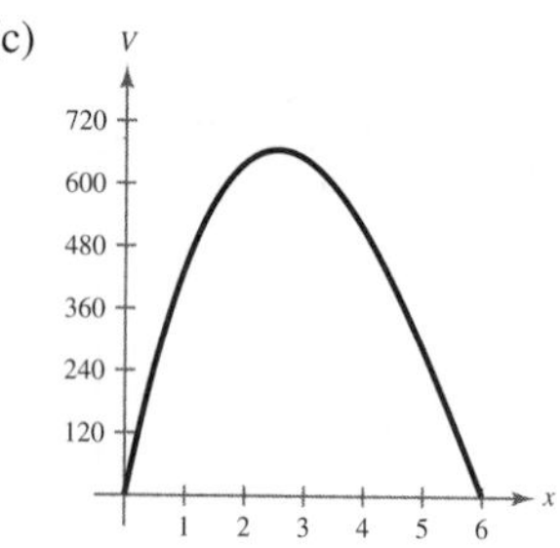

$x \approx 2.54$

77. (200, 160)

79. (a) $y_1 = -0.158t^3 + 2.850t^2 - 3.814t + 74.703$

(b) $y_2 = -0.007t^3 + 0.196t^2 + 2.533t + 60.844$

(c)

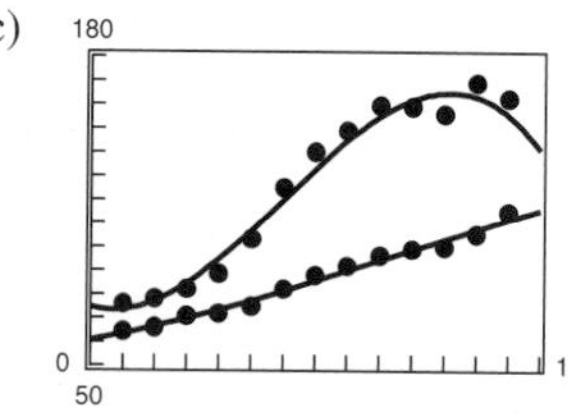

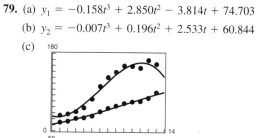

The median price of homes in the South is less than in the Northeast.

Section 3.3 *(page 277)*

1.

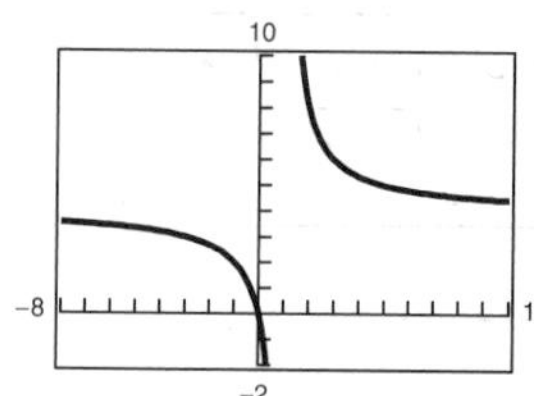

3.

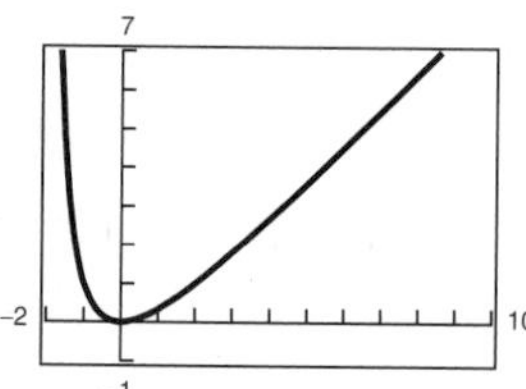

5.

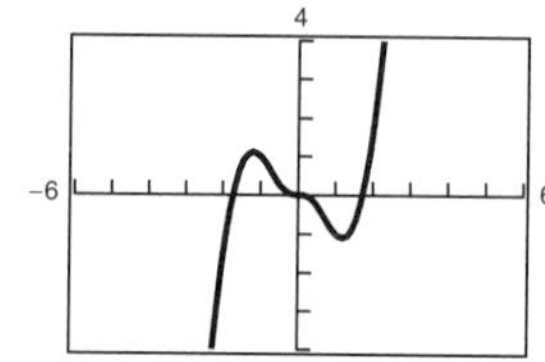

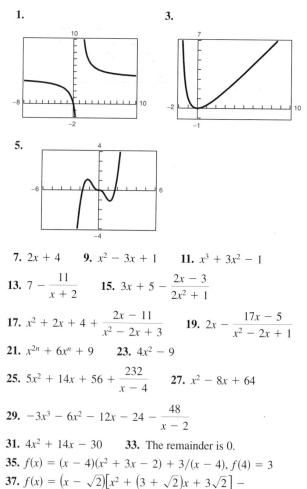

7. $2x + 4$ **9.** $x^2 - 3x + 1$ **11.** $x^3 + 3x^2 - 1$

13. $7 - \dfrac{11}{x + 2}$ **15.** $3x + 5 - \dfrac{2x - 3}{2x^2 + 1}$

17. $x^2 + 2x + 4 + \dfrac{2x - 11}{x^2 - 2x + 3}$ **19.** $2x - \dfrac{17x - 5}{x^2 - 2x + 1}$

21. $x^{2n} + 6x^n + 9$ **23.** $4x^2 - 9$

25. $5x^2 + 14x + 56 + \dfrac{232}{x - 4}$ **27.** $x^2 - 8x + 64$

29. $-3x^3 - 6x^2 - 12x - 24 - \dfrac{48}{x - 2}$

31. $4x^2 + 14x - 30$ **33.** The remainder is 0.

35. $f(x) = (x - 4)(x^2 + 3x - 2) + 3/(x - 4)$, $f(4) = 3$

37. $f(x) = (x - \sqrt{2})[x^2 + (3 + \sqrt{2})x + 3\sqrt{2}] - 8/(x - \sqrt{2})$, $f(\sqrt{2}) = -8$

39. (a) 1 (b) 4 (c) 4 (d) 1954

41. $(x - 2)(x + 3)(x - 1)$
Zeros: $2, -3, 1$

43. $(2x - 1)(x - 5)(x - 2)$
Zeros: $\frac{1}{2}, 5, 2$

45. $(x + \sqrt{3})(x - \sqrt{3})(x + 2)$
Zeros: $\pm\sqrt{3}, -2$

47. $(x - 1)(x - 1 - \sqrt{3})(x - 1 + \sqrt{3})$
Zeros: $1 \pm \sqrt{3}, 1$

49. (a) $R = 16.823 + 1.415t - 0.115t^2 - 0.023t^3$

(b)

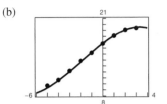

(c)

t	-5	-4	-3	-2	-1
R	9.75	10.80	12.16	13.72	15.32

t	0	1	2	3
R	16.82	18.10	19.01	19.41

(d) 16.205. No; the model will fall to the right and become negative.

51. $\pm 1, \pm 3$

53. $\pm 1, \pm 3, \pm 5, \pm 9, \pm 15, \pm 45, \pm\frac{1}{2}, \pm\frac{3}{2}, \pm\frac{5}{2}, \pm\frac{9}{2}, \pm\frac{15}{2}, \pm\frac{45}{2}$

55. (a) $\pm 1, \pm 2, \pm 4$

(b)

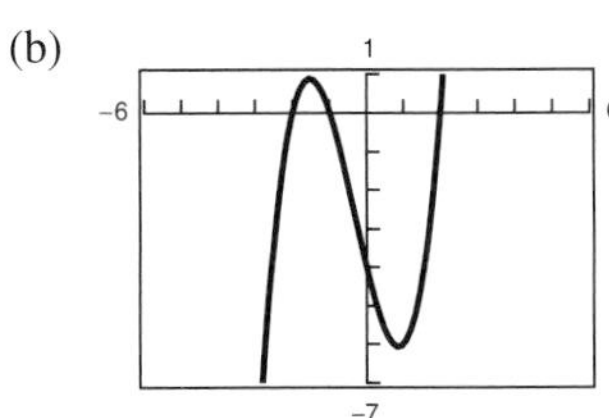

(c) $-2, -1, 2$

57. (a) $\pm 1, \pm 3, \pm\frac{1}{2}, \pm\frac{3}{2}, \pm\frac{1}{4}, \pm\frac{3}{4}$

(b)

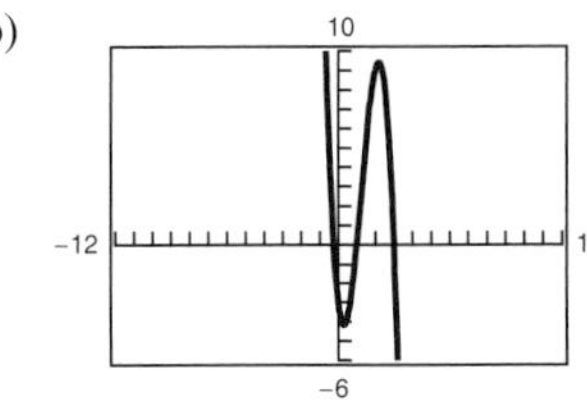

(c) $-\frac{1}{4}, 1, 3$

59. (a) $\pm 1, \pm 2, \pm 4, \pm 8, \pm\frac{1}{2}$

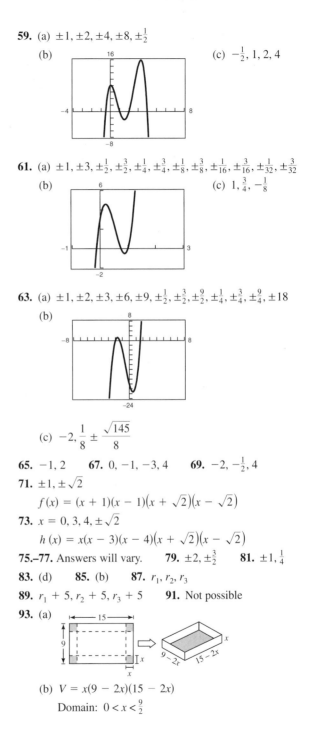

(b)

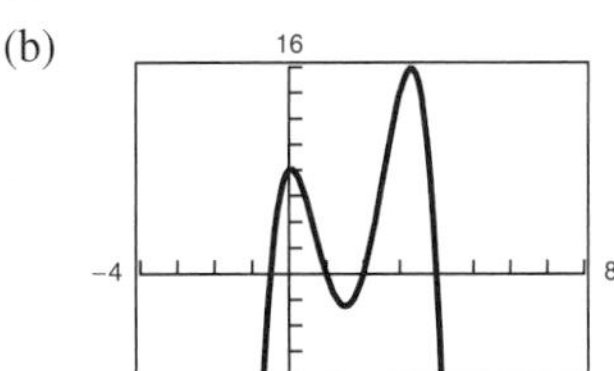

(c) $-\frac{1}{2}, 1, 2, 4$

61. (a) $\pm 1, \pm 3, \pm\frac{1}{2}, \pm\frac{3}{2}, \pm\frac{1}{4}, \pm\frac{3}{4}, \pm\frac{1}{8}, \pm\frac{3}{8}, \pm\frac{1}{16}, \pm\frac{3}{16}, \pm\frac{1}{32}, \pm\frac{3}{32}$

(b)

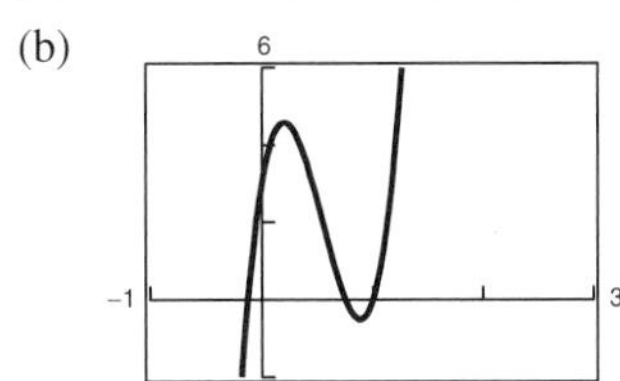

(c) $1, \frac{3}{4}, -\frac{1}{8}$

63. (a) $\pm 1, \pm 2, \pm 3, \pm 6, \pm 9, \pm\frac{1}{2}, \pm\frac{3}{2}, \pm\frac{9}{2}, \pm\frac{1}{4}, \pm\frac{3}{4}, \pm\frac{9}{4}, \pm 18$

(b)

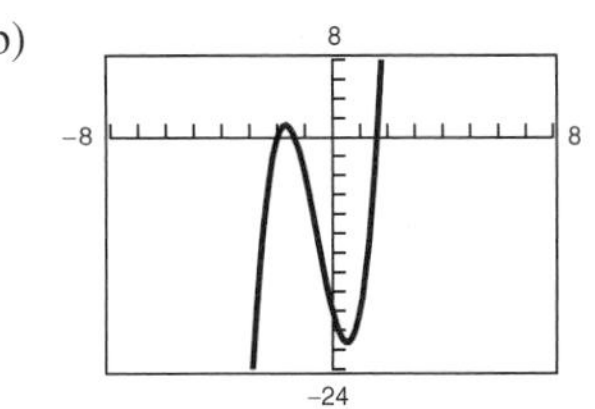

(c) $-2, \frac{1}{8} \pm \frac{\sqrt{145}}{8}$

65. $-1, 2$ **67.** $0, -1, -3, 4$ **69.** $-2, -\frac{1}{2}, 4$

71. $\pm 1, \pm\sqrt{2}$
$f(x) = (x + 1)(x - 1)(x + \sqrt{2})(x - \sqrt{2})$

73. $x = 0, 3, 4, \pm\sqrt{2}$
$h(x) = x(x - 3)(x - 4)(x + \sqrt{2})(x - \sqrt{2})$

75.–77. Answers will vary. **79.** $\pm 2, \pm\frac{3}{2}$ **81.** $\pm 1, \frac{1}{4}$

83. (d) **85.** (b) **87.** r_1, r_2, r_3

89. $r_1 + 5, r_2 + 5, r_3 + 5$ **91.** Not possible

93. (a)

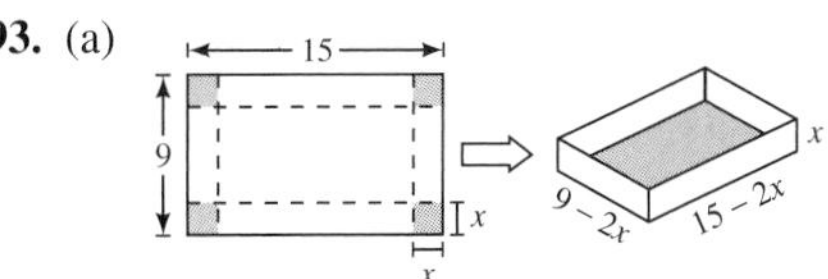

(b) $V = x(9 - 2x)(15 - 2x)$
Domain: $0 < x < \frac{9}{2}$

(c) 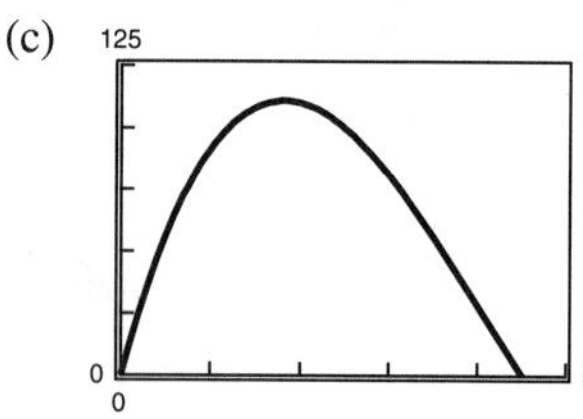

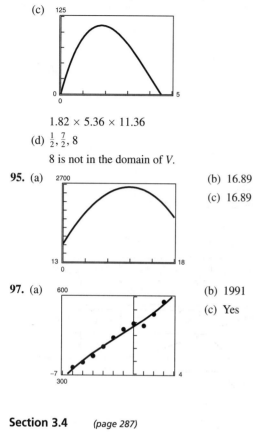

$1.82 \times 5.36 \times 11.36$

(d) $\frac{1}{2}, \frac{7}{2}, 8$

8 is not in the domain of V.

95. (a) 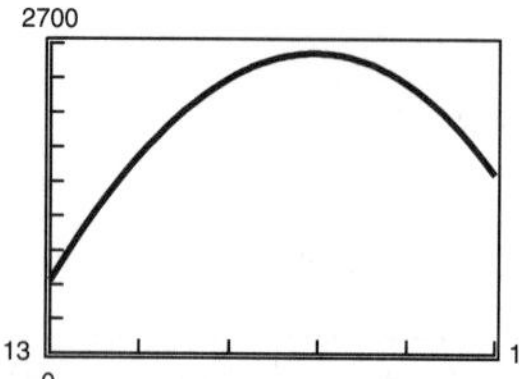

(b) 16.89
(c) 16.89

97. (a) 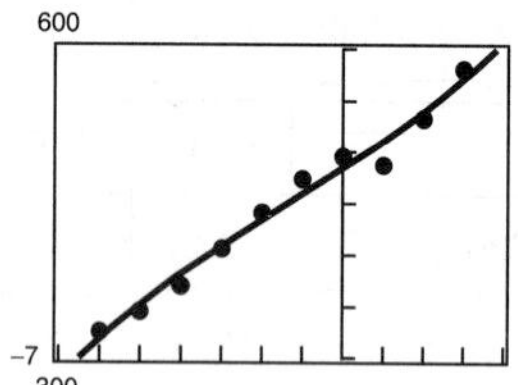

(b) 1991
(c) Yes

Section 3.4 *(page 287)*

1. 0, 6, 6 **3.** 3, 2, $\pm 3i$ **5.** (4, 0); same

7. No intercepts; same

9. $\pm 5i$
$(x + 5i)(x - 5i)$

11. $2 \pm \sqrt{3}$
$\left(x - 2 - \sqrt{3}\right)\left(x - 2 + \sqrt{3}\right)$

13. $\pm 3, \pm 3i$
$(x + 3)(x - 3)(x + 3i)(x - 3i)$

15. $1 \pm i$
$(z - 1 + i)(z - 1 - i)$

17. $-5, 4 \pm 3i$
$(t + 5)(t - 4 + 3i)(t - 4 - 3i)$

19. $-\frac{3}{4}, 1 \pm \frac{1}{2}i$
$(4x + 3)\left(x - 1 - \frac{1}{2}i\right)\left(x - 1 + \frac{1}{2}i\right)$

21. $-\frac{1}{5}, 1 \pm \sqrt{5}i$
$(5x + 1)\left(x - 1 + \sqrt{5}i\right)\left(x - 1 - \sqrt{5}i\right)$

23. $\pm i, \pm 3i$
$(x + i)(x - i)(x + 3i)(x - 3i)$

25. $2, 2, \pm 2i$
$(x - 2)^2(x + 2i)(x - 2i)$

27. $-2, -\frac{1}{2}, \pm i$
$(x + 2)(2x + 1)(x - i)(x + i)$

29. $x^3 - x^2 + 25x - 25$ **31.** $x^3 - 10x^2 + 33x - 34$

33. $x^4 + 37x^2 + 36$ **35.** $x^4 + 8x^3 + 9x^2 - 10x + 100$

37. (a) $(x^2 + 9)(x^2 - 3)$ (b) $(x^2 + 9)\left(x + \sqrt{3}\right)\left(x - \sqrt{3}\right)$
(c) $(x + 3i)(x - 3i)\left(x + \sqrt{3}\right)\left(x - \sqrt{3}\right)$

39. (a) $(x^2 - 2x - 2)(x^2 - 2x + 3)$
(b) $\left(x - 1 + \sqrt{3}\right)\left(x - 1 - \sqrt{3}\right)(x^2 - 2x + 3)$
(c) $\left(x - 1 + \sqrt{3}\right)\left(x - 1 - \sqrt{3}\right) \cdot$
$\left(x - 1 + \sqrt{2}i\right)\left(x - 1 - \sqrt{2}i\right)$

41. $-\frac{3}{2}, \pm 5i$ **43.** $\pm 2i, 1, -\frac{1}{2}$ **45.** $-3 \pm i, \frac{1}{4}$

47. (a) Answers will vary. (b) $1, 2, -3 \pm \sqrt{2}i$

49. (a) Answers will vary. (b) $\frac{3}{4}, \frac{1}{2} \pm \frac{\sqrt{5}}{2}i$

51. f does not have real coefficients **53.** (a) No (b) No

55. No. Setting $P = 9{,}000{,}000$ and solving the resulting equation yields imaginary roots.

57. $x^2 - 2ax + a^2 + b^2$

59. 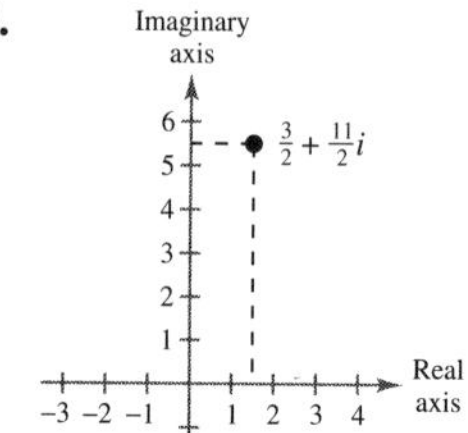

Section 3.5 *(page 295)*

1. (a)

x	$f(x)$	x	$f(x)$	x	$f(x)$
0.5	−2	1.5	2	5	0.25
0.9	−10	1.1	10	10	$0.\overline{1}$
0.99	−100	1.01	100	100	$0.\overline{01}$
0.999	−1000	1.001	1000	1000	$0.\overline{001}$

(b) Vertical asymptote: $x = 1$
Horizontal asymptote: $y = 0$

(c) Domain: all $x \neq 1$

3. (a)

x	$f(x)$
0.5	3
0.9	27
0.99	297
0.999	2997

x	$f(x)$
1.5	9
1.1	33
1.01	303
1.001	3003

x	$f(x)$
5	3.75
10	$3.\overline{33}$
100	$3.\overline{03}$
1000	$3.\overline{003}$

(b) Vertical asymptote: $x = 1$

Horizontal asymptotes: $y = \pm 3$

(c) Domain: all $x \neq 1$

5. (a)

x	$f(x)$
0.5	−1
0.9	−12.79
0.99	−147.8
0.999	−1498

x	$f(x)$
1.5	5.4
1.1	17.29
1.01	152.3
1.001	1502.3

x	$f(x)$
5	3.125
10	$3.\overline{03}$
100	$3.\overline{0003}$
1000	3

(b) Vertical asymptotes: $x = \pm 1$

Horizontal asymptote: $y = 3$

(c) Domain: all $x \neq \pm 1$

7. (a) **9.** (c) **11.** (b)

13. Domain: all $x \neq 0$

Vertical asymptote: $x = 0$

Horizontal asymptote: $y = 0$

15. Domain: all $x \neq 2$

Vertical asymptote: $x = 2$

Horizontal asymptote: $y = -1$

17. Domain: all $x \neq \pm 1$

Vertical asymptotes: $x = \pm 1$

19. Domain: all reals

Horizontal asymptote: $y = 3$

21. (a) Domain of f: all $x \neq -2$

Domain of g: all real numbers

(b) Vertical asymptote: None

(c)

x	−4	−3	−2.5	−2	−1.5	−1	0
$f(x)$	−6	−5	−4.5	Undef.	−3.5	−3	−2
$g(x)$	−6	−5	−4.5	−4	−3.5	−3	−2

(d) Differ only where f is undefined.

23. (a) Domain of f: all $x \neq 0, 3$; Domain of g: all $x \neq 0$

(b) Vertical asymptote: $x = 0$

(c)

x	−1	−0.5	0	0.5	2	3	4
$f(x)$	−1	−2	Undef.	2	$\frac{1}{2}$	Undef.	$\frac{1}{4}$
$g(x)$	−1	−2	Undef.	2	$\frac{1}{2}$	$\frac{1}{3}$	$\frac{1}{4}$

(d) Differ only where f is undefined and g is defined.

25. $f(x) = \dfrac{1}{x^2 + x - 2}$ **27.** $f(x) = \dfrac{2x^2}{1 + x^2}$

29. (a) 4 (b) Less than (c) Greater than

31. (a) 2 (b) Greater than (c) Less than

33. ± 2 **35.** 5

37. (a) \$28.33 million (b) \$170 million

(c) \$765 million

(d) No. The function is undefined.

39. (a)

M	200	400	600	800	1000
t	0.472	0.596	0.710	0.817	0.916

M	1200	1400	1600	1800	2000
t	1.009	1.096	1.178	1.255	1.328

The greater the mass, the more time required per oscillation.

(b) $M \approx 1306$ grams

41. (a) 333 deer, 500 deer, 800 deer (b) 1500

43. (a)

n	1	2	3	4	5
P	0.50	0.74	0.82	0.86	0.89

n	6	7	8	9	10
P	0.91	0.92	0.93	0.94	0.95

The percentage approaches 1 as n increases.

(b) 100%

45. $\frac{9}{2}$ **47.** $-\frac{7}{2}, 5$

Section 3.6 *(page 304)*

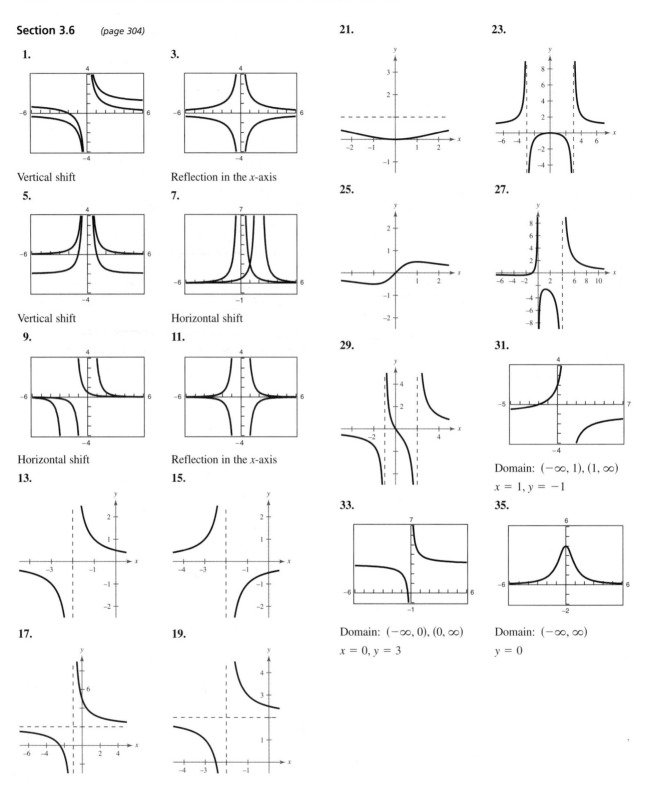

1.

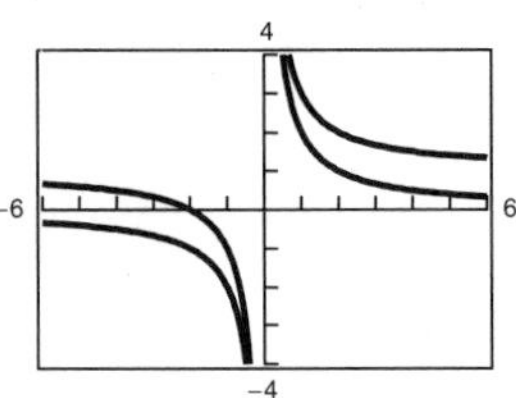

Vertical shift

3.

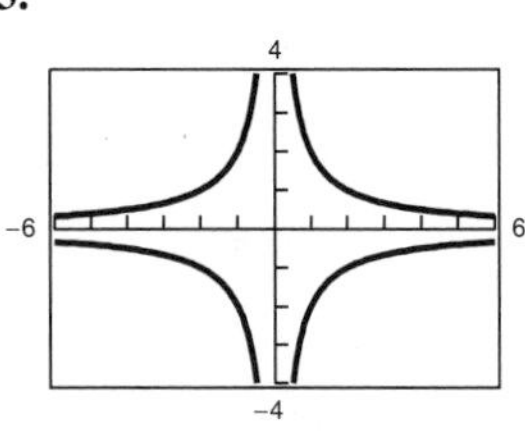

Reflection in the x-axis

5.

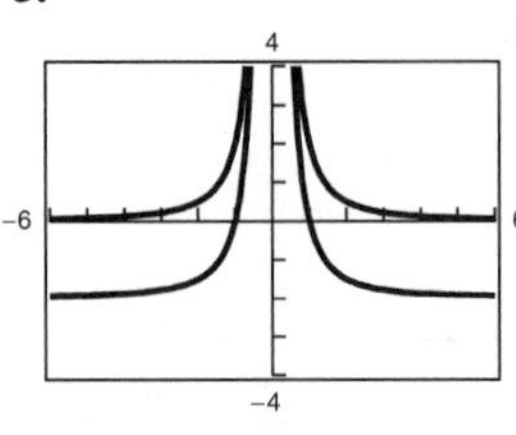

Vertical shift

7.

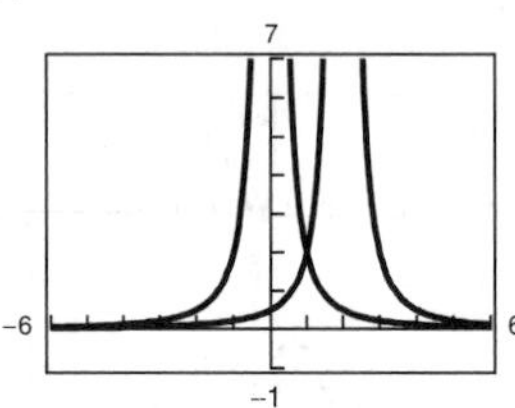

Horizontal shift

9.

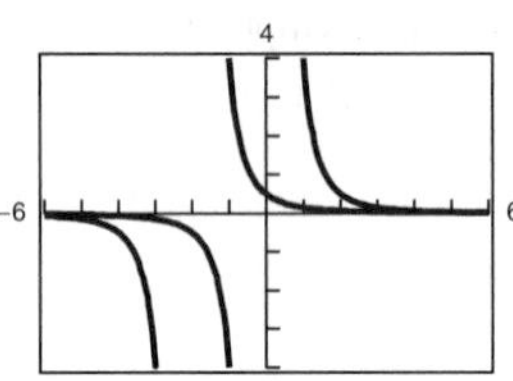

Horizontal shift

11.

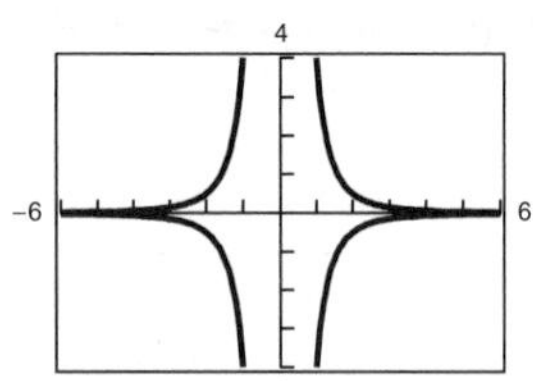

Reflection in the x-axis

13.

15.

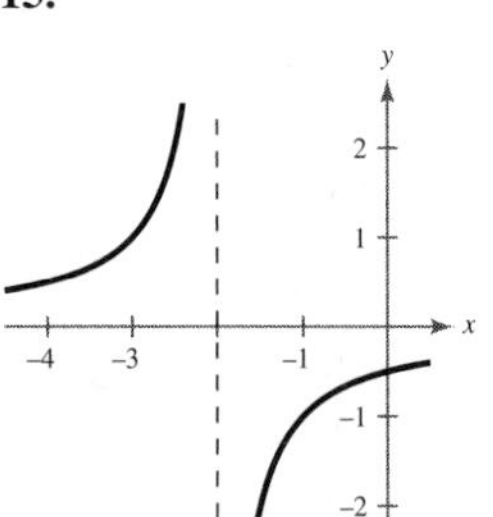

17.

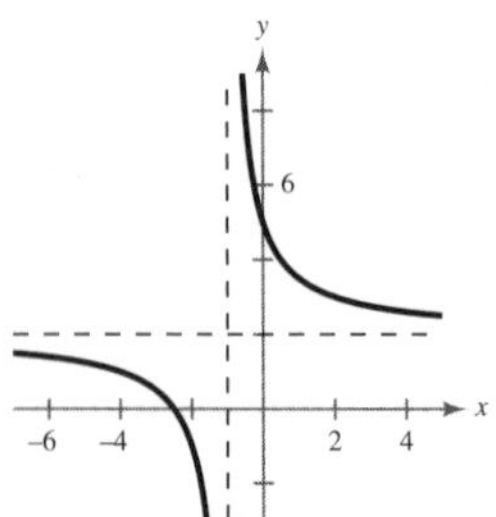

19.

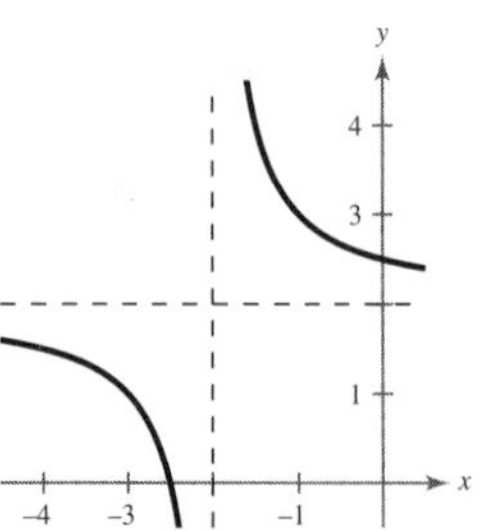

21.

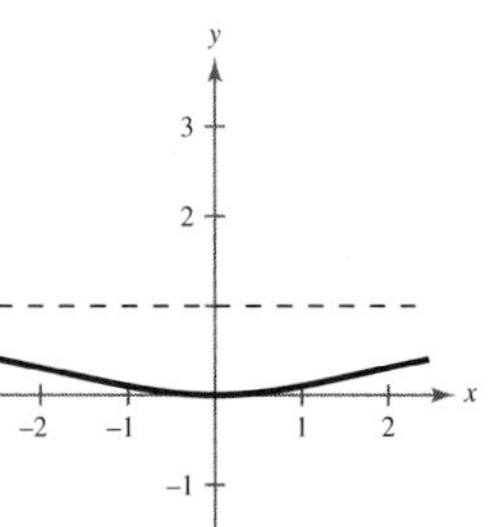

23.

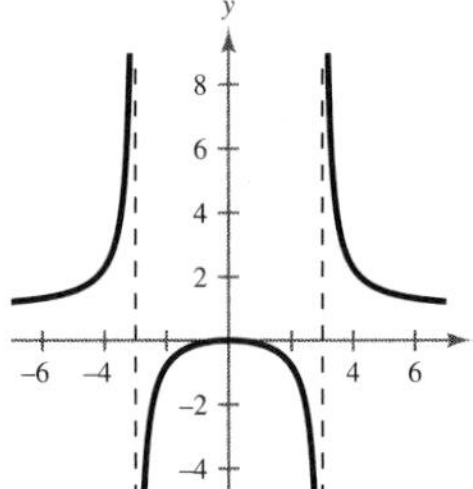

25.

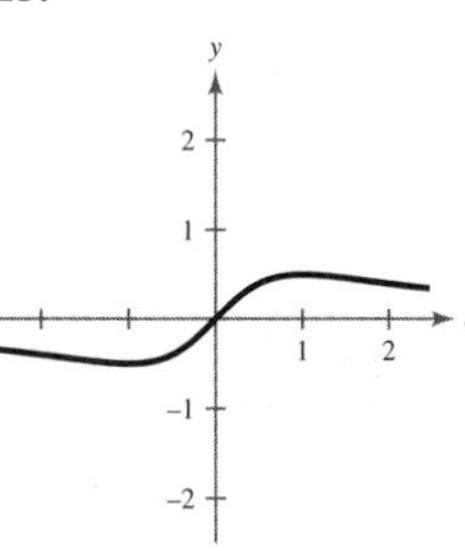

27.

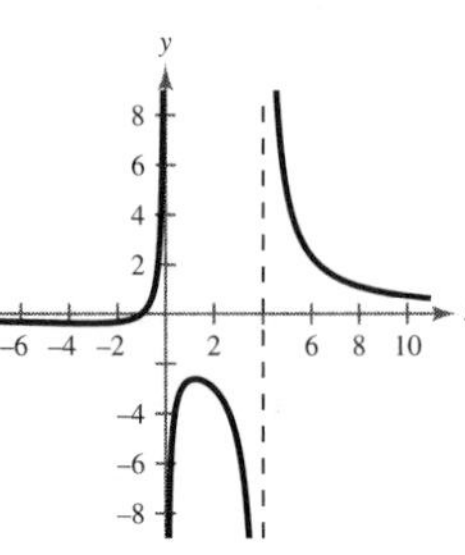

29.

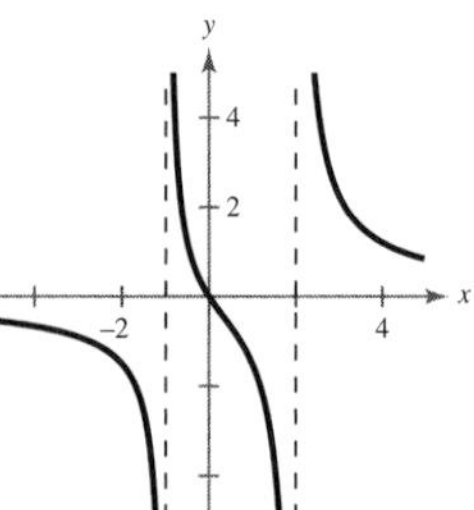

31.

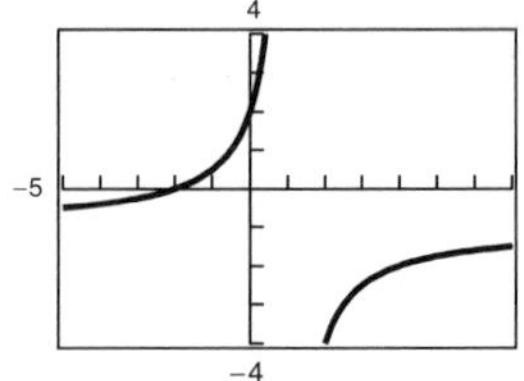

Domain: $(-\infty, 1), (1, \infty)$

$x = 1, y = -1$

33.

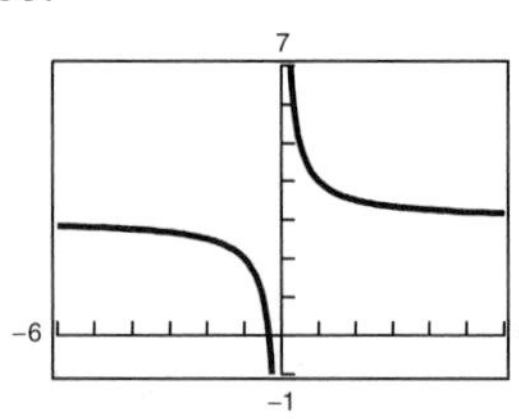

Domain: $(-\infty, 0), (0, \infty)$

$x = 0, y = 3$

35.

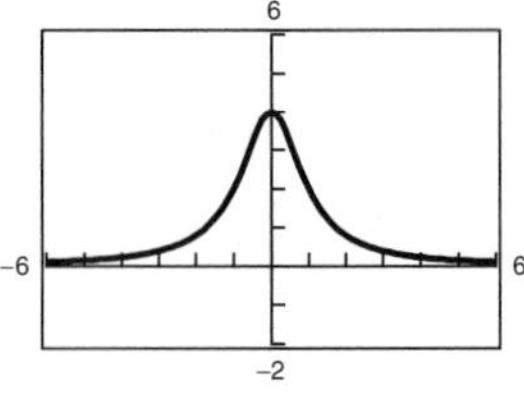

Domain: $(-\infty, \infty)$

$y = 0$

37.

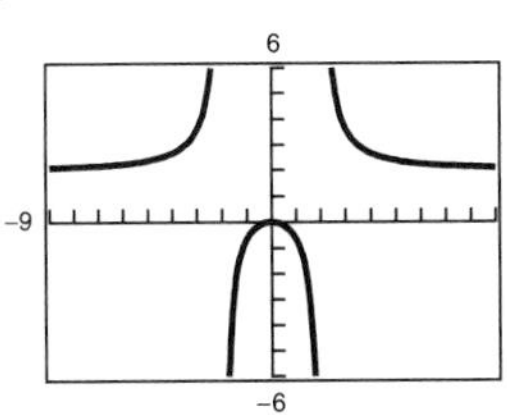

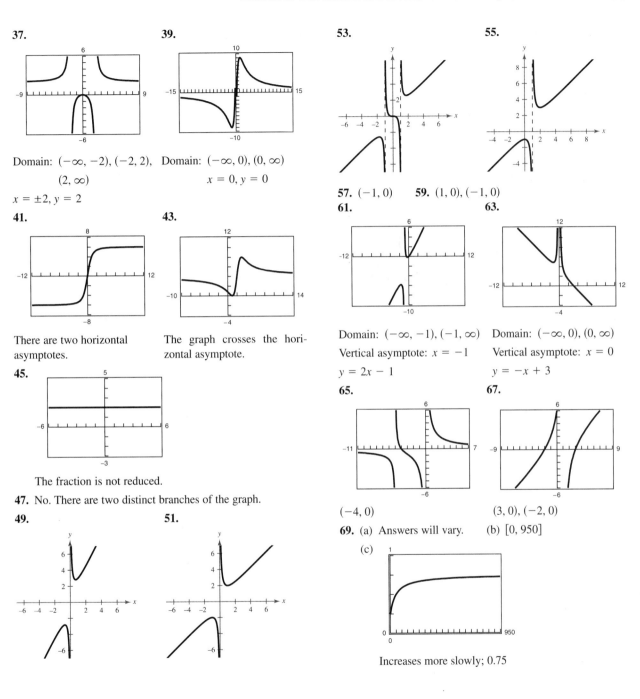

Domain: $(-\infty, -2), (-2, 2), (2, \infty)$

$x = \pm 2, y = 2$

39.

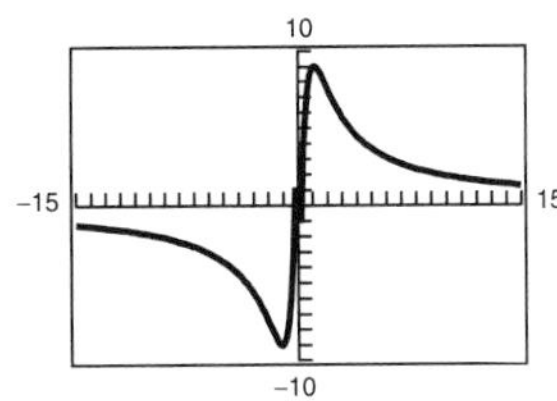

Domain: $(-\infty, 0), (0, \infty)$

$x = 0, y = 0$

41.

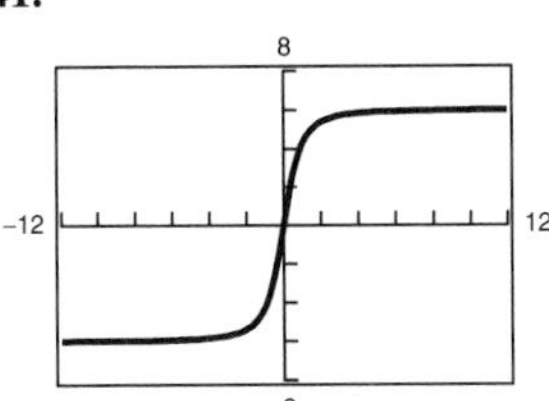

There are two horizontal asymptotes.

43.

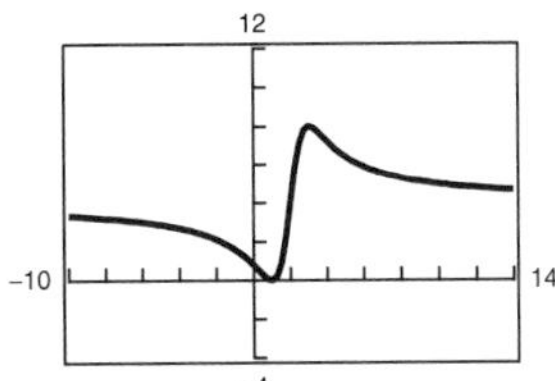

The graph crosses the horizontal asymptote.

45.

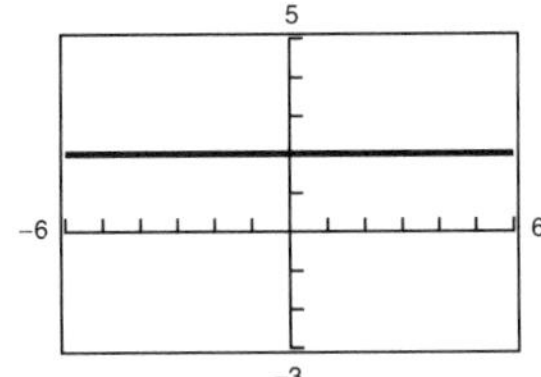

The fraction is not reduced.

47. No. There are two distinct branches of the graph.

49.

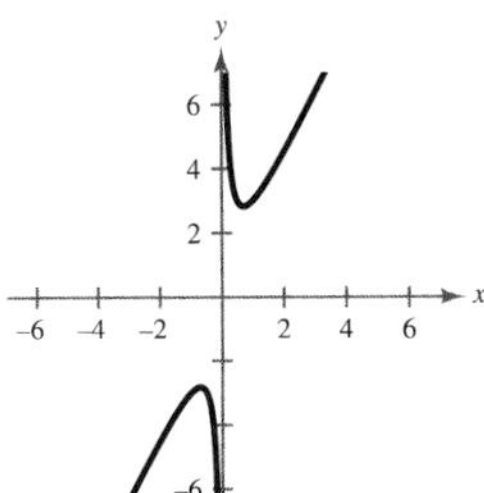

51.

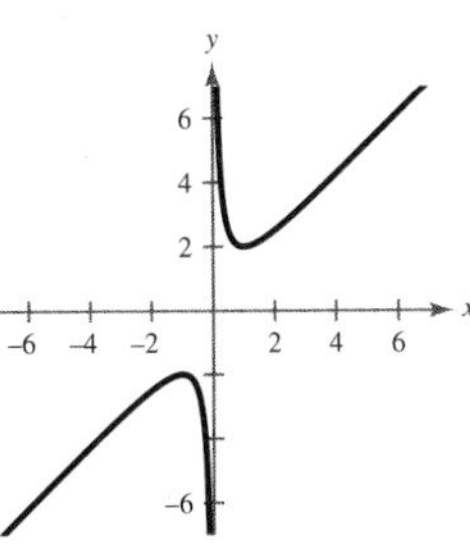

53.

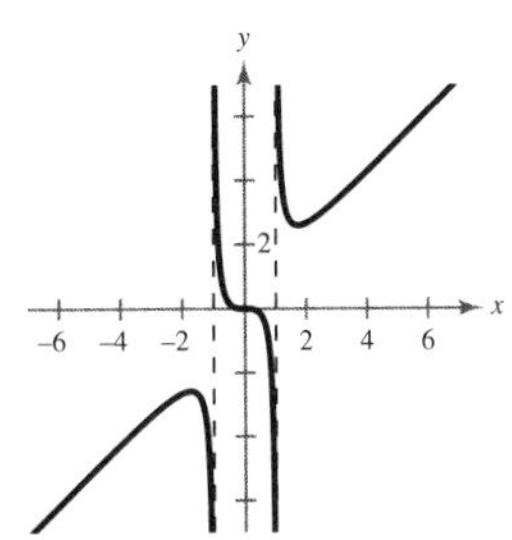

55.

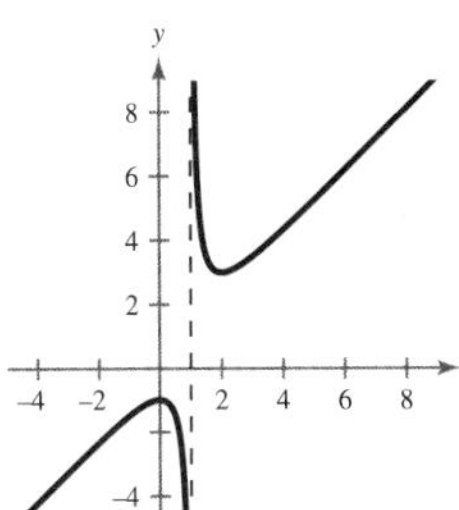

57. $(-1, 0)$ **59.** $(1, 0), (-1, 0)$

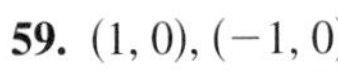

61.

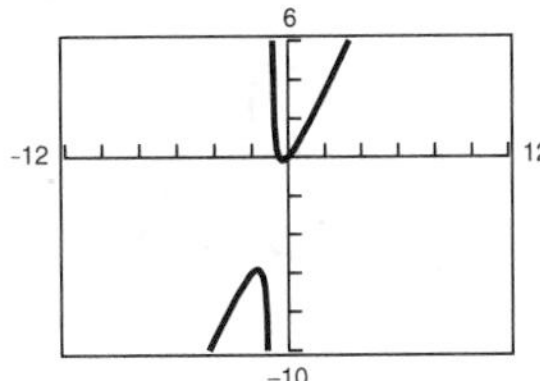

Domain: $(-\infty, -1), (-1, \infty)$

Vertical asymptote: $x = -1$

$y = 2x - 1$

63.

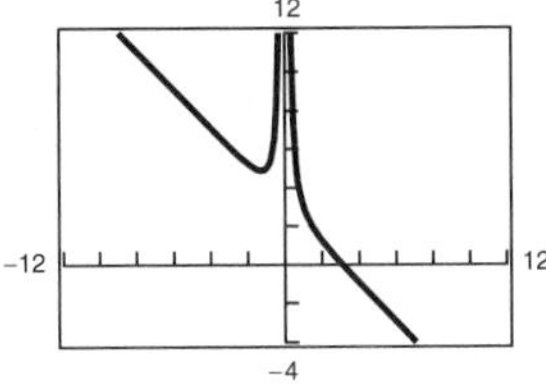

Domain: $(-\infty, 0), (0, \infty)$

Vertical asymptote: $x = 0$

$y = -x + 3$

65.

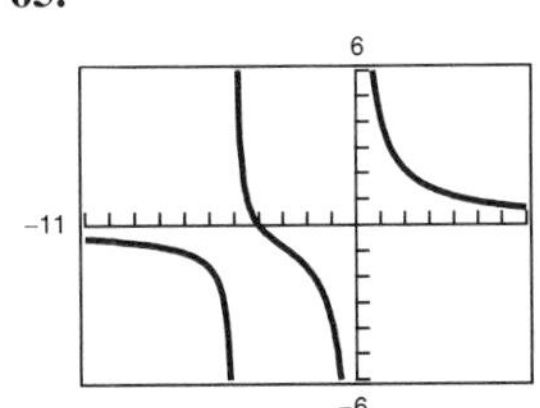

$(-4, 0)$

67.

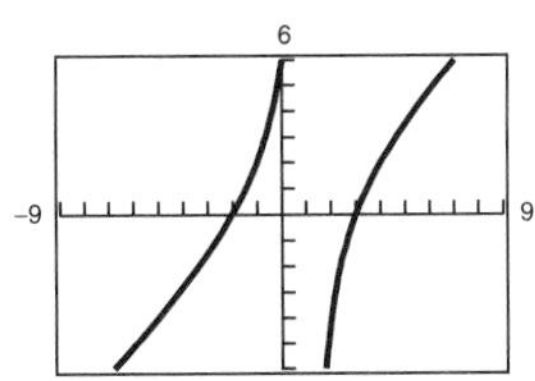

$(3, 0), (-2, 0)$

69. (a) Answers will vary. (b) $[0, 950]$

(c)

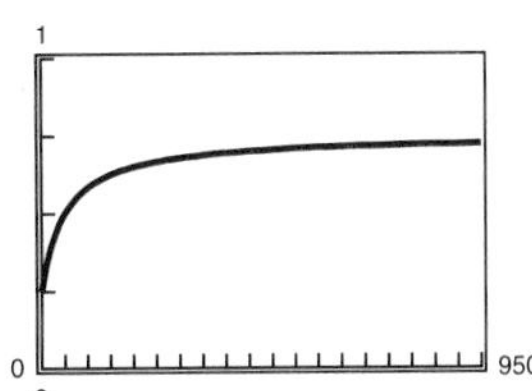

Increases more slowly; 0.75

71. 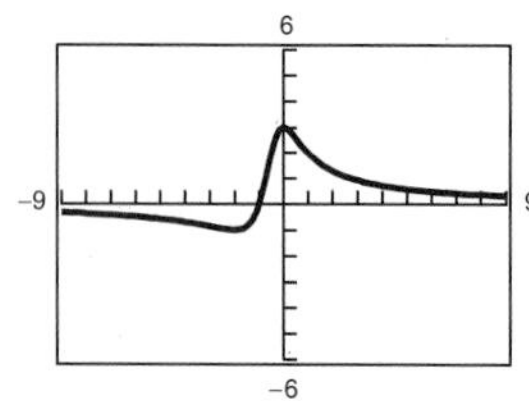

Minimum: $(-2, -1)$
Maximum: $(0, 3)$

73. (a) Answers will vary. (b) $(4, \infty)$ 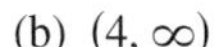

(c) 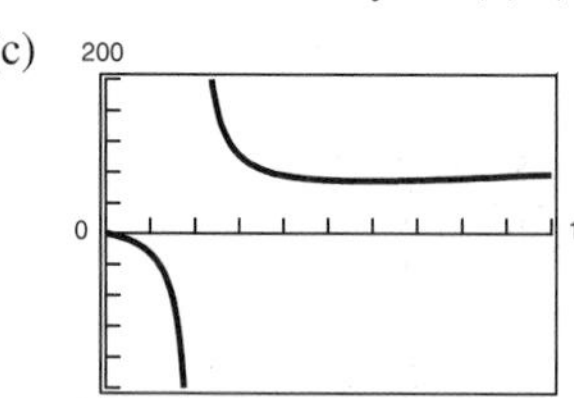

11.75×5.9 inches

75. 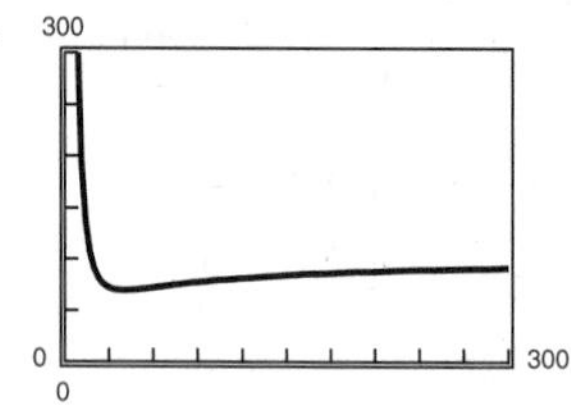

$x \approx 40$

77. (a) $C = 0$. The chemical will eventually dissipate.

(b) 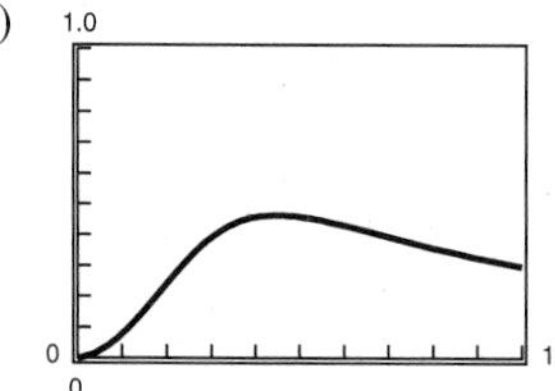

$t \approx 4.5$

79. (a) 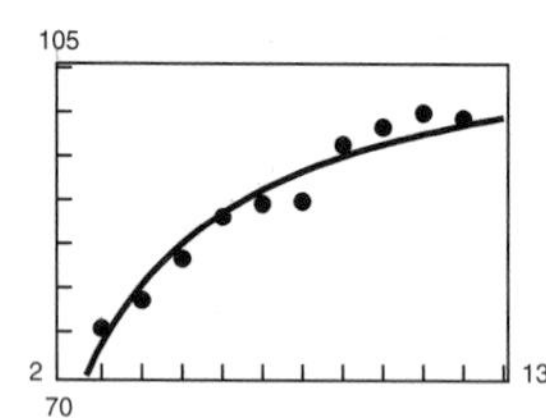

(b) $y = 2.81t + 68.77$

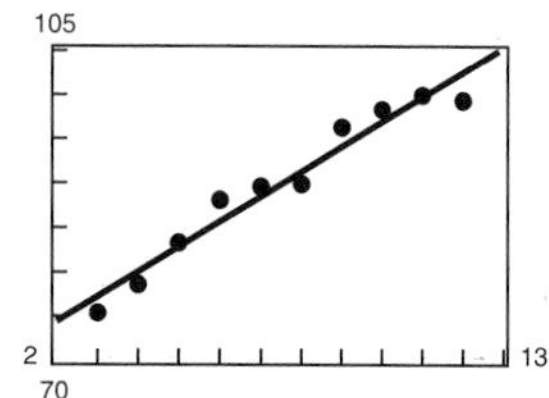

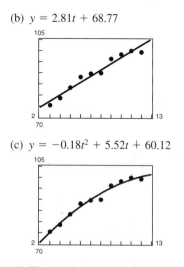

(c) $y = -0.18t^2 + 5.52t + 60.12$

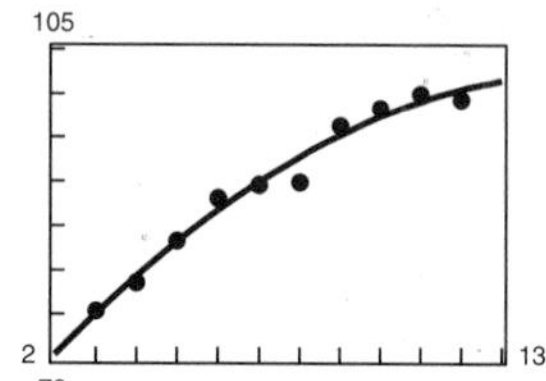

(d) The quadratic and rational models fit the data better than the line. The rational model may be a better predictor because the parabola is near its maximum.

81. $f(x) = \dfrac{x^2 - x - 6}{x - 2}$

FOCUS ON CONCEPTS *(page 308)*

1. Prefer the conditions (a) and (b) because profits would be increasing.

2. (a) Degree: 3; leading coefficient: positive
(b) Degree: 2; leading coefficient: positive
(c) Degree: 4; leading coefficient: positive
(d) Degree: 5; leading coefficient: positive

3. (a) No
(b) Yes (c) Yes

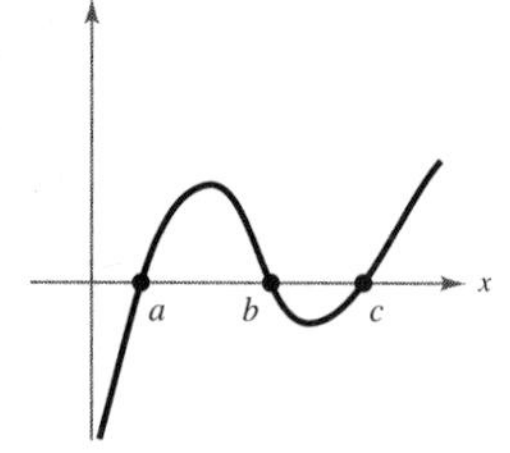

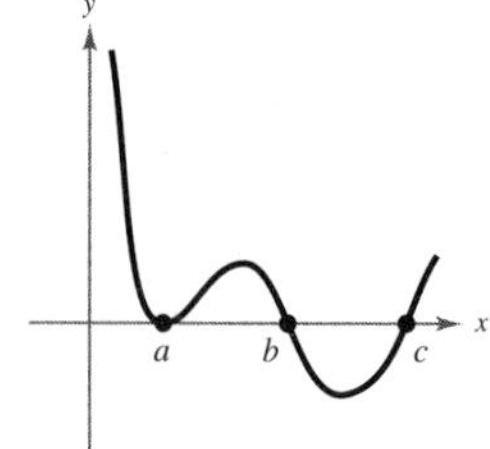

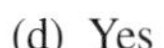
(d) Yes

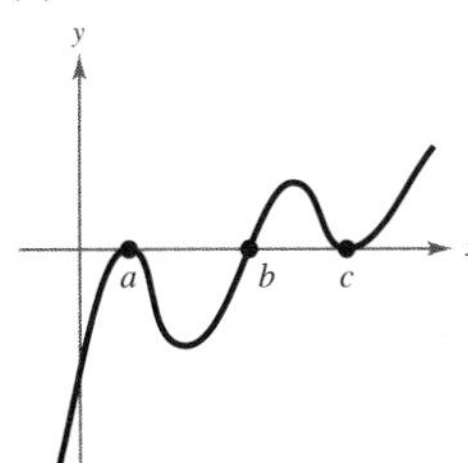

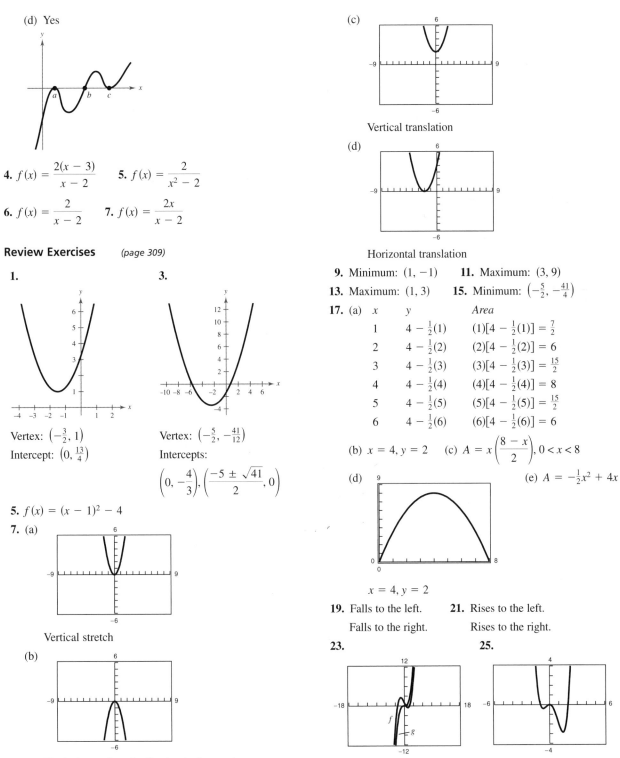

4. $f(x) = \dfrac{2(x-3)}{x-2}$ **5.** $f(x) = \dfrac{2}{x^2-2}$

6. $f(x) = \dfrac{2}{x-2}$ **7.** $f(x) = \dfrac{2x}{x-2}$

Review Exercises *(page 309)*

1.

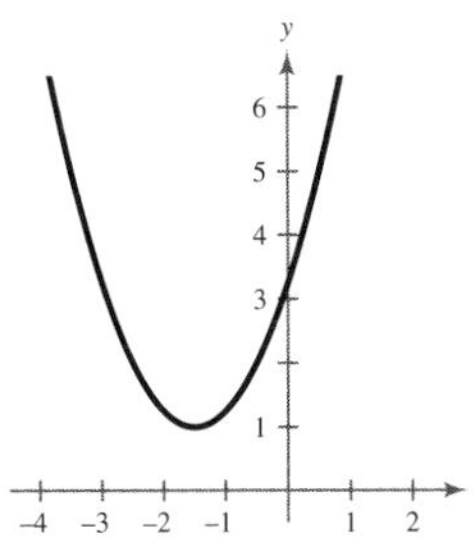

Vertex: $\left(-\frac{3}{2}, 1\right)$

Intercept: $\left(0, \frac{13}{4}\right)$

3.

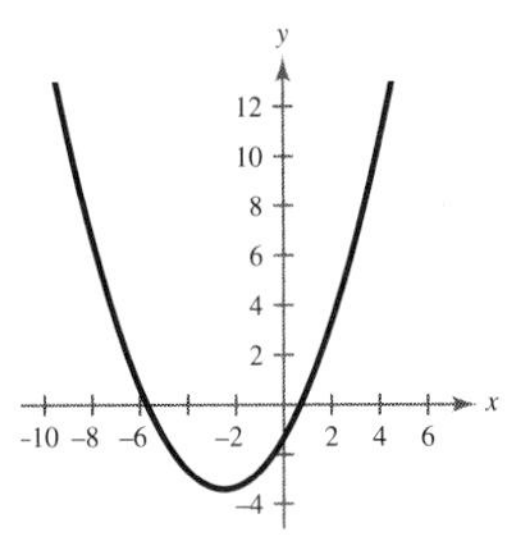

Vertex: $\left(-\frac{5}{2}, -\frac{41}{12}\right)$

Intercepts:

$\left(0, -\frac{4}{3}\right), \left(\frac{-5 \pm \sqrt{41}}{2}, 0\right)$

5. $f(x) = (x-1)^2 - 4$

7. (a)

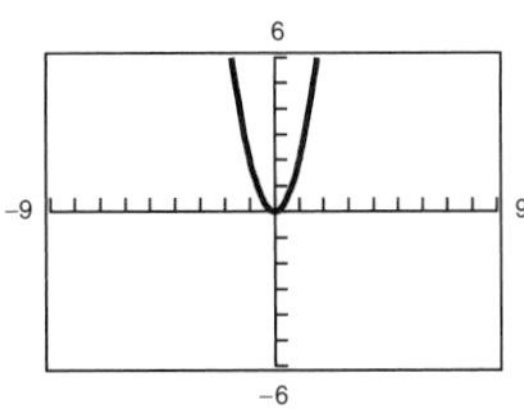

Vertical stretch

(b)

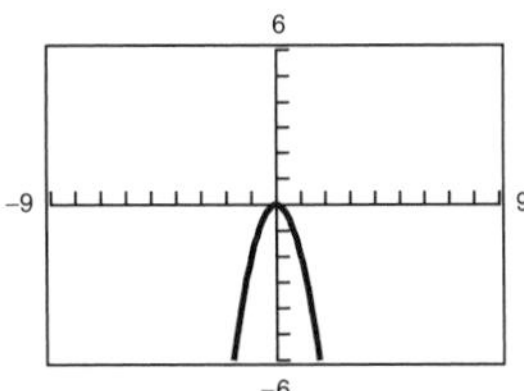

Vertical stretch and reflection in the x-axis

(c)

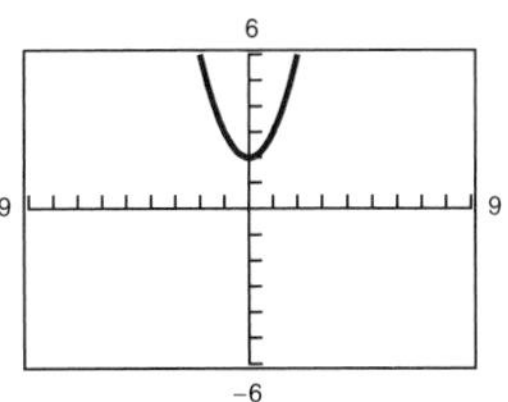

Vertical translation

(d)

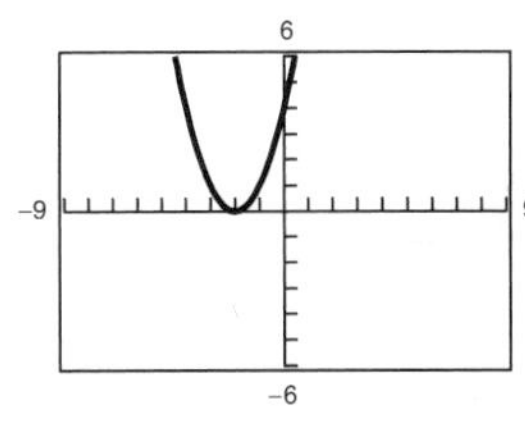

Horizontal translation

9. Minimum: $(1, -1)$ **11.** Maximum: $(3, 9)$

13. Maximum: $(1, 3)$ **15.** Minimum: $\left(-\frac{5}{2}, -\frac{41}{4}\right)$

17. (a)

x	y	*Area*
1	$4 - \frac{1}{2}(1)$	$(1)[4 - \frac{1}{2}(1)] = \frac{7}{2}$
2	$4 - \frac{1}{2}(2)$	$(2)[4 - \frac{1}{2}(2)] = 6$
3	$4 - \frac{1}{2}(3)$	$(3)[4 - \frac{1}{2}(3)] = \frac{15}{2}$
4	$4 - \frac{1}{2}(4)$	$(4)[4 - \frac{1}{2}(4)] = 8$
5	$4 - \frac{1}{2}(5)$	$(5)[4 - \frac{1}{2}(5)] = \frac{15}{2}$
6	$4 - \frac{1}{2}(6)$	$(6)[4 - \frac{1}{2}(6)] = 6$

(b) $x = 4, y = 2$ (c) $A = x\left(\dfrac{8-x}{2}\right), 0 < x < 8$

(d)

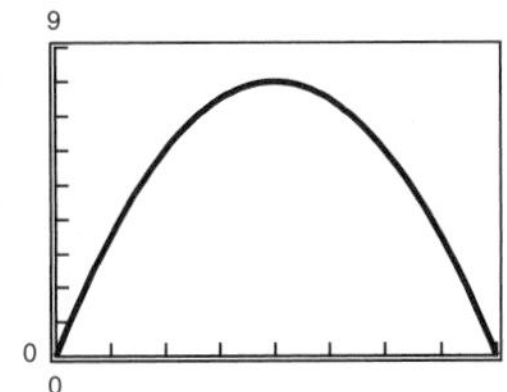

$x = 4, y = 2$

(e) $A = -\frac{1}{2}x^2 + 4x$

19. Falls to the left.
Falls to the right.

21. Rises to the left.
Rises to the right.

23.

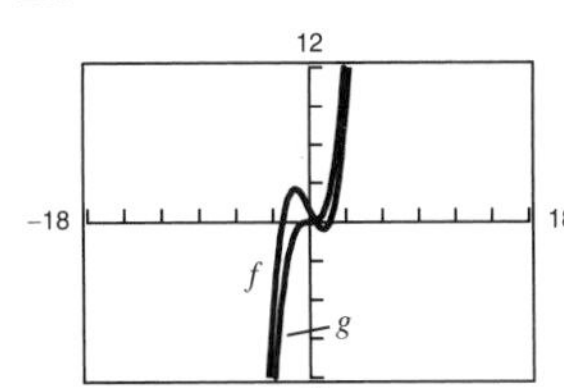

25.

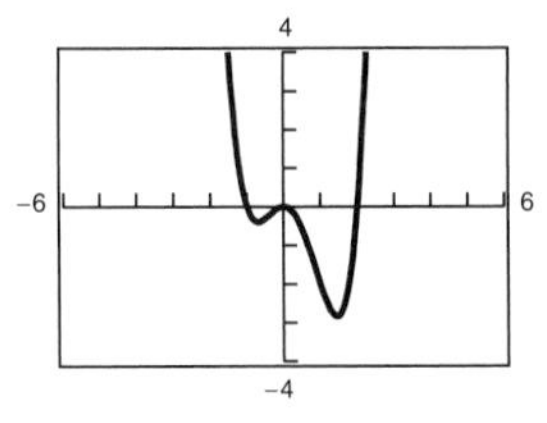

27.

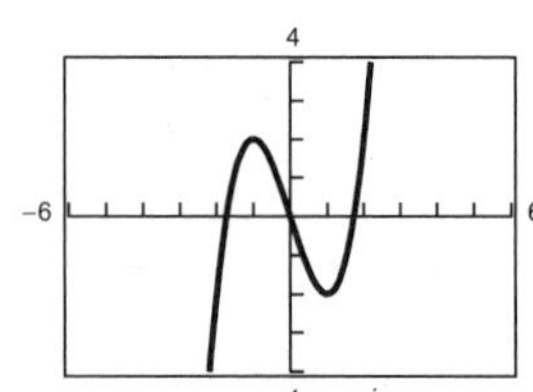

29.

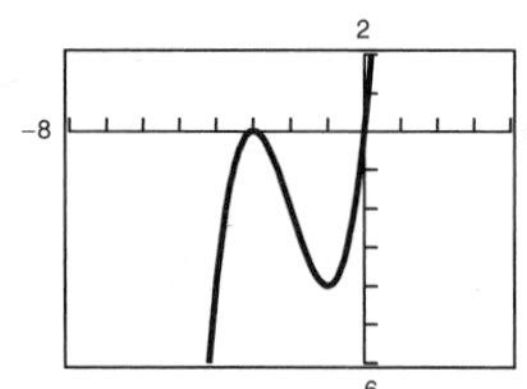

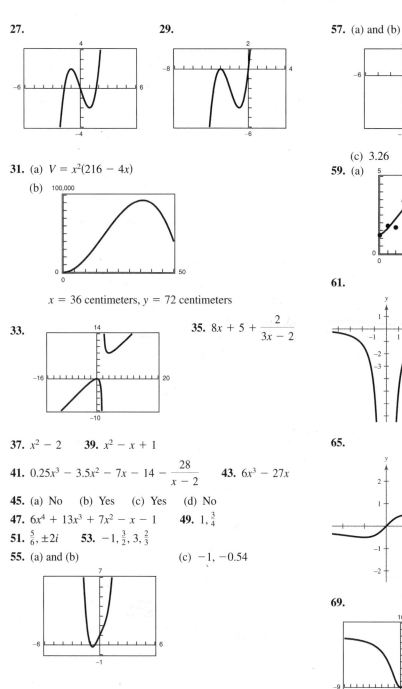

31. (a) $V = x^2(216 - 4x)$

(b)

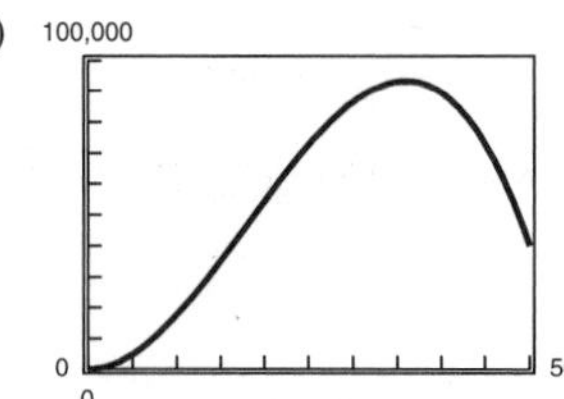

$x = 36$ centimeters, $y = 72$ centimeters

33.

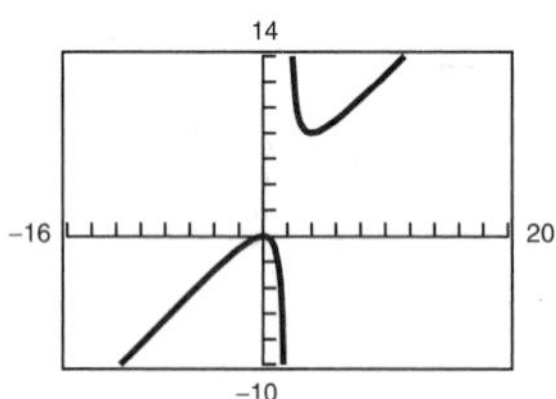

35. $8x + 5 + \dfrac{2}{3x - 2}$

37. $x^2 - 2$ **39.** $x^2 - x + 1$

41. $0.25x^3 - 3.5x^2 - 7x - 14 - \dfrac{28}{x - 2}$ **43.** $6x^3 - 27x$

45. (a) No (b) Yes (c) Yes (d) No

47. $6x^4 + 13x^3 + 7x^2 - x - 1$ **49.** $1, \frac{3}{4}$

51. $\frac{5}{6}, \pm 2i$ **53.** $-1, \frac{3}{2}, 3, \frac{2}{3}$

55. (a) and (b)

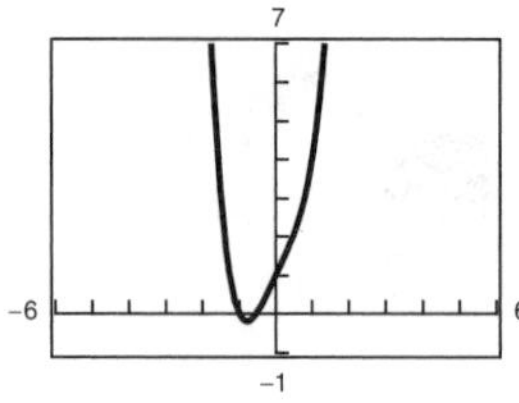

(c) $-1, -0.54$

57. (a) and (b)

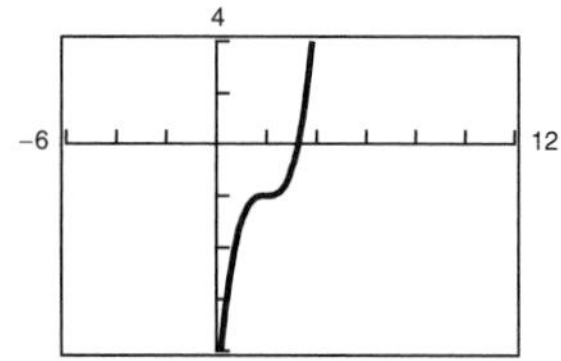

(c) 3.26

59. (a)

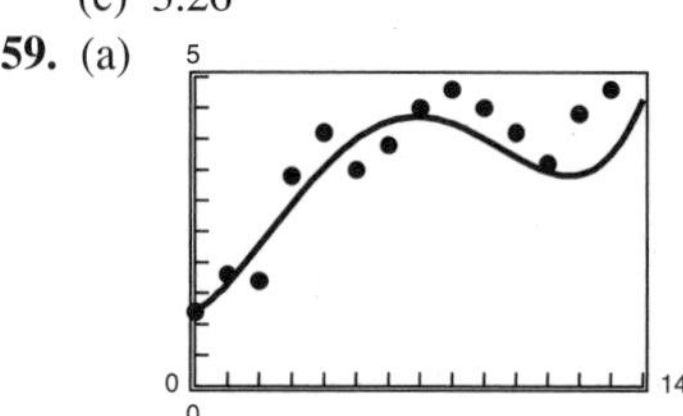

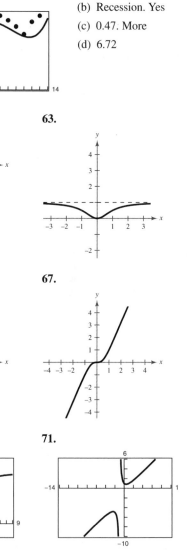

(b) Recession. Yes

(c) 0.47. More

(d) 6.72

61.

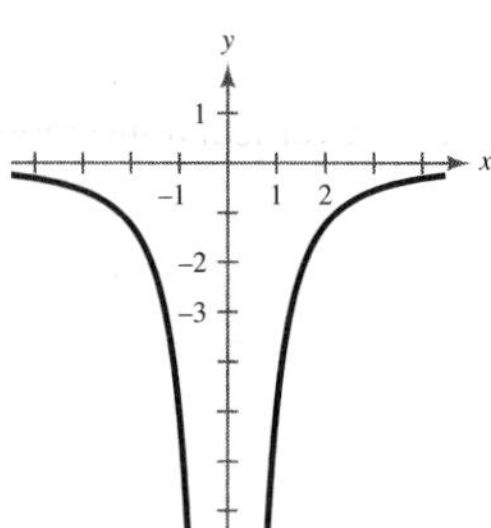

63.

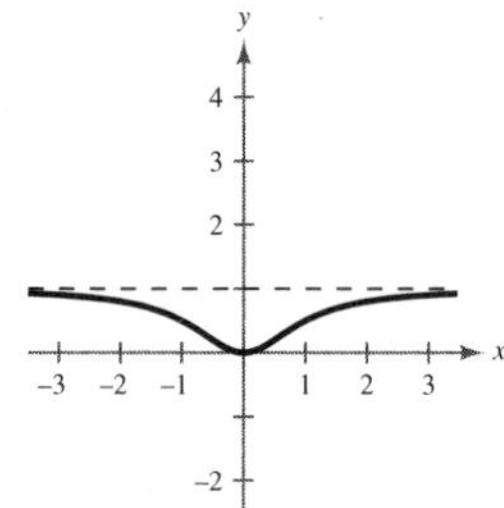

65.

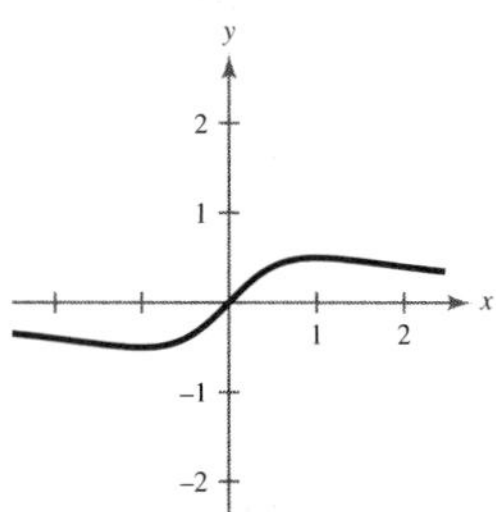

67.

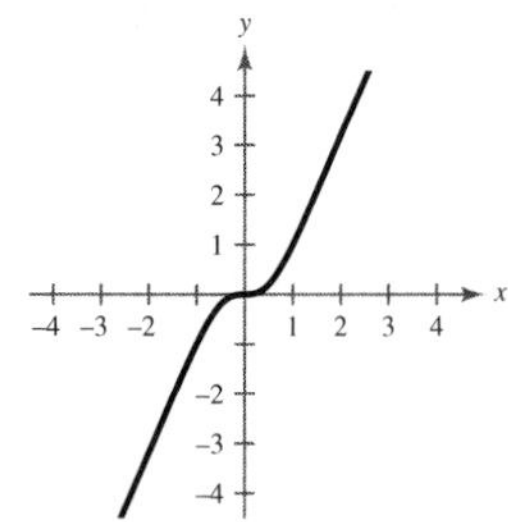

69.

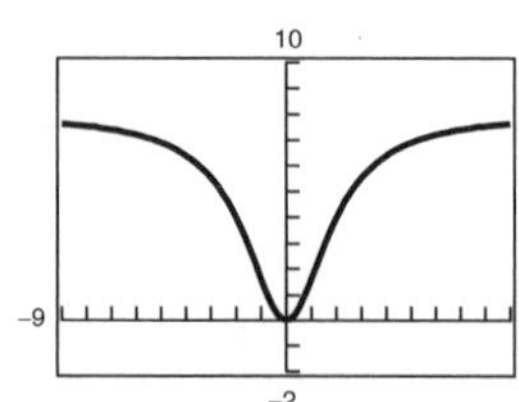

Horizontal asymptote: $y = 8$

71.

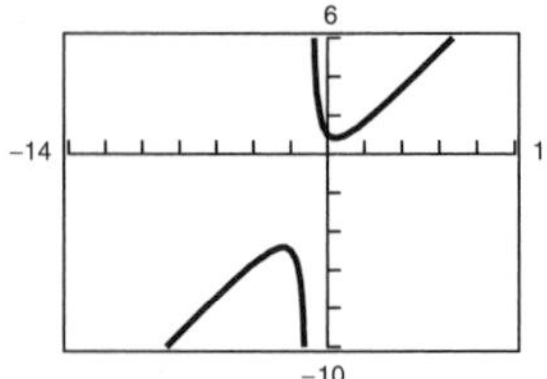

Vertical asymptote: $x = -1$

Slant asymptote: $y = x - 1$

73. $f(x) = \dfrac{2x^2}{x^2 - x - 12}$

75. As x increases, the cost approaches the horizontal asymptote, $\overline{C} = 0.5$.

77. (a)

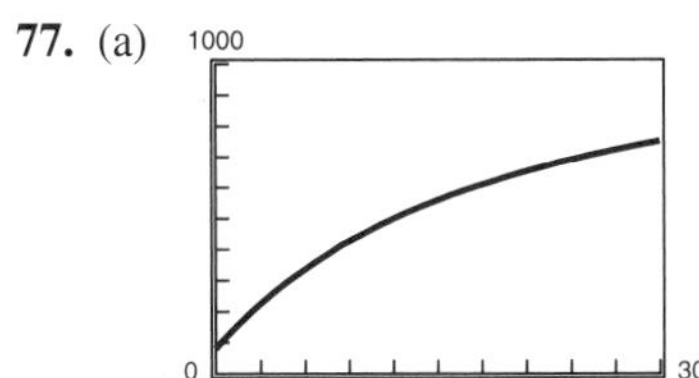

(b) 304,000; 453,333; 702,222 (c) 1,200,000

79. Because there are only a finite number of pixels, the utility may not attempt to evaluate the function where it does not exist.

Chapter Test *(page 314)*

1. (a) Reflection in the x-axis followed by a vertical translation

(b) Horizontal translation

2. Vertex: $(-2, -1)$

Intercepts: $(0, 3), (-3, 0), (-1, 0)$

3. $y = (x - 3)^2 - 6$

4. (a) 50 feet

(b) 5. Changing the constant term results in a vertical translation of the graph and therefore changes the maximum height.

5. $3x + \dfrac{x - 1}{x^2 + 1}$ **6.** $2x^3 + 4x^2 + 3x + 6 + \dfrac{9}{x - 2}$

7. $\pm 1, \pm 2, \pm 3, \pm 4, \pm 6, \pm 8, \pm 12, \pm 24, \pm\frac{1}{2}, \pm\frac{3}{2}$ **8.** $\pm 1, \pm 2, \pm\frac{1}{3}, \pm\frac{2}{3}$

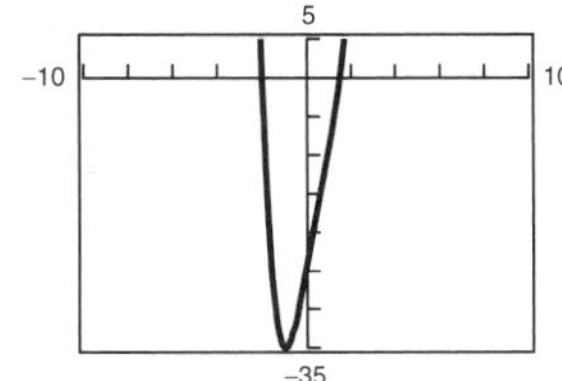

$-2, \frac{3}{2}$

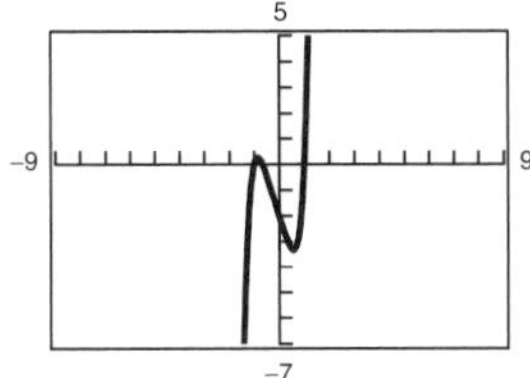

$\pm 1, -\frac{2}{3}$

9. $-0.819, 1.380$ **10.** $-1.414, -0.667, 1.414$

11. $f(x) = x^4 - 9x^3 + 28x^2 - 30x$

12. $f(x) = x^4 - 6x^3 + 16x^2 - 24x + 16$

13.

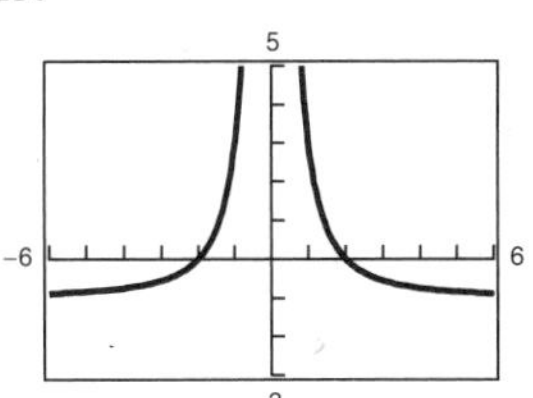

14.

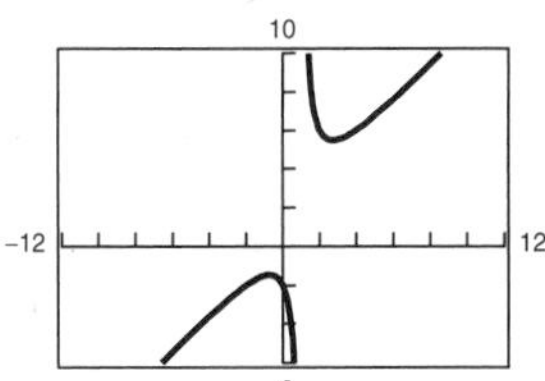

15. $f(x) = \dfrac{4x^2}{x^2 - 9}$

CHAPTER 4

Section 4.1 *(page 325)*

1. 946.852 **3.** 7.352 **5.** 0.006

7. 673.639 **9.** 0.472 **11.** $f(x) = h(x)$

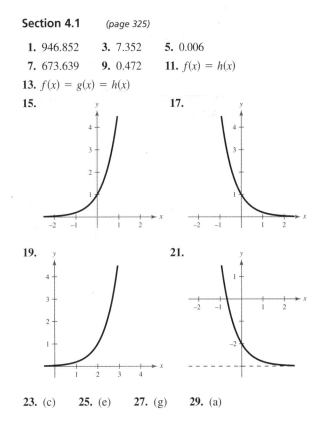

13. $f(x) = g(x) = h(x)$

15.

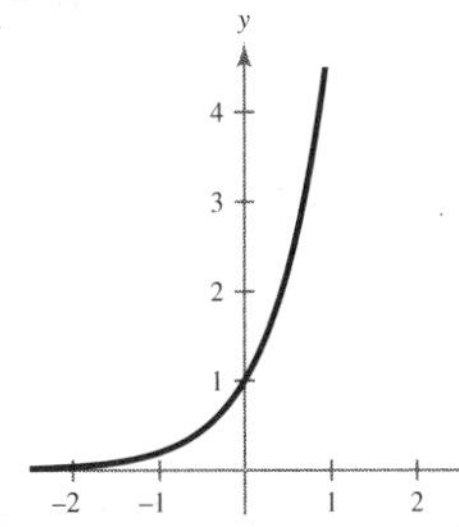

17.

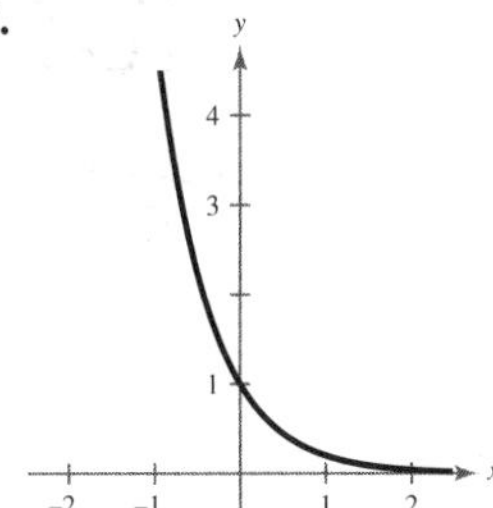

19.

21.

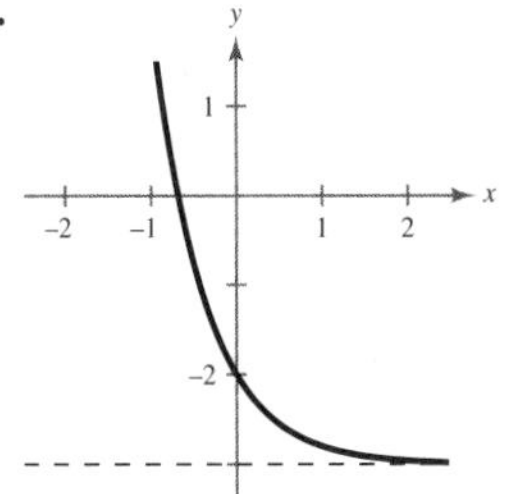

23. (c) **25.** (e) **27.** (g) **29.** (a)

31.

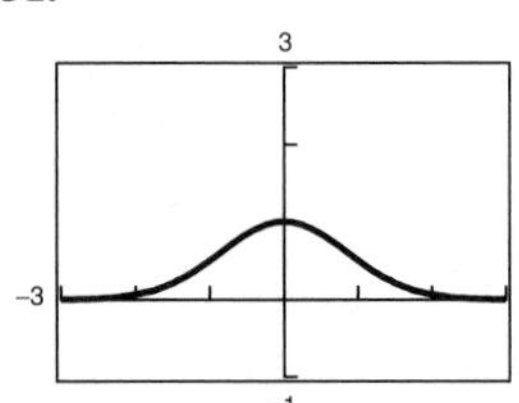

33.

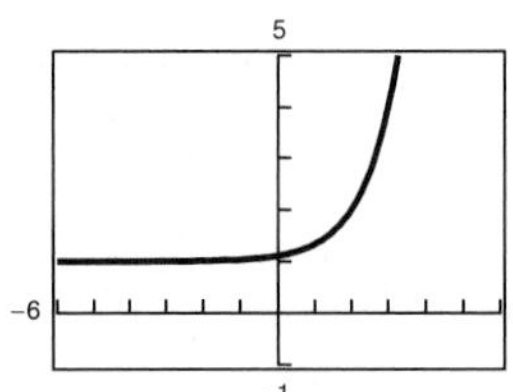

35.

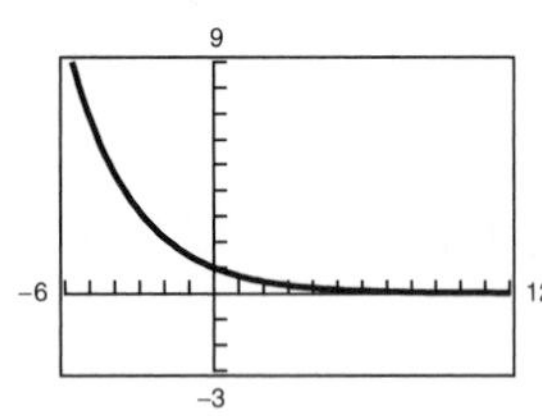

37.

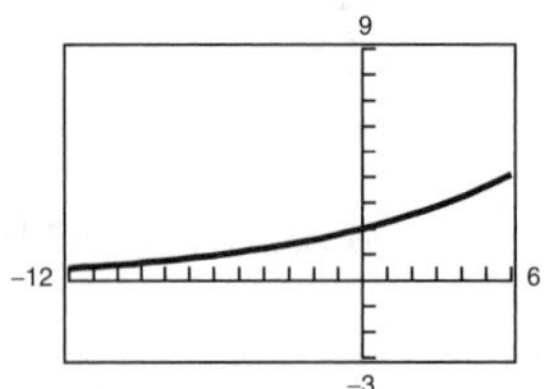

39.

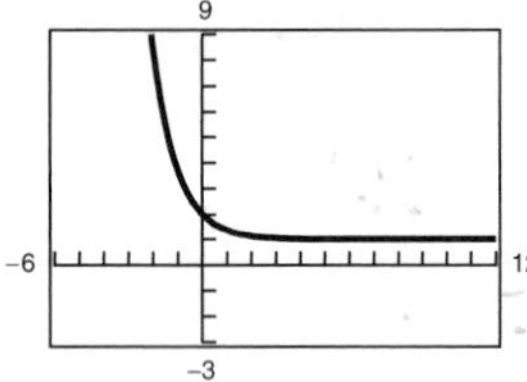

41. (a)

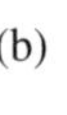

x	-1	-0.5	0	0.5	1
$f(x)$	0.3333	0.5774	1	1.7321	3
$g(x)$	0.25	0.5	1	2	4

$x < 0$

(b)

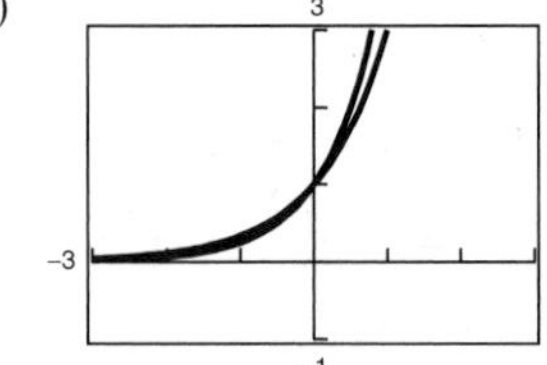

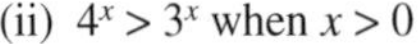

(i) $4^x < 3^x$ when $x < 0$ (ii) $4^x > 3^x$ when $x > 0$

43. (a)

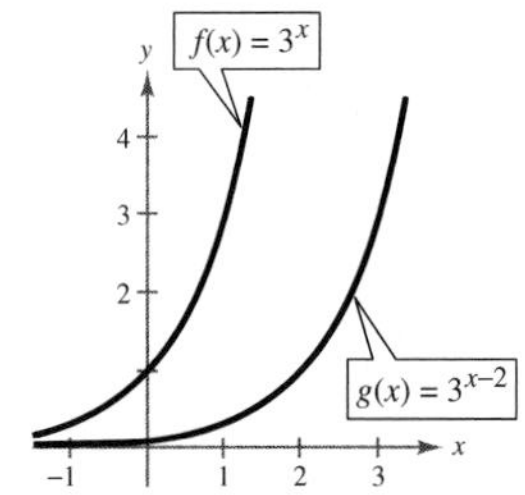

Horizontal shift two units to the right

(b)

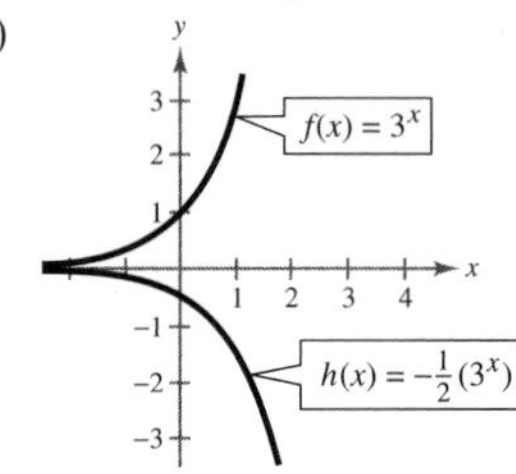

Vertical shrink and a reflection about the x-axis

(c)

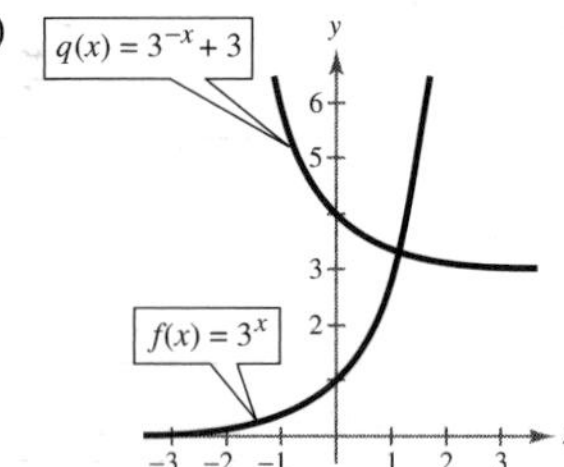

Reflection about the y-axis and a vertical translation

45. (a)

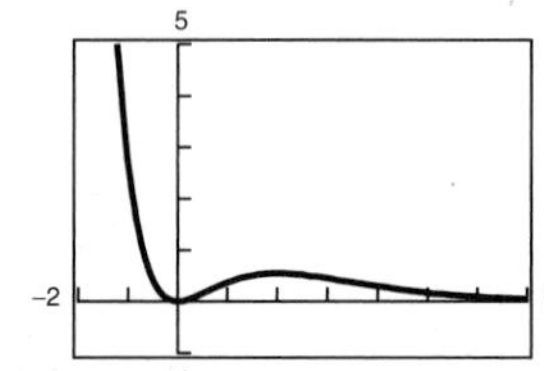

Decreasing: $(-\infty, 0)$, $(2, \infty)$

Increasing: $(0, 2)$

Relative maximum: $(2, 4e^{-2})$

Relative minimum: $(0, 0)$

(b)

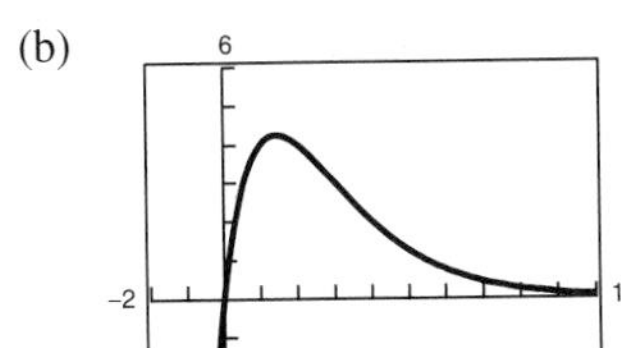

Decreasing: $(1.44, \infty)$

Increasing: $(-\infty, 1.44)$

Relative maximum: $(1.44, 4.25)$

47. The exponential function increases at a faster rate.

49.

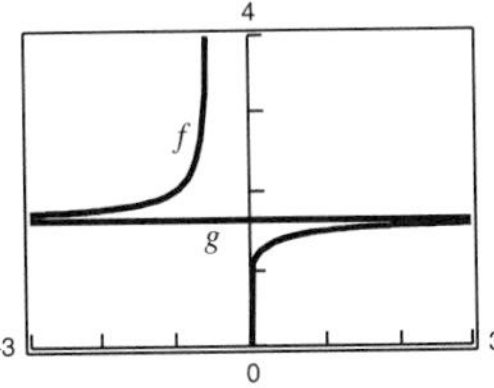

$f(x)$ approaches $g(x) = 1.6487$

51. \$212,605.41

53.

n	1	2	4
A	\$7764.62	\$8017.84	\$8155.09

n	12	365	Continuous
A	\$8250.97	\$8298.66	\$8300.29

55.

n	1	2	4
A	\$24,115.73	\$25,714.29	\$26,602.23

n	12	365	Continuous
A	\$27,231.38	\$27,547.07	\$27,557.94

57.

t	1	10	20
P	\$91,393.12	\$40,656.97	\$16,529.89

t	30	40	50
P	\$6720.55	\$2732.37	\$1110.90

59.

t	1	10	20
P	\$90,521.24	\$36,940.70	\$13,646.15

t	30	40	50
P	\$5040.98	\$1862.17	\$687.90

61. (a)

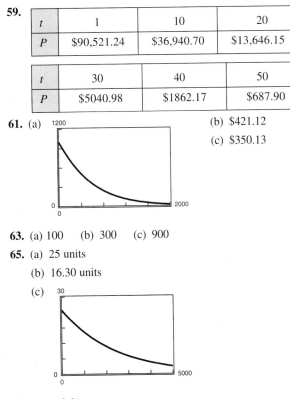

(b) \$421.12

(c) \$350.13

63. (a) 100 (b) 300 (c) 900

65. (a) 25 units

(b) 16.30 units

(c)

67. (a) and (b)

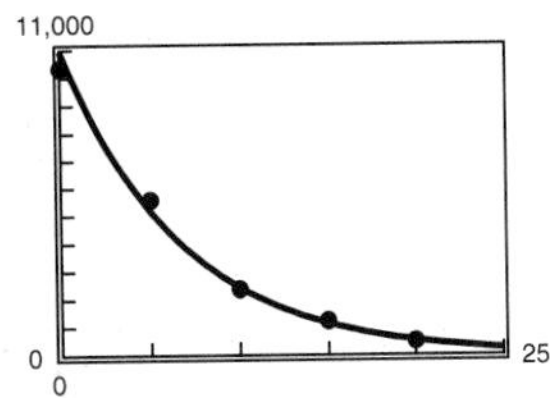

(c)

h	0	5	10	15	20
P	10,958	5176	2445	1155	546

(d) 3300 kilograms per square meter

(e) 11.3 kilometers

69. (a) $T = -1.239t + 73.021$

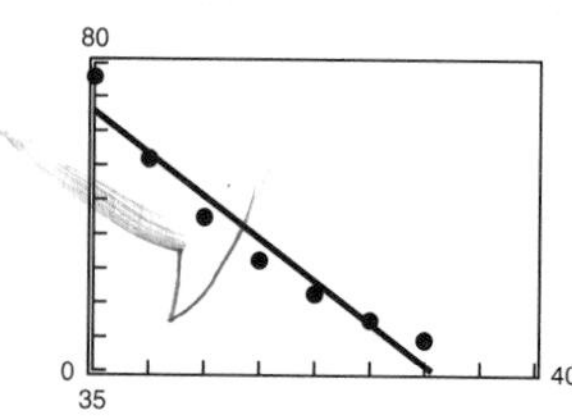

The temperature decreases at a slower rate as it approaches the room temperature.

(b) $T = 0.034t^2 - 2.264t + 77.295$

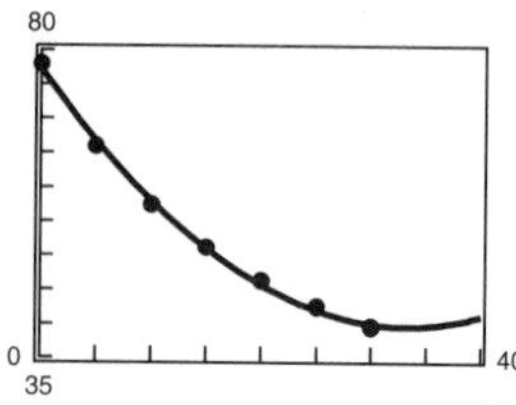

The parabola is increasing when $t = 60$.

(c) $T = 54.438(0.964)^t + 21$

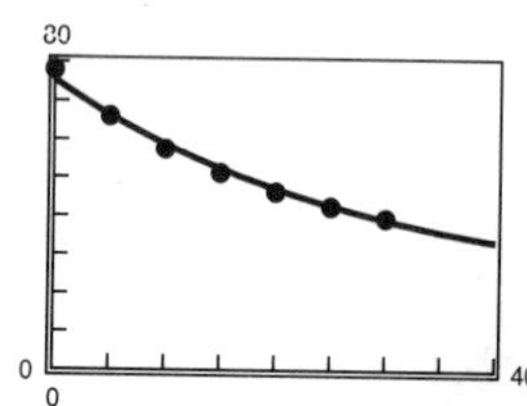

(d) The horizontal asymptote of the exponential model is $T = 0$.

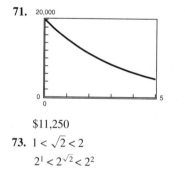

71.

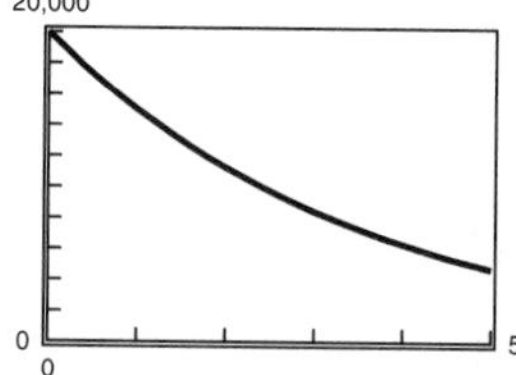

\$11,250

73. $1 < \sqrt{2} < 2$

$2^1 < 2^{\sqrt{2}} < 2^2$

75.

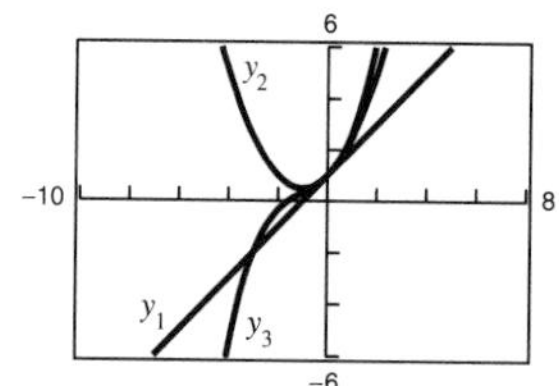

77. (a) $f(u + v) = a^{u+v} = a^u \cdot a^v = f(u) \cdot f(v)$

(b) $f(2x) = a^{2x} = (a^x)^2 = [f(x)]^2$

Section 4.2 *(page 337)*

1. $4^3 = 64$ **3.** $7^{-2} = \frac{1}{49}$ **5.** $32^{2/5} = 4$ **7.** $e^0 = 1$

9. $\log_5 125 = 3$ **11.** $\log_{81} 3 = \frac{1}{4}$

13. $\log_6 \frac{1}{36} = -2$ **15.** $\ln 20.0855 = 3$

17. $\ln 4 = x$ **19.** 4 **21.** $\frac{1}{2}$ **23.** 0

25. -2 **27.** 3 **29.** 2 **31.** 2.538 **33.** 2.161

35. 2.913 **37.** 1.005 **39.** -1.139

41.

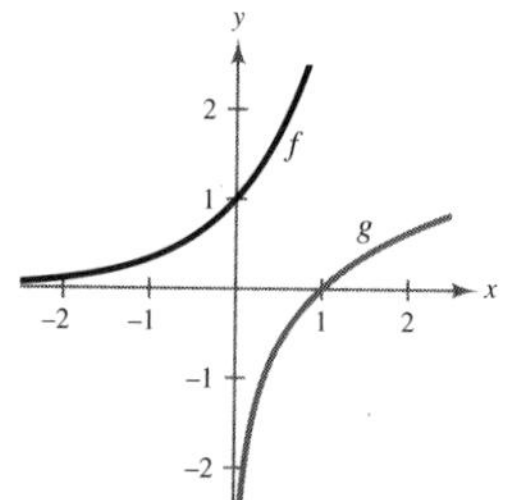

Reflections in the line $y = x$

$g = f^{-1}$

43.

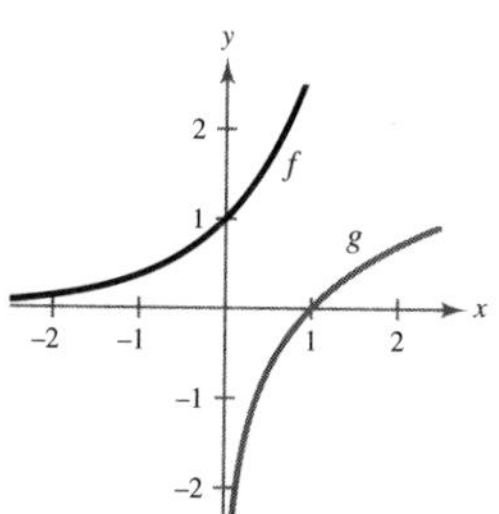

Reflections in the line $y = x$

$g = f^{-1}$

45. (c) **47.** (d) **49.** (b)

51. Domain: $(0, \infty)$

Vertical asymptote: $x = 0$

Intercept: $(1, 0)$

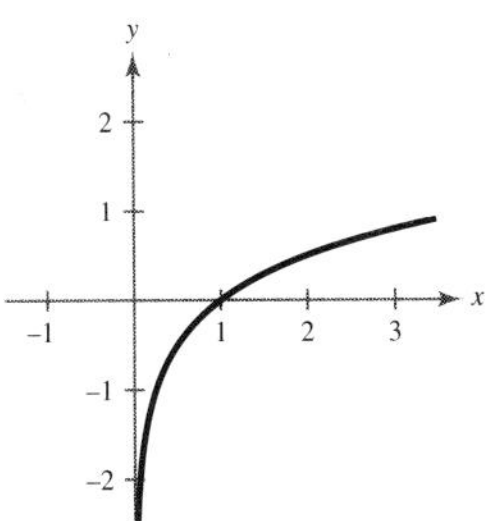

53. Domain: $(3, \infty)$

Vertical asymptote: $x = 3$

Intercept: $(4, 0)$

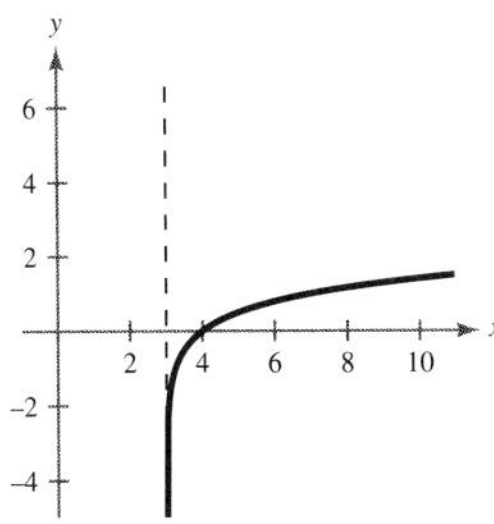

55. Domain: $(0, \infty)$

Vertical asymptote: $x = 0$

Intercept: $(9, 0)$

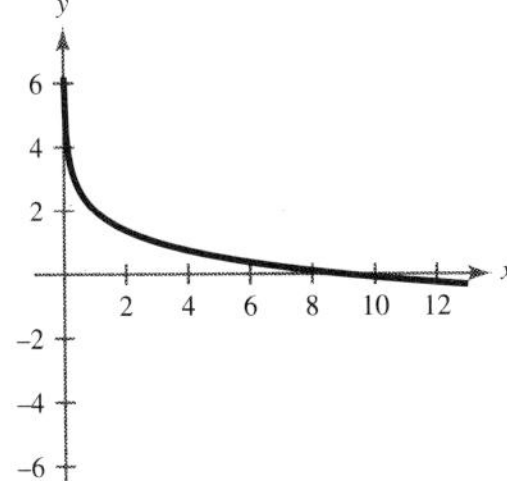

57. Domain: $(0, \infty)$

Vertical asymptote: $x = 0$

Intercept: $(5, 0)$

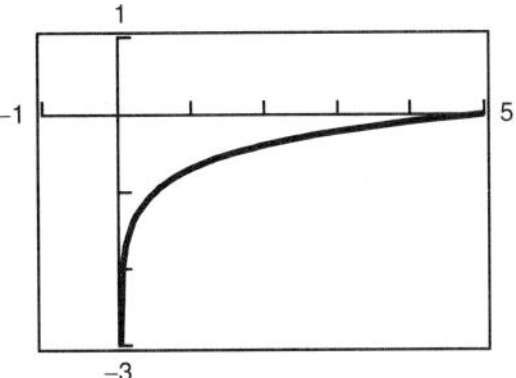

59. Domain: $(2, \infty)$

Vertical asymptote: $x = 2$

Intercept: $(3, 0)$

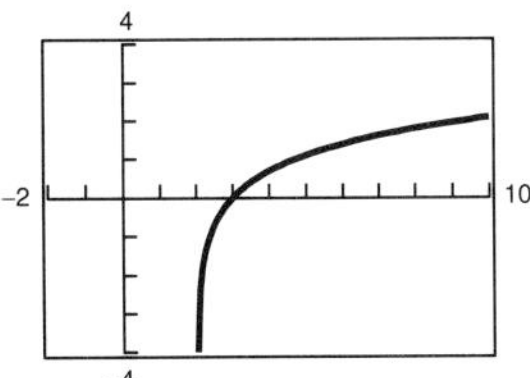

61. Domain: $(-\infty, 0)$

Vertical asymptote: $x = 0$

Intercept: $(-1, 0)$

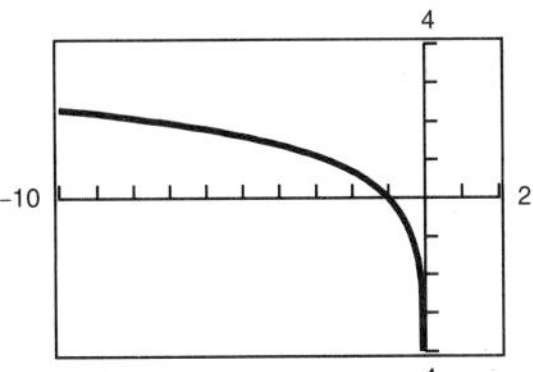

63. Domain: $(0, \infty)$

Decreasing: $(0, 2)$

Increasing: $(2, \infty)$

Relative minimum: $(2, 1.693)$

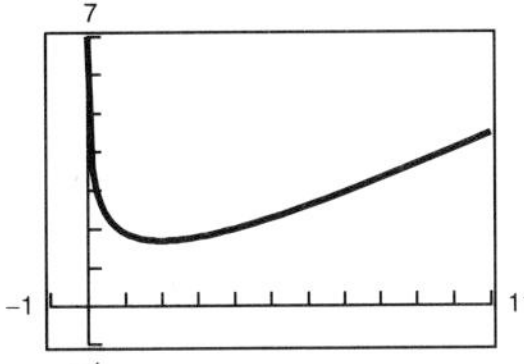

65. Domain: $(0, \infty)$

Decreasing: $(0, 0.37)$

Increasing: $(0.37, \infty)$

Relative minimum: $(0.37, -1.47)$

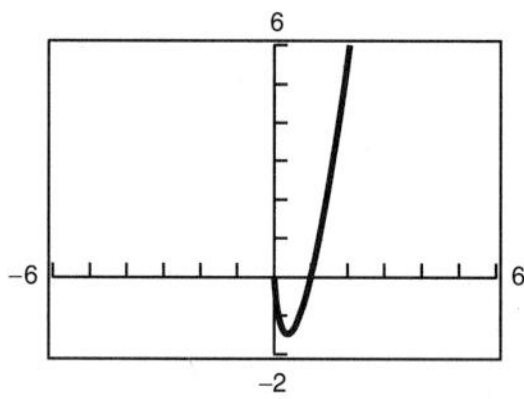

67. 23.68 years

69. (a) False (b) True (c) True (d) False

71. $y = (x - 1) - \frac{1}{2}(x - 1)^2 + \frac{1}{3}(x - 1)^3 - \frac{1}{4}(x - 1)^4$

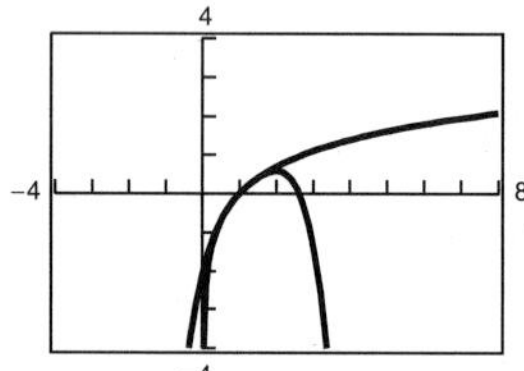

73. (a)

K	1	2	4	6	8	10	12
t	0	7.3	14.6	18.9	21.9	24.2	26.2

(b)

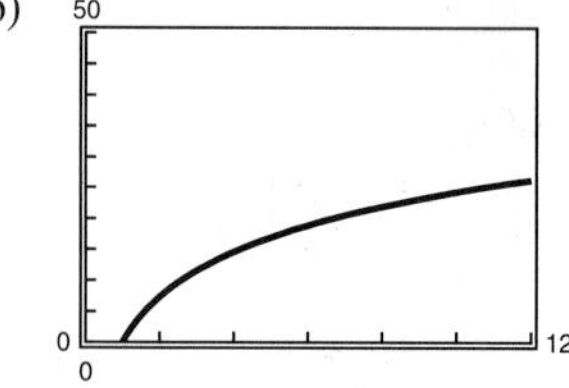

75. (a)

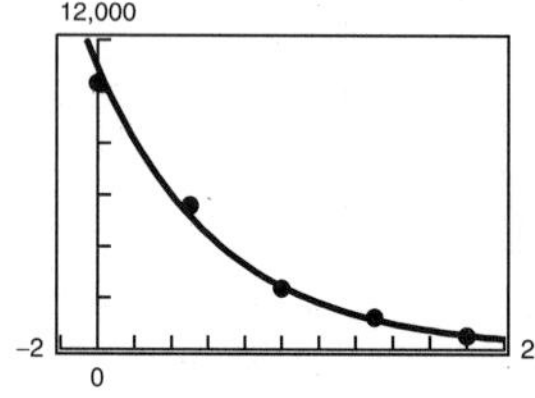

(b) $\ln P = -0.1499h + 9.3018$

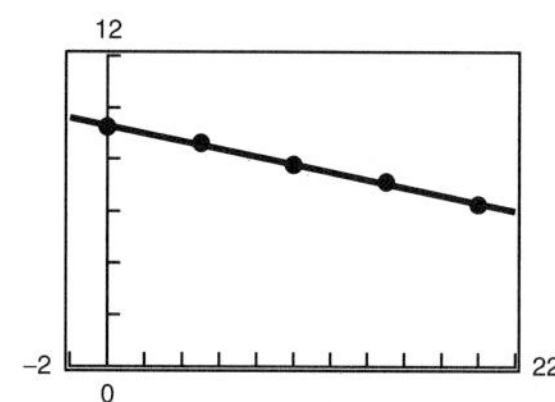

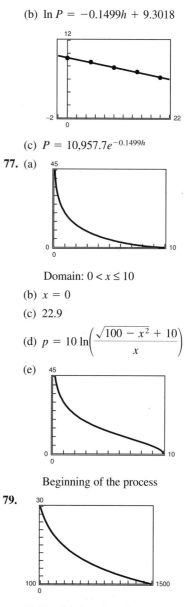

(c) $P = 10{,}957.7e^{-0.1499h}$

77. (a)

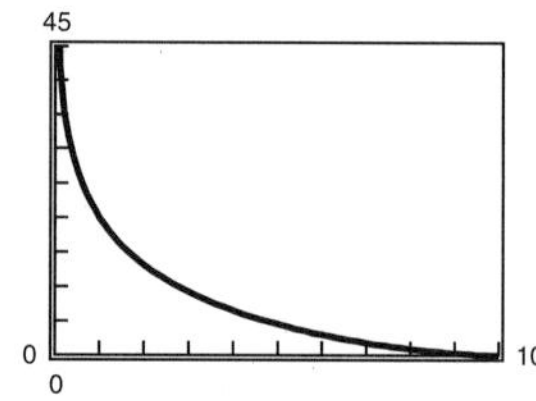

Domain: $0 < x \leq 10$

(b) $x = 0$

(c) 22.9

(d) $p = 10 \ln\left(\frac{\sqrt{100 - x^2} + 10}{x}\right)$

(e)

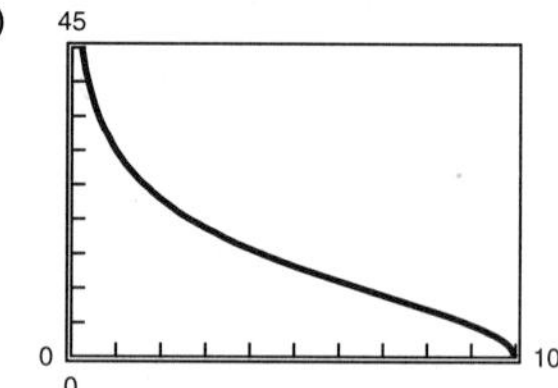

Beginning of the process

79.

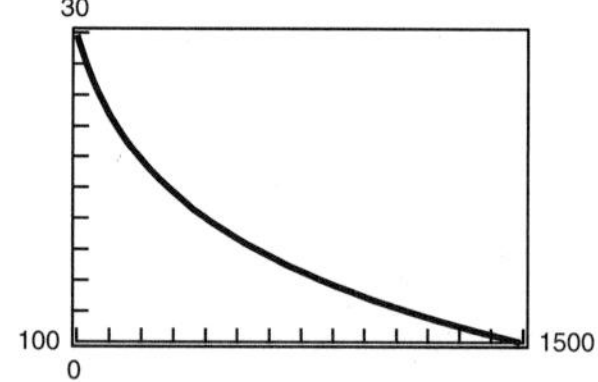

17.66 cubic feet per minute

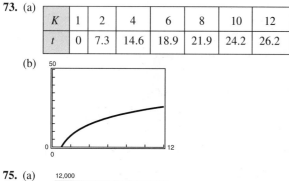

81. 21,357 foot-pounds **83.** 30 years

85. Total amount: \$473,886

Interest: \$323,886

87. (a) $(0, \infty)$ (b) $f^{-1} = 10^x$ (c) $3 < x < 4$

(d) $0 < x < 1$ (e) 10 (f) $10^{2n}:1$

Section 4.3 *(page 345)*

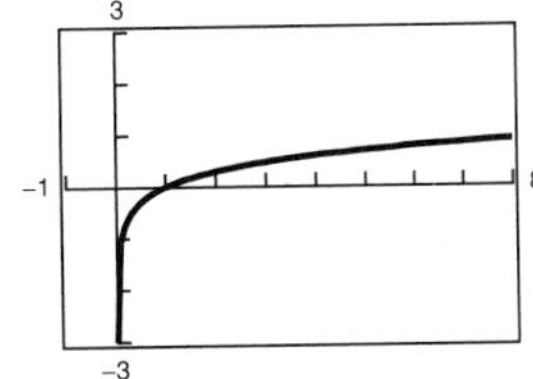

$f = g$

3. $\dfrac{\log_{10} 5}{\log_{10} 3}$ **5.** $\dfrac{\log_{10} x}{\log_{10} 2}$ **7.** $\dfrac{\ln 5}{\ln 3}$ **9.** $\dfrac{\ln x}{\ln 2}$

11. 1.771 **13.** -2.000 **15.** -0.417 **17.** 2.633

19. $\log_{10} 5 + \log_{10} x$ **21.** $\log_{10} 5 - \log_{10} x$

23. $4 \log_8 x$ **25.** $\frac{1}{2} \ln z$ **27.** $\ln x + \ln y + \ln z$

29. $\frac{1}{2} \ln(a - 1)$ **31.** $\ln z + 2 \ln(z - 1)$

33. $\frac{1}{3} \ln x - \frac{1}{3} \ln y$ **35.** $4 \ln x + \frac{1}{2} \ln y - 5 \ln z$

37. $2 \log_b x - 2 \log_b y - 3 \log_b z$

39.

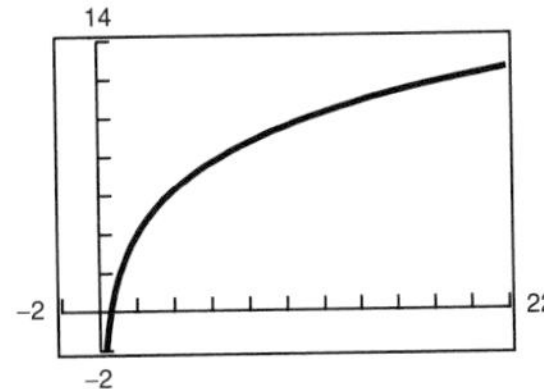

$y_1 = y_2$

41. $\ln 2x$ **43.** $\log_4 \dfrac{z}{y}$ **45.** $\log_2(x + 4)^2$

47. $\log_3 \sqrt[3]{5x}$ **49.** $\ln \dfrac{x}{(x + 1)^3}$ **51.** $\ln \dfrac{x - 2}{x + 2}$

53. $\ln \dfrac{x}{(x^2 - 4)^2}$ **55.** $\ln \sqrt[3]{\dfrac{x(x + 3)^2}{x^2 - 1}}$

57. $\ln \dfrac{\sqrt[3]{y(y + 4)^2}}{y - 1}$ **59.** $\ln \dfrac{9}{\sqrt{x^2 + 1}}$

61.

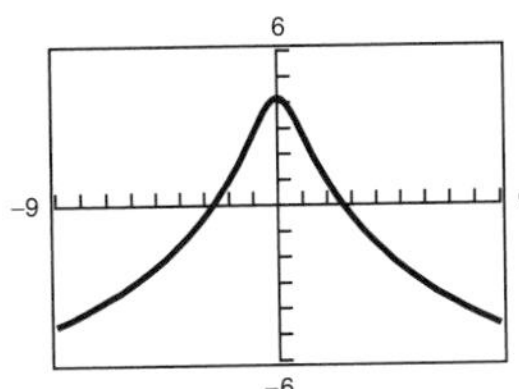

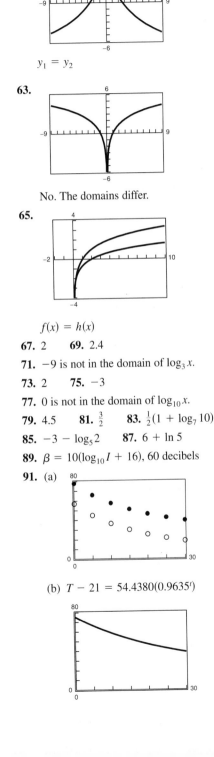

$y_1 = y_2$

63.

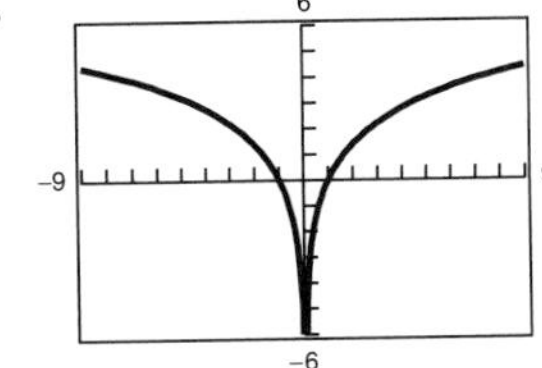

No. The domains differ.

65.

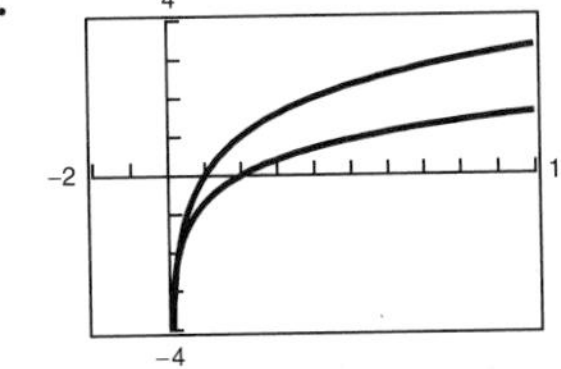

$f(x) = h(x)$

67. 2 **69.** 2.4

71. -9 is not in the domain of $\log_3 x$.

73. 2 **75.** -3

77. 0 is not in the domain of $\log_{10} x$.

79. 4.5 **81.** $\frac{3}{2}$ **83.** $\frac{1}{2}(1 + \log_7 10)$

85. $-3 - \log_5 2$ **87.** $6 + \ln 5$

89. $\beta = 10(\log_{10} I + 16)$, 60 decibels

91. (a)

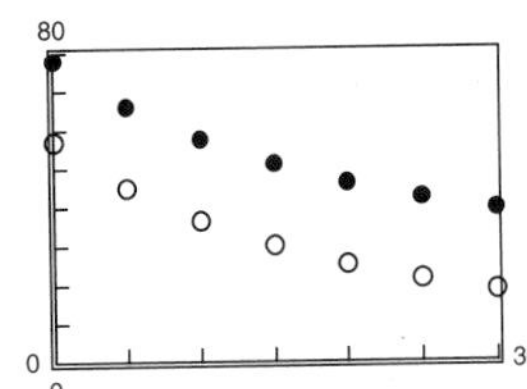

(b) $T - 21 = 54.4380(0.9635^t)$

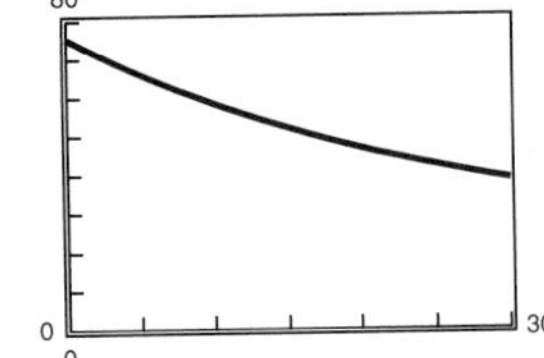

(c) $\ln(T - 21) = -0.0372t + 3.9971$

(d) $T = \dfrac{4960}{6t + 80} + 21$

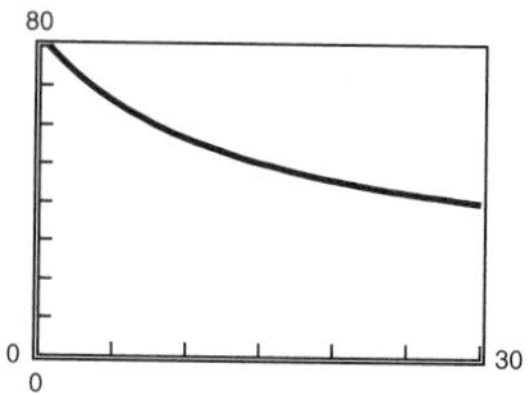

93.

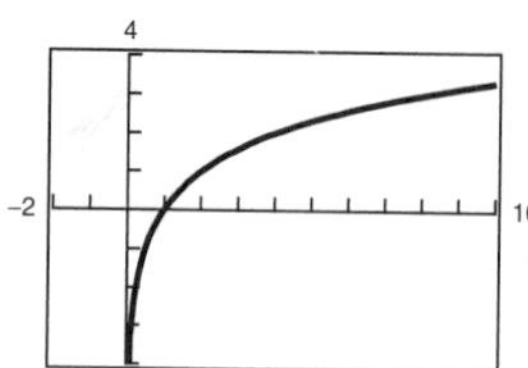

95.

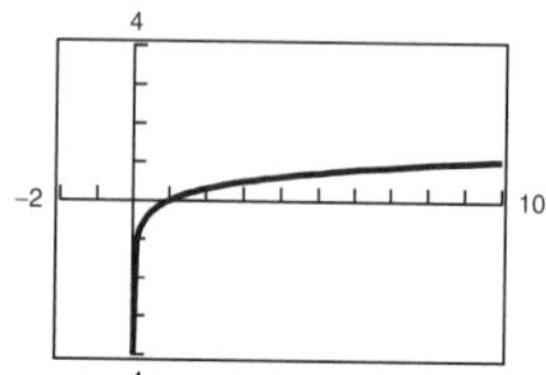

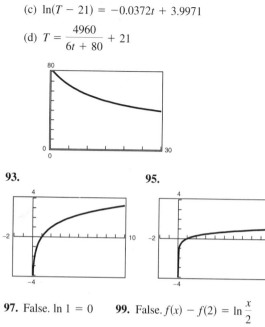

97. False. $\ln 1 = 0$ **99.** False. $f(x) - f(2) = \ln \dfrac{x}{2}$

101. False. $u = v^2$ **103.** Answers will vary.

105. $\dfrac{3x^4}{2y^3}$ **107.** 1

Section 4.4 *(page 355)*

1. (a) Yes (b) No **3.** (a) No (b) Yes (c) Yes

5. (a) No (b) No (c) Yes

7. (3, 8) **9.** (9, 2) **11.** 2 **13.** -2

15. 3 **17.** 64 **19.** $\frac{1}{10}$ **21.** x^2 **23.** $5x + 2$

25. x^2 **27.** $\ln 10 \approx 2.303$ **29.** 0

31. $\ln \frac{5}{3} \approx 0.511$ **33.** $\log_{10} 42 \approx 1.623$

35.

x	0.6	0.7	0.8	0.9	1.0
$f(x)$	6.05	8.17	11.02	14.88	20.09

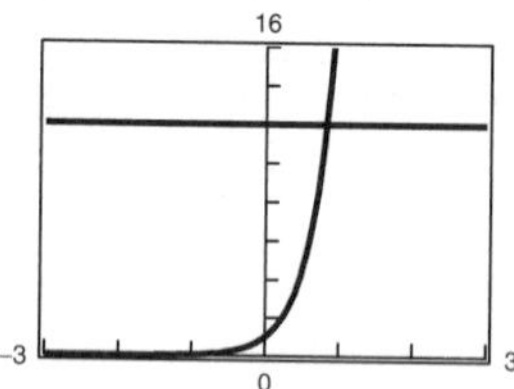

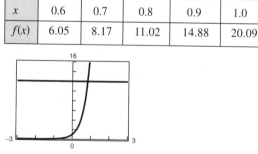

0.828

37.

x	5	6	7	8	9
$f(x)$	1756	1598	1338	908	200

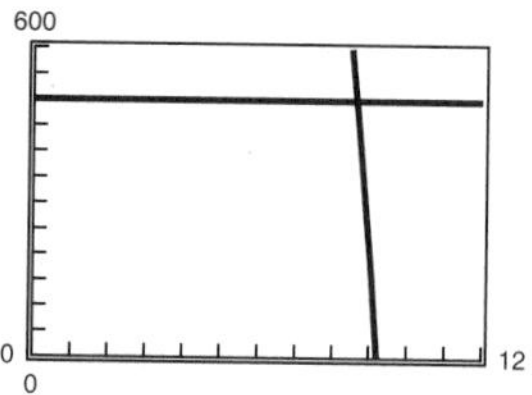

8.635

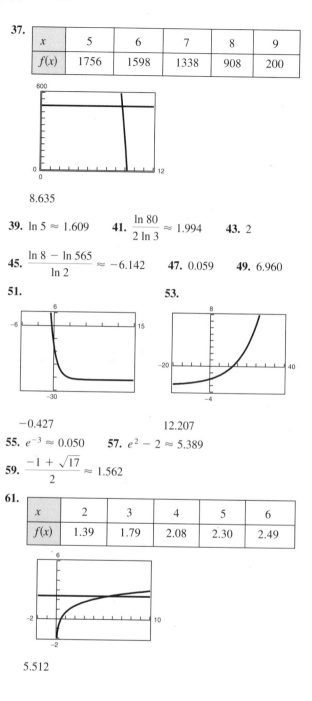

39. $\ln 5 \approx 1.609$ **41.** $\dfrac{\ln 80}{2 \ln 3} \approx 1.994$ **43.** 2

45. $\dfrac{\ln 8 - \ln 565}{\ln 2} \approx -6.142$ **47.** 0.059 **49.** 6.960

51.

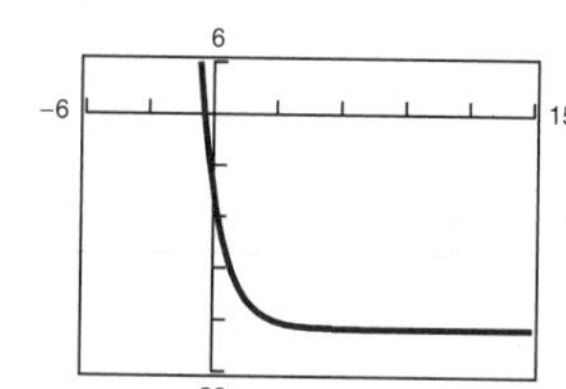

-0.427

53.

8
-20
40
-4

12.207

55. $e^{-3} \approx 0.050$ **57.** $e^2 - 2 \approx 5.389$

59. $\dfrac{-1 + \sqrt{17}}{2} \approx 1.562$

61.

x	2	3	4	5	6
$f(x)$	1.39	1.79	2.08	2.30	2.49

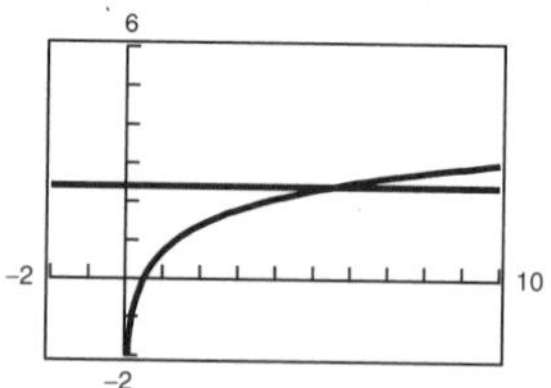

5.512

63.

x	12	13	14	15	16
$f(x)$	9.79	10.22	10.63	11.00	11.36

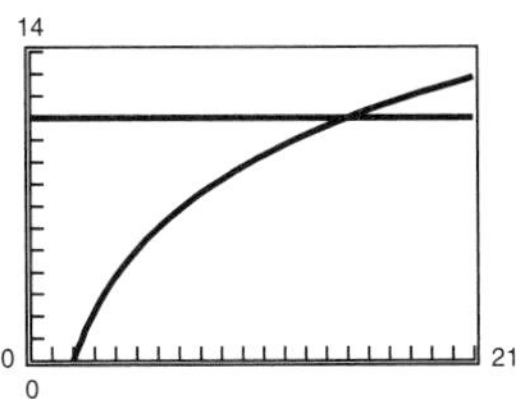

14.988

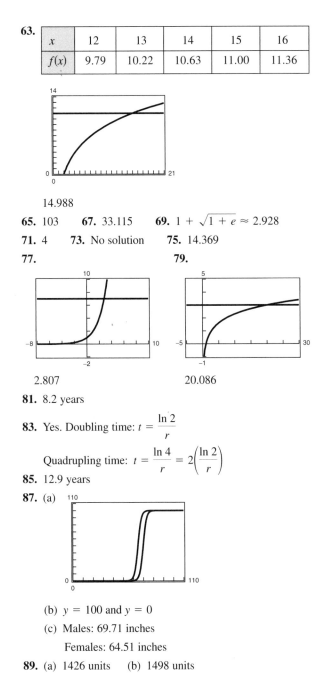

65. 103 **67.** 33.115 **69.** $1 + \sqrt{1 + e} \approx 2.928$

71. 4 **73.** No solution **75.** 14.369

77.

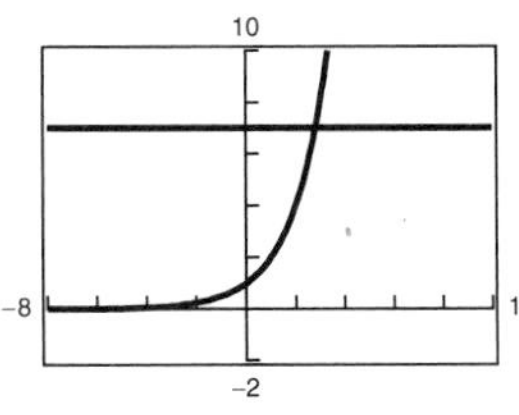

2.807

79.

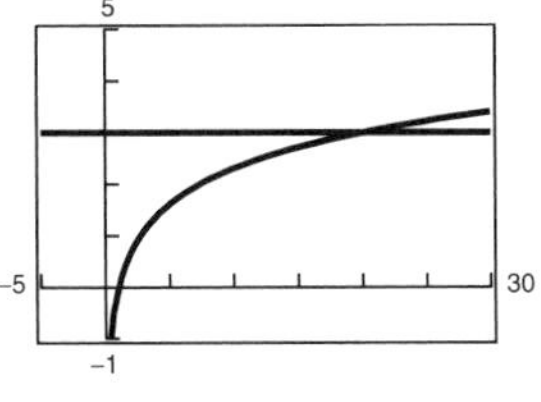

20.086

81. 8.2 years

83. Yes. Doubling time: $t = \dfrac{\ln 2}{r}$

Quadrupling time: $t = \dfrac{\ln 4}{r} = 2\left(\dfrac{\ln 2}{r}\right)$

85. 12.9 years

87. (a)

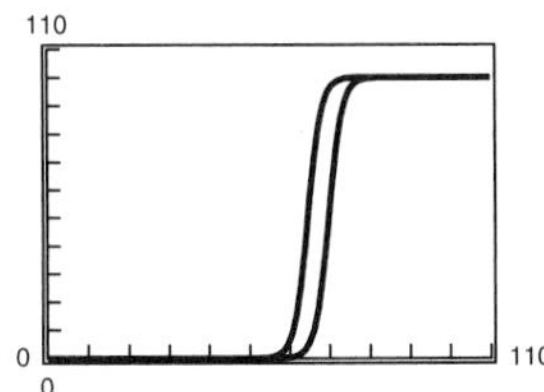

(b) $y = 100$ and $y = 0$

(c) Males: 69.71 inches

Females: 64.51 inches

89. (a) 1426 units (b) 1498 units

91. (a)

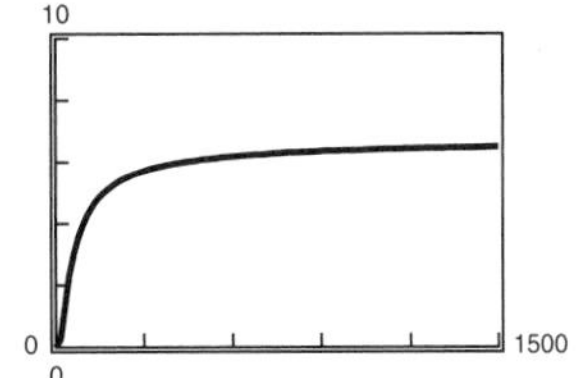

(b) $y = 6.7$. Yield will approach 6.7 million cubic feet per acre.

(c) 29.3 years

93. (a)

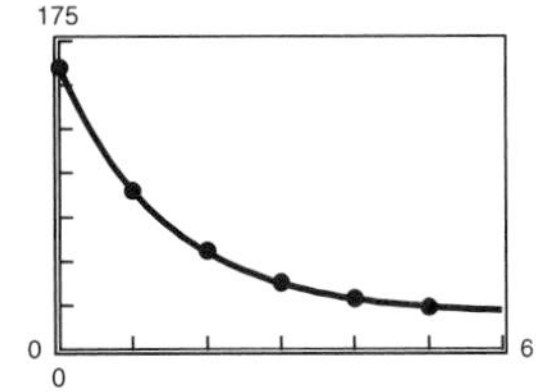

(b) $y = 20$. Room temperature

(c) 0.81 hour

95. (b) **97.** (f) **99.** (d)

Section 4.5 *(page 366)*

1. (b) **3.** (e) **5.** (f)

	Initial Investment	*Annual % Rate*	*Time to Double*	*Amount After 10 Years*
7.	\$1000	12%	5.78 yr	\$3320.12
9.	\$750	8.94%	7.75 yr	\$1833.67
11.	\$500	9.5%	7.30 yr	\$1292.85
13.	\$6376.28	4.5%	15.4 yr	\$10,000.00

15. \$112,087.09

17. (a) 6.642 years (b) 6.330 years

(c) 6.302 years (d) 6.301 years

19.

r	2%	4%	6%	8%	10%	12%
t	54.93	27.47	18.31	13.73	10.99	9.16

21.

r	2%	4%	6%	8%	10%	12%
t	55.47	28.01	18.85	14.27	11.53	9.69

23.

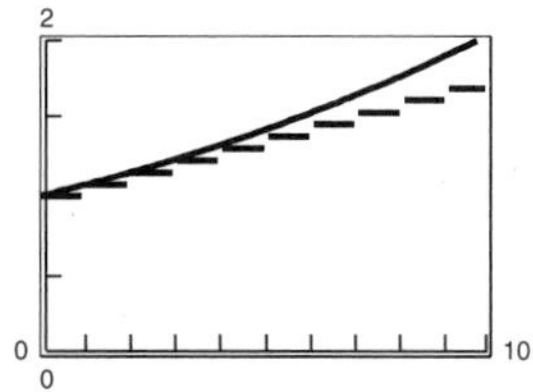

Continuous compounding

	Isotope	Half-Life (years)	Initial Quantity	Amount After 1000 Years
25.	Ra^{226}	1620	10 g	6.52 g
27.	C^{14}	5730	3 g	2.66 g

29. $y = e^{0.7675x}$ **31.** $y = e^{-0.4621x}$ **33.** 2013

35. $k = 0.0137$; 3288 **37.** $y = 4.22e^{0.0430t}$; 9.97 million

39. $y = 3e^{-0.0091t}$; 2.50 million

41. The greater the rate of growth, the greater the value of b.

43. 3.15 hours **45.** 95.8% **47.** $2423

49. (a) $S(t) = 100(1 - e^{-0.1625t})$

(b)

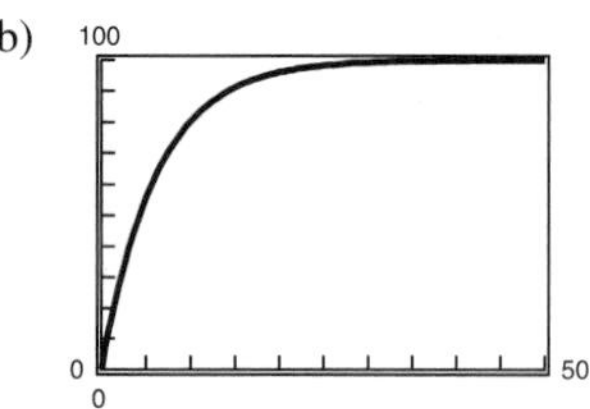

(c) 55,625

51. (a) $S = 10(1 - e^{-0.0575x})$ (b) 3314

53. (a) $N = 30(1 - e^{-0.050t})$

(b) 36 days

(c) No. It is not a linear function.

55. (a) 7.91 (b) 7.68

57. (a) 20 (b) 70 (c) 95 (d) 120

59. 95% **61.** 4.64 **63.** 1.58×10^{-6} moles per liter

65. 10,000,000 times

67. (a)

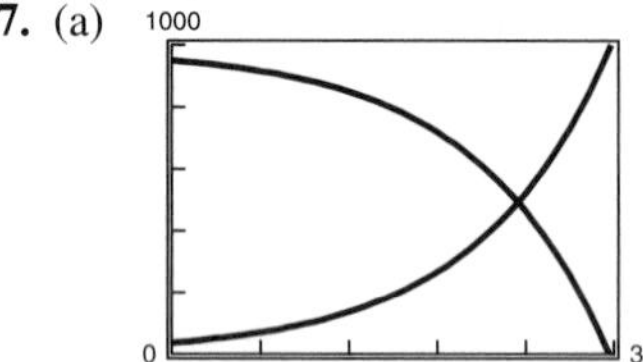

(b) Interest, $t \approx 28$ years

(c)

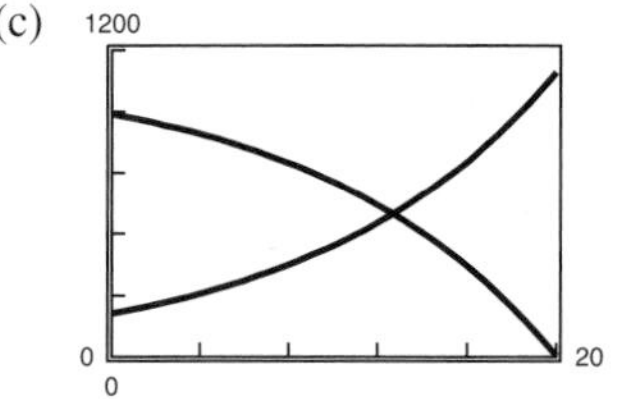

Interest, $t \approx 12.7$ years

69. (a) $t_3 = 0.2729s - 6.0143$

$t_4 = 1.5385e^{0.0291s}$

(b)

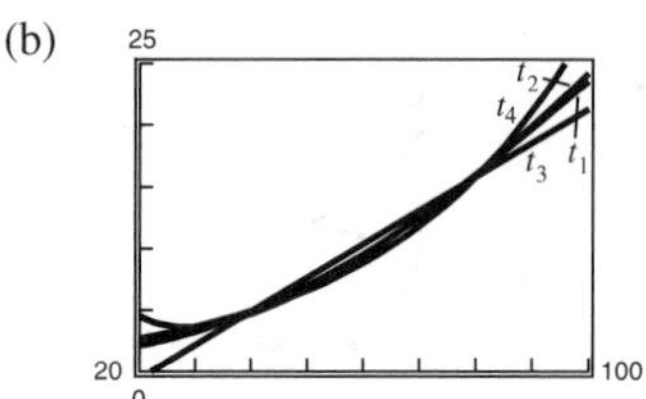

(c)

s	30	40	50	60	70	80	90
t_1	3.6	4.7	6.7	9.4	12.5	15.9	19.6
t_2	3.3	4.9	7.0	9.5	12.5	15.9	19.9
t_3	2.2	4.9	7.6	10.4	13.1	15.8	18.5
t_4	3.7	4.9	6.6	8.8	11.8	15.8	21.1

(d) Model: t_1; Sum ≈ 1.9

Model: t_2; Sum ≈ 1.2

Model: t_3; Sum ≈ 5.6

Model: t_4; Sum ≈ 2.6

Quadratic model fits best.

71. Answers will vary. **73.** 7:30 A.M.

75. $8x^2 - 24x + 18$

77. $x^3 - 5x^2 + 25x - 128 + \dfrac{641}{x + 5}$

Section 4.6 *(page 376)*

1. Logarithmic model **3.** Gaussian model

5. Exponential model **7.** Gaussian model

9. **11.**

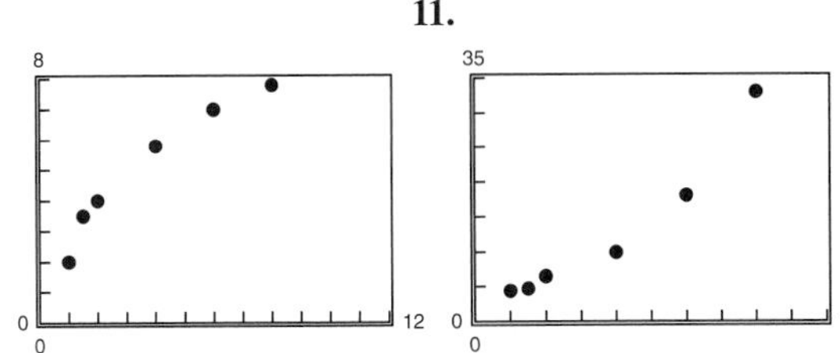

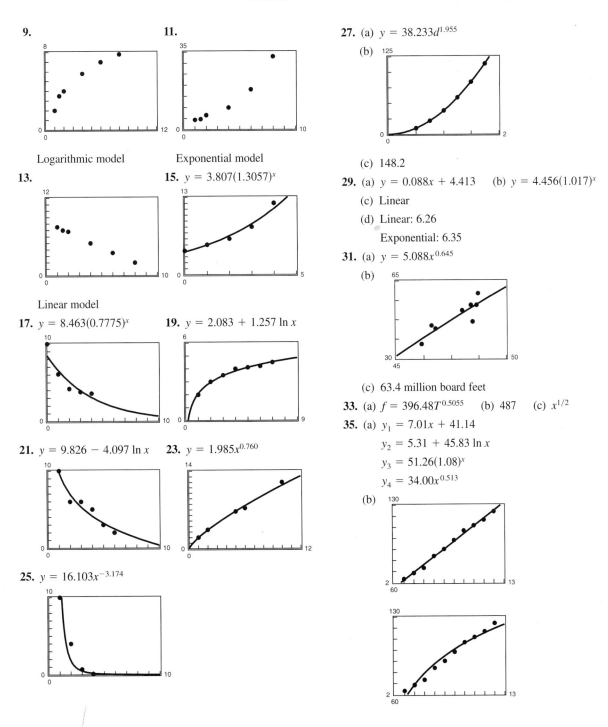

Logarithmic model   Exponential model

13. **15.** $y = 3.807(1.3057)^x$

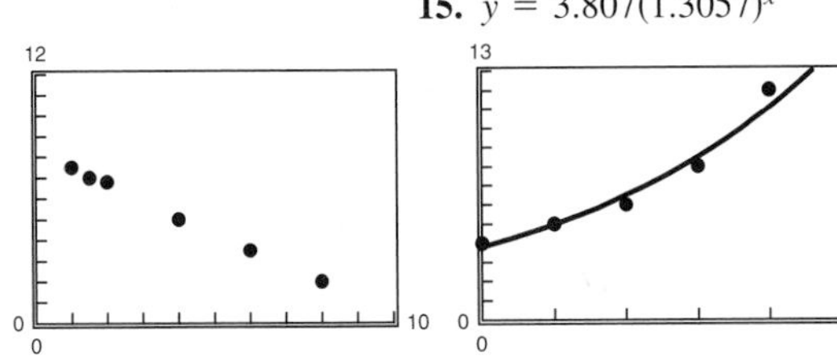

Linear model

17. $y = 8.463(0.7775)^x$ **19.** $y = 2.083 + 1.257 \ln x$

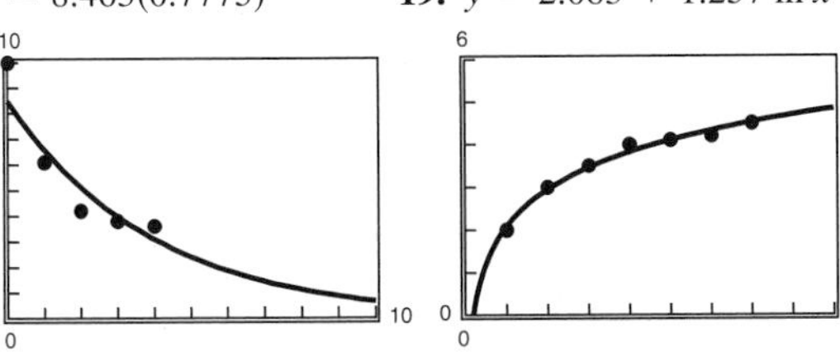

21. $y = 9.826 - 4.097 \ln x$ **23.** $y = 1.985x^{0.760}$

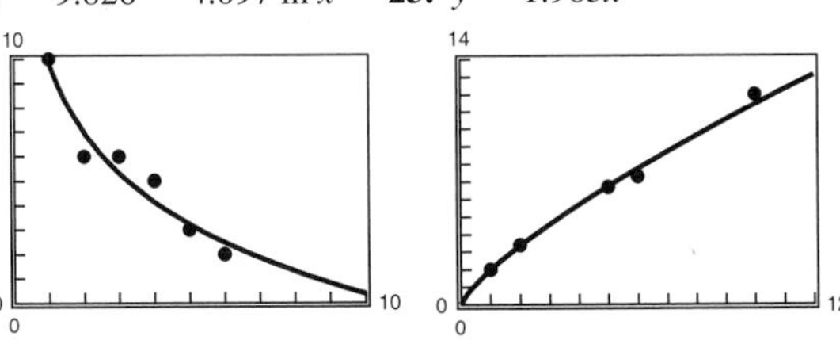

25. $y = 16.103x^{-3.174}$

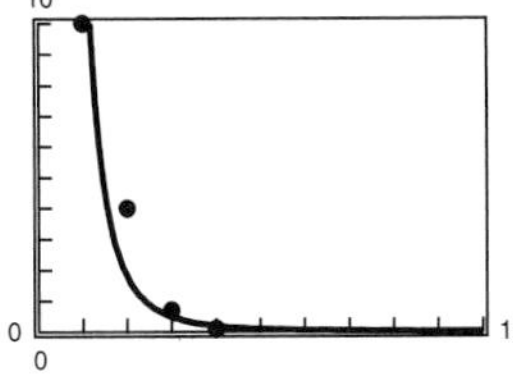

27. (a) $y = 38.233d^{1.955}$

(b)

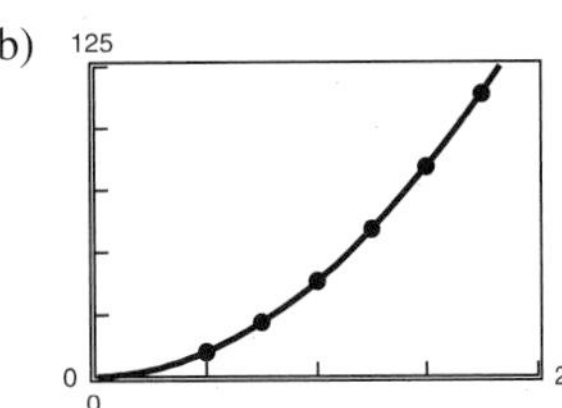

(c) 148.2

29. (a) $y = 0.088x + 4.413$ (b) $y = 4.456(1.017)^x$

(c) Linear

(d) Linear: 6.26

Exponential: 6.35

31. (a) $y = 5.088x^{0.645}$

(b)

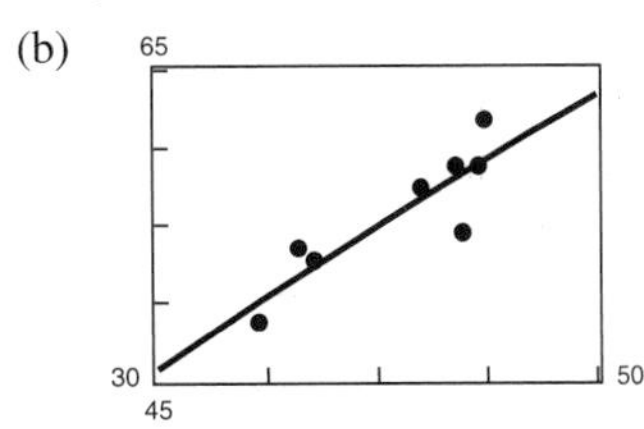

(c) 63.4 million board feet

33. (a) $f = 396.48T^{0.5055}$ (b) 487   (c) $x^{1/2}$

35. (a) $y_1 = 7.01x + 41.14$

$y_2 = 5.31 + 45.83 \ln x$

$y_3 = 51.26(1.08)^x$

$y_4 = 34.00x^{0.513}$

(b)

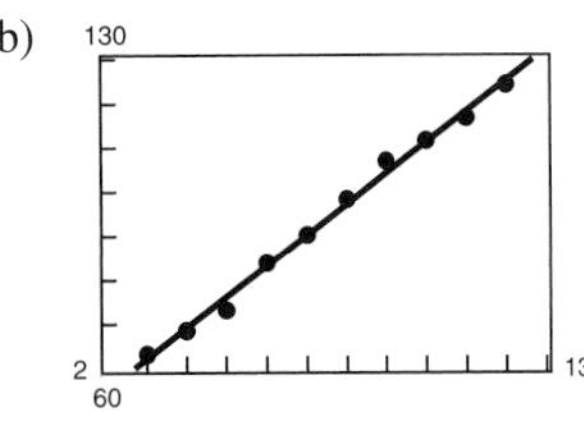

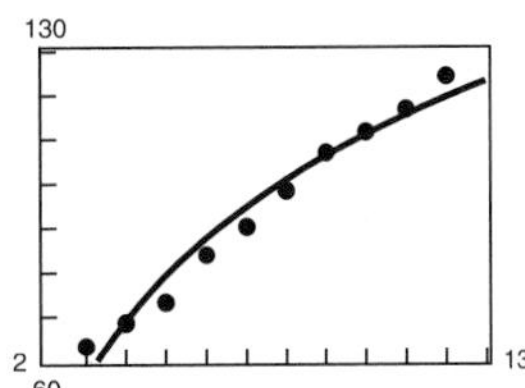

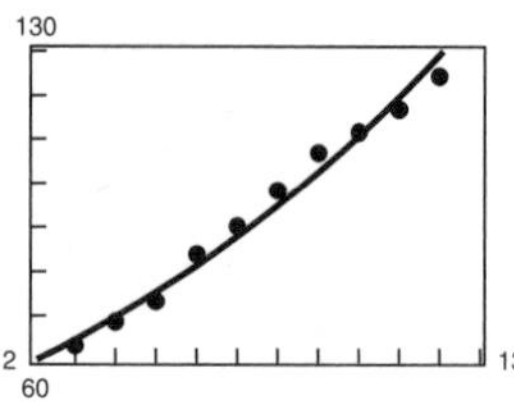

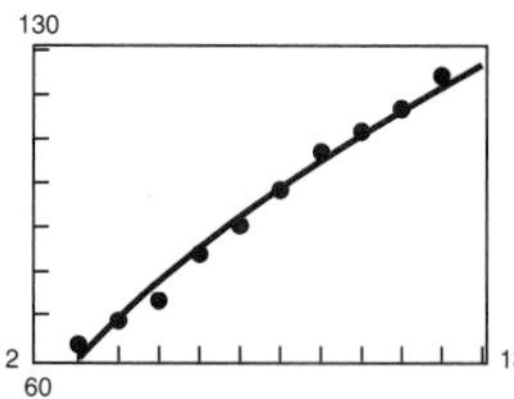

Linear model

(c)

x	y	$y - y_1$	$(y - y_1)^2$	$y - y_2$	$(y - y_2)^2$
3	63.2	1.03	1.06	7.54	56.86
4	68.6	−0.58	0.34	−0.24	0.06
5	73.2	−2.99	8.94	−5.87	34.46
6	83.9	0.70	0.49	−3.53	12.44
7	90.3	0.09	0.01	−4.19	17.57
8	98.4	1.18	1.39	−2.21	4.89
9	107.0	2.77	7.67	0.99	0.98
10	111.7	0.46	0.21	0.86	0.74
11	116.8	−1.45	2.10	1.59	2.54
12	124.3	−0.96	0.92	5.11	26.08

x	$y - y_3$	$(y - y_3)^2$	$y - y_4$	$(y - y_4)^2$
3	−1.37	1.88	3.46	11.99
4	−1.14	1.30	−0.64	0.41
5	−2.12	4.48	−4.43	19.66
6	2.56	6.54	−1.35	1.81
7	2.45	6.00	−1.96	3.84
8	3.52	12.40	−0.40	0.16
9	4.53	20.53	2.04	4.18
10	1.03	1.07	0.92	0.84
11	−2.72	7.40	0.46	0.22
12	−4.78	22.86	2.65	7.04

(d) Linear model

y_1: 23.14; y_2: 156.62; y_3: 84.46; y_4: 50.15

(e) Sum of the squares of the errors

FOCUS ON CONCEPTS *(page 380)*

1. $b < d < a < c$

b and d are negative.

2. (a) True. $\log_b uv = \log_b u + \log_b v$

(b) False. $2.04 \approx \log_{10}(10 + 100) \neq (\log_{10} 10)(\log_{10} 100) = 2$

(c) False. $1.95 \approx \log_{10}(100 - 10) \neq \log_{10} 100 - \log_{10} 10 = 1$

(d) True. $\log_b \frac{u}{v} = \log_b u - \log_b v$

3. Double the interest rate or time as it doubles the exponent in the exponential function.

4. (a) Logarithmic (b) Logistic (c) Exponential (d) Linear (e) None of the above (f) Exponential

Review Exercises *(page 381)*

1. (e) **3.** (b) **5.** (a)

7.

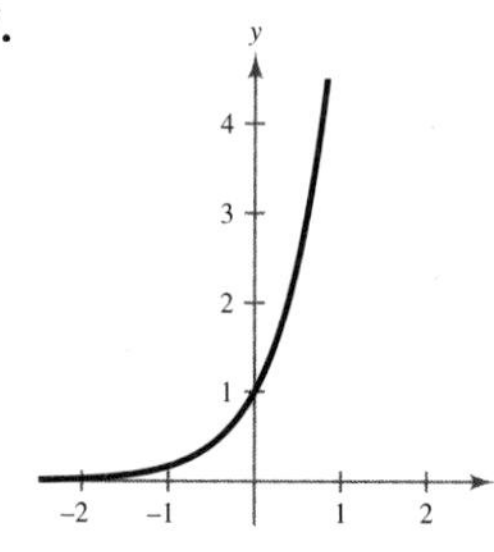

9.

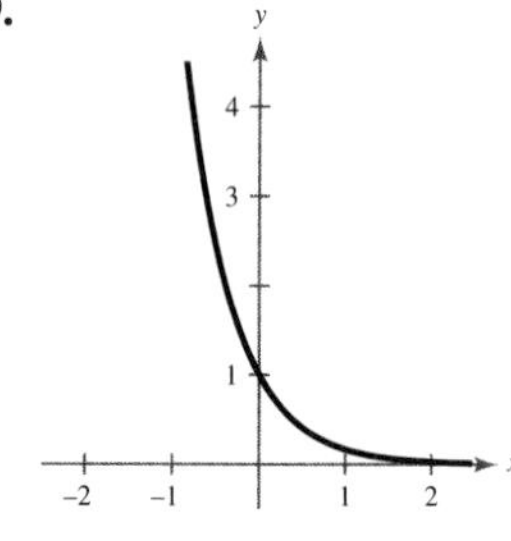

11.

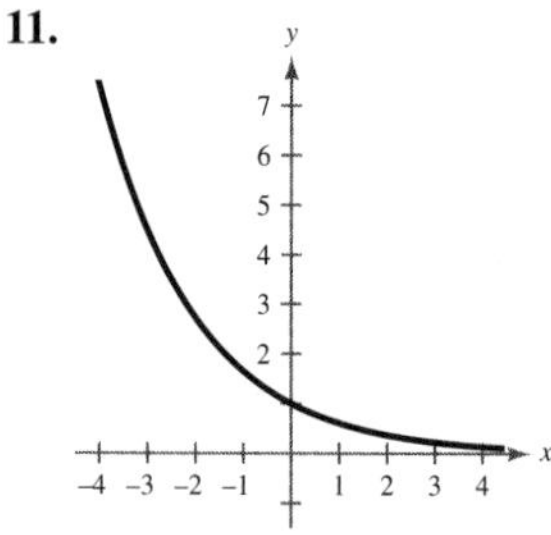

13.

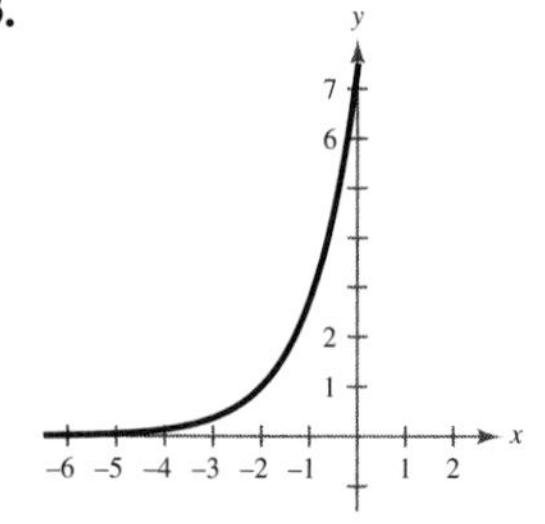

15.

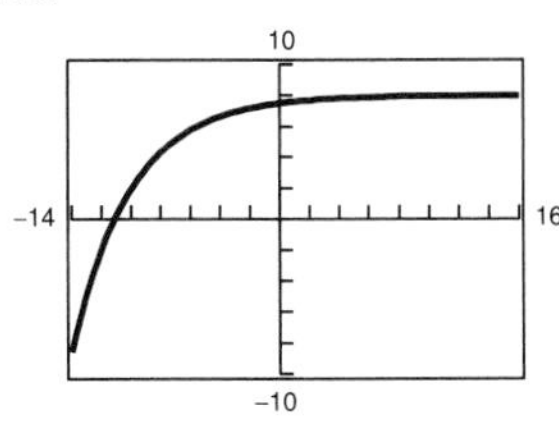

$y = 8$

17.

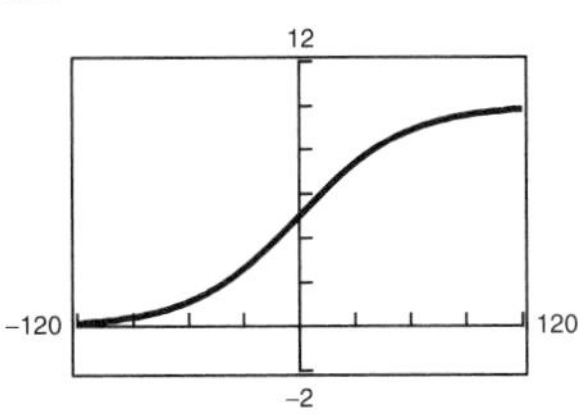

$y = 0, y = 10$

19.

n	1	2	4
A	\$9499.28	\$9738.91	\$9867.22

n	12	365	Continuous
A	\$9956.20	\$10,000.27	\$10,001.78

21.

t	1	10	20
P	\$184,623.27	\$89,865.79	\$40,379.30

t	30	40	50
P	\$18,143.59	\$8152.44	\$3663.13

23. (a)

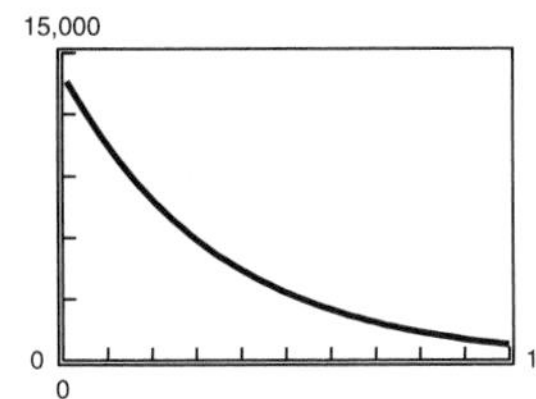

(b) \$7875

(c) At the beginning. Yes

25.

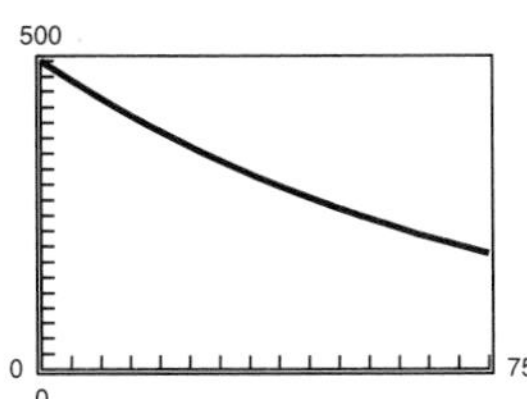

229.2

27.

Speed	50	55	60	65	70
Miles Per Gallon	28	26.4	24.8	23.4	22.0

29.

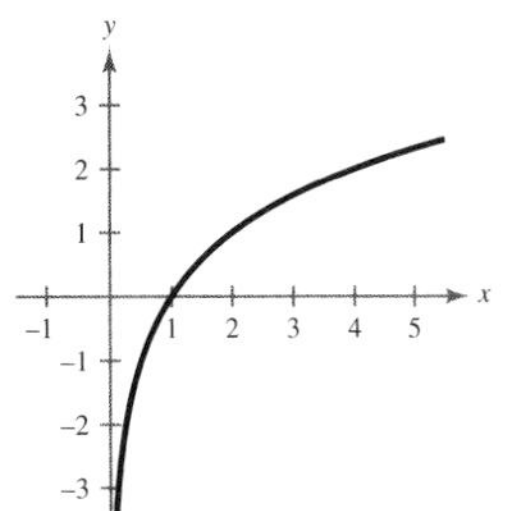

31.

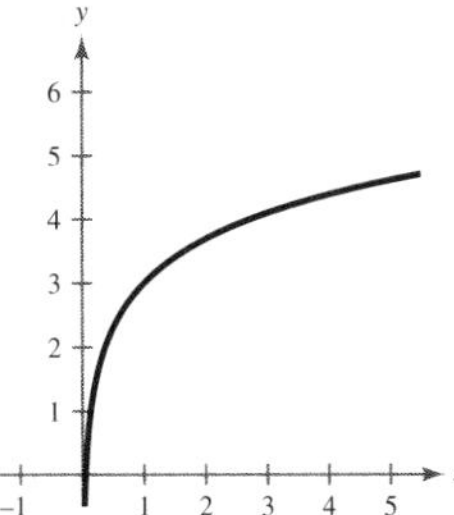

33.

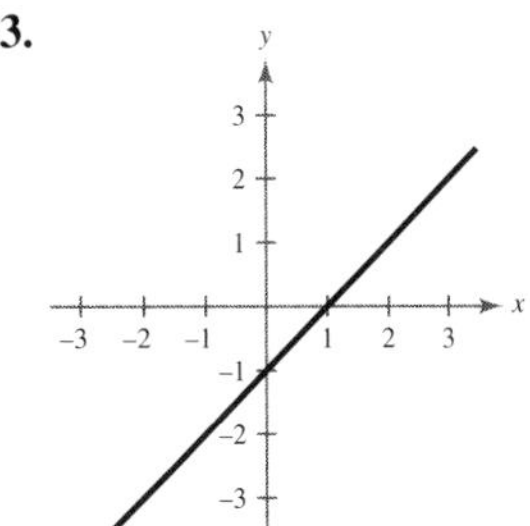

35.

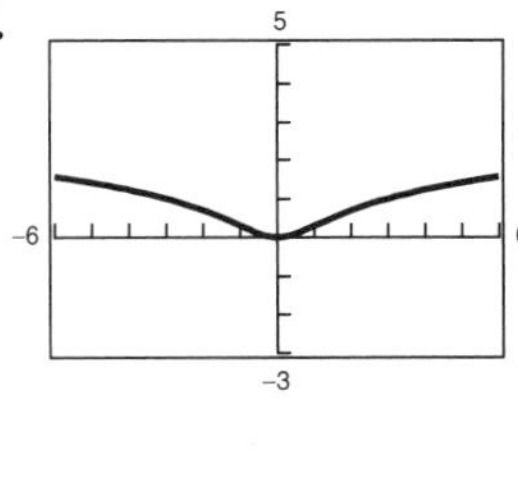

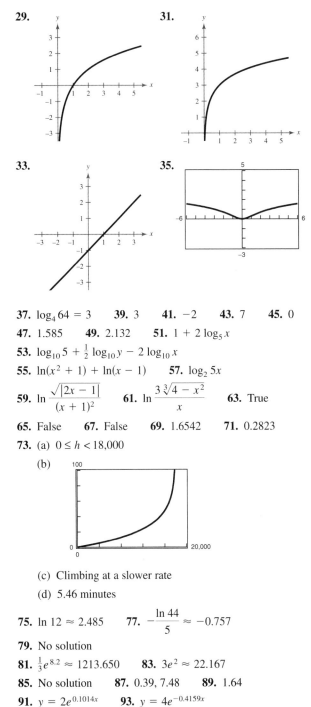

37. $\log_4 64 = 3$ **39.** 3 **41.** -2 **43.** 7 **45.** 0

47. 1.585 **49.** 2.132 **51.** $1 + 2\log_5 x$

53. $\log_{10} 5 + \frac{1}{2}\log_{10} y - 2\log_{10} x$

55. $\ln(x^2 + 1) + \ln(x - 1)$ **57.** $\log_2 5x$

59. $\ln \dfrac{\sqrt{|2x - 1|}}{(x + 1)^2}$ **61.** $\ln \dfrac{3\sqrt[3]{4 - x^2}}{x}$ **63.** True

65. False **67.** False **69.** 1.6542 **71.** 0.2823

73. (a) $0 \le h < 18{,}000$

(b)

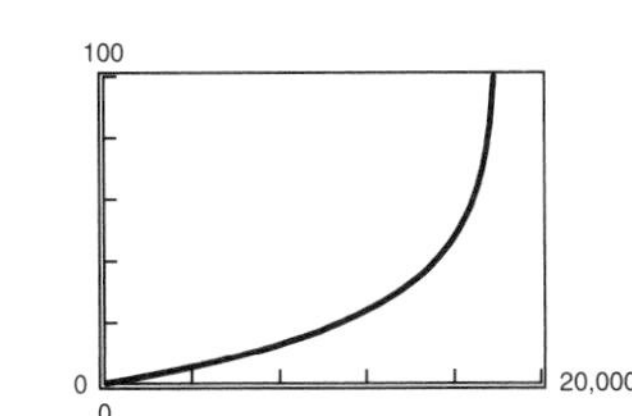

(c) Climbing at a slower rate

(d) 5.46 minutes

75. $\ln 12 \approx 2.485$ **77.** $-\dfrac{\ln 44}{5} \approx -0.757$

79. No solution

81. $\frac{1}{3}e^{8.2} \approx 1213.650$ **83.** $3e^2 \approx 22.167$

85. No solution **87.** 0.39, 7.48 **89.** 1.64

91. $y = 2e^{0.1014x}$ **93.** $y = 4e^{-0.4159x}$

95. (a) 1151 units (b) 1325 units

97. (a) 8.94% (b) \$1834.37 (c) 9.36% **99.** $10^{-3.5}$

101. $y = 234.684(0.8746)^x$

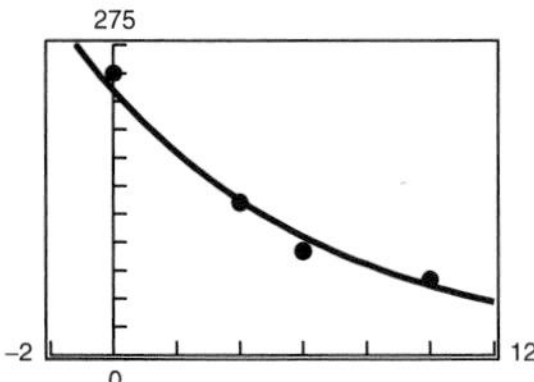

103. (a) $y = 12.907x^{0.5102}$

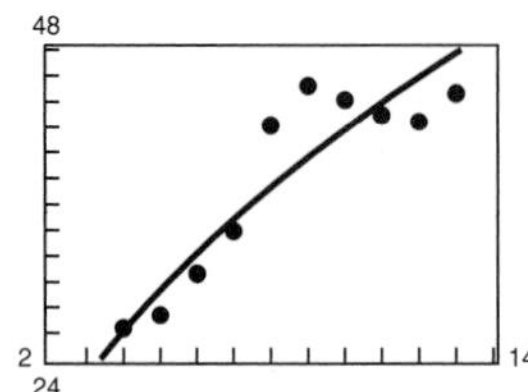

(b) $61.0 billion

Chapter Test (page 386)

1.

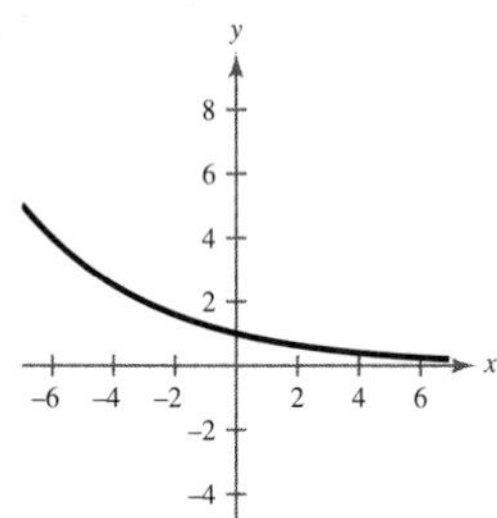

2.

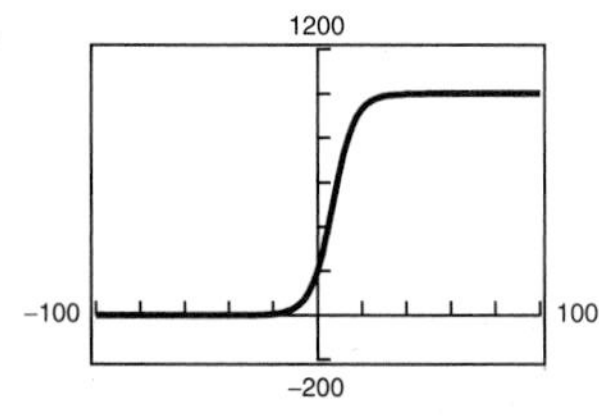

$y = 0, y = 1000$

3. $40,386.38 **4.** $4^3 = 64$

5.

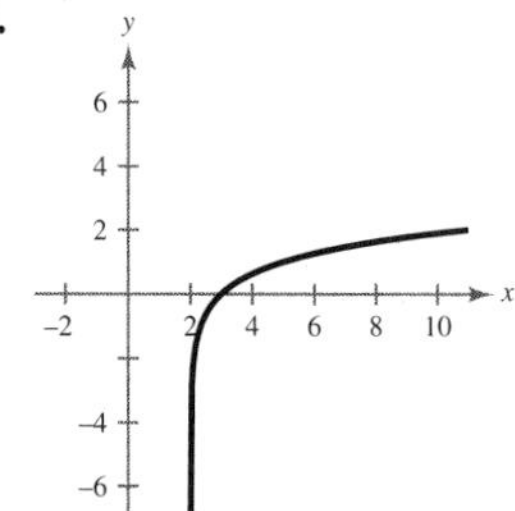

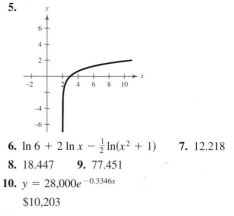

6. $\ln 6 + 2 \ln x - \frac{1}{2}\ln(x^2 + 1)$ **7.** 12.218

8. 18.447 **9.** 77.451

10. $y = 28{,}000e^{-0.3346x}$

$10,203

11. (a) 300 (b) 570 (c) 9 years

12. (c); it passes through the point (0, 0). Symmetric to the y-axis, and $y = 6$ is a horizontal asymptote.

13. $y = 5.280(1.4455)^x$

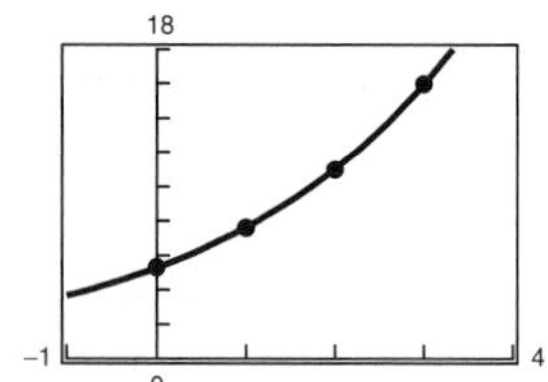

CHAPTER 5

Section 5.1 (page 395)

1. (2, 2) **3.** (2, 6), (−1, 3) **5.** (3, −4), (0, −5)
7. (0, 0), (2, −4) **9.** (−2, 1), (16, 4) **11.** (5, 5)
13. $\left(\frac{1}{2}, 3\right)$ **15.** (1, 1) **17.** $\left(\frac{20}{3}, \frac{40}{3}\right)$ **19.** No solution
21. (0, 0) **23.** (4, 3) **25.** $\left(\frac{5}{2}, \frac{3}{2}\right)$ **27.** (2, 2), (4, 0)

29.

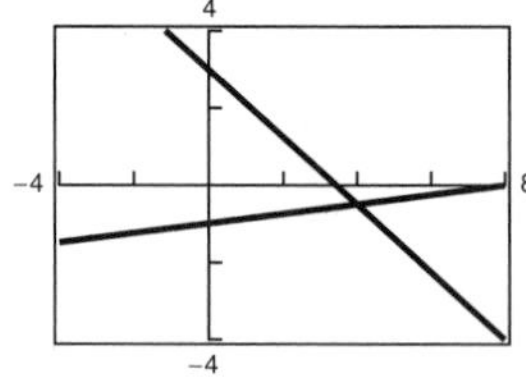

(4, −0.5)

31.

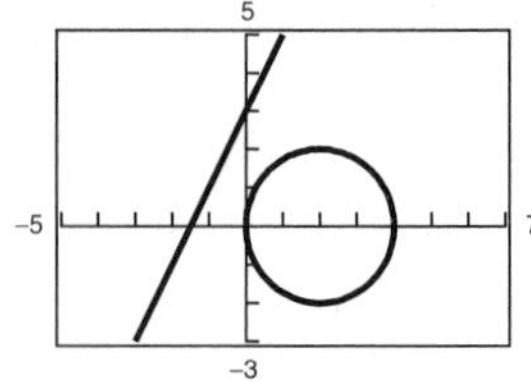

No points of intersection

33.

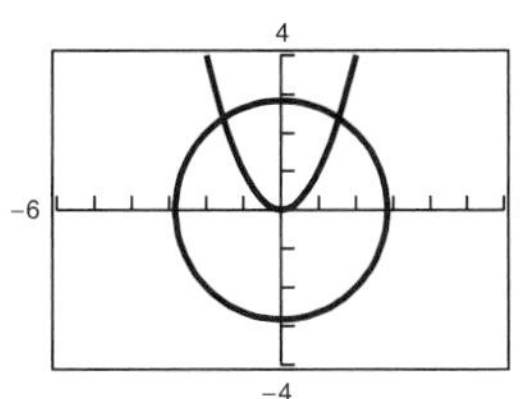

$(\pm 1.540,\ 2.372)$

35.

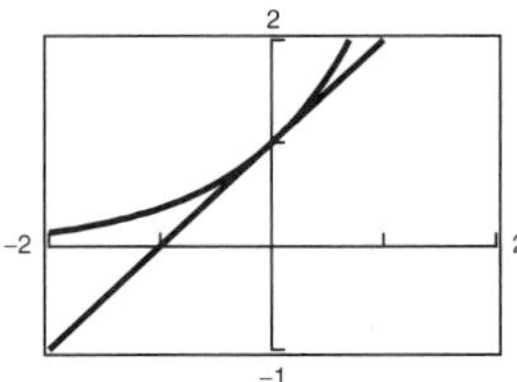

$(0, 1)$

37.

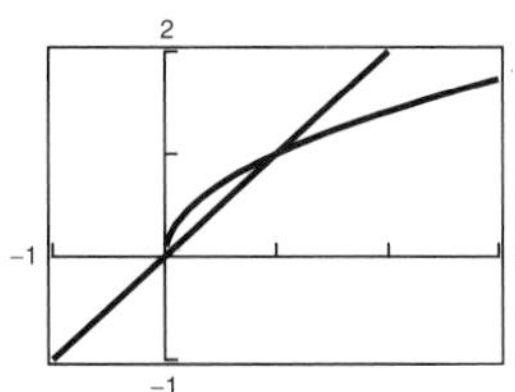

$(0, 0), (1, 1)$

39.

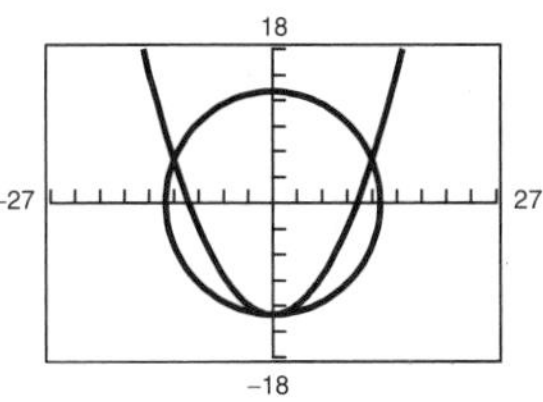

$(0, -13), (\pm 12, 5)$

41. $(1, 2)$ **43.** $(-2, 0), \left(\frac{29}{10}, \frac{21}{10}\right)$ **45.** No solution

47. $(0.287, 1.75)$ **49.** $(-1, 0), (0, 1), (1, 0)$

51. $\left(\frac{1}{2}, 2\right), \left(-4, -\frac{1}{4}\right)$

53. For a linear system the result will be a contradictory equation such as $0 = N$, where N is a nonzero real number. For nonlinear systems there may be an equation with imaginary roots.

55.

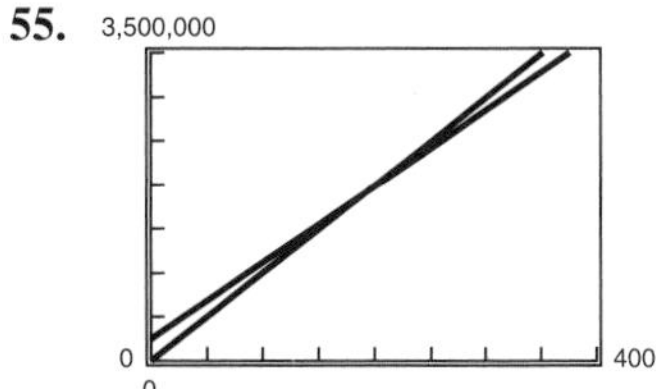

192 units

57.

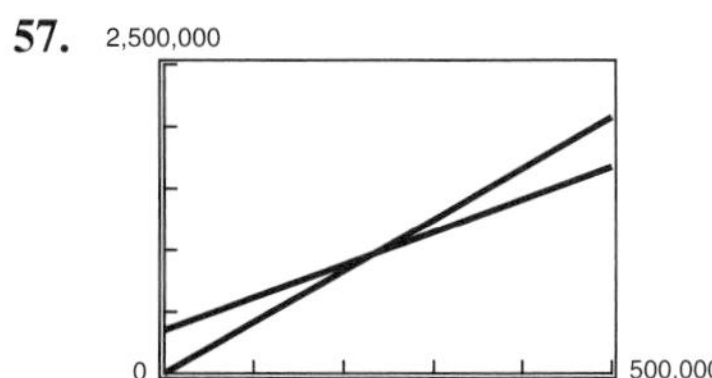

233,333 units

59. (a) $C = 3.45x + 16{,}000$

$R = 5.95x$

(b)

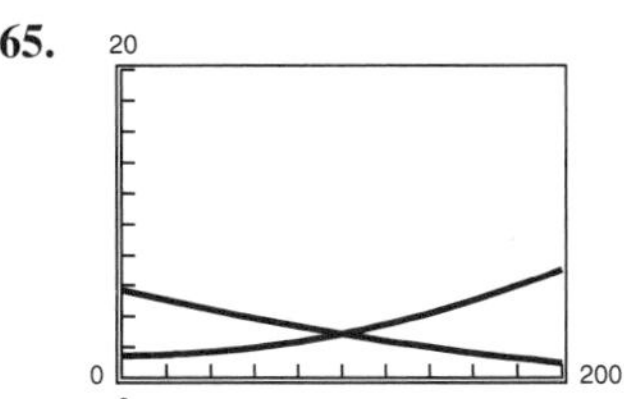

6400 units

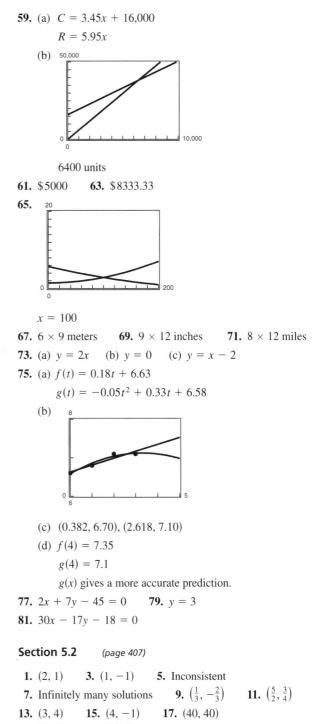

61. $5000 **63.** $8333.33

65.

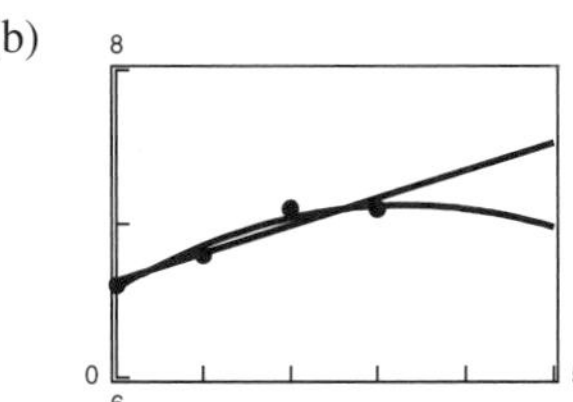

$x = 100$

67. 6×9 meters **69.** 9×12 inches **71.** 8×12 miles

73. (a) $y = 2x$ (b) $y = 0$ (c) $y = x - 2$

75. (a) $f(t) = 0.18t + 6.63$

$g(t) = -0.05t^2 + 0.33t + 6.58$

(b)

(c) $(0.382, 6.70), (2.618, 7.10)$

(d) $f(4) = 7.35$

$g(4) = 7.1$

$g(x)$ gives a more accurate prediction.

77. $2x + 7y - 45 = 0$ **79.** $y = 3$

81. $30x - 17y - 18 = 0$

Section 5.2 *(page 407)*

1. $(2, 1)$ **3.** $(1, -1)$ **5.** Inconsistent

7. Infinitely many solutions **9.** $\left(\frac{1}{3}, -\frac{2}{3}\right)$ **11.** $\left(\frac{5}{2}, \frac{3}{4}\right)$

13. $(3, 4)$ **15.** $(4, -1)$ **17.** $(40, 40)$

19. $\left(\frac{1}{2}, 0\right)$ **21.** $\left(\frac{18}{5}, \frac{3}{5}\right)$ **23.** $(5, -2)$

25. Infinitely many solutions **27.** $\left(\frac{90}{31}, -\frac{67}{31}\right)$ **29.** $\left(-\frac{6}{35}, \frac{43}{35}\right)$

31.

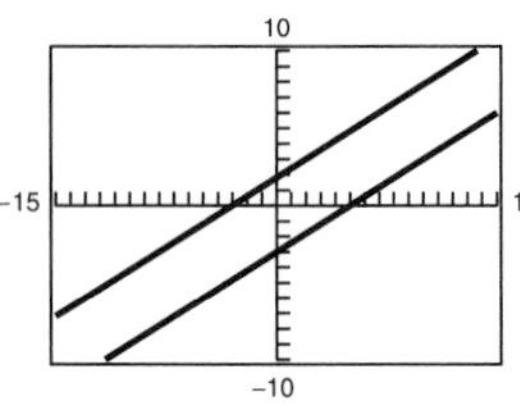

Inconsistent

33.

Consistent, one solution

35.

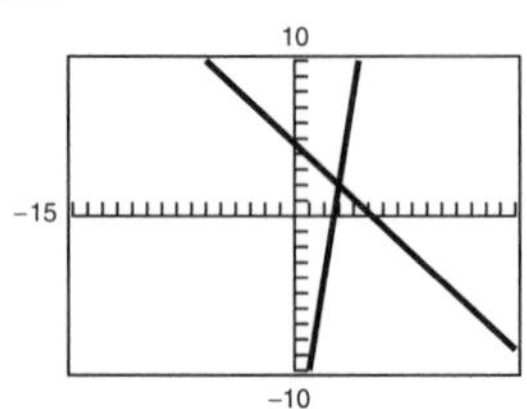

(3, 2)

37.

(3, −2)

39. (4, 1) **41.** (2, −1) **43.** $2x + 2y = 11$
$x - 4y = -7$

45. (39,600, 398). It is necessary to change the scale on the axes to see the point of intersection.

47. No. Two lines will intersect only once or coincide and the system will have infinitely many solutions.

49. $k = -4$ **51.** 550 miles per hour, 50 miles per hour

53. (a) $x + y = 10$
$0.2x + 0.5y = 3$

(b)

Decreases.

(c) 20% solution: $6\frac{2}{3}$ liters
50% solution: $3\frac{1}{3}$ liters

55. \$6000 **57.** 375 adults, 125 children

59. 225 kilometers and 75 kilometers

61. $y = 0.97x + 2.10$ **63.** $y = 0.318x + 4.061$

65. $y = \frac{3}{4}x + \frac{4}{3}$ **67.** $y = -2x + 4$

69. $y = 14x + 19$, 41.4 **71.** (80, 10)

73. (2,000,000, 100) **75.** $u = 1, \quad v = -\tan x$

77. $u = -x \ln x, \quad v = \ln x$ **79.** All real numbers

81. $-5 \le x \le 5$

Section 5.3 *(page 418)*

1. (1, −2, 4) **3.** (1, 2, −2) **5.** $\left(\frac{1}{2}, -2, 2\right)$

7. $x - 2y + 3z = 5$
$y - 2z = 9$
$2x - 3z = 0$

First step in putting the system in row-echelon form

9. (1, 2, 3) **11.** (−4, 8, 5) **13.** (5, −2, 0)

15. Inconsistent **17.** $\left(1, -\frac{3}{2}, \frac{1}{2}\right)$

19. $(-3a + 10, 5a - 7, a)$ **21.** $(-a + 3, a + 1, a)$

23. $(2a, 21a - 1, 8a)$ **25.** $\left(\frac{1}{2} - \frac{3}{2}a, 1 - \frac{2}{3}a, a\right)$

27. (1, 1, 1, 1) **29.** Inconsistent **31.** (0, 0, 0)

33. $(9a, -35a, 67a)$

35. No. There are two arithmetic errors. They are the constant in the second equation and the coefficient of z in the third equation.

37. $3x + y - z = 9$
$x + 2y - z = 0$
$-x + y + 3z = 1$

39. $y = \frac{1}{2}x^2 - 2x$

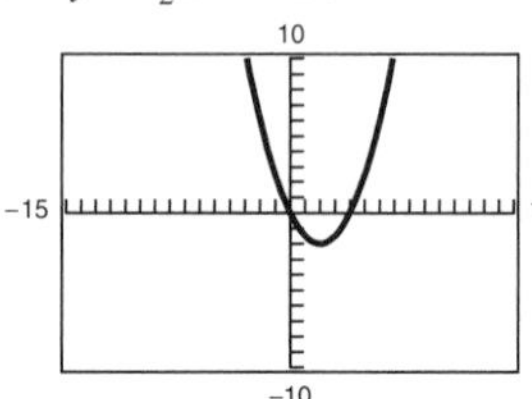

41. $y = x^2 - 6x + 8$

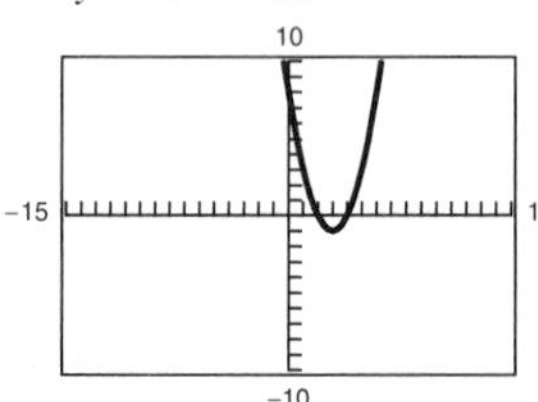

43. $x^2 + y^2 - 4x = 0$

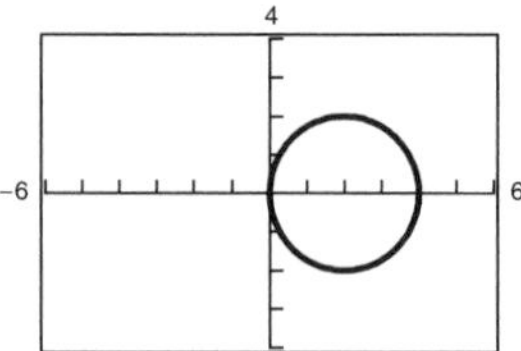

45. $x^2 + y^2 + 6x - 8y = 0$

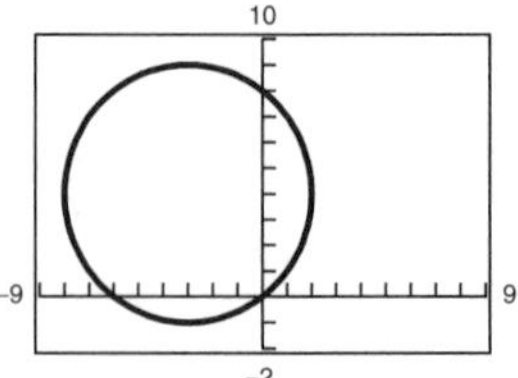

47. $s = -16t^2 + 144$ **49.** $s = 24t^2 - 264t + 692$

51. \$4000 at 5%
\$5000 at 6%
\$7000 at 7%

53. \$366,666.67 at 8%
\$316,666.67 at 9%
\$91,666.67 at 10%

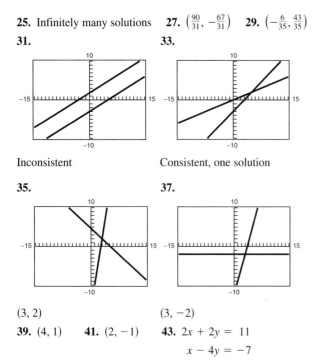

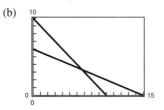

55. $250{,}000 - \frac{1}{2}s$ in certificates of deposit
$125{,}000 + \frac{1}{2}s$ in municipal bonds
$125{,}000 - s$ in blue-chip stocks
s in growth stocks

57. 20 liters of spray X
18 liters of spray Y
16 liters of spray Z

59. Use four medium trucks or two large trucks.
One medium truck and two small trucks

61. $t_1 = 96$ pounds
$t_2 = 48$ pounds
$a = -16$ feet per second squared

63.

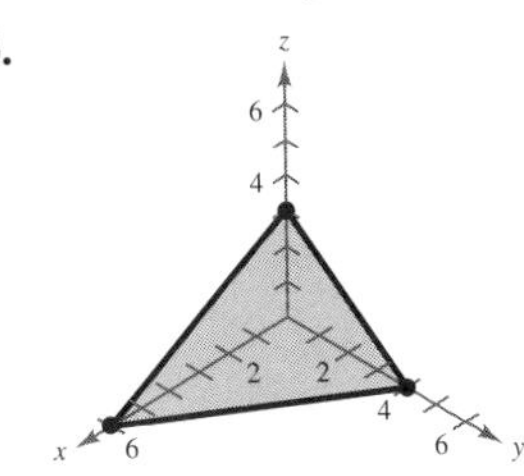

$(6, 0, 0)$, $(0, 4, 0)$, $(0, 0, 3)$, $(4, 0, 1)$

65.

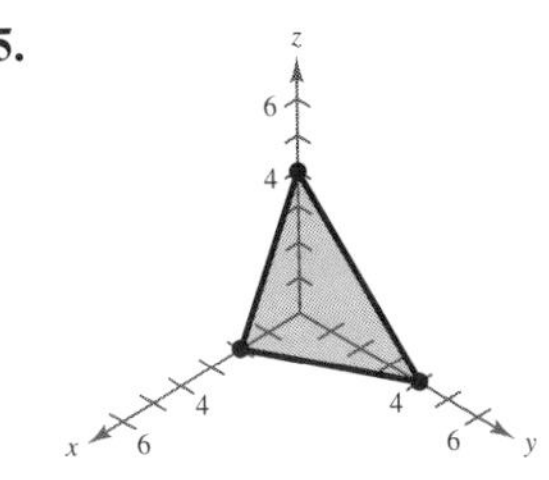

$(2, 0, 0)$, $(0, 4, 0)$, $(0, 0, 4)$, $(0, 2, 2)$

67. $y = -\frac{5}{24}x^2 - \frac{3}{10}x + \frac{41}{6}$ **69.** $y = x^2 - x$

71. (a) $y = 0.14x^2 - 4.43x + 58.40$

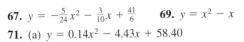

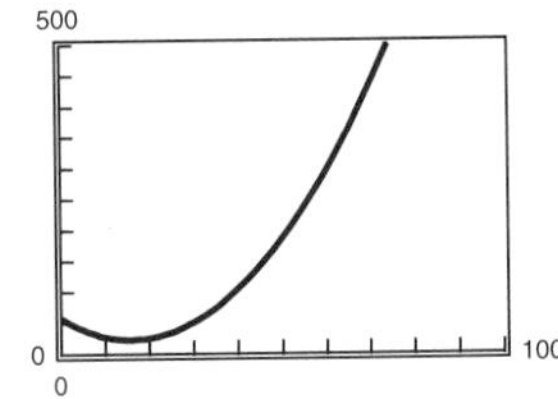

(b) 434.3

73. $x = 5$, $y = 5$, $\lambda = -5$

75. $x = \sqrt{2}/2$, $y = \frac{1}{2}$, $\lambda = 1$; $x = -\sqrt{2}/2$, $y = \frac{1}{2}$, $\lambda = 1$; $x = 0$, $y = 0$, $\lambda = 0$

77. The average increase in sales per year

79. 6.375 **81.** 80,000

Section 5.4 *(page 428)*

1. $\frac{A}{x} + \frac{B}{x - 14}$ **3.** $\frac{A}{x} + \frac{B}{x^2} + \frac{C}{x - 10}$

5. $\frac{A}{x} + \frac{Bx + C}{x^2 + 10}$ **7.** $\frac{1}{2}\left(\frac{1}{x - 1} - \frac{1}{x + 1}\right)$

9. $\frac{1}{x} - \frac{1}{x + 1}$ **11.** $\frac{1}{x} - \frac{2}{2x + 1}$ **13.** $\frac{1}{x - 1} - \frac{1}{x + 2}$

15. $-\frac{3}{x} - \frac{1}{x + 2} + \frac{5}{x - 2}$ **17.** $\frac{3}{x} - \frac{1}{x^2} + \frac{1}{x + 1}$

19. $\frac{3}{x - 3} + \frac{9}{(x - 3)^2}$ **21.** $-\frac{1}{x} + \frac{2x}{x^2 + 1}$

23. $\frac{1}{3(x^2 + 2)} - \frac{1}{6(x + 2)} + \frac{1}{6(x - 2)}$

25. $\frac{1}{8(2x + 1)} + \frac{1}{8(2x - 1)} - \frac{x}{2(4x^2 + 1)}$

27. $\frac{1}{x + 1} + \frac{2}{x^2 - 2x + 3}$

29. $x + 3 + \frac{6}{x - 1} + \frac{4}{(x - 1)^2} + \frac{1}{(x - 1)^3}$

31. $\frac{3}{2x - 1} - \frac{2}{x + 1}$ **33.** $\frac{2}{x} - \frac{1}{x^2} - \frac{2}{x + 1}$

35. $\frac{1}{x^2 + 2} + \frac{x}{(x^2 + 2)^2}$ **37.** $2x + \frac{1}{2}\left(\frac{3}{x - 4} - \frac{1}{x + 2}\right)$

39. $\frac{1}{2a}\left(\frac{1}{a + x} + \frac{1}{a - x}\right)$ **41.** $\frac{1}{a}\left(\frac{1}{y} + \frac{1}{a - y}\right)$

43. $\frac{3}{x} - \frac{2}{x - 4}$

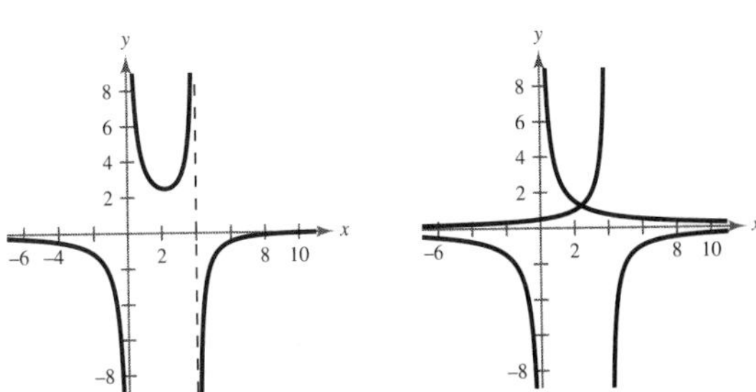

Vertical asymptotes are the same.

45. $\dfrac{3}{x-3} + \dfrac{5}{x+3}$

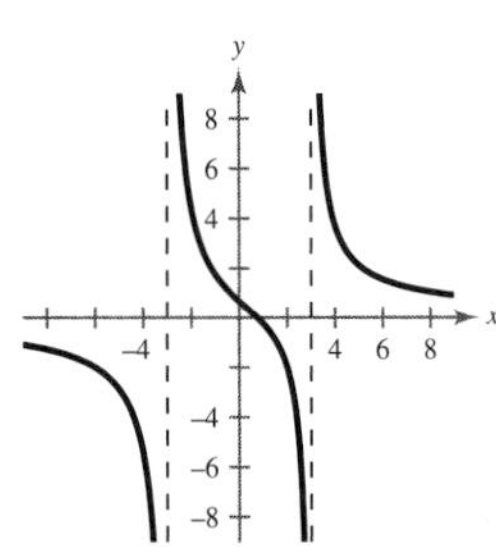

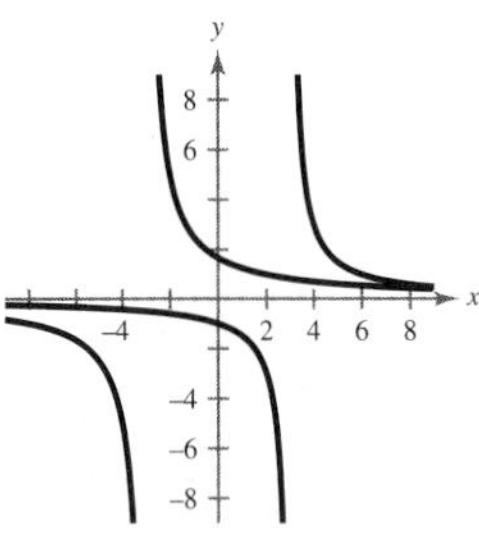

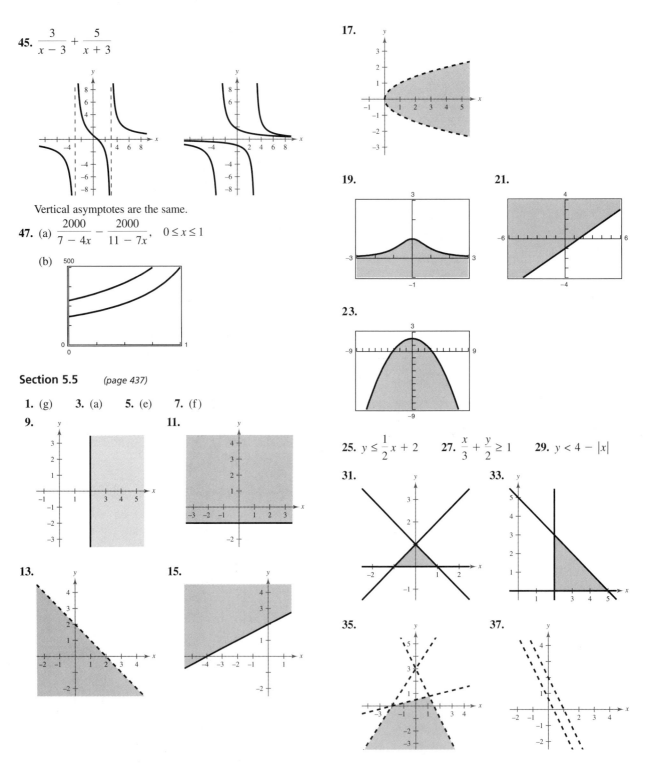

Vertical asymptotes are the same.

47. (a) $\dfrac{2000}{7-4x} - \dfrac{2000}{11-7x}, \quad 0 \le x \le 1$

(b)

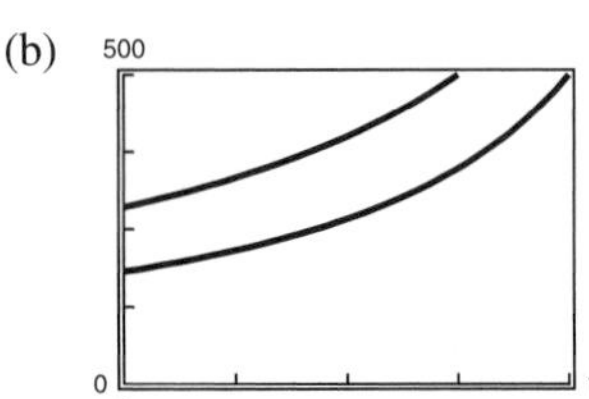

Section 5.5 *(page 437)*

1. (g) **3.** (a) **5.** (e) **7.** (f)

9.

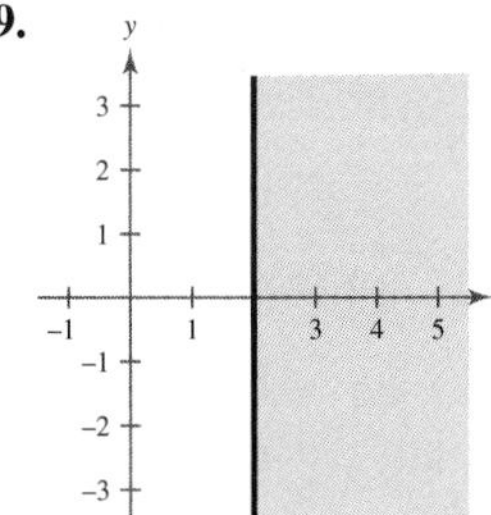

11.

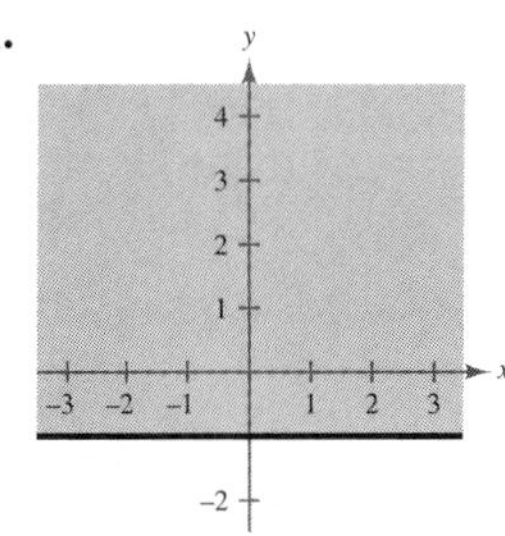

13.

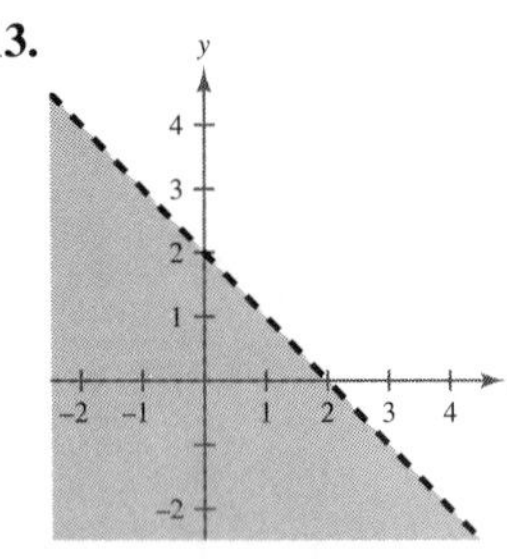

15.

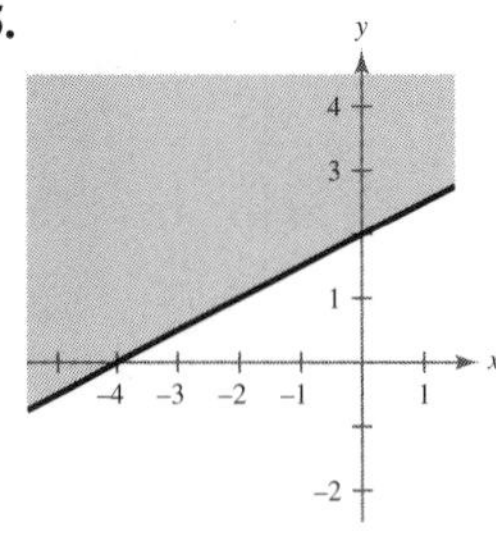

17.

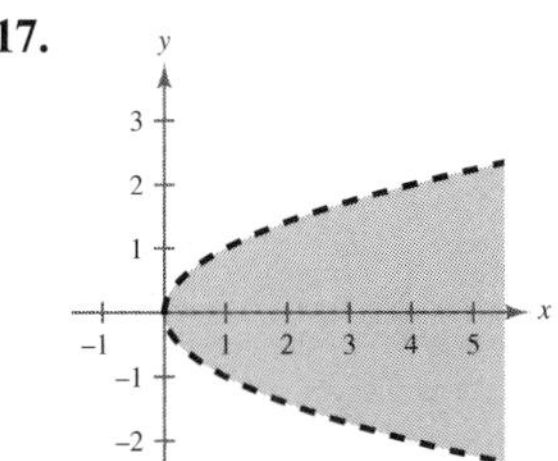

19.

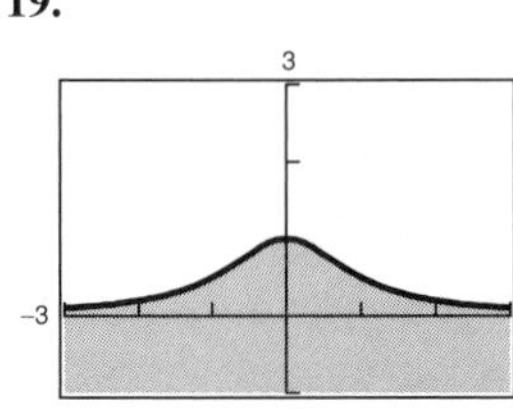

21.

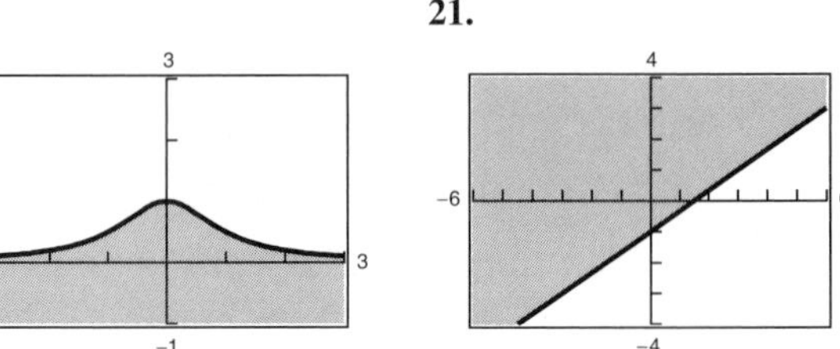

23.

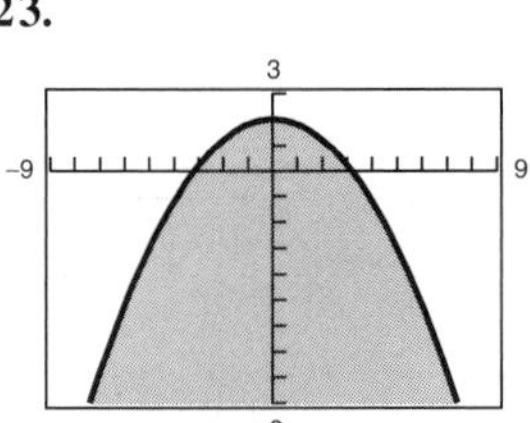

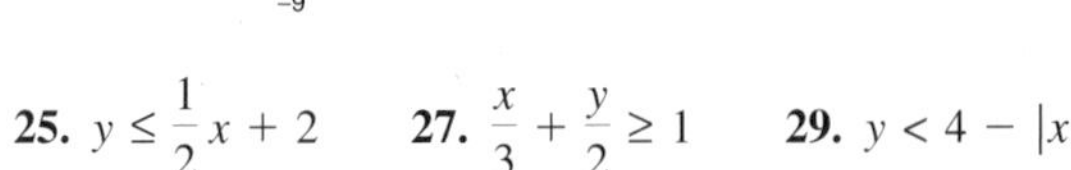

25. $y \le \dfrac{1}{2}x + 2$ **27.** $\dfrac{x}{3} + \dfrac{y}{2} \ge 1$ **29.** $y < 4 - |x|$

31.

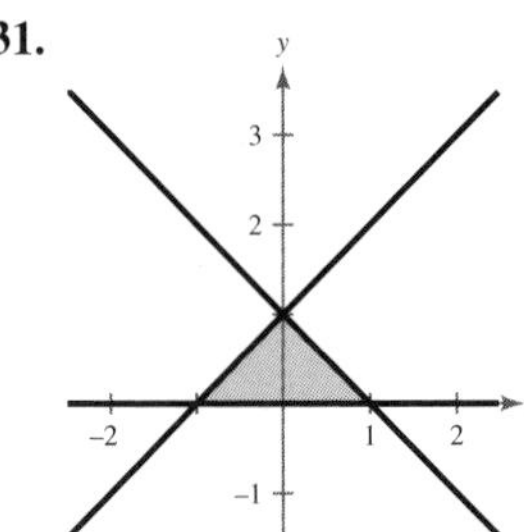

33.

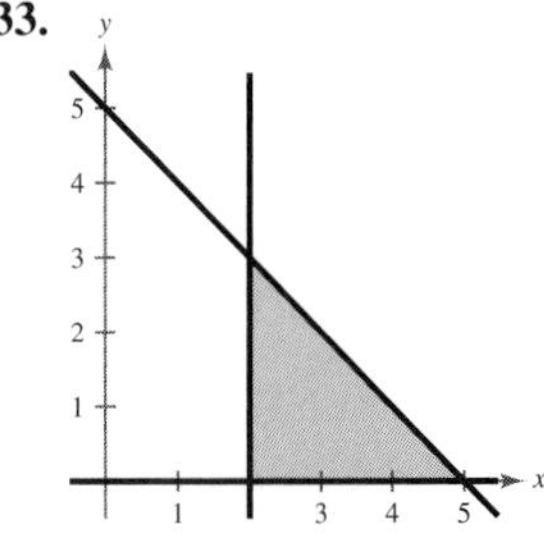

35.

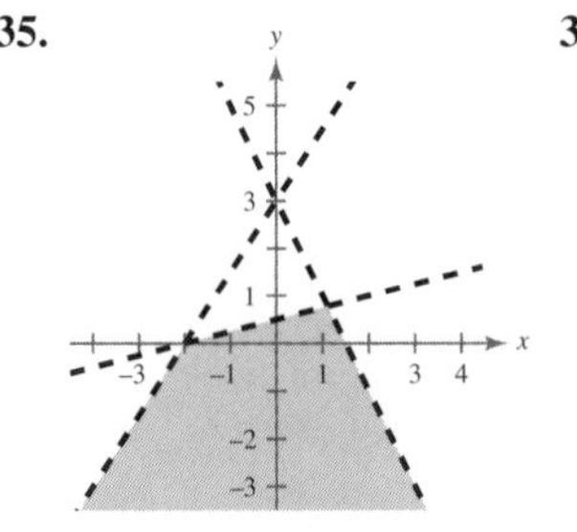

37.

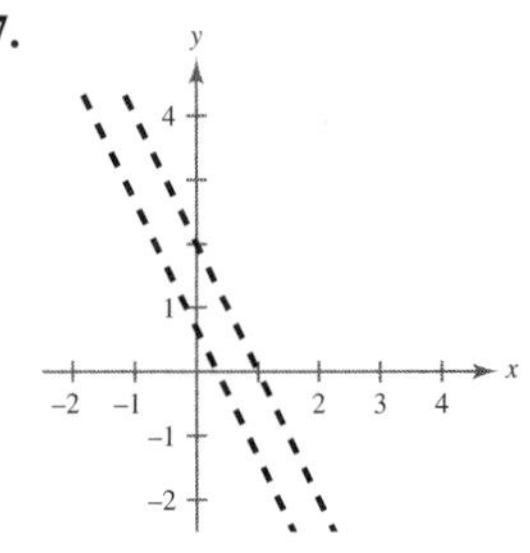

39.

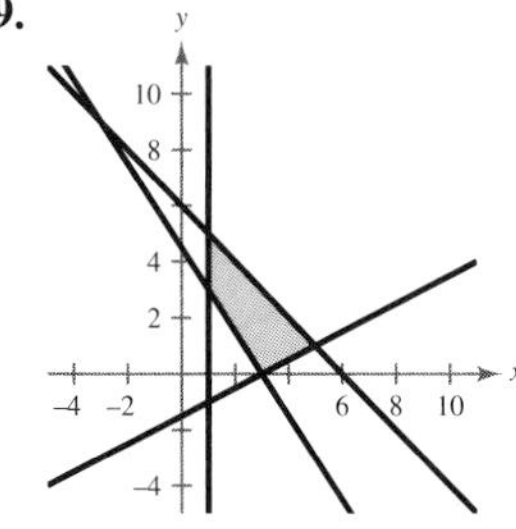

41.

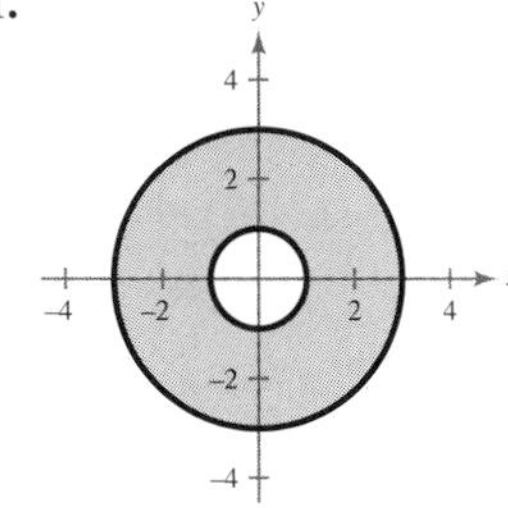

43.

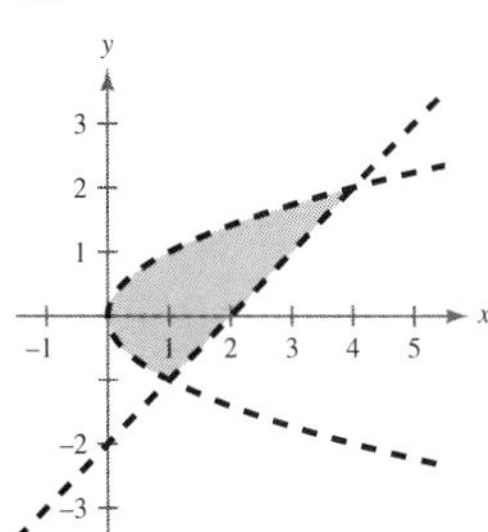

45.

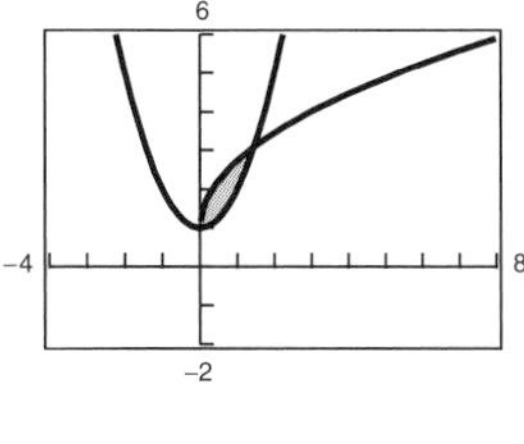

47.

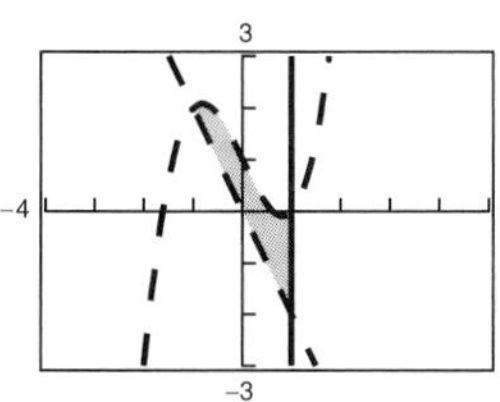

49.

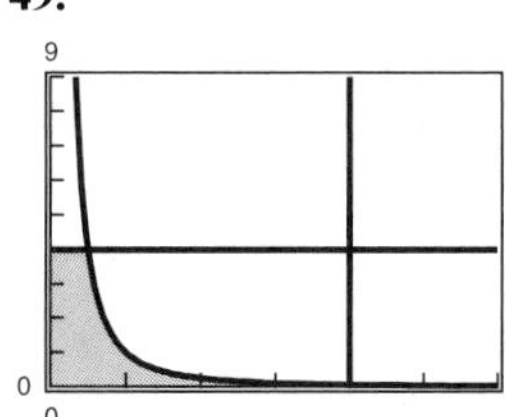

51. $\frac{1}{4}x + \frac{1}{4}y \le 1$
$x \ge 0$
$y \ge 0$

53. $y \ge 4 - x$
$y \ge 2 - \frac{1}{4}x$
$x \ge 0,\quad y \ge 0$

55. $x^2 + y^2 \le 16$
$x \ge 0,\quad y \ge 0$

57. $2 \le x \le 5$
$1 \le y \le 7$

59. $y \le \frac{3}{2}x$
$y \le -x + 5$
$y \ge 0$

61. (a) $x + y \le 20{,}000$
$y \ge 2x$
$x \ge 5000$
$y \ge 5000$

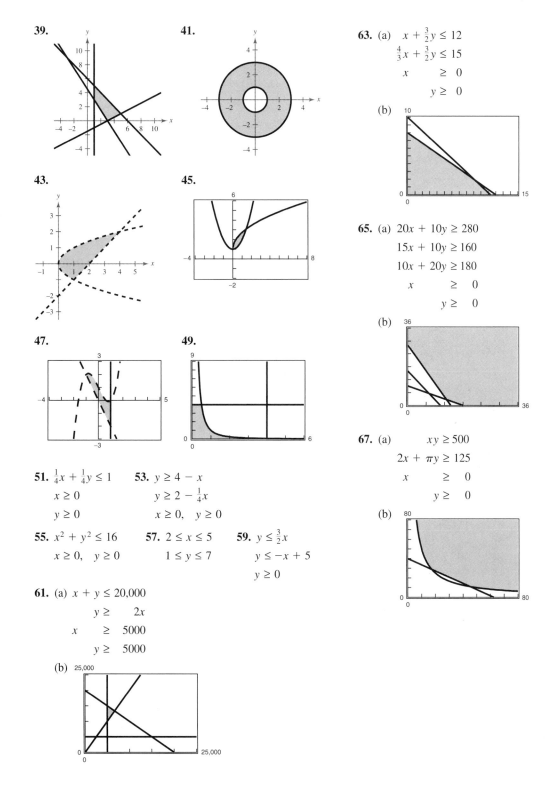

(b)

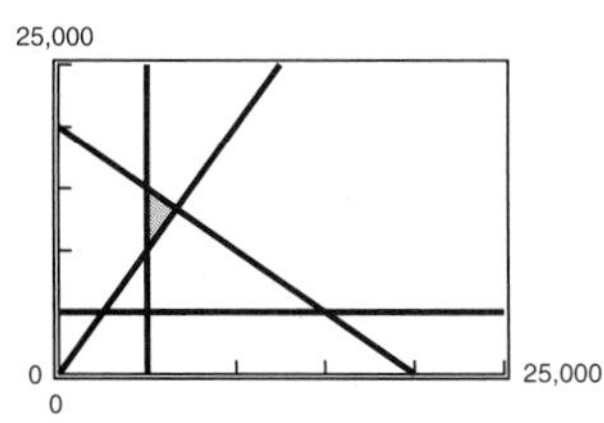

63. (a) $x + \frac{3}{2}y \le 12$
$\frac{4}{3}x + \frac{3}{2}y \le 15$
$x \ge 0$
$y \ge 0$

(b)

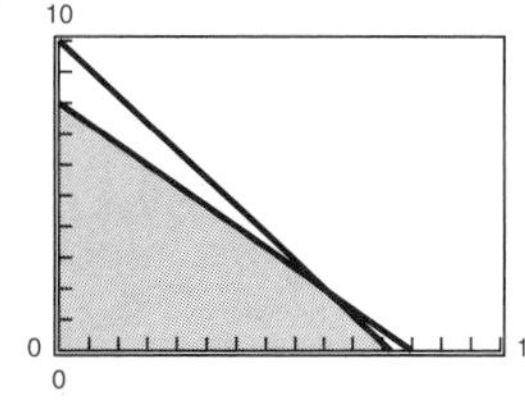

65. (a) $20x + 10y \ge 280$
$15x + 10y \ge 160$
$10x + 20y \ge 180$
$x \ge 0$
$y \ge 0$

(b)

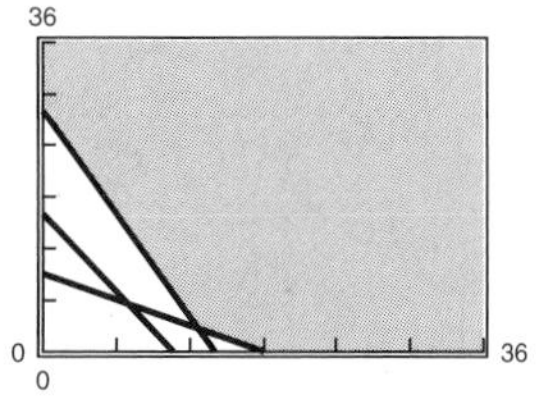

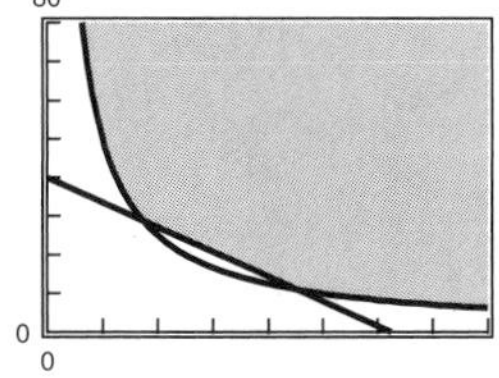

67. (a) $xy \ge 500$
$2x + \pi y \ge 125$
$x \ge 0$
$y \ge 0$

(b)

69.

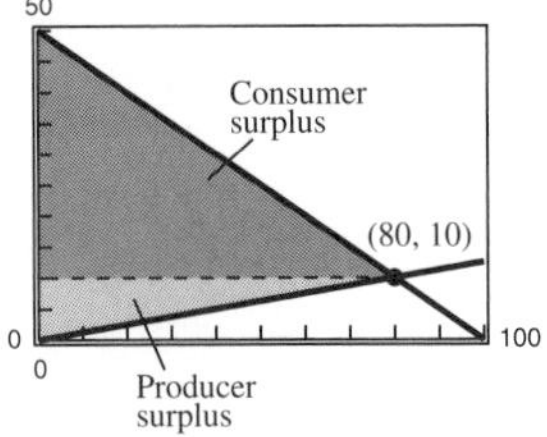

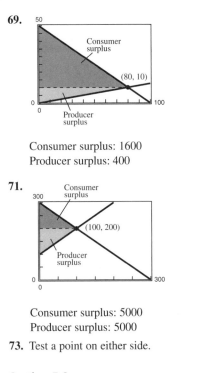

Consumer surplus: 1600
Producer surplus: 400

71.

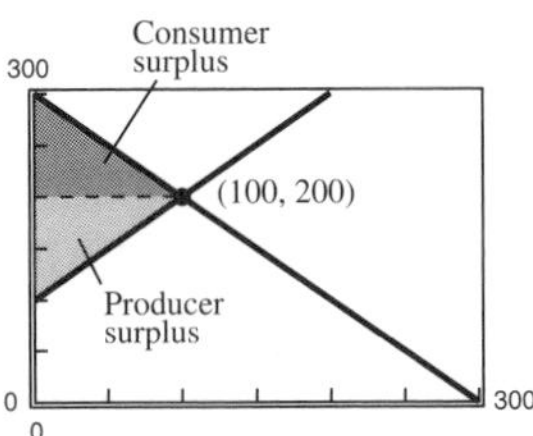

Consumer surplus: 5000
Producer surplus: 5000

73. Test a point on either side.

Section 5.6 *(page 447)*

1. Minimum at (0, 0): 0
Maximum at (0, 6): 30

3. Minimum at (0, 0): 0
Maximum at (6, 0): 60

5. Minimum at (0, 0): 0
Maximum at (3, 4): 17

7. Minimum at (0, 0): 0
Maximum at (4, 0): 20

9. Minimum at (0, 0): 0
Maximum at (60, 20): 740

11. Minimum at (0, 0): 0
Maximum at any point on the line segment connecting (60, 20) and (30, 45): 2100

13.

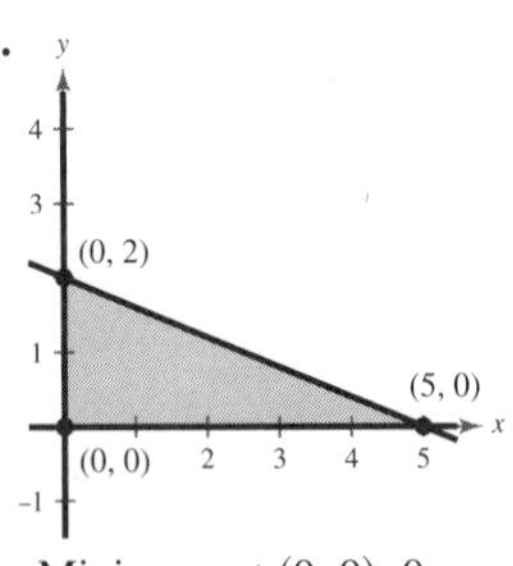

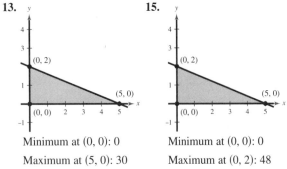

Minimum at (0, 0): 0
Maximum at (5, 0): 30

15.

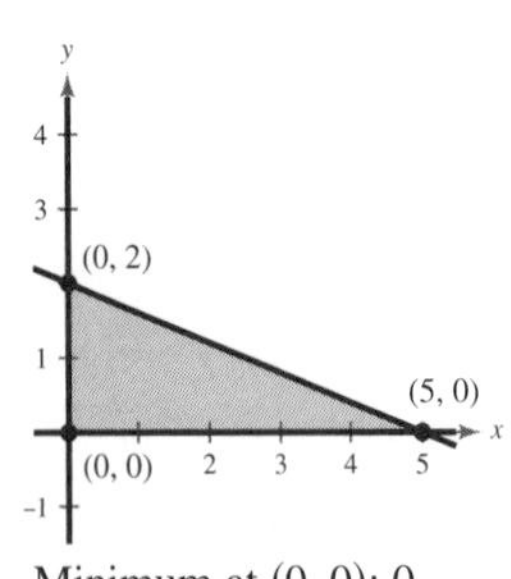

Minimum at (0, 0): 0
Maximum at (0, 2): 48

17.

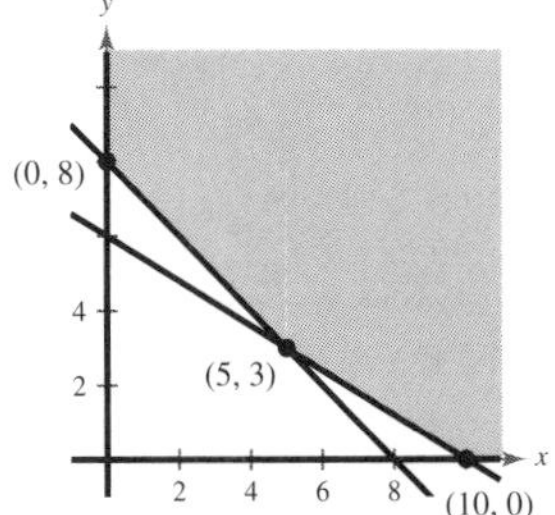

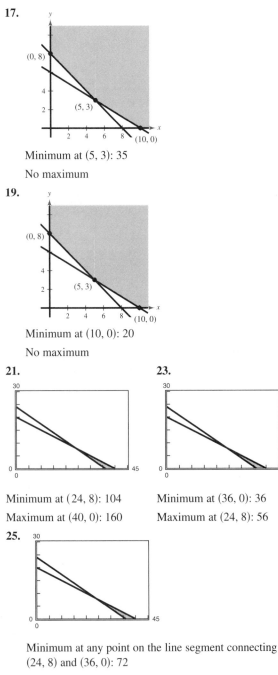

Minimum at (5, 3): 35
No maximum

19.

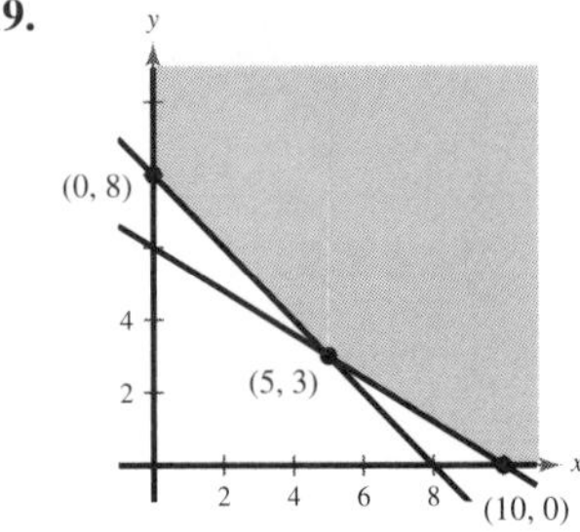

Minimum at (10, 0): 20
No maximum

21.

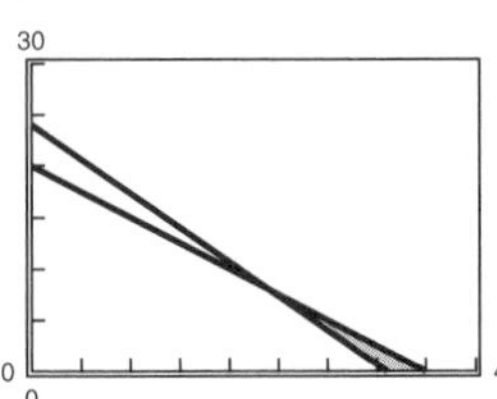

Minimum at (24, 8): 104
Maximum at (40, 0): 160

23.

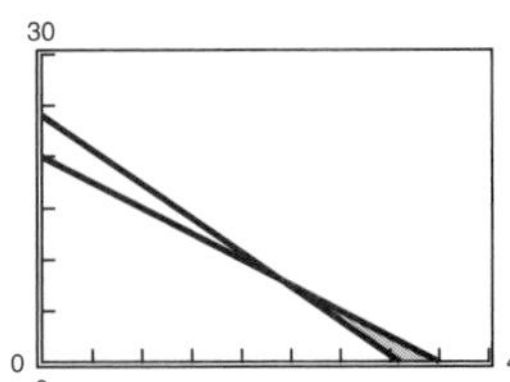

Minimum at (36, 0): 36
Maximum at (24, 8): 56

25.

30
0
0
45

Minimum at any point on the line segment connecting (24, 8) and (36, 0): 72
Maximum at (40, 0): 80

27. (a) and (b)

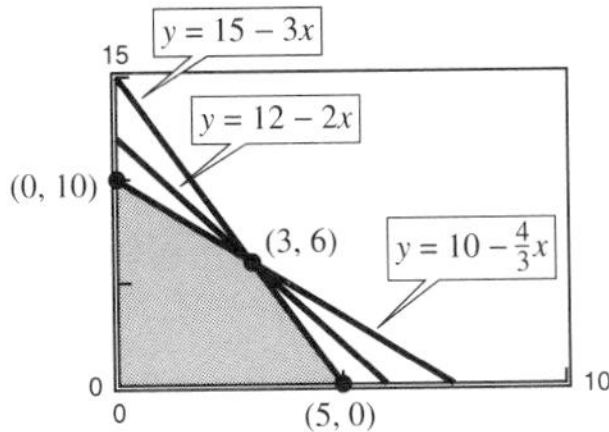

(c) (3, 6)

29. (a) and (b)

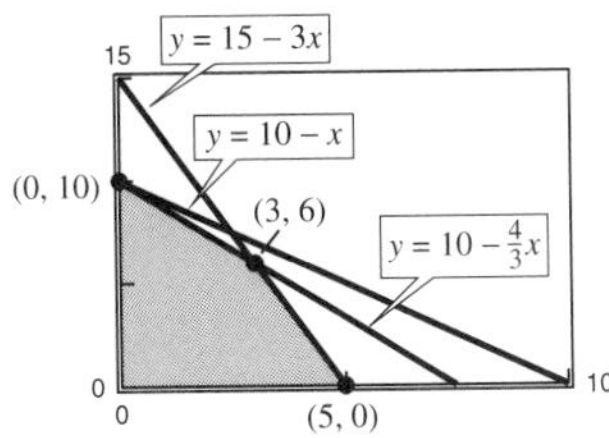

(c) (0, 10)

31. (a) and (b)

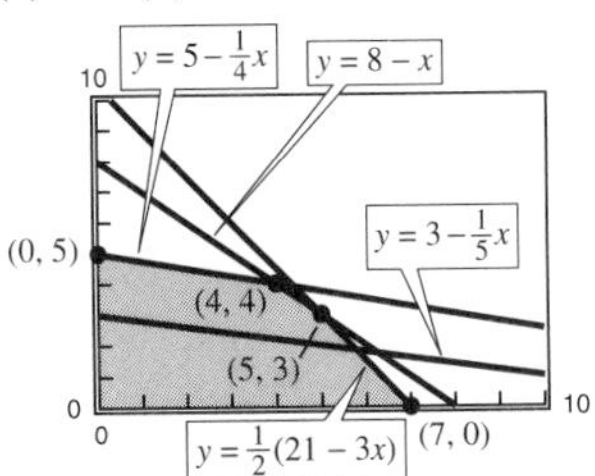

(c) (0, 5)

33. (a) and (b)

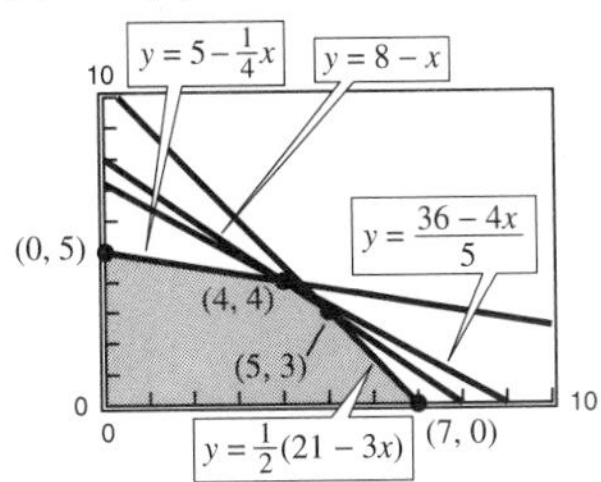

(c) (4, 4)

35. $z = x + 5y$ **37.** $z = 4x + y$

39. 200 units of the \$250 model

50 units of the \$400 model

Maximum profit: \$11,500

41. A: $\frac{1}{6}$ gallon

B: $\frac{5}{6}$ gallon

Minimum cost: \$1.255 per gallon

43. No audits

40 tax returns

Maximum revenue: \$12,000

45. 1000 units of Model A

500 units of Model B

Maximum profit: \$76,000

47.

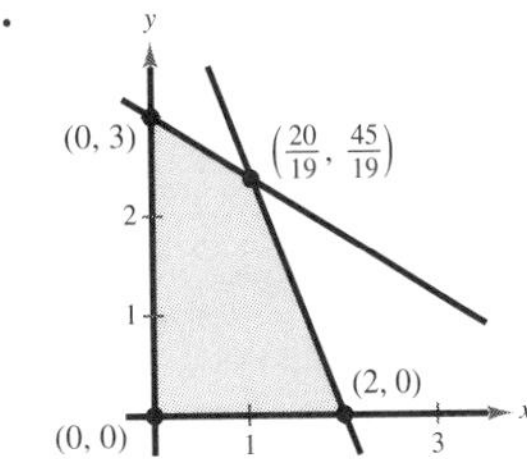

z is maximum at any point on the line segment connecting $(2, 0)$ and $\left(\frac{20}{19}, \frac{45}{19}\right)$.

49.

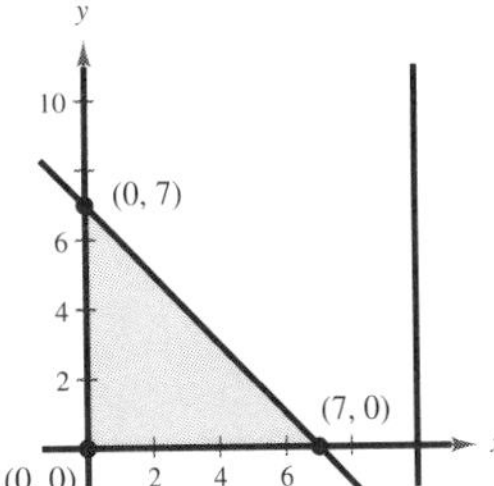

The constraint $x \le 10$ is extraneous. Maximum 14 at (0, 7)

51.

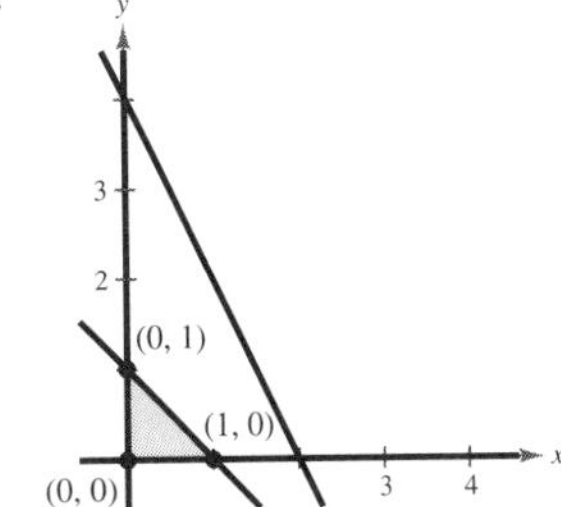

The constraint $2x + y \le 4$ is extraneous.

Maximum at (0, 1): 4

53. (a) $t > 9$ (b) $\frac{3}{4} < t < 9$ **55.** $\dfrac{9}{2(x + 3)}$

57. $\dfrac{x^2 + 2x - 13}{x(x - 2)}$

FOCUS ON CONCEPTS *(page 451)*

1. A solution of a system is an ordered pair that satisfies each equation in the system.

2. For a linear system the result will be a contradictory equation such as $0 = N$, where N is a nonzero real number. For nonlinear systems there may be an equation with imaginary roots.

3. There will be a contradictory equation of the form $0 = N$, where N is a nonzero real number.

4. The algebraic methods yield exact solutions.

5. (a) One (b) Two (c) Four

6. The system has no solution.

7. The lines are distinct and parallel.

$x + 2y = 3$
$2x + 4y = 9$

8. (a) Interchange any two equations.
(b) Multiply an equation by a nonzero constant.
(c) Add a multiple of one equation to any other equation in the system.

9. No. When -2 times Equation 1 is added to Equation 2, the constant is -11.

10. The first system is inconsistent because 4 times Equation 1 added to Equation 2 yields $0 = -20$.

11. (d) **12.** (b) **13.** (c) **14.** (a)

15. (a) The boundary would be included in the solution.
(b) The solution would be the half-plane on the opposite side of the boundary.

Review Exercises *(page 452)*

1. $(1, 1)$ **3.** $(5, 4)$ **5.** $(0, 0), (2, 8), (-2, 8)$

7. $(0, 0), (-3, 3)$ **9.** $(4, 4)$ **11.** $\left(\frac{5}{2}, 3\right)$

13. $(-0.5, 0.8)$ **15.** $(0, 0)$ **17.** $\left(\frac{14}{5} + \frac{8}{5}a, a\right)$

19. $3x + y = 7$
$-6x + 3y = 1$
21. 4762 units

23. 218.75 miles per hour, 193.75 miles per hour

25. $\left(\dfrac{500{,}000}{7}, \dfrac{159}{7}\right)$ **27.** $\left(\dfrac{38}{17}, \dfrac{40}{17}, -\dfrac{63}{17}\right)$

29. $(3a + 4, 2a + 5, a)$ **31.** $2x + y - 2z = 1$
$x + y - z = 0$
$2x - 3y - 2z = 5$

33. $y = 2x^2 + x - 5$

35. \$8000 at 7%
\$5000 at 9%
\$7000 at 11%

37. $\dfrac{3}{x + 2} - \dfrac{4}{x + 4}$ **39.** $1 - \dfrac{25}{8(x + 5)} + \dfrac{9}{8(x - 3)}$

41. $\dfrac{1}{2}\left(\dfrac{3}{x - 1} - \dfrac{x - 3}{x^2 + 1}\right)$

43. **45.**

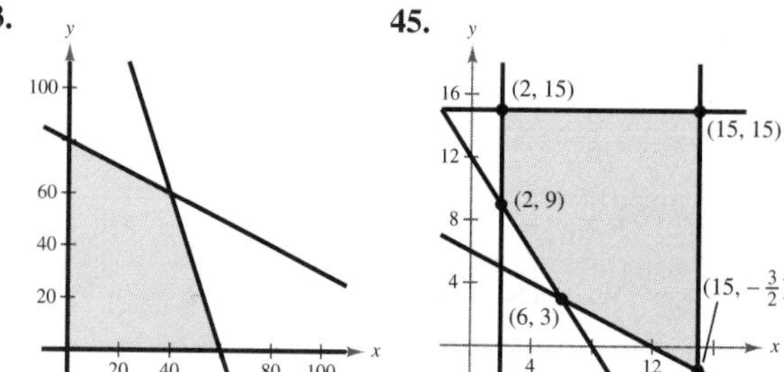

47. **49.**

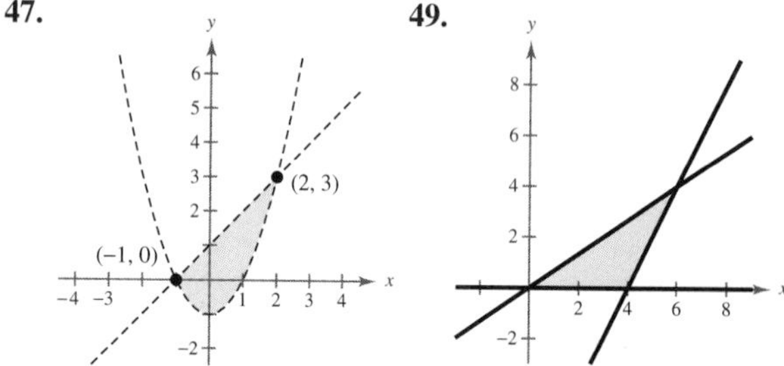

51. $-x + y \le 4$
$2x + y \le 22$
$-x + y \ge -2$
$2x + y \ge 7$

53. $x + y \le 1500$
$x \ge 400$
$y \ge 600$

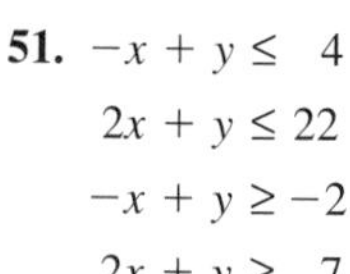

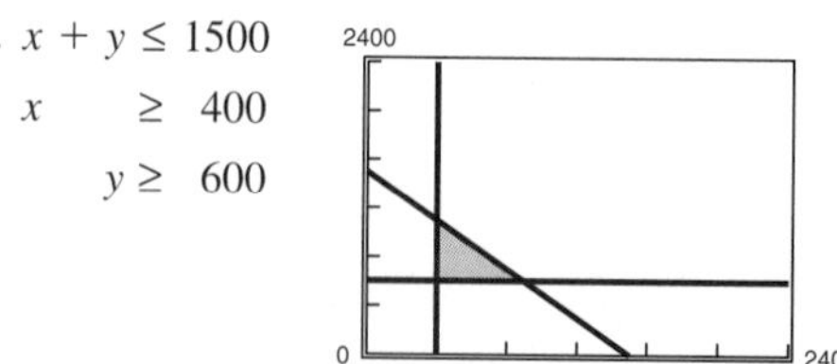

55. 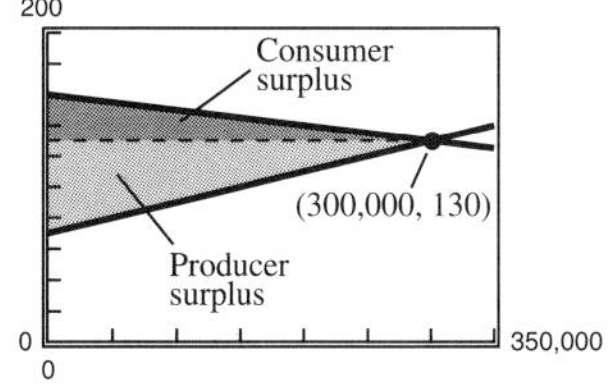

Consumer surplus: 4,500,000
Producer surplus: 9,000,000

57. 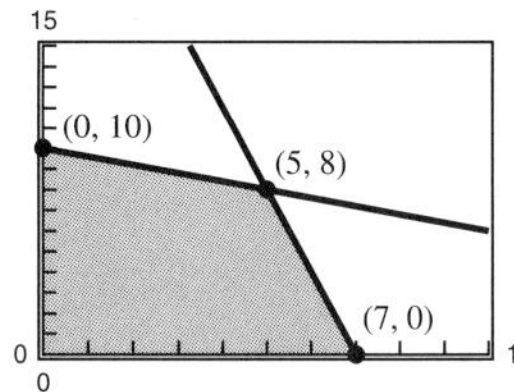

Maximum at (5, 8): 47

59. 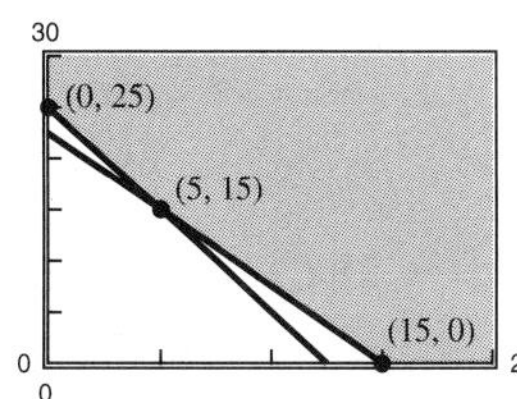

Minimum at (15, 0): 26.25

61. 20 perms **63.** Three bags of Brand X
Two bags of Brand Y
Minimum cost per bag: $21

65. $k_1 = -\frac{10}{3}$
$k_2 = \frac{16}{3}$

Cumulative Test for Chapters 3–5 *(page 456)*

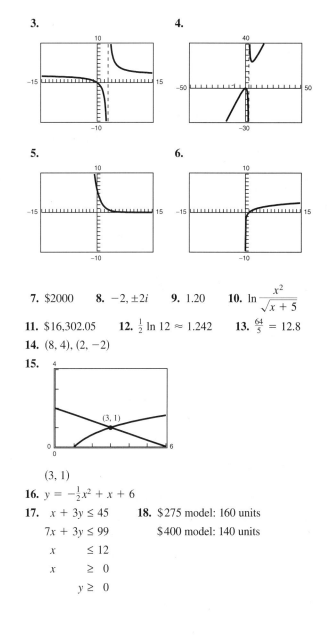

1. 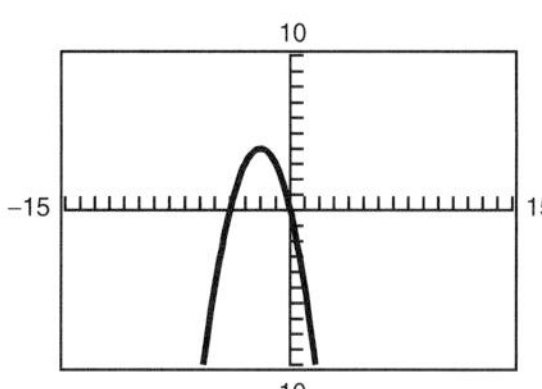

2. 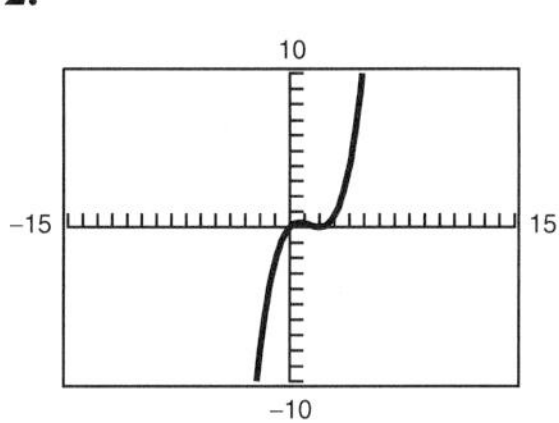

3. 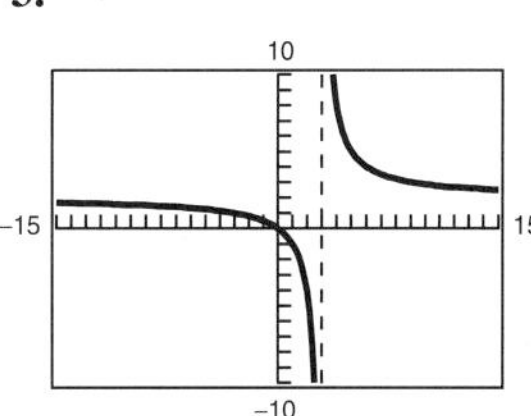

4. 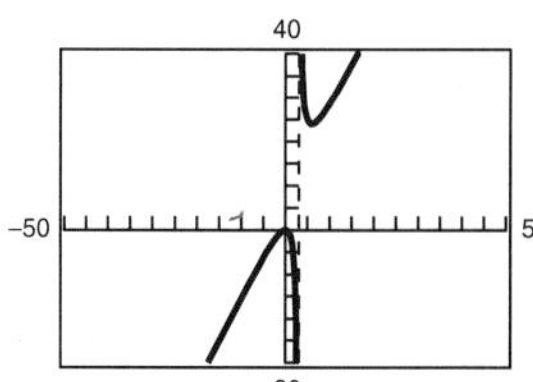

5. 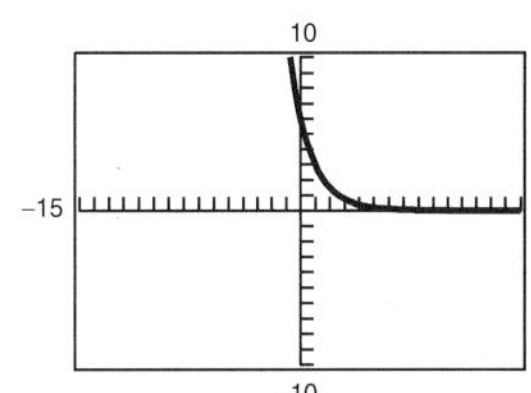

6. 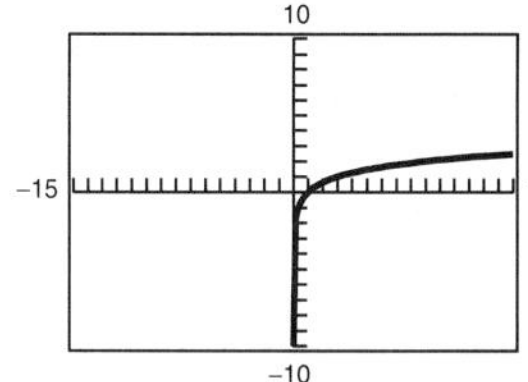

7. \$2000 **8.** $-2, \pm 2i$ **9.** 1.20 **10.** $\ln \frac{x^2}{\sqrt{x+5}}$

11. \$16,302.05 **12.** $\frac{1}{2}\ln 12 \approx 1.242$ **13.** $\frac{64}{5} = 12.8$

14. (8, 4), (2, −2)

15. 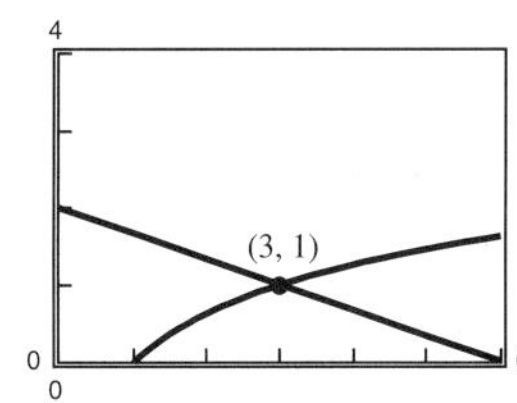

(3, 1)

16. $y = -\frac{1}{2}x^2 + x + 6$

17. $x + 3y \le 45$
$7x + 3y \le 99$
$x \le 12$
$x \ge 0$
$y \ge 0$

18. \$275 model: 160 units
\$400 model: 140 units

CHAPTER 6

Section 6.1 *(page 468)*

1. 3×2 **3.** 3×1 **5.** 2×2

7. $\begin{bmatrix} 4 & -3 & \vdots & -5 \\ -1 & 3 & \vdots & 12 \end{bmatrix}$ **9.** $\begin{bmatrix} 1 & 10 & -2 & \vdots & 2 \\ 5 & -3 & 4 & \vdots & 0 \\ 2 & 1 & 0 & \vdots & 6 \end{bmatrix}$

11. $x + 2y = 7$
$2x - 3y = 4$

13. $2x + 5z = -12$
$y - 2z = 7$
$6x + 3y = 2$

15. Reduced row-echelon form

17. Not in row-echelon form **19.** $\begin{bmatrix} 1 & 4 & 3 \\ 0 & 2 & -1 \end{bmatrix}$

21. $\begin{bmatrix} 1 & 1 & 4 & -1 \\ 0 & 5 & -2 & 6 \\ 0 & 3 & 20 & 4 \end{bmatrix}, \begin{bmatrix} 1 & 1 & 4 & -1 \\ 0 & 1 & -\frac{2}{5} & \frac{6}{5} \\ 0 & 3 & 20 & 4 \end{bmatrix}$

23. (a) $\begin{bmatrix} 1 & 2 & 3 \\ 0 & -5 & -10 \\ 3 & 1 & -1 \end{bmatrix}$ (b) $\begin{bmatrix} 1 & 2 & 3 \\ 0 & -5 & -10 \\ 0 & -5 & -10 \end{bmatrix}$ (c) $\begin{bmatrix} 1 & 2 & 3 \\ 0 & -5 & -10 \\ 0 & 0 & 0 \end{bmatrix}$

(d) $\begin{bmatrix} 1 & 2 & 3 \\ 0 & 1 & 2 \\ 0 & 0 & 0 \end{bmatrix}$ (e) $\begin{bmatrix} 1 & 0 & -1 \\ 0 & 1 & 2 \\ 0 & 0 & 0 \end{bmatrix}$

25. $\begin{bmatrix} 1 & 1 & 0 & 5 \\ 0 & 1 & 2 & 0 \\ 0 & 0 & 1 & -1 \end{bmatrix}$ **27.** $\begin{bmatrix} 1 & -1 & -1 & 1 \\ 0 & 1 & 6 & 3 \\ 0 & 0 & 0 & 0 \end{bmatrix}$

29. $\begin{bmatrix} 1 & 0 & 0 \\ 0 & 1 & 0 \\ 0 & 0 & 1 \end{bmatrix}$ **31.** $\begin{bmatrix} 1 & 2 & 0 & 0 \\ 0 & 0 & 1 & 0 \\ 0 & 0 & 0 & 1 \\ 0 & 0 & 0 & 0 \end{bmatrix}$

33. $x - 2y = 4$
$y = -3$
$(-2, -3)$

35. $x - y + 2z = 4$
$y - z = 2$
$z = -2$
$(8, 0, -2)$

37. $(7, -5)$ **39.** $(-4, -8, 2)$ **41.** $(3, 2)$

43. $(4, -2)$ **45.** $\left(\frac{1}{2}, -\frac{3}{4}\right)$ **47.** Inconsistent

49. $(4, -3, 2)$ **51.** $(2a + 1, 3a + 2, a)$

53. $(5a + 4, -3a + 2, a)$ **55.** $(0, 0)$

57. $(0, 2 - 4a, a)$ **59.** $(1, 0, 4, -2)$ **61.** $(-2a, a, a)$

63. $x + y + 7z = -1$
$x + 2y + 11z = 0$
$2x + y + 10z = -3$

65. \$800,000 at 8%
\$500,000 at 9%
\$200,000 at 12%

67. $\dfrac{4x^2}{(x+1)^2(x-1)} = \dfrac{1}{x-1} + \dfrac{3}{x+1} - \dfrac{2}{(x+1)^2}$

69. $y = x^2 + 2x + 5$

71. $y = \frac{1}{4}x^3 + x^2 - x - 2$ **73.** $y = -x^4 + 4x^2$

75. (a) $y = 170.5t^2 - 560.5t + 670$

(b)

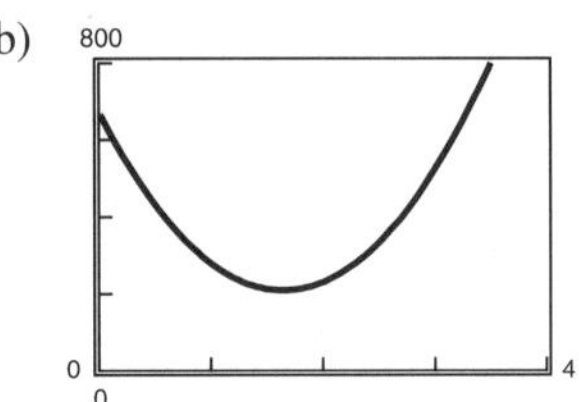

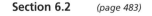

(c) 523

77. (a) $x_1 = s, x_2 = t, x_3 = 600 - s,$
$x_4 = s - t, x_5 = 500 - t, x_6 = s, x_7 = t$

(b) $x_1 = 0, x_2 = 0, x_3 = 600, x_4 = 0, x_5 = 500,$
$x_6 = 0, x_7 = 0$

(c) $x_1 = 0, x_2 = -500, x_3 = 600, x_4 = 500,$
$x_5 = 1000, x_6 = 0, x_7 = -500$

79. (a) $x_1 = 100 + t, x_2 = -100 + t, x_3 = 200 + t, x_4 = t$

(b) $x_1 = 100, x_2 = -100, x_3 = 200, x_4 = 0$

(c) $x_1 = 200, x_2 = 0, x_3 = 300, x_4 = 100$

81. Men: 1.171 minutes
Women: 1.239 minutes

83.

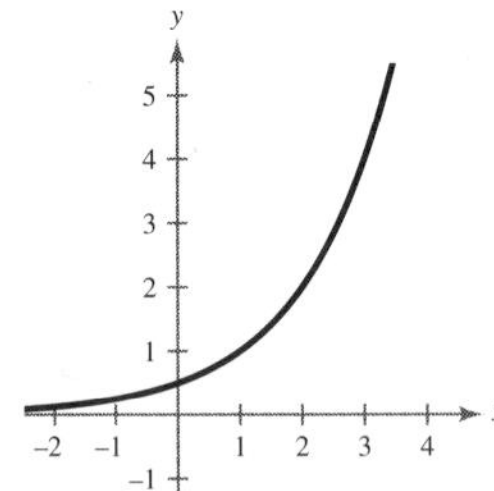

85.

Section 6.2 *(page 483)*

1. $x = -4, y = 22$ **3.** $x = 2, y = 3$

5. (a) $\begin{bmatrix} 3 & -2 \\ 1 & 7 \end{bmatrix}$ (b) $\begin{bmatrix} -1 & 0 \\ 3 & -9 \end{bmatrix}$

(c) $\begin{bmatrix} 3 & -3 \\ 6 & -3 \end{bmatrix}$ (d) $\begin{bmatrix} -1 & -1 \\ 8 & -19 \end{bmatrix}$

7. (a) $\begin{bmatrix} 7 & 3 \\ 1 & 9 \\ -2 & 15 \end{bmatrix}$ (b) $\begin{bmatrix} 5 & -5 \\ 3 & -1 \\ -4 & -5 \end{bmatrix}$

(c) $\begin{bmatrix} 18 & -3 \\ 6 & 12 \\ -9 & 15 \end{bmatrix}$ (d) $\begin{bmatrix} 16 & -11 \\ 8 & 2 \\ -11 & -5 \end{bmatrix}$

9. (a) $\begin{bmatrix} 3 & 3 & -2 & 1 & 1 \\ -2 & 5 & 7 & -6 & -8 \end{bmatrix}$

(b) $\begin{bmatrix} 1 & 1 & 0 & -1 & 1 \\ 4 & -3 & -11 & 6 & 6 \end{bmatrix}$

(c) $\begin{bmatrix} 6 & 6 & -3 & 0 & 3 \\ 3 & 3 & -6 & 0 & -3 \end{bmatrix}$

(d) $\begin{bmatrix} 4 & 4 & -1 & -2 & 3 \\ 9 & -5 & -24 & 12 & 11 \end{bmatrix}$

11. $\begin{bmatrix} -6 & -9 \\ -1 & 0 \\ 17 & -10 \end{bmatrix}$ **13.** $\begin{bmatrix} 3 & 3 \\ -\frac{1}{2} & 0 \\ -\frac{13}{2} & \frac{11}{2} \end{bmatrix}$

15. (a) $\begin{bmatrix} 0 & 15 \\ 6 & 12 \end{bmatrix}$ (b) $\begin{bmatrix} -2 & 2 \\ 31 & 14 \end{bmatrix}$ (c) $\begin{bmatrix} 9 & 6 \\ 12 & 12 \end{bmatrix}$

17. (a) $\begin{bmatrix} 0 & -10 \\ 10 & 0 \end{bmatrix}$ (b) $\begin{bmatrix} 0 & -10 \\ 10 & 0 \end{bmatrix}$ (c) $\begin{bmatrix} 8 & -6 \\ 6 & 8 \end{bmatrix}$

19. (a) $\begin{bmatrix} 6 & -21 & 15 \\ 8 & -23 & 19 \\ 4 & 7 & 5 \end{bmatrix}$ (b) $\begin{bmatrix} 9 & 0 & 13 \\ 7 & -2 & 21 \\ 1 & 4 & -19 \end{bmatrix}$

(c) $\begin{bmatrix} 20 & 7 & -8 \\ 24 & 7 & -2 \\ 2 & -5 & 30 \end{bmatrix}$

21. Not possible **23.** $\begin{bmatrix} -1 & 19 \\ 4 & -27 \\ 0 & 14 \end{bmatrix}$ **25.** $\begin{bmatrix} 1 & 0 & 0 \\ 0 & 1 & 0 \\ 0 & 0 & \frac{7}{2} \end{bmatrix}$

27. $\begin{bmatrix} 0 & 0 & 0 \\ 0 & 0 & 0 \\ 0 & 0 & 0 \end{bmatrix}$ **29.** $\begin{bmatrix} 41 & 7 & 7 \\ 42 & 5 & 25 \\ -10 & -25 & 45 \end{bmatrix}$

31. $\begin{bmatrix} 151 & 25 & 48 \\ 516 & 279 & 387 \\ 47 & -20 & 87 \end{bmatrix}$ **33.** Not possible

35. $A = \begin{bmatrix} -1 & 1 \\ -2 & 1 \end{bmatrix}$ **37.** $A = \begin{bmatrix} 2 & 3 \\ 1 & 4 \end{bmatrix}$

$X = \begin{bmatrix} x \\ y \end{bmatrix}$ $X = \begin{bmatrix} x \\ y \end{bmatrix}$

$B = \begin{bmatrix} 4 \\ 0 \end{bmatrix}$ $B = \begin{bmatrix} 5 \\ 10 \end{bmatrix}$

$x = 4, y = 8$ $x = -2, y = 3$

39. $\begin{bmatrix} -4 & 0 \\ 8 & 2 \end{bmatrix}$ **41.** $\begin{bmatrix} 0 & 0 & 0 \\ 0 & 0 & 0 \\ 0 & 0 & 0 \end{bmatrix}$

43. $AC = BC = \begin{bmatrix} 2 & 3 \\ 2 & 3 \end{bmatrix}$ **45.** Not possible

47. Not possible **49.** 2×2 **51.** Not possible

53. 2×3 **55.** $\begin{bmatrix} 72 & 48 & 24 \\ 36 & 108 & 72 \end{bmatrix}$

57. $BA = [\$1250 \quad \$1331.25 \quad \$981.25]$

The entries represent the profits from the two products at each of the three outlets.

59. $A^2 = \begin{bmatrix} -1 & 0 \\ 0 & -1 \end{bmatrix}$ **61.** $\begin{bmatrix} \$15{,}770 & \$18{,}300 \\ \$26{,}500 & \$29{,}250 \\ \$21{,}260 & \$24{,}150 \end{bmatrix}$

$A^3 = \begin{bmatrix} -i & 0 \\ 0 & -i \end{bmatrix}$

$A^4 = \begin{bmatrix} 1 & 0 \\ 0 & 1 \end{bmatrix}$

The entries are the wholesale and retail prices of the inventory at each outlet.

63. $\begin{bmatrix} 0.40 & 0.15 & 0.15 \\ 0.28 & 0.53 & 0.17 \\ 0.32 & 0.32 & 0.68 \end{bmatrix}$

65. Diagonal matrix whose entries are the products of the corresponding entries of A and B

67. $\frac{5}{4}$ **69.** $2 + \log_2 3 - \log_2 5$

Section 6.3 *(page 493)*

1.–7. Answers will vary.

9. $\begin{bmatrix} \frac{1}{2} & 0 \\ 0 & \frac{1}{3} \end{bmatrix}$ **11.** $\begin{bmatrix} -3 & 2 \\ -2 & 1 \end{bmatrix}$ **13.** $\begin{bmatrix} 1 & -1 \\ 2 & -1 \end{bmatrix}$

15. Does not exist **17.** Does not exist

19. $\begin{bmatrix} 1 & 1 & -1 \\ -3 & 2 & -1 \\ 3 & -3 & 2 \end{bmatrix}$ **21.** $\begin{bmatrix} 1 & 0 & 0 \\ -0.75 & 0.25 & 0 \\ 0.35 & -0.25 & 0.2 \end{bmatrix}$

23. $\begin{bmatrix} -\frac{1}{8} & 0 & 0 & 0 \\ 0 & 1 & 0 & 0 \\ 0 & 0 & \frac{1}{4} & 0 \\ 0 & 0 & 0 & -\frac{1}{5} \end{bmatrix}$ **25.** $\begin{bmatrix} -175 & 37 & -13 \\ 95 & -20 & 7 \\ 14 & -3 & 1 \end{bmatrix}$

27. $\frac{1}{2}\begin{bmatrix} -3 & 3 & 2 \\ 9 & -7 & -6 \\ -2 & 2 & 2 \end{bmatrix}$ **29.** $\frac{5}{11}\begin{bmatrix} 0 & -4 & 2 \\ -22 & 11 & 11 \\ 22 & -6 & -8 \end{bmatrix}$

31. Does not exist **33.** $\begin{bmatrix} -24 & 7 & 1 & -2 \\ -10 & 3 & 0 & -1 \\ -29 & 7 & 3 & -2 \\ 12 & -3 & -1 & 1 \end{bmatrix}$

35. Answers will vary. **37.** $(5, 0)$ **39.** $(-8, -6)$

41. $(3, 8, -11)$ **43.** $(2, 1, 0, 0)$ **45.** $(2, -2)$

47. No solution **49.** $\left(3, -\frac{1}{2}\right)$ **51.** $(-1, 3, 2)$

53. No solution **55.** $(5, 0, -2, 3)$

57. \$10,000 in AAA-rated bonds
\$5000 in A-rated bonds
\$10,000 in B-rated bonds

59. \$9000 in AAA-rated bonds
\$1000 in A-rated bonds
\$2000 in B-rated bonds

61. Answers will vary. **63.** $I_1 = -3$ amperes
$I_2 = 8$ amperes
$I_3 = 5$ amperes

65. $A^{-1} = \begin{bmatrix} \frac{1}{a_{11}} & 0 & 0 & 0 & \cdots & 0 \\ 0 & \frac{1}{a_{22}} & 0 & 0 & \cdots & 0 \\ 0 & 0 & \frac{1}{a_{33}} & 0 & \cdots & 0 \\ \vdots & \vdots & \vdots & \vdots & \cdots & \vdots \\ 0 & 0 & 0 & 0 & \cdots & \frac{1}{a_{nn}} \end{bmatrix}$

Section 6.4 *(page 502)*

1. 5 **3.** 5 **5.** 27 **7.** −24 **9.** 6 **11.** 0

13. 0 **15.** −9 **17.** −0.002 **19.** 0

21. (a) $M_{11} = -5, M_{12} = 2, M_{21} = 4, M_{22} = 3$
(b) $C_{11} = -5, C_{12} = -2, C_{21} = -4, C_{22} = 3$

23. (a) $M_{11} = 30, M_{12} = 12, M_{13} = 11, M_{21} = -36,$
$M_{22} = 26, M_{23} = 7, M_{31} = -4, M_{32} = -42, M_{33} = 12$
(b) $C_{11} = 30, C_{12} = -12, C_{13} = 11, C_{21} = 36, C_{22} = 26,$
$C_{23} = -7, C_{31} = -4, C_{32} = 42, C_{33} = 12$

25. −75 **27.** 96 **29.** 170 **31.** −58

33. −30 **35.** −168 **37.** 0 **39.** 412

41. −126 **43.** 0 **45.** −336 **47.** 410

49.–53. Answers will vary. **55.** −1, 4

57. $8uv - 1$ **59.** e^{5x} **61.** $1 - \ln x$

63. (a) −3 (b) −2
(c) $\begin{bmatrix} -2 & 0 \\ 0 & -3 \end{bmatrix}$ (d) 6

65. (a) 2 (b) −6
(c) $\begin{bmatrix} 1 & 4 & 3 \\ -1 & 0 & 3 \\ 0 & 2 & 0 \end{bmatrix}$ (d) −12

67. $A = \begin{bmatrix} 1 & 3 \\ -2 & 4 \end{bmatrix}, B = \begin{bmatrix} -4 & 0 \\ 3 & 5 \end{bmatrix}$
$|A + B| = -30, \; |A| + |B| = -10$

69. A square matrix is a square array of numbers. A determinant of a square matrix is a real number.

71. (a) Columns 2 and 3 are interchanged.
(b) Rows 1 and 3 are interchanged.

73. (a) 5 is factored from the first row of the matrix.
(b) 4 and 3 are factored from the second and third columns.

Section 6.5 *(page 514)*

1. 14 **3.** 7 **5.** $\frac{33}{8}$ **7.** 10 **9.** 28

11. $\frac{16}{5}$ or 0 **13.** $(2, -2)$ **15.** $(-1, 3, 2)$

17. $\left(0, -\frac{1}{2}, \frac{1}{2}\right)$ **19.** 250 square miles **21.** Collinear

23. Not collinear **25.** Collinear **27.** $3x - 5y = 0$

29. $x + 3y - 5 = 0$ **31.** $2x + 3y - 8 = 0$

33. $x = -3$

35. Uncoded: [20, 18, 15], [21, 2, 12], [5, 0, 9], [14, 0, 18], [9, 22, 5], [18, 0, 3], [9, 20, 25]
Encoded: [−52, 10, 27], [−49, 3, 34], [−49, 13, 27], [−94, 22, 54], [1, 1, −7], [0, −12, 9], [−121, 41, 55]

37. 1 −25 −65 17 15 −9 −12 −62 −119 27 51 48 43 67 48 57 111 117

39. −5 −41 −87 91 207 257 11 −5 −41 40 80 84 76 177 227

41. HAPPY NEW YEAR **43.** SEND PLANES
45. MEET ME TONIGHT RON

FOCUS ON CONCEPTS *(page 516)*

1. Interchange two rows.

Multiply a row by a nonzero constant.

Add a multiple of a row to another row.

2. They are the same.

3. A matrix in row-echelon form is in reduced row-echelon form if every column that has a leading 1 has zeros in every position above and below its leading 1.

4. Consistent—an infinite number of solutions

5. Inconsistent **6.** Consistent—a unique solution

7. Consistent—an infinite number of solutions

8. (a) The operation can be performed.
(b) The operation can be performed.

9. (a) The operation can be performed.
(b) The operation cannot be performed. The number of rows in *B* must be the same as the number of columns in *A*.

10. (a) The operation cannot be performed. The order of *A* and *B* must be the same to perform the operation.
(b) The operation can be performed.

11. The matrix must be square and its determinant nonzero.

12. A square matrix is a square array of numbers and a determinant is a real number associated with a square matrix.

13. No. The matrix must be square.

14. If *A* is a square matrix, the cofactor C_{ij} of the entry a_{ij} is $(-1)^{i+j}M_{ij}$, where M_{ij} is the determinant obtained by deleting the *i*th row and *j*th column of *A*. The determinant of *A* is the sum of the entries of any row or column of *A* multiplied by their respective cofactors.

15. No. The first two yield a unique solution to the system and the third yields an infinite number of solutions.

Review Exercises *(page 517)*

1. $\begin{bmatrix} 3 & -10 & \vdots & 15 \\ 5 & 4 & \vdots & 22 \end{bmatrix}$

3. $\begin{aligned} 5x + y + 7z &= -9 \\ 4x + 2y &= 10 \\ 9x + 4y + 2z &= 3 \end{aligned}$

5. $\begin{bmatrix} 1 & 0 & 0 \\ 0 & 1 & 0 \\ 0 & 0 & 1 \end{bmatrix}$ **7.** $\begin{bmatrix} 1 & 0 & 3 & -2 \\ 0 & 1 & 4 & -3 \end{bmatrix}$

9. $\begin{bmatrix} 1 & 0 & 1 \\ 0 & 1 & 1 \\ 0 & 0 & 0 \end{bmatrix}$ **11.** $(10, -12)$ **13.** $(-0.2, 0.7)$

15. $\left(-2a + \frac{3}{2}, 2a + 1, a\right)$ **17.** $\left(\frac{31}{42}, \frac{5}{14}, \frac{13}{84}\right)$

19. $(2, -3, 3)$ **21.** Inconsistent

23. The part of the matrix corresponding to the coefficients of the system reduces to a matrix in which the number of rows with nonzero entries is the same as the number of variables.

25. $(1, 1)$ **27.** $(3, 0, -4)$ **29.** $\begin{bmatrix} -13 & -8 & 18 \\ 0 & 11 & -19 \end{bmatrix}$

31. $\begin{bmatrix} 14 & -2 & 8 \\ 14 & -10 & 40 \\ 36 & -12 & 48 \end{bmatrix}$ **33.** $\begin{bmatrix} 44 & 4 \\ 20 & 8 \end{bmatrix}$

35. $\begin{bmatrix} 4 & 6 & 3 \\ 0 & 6 & -10 \\ 0 & 0 & 6 \end{bmatrix}$ **37.** $\begin{bmatrix} 48 & -18 & -3 \\ 15 & 51 & 33 \end{bmatrix}$

39. $\begin{bmatrix} 14 & -22 & 22 \\ 19 & -41 & 80 \\ 42 & -66 & 66 \end{bmatrix}$ **41.** $\begin{bmatrix} -14 & -4 \\ 7 & -17 \\ -17 & -2 \end{bmatrix}$

43. $\frac{1}{3}\begin{bmatrix} 9 & 2 \\ -4 & 11 \\ 10 & 0 \end{bmatrix}$ **45.** $\begin{aligned} 5x + 4y &= 2 \\ -x + y &= -22 \end{aligned}$

47. $\begin{bmatrix} \frac{1}{5} & \frac{1}{5} \\ \frac{1}{10} & -\frac{1}{15} \end{bmatrix}$ **49.** $\begin{bmatrix} \frac{1}{2} & -1 & -\frac{1}{2} \\ \frac{1}{2} & -\frac{2}{3} & -\frac{5}{6} \\ 0 & \frac{2}{3} & \frac{1}{3} \end{bmatrix}$ **51.** 550

53. 8 **55.** -3 **57.** 279 **59.** $(-3, 1)$

61. $(1, 1, -2)$ **63.** $(2. -4, 6)$ **65.** Inconsistent

67. $(1, 2)$ **69.** $\left(\frac{3}{4}, -\frac{1}{2}\right)$ **71.** $\left(\frac{2}{3}, \frac{1}{2}\right)$

73. Cramer's Rule does not apply.

75. Eight carnations, four roses **77.** $y = x^2 + 2x + 3$

79. (a) $a \approx 59.9, b \approx 3.6; y = 59.9 + 3.6t$
(b)

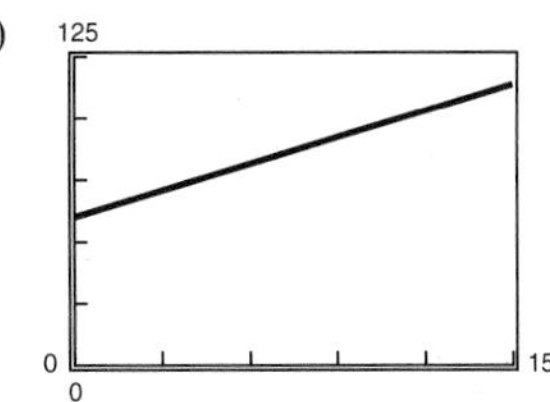

(c) The median price was increasing by an average of \$3600 per year. (d) \$113,900

81. 16 **83.** 7 **85.** $x - 2y + 4 = 0$

87. $2x + 6y - 13 = 0$ **89.** Answers will vary.

91. Because each of the three rows is multiplied by 4, $|4A| = 4^3|A| = 128$.

Chapter Test (page 522)

1. $\begin{bmatrix} 1 & 0 & 0 \\ 0 & 1 & 0 \\ 0 & 0 & 1 \end{bmatrix}$ **2.** $\begin{bmatrix} 1 & 0 & -1 & 2 \\ 0 & 1 & 0 & -1 \\ 0 & 0 & 0 & 0 \\ 0 & 0 & 0 & 0 \end{bmatrix}$

3. $\left(1, 3, -\frac{1}{2}\right)$ **4.** $y = -\frac{1}{2}x^2 + x + 2$

5. (a) $\begin{bmatrix} 1 & 5 & -2 \\ 0 & -4 & 3 \end{bmatrix}$ (b) $\begin{bmatrix} 15 & 12 & 12 \\ -12 & -12 & 0 \end{bmatrix}$

(c) $\begin{bmatrix} 7 & 14 & 0 \\ -4 & -12 & 6 \end{bmatrix}$

6. $\begin{bmatrix} 8 & -8 \\ 16 & -4 \\ 6 & 12 \end{bmatrix}$ **7.** $\begin{bmatrix} \frac{1}{2} & \frac{2}{5} \\ 1 & \frac{3}{5} \end{bmatrix}$ **8.** (13, 22)

9. -2 **10.** 7

CHAPTER 7

Section 7.1 (page 532)

1. 3, 5, 7, 9, 11 **3.** 2, 4, 8, 16, 32

5. $-2, 4, -8, 16, -32$ **7.** $2, \frac{3}{2}, \frac{4}{3}, \frac{5}{4}, \frac{6}{5}$

9. $3, \frac{12}{11}, \frac{9}{13}, \frac{24}{47}, \frac{15}{37}$ **11.** $0, 1, 0, \frac{1}{2}, 0$

13. $\frac{5}{2}, \frac{11}{4}, \frac{23}{8}, \frac{47}{16}, \frac{95}{32}$

15. $1, \frac{1}{2^{3/2}}, \frac{1}{3^{3/2}}, \frac{1}{4^{3/2}}, \frac{1}{5^{3/2}}$ **17.** $3, \frac{9}{2}, \frac{9}{2}, \frac{27}{8}, \frac{81}{40}$

19. $-1, \frac{1}{4}, -\frac{1}{9}, \frac{1}{16}, -\frac{1}{25}$ **21.** $\frac{2}{3}, \frac{2}{3}, \frac{2}{3}, \frac{2}{3}, \frac{2}{3}$

23. -73 **25.** 28, 24, 20, 16, 12 **27.** 3, 4, 6, 10, 18

29.

31.

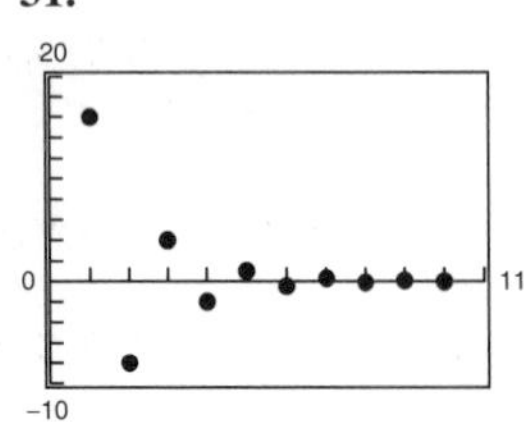

33.

35. (c) **37.** (d) **39.** $\frac{1}{30}$ **41.** 90 **43.** $n + 1$

45. $\dfrac{1}{2n(2n + 1)}$ **47.** $a_n = 3n - 2$

49. $a_n = n^2 - 1$ **51.** $a_n = \dfrac{n + 1}{n + 2}$

53. $a_n = \dfrac{(-1)^{n+1}}{2^n}$ **55.** $a_n = 1 + \dfrac{1}{n}$ **57.** $a_n = \dfrac{1}{n!}$

59. $a_n = (-1)^{n+1}$

61. 6, 8, 10, 12, 14

$a_n = 2n + 4$

63. 81, 27, 9, 3, 1

$a_n = \dfrac{243}{3^n}$

65. 35 **67.** 40 **69.** 30 **71.** $\frac{9}{5}$ **73.** 238

75. 30 **77.** 81 **79.** $\frac{47}{60}$ **81.** $\displaystyle\sum_{i=1}^{9} \frac{1}{3i}$

83. $\displaystyle\sum_{i=1}^{8} \left[2\left(\frac{i}{8}\right) + 3\right]$ **85.** $\displaystyle\sum_{i=1}^{6} (-1)^{i+1}3^i$

87. $\displaystyle\sum_{i=1}^{20} \frac{(-1)^{i+1}}{i^2}$ **89.** $\displaystyle\sum_{i=1}^{5} \frac{2^i - 1}{2^{i+1}}$

91. (a) $A_1 = \$5100.00$, $A_2 = \$5202.00$, $A_3 = \$5306.04$, $A_4 = \$5412.16$, $A_5 = \$5520.40$, $A_6 = \$5630.81$, $A_7 = \$5743.43$, $A_8 = \$5858.30$

(b) \$11,040.20

93. $A_1 = 529.73$, $A_2 = 555.79$, $A_3 = 588.31$, $A_4 = 627.29$, $A_5 = 672.73$, $A_6 = 724.63$, $A_7 = 782.99$, $A_8 = 847.81$, $A_9 = 919.09$, $A_{10} = 996.83$, $A_{11} = 1081.03$

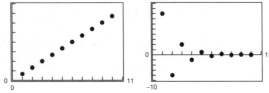

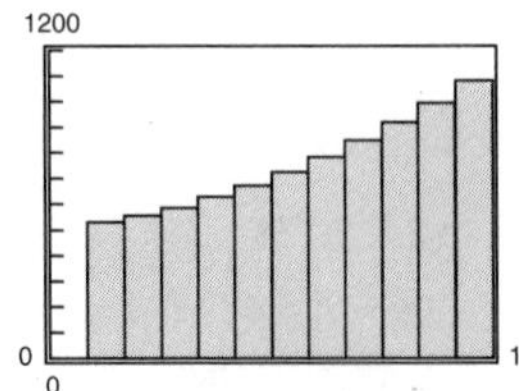

95. \$11,131.5 million

97. 1, 1, 2, 3, 5, 8, 13, 21, 34, 55, 89, 144

$1, 2, \frac{3}{2}, \frac{5}{3}, \frac{8}{5}, \frac{13}{8}, \frac{21}{13}, \frac{34}{21}, \frac{55}{34}, \frac{89}{55}$

Section 7.2 *(page 541)*

1. Arithmetic sequence, $d = -2$

3. Not an arithmetic sequence

5. Arithmetic sequence, $d = -\frac{1}{4}$

7. Arithmetic sequence, $d = 4$

9. Arithmetic sequence, $d = 0.4$

11. 8, 11, 14, 17, 20

Arithmetic sequence, $d = 3$

13. $\frac{1}{2}, \frac{1}{3}, \frac{1}{4}, \frac{1}{5}, \frac{1}{6}$

Not an arithmetic sequence

15. 97, 94, 91, 88, 85

Arithmetic sequence, $d = -3$

17. $1, 4, \frac{7}{3}, \frac{7}{2}, \frac{13}{5}$

Not an arithmetic sequence

19. 15, 19, 23, 27, 31

$a_n = 11 + 4n$

21. 200, 190, 180, 170, 160

$a_n = 210 - 10n$

23. $\frac{3}{2}, \frac{5}{4}, 1, \frac{3}{4}, \frac{1}{2}$

$a_n = \frac{7}{4} - \frac{1}{4}n$

25. 5, 11, 17, 23, 29 **27.** $-2.6, -3.0, -3.4, -3.8, -4.2$

29. 2, 6, 10, 14, 18 **31.** $-2, 2, 6, 10, 14$

33. $a_n = 1 + (n - 1)3$ **35.** $a_n = 100 + (n - 1)(-8)$

37. $a_n = x + (n - 1)(2x)$ **39.** $a_n = 4 + (n - 1)\left(-\frac{5}{2}\right)$

41. $a_n = 5 + (n - 1)\left(\frac{10}{3}\right)$ **43.** $a_n = 100 + (n - 1)(-3)$

45. (b) **47.** (c)

49.

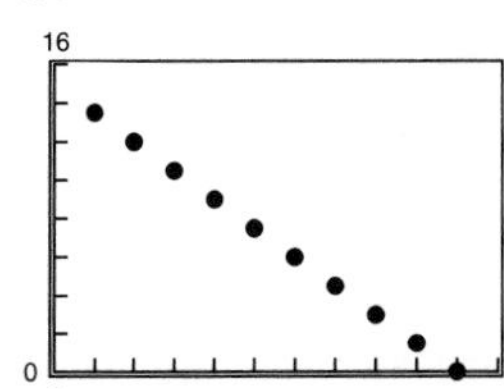

51.

53. Linear **55.** 620 **57.** 4600 **59.** 265

61. 4000 **63.** 1275 **65.** 25,250 **67.** 355

69. 126,750 **71.** 520 **73.** 44,625 **75.** 10,120

77. 10,000 **79.** (a) \$40,000 (b) \$217,500

81. 2340 **83.** 405 bricks **85.** 585 seats

87. 156 times

89. (a) 4, 9, 16, 25, 36 (b) n^2

(c) $\frac{n}{2}[1 + (2n - 1)] = n^2$

91. 4

Section 7.3 *(page 550)*

1. Geometric sequence, $r = 3$

3. Not a geometric sequence

5. Geometric sequence, $r = -\frac{1}{2}$

7. Not a geometric sequence **9.** Not a geometric sequence

11. 2, 6, 18, 54, 162 **13.** $1, \frac{1}{2}, \frac{1}{4}, \frac{1}{8}, \frac{1}{16}$

15. $5, -\frac{1}{2}, \frac{1}{20}, -\frac{1}{200}, \frac{1}{2000}$ **17.** $1, e, e^2, e^3, e^4$

19. $3, \frac{3x}{2}, \frac{3x^2}{4}, \frac{3x^3}{8}, \frac{3x^4}{16}$

21. 64, 32, 16, 8, 4 **23.** 4, 12, 36, 108, 324

$a_n = 128\left(\frac{1}{2}\right)^n$ $a_n = \frac{4}{3}(3)^n$

25. $6, -9, \frac{27}{2}, -\frac{81}{4}, \frac{243}{8}$

$a_n = 6\left(-\frac{3}{2}\right)^{n-1}$

27. $\left(\frac{1}{2}\right)^7$ **29.** $-\frac{2}{3^{10}}$ **31.** $100e^{8x}$ **33.** $500(1.02)^{39}$

35. 9 **37.** $-\frac{2}{9}$ **39.** (a) **41.** (b)

43.

45.

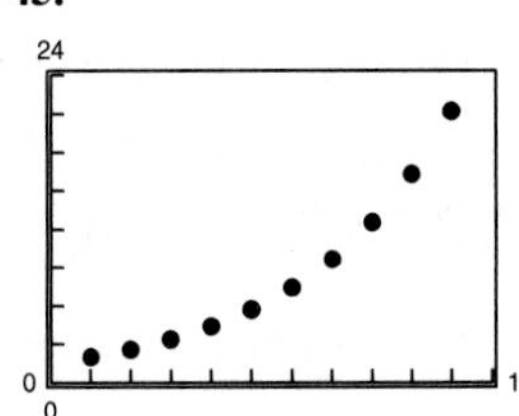

47. Increased powers of real numbers between -1 and 1 approach zero.

49. (a) \$2593.74 (b) \$2653.30 (c) \$2685.06

(d) \$2707.04 (e) \$2717.91

51. \$22,689.45 **53.** 8, 4, 6, 5 **55.** 511

57. 43 **59.** 29,921.31 **61.** 6.4 **63.** 2092.60

65. $\sum_{n=1}^{7} 5(3)^{n-1}$ **67.** \$7808.24 **69.** Answers will vary.

71. (a) \$26,198.27 (b) \$26,263.88

73. (a) \$637,678.02 (b) \$645,861.43

75. Answers will vary. **77.** 126 square inches

79. (a) \$5,368,709.11 (b) \$10,737,418.23
(c) \$21,474,836.47

81. 2 **83.** $\frac{2}{3}$ **85.** $\frac{16}{3}$ **87.** 32 **89.** $\frac{4}{11}$ **91.** $\frac{7}{22}$

93.

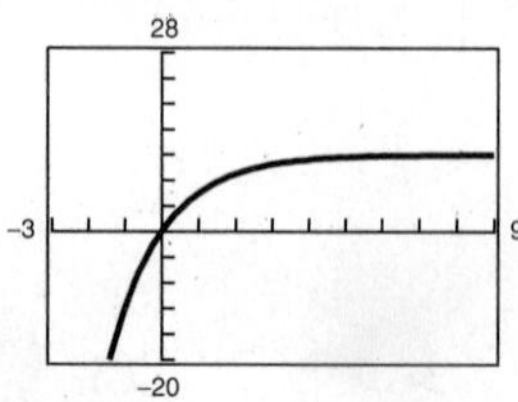

Horizontal asymptote: $y = 12$
Corresponds to the sum of the series

95. (a) 152.42 feet (b) 19 seconds

97. 45.65 miler per hour **99.** 2.4 hours

Section 7.4 *(page 563)*

1. $\dfrac{5}{(k+1)(k+2)}$ **3.** $\dfrac{(k+1)^2(k+2)^2}{4}$

5.–17. Answers will vary. **19.** 210 **21.** 91

23. 979 **25.** 70 **27.** -3402

29. $\sum_{n=0}^{\infty} (1 + 4n)$ **31.** $\sum_{n=1}^{\infty} \left(\frac{9}{10}\right)^{n-1}$

33. $\sum_{n=2}^{\infty} \dfrac{1}{2n(n-1)}$ **35.–47.** Answers will vary.

49. See the domino illustration and Figure 7.4.

51. 1, 3, 5, 7, 9 **53.** 4, 2, -2, -4, -2

55. 0, 3, 6, 9, 12
First differences: 3, 3, 3, 3
Second differences: 0, 0, 0
Linear

57. 3, 1, -2, -6, -11
First differences: -2, -3, -4, -5
Second differences: -1, -1, -1
Quadratic

59. 0, 1, 3, 6, 10
First differences: 1, 2, 3, 4
Second differences: 1, 1, 1
Quadratic

61. 2, 4, 6, 8, 10
First differences: 2, 2, 2, 2
Second differences: 0, 0, 0
Linear

63. 1, 2, 6, 15, 31
First differences: 1, 4, 9, 16
Second differences: 3, 5, 7
Neither

65. $a_n = n^2 - n + 3$ **67.** $a_n = \frac{1}{2}n^2 + n - 3$

69. $(2, 4), \left(-\frac{1}{2}, \frac{1}{4}\right)$ **71.** $(1, 2, 1)$

Section 7.5 *(page 570)*

1. 10 **3.** 1 **5.** 15,504 **7.** 4950 **9.** 4950

11. The first and last numbers in each row are 1. Every other number in each row is formed by adding the two numbers immediately above the number.

13. 35 **15.** 56 **17.** $x^4 + 4x^3 + 6x^2 + 4x + 1$

19. $a^3 + 6a^2 + 12a + 8$

21. $y^4 - 8y^3 + 24y^2 - 32y + 16$

23. $x^5 + 5x^4y + 10x^3y^2 + 10x^2y^3 + 5xy^4 + y^5$

25. $r^6 + 18r^5s + 135r^4s^2 + 540r^3s^3 + 1215r^2s^4 + 1458rs^5 + 729s^6$

27. $x^5 - 5x^4y + 10x^3y^2 - 10x^2y^3 + 5xy^4 - y^5$

29. $1 - 6x + 12x^2 - 8x^3$

31. $x^8 + 20x^6 + 150x^4 + 500x^2 + 625$

33. $\dfrac{1}{x^5} + \dfrac{5y}{x^4} + \dfrac{10y^2}{x^3} + \dfrac{10y^3}{x^2} + \dfrac{5y^4}{x} + y^5$

35. $2x^4 - 24x^3 + 113x^2 - 246x + 207$

37. $32t^5 - 80t^4s + 80t^3s^2 - 40t^2s^3 + 10ts^4 - s^5$

39. $81 - 216z + 216z^2 - 96z^3 + 16z^4$

41. 1,732,104 **43.** 180 **45.** $-326,592$ **47.** 210

49. $n + 1$ terms

51. $x^2 + 12x^{3/2} + 54x + 108x^{1/2} + 81$

53. $x^2 - 3x^{4/3}y^{1/3} + 3x^{2/3}y^{2/3} - y$

55. $3x^2 + 3xh + h^2$ **57.** $\dfrac{\sqrt{x+h} - \sqrt{x}}{h}$ **59.** -4

61. $2035 + 828i$ **63.** 1 **65.** 0.273 **67.** 0.171

69. 1.172 **71.** 510,568.785

73. $g(x) = x^3 + 12x^2 + 44x + 48$
Shifted four units to the left

75. $g(x) = -x^2 + 7x - 8$; Shifted two units to the right

77. Answers will vary. **79.** Answers will vary.

81. (a) 792 (b) 36 (c) 792 (d) 12

83.

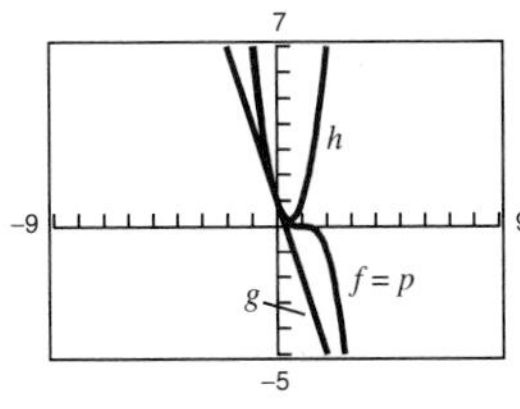

$p(x)$ is the expansion of $f(x)$.

85. (a) $g(t) = 0.1506t^2 + 3.7481t + 43.5584$

(b)

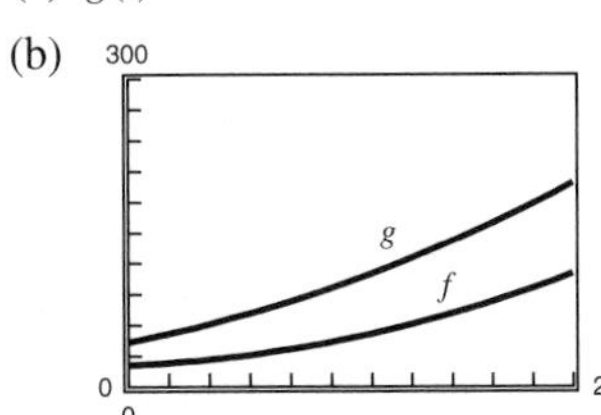

87. $g(x)$ is shifted eight units up from $f(x)$.

89. $g(x)$ is the reflection of $f(x)$ in the y-axis.

Section 7.6 (page 580)

1. 6 **3.** 5 **5.** 3 **7.** 7 **9.** 48 **11.** 12

13. 64 **15.** 12 **17.** 6,760,000

19. (a) 900 (b) 648 (c) 180 (d) 600

21. 64,000 **23.** (a) 720 (b) 48

25. 24 **27.** 336 **29.** 120

31. $n = 5$ or $n = 6$ **33.** 1,860,480

35. 970,200 **37.** 15,504

39. For some calculators the number is too great.

41. ABCD, ABDC, ACBD, ACDB, ADBC, ADCB, BACD, BADC, CABD, CADB, DABC, DACB

BCAD, BDAC, CBAD, CDAB, DBAC, DCAB, BCDA, BDCA, CBDA, CDBA, DBCA, DCBA

43. 120 **45.** 11,880 **47.** 420 **49.** 2520

51. AB, AC, AD, AE, AF, BC, BD, BE, BF, CD, CE, CF, DE, DF, EF

53. 4845 **55.** 3,838,380 **57.** 3,921,225 **59.** 21

61. (a) 70 (b) 30 **63.** (a) 70 (b) 54 (c) 16

65. 5 **67.** 20 **69.** Answers will vary.

71. Answers will vary. **73.** 8.30 **75.** 35

Section 7.7 (page 592)

1. $\{(H, 1), (H, 2), (H, 3), (H, 4), (H, 5), (H, 6), (T, 1), (T, 2), (T, 3), (T, 4), (T, 5), (T, 6)\}$

3. {ABC, ACB, BAC, BCA, CAB, CBA}

5. {(A, B), (A, C), (A, D), (A, E), (B, C), (B, D), (B, E), (C, D), (C, E), (D, E)}

7. $\frac{3}{8}$ **9.** $\frac{7}{8}$ **11.** $\frac{3}{13}$ **13.** $\frac{3}{26}$ **15.** $\frac{1}{12}$ **17.** $\frac{11}{12}$

19. $\frac{1}{3}$ **21.** $\frac{1}{5}$ **23.** $\frac{2}{5}$ **25.** 0.3 **27.** 0.85

29. (a) 925,000 (b) 0.18

31. (a) 0.58 (b) 0.956 (c) 0.004

33. (a) $\frac{672}{1254}$ (b) $\frac{582}{1254}$ (c) $\frac{548}{1254}$

35. $P(\{\text{Taylor wins}\}) = \frac{1}{2}$

$P(\{\text{Moore wins}\}) = P(\{\text{Jenkins wins}\}) = \frac{1}{4}$

37. (a) $\frac{21}{1292} \approx 0.016$ (b) $\frac{225}{646} \approx 0.348$ (c) $\frac{49}{323} \approx 0.152$

39. (a) $\frac{1}{3}$ (b) $\frac{5}{8}$ **41.** (a) $\frac{1}{120}$ (b) $\frac{1}{24}$

43. (a) $\frac{1}{169}$ (b) $\frac{1}{221}$ **45.** (a) $\frac{14}{55}$ (b) $\frac{12}{55}$ (c) $\frac{54}{55}$

47. (a) $\frac{1}{4}$ (b) $\frac{1}{2}$ (c) $\frac{9}{100}$ (d) $\frac{1}{30}$

49. (a) 0.9702 (b) 0.9998 (c) 0.0002

51. (a) $\frac{1}{1024}$ (b) $\frac{243}{1024}$ (c) $\frac{781}{1024}$

53. 0.1024 **55.** $\frac{7}{16}$

57. (a) As you consider successive people with distinct birthdays, the probabilities must decrease to take into account the birth dates already used. Because the birth dates of people are independent events, multiply the respective probabilities of distinct birthdays.

(b) $\frac{365}{365} \cdot \frac{364}{365} \cdot \frac{363}{365} \cdot \frac{362}{365}$

(c) Answers will vary.

(d) Q_n is the probability that the birthdays are *not* distinct, which is equivalent to at least two people having the same birthday.

(e)

n	10	15	20	23	30	40	50
P_n	0.88	0.75	0.59	0.49	0.29	0.11	0.03
Q_n	0.12	0.25	0.41	0.51	0.71	0.89	0.97

(f) 23

59. 0.454

FOCUS ON CONCEPTS (page 597)

1. Natural numbers

2. (a) Odd-numbered terms are negative.

(b) Even-numbered terms are negative.

3. True **4.** True **5.** True **6.** True

7. (a) Each term is obtained by adding the same constant (common difference) to the previous term.

(b) Each term is obtained by multiplying the same constant (common ratio) by the previous term.

8. (a) Arithmetic. There is a constant difference between consecutive terms.

(b) Geometric. Each term is a constant multiple of the previous term. In this case the common ratio is greater than 1.

9. Each term of the sequence is defined in terms of the previous term.

10. Increased powers of real numbers between 0 and 1 approach zero.

11. (d) **12.** (a) **13.** (b) **14.** (c)

15. The signs of the terms alternate in the expansion $(x - y)^n$.

16. Same

17. ${}_{10}P_6 > {}_{10}C_6$. Changing the order of any of the six elements selected results in a different permutation but the same combination.

18. $0 \le p \le 1$; closed

19. $\frac{1}{3}$. The probability that an event does not occur is 1 minus the probability that it does occur.

20. Meteorological records indicate that over an extended period of time with similar weather conditions it will rain 60% of the time.

Review Exercises (page 598)

1. $8, 5, 4, \frac{7}{2}, \frac{16}{5}$ **3.** $72, 36, 12, 3, \frac{3}{5}$

5.

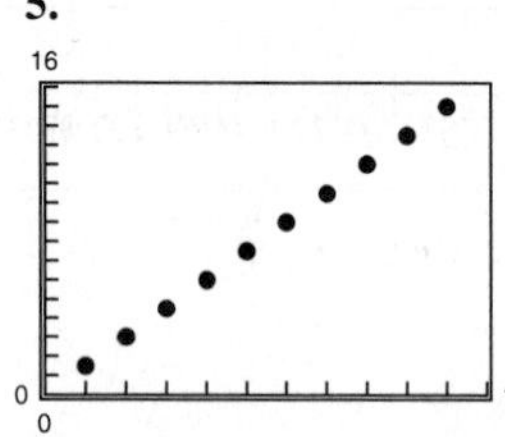

7.

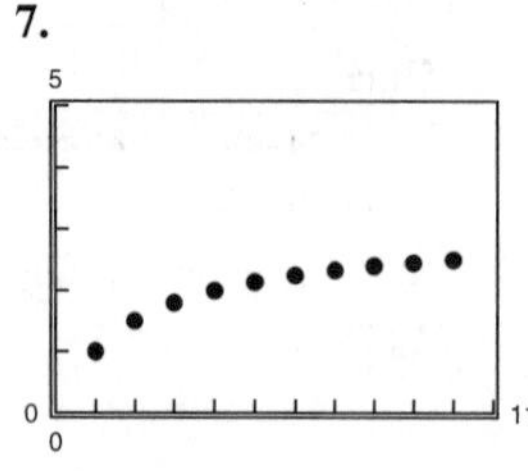

9. $\sum_{k=1}^{20} \frac{1}{2k}$ **11.** $\sum_{k=1}^{9} \frac{k}{k+1}$ **13.** 30 **15.** $\frac{205}{24}$

17. 6050 **19.** 418 **21.** 3, 7, 11, 15, 19

23. 1, 4, 7, 10, 13

25. 35, 32, 29, 26, 23
$a_n = 38 - 3n$

27. 9, 16, 23, 30, 37
$a_n = 2 + 7n$

29. $a_n = 103 - 3n$
1430

31. 80 **33.** 88 **35.** 25,250

37. (a) $43,000 (b) $192,500 **39.** $4, -1, \frac{1}{4}, -\frac{1}{16}, \frac{1}{64}$

41. $9, 6, 4, \frac{8}{3}, \frac{16}{9}$ or $9, -6, 4, -\frac{8}{3}, \frac{16}{9}$

43. $120, 40, \frac{40}{3}, \frac{40}{9}, \frac{40}{27}$
$a_n = 120\left(\frac{1}{3}\right)^{n-1}$

45. $25, -15, 9, -\frac{27}{5}, \frac{81}{25}$
$a_n = 25\left(-\frac{3}{5}\right)^{n-1}$

47. $a_n = 16\left(-\frac{1}{2}\right)^{n-1}$, 10.67 **49.** 127 **51.** 8 **53.** 12

55. 24.849 **57.** (a) $a_t = 120{,}000(0.7)^t$ (b) $20,168.40

59. $5111.82 **61.** Answers will vary.

63. Answers will vary. **65.** 15 **67.** 6720

69. $\frac{x^4}{16} + \frac{x^3y}{2} + \frac{3x^2y^2}{2} + 2xy^3 + y^4$

71. $\frac{64}{x^6} - \frac{576}{x^4} + \frac{2160}{x^2} - 4320 + 4860x^2 - 2916x^4 + 729x^6$

73. $41 + 840i$ **75.** 118,813,760 **77.** $\frac{1}{9}$

79. $P(\{3\}) = \frac{1}{6}$
$P(\{(1, 5), (5, 1), (2, 4), (4, 2), (3, 3)\}) = \frac{5}{36}$

81. $\frac{31}{32}$ **83.** 0.0475

Chapter Test (page 602)

1. $1, -\frac{2}{3}, \frac{4}{9}, -\frac{8}{27}, \frac{16}{81}$ **2.** 12, 16, 20, 24, 28

3. $a_n = 5000 - 100(n - 1)$ **4.** $a_n = 4\left(\frac{1}{2}\right)^{n-1}$

5. $\sum_{n=1}^{12} \frac{2}{3n+1}$ **6.** 3825 **7.** $-\frac{3}{2}$ **8.** 765

9. $47,868.33 **10.** 1140 **11.** 56 **12.** 26,000

13. 12,650 **14.** $\frac{3}{26}$ **15.** $\frac{1}{6}$ **16.** $\frac{1}{24}$

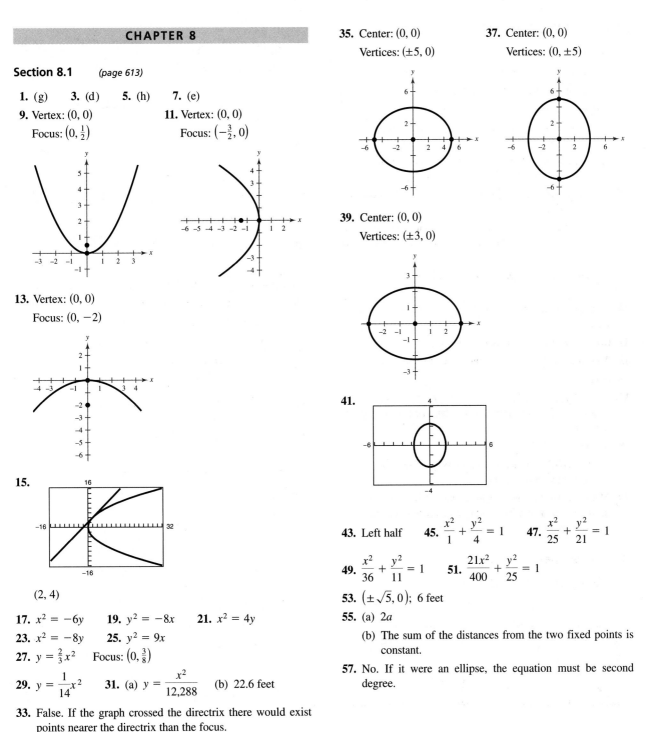

CHAPTER 8

Section 8.1 *(page 613)*

1. (g) **3.** (d) **5.** (h) **7.** (e)

9. Vertex: $(0, 0)$
Focus: $\left(0, \frac{1}{2}\right)$

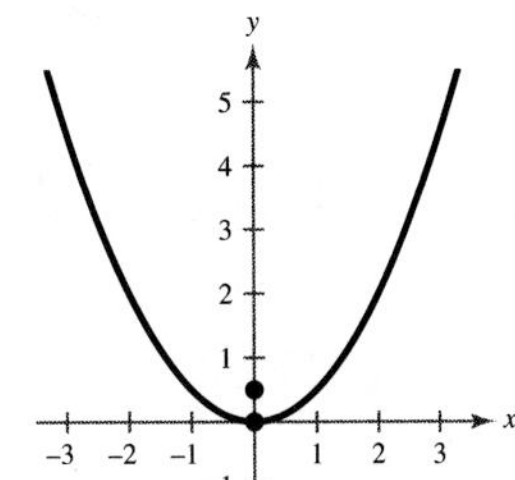

11. Vertex: $(0, 0)$
Focus: $\left(-\frac{3}{2}, 0\right)$

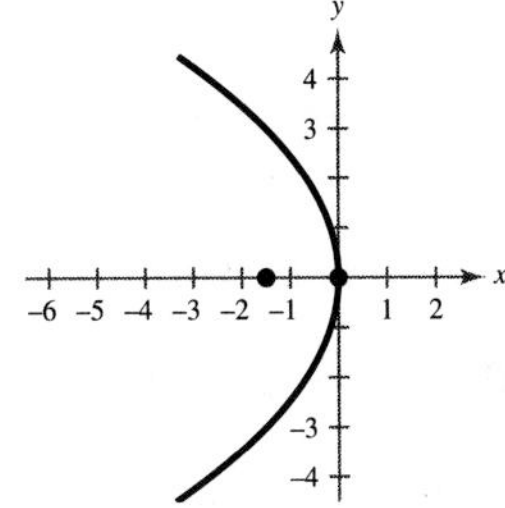

13. Vertex: $(0, 0)$
Focus: $(0, -2)$

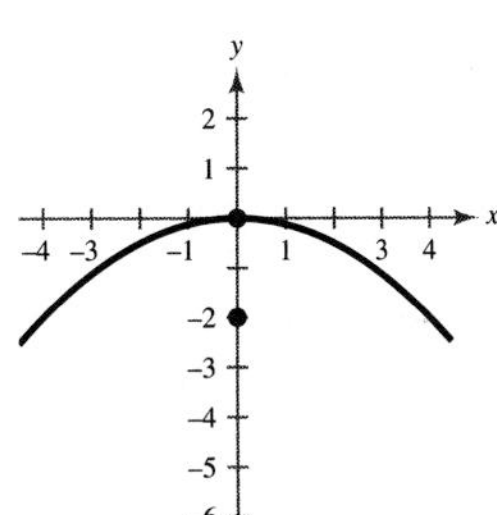

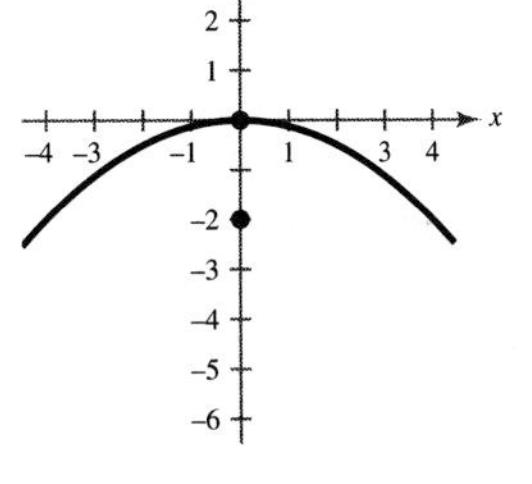

15.

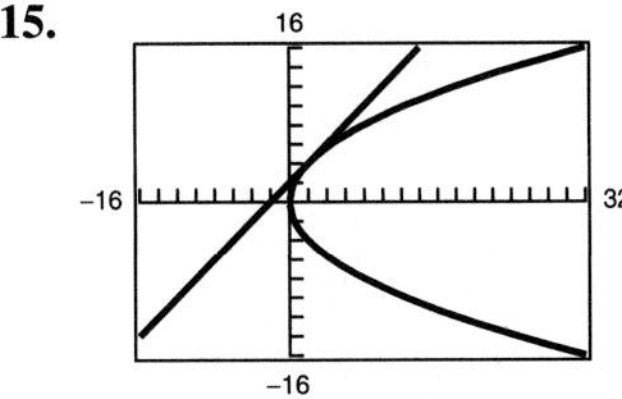

$(2, 4)$

17. $x^2 = -6y$ **19.** $y^2 = -8x$ **21.** $x^2 = 4y$

23. $x^2 = -8y$ **25.** $y^2 = 9x$

27. $y = \frac{2}{3}x^2$ Focus: $\left(0, \frac{3}{8}\right)$

29. $y = \dfrac{1}{14}x^2$ **31.** (a) $y = \dfrac{x^2}{12{,}288}$ (b) 22.6 feet

33. False. If the graph crossed the directrix there would exist points nearer the directrix than the focus.

35. Center: $(0, 0)$
Vertices: $(\pm 5, 0)$

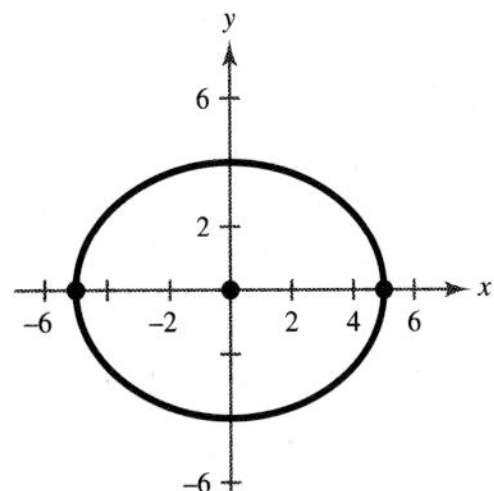

37. Center: $(0, 0)$
Vertices: $(0, \pm 5)$

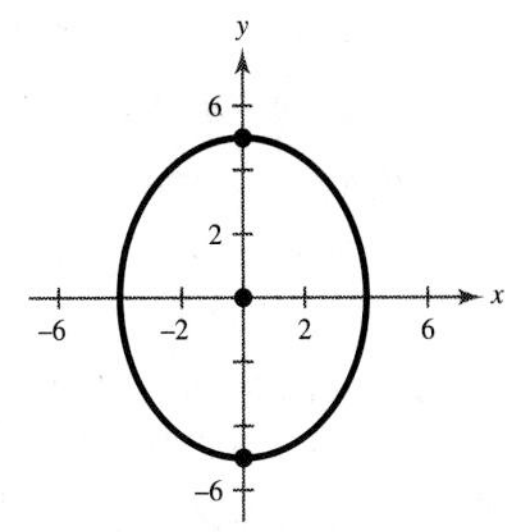

39. Center: $(0, 0)$
Vertices: $(\pm 3, 0)$

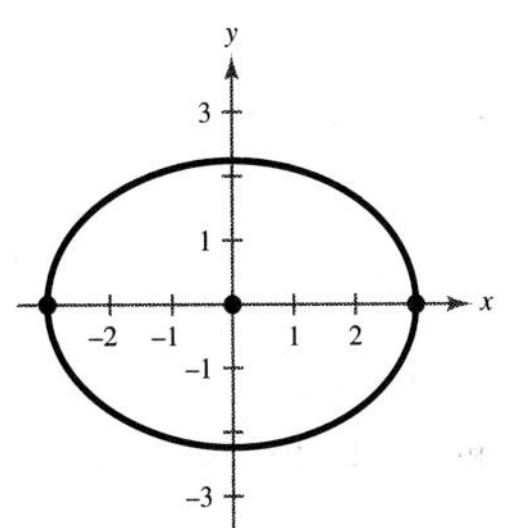

41.

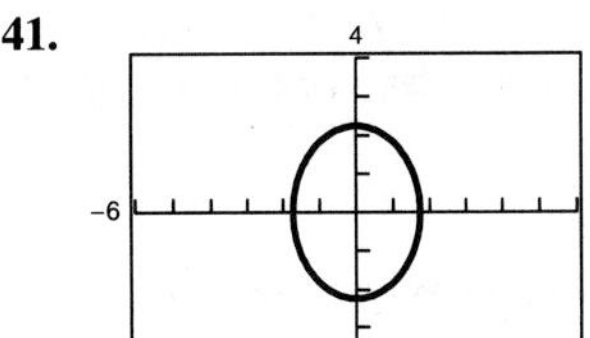

43. Left half **45.** $\dfrac{x^2}{1} + \dfrac{y^2}{4} = 1$ **47.** $\dfrac{x^2}{25} + \dfrac{y^2}{21} = 1$

49. $\dfrac{x^2}{36} + \dfrac{y^2}{11} = 1$ **51.** $\dfrac{21x^2}{400} + \dfrac{y^2}{25} = 1$

53. $\left(\pm\sqrt{5}, 0\right)$; 6 feet

55. (a) $2a$

(b) The sum of the distances from the two fixed points is constant.

57. No. If it were an ellipse, the equation must be second degree.

59. (a) $A = \pi a(20 - a)$

(b) $\dfrac{x^2}{196} + \dfrac{y^2}{36} = 1$

(c)

a	8	9	10	11	12	13
A	301.6	311.0	314.2	311.0	301.6	285.9

$a = 10$. Circle

(d) 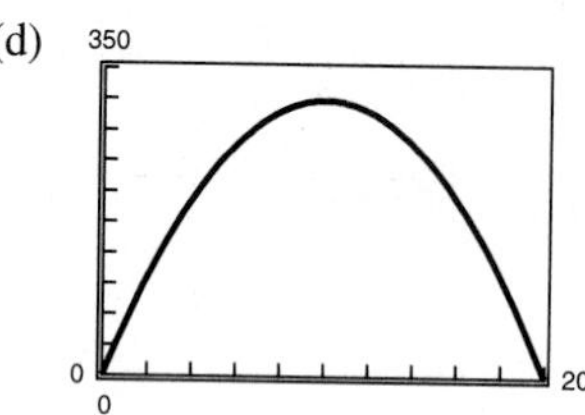

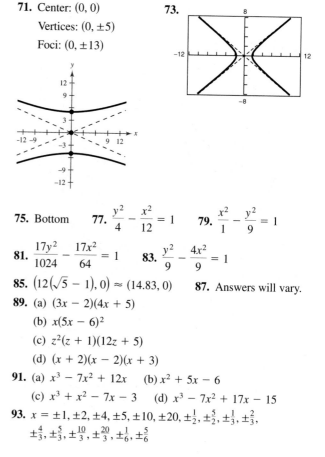

$a = 10$

61. The shape continuously changes from an ellipse with a vertical major axis of length 8 and minor axis of length 2 to a circle with a diameter of 8 and then to an ellipse with a horizontal major axis of length 16 and minor axis of length 8.

63. 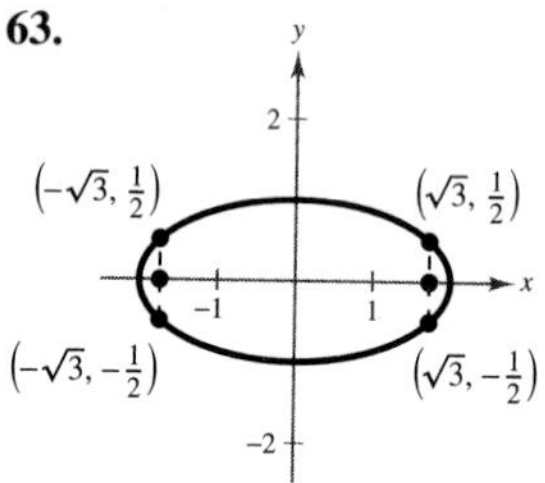

65. 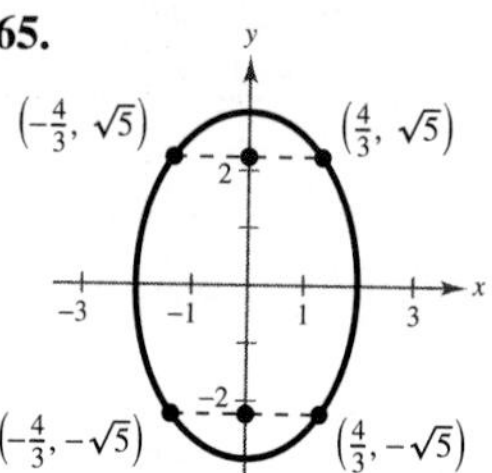

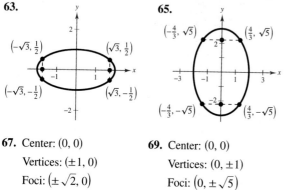

67. Center: $(0, 0)$
Vertices: $(\pm 1, 0)$
Foci: $(\pm\sqrt{2}, 0)$

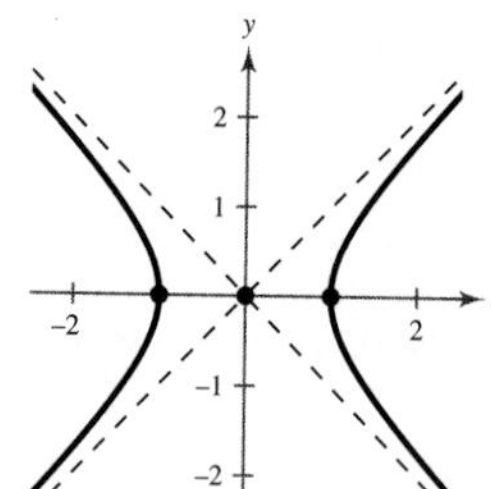

69. Center: $(0, 0)$
Vertices: $(0, \pm 1)$
Foci: $(0, \pm\sqrt{5})$

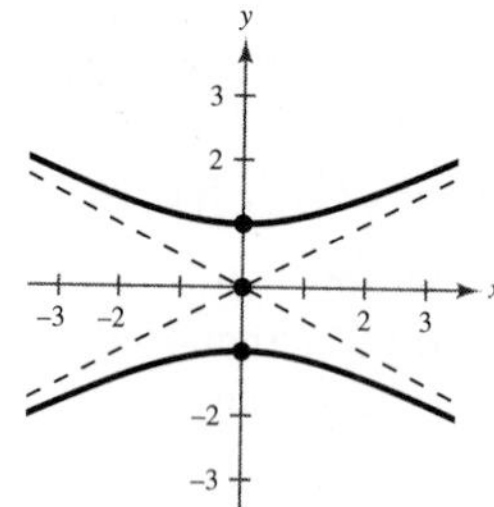

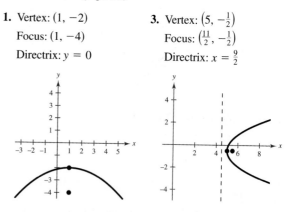

71. Center: $(0, 0)$
Vertices: $(0, \pm 5)$
Foci: $(0, \pm 13)$

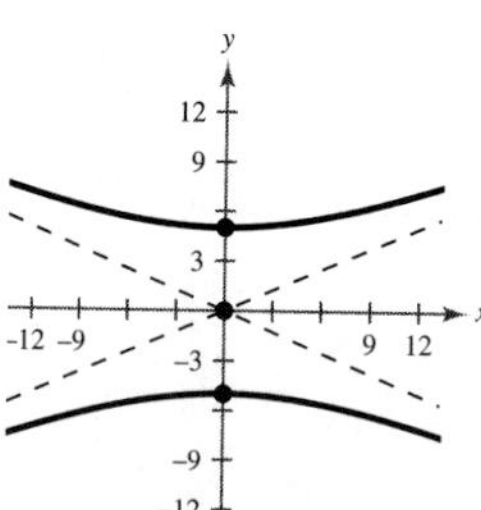

73. 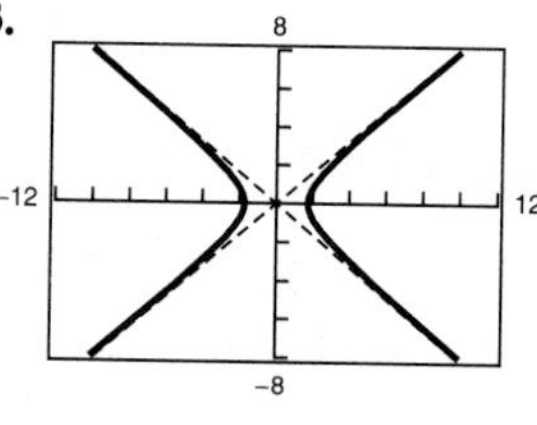

75. Bottom **77.** $\dfrac{y^2}{4} - \dfrac{x^2}{12} = 1$ **79.** $\dfrac{x^2}{1} - \dfrac{y^2}{9} = 1$

81. $\dfrac{17y^2}{1024} - \dfrac{17x^2}{64} = 1$ **83.** $\dfrac{y^2}{9} - \dfrac{4x^2}{9} = 1$

85. $\left(12(\sqrt{5} - 1), 0\right) \approx (14.83, 0)$ **87.** Answers will vary.

89. (a) $(3x - 2)(4x + 5)$
(b) $x(5x - 6)^2$
(c) $z^2(z + 1)(12z + 5)$
(d) $(x + 2)(x - 2)(x + 3)$

91. (a) $x^3 - 7x^2 + 12x$ (b) $x^2 + 5x - 6$
(c) $x^3 + x^2 - 7x - 3$ (d) $x^3 - 7x^2 + 17x - 15$

93. $x = \pm 1, \pm 2, \pm 4, \pm 5, \pm 10, \pm 20, \pm\frac{1}{2}, \pm\frac{5}{2}, \pm\frac{1}{3}, \pm\frac{2}{3}, \pm\frac{4}{3}, \pm\frac{5}{3}, \pm\frac{10}{3}, \pm\frac{20}{3}, \pm\frac{1}{6}, \pm\frac{5}{6}$

Section 8.2 *(page 623)*

1. Vertex: $(1, -2)$
Focus: $(1, -4)$
Directrix: $y = 0$

3. Vertex: $\left(5, -\frac{1}{2}\right)$
Focus: $\left(\frac{11}{2}, -\frac{1}{2}\right)$
Directrix: $x = \frac{9}{2}$

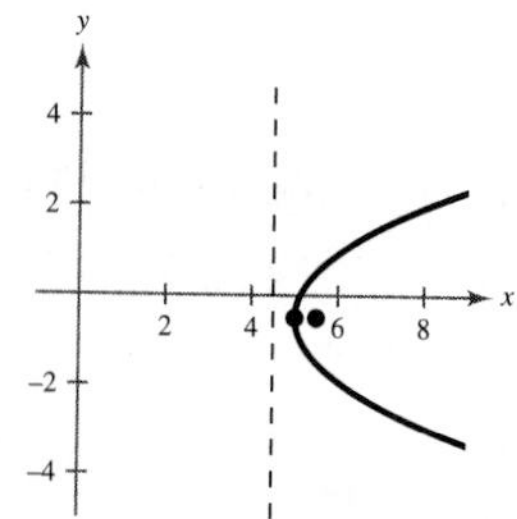

5. Vertex: $(1, 1)$
Focus: $(1, 2)$
Directrix: $y = 0$

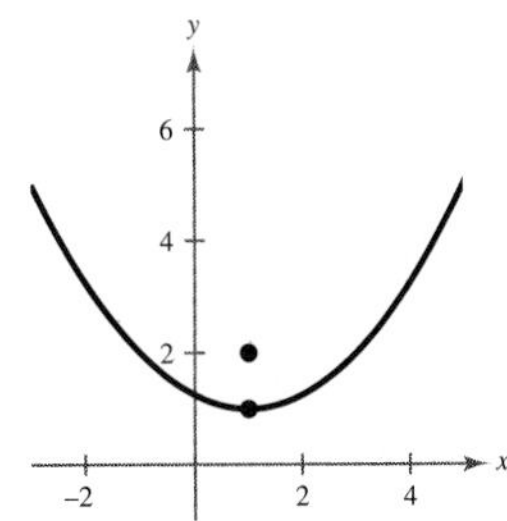

7. Vertex: $(-2, -3)$
Focus: $(-4, -3)$
Directrix: $x = 0$

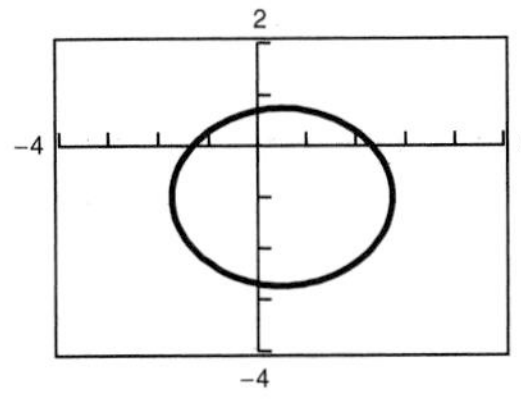

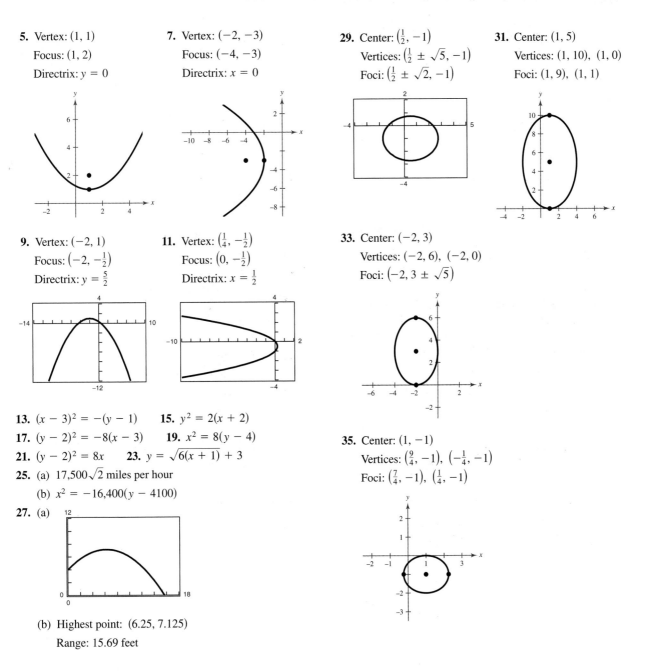

9. Vertex: $(-2, 1)$
Focus: $\left(-2, -\frac{1}{2}\right)$
Directrix: $y = \frac{5}{2}$

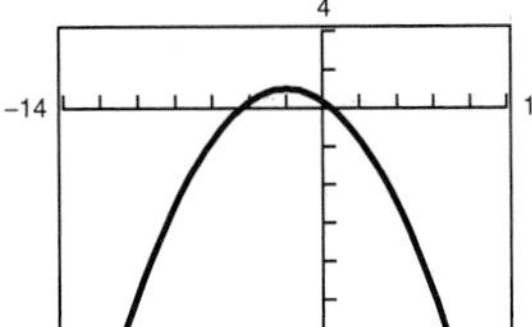

11. Vertex: $\left(\frac{1}{4}, -\frac{1}{2}\right)$
Focus: $\left(0, -\frac{1}{2}\right)$
Directrix: $x = \frac{1}{2}$

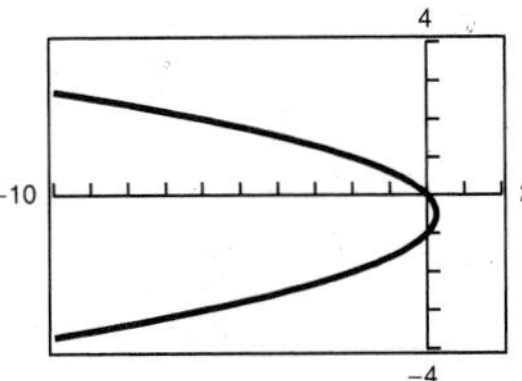

13. $(x - 3)^2 = -(y - 1)$ **15.** $y^2 = 2(x + 2)$
17. $(y - 2)^2 = -8(x - 3)$ **19.** $x^2 = 8(y - 4)$
21. $(y - 2)^2 = 8x$ **23.** $y = \sqrt{6(x + 1)} + 3$
25. (a) $17{,}500\sqrt{2}$ miles per hour
(b) $x^2 = -16{,}400(y - 4100)$
27. (a)

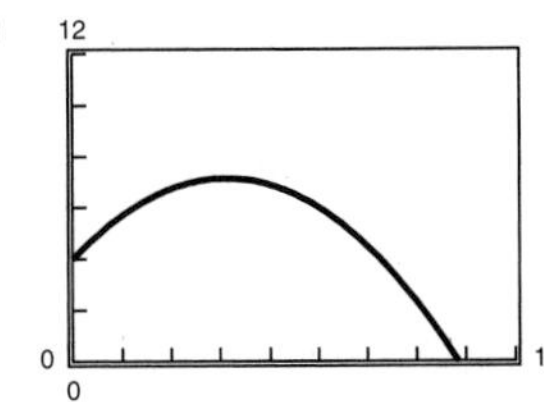

(b) Highest point: $(6.25, 7.125)$
Range: 15.69 feet

29. Center: $\left(\frac{1}{2}, -1\right)$
Vertices: $\left(\frac{1}{2} \pm \sqrt{5}, -1\right)$
Foci: $\left(\frac{1}{2} \pm \sqrt{2}, -1\right)$

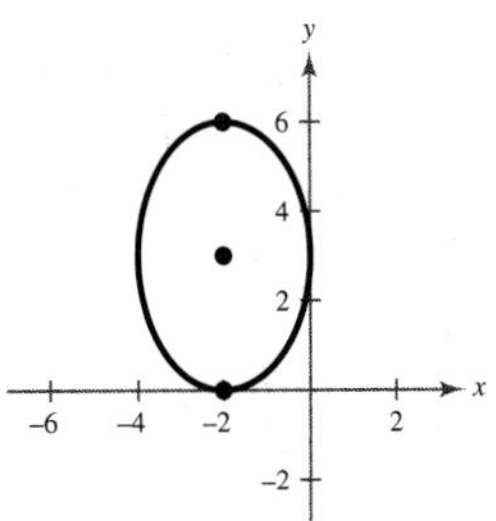

31. Center: $(1, 5)$
Vertices: $(1, 10)$, $(1, 0)$
Foci: $(1, 9)$, $(1, 1)$

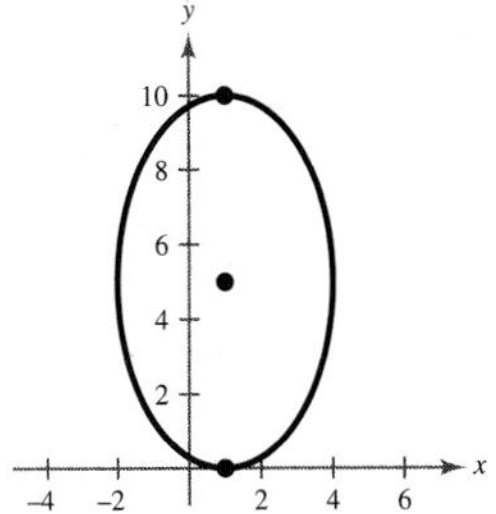

33. Center: $(-2, 3)$
Vertices: $(-2, 6)$, $(-2, 0)$
Foci: $\left(-2, 3 \pm \sqrt{5}\right)$

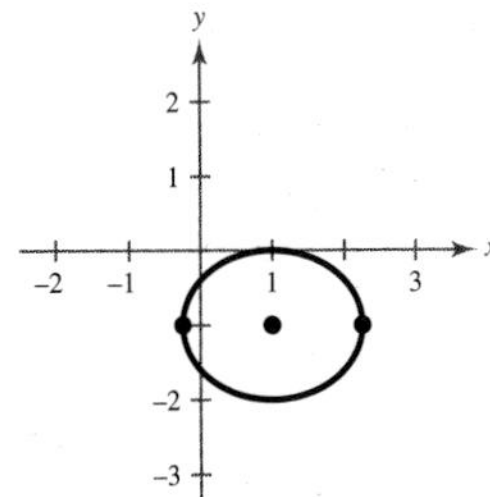

35. Center: $(1, -1)$
Vertices: $\left(\frac{9}{4}, -1\right)$, $\left(-\frac{1}{4}, -1\right)$
Foci: $\left(\frac{7}{4}, -1\right)$, $\left(\frac{1}{4}, -1\right)$

37. $\dfrac{(x-2)^2}{1} + \dfrac{(y-3)^2}{9} = 1$

39. $\dfrac{(x-2)^2}{9} + \dfrac{(y-2)^2}{4} = 1$

41. $\dfrac{(x-2)^2}{4} + \dfrac{(y-2)^2}{1} = 1$

43. $\dfrac{x^2}{48} + \dfrac{(y-4)^2}{64} = 1$ **45.** $\dfrac{(x-3)^2}{9} + \dfrac{(y-5)^2}{16} = 1$

47. $\dfrac{x^2}{16} + \dfrac{(y-4)^2}{12} = 1$ **49.** $x = \dfrac{3}{2}\left(2 + \sqrt{4 - y^2}\right)$

51. $\dfrac{x^2}{25} + \dfrac{y^2}{16} = 1$ **53.** 2,756,832,000; 4,575,168,000

55. (a) Answers will vary.

(b)

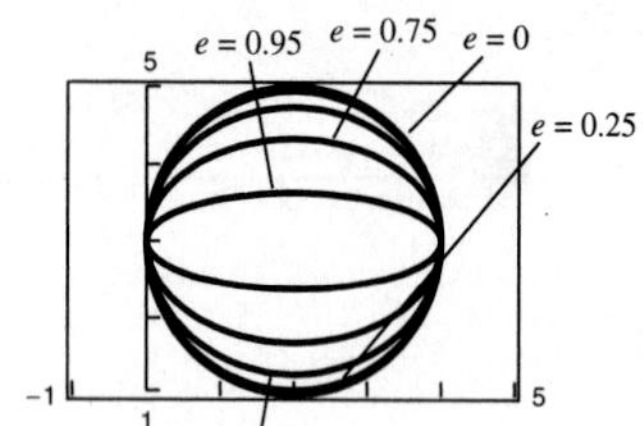

(c) Ellipse becomes more circular.

57. Center: $(1, -2)$

Vertices: $(3, -2)$, $(-1, -2)$

Foci: $\left(1 \pm \sqrt{5}, -2\right)$

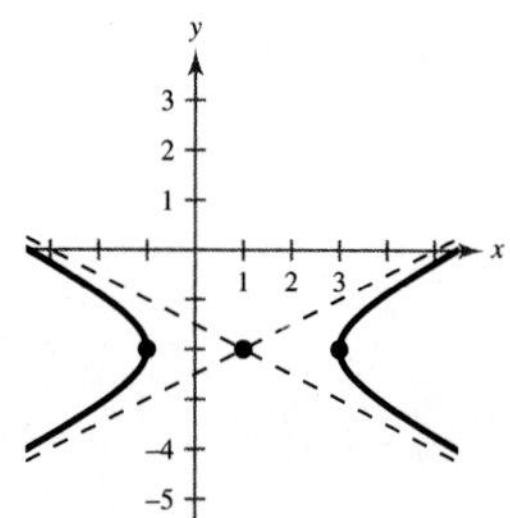

59. Center: $(2, -6)$

Vertices: $(2, -5)$, $(2, -7)$

Foci: $\left(2, -6 \pm \sqrt{2}\right)$

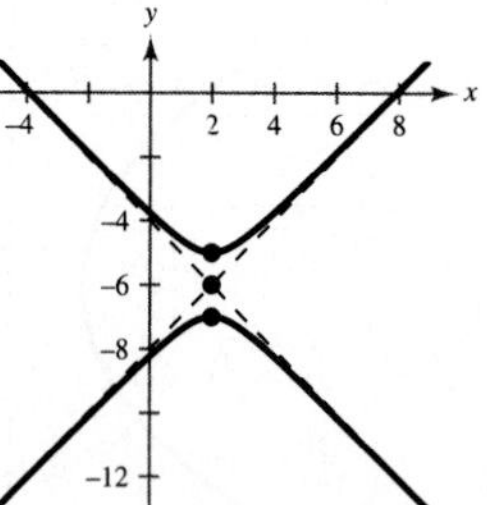

61. Center: $(2, -3)$

Vertices: $(3, -3)$, $(1, -3)$

Foci: $\left(2 \pm \sqrt{10}, -3\right)$

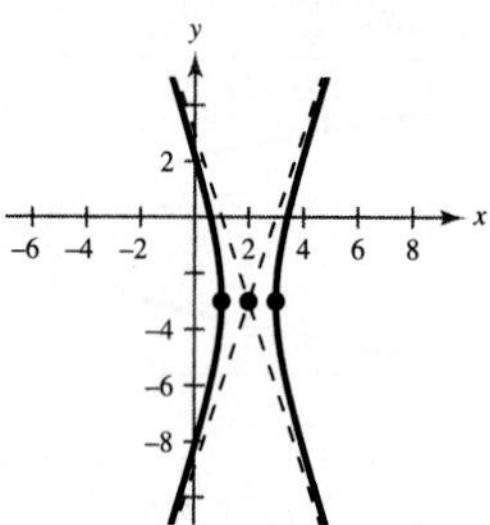

63. The graph of this equation is two lines intersecting at $(-1, -3)$.

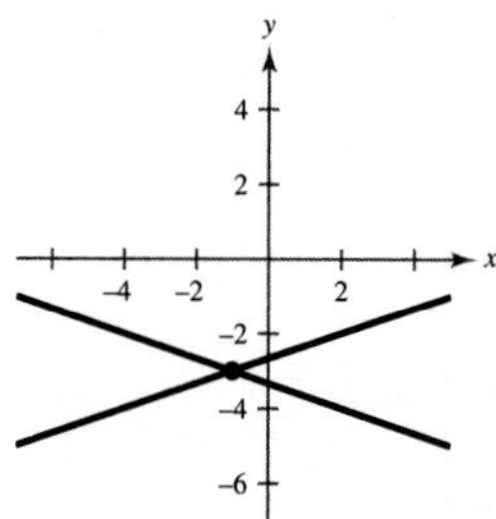

65. Center: $(1, -3)$

Vertices: $\left(1, -3 \pm \sqrt{2}\right)$

Foci: $\left(1, -3 \pm 2\sqrt{5}\right)$

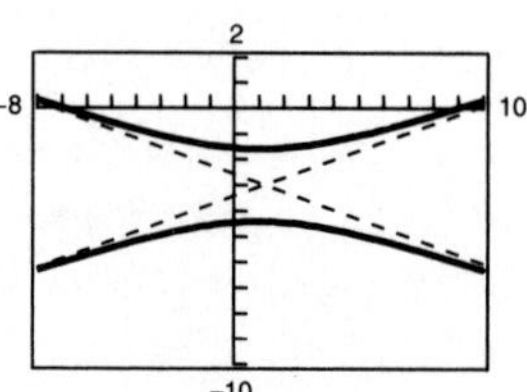

67. $(y-1)^2 - x^2 = 1$ **69.** $\dfrac{(x-3)^2}{4} - \dfrac{(y-2)^2}{16/5} = 1$

71. $\dfrac{(x-4)^2}{4} - \dfrac{y^2}{12} = 1$ **73.** $\dfrac{(y-5)^2}{16} - \dfrac{(x-4)^2}{9} = 1$

75. $\dfrac{y^2}{9} - \dfrac{4(x-2)^2}{9} = 1$ **77.** $\dfrac{(x-3)^2}{9} - \dfrac{(y-2)^2}{4} = 1$

79. Left half **81.** Circle **83.** Hyperbola

85. Ellipse **87.** Parabola

89.

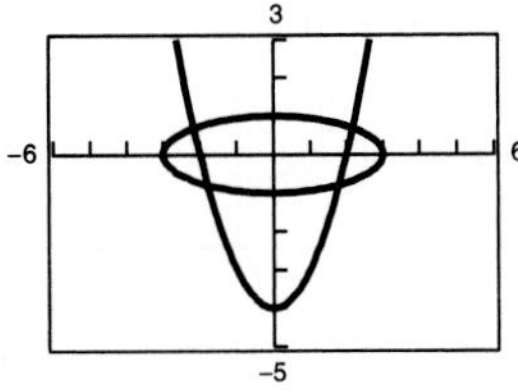

(±2.166, 0.692), (±1.788, −0.803)

91. $\dfrac{x^2}{328.15} + \dfrac{y^2}{19.39} = 1$

Section 8.3 *(page 632)*

1. (c) **3.** (b) **5.** (a) **7.** (f)

9. (a)

t	0	1	2	3	4
x	0	1	$\sqrt{2}$	$\sqrt{3}$	2
y	1	0	−1	−2	−3

(b)

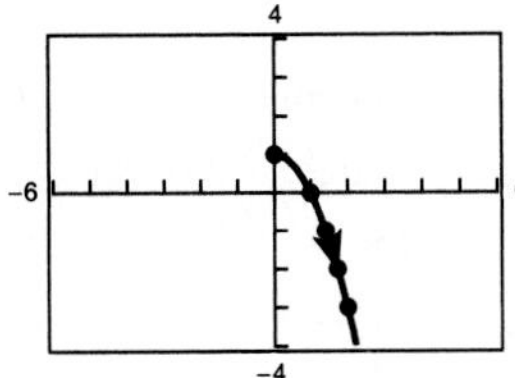

(c)

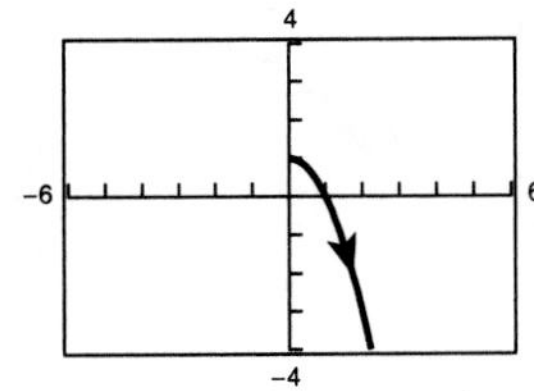

(d) $y = 1 - x^2$

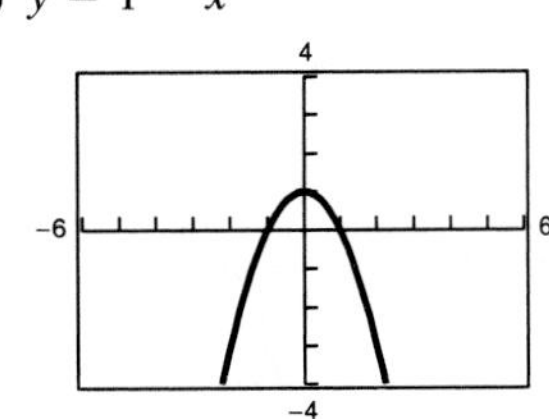

Entire parabola rather than just the right half

11.

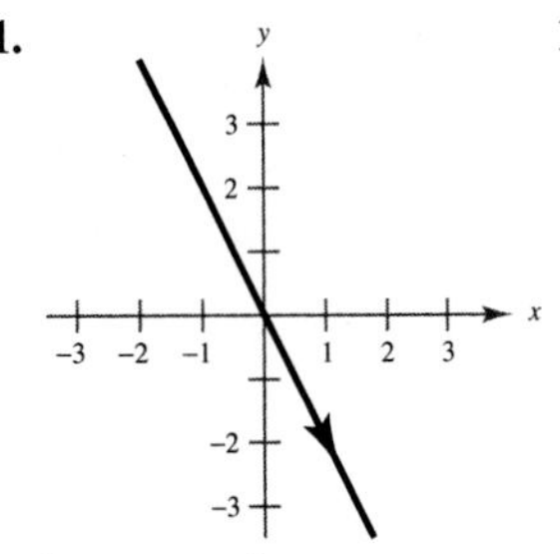

$2x + y = 0$

13.

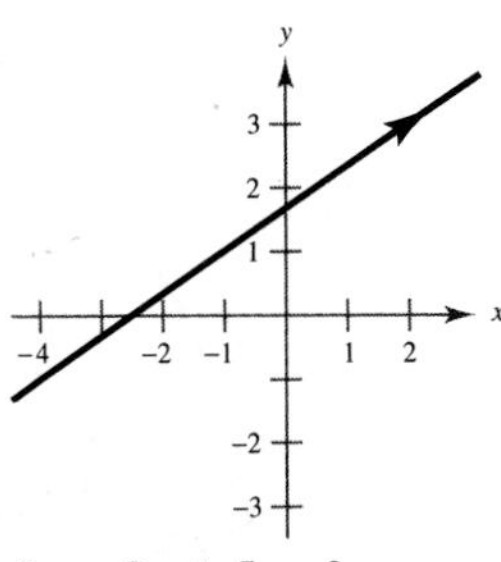

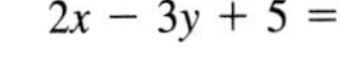

$2x - 3y + 5 = 0$

15.

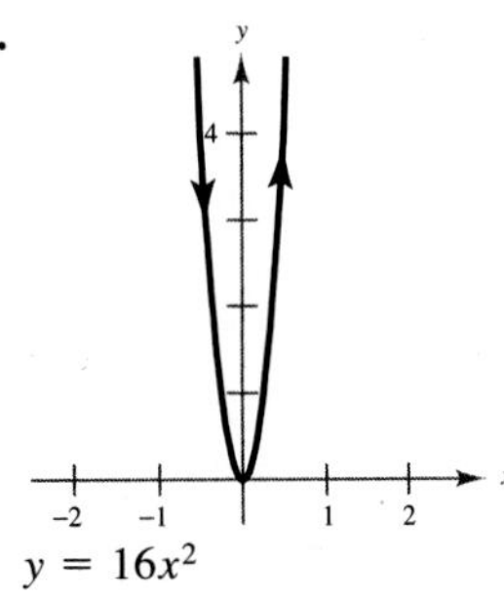

$y = 16x^2$

17.

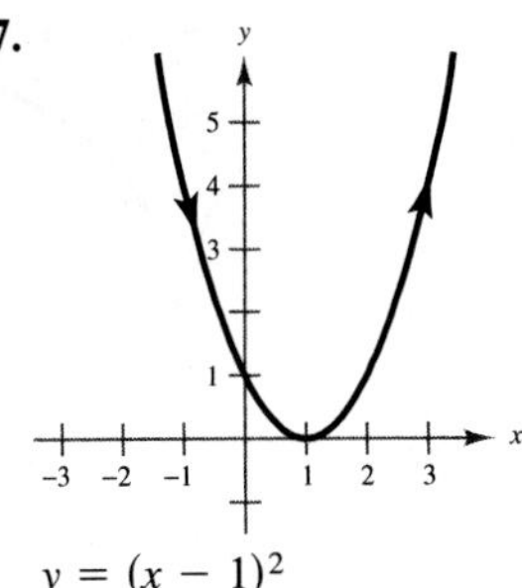

$y = (x-1)^2$

19.

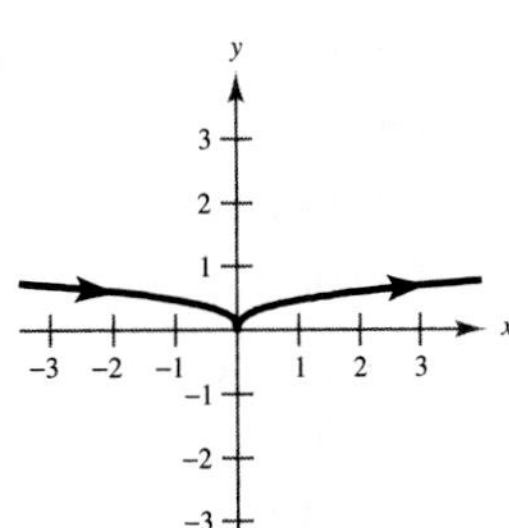

$y = \frac{1}{2}x^{2/3}$

21.

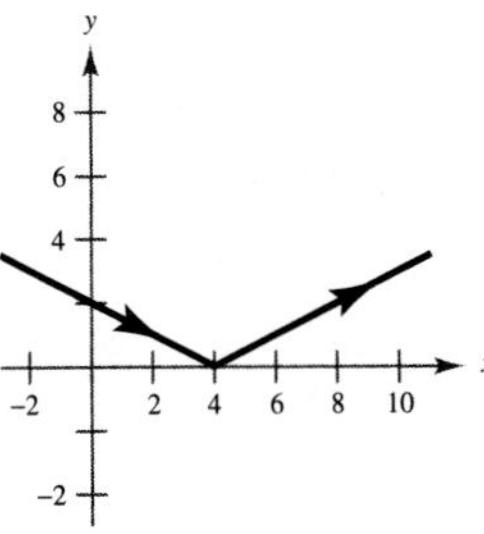

$y = \frac{1}{2}|x - 4|$

23.

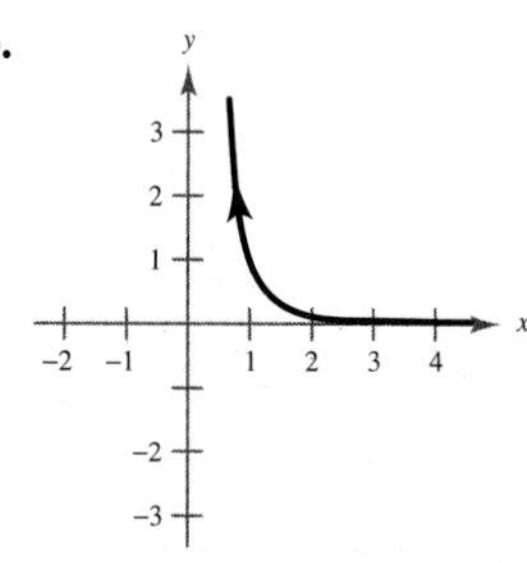

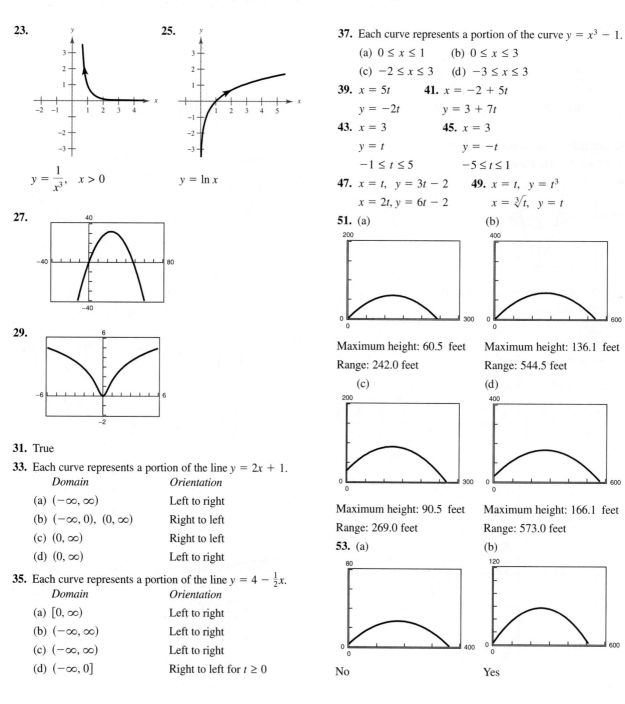

$y = \frac{1}{x^3}, \quad x > 0$

25.

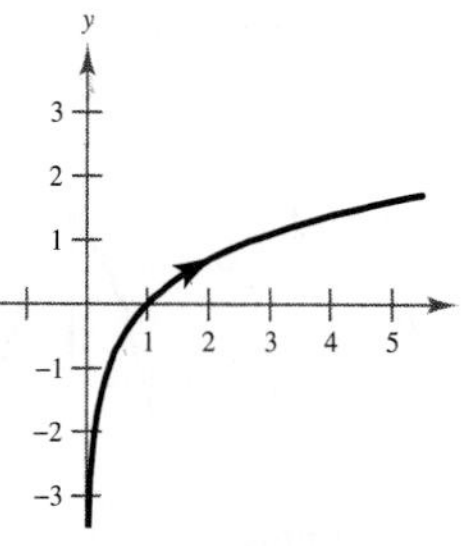

$y = \ln x$

27.

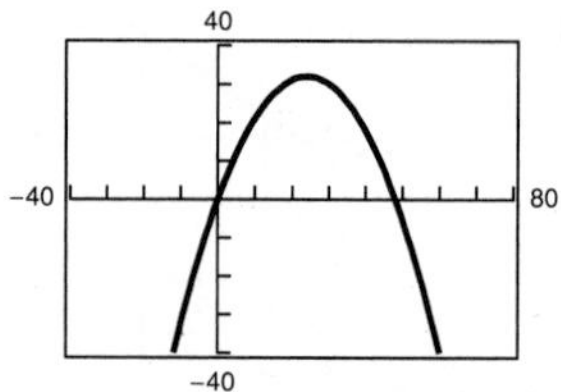

29.

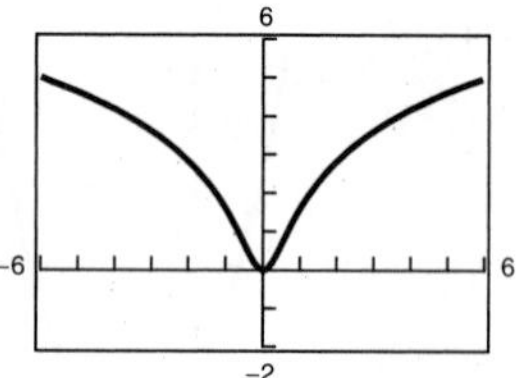

31. True

33. Each curve represents a portion of the line $y = 2x + 1$.

Domain	*Orientation*
(a) $(-\infty, \infty)$	Left to right
(b) $(-\infty, 0), \ (0, \infty)$	Right to left
(c) $(0, \infty)$	Right to left
(d) $(0, \infty)$	Left to right

35. Each curve represents a portion of the line $y = 4 - \frac{1}{2}x$.

Domain	*Orientation*
(a) $[0, \infty)$	Left to right
(b) $(-\infty, \infty)$	Left to right
(c) $(-\infty, \infty)$	Left to right
(d) $(-\infty, 0]$	Right to left for $t \geq 0$

37. Each curve represents a portion of the curve $y = x^3 - 1$.

(a) $0 \leq x \leq 1$ (b) $0 \leq x \leq 3$

(c) $-2 \leq x \leq 3$ (d) $-3 \leq x \leq 3$

39. $x = 5t$
$y = -2t$

41. $x = -2 + 5t$
$y = 3 + 7t$

43. $x = 3$
$y = t$
$-1 \leq t \leq 5$

45. $x = 3$
$y = -t$
$-5 \leq t \leq 1$

47. $x = t, \ y = 3t - 2$
$x = 2t, y = 6t - 2$

49. $x = t, \ y = t^3$
$x = \sqrt[3]{t}, \ y = t$

51. (a)

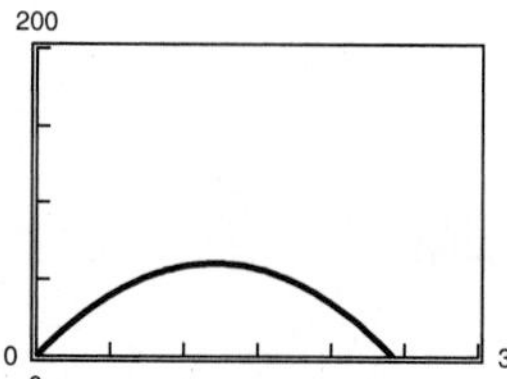

Maximum height: 60.5 feet
Range: 242.0 feet

(b)

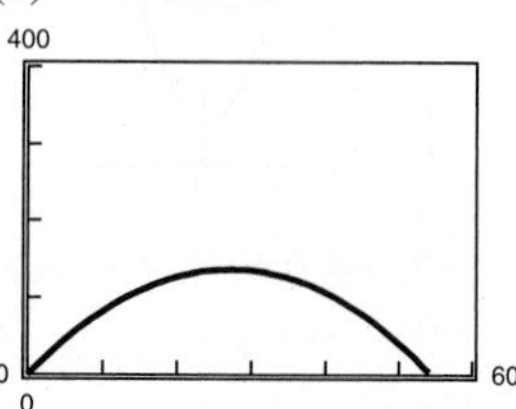

Maximum height: 136.1 feet
Range: 544.5 feet

(c)

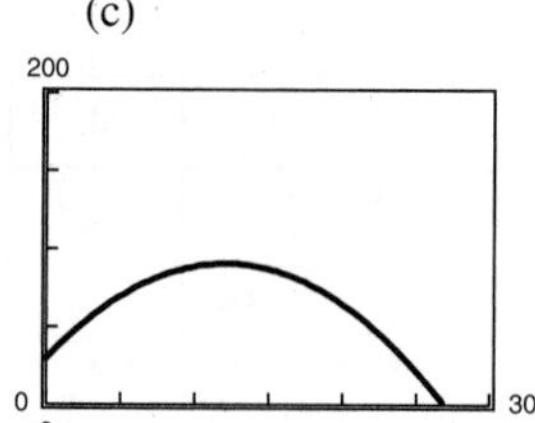

Maximum height: 90.5 feet
Range: 269.0 feet

(d)

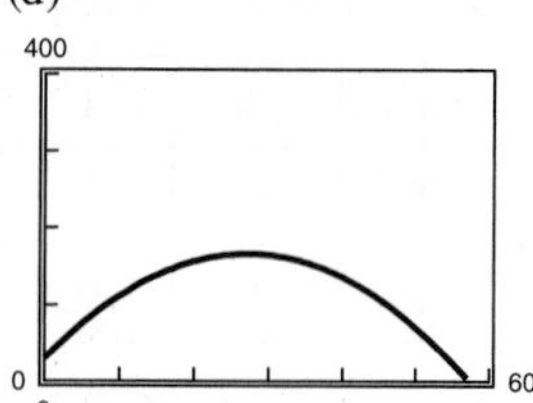

Maximum height: 166.1 feet
Range: 573.0 feet

53. (a)

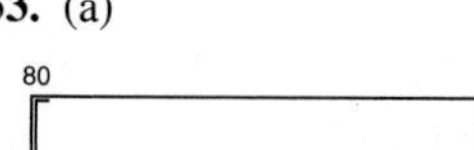

No

(b)

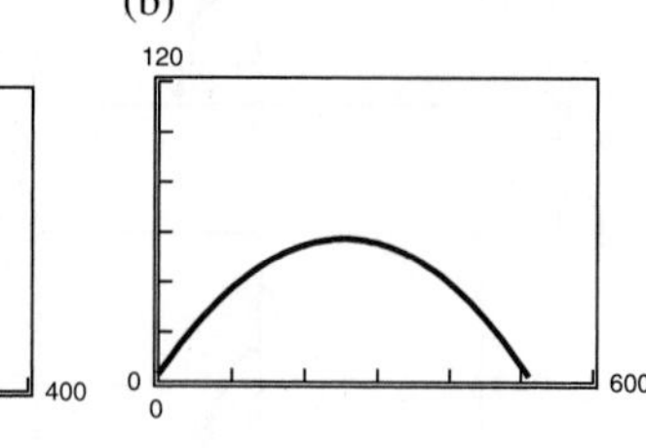

Yes

FOCUS ON CONCEPTS (page 635)

1. (a) Vertical translation
(b) Horizontal translation
(c) Reflection in the y-axis
(d) Parabola opens more slowly

2. (a) Major axis horizontal
(b) Circle
(c) Ellipse is flatter
(d) Horizontal translation

3. The extended diagonals of the central rectangle are asymptotes of the hyperbola.

4. 5. The ellipse becomes more circular and approaches a circle of radius 5.

5. The orientation would be reversed.

6. False. The following are two sets of parametric equations for the line.

$x = t, \quad y = 3 - 2t$

$x = 3t, \quad y = 3 - 6t$

7. (a) **8.** (c) **9.** (d) **10.** (b)

Review Exercises (page 636)

1. (e) **3.** (c) **5.** (f)

7. Parabola

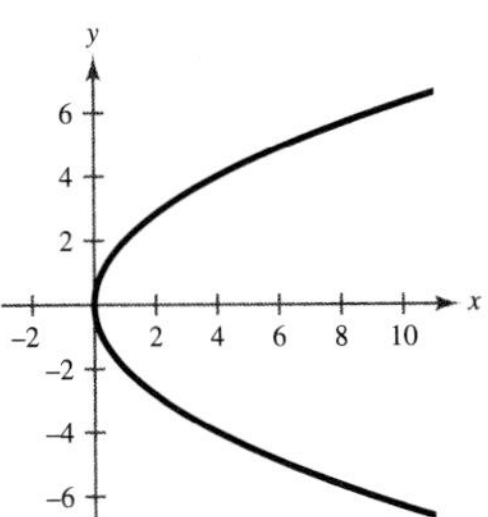

9. Ellipse

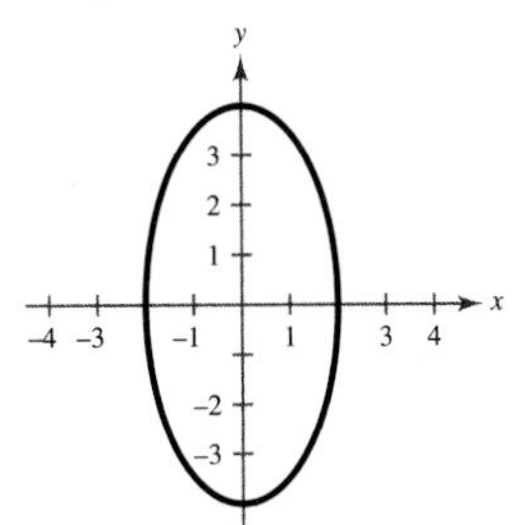

11. Hyperbola

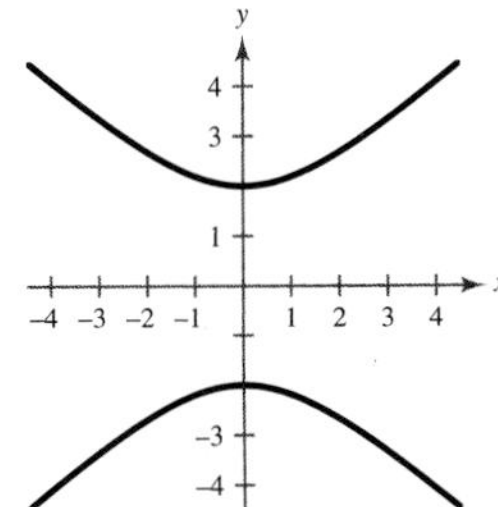

13. Degenerate circle (a point)

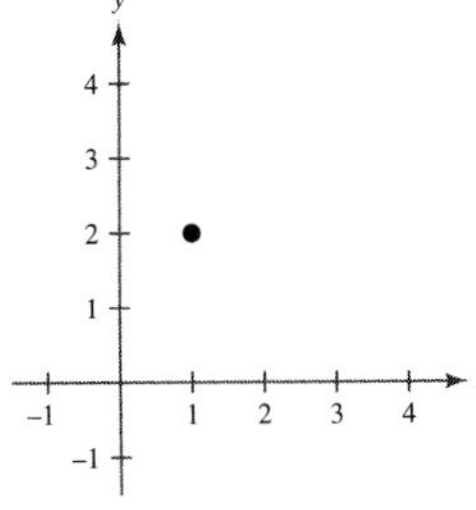

15. Parabola

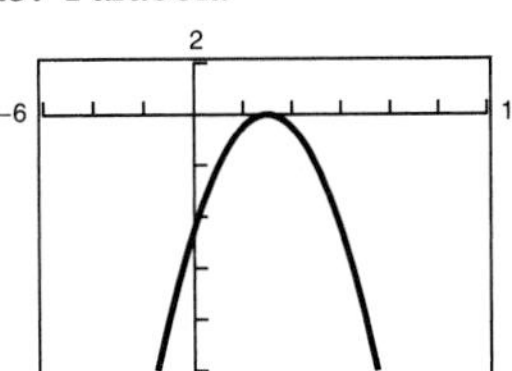

17. Ellipse

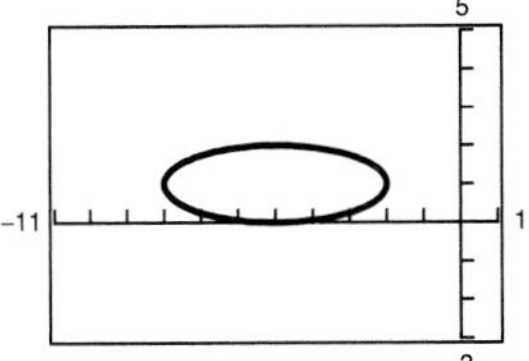

19. Hyperbola

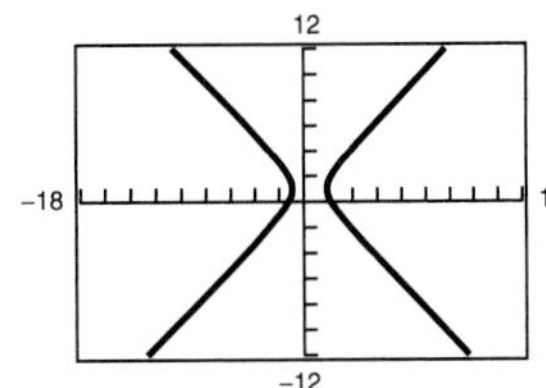

21. $y = 5x \pm \sqrt{24x^2 - 1}$
Hyperbola

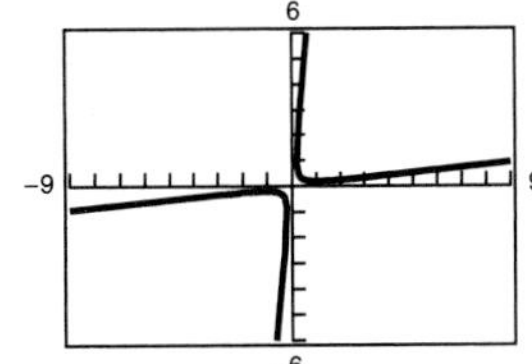

23. $(x + 6)^2 = -9(y - 4)$ **25.** $(x - 4)^2 = -8(y - 2)$

27. $(y - 2)^2 = -4x$ **29.** $\dfrac{(x - 5)^2}{25} + \dfrac{(y - 3)^2}{9} = 1$

31. $\dfrac{(x - 2)^2}{25} + \dfrac{y^2}{21} = 1$ **33.** $\dfrac{2x^2}{9} + \dfrac{y^2}{36} = 1$

35. $x^2 - \dfrac{y^2}{4} = 1$ **37.** $y^2 - \dfrac{x^2}{8} = 1$

39. $\dfrac{5(x - 4)^2}{16} - \dfrac{5y^2}{64} = 1$ **41.** $(0, 50)$

43. (a) $y = 4 - \frac{1}{4}x^2$
$x^2 + \left(y + 4\sqrt{3}\right)^2 = 64$

(b)

x	0	1	2	3	4
d	2.928	2.741	2.182	1.262	0

45. The foci should be placed 3 feet on either side of center and have the same height as the pillars.

47. $\dfrac{36x^2}{25} - \dfrac{36y^2}{11} = 1, \dfrac{64(x-2)^2}{25} - \dfrac{64y^2}{39} = 1$

49. (a) $-\dfrac{2x}{25} + \dfrac{3y}{25} = 1$

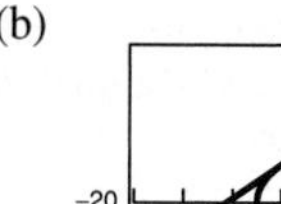

(b)

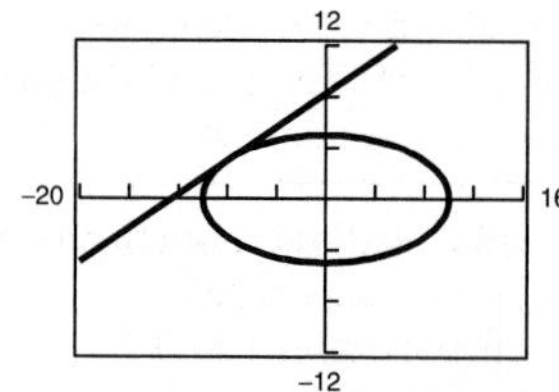

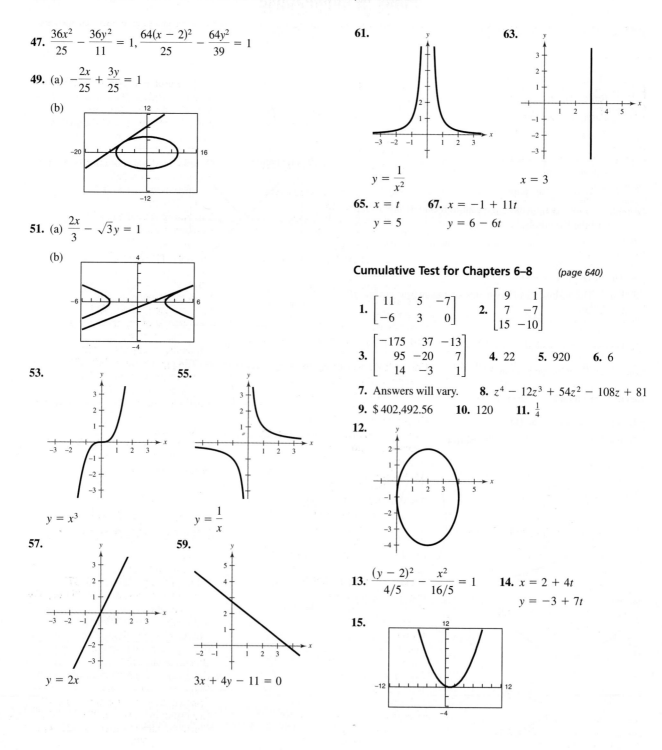

51. (a) $\dfrac{2x}{3} - \sqrt{3}y = 1$

(b)

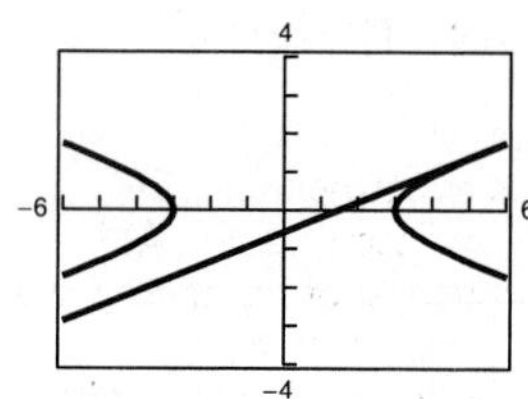

53.

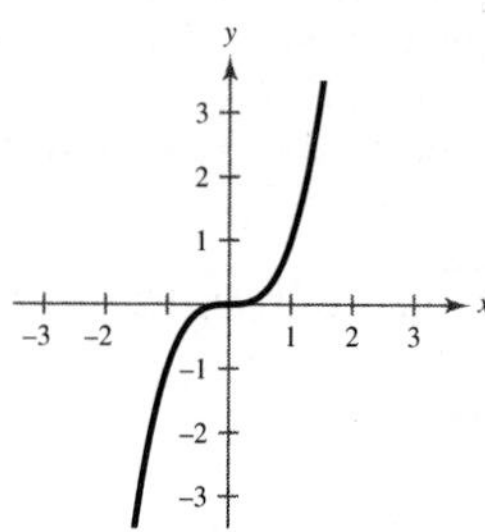

$y = x^3$

55.

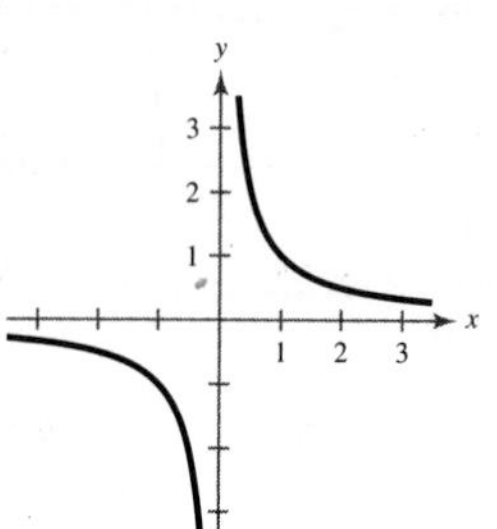

$y = \dfrac{1}{x}$

57.

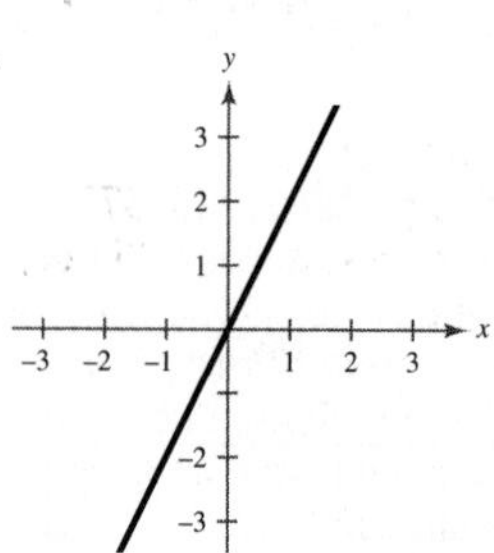

$y = 2x$

59.

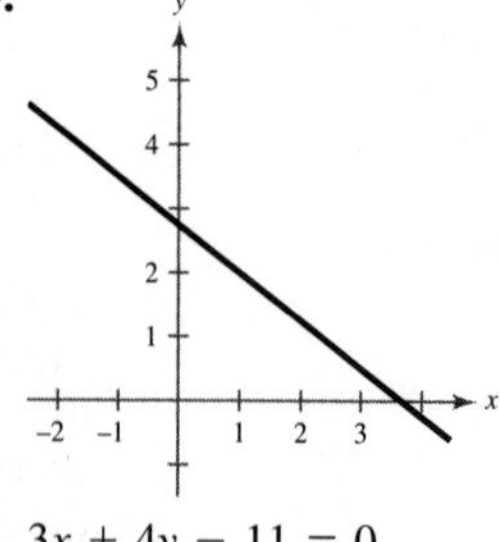

$3x + 4y - 11 = 0$

61.

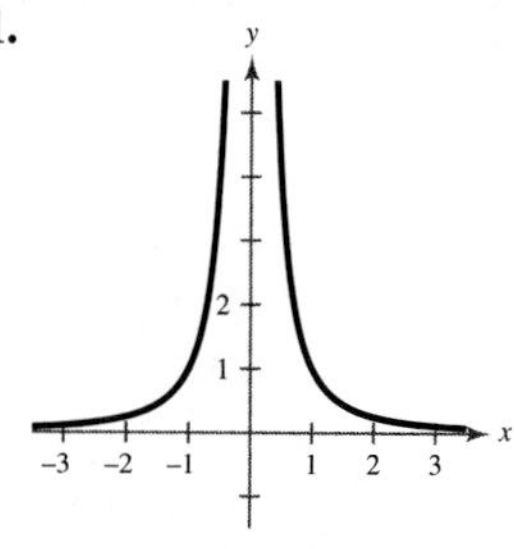

$y = \dfrac{1}{x^2}$

63.

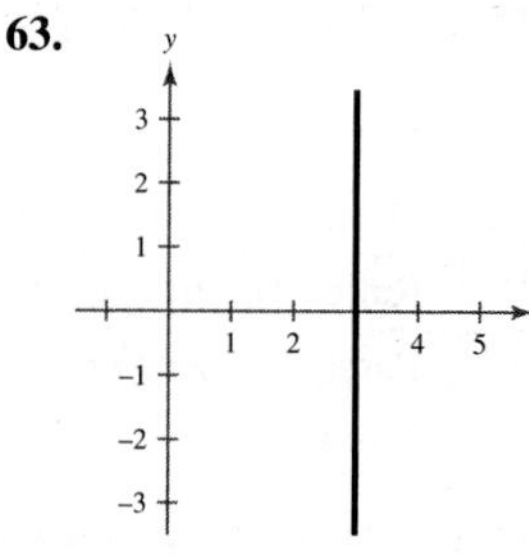

$x = 3$

65. $x = t$
$y = 5$

67. $x = -1 + 11t$
$y = 6 - 6t$

Cumulative Test for Chapters 6–8 *(page 640)*

1. $\begin{bmatrix} 11 & 5 & -7 \\ -6 & 3 & 0 \end{bmatrix}$ **2.** $\begin{bmatrix} 9 & 1 \\ 7 & -7 \\ 15 & -10 \end{bmatrix}$

3. $\begin{bmatrix} -175 & 37 & -13 \\ 95 & -20 & 7 \\ 14 & -3 & 1 \end{bmatrix}$ **4.** 22 **5.** 920 **6.** 6

7. Answers will vary. **8.** $z^4 - 12z^3 + 54z^2 - 108z + 81$

9. \$402,492.56 **10.** 120 **11.** $\frac{1}{4}$

12.

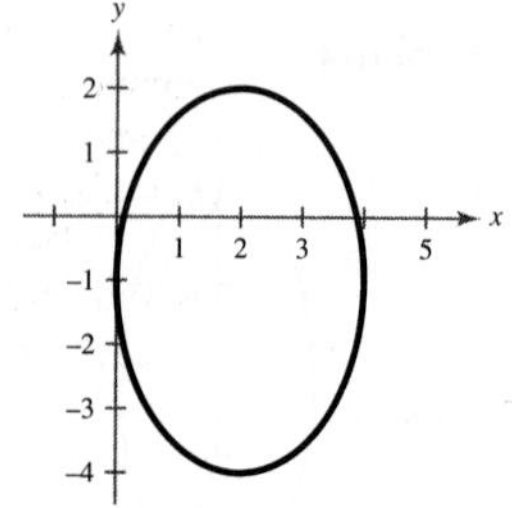

13. $\dfrac{(y-2)^2}{4/5} - \dfrac{x^2}{16/5} = 1$ **14.** $x = 2 + 4t$
$y = -3 + 7t$

15.

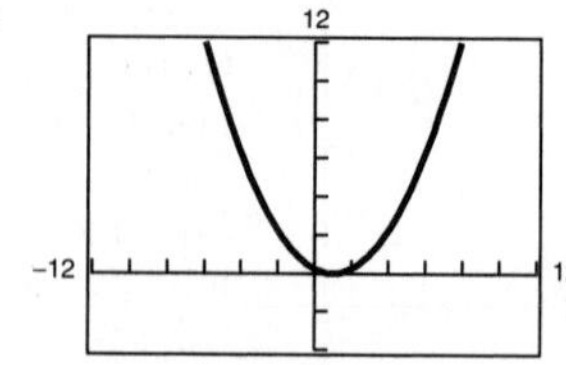

Index of Applications

Biology and Life Science Applications

Business Applications

Chemistry and Physics Applications

Construction Applications

Consumer Applications

Geometric Applications

Interest Rate Applications

Miscellaneous Applications

Time and Distance Applications

U.S. Demographics Applications

Index

G

H

I

O

P

Q

R

Formulas from Geometry

Triangle

$h = a \sin \theta$

Area $= \frac{1}{2}bh$

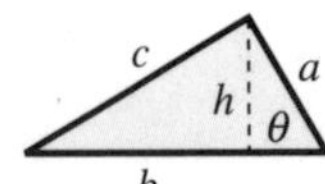

Law of Cosines:

$c^2 = a^2 + b^2 - 2ab \cos \theta$

Right Triangle

Pythagorean Theorem:

$c^2 = a^2 + b^2$

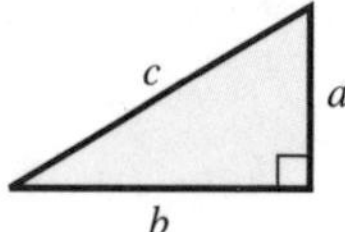

Equilateral Triangle

$h = \frac{\sqrt{3}s}{2}$

Area $= \frac{\sqrt{3}s^2}{4}$

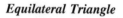

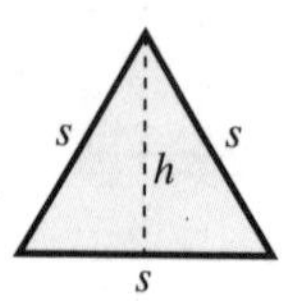

Parallelogram

Area $= bh$

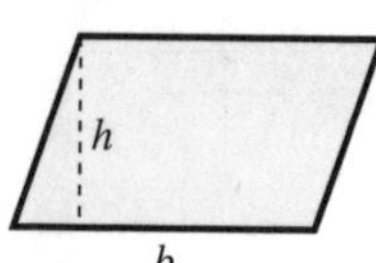

Trapezoid

Area $= \frac{h}{2}(a + b)$

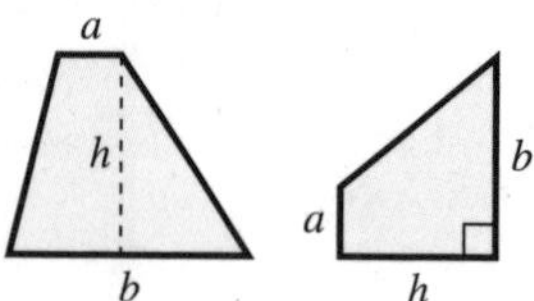

Circle

Area $= \pi r^2$

Circumference $= 2\pi r$

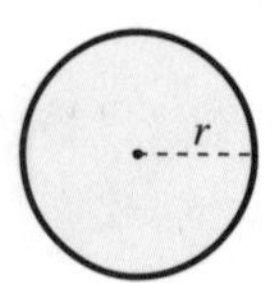

Sector of Circle

Area $= \frac{\theta r^2}{2}$

$s = r\theta$

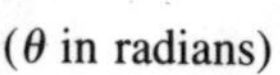

(θ in radians)

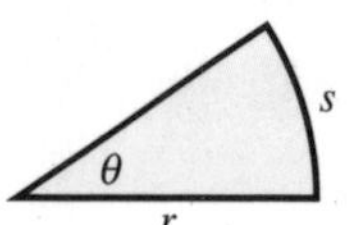

Circular Ring

(p = average radius,
w = width of ring)

Area $= \pi(R^2 - r^2)$
$= 2\pi pw$

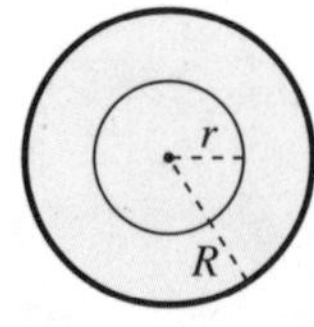

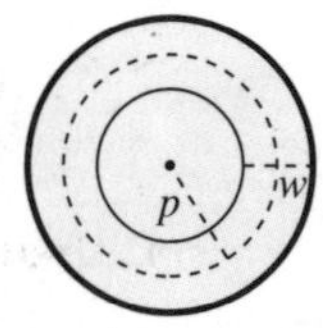

Ellipse

Area $= \pi ab$

Circumference $\approx 2\pi \sqrt{\frac{a^2 + b^2}{2}}$

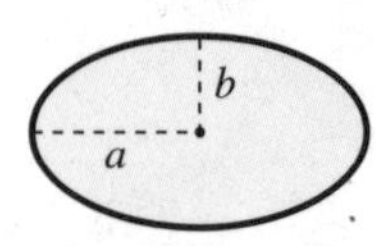

Cone

(A = area of base)

Volume $= \frac{Ah}{3}$

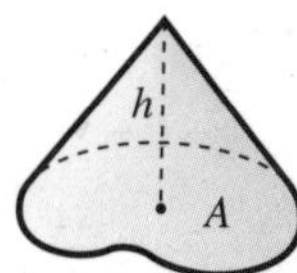

Right Circular Cone

Volume $= \frac{\pi r^2 h}{3}$

Lateral Surface Area $= \pi r\sqrt{r^2 + h^2}$

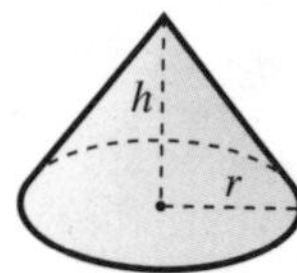

Frustrum of Right Circular Cone

Volume $= \frac{\pi(r^2 + rR + R^2)h}{3}$

Lateral Surface Area $= \pi s(R + r)$

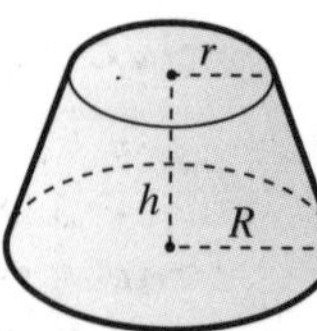

Right Circular Cylinder

Volume $= \pi r^2 h$

Lateral Surface Area $= 2\pi rh$

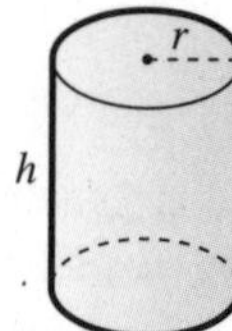

Sphere

Volume $= \frac{4}{3}\pi r^3$

Surface Area $= 4\pi r^2$

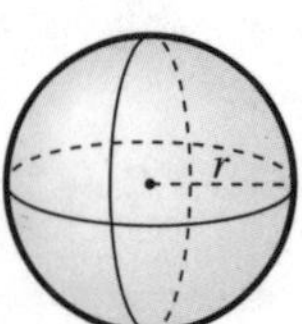